MANIPAL

UGET (MAHE) ENGINEERING ENTRANCE

15 Years'

SOLVED PAPERS

2024-2010

MANIPAL
UGET (MAHE) ENGINEERING ENTRANCE

15 Years'

SOLVED PAPERS
2024-2010

Authors
Atique, Sonam Jha, K Agarwal

arihant
SUCCESS AND NOTHING LESS

arihant
SUCCESS AND NOTHING LESS

Arihant Prakashan (Series), Meerut

卐 **Administrative & Production Offices**

Regd. Office
'Ramchhaya' 4577/15, Agarwal Road, Darya Ganj, New Delhi -110002
Tele: 011- 47630600, 43518550

Head Office
Kalindi, TP Nagar, Meerut (UP) - 250002, Tel: 0121-7156203, 7156204

卐 **Sales & Support Offices**
Agra, Ahmedabad, Bengaluru, Bareilly, Chennai, Delhi, Guwahati, Hyderabad, Jaipur, Jhansi, Kolkata, Lucknow, Nagpur & Pune.

卐 **PRICE** ₹510.00

卐 **PO No :** TXT-XX-XXXXXXX-X-XX

Published by Arihant Publications (India) Ltd.

For further information about the books published by Arihant, log on to www.arihantbooks.com or e-mail at info@arihantbooks.com

Follow us on

PREFACE

In the present world of tough competition you must learn not only how to accumulate exam-specific information, but also learn and master how to be mentally prepared to tackle a test, and use the knowledge gained through toils to your maximum advantage.

With this objective in mind, we present of this book containing Solved Papers (2010-24) for **Manipal Engineering Entrance Test** (UGET, MAHE).

Working on these mock tests in 'true spirit', you will be able to measure your preparedness for the 'real test', and would be also able to identify your weaknesses and take the corrective measures to rectify the lacunae before it is too late. Previous Years' Solved Papers will make you learn the exact pattern of the questions asked in Manipal Engineering Entrance Test, know how to manage your time effectively in the examination hall, and help you develop right exam strategies with positive mental framework.

With Best Wishes

Publisher

CONTENTS

SOLVED PAPERS

SYLLABUS

MATHEMATICS

Sets and Functions

Sets : Sets and their representations, The empty set, Finite and infinite sets, Equal sets, Subsets, Power set, Universal set, Venn Diagrams, Operation on Sets, Complement of a set, Union and intersection of two sets, **Relations and Functions:** Cartesian product of sets, relations and functions, **Trigonometric Functions:** Angles, Trigonometric functions, Trigonometric functions of sum and difference of two angles and trigonometric equations.

Algebra

Mathematical Induction : Motivation, Principle of mathematical induction, **Numbers and Quadratic Equations:** Complex numbers, Algebra of complex numbers, The modulus and the conjugate of a complex number, Argand plane and polar representation and Quadratic equations, **Linear Inequalities:** Inequalities, Algebraic solutions of linear inequalities, graphical solution and solution of system of linear inequalities in two variables, **Permutations and Combinations:** Permutations and Combinations, Derivation of formulae and their connections, **Binomial Theorem:** Binomial theorem for positive integral indices, General and middle terms, Applications, **Sequence and Series:** Sequences, Series, Arithmetic progression, Geometric progression, Relationship between AM and GM, sum to n terms of special series, **Matrices:** Matrix, types, operations, Transpose, Symmetric and skew symmetric matrices, elementary operation, Invertible matrices, **Determinants:** Properties, Area of a triangle, minors and cofactors, Adjoint and inverse of a matrix, applications.

Coordinate Geometry

Straight Lines : Slope of a line, Various forms of the equation of a line, General equation of a line, Distance of a point from a line, **Conic Sections:** Section of a cone, Circle, Parabola, Ellipse, Hyperbola, **Introduction to Three-dimensional Geometry:** Coordinate axes and coordinate planes in three-dimensional space, Coordinates of a point in space, Distance between two points, Section formula.

Calculus

Limits and Derivatives : Intuitive idea of derivatives, Limits of trigonometric functions, Derivatives, **Continuity and Differentiability:** Exponential and logarithmic functions, logarithmic differentiation, Derivatives of functions in parametric forms, Second order derivative, Mean value theorem, **Applications of Derivatives:** Rate of change of quantities, Increasing and decreasing functions, Tangents and normals, Approximations, Maxima and Minima, **Integrals and their Application:** Integration, methods of integration, Integrals of some particular functions, Integration by partial fractions, Integration by parts, Definite Integral, Fundamental theorem of calculus, Evaluation of definite integrals by substitution, Properties of definite integrals, Area under simple curves and two curves, **Differential equations:** Basic concepts, General and particular solutions of a differential equation, formation of a differential equation, methods of solving first order, first degree differential equations.

Mathematical Reasoning: Statements, New statements from old, Special words/phrases, Implications, Validating statements, **Statistics and Probability:** Measure of dispersion, Range, Mean deviation, Variance and standard deviation, Analysis of frequency distributions, Random experiments, Event, Axiomatic approach to probability, Conditional probability, multiplication theorem on probability, Independent events, Bayes' theorem, Random variables and its probability distributions, Bernoulli trials and binomial distribution, **Relations and Functions:** Types of relations and functions, Composition of functions and invertible function, Binary operations, **Inverse Trigonometric Functions:** Basic concepts, Properties of inverse trigonometric functions, **Vector algebra:** Basic concepts, types of vectors, Addition of vectors, Multiplication of a vector by a scalar, Product of two vectors, **Three-Dimensional Geometry:** Direction cosines and direction ratios of a line, Equation of a line in space, Angle between two lines, Shortest distance between two lines, Plane, Coplanarity of two lines, Angle between two planes, Distance of a point from a plane, Angle between a line and a plane, **Linear programming:** Problem and its mathematical formulation, different types of linear programming problems

PHYSICS

Physical World and Measurement

Physical World: Physics-scope and excitement; nature of physical laws; Physics, technology and society, **Units and Measurements:** Need for measurement: Units of measurement; systems of units; SI units, fundamental and derived units. Length, mass and time measurements; accuracy and precision of measuring instruments; errors in measurement; significant figures. Dimensions of physical quantities, dimensional analysis and its applications.

Kinematics

Motion in a Straight Line: Elementary concepts of differentiation and integration for describing motion, uniform and non- uniform motion, average speed and instantaneous velocity, uniformly accelerated motion, velocity - time and position-time graphs. Relations for uniformly accelerated motion, **Motion in a Plane:** Scalar and vector quantities; position and displacement vectors, general vectors and their notations; equality of vectors, multiplication of vectors by a real number; addition and subtraction of vectors, relative velocity, Unit vector; resolution of a vector in a plane, rectangular components, Scalar and Vector product of vectors. Motion in a plane, cases of uniform velocity and uniform acceleration-projectile motion, uniform circular motion.

Laws of Motion: Intuitive concept of force, Inertia, Newton's first law of motion; momentum and Newton's second law of motion; impulse, Newton's third law of motion, Law of conservation of linear momentum and its applications, Equilibrium of concurrent forces, Static and kinetic friction, laws of friction, rolling friction, lubrication, Dynamics of uniform circular motion: Centripetal force, examples of circular motion.

Work, Energy and Power: Work done by a constant force and a variable force; kinetic energy, work-energy, theorem, power, Notion of potential energy, potential energy of a spring, conservative forces: conservation of mechanical energy (kinetic and potential energies); **non-conservative forces:** motion in a vertical circle; elastic and inelastic collisions in one and two dimensions.

Motion of System of Particles and Rigid Body

System of Particles and Rotational Motion: Centre of mass of a two-particle system, momentum conservation and centre of mass motion. Centre of mass of a rigid body; centre of mass of a uniform rod. Moment of a force, torque, angular momentum, law of conservation of angular momentum and its applications, Equilibrium of rigid bodies, rigid body rotation and equations of rotational motion, comparison of linear and rotational motions, Moment of inertia, radius of gyration, values of moments of inertia for simple geometrical objects.

Gravitation: Universal law of gravitation. Acceleration due to gravity (recapitulation only) and its variation with altitude and depth, Gravitational potential energy and gravitational potential, escape velocity, orbital velocity of a satellite, Geo-stationary satellites.

Mechanical Properties of Solids: Stress-strain relationship, Hooke's law, Young's modulus, bulk modulus, **Mechanical Properties of Fluids:** Pressure due to a fluid column; Pascal's law and its applications (hydraulic lift and hydraulic brakes), effect of gravity on fluid pressure. Viscosity, Stokes' law, terminal velocity, streamline and turbulent flow, critical velocity, Bernoulli's theorem and its applications. Surface energy and surface tension, angle of contact, excess of pressure across a curved surface, application of surface tension ideas to drops, bubbles and capillary rise, **Thermal Properties of Matter:** Heat, temperature, (recapitulation only) thermal expansion; thermal expansion of solids, liquids and gases, anomalous expansion of water; specific heat capacity;

Cp, Cv- calorimetry; change of state - latent heat capacity. Heat transfer-conduction, convection and radiation, thermal conductivity, qualitative ideas of Blackbody radiation, Wein's displacement Law, Stefan's law, Greenhouse effect.

Thermodynamics: Thermal equilibrium and definition of temperature (zeroth law of thermodynamics), heat, work and internal energy. First law of thermodynamics, isothermal and adiabatic processes. Second law of thermodynamics: reversible and irreversible processes.

Behaviour of Perfect Gases and Kinetic Theory of Gases

Kinetic Theory: Equation of state of a perfect gas, work done in compressing a gas. Kinetic theory of gases - assumptions, concept of pressure. Kinetic interpretation of temperature; rms speed of gas molecules; degrees of freedom, law of equipartition of energy (statement only) and application to specific heat capacities of gases; concept of mean free path, Avogadro's number.

Oscillations: Periodic motion - time period, frequency, displacement as a function of time, periodic functions, Simple harmonic motion (S.H.M) and its equation; phase; oscillations of a loaded spring restoring force and force constant; energy in S.H.M. Kinetic and potential energies; simple pendulum derivation of expression for its time period. Free, forced and damped oscillations (qualitative ideas only), resonance, **Waves:** Wave motion: Transverse and longitudinal waves, speed of travelling wave, displacement relation for a progressive wave, principle of superposition of waves, reflection of waves, standing waves in strings and organ pipes, Beats.

Electrostatics

Electric Charges and Fields: Electric Charges; Conservation of charge, Coulomb's law-force between two-point charges, forces between multiple charges; superposition principle and continuous charge distribution. Electric field, electric field due to a point charge, electric field lines, electric dipole, electric field due to a dipole, torque on a dipole in uniform electric field. Electric flux, statement of Gauss's theorem and its applications to find field due to infinitely long straight wire, uniformly charged infinite plane sheet.

Electrostatic Potential and Capacitance: Electric potential, potential difference, electric potential due to a point charge, a dipole and system of charges; equipotential surfaces, electrical potential energy of a system of two-point charges and of electric dipole in an electrostatic field.

Conductors and insulators, free charges and bound charges inside a conductor. Dielectrics and electric polarisation, capacitors and capacitance, combination of capacitors in series and in parallel, capacitance of a parallel plate capacitor with and without dielectric medium between the plates, energy stored in a capacitor.

Current Electricity

Current Electricity: Electric current, flow of electric charges in a metallic conductor, drift velocity, mobility and their relation with electric current; Ohm's law, electrical resistance, V-I characteristics (linear and non-linear), electrical energy and power, electrical resistivity and conductivity; temperature dependence of resistance. Internal resistance of a cell, potential difference and emf of a cell, combination of cells in series and in parallel, Kirchhoff's laws and simple applications, Wheatstone bridge, metre bridge (qualitative ideas only) Potentiometer - principle and its applications to measure potential difference and for 14 comparing EMF of two cells; measurement of internal resistance of a cell (qualitative ideas only).

Magnetic Effects of Current and Magnetism

Moving Charges and Magnetism: Concept of magnetic field, Oersted's experiment. Biot - Savart law and its application to current carrying circular loop. Ampere's law and its applications to infinitely long straight wire. Straight and toroidal solenoids (only qualitative treatment), force on a moving charge in uniform magnetic and electric fields Force on a current-carrying conductor in a uniform magnetic field, force between two parallel current-carrying conductors-definition of ampere, torque experienced by a current loop in uniform magnetic field; moving coil galvanometer-its current sensitivity and conversion to ammeter and voltmeter.

Magnetism and Matter: Current loop as a magnetic dipole and its magnetic dipole moment, magnetic dipole moment of a revolving electron, bar magnet as an equivalent solenoid, magnetic field lines; earth's magnetic field and magnetic elements.

Electromagnetic Induction and Alternating Currents

Electromagnetic Induction: Electromagnetic induction; Faraday's laws, induced EMF and current; Lenz's Law, Eddy currents. Self and mutual induction. Alternating currents, peak and RMS value of alternating current/voltage; reactance and impedance; LC oscillations (qualitative treatment only), LCR series circuit, resonance; power in AC circuits AC generator and transformer.

Electromagnetic Waves: Electromagnetic waves, their characteristics, their Transverse nature (qualitative ideas only). Electromagnetic spectrum (radio waves, microwaves, infrared, visible, ultraviolet, X-rays, gamma rays) including elementary facts about their uses.

Optics: Ray Optics and Optical Instruments: Ray Optics, Refraction of light, total internal reflection and its applications, optical fibres, refraction at spherical surfaces, lenses, thin lens formula, lensmaker's formula, magnification, power of a lens, combination of thin lenses in contact, refraction of light through a prism. **Optical instruments:** Microscopes and astronomical telescopes (reflecting and refracting) and their magnifying powers.

Wave Optics Wave optics: Wave front and Huygen's principle, reflection and refraction of plane wave at a plane surface using wave fronts. Proof of laws of reflection and refraction using Huygen's principle. Interference, Young's double slit experiment and expression for fringe width, coherent sources and sustained interference of light, diffraction due to a single slit, width of central maximum.

Dual Nature of Radiation and Matter: Dual nature of radiation, Photoelectric effect, Hertz and Lenard's observations; Einstein's photoelectric equation-particle nature of light, **Experimental study of photoelectric effect:** Matter waves-wave nature of particles, de-Broglie relation.

Atoms: Alpha-particle scattering experiment; Rutherford's model of atom; Bohr model, energy levels, hydrogen spectrum, **Nuclei Composition and size of nucleus:** Nuclear force Mass-energy relation, mass defect, nuclear fission, nuclear fusion.

Electronic Devices

Semiconductor Electronics: Materials, Devices and Simple Circuits Energy bands in conductors, semiconductors and insulators (qualitative ideas only) Semiconductor diode - I-V characteristics in forward and reverse bias, diode as a rectifier; Special purpose p-n junction diodes: LED, photodiode, solar cell.

CHEMISTRY

Some Basic Concepts of Chemistry: General Introduction: Importance and scope of chemistry, Nature of matter, laws of chemical combination, **Dalton's atomic theory:** concept of elements, atoms and molecules, Atomic and molecular masses, mole concept and molar mass, percentage composition, empirical and molecular formula, chemical reactions, stoichiometry and calculations based on stoichiometry.

Structure of atom: Bohr's model and its limitations, concept of shells and subshells, dual nature of matter and light, de Broglie's relationship, Heisenberg uncertainty principle, concept of orbitals, quantum numbers, shapes of s, p and d orbitals, rules for filling electrons in orbitals - Aufbau principle, Pauli's exclusion principle and Hund's rule, electronic configuration of atoms, stability of half-filled and completely filled orbitals.

Classification of Elements and Periodicity in Properties: Modern periodic law and the present form of periodic table, periodic trends in properties of elements -atomic radii, ionic radii, inert gas radii, Ionization enthalpy, electron gain enthalpy, electronegativity, valency. Nomenclature of elements with atomic number greater than 100.

Chemical Bonding and Molecular structure: Valence electrons, ionic bond, covalent bond, bond parameters, Lewis structure, polar character of covalent bond, covalent character of ionic bond, valence bond theory, resonance, geometry of covalent molecules, VSEPR theory, concept of hybridization, involving s, p and d orbitals and shapes of some simple molecules, molecular orbital theory of homonuclear diatomic molecules (qualitative idea only), hydrogen bond.

States of Matter: Gases and Liquids: Three states of matter, intermolecular interactions, types of bonding, melting and boiling points, role of gas laws in elucidating the concept of the molecule, Boyle's law, Charles law, Gay Lussac's law, Avogadro's law, ideal behaviour, empirical derivation of gas equation, Avogadro's number, ideal gas equation. Deviation from ideal behaviour, liquefaction of gases, critical temperature, kinetic energy and molecular speeds (elementary idea) **Liquid State:** vapour pressure, viscosity and surface tension (qualitative idea only, no mathematical derivations).

Chemical Thermodynamics: Concepts of System and types of systems, surroundings, work, heat, energy, extensive and intensive properties, state functions. First law of thermodynamics -internal energy and enthalpy, heat capacity and specific heat, measurement of ΔU and ΔH, Hess's law of constant heat summation, enthalpy of bond dissociation, combustion, formation, atomization, sublimation, phase transition, ionization, solution and dilution. Second law of Thermodynamics (brief introduction). Introduction of entropy as a state function, Gibb's energy change for spontaneous and non- spontaneous processes, criteria for equilibrium. Third law of thermodynamics (brief introduction).

Equilibrium: Equilibrium in physical and chemical processes, dynamic nature of equilibrium, law of mass action, equilibrium constant, factors affecting equilibrium- Le Chatelier's principle, ionic equilibrium- ionization of acids and bases, strong and weak electrolytes, degree of ionization, ionization of poly basic acids, acid strength, concept of pH, Henderson Equation, hydrolysis of salts (elementary idea), buffer solution, solubility product, common ion effect (with illustrative examples).

Redox Reactions: Concept of oxidation and reduction, redox reactions, oxidation number, balancing redox reactions, in terms of loss and gain of electrons and change in oxidation number, applications of redox reactions.

Hydrogen: Position of hydrogen in periodic table, occurrence, isotopes, preparation, properties and uses of hydrogen, hydrides-ionic covalent and interstitial; physical and chemical properties of water, heavy water, hydrogen peroxide - preparation, reactions and structure and use; hydrogen as a fuel.

***s*-Block Elements (Alkali and Alkaline Earth Metals):** Group 1 and Group 2 Elements General introduction, electronic configuration, occurrence, anomalous properties of the first element of each group, diagonal relationship, trends in the variation of properties (such as ionization enthalpy, atomic and ionic radii), trends in chemical reactivity with oxygen, water, hydrogen and halogens, uses. Preparation and **Properties of Some Important Compounds:** Sodium Carbonate, Sodium Chloride, Sodium Hydroxide and Sodium Hydrogen carbonate, Biological importance of Sodium and Potassium. Calcium Oxide and Calcium Carbonate and their industrial uses, biological importance of Magnesium and Calcium.

***p*-Block Elements:** General Introduction to *p*-Block Elements **Group 13 Elements:** General introduction, electronic configuration, occurrence, variation of properties, oxidation states, trends in chemical reactivity, anomalous properties of first element of the group, Boron - physical and chemical properties, some important compounds, Borax, Boric acid, Boron Hydrides, **Aluminium:** Reactions with acids and alkalies, uses.
Group 14 Elements: General introduction, electronic configuration, occurrence, variation of properties, oxidation states, trends in chemical reactivity, anomalous behaviour of first elements. Carbon-catenation, allotropic forms, physical and chemical properties; uses of some important compounds: oxides. **Important compounds of Silicon and a few uses:** Silicon Tetrachloride, Silicones, Silicates and Zeolites, their uses.

Organic Chemistry - Some Basic Principles and Techniques: General introduction, methods of purification, qualitative and quantitative analysis, classification and IUPAC nomenclature of organic compounds. **Electronic displacements in a covalent bond:** inductive effect, electromeric effect, resonance and hyper conjugation. **Homolytic and heterolytic fission of a covalent bond:** free radicals, carbocations, carbanions, electrophiles and nucleophiles, types of organic reactions.

Hydrocarbons: Classification of Hydrocarbons **Aliphatic Hydrocarbons:** Alkanes - Nomenclature, isomerism, conformation (ethane only), physical properties, chemical reactions including free radical mechanism of halogenation, combustion and pyrolysis. Alkenes - Nomenclature, structure of double bond (ethene), geometrical isomerism, physical properties, methods of preparation, chemical reactions: addition of hydrogen, halogen, water, hydrogen halides (Markownikov's addition and peroxide effect), ozonolysis, oxidation, mechanism of electrophilic addition. Alkynes - Nomenclature, structure of triple bond (ethyne), physical properties, methods of preparation, **chemical reactions:** acidic character of alkynes, addition reaction of - hydrogen, halogens, hydrogen halides and water. **Aromatic Hydrocarbons:** Introduction,

IUPAC nomenclature, benzene: resonance, aromaticity, **chemical properties:** mechanism of electrophilic substitution. Nitration, sulphonation, halogenation, Friedel Craft's alkylation and acylation, directive influence of functional group in monosubstituted benzene, Carcinogenicity and toxicity.

Environmental Chemistry: Environmental pollution - air, water and soil pollution, chemical reactions in atmosphere, smog, major atmospheric pollutants, acid rain, ozone and its reactions, effects of depletion of ozone layer, greenhouse effect and global warming- pollution due to industrial wastes, green chemistry as an alternative tool for reducing pollution, strategies for control of environmental pollution.

Solutions: Types of solutions, expression of concentration of solutions of solids in liquids, solubility of gases in liquids, solid solutions, colligative properties - relative lowering of vapour pressure, Raoult's law, elevation of boiling point, depression of freezing point, osmotic pressure, determination of molecular masses using colligative properties, abnormal molecular mass, Van't Hoff factor.

Electrochemistry: Redox reactions, conductance in electrolytic solutions, specific and molar conductivity, variations of conductivity with concentration, Kohlrausch's Law, electrolysis and law of electrolysis (elementary idea), dry cell-electrolytic cells and Galvanic cells, lead accumulator, EMF of a cell, standard electrode potential, Nernst equation and its application to chemical cells, Relation between Gibbs energy change and EMF of a cell, fuel cells, corrosion.

Chemical Kinetics: Rate of a reaction (Average and instantaneous), **factors affecting rate of reaction:** concentration, temperature, catalyst; order and molecularity of a reaction, rate law and specific rate constant, integrated rate equations and half-life (only for zero and first order reactions), concept of collision theory (elementary idea, no mathematical treatment). Activation energy, Arrhenius equation.

Surface Chemistry: Adsorption - physisorption and chemisorption, factors affecting adsorption of gases on solids, catalysis, homogenous and heterogenous activity and selectivity; enzyme catalysis colloidal state distinction between true solutions, colloids and suspension; lyophilic, lyophobic multi-molecular and macromolecular colloids; properties of colloids; Tyndall effect, Brownian movement, electrophoresis, coagulation, emulsion - types of emulsions.

General Principles and Processes of Isolation of Elements: Principles and methods of extraction - concentration, oxidation, reduction - electrolytic method and refining; occurrence and principles of extraction of aluminium, copper, zinc and iron.

***p*-Block Elements: Group 16 Elements:** General introduction, electronic configuration, oxidation states, occurrence, trends in physical and chemical properties, **dioxygen:** Preparation, Properties and uses, classification of Oxides, Ozone, Sulphur - allotropic forms; **compounds of Sulphur:** Preparation Properties and uses of Sulphur-dioxide, **Sulphuric Acid:** industrial process of manufacture, properties and uses; Oxoacids of Sulphur

(Structures only). **Group 17 Elements:** General introduction, electronic configuration, oxidation states, occurrence, trends in physical and chemical properties; compounds of halogens, Preparation, properties and uses of Chlorine and Hydrochloric acid, interhalogen compounds, Oxoacids of halogens (structures only). **Group 18 Elements:** General introduction, electronic configuration, occurrence, trends in physical and chemical properties, uses.

***'d' and 'f'* Block Elements:** General introduction, electronic configuration, occurrence and characteristics of transition metals, general trends in properties of the first-row transition metals - metallic character, ionization enthalpy, oxidation states, ionic radii, colour, catalytic property, magnetic properties, interstitial compounds, alloy formation, preparation and properties of $K_2Cr_2O_7$ and $KMnO_4$. Lanthanoids - Electronic configuration, oxidation states, chemical reactivity and lanthanoid contraction and its consequences. Actinoids - Electronic configuration, oxidation states and comparison with lanthanoids.

Coordination Compounds: Coordination compounds - Introduction, ligands, coordination number, colour, magnetic properties and shapes, IUPAC nomenclature of mononuclear coordination compounds. Bonding, Werner's theory, VBT, and CFT; structure and stereoisomerism, importance of coordination compounds (in qualitative inclusion, extraction of metals and biological system).

Haloalkanes and Haloarenes: Haloalkanes: Nomenclature, nature of C-*X* bond, physical and chemical properties, mechanism of substitution reactions, optical rotation. **Haloarenes:** Nature of C-*X* bond, substitution reactions (Directive influence of halogen in monosubstituted compounds only). Uses and environmental effects of - dichloromethane, trichloromethane, tetrachloromethane, iodoform, freons, DDT.

Alcohols, Phenols and Ethers: Alcohols: Nomenclature, methods of preparation, physical and chemical properties (of primary alcohols only), identification of primary, secondary and tertiary alcohols, mechanism of dehydration, uses with special reference to methanol and ethanol.

Phenols: Nomenclature, methods of preparation, physical and chemical properties, acidic nature of phenol, electrophilic substitution reactions, uses of phenols. **Ethers:** Nomenclature, methods of preparation, physical and chemical properties, uses

Aldehydes, Ketones and Carboxylic Acids: Aldehydes and Ketones: Nomenclature, nature of carbonyl group, methods of preparation, physical and chemical properties, mechanism of nucleophilic addition, reactivity of alpha hydrogen in aldehydes, uses. **Carboxylic Acids:** Nomenclature, acidic nature, methods of preparation, physical and chemical properties; uses.

Organic compounds containing Nitrogen: Amines: Nomenclature, classification, structure, methods of preparation, physical and chemical properties, uses, identification of primary, secondary and tertiary amines. Cyanides and Isocyanides - will be mentioned at relevant places in text. **Diazonium salts:** Preparation, chemical reactions and importance in synthetic organic chemistry.

Biomolecules: Carbohydrates - Classification (aldoses and ketoses), monosaccharides (glucose and fructose), D-L configuration oligosaccharides (sucrose, lactose, maltose), polysaccharides (starch, cellulose, glycogen); Importance of carbohydrates. Proteins -Elementary idea of - amino acids, peptide bond, polypeptides, proteins, structure of proteins - primary, secondary, tertiary structure and quaternary structures (qualitative idea only), denaturation of proteins; enzymes. Hormones - Elementary idea excluding structure. Vitamins - Classification and functions. **Nucleic Acids:** DNA and RNA.

Polymers: Copolymerization, some important polymers: natural and synthetic like polythene, nylon polyesters, bakelite, rubber. Biodegradable and nonbiodegradable polymers.

Chemistry in Everyday life: Chemicals in medicines - analgesics, tranquilizers antiseptics, disinfectants, antimicrobials, antifertility drugs, antibiotics, antacids, antihistamines. Chemicals in food - preservatives, artificial sweetening agents, elementary idea of antioxidants, Cleansing agents- soaps and detergents, cleansing action.

ENGLISH

The test of English Language for the entrance examination covers 'General English' of the 10 + 2 Level. It is designed to test student's ability to grasp the fundamental concepts in English language, grammar and usage. Proficiency in English language is tested through objective-type multiple-choice questions on grammar (tenses, subject verb agreement, articles, prepositions, use of question tags, sentence structures, and use of linkers), and vocabulary (synonyms, commonly confused pair of words, one-word substitutions, idioms and phrases, and usage).

Students are not required to learn the glossary of complicated grammatical jargon, but should have the primary knowledge of word usage, matching of subject and verbs, correct usage of tenses, and grammatically correct sentences acceptable in standard written communication.

The syllabus will include the following areas:

- Tense – Use of proper tense and sequence of tense
- Subject – verb concord
- Pronouns and Determiners / Use of proper articles / Prepositions
- Linkers: Appropriate use of linkers and conjunctions in a sentence
- Sentence structure: Clauses of condition and time, Relative clauses, Fragments
- Usages in English: standard usages / Idioms and phrases
- Synonyms / Question tags
- One-word substitutions
- Commonly confused pair of words
- Sentence completion

MANIPAL
Engineering Entrance Exam
Solved Paper 2024*

Physics

Multiple Choice Questions (MCQs)

1. Two pendulums of time periods 3 s and 7 s, respectively start oscillating simultaneously from two opposite extreme positions. After how much time they will be in same phase?

(a) $\frac{21}{8}$ s (b) $\frac{21}{4}$ s

(c) $\frac{21}{2}$ s (d) $\frac{21}{10}$ s

2. The stress *versus* strain graphs for wires of two materials *A* and *B* are as shown in the figure. If Y_A and Y_B are the Young's modulii of the materials, then

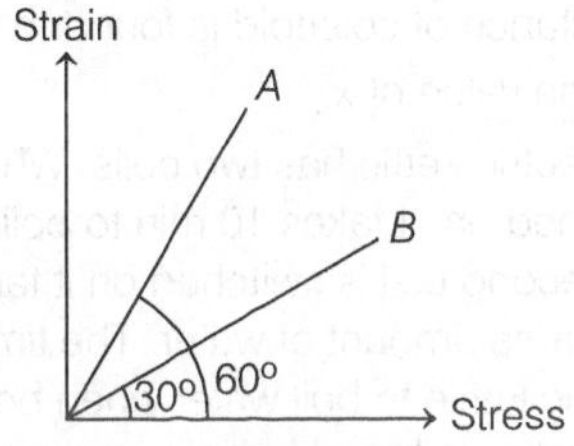

(a) $Y_B = 2Y_A$ (b) $Y_A = Y_B$

(c) $Y_B = 3Y_A$ (d) $Y_A = 3Y_B$

3. If the frequency of incident photon on metal surface is doubled, then stopping potential will become

(a) doubled

(b) less than double

(c) more than double

(d) less than existing value

4. In the following circuit diagram, potential difference across 4 μF capacitor is

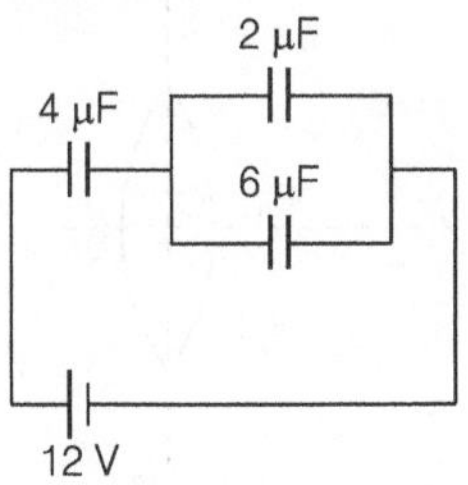

(a) 19 V (b) 14 V (c) 16 V (d) 8 V

5. Fundamental frequency of a sonometer wire is *n*, if the tension is made 3 times and length and diameter are also increased 3 times, what is the new frequency?

(a) $\frac{n}{3\sqrt{3}}$ (b) $3n$

(c) $\sqrt{3}\, n$ (d) $\frac{n}{\sqrt{3}}$ *

6. Two points move in the same straight line starting at the same moment from the same point in it. The first moves with constant velocity *u* and the second with constant acceleration *f*. During the time elapses before the second catches, the first greatest distance between the particle is

(a) $\frac{u}{f}$ (b) $\frac{u^2}{2f}$

(c) $\frac{f}{2u^2}$ (d) $\frac{f}{u^2}$

* *Based on memory*

7. A car turns on a banked road of radius 300 m. If the coefficient of friction between the tyres and the road is 0.3, then maximum speed of the car on the turn is
[Take, $g = 10\,m/s^2$]
(a) 10 m/s (b) 30 m/s
(c) 40 m/s (d) 50 m/s

8. If an electron in $n = 4$ orbit of hydrogen atom jumps down to $n = 3$ orbit, the amount of energy released and the wavelength of radiation emitted are
(a) 0.66 eV, 1.88×10^{6} m
(b) 1.89 eV, 1.98×10^{-7}m
(c) 0.29 eV, 1.78×10^{-5} m
(d) 0.98 eV, 0.93×10^{-6} m

9. When a lens is cut into two halves along XOX', then focal length of each half lens

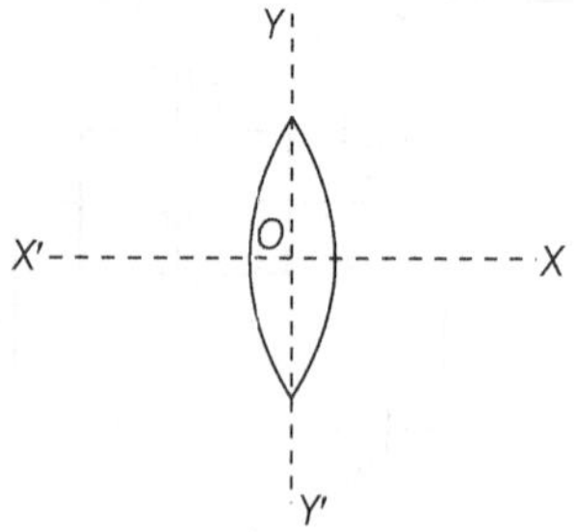

(a) increases
(b) decreases
(c) remains same
(d) None of the above

10. If E = energy, G = gravitational constant, I = impulse and M = mass, then dimensions of $\frac{GIM^2}{E^2}$ are same as that of
(a) time (b) mass
(c) length (d) force

Numerical Answer Type (NAT)

11. In the circuit shown below, the diode has 20 Ω in forward bias resistance. When V_i increases from 8 V to 12 V, then change in the current is found to be x mA. Find the value of x.

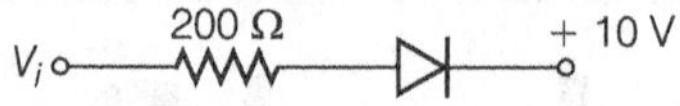

12. A particle is projected from the surface of the earth with an initial speed of 4.0 kms^{-1}. The maximum height (in km) attained by the particle is ....
[Radius of earth = 6400 km and $g = 9.8\ ms^{-2}$]

13. An electric dipole shown in the figure. Work done to move a charge particle of 1 μC from point Q to P is $x \times 10^{-7}$ J, then the value of x is ....

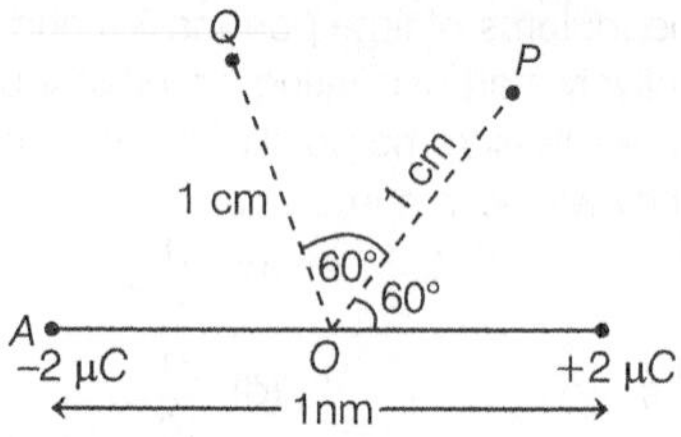

14. When 100 V DC is applied across a solenoid, a current of 1 A flows in it. When 100 V AC is applied across the same coil, the current drop is 0.5 A. If the frequency of the AC source is 50 Hz, then inductance of solenoid is found to be x mH, then find the value of x.

15. An electric kettle has two coils. When one coil is switched on, it takes 10 min to boil water and when the second coil is switched on it takes 20 min to boil same amount of water. The time taken by electric kettle to boil water when both coil are used in parallel, is found to be n second. Find the value of n.

Chemistry

Multiple Choice Questions (MCQs)

1. Consider the following amino acids.

(i) Lysine (ii) Glutamine
(ii) Arginine (iv) Leucine
(v) Serine (vi) Proline
(vii) Valine

Which of the given amino acids are basic in nature?

(a) (i) and (iii) (b) (i), (ii) and (iv)
(c) (iii) and (vii) (d) (iii), (v) and (vi)

2. The maximum work obtained from a reversible process is given as

(a) $-\Delta A$ (b) ΔA
(c) $-\Delta G$ (d) ΔG

3. Match List-I with List-II.

List-I (Compound)		List-II (Hybridisation)	
A.	$CuCl_5^{3-}$	I.	sp^3d^2
B.	$MnCl_5^{3-}$	II.	d^2sp^3
C.	$XeOF_4$	III.	dsp^3
D.	$Fe(CO)_5$	IV.	sp^3d

Choose the correct match from the options given below

(a) A-IV, B-III, C-I, D-II
(b) A-IV, B-III, C-II, D-I
(c) A-IV, B-I, C-III, D-II
(d) A-IV, B-II, C-III, D-I

4. The oxidation state of Cr in $[Cr(H_2O)_6]Cl_3$, $[Cr(C_6H_6)_2]$, $K_2[Cr(CN)_2(O)_2(O_2)(NH_3)]$ respectively are

(a) + 3, + 4, + 6 (b) + 3, + 2, + 4
(c) + 3, 0, + 6 (d) + 3, 0, + 4

5. Electronegativity of the following elements increases in the order

(a) C, N, Si, P (b) N, Si, C, P
(c) Si, P, C, N (d) P, Si, N, C

6. For the cell reaction,

$$Cu\ (s) + 2Ag^+(aq) \longrightarrow Cu^{2+}(aq) + 2Ag(s)$$

E°_{cell} is 0.46 V. The equilibrium constant of the reaction is

(a) 3.92×10^{12} (b) 3.92×10^{15}
(c) 8.92×10^{17} (d) 8.92×10^{10}

7. F_2 is formed by reacting K_2MnF_6 with

(a) SbF_5 (b) MnF_3
(c) $KSbF_6$ (d) MnF_4

8. Which of the following ion is colourless inspite of the presence of unpaired electrons?

(a) La^{3+} (b) Eu^{3+}
(c) Gd^{3+} (d) Lu^{3+}

9. In the following sequence of reactions,

$$CH_3—Br \xrightarrow{KCN} X \xrightarrow{H_3O^+} Y \xrightarrow[\text{Ether}]{LiAlH_4} Z$$

The final product Z is

(a) acetone
(b) methane
(c) acetaldehyde
(d) ethyl alcohol

10. The major product Y in the following reaction is

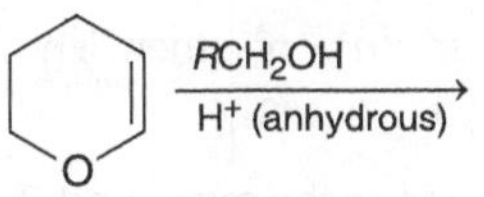

(a) hemiacetal (b) acetal
(c) an ether (d) an ester

Numerical Answer Type (NAT)

11. The wave number of the shortest wavelength of absorption spectrum of hydrogen atom is ...... (Rydberg constant = 109700 cm^{-1}).

12. If K_p for a reaction,

$$A(g) + 2B(g) \rightleftharpoons 3C(g) + D(g)$$

is 0.05 atm at 1000 K. Its K_c in term of R is $\frac{x \times 10^{-5}}{R}$.

The value of x is ....

13. Boiling point of water at 750 mm Hg is 99.63°C. The amount of sucrose that should be added to 500 g of water such that it boil at 100°C is .... g. (nearest integer)
(Molar elevation constant for water = 0.5 K kg/mol)

14. For a first order reaction, time required for 99% completion of a reaction is x times the time required for completion of 90%. The value of x is ....

15. The spin only magnetic moment of $[NiCl_4]^{2-}$ is .... (Nearest integer)

Mathematics

Multiple Choice Questions (MCQs)

1. If the number of terms in the expansion of $(x\sqrt{180} + \sqrt[3]{432})^{200}$ having integral coefficients is n, then the vaule of $\left[\frac{n}{6}\right]$ is (where $[\cdot]$ denotes greater integer function)

(a) 4 (b) 5 (c) 6 (d) 7

2. The solution of the equation $\log(\log_4(\sqrt{x+4} + \sqrt{x})) = 0$ is

(a) 2 (b) 4 (c) 6 (d) 8

3. If in a ΔABC, $\sin^2 A + \sin^2 B + \sin^2 C = 2$, then the triangle is always:

(a) isosceles triangle
(b) right angled
(c) acute angled
(d) obtuse angled

4. Let $D = \begin{vmatrix} n & n^2 & n^3 \\ n^2 & n^3 & n^5 \\ 1 & 2 & 3 \end{vmatrix}$, then $\lim\limits_{n \to \infty} \dfrac{M_{11} + C_{33}}{(M_{13})^2}$ is equal to (where M_{ij} is the minor and C_{ij} is the cofactor of ith row and j^{th} column's element)

(a) 0 (b) – 1 (c) – 2 (d) 3

5. $\dfrac{a}{b} = \dfrac{1}{3}$ and $\dfrac{b}{c} = \dfrac{3}{4}$, then the value of $\dfrac{a+2b}{b+2c}$ is

(a) $\dfrac{28}{33}$ (b) $\dfrac{7}{11}$
(c) $\dfrac{1}{2}$ (d) None of these

6. The equation of mirror image of the circle $x^2 + y^2 - 6x - 10y + 33 = 0$ about the line mirror $y = x$ is ......

(a) $x^2 + y^2 - 10x + 6y + 33 = 0$
(b) $x^2 + y^2 + 10x - 6y + 33 = 0$
(c) $x^2 + y^2 - 10x - 6y + 33 = 0$
(d) $x^2 + y^2 + 10x + 6y + 33 = 0$

7. Let $f : N \to N$ be defined as

$$f(n) = \begin{cases} \dfrac{n+1}{2}, & \text{if is } n \text{ odd} \\ \dfrac{n}{2}, & \text{if } n \text{ is even} \end{cases} \quad \text{for all } n \in N. \text{ Then } f \text{ is}$$

(a) injective but not surjective
(b) surjective but not injective
(c) both injective and surjective
(d) neither injective nor surjective

8. If $x = \sin(2\tan^{-1} 2)$, $y = \sin\left(\frac{1}{2}\tan^{-1}\frac{4}{3}\right)$, then

(a) $x = 1 - y$ (b) $x^2 = 1 - y$
(c) $x^2 = 1 + y$ (d) $y^2 = 1 - x$

9. In triangle ABC, $\sin A$, $\sin B$ and $\sin C$ are in AP $(C > 90°)$, then $\cos A$ is

(a) $\dfrac{3c - 4b}{2b}$ (b) $\dfrac{3c - 4b}{2c}$
(c) $\dfrac{4c - 3b}{2b}$ (d) $\dfrac{4c - 3b}{2c}$

10. A real valued differentiable function f satisfies $f(x) + f(y) + 2xy = f(x+y) \forall x, y \in R$. Given $f''(0) = 0$, then value of $\int_0^{\frac{\pi}{2}} f(\sin x) dx$ will be

(a) 0 (b) $\dfrac{\pi}{4}$ (c) $\dfrac{\pi}{2}$ (d) π

11. If two tangents drawn from the point (h, k) to the parabola $y^2 = 64x$ be such that the slope of one tangent is 8 times of the other, then the value of $\dfrac{k^2}{2n}$ is

(a) 9 (b) 27 (c) 81 (d) 162

12. If the angle between the pair of straight lines formed by joining the points of intersection of $x^2 + y^2 = 4$ and $y = 3x + c$ to the origin is a right angle, then c^2 is

(a) 20 (b) 13 (c) $\dfrac{1}{5}$ (d) 5

13. Total number of even divisions of 2079000 which are divisible by 15 are

(a) 54 (b) 128 (c) 108 (d) 72

14. If the coefficient of x^m, $m \in N$ in the expansion of $\left(\sqrt{2x} + \sqrt[3]{\dfrac{3}{x^2}}\right)^9$ is equal to k $(k \neq 0)$, the k is equal to

(a) 1008 (b) 2016
(c) 3024 (d) 1016

15. Let $f(x)$ is a polynomial function such that $f(x) + f\left(\dfrac{1}{x}\right) = f(x) \cdot f\left(\dfrac{1}{x}\right)$, $x > 0$ and $f(3) = 10$, if $\int f(x) dx = g(x) + c$ and $g(1) = \dfrac{4}{3}$, then the value of $g(3)$ is

(a) 10 (b) 9
(c) 8 (d) 12

16. If expression $x + \frac{1}{x}, |x > 0|$ attains is minimum value at $x = \alpha$, then α^3 is

17. If N denotes number of 8 digit numbers that contains exactly four nines, then unit digit of N is

18. Let $f(x) = \left[\frac{\sin x}{x}\right] + \left[\frac{2\sin x}{x}\right] + \dots + \left[\frac{10\sin x}{x}\right]$ (where $[\cdot]$ is the largest integer). then, find the value of $\lim_{x \to 0} f(x)$.

19. Given three vectors **a**, **b** and **c** each two of which are non collinear. Further if (**a** + **b**) is collinear with **c**, (**b** + **c**) is collinear with **a** and $|\mathbf{a}| = |\mathbf{b}| = |\mathbf{c}| = \sqrt{2}$. Then, the value of $|\mathbf{a}\cdot\mathbf{b} + \mathbf{b}\cdot\mathbf{c} + \mathbf{c}\cdot\mathbf{a}|$ is

20. Given $\frac{dy}{dx} + 2y \tan x = \sin x$ and $y = 0$ for $x = \frac{\pi}{3}$.
If maximum value of y is $\frac{1}{k}$, then 'k' is equal to

English

1. Fill in the blank with correct tense form.

He told me that he .... the movie.

(a) is finished
(b) was finished
(c) had finished
(d) not finished

2. Choose the sentence that follows the correct Subject Verb Agreement.

(a) One of the state in which Satyagraha were offered was Rajkot, where he has spent his youth.
(b) The average age at which people begins to need eye-glasses are increasing.
(c) The increase in consumption is directly proportional to the increase in income.
(d) The government have many new schemes for people who want to start businesses in rural areas.

3. Fill in the blanks with a suitable determiner.
I haven't .... butter but I have a lot of bread.

(a) much (b) a lot of
(c) little (d) some

4. Choose the appropriate conjunction from the options given below.

.... following all the instructions closely, he missed out an important guideline.

(a) Instead of (b) Although
(c) Inspite of (d) Otherwise

5. Determine the clause highlighted in the given sentence.

The bankers need to know **what they should do.**

(a) Noun clause (b) Adjective clause
(c) Adverb clause (d) Conditional clause

6. Choose the alternative which best expresses the meaning of the given underlined idiom/phrase.

Some colleagues at a workplace are **really hard nuts to crack.**

(a) hardcore traditions in a family.
(b) an inflexible person who is hard to convince.
(c) someone showing extra concern over little things.
(d) an easy-goer who can be easily convinced.

7. Choose the word that is nearest in meaning to the underlined word.

Ritu asked Rashmi not to **meddle** in her affairs.

(a) intercede (b) impose
(c) cross (d) interfere

8. Choose the word which can be substituted for the given word/ sentence.

Made in imitation so as to bepassed off fraudulently asgenuine

(a) Counterfeit (b) Staunch
(c) Authentic (d) Deliberate

9. We are ....to the true legacy of humanism, and we must never forget this.

(a) heirs (b) hair (c) hares (d) here

10. Fill in the blank with a suitable Question tag.
There's a cinema near the station,......... .

(a) aren't it (b) isn't there
(c) weren't it (d) isn't it

Answers

Physics

1. (a)	**2.** (c)	**3.** (c)	**4.** (d)	**5.** (a)	**6.** (b)	**7.** (b)	**8.** (a)	**9.** (c)	**10.** (a)
11. (9)	**12.** (935)	**13.** (1.8)	**14.** (550)	**15.** (400)					

Chemistry

1. (a)	**2.** (a)	**3.** (a)	**4.** (c)	**5.** (c)	**6.** (b)	**7.** (a)	**8.** (c)	**9.** (d)	**10.** (b)
11. (109700)	**12.** (5)	**13.** (127)	**14.** (2)	**15.** (3)					

Mathematics

1. (b)	**2.** (c)	**3.** (b)	**4.** (a)	**5.** (b)	**6.** (c)	**7.** (b)	**8.** (d)	**9.** (d)	**10.** (b)
11. (c)	**12.** (a)	**13.** (c)	**14.** (b)	**15.** (d)	**16.** (2)	**17.** (5)	**18.** (375)	**19.** (3)	**20.** (8)

English

1. (c)	**2.** (c)	**3.** (a)	**4.** (a)	**5.** (a)	**6.** (b)	**7.** (d)	**8.** (a)	**9.** (a)	**10.** (b)

SOLUTIONS

Physics

1. *(a)* General equation of oscillation are

$$y_1 = A\sin\left(\omega_1 t + \frac{\pi}{2}\right)$$

$$y_2 = A\sin\left(\omega_2 t + \frac{\pi}{2}\right)$$

Now, $\omega_1 t + \frac{\pi}{2} = \omega_2 t - \frac{\pi}{2}$

or $t = \frac{\pi}{\omega_2 - \omega_1}$

$$= \frac{\pi}{(2\pi/T_2) - (2\pi/T_1)}$$

$$= \frac{T_1 T_2}{2(T_1 - T_2)}$$

$$= \frac{3\times 7}{2(7-3)} = \frac{21}{8}\text{ s}$$

2. *(c)* As we know,

slope of strain – stress curve $= \frac{1}{Y}$

where, Y is Young's modulus.

or $\frac{(\text{Slope})_A}{(\text{Slope})_B} = \frac{Y_B}{Y_A} \Rightarrow \frac{\sqrt{3}}{\frac{1}{\sqrt{3}}} = \frac{Y_B}{Y_A}$

or $Y_B = 3Y_A$

3. *(c)* According to Einstein's photoelectric equation,

$$hf = W_0 + eV_0 \qquad \text{... (i)}$$

If frequency is doubled, then

$$h(2f) = W_0 + eV_0'$$

or $2hf = W_0 + eV_0'$ ...(ii)

Substituting hf from Eq. (i) in Eq. (ii), we have,

or $V_0' = 2V_0 + \frac{W_0}{e}$

or $V_0' > 2V_0$

4. *(d)* Equivalent capacitance of the given circuit diagram,

$$C_{eq} = \frac{(2+6)\times 4}{2+6+4} = \frac{32}{12} = \frac{8}{3}\mu\text{F}$$

Charge flowing through 4 μF capacitor

$$q = C_{eq}\cdot V$$

$$= \frac{8}{3}\times 10^{-6}\times 12 = 32\,\mu\text{C}$$

∴ Potential difference across 4 μC capacitor,

$$V = \frac{q}{C} = \frac{32}{4} = 8\text{ V}$$

5. *(a)* Fundamental frequency,

$$n = \frac{v}{2l} = \frac{\sqrt{T/\mu}}{2l} = \frac{\sqrt{T/\rho S}}{2l} = \frac{\sqrt{T/\rho\pi r^2}}{2l}$$

$$\therefore \quad n \propto \frac{\sqrt{T}}{2l}$$

Given that tension, diameter and length all are made 3 times.

∴ n will become $\frac{1}{3\sqrt{3}}$ times.

6. *(b)* The greatest distance, when velocities of both are equal.

or $ft = u$

$$\therefore \quad t = \frac{u}{f}$$

$$s_1 = ut = \frac{u^2}{f}$$

and $s_2 = \frac{1}{2}ft^2 = \frac{u^2}{2f}$

$$\therefore \quad s_{max} = s_1 - s_2 = \frac{u^2}{2f}$$

7. *(b)* Given, $r = 300\text{ m}, \mu = 0.3,$

$g = 10\text{ m/s}^2$

If v is the velocity of the vehicle while turning on the road, then

$$\frac{mv^2}{r} \le \mu mg$$

$$\Rightarrow \quad v \le \sqrt{\mu rg}$$

$$\Rightarrow \quad v_{max} = \sqrt{\mu rg}$$

$$= \sqrt{0.3\times 300\times 10} = 30\text{ m/s}$$

8. *(a)* $E = \frac{-13.6}{n^2}\text{ eV}$

$$E_4 - E_3 = 13.6\left[\frac{1}{(3)^2} - \frac{1}{(4)^2}\right]$$

$$= 13.6\left[\frac{7}{144}\right]\text{eV} = 0.66\text{ eV}$$

$$\lambda = \frac{1}{R\left[\frac{1}{(3)^2} - \frac{1}{(4)^2}\right]} = \frac{144}{1.09 \times 10^7 \times 7}\text{ m}$$

$$= 1.88 \times 10^{-6}\text{ m}$$

9. *(c)* If R be the magnitude of radius of curvature of equiconvex lens, then its focal length f is given by

$$\frac{1}{f} = (\mu - 1)\left(\frac{1}{R_1} - \frac{1}{R_2}\right)$$

$$= (\mu - 1)\left(\frac{1}{R} - \frac{1}{-R}\right) = \frac{2(\mu - 1)}{R}$$

$$\Rightarrow \quad \frac{1}{f} = \frac{2(\mu - 1)}{R} \qquad \text{...(i)}$$

When equiconvex lens is cut into two halves along XOX', then

$$R_1 = R, R_2 = -R$$

∴ Focal length f' is given as

$$\frac{1}{f'} = (\mu - 1)\left(\frac{1}{R} - \frac{1}{-R}\right) = \frac{2(\mu - 1)}{R} = \frac{1}{f} \qquad \text{[from (i)]}$$

$$\Rightarrow \quad f' = f$$

10. *(a)* The SI unit in which G is measured = $\text{N-m}^2/\text{kg}^2$

∴ Its dimensions will be $\frac{[MLT^{-2}][L^2]}{[M^2]} = [M^{-1}L^3T^{-2}]$

∵ Impulse = Force × Time = $[MLT^{-2}][T] = [MLT^{-1}]$

∵ Dimensions for energy = $[ML^2T^{-2}]$

Now put the dimensions of G, I, M and E in $\frac{GIM^2}{E^2}$

$$= \frac{[M^{-1}L^3T^{-2}][MLT^{-1}][M^2]}{[ML^2T^{-2}]^2}$$

$$= \frac{[M^{-1}L^3T^{-2}][MLT^{-1}][M^2]}{[M^2L^4T^{-4}]} = \frac{[M^2L^4T^{-3}]}{[M^2L^4T^{-4}]}$$

$$= [T^1] = [T]$$

The equation $\frac{GIM^2}{E^2}$ has dimension of time.

11. *(9)* Diode will be in forward bias when V_i is greater than 10 V and it conducts till V_i is 12 V.

Total circuit resistance

$= 200 + 20 = 220\,\Omega$

When V_i is less than 10 V, the diode will not conduct due to its reverse biasing. Thus, initial current, $I_1 = 0$.

When $V_i = 12\text{V}$, then diode will be in forward bias with effective voltage, $V' = 12 - 10 = 2\text{V}$

∴ Maximum current, $I_2 = \frac{V'}{R} = \frac{2}{220} = 0.009\text{A} = 9\text{ mA}$

∴ Change in current, $\Delta I = I_2 - I_1 = 9\text{ mA}$

12. *(935)* Using conservation of mechanical energy,

Decrease in kinetic energy = Increase in gravitational potential energy

$$\Rightarrow \quad \frac{1}{2}mv^2 = \Delta U$$

$$\Rightarrow \quad \frac{1}{2}mv^2 = \frac{mgh}{1 + \frac{h}{R}}$$

$$\Rightarrow \quad h = \frac{v^2}{2g - \frac{v^2}{R}}$$

Substituting the values, we get

$$h = \frac{(4.0 \times 10^3)^2}{2 \times 9.8 - \frac{(4 \times 10^3)^2}{6.4 \times 10^6}}$$

$$= 9.35 \times 10^5\text{ m} = 935\text{ km}$$

13. *(1.8)* From figure, electric dipole moment,

$$p = 2ql = q(2l)$$

$$= 2 \times 10^{-6} \times (1 \times 10^{-9})$$

$$\Rightarrow \quad p = 2 \times 10^{-15}\text{ cm}$$

Electric potential at point p due to dipole

$$V_p = \frac{1}{4\pi\varepsilon_0} \cdot \frac{p\cos\theta}{r^2}$$

$$V_p = 9 \times 10^9 \times \frac{2 \times 10^{-15} \times \cos 60^\circ}{(1 \times 10^{-2})^2}$$

$$= \frac{18 \times 10^{-6} \times \frac{1}{2}}{10^{-4}} = 9 \times 10^{-2} = 0.09\text{ V}$$

Similarly, electric potential at point Q due to electric dipole

$$V_Q = \frac{9 \times 10^9 \times (2 \times 10^{-15}) \times \cos 120^\circ}{(1 \times 10^{-2})^2}$$

$$= -0.09\text{ V}$$

∴ Potential difference between point Q and P.

$$V_{QP} = V_P - V_Q = 0.09 - (-0.09) = 0.18\text{ V}$$

∴ Work done to move the charge of 1 μC from point Q to P.

$$W = V_{QP} \cdot q = 0.18 \times 1 \times 10^{-6}$$

$$= 1.8 \times 10^{-7}\text{ J}$$

14. *(550)* When DC is applied across a solenoid, then

resistance, i.e. $R = \frac{V}{I}$

$$R = \frac{100}{1} = 100\,\Omega$$

When AC is applied, then

$$Z = \frac{100}{0.5} = 200\,\Omega$$

Net impedance,

$$Z = \sqrt{R^2 + X_L^2}$$

$$200 = \sqrt{(100)^2 + X_L^2}$$

$$\Rightarrow \quad (200)^2 = (100)^2 + X_L^2$$

$$\Rightarrow \quad 30000 = X_L^2 \Rightarrow X_L = 173.205\,\Omega$$

Now, $X_L = \omega L$

$$\Rightarrow \quad 173.205 = 2\pi f L \Rightarrow \frac{173.205}{2\pi \times f} = L$$

$$\Rightarrow \quad \frac{173.205}{2 \times 3.14 \times 50} = L \Rightarrow 0.55\,\Omega = L$$

$$L = 0.55\text{ H} = 550\text{ mH}$$

15. *(400)* Heat produced in a resistance R in time t is given by

$$H = \frac{V^2}{R}t \text{ [where, } V = \text{source voltage]}$$

For first coil,

$$H_1 = \frac{V^2}{R_1}t_1 = \frac{V^2}{R_1} \times 10 \times 60 \qquad [\because t_1 = 10 \times 60\text{s}]$$

For second coil, $H_2 = \dfrac{V^2}{R_2}t_2 \qquad [\because t_2 = 20 \times 60\text{s}]$

$$= \frac{V^2}{R_2} \times 20 \times 60$$

Given that, $H_1 = H_2$

$$\Rightarrow \quad \frac{V^2}{R_1} \times 10 \times 60 = \frac{V^2}{R_2} \times 20 \times 60$$

$$\Rightarrow \quad R_2 = 2R_1 \qquad \text{...(i)}$$

Let t be the time taken by electric kettle to boil water when both coil are used in parallel, then

$$H_P = \left(\frac{V^2}{R_1} + \frac{V^2}{R_2}\right) \times t = V^2 t \left(\frac{1}{R_1} + \frac{1}{2R_1}\right) \text{ [from Eq. (i)]}$$

$$H_P = \frac{3V^2 t}{2R_1}$$

But, $H_P = H_1$

$$\therefore \quad \frac{3V^2 t}{2R_1} = \frac{V^2}{R_1} \times 10 \times 60$$

$$\Rightarrow \quad t = 400\text{ s}$$

Chemistry

1. *(a)* Among the given amino acids, only two i.e. lysine and arginine are basic amino acid. While all other are neutral amino acid.

2. *(a)* Decrease in the value of Gibbs energy $(-\Delta G)$ gives the maximum useful work or net work done by the system for the given change whereas decrease in Helmholtz free energy or work function $(-\Delta A)$ gives the maximum work done for the given change.

3. *(a)* The correct match is A-IV, B-III, C-I, D-II.
Hybridisation involved in the given complexes are
$CuCl_5^{3-}$, Cu(II) = sp^3d $(4s, 4p, 4d)$
$MnCl_5^{3-}$, Mn(II) = dsp^3 $(3d, 4s, 4p)$
$XeOF_4$, Xe(VI) = sp^3d^2 $(5s, 5p, 5d)$
$Fe(CO)_5$, Fe(0) = d^2sp^3

4. *(c)* Oxidation state of Cr,

In $[Cr(H_2O)_6]Cl_3$

$$x + 0 \times 6 - 1 \times 3 = 0$$

$$x = 3 \qquad \text{...(i)}$$

In $[Cr(C_6H_6)_2]$

$$x + 0 \times 2 = 0$$

$$x = 0 \qquad \text{...(ii)}$$

In $K_2[Cr(CN)_2(O)_2(O_2)(NH_3)]$

$$(1 \times 2) + x - (1 \times 2) - (2 \times 2) - (1 \times 2) - (0 \times 1) = 0$$

$$x - 6 = 0$$

$$x = +6 \qquad \text{...(iii)}$$

Therefore, oxidation states are + 3, 0, + 6.

5. *(c)* As electronegativity increases across the period, therefore, electronegativity of nitrogen (N) is more than carbon (C) and electronegativity of P is more than Si.
Also electronegativity decreases down the group, therefore electronegativity of carbon is more than Si and electronegativity of N is more than P.
Hence, the correct increasing order will be Si, P, C, N.

6. *(b)* $Cu(s) + 2Ag^+(aq) \longrightarrow Cu^{2+}(aq) + 2Ag(s)$

$n = 2$, $E^\circ_{cell} = 0.46$ V

Now using the equation for E°_{cell}

$$E^\circ_{cell} = \frac{0.0591}{n}\log K_C$$

$$0.46 = \frac{0.0591}{n}\log K_C$$

$$K_C = \text{Antilog } 15.5668$$

$$K_C = 3.92 \times 10^{15}$$

7. *(a)* K_2MnF_6 react with SbF_5 to form F_2. In this reaction stronger Lewis acid SbF_5 displace the weaker one (MnF_4) from its salt. MnF_4 being unstable, readily decompose to give MnF_3 and fluorine.

The complete reaction is as follows,

$$K_2MnF_6 + 2SbF_5 \longrightarrow 2KSbF_6 + MnF_4 \text{ (unstable)}$$
$$2MnF_4 \longrightarrow 2MnF_3 + F_2$$

Overall reaction,

$$2K_2MnF_6 + 4SbF_5 \longrightarrow 4KSbF_6 + 2MnF_3 + F_2$$

8. *(c)* Gadolinium does not absorb light in the visible region of spectrum, hence it is colourless inspite of the presence of seven unpaired electrons.

9. *(d)* The complete reaction is as follows,

$$CH_3—Br \xrightarrow[—KBr]{KCN} \underset{\substack{(X)\\ \text{Acetonitrile}}}{CH_3—CN} \xrightarrow[\text{Hydrolysis}]{H_3O^+} \underset{\substack{(Y)\\ \text{Acetic acid}}}{CH_3COOH}$$

$$\xrightarrow[\text{Reduction}]{LiAlH_4/\text{ether}} \underset{\substack{(Z)\\ \text{Ethyl alcohol}}}{CH_3CH_2OH}$$

10. *(b)* Dihydropyran reacts with alcohols in presence of anhydrous H^+ to form acetal.

The complete reaction is as follows

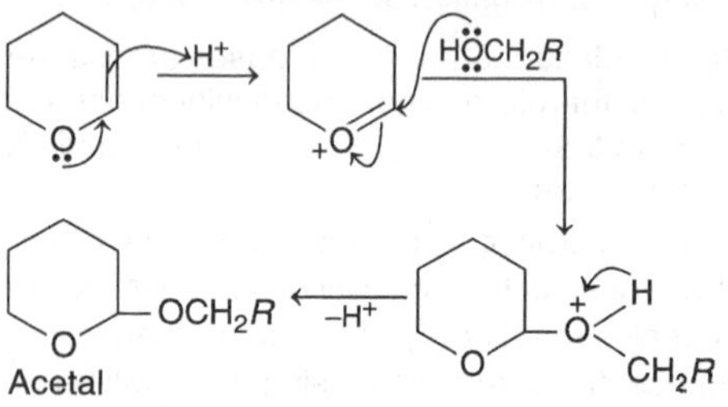

11. *(109700)*

Given, $R = 109700\,cm^{-1}$

Shortest wavelength mean transition is from infinity to $n = 1$.

Using the formula,

$$\bar{\nu} = R\left[\frac{1}{n_f^2} - \frac{1}{n_i^2}\right]$$

$$n_f = \infty, n_i = n = 1$$

$$\bar{\nu} = 109700\left[\frac{1}{1} - \frac{1}{\infty}\right]$$

$$\bar{\nu} = 109700\,cm^{-1}$$

12. *(5)* Given reaction,

$$A(g) + 2B(g) \rightleftharpoons 3C(g) + D(g)$$
$$\Delta n_g = 4 - 3 = 1$$

Formula for K_p,

$$K_p = K_C(RT)^{\Delta n}$$
$$0.05 = K_C(R \times 1000)^1$$
$$K_C = \frac{5 \times 10^{-5}}{R}$$

13. *(127)*

Elevation in boiling point of water required (ΔT_b).

$$= 100 - 99.63 = 0.37°C$$

Mass of solvent (water), $W_1 = 500\,g$

Molar mass of solvent, $M_1 = 18\,g/mol$

Molar mass of solute (sucrose) $(C_{12}H_{22}O_{11}) = 342\,g/mol$

$$M_2 = \frac{1000 \times K_b \times w_2}{w_1 \Delta T_b}$$

$$\text{or } w_2 = \frac{M_2 \times W_1 \times \Delta T_b}{1000 \times K_b} = \frac{342 \times 500 \times 0.37}{1000 \times 0.5}$$

$$w_2 = 126.54 \approx 127$$

14. *(2)*

$$t_{99\%} = \frac{2.303}{k} \log \frac{(a)}{a - 0.99a}$$

$$= \frac{2.303}{k} \log 10^2 \Rightarrow 2 \times \frac{2.303}{k} \quad \text{...(i)}$$

$$t_{90\%} = \frac{2.303}{k} \log \frac{a}{a - 0.90a}$$

$$= \frac{2.303}{k} \log 10 = \frac{2.303}{k} \quad \text{...(ii)}$$

Take the ratio of (i) and (ii)

$$\frac{t_{99\%}}{t_{90\%}} = 2$$

$$t_{99\%} = 2 \times t_{90\%}$$

Hence, 2 times is the final answer.

15. *(3)* Ni : $3d^8 4s^2$

for the compound $[NiCl_4]^{2-}$, Ni is +2 oxidation state. Hence, the electronic configuration of Ni^{2+} is given by

Ni^{2+} : $3d^8$

Since, Cl^- is a weak field ligand, so it will not cause pairing of electrons. So, for Ni^{2+} the configuration is

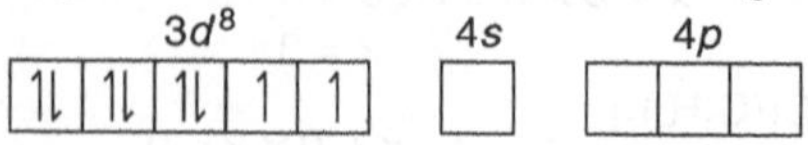

Hence, $n = 2$

Spin only magnetic moment is given by

$$\mu_s = \sqrt{n(n+2)} = \sqrt{2(2+2)}\text{ BM} = \sqrt{8} = 2.82\text{ BM}$$

$$\mu_s \simeq 3\text{ B M}$$

Mathematics

1. *(b)* $T_{r+1} = {}^{200}C_r (x \cdot \sqrt{180})^{200-r} \cdot (\sqrt[3]{432})^r$

$$= {}^{200}C_r\, 2^{200-r} \cdot 3^{200-r} \cdot 5^{\frac{200-r}{2}} \cdot x^{200-r}\, 2^r 3^r 2^{\frac{r}{3}}$$

$\therefore$ Coefficient $= {}^{200}C_r 2^{200} \cdot 3^{200} \cdot 5^{100} \cdot \dfrac{2^{\frac{r}{3}}}{5^{\frac{r}{2}}}$

r must be multiple of 6 and takes values between 0 to 200.

$\therefore$ 34 terms $= n$

$$\left[\frac{n}{6}\right] = \left[\frac{34}{6}\right] = 5$$

2. *(c)* $\because \log(\log_4(\sqrt{x+4} + \sqrt{x})) = 0$

$\Rightarrow \log_4(\sqrt{x+4} + \sqrt{x}) = 1$

$\Rightarrow \sqrt{x+4} + \sqrt{x} = 4^1 = 4$

From the given options, $x = 6$ satisfy the equation.

3. *(b)* $\sin^2 A + \sin^2 B + \sin^2 C = 2$

$= \dfrac{1-\cos 2A}{2} + \dfrac{1-\cos 2B}{2} + \dfrac{1-\cos 2C}{2} = 2$

$\Rightarrow \cos 2A + \cos 2B + \cos 2C = -1$

$\Rightarrow 2\cos(A+B)\cos(A-B) = -2\cos^2 C$

$\Rightarrow 2\cos(\pi - C)\cos(A-B) = -2\cos^2 C$ $\quad [\because A + B + C = \pi]$

$\Rightarrow 2\cos C[\cos C - \cos(A-B)] = 0$

$\Rightarrow -2\cos C[\cos(A+B) + \cos(A-B)] = 0$

$\Rightarrow -4\cos A \cdot \cos B \cdot \cos C = 0$

$\Rightarrow \cos A \cdot \cos B \cdot \cos C = 0$

Therefore, one of them should be equal to zero.

So, exactly one angle is 90°.

4. *(a)* $\because D = \begin{vmatrix} n & n^2 & n^3 \\ n^2 & n^3 & n^5 \\ 1 & 2 & 3 \end{vmatrix}$

Then, $M_{11} = \begin{vmatrix} n^3 & n^5 \\ 2 & 3 \end{vmatrix} = 3n^3 - 2n^5$

and $C_{33} = \begin{vmatrix} n & n^2 \\ n^2 & n^3 \end{vmatrix} = n^4 - n^4 = 0$

$M_{13} = \begin{vmatrix} n^2 & n^3 \\ 1 & 2 \end{vmatrix} = 2n^2 - n^3$

$$\therefore \lim_{n\to\infty} = \frac{M_{11} + C_{33}}{(M_{13})^2} = \lim_{n\to\infty} \frac{n^3(3-2n^2)+0}{n^4(2-n)^2}$$

$$= \lim_{n\to\infty} \frac{3-2n^2}{n[4+n^2-4n]} = \lim_{n\to\infty} \frac{n^2\left(\dfrac{3}{n^2} - 2\right)}{n^3\left(\dfrac{4}{n^2} + 1 - \dfrac{4}{n}\right)}$$

put $n = \infty$

$$= \frac{\left(\dfrac{3}{\infty^2} - 2\right)}{\infty\left(\dfrac{4}{\infty^2} + 1 - \dfrac{4}{\infty}\right)} = \frac{-2}{1 \times \infty} = 0$$

5. *(b)* $\because \dfrac{a}{b} = \dfrac{1}{3} \Rightarrow b = 3a$ and $\dfrac{b}{c} = \dfrac{3}{4} \Rightarrow \dfrac{3a}{c} = \dfrac{3}{4}$

$\Rightarrow \dfrac{a}{c} = \dfrac{1}{4} \Rightarrow c = 4a$

Then, $\dfrac{a+2b}{b+2c} = \dfrac{a+2(3a)}{(3a)+2(4a)}$

$$= \frac{a+6a}{3a+8a} = \frac{7a}{11a} = \frac{7}{11}$$

6. *(c)* Given circle is $x^2 + y^2 - 6x - 10y + 33 = 0$

Centre of the circle $= (3, 5)$

Radius of required circle is same as of given circle

$= \sqrt{(3)^2 + (5)^2 - 33} = 1$

Mirror image of centre about $y = x$ is $(5, 3)$

Hence, Required equation $(x-5)^2 + (y-3)^2 = 1$

$\Rightarrow x^2 + y^2 - 10x - 6y + 33 = 0$

7. *(b)* $\because \quad f(n) = \begin{cases} \dfrac{n+1}{2}, & n \text{ is odd} \\ \dfrac{n}{2}, & n \text{ is even} \end{cases}$

$\therefore \quad f(1) = \dfrac{1+1}{2} = 1$

and $\quad f(2) = \dfrac{2}{2} = 1$

$\because \quad f(1) = f(2) = 1$

So, f is not injective

and Range of $f(n)$ is all natural numbers.

So, it is surjective

Hence, f is surjective but not injective.

8. *(d)* $\because \sin(2\tan^{-1} 2) = 2\sin(\tan^{-1} 2) \cdot \cos(\tan^{-1} 2)$

$$= 2 \times \left(\frac{2}{\sqrt{5}}\right)\left(\frac{1}{\sqrt{5}}\right)$$

$$\left[\because \tan^{-1} 2 = \sin^{-1}\left(\frac{2}{\sqrt{5}}\right) = \cos^{-1}\left(\frac{1}{\sqrt{5}}\right)\right]$$

$$= \frac{4}{5}$$

$\Rightarrow \quad x = \dfrac{4}{5}$

and $\quad y = \sin\left(\dfrac{1}{2}\tan^{-1}\dfrac{4}{3}\right)$

$\Rightarrow \quad y^2 = \sin^2\left(\frac{1}{2}\tan^{-1}\frac{4}{3}\right)$

$= \frac{1-\cos\left(\tan^{-1}\frac{4}{3}\right)}{2}$

$\Rightarrow \quad y^2 = \frac{1-\frac{3}{5}}{2} \qquad \left[\because \tan^{-1}\frac{4}{3} = \cos^{-1}\frac{3}{5}\right]$

$\Rightarrow \quad y^2 = \frac{1}{5} = 1 - \frac{4}{5} = 1 - x$

Hence, $y^2 = 1 - x$

9. *(d)* As we know,

$$\frac{\sin A}{a} = \frac{\sin B}{b} = \frac{\sin C}{c}$$

and $\sin A$, $\sin B$ and $\sin C$ are in AP

So, a, b, c will also be in AP

$\therefore \quad 2b = a + c$

and $\cos A = \frac{b^2 + c^2 - a^2}{2bc} = \frac{b^2 + c^2 - (2b - c)^2}{2bc}$

$= \frac{b^2 + c^2 - (4b^2 + c^2 - 4bc)}{2bc}$

$= \frac{-3b^2 + 4bc}{2bc}$

$= \frac{b(4c - 3b)}{2bc} = \frac{4c - 3b}{2c}$

10. *(b)* $f(x) + f(y) + 2xy = f(x + y)$ ...(i)

differentiating the equation,

$\Rightarrow \quad f'(y) + 2x = f'(x + y)$

at $y = 0$

$f'(0) + 2x = f'(x)$

$\Rightarrow \quad f'(x) = 2x$

Integrating both sides.

$\int f'(x) = \int 2x \cdot dx$

$\Rightarrow \quad f(x) = x^2$

$\therefore \quad I = \int_0^{\frac{\pi}{2}} f(\sin x)\, dx$

$\Rightarrow \quad I = \int_0^{\frac{\pi}{2}} \sin^2 x \cdot dx$ ...(ii)

$\Rightarrow \quad I = \int_0^{\frac{\pi}{2}} \sin^2\left(\frac{\pi}{2} - x\right) dx = \int_0^{\frac{\pi}{2}} \cos^2 x \cdot dx$ ...(iii)

Adding Eqs. (ii) and (iii)

$2I = \int_0^{\frac{\pi}{2}} (\sin^2 x + \cos^2 x) dx = \int_0^{\frac{\pi}{2}} 1 \cdot dx$

$\Rightarrow \quad 2I = \frac{\pi}{2}$

$\Rightarrow \quad I = \frac{\pi}{4}$

11. *(c)* Let the equation of tangent is

$y = mx + \frac{16}{m} \qquad [\because a = 16]$

It passes through (h, k)

Then, $k = mh + \frac{16}{m} \Rightarrow m^2h + 16 - mk = 0$

According to the question,

roots of the equation will be m_1 and $8m_1$.

So, $m_1 + 8m_1 = \frac{k}{h} \Rightarrow 9m_1 = \frac{k}{h}$ ...(i)

and $m_1 \cdot 8m_1 = \frac{16}{h} \Rightarrow 8m_1^2 = \frac{16}{h}$ ...(ii)

By solving these equations, we get

$$\frac{k^2}{2h} = 81$$

12. *(a)* $\because x^2 + y^2 = 4 \Rightarrow x^2 + y^2 - 4(1)^2 = 0$ ...(i)

and $y = 3x + c \Rightarrow \frac{y - 3x}{c} = 1$

Put the value of '1' in Eq. (i)

$\Rightarrow \quad x^2 + y^2 - 4\left(\frac{y - 3x}{c}\right)^2 = 0$

$\Rightarrow \quad x^2 + y^2 - 4\left(\left(\frac{y}{c}\right)^2 + \left(\frac{3x}{c}\right)^2 - \frac{6xy}{c^2}\right) = 0$

$\Rightarrow \quad x^2\left(1 - \frac{36}{c^2}\right) + y^2\left(1 - \frac{4}{c^2}\right) + \frac{24xy}{c^2} = 0$

According to the question,

$\left(1 - \frac{36}{c^2}\right) + \left(1 - \frac{4}{c^2}\right) = 0 \Rightarrow 2 = \frac{40}{c^2}$

$\Rightarrow \quad c^2 = 20$

13. *(c)* $\because 2079000 = 2^3 \times 3^3 \times 5^3 \times 7 \times 11$

for the divisor to be even and divisible by 15; 2, 3 and 5 must be occur once.

There, the total number of required divisors are $3 \times 3 \times 3 \times 2 \times 2 = 108$

14. *(b)* $\because T_{r+1} = {}^9C_r (2x)^{\frac{9-r}{2}} \left(\frac{3}{x^2}\right)^{\frac{r}{3}}$

$= {}^9C_r\, 2^{\frac{9-r}{2}} \cdot 3^{\frac{r}{3}} \cdot x^{\left(\frac{9-r}{2} - \frac{2r}{3}\right)}$

$= {}^9C_r\, 2^{\frac{9-r}{2}} \cdot 3^{\frac{r}{3}} \cdot x^{\frac{27-7r}{6}}$

So, $\quad r = 3$

Then, coefficient $= {}^9C_3 \cdot 2^{\frac{9-3}{2}} \cdot 3^1$

$= \frac{9 \cdot 8 \cdot 7}{3 \cdot 2} \times 2^3 \times 3$

$= 2016$

15. *(d)* Let, $f(x) = 1 + x^n$ or $f(x) = 1 - x^n$

$\because \quad f(3) = 10$

So, $\quad f(x) = 1 + x^2$

Then, $\int f(x)dx = \int (1 + x^2)dx = x + \dfrac{x^3}{3} + c$

$\therefore \quad g(x) = x + \dfrac{x^3}{3} + c$

$\because \quad g(1) = \dfrac{4}{3}$

$\Rightarrow \quad \dfrac{4}{3} = 1 + \dfrac{1}{3} + c$

$\Rightarrow \quad C = 0$

Therefore,

$$g(x) = x + \frac{x^3}{3}$$

and $\quad g(3) = 3 + \dfrac{3^3}{3} = 12$

16. *(2)* $\because \quad x + \dfrac{1}{x^2} = \dfrac{x}{2} + \dfrac{x}{2} + \dfrac{1}{x^2}$

and AM $\geq$ GM

$\Rightarrow \quad \dfrac{x}{2} + \dfrac{x}{2} + \dfrac{1}{x^2} \geq \left(\dfrac{x}{2} \times \dfrac{x}{2} \times \dfrac{1}{x^2}\right)^{\frac{1}{3}}$

$\Rightarrow \quad \dfrac{x}{2} + \dfrac{x}{2} + \dfrac{1}{x^2} \geq \left(\dfrac{1}{4}\right)^{\frac{1}{3}}$

Minimum value of $x + \dfrac{1}{x^2} = \sqrt[3]{\dfrac{1}{4}}$

and for minimum value

$$\frac{x}{2} = \frac{1}{x^2} \Rightarrow x^3 = 2$$

Hence, $\alpha^3 = 2$

17. *(5)* N = Number of 8 digit number with exactly four 9.

$\Rightarrow N$ = Number with [{1st digit is 9} + {1st digit is not 9}]

$= {}^7C_3 \cdot 9^4 + 8 \cdot {}^7C_4 \cdot 9^3$

$= 9^3\left(9 \cdot \dfrac{7 \cdot 6 \cdot 5}{3 \cdot 2 \cdot 1} + 8 \cdot \dfrac{7 \cdot 6 \cdot 5}{3 \cdot 2 \cdot 1}\right)$

$= 729 \times (315 + 280) = 729 \times 595$

Hence, unit digit of N is 5.

18. *(375)* As we know,

$$\lim_{x \to 0} \frac{\sin x}{x} = 1$$

but it is actually less than 1.

So, $\lim_{x \to 0}\left[\dfrac{\sin x}{x}\right] = 0 = 1^2 - 1$

Similarly, $\lim_{x \to 0}\left[2\dfrac{\sin 2x}{x}\right] = \lim_{x \to 0}\left[4\dfrac{\sin 2x}{2x}\right]$

$= 3 = 2^2 - 1$

Hence, $\quad \lim_{x \to 0}\left[n\dfrac{\sin nx}{x}\right] = n^2 - 1$

Therefore,

$\lim_{x \to 0} f(x) = (1^2 - 1) + (2^2 - 1) + \ldots + (10^2 - (1)$

$= (1^2 + 2^2 + \ldots + 10^2) - (1 + 1 + \ldots + 1 \text{ times})$

$= \dfrac{10 \times 11 \times 21}{6} - 10$

$= 385 - 10 = 375$

19. *(3)* Let, $\mathbf{a} + \mathbf{b} = \lambda\mathbf{c}$ ...(i)

and $\mathbf{b} + \mathbf{c} = \mu\mathbf{a}$ ...(ii)

Subtracting Eq. (ii) from Eq. (i),

$$\mathbf{a} - \mathbf{c} = \lambda\mathbf{c} - \mu\mathbf{a}$$

$\Rightarrow \quad (1 + \mu)\mathbf{a} = (1 + \lambda)\mathbf{c}$

$\because$ **a** and **c** are non-collinear.

So, $1 + \mu = 0$ and $1 + \lambda = 0$

$\Rightarrow \lambda = \mu = -1$

Therefore, $\mathbf{a} + \mathbf{b} = -\mathbf{c}$

$\Rightarrow \mathbf{a} + \mathbf{b} + \mathbf{c} = 0 \Rightarrow |\mathbf{a} + \mathbf{b} + \mathbf{c}|^2 = 0$

$\Rightarrow \ |\mathbf{a}|^2 + |\mathbf{b}|^2 + |\mathbf{c}|^2 + 2(\mathbf{a} \cdot \mathbf{b} + \mathbf{b} \cdot \mathbf{c} + \mathbf{c} \cdot \mathbf{a}) = 0$

$\Rightarrow \ 2 + 2 + 2 + 2(\mathbf{a} \cdot \mathbf{b} + \mathbf{b} \cdot \mathbf{c} + \mathbf{c} \cdot \mathbf{a}) = 0$

$\Rightarrow \ \mathbf{a} \cdot \mathbf{b} + \mathbf{b} \cdot \mathbf{c} + \mathbf{c} \cdot \mathbf{a} = -3$

Hence, $|\mathbf{a} \cdot \mathbf{b} + \mathbf{b} \cdot \mathbf{c} + \mathbf{c} \cdot \mathbf{a}| = 3$

20. *(8)* $\dfrac{dy}{dx} + 2y \tan x = \sin x$

The given differential equation is of the form

$\dfrac{dy}{dx} + Py = Q$

By comparing, we get

$P = 2 \tan x$ and $Q = \sin x$

$\therefore$ I.F. $= e^{\int 2 \cdot \tan \cdot dx} = e^{2 \log \sec x} = e^{\log \sec^2 x} = \sec^2 x$

Then, solution of the differential equation

$$y \sec^2 x = \int \sin x \cdot \sec^2 x \cdot dx + c$$

$\Rightarrow \quad y \sec^2 x = \int \dfrac{\sin x}{\cos^2 x} dx + c$

$\Rightarrow \quad y \sec^2 x = \int \tan x \cdot \sec x \, dx + c$

$\Rightarrow \quad y \sec^2 x = \sec x + c$

$\Rightarrow \quad y = \dfrac{1}{\sec x} + \dfrac{c}{\sec^2 x} = \cos x + (\cos^2 x) \cdot c$

at $\quad x = \dfrac{\pi}{3}, y = 0$

$\therefore \quad 0 = \cos\dfrac{\pi}{3} + \left(\cos^2\dfrac{\pi}{3} \cdot c\right)$

$\Rightarrow \quad 0 = \dfrac{1}{2} + \dfrac{c}{4} \Rightarrow c = -2$

$\therefore \quad y = \cos x - 2\cos^2 x$

$$\Rightarrow \frac{y}{2} = -\cos^2 x + \frac{1}{2}\cos x = -\left(\cos^2 x - \frac{1}{2}\cos x\right)$$

$$\Rightarrow \frac{y}{2} = -\left[\left(\cos x - \frac{1}{4}\right)^2 - \left(\frac{1}{4}\right)^2\right] = -\left[\left(\cos x - \frac{1}{4}\right)^2\right] + \frac{1}{16}$$

So, maximum value of $\frac{y}{2} = \frac{1}{16}$

$\Rightarrow$ Maximum value of $y = \frac{1}{8} \Rightarrow k = 8$

English

1. *(c)* The correct filler for the given blank is 'had finished'.
2. *(c)* The Sentence that follows the correct Subject Verb Agreement is The increase in consumption is directly proportional to the increase in income.
3. *(a)* The correct filler for the given blank is 'much'.
4. *(a)* The correct filler for the given blank is 'Inspite of'.'Inspite of' is used to show that something happened or is true although something else might have happened toprevent it.
5. *(a)* The underlined expression is a noun clause expressing the future action.
6. *(b)* The idiom 'hard nut to crack' means a person who is very difficult to understand.
7. *(d)* The underlined word 'meddle' means to become involved in a situation that you have no right to be involved in, in a way that is annoying. Hence, 'interfere' is its synonym.
8. *(a)* Something made in exact imitation of something valuable with the intention to deceive or defraud is called 'Counterfeit'.
9. *(a)* The correct filler for the given blank is 'heirs'. An Heir means a person who inherits andcontinues the work of a predecessor.
10. *(b)* The correct filler for the given blank is 'isn't there'.

MANIPAL

Engineering Entrance Exam

Solved Paper 2023*

Physics

Multiple Choice Questions (MCQs)

1. A charged capacitor when filled with a dielectric $K = 3$ has charge Q_0, voltage V_0 and field E_0. If the dielectric is replaced with another one having $K = 9$, the new values of charge voltage and field will be respectively

(a) $3Q_0, 3V_0, 3E_0$

(b) $Q_0, 3V_0, 3E_0$

(c) $Q_0, \frac{V_0}{3}, 3E_0$

(d) $Q_0, \frac{V_0}{3}, \frac{E_0}{3}$

2. A proton accelerated through a potential difference of 100 V, has de-Broglie wavelength λ_0. The de-Broglie wavelength of an α-particle, accelerated through 800 V is

(a) $\frac{\lambda_0}{\sqrt{2}}$ (b) $\frac{\lambda_0}{2}$

(c) $\frac{\lambda_0}{4}$ (d) $\frac{\lambda_0}{8}$

3. A force $F = (2 + x)$ acts on a particle in x-direction, where F is in newton and x in metre. The work done by this force during a displacement from $x = 1.0$ m to $x = 2.0$ m is

(a) 2 J (b) 3.5 J

(c) 4.5 J (d) None of these

4. The acceleration due to gravity becomes $g/2$ (g = acceleration due to gravity on the surface of the earth) at a height equals to

(a) $\frac{R}{4}$ (b) $\frac{R}{2}$ (c) $\frac{R}{3}$ (d) $\frac{R}{5}$

5. A ray falls on a prism $ABC(AB = BC)$ and travels as shown in figure. The minimum refractive index of the prism material should be

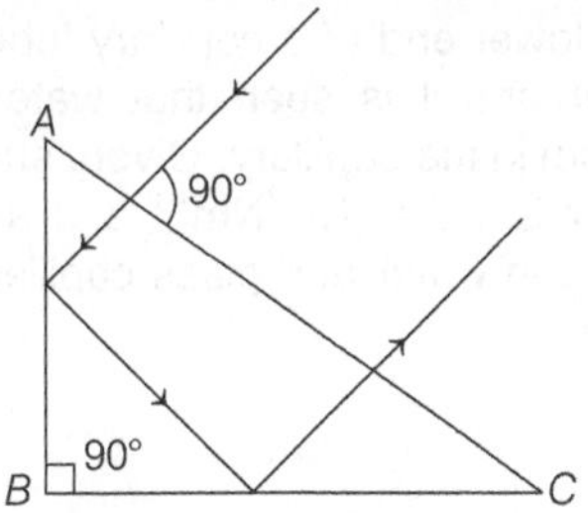

(a) 4/3 (b) $\sqrt{2}$ (c) 1.5 (d) $\sqrt{3}$

6. A function of time is represented as follows $\sin \omega t + \cos 2\omega t + \sin 4\omega t$. The motion represented by it is

(a) non-periodic

(b) periodic

(c) both non-periodic and periodic

(d) data insufficient

7. The resistance of a wire at 20°C is 20 Ω and at 500°C is 60Ω. At which temperature its resistance will be 25Ω?

(a) 50°C (b) 60°C

(c) 70°C (d) 80°C

8. Two stones having different masses m_1 and m_2 are projected at angles θ and (90° − θ) with same velocity from the same point. The ratio of their maximum heights is

(a) 1 : 1 (b) 1 : tanθ

(c) tan θ : 1 (d) $\tan^2 \theta : 1$

9. Ionisation potential of hydrogen atom is 13.6 V. Hydrogen atoms in the ground state are excited by monochromatic radiation of photon energy 12.1 eV. The spectral lines emitted by hydrogen atoms according to Bohr's theory will be

(a) one (b) two (c) three (d) four

10. In Young's double slit experiment, if the source of light changes from orange to blue, then

(a) the central bright fringe will become a dark fringe
(b) the distance between consecutive fringes will decrease
(c) the distance between consecutive fringes will increase.
(d) the intensity of the minima will increase.

Numerical Answer Type (NAT)

11. A disc of mass 5 kg and radius 50 cm rolls on the ground at the rate of $10 ms^{-1}$. Find the kinetic energy of the disc in J.

12. The lower end of a capillary tube is dipped into water and it is seen that water rises through 7.5 cm in the capillary. Given, surface tension of water is $7.5 \times 10^{-2} Nm^{-1}$ and angle of contact between water and glass capillary tube is zero. What will be the diameter (in mm) of the capillary tube? (given, $g = 10 ms^{-2}$)

13. Two steel wires having same length are suspended from a ceiling under the same load. If the ratio of their energy stored per unit volume is 1: 4, then the ratio of their diameters is found to be $\sqrt{k} : 1$. Find the value of k.

14. A vertical disc of diameter 10 cm makes 20 revolutions per second about a horizontal axis passing through its centre. A uniform magnetic field 10^{-1} T acts perpendicular to the plane of the disc. If the potential difference between its centre and rim is found to be $\frac{\pi}{x} \times 10^{-p}$ Volt, then find the ratio of x / p.

15. Light of wavelength 2475 Å is incident on barium. Photoelectrons emitted describe a circle of radius 100 cm by a magnetic field of flux density $\frac{1}{\sqrt{17}} \times 10^{-5}$ tesla. The value of work function of the barium is .... eV. $\left(\text{given, } \frac{e}{m} = 1.7 \times 10^{11} \right)$

Chemistry

Multiple Choice Questions (MCQs)

1. The number of Cl^- ions in 100 mL of 0.001 M NaCl solution is

(a) 6.022×10^{23} (b) 6.022×10^{20}
(c) 6.022×10^{19} (d) 6.022×10^{21}

2. Which of the following given below species have same bond order?

I. H_2 II. B_2 III. O_2^{2-}
IV. Be_2 V. N_2

Correct option is

(a) I and II (b) I, II and III
(c) I, II and IV (d) I, IV and V

3. Which of the following conclusion is correct regarding photoelectric effect?

(a) Energy of a photon depends upon frequency of light absorbed.
(b) Energy of a photon depends upon intensity of light.
(c) Kinetic energy of electrons increases due to increase in intensity of light.
(d) All the energy of photon is absorbed by metal surface.

4. Match the complex given in column-I with their hybridisation given in column-II.

	Column-I		Column-II
A.	$[Ni(CO)_4]$	1.	dsp^2
B.	$[Ni(CN)_4]^{2-}$	2.	sp^3
C.	$[Ni(NH_3)_6]^{2+}$	3.	d^2sp^3
D.	$[Fe(CN)_6]^{3-}$	4.	sp^3d^2

	A	B	C	D
(a)	1	2	3	4
(b)	2	1	4	3
(c)	1	3	2	4
(d)	2	3	1	4

5. Which of the following reagents are used to convert propene to propyne?

(a) $SOCl_2$ /Py, alc. KOH, H^+/H_2O
(b) Br_2/CCl_4/ alc. KOH/Δ
(c) alc. KCN, $SOCl_2$
(d) alc. KOH, B_2H_6/H_2O_2

6. Find the product Z in the series of reaction.

$$\text{Phenol} \xrightarrow{\text{Zn dust}} X \xrightarrow[\text{Anhyd. AlCl}_3]{\text{CH}_3\text{Cl}} Y \xrightarrow{\text{Alkaline KMnO}_4} Z$$

(a) benzaldehyde (b) benzene
(c) benzoic acid (d) benzophenone

7. Which of the given reaction doesn't able to form benzaldehyde as product?

(a) Rosenmund reaction
(b) Etard reaction
(c) Gattermann Koch reaction
(d) Cannizzaro reaction

8. Consider the graph for cyclic process.

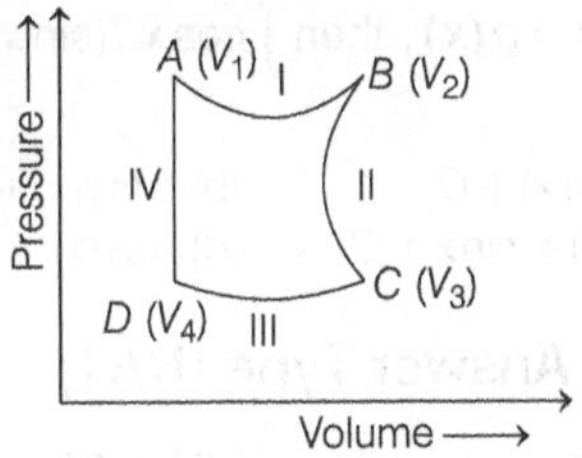

The isothermal compression process among them is

(a) I (b) II (c) III (d) IV

9. In which of the given form styrene exist at room temperature?

(a) Liquid (b) Solid
(c) Gas (d) Pseudo solid

10. The hydroxide of alkaline earth metal not soluble in water is

(a) magnesium hydroxide (b) calcium hydroxide
(c) beryllium hydroxide (d) barium hydroxide

Numerical Answer Type (NAT)

11. A 5% solution of a substance is isotonic with a 1.5% solution of urea (molar mass = 60 g mol^{-1}) in the same solvent. If the densities of both the solutions are assumed to be equal to 1 gcm^{-3} molar mass of the substance will be _____ .

12. The pK_a of acetic acid is 4.74. The concentration of CH_3COOH is 0.01 M. The pH of CH_3COOH is _____ .

13. Value of E_2° is _____ in the given diagram (ignore –ve sign)

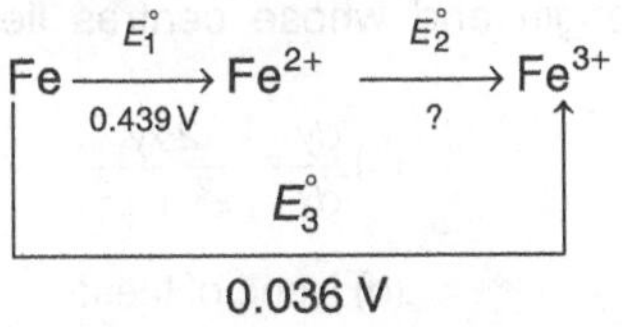

14. Total number of $2c-2e^-$ and $3c-3e^-$ bonds are present in diborane is _____.

15. Among Cu^{2+}, Fe^{3+}, Ti^{3+} and Zn^{2+} ions. The number of ions that impart colour is _____ .

Mathematics

Multiple Choice Questions (MCQs)

1. Let $A = \{1, 2, 3, 4, 5, 6, 7\}$ and $B = \{3, 6, 7, 9\}$. Then, the number of elements in the set $\{C \subseteq A : C \cap B \neq \phi\}$ is

(a) 112 (b) 120
(c) 108 (d) 96

2. Let $f(x) = \dfrac{1}{\sqrt{1+x^2}}$, then

(a) $f(x, y) = f(x) \cdot f(y)$ (b) $f(x, y) \geq f(x) \cdot f(y)$
(c) $f(x, y) \leq f(x) \cdot f(y)$ (d) $f(x, y) = f(x) - f(y)$

3. Set of values of x lying in $[0, 2\pi]$ satisfying the inequality $|\sin x| > 2\sin^2 x$ contains

(a) $\left(0, \dfrac{\pi}{6}\right) \cup \left(\pi, \dfrac{7\pi}{6}\right)$ (b) $\left(0, \dfrac{7\pi}{6}\right)$
(c) $\dfrac{\pi}{6}$ (d) None of these

4. If complex number z lies in the interior or on the boundary of circle of radius 3 units, then maximum and minimum values of $|z + 1|$ are

(a) (6, 0) (b) (3, 0) (c) (6, 3) (d) (4, 1)

5. If $A = \begin{bmatrix} 2 & -1 \\ -7 & 4 \end{bmatrix}$ and $B = \begin{bmatrix} 4 & 1 \\ 7 & 2 \end{bmatrix}$, then $B^T A^T$ is

(a) null matrix
(b) an identity matrix
(c) scalar but not an identity matrix
(d) such that $T_r(B^T A^T) = 4$

6. If the coefficients of x^3 and x^4 in the expansion of $(1 + ax + bx^2)(1 - 2x)^{18}$ in powers of x are both zero, then (a, b) is equal to

(a) $\left(14, \frac{272}{3}\right)$ (b) $\left(16, \frac{272}{3}\right)$
(c) $\left(16, \frac{251}{3}\right)$ (d) $\left(14, \frac{251}{3}\right)$

7. If $F : R \to R$ is a differentiable function and $f(3) = 6$, then $\lim_{x \to 3} \int_6^{f(x)} \frac{2t\,dt}{(x-2)}$ is equal to

(a) $18f'(3)$ (b) 0 (c) $24f'(3)$ (d) $3f'(3)$

8. The differential equation of all circles which pass through the origin and whose centres lies on Y-axis is

(a) $\frac{dy}{dx} = \frac{xy}{x^2 + y^2}$ (b) $\frac{dy}{dx} = \frac{2xy}{x^2 + y^2}$
(c) $\frac{dy}{dx} = \frac{2xy}{x^2 - y^2}$ (d) None of these

9. The area of the region bounded by the curves $y = |5 - x|$, $x = 1$, $x = 6$ and the X-axis

(a) 15 sq units (b) 13 sq units
(c) $\frac{17}{2}$ sq units (d) 16 sq units

10. If a^2, b^2 and c^2 are in AP, then $b + c, c + a$ and $a + b$ are in ....

(a) AP (b) GP
(c) HP (d) None of these

11. The area bounded by the curves $y = \tan x$, $-\frac{\pi}{3} \le x \le \frac{\pi}{3}$, $y = \cot x$, $\frac{\pi}{6} \le x \le \frac{\pi}{2}$ and the X-axis is

(a) $\ln\sqrt{3}$ (b) $\ln\sqrt{2}$ (c) $\ln 2$ (d) $\ln\left(\frac{3}{2}\right)$

12. If the line $\frac{x-1}{2} = \frac{y+1}{3} = \frac{z-1}{4}$ and $\frac{x-3}{1} = \frac{y-k}{2} = \frac{z}{1}$ intersect, then k is equal to

(a) -1 (b) $\frac{2}{9}$ (c) $\frac{9}{2}$ (d) 0

13. An unbaised coin is tossed n times. If the probability of getting 5 heads is equal to the probability of getting 6 heads, then the probability of getting 3 heads is

(a) ${}^{11}C_5\left(\frac{1}{2}\right)^5$ (b) ${}^{11}C_6\left(\frac{1}{2}\right)^6$
(c) ${}^{11}C_3\left(\frac{1}{2}\right)^{11}$ (d) $\frac{11}{1024}$

14. If $\sum_{i=1}^{9}(x_i - 5) = 9$ and $\sum_{i=1}^{9}(x_i - 5)^2 = 45$, then the standard deviation of the 9 items $x_1, x_2 \ldots x_9$ is

(a) 9 (b) 4 (c) 3 (d) 2

15. $\int f(x)\,dx = g(x)$, then $\int \cos x\, f(\sin x)\,dx$ is equal to

(a) $g(\cos x) + C$ (b) $g(\sin x) + C$
(c) $\int g(x) + \sin x + C$ (d) $f(\sin x) + g(\cos x) + C$

Numerical Answer Type (NAT)

16. The value of $f(x) = \lim_{x \to 0}\left(\frac{8^x + 2^x + 4^x}{3}\right)^{2/x}$, then $f(x)$ equals to

17. If the foot of the perpendicular drawn from the point (2, 0, 1) on a line passing through $(\alpha, 5, 1)$ is $\left(\frac{7}{3}, \frac{5}{3}, \frac{15}{4}\right)$, then α is equal to ....

18. The number of words (with or without meaning) that can be formed from all the letters of the word "LETTER" in which vowels never come together is ....

19. If the vectors $p = (a + 1)\hat{i} + a\hat{j} + a\hat{k}$, $q = a\hat{i} + (a + 1)\hat{j} + a\hat{k}$ and $r = a\hat{i} + a\hat{j} + (a + 1)\hat{k}$ $(a \in R)$ are coplanar, then the value of a is ....

20. Last two digits in $(9)^{50}$ be

English

1. Fill in the blank with correct tense form.

I ____(write) a letter when she knocked at the door.

(a) wrote (b) had writing
(c) have writing (d) was writing

2. Choose the correct sentence that follows the correct Subject-verb Agreements .

(a) Neither she laughed nor I did .
(b) Neither she laughed did nor I.
(c) Neither she laughed nor did I.
(d) Neither she laughed I did nor.

3. Fill in the blank with suitable determiner.

All the boys went to beach but only ____could swim.

(a) few (b) a few (c) little (d) the little

4. Choose the appropriate conjunction from the options given below.

I walked out____I wanted to breathe some fresh air.

(a) but (b) that (c) because (d) yet

5. Determine the clause highlighted in the given sentence.

The jury believed <u>that the man was guilty</u>.

(a) Adjective clause (b) Noun clause
(c) Adverb clause (d) Either (a) or (c)

6. Choose the alternative which best expresses the meaning of the given underlined idiom /phrase.

Chetan is very upset because the new manager always <u>picks on him</u>.

(a) advises (b) warns severely
(c) selects (d) treat badly

7. Choose the word nearest in meaning to the italicised part.

Many celebrities <u>encourage</u> young children to aspire for greatness.

(a) Daunt (b) Stimulate
(c) Undermine (d) Abate

8. Choose the one which can be substituted for the given words/sentence.

To keep a great person or event in people's memory.

(a) Communicate (b) Commensurate
(c) Commemorate (d) Commiserate

9. The government has the right to ____the stolen property.

(a) sees (b) seas (c) seize (d) size

10. Fill in the blank with suitable question tag.

There's a cinema near the station, ______ .

(a) aren't it (b) isn't there
(c) weren't it (d) isn't it

Answers

Physics

1. (d)	**2.** (d)	**3.** (b)	**4.** (a)	**5.** (b)	**6.** (b)	**7.** (d)	**8.** (d)	**9.** (c)	**10.** (b)
11. (375)	**12.** (0.4)	**13.** (2)	**14.** (1)	**15.** (4.5)					

Chemistry

1. (c)	**2.** (b)	**3.** (a)	**4.** (b)	**5.** (b)	**6.** (c)	**7.** (d)	**8.** (c)	**9.** (a)	**10.** (c)
11. (200)	**12.** (3.37)	**13.** (0.77)	**14.** (6)	**15.** (3)					

Mathematics

1. (a)	**2.** (b)	**3.** (d)	**4.** (d)	**5.** (b)	**6.** (b)	**7.** (a)	**8.** (c)	**9.** (c)	**10.** (c)
11. (c)	**12.** (c)	**13.** (c)	**14.** (d)	**15.** (b)	**16.** (16)	**17.** (4.372)	**18.** (120)	**19.** (–0.33)	**20.** (01)

English

1. (d)	**2.** (c)	**3.** (b)	**4.** (c)	**5.** (b)	**6.** (d)	**7.** (b)	**8.** (c)	**9.** (c)	**10.** (b)

SOLUTIONS

Physics

1. *(d)* When charging battery is disconnected then charge remains constant, V becomes $1/K$ times and E becomes $1/K$ times.
So, new value of charge is Q_0.
New value of $V = \frac{V_0 \times 3}{9} = \frac{V_0}{3}$
New value of $E = \frac{E_0 \times 3}{9} = \frac{E_0}{3}$

2. *(d)* As, $\lambda = \frac{h}{\sqrt{2qVm}} \propto \frac{1}{\sqrt{qVm}}$
$$\therefore \quad \frac{\lambda_\alpha}{\lambda_p} = \sqrt{\frac{q_p V_p m_p}{q_\alpha V_\alpha m_\alpha}} = \sqrt{\frac{1 \times 100 \times 1}{2 \times 800 \times 4}} = \frac{1}{8}$$
$$\therefore \quad \lambda_\alpha = \frac{\lambda_p}{8} = \frac{\lambda_0}{8}$$

3. *(b)* The work done in small displacement from x to $x + dx$ is
$$dW = Fdx = (2 + x)dx$$
Hence,
$$W = \int_1^2 dW = \int_1^2 (2 + x)\, dx = \int_1^2 2dx + \int_1^2 xdx$$
$$= \left[2x + \frac{x^2}{2}\right]_1^2 = 3.5 \text{ J}$$

4. *(a)* The acceleration due to gravity,
$$g = \frac{GM}{R^2} \quad \text{...(i)}$$
At a height h above the surface of the earth, the acceleration due to gravity is
$$g' = \frac{GM}{(R + h)^2} \quad \text{...(ii)}$$
From Eqs. (i) and (ii), we get
$$\frac{g}{g'} = \left(\frac{R + h}{R}\right)^2 = \left(1 + \frac{h}{R}\right)^2$$
$$\Rightarrow \quad \frac{g'}{g'} = \left(1 + \frac{h}{R}\right)^{-2} = \left(1 - \frac{2h}{R}\right)^2$$
Here, $g' = \frac{g}{2}$
$$\therefore \quad \frac{g/2}{g} = \left(1 - \frac{2h}{R}\right) \Rightarrow \frac{2h}{R} = \frac{1}{2} \text{ or } h = \frac{R}{4}$$

5. *(b)* Angle i at both focus will be 45°. For TIR to take place,
$$i > \theta_c \text{ or } \sin i > \sin\theta_c$$
$$\therefore \quad \frac{1}{\sqrt{2}} > \frac{1}{\mu} \text{ or } \mu > \sqrt{2}$$

6. *(b)* This is an example of a periodic motion. It can be noted that each term represents a periodic function with different angular frequency. Since, period is the least interval of time after which a function repeats its value, $\sin \omega t$ has a period $T_0 = \frac{2\pi}{\omega}$
$\cos 2\omega t$ has a period $\frac{\pi}{\omega} = \frac{T_0}{2}$
and $\sin 4\omega t$ has a period $\frac{2\pi}{4\omega} = \frac{T_0}{4}$
The period of the first term is a multiple of the periods of the last two terms. Therefore, the smallest interval of time after which the sum of the three terms repeats is T_0 and thus, the sum is a periodic function with a period $\frac{2\pi}{\omega}$.

7. *(d)* Resistance, $R_t = R_0(1 + \alpha t)$
$20 = R_0 (1 + 20\alpha)$ and $60 = R_0(1 + 500\alpha)$
From this equation, we can find
$$R_0 = 18.33\Omega$$
$$\alpha = 4.54 \times 10^{-3}$$
$$R_t = 25\Omega$$
Again, $R_t = R_0(1 + \alpha t)$
$$25 = 18.33(1 + 4.54 \times 10^{-3}\, t)$$
We find $t = 80°$ C

8. *(d)* Maximum height, $H_1 = \frac{u^2 \sin^2\theta}{2g}$ ...(i)
and $H_2 = \frac{u^2 \sin^2(90° - \theta)}{2g}$ ...(ii)
From Eqs. (i) and (ii), we get
$$\frac{H_1}{H_2} = \frac{\sin^2\theta}{\sin^2(90° - \theta)}, \frac{H_1}{H_2} = \frac{\tan^2\theta}{1}$$
$$\therefore \quad H_1 : H_2 = \tan^2\theta : 1$$

9. *(c)* Final energy of electron
$= -13.6 + 12.1 = -1.51\,\text{eV}$
This energy corresponds to third level *i.e.*, $n = 3$
Hence, number of spectral lines emitted
$$= \frac{n(n-1)}{2} = \frac{3(3-1)}{2} = 3$$

10. *(b)* We know that in Young's double slit experiment, fringe width $\beta = \frac{D\lambda}{d}, \beta \propto \lambda$

Since, $\lambda_{blue} < \lambda_{orange}$
Thus, $\beta_{blue} < \beta_{orange}$
$\therefore$ In Young's double slit experiment, if the source of light changes from orange to blue, then distance between the consecutive fringes will decrease.

11. *(375)* Here, mass of the disc, $M = 5\text{kg}$
Radius of the disc, $R = 50\,\text{cm} = 1/2\,\text{m}$
Linear velocity of the disc, $v = 10\text{ms}^{-1}$
As, $\quad v = R\omega$
$\therefore \quad 10 = \frac{1}{2}\omega$ or $\omega = 10 \times 2 = 20\,\text{rad s}^{-1}$

Also, moment of inertia of disc about an axis through its centre,
$$I = \frac{1}{2}MR^2 = \frac{1}{2} \times 5 \times \left(\frac{1}{2}\right)^2 = \frac{5}{8}\,\text{kg}^2\text{-m}$$
$$\therefore \text{KE of translation} = \frac{1}{2}mv^2 = \frac{1}{2} \times 5 \times (10)^2 = 250\,\text{J}$$
$$\text{Rotational kinetic energy} = \frac{1}{2}I\omega^2 = \frac{1}{2}\left(\frac{5}{8}\right)(20)^2 = 125\,\text{J}$$
Total kinetic energy = Translational + Rotational
$= (250 + 125) = 375\,\text{J}$

12. *(0.4)* Given, $h = 7.5\,\text{cm} = 7.5 \times 10^{-2}\,\text{m}$
$S = 7.5 \times 10^{-2}\,\text{Nm}^{-1}$
$\theta = 0^\circ, r = ?$
$$\text{As, } h = \frac{2S\cos\theta}{r\rho g} \Rightarrow 2r = \frac{4S\cos\theta}{h\rho g}$$
$$= \frac{4 \times 7.5 \times 10^{-2} \times \cos 0^\circ}{7.5 \times 10^{-2} \times 10^3 \times 10} = 4 \times 10^{-4} = 0.4\,\text{mm}$$

13. *(2)* Elastic potential energy stored in a loaded wire,
$$U = \frac{1}{2} \times (\text{Stress} \times \text{Strain} \times \text{Volume})$$
$\therefore$ Energy stored per unit volume,
$$U = \frac{U}{\text{Volume}} = \frac{1}{2} \times \text{Stress} \times \text{Strain} = \frac{1}{2}\left(\frac{F}{A}\right)^2 \times \frac{1}{Y}$$
Here, both wires are of same material and under same load, so the ratio of stored energies per unit volume, for both the wires will be
$$\frac{U_A}{U_B} = \frac{\frac{1}{2Y} \cdot \frac{F^2}{A_A^2}}{\frac{1}{2Y} \cdot \frac{F^2}{A_B^2}} = \frac{A_B^2}{A_A^2}, \frac{U_A}{U_B} = \frac{d_B^4}{d_A^4}$$
$$\text{Here, } \frac{U_A}{U_B} = \frac{1}{4}, \frac{d_B^4}{d_A^4} = \frac{1}{4} \text{ or } \frac{d_B}{d_A} = \frac{1}{\sqrt{2}}$$
$$\frac{d_A}{d_B} = \sqrt{2} : 1 = \sqrt{k} : 1 \quad \text{(given)}$$
$\therefore \quad k = 2$

14. *(1)* As we know, disc is equivalent to a rod of length *r*. Divide disc into rods, each rod can be replaced by a battery and these batteries are would be parallel in arrangement.

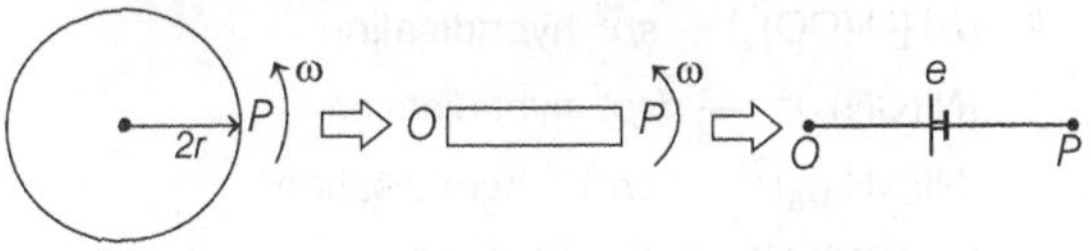

According to the question, radius of disc,
$$r = \frac{10}{2} = 5\,\text{cm} = 0.05\,\text{m}$$
Angular velocity, $\omega = 20 \times 2\pi = 40\pi$ rad/s
Magnetic field, $B = 10^{-1}$ T
The emf of a disc is given as
$$e = \frac{1}{2}B\omega r^2 = \frac{1}{2}(10^{-1})(40\pi)(0.05)^2$$
$$\Rightarrow \quad V_0 - V_A = 1.57 \times 10^{-2}\,\text{V}$$
$$= \frac{\pi}{2} \times 10^{-2}\,\text{V} = \frac{\pi}{x} \times 10^{-p}\,\text{V} \quad \text{(given)}$$
$$\therefore \quad \frac{x}{p} = \frac{2}{2} = 1$$

15. *(4.5)* Radius of circular path described by a charged particle in a magnetic field B is given by
$$r = \frac{\sqrt{2mK}}{qB} \Rightarrow K = \frac{q^2B^2r^2}{2m} = \left(\frac{e}{m}\right)\frac{eB^2r^2}{2}$$
where, K = kinetic energy of a particle
q = charge on a particle
r = radius of circular path
$$= \frac{1}{2} \times 1.7 \times 10^{11} \times 1.6 \times 10^{-19} \times \left(\frac{1}{\sqrt{17}} \times 10^{-5}\right)^2 \times (1)^2$$
$= 8 \times 10^{-20}\,\text{J} = 0.5\,\text{eV}$
Using the relation,
$$E = \phi + K_{max} \Rightarrow \phi = E - K_{max}$$
$$\text{where, } E = h\nu = \frac{hc}{\lambda} = \frac{12375}{\lambda(\text{in Å})}$$
$$= \left(\frac{12375}{2475}\right)\text{eV} - 0.5\,\text{eV} = 4.5\,\text{eV}$$

Chemistry

1. *(c)* 100 mL of 0.001 M NaCl contain Cl^- ions

$$= \frac{0.001}{1000} \times 100 = 10^{-4} \text{ mol}$$

$$= 10^{-4} \times (6.022 \times 10^{23}) = 6.022 \times 10^{19}\ Cl^- \text{ ions}$$

2. *(b)* H_2, B_2, O_2^{2-} have bond order equal which is to 1 and Be_2 and N_2 have 0 and 3 bond order respectively.

3. *(a)* Energy of photon depends upon frequency and not on intensity. As on increasing the intensity of light of a particular frequency *v*, the number of electrons ejected increases but not their kinetic energy.

4. *(b)* $[Ni(CO)_4]$ — sp^3-hybridisation

$[Ni(CN)_4]^{2-}$ — dsp^2-hybridisation

$[Ni(NH_3)_6]^{2+}$ — sp^3d^2-hybridisation

$[Fe(CN)_6]^{3-}$ — d^2sp^3-hybridisation

The correct answer is

A-2, B-1, C-4, D-3.

5. *(b)*

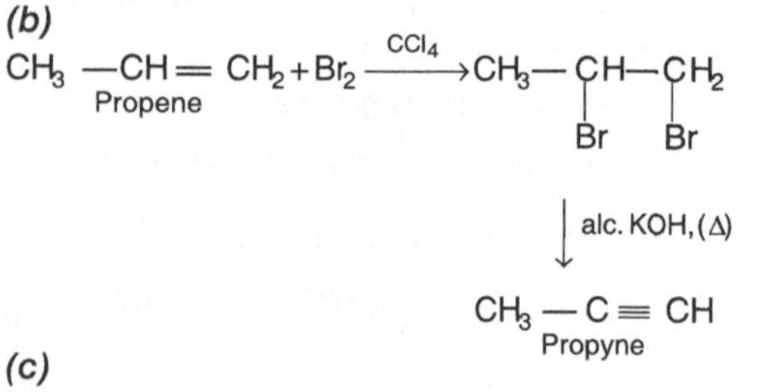

6. *(c)*

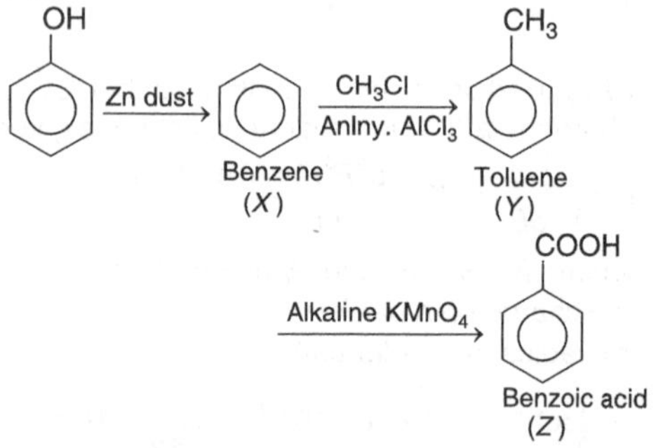

7. *(d)* In Cannizzaro reaction benzaldehyde acts as reactant not the product.

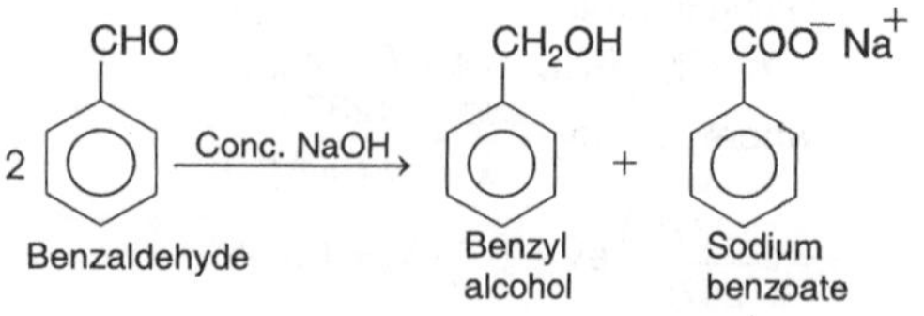

8. *(c)* I. → Isothermal expansion

II. → Adiabatic expansion

III. → Isothermal compression

IV. → Adiabatic compression

9. *(a)* Styrene exist as liquid at room temperature.

10. *(c)* As we move down the group, solubility increases. So, $Be(OH)_2$ is insoluble and $Ba(OH)_2$ is highly soluble in water.

11. *(200)* Solution with the same osmotic pressure are isotonic.

Let the molar mass of the substance be *M*.

$$\therefore \quad \pi_1 = \pi_2$$

$$C_1RT = C_2RT$$

$$C_1 = C_2$$

As density are also same

So, $$\frac{5}{M} = \frac{1.5}{60}$$

$$\Rightarrow \quad M = \frac{5 \times 60}{1.5} = 200 \text{ g mol}^{-1}$$

12. *(3.37)* For a weak acid,

$$pH = \frac{1}{2}[pK_a - \log C]$$

$$= \frac{1}{3}[4.74 - \log 10^{-2}]$$

$$= \frac{1}{2} \times 6.74 = 3.37$$

13. *(0.77)* $\Delta G_3 = \Delta G_1 + \Delta G_1$

$$n_3FE_3^\circ = n_1FE_1^\circ + n_2FE_2^\circ$$

$$3 \times 0.036 = 2 \times 0.439 + 1 \times E_2^\circ$$

$$\Rightarrow \quad E_2^\circ = 0.108 - 0.878$$

$$= -0.77 \text{ V}$$

14. *(6)*

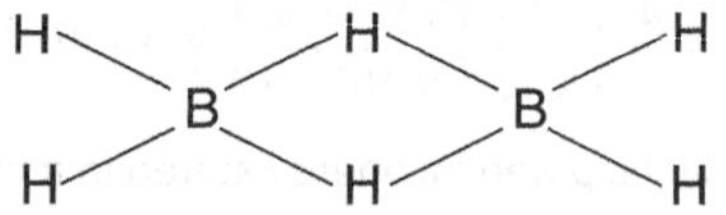

Number of $3c - 2e^{-1}$ and $2c - 2e^-$ bond present in diborane are 2 and 4, respectively.

Thus, total number of $3c - 2e^-$ and $2c - 2e^-$ bonds are 6.

15. *(3)* Cu^{2+}, Fe^{3+}, Ti^{3+} have unpaired electron which impart colour to the ions whereas Zn^{2+} has no unpaired, thus it does not impart any colour.

Mathematics

1. *(a)* $A = \{1, 2, 3, 4, 5, 6, 7\}$ and $B = \{3, 6, 7, 9\}$.

Total subset of $A = 2^7 = 128$

$C \cap B = \phi$, when set C contains the elements 1, 2, 4, 5.

$$S = \{C \subseteq A; C \cap B \neq \phi\}$$

Total $-(C \cap B = \phi) = 128 - 2^4 = 112$

2. *(b)* Given that

$$f(x) = \frac{1}{\sqrt{1+x^2}}$$

$$\therefore \quad f(y) = \frac{1}{\sqrt{1+y^2}}$$

and $f(x, y) = \dfrac{1}{\sqrt{1+x^2 y^2}}$

Now, $f(x, y) = f(x) \cdot f(y)$

$$\frac{1}{\sqrt{1+x^2y^2}} = \frac{1}{\sqrt{1+x^2}} \times \frac{1}{\sqrt{1+y^2}}$$

$$\frac{1}{\sqrt{1+x^2y^2}} = \frac{1}{\sqrt{(1+x^2)(1+y^2)}}$$

On squaring both sides, we get

$$\frac{1}{1+x^2y^2} = \frac{1}{(1+x^2)(1+y^2)}$$

$$\frac{1}{1+x^2y^2} \geq \frac{1}{(1+x^2)(1+y^2)}$$

$$[\because 1+x^2y^2 \leq 1+x^2+y^2+x^2y^2]$$

Hence, $f(x, y) \geq f(x) \cdot f(y)$

3. *(d)* $|\sin x| > 2\sin^2 x$

$$\Rightarrow |\sin x|(2|\sin x| - 1) < 0 \Rightarrow 0 < |\sin x| < \frac{1}{2}$$

$$\Rightarrow x \in \left(0, \frac{\pi}{6}\right) \cup \left(\frac{5\pi}{6}, \pi\right) \cup \left(\pi, \frac{7\pi}{6}\right) \cup \left(\frac{11\pi}{6}, 2\pi\right)$$

4. *(d)* Here, we have

Radius of circle = 3 units

we find the maximum and minimum value of $|z+1|$ for z lying in the interier or boundary of the circle $|z| = 3$.

For Max $|z+1|$, z must lies on the bondary of the circle, then $|z| = 3$ and $z = 3e^{i\theta}$, where θ is the argument of z.

Now, $|z+1| = |3e^{i\theta} + 1|$

$$= |(3\cos\theta + 1) + (3i \sin\theta)|$$

$$= \sqrt{(3\cos\theta + 1)^2 + (3\sin\theta)^2}$$

$$= \sqrt{9 + 1 + 6\cos\theta}$$

$$= \sqrt{10 + 6\cos\theta}$$

$\sin\theta, \cos\theta$ ranges from -1 to 1, the maximum value of $|z+1|$ occur, when $\cos\theta = 1$. Therefore,

Max $|z+1| = \sqrt{10+6} = \sqrt{16} = 4$

For minimum value The minimum value of $z+1$ occures, when z is at the centre of the circle $|z| = 3$. In this case $z = 0$.

On putting $z = 0$ into the expression, we have

Min $(|z+1|) = |0+1| = 1$

Hence, maximum value of $|z+1| = 4$

minimum value of $|z+1| = 1$

5. *(b)* $B^T A^T = (AB)^T$

$$= \left(\begin{bmatrix} 2 & -1 \\ -7 & 4 \end{bmatrix}\begin{bmatrix} 4 & 1 \\ 7 & 2 \end{bmatrix}\right)^T = \begin{bmatrix} 8-7 & 2-2 \\ -28+28 & -7+8 \end{bmatrix}^T$$

$$= \begin{bmatrix} 1 & 0 \\ 0 & 1 \end{bmatrix}^T = \begin{bmatrix} 1 & 0 \\ 0 & 1 \end{bmatrix} = I$$

6. *(b)* In the expansion of $(1 + ax + bx^2)(1 - 2x)^{18}$, coefficient of x^3 in $(1 + ax + bx^2)(1 - 2x)^{18}$

= Coefficient of x^3 in $(1-2x)^{18}$ + coefficient of x^2 in $a(1-2x)^{18}$ + coefficient of x in $b(1-2x)^{18}$.

$$= -{}^{18}C_3 \cdot 2^3 + a\,{}^{18}C_2 \cdot 2^2 - b\,{}^{18}C_1 \cdot 2$$

$$\because \quad -{}^{18}C_3 \cdot 2^3 + a\,{}^{18}C_2 \cdot 2^2 - b\,{}^{18}C_1 \cdot 2 = 0$$

$$\Rightarrow \frac{18 \times 17 \times 16}{3 \times 2} \cdot 8 + a \cdot \frac{18 \times 17}{2} 2^2 - b \cdot 18 \cdot 2 = 0$$

$$\Rightarrow 17a - b = \frac{34 \times 16}{3} \quad \ldots(i)$$

Similarly, coefficient of x^4

$${}^{18}C_4 \cdot 2^4 - a \cdot {}^{18}C_3\, 2^3 + b \cdot {}^{18}C_2 \cdot 2^2 = 0$$

$$\therefore \quad 32a - 32b = 240 \quad \ldots(ii)$$

From Eqs. (i) and (ii), we get

$$a = 16 \text{ and } b = \frac{272}{3}$$

7. *(a)* Let $I = \lim_{x \to 3} \int_6^{f(x)} \frac{2t}{x-2} dt \quad \left[\frac{0}{0} \text{ form, as } f(3) = 6\right]$

On applying the L' Hospital rule, we get

$$I = \lim_{x \to 3} \int_6^{f(x)} \frac{2f(x)\,f'(x)}{1}$$

So, $I = 3\,f(3) \cdot f'(3) = 18f'(3) \quad (\because f(3) = 6)$

$$\therefore \quad \lim_{x \to 3} \int_6^{f(x)} \frac{2t\,dt}{x-2} = 18f'(3)$$

8. *(c)* The equation of family of circles which passess through the origin and whose centrer lies on Y-axis is $(x-0)^2+(y-k)^2=k^2$

$\Rightarrow x^2+y^2+k^2-2yk=k^2$

$\Rightarrow x^2+y^2-2yk=0$

where, k is parameter.

On differentiating w.r.t . x, we get

$$2x+2y\frac{dy}{dx}-2\frac{dy}{dx}k=0$$

$$\Rightarrow \quad x+(y-k)\frac{dy}{dx}=0$$

$$\Rightarrow \quad x+\left\{y-\frac{x^2+y^2}{2y}\right\}\frac{dy}{dx}=0$$

$$\Rightarrow \quad x+\frac{1}{2y}(y^2-x^2)\frac{dy}{dx}=0$$

$$\Rightarrow \quad 2xy=(x^2-y^2)\frac{dy}{dx}$$

$$\Rightarrow \quad \frac{dy}{dx}=\frac{2xy}{x^2-y^2},$$

which is required equation.

9. *(c)* $\because$ Required area $=\left|\int_1^6 |5-x|dx\right|$

$$=\left|\int_1^5 -(5-x)dx+\int_5^6 (5-x)dx\right|$$

$$=\left|\left[-5x+\frac{x^2}{2}\right]_1^5+\left[5x-\frac{x^2}{2}\right]_5^6\right|$$

$$=\left|\left[-25+\frac{25}{2}+5-\frac{1}{2}\right]+\left[30-18-25+\frac{25}{2}\right]\right|$$

$$=\left[\frac{-50+25+10-1}{2}\right]+\left[\frac{-26+25}{2}\right]$$

$$=\left|-8-\frac{1}{2}\right|=\frac{17}{2}\text{ sq units}$$

10. *(c)* We have, a^2, b^2 and c^2 are in AP adding $ab+bc+ca$ to each of these terms, we get

$a^2+ab+bc+ca, b^2+ab+bc+ca$ and

$c^2+ab+bc+ca$ are in AP.

$\Rightarrow (a+b)(a+c), (b+c)(b+a)$ and $(c+a)(c+b)$ are in AP.

Dividing each term by $(b+c)(c+a)(a+b)$, we get

$\frac{1}{b+c}, \frac{1}{c+a}$ and $\frac{1}{a+b}$ are in AP

$\Rightarrow b+c, c+a$ and $a+b$ are in HP.

11. *(c)* The given curves, $y=\tan x$ and $y=\cot x$ meet at $x=\frac{\pi}{4}$.

$$\therefore \text{Required area}=\int_0^{\pi/4}\tan x\,dx+\int_{\pi/4}^{\pi/2}\cot x\,dx$$

$$=(\ln\sec x)_0^{\pi/4}+(\ln\sin x)_{\pi/4}^{\pi/2}$$

$$=\ln\sqrt{2}-\ln\frac{1}{\sqrt{2}}=2\ln\sqrt{2}=\ln 2.$$

12. *(c)* Given, two lines

$L_1:\frac{x-1}{2}=\frac{y+1}{3}=\frac{z-1}{4}$

and $L_2:\frac{x-3}{1}=\frac{y-k}{2}=\frac{z-0}{1}$

Let $L_1:\frac{x-1}{2}=\frac{y+1}{3}=\frac{z-1}{4}=p$

and $L_2:\frac{x-3}{1}=\frac{y-k}{2}=\frac{z-0}{1}=q$

Any point P on line L_1 is of type $P(2p+1, 3p-1$ and $4p+1)$ and any point Q on the line L_2 is of type $Q(q+3, 2q+k$ and $q)$

Since, L_1 and L_2 are intersecting each other, hence, both point P and Q should coincide at the point of intersection. Corresponding co-ordinates of P and Q should be same.

$2p+1=q+3, 3p-1=2q+k$ and $4p+1=q$

On solving, $2p+1=q+3$ and $4p+1=q$, we get $p=-\frac{3}{2}$ and $q=-5$.

On substituting the values of p and q in the third equation $3p-1=2q+k$, we get

$$3\left(-\frac{3}{2}\right)-1=2(-5)+k$$

$$\therefore \quad k=\frac{9}{2}$$

13. *(c)* The coin is tossed n times.

Let x denotes the number of heads in n trials and it is given that $p(x=5)=p(x=6)$

$$\Rightarrow {}^nC_5\left(\frac{1}{2}\right)^n={}^nC_6\left(\frac{1}{2}\right)^n \Rightarrow n=11$$

$$\therefore p(x=3)={}^{11}C_3\left(\frac{1}{2}\right)^{11}$$

14. *(d)* Given, $\sum_{i=1}^{9}(x_i-5)=9$

$$\sum x_i-45=9\Rightarrow \sum x_i=54$$

Also, $\sum_{i=1}^{9}(x_i - 5)^2 = 45$

$\Rightarrow \quad \sum x_i^2 - 10\times 54 + 25\times 9 = 45$

$\sum x_i^2 = 360$

$\therefore$ Standard deviation $(\sigma) = \sqrt{\dfrac{360}{9} - \left(\dfrac{54}{9}\right)^2} = 2.$

15. *(b)* Here, $\int f(x)dx = g(x)$

Now, $\int \cos x\, f(\sin x)dx$

Let, $\sin x = k \Rightarrow \cos x dx = dk$

$\therefore \int f(k)dk = g(k) + C = g(\sin x) + C$

16. *(16)* $y = \lim_{x\to 0}\left(\dfrac{8^x + 2^x + 4^x}{3}\right)^{2/x}$

$\Rightarrow \log y = \lim_{x\to 0}\dfrac{2}{x}\log\left(\dfrac{8^x + 2^x + 4^x}{3}\right)$

$= 2\lim_{x\to 0}\dfrac{1}{x}[\log(8^x + 2^x + 4^x) - \log 3]$

$= 2\lim_{x\to 0}\dfrac{[\log(8^x + 2^x + 4^x) - \log 3]}{x} \qquad \left[\dfrac{0}{0}\text{ form}\right]$

On applying L′ Hospital rule, we get

$= 2\lim_{x\to 0}\dfrac{[8^x\log 8 + 2^x\log 2 + 4^x\log 4]}{8^x + 2^x + 4^x}$

$= 2\left[\dfrac{\log 8 + \log 2 + \log 4}{3}\right]$

$= \dfrac{2}{3}[3\log 2 + \log 2 + 2\log 2] \qquad [\log(m)^n = n\log m]$

$= \dfrac{2}{3}\times 6\log 2$

$\log y = 4\log 2 = \log 16 \Rightarrow y = 16$

17. *(4.372)* It is given that $Q\left(\dfrac{7}{3}, \dfrac{5}{3}, \dfrac{15}{4}\right)$ is foot of perpendicular of point $P(2, 0, 1)$ to a line passes through point $A(\alpha, 5, 1)$. So $PQ \perp AQ$

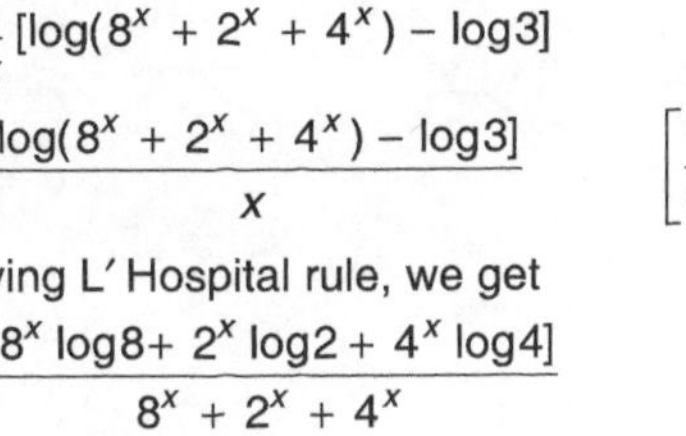

$\because$ Direction ratios of line segments PQ is $\left(\dfrac{7}{3}, \dfrac{5}{3}, \dfrac{15}{4}\right)$ and direction ratios of line AQ is $\left(\alpha - \dfrac{7}{3}, \dfrac{10}{3}, -\dfrac{11}{4}\right)$

Since, PQ and AQ is perpendicular to each other, so $\mathbf{PQ}\cdot\mathbf{AQ} = 0$

$\dfrac{7}{3}\left(\alpha - \dfrac{7}{3}\right) + \dfrac{5}{3}\times\dfrac{10}{3} + \dfrac{15}{4}\times\left(-\dfrac{11}{4}\right) = 0$

$\Rightarrow \quad \dfrac{7(3\alpha - 7)}{9} + \dfrac{50}{9} - \dfrac{165}{16} = 0$

$\Rightarrow \quad \dfrac{336\alpha - 784 + 800 - 1485}{144} = 0$

$\Rightarrow \quad 336\alpha = 1469 \Rightarrow \alpha = \dfrac{1469}{336} = 4.372$

18. *(120)* Given, word is LETTER, having vowels E, E and consonants L, T, T and R. Now, the number of ways to arrange the consonants are $\dfrac{4!}{2!} = 12$

Now, we have five place to put vowels E, E.

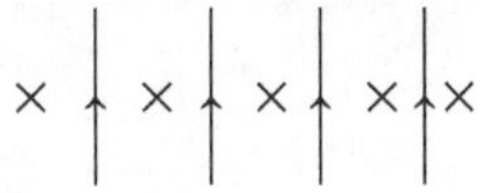

So, number of ways to arrange vowels is ${}^5C_2 = 10$

So, number of required words $= 12\times 10 = 120$

19. *(− 0.33)* Given, vectors are

$\mathbf{p} = (a+1)\hat{\mathbf{i}} + a\hat{\mathbf{j}} + a\hat{\mathbf{k}}$

$\mathbf{q} = a\hat{\mathbf{i}} + (a+1)\hat{\mathbf{j}} + a\hat{\mathbf{k}}$

and $\mathbf{r} = a\hat{\mathbf{i}} + a\hat{\mathbf{j}} + (a+1)\hat{\mathbf{k}}$

$[p\ q\ r] = 0$

$\Rightarrow \begin{vmatrix} a+1 & a & a \\ a & a+1 & a \\ a & a & a+1 \end{vmatrix} = 0$

$\Rightarrow (a+1)[(a+1)^2 - a^2] - a[a(a+1) - a^2] + a[a^2 - a(a+1)] = 0$

$\Rightarrow (a+1)[2a+1] - 2a[a] = 0$

$\Rightarrow 2a^2 + 3a + 1 - 2a^2 = 0$

$\Rightarrow a = -\dfrac{1}{3} = -0.33$

20. *(01)* $\because (9)^{50} = (10-1)^{50}$

$= [{}^{50}C_0(10)^{50}(-1)^0 - {}^{50}C_1(10)^{49}(1) + \ldots] - {}^{50}C_{49}(10)^1 + {}^{50}C_{50}(10)^0$

$\because$ All the terms in the bracket are multiples of 10^2.

So, we can write as 100λ

$= 100\lambda - 50\times 10 + 1 = 100\lambda - 500 + 1$

so, last 2 digits will be = 01.

English

1. *(d)* 'was writing' should be used in the given blank as the sentence belongs to past continuous tense.
2. *(c)* 'Neither she laughed nor did I' is the correct sentence that follows the Subject-Verb Agreement.
3. *(b)* 'a few' is the suitable determiner for the given blank.
4. *(c)* To show the reason or cause of an action, subordinating conjunction 'because' is used.
5. *(b)* The given underlined clause is Noun clause.
6. *(d)* 'Picking on someone' means to treat someone badly that hurts the emotion of that person.
7. *(b)* Encourage means to give support, confidence or hope to someone. So, 'Stimulate' would be its correct synonym as it also means to encourage someone.
8. *(c)* To remember officially and give respect to a great person or event, especially by a public ceremony is called 'Commemorate'.
9. *(c)* 'seize' is the correct word for the given blank.
10. *(b)* Use 'isn't there' because when there is a positive statement then the question tag is used in negative.

MANIPAL
Engineering Entrance Exam
Solved Paper 2022*

Physics

Multiple Choice Questions (MCQs)

1. The work done by all the forces (external and internal) on a system is equal to change in

(a) total energy (b) kinetic energy
(c) potential energy (d) None of these

2. The electric resistance of a wire is R. If the length of the wire is increased to doubled by stretching it, then the new resistance of the wire is

(a) $2R$ (b) $4R$
(c) R (d) $16R$

3. If the radius of a planet is three times the radius of the earth. Both have same mass-densities. v_P and v_E are the escape velocities of the planet and the earth respectively, then

(a) $v_P = 1.5\, v_E$ (b) $v_P = 3v_E$
(c) $v_E = 2v_P$ (d) $v_P = 2v_E$

4. At which temperature, magnitude of °C and °F are equal?

(a) 273 (b) 40
(c) – 40 (d) – 273

5. The potential at a point at a distance r from the centre of an electric dipole is proportional to

(a) $\frac{1}{r}$ (b) $\frac{1}{r^2}$
(c) $\frac{1}{r^3}$ (d) r^2

6. The time period of a mass suspended from a spring is T. If the spring is cut into three equal parts and connected in parallel. The same mass is suspended from these parallel springs, then the new time period of the mass will be

(a) $\frac{T}{4}$ (b) T
(c) $\frac{T}{3}$ (d) $3T$

7. The angle of a prism is A. One of its refracting surface is silvered. If light rays falling at an angle of incidence $2A$ on the first surface returns back through the same path after reflection from silvered surface. The refractive index μ of the prism is

(a) $2 \sin A$ (b) $2 \cos A$
(c) $\frac{1}{2} \cos A$ (d) $\tan A$

8. A particle of charge q and mass m is moving with a velocity $-2v\hat{\mathbf{i}}$ $(v \neq 0)$ towards a large screen placed in YZ-plane placed at a distance d. If there is a magnetic field $\mathbf{B} = B_0\hat{\mathbf{k}}$, the maximum value of v for which the particle will not strike the screen is

(a) $\frac{qdB_0}{m}$ (b) $\frac{qdB_0}{2m}$
(c) $\frac{2qdB_0}{m}$ (d) $\frac{qdB_0}{3m}$

9. A small object of uniform density rolls up a curved surface with an initial velocity v. It reaches upto a maximum height of $\frac{7v^2}{10g}$ with respect to initial position. Then the object is

*Based on memory

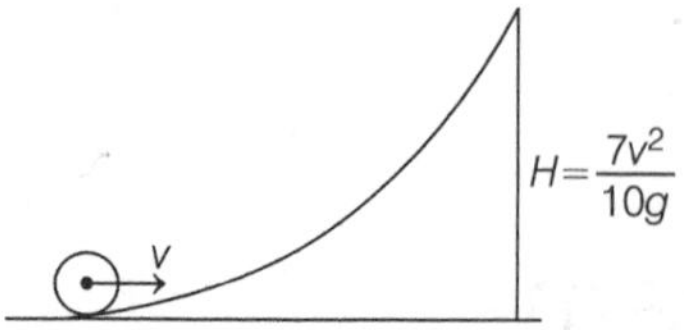

(a) ring (b) solid sphere
(c) hollow sphere (d) disc

10. An electron, helium ion (He^{++}) and proton having the same kinetic energy. The relation between their respective de-Broglie wavelengths λ_e, $\lambda_{He^{++}}$ and λ_p is

(a) $\lambda_e > \lambda_p > \lambda_{He^{++}}$
(b) $\lambda_e > \lambda_{He^{++}} > \lambda_p$
(c) $\lambda_e < \lambda_p < \lambda_{He^{++}}$
(d) $\lambda_e < \lambda_{He^{++}} = \lambda_p$

Numerical Answer Type (NAT)

11. A wire in the form of semi-circle of radius r rotates about its diameter with angular velocity ω in a magnetic field B. The axis of rotation is perpendicular to the field. The total resistance of the circuit is R. If the mean power generated per period of rotation is $\frac{(B\pi r^2\omega)^2}{xR}$, then the value of x is ....

12. A vessel contains oil (Density = 0.8 g/cm^3) over mercury (density = 13.6 g/cm^3). A homogeneous sphere floats with half of its volume immersed in mercury and the other half in oil. The density of the material of the sphere is x g/cm^3. The value of x is ....

13. The stress that has to be applied to the ends of a steel wire of length 20 cm to keep its length constant, when its temperature is raised by 100°C is 2.2×10^x Pa. The value of x is ....

(Given $Y = 2 \times 10^{11}$ Nm^{-2}, $\alpha = 1.1 \times 10^{-5}$ °C^{-1})

14. In a coil of resistance 150 Ω, a current is induced by changing the magnetic flux through it as shown by figure. The magnitude of flux through the coil is .... Wb.

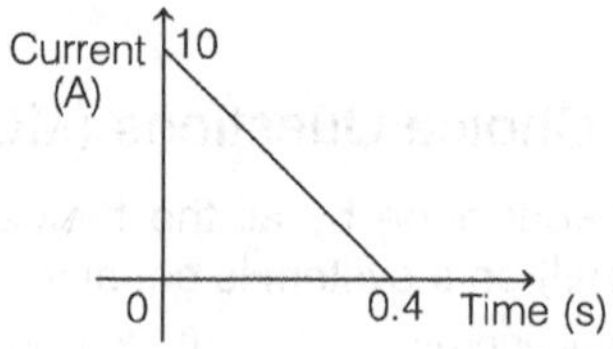

15. The electric field of light wave is given is

$$\mathbf{E} = 10^3 \cos\left(\frac{2\pi x}{5 \times 10^{-7}} - 2\pi \times 6 \times 10^{14} t\right)\hat{\mathbf{j}} \text{ N/C.}$$

This light falls on a metal plate of work function 1.5 eV. The stopping potential of the photoelectron is .... V.

(Energy of photon = $\frac{1240}{\lambda \text{ (in nm)}}$ eV).

Chemistry

Multiple Choice Questions (MCQs)

1. Arrange the following in increasing order of the volume (in L) occupied by them at STP

(i) 1.5 moles of CO_2
(ii) 14 g of N_2
(iii) 10^{21} molecules of oxygen

(a) (iii) < (ii) < (i) (b) (i) < (ii) < (iii)
(c) (ii) < (i) < (iii) (d) (i) < (iii) < (ii)

2. What is the first step and the final product formed in the reaction of H—Br with

$$CH_3-\underset{\underset{CH_3}{|}}{C}H-CH=CH_2\ ?$$

(a)
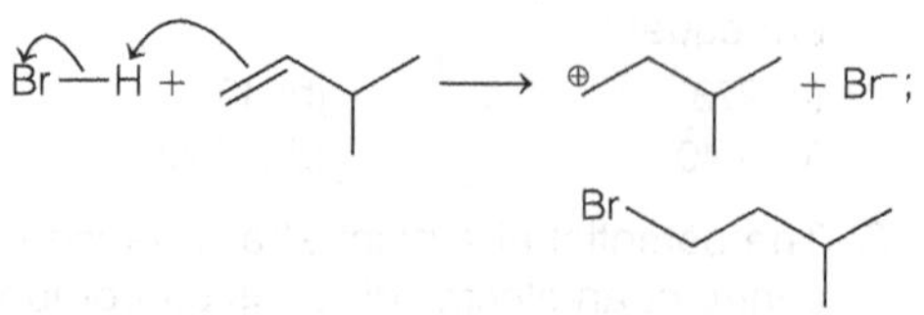

(b)
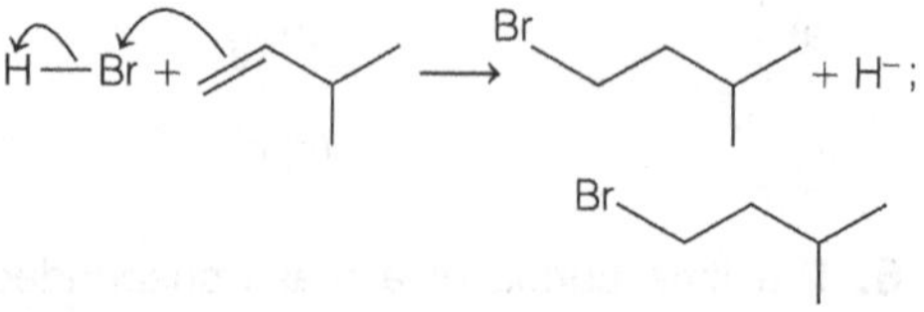

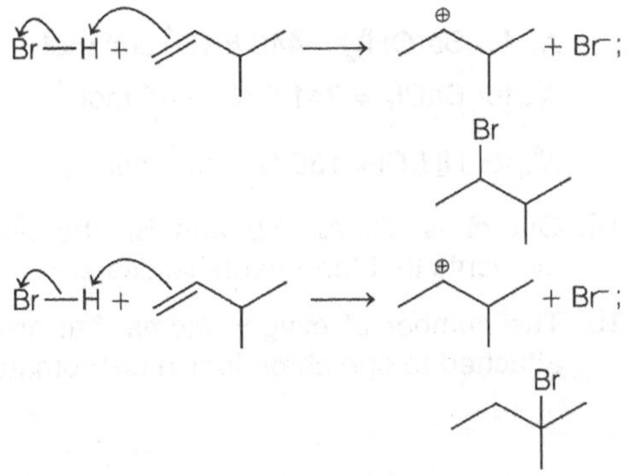

3. Alcoholic group of which position in the given molecule reacts fastest with the Lucas' reagent?

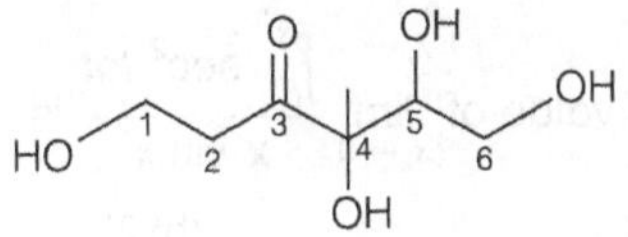

(a) 1 (b) 4 (c) 5 (d) 6

4. Select the correct statement.

(a) $[Ni(CN)_4]^{2-}$ is diamagnetic whereas $[Ni(CO)_4]$ is paramagnetic.

(b) $[Ni(CN)_4]^{2-}$ and $[Ni(CO)_4]$ both are diamagnetic.

(c) $[Ni(CN)_4]^{2-}$ is sp^3 hybridised and square planar in shape whereas $[Ni(CO)_4]$ is dsp^2 hybridised and tetrahedral in shape.

(d) $[Ni(CN)_4]^{2-}$ is paramagnetic and dsp^2 hybridised whereas $[Ni(CO)_4]$ is diamagnetic and sp^3 hybridised.

5. According to VSEPR theory, the molecular shapes of XeF_4, XeO_4, XeO_2F_2 and $XeOF_4$ respectively are

(a) square planar, square planar, *see-saw*, square pyramidal.

(b) square planar, tetrahedral, trigonal bipyramidal, octahedral.

(c) square planar, tetrahedral, *see-saw*, square pyramidal.

(d) octahedral, tetrahedral, trigonal bipyramidal, octahedral.

6. In the following sequence of reactions, the compound *C* formed would be

$$CH_3CH_2\overset{O}{\overset{\|}{C}}—NH_2 \xrightarrow{LiAlH_4} A \xrightarrow[\text{(ii) } H_2O]{\text{(i) } HNO_2} B \xrightarrow{CrO_3} C$$

(a) 2-propanol (b) propanol

(c) propanoic acid (d) propanal

7. Mond's process is used for refining .... Whereas van Arkel method is used for refining ....

(a) lead, zirconium (b) zirconium, nickel

(c) nickel, lead (d) nickel, titanium

8. If gas absorbs 150 J of heat and expands by 450 cm^3 against a constant pressure of 2×10^5 Nm^{-2}, then change in internal energy is

(a) – 60 J (b) 60 J (c) 240 J (d) – 240 J

9. In case of positive deviation from Raoult's law, the intermolecular attractive forces between the solute-solvent molecules as compared to those between the solute-solute and solvent-solvent molecules are

(a) weaker

(b) stronger

(c) same

(d) independent of intermolecular forces between solute-solvent molecules.

10. Polystyrene is a .... polymer whereas dacron is a .... polymer.

(a) step growth, chain growth

(b) chain growth, step growth

(c) condensation, addition

(d) thermoplastic, thermosetting

Numerical Answer Type (NAT)

11. Out of the given compounds, the number of compounds are weaker acids than benzoic acid are ....

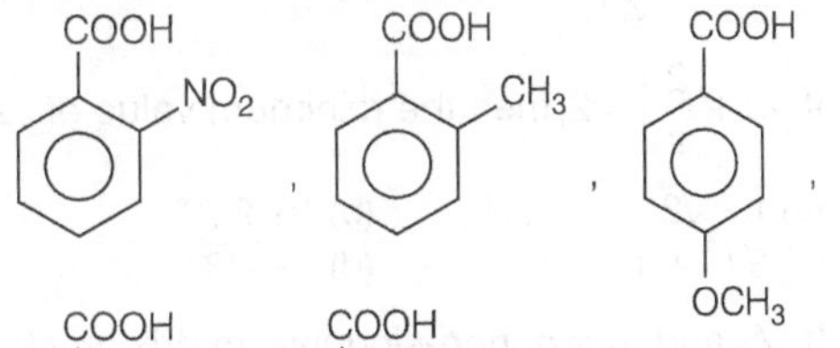

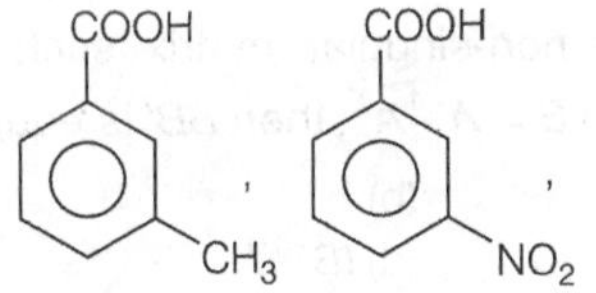

CH_3COOH,

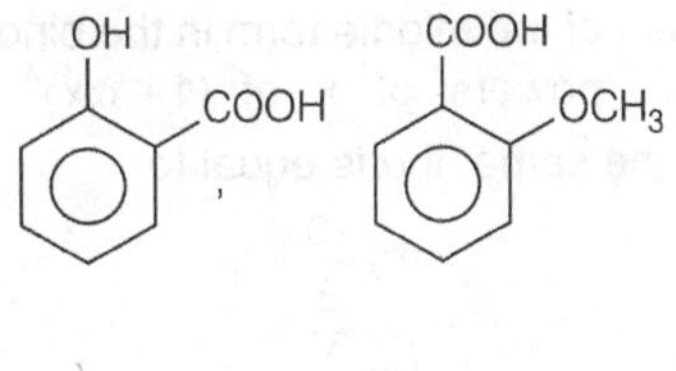

12. For the reaction,
$N_2O_5(g) \longrightarrow 2NO_2(g) + \frac{1}{2}O_2(g)$ the value of the rate of disappearance of N_2O_5 is given as 5.15×10^{-3} mol $L^{-1}s^{-1}$. The rate of formation of NO_2 is..........

13. From the following molar conductivities at infinite dilution, Λ_m° (in Ω^{-1} cm^2 mol^{-1}) for NH_4OH is ....

Λ_m° for $Ba(OH)_2 = 446.8\ \Omega^{-1}\ cm^2\ mol^{-1}$

Λ_m° for $BaCl_2 = 241.6\ \Omega^{-1}\ cm^2\ mol^{-1}$

Λ_m° for $NH_4Cl = 130\ \Omega^{-1}\ cm^2\ mol^{-1}$

14. Out of N, P, Ar, Sb and Bi, the number of elements that form pentahalides are ....

15. The number of oxygen atoms that are directly attached to one chromium in dichromate ion are ....

Mathematics

Multiple Choice Questions (MCQs)

1. If $A = \{1, 2\}$, $B = \{1, 2, 3, 4\}$, $C = \{5, 6\}$ and $D = \{5, 6, 7, 8\}$, then state which of the following is true?
(a) $(A \times C) \subset (B \times D)$ (b) $A \times B \subset C \times D$
(c) $(A \times B) \subset (A \times D)$ (d) $A \times C \subset B \times D$

2. Let $f(x) = \sqrt{1 + x^2}$, then
(a) $f(xy) = f(x) \cdot f(y)$ (b) $f(xy) \geq f(x) \cdot f(y)$
(c) $f(x, y) \leq f(x) \cdot f(y)$ (d) $f(xy) = f(x) - f(y)$

3. For what value of θ lying between 0 and π which satisfy inequality $\sin\theta\cos^3\theta > \sin^3\theta\cos\theta$
(a) $\theta \in \left(\frac{\pi}{4}, \frac{\pi}{2}\right)$ (b) $\theta \in \left(0, \frac{\pi}{4}\right)$
(c) $\theta \in \left(0, \frac{\pi}{2}\right)$ (d) None of these

4. If $\left| z + \frac{2}{z} \right| = 2$, then the minimum value of $|z|$ is
(a) $1 + \sqrt{2}$ (b) $1 + 2\sqrt{2}$
(c) $3\sqrt{3} + 1$ (d) $1 - \sqrt{3}$

5. If A and $n \times n$ non-singular matrix such that $AA^T = A^{-1}A$ and $B = A^{-1}A^T$, then BB' is equal to
(a) $I + B$ (b) I
(c) B^{-1} (d) $(B^{-1})'$

6. The coefficient of the middle term in the binomial expansion in powers of x of $(1 + \alpha x)^4$ and $(1 - \alpha x)^6$ is the same, if α is equal to
(a) $-\frac{5}{3}$ (b) $\frac{10}{3}$
(c) $-\frac{3}{10}$ (d) $\frac{3}{5}$

7. The value of $\lim_{x \to 0} \frac{\int_0^{x^2} \sec^2 t\,dt}{x \sin x}$ is
(a) 3 (b) 2
(c) 1 (d) -1

8. The differential equation of all circles passing through the origin and having their centre on the X-axis is
(a) $x^2 = y^2 + xy\frac{dy}{dx}$ (b) $x^2 = y^2 + 3xy\frac{dy}{dx}$
(c) $y^2 = x^2 + 2xy\frac{dy}{dx}$ (d) $y^2 = x^2 - 2xy\frac{dy}{dx}$

9. The area of the region bounded by the curves $y = |x - 2|$, $x = 1$, $x = 3$ and the X-axis is
(a) 1 sq unit (b) 2 sq units
(c) 3 sq units (d) 4 sq units

10. If α, β and γ are in AP and $\tan^{-1}\alpha$, $\tan^{-1}\beta$ and $\tan^{-1}\gamma$ are also in AP, then
(a) $\alpha - \beta - \gamma = 0$ (b) $\alpha = \beta = \gamma$
(c) $\alpha + \beta = \gamma$ (d) $2\alpha = 3\beta = \gamma$

11. The area of the region described by $A = \{(x, y) : x^2 + y^2 \leq 1 \text{ and } y^2 \leq 1 - x\}$ is
(a) $\frac{\pi}{2} + \frac{4}{3}$ (b) $\frac{\pi}{2} - \frac{1}{2}$
(c) $\frac{\pi}{4} + \frac{2}{3}$ (d) $\frac{\pi}{2} + \frac{2}{3}$

12. If the lines $\frac{x-2}{1} = \frac{y-3}{1} = \frac{z-4}{-k}$ and $\frac{x-1}{2k} = \frac{y-3}{2} = \frac{z-5}{1}$ are coplanar, then k can have

(a) exactly one value, $k = \frac{1}{2}$

(b) exactly one value, $k = \frac{1}{4}$

(c) exactly two values, $k = \frac{1}{2}, -\frac{3}{2}$

(d) any value

13. A multiple choice examination has 5 questions. Each question has 4 alternative answer of which exactly one is correct. The probability that a student will get 4 or more correct answer just by guessing is

(a) $\frac{1}{4^5}$ (b) $\left(\frac{3}{4}\right)^4$

(c) $\frac{1}{4^3}$ (d) $\frac{3}{4^5}$

14. If $\sum_{i=1}^{10}(x_i - 3) = 7$ and $\sum_{i=1}^{10}(x_i - 3)^2 = 27$, then the standard deviation of the 10 items

(a) 2.547 (b) 1.87 (c) 14.86 (d) 1.486

15. If $\int f(x)dx = g(x)$, then $\int x^9 \cdot f(x^5)\, dx$ is equal to

(a) $\frac{1}{5}(x^5 g(x^9) - 4\int g(x)\, dx + C$

(b) $\frac{1}{5}[x^9 g(x^5) - \frac{1}{5}\int x^4\, g(x^5)\, dx + C$

(c) $\frac{1}{5}[g(x^9) + \int g(x^5)\, dx]$

(d) $\frac{x^5}{5} g(x^5) - \int x^4 g(x^5)\, dx + C$

Numerical Answer Type (NAT)

16. $\lim_{x \to 2} \frac{x^2 + 2^{2-x} - 5}{2^{-x/2} - 2^{1-x}}$ is equal to ....

17. If the foot of the perpendicular drawn from the point (0, 2, 1) on a line passing through (α, 5, 1) is $\left(\frac{5}{3}, \frac{7}{3}, \frac{15}{4}\right)$, then α is equal to ....

18. The number of 4 letter word (with or without meaning) that can be formed from the eleven letters of the word 'EXAMINATION is ....

19. If $\begin{vmatrix} a & a^2 & 1+a^3 \\ b & b^2 & 1+b^3 \\ c & c^2 & 1+c^3 \end{vmatrix} = 0$ and vectors $(1, a, a^2)$, $(1, b, b^2)$ and $(1, c, c^2)$ are non-coplanar, then the product abc is ....

20. If the probability that a randomly chosen 6-digit number formed by using digits 1 and 8 only is a multiple of 21 is p, then 96 p is equal to ....

English

Multiple Choice Questions (MCQs)

1. Fill in the blank with the correct tense form.

Where is John? He .... his car in the garage.

(a) repairs (b) is repairing

(c) has repaired (d) repaired

2. Choose the correct sentence that follows the correct subject-verb agreement.

(a) One of my friends live in Russia.

(b) One of my friends living in Russia.

(c) One of my friends lives in Russia.

(d) None of the above

3. Fill in the blank with suitable determiner.

She showed .... concern for her step-daughter.

(a) a little (b) little

(c) few (d) some

4. Choose the appropriate conjunction from the options given below.

Both my father .... my wife are ill.

(a) and (b) or

(c) but (d) if

5. Determine the clause highlighted in the given sentence.

He wanted to ask her to come along.

(a) Finite clause (b) Non-finite clause

(c) Noun clause (d) Adverb clause

6. Choose the alternative which best expresses the meaning of the given underlined idiom/phrase.

She tries very hard to keep up with her rich neighbours.

(a) to imitate (b) to keep in touch

(c) to avoid (d) to be on par

7. Choose the word nearest in meaning to the italicised part.

The teacher felt that the student lacked *discrimination* in the study of his data.

(a) imagination (b) good taste
(c) objectivity (d) good judgement

8. Choose the one which can be substituted for the given words/sentence.
A person who maliciously destroys by fire.

(a) Antagonist (b) Activist
(c) Terrorist (d) Incendiary

9. Fill in the blank with correct alternative.
Avoid eating .... for healthy living.

(a) meet (b) meat
(c) mute (d) mate

10. Fill in the blank with suitable question tag.
Your father used to be the principal of this college, ... ?

(a) did he
(b) does he
(c) didn't he
(d) doesn't he

Answers

Physics

1. (b)	**2.** (b)	**3.** (b)	**4.** (c)	**5.** (b)	**6.** (c)	**7.** (b)	**8.** (b)	**9.** (b)	**10.** (a)
11. (8)	**12.** (7.2)	**13.** (8)	**14.** (300)	**15.** (0.98)					

Chemistry

1. (a)	**2.** (d)	**3.** (b)	**4.** (b)	**5.** (c)	**6.** (d)	**7.** (d)	**8.** (b)	**9.** (a)	**10.** (b)
11. (3)	**12.** (1.03×10^{-2})	**13.** (2.32.6)	**14.** (4)	**15.** (4)					

Mathematics

1. (a)	**2.** (c)	**3.** (b)	**4.** (d)	**5.** (b)	**6.** (c)	**7.** (c)	**8.** (c)	**9.** (a)	**10.** (b)
11. (a)	**12.** (b)	**13.** (c)	**14.** (d)	**15.** (d)	**16.** (12)	**17.** (10.893)	**18.** (2454)	**19.** (–1)	**20.** (33)

English

1. (b)	**2.** (c)	**3.** (b)	**4.** (a)	**5.** (b)	**6.** (d)	**7.** (d)	**8.** (d)	**9.** (b)	**10.** (c)

SOLUTIONS

Physics

1. *(b)* According to work-energy theorem,

$$W_{\text{external}} + W_{\text{internal}} = K_f - K_i = \Delta K$$

2. *(b)* As, $R = \rho \cdot \frac{L}{A} = \rho \cdot \frac{L^2}{A \cdot L} = \rho \cdot \frac{L^2}{V}$; ($V$ = Volume)

$$\therefore \quad R \propto L^2 \Rightarrow \frac{R_2}{R_1} = \left(\frac{L_2}{L_1}\right)^2 = \left(\frac{2L_1}{L_1}\right)^2 = 4$$

$$\therefore \quad R_2 = 4R_1 = 4R$$

3. *(b)* The escape velocity of a planet is given by,

$$v = \sqrt{\frac{2GM}{R}} = \sqrt{\frac{2G}{R} \times \frac{4}{3}\pi R^3 \cdot \rho}$$

$$\therefore \quad v \propto R\sqrt{\rho}$$

$$\therefore \quad \frac{v_P}{v_E} = \frac{R_P\sqrt{\rho_P}}{R_E\sqrt{\rho_E}} = \frac{R_P}{R_E} \quad (\because \rho_P = \rho_E)$$

$$\therefore \quad \frac{v_P}{v_E} = \frac{3R_E}{R_E} = 3 \quad (\because R_P = 3R_E)$$

$$\therefore \quad v_P = 3v_E$$

4. *(c)* As, $\frac{C}{100} = \frac{F - 32}{180}$

Now, $C = F$ (given)

$$\therefore \quad \frac{C}{100} = \frac{C - 32}{180}$$

$$\Rightarrow \quad 180C = 100C - 3200$$

$$\Rightarrow \quad 180C - 100C = -3200$$

$$\Rightarrow \quad 80C = -3200$$

$$\therefore \quad C = -\frac{3200}{80} = -40$$

$$\therefore \quad C = F = -40$$

5. *(b)* The electric potential due to an electric dipole at a distance r from its centre is given by

$$V = \frac{1}{4\pi\varepsilon_0} \cdot \frac{p\cos\theta}{r^2}$$

[where, p = electric dipole moment]

$$\therefore \quad V \propto \frac{1}{r^2}$$

6. *(c)* As, spring constant, $(k) \propto \frac{1}{\text{length of spring } (l)}$

$\because$ The spring cut into three equal parts

$\therefore$ $k' = 3k$ (For each spring)

We know that, time period of a body connected with spring.

$$T = 2\pi\sqrt{\frac{m}{k}} \quad \text{..(i)}$$

The parallel arrangement of the spring is given as

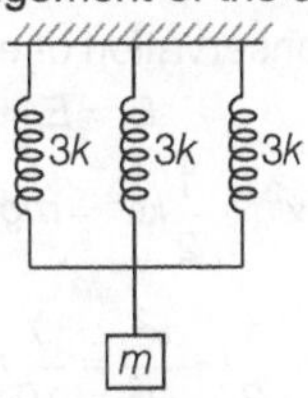

T' = Time period of springs ($k' = 3k$) connected in parallel

$$= 2\pi\sqrt{\frac{m}{3k'}} = 2\pi\sqrt{\frac{m}{3 \times 3k}}$$

$$= \frac{1}{3}\left(2\pi\sqrt{\frac{m}{k}}\right) = \frac{T}{3} \quad \text{[From Eq. (i)]}$$

7. *(b)* If reflected ray from silvered surface retrace its path, then it falls normally on the silvered surface.

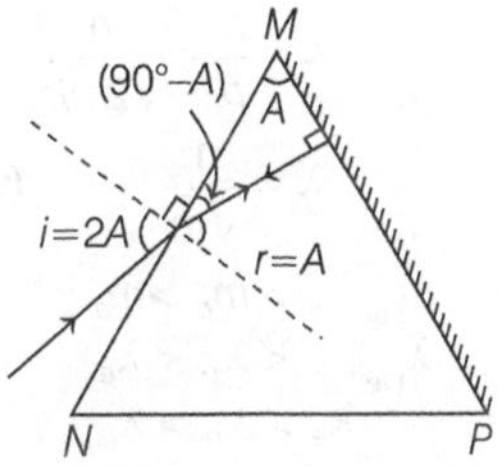

$\therefore$ From the geometry,

$$i = 2A, r = A$$

$$\therefore \quad \mu = \text{Refractive index} = \frac{\sin i}{\sin r} \text{ (From Snell's law)}$$

$$= \frac{\sin 2A}{\sin A} = \frac{2\sin A \cdot \cos A}{\sin A}$$

$$= 2\cos A$$

8. *(b)* The charged particle will not strike the screen, if $d > R$ (Radius of path)

$$\Rightarrow \quad d > \frac{m(2v)}{B_0 q} \Rightarrow v < \frac{qdB_0}{2m}$$

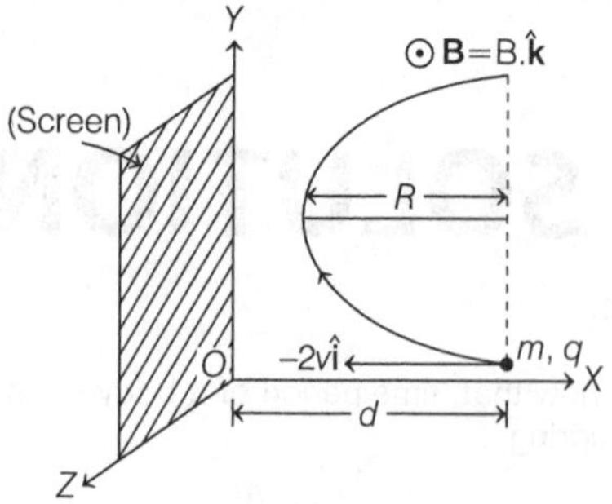

$\therefore \quad v_{max} = \dfrac{qdB_0}{2m}$

9. *(b)* By the conservation of energy,

$$E_i = E_f$$

$$\Rightarrow \quad \frac{1}{2}mv^2 + \frac{1}{2}I\omega^2 = mgH = mg\left(\frac{7v^2}{10g}\right)$$

$$\Rightarrow \quad \frac{1}{2}mv^2 + \frac{1}{2}I\cdot\frac{v^2}{R^2} = \frac{7}{10}mv^2 \qquad (\because v = \omega R)$$

$$\Rightarrow \quad \frac{1}{2}I\cdot\frac{v^2}{R^2} = \left(\frac{7}{10} - \frac{1}{2}\right)mv^2$$

$$= \frac{2}{10}mv^2 = \frac{1}{5}mv^2$$

$$\therefore \quad I = \frac{2}{5}mR^2$$

The above value of moment of inertia is equal to moment of inertia of solid sphere.

10. *(a)* de-Broglie wavelength is given by

$$\lambda = \frac{h}{p} = \frac{h}{\sqrt{2mk}}$$

$$\therefore \quad \lambda \propto \frac{1}{\sqrt{m}} \qquad (\because k_e = k_p = k_{He^{++}})$$

As, $\quad m_{He^{++}} > m_p > m_e$

$\therefore \quad \lambda_{He^{++}} < \lambda_p < \lambda_e$

or $\quad \lambda_e > \lambda_p > \lambda_{He^{++}}$

11. *(8)* $\phi = \mathbf{B}\cdot\mathbf{A} = BA\cos\theta = BA\cos\omega t$, $(\because \theta = \omega t)$

Now, emf $(e) = -\dfrac{d\phi}{dt} = -\dfrac{d}{dt}(BA\cos\omega t)$

$$= \omega BA\sin\omega t$$

$$i = \frac{e}{R} = \frac{\omega BA\sin\omega t}{R}$$

$$\therefore \quad P_{ins} = i^2R = \left(\frac{\omega^2B^2A^2}{R^2}\sin^2\omega t\right)\cdot R$$

$$= \frac{\omega^2B^2A^2}{R}\sin^2\omega t$$

$$\therefore \quad P_{avg} = \frac{\int_0^T P_{ins}\times dt}{\int_0^T dt}$$

$$= \frac{\omega^2B^2A^2}{R}\cdot\frac{\int_0^T \sin^2\omega t\,dt}{\int_0^T dt}$$

$$= \frac{1}{2}\cdot\frac{\omega^2B^2A^2}{R} \qquad \left(\because <\sin^2\omega t>_{t=T} = \frac{1}{2}\right)$$

$$= \frac{(B\pi r^2\omega)^2}{8R} \qquad \left(\because A = \frac{\pi r^2}{2}\right)$$

$$\therefore \quad x = 8$$

12. *(7.2)* For sphere to be float

weight $= F_{B_1} + F_{B_2}$

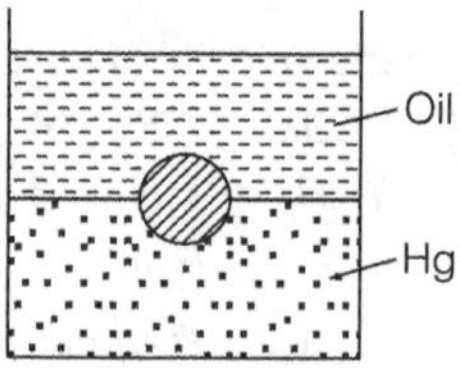

$$\Rightarrow \quad V\rho g = \rho_{oil}\times\frac{V}{2}\times g + \rho_{Hg}\times\frac{V}{2}\times g$$

$$\Rightarrow \quad \rho = \frac{\rho_{oil}}{2} + \frac{\rho_{Hg}}{2} = \frac{0.8}{2} + \frac{13.6}{2}$$

$$= 0.4 + 6.8$$

$\therefore \ \rho =$ density of sphere $= 7.2\ \text{g/cm}^3$

$\therefore \ x\ \text{g/cm}^3 = 7.2\ \text{g/cm}^3$

$\therefore \quad x = 7.2$

13. *(8)* From Hook's law,

$$Y = \frac{\text{Stress}}{\text{Strain}}$$

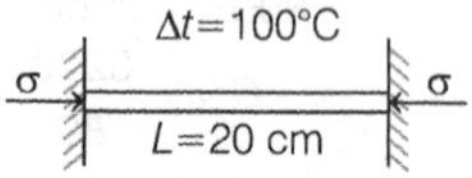

$\therefore \quad$ Stress $= Y \times$ Strain

Now, $\quad$ Strain $= \dfrac{\Delta L}{L} = \dfrac{L\alpha\Delta T}{L} = \alpha\Delta T$

$\therefore \quad$ Stress $= Y \times \alpha\Delta T$

$$= 2\times10^{11}\times1.1\times10^{-5}\times100$$

$$= 2.2\times10^8\ \text{Pa}$$

According to question,

$$2.2\times10^x = 2.2\times10^8$$

$$\therefore \quad x = 8$$

14. *(300)* By Faraday's law of electromagnetic induction,

$\varepsilon = \frac{d\phi}{dt}$, Also, $\varepsilon = iR$

$\therefore \quad \frac{d\phi}{dt} = iR \Rightarrow d\phi = iRdt$

$\Rightarrow \quad \int d\phi = R\int idt$

$\therefore \quad \phi = R \times$ Area of current *versus* time

$= 150 \times \frac{1}{2} \times 0.4 \times 10 = 300\text{ Wb}$

15. *(0.98)* $E = E_0 \cos(kx - \omega t)$

$\therefore \quad \omega = 2\pi \times 6 \times 10^{14}$

$\therefore \quad f = \frac{\omega}{2\pi} = \frac{2\pi \times 6 \times 10^{14}}{2\pi} = 6 \times 10^{14}\text{ Hz}$

$\phi = \text{Energy of photon} = \frac{1240}{\lambda\text{ (in nm)}}\text{ eV}$

$\therefore \quad \lambda = \frac{c}{f} = \frac{3 \times 10^8}{6 \times 10^{14}} = 0.5 \times 10^{-6}\text{ m} = 500\text{ nm}$

$\therefore \quad \phi = \frac{1240}{500}\text{ eV} = 2.48\text{ eV}$

Now, by Einstein photoelectric equation,

$\phi = \phi_0 + KE = \phi_0 + eV_s \quad (\because KE = eV_s)$

$\Rightarrow \quad 2.48 = 1.5 + eV_s$

$eV_s = 2.48 - 1.5 = 0.98\text{ eV}$

$\therefore \quad V_s = 0.98\text{ V}$

Chemistry

1. *(a)*

(i) Volume occupied by 1 mole of CO_2 = 22.4 L

Volume occupied by 1.5 moles of CO_2

= 1.5 × 22.4

= 33.6 L

(ii) Given, mass of nitrogen (m) = 14 g

Molar mass of N_2 (w) = 28 g

Number of moles of $N_2 = \frac{m}{w} = \frac{14}{28} = \frac{1}{2}$

= 0.5 mole

Volume occupied by 1 mole of N_2 at STP

= 22.4 L

Volume occupied by 0.5 mole of N_2 at STP

= 0.5 × 22.4

= 11.2 L

(iii) Given, number of molecules of $O_2 = 10^{21}$

Number of molecules of O_2 in 1 mole

$= 6.022 \times 10^{23}$ molecules

Number of moles $= \frac{10^{21}}{6.022 \times 10^{23}}$

$= 0.166 \times 10^{-2}$ moles

Volume occupied by 0.166×10^{-2} mole of $O_2 = 0.166 \times 10^{-2} \times 22.4$ L

$= 3.72 \times 10^{-2}$

= 0.0372 L

Therefore, correct increasing order of volume occupied by the given amount of molecules at STP is : (iii) < (ii) < (i).

2. *(d)*

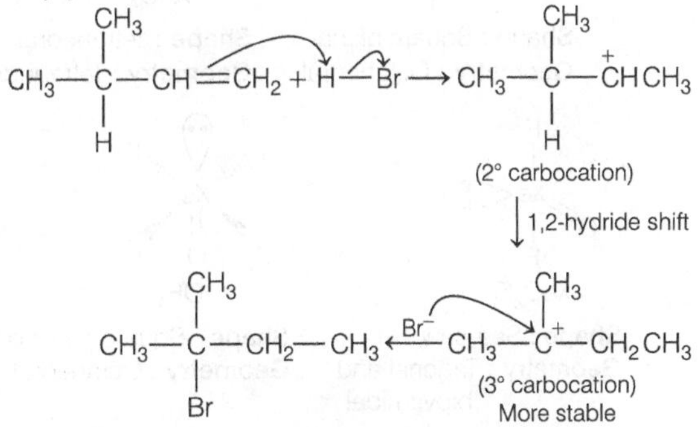

3. *(b)* Alcohols are soluble in Lucas' reagent (conc. HCl and $ZnCl_2$) while their halides are immiscible and produce turbidity in solution. Tertiary alcohols react the fastest and form halides easily. Primary alcohols do not produce turbidity at room temperature. Therefore, alcoholic group at the 4th position is tertiary, therefore it reacts the fastest.

4. *(b)* Both $[Ni(CN)_4]^{2-}$ and $[Ni(CO)_4]$ are diamagnetic.

$[Ni(CN)_4]^{2-}$

Electronic configuration of Ni^{2+} = [Ar] $3d^8$.

CN^- being strong field ligand cause pairing of electrons.

Hybridised orbitals of $[Ni(CN)_4]^{2-}$ =

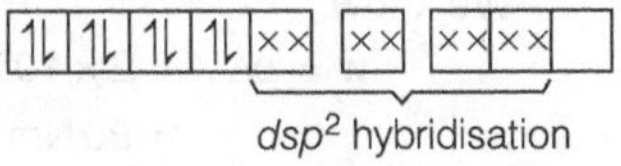

There are no unpaired electrons. Therefore, it is diamagnetic.

Shape : square planar

$[Ni(CO)_4]$

Electronic configuration of Ni = $[Ar]3d^8 4s^2$

CO being strong field ligand cause pairing of electrons.

Hybridised orbitals of $[Ni(CO)_4]$ 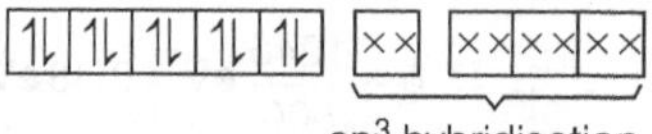

No unpaired electrons are there. Therefore, it is diamagnetic in nature.

Shape : Tetrahedral.

5. *(c)*

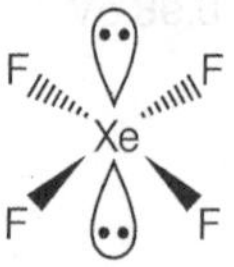

XeF_4

Shape : Square planar
Geometry : Octahedral

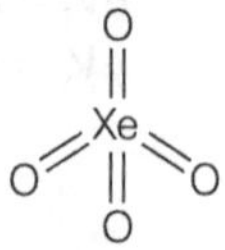

XeO_4

Shape : Tetrahedral
Geometry : Tetrahedral

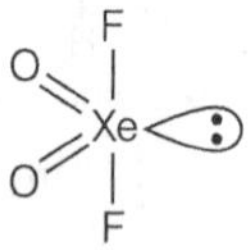

XeO_2F_2

Shape : See-saw
Geometry : Trigonal and bipyamidal

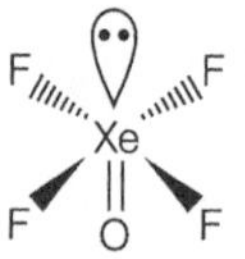

$XeOF_4$

Shape : Square pyramidal
Geometry : Octahedral

6. *(d)* $\underset{\text{Amide}}{CH_3CH_2\overset{O}{\overset{\|}{C}}—NH_2} \xrightarrow[\text{Reduction}]{LiAlH_4} \underset{\text{Amine}}{\underset{(A)}{CH_3CH_2CH_2NH_2}}$

$\xrightarrow[\text{(ii) } H_2O]{\text{(i) } HNO_2}$

$\underset{\text{Propanal}}{\underset{(C)}{CH_3CH_2CHO}} \xleftarrow[\text{Oxidation}]{CrO_3} \underset{\text{Propanol}}{\underset{(B)}{CH_3CH_2CH_2OH}}$

7. *(d)* Mond's process is used for refining nickel. Whereas, van Arkel method is used for refining zirconium and titanium.

8. *(b)* Given, $q = 150\,J$

We know,

$$w = p\Delta V = 2 \times 10^5\ Nm^{-2} \times 450 \times 10^{-6}\ m^3$$
$$= 90\,Nm = 90\,J$$

Also, $\Delta U = q - w = 150 - 90 = 60\,J$

9. *(a)* In case of positive deviation from Raoult's law, the intermolecular attractive forces between the solute-solvent molecules are weaker than those between the solute-solute and solvent-solvent molecules. Therefore, it becomes easy for the solute/solvent molecules to escape from the solution than in pure state. This will increase the vapour pressure and results in positive deviation.

10. *(b)* $nC_6H_5—CH{=}CH_2 \xrightarrow{\text{Polymerisation}}$

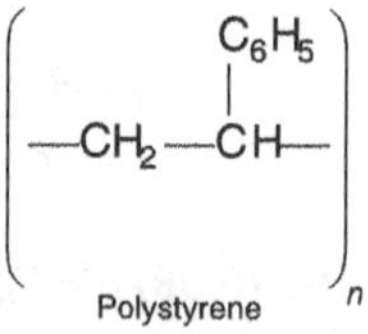

Polystyrene

Polystyrene is an addition and chain growth polymer. It is thermoplastic. Addition polymers have unsaturated monomers.

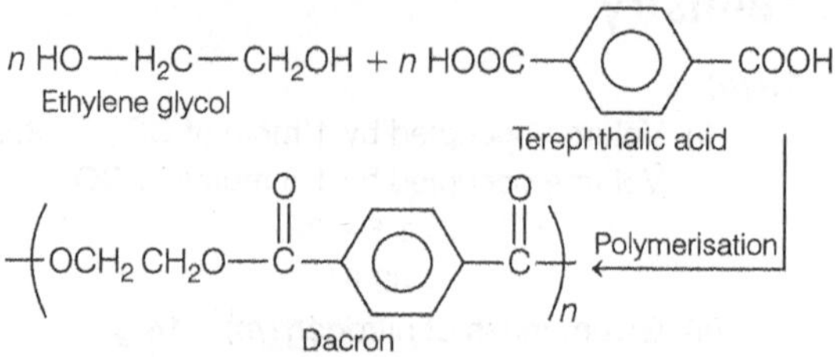

Dacron

Dacron is a condensation and step-growth polymer. It is a fibre.

Condensation polymerisation involves a repetitive condensation reaction between two bi-functional or tri-functional monomeric units.

11. *(3)* Attached electron withdrawing groups increase the acidity of benzoic acid by $-I$ and $-M$-effect (whenever applicable) whereas electron releasing groups decrease the acidity of benzoic acid.

Due to *othro*-effect, 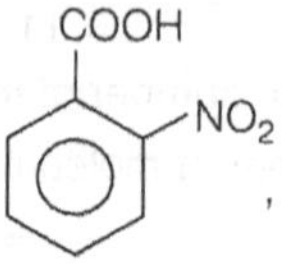

, 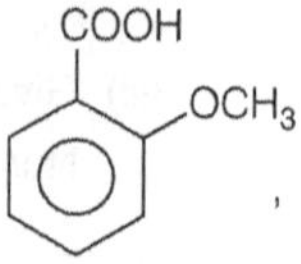

,

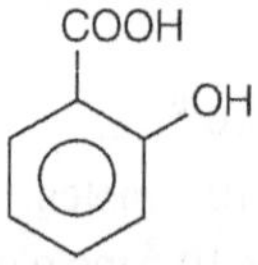

are stronger acids than 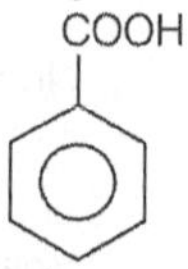

even though they have electron releasing groups (i.e. OCH_3) attached.

In 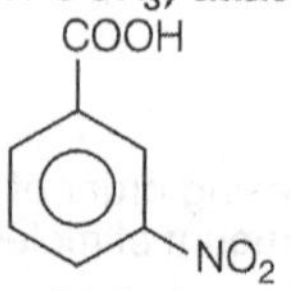

one electron withdrawing group is attached, due to this reason it is more acidic than benzoic acid.

Moreover, aliphatic carboxylic acid is a weaker acid than aromatic carboxylic acid because benzene itself is electron withdrawing in nature that makes the release of proton from —COOH easy.

Therefore, acids which are weaker than

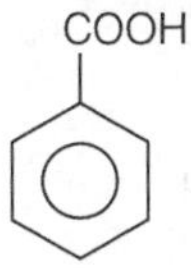

are

CH_3COOH, 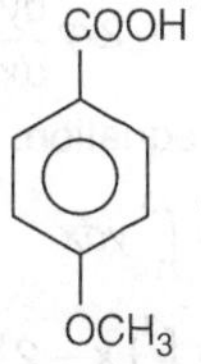, 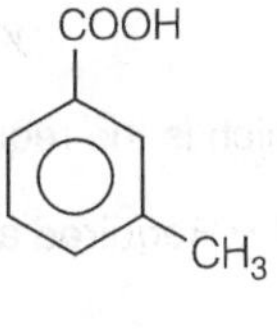

12. *(1.03×10^{-2})* For the given reaction,

$$N_2O_5(g) \longrightarrow 2NO_2(g) + \frac{1}{2}O_2(g)$$

$$-\frac{d[N_2O_5]}{dt} = \frac{1}{2}\frac{d[NO_2]}{dt} = \frac{2d[O_2]}{dt}$$

$$\therefore \quad \frac{d[NO_2]}{dt} = -2\frac{d[N_2O_5]}{dt} = 2 \times 5.15 \times 10^{-3}$$

$$= 1.03 \times 10^{-2} \text{ mol L}^{-1}\text{ s}^{-1}.$$

13. *(232.6)*

$$\overset{\circ}{\Lambda}_{m(NH_4OH)} = \overset{\circ}{\Lambda}_{NH_4^+} + \overset{\circ}{\Lambda}_{OH^-}$$

$$= (\overset{\circ}{\Lambda}_{NH_4^+} + \overset{\circ}{\Lambda}_{Cl^-}) + \frac{1}{2}(\overset{\circ}{\Lambda}_{Ba^{2+}} + 2\overset{\circ}{\Lambda}_{OH^-}) - \frac{1}{2}(\overset{\circ}{\Lambda}_{Ba^{2+}} + 2\overset{\circ}{\Lambda}_{Cl^-})$$

$$= \overset{\circ}{\Lambda}_{m\,(NH_4Cl)} + \frac{1}{2}\overset{\circ}{\Lambda}_{m\,[Ba(OH)_2]} - \frac{1}{2}\overset{\circ}{\Lambda}_{m\,(BaCl_2)}$$

$$= 130 + \frac{1}{2} + 446.8 - \frac{1}{2} \times 241.6$$

$$= 232.6\Omega^{-1}\text{ cm}^2\text{ mol}^{-1}.$$

14. *(4)* Nitrogen does not form pentahalide due to non-availability of the *d*-orbitals in its valence shell.

15. *(4)*

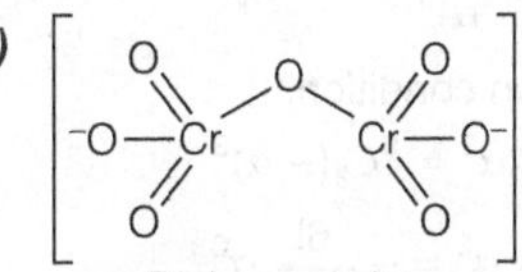

Dichromate ion

4 oxygen atoms are directly attached to one chromium ion.

Mathematics

1. *(a)* Here,

$$A \times C = \{1, 2\} \times \{5, 6\}$$
$$= \{(1, 5), (1, 6), (2, 5), (2, 6)\}$$

and $B \times D = \{1, 2, 3, 4\} \times \{5, 6, 7, 8\}$

$= \{(1, 5), (1, 6), (1, 7), (1, 8), (2, 5), (2, 6), (2, 7), (2, 8), (3, 5), (3, 6), (3, 7), (3, 8), (4, 5), (4, 6), (4, 7), (4, 8)\}$

$\therefore \quad (A \times C) \subset (B \times D)$

2. *(c)* Given that, $f(x) = \sqrt{1 + x^2}$

$\therefore \quad f(y) = \sqrt{1 + y^2}$

and $\quad f(xy) = \sqrt{1 + x^2y^2}$

Now, $\quad f(xy) = f(x)\,f(y)$

$\Rightarrow \quad \sqrt{1 + x^2y^2} = \sqrt{1 + x^2}\,\sqrt{1 + y^2}$

$= \sqrt{(1 + x^2)(1 + y^2)}$

On squaring both sides, we get

$$1 + x^2y^2 = (1 + x^2)(1 + y^2)$$

$\Rightarrow \quad 1 + x^2y^2 = 1 + y^2 + x^2 + x^2y^2$

$\therefore \quad 1 + x^2y^2 \le 1 + x^2 + y^2 + x^2y^2$

Hence, $\quad f(xy) \le f(x) \cdot f(y)$

3. *(b)* We have, $\sin\theta \cos^3\theta > \sin^3\theta \cos\theta$

$\Rightarrow \quad \sin\theta\cos^3\theta - \sin^3\theta\cos\theta > 0$

$\Rightarrow \quad \sin\theta\cos\theta(\cos^2\theta - \sin^2\theta) > 0$

$\Rightarrow \quad \sin 2\theta \cdot \cos 2\theta > 0$

$\Rightarrow \quad \sin 4\theta > 0$

$\Rightarrow \quad \theta \in \left(0, \frac{\pi}{4}\right) \quad [\because 4\theta \in (0, \pi)]$

4. *(d)* We can write

$$|z| = \left|z + \frac{2}{|z|} - \frac{2}{|z|}\right| \le \left|z + \frac{2}{z}\right| + \left|\frac{2}{|z|}\right|$$

$$= 2 + \frac{2}{|z|}$$

$\therefore \quad |z|^2 - 2|z| - 2 \le 0$

Here, $|z|$ lies between $\frac{2 \pm \sqrt{12}}{2} = 1 \pm \sqrt{3}$

$\therefore$ Minimum value of $|z| = 1 - \sqrt{3}$

5. *(b)* If A is non-singular matrix, then $|A| \neq 0$

$$AA^T = A^{-1}A \text{ and } B = A^{-1}A^T$$

$$\Rightarrow \quad BB' = (A^{-1}A^T)(A^{-1}A^T)^T$$
$$= A^{-1}A^T \cdot A(A^{-1})^T \qquad [\because (AI)' = A]$$
$$= A^{-1}AA^T(A^{-1})^T$$
$$= IA^T(A^{-1})^T$$
$$= A^T(A^{-1})^T$$
$$= (A^{-1}A)^T \qquad [\because (AB)^T = B^TA^T]$$
$$= I^T = I$$

6. *(c)* The coefficient of the middle term in power of $x(1+\alpha x)^4 = {}^4C_2\alpha^2$

The coefficient of the middle term in power of $x(1-\alpha x)^6 = {}^6C_3(-\alpha)^3$

According to given condition,

$${}^4C_2\alpha^2 = {}^6C_3(-\alpha)^3$$
$$\Rightarrow \quad \frac{4!}{2!2!}\alpha^2 = -\frac{6!}{3!3!}\alpha^3$$
$$\Rightarrow \quad 6\alpha^2 = -20\alpha^3$$
$$\Rightarrow \quad \alpha = -\frac{6}{20}$$
$$\therefore \quad \alpha = -\frac{3}{10}$$

7. *(c)* Let $I = \lim_{x \to 0} \dfrac{\int_0^{x^2} \sec^2 t\, dt}{x \sin x} \quad \left[\dfrac{0}{0} \text{ form}\right]$

Using L'Hospital's rule,

$$= \lim_{x \to 0} \frac{\sec^2 x^2 \cdot 2x}{x \cos x + \sin x} \quad \left[\frac{0}{0} \text{ form}\right]$$

Again using L'Hospital's rule,

$$\lim_{x \to 0} \frac{2x \cdot 2\sec^2 x^2 \cdot \tan x^2 \cdot 2x + 2\sec^2 x^2}{-x \sin x + \cos x + \cos x}$$
$$= \frac{0 + 2\sec^2 0}{0 + 2\cos 0} = 1$$

8. *(c)* General equation of all such circles which passing through the origin whose centre lie on X-axis is

$$x^2 + y^2 + 2gx = 0 \qquad \text{...(i)}$$

On differentiating w.r.t. 'x', we get

$$2x + 2y\frac{dy}{dx} + 2g = 0$$
$$\Rightarrow \quad 2g = -\left(2x + 2y\frac{dy}{dx}\right)$$

On putting the value of $2g$ in Eq. (i),

$$x^2 + y^2 + \left(-2x - 2y\frac{dy}{dx}\right)x = 0$$
$$\Rightarrow \quad x^2 + y^2 - 2x^2 - 2xy\frac{dy}{dx} = 0$$
$$\Rightarrow \quad y^2 = x^2 + 2xy\frac{dy}{dx}$$

which is the required equation.

9. *(a)* $\because$ Required area $= \int_1^3 y\,dx$

$$= \int_1^3 |x-2|\,dx$$
$$= \int_1^2 -(x-2)\,dx + \int_2^3 (x-2)\,dx$$
$$= \int_1^2 (2-x)\,dx + \int_2^3 (x-2)\,dx$$
$$= \left[2x - \frac{x^2}{2}\right]_1^2 + \left[\frac{x^2}{2} - 2x\right]_2^3$$
$$= \left[4 - 2 - \left(2 - \frac{1}{2}\right)\right] + \left[\frac{9}{2} - 6 - (2-4)\right]$$
$$= 2 - \frac{3}{2} - \frac{3}{2} + 2$$
$$= 1 \text{ sq unit}$$

10. (b) Since, α, β and γ are in AP.

$$\therefore \quad 2\beta = \alpha + \gamma \qquad \text{...(i)}$$

Also, $\tan^{-1}\alpha, \tan^{-1}\beta$ and $\tan^{-1}\gamma$ are also in AP.

$$\therefore \quad 2\tan^{-1}\beta = \tan^{-1}\alpha + \tan^{-1}\gamma$$
$$\Rightarrow \quad \tan^{-1}\left(\frac{2\beta}{1-\beta^2}\right) = \tan^{-1}\left(\frac{\alpha+\gamma}{1-\alpha\gamma}\right)$$
$$\Rightarrow \quad \frac{2\beta}{1-\beta^2} = \frac{\alpha+\gamma}{1-\alpha\gamma}$$
$$\Rightarrow \quad \frac{\alpha+\gamma}{1-\beta^2} = \frac{\alpha+\gamma}{1-\alpha\gamma} \qquad \text{[by using Eq. (i)]}$$
$$\Rightarrow \quad \beta^2 = \alpha\gamma$$

Since, α, β and γ are in AP as well as in GP.

$$\therefore \quad \alpha = \beta = \gamma$$

11. (a) Given, $A = \{(x, y) : x^2 + y^2 \le 1 \text{ and } y^2 \le 1 - x\}$

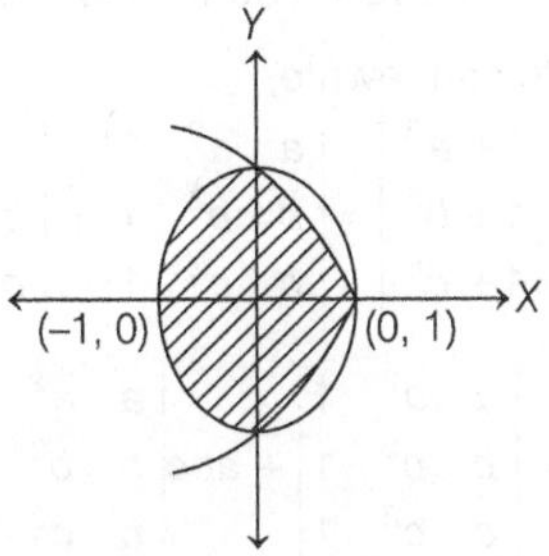

$\because$ Required area $= \frac{1}{2}\pi r^2 + 2\int_0^1 (1 - y^2)\, dy$

$= \frac{1}{2}\pi(1)^2 + 2\left[y - \frac{y^3}{3}\right]_0^1$

$= \frac{\pi}{2} + \frac{4}{3}$

12. *(b)* The given lines are

$$\frac{x-2}{1} = \frac{y-3}{1} = \frac{z-4}{-k} \quad \text{...(i)}$$

and $$\frac{x-1}{2k} = \frac{y-3}{2} = \frac{z-5}{1} \quad \text{...(ii)}$$

Condition for two lines are coplanar

$$\begin{vmatrix} x_1 - x_2 & y_1 - y_2 & z_1 - z_2 \\ l_1 & m_1 & n_1 \\ l_2 & m_2 & n_2 \end{vmatrix} = 0$$

where (x_1, y_1, z_1) and (x_2, y_2, z_2) are points on lines (i) and (ii) and (l_1, m_1, n_1) and (l_2, m_2, n_2) are direction cosines of lines (i) and (ii), respectively.

$$\therefore \quad \begin{vmatrix} 2-1 & 3-3 & 4-5 \\ 1 & 1 & -k \\ 2k & 2 & 1 \end{vmatrix} = 0$$

$$\Rightarrow \quad \begin{vmatrix} 1 & 0 & -1 \\ 1 & 1 & -k \\ 2k & 2 & 1 \end{vmatrix} = 0$$

$\Rightarrow 1(1 + 2k) - 0 - 1(2 - 2k) = 0$

$\Rightarrow 1 + 2k - 2 + 2k = 0$

$\Rightarrow 4k - 1 = 0 \Rightarrow k = \frac{1}{4}$

13. *(c)* Probability of guessing a correct answer, $p = \frac{1}{4}$

and probability of guessing a wrong answer $q = \frac{3}{4}$

So, the probability of guessing 4 or more correct answer

$P(X \ge 4) = {}^5C_4\left(\frac{1}{4}\right)^4 \cdot \left(\frac{3}{4}\right) + {}^5C_5\left(\frac{1}{4}\right)^5$

$= 5 \cdot \frac{1}{4^4} \cdot \frac{3}{4} + \frac{1}{4^5}$

$= \frac{15}{4^5} + \frac{1}{4^5} = \frac{16}{4^5} = \frac{4^2}{4^5} = \frac{1}{4^3}$

14. *(d)* We have,

$$\sum_{i=1}^{10}(x_i - 3) = 7 \text{ and } \sum_{i=1}^{10}(x_i - 3)^2 = 27$$

$\therefore$ Standard deviation is remain unchanged, if observation are added or subtracted by a fixed number.

$\therefore$ Standard deviation

$$= \sqrt{\frac{\sum_{i=1}^{10}(xi - 3)^2}{10} - \left(\frac{\sum_{i=1}^{10}(xi - 3)}{10}\right)^2}$$

$$= \sqrt{\frac{27}{10} - \left(\frac{7}{10}\right)^2}$$

$$= \sqrt{\frac{270}{100} - \frac{49}{100}} = \frac{\sqrt{221}}{10} = \frac{14.86}{10} = 1.486$$

15. *(d)* Given that, $\int f(x)\, dx = g(x)$

Let $I = \int x^9 \int (x^5)\, dx$

Put $x^5 = t \Rightarrow 5x^4\, dx = dt$

$= \frac{1}{5}\int t f(t)\, dt$

$= \frac{1}{5}[t \int f(t)\, dt - \int [f(t)\, dt]$

[integrating by part]

$= \frac{1}{5}[t\, g(t) - \int g(t)\, dt]$

$= \frac{1}{5}[x^2 g(x^5) - 5\int g(x^5)\, x^4\, dx] + C$

$= \frac{x^5}{5} g(x^5) - \int x^4 g(x^5)\, dx + C$

16. *(12)* We have,

$$\lim_{x \to 2} \frac{2^x + 2^{2-x} - 5}{2^{-x/2} - 2^{1-x}} \quad \left[\frac{0}{0} \text{ form}\right]$$

So, apply L' Hospital's rule

$$= \lim_{x \to 2} \frac{\frac{d}{dx}(2^x + 2^{2-x} - 5)}{\frac{d}{dx}(2^{-x/2} - 2^{1-x})}$$

$$= \lim_{x \to 2} \frac{2^x \ln 2 - 2^{2-x} \ln 2}{2^{1-x} \ln 2 - \frac{2^{-\frac{x}{2}} \ln 2}{2}}$$

$$= \lim_{x \to 2} \frac{2^{x+1} - 2^{3-x}}{2^{2-x} - 2^{-\frac{x}{2}}}$$

$$= \frac{2^3 - 2^1}{2^0 - 2^{-1}} = \frac{8-2}{2-\frac{1}{2}} = \frac{12}{2-1} = 12$$

17. *(10.893)* It is given that point $Q\left(\frac{5}{3}, \frac{7}{3}, \frac{15}{4}\right)$ is foot of perpendicular of point $P(0, 2, 1)$ to a line passes through point $A(\alpha, 5, 1)$. So, $PQ \perp AQ$.

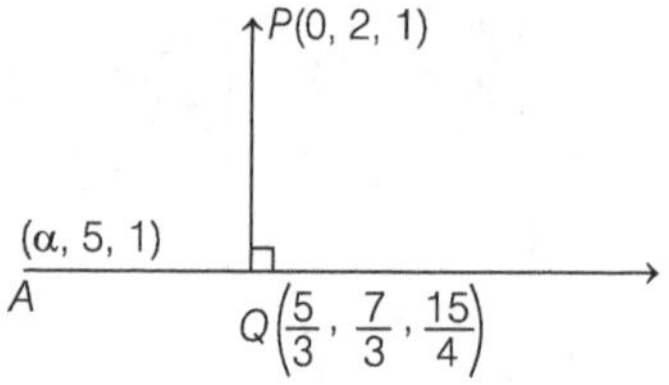

$\because$ Direction ratios of line segments PQ is $\left(\frac{5}{3}, \frac{1}{3}, \frac{11}{4}\right)$

and direction ratios of line AQ is $\left(\alpha - \frac{5}{3}, \frac{8}{3}, -\frac{11}{4}\right)$.

$$\therefore \quad \frac{5}{3}\left(\alpha - \frac{5}{3}\right) + \frac{1}{3} \times \frac{8}{3} + \frac{11}{4} \times \left(-\frac{11}{4}\right) = 0$$

$$\Rightarrow \quad 15\alpha - 25 + 8 + \frac{99(-11)}{16} = 0$$

$$\Rightarrow \quad 15\alpha - 17 = \frac{99 \times 11}{16}$$

$$\Rightarrow \quad 5\alpha = \frac{1089 - 817}{16}$$

$$\Rightarrow \quad 5\alpha = 54.466 \Rightarrow \alpha = 10.893$$

18. *(2454)* Given word is 'EXAMINATION' having letters (AA), (II), (NN), EXMOT, we have to form 4 letter words, then following cases are possible,

(I) 2 same, 2 same and number of words are
$${}^3C_2 \times \frac{4!}{2!\,2!} = 18$$

(II) 2 same, 2 different and number of words are
$${}^3C_1 \times {}^7C_2 \times \frac{4!}{2!} = 3 \times \frac{7 \times 6}{2} \times \frac{4 \times 3 \times 2}{2}$$
$$= 21 \times 36 = 756$$

(III) All are different and number of words are
$${}^8C_4 \times 4! = \frac{8 \times 7 \times 6 \times 5}{4!} 4! = 1680$$

So, total number of 4 letter words
$$18 + 756 + 1680 = 2454$$

19. *(–1)* We can rewrite,

$$\begin{vmatrix} a & a^2 & 1+a^3 \\ b & b^2 & 1+b^3 \\ c & c^2 & 1+c^3 \end{vmatrix} = \begin{vmatrix} a & a^2 & 1 \\ b & b^2 & 1 \\ c & c^2 & 1 \end{vmatrix} + \begin{vmatrix} a & a^2 & a^3 \\ b & b^2 & b^3 \\ c & c^2 & c^3 \end{vmatrix} = 0$$

$$= \begin{vmatrix} a & a^2 & 1 \\ b & b^2 & 1 \\ c & c^2 & 1 \end{vmatrix} + abc \begin{vmatrix} a & a^2 & 1 \\ b & b^2 & 1 \\ c & c^2 & 1 \end{vmatrix} = 0$$

$$= (1 + abc) \begin{vmatrix} a & a^2 & 1 \\ b & b^2 & 1 \\ c & c^2 & 1 \end{vmatrix} = 0$$

$$= (1 + abc) = 0 \qquad \left[\because \begin{vmatrix} a & a^2 & 1 \\ b & b^2 & 1 \\ c & c^2 & 1 \end{vmatrix} \neq 0\right]$$

$$\Rightarrow \quad abc = -1$$

20. *(33)* Total possibilities to make a 6-digit number using only 1 and 8

$\overline{2}\ \overline{2}\ \overline{2}\ \overline{2}\ \overline{2}\ \overline{2}$

Every position has 2 possibility either 1 or 8

$\therefore 2^6$

Number divisible by 21 :

1. for divisibility by 3

(sum of digits must be divisible by 3)

(a) Number of 1's = Number of 8

(b) Every digit is 1

(c) Every digit is 8.

2. for divisiblity by 7.

|2 (last digit) – remaining number| = 7k

$k \in z$

(a) all are 1s $\Rightarrow$ 1 no

(b) all are 8s $\Rightarrow$ 1 no

(c) 31's and 38s $\Rightarrow \frac{6!}{3!3!}$

Total number divisible by 21 = number divisible by 3 and 7

$$= \frac{6!}{3!3!} + 2 = \left(\frac{6 \times 5 \times 4 \times 3!}{3! \times 3 \times 2}\right) + 2 = 22$$

$$\therefore \text{Probability} = \frac{22}{2^6} = p$$

$$96p = 96 \times \frac{22}{2^6} = 96 \times \frac{22}{2 \times 2 \times 2 \times 2 \times 2 \times 2} = 33$$

English

1. *(b)* 'is repairing' is used in the given sentence as the sentence belongs to present continuous tense.
2. *(c)* 'One of my friends lives in Russia' is the correct sentence that follows the subject-verb agreement.
3. *(b)* For uncountables (concern) 'little' is used.
4. *(a)* 'and' is the suitable linker to join the given two phrases.
5. *(b)* The given underlined clause is belong to non-finite clause.
6. *(d)* 'To keep up' means to maintain something in good condition. So, option (d) would be its correct answer.
7. *(d)* Discrimination has the context of good judgement. Hence, option (d) is its correct answer.
8. *(d)* A person who maliciously destroys by fire is called incendiary.
9. *(b)* 'meat' is suitable word for the given blank.
10. *(c)* 'didn't he' is the suitable question tag for the given blank.

MANIPAL

Engineering Entrance Exam

Solved Paper 2021

Physics

1. A thin copper wire of length L increases in length by 1%, when heated from T_1 to T_2. What is the percentage change in area when a thin copper plate having dimensions $(10L \times 2L)$ is heated from T_1 to T_2 ?

(a) 2% (b) 20%
(c) 10% (d) 40%

2. A cyclist moving on a circular track of radius 40 m completes half a revolution in 40 s. Its average velocity is :

(a) zero (b) 4π m/s
(c) 2 m/s (d) 8π m/s

3. A particle is moving in a circular path with a constant speed v. If θ is the angular displacement. Then starting from $\theta = 0$, the maximum and minimum changes in the momentum will occur, when value of θ is respectively :

(a) 45° and 90° (b) 90° and 180°
(c) 180° and 360° (d) 90° and 270°

4. A particle of mass m is projected with a velocity v at an angle of 45° with horizontal. When the particle is at its maximum height, the magnitude of its angular momentum about the point of projection is :

(a) zero (b) $\dfrac{mv^3}{4\sqrt{2}g}$
(c) $\dfrac{mv^3}{\sqrt{2}g}$ (d) $\dfrac{mv^3}{\sqrt{2gh^3}}$

5. A 4 kg block A is placed on the top of a block B of mass 8 kg, which rests on a smooth table. A just slips on B when a force of 12 N is applied on A. Then, the maximum horizontal force required to make both A and B move together is :

(a) 12 N (b) 24 N
(c) 36 N (d) 48 N

6. A fireman weighing 80 kg slides down a pole. If the resisting force of friction is 720 N, his acceleration would be : (take $g = 10 \text{ m/s}^2$)

(a) 0.11 m/s^2 (b) 0.9 m/s^2
(c) 1 m/s^2 (d) zero

7. An object A moving horizontally with kinetic energy of 800 J experiences a constant opposing force of 100 N, while moving from a place x to a place y, where xy is 2m. What is the energy of A at y ?

(a) 700 J (b) 400 J
(c) 600 J (d) 300 J

8. A particle of mass m is moving in a horizontal circle of radius R under the centripetal force $= -k/R^2$ where k is a constant. What is the total energy of the particle ?

(a) $\dfrac{k}{2R}$ (b) $-\dfrac{k}{2R}$
(c) $\dfrac{k}{R}$ (d) $-\dfrac{k}{R}$

9. A solid sphere of radius R has moment of inertia I about its diameter. What will be moment of inertia of a shell of same mass and same radius about its diameter ?

(a) $\dfrac{3}{5}I$ (b) $\dfrac{5}{3}I$
(c) $\dfrac{2}{3}I$ (d) $\dfrac{2}{5}I$

10. A body is moving with constant velocity parallel to x-axis. Its angular momentum w.r.t., the origin will :

(a) be zero (b) decrease
(c) increase (d) remain constant

11. Imagine a light planet revolving around a very massive star in a circular orbit of radius R with a period of revolution T. If the gravitational force of attraction between the planet and the star is proportional to $R^{-5/R}$, then T^2 is proportional to :

(a) R^3 (b) $R^{7/2}$
(c) $R^{3/2}$ (d) $R^{3.75}$

12. A satellite goes along an elliptical path arounds earth. The rate of change of area swept by the same line joining earth and the satellite is proportional to :

(a) $r^{1/2}$ (b) r
(c) $r^{3/2}$ (d) r^2

13. A steel ring of radius r and cross-sectional area A is fitted on to a wooden disc of radius R ($R > r$). If Young's modulus be Y, then the force with which the steel ring is expanded, is

(a) $AY\dfrac{R}{r}$ (b) $AY\left(\dfrac{R-r}{r}\right)$
(c) $\dfrac{Y}{A}\left[\dfrac{R-r}{r}\right]$ (d) $\dfrac{Yr}{AR}$

14. To what height should a rectangular cylinder having square base of length 10 cm be filled, so that the total force on the bottom is equal to that on the sides ?

(a) 5 cm (b) 10 cm
(c) 20 cm (d) 6.67 cm

15. Two identical cylinders contain helium at 2.5 atm and argon at 1 atm respectively. If both the gases are filled in one of the cylinders, the pressure would be :

(a) 3.5 atm (b) 1.75 atm
(c) 1.50 atm (d) 1 atm

16. In the indicator diagram, AB is an isotherm and BC is an adiabat, because :

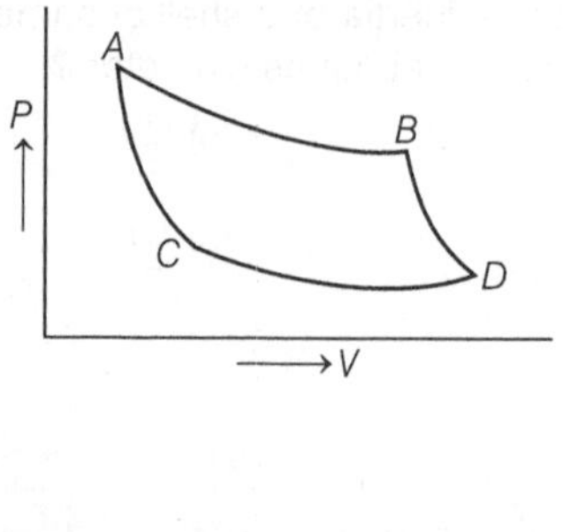

(a) AB and BC meet at B
(b) BC is shorter than AB
(c) slope of AB is less than slope of BC
(d) none of the above

17. The rays of sun are focussed on a piece of ice through a lens of diameter 5 cm, as a result of which 10 g ice melts in 10 min. The amount of heat received from sun, per unit area per minute is :

(a) 4 cal/cm^2 min
(b) 40 cal/cm^2 min
(c) 4 J/m^{2}min
(d) 400 cal/cm^2 min

18. The displacement y in centimetre is given in terms of time t second by the equation

$$y = 3\sin 314t + 4\cos 314t$$

The amplitude of SHM is :

(a) 7 cm (b) 3 cm
(c) 4 cm (d) 5 cm

19. A simple spring has length l and force constant k. It is cut into two springs of lengths l_1 and l_2 such that $l_1 = nl_2$ (n = an integer). The force constant of spring of length l_1 is :

(a) $k(1+n)$ (b) $\dfrac{k}{n}(1+n)$
(c) k (d) $\dfrac{k}{(n+1)}$

20. A wave travelling in a stretched string is described by the equation $y = A\sin(kx - \omega t)$. The maximum particle velocity is :

(a) $A\omega$ (b) ω/k
(c) $d\omega/dk$ (d) x/t

21. A stretched sonometer wire is in unison with a tuning fork. When length of wire is increase by 1%, the number of beats heard per second is 5. Then the frequency of the fork is :

(a) 500 Hz (b) 505 Hz
(c) 255 Hz (d) 250 Hz

22. If an electron has an initial velocity in a direction different from that of an electric field, the path of the electron is :

(a) a straight line
(b) a circle
(c) an ellipse
(d) a parabola

23. Due to the dipole shown in figure, the electric intensity will be parallel to the dipole axis at the point :

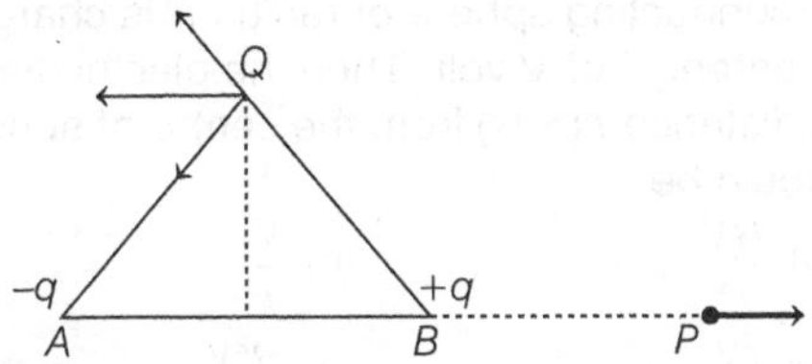

(a) Q
(b) P
(c) both P and Q
(d) neither P nor Q

24. Two capacitors of capacitance $2\mu F$ and $6\mu F$ are connected in series. A potential difference of 800 V is applied to the outer plates of the two capacitors system. The charge on each capacitor will be :

(a) 1200 C (b) 6000 C
(c) $6000\,\mu C$ (d) $1200\,\mu C$

25. A charge of 2×10^{-2} C moves at 30 rev/s in a circle of diameter 80 cm. The current linked with the circuit is :

(a) 0.02 A (b) 20 A
(c) 0.60 A (d) 60 A

26. The ratio of KE required to be given to the satellite to escape earth's gravitational field to the KE required to be given, so that the satellite moves in a circular orbit just above earth's atmosphere is :

(a) 1 (b) 1/2
(c) 2 (d) infinity

27. 8 g of O_2, 14 g of N_2, and 2 g of CO_2 is mixed in a container of 10 L capacity at 27°C. The pressure exerted by the mixture in terms of atmospheric pressure is ($R = 0.082$ L atm $k^{-1}mol^{-1}$) :

(a) 1.4 atm (b) 2.5 atm
(c) 3.7 atm (d) 8.7 atm

28. A rectangular vessel when full of water takes 10 min to be emptied through an orifice in its bottom. How much time will it take to be emptied, when half filled with water ?

(a) 4 min (b) 6 min
(c) 7 min (d) 8 min

29. During an adiabatic process, the pressure P of a fixed mass of an ideal gas changes by ΔP and its volume V changes by ΔV. If $\gamma = C_P/C_V$, then $\Delta V/V$ is given by :

(a) $-\Delta p/p$ (b) $-\gamma\,\Delta P/P$
(c) $-\Delta P/\gamma P$ (d) $\Delta P/\gamma^2 P$

30. An engine has an efficiency of 1/3. The amount of work this engine can perform per kcal of heat input is :

(a) 1400 cal (b) 700 cal
(c) 700 J (d) 1400 J

31. A mass M is attached to a string, oscillates with a period of 2 s. If the mass is increased by 4 kg. The time period increases by 1 s. Assuming, that Hooke's law is obeyed, the initial mass M was :

(a) 3.2 kg (b) 1 kg
(c) 2 kg (d) 8 kg

32. A heavy uniform rope hangs vertically from the ceiling, with its lower end free. A disturbance on the rope travelling upwards from the lower end has a velocity v at a distance x from the lower end such that :

(a) $v \propto x$ (b) $v \propto \sqrt{x}$
(c) $v \propto \frac{1}{x}$ (d) $v \propto \frac{1}{\sqrt{x}}$

33. The displacement of an oscillating particle is given by $y = A\sin[Bx + Ct + Dt]$. The dimensional formula for $[ABCD]$ is :

(a) $[M^0L^{-1}T^0]$ (b) $[M^0L^0T^{-1}]$
(c) $[M^0L^{-1}T^{-1}]$ (d) $[M^0L^0T^0]$

34. A point initially at rest moves along x-axis. Its acceleration varies with time as $a = (6t + 5)$ in m/s^2. If it starts from origin, the distance covered in 2 s is :

(a) 20 m (b) 18 m
(c) 16 m (d) 25 m

35. If $\vec{A}$, $\vec{B}$ and $\vec{C}$ are the unit vectors along the incident ray, reflected ray and outward normal to the reflecting surface, then :

(a) $\vec{B} = \vec{A} - \vec{C}$
(b) $\vec{B} = \vec{A} + (\vec{A} \cdot \vec{C})\vec{C}$
(c) $\vec{B} = \vec{A} + \vec{C}$
(d) $\vec{B} = \vec{A} - 2(\vec{A} \cdot \vec{C})\vec{C}$

36. A particle is projected from the ground with an initial speed of u at an angle θ with horizontal. The average velocity of the particle between its point of projection and highest point of trajectory is :

(a) $u\cos\theta$ (b) $\frac{u}{2}\sqrt{1+\cos^2\theta}$

(c) $\frac{u}{2}\sqrt{1+2\cos^2\theta}$ (d) $\frac{u}{2}\sqrt{1+3\cos^2\theta}$

37. A particle p is moving in a circle of radius r with a uniform speed v, C is the centre of the circle and AB is the diameter. The angular velocity of p about A and C is in the ratio :

(a) 1 : 1 (b) 1 : 2

(c) 2 : 1 (d) 4 : 1

38. A force of 5N gives a mass M_1 an acceleration equal to 8m/s^2 and M_2 an acceleration equal to 24 m/s^2. What is the acceleration, if both masses are tied together ?

(a) 16 m/s^2 (b) 6 m/s^2

(c) 12 m/s^2 (d) 4 m/s^2

39. A bullet is fired by a light rifle and the other with a heavy rifle by the same force. Which rifle will cause more injury to the shoulder ?

(a) Light rifle

(b) Heavy rifle

(c) Both will cause the same injury

(d) The information is insufficient

40. The displacement x and time t for a particle are related to each other as $t=\sqrt{x}+3$. What is work done in first 6 s of its motion ?

(a) 6 J (b) Zero

(c) 4 J (d) 2 J

41. A ball falls vertically on to a floor, with momentum p, and then bounces repeatedly, the coefficient of restitution is e. The total momentum imparted by the ball to the floor is

(a) $p(1+e)$ (b) $\frac{p}{(1-e)}$

(c) $p\left(\frac{1+e}{1-e}\right)$ (d) $p\left(1-\frac{1}{e}\right)$

42. A uniform thin bar of mass 6 m and length 12 L is bent to make a regular hexagon. Its moment of inertia about an axis passing through the centre of mass and perpendicular to the plane of hexagon is :

(a) $20\,mL^2$ (b) $30\,mL^2$

(c) $\frac{12}{5}mL^2$ (d) $6\,mL^2$

43. A conducting sphere of radius R is charged to a potential of V volt. Then the electric field at a distance $r(>R)$ from the centre of sphere would be :

(a) $\frac{RV}{r^2}$ (b) $\frac{V}{r}$

(c) $\frac{rV}{R^2}$ (d) $\frac{R^2V}{r^3}$

44. Given that $q_1+q_2=q$. For what ratio q_1/q_2 will the force between q_1 and q_2 be maximum ?

(a) 0.25 (b) 0.5

(c) 1 (d) 2

45. A current of 2A flows in the system of conductors as shown in the figure. The potential difference V_P-V_R will be :

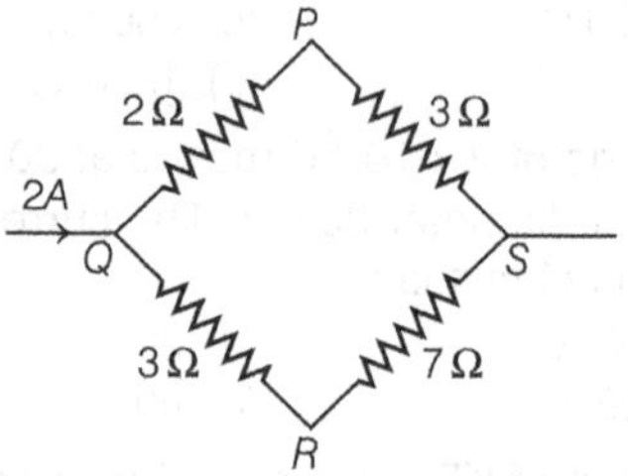

(a) − 2 V (b) − 1 V

(c) + 1 V (d) + 2 V

46. A coil of wire of resistance 50 Ω is embedded in a black of ice and a potential difference of 210 V is applied across it. The amount of ice which melts in 1 s is :

(a) 0.262 g (b) 2.62 g

(c) 26.2 g (d) 0.0262 g

47. When 1g hydrogen ($\text{ECE}=1.044\times10^{-8}\text{ kg C}^{-1}$) forms water, 34 kcal heat is liberated. The minimum voltage required to decompose water is :

(a) 0.75 V (b) 3 V

(c) 1.5 V (d) 4.5 V

48. A copper wire of diameter 1.6 mm carries a current I. The maximum magnetic field due to this wire is 5×10^{-4}T. The value of I is :

(a) 0.2 A (b) 0.5 A

(c) 2 A (d) 4 A

49. Two wires PQ and QR, carry equal currents I as shown in figure. One end of both the wires extends to infinity $\angle PQR = \theta$. The magnitude of the magnetic field at O on the bisector of these two wires at a distance r from point O is

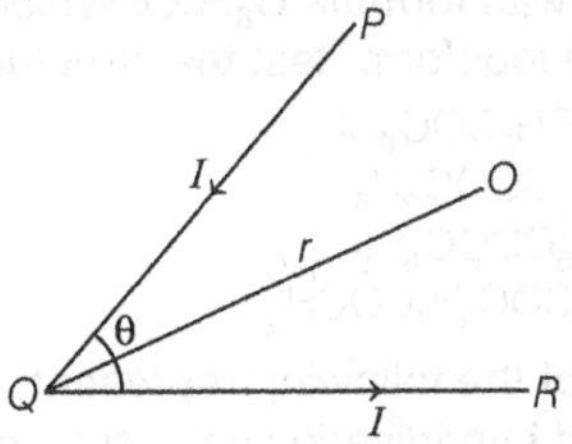

(a) $\frac{\mu_0 I}{4\pi r}\sin\frac{\theta}{2}$ (b) $\frac{\mu_0}{4\pi}\frac{I}{r}\cot\frac{\theta}{2}$

(c) $\frac{\mu_0}{4\pi}\frac{I}{r}\tan\frac{\theta}{2}$ (d) $\frac{\mu_0 I}{2\pi r}\frac{[1+\cos(\theta/2)]}{\sin(\theta/2)}$

50. A bar magnet has a magnetic moment equal to 5×10^{-5} Wb-m. It is suspended in a magnetising field equal to $8\pi \times 10^{-4}$ A-m^{-1}. The magnet vibrates with a period 15 s. The moment of inertia of the magnet is :

(a) 11.14 kg-m^2 (b) 0.57 kg-m^2

(c) 22.28 kg-m^2 (d) 0.057 kg-m^2

Chemistry

1. The structure for neo-heptyl alcohol is :

(a) 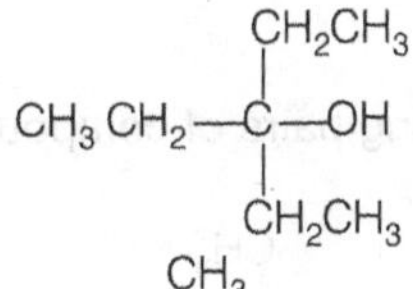

(b) 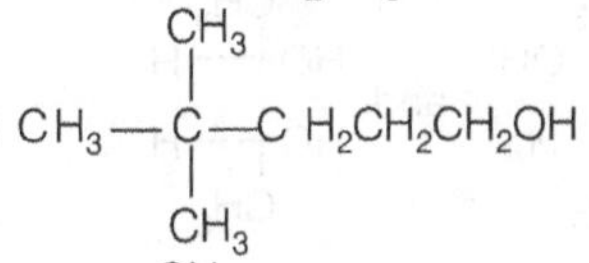

(c) 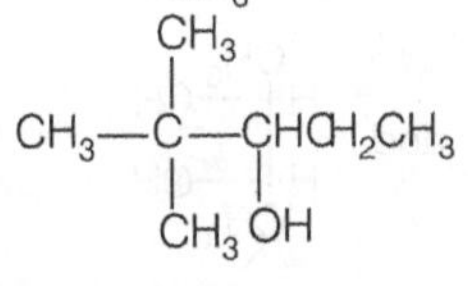

(d) $n-C_7H_{15}OH$

2. Which is the decreasing order of stability ?

(i) $CH_3—\overset{+}{C}H—CH_3$

(ii) $CH_3—\overset{+}{C}H—O—CH_3$

(iii) $CH_3\overset{+}{C}H\cdot CO\cdot CH_3$

(a) (i)<(ii)<(iii) (b) (i)> (ii)> (iii)

(c) (iii)> (i)> (ii) (d) (ii)> (iii)> (i)

3. Successive alkanes differ by :

(a) CH_2 (b) CH

(c) CH_3 (d) C_2H_4

4. The order of activity of the various *o* and *p*-director is :

(a) $—O^- > —OH > —OCOCH_3 > —COCH_3$

(b) $OH > —O^- > —OCOCH_3 > —COCH_3$

(c) $OH > —O^- > —COCH_3 > —OCOCH_3$

(d) $—O^- > —COCH_3 > —OCOCH_3 > —OH$

5. Number of π electrons present in naphthalene is

(a) 4 (b) 6

(c) 10 (d) 14

6. In the reaction $N_2 + 3H_2 \longrightarrow 2NH_3$, ratio by volume of N_2, H_2 and NH_3 is 1 : 3 : 2 . This illustrates law of :

(a) definite proportion (b) multiple proportion

(c) reciprocal proportion (d) gaseous volumes

7. Cyclohexanol is dehydrated to cyclohexane on heating with conc-H_2SO_4 . If the yield of this reaction is 75% how much cyclohexene will be obtained from 100 g of cyclohexanol ?

(a) 61.5 g (b) 75.0 g

(c) 20.0 g (d) 41.0 g

8. The order of increasing energies of the orbitals follows :

(a) $5p < 4f < 6s < 5d$ (b) $5p < 6s < 4f < 5d$

(c) $4f < 5p < 5d < 6s$ (d) $5p < 5d < 4f < 6s$

9. Which of the following is not possible ?

(a) $n = 3, l = 0, m = 0$ (b) $n = 3, l = 1, m = -1$

(c) $n = 2, l = 0, m = -1$ (d) $n = 2, l = 1, m = 0$

10. What is the wavelength associated with an electron moving with a velocity of 10^6 m/s ? ($h = 6.63 \times 10^{-34}$ Js)

(a) 72.7 nm (b) 7.27 nm

(c) 0.727 nm (d) 0.0727 nm

11. Which of the following is neutral refractorie material?

(a) SiO_2 (b) MgO

(c) CaO (d) SiC

12. What is the characteristic valence shell configuration of coinage metals ?

(a) ns^2np^6 (b) $(n-1)d^2ns^2$

(c) nd^9ns^1 (d) $(n-1)d^{10}ns^1$

13. Pure cold conc. HNO_3 makes iron passive, as the surface is covered with protective layer of

(a) Fe_2O_3 (b) FeO

(c) Fe_3O_4 (d) $Fe(NO_3)_3$

14. Which of the following is not correct for D_2O ?

(a) BVP is higher than H_2O

(b) D_2O reacts slowly than H_2O

(c) Viscosity is higher than H_2O at 25°

(d) Solubility of NaCl in it is more than H_2O

15. Ozone when react with potassium iodide solution liberates certain product, which turns starch paper blue. The liberated substance is :

(a) oxygen

(b) iodine

(c) hydrogen iodide

(d) potassium hydroxide

16. Red hot iron absorbs SO_2 giving the product.

(a) Fes + O_2 (b) Fe_2O_3 + FeS

(c) FeO + FeS (d) FeO + S

17. 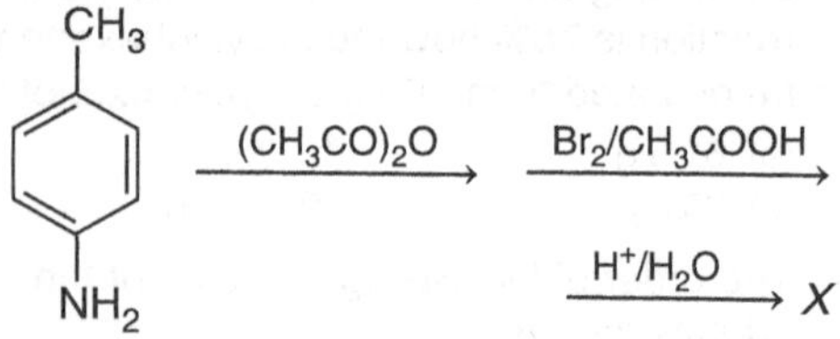

The product X is :

(a) 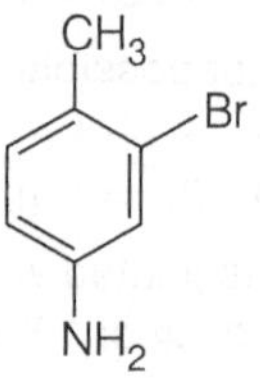

(b) 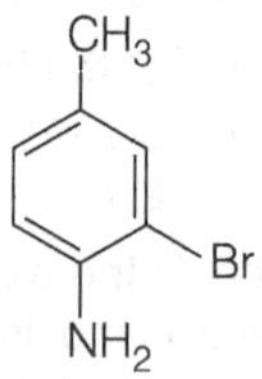

(c) 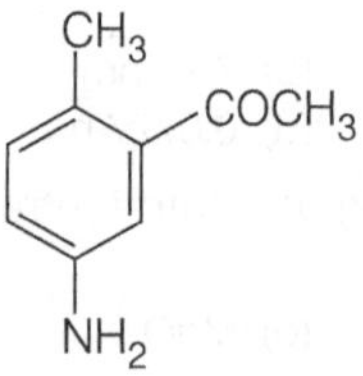

(d) 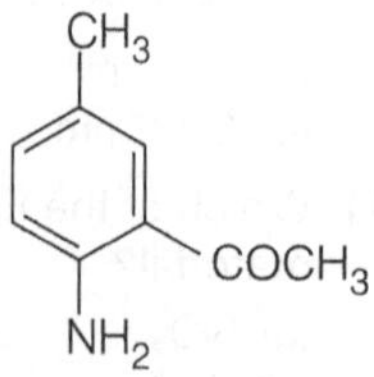

18. An ester (A) with molecular formula $C_9H_{10}O_2$ was treated with excess of CH_3MgBr and the compound so formed was treated with conc. H_2SO_4 to form olefin (B). Ozonolysis of B gave ketone with formula C_8H_8O. Which shows positive iodoform test the structure of A is :

(a) $CH_3CH_2COC_6CH_5$

(b) $C_6H_5COOC_2H_5$

(c) $C_6H_5COOC_6H_5$

(d) $CH_3COC_6H_4COCH_3$

19. Which of the following represents the given mode of hybridisation $sp^2 - sp^2 - sp - sp$ from left to right ?

(a) $H_2C=CH-C\equiv N$

(b) $CH\equiv C-C\equiv CH$

(c) $H_2C=C=C=CH_2$

(d) 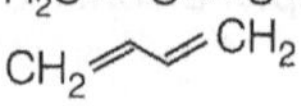

20. Which of the following pairs of compounds are enantiomers ?

(a) 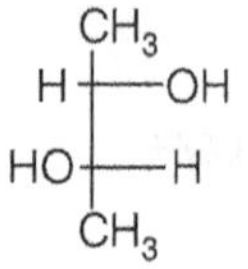 and 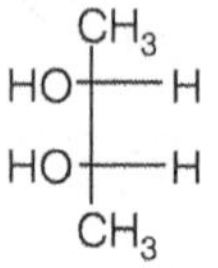

(b) 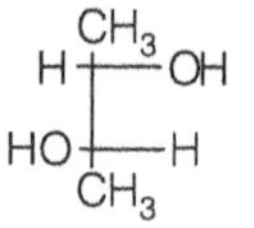 and 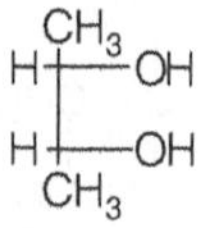

(c) 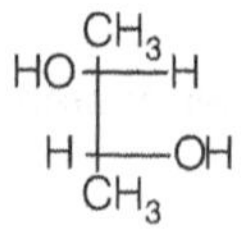 and 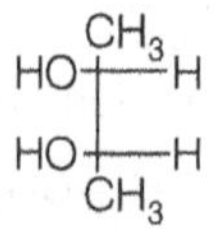

(d) 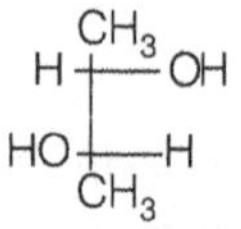 and 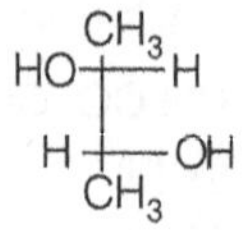

21. Name of the compound given below is :

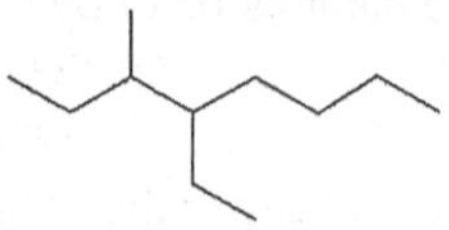

(a) 3-methyl-4-ethyloctane

(b) 2, 3-diethylheptane

(c) 5-ethyl-6-methyloct1ane

(d) 4-ethyl-3-methyloctane

22. Acrilan is a hard, horny and a high melting material. Its structure is :

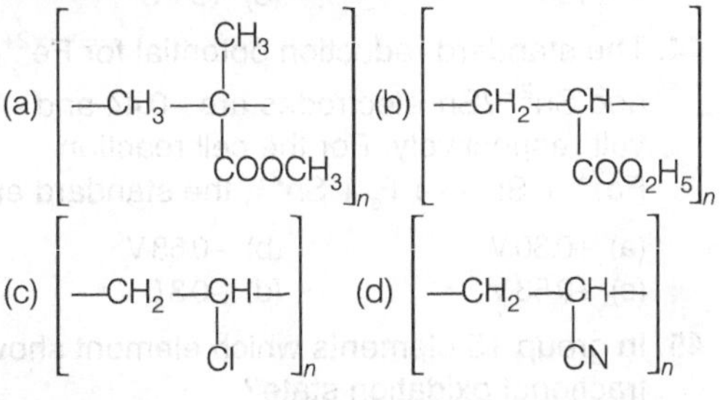

23. The low density of ice compared to water is due to :

(a) hydrogen bonding interactions
(b) dibble-dipole interactions
(c) dipole induced dipole interactions
(d) induced dipole induced dipole interactions.

24. 100 mL of 1.0 M HCl are mixed with 75 mL of 1.0 M Na_2CO_3. The resulting solution will be :

(a) acidic (b) basic
(c) neutral (d) amphoteric

25. Semiconductors are derived from compounds of

(a) *p*-block elements
(b) intrinsic semiconductor
(c) mixed conductor
(d) extrinsic semiconductor

26. $C(s) + O_2(g) \longrightarrow CO_2(g) \quad \Delta H = -94$ kcal

$2CO + O_2(g) \longrightarrow 2CO_2(g)\ \Delta H = -135.2$ kcal

The heat of formation of $CO(g)$ is :

(a) –26.4 kcal (b) 41.2 kcal
(c) 26.4 kcal (d) 229.2 kcal

27. The reagent NH_4Cl and aqueous NH_3 will precipitate :

(a) Ca^{2+} (b) Al^{3+}
(c) Mg^{2+} (d) Zn^{2+}

28. Whose name is not associated with the development of periodic table?

(a) Prout's (b) Newlands
(c) Rutherford (d) Lother Meyer

29. Which of the following set of quantum number is not applicable for an electron in an atom?

(a) $n = 1, l = 1, m = 1, s = +\frac{1}{2}$
(b) $n = 1, l = 0, m = 0, s = +\frac{1}{2}$
(c) $n = 2, l = 0, m = 0, s = +\frac{1}{2}$
(d) $n = 2, l = 0, m = 0, s = -\frac{1}{2}$

30. Sodium carbonate is manufactured by :

(a) Lowing process (b) Leblanc process
(c) Solvay process (d) Haber's process

31. $CH_3C \equiv CCH_3 \xrightarrow[(ii) H_2O/Zn]{(i) X} H_3C-\underset{\underset{O}{\|}}{C}-\underset{\underset{O}{\|}}{C}-CH_3$

in the above reaction *X* is :

(a) HNO_3 (b) O_2
(c) O_3 (d) $KMnO_4$

32. Which isomeric form of benzene hexachloride is used as insecticide?

(a) α-form (b) β-form
(c) γ-form (d) δ-form

33. Which one of the following is most reactive towards nucleophilic substitution reaction?

(a) $CH_2{=}CH{-}Cl$ (b) C_6H_5Cl
(c) $CH_3CH{=}CH{-}Cl$ (d) $ClCH_2{-}CH{=}CH_2$

34. A set of compound in which the reactivity of halogen atom in the ascending order is :

(a) chlorobenzene vinyl chloride, chloroethane
(b) chloroethane, chlorobenzene, vinyl chloride
(c) vinylchloride, chlorobenzene, chloroethane
(d) vinylchloride, chloroethane , chlorobenzene.

35.

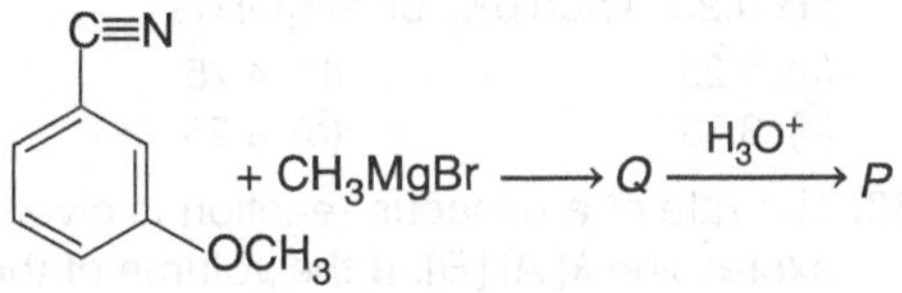

The product '*P*' in the above reaction is :

(a)

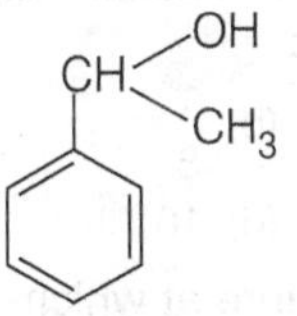

(b)

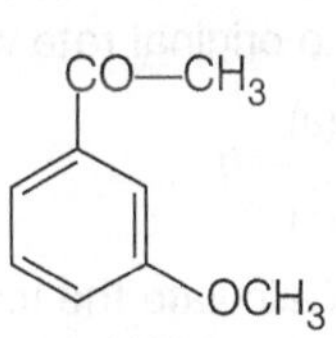

(c)

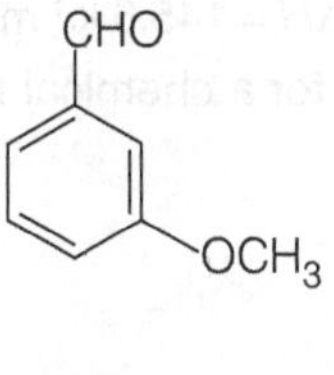

(d)

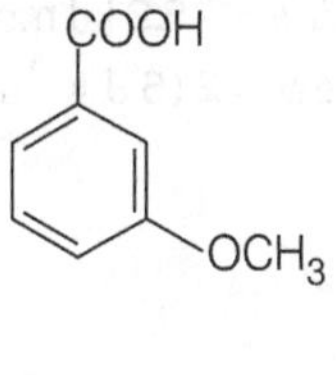

36. To become a carbohydrate a compound must contain at least :

(a) 2 carbons (b) 3-carbons
(c) 4 carbons (d) 6 carbons

37. For which of the following species Bohr's theory is not applicable?

(a) Be^{3+} (b) Li^{2+}
(c) He^{2+} (d) H

38. Diborane on reaction with olefins forms :

(a) mono acid derivatives
(b) diethyl borane
(c) triethyl borane
(d) ethyl borane

39. 3.92 g of ferrous ammonium sulphate crystals are dissolved in 100 mL of water, 20 mL of this solution requires 18 mL of $KMnO_4$ during titrating for complete oxidation. the weight of $KMnO_4$ present in one liter of solution is :

(a) 3.476 g (b) 12.38 g
(c) 34.76 g (d) 1.238 g

40. Addition of arsenic to germanium makes the latter :

(a) metallic conductor
(b) intrinsic semiconductor
(c) mixed conductor
(d) extrinsic semiconductor

41. Solution of 0.1 N NH_4OH and 0.1 NH_4Cl has pH 9.25. Then pK_b of NH_4OH is :

(a) 9.25 (b) 4.75
(c) 3.75 (d) 8.25

42. The rate of a gaseous reaction is given by the expression $k[A][B]$. If the volume of the reaction vessel is suddenly reduced to 1/4th of the initial volume, the reaction rate relating to original rate will be :

(a) $\frac{1}{10}$ (b) $\frac{1}{8}$
(c) 8 (d) 16

43. Calculate the temperature at which $\Delta G = -5.2$ kJ mol^{-1}, $\Delta H = 145.6$ kJ mol^{-1} and $\Delta S = 216$ J $k^{-1} mol^{-1}$ for a chemical reaction :

(a) 698.1°C (b) 698.1 K
(c) 130 K (d) 130°C

44. The standard reduction potential for Fe^{2+}/ Fe and Sn^{2+} /Sn electrodes are −0.44 and −0.14 volt respectively. For the cell reaction $Fe^{2+} + Sn \longrightarrow F_e + Sn^{2+}$, the standard emf is

(a) +0.30 V (b) −0.58 V
(c) +0.58 V (d) −0.3 0 V

45. In group 15 elements which element show a fractional oxidation state?

(a) N (b) As
(c) Sb (d) Bi

46. Ice floats on water because :

(a) its density is less than that of water
(b) crystal structure of ice has empty space
(c) both of the above
(d) none of the above

47. When the more electropositive metal displace less electropositive metals form their salt solution this process is called :

(a) auto reduction (b) electro reduction
(c) hydrometallurgy (d) none of these

48. When sodium hydroxide react with sand, it form :

(a) sodium silicate
(b) silicon oxide
(c) silicon hydroxide
(d) none of the above

49. A sample of 4.12 mg of unknown alcohol is added to CH_3MgBr then 1.56 mL of methane at STP was liberated. The alcohol is :

(a) C_2H_5OH (b) CH_3OH
(c) C_3H_7OH (d) C_4H_9OH

50. The IUPAC name for the formula

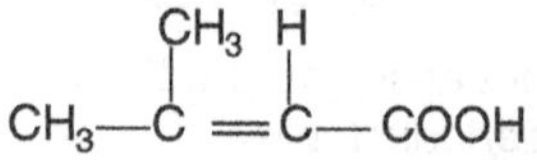

is

(a) 2-methyl 2-butanoic acid
(b) 3-methyl-3-butanoic acid
(c) 3-methyl-2-butanoic-acid
(d) 2-methyl-3-butanoic acid

Mathematics

1. $\dfrac{5^{\frac{3}{2}} - 2^{\frac{3}{2}}}{\sqrt{5} - \sqrt{2}} + \dfrac{5^{\frac{3}{2}} + 2^{\frac{3}{2}}}{\sqrt{5} + \sqrt{2}}$

(a) 7 (b) 14
(c) 12 (d) 8

2. If $9^{-x} = \dfrac{1}{27\sqrt{27}} = (81)^{-y}$, then (x, y) is equal to

(a) $\left(\dfrac{9}{4}, \dfrac{9}{8}\right)$ (b) $\left(\dfrac{3}{2}, \dfrac{3}{4}\right)$
(c) (3, 6) (d) (6, 3)

3. If $x^{4/3} + x^{-1/3} = 1$, $x^5 + 3x^2 - x$ is equal to

(a) 0 (b) 1
(c) – 1 (d) 2

4. If $x = \dfrac{2\sqrt{2} - \sqrt{7}}{2\sqrt{2} + \sqrt{7}}$, then $x + x^{-1}$ is equal to

(a) 28 (b) 32 (c) 30 (d) 24

5. $\left|\dfrac{x}{2} - 1\right| < 3$ implies that x lies in the interval

(a) (– 4, 8) (b) (– 3, 6)
(c) (– 4, 6) (d) (– 3, 8)

6. y is the sum of three numbers, one of which is a constant, the 2nd varies as x and the 3rd varies inversely as x. The values of y at $x = 1, -1$ and 3 are respectively 6, – 4 and 8. Then, y is equal to

(a) $1 + x - \dfrac{1}{x}$ (b) $1 + 2x + \dfrac{3}{x}$
(c) $2 + x + \dfrac{1}{x}$ (d) $2 - x + \dfrac{1}{x}$

7. $\log x + \log x^3 + \log x^5 + \dots + \log x^{2n-1}$ is equal to

(a) $2n \log x$ (b) $(2n - 1) \log x$
(c) $n^2 \log x$ (d) $(n^2 + 1) \log x$

8. $\dfrac{1}{\log_{25} 10} + \dfrac{1}{\log_4 10} + \dfrac{1}{\log_{\sqrt{2}} 10} + \dfrac{1}{\log_{\sqrt{5}} 10}$ is equal to

(a) 3/2 (b) 2
(c) 3 (d) 5/2

9. $1 + \dfrac{3}{1!} + \dfrac{5}{2!} + \dfrac{7}{3!} + \dots \infty$ is

(a) $5e$ (b) $2e$
(c) $3e$ (d) None of these

10. If the roots $x^2 + ax + 9 = 0$ are complex, then

(a) $a < 6$ (b) $a < -6$
(c) $|a| < 6$ (d) $|a| > 6$

11. The sum of the even multiples of 9 between 300 and 500 is

(a) 4356 (b) 5336
(c) 5346 (d) 3456

12. The sum of the first n terms of two AP's are in the ratio $(2n + 3) : (3n - 1)$. The ratio of the 5th terms is

(a) 11 : 6 (b) 21 : 26
(c) 13 : 16 (d) 8 : 5

13. The value of $\sum_{r=1}^{\infty} [3 \cdot 2^{-r} - 2 \cdot 3^{1-r}]$ is

(a) 2 (b) 1/2
(c) 1 (d) 0

14. The number of three digit numbers which have at least two identical digit, is

(a) 648 (b) 729
(c) 252 (d) 452

15. If $C(2n, 3) : C(n, 2) = 44 : 3$, then n is equal to

(a) 6 (b) 7
(c) 5 (d) 8

16. The middle term in the expansion of $\left(x - \dfrac{1}{2}y\right)^{10}$ is

(a) $-63\dfrac{x^5}{y^5}$ (b) $-\dfrac{63x^5y^5}{8}$
(c) $\dfrac{63x^5}{y^5}$ (d) $\dfrac{63x^5}{8y^5}$

17. The term independent of x in the expansion of $\left(2x^4 - \dfrac{1}{x^2}\right)^{12}$ is

(a) 6920 (b) 7920
(c) 7900 (d) 3960

18. The sum of the series $0.2 + 0.004 + 0.00006 + 0.0000008 + \dots \infty$ is

(a) $\dfrac{200}{891}$ (b) $\dfrac{2000}{9801}$
(c) $\dfrac{1000}{9801}$ (d) None of these

19. In the usual notation, $\frac{{}^nC_1}{2} + \frac{{}^nC_2}{3} + \ldots + \frac{{}^nC_n}{n+1}$ is equal to

(a) $\frac{2^{n+1}-1}{n+1}$ (b) $\frac{2^{n+1}-n-1}{n+1}$

(c) $\frac{2^{n+1}-n}{n+1}$ (d) None of these

20. Pick out the wrong statement. If A and B are square matrices of the same order, then

(a) $A + B = B + A$ (b) $(AB)' = B'A'$

(c) $A - B = I$ (d) $|AB| = |A||B|$

21. If $\begin{bmatrix} a & 2 & 3 \\ b & 5 & -1 \end{bmatrix} \times \begin{bmatrix} 1 & 2 \\ 3 & 4 \\ -1 & 1 \end{bmatrix} = \begin{bmatrix} 4 & 13 \\ 12 & 11 \end{bmatrix}$, (a, b) is

(a) (1, – 2) (b) (– 1, – 4)

(c) (1, 3) (d) (1, – 4)

22. If $a \neq \pm b$ and $\begin{bmatrix} a & a^2 & 1+a^3 \\ b & b^2 & 1+b^3 \\ 1 & 1 & 2 \end{bmatrix} = 0$, then ab is equal to

(a) – 1 (b) 1

(c) 2 (d) – 2

23. In the determinant $\begin{vmatrix} 3 & x & -1 \\ 2 & -1 & 4 \\ 1 & y & -3 \end{vmatrix}$, the sum of the cofactors of x and y is

(a) – 24 (b) 24

(c) – 4 (d) 4

24. If $\sin\theta = -\frac{24}{25}$ and θ is in the 4th quadrant, $7\tan\theta + 25\cos\theta$ is equal to

(a) 17 (b) – 17

(c) 14 (d) – 14

25. $\frac{\cos\theta}{1-\tan\theta} + \frac{\sin\theta}{1-\cot\theta}$ is equal to

(a) $\sec\theta + \operatorname{cosec}\theta$ (b) $\sin\theta + \cos\theta$

(c) $\tan\theta + \cot\theta$ (d) $\sin\theta - \cos\theta$

26. If $\sec A - \tan A + a = 0$, $\sin A$ is equal to

(a) $\frac{a^2-1}{a^2+1}$ (b) $\frac{1-a^2}{1+a^2}$

(c) $\frac{a^2+1}{a^2-1}$ (d) $\frac{1+a^2}{1-a^2}$

27. $\cos\frac{\pi}{12} + \cos\frac{17\pi}{12} + \cos\frac{11\pi}{12}$ is equal to

(a) 1 (b) – 1

(c) 0 (d) $\frac{1-\sqrt{3}}{2\sqrt{2}}$

28. $\cot^2\left(\frac{\pi}{4} + \frac{\theta}{2}\right)$ is equal to

(a) $\frac{1-\sin\theta}{1+\sin\theta}$ (b) $\frac{1-\cos\theta}{1+\cos\theta}$

(c) $\frac{1+\sin\theta}{1-\sin\theta}$ (d) $\frac{2-\sin\theta}{2+\sin\theta}$

29. If A, B, C are the angles of a triangle, then $\cos B + \cos C - \cos A + 1$ is equal to

(a) $4\sin\frac{A}{2}\cos\frac{B}{2}\cos\frac{C}{2}$

(b) $-4\sin\frac{A}{2}\cos\frac{B}{2}\cos\frac{C}{2}$

(c) $4\cos\frac{A}{2}\sin\frac{B}{2}\sin\frac{C}{2}$

(d) $4\sin\frac{A}{2}\sin\frac{B}{2}\sin\frac{C}{2}$

30. $\sin^{-1}\frac{1}{\sqrt{5}} + \cos^{-1}\frac{3}{\sqrt{10}}$ is equal to

(a) $\pi/6$ (b) $\pi/4$

(c) $\pi/3$ (d) $2\pi/3$

31. The point on the line $y = x$ equidistant from (4, 0) and (5, 1) is

(a) (2, 2) (b) (3, 3)

(c) $\left(\frac{5}{2}, \frac{5}{2}\right)$ (d) $\left(\frac{1}{2}, \frac{1}{2}\right)$

32. If $A(-1, 2)$, $B(5, 1)$, $C(6, 5)$ are the vertices of a parallelogram $ABCD$. The equation to the diagonal through B is

(a) $x + y + 6 = 0$ (b) $x + y - 6 = 0$

(c) $x - y - 4 = 0$ (d) $x - 2y - 1 = 0$

33. A line cuts off on the coordinate axes positive intercepts whose sum is 4. If it passes through $\left(\frac{9}{2}, -5\right)$, its equation is

(a) $10x + 6y = 15$ (b) $2x - y = 14$

(c) $4x + y = 13$ (d) None of these

34. The ratio in which the segment joining (2, 1) and (0, – 2) is divided by the line $2x - 3y + 4 = 0$ is

(a) 1 : 2 (b) 2 : 1

(c) – 1 : 2 (d) – 2 : 1

35. The equation to the line through the point of intersection of $x - y + 1 = 0, 3x + 2y + 4 = 0$ and perpendicular to $x - 4y = 0$ is

(a) $4x + y + 5 = 0$ (b) $4x + y + 3 = 0$
(c) $4x + y - 5 = 0$ (d) $4x + y - 3 = 0$

36. If ω is a cube root of unity, then $(1 + \omega - \omega^2)(1 - \omega + \omega^2)$ is

(a) 1 (b) 0
(c) 2 (d) 4

37. If $y = \sqrt{\sin x + \sqrt{\sin x + \ldots}}$, then $\frac{dy}{dx}$ is

(a) $\frac{\sin x}{2y - 1}$ (b) $\frac{\cos x}{1 - 2y}$
(c) $\frac{\cos x}{2y - 1}$ (d) 0

38. $\int_0^1 x(1-x)^{12}\, dx$ is equal to

(a) $\frac{1}{132}$ (b) $\frac{1}{156}$
(c) $\frac{1}{182}$ (d) None of these

39. The eccentricity of the conic $9x^2 - 16y^2 = 144$ is

(a) $\frac{4}{5}$ (b) $\frac{4}{3}$ (c) $\frac{5}{4}$ (d) $\sqrt{7}$

40. $\sum_{n=1}^{\infty} \frac{(-1)^n}{n+1}$ is equal to

(a) e^{-1} (b) $\log 2 - 1$
(c) 1 (d) 0

41. The function and its derivative are same for

(a) $\sin x$ (b) $\cos x$
(c) $\log x$ (d) e^x

42. The equation $(\cos p - 1)x^2 + \cos px + \sin p = 0$ in the variable x, has real roots. Then, p can take any value in the interval

(a) $(0, \pi)$ (b) $(-\pi, 0)$
(c) $\left(-\frac{\pi}{2}, \frac{\pi}{2}\right)$ (d) $(-\pi, \pi)$

43. If $\log_3 2, \log_3(2^x - 5)$ and $\log_3\left(2^x - \frac{7}{2}\right)$ are in AP, the value of x is

(a) 2 (b) 3
(c) 0 (d) $\frac{1}{3}$

44. $\int_{-\frac{\pi}{2}}^{\frac{\pi}{2}} |\sin x|\, dx$ is

(a) 2 (b) 0 (c) $\frac{\pi}{2}$ (d) 1

45. The value of $|\vec{\mathbf{a}} \times \vec{\mathbf{b}} + \vec{\mathbf{b}} \times \vec{\mathbf{a}}|$ is

(a) 1 (b) $2|\vec{\mathbf{a}} \times \vec{\mathbf{b}}|$
(c) 0 (d) None of these

46. The complex number z which satisfy the equation $\left|\frac{i+z}{i-z}\right| = 1$ lies on

(a) a circle $x^2 + y^2 = 1$ (b) the x-axis
(c) the y-axis (d) the line $x + y = 1$

47. If $a^2 + b^2 + c^2 = 1$, then $ab + bc + ca$ lies in the interval

(a) $\left[\frac{1}{2}, 2\right]$ (b) $\left[-1, \frac{1}{2}\right]$
(c) $\left[-\frac{1}{2}, 1\right]$ (d) $[-1, 1]$

48. In a simple regular graph, total degree is 28. If the graph has more than one cycle in it, then the degree of each vertex is

(a) 2 (b) 4
(c) 7 (d) 14

49. The middle term of $\left(\sqrt{x} - \frac{1}{\sqrt{x}}\right)^6$ is

(a) -20 (b) -1
(c) 1 (d) None of these

50. If α and β are the roots of $4x^2 + 3x + 7 = 0$, then the value of $\frac{1}{\alpha} + \frac{1}{\beta}$ is

(a) $-\frac{3}{4}$ (b) $\frac{3}{7}$ (c) $-\frac{3}{7}$ (d) $\frac{4}{7}$

51. If $A + B + C = \pi$, then
$$\begin{vmatrix} \sin(A+B+C) & \sin B & \cos C \\ -\sin B & 0 & \tan A \\ \cos(A+B) & -\tan A & 0 \end{vmatrix}$$ is equal to

(a) $\sin A$ (b) $\sin A \cos B$
(c) 0 (d) None of these

52. The coefficient of x in the expansion of $(1 + x + x^2 + x^3)^{-3}$ is

(a) 6 (b) 9
(c) 5 (d) -3

53. If $a < b$, then $a < \frac{a+b}{2} < .....$

(a) $2a$ (b) $2b$
(c) b (d) None of these

54. If $\tan^{-1} x + \tan^{-1} y + \tan^{-1} z = \pi$, then the value of $x + y + z - xyz$ is

(a) 1 (b) 0 (c) -1 (d) $\frac{1}{2}$

55. The value of $\frac{C(n, 2)}{(n+1)!}$ is

(a) $\frac{1}{2}e + 1$ (b) $e + 1$ (c) $\frac{1}{2}e - 1$ (d) e

56. Differential equations of the family of curves $y = a \cos \mu x + b \sin \mu x$, where a and b are arbitrary constants is given by

(a) $\frac{d^2y}{dx^2} + \mu y = 0$ (b) $\frac{d^2y}{dx^2} - \mu^2 y = 0$
(c) $\frac{d^2y}{dx^2} + \mu^2 y = 0$ (d) None of these

57. The value of $\lim_{\theta \to 0} \frac{\tan \theta}{\theta}$ is

(a) 0 (b) 1
(c) ∞ (d) None of these

58. The area included between the curves $y^2 = 2x$ and $x^2 = 2y$ is

(a) $\frac{3}{4}$ sq unit (b) $\frac{4}{3}$ sq unit
(c) $\frac{1}{2}$ sq unit (d) $\frac{4}{3}$ sq unit

59. $\int_1^e \frac{\log x}{x} dx$ is

(a) $\frac{1}{2}$ (b) 1
(c) e (d) None of these

60. $\int_0^1 \sin^{-1} x \, dx$ is

(a) $\frac{\pi}{2} - 1$ (b) 1
(c) π (d) 0

61. The function $\sin x (1 + \cos x)$, $0 \le x \le \frac{\pi}{2}$ has maximum value, when x is equal to

(a) 0 (b) $\frac{\pi}{2}$
(c) $\frac{\pi}{6}$ (d) None of these

62. If $A = \begin{bmatrix} 1 & 0 & 0 \\ 0 & 1 & 0 \\ a & b & -1 \end{bmatrix}$, then A^2 is equal to

(a) unit matrix (b) null matrix
(c) A (d) $-A$

63. The value of 'a' for which the system of equations
$ax + y + z = 0, x + ay + z = 0, x + y + z = 0$
possesses non-zero solutions are given by

(a) 1, 2 (b) 1, -1
(c) 1 (d) None of these

64. $\vec{\mathbf{a}} = \frac{1}{7}(2\hat{\mathbf{i}} + 3\hat{\mathbf{j}} + 6\hat{\mathbf{k}})$, $\vec{\mathbf{b}} = \frac{1}{7}(3\hat{\mathbf{i}} - \lambda\hat{\mathbf{j}} + 2\hat{\mathbf{k}})$. If $\vec{\mathbf{a}}$ and $\vec{\mathbf{b}}$ are mutually perpendicular, then value of λ is

(a) 2 (b) -1 (c) 6 (d) -6

65. The value of $\{(\vec{\mathbf{a}} \times \vec{\mathbf{b}})^2 + (\vec{\mathbf{a}} \cdot \vec{\mathbf{b}})^2\} \div a^2 b^2$ is

(a) 0 (b) 1
(c) 2 (d) None of these

66. The system of linear equations $x + y + z = 0$, $2x + y - z = 0$, $3x + 2y = 0$ has

(a) no solution (b) a unique solution
(c) infinitely many solutions (d) None of these

67. $[\vec{\mathbf{a}} + \vec{\mathbf{b}}\ \vec{\mathbf{b}} + \vec{\mathbf{c}}\ \vec{\mathbf{c}} + \vec{\mathbf{a}}]$ is equal to

(a) $[\vec{\mathbf{a}}\ \vec{\mathbf{b}}\ \vec{\mathbf{c}}]$ (b) $\Sigma (\vec{\mathbf{a}} \cdot \vec{\mathbf{b}}) \vec{\mathbf{c}}$
(c) $2 [\vec{\mathbf{a}}\ \vec{\mathbf{b}}\ \vec{\mathbf{c}}]$ (d) $|\vec{\mathbf{a}}|\ |\vec{\mathbf{b}}|\ |\vec{\mathbf{c}}|$

68. If α, β, γ are the angles which a half ray makes with the positive direction of the axes, then $\sin^2 \alpha + \sin^2 \beta + \sin^2 \gamma$ is equal to

(a) 1 (b) 2 (c) 0 (d) -1

69. A parallelogram is constructed on the vectors $\vec{\mathbf{a}} = 3\vec{\alpha} - \vec{\beta}$, $\vec{\mathbf{b}} = \vec{\alpha} + 3\vec{\beta}$. If $|\vec{\alpha}| = |\vec{\beta}| = 2$ and the angle between $\vec{\alpha}$ and $\vec{\beta}$ is $\frac{\pi}{3}$, then length of a diagonal of the parallelogram is

(a) $4\sqrt{3}$ (b) $4\sqrt{5}$
(c) $4\sqrt{7}$ (d) None of these

70. $\lim_{x \to 0} \frac{\sin 3x - \sin x}{\sin x}$ is

(a) -2 (b) 2
(c) 0 (d) None of these

English and General Aptitude

Direction (Q. Nos. 1-5) *Read the passage and choose the correct answer from the given options.*

Sariska National Park is a wildlife sanctuary, located at a distance of 107 km from Jaipur. The park possesses historical monuments and temples, which reflect the legacy of the maharajas of Alwar. The park is bigger than Ranthambore but has a similar topography. Though this sanctuary does not have many tigers yet many wild animals dwell on the grounds of this park. Apart from other carnivores and herbivores, you can also trace Rhesus monkeys, which can usually be seen playing tricks upon one another around the Tad Vriksh(palm tree).

The park also shelters many different species of birds including Bush Quails, Sand Grouses, Tree Pies, Golden Backed Woodpeckers etc. Anima lovers would definitely love this place as it will offer them some of the best scenes to remember. The best place to spot wildlife is at the waterholes where animals come to satisfy their thirst. You can trace hundreds of birds at the Kalighati waterhole and Nilgai can be spotted at Salupka waterhole.

1. In the Sariska National Park, there are ____________

(a) only herbivores
(b) only carnivores
(c) both carnivores and herbivores
(d) no carnivores or herbivores

2. The Sariska National Park is not far from

(a) Jodhpur (b) Raipur
(c) Mewar (d) Jaipur

3. The park possesses

(a) ramparts and monuments
(b) monuments and temples
(c) temples and ramparts
(d) monuments and forts

4. Animals come to satisfy their thirst at a

(a) borehole (b) deepwell
(c) waterhole (d) borewell

5. The Sand Grouse is a species of

(a) birds (b) plants
(c) animals (d) trees

Direction (Q. Nos. 6-8) *Sentences are given with blank to be filled with appropriate word out of four alternatives given:*

6. I am vexed __________ his silence

(a) about (b) for
(c) by (d) at

7. He fell a victim __________ his own avarice.

(a) by (b) to
(c) of (d) with

8. We should abide __________ the laws of our country.

(a) with (b) in
(c) by (d) to

Direction (Q. Nos. 9-12) *Choose the word that is opposite in meaning to the given words.*

9. Collision

(a) Compassion (b) Agreement
(c) Perfection (d) Conflict

10. Grudge

(a) hatred (b) modest
(c) eternal (d) affection

11. Insolvent

(a) irrational (b) alluring
(c) affluent (d) defaulter

12. Scanty

(a) Lavish (b) skillful
(c) dirty (d) Insufficient

Direction (Q. Nos. 13-15) *Out of four alternatives, choose the one which best expresses the meaning of the given word.*

13. Emphatic

(a) hesitant (b) extinct
(c) gigantic (d) definite

14. Compatible

(a) liable (b) viable
(c) eligible (d) consistent

15. Pandemonium

(a) a kind of harmonium (b) repercussion
(c) chaos (d) symposium

Directions (Q. Nos. 16-18) *In the questions given below, select the one which is different from the other three responses.*

16. (a) Petrol (b) Diesel (c) Kerosene (d) LPG

17. (a) ACF (b) EGI (c) IKN (d) MOR

18. (a) 126 (b) 215 (c) 342 (d) 511

19. If 'parrot' is known as 'peacock', 'peacock' is known as 'swallow', 'swallow' is known as 'pigeon' and 'pigeon' is known as 'sparrow', then what would be the name of Indian National Bird?

(a) Parrot (b) Swallow (c) Peacock (d) Pigeon

20. Shyama says that father of Rajiv's father is my father. How Shyama is related to Rajiv?

(a) Sister (b) Mother (c) Father's sister (aunt) (d) Niece

21. Which one set of letters when sequentially placed at the gaps in the given letter series shall complete it?

g_ _ g_ _ gh_ g_i

(a) h i h i i h (b) h h i i h h (c) j k l j k l (d) h i j k l i

22. If 'EARN' is written as 'GCTP', how 'NEAR' can be written in that code?

(a) CTGP (b) GPTC (c) PGCT (d) PCGT

23. If 20 + 3 = 15 and 15 + 4 = 30, then 12 + 6 = ?

(a) 36 (b) 27 (c) 60 (d) 9

24. Which of the following answer figures indicates best relation between aeroplanes, train, travel?

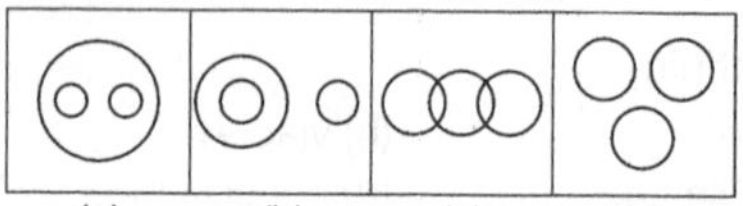

(a) (b) (c) (d)

25. If '+' and '–' signs are interchanged, similarly '×' and '÷' are interchanged, then find the answer of the following equation.

$10 - 2 + 12 \times 1 \div 0$

(a) 8 (b) 12 (c) 1 (d) 0

26. In the morning, I walked 2 km facing the Sun and stopped. From there, I walked 4 km on my right hand side, from there I walked 1 km again facing the Sun, from there I walked on my right hand side. In which direction I shall be walking now?

(a) East (b) South-East (c) South-West (d) Right hand side

Directions (Q. Nos. 27-28) *In the questions given below, select the missing number / letter from the given responses.*

27. 23, 24, 27, 28, 31, 32,?

(a) 30 (b) 33 (c) 29 (d) 35

28.

49	100	64
9	36	4
81	1	25
19	?	15

(a) 14 (b) 16 (c) 17 (d) 18

Directions (Q. Nos. 29 and 30) *In the question given below, select the related letters/ words/ numbers from the given alternatives.*

29. Sapling : Tree : : Girl : ?

(a) Woman (b) Wife (c) Man (d) Child

30. Pigeon : Peace : : ?

(a) Crown : Head
(b) White Flag : Surrender
(c) Laurels : Victory
(d) War : Liberty

Answers

Physics

1. (a)	**2.** (c)	**3.** (c)	**4.** (b)	**5.** (c)	**6.** (c)	**7.** (c)	**8.** (b)	**9.** (b)	**10.** (d)
11. (b)	**12.** (a)	**13.** (b)	**14.** (a)	**15.** (a)	**16.** (c)	**17.** (a)	**18.** (d)	**19.** (b)	**20.** (a)
21. (b)	**22.** (d)	**23.** (c)	**24.** (d)	**25.** (c)	**26.** (c)	**27.** (c)	**28.** (c)	**29.** (c)	**30.** (d)
31. (a)	**32.** (b)	**33.** (b)	**34.** (b)	**35.** (d)	**36.** (d)	**37.** (b)	**38.** (b)	**39.** (a)	**40.** (b)
41. (c)	**42.** (a)	**43.** (a)	**44.** (b)	**45.** (b)	**46.** (b)	**47.** (c)	**48.** (c)	**49.** (d)	**50.** (b)

Chemistry

1. (b)	**2.** (b)	**3.** (a)	**4.** (a)	**5.** (c)	**6.** (d)	**7.** (a)	**8.** (b)	**9.** (c)	**10.** (c)
11. (d)	**12.** (d)	**13.** (c)	**14.** (d)	**15.** (b)	**16.** (c)	**17.** (b)	**18.** (b)	**19.** (a)	**20.** (d)
21. (d)	**22.** (d)	**23.** (a)	**24.** (b)	**25.** (a)	**26.** (a)	**27.** (b)	**28.** (c)	**29.** (a)	**30.** (c)
31. (c)	**32.** (c)	**33.** (d)	**34.** (a)	**35.** (b)	**36.** (b)	**37.** (c)	**38.** (c)	**39.** (a)	**40.** (d)
41. (b)	**42.** (d)	**43.** (b)	**44.** (d)	**45.** (a)	**46.** (c)	**47.** (c)	**48.** (a)	**49.** (c)	**50.** (c)

Mathematics

1. (b)	**2.** (a)	**3.** (c)	**4.** (c)	**5.** (a)	**6.** (b)	**7.** (c)	**8.** (d)	**9.** (c)	**10.** (c)
11. (a)	**12.** (b)	**13.** (d)	**14.** (c)	**15.** (a)	**16.** (b)	**17.** (b)	**18.** (b)	**19.** (d)	**20.** (c)
21. (d)	**22.** (a)	**23.** (c)	**24.** (b)	**25.** (b)	**26.** (b)	**27.** (d)	**28.** (a)	**29.** (a)	**30.** (b)
31. (c)	**32.** (b)	**33.** (a)	**34.** (c)	**35.** (a)	**36.** (d)	**37.** (c)	**38.** (c)	**39.** (c)	**40.** (b)
41. (d)	**42.** (a)	**43.** (b)	**44.** (a)	**45.** (c)	**46.** (b)	**47.** (c)	**48.** (b)	**49.** (a)	**50.** (c)
51. (c)	**52.** (d)	**53.** (c)	**54.** (b)	**55.** (c)	**56.** (c)	**57.** (b)	**58.** (b)	**59.** (a)	**60.** (a)
61. (d)	**62.** (a)	**63.** (c)	**64.** (c)	**65.** (b)	**66.** (c)	**67.** (c)	**68.** (b)	**69.** (c)	**70.** (b)

English and General Aptitude

1. (c)	**2.** (d)	**3.** (b)	**4.** (c)	**5.** (a)	**6.** (d)	**7.** (b)	**8.** (c)	**9.** (b)	**10.** (d)
11. (c)	**12.** (a)	**13.** (d)	**14.** (d)	**15.** (c)	**16.** (d)	**17.** (b)	**18.** (a)	**19.** (b)	**20.** (c)
21. (a)	**22.** (c)	**23.** (b)	**24.** (a)	**25.** (b)	**26.** (d)	**27.** (d)	**28.** (c)	**29.** (a)	**30.** (b)

SOLUTIONS

Physics

1. *(a)* $A = 10L \times 2L = 20L^2$

$$\frac{\Delta A}{A} \times 100 = 2\frac{\Delta L}{L} \times 100 = 2 \times 1 = 2\%$$

2. *(c)* Average velocity $= \frac{\text{displacement}}{\text{time}} = \frac{2r}{t} = \frac{2 \times 40}{40}$

$$= 2 \text{ m/s}$$

3. *(c)* When $\theta = 180°$, the particle will be at diametrically opposite point. Where its velocity is opposite to the direction of motion.

The change in momentum $= mv - (-mv) + 2mv$ (maximum).

When $\theta = 360°$, the particle is at the initial position with momentum mv change in momentum $= mv - mv = 0$ (minimum).

4. *(b)* Maximum height

$$H = \frac{v^2 \sin^2 45°}{2g} = \frac{v^2}{2g} \times \frac{1}{2} = \frac{v^2}{4g}$$

Momentum of particle at the highest point

$$p = mv \cos 45° = mv/\sqrt{2}$$

Angular momentum = pH

$$= \frac{mv}{\sqrt{2}} \times \frac{v^2}{4g} = \frac{mv^3}{4\sqrt{2}g}$$

5. *(c)* The mass of block $A = 4$ kg

Force of friction between A and $B = 12$ N

Mass of block A and $B = 4 + 8 = 12$ kg

i.e, mass is increased 3 times.

The force of friction also increases 3 times

$$= 12 \times 3 = 36 \text{ N}$$

Hence, maximum horizontal force required = 36 N

6. *(c)* Force of friction = 720 N

If a is the acceleration of fireman, sliding down the pole, then

$$mg - F = ma$$

or
$$a = \frac{mg - F}{m}$$

$$= \frac{(80 \times 10 - 720)}{80}$$

$$= 1 \text{ m/s}^2$$

7. *(c)* Energy at y = total KE − work done from x to y

$$= 800 - 100 \times 2 = 600 \text{ J}$$

8. *(b)* Centripetal force $= \frac{mv^2}{R} = \frac{k}{R^2}$ (in magnitude)

$$\therefore \quad \text{KE} = \frac{1}{2}mv^2 = \frac{1}{2}\frac{k}{R}$$

$$\text{PE} = -\int F dR = -\int \frac{-k}{R^2} dR = -\frac{k}{R}$$

$\therefore$ Total energy = KE + PE

$$= \frac{k}{2R} - \frac{k}{R} = -\frac{k}{2R}$$

9. *(b)* For a solid sphere,

$$I = \frac{2}{5}MR^2$$

For a spherical shell,

$$I' = \frac{2}{3}MR^2$$

$$= \frac{5}{2} \times \frac{2}{5}MR^2 = \frac{5}{3}I$$

10. *(d)* Angular momentum w.r.t. origin ($L = mvr$) will remain constant.

11. *(b)* Gravitational force $\left(= \frac{GMm}{R^{5/2}}\right)$ provides the necessary centripetal force (*i.e*, $mR\omega^2$). So,

$$\frac{GMm}{R^{5/2}} = mR\omega^2 = mR\left(\frac{2\lambda}{T}\right)^2 = \frac{4\pi^2 mR}{T^2}$$

or
$$T^2 = \frac{4\pi^2 R^{7/2}}{GM}$$

ie,
$$T^2 \propto R^{7/2}$$

12. *(a)* Areal velocity $= \frac{dA}{dt} = \frac{L}{2m}$

$$= \frac{2vr}{2m} = \frac{vr}{2} = \frac{r}{2}\sqrt{\frac{GM}{r}} = \frac{1}{2}\sqrt{GMr}$$

So,
$$\frac{dA}{dt} \propto \sqrt{r}$$

13. *(b)* Strain $= \frac{\Delta l}{l} = \frac{2\pi R - 2\pi r}{2\pi r} = \left(\frac{R - r}{r}\right)$

$$\text{Stress} = Y \times \text{strain} = \frac{F}{A}$$

$$\therefore \quad F = Y \times A \times \text{strain} = AY \times \left(\frac{R-r}{r}\right)$$

14. *(a)* If h is the height of liquid column in cylinder and l is the length of each side of base. Then

$$F_B = (h\rho g)(l \times l) = h\rho g l^2$$

$$F_S = \left(\frac{h\rho g}{2}\right)(4l \times h) = 2h^2\rho g l$$

$$\therefore \quad h\rho g l^2 = 2h^2 \rho g l$$

or $$h = \frac{l}{2} = \frac{10}{2} = 5 \text{ cm}$$

15. *(a)* Total pressure, when the two gases are filled in one cylinder is equal to sum of the pressures of the individual gases.

Thus, total pressure = 2.5 + 1 = 3.5 atm

16. *(c)* Slope of isothermal curve (AB) is smaller than slope of adiabatic curve (BC).

17. *(a)* Mass of ice melted per minute $= \frac{10}{10} = 1$ g

Quantity of heat used = 1 × 80 cal

$$\text{Area of lens} = \pi r^2 = 3.14 \times (2.5)^2 = 19.625 \text{ cm}^2$$

∴ Amount of heat received from sun

$$= \frac{80}{19.625} = 4 \text{ cal / cm}^2\text{min.}$$

18. *(d)* $y = 3 \sin 314\, t + 4 \cos 314\, t$

$$= r \cos\theta \sin 314\, t + r \sin\theta \cos 314\, t$$

$$= r \sin(314\, t + \theta) \qquad \text{...(i)}$$

where, $r\cos\theta = 3$ and $r \sin\theta = 4$

$$\therefore \quad r^2(\cos^2\theta + \sin^2\theta) = 3^2 + 4^2 = 25$$

or $$r = 5 \text{ cm}$$

19. *(b)* Since $l_1 + nl_2$ where n is an integer, so the spring is made of $(n+1)$ equal parts in length each of length l_2

$$\therefore \quad \frac{1}{k} = \frac{(n+1)}{k'}$$

or $$k' = (n+1)\,k$$

The spring of length $l_1 (= nl_2)$ will be equivalent to n springs connected in series where spring constant

$$k'' = \frac{k'}{n} = \frac{(n+1)\,k}{n}$$

20. *(a)* $$y = A \sin(kx - \omega t)$$

$$\frac{dy}{dt} = A \cos(kx - \omega t) \times (\omega)$$

$$\left(\frac{dy}{dt}\right)_{\max} = A(1)(\omega) = A\omega$$

21. *(b)* Let frequency of tuning fork be n_1

$$\frac{n_2}{n_1} = \frac{100}{101}$$

As $$n_1 - n_2 = 5$$

So, $$n_1 = 505 \text{ Hz}$$

22. *(d)* The path is a parabola, because initial velocity can be resolved into two rectangular components, one along $\vec{\mathbf{E}}$ and other perpendicular to $\vec{\mathbf{E}}$. The former decreases at a constant rate and later is unaffected. The resultant path is, therefore a parabola.

23. *(c)* $\vec{\mathbf{E}}$ is parallel to dipole axis, at both the points P and Q, i.e. on axial line as well as equatorial line.

24. *(d)* $$\frac{1}{C_S} = \frac{1}{2} + \frac{1}{6} = \frac{3+1}{6} = \frac{4}{6} = \frac{2}{3}$$

$$\therefore \quad C_S = \frac{3}{2}\mu\text{F}$$

In combination charge on each capacitor in same.

So, $$Q = C_S V = \frac{3}{2} \times 800 = 1200\,\mu\text{C}$$

25. *(c)* $$I = \frac{q}{T} = qv$$

$$= 2 \times 10^{-2} \times 30 = 0.60 \text{ A}$$

26. *(c)* $$v_e = \sqrt{2gR} \text{ and } v_o = \sqrt{gR}$$

$$\therefore \quad \frac{\text{KE}_1}{\text{KE}_2} = \frac{\frac{1}{2}mv_e^2}{\frac{1}{2}mv_o^2} = \frac{v_e^2}{v_o^2} = 2$$

27. *(c)* $$P = \frac{n_1RT + n_2RT + n_3RT}{V}$$

$$= (n_1 + n_2 + n_3)\frac{RT}{V}$$

$$= \left(\frac{8}{16} + \frac{14}{28} + \frac{22}{44}\right) \times \frac{0.082 \times 300}{10}$$

$$= \left(\frac{1}{2} + \frac{1}{2} + \frac{1}{2}\right) \times 2.46$$

$$= \frac{3}{2} \times 2.46 = 3.69 \text{ atm} \approx 3.7 \text{ atm}$$

28. *(c)* Velocity of liquid flowing out of orifice,

$$v = \sqrt{2gh}$$

Average velocity of liquid flowing out of orifice till the whole vessel is emptied

$$v_{av} = \frac{v+0}{2} = \frac{1}{2}\sqrt{2gh}$$

If A is area of cross-section of the orifice, then the time taken by liquid of volume V to flow out of tank through orifice is

$$t = \frac{V}{A \times \frac{1}{2}\sqrt{2gh}}$$

In the second case, volume is $V/2$ and height of liquid in vessel is $h/2$. Therefore, time t is :

$$t' = \frac{V/2}{A \times \frac{1}{2}\sqrt{2gh/2}} = \frac{t}{\sqrt{2}} = \frac{10}{\sqrt{2}} = 7 \text{ min}$$

29. *(c)* $k_{adia} = \gamma P = -\frac{\Delta P}{\Delta V / V}$

$\therefore \quad \frac{\Delta V}{V} = -\frac{\Delta P}{\gamma P}$

30. *(d)* $\eta = \frac{W}{Q_1}$

$\Rightarrow \quad w = \eta Q_1 = \frac{1}{3} \times 1000 \text{ cal}$

$= \frac{1000}{3} \times 4.2 \text{ J} = 1400 \text{ J}$

31. *(a)* $T = 2 = 2\pi\sqrt{\frac{M}{k}}$

and $2 + 1 = 2\pi\sqrt{\frac{M+4}{k}}$

or $3 = 2\pi\sqrt{\frac{M+4}{k}}$

$\therefore \quad \frac{4}{9} = \frac{M}{M+4}$

or $4M + 16 = 9M$

or $M = \frac{16}{5} = 3.2 \text{ kg}$

32. *(b)* Let m = mass per unit length of rope

T = tension in the rope at a distance x from the lower end

$\therefore T = (mg)\, x$ = weight of x metre of rope

As $v = \sqrt{\frac{T}{m}}$

$\therefore \quad v = \sqrt{\frac{mgx}{m}} = \sqrt{gx}$

i.e., $v \propto \sqrt{x}$

33. *(b)* $y = A \sin[Bx + Ct + D]$

As each term inside the bracket is dimension less,

$\therefore \quad [A] = [y] = [L], \quad [B] = \frac{1}{[x]} = [L^{-1}]$

$[C] = \frac{1}{[t]} = [T^{-1}]$ and D is dimensionless.

$\therefore \quad [ABCD] = [L][L^{-1}][T^{-1}][1] = [M^0L^0T^{-1}]$

34. *(b)* Given, $a = \frac{dv}{dt} = 6t + 5$

Integrating it, $\int_0^v dv = \int_0^t (6t+5)\, dt$

$$v = \frac{6t^2}{2} + 5t$$

As $v = \frac{ds}{dt}$, so $ds = \left(\frac{6t^2}{2} + 5t\right) dt$

Integrating it, $\int_0^s ds = \int_0^t \left(\frac{6t^2}{2} + 5t\right) dt$

$$s = \frac{3t^3}{3} + \frac{5t^2}{2}$$

When $t = 2\text{s}, s = 3\frac{(2)^3}{3} + \frac{5(2)^2}{2}$

$= 8 + 10 = 18 \text{ m}$

35. *(d)* Let $\vec{A}$, $\vec{B}$ and $\vec{C}$ be as shown in figure. Let θ be the angle of incidence, which is equal to the angle of reflection. Resolving these vectors in rectangular components, we have

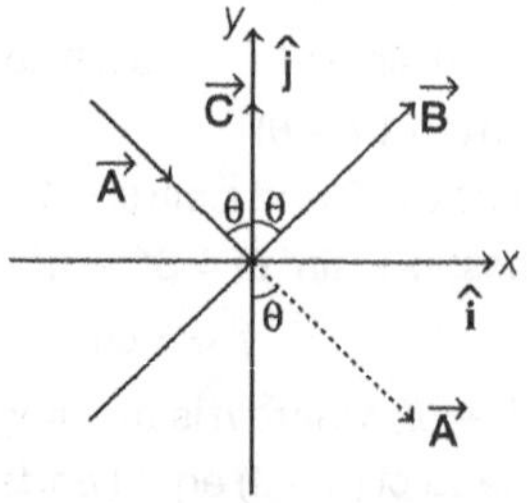

$$\vec{A} = \sin\theta\,\hat{i} - \cos\theta\,\hat{j}$$

$$\vec{B} = \sin\theta\,\hat{i} + \cos\theta\,\hat{j}$$

$$\vec{B} - \vec{A} = 2\cos\theta\,\hat{j}$$

or $\vec{B} = \vec{A} + 2\cos\theta\,\hat{j}$

Now, $\vec{A} \cdot \vec{C} = (1)(1)\cos(180° - \theta) = -\cos\theta$

$\therefore \quad \vec{B} = \vec{A} - 2(\vec{A} \cdot \vec{C})\,\hat{j}$

or $\vec{B} = \vec{A} - 2(\vec{A} \cdot \vec{C})\,\vec{C}$ (as $\hat{j} = \vec{C}$)

36. *(d)* Refer figure as shown, the average velocity

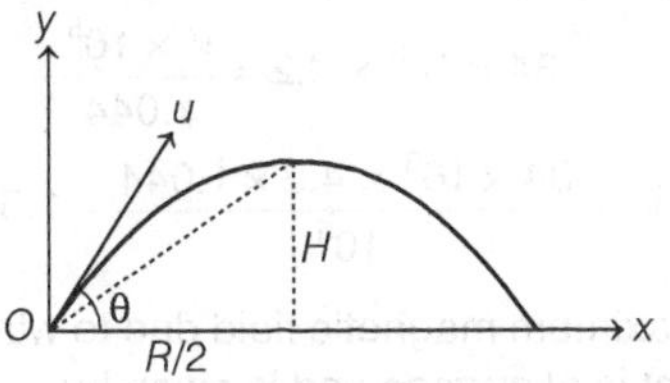

$$= \frac{\text{displacement}}{\text{time taken}}$$

$$v_{av} = \frac{\sqrt{H^2 + R^2/4}}{T/2} \quad \text{...(i)}$$

Here, $H = \frac{u^2 \sin^2 \theta}{2g}$

$$R = \frac{u^2 \sin 2\theta}{g}$$

and $T = \frac{2u \sin \theta}{g}$

Putting these values in Eq. (i), we get

$$v_{av} = \frac{u}{2}\sqrt{1 + 3\cos^2 \theta}$$

37. *(b)* Angular velocity about *A*,

$$\omega_1 = \frac{v}{2r}$$

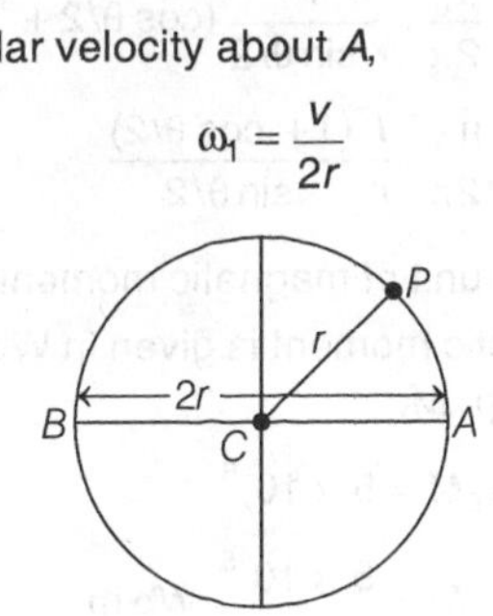

Angular velocity about *C*,

$$\omega_2 = \frac{v}{r}$$

$$\therefore \quad \frac{\omega_1}{\omega_2} = \frac{v/2r}{v/r} = 1:2$$

38. *(b)* Mass $M_1 = \frac{5}{8}$ kg, Mass $M_2 = \frac{5}{24}$ kg

Total mass $= M_1 + M_2 = \frac{5}{8} + \frac{5}{24} = \frac{20}{24}$ kg

Therefore, total acceleration in two masses

$$= \frac{F}{M_1 + M_2} = \frac{5}{20/24}$$

$$= \frac{5 \times 24}{20} = 6\,\text{ms}^{-2}$$

39. *(a)* While firing a bullet from a rifle, there will be equal back thrust on light rifle as well as heavy rifle. Due to this, the light rifle will move backwards with greater speed than heavy rifle. That is why light rifle will cause more injury.

40. *(b)* $t = \sqrt{x} + 3$

or $x = (t - 3)^2$

Now, $v = \frac{dx}{dt} = 2(t - 3)$

At $t = 0, v_1 = 2(-3) = -6$

At $t = 6, v_2 = 2(6 - 3) = 6$

Work done = change in KE

$$= \frac{1}{2}mv_2^2 - \frac{1}{2}mv_1^2 = 0$$

41. *(c)* When a particle undergoes normal collision with a floor or a wall, with coefficient of restitution *e*, the speed after collision is *e* times the speed before collision. Therefore, change in momentum after Ist impact is

$$ep - (-p) = p(1 + e).$$

After the second impact, change in momentum would be $e(ep) - (-ep) = ep(1 + e)$ and soon.

Therefore, total change in momentum of ball

$$= p(1 + e)[1 + e + e^2 + \ldots]$$

$$= p\left(\frac{1 + e}{1 - e}\right)$$

42. *(a)* Length of each side of hexagon = 2*L*

and mass of each side = *m*

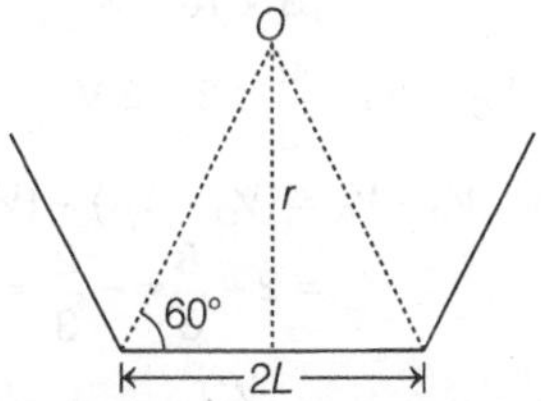

Let *O* be the centre of mass of hexagon. Therefore, perpendicular distance of *O* from each side, $r = L \tan 60° = L\sqrt{3}$

The desired moment of inertia of hexagon about *O* is

$$I = 6[I_{\text{one side}}] = 6\left[\frac{m(2L)^2}{12} + mr^2\right]$$

$$= 6\left[\frac{mL^2}{3} + m(L\sqrt{3})^2\right]$$

$$= 20\,mL^2$$

43. *(a)* As $V = \dfrac{q}{4\pi\varepsilon_0 R}$

$\therefore \quad \dfrac{q}{4\pi\varepsilon_0} = VR$ and $E = \dfrac{q}{4\pi\varepsilon_0 r^2}$

$$\therefore \quad E = \frac{VR}{r^2}$$

44. *(b)* $$F = \frac{q_1 q_2}{4\pi\varepsilon_0 r^2} = \frac{q_1(q - q_1)}{4\pi\varepsilon_0 r^2}$$

For maximum force between q_1 and q_2

$$\frac{dF}{dq_1} = 0$$

$$\therefore \quad q - 2q_1 = 0$$

or $$q_1 = \frac{q}{2} \text{ or } \frac{q_1}{q} = 0.5$$

45. *(b)* Resistance between Q and S,

$$R' = \frac{5 \times 10}{5 + 10} = \frac{50}{15} = \frac{10}{3}\Omega$$

Potential difference across Q and S,

$$V_Q - V_S = 2 \times \frac{10}{3} = \frac{20}{3}\text{ V}$$

Current through arm QPS,

$$I_1 = \frac{20}{3 \times 5} = \frac{4}{3}\text{ A}$$

$$V_Q - V_P = \frac{4}{3} \times 2 = \frac{8}{3}\text{ V}$$

Current through arm QRS,

$$I_2 = \frac{20}{3 \times 10} = \frac{2}{3}\text{ A}$$

$$V_Q - V_R = \frac{2}{3} \times 3 = 2\text{ V}$$

Hence, $V_P - V_R = (V_Q - V_R) - (V_Q - V_P)$

$$= 2 - \frac{8}{3} = -\frac{2}{3} \approx -1\text{ V}$$

46. *(b)* Heat produced $= \dfrac{V^2 t}{4.2R} = mL$

or $$m = \frac{V^2 t}{4.2\,RL} = \frac{(210)^2 \times 1)}{4.2 \times 50 \times 80} = 2.62\text{ g}$$

47. *(c)* $q = \dfrac{m}{z} = \dfrac{1 \times 10^{-3}}{1.044 \times 10^{-8}} = \dfrac{10^5}{1.044}$ C

Given, H= 34 k-cal = $34 \times 10^3 \times 4.2$ J

Heat produced, $H = Vq$

$$\therefore \quad 34 \times 10^3 \times 4.2 = \frac{V \times 10^5}{1.044}$$

$$\therefore \quad V = \frac{34 \times 10^3 \times 4.2 \times 1.044}{10^5} = 1.5\text{ V}$$

48. *(c)* Maximum magnetic field due to wire carrying current is at surface and is given by

$$B = \frac{\mu_0}{4\pi}\frac{2I}{R}$$

or $$I = \frac{BR}{2 \times (\mu_0 / 4\pi)}$$

$$\therefore \quad I = \frac{5 \times 10^{-4} \times \left(\dfrac{1.6 \times 10^{-3}}{2}\right)}{2 \times 10^{-7}} = 2\text{ A}$$

49. *(d)* Perpendicular of O from PQ or QR, $a = r \sin\theta/2$ magnetic field induction at O due to current through PQ and QR is

$$B = \frac{\mu_0}{4\pi} \cdot \frac{I}{a}[\sin(90° - \theta/2) + \sin 90°] \times 2$$

$$= \frac{\mu_0}{2\pi} \cdot \frac{I}{r \sin\theta/2}(\cos\theta/2 + 1)$$

$$= \frac{\mu_0}{2\pi} \cdot \frac{I}{r}\frac{(1 + \cos\theta/2)}{\sin\theta/2}$$

50. *(b)* The SI unit of magnetic moment is A-m^2, but the magnetic moment is given in Wb-m^2, which is the unit of $\mu_0 M$.

$$\therefore \quad \mu_0 M = 5 \times 10^{-5}$$

$$M = \frac{5 \times 10^{-5}}{\mu_0}\text{ Wb-m}$$

$$= \frac{5 \times 10^{-5}}{4\pi \times 10^{-7}} = \frac{125}{\pi}$$

As $$T = 2\pi\sqrt{\frac{I}{MB}}$$

$$\therefore \quad I = \frac{T^2 MB}{4\pi^2} = \frac{(15)^2\left(\dfrac{125}{\pi}\right) \times 8\pi \times 10^{-4}}{4\pi^2}$$

$$= 0.57\text{ kg m}^2$$

Chemistry

1. *(b)* The prefix neo means a carbon atom having four alkyl groups, position of alcohol group should be at C_1.

2. *(b)* Presence of $-I$ group in a carbocation decreases delocalisation of the positive charge and hence decreases stability of the carbocation and *vice versa*. Thus the carbocation (iii) having $—COCH_3$ grouping is the least stable, followed by (ii) having $—OCH_3$ grouping, followed by (i) having only alkyl group.

3. *(a)* Alkanes have a general formula C_nH_{2n+2}

So, when $n = 1$, $C_1H_{2\times 1+2} = CH_4$

$n = 2$, $C_2H_{2\times 2}+2 = C_2H_6$

$n = 3$, $C_3H_{2\times 3+2} = C_3H_8$

So, the all have a difference of CH_2.

4. *(a)*

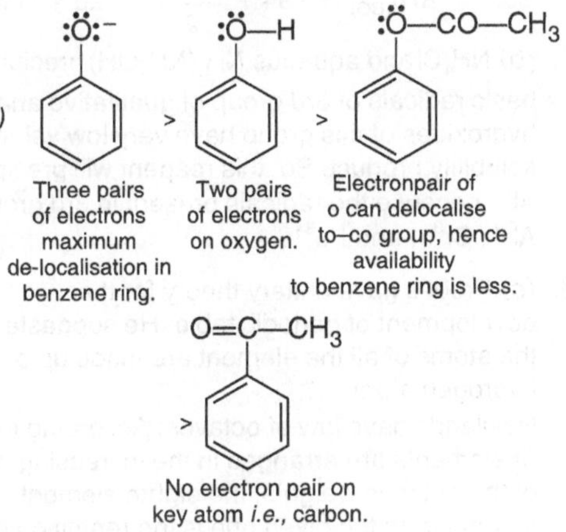

5. *(c)* Naphthalene has a structure,

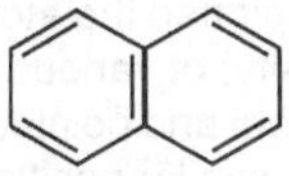

It has 5π bonds and thus the number of π electrons is $5 \times 2 = 10$.

6. *(d)* The volumes of the reacting gases and those of gaseous products bear the simple ratio. This is called the law of gaseous volumes or Gay-Lussac's law.

7. *(a)*

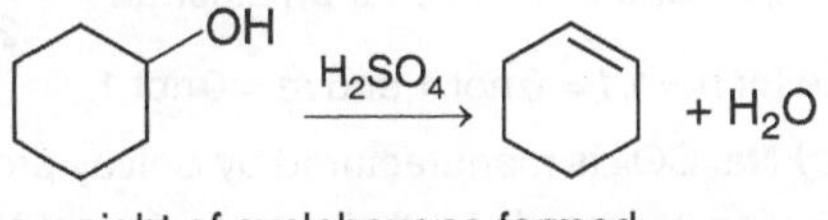

The weight of cyclohexene formed

$$= \frac{82 \times 100}{100} \times \frac{75}{100} = 61.5 \text{ g}$$

8. *(b)* According to aufbau principle. The order of increasing energies may be summed up as below.

1*s*, 2*s*, 2*p*, 3*s*, 3*p*, 4*s*, 3*d*, 4*p*. 5*s*, 4*d*, 5*p*, 6*s*, 4*f*, 5*d*, 6*p*, 7*s*, 5*f*, 6*d* and 7*p*

So, the correct order is $5p < 6s < 4f < 5d$

9. *(c)* When $n = 3$, $l = 0, 1, 2$ and

For $l = 0, m = 0$

$l = 1, m = -1, 0, +1$

$l = 2, m = -2, -1, 0, +1, +2$

So, for the value $l = 0$, m cannot be -1

10. *(c)* $\lambda = \frac{h}{mv} = \frac{6.63 \times 10^{-34}}{9.11 \times 10^{-31} \times 10^6}$

$= 0.727 \times 10^{-9} = 0.727$ nm.

11. *(d)* Substances which resist high temperatures and do not become soft are called refractories material. The neutral refractories are graphite, chromite and carboren, dum (SiC). Acidic refractories are SiO_2, basic refractories are CaO and MgO.

12. *(d)* Coinage metals are Cu, Ag and Au. They are transition metal and present in group II or I B. The valence shell configuration is $(n-1)d^{10}ns^1$.

13. *(c)* Cold conc. HNO_3 makes it passive in which state it fails to exhibit the properties of normal iron. Passivity is said to be due to the deposit of a thin layer of iron oxide (Fe_3O_4) on the surface.

14. *(d)* The b.p. and viscosity of D_2O is higher than water. Solubility of NaCl in it is less than H_2O because reaction velocity is slightly less in D_2O due to the that deuterium bond is strong than protium bond.

15. *(b)* When potassium iodide react with ozone it oxidise KI and liberate I_2 it turns starch paper blue.

16. *(c)* Red hot iron absorbs SO_2 and give FeO and FeS.

$$Fe + SO_2 \longrightarrow FeO + FeS$$

17. *(b)*

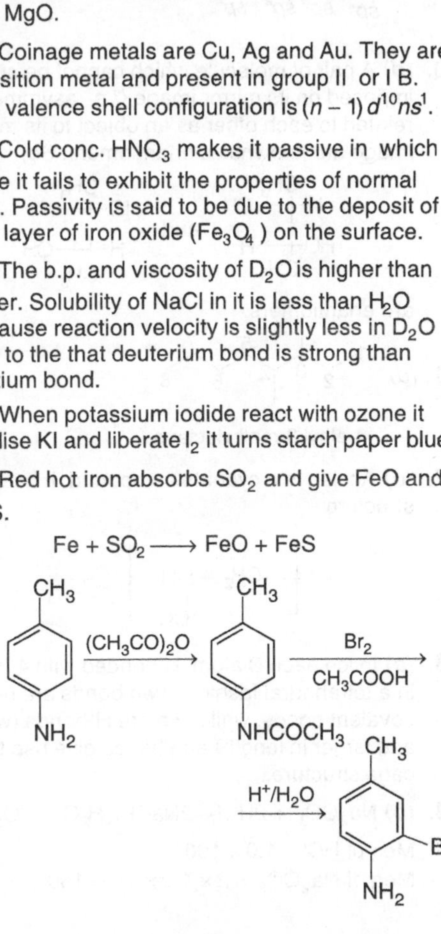

18. *(b)* Only $C_6H_5COOC_2H_5$ corresponds with the molecular formula of the ester *A*. Further it explains all the given reactions.

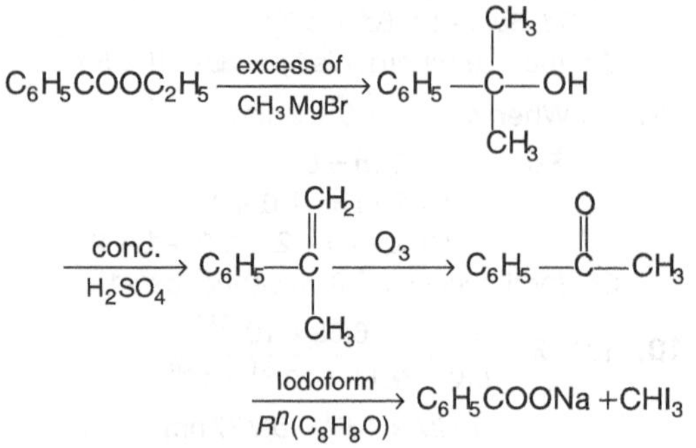

19. *(a)*

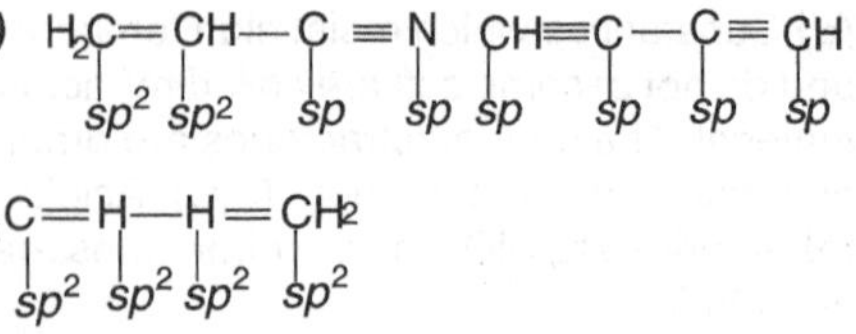

20. *(d)* A pair of molecule which cannot be super imposed on its mirror image (*i.e.*, asymmetrical) related to each other as an object to its mirror image are knowns as enantiomers, So,

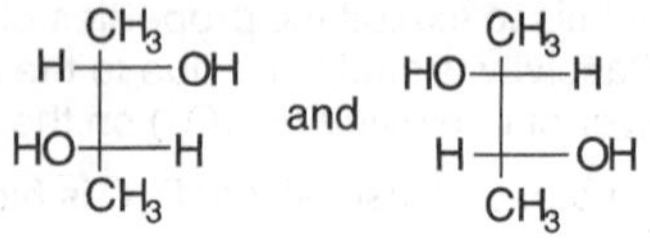

are enantiomers.

21. *(d)*

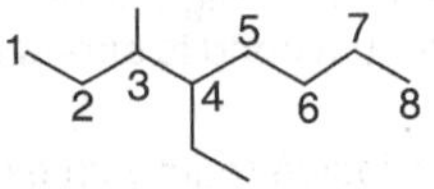

4-ethyl 3-methyloctane

22. *(d)* Acrilan is a cyanide compound. It has a structure

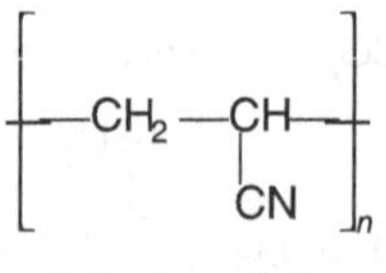

23. *(a)* In ice each O-atom is bonded with 4 H-atoms in a tetrahedral fashion. Two bonds are normal covalent bonds, while two are H-bonds (weaker and larger in length) and hence give rise to loose cage structure.

24. *(b)* $Na_2CO_3 + 2HCl — 2NaCl + H_2O + CO_2$

Meq of HCl $= 1.0 \times 100$

Meq of $Na_2CO_3 = 2 \times 1.0 \times 75 = 150$

Thus, Na_2CO_3 is in excess, and being the salt of weak acid (H_2CO_3) and strong base (NaOH), undergoes hydrolysis to produce basic character.

$$Na_2CO_3 + 2H_2O \longrightarrow 2Na^+ + 2OH^- + H_2CO_3$$

25. *(a)* Semiconductor are derived from compounds of *p*-block elements.

26. *(a)* The required reaction is

$$C(s) + \frac{1}{2}O_2(g) \longrightarrow CO(g)$$

$$CO_2(g) \longrightarrow CO(g) + \frac{1}{2}O_2(g)$$

$$\Delta H = \frac{+135.2}{2} \text{ kcal} \quad \ldots(i)$$

$$C(s) + O_2(g) \longrightarrow CO_2(g)$$

$$\Delta H = -94 \text{ kcal} \quad \ldots(ii)$$

The required equation is come out by adding Eq. (i) and (ii)

So, $\Delta H_{f(CO)} = -94 + \frac{135.2}{2} = -26.4$ kcal

27. *(b)* NH_4Cl and aqueous $NH_3(NH_4OH)$ precipitates basic radicals of 3rd group of qualitative analysis hydroxides of this group have very low value of solubility product. So, this reagent will precipitate Al^{3+} because the radicals present in 3rd group are Al^{3+}, Cr^{3+}, and Fe^{3+}.

28. *(c)* Prout's gave unitary theory for the development of periodic table. He suggested that the atoms of all the element are made up of hydrogen atom.

Newlands gave law of octaves: According to him, "If elements are arranged in the increasing order of their atomic weights, the eighth element starting from the given one is the repetition of the first like the eighth node in an octave of music.

Lother Meyer's plotted the atomic volumes (*i.e.*, at. wt/density) of various elements against their atomic weight and pointed out that the similar element occupy similar position on the curve.

Rutherford is associated with structure of atom. He bombarded high speed α-particles on nitrogen and proved that nuclei of all atoms have protons.

29. *(a)* For a given *n* value possible *l* values are 0 to $(n-1)$ possible *m* values for a given *l* value is $-l, 0, +l$ and value of *s* for an electron $= +\frac{1}{2}$ or $-\frac{1}{2}$.

So for $n = 1$, $l = 0$ not 1 and $m = 0$ not 1.

30. *(c)* Na_2CO_3 is manufactured by Solvay process.

Lowing process is for the manufacture of NaOH

Leblanc process is for the manufacture of K_2CO_3

Haber's process is for the manufacture of NH_3

31. *(c)* This reaction is ozonolysis. O_3 is added to triple bond and then on hydrolysis form aldehyde or ketones

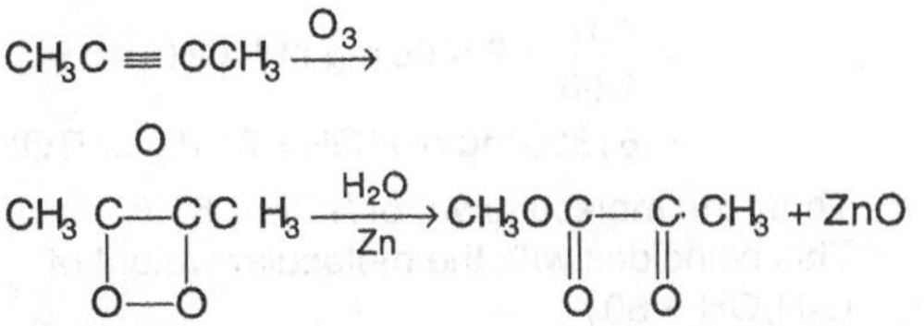

32. *(c)* Benzene-hexachloride exists in eight isomeric forms namely α, β, γ and δ etc. Out of which γ-isomer is most active and used as insecticide under the name of gammexene.

33. *(d)* Stability of corresponding carbocation

$$CH_2{=}CH{-}\overset{+}{C}H_2 > CH_2 = \overset{+}{C}H > C_6H_5^+$$

So, $ClCH_2{-}CH{=}CH_2$ is most reactive towards nucleophilic substitution reaction.

34. *(a)* Stability of corresponding carbocation

$$CH_3CH_2 > CH_2{=}\overset{+}{C}H > C_6H_5^+$$

So, the correct order of reactivity of halogen atom is chlorobenzene, vinyl chloride, chloroethane.

35. *(b)* Cyanide with Grignard reagent forms ketone.

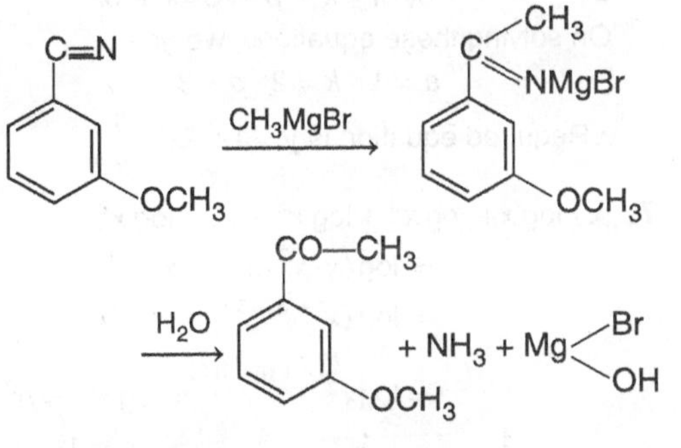

36. *(b)* Carbohydrates are optically active polyhydroxy aldehyde ketones it must contain at least three carbon atom. Hydroxyaldehyde with 2 carbon atom *i.e.*, $CH_2OH \cdot CHO$ is not carbohydrate as it is optically inactive.

37. *(c)* Bohr's atom model is applicable only to one electron species, He^{2+} does not contain any electron all other species have one electron each.

38. *(c)* Diborane readily adds at 0°C to 25°C to the olefin or acetylenic compound to form trialkyl boranes. This reaction is known as hydro-boration.

$$\underset{\text{Ethylene}}{6CH_2{=}CH_2} + B_2H_6 \xrightarrow{\text{ether}}$$

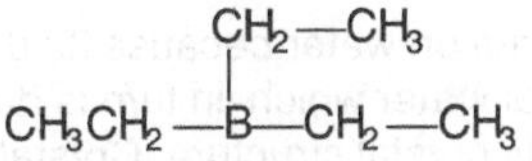

Triethyl borane

39. *(a)* Normality of salt solution $= \frac{3.92}{392} \times \frac{1}{100} \times 1000$

$$= 0.1\,N$$

20 mL of 0.1 N salt solution $\equiv$ 18 mL of $KMnO_4$ solution

$\therefore$ Normality of $KMnO_4$ solution $= \frac{20 \times 0.1}{18}$

$$= \frac{1}{9}\,N$$

$\therefore$ Strength of $KMnO_4$ solution $= \frac{1}{9} \times 31.6$

$$= 3.5\ g\ L^{-1}$$

40. *(d)* Since, As is added from outside, it is an extrinsic semiconductor

41. *(b)* $pOH = pK_b + \log\frac{[\text{salt}]}{[\text{base}]} = pK_b + 0$

$$pK_b = pOH = 14 - pH = 14 - 9.25 = 4.75$$

42. *(d)* On reducing the volume $\frac{1}{4}$th of the initial volume, each concentration will be increased by 4 times.

$$(\text{Rate})_1 = k[A]\,[B]$$

$$(\text{Rate})_2 = k\,[4\,[A]\,[4\,[B]]$$

Hence, $\frac{[\text{Rate}]_2}{[\text{Rate}]} = 4 \times 4 = 16$

43. *(b)* $\Delta G_1 = \Delta H - T\Delta S$ or $T = \frac{\Delta H - \Delta G}{\Delta S}$

$$= \frac{[145.6 - (-5.2)] \times 10^3}{216} = 698.1\ K$$

44. *(d)* For the cell reaction, Fe acts as cathode and Sn as anode

Hence, $E°_{cell} = E°_{cathode} - E°_{anode}$

$$= -0.44 - (0.14) = -0.30\ V$$

The negative emf suggests that the reaction goes spontaneously in reversed direction.

45. *(a)* In group number 15 elements shows –III and III oxidation state (by gain or loss of three *p* electrons). These shows V oxidation state due to loss of $np^3 + ns^2$ electrons. But nitrogen is unique in having a large number of oxidation states including a fractional one (*i.e.*, $-\frac{1}{3}$ in N_3H)

46. *(c)* Ice floats on water because its density is less than that of water which in turn is due to empty space in its crystal structure. Crystal structure of ice is regular hexagon with empty space at the centre.

47. *(c)* When more electropositive metal displace less electropositive metals from their salt solution, this is reduction by precipitation and the process is called hydrometallurgy.

48. *(a)* NaOH when react with sand SiO_2, it forms, sod. silicate which is generally called glass.

$$2NaOH + SiO_2 \longrightarrow \underset{\text{Sod. silicate (glass)}}{Na_2SiO_3} + H_2O$$

49. *(c)* 1.56 mL of CH_4 at STP is liberated by 4.12 mg of alcohol. 22400 mL of CH_4 at STP is liberated by

$$= \frac{4.12}{1.56} \times 22400 \text{ mg of } ROH$$

$$= 61800 \text{ mg of } ROH = 61.8 \text{ g of } ROH$$

Thus, the molecular wt. of $ROH = 61.8$.
This coincides with the molecular weight of $C_3H_7OH = 60$.

50. *(c)* 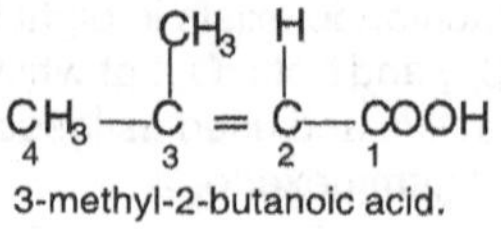

3-methyl-2-butanoic acid.

Mathematics

1. *(b)* Now, $\dfrac{5^{\frac{3}{2}} - 2^{\frac{3}{2}}}{\sqrt{5} - \sqrt{2}} + \dfrac{5^{\frac{3}{2}} + 2^{\frac{3}{2}}}{\sqrt{5} + \sqrt{2}}$

$$= \frac{\sqrt{5}^3 - \sqrt{2}^3}{\sqrt{5} - \sqrt{2}} + \frac{\sqrt{5}^3 + \sqrt{2}^3}{\sqrt{5} + \sqrt{2}}$$

$$= (5 + 2 + \sqrt{10}) + (5 + 2 - \sqrt{10})$$

$$= 7 + 7 = 14$$

2. *(a)* Given, $9^{-x} = \dfrac{1}{27\sqrt{27}} = (81)^{-y}$

$$\Rightarrow \quad (3)^{-2x} = (3)^{-3}\,(3)^{-\frac{3}{2}} = (3)^{-4y}$$

$$\Rightarrow \quad -2x = -\frac{9}{2} \text{ and } -4y = -\frac{9}{2}$$

$$\Rightarrow \quad x = \frac{9}{4} \text{ and } y = \frac{9}{8}$$

3. *(c)* $x^{4/3} + x^{-1/3} = 1 \Rightarrow x^{4/3} \cdot x^{1/3} + 1 = x^{1/3}$

$$\Rightarrow \quad x^{5/3} - x^{1/3} = -1$$

On cubing both sides, we get

$$x^5 - x - 3x^2\,(x^{5/3} - x^{1/3}) = -1$$

$$\Rightarrow \quad x^5 - x + 3x^2 = -1$$

4. *(c)* Given, $x = \dfrac{2\sqrt{2} - \sqrt{7}}{2\sqrt{2} + \sqrt{7}}$

$$\therefore \quad x + \frac{1}{x} = \frac{2\sqrt{2} - \sqrt{7}}{2\sqrt{2} + \sqrt{7}} + \frac{2\sqrt{2} + \sqrt{7}}{2\sqrt{2} - \sqrt{7}}$$

$$= \frac{8 + 7 - 4\sqrt{14} + 8 + 7 + 4\sqrt{14}}{8 - 7} = 30$$

5. *(a)* $\left|\dfrac{x}{2} - 1\right| < 3 \Rightarrow -3 < \dfrac{x}{2} - 1 < 3$

$$\Rightarrow \quad -3 + 1 < \frac{x}{2} < 3 + 1$$

$$\Rightarrow \quad -2 < \frac{x}{2} < 4$$

$$\Rightarrow \quad -4 < x < 8$$

$$\therefore \quad x \in (-4, 8)$$

6. *(b)* $y = a + kx + p/x$

$$\therefore \quad 6 = a + k + p$$

$$-4 = a - k - p$$

and $\quad 8 = a + 3k + p/3$

$$\Rightarrow \quad 6 - 1 = k + p \Rightarrow 5 = k + p$$

On solving these equations, we get

$$a = 1, \quad k = 2, \quad p = 3$$

$\therefore$ Required equation is $y = 1 + 2x + \dfrac{3}{x}$

7. *(c)* $\log x + \log x^3 + \log x^5 + \ldots + \log x^{2n-1}$

$$= \log (x \cdot x^3 \cdot x^5 \; \ldots x^{2n-1})$$

$$= \log (x^{1 + 3 + 5 + \ldots + 2n - 1})$$

$$= \log x^{\frac{n}{2}[2 + (n-1)2]} = \log x^{n^2} = n^2 \log x$$

8. *(d)* $\dfrac{1}{\log_{25} 10} + \dfrac{1}{\log_4 10} + \dfrac{1}{\log_{\sqrt{2}} 10} + \dfrac{1}{\log_{\sqrt{5}} 10}$

$$= \frac{\log 25}{\log 10} + \frac{\log 4}{\log 10} + \frac{\log \sqrt{2}}{\log 10} + \frac{\log \sqrt{5}}{\log 10}$$

$$= \frac{2 \log 5 + \frac{1}{2} \log 5}{\log 10} + \frac{2 \log 2 + \frac{1}{2} \log 2}{\log 10}$$

$$= \frac{5}{2}\left(\frac{\log 5 + \log 2}{\log 10}\right) = \frac{5}{2}\,\frac{\log 10}{\log 10} = \frac{5}{2}$$

9. *(c)* Let T_n be the nth term of a given series :

$$\therefore \quad T_n = \frac{(2n-1)}{(n-1)!};\; n = 1, 2, 3, \ldots, \infty$$

$$\Rightarrow \quad T_n = \frac{2n}{(n-1)!} - \frac{1}{(n-1)!}$$

$$= 2\left[\frac{(n-1)}{(n-1)!} + \frac{1}{(n-1)!}\right] - \frac{1}{(n-1)!}$$

$$= \frac{2}{(n-2)!} + \frac{1}{(n-1)!}$$

$$\therefore \sum_{n=1}^{\infty} T_n = \sum_{n=1}^{\infty} \frac{2}{(n-2)!} + \sum_{n=1}^{\infty} \frac{1}{(n-1)!}$$

$$= 2e + e = 3e$$

10. *(c)* Since, roots are complex.

$$\therefore \quad a^2 - 36 < 0 \quad (\because \text{ discriminant} < 0)$$

$$\Rightarrow \quad a^2 < 36$$

$$\Rightarrow \quad |a| < 6$$

11. *(a)* Even multiples of 9 are of the form $18n$.

$18n$ for $n = 1, \ldots, 27$ lies between 1 to 500

and $18n$ for $n = 1, \ldots, 16$ lies between 1 to 300.

$$\therefore \text{Required sum} = 18\left[\sum_{n=1}^{27} n - \sum_{n=1}^{16} n\right]$$

$$= 18\left[\frac{27 \times 28}{2} - \frac{16 \times 17}{2}\right] = 4356$$

12. *(b)* Let $S_{n_1} = \frac{n}{2}[2a_1 + (n-1)d_1]$

and $S_{n_2} = \frac{n}{2}[2a_2 + (n-1)d_2]$

$$\Rightarrow \quad \frac{\frac{n}{2}[2a_1 + (n-1)d_1]}{\frac{n}{2}[2a_2 + (n-1)d_2]} = \frac{2n+3}{3n-1} \quad \text{(given)}$$

For 5th term, replace n by $2 \times 5 - 1 = 9$

$$\Rightarrow \quad \frac{2a_1 + 8d_1}{2a_2 + 8d_2} = \frac{18+3}{27-1}$$

$$\Rightarrow \quad \frac{a_1 + 4d_1}{a_2 + 4d_2} = \frac{21}{26}$$

13. *(d)* $\sum_{r=1}^{\infty} [3 \cdot 2^{-r} - 2 \cdot 3^{1-r}] = 3\sum_{r=1}^{\infty} \frac{1}{2^r} - 2\sum_{r=1}^{n} 3^{1-r}$

$$= 3\left[\frac{1}{2} + \frac{1}{2^2} + \ldots \infty\right] - 2\left[1 + \frac{1}{3} + \frac{1}{3^2} + \ldots \infty\right]$$

$$= \frac{3}{2}\left[1 + \frac{1}{2} + \frac{1}{2^2} + \ldots\right] - 2\left[1 + \frac{1}{3} + \frac{1}{3^2} + \ldots \infty\right]$$

$$= \frac{3}{2}\left(\frac{1}{1-\frac{1}{2}}\right) - 2\left(\frac{1}{1-\frac{1}{3}}\right) = \frac{3}{2} \times 2 - \frac{2 \times 3}{2} = 0$$

14. *(c)* Number of three digits number when two 1's occur $= 9 + 8 + 9 = 26$.

Similarly, for two 2's ... 9's

We have, 26×9 numbers $= 234$ numbers

Number of numbers when two zero's occur $= 9$

Number of numbers when all three digits are same $= 9$

$\therefore$ Total three digit numbers having at least two identical digits $= 234 + 9 + 9 = 252$

15. *(a)* Given, ${}^{2n}C_3 : {}^nC_2 = \frac{44}{3}$

$$\Rightarrow \quad \frac{2n!}{(2n-3)!\,3!} \times \frac{2!(n-2)!}{n!} = \frac{44}{3}$$

$$\Rightarrow \frac{2n(2n-1)(2n-2)}{3} \times \frac{1}{n(n-1)} = \frac{44}{3}$$

$$\Rightarrow \quad \frac{2}{3} \times 2(2n-1) = \frac{44}{3}$$

$$\Rightarrow \quad 2n - 1 = 11 \Rightarrow 2n = 12 \Rightarrow n = 6$$

16. *(b)* Middle term $= T_6 = T_{5+1}$

$$= {}^{10}C_5\,(x)^5\left(-\frac{1}{2}y\right)^5 = \frac{10!}{5! \times 5!}\frac{(x)^5}{(-2)^5}y^5$$

$$= -\frac{10 \times 9 \times 8 \times 7 \times 6}{5 \times 4 \times 2 \times 3 \times 2^5}x^5y^5$$

$$= -\frac{63}{8}x^5y^5$$

17. *(b)* The general term is

$$T_{r+1} = {}^{12}C_r\,(2x^4)^r\left(-\frac{1}{x^2}\right)^{12-r}$$

$$= {}^{12}C_r\,(2)^r\,x^{4r}\,(-1)^{12-r}\,(x)^{-2(12-r)}$$

For term independent of x, we put

$4r - 24 + 2r = 0 \Rightarrow 6r = 24 \Rightarrow r = 4$

$$\therefore \quad T_5 = {}^{12}C_4\,(2)^4\,(-1)^{12-4}$$

$$= \frac{12 \times 11 \times 10 \times 9}{4 \times 3 \times 2 \times 1} \times (2)^4 = 7920$$

18. *(b)* Sum $= \frac{2}{10} + \frac{4}{10^3} + \frac{6}{10^5} + \frac{8}{10^7} + \ldots \infty$

It is an arithmetic-geometric series.

$$\therefore \quad S - \frac{S}{10^2} = \frac{2}{10} + \frac{2}{10^3} + \frac{2}{10^5} + \ldots \infty$$

$$\Rightarrow \quad \frac{99S}{100} = \frac{20}{99} \Rightarrow S = \frac{2000}{9801}$$

19. *(d)* $\frac{{}^nC_1}{2} + \frac{{}^nC_2}{3} + \ldots + \frac{{}^nC_n}{n+1} = \sum_{r=0}^{n} \frac{{}^nC_r}{r+1} - {}^nC_0$

$$= \sum_{r=0}^{n} \frac{1}{n+1}\frac{n+1}{r+1}{}^nC_r - {}^nC_0$$

$$= \frac{1}{n+1}\sum_{r=0}^{n} {}^{n+1}C_{r+1} - {}^nC_0$$

$= \frac{1}{n+1}[({}^{n+1}C_0 + {}^{n+1}C_1 + \dots + {}^{n+1}C_{n+1}) - {}^{n+1}C_0] - C_0$

$= \frac{1}{n+1}(2^{n+1} - 1) - 1$

$= \frac{2^{n+1} - n - 1 - 1}{n+1} = \frac{2^{n+1} - n - 2}{n+1}$

20. *(c)* The only wrong statement is option(c).

$\because$ It is not necessary that $A - B = I, \forall A, B$ square matrix of same order.

21. *(d)* $\begin{bmatrix} a & 2 & 3 \\ b & 5 & -1 \end{bmatrix}\begin{bmatrix} 1 & 2 \\ 3 & 4 \\ -1 & 1 \end{bmatrix} = \begin{bmatrix} a+6-3 & 2a+8+3 \\ b+15+1 & 2b+20-1 \end{bmatrix}$

$\Rightarrow \begin{bmatrix} a+3 & 2a+11 \\ b+16 & 2b+19 \end{bmatrix} = \begin{bmatrix} 4 & 13 \\ 12 & 11 \end{bmatrix}$ (given)

$\Rightarrow \quad a + 3 = 4 \Rightarrow a = 1$

and $\quad b + 16 = 12 \Rightarrow b = 12 - 16 = -4$

$\therefore \quad (a, b) = (1, -4)$

22. *(a)* Given that, $\begin{vmatrix} a & a^2 & 1+a^3 \\ b & b^2 & 1+b^3 \\ 1 & 1 & 1+1 \end{vmatrix} = 0$

$\Rightarrow \begin{vmatrix} a & a^2 & 1 \\ b & b^2 & 1 \\ 1 & 1 & 1 \end{vmatrix} + \begin{vmatrix} a & a^2 & a^3 \\ b & b^2 & b^3 \\ 1 & 1 & 1 \end{vmatrix} = 0$

$\Rightarrow \begin{vmatrix} a & a^2 & 1 \\ b & b^2 & 1 \\ 1 & 1 & 1 \end{vmatrix} + ab\begin{vmatrix} 1 & a & a^2 \\ 1 & b & b^2 \\ 1 & 1 & 1 \end{vmatrix} = 0$

$\Rightarrow \quad (ab+1)\begin{vmatrix} a & a^2 & 1 \\ b & b^2 & 1 \\ 1 & 1 & 1 \end{vmatrix} = 0$

$\Rightarrow (ab+1)[a(b^2 - 1) - a^2(b-1) + 1(b - b^2)] = 0$

$\Rightarrow \quad (ab+1)(b-1)[a(b+1) - a - b)] = 0$

$\Rightarrow \quad (ab+1)(b-1)[ab + a - a - b] = 0$

$\Rightarrow \quad b(ab+1)(b-1)(a-1) = 0$

$\Rightarrow \quad b = 0, 1; a = 1, ab = -1$

23. *(c)* Given, determinant $\begin{vmatrix} 3 & x & -1 \\ 2 & -1 & 4 \\ 1 & y & -3 \end{vmatrix}$

Here, cofactor of $x = 10$

and cofactor of $y = -14$

$\therefore$ Their sum $= -4$

24. *(b)* Since, $\sin\theta = -\frac{24}{25}$

$\therefore$ Base $AB = \sqrt{(25)^2 - (24)^2} = 7$

$\therefore \quad \cos\theta = \frac{7}{25}$

and $\quad \tan\theta = -\frac{24}{7} \quad (\because \theta \in \text{4th quadrant})$

$\therefore 7\tan\theta + 25\cos\theta = -24 + 7 = -17$

25. *(b)* $\frac{\cos\theta}{1 - \tan\theta} + \frac{\sin\theta}{1 - \cot\theta}$

$= \frac{\cos^2\theta}{\cos\theta - \sin\theta} + \frac{\sin^2\theta}{\sin\theta - \cos\theta}$

$= \frac{\cos^2\theta - \sin^2\theta}{\cos\theta - \sin\theta} = \cos\theta + \sin\theta$

26. *(b)* Given, $\sec A - \tan A = -a$

$\Rightarrow \quad \frac{1 - \sin A}{\cos A} = -a$

$\Rightarrow \quad \frac{1 - \sin A}{\sqrt{1 - \sin^2 A}} = -a$

$\Rightarrow \quad 1 + \sin^2 A - 2\sin A = a^2(1 - \sin^2 A)$

$\Rightarrow (1 + a^2)\sin^2 A - 2\sin A + (1 - a^2) = 0$

$\Rightarrow \quad (\sin A - 1)\left[\sin A - \frac{(1 - a^2)}{(1 + a^2)}\right] = 0$

$\Rightarrow \quad \sin A = 1 \text{ or } \sin A = \frac{1 - a^2}{1 + a^2}$

27. *(d)* $\cos\frac{\pi}{12} + \cos\frac{17\pi}{12} + \cos\frac{11\pi}{12}$

$= \cos\frac{\pi}{12} + \cos\left(\frac{3\pi}{2} - \frac{\pi}{12}\right) + \cos\left(\pi - \frac{\pi}{12}\right)$

$= \cos\frac{\pi}{12} - \cos\frac{\pi}{12} - \sin\frac{\pi}{12}$

$= -\sin\frac{\pi}{12} = \frac{1 - \sqrt{3}}{2\sqrt{2}}$

28. *(a)* $\cot^2\left(\frac{\pi}{4} + \frac{\theta}{2}\right)$

$= \left[\cot\left(\frac{\pi}{4} + \frac{\theta}{2}\right)\right]^2 = \left[\frac{\cot\frac{\theta}{2}\cot\frac{\pi}{4} - 1}{\cot\frac{\theta}{2} + \cot\frac{\pi}{4}}\right]^2$

$= \left[\frac{\cos\frac{\theta}{2} - \sin\frac{\theta}{2}}{\cos\frac{\theta}{2} + \sin\frac{\theta}{2}}\right]^2 = \frac{1 - \sin\theta}{1 + \sin\theta}$

29. *(a)* Now, $\cos B + \cos C + (1 - \cos A)$

$$= 2\cos\left(\frac{B+C}{2}\right)\cos\left(\frac{B-C}{2}\right) + 2\sin^2\frac{A}{2}$$

$$= 2\cos\left(\frac{\pi}{2} - \frac{A}{2}\right)\cos\left(\frac{B-C}{2}\right) + 2\sin^2\frac{A}{2}$$

$$(\because B + C = \pi - A)$$

$$= 2\sin\frac{A}{2}\left[\cos\left(\frac{B-C}{2}\right) + \sin\frac{A}{2}\right]$$

$$= 2\sin\frac{A}{2}\left[\cos\left(\frac{B-C}{2}\right) + \cos\left(\frac{B+C}{2}\right)\right]$$

$$[\because A = \pi - (B+C)]$$

$$= 2\sin\frac{A}{2} \cdot 2\cos\frac{B}{2} \cdot \cos\frac{C}{2}$$

$$= 4\sin\frac{A}{2}\cos\frac{B}{2}\cos\frac{C}{2}$$

30. *(b)* $\sin^{-1}\frac{1}{\sqrt{5}} + \cos^{-1}\frac{3}{\sqrt{10}}$

Now, $\sin^{-1}\frac{1}{\sqrt{5}} = \tan^{-1}\left(\frac{1/\sqrt{5}}{\sqrt{1-\frac{1}{5}}}\right) = \tan^{-1}\frac{1}{2}$

and $\cos^{-1}\frac{3}{\sqrt{10}} = \tan^{-1}\left(\frac{\sqrt{1-\frac{9}{10}}}{\frac{3}{\sqrt{10}}}\right) = \tan^{-1}\frac{1}{3}$

$$\therefore \tan^{-1}\frac{1}{2} + \tan^{-1}\frac{1}{3} = \tan^{-1}\left(\frac{\frac{1}{2}+\frac{1}{3}}{1-\frac{1}{6}}\right) = \tan^{-1}(1) = \frac{\pi}{4}$$

31. *(c)* Let the required point is (a, a).

$$\therefore (4-a)^2 + a^2 = (5-a)^2 + (1-a)^2$$

$$\Rightarrow 16 + a^2 - 8a + a^2 = 25 + a^2 - 10a + 1 + a^2 - 2a$$

$$\Rightarrow 16 - 25 - 1 + a^2 + a^2 - a^2 - a^2 - 8a + 10a + 2a = 0$$

$$\Rightarrow 4a = 10$$

$$\Rightarrow a = \frac{10}{4} = \frac{5}{2}$$

$\therefore$ Required point is $\left(\frac{5}{2}, \frac{5}{2}\right)$.

32. *(b)* Diagonals of parallelogram bisects each other.

$\Rightarrow$ Diagonal BD passes through mid point of AC.

ie, $\left(\frac{5}{2}, \frac{7}{2}\right)$.

$\therefore$ Equation of diagonal passing through $(5, 1)$ and $\left(\frac{5}{2}, \frac{7}{2}\right)$ is

$$y - 1 = \frac{\frac{7}{2} - 1}{\frac{5}{2} - 5}(x - 5)$$

$$\Rightarrow y - 1 = -1(x - 5)$$

$$\Rightarrow y + x - 6 = 0$$

33. *(a)* Let the equation of line be $\frac{x}{a} + \frac{y}{b} = 1$.

Then $a + b = 4$

$$\Rightarrow b = 4 - a$$

$$\therefore \frac{x}{a} + \frac{y}{4-a} = 1$$

$$\Rightarrow x(4-a) + ya = a(4-a)$$

This passes through $\left(\frac{9}{2}, -5\right)$.

$$\therefore \frac{9}{2}(4-a) - 5a = a(4-a)$$

$$\Rightarrow 36 - 9a - 10a = 8a - 2a^2$$

$$\Rightarrow 2a^2 - 27a + 36 = 0$$

$$\Rightarrow a = 12 \text{ or } \frac{3}{2} \text{ but } a \neq 12$$

$$\therefore a = \frac{3}{2} \Rightarrow b = \frac{5}{2}$$

$\therefore$ Required equation is

$$\frac{2x}{3} + \frac{2y}{5} = 1$$

$$\Rightarrow 10x + 6y = 15$$

34. *(c)* Suppose the segment $2x - 3y + 4 = 0$ divides line segment joining $A(2, 1)$ and $B(0, -2)$ in the ratio $k : 1$ at point C.

$$\therefore C \text{ is } \left(\frac{2}{k+1}, \frac{-2k+1}{k+1}\right).$$

Since, C lies on $2x - 3y + 4 = 0$

$$\Rightarrow \frac{4}{k+1} - 3\frac{(-2k+1)}{k+1} + 4 = 0$$

$$\Rightarrow 4 + 6k - 3 + 4k + 4 = 0$$

$$\Rightarrow 10k + 5 = 0$$

$$\Rightarrow k = -\frac{1}{2}$$

$\therefore$ Ratio is $-1 : 2$.

35. *(a)* Point of intersection of $x - y + 1 = 0$ and $3x + 2y + 4 = 0$ is

Slope of required line is -4.

$\therefore$ Required line is $\left(y + \frac{1}{5}\right) = -4\left(x + \frac{6}{5}\right)$

$\Rightarrow \quad 5y + 1 = -20x - 24$

$\Rightarrow \quad 5y + 20x = -25$

$\Rightarrow \quad y + 4x + 5 = 0$

36. *(d)* If ω is a cube root of unity, then $1 + \omega + \omega^2 = 0$ and $\omega^3 = 1$.

$\therefore (1 + \omega - \omega^2)(1 - \omega + \omega^2)$

$= (-\omega^2 - \omega^2)(-\omega - \omega)$

$= 2\omega^2 \cdot 2\omega = 4$

37. *(c)* $y = \sqrt{\sin x + \sqrt{\sin x + \sqrt{\sin x + \ldots}}}$

$\Rightarrow \quad y = \sqrt{\sin x + y}$

$\Rightarrow \quad y^2 = \sin x + y$

On differentiating both sides w.r.t. x, we get

$$2y\frac{dy}{dx} = \cos x + \frac{dy}{dx} \Rightarrow \frac{dy}{dx}(2y - 1) = \cos x$$

$$\Rightarrow \quad \frac{dy}{dx} = \frac{\cos x}{2y - 1}$$

38. *(c)* $\int_0^1 x(1-x)^{12}\,dx = \left[\dfrac{x(1-x)^{13}}{-13} + \int_0^1 \dfrac{(1-x)^{13}}{13}\,dx\right]_0^1$

$$= \left[\frac{x(1-x)^{13}}{-13}\right]_0^1 + \left[\frac{(1-x)^{14}}{-14 \times 13}\right]_0^1$$

$$= [0 - 0] + \left[0 - \frac{1}{-14 \times 13}\right] = \frac{1}{182}$$

39. *(c)* Given conic is $9x^2 - 16y^2 = 144$

$$\Rightarrow \quad \frac{x^2}{16} - \frac{y^2}{9} = 1$$

$$\therefore \quad e = \sqrt{1 + \frac{b^2}{a^2}} = \sqrt{1 + \frac{9}{16}} = \frac{5}{4}$$

40. *(b)* $\sum_{n=1}^{\infty} \dfrac{(-1)^n}{n+1} = -\dfrac{1}{2} + \dfrac{1}{3} - \dfrac{1}{4} + \ldots \infty$

$$= 1 - \frac{1}{2} + \frac{1}{3} - \frac{1}{4} + \ldots \infty - 1$$

$$= \log 2 - 1$$

41. *(d)* Let $y = e^x$

$$\frac{d}{dx}(y) = \frac{d}{dx}(e^x) \Rightarrow \frac{dy}{dx} = e^x$$

42. *(a)* Since, the given equation has real roots.

$\therefore$ Discriminant,

$$D = \cos^2 p - 4(\cos p - 1)\sin p \geq 0$$

$\Rightarrow \quad D = (\cos p - 2\sin p)^2 + 4\sin p(1 - \sin p) \geq 0$

$\Rightarrow \quad 0 < p < \pi$

43. *(b)* $\log_3 2, \log_3(2^x - 5), \log_3\left(2^x - \dfrac{7}{2}\right)$ are in AP.

$$\Rightarrow \quad \log_3(2^x - 5) = \frac{\log_3\left(2^x - \frac{7}{2}\right) + \log_3 2}{2}$$

$$\Rightarrow \quad (2^x - 5)^2 = 2 \cdot \left(2^x - \frac{7}{2}\right)$$

$\Rightarrow \quad (2^x)^2 + 25 - 10 \cdot 2^x = 2 \cdot 2^x - 7$

$\Rightarrow \quad (2^x)^2 - 12 \cdot 2^x + 25 + 7 = 0$

$\Rightarrow \quad (2^x)^2 - 12 \cdot 2^x + 32 = 0$

$\Rightarrow \quad 2^x = 8$ or $2^x = 4$

$\Rightarrow \quad x = 3$ or 2

$\Rightarrow \quad x = 3$ ($\because x = 2$ does not satisfy the given series)

44. *(a)* $\int_{-\pi/2}^{\pi/2} |\sin x|\,dx = 2\int_0^{\pi/2} \sin x\,dx$

$$= 2[-\cos x]_0^{\pi/2} = 2$$

45. *(c)* $|\vec{\mathbf{a}} \times \vec{\mathbf{b}} + \vec{\mathbf{b}} \times \vec{\mathbf{a}}| = |\vec{\mathbf{a}} \times \vec{\mathbf{b}} - \vec{\mathbf{a}} \times \vec{\mathbf{b}}| = 0$

46. *(b)* Let $z = x + iy$

$\therefore \quad i + z = x + i(1 + y)$

and $\quad i - z = -x + i(1 - y)$

$\because \quad \left|\dfrac{i + z}{i - z}\right| = 1$

$\Rightarrow \quad |i + z| = |i - z|$

$\Rightarrow \quad x^2 + (1 + y)^2 = x^2 + (1 - y)^2$

$\Rightarrow \quad x^2 + y^2 + 2y + 1 = x^2 + y^2 + 1 - 2y$

$\Rightarrow \quad 4y = 0$

$\Rightarrow \quad y = 0$

ie, x-axis.

47. *(c)* We know, $(a + b + c)^2 \geq 0$

$\Rightarrow \quad a^2 + b^2 + c^2 + 2ab + 2bc + 2ca \geq 0$

$\Rightarrow \quad 1 + 2(ab + bc + ca) \geq 0$

$\Rightarrow \quad (ab + bc + ca) \geq \dfrac{-1}{2} \quad \ldots$(i)

Also, $\quad a^2 + b^2 + c^2 = 1$

$\Rightarrow \quad a^2 + b^2 + c^2 - (ab + bc + ca)$

$$= \frac{1}{2}[(a - b)^2 + (b - c)^2 + (c - a)^2] \geq 0$$

$\Rightarrow \quad 1 - (ab + bc + ca) \geq 0$

$\Rightarrow \quad ab + bc + ca \leq 1 \quad \ldots$(ii)

Hence, from Eqs. (i) and (ii),

$$-\frac{1}{2} \leq ab + bc + ca \leq 1$$

48. *(b)* Total degree = number of vertices × degree

$= 4 \times 7$ or 2×14

If there are 14 vertices each of degree 2, then graph is a cycle.

Hence, there are 7 vertices and each of degree 4.

49. *(a)* Middle term is $\left(\frac{n}{2}+1\right)$th term for n even.

$\therefore$ Middle term $= \left(\frac{6}{2}+1\right) = 4$ th term

$$\therefore \quad T_4 = {}^6C_3 (\sqrt{x})^3 \left(-\frac{1}{\sqrt{x}}\right)^{6-3}$$

$$= \frac{6!}{3!\,3!}(\sqrt{x})^3 \frac{(-1)^3}{(\sqrt{x})^3}$$

$$= \frac{-6 \times 5 \times 4}{6}$$

$$= -20$$

50. *(c)* Since, α and β are the roots of the equation

$$4x^2 + 3x + 7 = 0$$

$$\therefore \quad \alpha + \beta = -3/4, \alpha\beta = 7/4$$

Now, $\frac{1}{\alpha} + \frac{1}{\beta} = \frac{\alpha+\beta}{\alpha\beta}$

$$= -\frac{3 \cdot 4}{4 \cdot 7} = -\frac{3}{7}$$

51. *(c)* In a triangle ABC,

$$A + B + C = \pi,$$

$\Rightarrow \sin(A+B+C) = 0, \cos(A+B) = -\cos C$

$$\therefore \quad \begin{vmatrix} \sin(A+B+C) & \sin B & \cos C \\ -\sin B & 0 & \tan A \\ \cos(A+B) & -\tan A & 0 \end{vmatrix}$$

$$= \begin{vmatrix} 0 & \sin B & \cos C \\ -\sin B & 0 & \tan A \\ -\cos C & -\tan A & 0 \end{vmatrix}$$

$= 0$ $\left(\because \text{ It is the determinant of a skew symmetric matrix of odd order.}\right)$

52. *(d)* $(1 + x + x^2 + x^3)^{-3} = \left[\frac{(x^4-1)}{x-1}\right]^{-3}$

$$= (x^4-1)^{-3}(x-1)^3$$

$$= (-1)^{-3}(1-x^4)^{-3}(x-1)^3$$

$$= (-1)(1 + 3x^4 + \dots)(x^3 - 3x^2 + 3x - 1)$$

$\therefore$ Coefficient of x in $(1+x+x^2+x^3)^{-3}$ is -3.

53. *(c)* $a < b \Rightarrow \frac{a+b}{2} < \frac{b+b}{2}$

$$\Rightarrow \quad \frac{a+b}{2} < b$$

54. *(b)* Given, $\tan^{-1}x + \tan^{-1}y + \tan^{-1}z = \pi$

$$\Rightarrow \quad \tan^{-1}\left(\frac{x+y+z-xyz}{1-xy-yz-zx}\right) = \pi$$

$$\Rightarrow \quad x + y + z - xyz = 0$$

55. *(c)* Now, $\frac{{}^nC_2}{(n+1)!} = \frac{\frac{n!}{(n-2)!\,2!}}{(n+1)!}$

$$= \frac{n!}{2(n+1)!\,(n-2)!} = \frac{n(n-1)}{2(n+1)!}$$

$$= \frac{n^2 - 1 + 1 - n}{2(n+1)!}$$

$$= \frac{1}{2}\left[\frac{(n-1)}{n!} - \frac{1}{n!} + \frac{2}{(n+1)!}\right]$$

$$= \frac{1}{2}\left[\frac{1}{(n-1)!} - \frac{2}{n!} + \frac{2}{(n+1)!}\right]$$

Put $n = 0, 1, 2, 3, \dots$, we get

$$= \frac{1}{2}[e - 2(e) + 2(e-1)]$$

$$= \frac{1}{2}[e-2] = \frac{e}{2} - 1$$

56. *(c)* Given, $y = a \cos \mu x + b \sin \mu x$ ...(i)

On differentiating w.r.t. x, we get

$$\frac{dy}{dx} = -a\mu \sin \mu x + b\mu \cos \mu x$$

Again, differentiating, we get

$$\frac{d^2y}{dx^2} = -a\mu^2 \cos \mu x - b^2\mu \sin \mu x$$

$$= -\mu^2 y \quad \text{[from Eq. (i)]}$$

$$\Rightarrow \quad \frac{d^2y}{dx^2} + \mu^2 y = 0$$

57. *(b)* $\lim\limits_{\theta \to 0} \frac{\tan\theta}{\theta}$. Applying L'Hospital's rule,

$$= \lim_{\theta \to 0} \frac{\sec^2\theta}{1} = 1$$

58. *(b)* The point of intersection of the curves $y^2 = 2x$ and $x^2 = 2y$ are $O(0, 0)$, $A(2, 2)$.

$\therefore$ Required area $= \int_0^2 \left(\sqrt{2x} - \frac{x^2}{2}\right) dx$

$$= \left[\sqrt{2}\,\frac{x^{3/2}}{3/2} - \frac{x^3}{6}\right]_0^2$$

$$= \left[\frac{2}{3}\cdot 2^2 - \frac{2^3}{6}\right] = \left[\frac{8}{3} - \frac{4}{3}\right]$$

$$= \frac{4}{3} \text{ sq unit}$$

59. *(a)* Let $I = \int_1^e \frac{\log x}{x}\,dx$

Put $\log x = t \Rightarrow \frac{1}{x}\,dx = dt$

$\therefore \quad I = \int_0^1 t\,dt = \left[\frac{t^2}{2}\right]_0^1 = \frac{1}{2}$

60. *(a)* Let $I = \int_0^1 \sin^{-1} x\,dx$

$$= [x \sin^{-1} x]_0^1 - \int_0^1 \frac{x}{\sqrt{1-x^2}}\,dx$$

Put $1 - x^2 = t^2 \Rightarrow -2x\,dx = 2t\,dt$

$$\therefore \quad I = \left[\frac{\pi}{2} - 0\right] + \int_1^0 \frac{t\,dt}{t}$$

$$= \frac{\pi}{2} + \int_1^0 dt = \frac{\pi}{2} + [t]_1^0 = \frac{\pi}{2} - 1$$

61. *(d)* Let $z = \sin x\,(1 + \cos x)$

or $z = \sin x + \frac{1}{2}\sin 2x$

On differentiating w.r.t. x, we get

$$\frac{dz}{dx} = \cos x + \cos 2x$$

For maximum or minimum, put $\frac{dz}{dx} = 0$

$\Rightarrow \quad \cos x + \cos 2x = 0$

$\Rightarrow \quad 2\cos^2 x - 1 + \cos x = 0$

$\Rightarrow \quad (\cos x + 1)(2\cos x - 1) = 0$

$\Rightarrow \quad \cos x = -1, \frac{1}{2}$

$\Rightarrow \quad x = \frac{\pi}{3}, \pi$

But $0 \le x \le \frac{\pi}{2}$, therefore we take only $x = \frac{\pi}{3}$.

$\therefore \quad \frac{d^2z}{dx^2} = -\sin x - 2\sin 2x = -$ ve, for $x = \pi/3$

$\therefore$ It is maximum at $x = \pi/3$.

62. *(a)* Given, $A = \begin{bmatrix} 1 & 0 & 0 \\ 0 & 1 & 0 \\ a & b & -1 \end{bmatrix}$

$$\therefore A^2 = A \cdot A = \begin{bmatrix} 1 & 0 & 0 \\ 0 & 1 & 0 \\ a & b & -1 \end{bmatrix}\begin{bmatrix} 1 & 0 & 0 \\ 0 & 1 & 0 \\ a & b & -1 \end{bmatrix}$$

$$= \begin{bmatrix} 1 & 0 & 0 \\ 0 & 1 & 0 \\ a-a & b-b & 1 \end{bmatrix} = \begin{bmatrix} 1 & 0 & 0 \\ 0 & 1 & 0 \\ 0 & 0 & 1 \end{bmatrix}$$

63. *(c)* Given, system of equations are

$ax + y + z = 0, x + ay + z = 0$

and $x + y + z = 0$ possesses non-zero solution.

$$\therefore \quad \begin{vmatrix} a & 1 & 1 \\ 1 & a & 1 \\ 1 & 1 & 1 \end{vmatrix} = 0$$

$\Rightarrow \quad a(a-1) - 1(1-1) + 1(1-a) = 0$

$\Rightarrow \quad a^2 - a + 0 + 1 - a = 0$

$\Rightarrow \quad a^2 - 2a + 1 = 0$

$\Rightarrow \quad (a-1)^2 = 0$

$\Rightarrow \quad a = 1$

64. *(c)* Given, $\vec{\mathbf{a}} = \frac{1}{7}(2\hat{\mathbf{i}} + 3\hat{\mathbf{j}} + 6\hat{\mathbf{k}}), \vec{\mathbf{b}} = \frac{1}{7}(3\hat{\mathbf{i}} - \lambda\hat{\mathbf{j}} + 2\hat{\mathbf{k}})$

Since, the vectors $\vec{\mathbf{a}}$ and $\vec{\mathbf{b}}$ are perpendicular.

$\therefore \quad \vec{\mathbf{a}} \cdot \vec{\mathbf{b}} = 0$

$\Rightarrow \frac{1}{7}(2\hat{\mathbf{i}} + 3\hat{\mathbf{j}} + 6\hat{\mathbf{k}}) \cdot \frac{1}{7}(3\hat{\mathbf{i}} - \lambda\hat{\mathbf{j}} + 2\hat{\mathbf{k}}) = 0$

$\Rightarrow \quad \frac{1}{49}(6 - 3\lambda + 12) = 0$

$\Rightarrow \quad 3\lambda = 18 \Rightarrow \lambda = 6$

65. *(b)* $\{(\vec{\mathbf{a}} \times \vec{\mathbf{b}})^2 + (\vec{\mathbf{a}} \cdot \vec{\mathbf{b}})^2\} \div a^2b^2$

$= \{(|\vec{\mathbf{a}}||\vec{\mathbf{b}}|\sin\theta\,\hat{\mathbf{n}})^2 + (|\vec{\mathbf{a}}||\vec{\mathbf{b}}|\cos\theta)^2\} \div a^2b^2$

$= a^2b^2(\sin^2\theta + \cos^2\theta) \div a^2b^2 = 1$

66. *(c)* Given system of equations are

$x + y + z = 0, 2x + y - z = 0, 3x + 2y = 0$

$$\text{Let } \Delta = \begin{vmatrix} 1 & 1 & 1 \\ 2 & 1 & -1 \\ 3 & 2 & 0 \end{vmatrix} = 1(0+2) - 1(0+3) + 1(4-3)$$

$$= +2 - 3 + 1 = 0$$

Hence, it has infinitely many solutions.

67. *(c)* $[\vec{\mathbf{a}} + \vec{\mathbf{b}}\;\; \vec{\mathbf{b}} + \vec{\mathbf{c}}\;\; \vec{\mathbf{c}} + \vec{\mathbf{a}}]$

$= (\vec{\mathbf{a}} + \vec{\mathbf{b}}) \cdot [(\vec{\mathbf{b}} + \vec{\mathbf{c}}) \times (\vec{\mathbf{c}} + \vec{\mathbf{a}})]$

$= (\vec{\mathbf{a}} + \vec{\mathbf{b}}) \cdot [(\vec{\mathbf{b}} \times \vec{\mathbf{c}} + \vec{\mathbf{b}} \times \vec{\mathbf{a}} + 0 + \vec{\mathbf{c}} \times \vec{\mathbf{a}})]$

$= [\vec{a}\ \vec{b}\ \vec{c}] + 0 + 0 + 0 + 0 + [\vec{b}\ \vec{c}\ \vec{a}]$

$= [\vec{a}\ \vec{b}\ \vec{c}] + [\vec{a}\ \vec{b}\ \vec{c}]$

$= 2[\vec{a}\ \vec{b}\ \vec{c}]$

68. *(b)* Since, a ray makes angles α, β, γ with the positive direction of the axes.

$\therefore \quad \cos^2\alpha + \cos^2\beta + \cos^2\gamma = 1$

$\Rightarrow 1 - \sin^2\alpha + 1 - \sin^2\beta + 1 - \sin^2\gamma = 1$

$\Rightarrow \quad \sin^2\alpha + \sin^2\beta + \sin^2\gamma = 2$

69. *(c)* Given, $\vec{a} = 3\vec{\alpha} - \vec{\beta}, \vec{b} = \vec{\alpha} + 3\vec{\beta}$

and $\quad |\vec{\alpha}| = |\vec{\beta}| = 2$

$\therefore \quad |\vec{a} + \vec{b}|^2 = |4\vec{\alpha} + 2\vec{\beta}|^2$

$= 16|\vec{\alpha}|^2 + 4|\vec{\beta}|^2 + 16|\vec{\alpha}||\vec{\beta}|\cos\frac{\pi}{3}$

$= 16(4) + 4(4) + 16(2 \times 2) \times \frac{1}{2}$

$= 64 + 16 + 32 = 112$

$\Rightarrow |\vec{a} + \vec{b}| = 4\sqrt{7}$

70. *(b)* $\lim_{x \to 0} \frac{\sin 3x - \sin x}{\sin x}$

$= \lim_{x \to 0} \frac{2\cos 2x \sin x}{\sin x} = 2$

English and General Aptitude

1. *(c)* In the Sariska National Park, there are both carnivores and herbivores.

2. *(d)* The Sariska National Park is not far from Jaipur.

3. *(b)* The park possesses monuments and temples.

4. *(c)* Animals come to satisfy their thirst at a waterhole.

5. *(a)* The Sand Grouse is a species of bird.

6. *(d)* "vex" is followed by preposition "at".

7. *(b)* The phrase will read as "to fall a victim to ________". Hence, "to" is suitable.

8. *(c)* "Abide" is followed by preposition "by".

9. *(b)* "Collision" is a kind of mental or physical conflict so, the antonym will be "agreement".

10. *(d)* "Grudge" is dislike or apathy for a thing/ person. Hence, "affection" is its correct antonym.

11. *(c)* "Insolvent" is a condition of being scarce in terms of money. Hence, the opposite will be "affluent".

12. *(a)* "Scanty" means lack of something so, the antonym will be "Lavish" which means "having something in plenty".

13. *(d)* "Emphatic" is a situation of extreme pressure or substance. So, "definite" is its synonym.

14. *(d)* "Compatible" and "consistent" are the words with similar meaning-suitable with other things.

15. *(c)* "Pandemonium" and "chaos" both are similar in meaning–a complete disorder. Hence, these are similar.

16. *(d)* Except 'LPG', all others are liquids while 'LPG' is a Liquified Petroleum Gas.

17. *(b)*

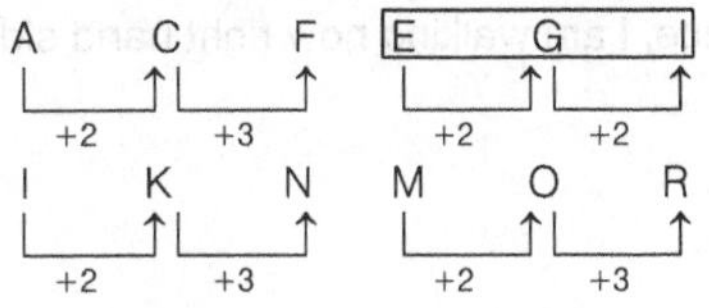

Hence, 'EGI' is different from the other three.

18. *(a)* $126 = 5^3 + 1 \quad 215 = 6^3 - 1$

$342 = 7^3 - 1 \quad 511 = 8^3 - 1$

Hence, '126' is different from the other three.

19. *(b)* We know that, peacock is the Indian National Bird but here peacock is known as swallow. So, the answer is swallow.

20. *(c)*

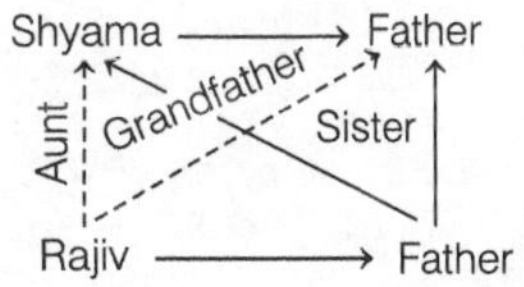

Hence, from the above diagram it is clear that Shyama is father's sister (aunt) of Rajiv.

21. *(a)* ghi/ghi/ghi/ghi $\Rightarrow$ hihiih

22. *(c)* As, Similarly,

$E \xrightarrow{+2} G \qquad N \xrightarrow{+2} P$

$A \xrightarrow{+2} C \qquad E \xrightarrow{+2} G$

$R \xrightarrow{+2} T \qquad A \xrightarrow{+2} C$

$N \xrightarrow{+2} P \qquad R \xrightarrow{+2} T$

23. *(b)* As, $20 + 3 \Rightarrow (2 + 0 + 3) \times 3 = 15$

and $15 + 4 \Rightarrow (1 + 5 + 4) \times 3 = 30$

Similarly, $12 + 6 \Rightarrow (1 + 2 + 6) \times 3 = \boxed{27}$

24. *(a)*

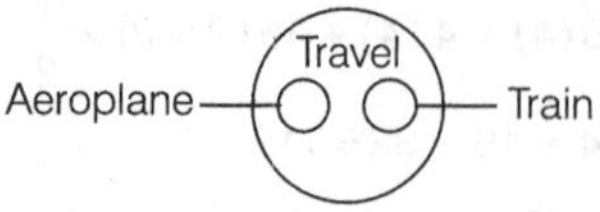

Both aeroplanes and train are medium of travel.

25. *(b)* Given, equation $= 10 - 2 + 12 \times 1 \div 0$

According to the question, after interchanging the signs

$$10 + 2 - 12 \div 1 \times 0 = 12 - 12 \times 0$$
$$= 12 - 0 = 12$$

26. *(d)*

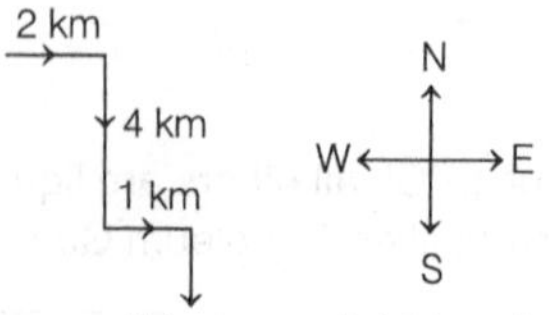

Hence, I am walking now right hand side direction.

27. *(d)* The pattern of series are as follows

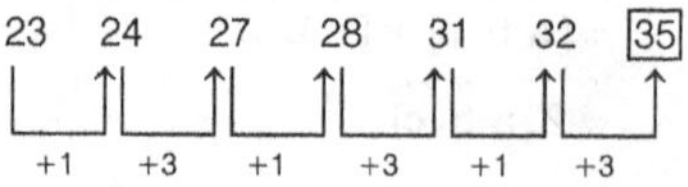

28. *(c)* As,

$\sqrt{49} + \sqrt{9} + \sqrt{81} = 7 + 3 + 9 = 19$

(in first column)

and $\sqrt{64} + \sqrt{4} + \sqrt{25}$

$= 8 + 2 + 5 = 15$

(in third column)

Similarly, $\sqrt{100} + \sqrt{36} + \sqrt{1}$

$= 10 + 6 + 1 = \boxed{17}$

(in second column)

29. *(a)* As, 'Sapling' is related to 'Tree'. Similarly, 'Girl' is related to 'Woman'.

30. *(b)* As, 'Pigeon' is a symbol of 'Peace'. Similarly, 'White Flag' is a symbol of 'Surrender'.

MANIPAL

Engineering Entrance Exam

Solved Paper 2020

Physics

1. A body of mass m rises to a height $h = \frac{R}{5}$ from the earth's surface, where R is the earth's radius. If g is acceleration due to gravity at the earth's surface, the increase in potential energy will be

(a) mgh (b) $\frac{4}{5}mgh$ (c) $\frac{5}{6}mgh$ (d) $\frac{6}{7}mgh$

2. The energy required to break the covalent bond in a semiconductor is

(a) always 1 eV
(b) equal to the forbidden energy gap of semiconductor
(c) equal to fermi energy
(d) much less than fermi energy

3. When a glass prism of refracting angle 60° is immersed in a liquid its angle of minimum deviation is 30°. The critical angle of glass with respect to the liquid medium is

(a) 45° (b) 42° (c) 50° (d) 52°

4. Faraday constant

(a) depends on the amount of the electrolyte
(b) depends on the current in the electrolyte
(c) is a universal constant
(d) depends on the amount of charge passed through the electrolyte

5. A comb is run through wet hair on a rainy day, then

(a) it will attract large number of small bits of paper
(b) it will not go through the hair
(c) it will not attract small bits of paper
(d) None of the above

6. The displacement-time graph of a particle moving along a straight line is shown in the figure.

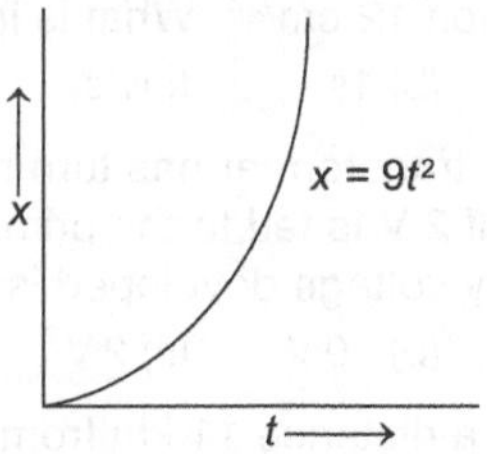

The acceleration-time graph of this particle is

(a)
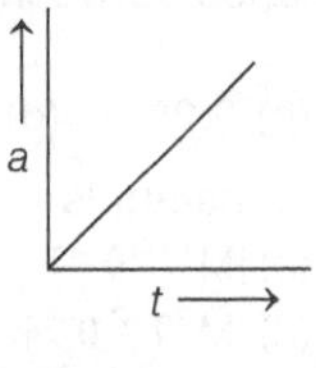

(b)
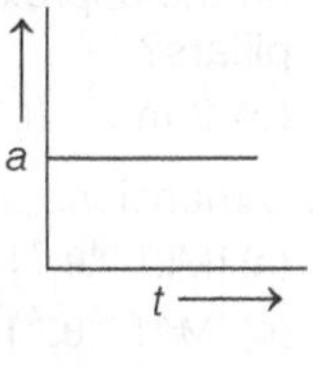

(c)
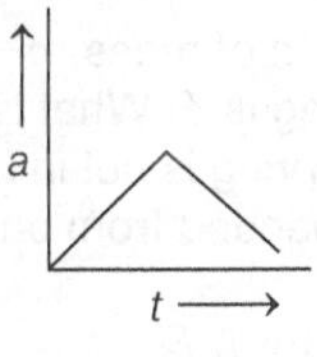

(d)
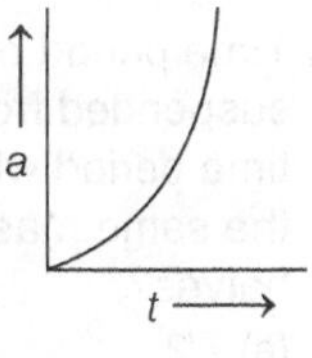

7. The maximum kinetic energy of photoelectrons coming out of a metal surface is 10 eV. The minimum voltage required to stop, the emission of electrons from this metal surface is

(a) 10 V (b) 5 V
(c) −5 V (d) −10 V

8. Thermal radiation exist in which part of electromagnetic spectrum?
(a) Ultraviolet (b) Infrared
(c) Visible (d) Violet

9. A rod of length L is composed of a uniform length $\frac{L}{2}$ of wood whose mass in m_w and a uniform length $\frac{L}{2}$ of brass whose mass is m_b. The moment of inertia I of the rod about an axis perpendicular to the rod and through its centre is equal to
(a) $(m_w + m_b)\frac{L^2}{6}$ (b) $(m_w + m_b)\frac{L^2}{2}$
(c) $(m_w + m_b)\frac{L^2}{12}$ (d) $(m_w + m_b)\frac{L^2}{3}$

10. A particle, doing simple harmonic motion, at a distance 3 cm from mean position has acceleration 12 cm/s^2. What is its time period?
(a) 0.5 s (b) 1s (c) 2s (d) 3.14 s

11. A step up transformer has turn ratio 10 : 1. A cell of emf 2 V is fed to the primary, then the secondary voltage developed is
(a) 20 V (b) 10 V (c) 2 V (d) zero

12. A man at a distance 11 km from two pillars wants to see two pillars separately. What will be the approximate distance between the pillars?
(a) 3 m (b) 1 m (c) 0.25 m (d) 0.5 m

13. Dimensions of Stefan's constant is
(a) $[MLT^{-3}\theta^{-4}]$ (b) $[MT^{-3}\theta^{-4}]$
(c) $[M^2T^{-3}\theta^{-4}]$ (d) $[M^2T^{-2}\theta^{-4}]$

14. Time period of oscillation of mass m suspended from a spring is T. What is the time period when the spring is cut in half and the same mass is suspended from one of the halves?
(a) $T/2$ (b) $T/\sqrt{2}$
(c) $\sqrt{2}\,T$ (d) $2T$

15. An electric current of 2 A passes through a wire of resistance 25 Ω. How much heat will be generated in 1 min?
(a) 6×10^3 J (b) 3.6×10^3 J
(c) 0.6×10^3 J (d) 0.36×10^3 J

16. A mass m hanging from a spring is doing simple harmonic motion with frequency f. If the mass is increased by 4 times, then frequency will be
(a) $2f$ (b) $f/2$ (c) $4f$ (d) $f/4$

17. Internal energy of a gas remains unchanged in
I. an isothermal process
II. an adiabatic process
III. a reversible process
IV. a cyclic process
Which of these are true?
(a) I and IV (b) I, III and IV
(c) III and IV (d) II and III

18. An electric charge in uniform motion produces
(a) only electric field
(b) only magnetic field
(c) Both electric and magnetic field
(d) Neither electric nor magnetic field

19. If λ is the incident wavelength and λ_0 is the threshold wavelength for a metal surface, photoelectric effect takes place only, if
(a) $\lambda \le \lambda_0$ (b) $\lambda \ge \lambda_0$
(c) $\lambda \ge 2\lambda_0$ (d) None of these

20. The mass number of an atom is 15 and its atomic number is 7. Now, this atom absorbs an α-particle and emits a proton. What will be the mass number of changed atom?
(a) 16 (b) 18 (c) 17 (d) 15

21. What is the direction of the electric field at the centre O of the square in the figure shown below? Given that, $q = 10$ nC and the side of the square is 5 cm.

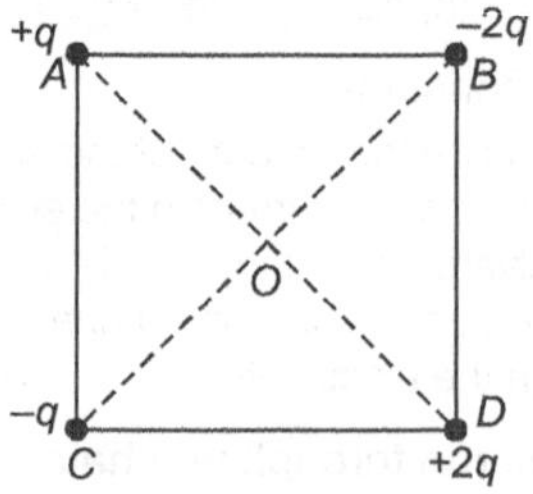

(a) at 45° to OA upward
(b) at 135° to OA towards BD
(c) no direction, because $E = 0$
(d) None of the above

22. Which equation is valid for adiabatic process?

(a) $TV^{\gamma-1}$ = constant (b) $pV^{\gamma-1}$ = constant

(c) $T^{\gamma}V^{\gamma-1}$ = constant (d) $\frac{p^{\gamma-1}}{T^{\gamma-1}}$ = constant

23. Which of the following relation correct? (v_{rms}-root mean square velocity, $\bar{v}$-mean velocity and v_{mp}-most probable velocity)

(a) $v_{rms} > \bar{v} < v_{mp}$ (b) $v_{rms} < \bar{v} > v_{mp}$

(c) $v_{rms} > \bar{v} > v_{mp}$ (d) None of these

24. The effect of reverse bias in a junction diode on its potential barrier is

(a) increases (b) decreases

(c) remains same (d) None of these

25. During an experiment, an ideal gas is found to obey an additional law Vp^2 = constant. The gas is initially at temperature T and volume V. The temperature of the gas will be following, when it expands to a volume 2 V?

(a) $\sqrt{2}\,T$ (b) $\sqrt{4}\,T$ (c) $\sqrt{6}\,T$ (d) $\sqrt{5}\,T$

26. In which of the following process, convection does not take place primarilly?

(a) Sea and land breeze

(b) Boiling of water

(c) Warming of glass bulb due to filament

(d) Heating air around furnace

27. The horizontal component of earth's magnetic field at a place is 0.4×10^4 T. If the angle of dip is 45° the value of total intensity of earth's magnetic field is

(a) 0.5×10^{-4} T (b) 0.4×10^{-4} T

(c) 0.5×10^{-6} T (d) 0.4×10^{-6} T

28. If an observer moves towards stationary source, then the apparent frequency is given by

(a) $f' = f\left(\frac{v + v_0}{v}\right)$ (b) $f' = f\left(\frac{v - v_0}{v}\right)$

(c) $f' = f\left(\frac{v}{v + v_0}\right)$ (d) $f' = f\left(\frac{v}{v - v_0}\right)$

29. Two coils have mutual inductance of 1.5 H. If current in primary coil is suddenly raised to 5 A in one milli second, the induced emf in the secondary coil is

(a) 75 V (b) 750 V

(c) 7500 V (d) 75000 V

30. The formula for magnetic field of a certain component is given by $B = \frac{\mu_0 Ni}{2\pi r}$,

where, N = total number of turns

i = current through it

r = radius.

The component is

(a) ring (b) solenoid

(c) toroid (d) None of these

31. A convex lens of focal length 0.12 m produces an image, which is three times as long as the object. The distance between the object and the lens for a real image

(a) 0.16 m (b) – 0.16 m

(c) 1.6 m (d) – 1.6 m

32. An object is moving in a circle at constant speed v. The magnitude of the rate of change of momentum of the object is

(a) zero (b) proportional to v^2

(c) proportional to v^3 (d) proportional to v

33. Five equal capacitors each with capacitance C are connected as shown in figure. Then, the equivalent capacitance between A and B is

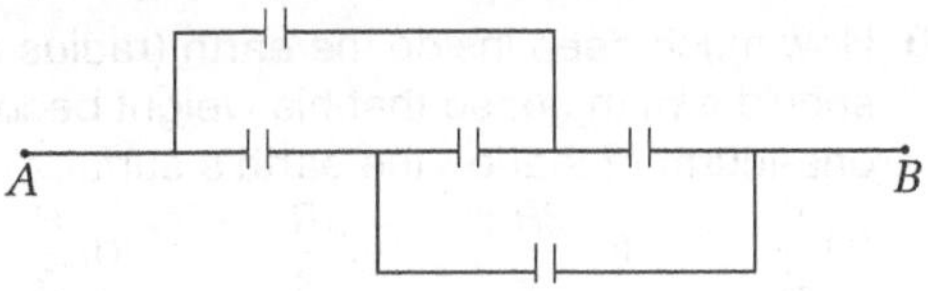

(a) 5C (b) $\frac{C}{5}$ (c) 3C (d) C

34. The time period of a freely suspended magnet is 4 s. If it is broken in length into two equal parts and one part is suspended in the same way, then the time period will be

(a) 4 s (b) 2 s (c) 0.5 s (d) 0.25 s

35. A spring of force constant k is cut into two pieces such that one piece is double the length of other. Then, the long piece will have a force constant of

(a) $\frac{2}{3}k$ (b) $\frac{3}{2}k$ (c) 3k (d) 6 k

36. Half-life period of a radiactive substance is 10 min, then amount of substance decayed in 40 min will be

(a) 25% (b) 50%

(c) 75% (d) None of these

37. Two identical containers A and B with frictionless pistons contain the same ideal gas at the same temperature and the same volume V. The mass of the gas in A is m_A and that in B is m_B. The gas in each container is now allowed to expand isothermally to the same final volume $2V$. The changes in the pressure in A and B are to be found Δp and $1.5\ \Delta p$ respectively, then relation for masses will be

(a) $4m_A = 9m_B$ (b) $2m_A = 3m_B$
(c) $3m_A = 2m_B$ (d) $9m_a = 4m_B$

38. If one mole of a monoatomic gas $\left(\gamma = \frac{5}{3}\right)$ is mixed with one mole diatomic gas $\left(\gamma = \frac{7}{5}\right)$, the value of γ for the mixture is

(a) 1.40 (b) 1.50 (c) 1.53 (d) 3.07

39. A solid sphere of mass 2 kg rolls on a smooth horizontal surface at 10 m/s and then rolls up a smooth 30° incline. The maximum height reached by the sphere is ($g = 9.8\ m/s^2$)

(a) 10 m (b) 4.9 m (c) 14.2 m (d) 7.1 m

40. How much deep inside the earth (radius R) should a man go, so that his weight becomes one-fourth of that on the earth's surface?

(a) $\frac{R}{2}$ (b) $\frac{3R}{4}$ (c) $\frac{R}{4}$ (d) $\frac{R}{3}$

41. Four projectiles are thrown with the same initial speed making angles 27°, 36°, 43°, 51°, with the horizontal. The range of projectiles will be largest for the projectile fired at angle

(a) 27° (b) 36° (c) 43° (d) 51°

42. A wire of resistance 5 Ω is drawn out so that its length is increased by twice its original length, its new resistance is

(a) 45 Ω (b) 54 Ω (c) 20 Ω (d) 5 Ω

43. The magnifying power of a telescope is m'. If the focal length of the eye piece is doubled, then its magnifying power will become

(a) $\sqrt{2}\ m$ (b) $3\ m$ (c) $2\ m$ (d) $\frac{m}{2}$

44. On a heater coil it is written that, 250 V, 500 W. What is the resistance of this coil?

(a) 62.5 Ω (b) 100 Ω (c) 200 Ω (d) 125 Ω

45. When a dielectric slab is introduced between the plates of a capacitor connected to a battery, then

(a) charge on capacitor increases
(b) potential difference across the capacitor increases
(c) energy stored increases
(d) capacity remains the same

46. A concave lens is kept in contact with a convex lens of focal length 20 cm. The combination works as convex lens of focal length 50 cm. The power of concave lens is

(a) $P = -3.0$ D (b) $P = +3.0$ D
(c) $P = -0.3$ D (d) $P = +0.3$ D

47. The critical angle for glass water interface $\left(\text{if } {}_a\mu_g = \frac{3}{2},\ {}_a\mu_w = \frac{4}{3}\right)$

(a) $\sin^{-1}\left(\frac{8}{9}\right)$ (b) $\sin^{-1}\left(\frac{9}{8}\right)$
(c) $\sin^{-1}\left(\frac{3}{2}\right)$ (d) $\sin^{-1}\left(\frac{4}{3}\right)$

48. What is the current through the circuit and the pontential difference across the diode shown in figure. The drift current for the diode is 30μ A.

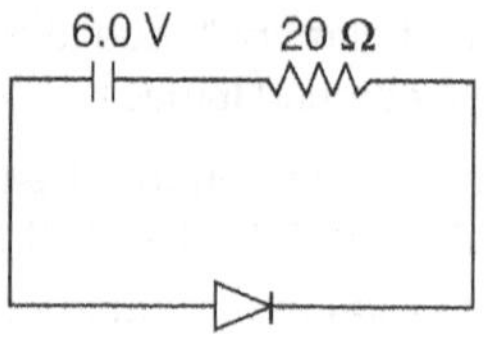

(a) 30 μ A , 5.99 V (b) 30 μ A, 5 V
(c) 20 μA , 6 V (d) 20 μA , 5.99 V

49. A man who wears glasses of power 3 D must holds a newspaper at least 25 cm away to see the print clearly. How far away would the newspaper have to be if he took off the glasses and still wanted clear vision?

(a) 10 cm (b) 25 cm
(c) 1 m (d) – 1 m

50. The force per unit length between two parallel current carrying straight conductors separated by $2d$ is given by the formula

(a) $\frac{\mu_0}{4\pi}\frac{i_1 i_2}{d}$ (b) $\frac{\mu_0}{4\pi}\frac{i_1 i_2}{2d}$
(c) $\frac{\mu_0}{\pi}\frac{i_1 i_2}{2d}$ (d) None of these

Chemistry

1. Which of the following is not basic amino acid?

(a) Leucine (b) Lysine
(c) Arginine (d) Histidine

2. In qualitative analysis, NH_4Cl is added before NH_4OH

(a) to increase $[OH^-]$ concentration
(b) for making HCl
(c) to decrease $[OH^-]$ concentration
(d) statement is wrong

3. Certain electric current for half an hour can collect 11.2 L of hydrogen at NTP. Same current when passed through an electrolytic solution for one hour, can deposite how much silver?

(a) 216 g (b) 108 g (c) 47 g (d) 60 g

4. Which of the following complexes is an outer orbital complex?

(a) $[Co(NH_3)_6]^{3+}$ (b) $[Fe(CN)_6]^{4-}$
(c) $[Ni(NH_3)_6]^{2+}$ (d) $[Mn(CN)_6]^{4-}$

5. Which is true for a cyclic process?

(a) $\Delta E = 0$ (b) $\Delta E = q - W$
(c) $q = W$ (d) All of these

6. In the Cannizzaro reaction given below

$$2C_6H_5CHO \xrightarrow{OH^-} C_6H_5CH_2OH + C_6H_5COO^-$$

The slowest step is

(a) The attack of OH^- at the carbonyl group
(b) The transfer of hydride ion to the carbonyl group
(c) The abstraction of proton from the carboxylic acid
(d) The deprotonation of benzyl alcohol

7. The rate of a reaction doubles when the initial concentration of the reactant is made four fold. If the initial concentration is made 400 fold, then the rate will become

(a) 400 times (b) 200 times
(c) 40 times (d) 20 times

8. Which compound present in diesel?

(a) Cetane
(b) $TiCl_4$
(c) Cyclo pentandienyl manganese carbonyl
(d) Iso octane

9. Which of the following is an organometallic compound?

(a) Lithium acetate
(b) Methyl lithium
(c) Lithium dimethyl amide
(d) Lithium methoxide

10. If in the reaction, $N_2O_4 \rightleftharpoons 2NO_2$; α is the degree of dissociation of N_2O_4, then total number of moles at equilibrium is

(a) $(1-\alpha)$ (b) $(1+\alpha)$ (c) $(1-\alpha)^2$ (d) $(1-\alpha)^2$

11. $CHCl_3 \xrightarrow[h\nu]{[O]} A \xrightarrow{NH_3} B \xrightarrow[PoCl_3]{\text{Malonic acid}} C$

The end product C in the following reaction is used as

(a) explosive (b) hypnotic
(c) tear gas (d) analgesic

12. Which statement is false for white phosphorus (P_4)?

(a) It has six P—P single bonds
(b) It has four P—P single bonds
(c) It has four lone pairs of electrons
(d) It has PPP angle 60°

13. Natural rubber is a polymer of monomer isoprene. During polymerisaton

(a) 1, 4 addition takes place
(b) 1, 2 addition takes place
(c) 1, 3 addition takes place
(d) both double bonds are converted into single bond

14. Match List I (species) with List II (hybridisation) and select the correct code given below

	List I		List II
A.	XeF_4	(i)	dsp^2
B.	H_2O	(ii)	sp^3
C.	PCl_5	(iii)	sp^3d^2
D.	$[Pt(NH_3)_4]^{2+}$	(iv)	sp^3d

Codes

	A	B	C	D
(a)	(iii)	(ii)	(iv)	(i)
(b)	(i)	(iv)	(ii)	(iii)
(c)	(iii)	(iv)	(ii)	(i)
(d)	(i)	(ii)	(iv)	(iii)

15. A new carbon-carbon bond formation is possible in :

I. Cannizzaro reaction
II. Friedel-Craft reaction
III. Clemmensen reduction
IV. Reimer-Tiemann reaction

(a) I, II and III (b) II, III and IV
(c) I and III (d) II and IV

16. Which of the following compounds formed during Perkin's reaction?

(a) Resorcinol (b) Cinnamic acid
(c) Benzaldehyde (d) Benzoin

17. The ratio of de-Broglie wavelengths for electron accelerated through 200 V and 50 V is

(a) 1 : 2 (b) 2 : 1 (c) 3 : 10 (d) 10 : 3

18. Malachite decomposed to give $A + CO_2 + H_2O$ and compound *A* on reduction with carbon gives $CO + B$. Here, *A* and *B* are

(a) CuO, Cu (b) Cu_2O, CuO
(c) Cu_2O, Cu (d) $CuCO_3$, Cu

19. Match List I with List II and choose correct answer from the codes given below

	List I		List II
A.	$NaNO_3$	(i)	Baking soda
B.	$Na(NH_4)HPO_4$	(ii)	Chile salt petre
C.	$NaHCO_3$	(iii)	Microcosmic salt
D.	$Na_2CO_3 \cdot 10H_2O$	(iv)	Washing soda

	A	B	C	D
(a)	(i)	(ii)	(iii)	(iv)
(b)	(ii)	(iii)	(i)	(iv)
(c)	(iii)	(i)	(ii)	(iv)
(d)	(iv)	(i)	(ii)	(iii)

20. When MnO_2 is heated with PbO_2 and conc. HNO_3, pink colour is obtained due to formation of

(a) $KMnO_4$ (b) $HMnO_4$
(c) $Pb(MnO_4)_2$ (d) $PbMnO_4$

21. Which is mismatched for NaCl crystal ?

(a) $\frac{r^+}{r^-} = 0.414$ to 0.732
(b) Coordination number = 6:6
(c) Edge of unit cell = $(r^+ + r^-)$
(d) Crystal structure = fcc

22. Which of the following ions has the highest magnetic moment ?

(a) Zn^{2+} (b) Ti^{3+} (c) Sc^{3+} (d) Mn^{2+}

23. Correct order for solubility of alkaline earth metals in water is

(a) $MgF_2 > CaF_2 > SrF_2 > BaF_2$
(b) $MgF_2 < CaF_2 < SrF_2 < BaF_2$
(c) $MgF_2 > CaF_2 < SrF_2 < BaF_2$
(d) $BaF_2 > MgF_2 > SrF_2 > CaF_2$

24. An organic compound *A* contains 20% C, 46.66% N and 6.66% H. It gives NH_3 gas on heating with NaOH. *A* can be

(a) CH_3CONH_2 (b) $C_6H_5CONH_2$
(c) NH_2CONH_2 (d) $CH_3NHCONH_2$

25. In the equilibrium mixture, $KI + I_2 \rightleftharpoons KI_3$; the concentration of KI and I_2 is made two fold and three fold respectively. The concentration of KI_3 becomes

(a) two fold (b) three fold
(c) five fold (d) six fold

26. $A \xrightarrow{HOH/H^+}$ glucose + fructose

$B \xrightarrow{HOH/H^+}$ glucose + glucose

$C \xrightarrow{HOH/H^+}$ glucose + galactose

The disaccharides *A*, *B* and *C* respectively are

(a) lactose, sucrose and maltose
(b) sucrose, maltose and lactose
(c) sucrose, lactose and maltose
(d) maltose, sucrose and lactose

27. *n*/*p* ratio during positron decay

(a) increases (b) decreases
(c) remains constant (d) All of these

28. Number of carbon atoms in kerosene is

(a) C_{17-20} (b) C_{12-16} (c) C_{20-25} (d) C_{25-30}

29. Which of the following polymers can be used for lubrication and as an insulator?

(a) SBR (b) PVC (c) PTFE (d) PAN

30. The IUPAC name of

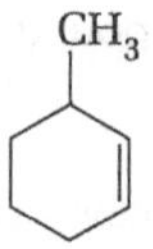

is

(a) 3-methyl cyclohexene
(b) 1-methyl cyclohex-2-ene

(c) 6-methyl cyclohexene
(d) 1-methyl cyclohex-5-ene

31. The correct representation of a complex ion is
(a) $[Co(H_2O)(NH_3)_4Cl]^{2+}$
(b) $[CoCl(H_2O)(NH_3)_4]^{2+}$
(c) $[Co(NH_3)_4Cl(H_2O)]^{2+}$
(d) $[Co(NH_3)_4(H_2O)Cl]^{2+}$

32. Which element has maximum first ionisation potential?
(a) Cs (b) F (c) Na (d) He

33. In which of the following $\Delta E = \Delta H$?
(a) $N_2O_4(g) \rightleftharpoons 2NO_2(g)$
(b) $2SO_2(g) + O_2(g) \rightleftharpoons 2SO_3(g)$
(c) $H_2(g) + I_2(g) \rightleftharpoons 2HI(g)$
(d) $H_2(g) + \frac{1}{2}O_2(g) \rightleftharpoons H_2O(l)$

34. Calcination is used in metallurgy for removal of
(a) water and sulphide (b) water and CO_2
(c) CO_2 and H_2S (d) H_2O and H_2S

35. What is the value of x on the $[Ni(CN)_4]^x$ complex ion?
(a) +2 (b) − 2 (c) 0 (zero) (d) + 4

36. The activation energy of a reaction is zero. The rate constant of this reaction
(a) increases with an increase of temperature
(b) decreases with an increase of temperature
(c) decreases with decrease of temperature
(d) is independent of temperature

37. Which of the following does not exist?
I. $HO_3S—S—SO_3H$ II. HO—Te—OH
III. $HS—S_2—SH$ IV. HS—Po—OH
(a) Only II (b) Only III
(c) II and IV (d) I, III and IV

38. Which of the following notations shows the product incorrectly?
(a) ${}^{10}_{5}B\ (\alpha, n)\ {}^{13}_{7}N$ (b) ${}^{242}_{96}Cm\ (\alpha, 2n)\ {}^{243}_{97}Bk$
(c) ${}^{14}_{7}N\ (n, p)\ {}^{14}_{6}C$ (d) ${}^{28}_{14}Si\ (d, \gamma)\ {}^{30}_{15}P$

39. Acidic dichromate ion reacts with hydrogen peroxide to give deep blue colour. This is due to the formation of
(a) $CrO(O)_2$ (b) CrO_5
(c) Both (a) and (b) (d) None of (a) and (b)

40. In blast furnace, the highest temperature is in
(a) reduction zone (b) slag zone
(c) fusion zone (d) combustion zone

41. The reagent which distinguishes formic acid and acetic acid is
(a) 2, 4-dinitrophenyl hydrazine
(b) $HgCl_2$
(c) C_2H_5ONa
(d) Hg_2Cl_2

42. Neon atom requires energy to remove one electron from its outermost orbit, so also to add one electron into the outermost orbit.
Which of the following is correct about fluorine?
I. Fluorine releases energy when an electron is removed.
II. Fluorine requires energy to add one electron.
(a) (I) and (II) are correct
(b) (I) is correct, (II) is false
(c) (I) is false, (II) is correct
(d) (I) and (II) are false

43. Liquid hydrocarbon is converted into mixture of gaseous hydrocarbons by
(a) cracking
(b) oxidation
(c) hydrolysis
(d) distillation under reduced pressure

44. Match List-I with List-II and select the correct answer from the given codes.

	List - I (Reaction)		List - II (Reagent/Catalyst)
A.	Cannizaro reaction	1.	$SnCl_2$ / HCl
B.	Stephen's reaction	2.	NaOH
C.	Clemmensen reduction	3.	Zn / Hg—conc.HCl
D.	Rosenmund's method	4.	$\frac{Pd / BaSO_4}{\text{Boiling xylene}}$

Codes

	A	B	C	D
(a)	1	2	3	4
(b)	2	1	3	4
(c)	4	3	2	1
(d)	1	4	2	3

45. The correct order of ionic radius of nitrogen family is
(a) $N^{3-} < P^{3-} < As^{3-} < Sb^{3-} < Bi^{3-}$
(b) $N^{3-} < P^{3-} < Sb^{3-}$
(c) $P^{3-} = As^{3-} > Bi^{3-}$
(d) $N^{3-} > Bi^{3-} > Sb^{3-}$

46. Which of the following statement is false?
(a) Chlorophyll is responsible for the synthesis of carbohydrate in plants
(b) In presence of oxygen, haemoglobin forms oxyhaemoglobin
(c) Acetyl salicylic acid is known as aspirin
(d) Vitamin B_{12} contains Mg^{2+} ion

47. Among following, compound with the lowest pK_a value, is

(a) 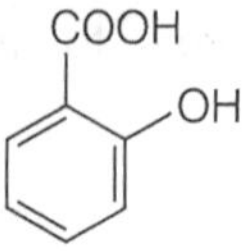

(b) 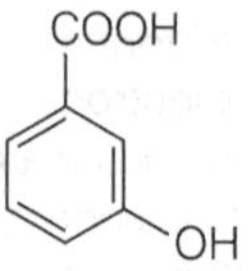

(c) 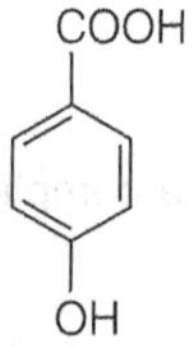

(d) 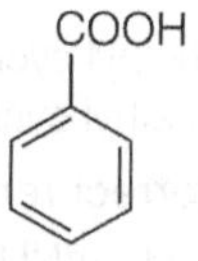

48. Which one of the following will most readily be dehydrated in acidic solution?

(a) 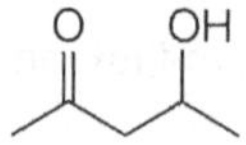

(b) 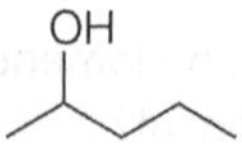

(c) 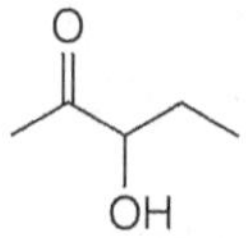

(d) 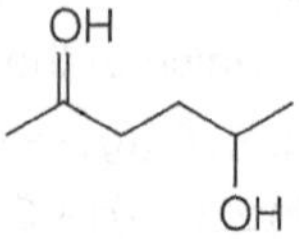

49. Paracetamol is a following drug
(a) antipyretic (b) antiseptic
(c) antibiotic (d) anaesthetic

50. $K_2[HgI_4]$ detects the following ion
(a) Cl^- (b) NO_2^- (c) NO_3^- (d) NH_4^+

Mathematics

1. If A and B are two such events that $P(A \cup B) = P(A \cap B)$, then which of the following is true?
(a) $P(A) + P(B) = 0$
(b) $P(A) + P(B) = P(A)\,P(B/A)$
(c) $P(A) + P(B) = 2P(A)\,P(B/A)$
(d) None of the above

2. Equation of tangent to the circle $x^2 + y^2 - 2x - 2y + 1 = 0$ perpendicular to $y = x$ is given by
(a) $x + y \pm 1 = 0$ (b) $x + y = 2 \pm \sqrt{3}$
(c) $x - y \pm 3 = 0$ (d) $x - y \pm 1 = 0$

3. If point D divides the base BC of a ΔABC in the ratio $n : m$, then the value of $mBD^2 + nCD^2 + (m+n)AD^2$ is
(a) $mAC^2 + nAB^2$
(b) $(m+n)(AC^2 + AB^2)$
(c) $nAC^2 + mAB^2$
(d) None of the above

4. A man is standing on the horizontal plane. The angle of elevation of top of the pole is α. If he walks a distance double the height of the pole, then the elevation of the pole is 2α. The value of α is
(a) $\frac{\pi}{12}$ (b) $\frac{\pi}{4}$ (c) $\frac{\pi}{3}$ (d) $\frac{\pi}{6}$

5. If $f(x) = \begin{cases} \dfrac{\sin[x]}{[x]} & \text{for } [x] \neq 0 \\ 0 & \text{for } [x] = 0 \end{cases}$

where, $[x]$ denotes the greatest integer less than or equal to x, then $\lim_{x \to 0} f(x)$ is equal to
(a) 1 (b) –1
(c) 0 (d) Does not exist

6. The value of the angle between two straight lines $y = (2 - \sqrt{3})x + 5$ and $y = (2 + \sqrt{3})x - 7$ is
(a) 30° (b) 60°
(c) 45° (d) 90°

7. Equation of a plane passing through $(-1, 1, 1)$ and $(1, -1, 1)$ and perpendicular to $x + 2y + 2z = 5$ is

(a) $2x + 3y - 3z + 3 = 0$
(b) $x + y + 3z - 5 = 0$
(c) $2x + 2y - 3z + 3 = 0$
(d) $x + y + z - 3 = 0$

8. If $f'(x) > 0 \ \forall x \in R, f'(3) = 0$ and $g(x) = f(\tan^2 x - 2\tan x + 4), 0 < x < \frac{\pi}{2}$, then $g(x)$ is increasing in

(a) $\left(0, \frac{\pi}{4}\right)$ (b) $\left(\frac{\pi}{6}, \frac{\pi}{3}\right)$ (c) $\left(0, \frac{\pi}{3}\right)$ (d) $\left(\frac{\pi}{4}, \frac{\pi}{2}\right)$

9. Probability of getting a total of 7 or 9 in a single throw of two dice is

(a) $\frac{5}{18}$ (b) $\frac{1}{6}$
(c) $\frac{1}{9}$ (d) None of these

10. What is compiler?

(a) Application software (b) System software
(c) Utility software (d) All of these

11. Let $f(x)$ be differentiable on the interval $(0, \infty)$ such that $f(1) = 1$ and

$$\lim_{t \to x} \frac{t^2 f(x) - x^2 f(t)}{t - x} = 1 \text{ for each } x > 0.$$

Then, $f(x)$ is equal to

(a) $\frac{1}{3x} + \frac{2}{3}x^2$ (b) $-\frac{x}{3} + \frac{4x^2}{3}$
(c) $-\frac{1}{x}$ (d) $-\frac{1}{x} + \frac{2}{x^2}$

12. If $\int \sin\left\{2\tan^{-1}\sqrt{\frac{1-x}{1+x}}\right\}dx$

$$= A\sin^{-1}x + Bx\sqrt{1-x^2} + C,$$

then $A + B$ is equal to

(a) 10 (b) $\frac{1}{2}$
(c) 1 (d) $-\frac{1}{2}$

13. The coefficient of x^4 in $(1 + x + x^3 + x^4)^{10}$ is

(a) 210 (b) 100
(c) 310 (d) 110

14. The locus of centre of circles which cuts orthogonally the circle $x^2 + y^2 - 4x + 8 = 0$ and touches $x + 1 = 0$, is

(a) $y^2 + 6x + 7 = 0$ (b) $x^2 + y^2 + 2x + 3 = 0$
(c) $x^2 + 3y + 4 = 0$ (d) None of the above

15. Let $f(x) = \begin{vmatrix} \sin 3x & 1 & 2\left(\cos\frac{3x}{2} + \sin\frac{3x}{2}\right)^2 \\ \cos 3x & -1 & 2\left(\cos^2\frac{3x}{2} - \sin^2\frac{3x}{2}\right) \\ \tan 3x & 4 & 1 + 2\tan 3x \end{vmatrix}$

Then, the value of $f'(x)$ at $x = (2n + 1)\pi, n \in I$ (the set of integers) is equal to

(a) $(-1)^n$ (b) $(-1)^{n+1}$
(c) 3 (d) 9

16. The condition for the line $lx + my + n = 0$ to be a normal to $\frac{x^2}{25} + \frac{y^2}{9} = 1$ is

(a) $\frac{l^2}{9} + \frac{m^2}{25} = \frac{n^2}{256}$ (b) $\frac{9}{m^2} + \frac{25}{l^2} = \frac{256}{n^2}$
(c) $\frac{l^2}{9} - \frac{m^2}{25} = \frac{n^2}{256}$ (d) None of these

17. The least value of a, for which the function $\frac{4}{\sin x} + \frac{1}{1 - \sin x} = a$ has atleast one solution in the interval $\left(0, \frac{\pi}{2}\right)$, is

(a) 9 (b) 4
(c) 5 (d) 1

18. If one line of regression coefficient is less than unity, then the other will be

(a) less than unity (b) equal to unity
(c) greater than unity (d) All of these

19. Three concurrent edges of a parallelopiped are given by

$$\mathbf{a} = 2\hat{\mathbf{i}} - 3\hat{\mathbf{j}} + \hat{\mathbf{k}},$$
$$\mathbf{b} = \hat{\mathbf{i}} - \hat{\mathbf{j}} + 2\hat{\mathbf{k}}$$
and $$\mathbf{c} = 2\hat{\mathbf{i}} + \hat{\mathbf{j}} - \hat{\mathbf{k}}.$$

The volume of the parallelopiped is

(a) 14 cu units (b) 20 cu units
(c) 25 cu units (d) 60 cu units

20. Roots of equation $x^3 - 6x + 1 = 0$ lie in the interval

(a) (2, 3) (b) (3, 4) (c) (3, 5) (d) (4, 6)

21. If $\lim_{x \to 0} \frac{\sin(\sin x) - \sin x}{ax^3 + bx^5 + c} = \frac{-1}{12}$, then

(a) $a = 2, b \in R, c = 0$ (b) $a = -2, b \in R, c = 0$
(c) $a = 1, b \in R, c = 0$ (d) $a = -1, b \in R, c = 0$

22. According to Newton-Raphson method, the value of $\sqrt{12}$ upto three places of decimal will be

(a) 3.463 (b) 3.462
(c) 3.467 (d) None of these

23. If $\frac{(3-i)^2}{2+i} = A + iB$, where A and B are real numbers, then A and B are equal to

(a) $A = -4, B = 2$ (b) $A = 2, B = -4$
(c) $A = 2, B = 4$ (d) None of these

24. The radical centre of the system of circles,

$$x^2 + y^2 + 4x + 7 = 0,$$
$$2(x^2 + y^2) + 3x + 5y + 9 = 0$$
and $$x^2 + y^2 + y = 0 \text{ is}$$

(a) (−2, −1) (b) (1, −2)
(c) (−1, −2) (d) None of these

25. The curve, for which the area of the triangle formed by X-axis, the tangent line at any point P and line OP is equal to a^2, is given by

(a) $y = x - Cx^2$ (b) $x = Cy \pm \frac{a^2}{y}$
(c) $y = Cx \pm \frac{a^2}{x}$ (d) None of these

26. Solution of the equation $\cos^2 x \frac{dy}{dx} - (\tan 2x)y = \cos^4 x, |x| < \frac{\pi}{4}$, where $y\left(\frac{\pi}{6}\right) = \frac{3\sqrt{3}}{8}$, is given by

(a) $y \frac{\tan 2x}{1 - \tan^2 x} = 0$ (b) $y(1 - \tan^2 x) = C$
(c) $y = \sin 2x + C$ (d) $y = \frac{1}{2} \cdot \frac{\sin 2x}{1 - \tan^2 x}$

27. If 1, ω and ω^2 are the cube roots of unity, then the value of $(1 - \omega + \omega^2)(1 + \omega - \omega^2)$ is equal to

(a) 4 (b) 0
(c) 2 (d) 3

28. The equation of the curve through the point (1, 0), whose slope is $\frac{y-1}{x^2 + x}$, is

(a) $2x(y - 1) + x + 1 = 0$
(b) $(x + 1)(y - 1) + 2x = 0$
(c) $x(y - 1)(x + 1) + 2 = 0$
(d) $x(y + 1) + y(x + 1) = 0$

29. The number of points, where $f(x) = [\sin x + \cos x]$ (where $[\cdot]$ denotes the greatest integer function) and $x \in (0, 2\pi)$ is not continuous, is

(a) 3 (b) 4 (c) 5 (d) 6

30. The value of $\cot^{-1}\left(\frac{\sqrt{1+\sin x} + \sqrt{1-\sin x}}{\sqrt{1+\sin x} - \sqrt{1-\sin x}}\right)$ is equal to

(a) $\frac{x}{3}$ (b) $\frac{x}{4}$ (c) 1 (d) $\frac{x}{2}$

31. If $A(-1, 3, 2)$, $B(2, 3, 5)$ and $C(3, 5, -2)$ are vertices of a ΔABC, then angles of ΔABC are

(a) $\angle A = 90°, \angle B = 30°, \angle C = 60°$
(b) $\angle A = \angle B = \angle C = 60°$
(c) $\angle A = \angle B = 45°, \angle C = 90°$
(d) None of the above

32. $\lim_{x \to 0} \frac{\int_0^{x^2} \sin \sqrt{t}\, dt}{x^3}$ is equal to

(a) $\frac{2}{3}$ (b) $\frac{1}{3}$ (c) 0 (d) ∞

33. If $\mathbf{a}$, $\mathbf{b}$ and $\mathbf{c}$ are three non-coplanar vectors, then $[\mathbf{a} \times \mathbf{b}\ \mathbf{b} \times \mathbf{c}\ \mathbf{c} \times \mathbf{a}]$ is equal to

(a) $[\mathbf{a\ b\ c}]^3$ (b) $[\mathbf{a\ b\ c}]^2$
(c) 0 (d) None of these

34. If geometric mean and harmonic mean of two numbers a and b are 16 and 64/5 respectively, then the value of $a : b$ is

(a) 4 : 1 (b) 3 : 2 (c) 2 : 3 (d) 1 : 4

35. If the sum of four numbers in GP is 60 and the arithmetic mean of the first and last numbers is 18, then the numbers are

(a) 3, 9, 27, 81 (b) 4, 8, 16, 32
(c) 2, 6, 18, 54 (d) None of these

36. The sum of the real solutions of equation $2|x|^2 + 51 = |1 + 20x|$ is

(a) 5 (b) 24
(c) 0 (d) None of these

37. The quadratic equation whose roots are $\frac{1}{3+\sqrt{2}}$ and $\frac{1}{3-\sqrt{2}}$, will be

(a) $7x^2 - 6x + 1 = 0$ (b) $6x^2 - 7x + 1 = 0$
(c) $x^2 - 6x + 7 = 0$ (d) $x^2 - 7x + 6 = 0$

38. Who said, "Number of transistors per square inch on integrated circuits double every year, since their invention will continue to do so in the foreseable future"?

(a) Alan Turing (b) Jon Von Neumann
(c) Herbert Simon (d) Gorden Moore

39. The number of vectors of unit length perpendicular to vectors $\mathbf{a} = \hat{\mathbf{i}} + \hat{\mathbf{j}}$ and $\mathbf{b} = \hat{\mathbf{j}} + \hat{\mathbf{k}}$, is

(a) infinite (b) one (c) two (d) three

40. By trapezoidal rule, the approximate value of the integral $\int_0^6 \frac{dx}{1+x^2}$ is

(a) 1.3128 (b) 1.4108
(c) 1.4218 (d) None of these

41. The radius of a cylinder is increasing at the rate of 2 m/s and its height is decreasing at the rate of 3 m/s. When the radius is 3 m and height is 5 m, then the volume of the cylinder would change at the rate of

(a) $87\pi\ m^3/s$ (b) $33\ \pi\ m^3/s$
(c) $27\pi\ m^3/s$ (d) $15\ \pi m^3/s$

42. The point on the straight line $y = 2x + 11$ which is nearest to the circle $16(x^2 + y^2) + 32x - 8y - 50 = 0$, is

(a) $\left(\frac{9}{2}, 2\right)$ (b) $\left(\frac{9}{2}, -2\right)$ (c) $\left(-\frac{9}{2}, 2\right)$ (d) $\left(-\frac{9}{2}, -2\right)$

43. The locus of the extrimities of the latusrectum of the family of ellipses $b^2x^2 + y^2 = a^2b^2$ having a given major axis, is

(a) $x^2 \pm ay = a^2$ (b) $y^2 \pm bx = a^2$
(c) $x^2 \pm by = a^2$ (d) $y^2 \pm ax = b^2$

44. The solution of differential equation $(y \log x - 1)ydx = xdy$ is

(a) $y(\log e^x + Cx) = 1$ (b) $\left(\log\frac{x}{e} + Cx\right)x = y$
(c) $(\log Cx^2 + ex^2)y = x$ (d) None of these

45. If m things are distributed among a men and b women. Then, the chance that the number of things received by men is odd, is

(a) $\frac{(b-a)^m - (b+a)^m}{2(b+a)^m}$ (b) $\frac{(b+a)^m - (b-a)^m}{2(b+a)^m}$
(c) $\frac{(b+a)^m - (b-a)^m}{(b+a)^m}$ (d) None of these

46. The solution of the differential equation

$$\sqrt{a+x}\,\frac{dy}{dx} + xy = 0 \text{ is}$$

(a) $y = Ce^{\frac{2}{3}(2a-x)\sqrt{x+a}}$ (b) $y = Ce^{\frac{2}{3}(a-x)\sqrt{x+a}}$
(c) $y = Ce^{\frac{2}{3}(2a+x)\sqrt{x+a}}$ (d) $y = Ce^{-\frac{2}{3}(2a-x)\sqrt{x+a}}$

47. The values of a, if $f(x) = 2e^x - ae^{-x} + (2a+1)x - 3$ increases x, are in

(a) $[0, \infty)$ (b) $(-\infty, 0]$
(c) $(-\infty, \infty)$ (d) $(1, \infty)$

48. By Newton-Raphson method, the positive root of the equation $x^4 - x - 10 = 0$ is

(a) 1.871 (b) 1.868
(c) 1.856 (d) None of these

49. In ΔABC, if $3a = b + c$, then value of $\cot\frac{B}{2}\cot\frac{C}{2}$ will be

(a) 1 (b) 2 (c) $\sqrt{3}$ (d) $\sqrt{2}$

50. If $\sin^{-1} x + \cot^{-1}\left(\frac{1}{2}\right) = \frac{\pi}{2}$, then value of x will be

(a) 0 (b) $\frac{1}{\sqrt{5}}$ (c) $\frac{2}{\sqrt{5}}$ (d) $\frac{\sqrt{3}}{2}$

51. The value of $\int_0^{\sqrt{\ln(\pi/2)}} \cos(e^{x^2})2x\,e^{x^2}dx$ is

(a) 1 (b) $1 + \sin 1$
(c) $1 - \sin 1$ (d) $(\sin 1) - 1$

52. API stands for

(a) Access Programming Interface
(b) Android Programming Interface
(c) Application Programming Interface
(d) None of the above

53. If $Z = i\log(2 - \sqrt{3})$, then the value of $\cos Z$ will be

(a) i (b) $2i$ (c) 1 (d) 2

54. $\lim_{x \to 2} \frac{2 - \sqrt{2+x}}{2^{1/3} - (4-x)^{1/3}}$ is equal to

(a) $2 \cdot 3^{-1/2}$ (b) $3 \cdot 2^{-4/3}$
(c) $-3 \cdot 2^{-4/3}$ (d) None of these

55. The three lines of a triangle are given by $(x^2 - y^2)(2x + 3y - 6) = 0$. If the point $(-2, \lambda)$ lies inside and $(\mu, 1)$ lies outside the triangle, then

(a) $\lambda \in \left(1, \frac{10}{3}\right); \mu \in (-3, 5)$
(b) $\lambda \in \left(2, \frac{10}{3}\right); \mu \in (-1, 1)$
(c) $\lambda \in \left(-1, \frac{9}{2}\right); \mu \in \left(-2, \frac{10}{3}\right)$
(d) None of the above

56. System software : Utility software : :

(a) Operating system: Anti-virus
(b) Anti-virus : Operating system
(c) Anti-virus : MS office
(d) MS office : Anti-virus

57. The number of common tangents to two circles $x^2 + y^2 = 4$ and $x^2 + y^2 - 8x + 12 = 0$ is

(a) 1 (b) 2 (c) 3 (d) 4

58. If $2a + 3b + 6c = 0$, then the equation $ax^2 + bx + c = 0$ has atleast one real root in

(a) (0, 1) (b) $\left(0, \frac{1}{2}\right)$
(c) $\left(\frac{1}{4}, \frac{1}{2}\right)$ (d) None of these

59. If for all $x, y \in N$, there exists a function $f(x)$ satisfying $f(x + y) = f(x) \cdot f(y)$ such that $f(1) = 3$ and $\sum_{x=1}^{n} f(x) = 120$, then value of n will be

(a) 4 (b) 5
(c) 6 (d) None of these

60. If $f(x) = \begin{cases} \sin\left(\frac{\pi x}{2}\right), & \text{if } x < 1 \\ 3 - 2x, & \text{if } x \geq 1 \end{cases}$, then $f(x)$ has

(a) local minimum at $x = 1$
(b) local maximum at $x = 1$
(c) Both local maximum and local minimum at $x = 1$
(d) None of the above

61. The general solution of the differential equation $\frac{dy}{dx} = y \tan x - y^2 \sec x$ is

(a) $\tan x = (C + \sec x)y$ (b) $\sec y = (C + \tan y)x$
(c) $\sec x = (C + \tan x)y$ (d) $\tan y = (C + \sec x)x$

62. The values of λ such that $(x, y, z) \neq (0, 0, 0)$ and $(\hat{\mathbf{i}} + \hat{\mathbf{j}} + 3\hat{\mathbf{k}})x + (3\hat{\mathbf{i}} - 3\hat{\mathbf{j}} + \hat{\mathbf{k}})y + (-4\hat{\mathbf{i}} + 5\hat{\mathbf{j}})z = \lambda(\hat{\mathbf{i}}x + \hat{\mathbf{j}}y + \hat{\mathbf{k}}z)$ are

(a) 0, 1 (b) −1, 1 (c) −1, 0 (d) −2, 0

63. In ΔABC, if $\cot A$, $\cot B$ and $\cot C$ are in AP, then a^2, b^2 and c^2 are in

(a) HP (b) AP
(c) GP (d) None of these

64. By Simpson's $\frac{1}{3}$rd rule, the approximate value of the integral $\int_1^2 e^{-x/2} dx$ using four intervals, is

(a) 0.377 (b) 0.487 (c) 0.477 (d) 0.387

65. A die is rolled three times. The probability of getting a larger number than the previous number is

(a) $\frac{5}{216}$ (b) $\frac{5}{54}$ (c) $\frac{1}{6}$ (d) $\frac{5}{36}$

66. If $f(x) = \log_e(6 - |x^2 + x - 6|)$, then domain of $f(x)$ has how many integral values of x?

(a) 5 (b) 4
(c) Infinite (d) None of these

67. The graph of the equation $y^2 + z^2 = 0$ in three dimensional space is

(a) YZ-plane (b) Z-axis (c) Y-axis (d) X-axis

68. If $Z_1 = 1 + i$, $Z_2 = -2 + 3i$ and $Z_3 = \frac{ai}{3}$ are collinear, where $i^2 = -1$, then the value of a will be

(a) −1 (b) 3 (c) 4 (d) 5

69. If a, b and c are in HP, then for any $n \in N$, which one of the following is true?

(a) $a^n + c^n < 2b^n$ (b) $a^n + c^n > 2b^n$
(c) $a^n + c^n = 2b^n$ (d) None of these

70. The value of $\tan\left[2\tan^{-1}\left(\frac{1}{5}\right) - \frac{\pi}{4}\right]$ is

(a) $\frac{17}{7}$ (b) $\frac{-17}{7}$ (c) $\frac{7}{17}$ (d) $\frac{-7}{17}$

English and General Aptitude

Directions (Q. Nos. 1-5) *Read the given passage carefully and answer the questions that follow by selecting the most appropriate option.*

No one knows when or by whom rockets were invented. In all probability the rocket was not suddenly invented but evolved gradually over a long period of time, perhaps in different parts of the world at the same time. Some historians of rocketry, notably Willy Ley, trace the development of rockets to the 13th century China, a land noted in ancient times for its fire work display. In the year 1232 AD when the Mongols laid siege to the city of Kai-Feng Fu, the capital of Honan province, the Chinese defenders used weapons that were described as "arrows of flying fire". There is no explicit statement that these arrows were rockets, but some students have concluded that they were because the record does not mentioned bows or other means of shooting the arrows. In the same battle, we read, the defenders dropped from the walls of the city a kind of bomb described as *"heaven-shaking thunder*". From these meagre references some students have concluded that the Chinese, by the year 1232 had discovered gunpowder and had learned to use it to make explosive bombs as well as propulsive charges for rockets.

1. The passage gives primarily a history of
(a) the bravery of the Chinese
(b) the invention of rockets
(c) the attack on China by the Mongols
(d) the battle against the Chinese wall

2. According to this passage, rockets were invented by
(a) Willy Ley
(b) unknown people
(c) the Mongols
(d) the ruler of Honan province

3. According to this passage, rockets were
(a) a gift of God to the Chinese
(b) invented in the twentieth century
(c) invented in 1232 AD
(d) developed over many centuries

4. The phrase 'arrows of flying fire'
(a) means some ancient phenomenon in the skies
(b) refers to lightning and thunder
(c) is another name for rockets
(d) is assumed to refer to rockets

5. The bombs have been referred to as "heaven shaking thunder" because they
(a) contain gunpowder
(b) make thunderous noise
(c) are propelled by rockets
(d) seem to fall from heaven

Directions (Q. Nos. 6-8) *Complete the following sentence(s) with an appropriate word.*

6. The lawyer has plenty of ....
(a) criminals (b) buyers
(c) customers (d) clients

7. "I have brought the book. It's ..... !" Ravi said assertively to all the boys present.
(a) mine (b) my (c) me (d) myself

8. "The project is good, but there is .... missing to make it an excellent work," the engineer commented.
(a) everything (b) anything
(c) something (d) nothing

Directions (Q. Nos. 9-12) *In the following questions, select the antonym of the given words.*

9. Stingy
(a) Clean (b) Tight
(c) Generous (d) Cheap

10. Barren
(a) Fertile (b) Rich
(c) Prosperous (d) Positive

11. Virtue
(a) Vice (b) Failure (c) Fault (d) Offence

12. Nervous
(a) Flawless (b) Immature
(c) Smooth (d) Composed

Directions (Q. Nos. 13-15) *In the following questions, choose the alternative which best expresses the meaning of the word given in capital letters.*

13. Savour
(a) Taste (b) Protector
(c) Sour (d) Flavour

14. Rivalled
(a) Hatred (b) Revised
(c) Competed (d) Contradicted

15. Trimming
(a) Skimming (b) Arranging
(c) Planning (d) Cutting

Directions (Q. Nos. 16-20) *In each of the following questions, there is a certain relationship between two given words on side of (::) and one word is given on another side (::) while another word is to be found from the given alternatives, having the same relation with this word as the words of the given pair bear. Choose the correct alternative.*

16. Donkey : Brays :: Wolf : ?
(a) Bellows (b) Howls
(c) Whimpers (d) Roars

17. Astronauts : Space :: Argonauts : ?
(a) Fire (b) Ship
(c) Treasure (d) Sea

18. Orthopaedic : Bones :: Dermatologist : ?
(a) Feet (b) Skin
(c) Heart (d) Lungs

19. Basilica : Church :: Dormer : ?
(a) Window (b) Chapel
(c) Movie (a) Servant

20. Although : Nevertheless :: Though : ?
(a) Therefore (b) Yet
(c) However (d) Simultaneously

Directions (Q. Nos. 21-23) *In each of the following questions, a group of four words are given. Choose the word which is odd.*

21. Find the odd word.
(a) Sun (b) Mercury (c) Mars (d) Venus

22. Find the odd word.
(a) Stneng (b) Hesitant
(c) Daring (b) Brave

23. Find the odd word.
(a) Marigold (b) Tulip
(c) Lotus (d) Rose

24. In the English alphabet, find the position of *P* from right.
(a) 12 (b) 13 (c) 10 (d) 11

25. Which of the following letters is 14th to the right of 6th letter from the left in the English alphabet?
(a) R (b) P (c) W (d) T

26. If English alphabet is written in backward order, then what will be the 13th letter to the left of the 3rd letter from right?
(a) P (b) N (c) R (d) Q

Directions (Q. Nos. 27-28) *In the following series, replace the question mark (?) with the suitable option.*

27. 23, 28, 34, 41, 49, ?
(a) 57 (b) 59 (c) 56 (d) 58

28. 11, 13, 17, 19, 23, 25, ?
(a) 29 (b) 27 (c) 31 (d) 37

Directions (Q. Nos. 29-30) *In each of the following questions, a set of figures carrying certain characters is given. Assuming that the characters in each set follow a similar pattern, find the missing character in each case.*

29.
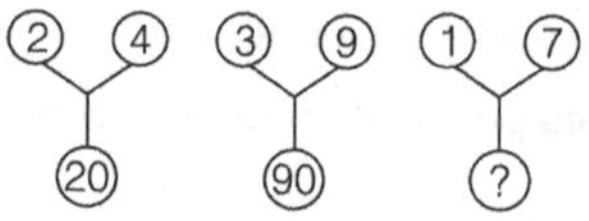

(a) 160 (b) 100 (c) 50 (d) 75

30.
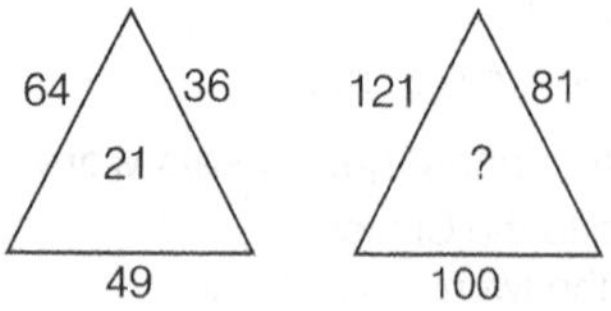

(a) 40 (b) 30 (c) 20 (d) 10

Answers

Physics

1. (c)	**2.** (b)	**3.** (a)	**4.** (c)	**5.** (c)	**6.** (b)	**7.** (d)	**8.** (b)	**9.** (c)	**10.** (d)
11. (d)	**12.** (a)	**13.** (b)	**14.** (b)	**15.** (a)	**16.** (b)	**17.** (a)	**18.** (c)	**19.** (a)	**20.** (b)
21. (a)	**22.** (a)	**23.** (c)	**24.** (a)	**25.** (a)	**26.** (c)	**27.** (a)	**28.** (a)	**29.** (c)	**30.** (c)
31. (a)	**32.** (b)	**33.** (d)	**34.** (b)	**35.** (b)	**36.** (d)	**37.** (c)	**38.** (b)	**39.** (d)	**40.** (b)
41. (c)	**42.** (c)	**43.** (d)	**44.** (d)	**45.** (a)	**46.** (a)	**47.** (a)	**48.** (a)	**49.** (d)	**50.** (a)

Chemistry

1. (a)	**2.** (c)	**3.** (a)	**4.** (c)	**5.** (a)	**6.** (b)	**7.** (d)	**8.** (a)	**9.** (b)	**10.** (b)
11. (b)	**12.** (a)	**13.** (a)	**14.** (a)	**15.** (d)	**16.** (b)	**17.** (a)	**18.** (c)	**19.** (b)	**20.** (b)
21. (c)	**22.** (d)	**23.** (a)	**24.** (c)	**25.** (a)	**26.** (b)	**27.** (a)	**28.** (b)	**29.** (c)	**30.** (a)
31. (d)	**32.** (d)	**33.** (c)	**34.** (b)	**35.** (b)	**36.** (d)	**37.** (c)	**38.** (b)	**39.** (b)	**40.** (d)
41. (b)	**42.** (d)	**43.** (b)	**44.** (b)	**45.** (a)	**46.** (d)	**47.** (a)	**48.** (a)	**49.** (a)	**50.** (d)

Mathematics

1. (c)	**2.** (*)	**3.** (c)	**4.** (a)	**5.** (d)	**6.** (b)	**7.** (c)	**8.** (d)	**9.** (a)	**10.** (b)
11. (a)	**12.** (c)	**13.** (c)	**14.** (a)	**15.** (c)	**16.** (b)	**17.** (a)	**18.** (c)	**19.** (a)	**20.** (a)
21. (b)	**22.** (d)	**23.** (b)	**24.** (a)	**25.** (b)	**26.** (d)	**27.** (a)	**28.** (*)	**29.** (c)	**30.** (d)
31. (d)	**32.** (a)	**33.** (b)	**34.** (a)	**35.** (b)	**36.** (d)	**37.** (a)	**38.** (d)	**39.** (c)	**40.** (b)
41. (b)	**42.** (c)	**43.** (a)	**44.** (d)	**45.** (b)	**46.** (a)	**47.** (a)	**48.** (a)	**49.** (b)	**50.** (b)
51. (c)	**52.** (c)	**53.** (d)	**54.** (c)	**55.** (d)	**56.** (a)	**57.** (c)	**58.** (a)	**59.** (a)	**60.** (b)
61. (c)	**62.** (c)	**63.** (b)	**64.** (c)	**65.** (b)	**66.** (d)	**67.** (a)	**68.** (d)	**69.** (b)	**70.** (d)

English and General Aptitude

1. (b)	**2.** (b)	**3.** (d)	**4.** (d)	**5.** (b)	**6.** (d)	**7.** (a)	**8.** (c)	**9.** (c)	**10.** (a)
11. (a)	**12.** (a)	**13.** (a)	**14.** (c)	**15.** (d)	**16.** (b)	**17.** (d)	**18.** (b)	**19.** (a)	**20.** (c)
21. (a)	**22.** (b)	**23.** (b)	**24.** (d)	**25.** (d)	**26.** (a)	**27.** (d)	**28.** (a)	**29.** (c)	**30.** (b)

Note (*) None of the option is correct.

SOLUTIONS

Physics

1. *(c)* Increase in potential energy

= Final potential energy – Initial potential energy

$$= -\frac{GMm}{\left(R+\frac{R}{5}\right)} - \left(-\frac{GMm}{R}\right)$$

$$= -\frac{5}{6}\frac{GMm}{R} + \frac{GMm}{R}$$

$$= \frac{1}{6}\frac{GMm}{R} = \frac{1}{6}\frac{GMmR}{R^2}$$

$$= \frac{1}{6}mgR \qquad \left[\because g = \frac{GM}{R^2}\right]$$

$$= \frac{5}{6}\,mgh \qquad \left[\because h = \frac{R}{5}\right]$$

2. *(b)* To break the covalent bond in a semiconductor, energy equal to forbidden energy gap is required.

3. *(a)* Given, $A = 60°$ and $\delta_m = 30°$

$$\text{So, } \mu = \frac{\sin\left(\frac{A+\delta_m}{2}\right)}{\sin\frac{A}{2}} = \frac{\sin\left(\frac{60°+30°}{2}\right)}{\sin\frac{60°}{2}}$$

$$= \frac{\sin 45°}{\sin 30°} = \frac{1}{\sqrt{2}} \times 2 = \frac{2\times\sqrt{2}}{\sqrt{2}\times\sqrt{2}}$$

$$= \frac{2\sqrt{2}}{2} = \sqrt{2}$$

We have, $\mu = \dfrac{1}{\sin C}$

or $\sqrt{2} = \dfrac{1}{\sin C}$

or $C = \sin^{-1}\left(\dfrac{1}{\sqrt{2}}\right) = 45°$

4. *(c)* Faraday constant is a universal constant.

5. *(c)* If a comb is run through wet hair on rainy day, then it will not attract small bits of paper.

6. *(b)* Given, $x = 9t^2$

$$\therefore \quad v = \frac{dx}{dt} = \frac{d}{dt}(9t^2) = \frac{9d}{dt}(t^2) = 18\,t$$

also, $a = \dfrac{d^2x}{dt^2} = \dfrac{dv}{dt} = \dfrac{d}{dt}(18t) = 18$

So, the graph between acceleration and time is a line parallel to time axis as acceleration is constant.

7. *(d)* The minimum negative potential V_0 is called stopping potential.

Here, $K_{max} = 10$ eV

So, the stopping potential, $eV_0 = K_{max}$

$\Rightarrow \quad eV_0 = 10$ eV

or $\quad V_0 = -10$ V

8. *(b)* Thermal radiations exists in infrared part of the electromagnetic spectrum.

9. *(c)* For a thin uniform rod, moment of inertia about an axis through its centre perpendicular to length of rod, $I = \dfrac{1}{12}ML^2$

Here, $M = (m_w + m_b)$

$\therefore \quad I = \dfrac{1}{12}(m_w + m_b)\,L^2$

10. *(d)* The time period of a particle executing SHM is given as

$$T = 2\pi\sqrt{\frac{x}{a}} = 2\pi\sqrt{\frac{\text{Displacement}}{\text{Acceleration}}}$$

$$= 2\pi\sqrt{\frac{3\times10^{-2}}{12\times10^{-2}}} = 2\pi\times\frac{1}{2}$$

$$= \pi = 3.14 \text{ s}$$

11. *(d)* A transformer is essentially an AC device, it does not work on DC. So, voltage developed across secondary is zero.

12. *(a)* Resolving power of eye $= \left(\dfrac{1}{60}\right)^{\circ} = \dfrac{1}{60}\times\dfrac{\pi}{180}$

Let the minimum distance between the poles be d,

then, $\dfrac{d}{11000} = \dfrac{1}{60}\times\dfrac{\pi}{180}$

or $\quad d = 11000 \times \dfrac{1}{60}\times\dfrac{\pi}{180} = 3$ m

13. *(b)* Stefan's constant, $\sigma = \frac{E}{T^4}$ (where, E is energy/s area)

$$= \frac{[ML^2 T^{-2}]}{[T][L^2][\theta^4]} = [MT^{-3}\theta^{-4}]$$

14. *(b)* Time period of oscillation of mass m suspended from a spring

$$T = 2\pi\sqrt{\frac{m}{k}}$$

If the spring is cut into two halves, then the new time period.

$$T' = 2\pi\sqrt{\frac{m}{2k}} = 2\frac{\pi}{\sqrt{2}}\sqrt{\frac{m}{k}} = \frac{T}{\sqrt{2}}$$

15. *(a)* Given, $I = 2.0$ A

$$R = 25\,\Omega$$
$$t = 1\text{ min} = 60\text{ s}$$

$\therefore$ Heat produced, $H = I^2 Rt$

$$= (2)^2 \times 25 \times 60$$
$$= 4 \times 25 \times 60\text{ J}$$
$$= 6 \times 10^3\text{ J}$$

16. *(b)* Frequency for a mass m executing SHM

$$f = \frac{1}{2\pi}\sqrt{\frac{k}{m}} \quad \text{...(i)}$$

When the mass is increased by 4 times, then the new frequency

$$f' = \frac{1}{2\pi}\sqrt{\frac{k}{4m}} = \frac{1}{2}\cdot\frac{1}{2\pi}\sqrt{\frac{k}{m}} = \frac{f}{2}$$

17. *(a)* For isothermal process, $dU = 0$,

i.e., internal energy remains unchanged.

For a cyclic process, $dU = 0$,

i.e., internal energy remains unchanged.

18. *(c)* An electric charge in uniform motion produces both electric field and magnetic field.

19. *(a)* For photoelectric effect to take place, the incident wavelength (λ) should be equal to or smaller than the threshold wavelength (λ_0) *i.e.*,

$$\lambda \le \lambda_0$$

20. *(b)* ${}_7X^{15} + {}_2\text{He}^4 \longrightarrow {}_9Y^{19}$

$${}_9Y^{19} \longrightarrow {}_8Z^{18} + {}_1\text{H}^1$$

According to the conservation of mass number

$$19 = 18 + 1$$

So, the mass number of changed atom = 18

21. *(a)* $AD = BC = \sqrt{(5)^2 + (5)^2} = \sqrt{25 + 25} = \sqrt{2} \times 5$ cm

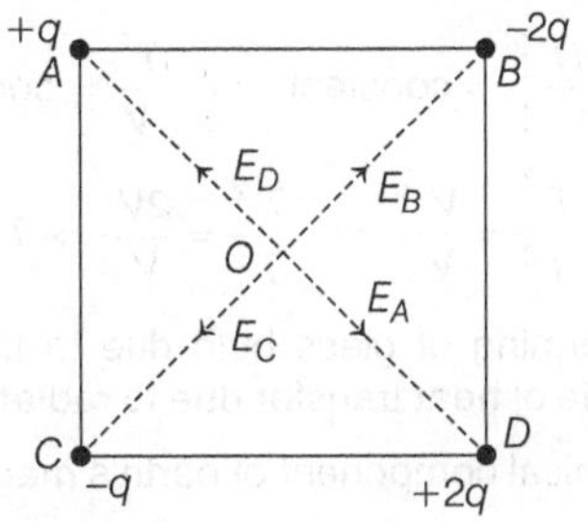

$$\Rightarrow AO = BO = CO = OD = \frac{5\sqrt{2}}{2}\text{ cm}$$

The electric field,

$$E = \frac{1}{4\pi\varepsilon_0}\cdot\frac{q}{r^2}$$

So, $$E_A = \frac{9 \times 10^9 \times 10 \times 10^{-9} \times 4}{25 \times 2}$$

$= 7.2\ \text{NC}^{-1}$ along OD

$$E_B = \frac{9 \times 10^9 \times 2 \times 10 \times 10^{-9} \times 4}{25 \times 2}$$

$= 14.4\ \text{NC}^{-1}$ along OB

$$E_C = \frac{9 \times 10^9 \times 10 \times 10^{-9} \times 4}{25 \times 2}$$

$= 7.2$ along OC

$$E_D = \frac{9 \times 10^9 \times 2 \times 10 \times 10^{-9} \times 4}{25 \times 2}$$

$= 14.4$ along OA

Resultant of E_A and E_D, $E_1 = (14.4 - 7.2)$

$= 7.2\ \text{NC}^{-1}$ along OA.

Resultant of E_B and E_C, $E_2 = (14.4 - 7.2)$

$= 7.2\ \text{NC}^{-1}$ along OB.

Since, E_1 and E_2 are perpendicular to each other.

$\therefore E = \sqrt{E_1^2 + E_2^2}$ is along 45° to OA upward.

22. *(a)* For an adiabatic process, $TV^{\gamma - 1} =$ constant.

23. *(c)* The correct relation is $v_{rms} > \bar{v} > v_{mp}$

As $\bar{v} = 1.6\sqrt{\frac{RT}{M}}$, $v_{rms} = 1.73\sqrt{\frac{RT}{M}}$

and $v_{mp} = 1.41\sqrt{\frac{RT}{M}}$

24. *(a)* Due to reverse biasing, in a junction diode the potential barrier increases.

25. *(a)* Given, $Vp^2 =$ constant

or $V\left[\frac{RT}{V}\right]^2 =$ constant or $\frac{T^2}{V} =$ constant

or $\frac{T'^2}{T^2} = \frac{V'}{V} \Rightarrow \frac{T'^2}{T^2} = \frac{2V}{V} \Rightarrow T' = \sqrt{2}T$

26. *(c)* Warming of glass bulb due to filament is an example of heat transfer due to radiation.

27. *(a)* Vertical component of earth's magnetic field is

$$B_V = B_H \tan\delta$$

where, B_H = horizontal component of earth's magnetic field

δ = angle of dip.

Total intensity of earth's magnetic field

$$B = \sqrt{B_H^2 + B_V^2}$$

$$= \sqrt{B_H^2 + B_H^2} \quad [\because \tan\delta = \tan 45° = 1]$$

$$= \sqrt{2}\, B_H$$

$$= 1.44 \times 0.4 \times 10^{-4}\text{ T}$$

$$= 5.656 \times 10^{-5}\text{ T}$$

$$= 0.5 \times 10^{-4}\text{ T}$$

28. *(a)* Generalised formula for apparent frequency is

$$f_a = \left(\frac{v \pm v_o}{v \mp v_s}\right) f_0$$

According to the question source is stationary,

$$v_s = 0$$

$$f_a = f' = \left(\frac{v \pm v_o}{v}\right) f_0$$

Observer is moving towards the source

$\Rightarrow \quad f_a = f'\left(\frac{v + v_o}{v}\right) f$

where, v = speed of the wave

v_o = speed of the observer

f = original frequency of the source

29. *(c)* Magnetic flux linked with the secondary due to current in the primary.

$$\phi_2 = MI_1$$

$\Rightarrow \quad \frac{d\phi_2}{dt} = M\frac{dI_1}{dt}$

$\Rightarrow$ Induced emf in the secondary

$$\varepsilon_2 = \frac{d\phi_2}{dt} = M\frac{dI_1}{dt}$$

where, M = mutual inductance = 1.5 H

$$\frac{dI_1}{dt} = \frac{5}{1 \times 10^{-3}} = 5000\text{ A/s}$$

$\therefore \quad \varepsilon_2 = -1.5 \times 5000\text{ V}$

$$= 7.5 \times 10^3 V = 7500\text{ V}$$

30. *(c)* Magnetic field at the centre of a toroid is given by

$$B = \frac{\mu_0 Ni}{2\pi r}$$

where, $n = \frac{N}{2\pi r}$ = number of turns per unit length of the toroid.

31. *(a)* Magnification produced by the lens

$$m = \frac{v}{u} = -3 \Rightarrow v = -3u$$

Applying lens of formula, we have

$$\frac{1}{v} - \frac{1}{u} = \frac{1}{f}$$

$\Rightarrow \quad \frac{1}{-3u} - \frac{1}{u} = \frac{1}{0.12}$

or $\quad \frac{-1-3}{3u} = \frac{100}{12} = \frac{25}{3}$

or $\quad -\frac{4}{3u} = \frac{25}{3}$

or $\quad u = -\frac{4}{25}\text{ m} = 0.16\text{ m}$

32. *(b)* Rate of change of momentum of the object,

$$\frac{\Delta p}{\Delta t} = F_{\text{ext}} = \frac{mv^2}{r}$$

$\frac{\Delta p}{\Delta t} \propto v^2$ (where, m and r are constant)

33. *(d)* Consider the given circuit,

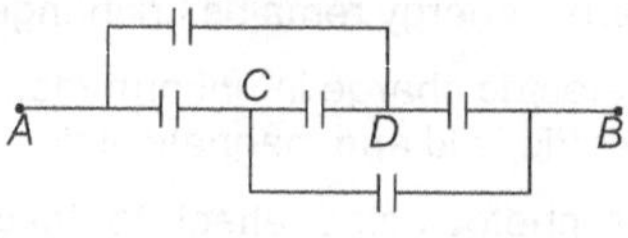

The circuit can be redrawn as shown below.

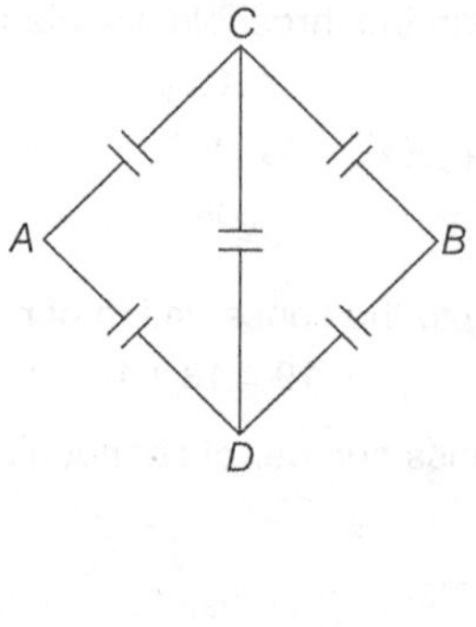

The circuit is a balanced bridge. We can remove the capacitor between C and D.

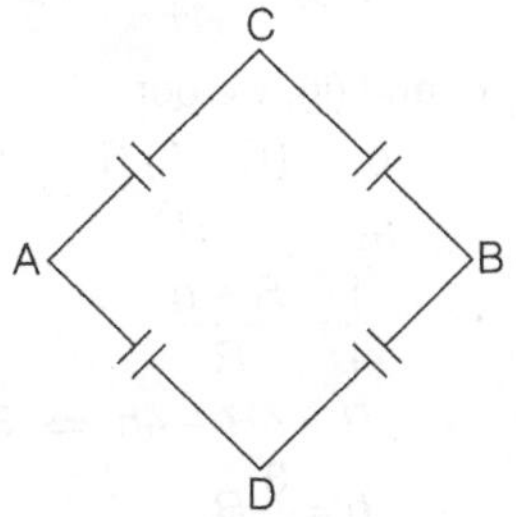

Thus, equivalent capacitance between A and B is

$$C_{AB} = \frac{C}{2} + \frac{C}{2} = C$$

34. *(b)* Time period of a suspended magnet is given by

$$T = 2\pi\sqrt{\frac{I}{MB}}$$

where, I = moment of inertia of the magnet about the point of suspension.

M = magnetic moment

B = magnetic field

When the magnet is broken in length

$$T^{2'} = 2\pi\sqrt{\frac{I'}{M'B}}$$

$$\Rightarrow \quad \frac{T}{T^{2'}} = \sqrt{\frac{I}{I'}} \times \sqrt{\frac{M'}{M}} = \sqrt{\frac{mL^2}{m'L'^2}} \times \sqrt{\frac{L'}{L}}$$

Here, m and m' are the masses of the magnet and L and L's are their lengths.

$$= \sqrt{\frac{m}{m'}} \times \sqrt{\frac{L}{L'}} = \sqrt{\frac{m}{m/2}} \times \sqrt{\frac{L}{L/2}} = \sqrt{2} \times \sqrt{2} = 2$$

$$\Rightarrow \quad T' = \frac{T}{2} = \frac{4}{2} = 2 \text{ s}$$

35. *(b)* Let length of the original spring is L.

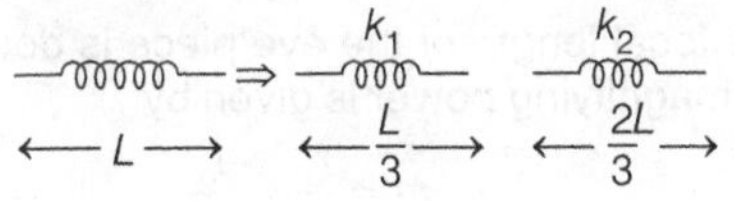

As we know that,

Spring constant, $k \propto \frac{1}{\text{length}}$

$$\Rightarrow \quad k \propto \frac{1}{L}$$

$$k_1 \propto \frac{3}{L}$$

$$k_2 \propto \frac{3}{2L}$$

$$\Rightarrow \quad \frac{k_2}{k} = \frac{3/2L}{1/L} = \frac{3}{2}$$

$$\Rightarrow \quad k_2 = \frac{3}{2}k$$

36. *(d)* Number of half-lives in 40 min

$$n = \frac{40}{T_{1/2}} = \frac{40}{10} = 4$$

Amount of the substance remaining after n half- lives

$$N = N_0\left(\frac{1}{2}\right)^n = N_0\left(\frac{1}{2}\right)^4 = \frac{N_0}{16}$$

where, N_0 is original amount of the substance.

Amount of the substance decayed in 40 min.

$$N' = N_0 - N = N_0 - \frac{N_0}{16} = \frac{15}{16}N_0$$

$$= \left(\frac{15}{16} \times 100\%\right) \text{ of } N_0$$

$$= 15 \times 6.25\ \% \text{ of } N_0$$

$$= 93.75\ \% \text{ of } N_0$$

37. *(c)* For isothermal expansion of an ideal gas

$$pV = \text{constant}$$

$$\Rightarrow \quad p\Delta V + V\Delta p = 0$$

For container A,

$$p_{iA}(V) + V(\Delta p) = 0$$

$$\Rightarrow \quad p_{iA} = -\frac{V\Delta P}{V} = -\Delta p \qquad \ldots(\text{i})$$

where, p_i is initial pressure.

For container B,

$$p_{iB}(V) + V(1.5\,\Delta p) = V$$

$$\Rightarrow \quad p_{iB} = -1.5\,\Delta p \qquad \ldots(\text{ii})$$

From Eqs. (i) and (ii), we have

$$p_{iA} = \frac{p_{iB}}{1.5} = \frac{2}{3}p_{iB}$$

$$\Rightarrow \quad m_A = \frac{2}{3}mB$$

$$\Rightarrow \quad 3m_A = 2mB \qquad [\because p \propto m]$$

38. *(b)* For monoatomic gas,

$$\gamma_1 = \frac{C_{p_1}}{C_{V_1}} = \frac{5}{3} \text{ and } C_{p_1} - C_{V_1} = R$$

this gives $C_{p_1} = \frac{5}{2}R$ and $C_{V_1} = \frac{3}{2}R$

For diatomic gas,

$$\gamma_2 = \frac{C_{p_2}}{C_{V_2}} = 7/5 \text{ and } C_{p_2} - C_{V_2} = R$$

$$\Rightarrow \quad C_{p_2} = \frac{7}{2}R \text{ and } C_{V_2} = \frac{5}{2}R$$

Now, we can write,

$$\gamma_{mix} = \frac{n_1 C_{p_1} + n_2 C_{p_2}}{n_1 C_{V_1} + n_2 C_{V_2}}$$

$$= \frac{C_{p_1} + C_{p_2}}{C_{V_1} + C_{V_2}} = \frac{\frac{5}{2}R + \frac{7}{2}R}{\frac{3}{2}R + \frac{5}{2}R} = \frac{12}{8} = \frac{3}{2} = 1.5$$

39. *(d)* Applying conservation of mechanical energy, we can write

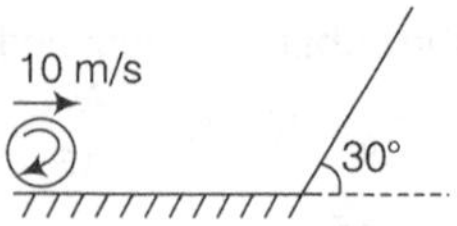

$$\frac{1}{2}I\omega^2 + \frac{1}{2}mv^2 = mgh$$

where, h is maximum height reached by the sphere, v is linear velocity and ω is angular velocity of the sphere.

$$\frac{1}{2}I\left(\frac{v}{r}\right)^2 + \frac{1}{2}mv^2 = mgh \qquad [\because v = \omega r]$$

$$\Rightarrow \quad \frac{1}{2} \times \frac{2}{5}mr^2 \times \frac{v^2}{r^2} + \frac{1}{2}mv^2 = mgh$$

$$\frac{1}{5}mv^2 + \frac{1}{2}mv^2 = mgh$$

$$\frac{7}{10}mv^2 = mgh$$

$$h = \frac{7}{10} \times \frac{v^2}{g} = \frac{7}{10} \times \frac{10^2}{9.8} = \frac{70}{9.8}$$

$$= \frac{100}{14} = \frac{50}{7} = 7.1 \text{ m.}$$

40. *(b)* Suppose depth inside the earth is h.

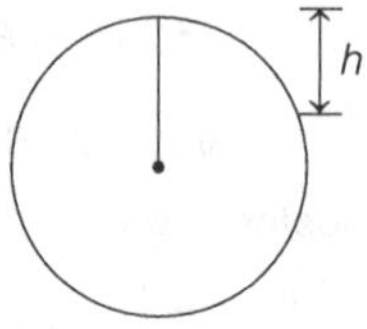

As we know that, inside the earth

$$g_i \propto \left(\frac{R-h}{R^3}\right) \qquad \ldots\text{(i)}$$

On the surface of the earth

$$g_s \propto \frac{1}{R^2} \qquad \ldots\text{(ii)}$$

From Eqs. (i) and (ii), we get

$$\frac{g_i}{g_s} = \frac{(R-h)R^2}{R^3} = \frac{R-h}{R}$$

$$\Rightarrow \quad \frac{1}{4} = \frac{R-h}{R}$$

$$\Rightarrow \quad R = 4R - 4h \Rightarrow 3R = 4h$$

$$\Rightarrow \quad h = \frac{3}{4}R.$$

41. *(c)* Range of a projectile is maximum for angle of projection of 45°.

The nearest value is 43°. Thus, among given values of angles of projection, range will be maximum for 43°.

42. *(c)* Resistance of a wire is given by

$$R = \frac{\rho L}{A}$$

where, ρ = resistivity, L = length of the wire

A = area of cross- section of the wire

Equating initial and finial volumes of the wire, we can write

$$AL = A'L' = A' \times (2L) \Rightarrow A' = \frac{A}{2}$$

$$\therefore \quad \frac{R'}{R} = \frac{\rho L'/A'}{\rho L/A} = \left(\frac{L'}{L}\right)\left(\frac{A}{A'}\right)$$

$$= \left(\frac{2L}{L}\right)\left(\frac{A}{A/2}\right) = 4$$

$$\Rightarrow \quad R' = 4R = 4(5) = 20\,\Omega.$$

43. *(d)* Magnifying power of a telescope is given by

$$m = \frac{f_o}{f_e}$$

If the focal length of the eye piece is doubled, the new magnifying power is given by

$$m' = \frac{f_o}{f_e'} = \frac{f_o}{2f_e}$$

$$\therefore \quad m' = \frac{m}{2}$$

44. *(d)* Given $V = 250$ V and $P = 500$ W

As we know that,

$$P = \frac{V^2}{R}$$

$$\Rightarrow \quad R = V^2/P = \frac{250 \times 250}{500}$$

$$= \frac{625}{5} = 125\,\Omega.$$

45. *(a)* As the battery remains connected. potential difference (V) of the capacitor remains constant. Suppose dielectric of dielectric constant K is introduced between the plates. Now, capcitance of the capacitor is

$$C' = KC_0$$

where, C_0 is original capacitance.

Initial energy stored in the capacitor

$$U_i = \frac{1}{2} C_0 V^2$$

Final energy stored in the capacitor

$$U_f = \frac{1}{2} C' V^2 = \frac{1}{2} (K C_0) V^2$$

Clearly, $\frac{U_f}{U_i} = K > 1 \Rightarrow U_f > U_i$

As, $Q = CV$

$\therefore$ V remains constant and C increases.

$\therefore$ Q increases .

46. *(a)* Power of the combination

$$P = P_1 + P_2$$

$$= P_{\text{concave}} + P_{\text{convex}}$$

$$\Rightarrow \quad \frac{100}{50} = P_{\text{concave}} + \frac{100}{20}$$

or $\quad 2 = P_{\text{concave}} + 5$

or $\quad P_{\text{concave}} = 2 - 5 = -3.0\text{ D}$

47. *(a)* Critical angle (θ_c) for glass water interface can be written as

$$\text{Sin } \theta_c = \frac{{}_a\mu_w}{{}_a\mu_g} = \frac{4/3}{3/2} = \frac{8}{9}$$

$$\Rightarrow \quad \theta_c = \sin^{-1}(8/9)$$

48. *(a)* In the given circuit, the diode is reverse biased. Only drift currant flows through the diode.

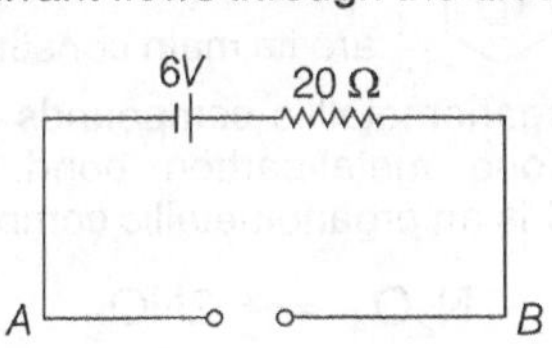

Consider the circuit in which by applying KVL, we can write

$$V_A - 6 + I \times 20 = V_B$$

$$\Rightarrow \quad V_A - V_B = 6 - I \times 20 = 6 - 30 \times 10^{-6} \times 20$$

$$= 6 - 0.006 = -5.9994\text{ V}$$

49. *(d)* Given, power $P = 3\text{D}$

$$f = \frac{1}{P} = \frac{1}{3}\text{ m} = \frac{100}{3}\text{ cm.}$$

$$u = -25\text{ cm.}$$

Applying lens formula, we can write

$$\frac{1}{v} - \frac{1}{u} = \frac{1}{f}$$

$$\Rightarrow \quad \frac{1}{v} + \frac{1}{25} = \frac{3}{100}$$

$$\Rightarrow \quad \frac{1}{v} = \frac{3}{100} - \frac{1}{25} = \frac{3-4}{100} = \frac{-1}{100}$$

$$v = -100\text{ cm.}$$

Now, again using lens formula we can write

$$\frac{1}{v} - \frac{1}{u} = \frac{1}{f}$$

As, focal length of a far sighted person deceases

$$\Rightarrow \quad \frac{1}{-100} - \frac{1}{u} > \frac{1}{1.7} = \frac{10}{17} \qquad [\because f < 1.7\text{ cm}]$$

$$\Rightarrow \quad \frac{1}{u} < \frac{-1}{100} - \frac{10}{17}$$

$$= \frac{-17 - 1000}{1700} = \frac{-1017}{1700}$$

$$\Rightarrow u > \frac{1700}{1000} \approx -1.7 = -1\text{ m (nearest option)}$$

50. *(a)* Force per unit length between the two parallel current carrying wires

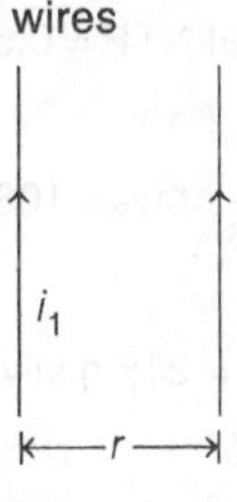

$$\frac{F}{L} = \frac{\mu_0}{2\pi r} i_1 i_2$$

$$= \frac{\mu_0}{2\pi\,(2d)} i_1 i_2 = \frac{\mu_0 i_1 i_2}{4\pi d} \qquad [\because r = 2d]$$

Chemistry

1. *(a)*

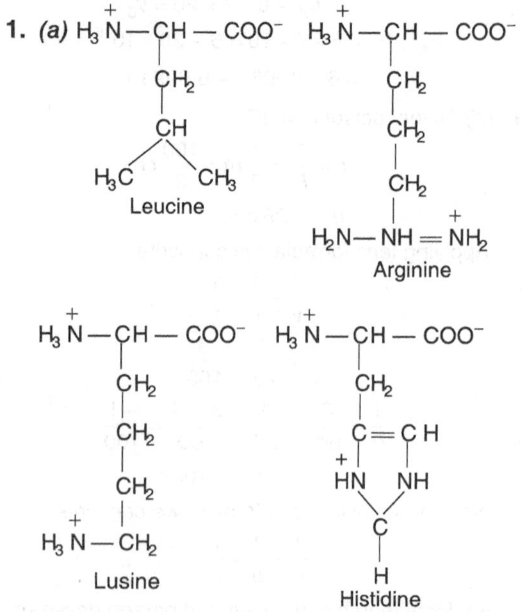

Leucine, Arginine, Lusine, Histidine

Leucine is not a basic amino acid while others three i.e., arginine, lysine and histidine are basic amino acids as they have more number of amino groups than carboxyl groups.

2. *(c)* In qualitative analysis, NH_4Cl is added before NH_4OH to decrease $[OH^-]$ concentration. NH_4Cl suppresses the ionisation of NH_4OH due to common ion effect. Otherwise basic radicals of group V and group VI will be precipitated along with IIIrd group basic radicals.

3. *(a)* 11.2 L H_2 gas at NTP = 0.5 mole of H_2 gas =1 g of H_2 g

$$\frac{w_1}{w_2}=\frac{E_1}{E_2}=\frac{1}{w_2}=\frac{1}{108} \text{ or } w_2 = 108\text{g silver is deposited}$$

in half an hour.

Therefore, $108 \times 2 = 216$ g silver will be deposited in one hour.

4. *(c)* A. $[Co(NH_3)_6]^{+3}$ inner orbital (d^2sp^3) complex

B. $[Fe(CN)_6]^{4-}$ inner orbital (d^2sp^3) complex

C. $[Ni(NH_3)_6]^{2+}$ outer orbital (sp^3d^2) complex

D. $[Mn(CN)_6]^{2-}$ inner orbital (d^2sp^3) complex

5. *(a)* For a cyclic process, $\Delta E = 0$. In cyclic process, system returns to its original state after a number of reactions.

6. *(b)* **Step I.** **Step II.**

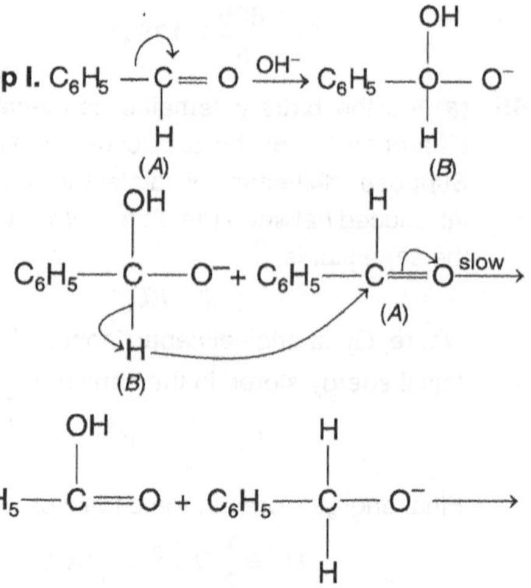

$C_6H_5COO^- + C_6H_5CH_2OH$

H^- is obtained by fission of C — H bond of (*B*). It is transferred to (II). This is the slowest step.

7. *(d)* $R = k[A]^x$ ...(i)

$$2R = k[4A]^x \quad \text{...(ii)}$$

Divide Eq. (ii) by (i)

$$2 = [4]^x$$

$$R = k[A]^{1/2}$$

$$x = \frac{1}{2} = 0.5$$

$$R = k(400\ A)^{1/2}$$

$$\text{Rate} = 20k[A]^{1/2}$$

Rate becomes 20 times.

8. *(a)* Diesel is fraction of petroleum having $C_{15} - C_{18}$ composition and boiling point in the range of 573 – 673 K. Generally cetane $[CH_3 - (CH_2)_{14} - CH_3]$ and α-methyl naphthalene

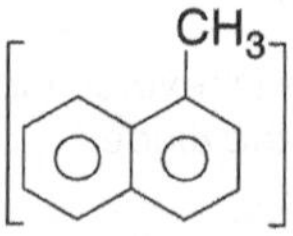

are its main constituents.

9. *(b)* Organometalllic compounds must contain, at least one metal-carbon bond. Methyl lithium, CH_3 Li is an organometallic compound.

10. *(b)* $N_2O_4 \rightleftharpoons 2NO_2$

	N_2O_4	$2NO_2$
Initial	1	0
At. equil.	1	0

Total number of moles at equilibrium

$$= (1 - \alpha) + 2\alpha = 1 + \alpha$$

11. *(b)*

$$CHCl_3 \xrightarrow[h\nu]{[O]} COCl_2 \xrightarrow{NH_3} NH_2CONH_2$$

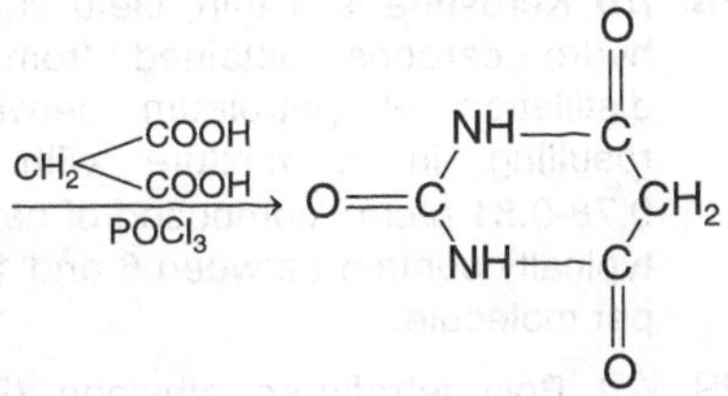

Barbituric acid

Barbituric acid and its derivatives are used in medicines as hypnotics and sedatives.

12. *(a)*

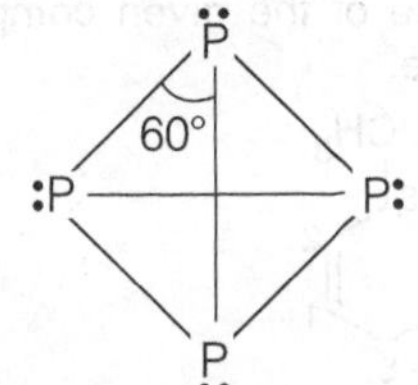

White phosphorus (P_4)
It has six P — P single bonds.

13. *(a)* Natural rubber is a linear polymers of isoprene and is also called as *cis*-1, 4-polyisoprene.

It is a linear 1, 4-polymer of isoprene (2-methyl-1,3-butadiene).

$$nCH_2=\underset{|}{\overset{CH_3}{C}}-CH=CH_2 \xrightarrow[\text{polymerisation}]{\text{1, 4 addition}}$$

isoprene

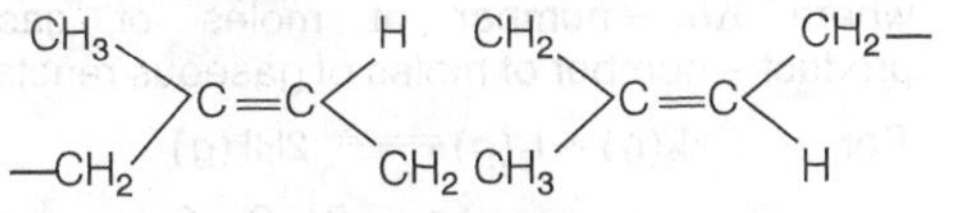

Natural rubber

14. *(a)*

A.	XeF_4	sp^3d^2
B.	H_2O	sp^3
C.	PCl_5	sp^3d
D.	$[Pt(NH_3)_4]^{2+}$	dsp^2

15. *(d)* A new carbon- carbon bond formation takes place in Friedel-Craft reaction and Reimer-Tiemann reaction.

16. *(b)* $C_6H_5CHO + (CH_3CO)_2O \xrightarrow[453\ K]{CH_3COONa}$

$$C_6H_5CH=CH\overset{O}{\overset{\|}{C}}-O-\overset{O}{\overset{\|}{C}}-CH_3 \xrightarrow[-CH_3COOH]{H_2O/H^+}$$

$$C_6H_5\ CH=CHCOOH$$

Cinnamic acid

17. *(a)* $\lambda = \dfrac{h}{mv} = \dfrac{h}{\sqrt{2m \times (KE)}}$

$$\lambda_1 = \frac{h}{\sqrt{2m \times 200}}$$

$$\lambda_2 = \frac{h}{\sqrt{2m \times 50}}$$

$$\frac{\lambda_1}{\lambda_2} = \frac{1 \times \sqrt{50}}{\sqrt{200} \times 1} = \frac{1}{2}$$

$$\lambda_1 : \lambda_2 = 1:2$$

18. *(c)* $CuCO_3 \cdot Cu(OH)_2 \longrightarrow \underset{(A)}{Cu_2O} + CO_2 + H_2O$

$$Cu_2O + C \longrightarrow \underset{(B)}{2Cu} + CO$$

19. *(b)*

	Formula	Name of the compound
A.	$NaNO_3$	Chile salt petre
B.	$Na(NH_4)HPO_4$	Microcosmic salt
C.	$NaHCO_3$	Baking soda
D.	$Na_2CO_3 \cdot 10H_2O$	Washing soda

20. *(b)* $2MnO_2 + 4HNO_3 \longrightarrow 2Mn(NO_3)_2 + 2H_2O + O_2$

$$2Mn(NO_3)_2 + 5PbO_2 + 6HNO_3 \longrightarrow \underset{\text{Permanganic acid (violet pink)}}{2HMnO_4} + 5Pb(NO_3)_2 + 2H_2O$$

21. *(c)* Edge length of (NaCl crystal) unit cell

$$= 2(r_{Na^+} + r_{Br^-})$$

22. *(d)* Magnetic moment $\propto$ no. of unpaired electron.

$Zn^{2+} = 3d^{10}4s^0$ unpaired electron=0;

$$\mu = 0$$

$Ti^{3+} = 3d^{10}4s^0$ unpaired electron=1

$$\mu = \sqrt{n(n+2)} = 1.73$$

$Sc^{3+} = 3d^0 4s^0$ unpaired electron=0;

$\mu = 0$

$Mn^{2+} = 3d^5 4s^0$ unpaired electrons =5

$\mu = \sqrt{5(5+2)} = 5.91$

23. *(a)* Order of solubility of fluorides of alkaline earth metals in water is

Halides of alkaline earth metals (except Be) are ionic solids and are therefore are water soluble and their solubility in water decreases from Mg to Ba due to the decrease in the hydration energy.

However, fluorides of alkanline earth metals excepts BeF_2 are almost insoluble in water.

24. *(c)*

Element	% amount	Atomic mass	Mole ratio	Simplest ratio
C	20	12	$\frac{20}{12} = 1.66$	$\frac{1.66}{1.66} = 1$
N	46.66	14	$\frac{46.66}{14} = 3.33$	$\frac{3.33}{1.66} = 2$
H	6.66	1	$\frac{6.66}{1} = 6.66$	$\frac{6.66}{1.66} = 4$
O	26.68	16	$\frac{26.68}{16} = 1.66$	$\frac{1.66}{1.66} = 1$

$\therefore$ Empirical formula $= CN_2H_4O$

25. *(a)* $KI + I_2 \longrightarrow KI_3$ or $I^- + I_2 \longrightarrow I_3^-$
from KI

On increasing concentration of KI and I_2 two fold and three fold respectively, the concentration of KI_3 becomes two fold.

26. *(b)* $\underset{(C_{12}H_{22}O_{11})}{(+)\text{-Sucrose}} \xrightarrow{HOH/H^+} \underset{(C_6H_{12}O_6)}{D(+)\text{-glucose}} + \underset{(C_6H_{12}O_6)}{D(-)\text{-fructose}}$

$\underset{(C_{12}H_{22}O_{11})}{\text{Maltose}} \xrightarrow{HOH/H^+} \text{Glucose} + \text{Glucose}$

$\underset{(C_{12}H_{22}O_{11})}{\text{Lactose}} \xrightarrow{HOH/H^+} D(+)\text{-glucose}$
$+ D(+)\text{-galactose}$

Hence, A = Sucrose, B = Maltose, C = Lactose

27. *(a)* A decay process is possible only if accompanied by the release of energy. There are three different decay processes that increase the n/p ratio. These are : (i) electron capture, (ii) position emission (β^+- emission) and (iii) alpha particle decay.

28. *(b)* Kerosene is a thin, clear liquid formed from hydro carbons obtained from the fractional distillation of petroleum between 150-275°C resulting in a mixture with a density of 0.78-0.81 g/cm^3 composed of carbon chains that typically contain between 6 and 16 carbon atoms per molecule.

29. *(c)* Poly tetrafluoro ethylene (PTFE) or teflon $-\!\!\left(CF_2 - CF_2\right)\!\!_n-$ is an addition homopolymer. It is used as lubricant, insulator and in making cooking wares.

30. *(a)* The IUPAC name of the given compound is 3-methyl cyclohexene.

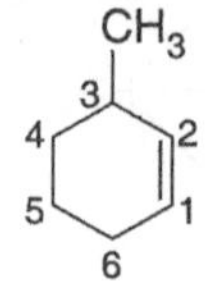

31. *(d)* The correct representation of a complexionian is $[Co(NH_3)_4(H_2O)Cl]^{2+}$.

32. *(d)* Atoms and ions having inert gas configuration (such as helium), i.e. stable (half-filled) configuration have high ionisation potential, i.e. it is relatively easy to remove an electron from a partially filled valence shell, where Z_{eff} is lower but it is relatively difficult to remove an electron from an atom or ion that has a filled valence shell, where Z_{eff} is higher.

33. *(c)* $\Delta H = \Delta E + \Delta n_g RT$

where Δn_g = number of moles of gaseous product – number of moles of gaseous reactant

For $\quad H_2(g) + I_2(g) \rightleftharpoons 2HI(g)$

$\Delta n_g = 2 - 2 = 0$

Thus, $\quad \Delta H = \Delta E + 0 \times RT$ or $\Delta H = \Delta E$

34. *(b)* Calcination is used in metallurgy proceeds only with the expulsion of some small molecules like water, CO_2, SO_2 etc. without any chemical changes. e.g.

$2Al(OH)_3 \xrightarrow{1500°C} Al_2O_3 + 3H_2O \uparrow$

$FeCO_3 \xrightarrow{\Delta} FeO + CO_2 \uparrow$

35. *(b)* $[Ni(CN)_4)]^x \equiv [Ni(CN)_4]^{2-}$

In this complex, the oxidation state of nickel is +2 and that of (CN) is – 1.

$\therefore \quad x = +2 + y(-1) = -2$

36. *(d)* From Arrhenius equation,

$$\log k = \log A - \frac{E_a}{2.303\,RT}$$

As the activation energy of a reaction is zero, the rate constant of this reaction becomes zero. It shows that the reaction is independent of temperature.

37. *(c)* I. $H_2S_3O_6$– trithionic acid is an example of polythionic acid.

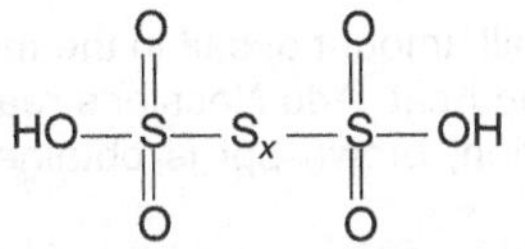

III. H_2S_4 or HS—S—S—SH is an hydrosulfane.

The compounds (II) and (IV) do not exist.

38. *(b)* ${}^{13}_{7}N \xrightarrow{-_2He^4} {}^{10}_{5}B$

${}^{242}_{96}Cm \xrightarrow{-_0n^1} {}^{244}_{94}Pu$

${}^{14}_{7}N \xrightarrow{-_0n^1} {}^{14}_{6}C$

${}^{30}_{15}P \xrightarrow[-g]{-_1H^1} {}^{28}_{14}Si$

39. *(b)* When cold H_2O_2 is added in cold mixture of $K_2Cr_2O_7$ and conc. H_2SO_4 (i.e. $K_2Cr_2O_7$ in acidic medium), a deep blue solution of blue perchromate is obtained as,

$$K_2Cr_2O_7 + H_2SO_4 \longrightarrow K_2SO_4 + H_2Cr_2O_7$$

$$H_2Cr_2O_7 + 4\,H_2O_2 \longrightarrow \underset{\text{Blue perchromate}}{2CrO_5} + 5\,H_2O$$

40. *(d)* In blast furnace the highest temperature is in combustion zone. In the first stage, coke burns to produce CO_2 which rises temperature.

$$C + O_2 \longrightarrow CO_2 + \text{Heat}$$

This reaction is exothermic and hence temperature is 2170 K. It is known as combustion zone.

41. *(b)* $HgCl_2$ distinguishes formic acid and acetic acid. Formic acid produces a white precipitate with $HgCl_2$ solution. Acetic acid does not give this test.

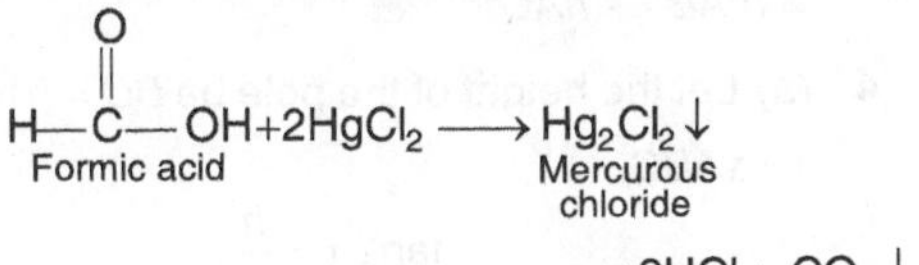

$+2HCl + CO_2\downarrow$

42. *(d)* (i) Fluorine required energy when an electron is removed, i.e. ionisation energy.

(ii) Fluorine releases energy to add one electron, i.e. electron affinity ($-$ 328 kJ mol^{-1})

43. *(b)* Liquid hydrocarbon is converted into mixture of gaseous hydrocarbons by oxidation oxidation can be complete or partial. Complete oxidation is called combustion which is shown by the generalised reaction as

$$C_nH_{2n+2} + \left(\frac{3n+1}{2}\right) O_2 \xrightarrow{\text{Flame}} nCO_2 + (n+1)\,H_2O + \text{Heat}$$

44. *(b)* A $\rightarrow$ 2 B $\rightarrow$ 1 C $\rightarrow$ 3 D $\rightarrow$ 4

Cannizaro reaction – NaOH

Stephan's reaction – $SnCl_2$ / HCl

Clemmensen reduction – Zn/Hg –Conc. HCl

Rosenmund's method – $\frac{\text{Pd - BaSO}_4}{\text{Boiling xylene}}$

45. *(a)* Atomic raddii as well as ionic radii increases down the group. All the element of the nitrogen family contain same charge, i.e. – 3.

Thus, the correct order of ionic radius of nitrogen family is $N^{3-} < P^{3-} < As^{3-} < Sb^{3-} < Bi^{3-}$

46. *(d)* Chlorophyll is responsible for the synthesis of glucose (carbohydrate) in plants.

In presence of oxygen, haemoglobin forms oxyhemoglobin.

Acetyl salicylic acid is known as aspirin

Vitamin B_{12} (Cyanocobalamin, $C_{63}H_{88}O_4N_{14}PCo$) contains Co^{2+} ion.

47. *(a)* Strength of acid is indicated by pK_a value. Higher the value of K_a or lower the value of pK_a, stronger is the acid. Among the given aromatic acids, the strength decreases as follows

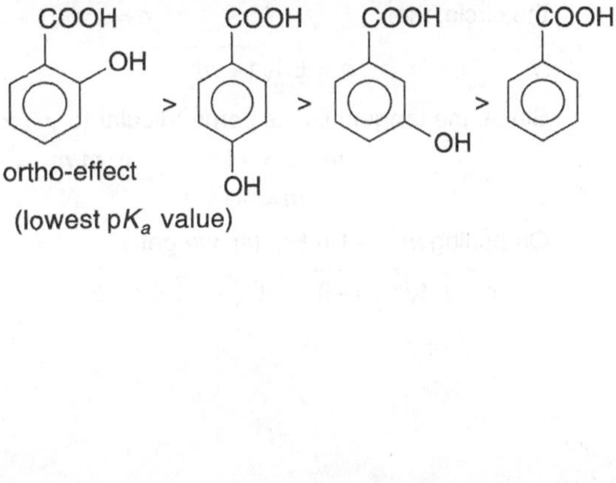

ortho-effect

(lowest pK_a value)

***Ortho*-effect** The effect of a group is maximum at *ortho*-position due to nearness, is called *ortho* effect.

48. *(a)*

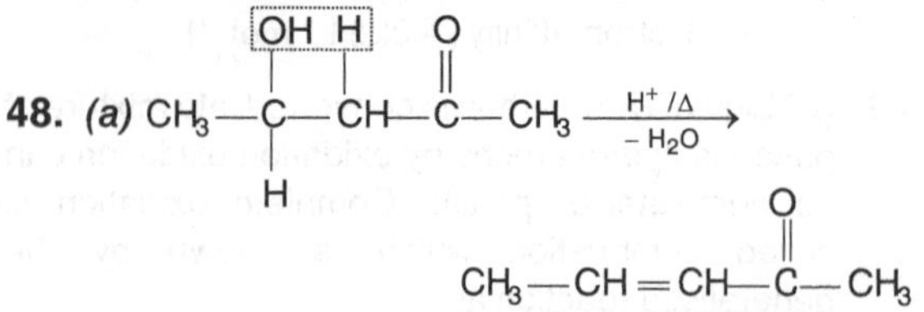

α, β-unsaturated ketones

Among the given compounds, the compound present in option (a) is readily dehydrated in acidic medium due to presence of acidic hydrogen with — OH group, In option (d), the strength of acidic hydrogen is less.

49. *(a)* The chemical substances which are used to bring down body temperature during high fever are called antipyretics,

e.g. paracetamol, aspirin, phenacetin etc.

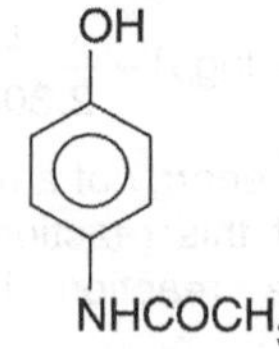

Paracetamol
(4-acetamidophenol)

50. *(d)* Analysis of NH_4^+ cation

Take small amount of salt in the test tube and add NaOH and heat. Add Nessler's reagent $[K_2H_gI_4]$ to the solution, brown ppt is obtained. It confirmed NH_4^+ ion.

$NH_3 + 3NaOH + 2K_2[HgI_4] \longrightarrow$

Nessler's reagent

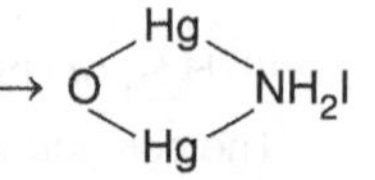

Iodide of million'sbase
(brown ppt.)

$+4KI + 3NaI + 2H_2O$

Mathematics

1. *(c)* Given, $P(A \cup B) = P(A \cap B)$

$\Rightarrow \quad P(A) + P(B) - P(A \cap B) = P(A \cap B)$

$\Rightarrow \quad P(A) + P(B) = 2P(A \cap B)$

$\Rightarrow \quad P(A) + P(B) = 2 \times P(A)\, P(B/A)$

$$\left[\because P(B/A) = \frac{P(A \cap B)}{P(A)}\right]$$

2. *(*)* Let the equation of tangent be

$$y = mx + c \qquad \ldots(i)$$

Given equation of circle is

$$x^2 + y^2 - 2x - 2y + 1 = 0$$

Its centre is (1, 1)

and radius, $r = \sqrt{1^2 + 1^2 - 1} = \sqrt{1 + 1 - 1} = 1$

We know that, if line $y = mx + c$ be a tangent to the circle, then $\quad c = \pm r\sqrt{1 + m^2}$

$\therefore \quad c = \pm 1\sqrt{1 + m^2} \qquad \ldots(ii)$

Since, the tangent line is perpendicular to $y = x$.

$\therefore \quad m \times 1 = -1 \qquad (\because m_1 m_2 = -1)$

$\Rightarrow \quad m = -1$

On putting $m = -1$ in Eq. (ii), we get

$$c = \pm 1\sqrt{1 + (-1)^2} = \pm\sqrt{1 + 1} = \pm\sqrt{2}$$

On putting the values of $m = -1$ and $c = \pm\sqrt{2}$ in Eq. (i), we get

$$y = -x \pm \sqrt{2}$$

3. *(c)* $mBD^2 + nCD^2 + (m + n)AD^2$

$= mBD^2 + nCD^2 + mAD^2 + nAD^2$

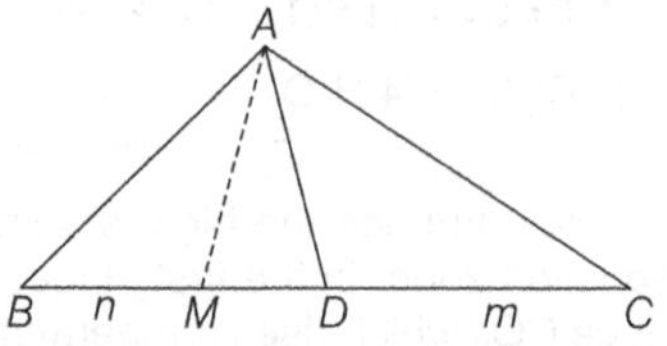

$= m(BD^2 + AD^2) + n(CD^2 + AD^2)$

$= m(AB^2 + 2DM \cdot DB) + n(AC^2 - 2DM \cdot DC)$

$= mAB^2 + m \cdot 2DM \cdot DB + nAC^2 - n \cdot 2DM \cdot DC$

$= mAB^2 + nAC^2 + 2DM(mDB - nDC)$

$= mAB^2 + nAC^2 + 2DM(mn - nm)$

$= mAB^2 + nAC^2$

4. *(a)* Let the height of the pole be $BC = h$ m.

In ΔABC,

$$\tan 2\alpha = \frac{h}{x} \qquad \ldots(i)$$

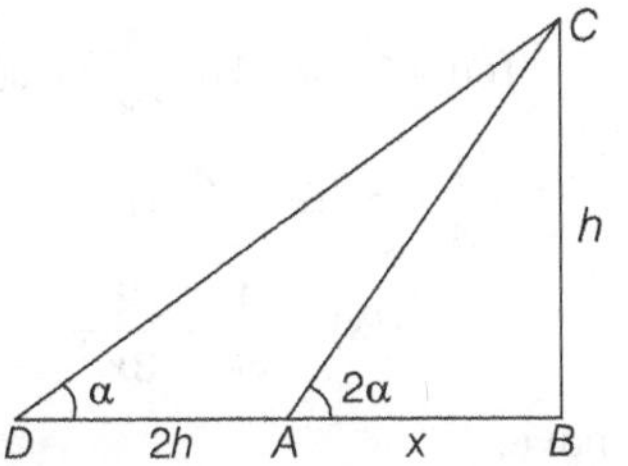

and in ΔDBC,

$$\tan\alpha = \frac{h}{2h + x}$$

$$\Rightarrow \quad \tan\alpha = \frac{\frac{h}{x}}{2\left(\frac{h}{x}\right) + 1}$$

$$\Rightarrow \quad \tan\alpha = \frac{\tan 2\alpha}{2\tan 2\alpha + 1}$$

$$\Rightarrow \quad \tan\alpha\,(2\tan 2\alpha + 1) = \tan 2\alpha$$

$$\Rightarrow \tan\alpha\left[2 \times \frac{2\tan\alpha}{1 - \tan^2\alpha} + 1\right] = \frac{2\tan\alpha}{1 - \tan^2\alpha}$$

$$\Rightarrow \quad [4\tan\alpha + 1 - \tan^2\alpha] = 2$$

$$\Rightarrow \quad \tan^2\alpha - 4\tan\alpha + 1 = 0$$

$$\Rightarrow \quad \tan\alpha = \frac{+4 \pm \sqrt{16 - 4}}{2 \times 1}$$

$$= \frac{4 \pm \sqrt{12}}{2} = \frac{4 \pm 2\sqrt{3}}{2}$$

$$\Rightarrow \quad \tan\alpha = 2 \pm \sqrt{3}$$

Taking '–' sign, we get

$$\Rightarrow \quad \tan\alpha = 2 - \sqrt{3}$$

$$\Rightarrow \quad \tan\alpha = \tan 15°$$

$$\Rightarrow \quad \alpha = 15° \text{ or } \frac{\pi}{12}$$

5. *(d)* Given, $f(x) = \begin{cases} \frac{\sin[x]}{[x]} & \text{for}[x] \neq 0 \\ 0 & \text{for}[x] = 0 \end{cases}$

$$\text{LHL} = \lim_{x \to 0^-} f(x) = \lim_{x \to 0^-} \frac{\sin(1 + [x])}{[x]}$$

$$= \frac{\sin[1 - 1]}{-1} = \frac{\sin 0}{-1} = 0$$

$$\text{RHL} = \lim_{x \to 0^+} f(x) = \lim_{x \to 0^+} \frac{\sin(1 + [x])}{[x]}$$

$$= \frac{\sin(1 + 0)}{0} = \infty$$

Hence, limit does not exist.

6. *(b)* Given equation of straight lines are

$$y = (2 - \sqrt{3})x + 5$$

and $$y = (2 + \sqrt{3})x - 7$$

On comparing with $y = mx + c$, we get

$$m_1 = 2 - \sqrt{3}$$

and $$m_2 = 2 + \sqrt{3}$$

$$\therefore \quad \tan\theta = \frac{m_2 - m_1}{1 + m_1 m_2}$$

$$= \frac{2 + \sqrt{3} - (2 - \sqrt{3})}{1 + (2 - \sqrt{3})(2 + \sqrt{3})}$$

$$= \frac{2\sqrt{3}}{1 + (4 - 3)} = \frac{2\sqrt{3}}{1 + 4 - 3}$$

$$\Rightarrow \quad \tan\theta = \frac{2\sqrt{3}}{2} = \sqrt{3}$$

$$\Rightarrow \quad \theta = 60°$$

7. *(c)* Equation of plane passing through $(-1, 1, 1)$ is

$$a(x + 1) + b(y - 1) + c(z - 1) = 0 \qquad \text{...(i)}$$

Also, it is passing through $(1, -1, 1)$.

$$\therefore \quad a(1 + 1) + b(-1 - 1) + c(1 - 1) = 0$$

$$\Rightarrow \quad 2a - 2b + 0c = 0 \qquad \text{...(ii)}$$

Also, required equation of plane (i) is perpendicular to $x + 2y + 2z = 5$.

$$\therefore \quad a \times 1 + b \times 2 + c \times 2 = 0$$

$$\Rightarrow \quad a + 2b + 2c = 0 \qquad \text{...(iii)}$$

Eqs. (ii) and (iii) are identical.

$$\therefore \quad \frac{a}{-4 - 0} = \frac{-b}{4 - 0} = \frac{c}{4 + 2}$$

$$\Rightarrow \quad \frac{a}{-4} = \frac{b}{-4} = \frac{c}{6}$$

$$\Rightarrow \quad \frac{a}{-2} = \frac{b}{-2} = \frac{c}{3} = \lambda \qquad \text{(say)}$$

$$\Rightarrow \quad a = -2\lambda, b = -2\lambda, c = 3\lambda$$

On putting the values of a, b and c in Eq. (i), we get

$$-2\lambda(x + 1) - 2\lambda(y - 1) + 3\lambda(z - 1) = 0$$

$$\Rightarrow \quad \lambda[-2x - 2 - 2y + 2 + 3z - 3] = 0$$

$$\Rightarrow \quad -2x - 2y + 3z - 3 = 0$$

$$\Rightarrow \quad 2x + 2y - 3z + 3 = 0$$

8. *(d)* Given, $g(x) = f(\tan^2 x - 2\tan x + 4)$

On differentiating w.r.t. x, we get

$$g'(x) = f'(\tan^2 x - 2\tan x + 4) \times (2\tan x \sec^2 x - 2\sec^2 x)$$

$$= f'(\tan^2 x - 2\tan x + 4)\, 2\sec^2 x\,(\tan x - 1)$$

$\because \quad f'(x) > 0 \; \forall x \in \left(0, \frac{\pi}{2}\right)$

Also, $2\sec^2 x > 0 \; \forall x \in \left(0, \frac{\pi}{2}\right) \left[\because x \in \left(0, \frac{\pi}{2}\right), \text{given}\right]$

But $\tan x - 1 > 0 \; \forall x \in \left(\frac{\pi}{4}, \frac{\pi}{2}\right)$

$\therefore \quad g'(x) > 0 \; \forall x \in \left(\frac{\pi}{4}, \frac{\pi}{2}\right)$

Hence, $g(x)$ is increasing in $\left(\frac{\pi}{4}, \frac{\pi}{2}\right)$.

9. *(a)* The total number of sample points in a sample space of single throw of two dice, $n(S) = 36$

Let E_1 = Event of getting a sum 7

E_2 = Event of getting a sum 9

$\therefore \quad E_1 = \{(1, 6), (6, 1), (2, 5), (5, 2), (3, 4), (4, 3)\}$

$\Rightarrow \quad n(E_1) = 6$

and $\quad E_2 = \{(3, 6), (6, 3), (4, 5), (5, 4)\}$

$\Rightarrow \quad n(E_2) = 4$

$\therefore \quad P(E_1) = \frac{n(E_1)}{n(S)} = \frac{6}{36} = \frac{1}{6}$

and $\quad P(E_2) = \frac{n(E_2)}{n(S)} = \frac{4}{36} = \frac{1}{9}$

Now, $P(E_1 \cup E_2) = P(E_1) + P(E_2)$

$= \frac{1}{6} + \frac{1}{9} = \frac{6+4}{36} = \frac{10}{36} = \frac{5}{18}$

10. *(b)* Compiler is a system software which transforms source code written in a programming language into another computer language.

11. *(a)* Given, $\lim_{t \to x} \frac{t^2 f(x) - x^2 f(t)}{t - x} = 1$

Using L' Hospital's rule, we get

$$\lim_{t \to x} \frac{t^2 f'(x) - 2xf(t)}{-1} = 1$$

$$\Rightarrow \quad x^2 f'(x) - 2x f(x) + 1 = 0$$

$$\Rightarrow \quad \frac{x^2 f'(x) - 2xf(x)}{(x^2)^2} + \frac{1}{x^4} = 0$$

$$\Rightarrow \quad \frac{d}{dx}\left(\frac{f(x)}{x^2}\right) = -\frac{1}{x^4}$$

On integrating both sides, we get

$$\frac{f(x)}{x^2} = +\frac{1}{3x^3} + c \Rightarrow f(x) = \frac{1}{3x} + cx^2$$

Also, $\quad f(1) = 1 \Rightarrow 1 = \frac{1}{3 \times 1} + c(1)^2$

$$\Rightarrow \quad 1 = \frac{1}{3} + c \Rightarrow \frac{2}{3} = c$$

$$\therefore \quad f(x) = \frac{1}{3x} + \frac{2}{3x^2}$$

12. *(c)* We have,

$$\int \sin\left\{2\tan^{-1}\sqrt{\frac{1-x}{1+x}}\right\}dx = A\sin^{-1}x + Bx\sqrt{1-x^2} + C$$

Let $\quad I = \int \sin\left\{2\tan^{-1}\sqrt{\frac{1-x}{1+x}}\right\}dx$

Put $x = \cos 2\theta \Rightarrow dx = -2\sin 2\theta \, d\theta$

$$\therefore \quad I = -\int \sin\left\{2\tan^{-1}\sqrt{\frac{1-\cos 2\theta}{1+\cos 2\theta}}\right\} 2\sin 2\theta \, d\theta$$

$$= -\int \sin\left\{2\tan^{-1}\sqrt{\frac{2\sin^2\theta}{2\cos^2\theta}}\right\} 2\sin 2\theta \, d\theta$$

$$= -\int \sin\{2\tan^{-1}\sqrt{\tan^2\theta}\} 2\sin 2\theta \, d\theta$$

$$= -\int \sin\{2\tan^{-1}\tan\theta\} 2\sin 2\theta \, d\theta$$

$$= -\int [\sin(2\theta)] \, 2\sin 2\theta \, d\theta$$

$$= -\int 2\sin^2 2\theta \, d\theta = -\int (1 - \cos 4\theta) \, d\theta$$

$$= -\left[\theta - \frac{\sin 4\theta}{4}\right] + C$$

$$= \left[-\frac{1}{2}\cos^{-1}x + \frac{1}{4} \times 2\sin 2\theta \cos 2\theta\right] + C$$

$$= \left[-\frac{1}{2}\left(\frac{\pi}{2} - \sin^{-1}x\right) + \frac{1}{2}\sqrt{1-x^2} \times x\right] + C$$

$$= \frac{1}{2}\sin^{-1}x + \frac{x}{2}\sqrt{1-x^2} + \left(C - \frac{\pi}{4}\right)$$

But $I = A\sin^{-1}x + Bx\sqrt{1-x^2} + C$

$$\therefore \quad A = \frac{1}{2} \text{ and } B = \frac{1}{2}$$

Hence, $\quad A + B = \frac{1}{2} + \frac{1}{2} = 1$

13. *(c)* Let $E = (1 + x + x^3 + x^4)^{10}$

$$= [1 + x + x^3(1 + x)]^{10} = (1 + x)^{10}(1 + x^3)^{10}$$

$$= [{}^{10}C_0 + {}^{10}C_1 x^1 + {}^{10}C_2 x^2 + \ldots + {}^{10}C_9 x^9$$

$$+ {}^{10}C_{10} x^{10}] \times [{}^{10}C_0 1 + {}^{10}C_1 1 \times (x^3)$$

$$+ {}^{10}C_2 (x^3)^2 + {}^{10}C_3 (x^3)^3 + \ldots]$$

$\therefore$ Coefficient of x^4 in $E = {}^{10}C_4 \times {}^{10}C_0 + {}^{10}C_1 \times {}^{10}C_1$

$$= \frac{10!}{6! \times 4!} \times 1 + 10 \times 10$$

$$= \frac{10 \times 9 \times 8 \times 7}{4 \times 3 \times 2 \times 1} + 100$$

$$= 210 + 100 = 310$$

14. *(a)* Let equation of circle be

$$x^2 + y^2 + 2gx + 2fy + c = 0$$

Given equation of circle is

$$x^2 + y^2 - 4x + 8 = 0$$

The centres of above circles are $(-g, -f)$ and $(2, 0)$.

Condition of orthogonality is

$$2(g_1 g_2 + f_1 f_2) = c_1 + c_2$$

$\therefore \quad 2(g \times (-2) + (f) \times 0) = c + 8$

$\Rightarrow \quad -4g = c + 8 \quad \ldots\text{(i)}$

Also, the assume circle touch the line $x + 1 = 0$.

$\therefore$ The perpendicular drawn from centre to the line is equal to radius.

$\therefore \quad \dfrac{-g + 1}{\sqrt{1^2}} = \sqrt{g^2 + f^2 - c}$

$\Rightarrow \quad -g + 1 = \sqrt{g^2 + f^2 - c}$

On squaring both sides, we get

$$g^2 + 1 - 2g = g^2 + f^2 - c$$

$\Rightarrow \quad c = f^2 + 2g - 1$

Putting the value of c in Eq. (i), we get

$$-4g = f^2 + 2g - 1 + 8$$

$\Rightarrow f^2 + 2g + 4g + 7 = 0$

$\Rightarrow \quad f^2 + 6g + 7 = 0$

$\therefore$ Locus of centre of circle is $y^2 + 6x + 7 = 0$.

15. *(c)* Given,

$$f(x) = \begin{vmatrix} \sin 3x & 1 & 2\left(\cos\frac{3x}{2} + \sin\frac{3x}{2}\right)^2 \\ \cos 3x & -1 & 2\left(\cos^2\frac{3x}{2} - \sin^2\frac{3x}{2}\right) \\ \tan 3x & 4 & 1 + 2\tan 3x \end{vmatrix}$$

On differentiating w.r.t. x, we get

$$f'(x) = \begin{vmatrix} \frac{d}{dx}(\sin 3x) & 1 & 2\left(\cos\frac{3x}{2} + \sin\frac{3x}{2}\right)^2 \\ \frac{d}{dx}(\cos 3x) & -1 & 2\left(\cos^2\frac{3x}{2} - \sin^2\frac{3x}{2}\right) \\ \frac{d}{dx}(\tan 3x) & 4 & 1 + 2\tan 3x \end{vmatrix}$$

$$+ \begin{vmatrix} \sin 3x & \frac{d}{dx}(1) & 2\left(\cos\frac{3x}{2} + \sin\frac{3x}{2}\right)^2 \\ \cos 3x & \frac{d}{dx}(-1) & 2\left(\cos^2\frac{3x}{2} - \sin^2\frac{3x}{2}\right) \\ \tan 3x & \frac{d}{dx}(4) & 1 + 2\tan 3x \end{vmatrix}$$

$$+ \begin{vmatrix} \sin 3x & 1 & 2\frac{d}{dx}\left(\cos\frac{3x}{2} + \sin\frac{3x}{2}\right)^2 \\ \cos 3x & -1 & 2\frac{d}{dx}\left(\cos^2\frac{3x}{2} - \sin^2\frac{3x}{2}\right) \\ \tan 3x & 4 & \frac{d}{dx}(1 + 2\tan 3x) \end{vmatrix}$$

$$= \begin{vmatrix} 3\cos 3x & 1 & 2\left(\cos\frac{3x}{2} + \sin\frac{3x}{2}\right)^2 \\ 3\sin 3x & -1 & 2\left(\cos^2\frac{3x}{2} - \sin^2\frac{3x}{2}\right) \\ 3\sec^2 3x & 4 & 1 + 2\tan 3x \end{vmatrix}$$

$$+ \begin{vmatrix} \sin 3x & 0 & 2\left(\cos\frac{3x}{2} + \sin\frac{3x}{2}\right)^2 \\ \cos 3x & 0 & 2\left(\cos^2\frac{3x}{2} - \sin^2\frac{3x}{2}\right) \\ \tan 3x & 0 & 1 + 2\tan 3x \end{vmatrix}$$

$$+ \begin{vmatrix} \sin 3x & 1 & 2 \times 2\left(\cos\frac{3x}{2} + \sin\frac{3x}{2}\right) \times \left(\frac{-3}{2}\sin\frac{3x}{2} + \frac{3}{2}\cos\frac{3x}{2}\right) \\ \cos 3x & -1 & 2\left(-2\cos\frac{3x}{2} \times \frac{3}{2}\sin\frac{3x}{2} - 2\sin\frac{3x}{2} \times \frac{3}{2}\cos\frac{3x}{2}\right) \\ \tan 3x & 4 & (0 + 2 \times 3\sec^2 3x) \end{vmatrix}$$

At $x = (2n+1)\pi$,

$$f'(x) = \begin{vmatrix} 3(-1) & 1 & 2(1) \\ 0 & -1 & 2(-1) \\ 3 & 4 & 1+0 \end{vmatrix} + 0 + \begin{vmatrix} 0 & 1 & 4(0-1) \times \left[-\frac{3}{2}(-1) + \frac{3}{2} \times 0\right] \\ -1 & -1 & 0 \\ 0 & 4 & 0 \end{vmatrix}$$

$$= \begin{vmatrix} -3 & 1 & 2 \\ 0 & -1 & -2 \\ 3 & 4 & 1 \end{vmatrix} + 0 + \begin{vmatrix} 0 & 1 & -6 \\ -1 & -1 & 0 \\ 0 & 4 & 0 \end{vmatrix}$$

$$= [-3(-1+8) - 1(0+6) + 2(0+3)] + [0 - 1(0-0) - 6(-4)]$$

$$= -21 + 24 = 3$$

16. *(b)* Given equation of line is $lx + my + n = 0$

and equation of ellipse is $\frac{x^2}{25} + \frac{y^2}{9} = 1$.

$\therefore$ The equation of any normal to the ellipse is

$$5x\sec\theta - 3y\ \text{cosec}\ \theta = 25 - 9$$

$\left(\because \text{the equation of any normal to the ellipse } \frac{x^2}{a^2} + \frac{y^2}{b^2} = 1 \text{ is } ax\sec\theta - by\ \text{cosec}\theta = a^2 - b^2\right)$

$$\Rightarrow \quad 5x\sec\theta - 3y\text{cosec}\ \theta - 16 = 0 \quad \text{...(i)}$$

As the Eq. (i) is the normal to the ellipse.

$$\therefore \quad \frac{5\sec\theta}{l} = \frac{-3\ \text{cosec}\ \theta}{m} = \frac{-16}{n}$$

$$\Rightarrow \quad \cos\theta = \frac{5n}{-16l} \quad \text{and} \quad \sin\theta = \frac{3n}{16m}$$

$$\because \quad \cos^2\theta + \sin^2\theta = 1$$

$$\therefore \quad \left(\frac{5n}{16l}\right)^2 + \left(\frac{3n}{16m}\right)^2 = 1$$

$$\Rightarrow \quad \frac{25n^2}{256l^2} + \frac{9n^2}{256m^2} = 1$$

$$\Rightarrow \quad \frac{25}{l^2} + \frac{9}{m^2} = \frac{256}{n^2}$$

17. *(a)* Given, $\frac{4}{\sin x} + \frac{1}{1 - \sin x} = a$

$$\Rightarrow \quad 4(1 - \sin x) + \sin x = a\sin x(1 - \sin x)$$

$$\Rightarrow \quad 4 - 4\sin x + \sin x = a\sin x - a\sin^2 x$$

$$\Rightarrow \quad a\sin^2 x - (3 + a)\sin x + 4 = 0 \quad \text{...(i)}$$

It is a quadratic equation in $\sin x$, so

$$D \geq 0$$

$$\Rightarrow \quad (3+a)^2 - 4 \times 4a \geq 0$$

$$\Rightarrow \quad 9 + a^2 + 6a - 16a \geq 0$$

$$\Rightarrow \quad a^2 - 10a + 9 \geq 0$$

$$\Rightarrow \quad (a-1)(a-9) \geq 0$$

$$\Rightarrow \quad a \geq 9 \text{ or } a \leq -9$$

Now, at $a = 9$, Eq. (i) becomes

$$9\sin^2 x - 12\sin x + 4 = 0$$

$$\Rightarrow \quad (3\sin x - 2)^2 = 0$$

$$\Rightarrow \quad \sin x = \frac{2}{3} < 0 \Rightarrow x \in \left(0, \frac{\pi}{2}\right)$$

Hence, least value of a is 9.

18. *(c)* If one line of regression coefficient is less than unity, then other will be greater than unity.

19. *(a)* Given edges of a parallelopiped are $\mathbf{a} = 2\hat{i} - 3\hat{j} + \hat{k}$, $\mathbf{b} = \hat{i} - \hat{j} + 2\hat{k}$ and $\mathbf{c} = 2\hat{i} + \hat{j} - \hat{k}$

$\therefore$ Volume of parallelopiped $= [\mathbf{a}\,\mathbf{b}\,\mathbf{c}]$

$$= \left\| \begin{matrix} 2 & -3 & 1 \\ 1 & -1 & 2 \\ 2 & 1 & -1 \end{matrix} \right\|$$

$$= |2(1-2) + 3(-1-4) + 1(1+2)|$$

$$= |-2 - 15 + 3| = 14 \text{ cu units}$$

20. *(a)* Let $f(x) = x^3 - 6x + 1$

Now, $f(2) = (2)^3 - 6 \times 2 + 1$
$= 8 - 12 + 1 = -3$

and $f(3) = (3)^3 - 6 \times 3 + 1$
$= 27 - 18 + 1 = 10$

Here, we see that $f(2)$ and $f(3)$ have opposite signs, so one of the roots lies in $(2, 3)$.

21. *(b)* Given, $\lim_{x \to 0} \frac{\sin(\sin x) - \sin x}{ax^3 + bx^5 + c} = -\frac{1}{12}$ ...(i)

$$\Rightarrow \quad \frac{\sin\sin 0 - \sin 0}{a(0)^3 + b(0)^5 + c} = -\frac{1}{12}$$

$$\Rightarrow \quad \frac{0}{c} = -\frac{1}{12} \Rightarrow c = 0$$

Applying L' Hospital's rule in Eq.(i),we get

$$\lim_{x \to 0} \frac{\cos(\sin x)\cos x - \cos x}{3ax^2 + 5bx^4 + 0} = -\frac{1}{12}$$

$$\Rightarrow \quad \lim_{x \to 0} \frac{\cos x(\cos(\sin x) - 1)}{x^2(3a + 5bx)} = -\frac{1}{12}$$

$$\Rightarrow \quad \lim_{x \to 0} \cos x \frac{2\sin^2\left(\frac{\sin x}{2}\right)}{x^2(3a + 5bx)} = -\frac{1}{12}$$

$$\Rightarrow \frac{\cos 0}{3a+5b\times 0}\times \lim_{x\to 0}\frac{2\sin^2\left(\frac{\sin x}{2}\right)}{4\left(\frac{\sin x}{2}\right)^2\times\left(\frac{x}{\sin x}\right)^2}=-\frac{1}{12}$$

$$\Rightarrow \frac{1}{3a+0}\times\frac{1}{2}\times\frac{1}{1}=-\frac{1}{12}\quad\left(\because \lim_{\theta\to 0}\frac{\sin\theta}{\theta}=1\right)$$

$$\Rightarrow \frac{1}{6a}=-\frac{1}{12}\Rightarrow 6a=-12$$

$$\therefore \quad a=-2$$

Hence, $a=-2$, $b\in R$ and $c=0$

22. *(d)* Let $x=\sqrt{12}$

$$\Rightarrow x^2=12$$

$$\Rightarrow x^2-12=0$$

Let $f(x)=x^2-12$

The first approximation in the Newton-Raphson method is given by

$$x_1=x_0-\frac{f(x_0)}{f'(x_0)}$$

$$=x_0-\frac{x_0^2-12}{2x_0}=\frac{x_0{}^2+12}{2x_0}$$

$$\because \quad 3<\sqrt{12}<3.5$$

We can take $x_0=3.5$

$$\therefore \quad x_1=\frac{(3.5)^2+12}{2\times 3.5}=\frac{12.25+12}{7}$$

$$=\frac{24.25}{7}=3.464$$

23. *(b)* Given equation is

$$\frac{(3-i)^2}{2+i}=A+iB$$

$$\Rightarrow \frac{9-1-6i}{2+i}=A+iB$$

$$\Rightarrow \frac{8-6i}{2+i}=A+iB$$

$$\Rightarrow \frac{2(4-3i)}{2+i}\times\frac{2-i}{2-i}=A+iB$$

$$\Rightarrow \frac{2[8-4i-6i-3]}{4+1}=A+iB$$

$$\Rightarrow \frac{2[5-10i]}{5}=A+iB$$

$$\Rightarrow 2-4i=A+iB$$

On equating the real and imginary parts from both sides, we get

$$A=2 \text{ and } B=-4$$

24. *(a)* Given system of circles is

$$x^2+y^2+4x+7=0 \quad \ldots(i)$$

$$2(x^2+y^2)+3x+5y+9=0$$

or $$x^2+y^2+\frac{3}{2}x+\frac{5}{2}y+\frac{9}{2}=0 \quad \ldots(ii)$$

and $$x^2+y^2+y=0 \quad \ldots(iii)$$

The radical centre can be obtained by solving the Eqs. (i), (ii) and (iii).

On subtracting Eq. (ii) from Eq. (i), we get

$$4x-\frac{3}{2}x-\frac{5}{2}y+7-\frac{9}{2}=0$$

$$\Rightarrow \frac{5}{2}x-\frac{5}{2}y+\frac{5}{2}=0$$

$$\Rightarrow x-y+1=0 \quad \ldots(iv)$$

On subtracting Eq. (iii) from Eq. (i), we get

$$4x-y+7=0 \quad \ldots(v)$$

On solving Eqs. (iv) and (v), we get

$$x=-2 \text{ and } y=-1$$

Hence, radical centre is $(-2,-1)$.

25. *(b)* Tangent drawn at any point (x, y) is

$$Y-y=\frac{dy}{dx}(X-x)$$

When $Y=0$, $X=x-y\frac{dx}{dy}$

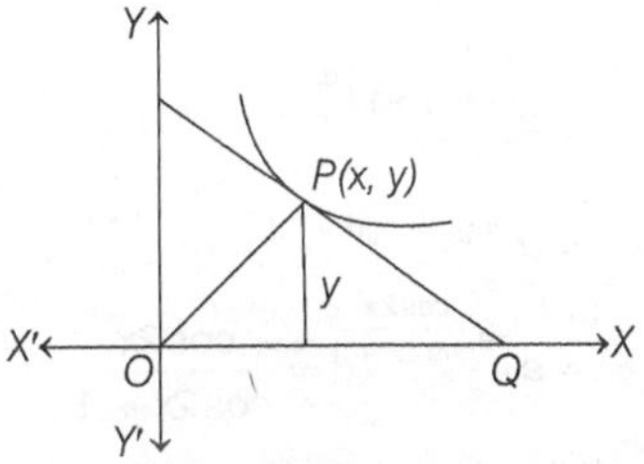

$\because$ Area of $\Delta OPQ=a^2$ (given)

$$\therefore \left|\frac{1}{2}X\cdot y\right|=a^2$$

$$\Rightarrow \left|\left(x-y\frac{dx}{dy}\right)y\right|=2a^2$$

$$\Rightarrow xy-y^2\frac{dx}{dy}=\pm 2a^2$$

$$\Rightarrow \frac{dx}{dy}-\frac{x}{y}=\pm\frac{2a^2}{y^2}$$

Here, $P=-\frac{1}{y}$ and $Q=\pm\frac{2a^2}{y^2}$

$$\therefore \quad \text{IF} = e^{\int P\,dy} = e^{\int -\frac{1}{y}dy}$$

$$= e^{-\log y} = e^{\log\frac{1}{y}} = \frac{1}{y}$$

Hence, required solution is

$$x \times \frac{1}{y} = \int \pm \frac{2a^2}{y^2} \times \frac{1}{y} dy$$

$$\Rightarrow \quad \frac{x}{y} = \pm \frac{2a^2 y^{-2}}{-2} + C$$

$$\Rightarrow \quad x = Cy \pm \frac{a^2}{y}$$

which is the required curve.

26. *(d)* Given differential equation can be written as

$$\frac{dy}{dx} - \left(\frac{\tan 2x}{\cos^2 x}\right) y = \cos^2 x$$

Here, $P = -\dfrac{\tan 2x}{\cos^2 x} = \dfrac{-\sin 2x}{\cos 2x\left(\dfrac{\cos 2x + 1}{2}\right)}$

and $\qquad Q = \cos^2 x$

$$\because \quad \text{IF} = e^{\int P\,dx} = e^{-\int \frac{2\sin 2x}{\cos 2x(\cos 2x + 1)}dx}$$

Put $\cos 2x = t \Rightarrow -2\sin 2x\, dx = dt$

$$\therefore \quad \text{IF} = e^{\int \frac{1}{t}\left(\frac{1}{t+1}\right)dt}$$

$$= e^{\int \left(\frac{1}{t} - \frac{1}{t+1}\right)dt}$$

$$= e^{[\log t - \log(t+1)]} = e^{\log\frac{t}{t+1}}$$

$$= e^{\log\frac{\cos 2x}{\cos 2x + 1}} = \frac{\cos 2x}{\cos 2x + 1}$$

Now, solution is

$$y \times \frac{\cos 2x}{\cos 2x + 1} = \int \frac{\cos 2x}{\cos 2x + 1} \times \cos^2 x\, dx + C$$

$$= \int \frac{\cos 2x}{2\cos^2 x} \times \cos^2 x\, dx + C$$

$$= \frac{1}{2}\int \cos 2x\, dx + C$$

$$= \frac{1}{2} \times \frac{\sin 2x}{2} + C$$

$$\Rightarrow \quad y\frac{\cos 2x}{\cos 2x + 1} = \frac{1}{4}\sin 2x + C \qquad \text{...(i)}$$

But $\qquad y\left(\dfrac{\pi}{6}\right) = \dfrac{3\sqrt{3}}{8}$

$$\therefore \quad \frac{3\sqrt{3}}{8} \times \frac{\cos\left(2 \times \frac{\pi}{6}\right)}{\cos\left(2 \times \frac{\pi}{6}\right) + 1} = \frac{1}{4}\sin 2\left(\frac{\pi}{6}\right) + C$$

$$\Rightarrow \quad \frac{\frac{3\sqrt{3}}{8} \times \frac{1}{2}}{\frac{1}{2} + 1} = \frac{1}{4} \times \frac{\sqrt{3}}{2} + C$$

$$\Rightarrow \quad \frac{3\sqrt{3}}{2 \times 8 \times \frac{3}{2}} = \frac{\sqrt{3}}{8} + C$$

$$\Rightarrow \quad \frac{\sqrt{3}}{8} = \frac{\sqrt{3}}{8} + C \Rightarrow C = 0$$

From Eq. (i), we get

$$y\frac{\cos 2x}{\cos 2x + 1} = \frac{1}{4}\sin 2x + 0$$

$$\Rightarrow \quad y = \frac{1}{4}\frac{\sin 2x}{\frac{\cos 2x}{\cos 2x + 1}} = \frac{1}{4}\frac{\sin 2x}{\frac{\cos^2 x - \sin^2 x}{2\cos^2 x}}$$

$$= \frac{1}{2} \cdot \frac{\sin 2x}{1 - \tan^2 x}$$

27. *(a)* $(1 - \omega + \omega^2)(1 + \omega - \omega^2)$

$$= (1 + \omega^2 - \omega)(1 + \omega - \omega^2)$$

$$= (-\omega - \omega)(-\omega^2 - \omega^2) \quad (\because 1 + \omega + \omega^2 = 0)$$

$$= (-2\omega)(-2\omega^2)$$

$$= 4(\omega^3) = 4 \times 1 \qquad (\because \omega^3 = 1)$$

$$= 4$$

28. *(*)* Given, $\dfrac{dy}{dx} = \dfrac{y-1}{x^2 + x}$

$$\Rightarrow \quad \frac{1}{y-1}dy = \frac{dx}{x(x+1)}$$

On integrating both sides, we get

$$\int \frac{1}{(y-1)}dy = \int \left(\frac{1}{x} - \frac{1}{x+1}\right)dx$$

$$\Rightarrow \quad \log|y - 1| = [\log|x| - \log|x + 1|] + C$$

$$\Rightarrow \quad \log|y - 1| = \log\left|\frac{x}{x+1}\right| + C \qquad \text{...(i)}$$

At point (1, 0), we get

$$\log|0 - 1| = \log\left(\frac{1}{2}\right) + C$$

$$\Rightarrow \quad C = -\log\frac{1}{2} + 0$$

From Eq. (i), we get

$$\log|y-1| = \log\left|\frac{x}{x+1}\right| - \log\frac{1}{2}$$

$$\Rightarrow \quad \log(y-1) = \log 2\left|\frac{x}{x+1}\right|$$

$$\Rightarrow \quad (y-1) = \frac{2x}{x+1}$$

$$\Rightarrow \quad (y-1)(x+1) = 2x$$

$$\Rightarrow \quad (y-1)(x+1) - 2x = 0$$

29. *(c)* Given, $f(x) = [\sin x + \cos x]$

$$= \left[\sqrt{2}\left(\frac{1}{\sqrt{2}}\sin x + \frac{1}{\sqrt{2}}\cos x\right)\right]$$

$$= \left[\sqrt{2}\sin\left(x + \frac{\pi}{4}\right)\right]$$

We know that, greatest integer function is discontinuous on integer values.

Function $\sqrt{2}\sin\left(x + \frac{\pi}{4}\right)$ will gives integer values at

$x = 90°, 135°, 180°, 270°, 315°$,

Hence, there are five points in the given interval, in which $f(x)$ is not continuous.

30. *(d)* Let $I = \cot^{-1}\left(\frac{\sqrt{1+\sin x} + \sqrt{1-\sin x}}{\sqrt{1+\sin x} - \sqrt{1-\sin x}}\right)$

$$= \cot^{-1}\left[\frac{\sqrt{\left(\sin\frac{x}{2} + \cos\frac{x}{2}\right)^2} + \sqrt{\left(\cos\frac{x}{2} - \sin\frac{x}{2}\right)^2}}{\sqrt{\left(\sin\frac{x}{2} + \cos\frac{x}{2}\right)^2} - \sqrt{\left(\cos\frac{x}{2} - \sin\frac{x}{2}\right)^2}}\right]$$

$$= \cot^{-1}\left(\frac{\sin\frac{x}{2} + \cos\frac{x}{2} + \cos\frac{x}{2} - \sin\frac{x}{2}}{\sin\frac{x}{2} + \cos\frac{x}{2} - \cos\frac{x}{2} + \sin\frac{x}{2}}\right)$$

$$= \cot^{-1}\left(\frac{2\cos\frac{x}{2}}{2\sin\frac{x}{2}}\right) = \cot^{-1}\left(\cot\frac{x}{2}\right) = \frac{x}{2}$$

31. *(d)* Given vertices of a ΔABC are $A(-1, 3, 2)$, $B(2, 3, 5)$ and $C(3, 5, -2)$.

Now DR's of $AB = (2+1, 3-3, 5-2) = (3, 0, 3)$

DR's of $BC = (3-2, 5-3, -2-5) = (1, 2, -7)$

and DR's of $CA = (-1-3, 3-5, 2+2) = (-4, -2, 4)$

Now, the angle between AB and BC,

$$\cos B = \frac{|3\times1 + 0\times2 + 3\times(-7)|}{\sqrt{3^2+0^2+3^2}\sqrt{1^2+2^2+(-7)^2}}$$

$$= \frac{|3+0-21|}{\sqrt{9+0+9}\sqrt{1+4+49}}$$

$$= \frac{18}{3\sqrt{2}\times3\sqrt{6}} = \frac{2}{2\sqrt{3}} = \frac{1}{\sqrt{3}}$$

angle between BC and CA,

$$\cos C = \frac{|1\times(-4) + 2(-2) + (-7)(4)|}{\sqrt{1^2+2^2+(-7)^2}\sqrt{(-4)^2+(-2)^2+(4)^2}}$$

$$= \frac{|-4-4-28|}{\sqrt{1+4+49}\sqrt{16+4+16}}$$

$$= \frac{36}{\sqrt{54}\sqrt{36}} = \frac{36}{3\sqrt{6}\times6}$$

$$= \frac{2}{\sqrt{2}\sqrt{3}} = \frac{\sqrt{2}}{\sqrt{3}}$$

and angle between AC and AB,

$$\cos A = \frac{|-4\times3 + (-2)\times0 + 4\times3|}{\sqrt{(-4)^2+(-2)^2+(4)^2}\sqrt{3^2+0^2+3^2}}$$

$$= |0|$$

$$\Rightarrow \quad A = 90°$$

32. *(a)* $\lim\limits_{x\to0}\frac{\int_0^{x^2}\sin\sqrt{t}\,dt}{x^3}$ $\left(\text{form}\frac{0}{0}\right)$

$$= \lim_{x\to0}\frac{\sin\sqrt{x^2}\times2x}{3x^2}$$

(using L' Hospital's rule)

$$= \lim_{x\to0}\frac{2}{3}\frac{\sin x}{x} = \frac{2}{3}\times1 = \frac{2}{3}$$

33. *(b)* $[\mathbf{a}\times\mathbf{b}\ \ \mathbf{b}\times\mathbf{c}\ \ \mathbf{c}\times\mathbf{a}]$

$$= (\mathbf{a}\times\mathbf{b})\cdot[(\mathbf{b}\times\mathbf{c})\times(\mathbf{c}\times\mathbf{a})]$$

$$= (\mathbf{a}\times\mathbf{b})\cdot[((\mathbf{b}\times\mathbf{c})\cdot\mathbf{a})\,\mathbf{c} - ((\mathbf{b}\times\mathbf{c})\cdot\mathbf{c})\,\mathbf{a}]$$

$$= (\mathbf{a}\times\mathbf{b})\cdot([\mathbf{b}\ \mathbf{c}\ \mathbf{a}]\,\mathbf{c} - [\mathbf{b}\ \mathbf{c}\ \mathbf{c}]\,\mathbf{a})$$

$$= (\mathbf{a}\times\mathbf{b}\cdot\mathbf{c})[\mathbf{b}\ \mathbf{c}\ \mathbf{a}] - [\mathbf{a}\times\mathbf{b}\cdot\mathbf{a}]\ 0$$

$$= [\mathbf{a}\ \mathbf{b}\ \mathbf{c}]\,[\mathbf{a}\ \mathbf{b}\ \mathbf{c}] - 0 = [\mathbf{a}\ \mathbf{b}\ \mathbf{c}]^2$$

34. *(a)* Geometric mean of a and $b = \sqrt{ab}$

$$\Rightarrow \quad \sqrt{ab} = 16 \quad \text{(given)}$$

$$\Rightarrow \quad ab = 256 \quad \ldots\text{(i)}$$

And harmonic mean of a and $b = \frac{2ab}{a+b}$

$$\therefore \quad \frac{2ab}{a+b} = \frac{64}{5} \quad \text{(given)}$$

$$\Rightarrow \quad \frac{2 \times 256}{a+b} = \frac{64}{5} \quad \text{[from Eq. (i)]}$$

$$\Rightarrow \quad a + b = 40 \quad \ldots\text{(ii)}$$

Now, $(a-b) = \sqrt{(a+b)^2 - 4ab}$

$$= \sqrt{(40)^2 - 4 \times 256}$$

$$= \sqrt{1600 - 1024}$$

$$= \sqrt{576}$$

$$\Rightarrow \quad a - b = 24 \quad \ldots\text{(iii)}$$

On solving Eqs. (ii) and (iii), we get

$$a = 32 \quad \text{and} \quad b = 8$$

$$\therefore \quad a : b = 32 : 8 = 4 : 1$$

35. *(b)* Let four terms in a GP be ar^3, ar, $\frac{a}{r}$ and $\frac{a}{r^3}$.

According to the given condition,

$$ar^3 + ar + \frac{a}{r} + \frac{a}{r^3} = 60 \quad \ldots\text{(i)}$$

and $\frac{ar^3 + \frac{a}{r^3}}{2} = 18$

$$\Rightarrow \quad ar^3 + \frac{a}{r^3} = 36 \quad \ldots\text{(ii)}$$

Now, from Eq. (i), we have

$$\left(ar + \frac{a}{r}\right) + ar^3 + \frac{a}{r^3} = 60$$

$$\Rightarrow \quad a\left(r + \frac{1}{r}\right) + 36 = 60 \quad \text{[from Eq. (ii)]}$$

$$\Rightarrow \quad a\left(r + \frac{1}{r}\right) = 24 \quad \ldots\text{(iii)}$$

On dividing Eq. (iii) by Eq. (ii), we get

$$\frac{a\left(r^3 + \frac{1}{r^3}\right)}{a\left(r + \frac{1}{r}\right)} = \frac{36}{24}$$

$$\Rightarrow \quad \frac{\left(r + \frac{1}{r}\right)\left(r^2 + \frac{1}{r^2} - 1\right)}{r + \frac{1}{r}} = \frac{3}{2}$$

$$\Rightarrow \quad 2\left(r^2 + \frac{1}{r^2} - 1\right) = 3$$

$$\Rightarrow \quad \frac{2(r^4 + 1 - r^2)}{r^2} = 3$$

$$\Rightarrow \quad 2r^4 + 2 - 2r^2 = 3r^2$$

$$\Rightarrow \quad 2r^4 - 5r^2 + 2 = 0$$

$$\Rightarrow \quad 2r^4 - 4r^2 - r^2 + 2 = 0$$

$$\Rightarrow \quad 2r^2(r^2 - 2) - 1(r^2 - 2) = 0$$

$$\Rightarrow \quad (r^2 - 2)(2r^2 - 1) = 0$$

$$\Rightarrow \quad r^2 = 2, 2r^2 = 1$$

$$\Rightarrow \quad r = \pm\sqrt{2}, r = \pm\frac{1}{\sqrt{2}}$$

On putting $r = \sqrt{2}$ in Eq. (iii), we get

$$a\left(\sqrt{2} + \frac{1}{\sqrt{2}}\right) = 24$$

$$\Rightarrow \quad a\left(\frac{2+1}{\sqrt{2}}\right) = 24$$

$$\Rightarrow \quad 3a = 24\sqrt{2}$$

$$\Rightarrow \quad a = 8\sqrt{2}$$

$\therefore$ Series becomes $8\sqrt{2}(\sqrt{2})^3$, $8\sqrt{2}(\sqrt{2})$, $\frac{8\sqrt{2}}{\sqrt{2}}$ and $\frac{8\sqrt{2}}{(\sqrt{2})^3}$ *i.e.*, 32, 16, 8 and 4.

If we take $r = \frac{1}{\sqrt{2}}$, we get the series 4, 8, 16 and 32.

36. *(d)* Given, $2|x|^2 + 51 = |1 + 20x|$

Then, following two possible cases arise:

***Case* I** When $1 + 20x > 0$

$$\Rightarrow \quad x > -\frac{1}{20}$$

$$\therefore \quad 2x^2 + 51 = (1 + 20x)$$

$$\Rightarrow \quad 2x^2 - 20x + 50 = 0$$

$$\Rightarrow \quad x^2 - 10x + 25 = 0 \quad [\because 2 \neq 0]$$

$$\Rightarrow \quad (x-5)^2 = 0 \Rightarrow x = 5, 5$$

***Case* II** When $1 + 20x < 0$

$$\Rightarrow \quad x < -\frac{1}{20}$$

$$\therefore \quad 2x^2 + 51 = -(1 + 20x)$$

$$\Rightarrow \quad 2x^2 - 20x + 52 = 0$$

$$\Rightarrow \quad x^2 - 10x + 26 = 0$$

Here, $D < 0$

Thus, roots are imaginary.

Hence, sum of real roots $= 5 + 5 = 10$

37. *(a)* Given roots are $\frac{1}{3+\sqrt{2}}$ and $\frac{1}{3-\sqrt{2}}$

i.e. $\frac{1}{(3+\sqrt{2})} \times \frac{(3-\sqrt{2})}{(3-\sqrt{2})}$

and $\frac{1}{(3-\sqrt{2})} \times \frac{(3+\sqrt{2})}{(3+\sqrt{2})}$

$\Rightarrow \frac{3-\sqrt{2}}{7}$ and $\frac{3+\sqrt{2}}{7}$

So, the required quadratic equation is

x^2 – (Sum of the roots) x

+ Product of the roots = 0

$$\Rightarrow x^2 - \left[\frac{3-\sqrt{2}}{7} + \frac{3+\sqrt{2}}{7}\right]x + \left(\frac{3-\sqrt{2}}{7}\right)\left(\frac{3+\sqrt{2}}{7}\right) = 0$$

$$\Rightarrow x^2 - \left(\frac{6}{7}\right)x + \frac{9-2}{49} = 0$$

$$\Rightarrow x^2 - \frac{6}{7}x + \frac{7}{49} = 0 \Rightarrow 7x^2 - 6x + 1 = 0$$

38. *(d)* Gorden Moore

39. *(c)* Vector which is perpendicular to vertors **a** and **b**, is **a** × **b** or **b** × **a**.

Hence, two vectors are possible which are perpendicular to **a** and **b**.

40. *(b)* Given that,

$$I = \int_0^6 \frac{dx}{1+x^2} \quad \ldots\text{(i)}$$

From Eq. (i), $f(x) = \frac{1}{1+x^2}$

Now, divide the interval [0, 6] into six parts each of width

$$h = \frac{6-0}{6} = 1.$$

The value of $f(x)$ are given below

x	0	1	2	3	4	5	6
$f(x)$	1	0.5	0.2	0.1	0.0588	0.0385	0.027

The trapezoidal rule is

$$\int_{x_0}^{x_0+nh} y\,dx = \frac{h}{2}[(y_0 + y_n) + 2(y_1 + y_2 + y_3 + \ldots + y_{n-1})]$$

$$\therefore \int_0^6 \frac{1}{1+x^2}\,dx = \frac{1}{2}[(1+0.027) + 2\{0.5 + 0.2 + 0.1 + 0.0588 + 0.0385\}]$$

$$= \frac{1}{2}[1.027 + 2(0.8973)]$$

$$= \frac{1}{2}[1.027 + 1.7946]$$

$$= \frac{1}{2}[2.8216] = 1.4108$$

41. *(b)* Given,

$$\frac{dr}{dt} = 2 \text{ m/s} \quad \ldots\text{(i)}$$

$$\frac{dh}{dt} = -3 \text{ m/s}$$

∵ Volume of cylinder, $V = \pi r^2 h$

$$\therefore \frac{dV}{dt} = \pi\left[2rh\frac{dr}{dt} + r^2\frac{dh}{dt}\right]$$

At $r = 3$ m and $h = 5$ m,

$$\frac{dV}{dt} = \pi[2 \times 3 \times 5 \times 2 - 9 \times 3] \quad \text{[using Eq. (i)]}$$

$$\Rightarrow \frac{dV}{dt} = \pi[60 - 27] = 33\pi$$

$$\Rightarrow \frac{dV}{dt} = 33\pi \text{ m}^3\text{/s}$$

42. *(c)* Let required point be (α, β) on the straight line $y = 2x + 11$, which is nearest to the circle

$$16(x^2 + y^2) + 32x - 8y - 50 = 0$$

$$\Rightarrow x^2 + y^2 + 2x - \frac{1}{2}y - \frac{50}{16} = 0$$

$\therefore$ Centre of circle = $\left(-1, \frac{1}{4}\right)$

and radius, $r = \sqrt{1 + \frac{1}{16} + \frac{50}{16}} = \frac{\sqrt{67}}{4}$

Now, equation of straight line passing through centre $\left(-1, \frac{1}{4}\right)$ and (α, β) is

$$y - \frac{1}{4} = \left(\frac{\beta - \frac{1}{4}}{\alpha + 1}\right)(x+1) \quad \ldots\text{(i)}$$

Now, gradient of this straight line = $\left(\frac{\beta - \frac{1}{4}}{\alpha + 1}\right)$

Since, straight line (i) is perpendicular to the line

$$y = 2x + 11$$

$$\therefore \quad \left(\frac{\beta - \frac{1}{4}}{\alpha + 1}\right) \times 2 = -1 \qquad [\because m_1 \cdot m_2 = -1]$$

$$\Rightarrow \quad 2\beta - \frac{1}{2} = -\alpha - 1$$

$$\Rightarrow \quad 2\beta + \alpha = -1 + \frac{1}{2} = \frac{-1}{2}$$

$$\Rightarrow \quad 4\beta + 2\alpha = -1 \qquad \ldots\text{(ii)}$$

$\because$ Point (α, β) lies on straight line

$$y = 2x + 11$$

$$\therefore \quad \beta = 2\alpha + 11$$

$$\Rightarrow \quad \beta - 2\alpha = 11 \qquad \ldots\text{(iii)}$$

On solving Eqs. (ii) and (iii), we get

$$5\beta = 10 \Rightarrow \beta = 2$$

and $2\alpha = 2 - 11 = -9 \Rightarrow \alpha = \frac{-9}{2}$

$\therefore$ Required point is $\left(\frac{-9}{2}, 2\right)$.

43. *(a)* Given equation is

$$b^2x^2 + y^2 = a^2b^2$$

$$\Rightarrow \quad \frac{x^2}{a^2} + \frac{y^2}{a^2b^2} = 1 \qquad \ldots\text{(i)}$$

Above equation of an ellipse with semi-major axis (a) and semi-minor axis (ab).

Now, ecentricity,

$$e = 1 - \frac{a^2b^2}{a^2} \Rightarrow b^2 = 1 - e^2 \qquad \ldots\text{(ii)}$$

Let (x, y) be extrimities of latusrectum, then

$$x = ae \text{ and } y = \pm\frac{a^2b^2}{a}$$

$$\Rightarrow \quad \frac{x}{a} = e \text{ and } \frac{y}{a} = \pm b^2$$

From Eq. (ii), we get

$$\pm\frac{y}{a} = 1 - \frac{x^2}{a^2} \Rightarrow \pm ay + x^2 = a^2$$

Hence, locus of latusrectum is $x^2 \pm ay = a^2$.

44. *(d)* Given differential equation is

$$(y\log x - 1)\, y\, dx = x dy$$

$$\Rightarrow \quad \frac{dy}{dx} = \frac{(y\log x - 1)y}{x} = \frac{y^2 \log x}{x} - \frac{y}{x}$$

$$\Rightarrow \quad \frac{dy}{dx} + \frac{y}{x} = \frac{y^2 \log x}{x}$$

$$\Rightarrow \quad \frac{1}{y^2}\frac{dy}{dx} + \frac{y^{-1}}{x} = \frac{\log x}{x} \qquad \ldots\text{(i)}$$

Put $\quad y^{-1} = v \Rightarrow -y^{-2}\frac{dy}{dx} = \frac{dv}{dx}$

$$\Rightarrow \quad y^{-2}\frac{dy}{dx} = -\frac{dv}{dx}$$

From Eq. (i), we have

$$-\frac{dv}{dx} + \frac{v}{x} = \frac{\log x}{x}$$

$$\Rightarrow \quad \frac{dv}{dx} - \frac{v}{x} = -\frac{\log x}{x} \qquad \ldots\text{(ii)}$$

This is linear differential equation.

Here, $\text{IF} = e^{\int\left(-\frac{1}{x}\right)dx} = e^{-\log x} = \frac{1}{x}$

So, solution is $v \cdot \text{IF} = \int \text{IF} \cdot Q\, dx + C$

$$v \cdot \frac{1}{x} = \int \frac{1}{x}\left(-\frac{\log x}{x}\right) dx + C$$

$$\Rightarrow \quad v \cdot \frac{1}{x} = -\int \frac{\log x}{x^2}\, dx + C$$

$$= -\left[\log x\left(\frac{-1}{x}\right) + \int \frac{1}{x}\cdot\frac{1}{x}\, dx\right] + C$$

$$\Rightarrow \quad \frac{1}{x \cdot y} = \frac{\log x}{x} + \frac{1}{x} + C \qquad \left[\because v = \frac{1}{y}\right]$$

$$\Rightarrow \quad 1 = y[\log x + 1 + Cx]$$

$$\Rightarrow \quad 1 = y[\log x + \log e + Cx] \qquad [\because 1 = \log e]$$

$$\Rightarrow \quad 1 = y[\log e \cdot x + Cx]$$

45. *(b)* A particular thing is received by a man with probability, $p = \frac{a}{a+b}$ and by a woman with probability, $q = \frac{b}{a+b}$.

Now, this experiment is repeated m times. Since, probability in each trial remains same for the men or women. Thus, we can apply binomial distribution.

Since, men should get odd number of things, so the random variable x would have $x = 1, 3, 5, \ldots$

$\therefore$ Required probability

$$= P(X = 1) + P(X = 3) + P(X = 5) + \ldots$$

$$= {}^mC_1 pq^{m-1} + {}^mC_3 p^3q^{m-3} + {}^mC_5 p^5q^{m-5} + \ldots$$

$$= \left[\frac{(q+p)^m - (q-p)^m}{2}\right]$$

$$= \left[\frac{1 - \left(\frac{b}{a+b} - \frac{a}{a+b}\right)^m}{2}\right] \quad [\because p + q = 1]$$

$$= \frac{(a+b)^m - (b-a)^m}{2(a+b)^m}$$

46. *(a)* Given differential equation is

$$\sqrt{a+x}\,\frac{dy}{dx} + xy = 0$$

$$\Rightarrow \quad \sqrt{a+x}\,\frac{dy}{dx} = -xy$$

$$\Rightarrow \quad \frac{dy}{y} = \frac{-x}{\sqrt{a+x}}\,dx$$

On integrating both sides, we get

$$\log y - \log C = \int\left[\frac{-x-a}{\sqrt{a+x}} + \frac{a}{\sqrt{a+x}}\right]dx$$

$$\Rightarrow \quad \log\left(\frac{y}{C}\right) = -\int \sqrt{x+a}\,dx + a\int \frac{1}{\sqrt{a+x}}\,dx$$

$$= -\frac{(x+a)^{3/2}}{3/2} + a\,\frac{(a+x)^{1/2}}{1/2}$$

$$\Rightarrow \quad \log\left(\frac{y}{C}\right) = -\frac{2}{3}(x+a)\sqrt{x+a} + 2a\sqrt{x+a}$$

$$= \frac{2}{3}(2a - x)\sqrt{x+a}$$

$$\Rightarrow \quad \frac{y}{C} = e^{\frac{2}{3}(2a-x)\sqrt{x+a}}$$

$$\Rightarrow \quad y = C\cdot e^{\frac{2}{3}(2a-x)\sqrt{x+a}}$$

which is the required solution.

47. *(a)* Given function is

$$f(x) = 2e^x - ae^{-x} + (2a+1)x - 3$$

For increasing function, $f'(x) > 0, \forall x$

$$\Rightarrow f'(x) = 2e^x + ae^{-x} + (2a+1) > 0$$

$$\Rightarrow \quad (2e^x + 1) + a(e^{-x} + 2) > 0$$

$$\Rightarrow \quad a(e^{-x} + 2) > -(1 + 2e^x)$$

$$\Rightarrow \quad a > \frac{-(1+2e^x)}{(e^{-x}+2)}$$

$$\Rightarrow \quad a > \frac{-(1+2e^x)}{(1+2e^x)}\cdot e^x$$

$$\Rightarrow \quad a > -e^x$$

$\because \quad e^x \in (0, \infty) \Rightarrow -e^x \in (-\infty, 0)$

and $\quad a > -e^x$

$\therefore \quad a \in [0, \infty)$

48. *(a)* Given, $f(x) = x^4 - x - 10$

We assume $x_0 = 2$ is the approximate root of $f(x)$.

Then, $\quad h = -\frac{f(x_0)}{f'(x_0)} = -\frac{f(2)}{f'(2)}$

$$\Rightarrow \quad h = -\left[\frac{(2)^4 - 2 - 10}{4(2)^3 - 1}\right]$$

$$\Rightarrow \quad h = -\left[\frac{16-12}{31}\right] = \frac{-4}{31}$$

$$\Rightarrow \quad h = -0.129$$

$\therefore$ Positive square root of $f(x)$ by Newton-Raphson method,

$$x_1 = x_0 + h = 2 + (-0.129)$$

$$= 2 - 0.129 = 1.871$$

49. *(b)* Given, $3a = b + c$

$\because \quad \cot\left(\frac{B}{2}\right) = \sqrt{\frac{s(s-b)}{(s-a)(s-c)}}$

and $\quad \cot\left(\frac{C}{2}\right) = \sqrt{\frac{s(s-c)}{(s-a)(s-b)}}$

$$\therefore \cot\left(\frac{B}{2}\right)\cdot\cot\left(\frac{C}{2}\right) = \sqrt{\frac{s(s-b)}{(s-a)(s-c)}} \times \sqrt{\frac{s(s-c)}{(s-a)(s-b)}}$$

$$= \frac{s}{(s-a)} = \frac{2s}{2s-2a}$$

On putting $2s = a + b + c$, we get

$$\cot\left(\frac{B}{2}\right)\cdot\cot\left(\frac{C}{2}\right) = \frac{a+b+c}{a+b+c-2a}$$

$$\Rightarrow \quad \cot\left(\frac{B}{2}\right)\cdot\cot\left(\frac{C}{2}\right) = \frac{a+3a}{a+3a-2a} = \frac{4a}{2a} = 2$$

50. *(b)* Given, $\sin^{-1}x + \cot^{-1}\left(\frac{1}{2}\right) = \frac{\pi}{2}$

$$\Rightarrow \quad \sin^{-1}x = \frac{\pi}{2} - \cot^{-1}\left(\frac{1}{2}\right)$$

$$\Rightarrow \quad \sin^{-1}x = \tan^{-1}\left(\frac{1}{2}\right)$$

$$\left[\because \tan^{-1}\theta + \cot^{-1}\theta = \frac{\pi}{2}\right]$$

$$\Rightarrow \quad x = \sin\left\{\tan^{-1}\left(\frac{1}{2}\right)\right\}$$

$$\Rightarrow \quad x = \sin\left\{\sin^{-1}\left(\frac{1}{\sqrt{5}}\right)\right\}$$

$$\Rightarrow \quad x = \frac{1}{\sqrt{5}}$$

51. *(c)* Given, $I = \int_0^{\sqrt{\ln(\pi/2)}} \cos(e^{x^2}) \cdot 2xe^{x^2}\, dx$

Put $e^{x^2} = t \Rightarrow 2xe^{x^2}\, dx = dt$

Now, lower limit, $t = 1$

Upper limit, $t = e^{\ln \pi/2} = \frac{\pi}{2}$

$$\therefore \quad I = \int_1^{\pi/2} \cos(t)\, dt = [\sin t]_1^{\pi/2}$$

$$= \sin\left(\frac{\pi}{2}\right) - \sin 1$$

$$\Rightarrow \quad I = 1 - \sin 1$$

52. *(c)* Application Programming Interface.

53. *(d)* Given, $Z = i\log(2-\sqrt{3})$

$$\Rightarrow \cos Z = \cos[i\log(2-\sqrt{3})]$$

$$= \cos h\{\log(2-\sqrt{3})\}$$

$$[\because \cos ix = \cosh x, \text{ for } x \in R]$$

$$= \frac{e^{\log(2-\sqrt{3})} + e^{-\log(2-\sqrt{3})}}{2}$$

$$\left[\because \cos hx = \frac{e^x + e^{-x}}{2}\right]$$

$$= \frac{e^{\log(2-\sqrt{3})} + e^{\log\left(\frac{1}{2-\sqrt{3}}\right)}}{2}$$

$$= \frac{2-\sqrt{3} + \frac{1}{2-\sqrt{3}}}{2}$$

$$= \frac{(2-\sqrt{3})^2 + 1}{2(2-\sqrt{3})} = \frac{4 + 3 - 4\sqrt{3} + 1}{2(2-\sqrt{3})}$$

$$= \frac{8 - 4\sqrt{3}}{2(2-\sqrt{3})} = \frac{4(2-\sqrt{3})}{2(2-\sqrt{3})} = 2$$

54. *(c)* Given, $\lim_{x \to 2} \frac{2 - \sqrt{2+x}}{2^{1/3} - (4-x)^{1/3}}$ $\left[\frac{0}{0}\text{ form}\right]$

$$= \lim_{x \to 2} \frac{0 - \frac{1}{2\sqrt{2+x}}}{0 - \frac{1}{3} \cdot \frac{(-1)}{(4-x)^{2/3}}}$$

$$= \lim_{x \to 2} \frac{-3(4-x)^{2/3}}{2\sqrt{2+x}}$$

$$= \frac{-3}{2} \times \frac{(4-2)^{2/3}}{\sqrt{2+2}} = \frac{-3}{2} \times \frac{2^{2/3}}{2}$$

$$= -3 \cdot 2^{\frac{2}{3}-2} = -3 \cdot 2^{-4/3}$$

55. *(d)* Three lines of triangle are given by

$$(x^2 - y^2)(2x + 3y - 6) = 0$$

$$\Rightarrow (x - y)(x + y)(2x + 3y - 6) = 0$$

$\therefore$ The three lines of triangle are

$$x - y = 0, x + y = 0$$

and $\quad 2x + 3y - 6 = 0$

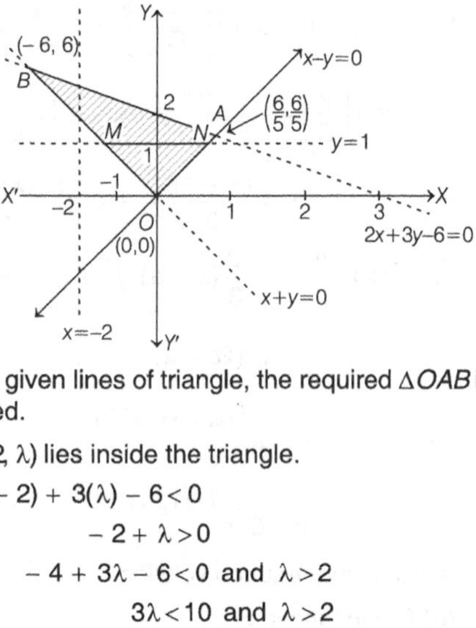

From given lines of triangle, the required ΔOAB is formed.

$\because (-2, \lambda)$ lies inside the triangle.

$\therefore \quad 2(-2) + 3(\lambda) - 6 < 0$

and $\quad -2 + \lambda > 0$

$$\Rightarrow \quad -4 + 3\lambda - 6 < 0 \text{ and } \lambda > 2$$

$$\Rightarrow \quad 3\lambda < 10 \text{ and } \lambda > 2$$

$$\Rightarrow \quad \lambda < \frac{10}{3} \text{ and } \lambda > 2$$

$$\therefore \quad \lambda \in \left(2, \frac{10}{3}\right) \quad \ldots\text{(i)}$$

Now, $(\mu, 1)$ lies outside the triangle.

To find value of μ, we find the interval $[M, N]$ for values of x.

$$x + 1 \geq 0 \text{ and } x - 1 \leq 0$$

$\Rightarrow \qquad x \geq -1$ and $x \leq 1$

$\therefore \qquad x \in [-1, 1]$

$\because$ $(\mu, 1)$ lies outside the triangle.

$\therefore \qquad \mu \in (-\infty, -1) \cup (1, \infty)$

or $\qquad \mu \in R - [-1, 1]$

56. *(a)* System software : Utility software :: Operating system : Arti-virus

Utility software-Anti-virus

57. *(c)* Given equations of circles are

$$x^2 + y^2 = 4$$

and $\qquad x^2 + y^2 - 8x + 12 = 0$

or $\qquad x^2 + y^2 = (2)^2 \qquad \ldots(i)$

$$(x-4)^2 + (y-0)^2 = (2)^2 \qquad \ldots(ii)$$

The figure of both the circles are shown below :

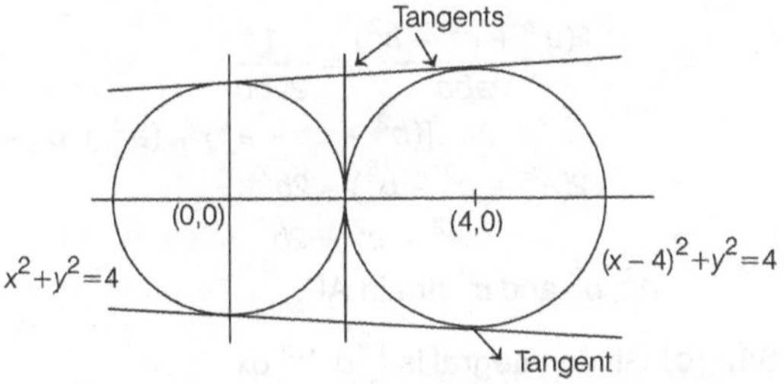

From figure, we see that there is exactly three common tangents.

58. *(a)* Let $f(x) = \frac{1}{3}ax^3 + \frac{1}{2}bx^2 + cx$

Then, $f(x)$ is a polynomial.

So, it is continuous in R.

Now, $\qquad f(0) = 0$

and $\qquad f(1) = \frac{a}{3} + \frac{b}{2} + c = \frac{2a + 3b + 6c}{6}$

$\Rightarrow \qquad f(1) = 0 \qquad [\because 2a + 3b + 6c = 0$, given$]$

$\because f(x)$ is a polynomial, so it is differentiable in R, so in $(0, 1)$.

Hence, by Rolle's theorem, there exists atleast one point $x \in (0, 1)$, there exists such that

$$f'(x) = 0$$

$\Rightarrow \qquad ax^2 + bx + c = 0$

Hence, required interval is $(0, 1)$.

59. *(a)* Given, for $x, y \in N$,

$$f(x + y) = f(x) \cdot f(y)$$

Then, function will be of the form

$$f(x) = a^x, \text{ where } a \in N \qquad [\because a \neq 1]$$

$\because \qquad f(1) = 3$

$\Rightarrow \qquad f(1) = a^1 = 3$

$\Rightarrow \qquad a = 3$

$\therefore$ Function is $f(x) = 3^x$.

Now, $\qquad \sum_{x=1}^{n} f(x) = 120$

$\Rightarrow \qquad \sum_{x=1}^{n} 3^x = 120$

$\Rightarrow 3 + 3^2 + 3^3 + \ldots + 3^n = 120$

$\Rightarrow \qquad \frac{3(3^n - 1)}{3 - 1} = 120$

$\Rightarrow \qquad 3^n = 1 + \frac{120 \times 2}{3}$

$\Rightarrow \qquad 3^n = 81 = 3^4$

So, $\qquad n = 4$

60. *(b)* Given function is

$$f(x) = \begin{cases} \sin\left(\frac{\pi x}{2}\right), & \text{if } x < 1 \\ 3 - 2x, & \text{if } x \leq 1 \end{cases}$$

$\because$ LHL and RHL of $f(x)$ is 1, when $x \to 1$

and $\qquad f(1) = 3 - 2 = 1$

$\therefore \quad f(x)$ is continuous at $x = 1$.

Now, graph of function $f(x)$ is

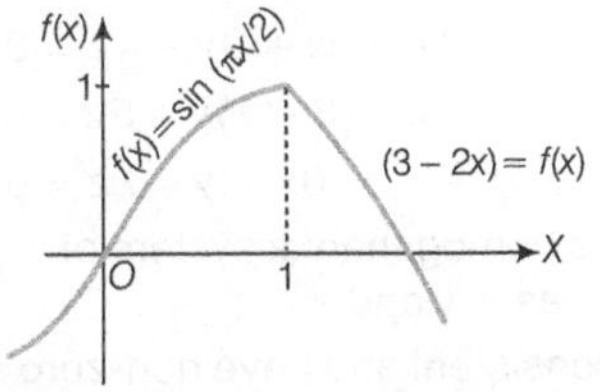

From the above graph of function, we see that $f(x)$ has local maxima at $x = 1$.

61. *(c)* Given differential equation is

$$\frac{dy}{dx} = y\tan x - y^2 \sec x$$

$\Rightarrow \qquad \frac{dy}{dx} - y\tan x = -y^2 \sec x$

$\Rightarrow \qquad \frac{1}{y^2}\frac{dy}{dx} - \frac{\tan x}{y} = -\sec x \qquad \ldots(i)$

Put $\qquad \frac{-1}{y} = u \Rightarrow \frac{1}{y^2}\frac{dy}{dx} = \frac{du}{dx}$

From Eq. (i),

$$\frac{du}{dx} + \tan x \cdot u = -\sec x \quad \text{...(ii)}$$

This is linear differential equation of the form $\frac{du}{dx} + P \cdot u = Q$, where $P = \tan x$, $Q = -\sec x$

$$\therefore \quad IF = e^{\int \tan x\, dx} = e^{\log \sec x} = \sec x$$

Hence, general solution is

$$u \cdot IF = \int IF \cdot Q\, dx + C_1$$

$$\Rightarrow \quad u \cdot \sec x = \int (\sec x) \cdot (-\sec x)\, dx + C_1$$

$$\Rightarrow \quad u \cdot \sec x = -\int \sec^2 x\, dx + C_1$$

$$\Rightarrow \quad u \sec x = -\tan x + C_1$$

$$\Rightarrow \quad \frac{-\sec x}{y} = -\tan x + C_1 \quad \left[\because u = \frac{-1}{y}\right]$$

$$\Rightarrow \quad \sec x = y(\tan x - C_1)$$

$$\Rightarrow \quad \sec x = y(\tan x + C) \quad [\text{where, } C = -C_1]$$

62. *(c)* Given,

$$(\hat{i} + \hat{j} + 3\hat{k})x + (3\hat{i} - 3\hat{j} + \hat{k})y + (-4\hat{i} + 5\hat{j})z = \lambda(\hat{i}x + \hat{j}y + \hat{k}z)$$

On equating the coefficients of $\hat{i}$, $\hat{j}$ and $\hat{k}$ both sides, we have

$$x + 3y - 4z = \lambda x,$$
$$x - 3y + 5z = \lambda y$$
and
$$3x + y + 0 = \lambda z$$

Above three equations can be rewritten as

$$(1 - \lambda)x + 3y - 4z = 0$$
$$x - (3 + \lambda)y + 5z = 0$$
$$3x + y - \lambda z = 0$$

This is homogeneous system of equations in three variables x, y and z.

It is consistent and have non-zero solution.

i.e. $(x, y, z) \neq (0, 0, 0)$, if determinant of coefficient matrix is zero.

$$\Rightarrow \quad \begin{vmatrix} 1-\lambda & 3 & -4 \\ 1 & -(3+\lambda) & 5 \\ 3 & 1 & -\lambda \end{vmatrix} = 0$$

On expanding along first row, we have

$$(1 - \lambda)[\lambda(3 + \lambda) - 5] - 3(-\lambda - 15) - 4(1 + 9 + 3\lambda) = 0$$

$$\Rightarrow (1 - \lambda)(\lambda^2 + 3\lambda - 5) + 3\lambda + 45 - 40 - 12\lambda = 0$$

$$\Rightarrow \lambda^2 + 3\lambda - 5 - \lambda^3 - 3\lambda^2 + 5\lambda - 9\lambda + 5 = 0$$

$$\Rightarrow \quad -\lambda^3 - 2\lambda^2 - \lambda = 0$$

$$\Rightarrow \quad \lambda(\lambda^2 + 2\lambda + 1) = 0$$

$$\Rightarrow \quad \lambda(\lambda + 1)^2 = 0$$

$$\Rightarrow \quad \lambda = 0, -1$$

63. *(b)* It is given, $\cot A$, $\cot B$ and $\cot C$ are in AP.

$$\Rightarrow \quad 2\cot B = \cot A + \cot C$$

$$\Rightarrow \quad 2\frac{\cos B}{\sin B} = \frac{\cos A}{\sin A} + \frac{\cos C}{\sin C}$$

$$\because \quad \frac{\sin A}{a} = \frac{\sin B}{b} = \frac{\sin C}{c} = k$$

$$\Rightarrow \quad \frac{2\cos B}{kb} = \frac{\cos A}{ak} + \frac{\cos C}{ck}$$

$$\Rightarrow \quad \frac{2\cos B}{b} = \frac{\cos A}{a} + \frac{\cos C}{c}$$

$$\Rightarrow \quad \frac{2}{b}\left[\frac{a^2 + c^2 - b^2}{2ac}\right] = \frac{1}{a}\left[\frac{b^2 + c^2 - a^2}{2bc}\right] + \frac{1}{c}\left[\frac{a^2 + b^2 - c^2}{2ab}\right]$$

$$\Rightarrow \quad \frac{2(a^2 + c^2 - b^2)}{2abc} = \frac{1}{2abc}[(b^2 + c^2 - a^2) + (a^2 + b^2 - c^2)]$$

$$\Rightarrow \quad 2(a^2 + c^2 - b^2) = 2b^2$$

$$\Rightarrow \quad a^2 + c^2 = 2b^2$$

$\therefore$ a^2, b^2 and c^2 are in AP.

64. *(c)* Given integral is $\int_1^2 e^{-x/2}\, dx$

On dividing interval [1, 2] in four parts, we have

$$h = \frac{2-1}{4} = \frac{1}{4}$$

Now, value of $f(x) = e^{-x/2}$ is given below

x	1	$\frac{5}{4}$	$\frac{3}{2}$	$\frac{7}{4}$	e
$f(x)$	0.6065	0.5352	0.4724	0.4168	0.3679

Now, Simpson's $\frac{1}{3}$ rule is

$$\int_{x_0}^{x_0 + nh} f(x)dx = \frac{h}{3}[(y_0 + y_n) + 4(y_1 + y_3 + \dots + y_{n-1}) + 2(y_2 + y_4 + \dots + y_{n-2})]$$

$$\therefore \int_1^2 e^{-x/2} dx = \frac{1}{12}[(0.6065 + 0.3679) + 4(0.5352 + 0.4168) + 2(0.4724)]$$

$$= \frac{1}{12}[0.9744 + 3.808 + 0.9448]$$

$$= \frac{1}{12} \times 5.7272 = 0.477$$

65. *(b)* When a die is rolled three times.

Then, cardinality of sample space = 6^3

Now, according to question, the favourable outcomes

= {(1, 2, 3), (1, 2, 4), (1, 2, 5), (1, 2, 6), (1, 3, 4), (1, 3, 5), (1, 3, 6), (1, 4, 5), (1, 4, 6), (1, 5, 6), (2, 3, 4), (2, 3, 5), (2, 3, 6), (2, 4, 5), (2, 4, 6), (3, 4, 5), (3, 4, 6), (4, 5, 6), (2, 5, 6), (3, 5, 6)}

$\therefore$ Required probability $= \dfrac{20}{6^3} = \dfrac{5}{54}$

66. *(d)* Given, $f(x) = \log_e (6 - |x^2 + x - 6|)$

The function $f(x)$ is defined, if

$$(6 - |x^2 + x - 6|) > 0$$
$$\Rightarrow \quad |x^2 + x - 6| < 6$$
$$\Rightarrow \quad -6 < x^2 + x - 6 < 6$$

If $\quad x^2 + x - 6 < 6$

$$\Rightarrow \quad x^2 + x - 12 < 0$$
$$\Rightarrow \quad (x + 4)(x - 3) < 0$$

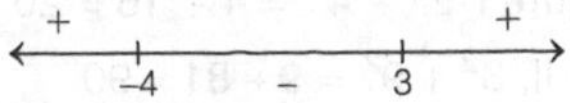

$$\therefore \quad x \in (-4, 3) \quad \ldots\text{(i)}$$

Now, if $-6 < x^2 + x - 6$

$$\Rightarrow \quad x^2 + x > 0 \Rightarrow x(x + 1) > 0$$

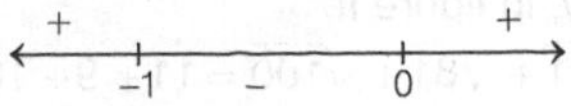

$$\therefore \quad x \in (0, \infty) \quad \ldots\text{(ii)}$$

From Eqs. (i) and (ii), we get

$$x \in (0, 3)$$

$\because f(x)$ has only two integral values.

$$\therefore \quad x = 1, 2$$

67. *(a)* The given equation is

$$y^2 + z^2 = 0$$

Above equation represents a point circle in the *YZ*-plane.

68. *(d)* Given complex numbers are

$$Z_1 = (1 + i) \equiv (1, 1) \equiv (a_1, b_1)$$
$$Z_2 = (-2 + 3i) \equiv (-2, 3) \equiv (a_2, b_2)$$

and $\quad Z_3 = \dfrac{ai}{3} \equiv \left(0, \dfrac{a}{3}\right) \equiv (a_3, b_3)$

These three points will be collinear, if

$$\frac{1}{2}[a_1(b_2 - b_3) + a_2(b_3 - b_1) + a_3(b_1 - b_2)] = 0$$

$$\Rightarrow \quad 1\left(3 - \frac{a}{3}\right) - 2\left(\frac{a}{3} - 1\right) + 0(1 - 3) = 0$$
$$\Rightarrow \quad \frac{9 - a}{3} - 2\left(\frac{a - 3}{3}\right) + 0 = 0$$
$$\Rightarrow \quad 9 - a - 2a + 6 = 0$$
$$\Rightarrow \quad 3a = 15$$
$$\Rightarrow \quad a = 5$$

69. *(b)* Given that, *a*, *b* and *c* are in HP.

$\therefore$ Harmonic mean of *a* and *c* is *b*.

and geometric mean of *a* and *c* is $\sqrt{ac}$.

$\because$ Geometric mean > Harmonic mean

$$\Rightarrow \quad \sqrt{ac} > b \quad \ldots\text{(i)}$$

Now, for the positive numbers a^n and c^n, we have

$$\text{Geometric mean} = \sqrt{a^n \cdot c^n}$$
$$\text{and arithmetic mean} = \frac{a^n + c^n}{2}$$
$$\because \quad \text{AM} > \text{GM}$$
$$\therefore \quad \frac{a^n + c^n}{2} > \sqrt{a^n \cdot c^n}$$
$$\Rightarrow \quad a^n + c^n > 2\sqrt{a^n \cdot c^n}$$
$$\Rightarrow \quad a^n + c^n > 2(\sqrt{ac})^n > 2b^n \quad \text{[using Eq. (i)]}$$
$$\Rightarrow \quad a^n + c^n > 2b^n$$

70. *(d)* $\tan\left[2\tan^{-1}\left(\dfrac{1}{5}\right) - \dfrac{\pi}{4}\right] = \tan\left[\tan^{-1}\left(\dfrac{\frac{2}{5}}{1 - \frac{1}{25}}\right) - \dfrac{\pi}{4}\right]$

$$\left[\because 2\tan^{-1}\theta = \tan^{-1}\left(\frac{2\theta}{1 - \theta^2}\right)\right]$$

$$= \tan\left[\tan^{-1}\left(\frac{10}{24}\right) - \frac{\pi}{4}\right]$$

$$= \frac{\tan\left\{\tan^{-1}\left(\frac{10}{24}\right)\right\} - \tan\left(\frac{\pi}{4}\right)}{1 + \tan\left\{\tan^{-1}\left(\frac{10}{24}\right)\right\} \cdot \tan\left(\frac{\pi}{4}\right)}$$

$$= \frac{\frac{10}{24} - 1}{1 + \frac{10}{24} \cdot 1}$$

$$= \frac{10 - 24}{24 + 10}$$

$$= \frac{-14}{34} = \frac{-7}{17}$$

General Aptitude

16. *(b)* 'Brays' is the sound produced by Donkey, similarly howls is the sound produced by 'Wolf'.

17. *(d)* As 'astronauts' related to space, in the same manner Argonauts is related to 'sea'.

18. *(b)* As orthopaedic is the specialist of Bones, in the same manner dermatologist is the specialist of Skin.

19. *(a)* 'Basilica' is an early Christian church building, in the same manner, 'dormer' is a window that projects vertically from a sloping roof.

20. *(c)* As 'although' means nevertheless in the same manner, 'Though' means 'however'.

21. *(a)* Except Sun, all other are planets.

22. *(b)* Hesitant is different as all others are synonyms of each other.

23. *(b)* Except Tulip, all others are flowers whereas Tulip is a kind of plant.

24. *(d)* Position of P from left = 16

$\therefore$ Position of P from right = 27 − 16 = 11

25. *(d)* 14th to the right of 6th letter from the left

= (6 + 14)th letter from the left

= 20th letter from the left

= T

26. *(a)* In backward order of alphabet,

3rd letter from right = C

13th letter to the left of 3rd letter from right = (13 + 3)th letter from the right

= 16th letter from the right

= P

27. *(d)* The pattern is as follows

23 28 34 41 49 (58)

+5 +6 +7 +8 +9

$\therefore ? = 58$

28. *(a)* The pattern is as follows

+6 +6 +6

11 13 17 19 23 25 (29)

+6 +6

$\therefore ? = 29$

29. *(c)* In figure I, $2^2 + 4^2 = 4 + 16 = 20$

In figure II, $3^2 + 9^2 = 9 + 81 = 90$

Similarly in figure III, $1^2 + 7^2 = 1 + 49 = 50$

30. *(b)* In figure-I

$$\sqrt{64} + \sqrt{36} + \sqrt{49} = 8 + 6 + 7 = 21$$

Similarly, in figure II.

$$\sqrt{121} + \sqrt{81} + \sqrt{100} = 11 + 9 + 10 = 30$$

MANIPAL

Engineering Entrance Exam

Solved Paper 2019

PHYSICS

1. Three blocks of masses 2 kg, 3 kg and 5 kg are connected to each other with light string and are then placed on a frictionless surface as shown in the figure. The system is pulled by a force $F = 10$ N, then tension T_1 is

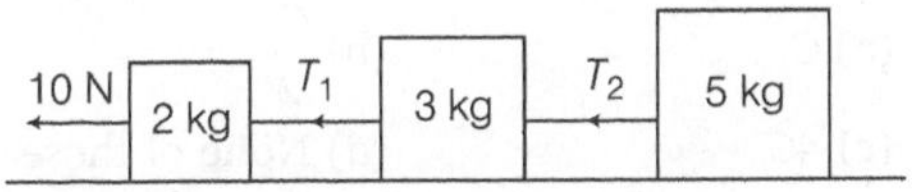

(a) 1 N (b) 5 N
(c) 8 N (d) 10 N

2. Two masses m_1 and m_2 are attached to a string which passes over a frictionless smooth pulley. When $m_1 = 10$ kg, $m_2 = 6$ kg, the acceleration of masses is

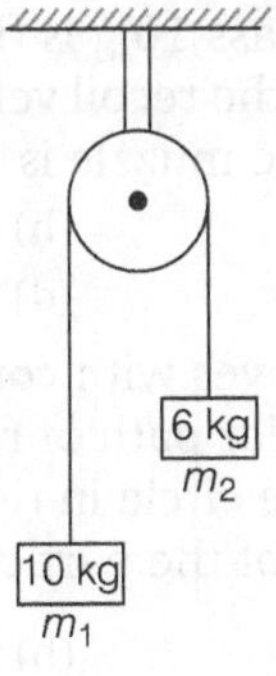

(a) 20 m/s^2 (b) 5 m/s^2
(c) 2.5 m/s^2 (d) 10 m/s^2

3. The unit of L/R is (where L = inductance and R = resistance)
(a) sec (b) sec^{-1}
(c) volt (d) ampere

4. If force is proportional to square of velocity, then the dimensions of proportionality constant is
(a) $[ML^{-1}T]$ (b) $[ML^{-1}T^0]$
(c) $[MLT^0]$ (d) $[M^0LT^{-1}]$

5. Minimum energy required to take out the only one electron from ground state of He^+ is
(a) 13.6 eV (b) 54.4 eV
(c) 27.2 eV (d) 6.8 eV

6. Photons of 5.5 eV energy fall on the surface of the metal emitting photoelectrons of maximum kinetic energy 4.0 eV. The stopping voltage required for these electrons is
(a) 5.5 V
(b) 1.5 V
(c) 9.5 V
(d) 4.0 V

7. Which is different from others in units?
(a) Phase difference
(b) Mechanical equivalent
(c) Loudness of sound
(d) Poisson's ratio

8. Continuous emission spectrum is produced by
(a) in candescent electric lamp
(b) mercury vapour lamp
(c) sodium vapour lamp
(d) polyatomic substances

9. A ball is dropped from top of a tower of 100 m height. Simultaneously another ball was thrown upward from bottom of the tower with a speed of 50 m/s. They will cross each other after ($g = 10\,m/s^2$)
(a) 1 s (b) 2 s (c) 3 s (d) 4 s

10. The driver of a car moving towards a rocket launching pad with a speed of 6 m/s observes that the rocket is moving with speed of 10 m/s. The upward speed of the rocket as seen by the stationary observer is
(a) 4 m/s (b) 6 m/s (c) 8 m/s (d) 11 m/s

11. A satellite revolves very near to the earth surface. Its speed should be around
(a) 5 km/s (b) 8 km/s
(c) 2 km/s (d) 11 km/s

12. If the density of earth is doubled keeping its radius constant, then acceleration due to gravity g is
(a) 20 m/s^2 (b) 10 m/s^2
(c) 5 m/s^2 (d) 2.5 m/s^2

13. A simple pendulum oscillates in a vertical plane. When it passes through the mean position, the tension in the string is 3 times the weight of the pendulum bob. What is the maximum displacement of the pendulum of the string with respect to the vertical ?
(a) 30° (b) 45°
(c) 60° (d) 90°

14. A body of mass 8 kg is moved by a force $F = 3\,x$ N, where x is the distance covered. Initial position is $x = 2$ m and the final position is $x = 10$ m. The initial speed is zero. The final speed is
(a) 6 m/s (b) 12 m/s
(c) 18 m/s (d) 14 m/s

15. Find the equivalent resistance across AB

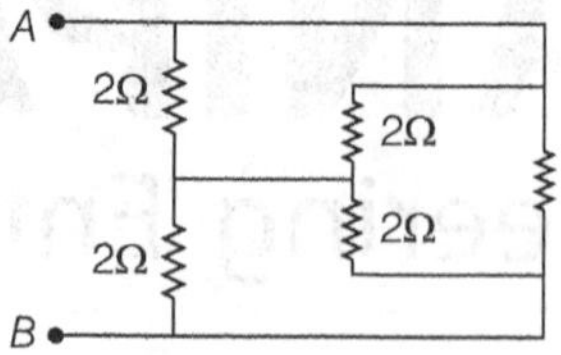

(a) 1Ω (b) 2Ω (c) 3Ω (d) 4Ω

16. The nuclear reaction ${}_1H^1 + {}_1H^1 \rightarrow {}_2He^4$ (mass of deuteron = 2.0141 amu and of He = 4.0024 amu) is
(a) fusion reaction releasing 24 MeV energy
(b) fusion reaction absorbing 24 MeV energy
(c) fission reaction releasing 0.0258 MeV energy
(d) fission reaction absorbing 0.0258 MeV energy

17. A thin metal plate P is inserted half way between the plates of a parallel plate capacitor of capacitance C in such a way that it is parallel to the two plates. The capacitance now becomes
(a) C (b) $\frac{C}{2}$
(c) $4C$ (d) None of these

18. An inclined plane makes an angle 30° with horizontal. A solid sphere rolling down this inclined plane has a linear acceleration of
(a) $\frac{5g}{14}$ (b) $\frac{2g}{3}$
(c) $\frac{g}{3}$ (d) $\frac{5g}{7}$

19. A bullet of mass 10 g is fired from a gun of mass 1 kg. If the recoil velocity is 5 m/s, the velocity of the muzzle is
(a) 0.05 m/s (b) 5 m/s
(c) 50 m/s (d) 500 m/s

20. A particle moves with constant speed v along a circular path of radius r and completes the circle in time T. The acceleration of the particle is
(a) $\frac{2\pi v}{T}$ (b) $\frac{2\pi r}{T}$
(c) $\frac{2\pi r^2}{T}$ (d) $\frac{2\pi v^2}{T}$

21. The separation between C and O-atoms in CO is 1.2 Å. The distance of carbon atom from the centre of mass is
(a) 0.3 Å (b) 0.7 Å (c) 0.5 Å (d) 0.9 Å

22. A body moves a distance of 10 m under the action of force $F = 10\,N$. If the work done is 25 J, the angle which the force makes with the direction of motion is
(a) 0° (b) 30°
(c) 60° (d) None of these

23. When a spring is stretched by 2 cm, it stores 100 J of energy. If it is stretched further by 2 cm, the stored energy will be increased by
(a) 100 J (b) 200 J (c) 300 J (d) 400 J

24. Two wires A and B are of same materials. Their lengths are in the ratio 1 : 2 and diameters are in the ratio 2 : 1. When stretched by force F_A and F_B, respectively, they get equal increase in their lengths. Then the ratio $\frac{F_A}{F_B}$ should be
(a) 1 : 2 (b) 1 : 1 (c) 2 : 1 (d) 8 : 1

25. Mixed He^+ and O^{2+} ions (mass of $He^+ = 4$ amu and that of $O^{2+} = 16$ amu) beam passes a region of constant perpendicular magnetic field. If kinetic energy of all the ions is same, then
(a) He^+ ions will be deflected more than those of O^{2+}
(b) He^+ ions will be deflected less than those of O^{2+}
(c) all the ions will be deflected equally
(d) no ions will be deflected

26. In Young's double slit experiment the wavelength of light was changed from 7000 Å to 3500 Å. While doubling the separation between the slits which of the following is not true for this experiment?
(a) The width of the fringes changes
(b) The colour of bright fringes changes
(c) The separation between successive bright fringes changes
(d) The separation between successive dark fringes remains unchanged

27. The coherence of two light sources means that the light waves emitted have
(a) same frequency
(b) same intensity
(c) constant phase difference
(d) same velocity

28. The valence band and conduction band of a solid overlap at low temperature, the solid may be
(a) a metal
(b) a semiconductor
(c) an insulator
(d) None of the above

29. The dominant contribution to current comes from holes in case of
(a) metals
(b) intrinsic semiconductors
(c) p-type extrinsic semiconductors
(d) n-type extrinsic semiconductors

30. A laser device produces amplification in the
(a) microwave region
(b) ultraviolet or visible region
(c) infrared region
(d) None of the above

31. A point object O is placed in front of a glass rod having spherical end of radius of curvature 30 cm. The image would be formed at

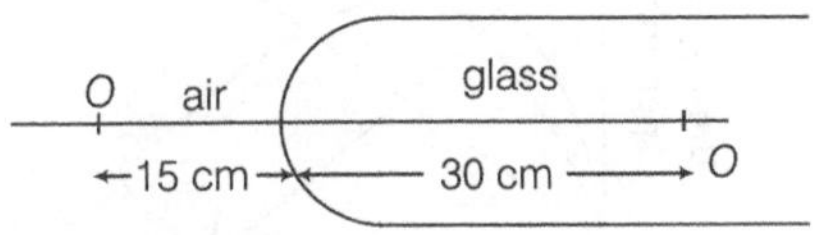

(a) 30 cm left
(b) infinity
(c) 1 cm to the right
(d) 18 cm to the left

32. In the formation of a rainbow light from the sun on water droplets undergoes
(a) dispersion only
(b) only total internal reflection
(c) dispersion and total internal reflection
(d) None of the above

33. If no external voltage is applied across p-n junction, there would be
(a) no electric field across the junction
(b) an electric field pointing from n-type to p-type side across the junction
(c) an electric field pointing from p-type to n-type side across the junction
(d) a temporary electric field during formation of p-n junction that would subsequently disappear

34. Light travelling from a transparent medium to air undergoes total internal reflection at an angle of incidence of 45°. Then refractive index of the medium may be
(a) 1.5 (b) 1.3
(c) 1.1 (d) $1/\sqrt{2}$

35. A force F acting on a body depends on its displacement S as $F \propto S^{-1/3}$. The power delivered by F will depend on displacement as
(a) $S^{2/3}$ (b) $S^{-5/3}$
(c) $S^{1/2}$ (d) S^{0}

36. Binding energy per nucleon plot against the mass number for stable nuclei is shown in the figure. Which curve is correct?

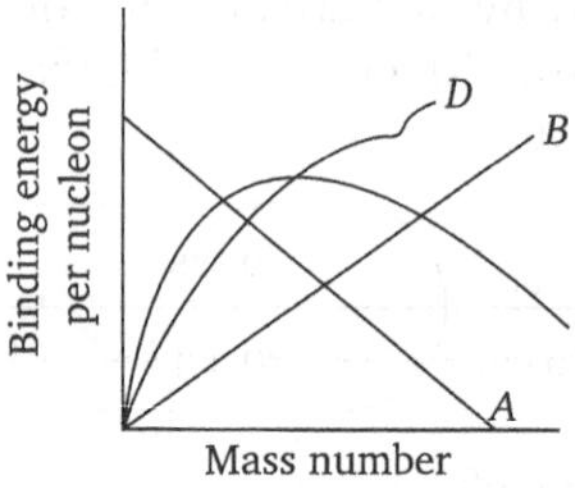

(a) A (b) B
(c) C (d) D

37. A light ray from air is incident (as shown in figure) at one end of a glass fiber (refractive index, $\mu = 1.5$) making an incidence angle of 60° on the lateral surface, so that it undergoes a total internal reflection. How much time would it take to traverse the straight fiber of length 1 km ?

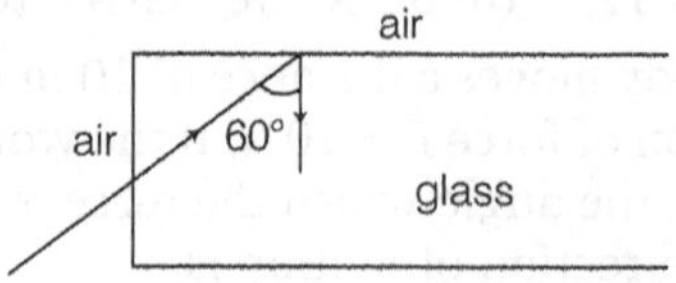

(a) 3.33 μs (b) 6.67 μs
(c) 5.77 μs (d) 3.85 μs

38. A thermodynamic system is taken through the cycle $PQRSP$ process. The net work done by the system is

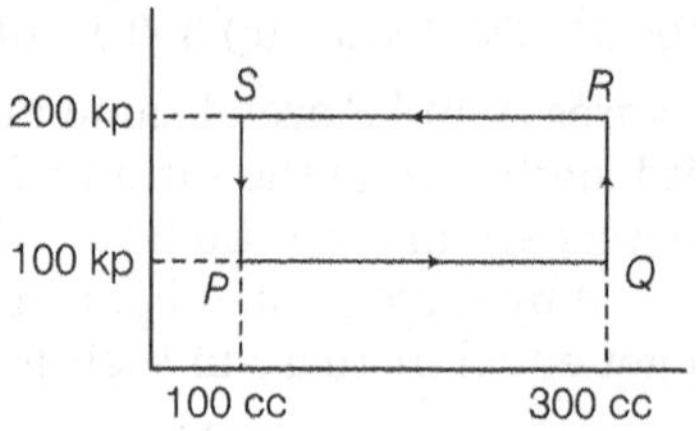

(a) 20 J (b) – 20 J (c) 400 J (d) – 374 J

39. Consider four circuits shown in the figure below. In which circuit power dissipated is greatest?

(Neglect the internal resistance of the power supply).

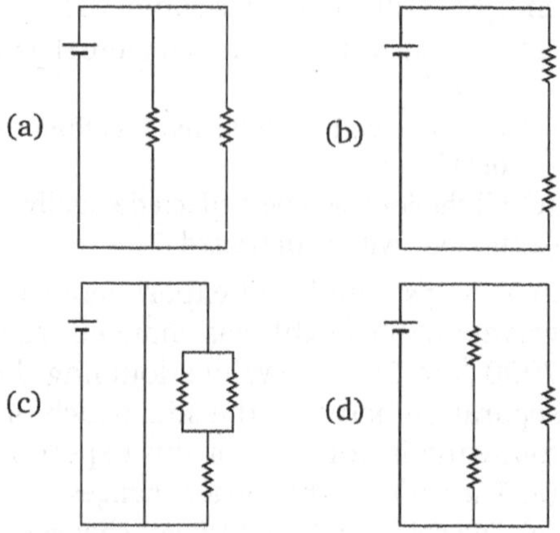

40. Two capacitors C_1 and $C_2 = 2C_1$ are connected in a circuit with a switch between them as shown in the figure. Initially the switch is open and C_1 holds charge Q. The switch is closed. At steady state, the charge on each capacitors will be

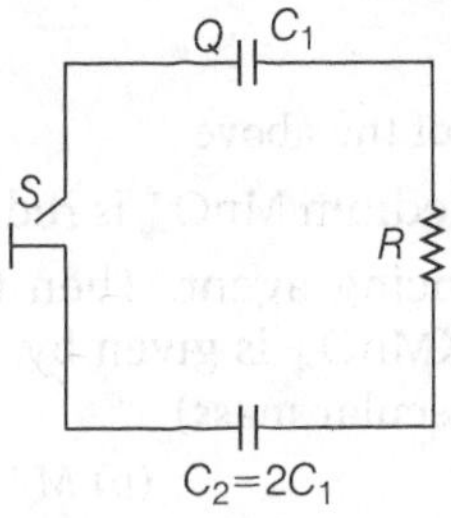

(a) $Q, 2Q$ (b) $\frac{Q}{3}, \frac{2Q}{3}$

(c) $\frac{3Q}{2}, 3Q$ (d) $\frac{2Q}{3}, \frac{4Q}{3}$

41. A particle is moving in a vertical circle. The tensions in the string when passing through two positions at angles 30° and 60° from vertical (lowest positions) are T_1 and T_2, respectively. Then

(a) $T_1 = T_2$

(b) $T_2 > T_1$

(c) $T_1 > T_2$

(d) tension in the string always remains the same

42. A coil of 100 turns carries a current of 5 mA and creates a magnetic flux of 10^{-5} wb. The inductance is

(a) 0.2 mH (b) 2.0 mH

(c) 0.02 mH (d) None of these

43. The starter motor of a car draw a current $I = 300$ A from the battery of voltage 12 V. If the car starts only after 2 minutes, what is the energy drawn from the battery ?

(a) 3 kJ (b) 30 kJ

(c) 7.2 kJ (d) 432 kJ

44. Surface of the lake is at 2°C and depth of the lake is 20 m. Find the temperature of the bottom of the lake

(a) 2°C (b) 3°C

(c) 4°C (d) None of these

45. $y_1 = 4\sin(\omega t + kx)$, $y_2 = -4\cos(\omega t + kx)$, the phase difference is

(a) $\frac{\pi}{2}$ (b) $\frac{3\pi}{2}$ (c) π (d) zero

46. A charged particle of mass 0.003 g is held stationary in space by placing it in a downward direction of electric field of 6×10^4 N/C. Then the magnitude of the charge is

(a) 5×10^{-4} C (b) 5×10^{-10} C

(c) -18×10^{-6} C (d) -5×10^{-9} C

47. A parallel plate capacitor has an electric field of 10^5 V / m between the plates. If the charge on the capacitor plate is 1 μC, the force on each capacitor plate is

(a) 0.5 N (b) 0.05 N

(c) 0.005 N (d) None of these

48. 1 g of water at atmospheric pressure has volume of 1 cc and when boiled it becomes 1681 cc of steam. The heat of vaporisation of water is 540 cal/g. Then the change in its internal energy in this process is

(a) 540 cal

(b) 500 cal

(c) 1681 cal

(d) None of the above

49. A physicist works in a laboratory where the magnetic field is 2 T. She wears a necklace enclosing area 0.01 m^2 in such a way that the plane of the necklace is normal to the field and is having a resistance, $R = 0.01\Omega$. Because of power failure, the field decays to 1 T in time 10^{-3} s. Then what is the total heat produced in her necklace?

(a) 10 J (b) 20 J

(c) 30 J (d) 40 J

50. The temperature coefficient of resistance of a wire is 0.00125° C^{-1}. At 300 K, its resistance is 1Ω. At what temperature the resistance of the wire will be 2Ω?

(a) 800 K

(b) 1100 K

(c) 600 K

(d) None of the above

CHEMISTRY

1. The solubility CaF_2 is s moles/litre. Then solubility product is
(a) s^2 (b) $4s^3$ (c) $3s^2$ (d) s^3

2. If p, T, ρ and R represents pressure, temperature, density and universal gas constant respectively, then the molar mass of the ideal gas is given by
(a) $\frac{\rho RT}{p}$ (b) $\frac{\rho T}{pR}$ (c) $\frac{p}{\rho RT}$ (d) $\frac{RT}{\rho p}$

3. The kinetic energy of a gas molecule is ..... temperature
(a) independent of
(b) directly proportional to
(c) inversely proportional to
(d) directly proportional to square root of

4. The ratio of rate of diffusion of $SO_2 (M = 64)$ and oxygen $(M = 32)$ is
(a) 1 : 1 (b) 2 : 1
(c) 1 : 2 (d) 1 : 1.414

5. An aqueous solution freezes at – 0.186° C, then elevation in boiling point is $(K_b = 0.512, K_f = 1.86)$
(a) 0.0512° C (b) 100.0512° C
(c) – 0.0512° C (d) None of these

6. Arrange the following in the increasing order of their bond order

$$O_2, O_2^+, O_2^- \text{ and } O_2^{2-}$$

(a) $O_2^{2-}, O_2^-, O_2, O_2^+$ (b) $O_2^{2-}, O_2^-, O_2^+, O_2$
(c) $O_2^+, O_2, O_2^-, O_2^{2-}$ (d) $O_2, O_2^+, O_2^-, O_2^{2-}$

7. In hydrolysis of a salt of weak acid and strong base, $A^- + H_2O \rightleftharpoons HA + OH^-$, the hydrolysis constant (K_h) is equal to
(a) $\frac{K_w}{K_a}$ (b) $\frac{K_w}{K_b}$ (c) $\sqrt{\frac{K_a}{C}}$ (d) $\frac{K_w}{K_a \times K_b}$

8. For a reaction of the type $aA + bB \longrightarrow$ Products, the $-\frac{d[A]}{dt}$ is equal to
(a) $-\frac{d[B]}{dt}$ (b) $-\frac{1}{b} \times \frac{d[B]}{dt}$
(c) $-\frac{a}{b} \times \frac{d[B]}{dt}$ (d) $-\frac{b}{a} \times \frac{d[B]}{dt}$

9. In a mixture of 1 g H_2 and 8 g O_2 the mole fraction of hydrogen is
(a) 0.667
(b) 0.5
(c) 0.33
(d) None of the above

10. In acid medium MnO_4^- is reduced to Mn^{2+}, by a reducing agent. Then the equivalent mass of $KMnO_4$ is given by
(M = molecular mass)
(a) $M/2$ (b) M
(c) $M/5$ (d) $M/3$

11. For the reaction
$$CH_3COOH_{(l)} + 2O_{2(g)} \rightleftharpoons 2CO_{2(g)} + 2H_2O_{(l)}$$
at 25°C and 1 atm pressure, $\Delta H = -874$ kJ. Then the change in internal energy (ΔE) is
(a) – 874 kJ (b) – 871.53 kJ
(c) – 876.47 kJ (d) + 874 kJ

12. The electrophile involved in the sulphonation of benzene is
(a) SO_3^+ (b) SO_3^{2-}
(c) H_3^+O (d) SO_3

13. According to the adsorption theory of catalysis, the speed of reaction increases because
(a) adsorption lowers the activation enegry of the reaction
(b) the concentration of reactant molecules at the active centres of the catalyst becomes high due to adsorption
(c) in the process of adsorption, the activation energy of the molecules becomes large
(d) adsorption produces heat which increases the speed of the reaction

14. The product of reaction between alcohalic silver nitrite with etyle bromide is
(a) ethene
(b) ethane
(c) ethyle nitrile
(d) nitro ethane

15. The specific conductivity of 0.1 N KCl solution is 0.0129 $ohm^{-1}cm^{-1}$. The resistance of the solution in the cell is 100 ohm. The cell constant of the cell will be
(a) 1.10 (b) 1.29
(c) 0.56 (d) 2.80

16. The standard emf of a cell $Zn / Zn^{2+} || Fe^{2+}/Fe$, if electrode potentials for (Zn/Zn^{2+}) and (Fe^{2+}/Fe) and 0.763 V and – 0.44 V respectively is
(a) + 0.323 V (b) – 1.203 V
(c) + 1.203 V (d) – 0.323 V

17. Which of the following has highest second ionisation energy
(a) Chromium
(b) Calcium
(c) Iron
(d) Cobalt

18. Which of the following plot represents the graph of pH against volume of alkali added in the titration of NaOH and HCl?

(a) 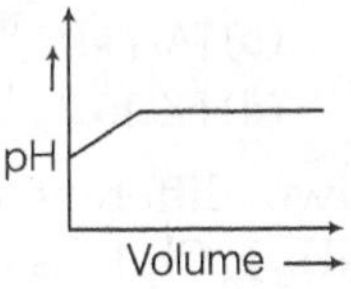

(b) 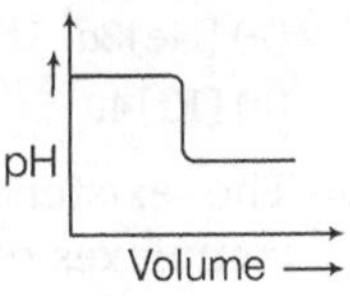

(c) 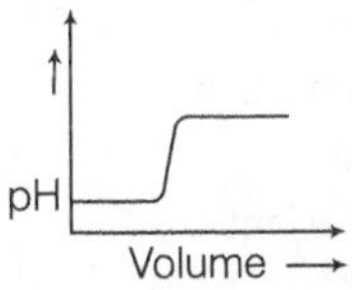

(d) 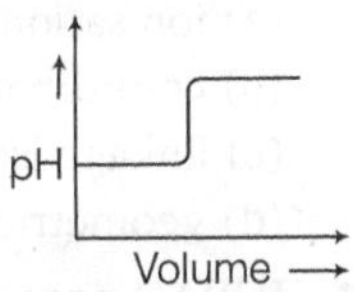

19. CO_2 is a gas, while SiO_2 is a solid, but both are
(a) acidic
(b) ionic
(c) discrete molecules
(d) covalent containing π-bonds

20. Pure conc. HNO_3 make iron passive, as the surface is covered with protective layer of
(a) Fe_2O_3
(b) FeO
(c) Fe_3O_4
(d) $Fe(NO_3)_3$

21. Ozone when reacts with potassium iodide solution liberates certain product, which turns starch paper blue. The liberated substance is
(a) oxygen
(b) iodine
(c) hydrogen iodide
(d) potassium hydroxide

22. Red hot iron absorbs SO_2 giving the product
(a) $FeS + O_2$ (b) $Fe_2O_3 + FeS$
(c) FeO + FeS (d) FeO + S

23. Ethyl iodide when heated with sodium in dry ether gives pure
(a) C_4H_{10} (b) C_2H_6
(c) C_3H_8 (d) C_2H_5OH

24. For the reaction,

$$CH_3—CH{=}CH_2 + HOCl \longrightarrow A$$

The product A is :
(a) $CH_3—CHCl—CH_2OH$
(b) $CH_3—CH(OH)—CH_2—Cl$
(c) $CH_3—CH_2—CH_2—COCl$
(d) $CH_3—C(Cl)(OH)—CH_3$

25. Which of the following is not correct for ionic crystals?
(a) They possess high melting point and boiling point
(b) All are electrolyte
(c) Exhibit the property of isomorphism
(d) Exhibit directional properties of the bond

26. Which of the following is not true in Rutherford's nuclear' model of atom?
(a) Protons and neutrons are present inside nucleus
(b) Volume of nucleus is very small as compared to volume of atom
(c) The number of protons and neutrons are always equal
(d) The number of electrons and protons are always equal

27. All the s-block elements of the periodic table are placed in the groups .....
(a) IA and IIA (b) IIIA and IVA
(c) B sub groups (d) VA to VIIA

28. The magnetic quantum number for *d*-orbital is given by
(a) 2 (b) 0, ± 1, ± 2
(c) 0, 1, 2 (d) 5

29. XeF_2 is isostructural with
(a) ICL_2 (b) $SbCl_3$
(c) $BaCl_2$ (d) TeF_2

30. In the process, $O_2^+ \longrightarrow O_2^{2+} + e^-$ the electron lost is from
(a) bonding π-orbital
(b) antibonding π-orbital
(c) $2p_z$ orbital (d) $2p_x$ orbital

31. Bond between A and B can be represented by

$$\underset{(I)}{A—B}, \underset{(II)}{A^+B^-}, \underset{(III)}{A^-B^+}$$

If A is more electronegative than B, then least contribution to the actual structure comes from
(a) I
(b) II
(c) III
(d) All the structures have equal contribution

32. In the following reaction
$NaOH + S \rightarrow A + Na_2S + H_2O$; A is
(a) Na_2SO_4
(b) Na_2SO_3
(c) Na_2S
(d) $Na_2S_2O_3$

33. SiF_4 gets hydrolysed giving
(a) SiO_2 (b) $Si(OH)_2F_2$
(c) H_2SiF_6 (d) $Si(OH)_4$

34. All ores are minerals, while all minerals are not ores, because
(a) the metal cannot be extracted economically from all the minerals
(b) minerals are complex compounds
(c) the minerals are obtained from mines
(d) All of these are correct

35. In the reaction,
$P_2O_5 + 3CaO \longrightarrow Ca_3(PO_4)_2$, P_2O_5 acts as
(a) acidic flux (b) basic flux
(c) basic impurity (d) acidic impurity

36. In the given reaction, the oxide of sodium is

$$\begin{pmatrix} 4Na + O_2 \longrightarrow 2Na_2O \\ Na_2O + H_2O \longrightarrow 2NaOH \end{pmatrix}$$

(a) acidic (b) basic
(c) amphoteric (d) neutral

37. When CO_2 is passed through solution of calcium hydroxide, which one of the following compound is precipitated?
(a) $Ca(HCO_3)_2$ (b) CaO
(c) $CaCO_3$ (d) $Ca(OH)_2$

38. Ferric alum has the composition
$(NH_4)_2SO_4 \cdot Fe_2(SO_4)_3 \cdot xH_2O$
The value of x is
(a) 7 (b) 24
(c) 6 (d) 15

39. What is the general electronic configuration for 2nd row transition series?
(a) $[Ne]3d^{1-10}, 4s^2$ (b) $[Ar]3d^{1-10}, 4s^{1-2}$
(c) $[Kr]4d^{1-10}, 5s^{1-2}$ (d) $[Xe]5d^{1-10}, 5s^{1-2}$

40. The existence of two different coloured complexes of $[Co(NH_3)_4Cl_2]^+$ is due to
(a) ionisation isomerism
(b) co-ordination isomerism
(c) linkage isomerism
(d) geometrical isomerism

41. IUPAC name of the compound

$$CH_3—\underset{\underset{OH}{|}}{C}H—CH_2—\underset{\underset{CH_3}{|}}{C}H—CH_3$$ is

(a) 4-methyl pentene-2-ol
(b) 2-methyl pentanol-4
(c) 4, 4-dimethyl-butane-2-ol
(d) 4-methyl pentane-2-ol

42. Alkyl halide on heating with alc.NH_3 in a sealed tube results .....
(a) 1° amine (b) 2° amine
(c) 3° amine (d) all of these

43. Among H—CHO, CH_3CHO and C_6H_5CHO, which will undergo Cannizaro's reaction?
(a) HCHO and CH_3—CHO
(b) CH_3—CHO and C_6H_5CHO
(c) C_6H_5CHO and HCHO
(d) All of the above

44. The main product of the reaction of CH_3CONH_2 with Br_2 in aqueous potassium hydroxide medium is
(a) CH_3—CH_2—NH_2 (b) CH_3Br
(c) $CH_3CONHBr$ (d) CH_3NH_2

45. In the reaction,
$$HCHO + CH_3MgI \longrightarrow A \xrightarrow{H_2O} B + Mg(OH)I$$
What are A and B?
(a) CH_3OMgI and CH_3—OH
(b) CH_3CH_2OMgI and C_2H_5—O—C_2H_5
(c) CH_3CH_2OMgI and CH_3—CH_2—OH
(d) CH_3—CH_2—I and CH_3—CH_2—OH

46. Acetylation of a secondary amine in alkaline medium yields
(a) N, N-dialkyl acetamide
(b) N, N-dialkyl amine
(c) N, N-dialkyl amide
(d) acetyl dialkyl amine

47. In acid medium nitrobenzene is reduced to aniline as shown in the reaction
$$C_6H_5—NO_2 + 6[H] \longrightarrow C_6H_5—NH_2 + 2H_2O$$
The reducing agent used in this reaction is :
(a) $LiAlH_4$ (b) Sn / HCl
(c) Na/Alcohol (d) H_2/Ni

48. Acetyl salicylic acid is used as
(a) anti oxidant (b) analgesic drug
(c) anti biotic drug (d) anaesthetic

49. C_6H_6 consists of one ring, while naphthlene consists of two rings. Both of them are aromatic and obey the $(4n + 2)$ rule. Thus the number of π-electrons inside rings of C_6H_6 and naphthalene are respectively
(a) 3, 5
(b) 5, 10
(c) 6, 10
(d) 6, 12

50. In the oxidation of C_6H_5—CH_2—CH_3 by $KMnO_4$ the product formed is
(a) C_6H_5—CH_2—CHO
(b) C_6H_5—CH_2—COOH
(c) C_6H_6—COOH
(d) C_6H_5—CH_2—OH

MATHEMATICS

1. If the angles between the pair of straight lines represented by the equation $x^2 - 3xy + \lambda y^2 + 3x - 5y + 2 = 0$ is $\tan^{-1}\frac{1}{3}$.
Where λ is a non-negative real number, then λ is
(a) 2 (b) 0
(c) 3 (d) 1

2. The distance of the line $2x - 3y = 4$ from the point (1, 1) measured parallel to the line $x + y = 1$ is
(a) $\sqrt{2}$ (b) $5/\sqrt{2}$
(c) $1/\sqrt{2}$ (d) 6

3. The equation of bisectors of the angles between the lines $|x| = |y|$ are
(a) $y = \pm x$ and $x = 0$
(b) $x = \frac{1}{2}$ and $y = \frac{1}{2}$
(c) $y = 0$ and $x = 0$
(d) None of the above

4. The base of vertices of an isosceles triangle PQR are $Q(1, 3)$ and $R(-2, 7)$. The vertex P can be
(a) (1, 6) (b) $\left(\frac{1}{2}, 5\right)$
(c) $\left(\frac{5}{6}, 6\right)$ (d) None of these

5. The normal at the point (3, 4) on a circle cuts the circle at the point (– 1, – 2). Then the equation of the circle is
(a) $x^2 + y^2 + 2x - 2y - 13 = 0$
(b) $x^2 + y^2 - 2x - 2y - 11 = 0$
(c) $x^2 + y^2 - 2x + 2y + 12 = 0$
(d) $x^2 + y^2 - 2x - 2y + 14 = 0$

6. If $\cos P = \frac{1}{7}$ and $\cos Q = \frac{13}{14}$, where P and Q both are acute angles. Then the value of $P - Q$ is
(a) 30° (b) 60°
(c) 45° (d) 75°

7. The equation $3 \cos x + 4 \sin x = 6$ has ..... solution.
(a) finite (b) infinite
(c) one (d) no

8. If $\sec^{-1} x = \text{cosec}^{-1} y$, then $\cos^{-1}\frac{1}{x} + \cos^{-1}\frac{1}{y}$ is equal to
(a) π (b) $\frac{\pi}{4}$ (c) $-\frac{\pi}{2}$ (d) $\frac{\pi}{2}$

9. If n be any integer, then $n(n + 1)(2n + 1)$ is:
(a) odd number
(b) integral multiple of 6
(c) perfect square
(d) does not necessarily have any of the foregoing proof

10. If $\tan\theta = -\frac{4}{3}$, then the value of $\sin\theta$ is
(a) $-\frac{4}{5}$ but $\neq \frac{4}{5}$ (b) $-\frac{4}{5}$ or $\frac{4}{5}$
(c) $\frac{4}{5}$ but $\neq -\frac{4}{5}$ (d) $\frac{1}{5}$

11. If $c = 2\cos\theta$, then the value of the determinant $\Delta = \begin{vmatrix} c & 1 & 0 \\ 1 & c & 1 \\ 6 & 1 & c \end{vmatrix}$ is
(a) $\frac{\sin 4\theta}{\sin\theta}$ (b) $\frac{2\sin^2 2\theta}{\sin\theta}$
(c) $4\cos^2\theta(2\cos\theta - 1)$ (d) None of these

12. The equation of the parabola whose vertex is (– 1, – 2), axis is vertical and which passes through the point (3, 6), is
(a) $x^2 + 2x - 2y - 3 = 0$
(b) $2x^2 = 3y$
(c) $x^2 - 2x + 2y - 3 = 0$
(d) $x^2 - 2x - 2y - 3 = 0$

13. The length of the axis of the conic $9x^2 + 4y^2 - 6x + 4y + 1 = 0$ are
(a) $\frac{1}{2}, 9$ (b) $3, \frac{2}{5}$ (c) $1, \frac{2}{3}$ (d) 3, 2

14. If $f(x) = \cot^{-1}\left(\frac{3x - x^3}{1 - 3x^2}\right)$ and $g(x) = \cos^{-1}\left(\frac{1 - x^2}{1 + x^2}\right)$, then $\lim_{x \to a}\frac{f(x) - f(a)}{g(x) - g(a)}, 0 < a < \frac{1}{2}$, is
(a) $\frac{3}{2(1 + a^2)}$ (b) $\frac{3}{2(1 + x^2)}$
(c) $\frac{3}{2}$ (d) $-\frac{3}{2}$

15. If $f(x) = \begin{cases} x, & 0 \le x \le 1 \\ 2x - 1, & 1 < x \end{cases}$, then
(a) f is discontinuous at $x = 1$
(b) f is differentiable at $x = 1$
(c) f is continuous but not differentiable at $x = 1$
(d) None of the above

16. $\lim_{x \to -2}\frac{\sin^{-1}(x + 2)}{x^2 + 2x}$ is equal to
(a) 0 (b) ∞
(c) $-\frac{1}{2}$ (d) None of these

17. Let $f(x) = x^p \cos\left(\frac{1}{x}\right)$, when $x \neq 0$ and $f(x) = 0$, when $x = 0$. Then $f(x)$ will be differentiable at $x = 0$, if
(a) $p > 0$ (b) $p > 1$
(c) $0 < p < 1$ (d) $\frac{1}{2} < p < 1$

18. The derivative of $f(x) = 3|2 + x|$ at the point $x_0 = -3$ is
(a) 3
(b) -3
(c) 0
(d) None of the above

19. Derivative of the function $f(x) = \log_5(\log_7 x), x > 7$ is
(a) $\frac{1}{x(\log 5)(\log 7)(\log_7 x)}$
(b) $\frac{1}{x(\log 5)(\log 7)}$
(c) $\frac{1}{x(\log x)}$
(d) None of the above

20. If $z = x + iy, z^{1/3} = a - ib$, then $\frac{x}{a} - \frac{y}{b} = k(a^2 - b^2)$, where k is equal to :
(a) 1 (b) 2
(c) 3 (d) 4

21. The number of real solutions of the equation $1 + |e^x - 1| = e^x(e^x - 2)$ is
(a) 1 (b) 2
(c) 4 (d) 8

22. If $u = x^2 + y^2$ and $x = s + 3t, y = 2s - t$, then $\frac{d^2u}{ds^2}$ is equal to
(a) 12 (b) 10
(c) 32 (d) 36

23. If the equation $x^2 + px + q = 0$ and $x^2 + qx + p = 0$ have a common root then $p + q + 1$ is equal to
(a) 0 (b) 1
(c) 2 (d) -1

24. The value of a $(a \geq b)$ for which the sum of the cubes of the roots of $x^2 - (a-2)x + (a-3) = 0$ assumes the least value, is
(a) 3
(b) 4
(c) 5
(d) None of the above

25. Let z_1, z_2, z_3 be three vertices of an equilateral triangle circumscribing the circle $|z| = \frac{1}{2}$. If $z_1 = \frac{1}{2} + \frac{i\sqrt{3}}{2}$ and z_1, z_2, z_3 were in anticlockwise sense, then z_2 is
(a) $1 + \sqrt{3}i$ (b) $1 - \sqrt{3}i$ (c) 1 (d) -1

26. If $z = \frac{-2}{1 + \sqrt{3}i}$, then the value of arg (z) is
(a) π (b) $\frac{\pi}{3}$ (c) $\frac{2\pi}{3}$ (d) $\frac{\pi}{4}$

27. Let ω is an imaginary cube roots of unity, then the value of
$2(1 + \omega)(1 + \omega^2) + 3(2\omega + 1)(2\omega^2 + 1) + \ldots + (n + 1)(n\omega + 1)(n\omega^2 + 1)$ is
(a) $\left[\frac{n(n+1)}{2}\right]^2 + n$ (b) $\left[\frac{n^2(n+1)^2}{4}\right]$
(c) $\left[\frac{n(n+1)}{2}\right]^2 - n$ (d) None of these

28. The locus of the point z satisfying $\arg\left(\frac{z-1}{z+1}\right) = k$, (where k is non-zero) is
(a) a circle with centre on y-axis
(b) circle with centre on x-axis
(c) a straight line parallel to x-axis
(d) a straight line making an angle 60° with the x-axis

29. If (3,4,5), $Q(4, 6, 3)$, $R(-1, 2, 4)$, $S(1, 0, 5)$, then the projection of RS on PQ is
(a) $-\frac{2}{3}$ (b) $\frac{4}{3}$ (c) $\frac{1}{2}$ (d) 2

30. If a line makes α, β, γ with the positive direction of x, y and z-axes respectively. Then, $\cos^2\alpha + \cos^2\beta + \cos^2\gamma$ is equal to
(a) $\frac{1}{2}$ (b) $-\frac{1}{2}$ (c) -1 (d) 1

31. The projection of a line on a co-ordinate axes are 2, 3, 6. Then the length of the line is
(a) 7 (b) 5 (c) 1 (d) 11

32. The decimal equivalent of the binary number 10011.1 is
(a) 19.50 (b) 11001.11
(c) 5005.55 (d) 19.10

33. The binary represents of 60 is
(a) 101110 (b) 111100
(c) 110011 (d) 110000

34. Which of the following statement is not tautology?
(a) $\sim(p \wedge q) \vee p$
(b) $(p \wedge q) \Rightarrow p$
(c) $q \vee \sim(p \wedge q)$
(d) $(\sim p \wedge q) \cap (\sim p \vee p)$

35. The period of
$$f(x) = \sin\left(\frac{\pi x}{n-1}\right) + \cos\left(\frac{\pi x}{n}\right),\ n \in Z,\ n > 2$$
is
(a) $2\pi n(n-1)$
(b) $4n(n-1)$
(c) $2n(n-1)$
(d) None of the above

36. For $\theta > \frac{\pi}{3}$, the value of $f(\theta) = \sec^2\theta + \cos^2\theta$ always lies in the interval
(a) $(0, 2)$ (b) $[0, 1]$
(c) $(1, 2)$ (d) $[2, \infty)$

37. The radius of the circle whose arc of length 15 cm makes an angle of $\frac{3}{4}$ radian at the centre, is
(a) 10 cm (b) 20 cm
(c) $11\frac{1}{4}$ cm (d) $22\frac{1}{2}$ cm

38. If $f_n(x) = e^{f_{(n-1)}(x)}$, for all $n \in N$ and $f_0(x) = x$, then $\frac{d}{dx}\{f_n(x)\}$ is equal to
(a) $f_n(x)\, f_{n-1}(x)$
(b) $f_n(x)\frac{d}{dx}\{f_{n+1}(x)\}$
(c) $f_n(x) \cdot f_{n-1}(x) \ldots f_2(x)\, f_1(x)$
(d) None of the above

39. If $3^x + 2^{2x} \geq 5^x$, then the solution set for x is
(a) $(-\infty, 2]$ (b) $[2, \infty)$
(c) $[0, 2]$ (d) $\{2\}$

40. The number of integral solution of $\frac{x+1}{x^2+2} > \frac{1}{4}$ is
(a) 1 (b) 2
(c) 5 (d) None of these

41. The triangle PQR of which the angles P, Q, R satisfy $\cos P = \frac{\sin Q}{2 \sin R}$ is
(a) equilateral (b) right angled
(c) any triangle (d) isosceles

42. If $f(x) = (a - x^n)^{1/n}$, where $a > 0$ and n is a positive integer, then $f[f(x)]$ is equal to
(a) x^3 (b) x^2
(c) x (d) None of these

43. A function $f(x) = \frac{x^2 - 3x + 2}{x^2 + 2x - 3}$ is
(a) maximum at $x = -3$
(b) maximum at $x = -3$ and maximum at $x = 1$
(c) maximum at $x = 1$
(d) function is increasing in its domain

44. The locus of the point $P(x, y)$ satisfying the relation
$$\sqrt{(x-3)^2 + (y-1)^2} + \sqrt{(x+3)^2 + (y-1)^2} = 6$$ is
(a) straight line
(b) pair of straight lines
(c) circle
(d) ellipse

45. If z_1, z_2 and z_3 are complex number such that $|z_1| = |z_2| = |z_3| = \left|\frac{1}{z_1} + \frac{1}{z_2} + \frac{1}{z_3}\right| = 1$ then $|z_1 + z_2 + z_3|$ is
(a) equal to 1 (b) less than 1
(c) greater than 3 (d) equal to 3

46. Equation $x * a = b$ has ijn group $(G, \forall)$
(a) unique solution $b * a^{-1}$
(b) unique solution $a^{-1} * b$
(c) unique solution $a^{-1} * b^{-1}$
(d) many solutions

47. If $f(x) = \cos[\pi^2]x + \cos[-\pi^2]x$, then

(a) $f\left(\frac{\pi}{4}\right) = 2$ (b) $f(-\pi) = 2$

(c) $f(\pi) = 1$ (d) $f\left(\frac{\pi}{2}\right) = -1$

48. The range of
$f(x) = \sec\left(\frac{\pi}{4}\cos^2 x\right), -\infty < x < \infty$ is

(a) $[1, \sqrt{2}]$
(b) $[1, \infty)$
(c) $[-\sqrt{2}, -1] \cup [1, \sqrt{2}]$
(d) $(-\infty, 1] \cup [1, \infty)$

49. The domain of the function
$f(x) = \frac{\sin^{-1}(3-x)}{\log(|x|-2)}$ is :

(a) $[2, 4]$
(b) $(3, 4]$
(c) $[2, \infty)$
(d) $(-\infty, 3) \cup [2, \infty)$

50. The remainder obtained when $1! + 2! + \ldots + 200!$ is divided by 14 is
(a) 3
(b) 4
(c) 5
(d) None of the above

51. $\cos^{-1}\{\cos 2\cot^{-1}(\sqrt{2}-1)\}$ is equal to

(a) $\sqrt{2}-1$ (b) $\frac{\pi}{4}$

(c) $\frac{3\pi}{4}$ (d) 0

52. The function $f(x) = [x]\cos\left[\frac{2x-1}{2}\right]\pi$, where [·] denotes the greatest integer function, is discontinuous at
(a) all x
(b) no x
(c) all integral points
(d) x which is not an integer

53. A perpendicular is drawn from the point $P(2, 4, -1)$ to the line $\frac{x+5}{1} = \frac{y+3}{4} = \frac{z-6}{-9}$. The equation of the perpendicular from P to the given line is

(a) $\frac{x-2}{6} = \frac{y-4}{3} = \frac{z+1}{2}$
(b) $\frac{x+2}{6} = \frac{y-4}{3} = \frac{z+1}{2}$
(c) $\frac{x+2}{-6} = \frac{y-4}{3} = \frac{z+1}{2}$
(d) $\frac{x+2}{6} = \frac{y+4}{3} = \frac{z+1}{2}$

54. One mapping (function) is selected at random from all the mappings of the set $A = \{1, 2, 3, \ldots, n\}$ into itself. The probability that the mapping selected is one-one, is

(a) $\frac{n!}{n^{n-1}}$ (b) $\frac{n!}{n^n}$

(c) $\frac{n!}{2n^n}$ (d) None of these

55. If $\mathbf{a} = -\hat{\mathbf{i}} + 2\hat{\mathbf{j}} - \hat{\mathbf{k}}$, $\mathbf{b} = \hat{\mathbf{i}} + \hat{\mathbf{j}} - 3\hat{\mathbf{k}}$ and $\mathbf{c} = -4\hat{\mathbf{i}} - \hat{\mathbf{k}}$, then $\mathbf{a} \times (\mathbf{b} \times \mathbf{c}) + (\mathbf{a} \cdot \mathbf{b})\mathbf{c}$

(a) $5\hat{\mathbf{i}} + 5\hat{\mathbf{j}} - 15\hat{\mathbf{k}}$ (b) 0

(c) $12\hat{\mathbf{j}} + 4\hat{\mathbf{k}}$ (d) $-3\hat{\mathbf{i}} + 6\hat{\mathbf{j}} - 3\hat{\mathbf{k}}$

56. The number of five digits numbers that can be formed without any restriction is
(a) 990000 (b) 10000
(c) 90000 (d) None of these

57. The maximum value of $3\cos\theta + 4\sin\theta$ is
(a) -5 (b) 5
(c) 25 (d) None of these

58. The vertex connectivity of any tree is
(a) one (b) two
(c) three (d) None of these

59. If $\mathbf{a}$, $\mathbf{b}$ and $\mathbf{c}$ are non-collinear vectors such that for some scalars, x, y, z, $x\mathbf{a} + x\mathbf{b} + x\mathbf{c} = 0$, then
(a) $x = 0, y = 0, z = 0$
(b) $x \neq 0, y \neq 0, z = 0$
(c) $x = 0, y \neq 0, z \neq 0$
(d) $x \neq 0, y \neq 0, z \neq 0$

60. Which is the correct order for a given number α in increasing order?
(a) $\log_2 \alpha, \log_3 \alpha, \log_e \alpha, \log_{10} \alpha$
(b) $\log_{10} \alpha, \log_3 \alpha, \log_e \alpha, \log_2 \alpha$
(c) $\log_{10} \alpha, \log_e \alpha, \log_2 \alpha, \alpha \log_3 \alpha$
(d) None of the above

61. In ΔABC, $\frac{b-c}{r_1}+\frac{c-a}{r_2}+\frac{a-b}{r_3}$ is equal to
(a) 0 (b) abc
(c) $a+b+c$ (d) $ab+bc+ca$

62. If $n \in N$, then $|\sin nx|$
(a) $\le n|\sin x|$ (b) $\ge n|\sin x|$
(c) $= n|\sin x|$ (d) None of these

63. If x follows a binomial distribution with parameters $n = 100$ and $p = \frac{1}{3}$, then $p(X = r)$ is maximum when r equals
(a) 16 (b) 32
(c) 33 (d) None of these

64. The direction cosines of any normal to the xy-plane are
(a) 1, 0, 0 (b) 0, 1, 0
(c) 1, 1, 0 (d) 0, 0, 1

65. If $\int f(x)\,dx = F(x)$, then $\int x^3 f(x^2)\,dx$ is equal to
(a) $\frac{1}{2}[x^2\{F(x)\}^2 - \int \{F(x)\}^2\,dx]$
(b) $\frac{1}{2}[x^2 F(x^2) - \int F(x^2)\,d(x^2)]$
(c) $\frac{1}{2}[x^2F(x) - \frac{1}{2}\int \{F(x)\}^2\,dx]$
(d) None of the above

66. The approximate value of $\int_1^5 x^2\,dx$ using trapezoidal rule with $n = 4$ is
(a) 41 (b) 41.5 (c) 41.75 (d) 42

67. If the distance between the foci of an ellipse is 6 and the length of the minor axis is 8, then the eccentricity is
(a) $\frac{1}{\sqrt{5}}$ (b) $\frac{1}{2}$ (c) $\frac{3}{5}$ (d) $\frac{4}{5}$

68. If $I_n = \int \sin^n x\,dx$, then $nI_n - (n-1)I_{n-2}$ equals
(a) $\sin^{n-1} x \cos x$ (b) $\cos^{n-1} x \sin x$
(c) $-\sin^{n-1} x \cos x$ (d) $-\cos^{n-1} x \sin x$

69. The adjoining graph

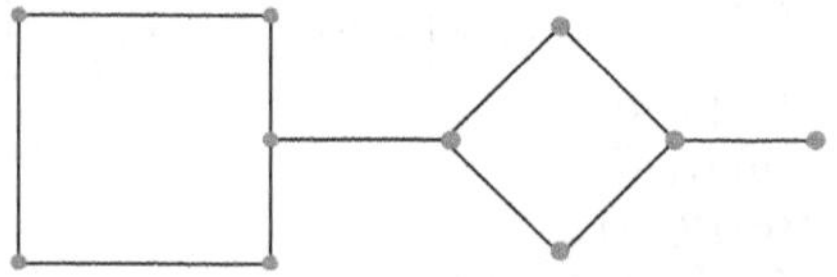

(a) connected (b) disconnected
(c) Neither connected nor disconnected
(d) None of the above

70. The remainder obtained when 5^{124} is divided by 124 is
(a) 5 (b) 0 (c) 2 (d) 1

ENGLISH AND GENERAL APTITUDE

Directions (Q. Nos. 1-5) *Read the given passage carefully and answer the questions that follow by selecting the most appropriate option.*

Helicopters are very different from airplanes. They can do three things that airplanes cannot do. First, when airplanes move upward, they must also move forward, but helicopters can move straight up without moving ahead. Second, helicopters can fly backward, which airplanes cannot do. Third, helicopters can use their rotors to hover in the air, which is impossible for airplanes. Helicopters can **perform** actions that airplanes cannot, they are used for different tasks. Since, helicopters can take off without moving forward, they do not need a runway for take off. They are used in **congested** areas where there is no room for airplanes or in isolated areas, which do not have airports. Because they can hover, they are used on firefighting missions to drop water on fires. They are used in logging operations to lift trees out of forests. Helicopters are used as air ambulances to airlift patients out of situations, which are difficult to reach by conventional ambulances.

The police used helicopters to follow suspects on the ground or to search for cars on the ground. Of course, helicopters have military uses because of their design and capabilities.

1. The word 'congested' in the passage means
   (a) the place is roomy
   (b) the place has no place
   (c) the place can be accessed by police vans
   (d) the place has so much of smoke due to fire
2. A helicopter can hover while an airplane cannot, according to the passage. 'Hover' in the passage means
   (a) stay at one place in the air
   (b) move straight up in the air
   (c) go backward in the air
   (d) fly sideways
3. Point out the grammatical category of the word 'perform' in the passage.
   (a) Noun (b) Verb
   (c) Adverb (d) Helping verb
4. Why is a helicopter used as an ambulance?
   (a) Its movement can be manoeuvred easily
   (b) It crosses all difficulties of traffic
   (c) It reaches the inaccessible places easily
   (d) It can fire-fight
5. Point out the grammatical category of the word 'very' in the passage.
   (a) Adverb (b) Adjective
   (c) Noun (d) Verb

Directions (Q. Nos. 6-8) *Complete the sentence in the following questions with most appropriate choice.*

6. He was very much ashamed .... his mischief.
   (a) for (b) of (c) with (d) over
7. .... he stood first, he would have gone to America.
   (a) If (b) Had
   (c) Have (d) Anyways
8. Of the sword and pen, the .... is more powerful.
   (a) last one (b) later (c) latter (d) last

Directions (Q. Nos. 9-11) *Each of the following question consist of a word, followed by four words or group of words. Select the antonym of the words given below.*

9. Adverse
   (a) Hostile (b) Friendly
   (c) Inimical (d) Unwilling
10. Expansion
    (a) Depression (b) Negligent
    (c) Precise (d) Enlargement
11. Thrift
    (a) Parsimony (b) Dismantle
    (c) Prodigality (d) Essential
12. Pick out the correct synonym of the word 'cantankerous'.
    (a) Rash (b) Quarrelsome
    (c) Noisy (d) Disrespectful

Directions (Q. Nos. 13-15) *In each of the following questions, choose the alternative which best expresses the meaning of the word given in capital letters.*

13. INTREPID
    (a) Coward (b) Fearless
    (c) Selfish (d) Uqly
14. LAMENTABLE
    (a) Deplorable (b) Contemptible
    (c) Remorseful (d) Scornful
15. DEFT
    (a) Skilful (b) Vigorous
    (c) Swift (d) Store

Directions (Q. Nos. 16-20) *In each of the following questions, there is a certain relationship between two given words on one side of (::) and one word is given on another side (::) while another word is to be found from the given alternatives, having the same relation with this word as the words of the given pair bear. Choose the correct alternative.*

16. King : Throne :: Judge : ?
    (a) Lawyer (b) Bench
    (c) Court (d) Trial
17. Lion : Roar :: Ass : ?
    (a) Bark
    (b) Trumpet
    (c) Howl
    (d) Bray
18. Lamb : Frisk :: Mouse : ?
    (a) Trot (b) Scamper
    (c) Gallop (d) Flit
19. Circle : Arc :: Square : ?
    (a) Line (b) Triangle
    (c) Sphere (d) Rectangle

20. Brinjal : Vegetable :: Orange : ?
(a) Fruit (b) Stem
(c) Leaf (d) Root

Direction (Q. Nos. 21-23) *In each of the following questions, a group of four words are given. Choose the word which is odd.*

21. Find the odd word.
(a) Walking
(b) Running
(c) Moving
(d) Reading

22. Find the odd word.
(a) Ghaziabad (b) Varanasi
(c) Gorakhpur (d) Bhagalpur

23. Find the odd word.
(a) Orange
(b) Apple
(c) Guava
(d) Grapes

24. In the English alphabet, find the position of S from right.
(a) 8 (b) 5
(c) 4 (d) 9

25. In the English alphabet, find the position of L from left.
(a) 12
(b) 16
(c) 11
(d) 15

26. Which letter is 7th from right in the English alphabet?
(a) P (b) C
(c) T (d) V

Directions (Q. Nos. 27-28) *In the following series, replace the question mark (?) with the suitable option.*

27. 6, 9, 12, 15, 18, ?
(a) 21
(b) 20
(c) 19
(d) 22

28. 360, 180, 90, 45, 22.5, ?
(a) 15
(b) 11.25
(c) 20
(d) 10.5

Directions (Q. Nos. 29-30) *In each of the following questions, a set of figures carrying certain characters is given. Assuming that the characters in each set follow a similar pattern, find the missing character in each case.*

29.

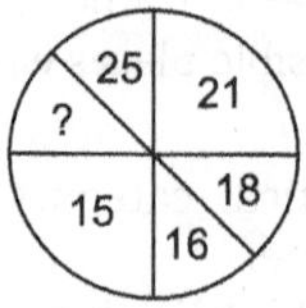

(a) 28 (b) 30
(c) 35 (d) 27

30.

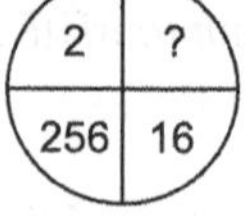

(a) 8
(b) 4
(c) 32
(d) 16

Answers

PHYSICS

1. (c)	**2.** (c)	**3.** (a)	**4.** (b)	**5.** (b)	**6.** (d)	**7.** (d)	**8.** (a)	**9.** (b)	**10.** (d)
11. (b)	**12.** (a)	**13.** (d)	**14.** (a)	**15.** (a)	**16.** (a)	**17.** (c)	**18.** (a)	**19.** (d)	**20.** (a)
21. (b)	**22.** (d)	**23.** (c)	**24.** (d)	**25.** (c)	**26.** (a)	**27.** (c)	**28.** (a)	**29.** (c)	**30.** (b)
31. (a)	**32.** (c)	**33.** (b)	**34.** (a)	**35.** (d)	**36.** (c)	**37.** (d)	**38.** (b)	**39.** (a)	**40.** (b)
41. (c)	**42.** (c)	**43.** (d)	**44.** (d)	**45.** (b)	**46.** (b)	**47.** (b)	**48.** (b)	**49.** (a)	**50.** (b)

CHEMISTRY

1. (b)	**2.** (a)	**3.** (b)	**4.** (d)	**5.** (a)	**6.** (a)	**7.** (a)	**8.** (c)	**9.** (a)	**10.** (c)
11. (a)	**12.** (d)	**13.** (a)	**14.** (d)	**15.** (b)	**16.** (a)	**17.** (a)	**18.** (a)	**19.** (a)	**20.** (c)
21. (b)	**22.** (c)	**23.** (a)	**24.** (b)	**25.** (d)	**26.** (c)	**27.** (a)	**28.** (b)	**29.** (a)	**30.** (a)
31. (b)	**32.** (b)	**33.** (d)	**34.** (a)	**35.** (a)	**36.** (b)	**37.** (c)	**38.** (b)	**39.** (c)	**40.** (d)
41. (d)	**42.** (d)	**43.** (c)	**44.** (d)	**45.** (c)	**46.** (a)	**47.** (b)	**48.** (b)	**49.** (c)	**50.** (c)

MATHEMATICS

1. (a)	**2.** (a)	**3.** (c)	**4.** (c)	**5.** (b)	**6.** (b)	**7.** (d)	**8.** (d)	**9.** (b)	**10.** (b)
11. (d)	**12.** (a)	**13.** (c)	**14.** (d)	**15.** (c)	**16.** (c)	**17.** (b)	**18.** (b)	**19.** (a)	**20.** (d)
21. (a)	**22.** (b)	**23.** (a)	**24.** (a)	**25.** (d)	**26.** (c)	**27.** (a)	**28.** (a)	**29.** (b)	**30.** (d)
31. (a)	**32.** (a)	**33.** (b)	**34.** (d)	**35.** (c)	**36.** (d)	**37.** (b)	**38.** (c)	**39.** (a)	**40.** (c)
41. (d)	**42.** (c)	**43.** (d)	**44.** (b)	**45.** (a)	**46.** (a)	**47.** (d)	**48.** (a)	**49.** (b)	**50.** (c)
51. (c)	**52.** (c)	**53.** (a)	**54.** (b)	**55.** (a)	**56.** (c)	**57.** (b)	**58.** (a)	**59.** (a)	**60.** (b)
61. (a)	**62.** (d)	**63.** (c)	**64.** (d)	**65.** (b)	**66.** (d)	**67.** (c)	**68.** (c)	**69.** (a)	**70.** (a)

ENGLISH AND GENERAL APTITUDE

1. (b)	**2.** (a)	**3.** (b)	**4.** (c)	**5.** (a)	**6.** (b)	**7.** (b)	**8.** (c)	**9.** (b)	**10.** (b)
11. (c)	**12.** (c)	**13.** (b)	**14.** (a)	**15.** (a)	**16.** (b)	**17.** (d)	**18.** (b)	**19.** (a)	**20.** (a)
21. (d)	**22.** (d)	**23.** (d)	**24.** (a)	**25.** (a)	**26.** (c)	**27.** (a)	**28.** (b)	**29.** (b)	**30.** (b)

Hints & Solutions

PHYSICS

1. (c) Acceleration of the system

$$a = \frac{F}{m} = \frac{10}{2+3+5}$$

$$= \frac{10}{10} = 1\ \text{m/s}^2$$

$\therefore$ $T_1 = (3+5) \times a = 8 \times 1 = 8\ \text{N}$

2. (c) Acceleration of the system

$$a = \frac{m_1 - m_2}{m_1 + m_2} g = \frac{10-6}{10+6} \times g$$

$$= \frac{4}{16} \times 10 = 2.5\ \text{m/s}^2$$

3. (a) $\because L/R$ is a time constant so the unit of L/R is second.

4. (b) According to the question, $F \propto v^2$

or $F = kv^2$

$$k = \frac{F}{v^2} \Rightarrow \frac{[\text{MLT}^{-2}]}{[\text{L}^2\text{T}^{-2}]} = [\text{ML}^{-1}\ \text{T}^0]$$

5. (b) Ionisation energy

$$E = 13.6\, Z^2$$

For He^+, $Z = 2$

$$E = 13.6\,(2)^2 = 13.6 \times 4 = 54.4\ \text{eV}$$

6. (d) Max. kinetic energy of photoelectrons $= 4\ \text{eV}$

If stopping potential is V_0, then

$$eV_0 = 4\ \text{eV}$$

$\therefore$ $V_0 = 4$ volt

7. (d) Out of the given choices, Poisson's ratio has not unit.

8. (a) In candescent electric lamp produces continuous emission spectrum whereas mercury and sodium vapour given line emission spectrum. Polyatomic substance such as H_2, CO_2 and $KMNO_4$ produces band absorption spectrum.

9. (b) Let the two balls cross each other after time t.

Height covered by first ball

$$h_1 = \frac{1}{2} gt^2$$

Height covered by second ball

$$h_2 = ut - \frac{1}{2} gt^2$$

but $h_1 + h_2 = 100$

$$\frac{1}{2} gt^2 + ut - \frac{1}{2} gt^2 = 100$$

$$ut = 100$$

$$50 \times t = 100 \Rightarrow t = 2\ \text{s}$$

10. (d) Observed speed, $v = \sqrt{(10)^2 + (6)^2}$

$$= \sqrt{100 + 36} = \sqrt{136} \simeq 11\ \text{m/s}$$

11. (b) Orbital speed of satellite near the earth surface is approximately 8 km/s.

12. (a) $g = \dfrac{GM_e}{R_e^2} = G\dfrac{\frac{4}{3}\pi R_e^3 d}{R_e^2} = \dfrac{4}{3}\pi G R_e d$

$g \propto d$

So, if density is doubled, then acceleration due to gravity will also be doubled.

13. (d) In mean position,

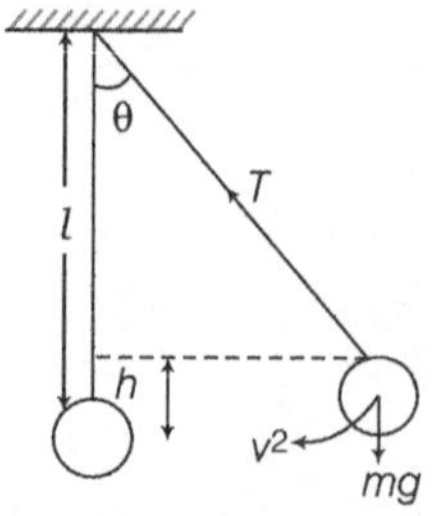

$$T = 3\ mg$$

Net force, $T - mg = \dfrac{mv^2}{l}$

$$3mg - mg = \frac{mv^2}{l}$$

$$\Rightarrow \quad v = \sqrt{2\,gl}$$

Let maximum displacement with respect to vertical is θ, then

$$\frac{1}{2}mv^2 = mgh$$

$$= mg(l - l\cos\theta)$$

$$\frac{1}{2}m(2\,g\,l) = mgl\,(1 - \cos\theta)$$

$$\cos\theta = 0$$

$$\Rightarrow \quad \theta = 90°$$

14. (a) Force, $F = m\,a = m\dfrac{dv}{dt}$

$$3x = 8\frac{dv}{dx}\cdot\frac{dx}{dt}$$

$$3x = 8\,v\frac{dv}{dx} \qquad \left[\because \frac{dx}{dt} = v\right]$$

$$3x\,dx = 8v\,dv$$

Integrating both sides

$$3\left[\frac{x^2}{2}\right]_2^{10} = 8\left[\frac{v^2}{2}\right]_0^{v}$$

$$\frac{3}{2}[10^2 - 2^2] = 4v^2$$

$$\frac{3}{2}\times 96 = 4v^2$$

$$v = 6\text{ m/s}$$

15. (a) R_1 and R_2 are in parallel order, so their equivalent resistance

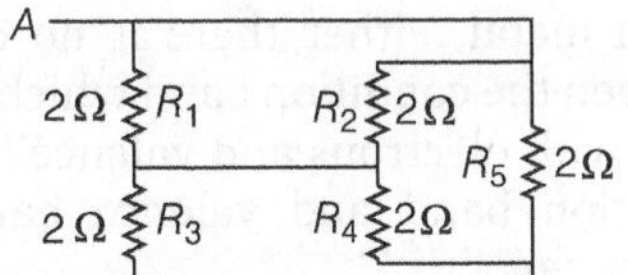

$$R' = \frac{2\times 2}{2+2} = 1\,\Omega$$

Similarly, equivalent resistance of R_3 and R_4

$$R'' = 1\,\Omega$$

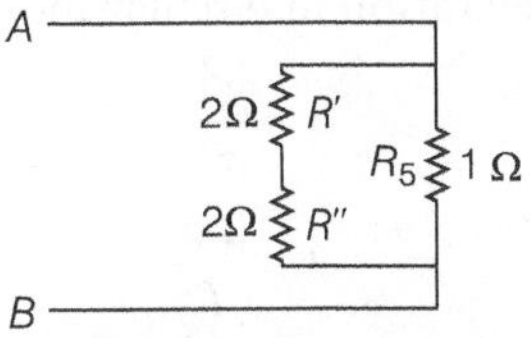

R' and R'' are in series order, so their equivalent resistance

$$R''' = R' + R''$$

$$= 1 + 1 = 2\,\Omega$$

R''' and R_5 are in parallel.

So, $\quad R_{AB} = \dfrac{2\times 2}{2+2} = 1\,\Omega$

16. (a) ${}_1H^1 + {}_1H^1 \longrightarrow {}_2He^4$

Mass defect

$$\Delta m = 2\times \text{mass of } {}_1H^1 - \text{mass of } {}_2He^4$$

$$= 2\times 2.0141 - 4.0024$$

$$= 0.0258$$

Energy released $= \Delta m \times 931$ MeV

$$= 0.0258\times 931 = 24\text{ MeV}$$

17. (c) When a thin metal plate is placed between the plates of a parallel plate capacitor, then its capacitance increases so, out of the given choices, 4 C will be the right answer.

18. (a) Linear acceleration, $a = \dfrac{g\sin\theta}{1 + \dfrac{k^2}{r^2}}$

For solid sphere $\dfrac{k^2}{r^2} = \dfrac{2}{5}$

$$\therefore \quad a = \frac{g\sin\theta}{1 + \dfrac{2}{5}}$$

$$= \frac{5}{7}g\sin\theta = \frac{5}{7}g\sin 30°$$

$$= \frac{5}{7}g\times\frac{1}{2} = \frac{5g}{14}$$

19. (d) From law of conservation of momentum

$$MV = mv$$

$$1\times 5 = 10\times 10^{-3}\times v$$

$$v = \frac{5}{10\times 10^{-3}} = 500\text{ m/s}$$

20. (a) Acceleration in circular motion

$$a = \frac{v^2}{r}$$

But, $$v = \frac{2\pi r}{T}$$

$$r = \frac{vT}{2\pi}$$

$\therefore$ $$a = \frac{v^2}{vT / 2\pi}$$

$$= \frac{2\pi v}{T}$$

21. (b) The centre of mass will be slightly farther from C-atom (because it is lighter), so correct choice will be (b).

22. (d) Work done, $W = Fd \cos \theta$

$$25 = 10 \times 10 \cos \theta$$

$$\cos \theta = \frac{1}{4} \Rightarrow \theta = \cos^{-1}\left(\frac{1}{4}\right)$$

23. (c) Ist case,

$$U = \frac{1}{2} kx_1^2$$

$$100 = \frac{1}{2} k (2 \times 10^{-2})^2$$

$$k = \frac{200}{4 \times 10^{-4}}$$

$$= 50 \times 10^4$$

IInd case

$$U = \frac{1}{2} kx_2^2$$

$$= \frac{1}{2} \times 50 \times 10^4 \, (4 \times 10^{-2})^2$$

$$= 25 \times 10^4 \times 16 \times 10^{-4}$$

$$= 400 \text{ J}$$

So, stored energy increases by

$$400 - 100 = 300 \text{ J}$$

24. (d) Young's modulus, $Y = \frac{F \cdot L}{\pi r^2 \Delta L}$

Both wires are of same material.

So, $$Y_1 = Y_2$$

$$\frac{F_A \cdot L_A}{r_A^2 \Delta L_A} = \frac{F_B L_B}{r_B^2 \cdot \Delta L_B}$$

Here, $$\Delta L_A = \Delta L_B$$

$\therefore$ $$\frac{F_A}{F_B} = \frac{L_B}{L_A} \times \left(\frac{r_A}{r_B}\right)^2$$

$$= \frac{2}{1} \times \left(\frac{2}{1}\right)^2 = \frac{8}{1}$$

$$F_A : F_B = 8 : 1$$

25. (c) Radius of path of charged particle

$$r = \frac{\sqrt{2m\,E_k}}{Bq}$$

E_k remains same, so

$$\frac{r_1}{r_2} = \sqrt{\frac{m_1}{m_2}} \cdot \frac{q_2}{q_1}$$

$$= \sqrt{\frac{4}{16}} \times \frac{2}{1} = \frac{1}{2} \times \frac{2}{1} = \frac{1}{1}$$

So, correct choice is (c).

26. (a) $\beta = \frac{\lambda D}{d}$

$$\frac{\beta_1}{\beta_2} = \frac{\lambda_1}{\lambda_2} \cdot \frac{d_2}{d_1} = \frac{7000}{3500} \times \frac{2d}{d}$$

$$\beta_2 = \frac{\beta_1}{4}$$

So, the width of fringes changes.

27. (c) Coherent sources of light are those sources of light which emit light waves of same wavelength, same frequency and are in same phase or having constant phase difference.

28. (a) In metal, either there is no energy gap between the condition band which is partially filled with electrons and valence band or the condition band and valence band overlap each other.

29. (c) The semiconductors in which majority charge carriers are holes i.e. positinely charged and minority charge are holes, so, in p-type semiconductor, the holes movement results in the formation of the current.

30. (b) A laser device produces amplification in the ultraviolet or visible region.

31. (a) Focal length of glass rod, $f = \frac{R}{2}$

$$= \frac{30}{2} = 15 \text{ cm}$$

Distance of object, $u = -15$ cm

From formula,

$$\frac{\mu_2}{v} - \frac{\mu_1}{u} = \frac{\mu_2 - \mu_1}{R}$$

$$\mu_2 = 1.5 \quad \text{(for glass)}$$

$$\mu_1 = 1 \quad \text{(for air)}$$

$$\frac{1.3}{v} = -\frac{1}{-15} = \frac{1.5-1}{30}$$

$$\frac{1.5}{v} + \frac{1}{15} = \frac{0.5}{30}$$

$$\frac{1.5}{v} = \frac{1}{60} - \frac{1}{15}$$

$$\frac{1.5}{v} = \frac{-1}{20}$$

$$\Rightarrow \quad v = -30 \text{ cm}$$

32. (c) Rainbow occurs in the form of concentric colourd circular arcs in the sky, when the sun shines on rain drops during or after the shower. This phenomenon occurs due to combined effect of dispersion, refraction and total internal reflection of sunlight by spherical drop of water.

33. (b) Near the junction positive charge is built on n-side and negative charge on P-side. This sets up potential difference across the junction and internal electric field E_i directed from n-side to P-side. The equilibrium is established when the field E_i becomes strong enough to stop further diffusion of the majority charge carriers.

34. (a) At 45° total internal reflection takes place so, critical angle, $C = 45°$.

Refractive index, $\mu = \frac{1}{\sin C} = \frac{1}{\sin 45°}$

$$= \sqrt{2} = 1.414$$

$$\approx 1.5$$

35. (d) $\because\ F \propto S^{-1/3}$ or $a \propto S^{-1/3}$

or $Vdv \propto S^{-1/3} ds \left(a = V \cdot \frac{dV}{ds} \right)$

Integrating we have, $V^2 \propto S^{2/3}$

or $\quad V \propto S^{1/3}$

Now, power $P \propto F \cdot V \propto S^{-1/3} \cdot S^{1/3}$

or $\quad P \propto S^0$

36. (c) Binding Energy per nuclear

$$= \frac{\text{Total binding energy}}{\text{Number of nucleon } (A)}$$

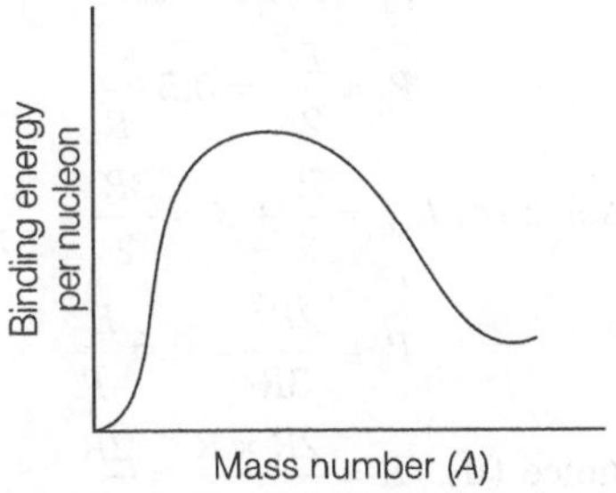

37. (d) Velocity of light in fiber

$$v = \frac{c}{\mu} = \frac{3 \times 10^8}{1.5} = 2 \times 10^8 \text{ m/s}$$

Straight length is 1 km,

Distance covered by light,

$$s = \frac{1 \text{ km}}{\sin 60°} = \frac{1000}{\sqrt{3}/2} = \frac{2000}{\sqrt{3}}$$

So, time taken $= \frac{s}{v} = \frac{2000/\sqrt{3}}{2 \times 10^8} = 3.85\ \mu\text{s}$

38. (b) $W_{PQ} = P\Delta V$

$$= 100 \times 10^3 (300 - 100) \times 10^{-6}$$

$$= 100 \times 10^{-3} \times 200 = 20 \text{ J}$$

$W_{QR} = 0$ ($\because V$ is constant)

$W_{RS} = P\Delta V$

$$= 200 \times 10^3 (100 - 300) \times 10^{-6}$$

$$= -200 \times 10^{-3} \times 200 = -40 \text{ J}$$

$W_{SP} = 0$

Net work done

$$= 20 + (-40) = -20 \text{ J}$$

39. (a) Let each resistance is R

$$P = \frac{E^2}{R_{eq}}$$

In choice (a) $R_{eq} = \frac{R \times R}{R + R} = \frac{R}{2}$

$$\Rightarrow \quad P_1 = \frac{E^2}{\frac{R}{2}} = \frac{2E^2}{R}$$

In choice (b)

$$R_{eq} = R + R = 2R$$

$$P_2 = \frac{E^2}{2R} = 0.5\frac{E^2}{R}$$

In choice (c) $R_{eq} = \frac{R}{2} + R = \frac{3R}{2}$

$$P_3 = \frac{2E^2}{3R} = 0.6\frac{E^2}{R}$$

In choice (d) $R = \frac{2R \times R}{2R + R} = \frac{2R}{3}$

$$P_4 = \frac{3E^2}{2R} = 1.5\frac{E^2}{R}$$

So, P_1 is maximum.

40. (b) Charge redistributes on the capacitors in the ratio of their capacitance.

$$\frac{Q_1}{Q_2} = \frac{C_1}{C_2} = \frac{C_1}{2C_1} = \frac{1}{2}$$

$$Q_1 : Q_2 = 1 : 2$$

$$\therefore \quad Q_1 = \frac{1}{1+2} \times Q = \frac{Q}{3}$$

$$Q_2 = \frac{2}{1+2} Q = \frac{2}{3} Q$$

41. (c) Tension in the string

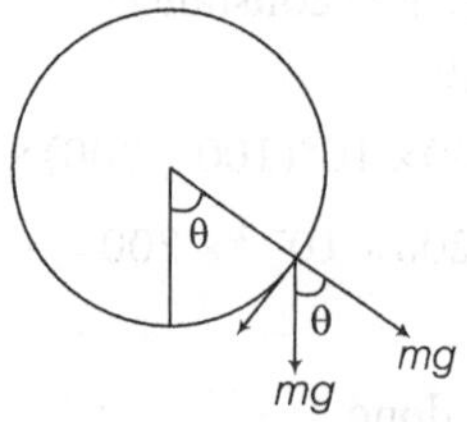

$$T = \frac{mv^2}{l} + mg\cos\theta$$

So, $\quad T \propto \cos\theta$

$$\frac{T_1}{T_2} = \frac{\cos\theta_1}{\cos\theta_2} = \frac{\cos 30°}{\cos 60°}$$

$$= \frac{\sqrt{3}/2}{1/2} = \frac{\sqrt{3}}{1}$$

$$\therefore \quad T_1 > T_2$$

42. (c) $L = \frac{\phi}{nI} = \frac{10^{-5}}{100 \times 5 \times 10^{-3}} = \frac{10^{-2}}{100 \times 5}$

$$= \frac{10^{-4}}{5}\,\text{H} = \frac{10^{-1}}{5}\,\text{mH}$$

$$= 0.02\,\text{mH}$$

43. (d) Energy drawn from the battery

$$E = VIt$$

$$= 12 \times 300 \times 2 \times 60$$

$$= 432 \times 10^3\,\text{J}$$

$$= 432\,\text{kJ}$$

44. (d) From $\frac{P_1}{T_1} = \frac{P_2}{T_2}$

$$\frac{1 \times 10^5}{2 + 273} = \frac{hdg + 1 \times 10^5}{T_2}$$

[$\because P_1$ = atmospheric pressure]

$h = 20$ m, d = density of water

$= 1 \times 10^3\,\text{kg/m}^3$

$$\therefore \quad \frac{1 \times 10^5}{275} = \frac{20 \times 10^3 \times 10 + 1 \times 10^5}{T_2}$$

$$\frac{1 \times 10^5}{275} = \frac{3 \times 10^5}{T_2}$$

$$T_2 = 275 \times 3 = 825\,\text{K}$$

$$= 825 - 273$$

$$= 552°\,\text{C}$$

45. (b) $y_1 = 4\sin(\omega t + kx)$ ...(i)

$y_2 = -4\cos(\omega t + kx)$

$= -4\sin(\omega t + kx + \pi/2)$

$= 4\sin\left(\omega t + kx + \frac{3\pi}{2}\right)$ ...(ii)

From Eqs. (i) and (ii), we get

The phase difference, $\Delta\phi = \frac{3\pi}{2}$

46. (b) $m = 0.003\text{ g} = 0.003 \times 10^{-3}\text{ kg}$

$$E = 6 \times 10^4\text{ N/C}$$

The particle is stationary, so

Electric force = Weight of the particle

$$qE = mg$$

$$q = \frac{mg}{E} = \frac{0.003 \times 10^{-3} \times 10}{6 \times 10^4}$$

$$= \frac{3 \times 10^{-10} \times 10}{6} = 5 \times 10^{-10}\text{C}$$

47. (b) Force on each plate, $F = \frac{qE}{2}$

$$= \frac{1 \times 10^{-6} \times 10^5}{2} = \frac{10^{-1}}{2}$$

$$= 0.05\text{ N}$$

48. (b) Given heat, $Q = mL = 1 \times 540$ cal

$$= 540 \times 4.18 \times 10^7\text{ J}$$

Work done, $W = P\Delta V$

$P = 1\text{ atm} = 1 \times 10^5\text{ N/m}^2 = 10^6\text{ dyne/cm}^2$

$W = 10^6 \times 1680$ J

Change in internal energy

$\Delta U = Q - W$

$$= 540 \times 4.18 \times 10^7 - 10^6 \times 1680$$

$$= \frac{540 \times 4.18 \times 10^7 - 10^6 \times 1680}{4.18 \times 10^7}\text{ cal}$$

$$= 500\text{ cal}$$

49. (a) Induced emf

$$E = -\frac{\Delta\phi}{\Delta t} = -A\frac{\Delta B}{\Delta t}$$

$$= -0.01 \times \frac{1-2}{10^{-3}} = \frac{0.01}{10^{-3}}$$

$$= 10\text{ V}$$

$\therefore$ Heat produced, $H = \frac{E^2}{R}t$

$$= \frac{(10)^2}{0.01} \times 10^{-3}$$

$$= \frac{100 \times 10^{-3}}{0.01}$$

$$= 10\text{ J}$$

50. (b) $R_t = R_0(1 + \alpha\,\Delta T)$

$$2 = 1(1 + 0.00125 \times \Delta T)$$

$$\Delta T = \frac{1}{0.00125} \quad \text{(neglecting – ve sign)}$$

$$T = 800 + 300 = 1100\text{ K}$$

CHEMISTRY

1. (b) $\underset{s}{CaF_2} \rightleftharpoons \underset{s}{Ca^{2+}} + \underset{2s}{2F^-}$

$$[Ca^{2+}] = s$$

$$[F^-] = 2s$$

$$K_{sp} = [Ca^{2+}] \cdot [F^-]^2 = s \cdot (2s)^2$$

$$K_{sp} = 4s^3$$

2. (a) If p, T, ρ and R represents pressure, temperature, density and universal gas constant then ideal gas equation

$pV = nRT$ for n mole of gas

$pV = RT$ for 1 mole of gas

We know that

$$M = V \times \rho$$

$$V = \frac{M}{\rho}$$

Put the value of V in gas equation

$$p \times \frac{M}{\rho} = RT$$

then molar mass of gas

$$M = \frac{\rho RT}{p}$$

3. (b) The kinetic energy of a gas molecule is directly proportional to absolute temperature. According to kinetic theory of gases, the pressure P exerted by a gas is given by

$$p = \frac{1}{3}\frac{Mu^2}{V}$$

or $$pV = \frac{1}{3} Mu^2$$

According to gas equation $pV = RT$

$$\therefore \quad \frac{1}{3} Mu^2 = RT$$

or $$\frac{1}{3} \frac{Mu^2}{N} = \frac{RT}{N}$$

We know that

$$\frac{R}{N} = K$$

K = Boltzmann constant

or $$\frac{1}{2} Mu^2 = \frac{3}{2} KT$$

But $\frac{1}{2} Mu^2$ is average kinetic energy per molecule of the gas.

$$\therefore \quad \text{Average K.E.} = \frac{3}{2} KT$$

$$\text{K.E.} \propto T$$

Thus average kinetic energy of one mole of any gas is directly proportional to its absolute temperature.

4. (d) Molecular weight of $SO_2 (M_1) = 64$

Molecular weight of $O_2 (M_2) = 32$

According to Graham's law of diffusion

$$\frac{r_1}{r_2} = \sqrt{\frac{M_2}{M_1}}$$

$$\frac{r_1}{r_2} = \sqrt{\frac{32}{64}}$$

$$= \sqrt{\frac{1}{2}} = \frac{1}{\sqrt{2}}$$

or $$= \frac{1}{1.414}$$

$$r_1 : r_2 = 1 : 1.414$$

5. (a) Elevation in boiling point $\Delta T_b = K_b \times m$

Dipression in freezing point

$$\Delta T_f = K_f \times m$$

$$\frac{\Delta T_b}{\Delta T_f} = \frac{K_b}{K_f}$$

$$\frac{\Delta T_b}{0.186} = \frac{0.512}{1.86}$$

$$\Delta T_b = 0.0512°C$$

6. (a) Species : $O_2^{2-}, O_2^-, O_2, O_2^+$

Bond Order : 1 1.5 2 2.5

7. (a) In hydrolysis of a salt of weak acid and strong base.

$$A^- + H_2O \rightleftharpoons HA + OH^-$$

the hydrolysis constant K_h

$$K_h = \frac{[HA][OH^-]}{[A^-]}$$

(According to law of mass action)

($\because$ conc. of water remains constant)

For the ionisation of weak acid HA,

we has $HA \rightleftharpoons H^+ + A^-$

$$K_a = \frac{[H^+][A^-]}{[HA]}$$

Also we know that

$$\frac{K_w}{K_a} = \frac{[H^+][OH^-][HA]}{[H^+][A^-]}$$

$$\frac{K_w}{K_a} = \frac{[OH^-][HA]}{[A^-]}$$

or $$\frac{K_w}{K_a} = K_h$$

8. (c) $aA + bB \longrightarrow$ Products

Rate of reactant for $A = -\frac{1}{a}\frac{d[A]}{dt}$

Rate of reactant for $B = -\frac{1}{b}\frac{d[B]}{dt}$

the $-\frac{d[A]}{dt}$ is equal to $= -\frac{a}{b}\frac{d[B]}{dt}$

9. (a) Number of moles of hydrogen

$$(H_2) = \frac{\text{Weight in gram of } H_2}{\text{Molecular weight of } H_2}$$

Weight of H_2 in gram = 1g

molecular weight of $H_2 = 2$

Number of moles of $H_2 = \frac{1}{2} = 0.5$ moles

Number of moles of

$$O_2 = \frac{\text{Weight in g of } O_2}{\text{Molecular weight of } O_2} = \frac{8}{32}$$

$$= 0.25 \text{ moles}$$

Mole fraction of

$$H_2 = \frac{\text{Number of moles of } H_2}{\text{Total number of moles}}$$

$$= \frac{0.5}{0.25 + 0.5} = 0.667$$

10. (c) $\overset{+7}{MnO_4^-} \rightarrow Mn^{+2}$ (Acid medium). Change in oxidation number = + 5 Equivalent weight of $KMnO_4 = \frac{\text{Molecular weight of } KMnO_4}{\text{Change in oxidation number}}$

$$= \frac{M}{5}$$

11. (a) For the reaction

$$CH_3COOH\,(l) + 2O_2(g) \rightleftharpoons 2CO_2\,(g) + 2H_2O\,(l)$$

$$\Delta H = -874 \text{ kJ}$$

We know that $\Delta H = \Delta E + P\Delta V$

From ideal gas equation $P\Delta V = \Delta nRT$

Then $\Delta H = \Delta E + \Delta nRT$

Number of moles of reactant

= Number of moles of products

$\therefore \quad \Delta n = 0$

Then $\quad \Delta H = \Delta E$

$\Delta E = -874 \text{ kJ}$

12. (d) The electrophile involved in the sulphonation of benzene is SO_3

$$2H_2SO_4 \rightarrow SO_3 + H_3O^+ + HSO_4^-$$

13. (a) In heterogeneous catalysis, adsorption lowers the activation energy of the reaction.

14. (d) $C_2H_5Br + AgNO_2(\text{alc.}) \rightarrow \underset{\text{Nitro ethane}}{C_2H_5NO_2} + AgBr$

15. (b) Specific conductivity

$$(K) = \frac{1}{R} \times \text{Cell constant}$$

Cell constant $= K \times R$

$$= 0.0129 \times 100 = 1.29$$

16. (a) Cell $Zn \,|\, Zn^{2+} \,||\, Fe^{2+} \,|\, Fe$

Electrode potential for

$$Zn / Zn^{2+} = 0.763 \text{ V}$$

Electrode potential for

$$Fe^{2+}/Fe = -0.44 \text{ V}$$

EMF of cell $= E^\circ_{\text{oxidation (Fe)}} + E^\circ_{\text{reduction (Zn)}}$

$$= -0.44 + 0.763$$

$$= 0.323 \text{ V}$$

17. (a) The electronic configuration of $_{24}C_r$ is $[Ar]_{18}\, 3d_1^5 4s^1$. Chromium after loss of one electron gain stable configuration due to presence of half filled d-orbitals, therefore its second ionisation enthalphy is highest.

18. (a) If pH is less then 7, the solution is acidic but if the pH is more than 7 the solution becomes basic.

0 — Acid — 7 — Base — 14

If alkali is added in acid solution the pH increases.

19. (a) CO_2 is a gas, while SiO_2 is a solid but both are acidic in nature, both reacts with base and forms carbonate and silicates.

$$CO_2 + \underset{\text{Base}}{NaOH} \longrightarrow NaHCO_3$$

$$SiO_2 + \underset{\text{Base}}{2NaOH} \longrightarrow Na_2SiO_3 + H_2O$$

20. (c) HNO_3 is a strong oxidising agent. So it oxidise iron and form a layer of Fe_3O_4 that works as protective layer.

21. (b) O_3 (Ozone) works as oxidising agent. It reacts with potassium iodide (KI) solution and liberate iodine vapours that turns starch paper blue.

$$O_3 \longrightarrow O_2 + O$$

$$2KI + H_2O + O \longrightarrow 2KOH + I_2$$

$$2KI + H_2O + O_3 \longrightarrow 2KOH + I_2 + O_2$$

22. (c) It is the oxidation process.

$$3Fe + SO_2 \longrightarrow 2FeO + FeS$$

Fe is oxidised to form FeO and FeS.

23. (a) It is Wurtz reaction.

$$C_2H_5I + 2Na + IC_2H_5 \xrightarrow{\text{Dry ether}} C_2H_5—C_2H_5 + 2NaI$$

So, C_4H_{10} is formed.

24. (b) For the reaction

$$CH_3—CH{=}CH_2 + HOCl \longrightarrow A$$

This is addition reaction, this reaction involves in unsaturated compound or that compound which has multiple bonds (= or ≡ bonds)

$$CH_3—CH{=}CH_2 + HO\overset{+}{Cl} \longrightarrow \underset{\substack{|\\OH}}{CH_3—CH}—\underset{\substack{|\\Cl}}{CH_2}$$

According to Markownfikoff's rule, the –ve part of reagent is added to that carbon atom of the double bond which bears the least number of hydrogen atom.

25. (d) Ionic crystals have no directional bonds as the oppositely charged ions are attracted by electrostatic forces and arranged in a definite ratio.

26. (c) The sum of protons and neutrons give the mass number. The number of neutrons and protons are not equal.

27. (a) According to the modern periodic table. If the last electron of element is placed in *s*-orbital, then elements are known as *s*-block element. Group IA has elements with ns^1 configuration and IIA has elements with ns^2 configuration.

28. (b) Azimuthal quantum number l for d orbitals is 2. so the value of magnetic quantum number is $-l$ to $+l$

then -2 to $+2$ $-2, -1, 0, +1, +2$

29. (a) $F—\ddot{\underset{..}{Xe}}—F$ sp^3d and linear

$Cl—\ddot{\underset{..}{I}}—F$ sp^3d and linear

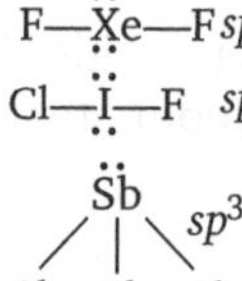
sp^3 and pyramidal

Cl—Ba—Cl sp and linear

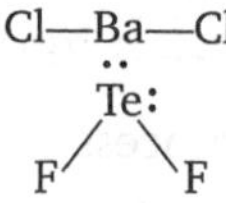
sp^3 and Y-shape

30. (a) $O_2^+ \longrightarrow O_2^{2+} + e^-$

O_2^+ has $(15e^-) = \sigma 1s^2, \sigma^* 1s^2, \sigma 2s^2, \sigma^* 2s^2$
$\sigma 2p_x^2, \pi 2p_y^2, \pi 2p_z^2, \pi^* 2p_y^1$

So O_2^+ lost a electron to form O_2^{2+}, the e^- is from antibonding π orbital.

31. (b) Structure II shows *A* to be electropositive. Hence, It will have least contribution.

32. (b) $NaOH + 2S \longrightarrow \underset{\text{Sodium sulphite}}{Na_2SO_3} + Na_2S + H_2O$

33. (d) $SiF_4 + 4H_2O \longrightarrow \underset{\text{Silicic acid}}{Si(OH)_4} + 4HCl$

The trend towards hydrolysis decreases down the group. Silicon have vacant orbital to which water molecules can co-ordinate and hence their halides are hydrolysed by water.

34. (a) For extraction of metal only those minerals are used (ores) for which the process is economic.

35. (a) P_2O_5 is an acidic oxide, acting as an acidic flux.

36. (b) Basic Oxide of metals are basic in nature and form basic hydroxide with water.

37. (c) CO_2 gas is passed in $Ca(OH)_2$ solution the solution becomes milky white to form $CaCO_3$ and if the CO_2 is passed in excess the milky white solution become colourless due to the formation of $Ca(HCO_3)_2$.

$$Ca(OH)_2 + CO_2 \longrightarrow CaCO_3 \downarrow + H_2O$$

38. (b) $(NH_4)_2SO_4 \cdot Fe_2(SO_4)_3 \cdot 24H_2O$

39. (c) Transition elements have three series 3*d*, 4*d* and 5*d*.

3*d* series has elements 21 (Sc) to 30 (Zn) atomic number. 4*d* series has elements 39 (Y) to 48 (Cd) atomic number.

$[Kr]4d^{1-10}, 5s^{1-2}$ has IInd row series.

40. (d) This complex is of the type MA_4B_2 with co-ordination number 6. It shows geometrical isomerism. *Cis* form is violet coloured and *trans* form is green.

41. (d) $\overset{1}{CH_3}-\underset{\underset{OH}{|}}{\overset{2}{CH}}-\overset{3}{CH_2}-\underset{\underset{CH_3}{|}}{\overset{4}{CH}}-\overset{5}{CH_3}$

4-methyl pentane-2-ol

42. (d) $R-X \xrightarrow[\text{Sealed tube } 100°C]{NH_3} \underset{\text{Primary amine 1°}}{R \cdot NH_2} + HX$

$RX + R \cdot NH_2 \longrightarrow \underset{\text{2° amine}}{R_2NH} + HX$

$R_2NH + RX \longrightarrow \underset{\text{3° amine}}{R_3N} + HX$

43. (c) Cannizaro reaction is possible in which aldehyde where the α-hydrogen atoms are absent.

e.g. HCHO, C_6H_5CHO

$2HCHO \xrightarrow[\Delta]{NaOH} HCOONa + CH_3OH$

$2C_6H_5CHO \xrightarrow[\Delta]{NaOH} C_6H_5CH_2OH + C_6H_5COONa$

44. (d) This is Hoffmann's bromide reaction. In this reaction, product compound has one less 'C' atom than reactant.

So it causes shortening of 'C' chain.

$-CONH_2 \longrightarrow -NH_2$

$CH_3CONH_2 + Br_2 + 4KOH \longrightarrow CH_3NH_2 + 2KBr + 2H_2O + K_2CO_3$

45. (c) $HCHO + CH_3MgI \longrightarrow A \xrightarrow{H_2O} B + Mg(OH)I$

$H-\overset{\overset{H}{|}}{C}=O + CH_3MgI \longrightarrow \underset{A}{H-\overset{\overset{H}{|}}{\underset{\underset{CH_3}{|}}{C}}-OMgI}$

$\xrightarrow{H_2O} \underset{B}{H-\overset{\overset{H}{|}}{\underset{\underset{CH_3}{|}}{C}}-OH} + Mg(OH)I$

46. (a) $\underset{\text{Secondary amine}}{R_2NH} + \underset{\text{Acetyl chloride}}{CH_3COCl} \longrightarrow \underset{\text{N, N, dialkyl acetamide}}{R_2NCO \cdot CH_3}$

47. (b) Reducing agent Sn/HCl is used.

It changes $-NO_2$ group into $-NH_2$.

48. (b) Acetyl salicylic acid is used as analgesic drug. It is also known as aspirin.

49. (c) Benzene, $(4n + 2)\pi$ electrons

$n = 1$ (6π electrons)

Naphthalene, $(4n + 2)\pi$; $n = 2$, 10π electrons

Both obey Huckel's rule.

50. (c)

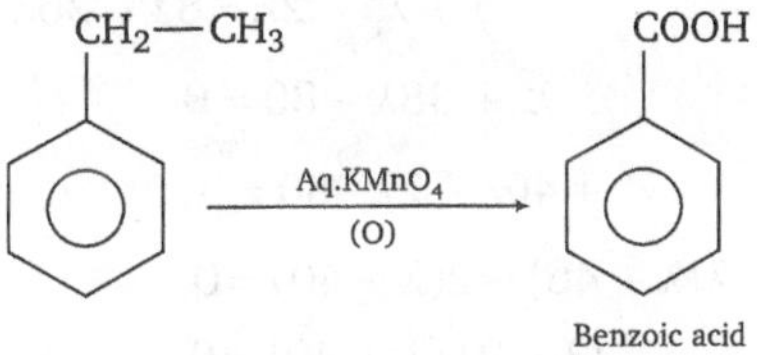

MATHEMATICS

1. (a) Let θ be the angle between the given pair of straight lines

$$x^2 - 3xy + \lambda y^2 + 3x - 5y + 2 = 0 \quad \text{...(i)}$$

$$\therefore \quad \theta = \tan^{-1}\frac{1}{3} \quad \text{(given)}$$

or $\tan\theta = \frac{1}{3}$

We know that,

$$\tan\theta = \frac{2\sqrt{h^2 - ab}}{a+b} \quad \text{...(ii)}$$

where, $h = -\frac{3}{2}, a = 1, b = \lambda$

put these value in Eq. (ii), we get

$$\frac{1}{3} = \frac{2\sqrt{\frac{9}{4} - \lambda \cdot 1}}{1+\lambda}$$

$$\Rightarrow \quad (1+\lambda) = 3\sqrt{9 - 4\lambda}$$

On squaring both sides, we get

$$(1+\lambda)^2 = (3\sqrt{9-4\lambda})^2$$

$$1 + \lambda^2 + 2\lambda = 81 - 36\lambda$$

$$\Rightarrow \quad \lambda^2 + 38\lambda - 80 = 0$$

$$\Rightarrow \quad \lambda^2 + 40\lambda - 2\lambda - 80 = 0$$

$$\Rightarrow \quad \lambda(\lambda + 40) - 2(\lambda + 40) = 0$$

$$\Rightarrow \quad (\lambda - 2)(\lambda + 40) = 0$$

$$\Rightarrow \quad \lambda = 2, -40 \Rightarrow \lambda = 2$$

($\because$ 'λ' is non-negative real number)

2. (a) Given that equation of lines are

$$2x - 3y = 4 \quad \text{...(i)}$$

and $x + y = 1$

or $x + y - 1 = 0 \quad$ ...(ii)

Now, let any line parallel to $x + y - 1 = 0$ is given by $x + y + \lambda = 0$ which passes through (1, 1).

$$\therefore \quad 1 + 1 + \lambda = 0 \Rightarrow \lambda = -2$$

$\therefore$ Required line is $x + y - 2 = 0$

or $x + y = 2 \quad$ ...(iii)

$\therefore$ Point of intersection of lines (i) and (iii) is (2, 0).

$\therefore$ Required distance

$$= \sqrt{(2-1)^2 + (0-1)^2} = \sqrt{1+1} = \sqrt{2}$$

3. (c) The equation of lines are $x + y = 0$ and $x - y = 0$

$\therefore$ The equation of bisectors of the angles between these lines are

$$\frac{x+y}{\sqrt{1+1}} = \pm\frac{x-y}{\sqrt{1+1}}$$

$$\Rightarrow \quad x + y = \pm(x - y) \quad \text{...(ii)}$$

Taking + ve sign, we get $y = 0$.

Taking – ve sign, we get $x = 0$.

Hence the equation of bisectors are

$$x = 0, y = 0.$$

4. (c) Given that, the base of an isosceles triangle PQR are $Q(1, 3)$ and $R(-2, 7)$. Let the co-ordinates of point P are (x, y).

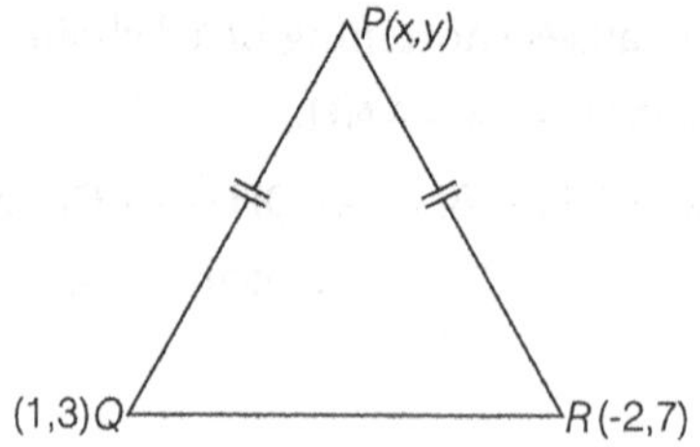

$\because$ In an isosceles triangle $PQ = PR$

i.e. $\sqrt{(x-1)^2 + (y-3)^2}$

$$= \sqrt{(x+2)^2 + (y-7)^2}$$

$$\Rightarrow \quad (x-1)^2 + (y-3)^2 = (x+2)^2 + (y-7)^2$$

$$\Rightarrow \quad x^2 - 2x + 1 + y^2 - 6y + 9$$

$$= x^2 + 4x + 4 + y^2 - 14y + 49$$

$$\Rightarrow -2x - 6y + 10 = 4x - 14y + 49 + 4$$

$$\Rightarrow \quad -6x + 8y = 43$$

Here, option 'c' (5/6, 6) satisfies the equation.

$\therefore$ Co-ordinate of P are (5/6, 6).

5. (b) Given that normal cuts the circle at points $A(3, 4)$ and $B(-1, -2)$. As we know, normal line is always passing through the centre of the circle and AB is the diameter of the circle. Therefore, A and B are the end points of a diameter.

$\therefore$ Required equation of circle is

$$(x-3)(x+1)+(y-4)(y+2)=0$$

$$\Rightarrow x^2-2x-3+y^2-2y-8=0$$

$$\Rightarrow \quad x^2+y^2-2x-2y-11=0$$

6. (b) Given that, $\cos P=\frac{1}{7}, \cos Q=\frac{13}{14}$

$$\because \sin P=\sqrt{1-\cos^2 P}=\sqrt{1-\frac{1}{49}}=\frac{1}{7}\sqrt{48}$$

Now, $\sin Q=\sqrt{1-\cos^2 Q}=\sqrt{1-\frac{169}{196}}=\frac{\sqrt{27}}{14}$

Now,

$$\cos(P-Q)=\cos P\cos Q+\sin P\sin Q$$

$$\cos(P-Q)=\frac{1}{7}\times\frac{13}{14}+\frac{1}{7}\sqrt{48}\times\frac{1}{14}\sqrt{27}$$

$$=\frac{13}{98}+\frac{36}{98}=\frac{49}{98}=\frac{1}{2}$$

$$\Rightarrow \quad \cos(P-Q)=\frac{1}{2}$$

$$\Rightarrow \quad \cos(P-Q)=\cos 60^\circ$$

$$\therefore \quad P-Q=60^\circ$$

7. (d) Given that $3\cos x+4\sin x=6$

Since the maximum value of $(3\cos x+4\sin x)=+\sqrt{3^2+4^2}=5$. Hence there is no '$x$' satisfying $3\cos x+4\sin x=6$.

$\therefore$ No solution, as $|\sin x|\le 1, |\cos x|\le 1$ and both of them do not attain their maximum value for the same angle.

Alternate Method, $3\cos x+4\sin x=6$

Let $\quad r\cos\theta=3$

and $\quad r\sin\theta=4$

$$\Rightarrow \quad \theta=\tan^{-1}\left(\frac{4}{3}\right)$$

$$\Rightarrow \quad 5(\cos\theta\cos x+\sin\theta\sin x)=6$$

$$\Rightarrow \cos(\theta-x)=\frac{6}{5} \text{ or } \cos(x-\theta)=\frac{6}{5}$$

or $\cos(x-\theta)>1$ but $-1\le\cos x\le 1$.

$\therefore$ There exist no solution in $[-1, 1]$.

8. (d) Given that, $\sec^{-1}x=\operatorname{cosec}^{-1}y$

or $\quad \cos^{-1}\left(\frac{1}{x}\right)=\sin^{-1}\left(\frac{1}{y}\right)$

On adding $\cos^{-1}\left(\frac{1}{y}\right)$ on both sides, we get

$$\cos^{-1}\left(\frac{1}{x}\right)+\cos^{-1}\left(\frac{1}{y}\right)=\sin^{-1}\left(\frac{1}{y}\right)+\cos^{-1}\left(\frac{1}{y}\right)$$

$$\Rightarrow \cos^{-1}\left(\frac{1}{x}\right)+\cos^{-1}\left(\frac{1}{y}\right)=\frac{\pi}{2}$$

$$\left[\because \sin^{-1}\theta+\cos^{-1}\theta=\frac{\pi}{2}\right]$$

9. (b) Given that n be any integer, then $n(n+1)(2n+1)$ always an integral multiple of even numbers (i.e. 6, 30, 84, ...) when

$n=1, n(n+1)(2n+1)=1(2)(3)=6$

$n=2, n(n+1)(2n+1)=2(3)(5)=30$

and so on.

10. (b) We have, $\tan\theta=-\frac{4}{3}$

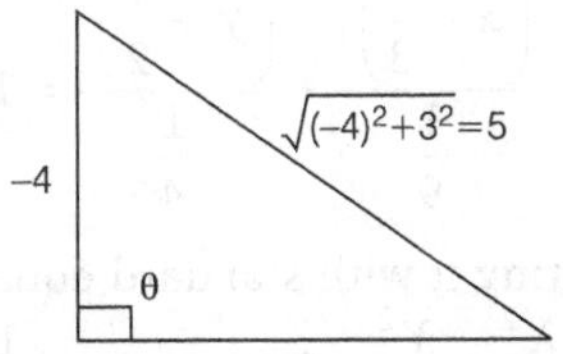

$\because \tan\theta$ is less than 0 and $0<\theta<\frac{\pi}{2}$

or $\quad \frac{3\pi}{2}<\theta<2\pi$.

$$\therefore \sin\theta=\frac{\text{perpendicular}}{\text{hypotenuse}}=-\frac{4}{5} \text{ (i.e. – ve)}$$

But, θ lies also between $\frac{3\pi}{2}$ and 2π

$$\therefore \quad \sin\theta=\frac{4}{5} \text{ also.}$$

11. (d) Given that $c = 2\cos\theta$

$$\therefore \Delta = \begin{vmatrix} c & 1 & 0 \\ 1 & c & 1 \\ 6 & 1 & c \end{vmatrix} = \begin{vmatrix} 2\cos\theta & 1 & 0 \\ 1 & 2\cos\theta & 1 \\ 6 & 1 & 2\cos\theta \end{vmatrix}$$

On expanding, we get

$$\Delta = 2\cos\theta[(4\cos^2\theta - 1)]$$
$$-1[(2\cos\theta - 6)] + 0[(1 - 12\cos\theta)]$$
$$= 8\cos^3\theta - 2\cos\theta - 2\cos\theta + 6$$
$$= 8\cos^3\theta - 4\cos\theta + 6$$

12. (a) The equation of parabola whose vertex is $(-1, -2)$ is

$$(x+1)^2 = 4a(y+2)$$

which passes through (3, 6)

i.e. $16 = 4a \times 8 \Rightarrow a = \frac{1}{2}$

$\therefore$Required equation is

$(x+1)^2 = 2(y+2)$ or $x^2 + 2x - 2y - 3 = 0$

13. (c) Given that equation of conic is

$$9x^2 + 4y^2 - 6x + 4y + 1 = 0$$
$$\Rightarrow \quad 9x^2 - 6x + 4y^2 + 4y + 1 = 0$$
$$\Rightarrow \quad 9\left(x - \frac{1}{3}\right)^2 + 4\left(y + \frac{1}{2}\right)^2 = 1$$

or, $$\frac{\left(x - \frac{1}{3}\right)^2}{\frac{1}{9}} + \frac{\left(y + \frac{1}{2}\right)^2}{\frac{1}{4}} = 1$$

Comparing it with standard equation of ellipse $\frac{X^2}{a^2} + \frac{Y^2}{b^2} = 1$, we get $a = \frac{1}{3}, b = \frac{1}{2}$

Length of major axis $= 2b = 2 \times \frac{1}{2} = 1$

and length of minor axis $= 2a = 2 \times \frac{1}{3} = \frac{2}{3}$

14. (d) Given that, $f(x) = \cot^{-1}\left(\frac{3x - x^3}{1 - 3x^2}\right)$

and $$g(x) = \cos^{-1}\left(\frac{1 - x^2}{1 + x^2}\right)$$

Put $x = \tan\theta$ in $f(x)$, we get

$$f(x) = \cot^{-1}\left(\frac{3\tan\theta - \tan^3\theta}{1 - 3\tan^2\theta}\right)$$
$$\Rightarrow \quad f(x) = \frac{\pi}{2} - 3\tan^{-1}x$$

Again put $x = \tan\theta$ in $g(x)$, we get

$$g(x) = \cos^{-1}\left(\frac{1 - \tan^2\theta}{1 + \tan^2\theta}\right)$$
$$g(x) = 2\tan^{-1}x$$
$$\therefore \lim_{x \to a} \frac{f(x) - f(a)}{g(x) - g(a)}$$
$$= \lim_{x \to a} \frac{\left(\frac{\pi}{2} - 3\tan^{-1}x\right) - \frac{\pi}{2} + 3\tan^{-1}a}{2\tan^{-1}x - 2\tan^{-1}a}$$
$$= \lim_{x \to a} \frac{-3(\tan^{-1}a - \tan^{-1}x)}{2(\tan^{-1}a - \tan^{-1}x)} = -\frac{3}{2}$$

15. (c) Given that, $f(x) = \begin{cases} x, & 0 \le x \le 1 \\ 2x - 1, & 1 < x \end{cases}$

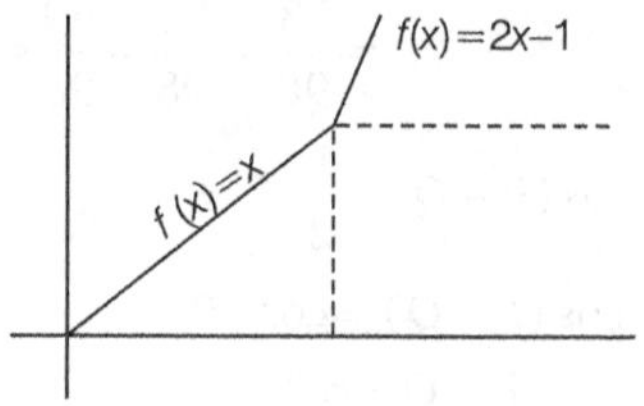

From graph $f(x)$ is continuous, $\forall\ x \in R$.

i.e. $f'(x) = 1$ when $0 \le x \le 1$

and $f'(x) = 2 - 0$ when $x > 1$.

$\therefore$Left hand limit $\ne$ Right hand limit

$\therefore$At $x = 1$, $f(x)$ is continuous but not differentiable.

16. (c) We have, $y = \lim_{x \to -2} \frac{\sin^{-1}(x+2)}{x^2 + 2x}$

$$\Rightarrow \quad y = \lim_{x \to -2} \frac{\sin^{-1}(x+2)}{(x+2)} \times \frac{1}{x}$$
$$\Rightarrow \quad y = \lim_{x \to -2} (1) \times \frac{1}{x} \quad \left[\because \lim_{x \to \theta} \frac{\sin\theta}{\theta} = 1\right]$$
$$= -\frac{1}{2}$$

17. (b) Given that, $f(x)=\begin{cases} x^p \cos\left(\frac{1}{x}\right), & x \neq 0 \\ 0, & x=0 \end{cases}$

$f(x)$ will be differentiable at $x=0$, if $\lim_{x \to 0} \frac{f(x)-f(0)}{x-0}$ exists finitely.

$$\Rightarrow \lim_{x \to 0} \frac{x^p \cos\left(\frac{1}{x}\right)}{x} \text{ exists finitely}$$

$$\Rightarrow \lim_{x \to 0} x^{p-1} \cos\left(\frac{1}{x}\right) \text{ exists finitely}$$

$$\Rightarrow p-1>0 \Rightarrow p>1$$

18. (b) We have, $f(x)=3|2+x|$

as $f(x)=|x| \Rightarrow f'(x)=\frac{x}{|x|}$

Now, $f(x)=3|x+2|$

$$\Rightarrow f(x)=3\frac{x+2}{|x+2|}=-3 \quad (\text{as } x_0=-3)$$

19. (a) Given that, $f(x)=\log_5(\log_7 x)$

or it can be written as

$$f(x)=\frac{\log_e(\log_7 x)}{\log_e 5} \quad \left[\because \log_a b=\frac{\log b}{\log a}\right]$$

On differentiating with respect to x, we get

$$\text{Now, } f'(x)=\frac{1}{\log_e 5}\left[\frac{1}{\log_7 x}\times\frac{d}{dx}(\log_7 x)\right]$$

$$=\frac{1}{\log_e 5}\times\frac{1}{\log_7 x}\times\frac{1}{\log_e 7}\cdot\frac{1}{x}$$

$$=\frac{1}{x(\log 5)(\log 7)(\log_7 x)}$$

20. (d) Here, $z^{1/3}=a-ib$

$\Rightarrow z=(a-ib)^3$

$\Rightarrow z=a^3-3a^2(ib)+3a(ib)^2-(ib)^3$

$\Rightarrow x+iy=(a^3-3b^2a)+i(-3a^2b+b^3)$

Equating real and imaginary parts, we get

$$x=a(a^2-3b^2) \text{ and } y=b(b^2-3a^2)$$

$$\text{Now, } \frac{a(a^2-3b^2)}{a}-\frac{b(b^2-3a^2)}{b}=k(a^2-b^2)$$

$$\Rightarrow 4a^2-4b^2=k(a^2-b^2) \Rightarrow k=4$$

21. (a) We have, $1+|e^x-1|=e^x(e^x-2)$

$$1+1+|e^x-1|=e^{2x}-2e^x+1$$

$$=(e^x-1)^2$$

$$1+1+|e^x-1|=|e^x-1|^2$$

Let $(e^x-1)=y$, then

$$y^2-y-2=0 \Rightarrow y=2,-1$$

$$\Rightarrow |e^x-1|=2 \Rightarrow e^x-1=\pm 2$$

$$\Rightarrow e^x=1\pm 2 \Rightarrow e^x=3,-1$$

or $\quad e^x=3 \quad \Rightarrow x=\log_e 3$

$\therefore$ There is one real solution of the equation.

22. (b) Given that,

$$u=x^2+y^2$$

and $\quad x=s+3t,\ y=2s-t$

Now, $\quad u=x^2+y^2$

$$=(s+3t)^2+(2s-t)^2$$

$$=s^2+9t^2+6st+4s^2+t^2-4st$$

$$\Rightarrow u=5s^2+2st+10t^2$$

On differentiating with respect to s, we get

$$\frac{du}{ds}=10s+2t$$

Again differentiating $\frac{d^2u}{ds^2}=10$

23. (a) Given equations are

$$x^2+px+q=0 \quad \text{...(i)}$$

$$x^2+qx+p=0 \quad \text{...(ii)}$$

Let α be the common root, then

$$\alpha^2+p\alpha+q=0 \text{ and } \alpha^2+q\alpha+p=0$$

$$\Rightarrow \frac{\alpha^2}{p^2-q^2}=\frac{\alpha}{q-p}=\frac{1}{q-p}$$

$$\Rightarrow \alpha^2=\frac{(p-q)(p+q)}{-(p-q)}=-(p+q) \quad \text{...(iii)}$$

and $\quad \alpha=\frac{q-p}{q-p}=1 \quad$ ...(iv)

$\therefore$ From Eq. (iii)

$$\alpha^2 = -p + q$$

$$\Rightarrow \quad (1)^2 = -(p+q) \quad \text{[from Eq. (iv)]}$$

$$\Rightarrow \quad p + q = -1 \Rightarrow p + q + 1 = 0$$

24. (a) Let α and β be the roots of the equation $x^2 - (a-2)x + (a-3) = 0$

$\therefore \ \alpha + \beta = (a-2)$ and $\alpha\beta = a - 3$

Now, $\alpha^3 + \beta^3 = (\alpha+\beta)^3 - 3\alpha\beta(\alpha+\beta)$

$$= (a-2)^3 - 3(a-3)(a-2)$$

$$= a^3 - 9a^2 + 27a - 26 = (a-3)^3 + 1$$

It assumes the least value, if $(a-3)^3 = 0$

$$\therefore \quad a = 3$$

25. (d) Given that, $|z| = \frac{1}{2}$ and $z_1 = \frac{1}{2} + \frac{\sqrt{3}i}{2}$

Here, one of the number must be a conjugate of $z_1 = \frac{1}{2} + \frac{\sqrt{3}i}{2}$ i.e. $z_2 = \frac{1}{2} - \frac{\sqrt{3}i}{2}$

and $z_2 = z_1 e^{-i2\pi/3}$

$$\Rightarrow \quad z_2 = \left(\frac{1}{2} + \frac{i\sqrt{3}}{2}\right)\left[\cos\frac{2\pi}{3} + i\sin\frac{2\pi}{3}\right]$$

$$= -2 \times \frac{1}{2} = -1$$

$$\left[\because (1 + i\sqrt{3})\left(\cos\frac{2\pi}{3} + i\sin\frac{2\pi}{3}\right) = -2\right]$$

26. (c) Given that, $z = \frac{-2}{1+\sqrt{3}i} \times \frac{1-\sqrt{3}i}{1-\sqrt{3}i}$

$$= -\left(\frac{1 - i\sqrt{3}}{2}\right) = -\frac{1}{2} + \frac{i\sqrt{3}}{2}$$

$$\therefore \quad \theta = \tan^{-1}\left(\frac{y}{x}\right) = \tan^{-1}\left(\frac{\sqrt{3}}{2} \times 2\right)$$

$$= \tan^{-1}(\sqrt{3}) = \frac{\pi}{3}$$

$$\therefore \quad \arg(z) = \pi - \theta = \pi - \frac{\pi}{3} = \frac{2\pi}{3}.$$

27. (a) Given that,

$$1 + \omega + \omega^2 = 0 \text{ and } \omega^3 = 1$$

$$\therefore 2(1+\omega)(1+\omega^2) + 3(2\omega+1)(2\omega^2+1) + \ldots + (n+1)(n\omega+1)(n\omega^2+1)$$

$$= 2(1) + 3(3) + 4(7) + \ldots + (n^3+1)$$

Let $\quad T_n = n^3 + 1$

$$\Rightarrow \quad \Sigma T_n = \left[\frac{n(n+1)}{2}\right]^2 + n$$

28. (a) We have, $\arg\left(\frac{z-1}{z+1}\right) = k, \ (k \neq 0)$

Put $\ z = x + iy$, we get

$$\arg\left(\frac{x - 1 + iy}{x + 1 + iy}\right) = k$$

$$\Rightarrow \arg\left[\frac{(x-1+iy)(x+1-iy)}{(x+1+iy)(x+1-iy)}\right] = k$$

$$\Rightarrow \quad \arg\left[\frac{(x^2-1) + iy(2) + y^2}{(x+1)^2 + y^2}\right] = k$$

$$\Rightarrow \quad \arg\left[\frac{x^2 + y^2 - 1 + 2iy}{(x+1)^2 + y^2}\right] = k$$

$$\Rightarrow \quad \frac{2y}{x^2 + y^2 - 1} = k$$

$$\left[\because \arg(z) = \tan^{-1}\frac{y}{x}\right]$$

$$\Rightarrow \quad (x^2 + y^2)k - k - 2y = 0$$

$$\Rightarrow \quad kx^2 + ky^2 - 2y - k = 0$$

$$\Rightarrow \quad x^2 + y^2 - \frac{2y}{k} - 1 = 0$$

Which represent a circle of centre $(0, 1/k)$ i.e. on y-axis.

29. (b) We have,

$P(3, 4, 5), Q(4, 6, 3), R(-1, 2, 4)$ and $S(1, 0, 5)$.

$\therefore$ Direction ratios of PQ are $(1, 2, -2)$ and direction cosines of PQ are

$$\frac{1}{\sqrt{1^2 + 2^2 + (-2)^2}}, \frac{2}{\sqrt{1^2 + 2^2 + (-2)^2}}, \frac{-2}{\sqrt{1^2 + 2^2 + (-2)^2}}$$

$$= \frac{1}{\sqrt{9}}, \frac{2}{\sqrt{9}}, -\frac{2}{\sqrt{9}} = \frac{1}{3}, \frac{2}{3}, -\frac{2}{3}$$

$\therefore$ Required projection of RS on PQ is $l(x_2 - x_1) + m(y_2 - y_1) + n(z_2 - z_1)$

30. (d) Let α, β, γ be the angles which the line makes with the positive directions at the axes of x, y and z respectively, then $\cos\alpha$, $\cos\beta$ and $\cos\gamma$ are called the direction cosines of the line and these are usually denoted by

$$l = \cos\alpha,\ m = \cos\beta,\ n = \cos\gamma$$

$$\therefore \cos^2\alpha + \cos^2\beta + \cos^2\gamma = l^2 + m^2 + n^2 = 1$$

31. (a) Let $\mathbf{a} = a_1\hat{\mathbf{i}} + a_2\hat{\mathbf{j}} + a_3\hat{\mathbf{k}}$

Here, $(a_1\hat{\mathbf{i}})\cdot\hat{\mathbf{i}} = a_1 = 2$; $(a_2\hat{\mathbf{j}})\cdot\hat{\mathbf{j}} = a_2 = 3$ and $(a_3\hat{\mathbf{k}})\cdot\hat{\mathbf{k}} = a_3 = 6$

$\therefore\ \mathbf{a} = 2\hat{\mathbf{i}} + 3\hat{\mathbf{j}} + 6\hat{\mathbf{k}}$

Now, required length $= |\mathbf{a}| = \sqrt{2^2 + 3^2 + 6^2}$

$= \sqrt{4 + 9 + 36} = \sqrt{49} = 7$

32. (a) $(10011.1)_2 = 2^4 + 0\times 2^3 + 0\times 2^2 + 1 \times 2^1 + 1\times 2^0 + 1\times 2^{-1}$

$= 16 + 2 + 1 + 0.5 = (19.5)_{10}$

33. (b)

2	60	
2	30	0
2	15	0
2	7	1
2	3	1
	1	1

$\therefore\ (60)_{10} = (111100)_2$

34. (d) $(\sim p \wedge q) \cap (\sim p \vee p)$ is true.

35. (c) Given that, $f(x) = \sin\left(\dfrac{\pi x}{n-1}\right) + \cos\left(\dfrac{\pi x}{n}\right)$

Let $f(x) = g(x) + p(x),\ \forall\ n > 2$

$\therefore$ Period of $g(x) = \dfrac{2\pi\times(n-1)}{\pi} = 2(n-1)$

and period of $p(x) = \dfrac{2\pi n}{\pi} = 2n$

Period of $f(x) =$ LCM of $p(x)$ and $q(x)$

$= 2n(n-1)$

36. (d) We have, $f(\theta) = \sec^2\theta + \cos^2\theta$

Since AM $\geq$ GM

$$\frac{\sec^2\theta + \cos^2\theta}{2} \geq (\sec^2\theta\cdot\cos^2\theta)^{1/2}$$

$$\Rightarrow \frac{\sec^2\theta + \cos^2\theta}{2} \geq \sqrt{\sec^2\theta\cdot\cos^2\theta}$$

$$\Rightarrow \frac{\sec^2\theta + \cos^2\theta}{2} \geq 1 \left[\because \sec\theta = \frac{1}{\cos\theta}\right]$$

$$\Rightarrow \sec^2\theta + \cos^2\theta \geq 2 \Rightarrow f(\theta) \geq 2$$

$\therefore$ The value of $f(\theta)$ always lies in the interval $[2, \infty)$.

37. (b) Given that, arc = 15 cm.

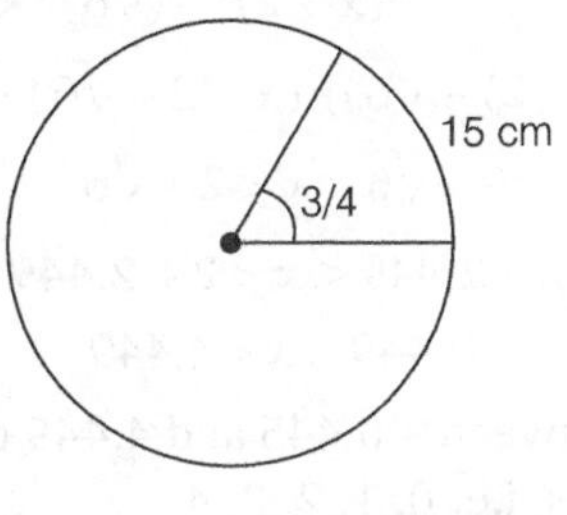

$$\theta = \frac{3}{4} \text{ radian}$$

We know that, Angle $= \dfrac{\text{Arc}}{\text{Radius}}$

$$\Rightarrow \text{Radius} = \frac{\text{Arc}}{\text{Angle}} = \frac{15\times 4}{3} = 20 \text{ cm}$$

38. (c) $\dfrac{d}{dx}\{f_n(x)\} = \dfrac{d}{dx}\{e^{f_{n-1}(x)}\}$

$= e^{f_{n-1}(x)}\dfrac{d}{dx}\{f_{n-1}(x)\} = f_n(x)\cdot\dfrac{d}{dx}\{e^{f_{n-1}(x)}\}$

$= f_n(x)\cdot e^{f_{n-2}(x)}\cdot\dfrac{d}{dx}\{f_{n-2}(x)\}$

$= f_n(x)\cdot f_{n-1}(x)\ldots f_2(x)\dfrac{d}{dx}\{f_1(x)\}$

$= f_n(x)\cdot f_{n-1}(x)\ldots f_2(x)\cdot f_1(x)\cdot\dfrac{d}{dx}e^{f_0(x)}$

$= f_n(x)\cdot f_{n-1}(x)\ldots f_2(x)\cdot f_1(x)\cdot\dfrac{d}{dx}(f_0(x))$

$= f_n(x)\cdot f_{n-1}(x)\ldots f_2(x)\cdot f_1(x)\cdot\dfrac{d}{dx}(x)$

$[\because f_0(x) = x \text{ (given)}]$

$= f_n(x)\cdot f_{n-1}(x)\ldots f_2(x)\cdot f_1(x).$

39. (a) We have, $3^x + 2^{2x} \geq 5^x$

$3^x + (2^2)^x \geq 5^x$; $3^x + 4^x \geq 5^x$

$$\Rightarrow \left(\frac{3}{5}\right)^x + \left(\frac{4}{5}\right)^x \geq 1$$

$\Rightarrow \quad (\sin\theta)^x + (\cos\theta)^x \geq 1$

[by triangle inequality]

$\Rightarrow \quad x \leq 2$

$\therefore$ Solution set is $(-\infty, 2]$

40. (c) We have, $\dfrac{x+1}{x^2+2} > \dfrac{1}{4}$

$\Rightarrow \quad x^2+2-4x-4<0 \Rightarrow x^2-4x-2<0$

$\Rightarrow \quad (x-2)^2-(\sqrt{6})^2<0$

$\Rightarrow \{(x-2)-\sqrt{6})\}\,(x-2+\sqrt{6})<0$

$\Rightarrow \quad 2-\sqrt{6}<x<2+\sqrt{6}$

$\Rightarrow \quad 2-2.449<x<2+2.449$

$\Rightarrow \quad -0.449<x<4.449$

$\therefore$ In between -0.445 and 4.449 five values of x lies, i.e. 0, 1, 2, 3, 4.

$\therefore$ There are 5 integral solution of inequality.

41. (d) Let sides of a triangle are p, q, r.

Applying sine rule $\dfrac{p}{\sin P} = \dfrac{q}{\sin Q} = \dfrac{r}{\sin R} = \lambda$

$\because \quad \cos P = \dfrac{\sin Q}{2\sin R}$

$\Rightarrow \quad \dfrac{q^2+r^2-p^2}{2qr} = \dfrac{q}{2r}$

$\Rightarrow \quad q^2+r^2-p^2 = q^2 \Rightarrow r^2 = p^2 \Rightarrow r = p$

$\therefore$ Triangle is isosceles.

42. (c) Given that, $f(x) = (a-x^n)^{1/n}$

$\therefore \quad f[f(x)] = [a-\{f(x)\}^n]^{1/n}$

$= [a-(a-x^n)]^{1/n} = x$

43. (d) $f(x) = \dfrac{x^2-3x+2}{x^2+2x-3}$

On differentiating with respect to x, we get

$$f'(x) = \frac{(x^2+2x-3)(2x-3)-(x^2-3x+2)(2x+2)}{(x^2+2x-3)^2}$$

$$= \frac{(2x^3+4x^2-6x-3x^2-6x+9)-(2x^3-6x^2+4x+2x^2-6x+4)}{(x^2+2x-3)^2}$$

$$= \frac{5x^2-10x+5}{(x^2+2x-3)^2} > 0$$

$$= \frac{5(x-1)^2}{(x+3)^2(x-1)^2} > 0 = \frac{5}{(x+3)^2} > 0$$

$\therefore$ $f(x)$ is increasing in its domain.

44. (b) Given that,

$\sqrt{(x-3)^2+(y-1)^2} + \sqrt{(x+3)^2+(y-1)^2} = 6$

$\Rightarrow \sqrt{(x-3)^2+(y-1)^2} = 6 - \sqrt{(x+3)^2+(y-1)^2}$

On squaring both sides, we get

$(x-3)^2+(y-1)^2 = 36+(x+3)^2+(y-1)^2 - 12\sqrt{(x+3)^2+(y-1)^2}$

$\Rightarrow x^2-6x+9+(y^2-2y+1)$

$= 36+x^2+6x+9+y^2-2y+1 - 12\sqrt{(x+3)^2+(y-1)^2}$

$\Rightarrow (x+3) = \sqrt{(x+3)^2+(y-1)^2}$

On squaring both sides of above equation

$\Rightarrow (x+3)^2 = [(x+3)^2+(y-1)^2]$

$\Rightarrow (y-1)^2 = 0$

$\Rightarrow$ Which represents a pair of straight lines.

45. (a) Since, $|z_1| = |z_2| = |z_3| = 1$

we get $z_1\bar{z}_1 = z_2\bar{z}_2 = z_3\bar{z}_3 = 1$

Now, $\quad 1 = \left|\dfrac{1}{z_1}+\dfrac{1}{z_2}+\dfrac{1}{z_3}\right|$

$= |\bar{z}_1+\bar{z}_2+\bar{z}_3| = |\overline{z_1+z_2+z_3}|$

$= |z_1+z_2+z_3|$

46. (a) Unique solution of $x * a = b$ is taken by $x = b * a^{-1}$.

47. (d) We have, $f(x) = \cos[\pi^2]x + \cos[-\pi^2]x$ ...(i)

Put the value of $\pi = 3.141$

$\Rightarrow \quad \pi^2 =$ app. 9.6 in (i)

$\Rightarrow f(x) = \cos 9x + \cos 10x \quad [\because \pi = 3.141]$

Put $\quad x = \dfrac{\pi}{4}$,

$f\left(\dfrac{\pi}{4}\right) = \cos\dfrac{9\pi}{4} + \cos\dfrac{10\pi}{4} = \dfrac{1}{\sqrt{2}} + 0 = \dfrac{1}{\sqrt{2}}$

Now, $f(-\pi) = \cos 9\pi + \cos 10\pi = -1+1 = 0$

and $f(\pi) = \cos 9\pi + \cos 10\pi = 0$

$\therefore \quad f\left(\frac{\pi}{2}\right) = \cos\frac{9\pi}{2} + \cos\frac{10\pi}{2}$

$= \cos\frac{9\pi}{2} + \cos 5\pi = 0 + (-1) = -1$

48. (a) Given that, $f(x) = \sec\left(\frac{\pi}{4}\cos^2 x\right)$

Since, $\cos^2 x$ lies between 0 to 1.

i.e. $0 \le \cos^2 x \le 1 \;\forall\; x \in (-\infty, \infty)$

$\Rightarrow \quad 0 \le \frac{\pi}{4}\cos^2 x \le \frac{\pi}{4}$

$\Rightarrow \quad \sec 0 \le \sec\left(\frac{\pi}{4}\cos^2 x\right) \le \sec\frac{\pi}{4}$

$\Rightarrow \quad 1 \le \sec\left(\frac{\pi}{4}\cos^2 x\right) \le \sqrt{2}$

49. (b) Given that, $f(x) = \frac{\sin^{-1}(3-x)}{\log(|x|-2)}$

From numerator of the function, we have

$\sin^{-1}(3-x)$, here $-1 \le 3-x \le 1$

$\Rightarrow \quad -4 \le -x \le -2$

$\Rightarrow \quad 2 \le x \le 4$ ...(i)

From denominator of the function, we have

$|x| - 2 > 1$ or $|x| > 3$

$\Rightarrow \quad x < -3 \cup x > 3$ ...(ii)

From both cases, we get the domain of the given function.

i.e. $3 < x \le 4$ or $x \in (3, 4]$.

50. (c) $(1! + 2! + 3! + 4! + 5! + \ldots + 200!) \div 14$

$= 1 + 2 + 6 + 24 + 120 + 720 + (14M)$

$= 873 + 14M = 5 + 868 + 14M$

$= 5 + 14 \times 62 + 14M = 5 + 14(62 + M)$

Hence, when $1! + 2! + \ldots + 200!$ is divided by 14, remainder is 5.

51. (c) $\cos^{-1}[\cos\{2 \cdot \cot^{-1}(\sqrt{2} - 1\}]$

$= \cos^{-1}[\cos\{2 \times 67.5\}]]$

$= \cos^{-1}[\cos 135°] = 135° = \frac{3\pi}{4}$

52. (c) We have $f(x) = [x]\cos\left[\frac{2x-1}{2}\right]\pi$

Since, $g(x) = [x]$ is always discontinuous at all integral values of points. Hence, $f(x)$ is discontinuous for all integral points.

53. (a) Obtain the foot of the perpendicular $N(-4, 1, -3)$.

Find the equations of the line passing thorugh $P(2, 4, -1)$ and $N(-4, 1, -3)$.

$\frac{x-2}{6} = \frac{y-4}{3} = \frac{z+1}{2}$

54. (b) Total number of mappings from a set A having elements into itself is n^n. And the total number of one to one mapping is $n!$

$\therefore$ Required probability $= \frac{n!}{n^n}$.

55. (a) $\mathbf{a} = -\hat{\mathbf{i}} + 2\hat{\mathbf{j}} - \hat{\mathbf{k}}$, $\mathbf{b} = \hat{\mathbf{i}} + \hat{\mathbf{j}} - 3\hat{\mathbf{k}}$,

$\mathbf{c} = -4\hat{\mathbf{i}} - \hat{\mathbf{k}}$

$\therefore \mathbf{a} \times (\mathbf{b} \times \mathbf{c}) + (\mathbf{a} \cdot \mathbf{b})\mathbf{c} = (\mathbf{a} \cdot \mathbf{c})\mathbf{b}$

$= (4 + 1)(\hat{\mathbf{i}} + \hat{\mathbf{j}} - 3\hat{\mathbf{k}}) = 5\hat{\mathbf{i}} + 5\hat{\mathbf{j}} - 15\hat{\mathbf{k}}$

56. (c) Since, total number of numbers 1 to 5 digits are 99999 and total number of numbers 1 to 4 digits are 9999.

Hence, the total number of numbers of exactly 5 digits $= 99999 - 9999 = 90000$.

57. (b) The maximum value of $3\cos\theta + 4\sin\theta$

$= \sqrt{3^2 + 4^2} = \sqrt{25} = 5$

58. (a) The vertex connectivity of any tree is one.

59. (a) Since, $x\mathbf{a} + y\mathbf{b} + z\mathbf{c} = 0$

Also, **a**, **b** and **c** are non-collinear vectors, then $x = y = z = 0$.

60. (b) Since, 10, 3, e, 2 are in decreasing order.

$\therefore \log_{10}\alpha, \log_3\alpha, \log_e\alpha, \log_2\alpha$ are in increasing order.

61. (a) Now, $\frac{b-c}{r_1} + \frac{c\ a}{r_2} + \frac{a-b}{r_3}$

$= \frac{[(b-c)(s-a) + (c-a)(s-b) + (a-b)(s-c)]}{\Delta} = 0$

62. (d) Let $T = |\sin nx|$

Put $n = 2$, then

$T = |\sin 2x| = |2\sin x\cos x| \le 2|\sin x|$

63. (c) Since, $(n+1)\,p = \frac{101}{3}$ is not an integer.

Therefore, $p(X = r)$ is maximum when

$$r = \left[\frac{101}{3}\right] = 33.$$

64. (d) The equation of xy-plane is $z = 0$

$\therefore$ Direction cosines of its normal are 0, 0, 1.

65. (b) We have, $\int f(x)\,dx = F(x)$

$$\therefore \int x^3 f(x^2)dx = \frac{1}{2}\int x^2 f(x^2)\,d(x^2)$$

66. (d) $\int_1^5 x^2\,dx$

$$= \frac{4}{4}\left[\frac{1}{2}f(1) + f(2) + f(3) + f(4) + \frac{1}{2}f(5)\right]$$

$$= \frac{1}{2}[x^2F(x^2) - \int F(x^2)\,d(x^2)]$$

$$= \left[\frac{1}{2} + 4 + 9 + 16 + \frac{25}{2}\right] = 42 \text{ [using } f(x) = x^2]$$

67. (c) Here, $2ae = 6$ and $2b = 8 \Rightarrow b = 4$

$$\Rightarrow b^2 = 16 \Rightarrow a^2(1-e^2) = 16$$

$$\Rightarrow \frac{9}{e^2}(1-e^2) = 16 \Rightarrow 9 - 9e^2 = 16e^2$$

$$\Rightarrow 25e^2 = 9 \Rightarrow e = \frac{3}{5}$$

68. (c) We know that, if $I_n = \int \sin^n x\,dx$, then

$$I_n = \frac{\sin^{n-1}x\cos x}{n} + \frac{n-1}{n}I_{n-2}$$

where n is a positive integer.

$$\Rightarrow nI_n - (n-1)\,I_{n-2} = -\sin^{n-1}x\cos x$$

69. (a) The given graph is a connected graph.

70. (a) We have, $5^{124} = (5^3)^{41} \cdot 5$

Now, $5^3 \equiv 1 \pmod{124}$

$\therefore$ $(5^3)^{41} \equiv 1 \pmod{124}$

$(5^3)^{41} \cdot 5 \equiv 1 \pmod{124}$

$\Rightarrow$ $5^{124} \equiv 5 \pmod{124}$

GENERAL APTITUDE

16. (a) As 'kind' is related to 'judge', in the same way 'Throne' is related to 'Bench'.

17. (b) 'Roar' is the sound produced by 'Lion', similarly 'Bray' is the sound produced by 'Ass'.

18. (a) 'Frisk' is the name given to the movement of 'Lamb' similarly 'Scamper' is the name given to the movement of 'Mouse'.

19. (d) As 'Arc' is a part of 'Circle', in the same way 'Line' is a part of 'Square'.

20. (d) 'Brinjal' is a 'Vegetable', in the same way 'Orange' is a 'Fruit'.

21. (d) All except 'Reading' are related to motion.

22. (d) All except 'Bhagalpur' are in the state of Uttar Pradesh.

23. (d) Only Grapes grow in bunches and so it is odd one among others.

24. (a) Position of S from left = 19

$\therefore$ Position of S from right = 27 − 19 = 8

25. (a) From CFILORUX, Position of L from left = 12

26. (c) 7th letter from right = (27 − 7) th = 20th letter from left = T

27. (a) The pattern is as follows

6 →(+3) 9 →(+3) 12 →(+3) 15 →(+3) 18 →(+3) (21)

$\therefore$? = 20

28. (b) The pattern is as follows

360 →(÷2) 180 →(÷2) 90 →(÷2) 45 →(÷2) 22.5 →(÷2) (11.25)

$\therefore$? = 11.25

29. (b) Moving anti-clockwise, the pattern is as follows 15 + 1 = 16; 16 + 2 = 18

18 + 3 = 21 and 21 + 4 = 25

So, missing number 25 + 5 = 30

30. (b) The pattern in clockwise direction

$$2 \xrightarrow{\times 2} \boxed{4} \xrightarrow{\times 4} 16 \xrightarrow{\times 16} 256 \;\therefore\; ? = 4$$

MANIPAL
Engineering Entrance Exam
Solved Paper 2018

Physics

1. At great distances from an electric dipole, the electric field strength due to the dipole varies with the distance as

(a) $\frac{1}{r}$ (b) $\frac{1}{r^2}$

(c) $\frac{1}{r^3}$ (d) $\frac{1}{r^4}$

2. A charged particle travels along a straight line with a speed v in a region where both electric field **E** and magnetic fields **B** are present. It follows that

(a) $|\mathbf{E}| = v|\mathbf{B}|$ and the two fields are parallel

(b) $|\mathbf{E}| = v|\mathbf{B}|$ and the two fields are perpendicular

(c) $|\mathbf{B}| = v|\mathbf{E}|$ and the two fields are parallel

(d) $|\mathbf{B}| = v|\mathbf{E}|$ and the two fields are perpendicular

3. A solenoid of length 50 cm and a radius of cross-section 1 cm has 1000 turns of wire wound over it. If the current carried is 5A, the magnetic field on its axis, near the centre of the solenoid is approximately (permeability of free space $\mu_0 = 4\pi \times 10^{-7}$ T-m/A)

(a) 0.63×10^{-2} T (b) 1.26×10^{-2} T

(c) 2.51×10^{-2} T (d) 6.3 T

4. A transformer of 100% efficiency has 200 turns in the primary and 40000 turns in secondary. It is connected to a 220 V main supply and secondary feeds to a 100 kΩ resistance. The potential difference per turn is

(a) 1.1 V (b) 25 V

(c) 18 V (d) 11 V

5. In a single slit diffraction pattern, the distance between the first minimum on the left and the first minimum on the right is 5 mm. The screen on which the diffraction pattern is displayed is at a distance of 80 cm from the slit. The wavelength is 6000 Å.The slit width in (mm) is about

(a) 0.576 (b) 0.348 (c) 0.192 (d) 0.096

6. When a ray is incident on a medium of refractive index n at Brewster's angle, it gets

(a) totally reflected (b) totally absorbed

(c) circularly polarised (d) plane polarised

7. The binding energy per nucleon is maximum in the case of

(a) ${}_{2}^{4}\text{He}$ (b) ${}_{26}\text{Fe}^{56}$

(c) ${}_{56}\text{Ba}^{141}$ (d) ${}_{92}\text{U}^{225}$

8. A thin convex lens of refractive index 1.5 has 20 cm focal length in air. If the lens is completely immersed in a liquid of refractive index 1.6, its focal length will be

(a) − 160 cm (b) − 100 cm

(c) + 10 cm (d) + 100 cm

9. A mass $3m$, initially at rest at the origin explodes into three fragments of equal mass. Two of the fragments have speed v each and move perpendicular to each other. The third fragment will move with a speed

(a) $\frac{v}{\sqrt{2}}$ (b) $\frac{v}{2}$

(c) v (d) $v\sqrt{2}$

10. Standing waves are formed on a string when interference occurs between two waves having

(a) the same amplitude travelling in the same direction with no phase difference between them
(b) the same amplitude, travelling in the opposite direction with no phase difference between them
(c) different amplitudes travelling in the same direction
(d) different amplitudes travelling in the opposite direction

11. An open tank filled with water (density) ρ has a narrow hole at a depth of h below the water surface. The velocity of water flowing out is

(a) $h\rho g$ (b) $2gh$
(c) $\sqrt{2gh}$ (d) gh

12. Three concurrent forces of the same magnitude are in equilibrium. What is the angle between the forces ? Also, name the triangle formed by the forces as sides

(a) 60° equilateral triangle
(b) 120° equilateral triangle
(c) 120°, 30°, 30° an isosceles triangle
(d) 120° an obtuse triangle

13. A straight conductor carries a current of 5 A. An electron travelling with a speed of 5×10^6 ms^{-1} parallel to the wire at a distance of 0.1 m from the conductor, experiences a force of

(a) 8×10^{-20} N (b) 3.2×10^{-19} N
(c) 8×10^{-18} N (d) 1.6×10^{-19} N

14. A uniform wire of resistance 9Ω is cut into 3 equal parts. They are connected in the form of equilateral triangle ABC. A cell of emf 2V and negligible inernal resistance is connected across B and C. Potential difference across AB is

(a) 1 V (b) 2 V
(c) 3 V (d) 0.5 V

15. In a potentiometer experiment two cells of emf E_1 and E_2 are used in series and in conjunction and the balancing length is found to be 58 cm of the wire. If the polarity of E_2 is reversed, then the balancing length becomes 29 cm. The ratio $\frac{E_1}{E_2}$ of the emfs of the two cells is

(a) 1 : 1 (b) 2 : 1 (c) 3 : 1 (d) 4 : 1

16. A spring balance is attached to the ceiling of a lift. A man hangs his bag on the spring and the spring reads 49 N, when the lift is stationary. If the lift moves downward with an acceleration of 5 m/s^2, the reading of the spring balance will be

(a) 49 N (b) 24 N
(c) 74 N (d) 15 N

17. A body cools in 7 min from 60°C to 40°C. What times (in min) does it take to cool from 40°C to 28°C, if surrounding temperature is 10°C ? (Assume Newton's law of cooling)

(a) 3.5 (b) 14 (c) 7 (d) 10

18. The stopping potential for photoelectric emission from a metal surface is plotted along Y-axis and frequency ν of incident light along X-axis. A straight line is obtained as shown. Planck's constant is given by

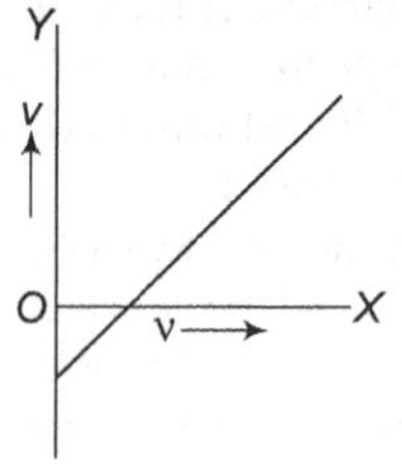

(a) slope of the line
(b) product of slope of the line and charge on the electron
(c) intercept along Y-axis divided by charge on the electron
(d) product of intercept along X-axis and mass of the electron

19. A parallel beam of monochromatic light is incident normally on a slit. The diffraction pattern is observed on a screen placed at the focal plane of a convex lens. If the slit width is increased, the central maximum of the diffraction pattern will

(a) become broader and fainter
(b) become broader and brighter
(c) become narrower and fainter
(d) become narrower and brighter

20. The bob A of a simple pendulum is released when the string makes an angle of 45° with the vertical. It hits an other bob B of the same material and same mass kept at rest on the table. If the collision is elastic

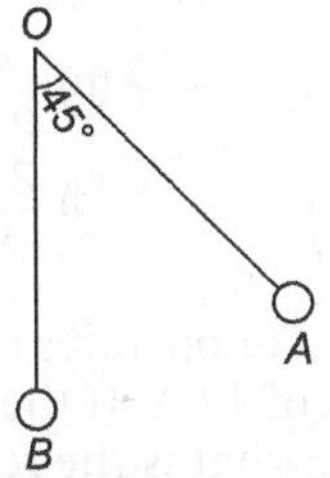

(a) both A and B rise to the same height
(b) both A and B come to rest at B
(c) both A and B move with the same velocity of A
(d) A comes to rest and B moves with the velocity of A

21. A battery of emf 12 V and internal resistance 2 Ω is connected in series with a tangent galvanometer of resistance 4 Ω. The deflection is 60° when the plane of the coil is along the magnetic meridian. To get a deflection of 30°, the resistance to be connected in series with the tangent galvanometer is

(a) 12 Ω (b) 20 Ω
(c) 10 Ω (d) 5 Ω

22. A transistor is used in common-emitter configuration. Given its $\alpha = 0.9$, calculate the change in collector current when the base current changes by 2 μA.

(a) 1 μA (b) 0.9 μF
(c) 30 μA (d) 18 μA

23. Energy needed in breaking a drop of radius R into n drops of radii r is given by

(a) $4\pi T(nr^2 - R^2)$
(b) $\frac{4}{3}\pi(r^3 n - R^2)$
(c) $4\pi T(R^2 - nr^2)$
(d) $4\pi T(nr^2 + R^2)$

24. A gas at the temperature 250 K is contained in a closed vessel. If the gas is heated through 1 K, then percentage increase in its pressure will be

(a) 0.4% (b) 0.2%
(c) 0.1% (d) 0.8%

25. Voltage and current in an AC circuit are given by $V = 5\sin\left(100\pi t - \frac{\pi}{6}\right)$ and $I = 4\sin\left(100\pi t + \frac{\pi}{6}\right)$

(a) voltage leads the current by 30°
(b) current leads the voltage by 30°
(c) current leads the voltage by 60°
(d) voltage leads the current by 60°

26. A bar magnet is released into a copper ring directly below it. The acceleration of the magnet will be

(a) equal to the acceleration due to gravity at that place
(b) less than the acceleration due to gravity at that place
(c) greater than the acceleration due to gravity at that place
(d) twice the acceleration due to gravity at that place

27. Two different isotherms representing the relationship between pressure p and volume V at a given temperature of the same ideal gas are shown for masses m_1 and m_2 of the gas respectively in the figure given, then

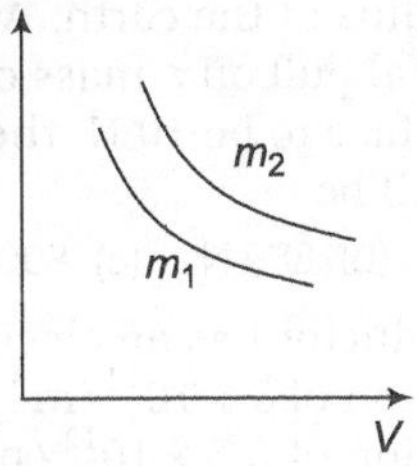

(a) $m_1 > m_2$ (b) $m_1 = m_2$ (c) $m_1 < m_2$ (d) $m_1 \gtreqless m_2$

28. A, B and C are parallel conductors of equal lengths carrying currents I, I and $2I$ respectively. Distance between A and B is x.

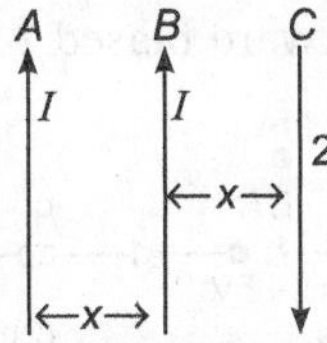

Distance between B and C is also x. F_1 is the force exerted by B on A. F_2 is the force exerted by C on A. Choose the correct answer.

(a) $F_1 = 2F_2$ (b) $F_2 = 2F_1$
(c) $F_1 = F_2$ (d) $F_1 = -F_2$

29. The light ray is incident at angle of 60° on a prism of angle 45°. When the light rays falls on the other surface at 90°, the refractive index of the material of prism μ and the angle of deviation δ are given by

(a) $\mu = \sqrt{2}, \delta = 30°$ (b) $\mu = 1.5, \delta = 15°$

(c) $\mu = \frac{\sqrt{3}}{2}, \delta = 30°$ (d) $\mu = \sqrt{\frac{3}{2}}, \delta = 15°$

30. A plano-convex lens is made of refractive index of 1.6. The radius of curvature of the curved surface is 60 cm. The focal length of the lens is

(a) 400 cm (b) 200 cm (c) 100 cm (d) 50 cm

31. A body of mass 2 kg makes an elastic collision with another body at rest and continues to move in the original direction with one-fourth its original speed. The mass of the second body which collides with the first body is

(a) 2 kg (b) 1.2 kg (c) 3 kg (d) 1.5 kg

32. A research satellite of mass 200 kg circle the earth in an orbit of average radius $\frac{3R}{2}$, where R is the radius of the earth. Assuming the gravitational pull on a mass of 1 kg on the earth's surface to be 10 N, the pull on the satellite will be

(a) 880 N (b) 889 N (c) 890 N (d) 892 N

33. A semiconductor has an electron concentration of $8 \times 10^{13}/\text{m}^3$ and hole concentration of $5.5 \times 10^{12}/\text{m}^3$. The semiconductor is

(a) *n*-type (b) *p*-type

(c) intrinsic semiconductor (d) *p*-*n* junction

34. In the figure shown below, which of the diodes are forward biased ?

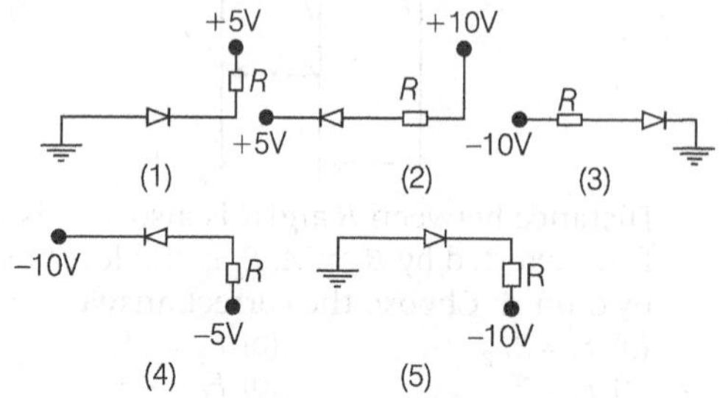

(a) 1, 2, 3 (b) 2, 4, 5 (c) 1, 3, 4 (d) 2, 3, 4

35. The path difference between the two waves

$$y_1 = a_1 \sin\left(\omega t - \frac{2\pi x}{\lambda}\right) \text{ and}$$

$$y_2 = a_2 \cos\left(\omega t - \frac{2\pi x}{\lambda} + \phi\right) \text{ is}$$

(a) $\frac{\lambda}{2\pi}\phi$ (b) $\frac{\lambda}{2\pi}\left(\phi + \frac{\pi}{2}\right)$

(c) $\frac{2\pi}{\lambda}\left(\phi - \frac{\pi}{2}\right)$ (d) $\frac{2\pi}{\lambda}\phi$

36. An electric motor operates on a 50 V supply and a current of 12 A. If the efficiency of the motor is 30%, what is the resistance of the winding of the motor ?

(a) 6 Ω (b) 4 Ω

(c) 2.9 Ω (d) 3.1 Ω

37. Two long straight wires are set parallel to each other. Each carries a current in the same direction and the separation between them is $2r$. The intensity of the magnetic field mid-way between them is

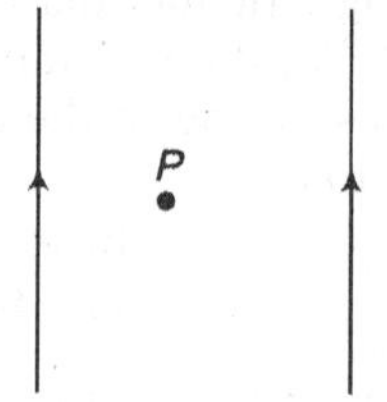

(a) $\frac{\mu_0 i}{r}$ (b) $\frac{4\mu_0 i}{r}$

(c) zero (d) $\frac{\mu_0 i}{4r}$

38. A galvanometer of resistance 20 Ω is to be converted into an ammeter of range 1 A. If a current of 1 mA produces full scale deflection, the shunt required for the purpose is

(a) 0.01 Ω (b) 0.05 Ω

(c) 0.02 Ω (d) 0.04 Ω

39. In a step-up transformer the voltage in the primary is 220 V and the current is 5 A. The secondary voltage is found to be 22000 V. The current in the secondary (neglect losses) is

(a) 5 A (b) 50 A

(c) 500 A (d) 0.05 A

40. In Young's double slit experiment, the intensity of light coming from one of the slits is double the intensity from the other slit.

The ratio of the maximum intensity to the minimum intensity in the interference fringe pattern observed is

(a) 34 (b) 40 (c) 25 (d) 38

41. The distance between charges 5×10^{-11} C and -2.7×10^{-11}C is 0.2 m. The distance at which is third charge should be placed from $4e$ in order that it will not experience any force along the line joining the two charges is

(a) 0.44 m (b) 0.65 m
(c) 0.556 m (d) 0.350 m

42. Small rain drops of the same size are charged to potential V volt each. If n such drops coalesce to form a single drop, then the potential of the bigger drop is

(a) $n^{1/3}$ V (b) $n^{2/3}$ V
(c) nV (d) $n^{3/2}$ V

43. A small sphere is charged to a potential of 50 V and a big hollow sphere is charged to potential of 100 V. Charge will flow from the smaller sphere to the bigger one when

(a) the smaller one is placed inside the bigger one and connected by a wire
(b) the bigger one is placed inside the smaller one and connected by means of a wire
(c) bigger one placed by the side of the smaller one and connected by a wire
(d) smaller one placed by the side of the bigger one

44. If refractive index of a material of equilateral prism is $\sqrt{3}$, then angle of minimum deviation of the prism is

(a) 30° (b) 45°
(c) 60° (d) 75°

45. A long solenoid of length L has a mean diameter D. It has n layers of winding of N turns each. If it carries a current i, the magnetic field at its centre will be

(a) proportional to D
(b) inversely proportional to D
(c) independent of D
(d) proportional to L

46. A wire of length 100 cm is connected to a cell of emf 2 V and negligible internal resistance. The resistance of the wire is 3 Ω, the additional resistance required to produce a potential difference of 1 mV/cm is

(a) 60 Ω (b) 47 Ω (c) 57 Ω (d) 35 Ω

47. In the figure a carbon resistor has bands of different colours on its body as mentioned in the figure. The value of the resistance is

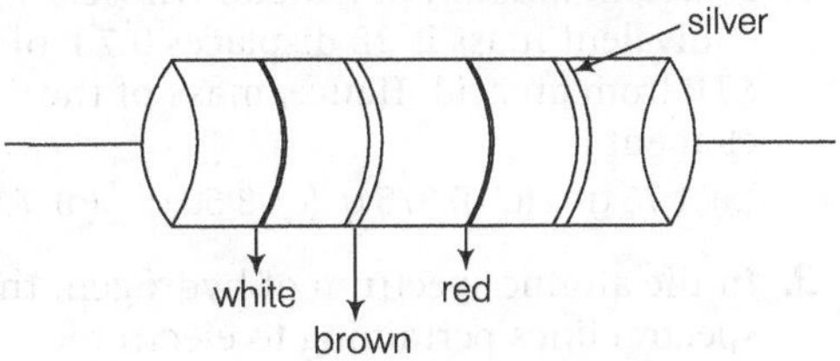

(a) 2.2 kΩ (b) 3.3 kΩ (c) 5.6 kΩ (d) 9.1 kΩ

48. The mass defect in a particular nuclear reaction is 0.3 g. The amount of energy liberated in killowatt hour is (velocity of light $= 3 \times 10^8$ m / s)

(a) 1.5×10^6 (b) 2.5×10^6
(c) 3×10^6 (d) 7.5×10^6

49. A photon and an electron have equal energy $E \cdot \lambda_{photon} / \lambda_{electron}$ is proportional to

(a) $\sqrt{E}$ (b) $\frac{1}{\sqrt{E}}$
(c) $\frac{1}{E}$ (d) does not depend upon E

50. The magnitude of current i in ampere unit is

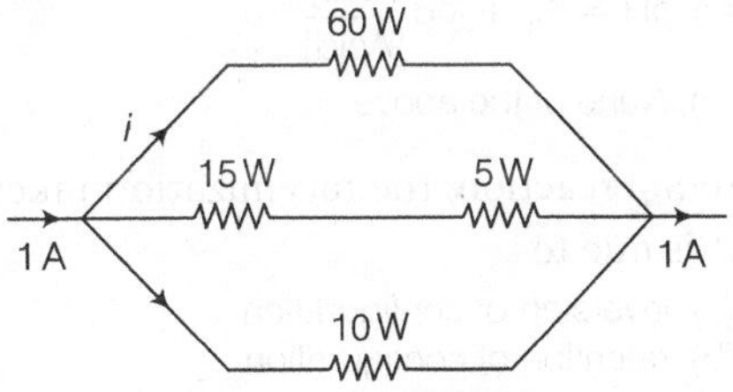

(a) 0.1 (b) 0.3
(c) 0.6 (d) None of these

Chemistry

1. If 20 g of $CaCO_3$ is treated with 100 mL of 20% HCl solution, the amount of CO_2 produced is
(a) 22.4 L (b) 8.80 g (c) 4.40 g (d) 2.24 L

2. A certain amount of a metal whose equivalent mass is 28 displaces 0.7 L of H_2 at STP from an acid. Hence, mass of the element is
(a) 1.75 g (b) 0.875 g (c) 3.50 g (d) 7.00 g

3. In the atomic spectrum of hydrogen, the spectral lines pertaining to electronic transition of $n = 4$ to $n = 2$ refers to
(a) Lyman series (b) Balmer series
(c) Paschen series (d) Brackett series

4. The electron affinity of Be is almost similar to that of
(a) Li (b) B
(c) Na (d) Ne

5. For a process to be spontaneous
(a) ΔG must be –ve (b) ΔG should be +ve
(c) ΔH must be –ve (d) ΔS must be –ve

6. pH of an acid buffer is given by
(a) $pH = pK_a + \log \frac{[Salt]}{[Acid]}$
(b) $pH = pK_a - \log \frac{[Salt]}{[Acid]}$
(c) $pH = K_a + \log \frac{[Salt]}{[Acid]}$
(d) None of the above

7. In S_N1 reaction, the racemization takes place. It is due to
(a) inversion of configuration
(b) retention of configuration
(c) conversion of configuration
(d) Both (a) and (b)

8. When phenol is treated with excess of bromine water it gives
(a) *m*-bromophenol
(b) *o*- and *p*-bromophenol
(c) 2, 4-dibromophenol
(d) 2, 4, 6-tribromophenol

9. $CH_3COOH + NH_3 \longrightarrow X \xrightarrow{\Delta} Y$; *Y* is
(a) CH_3NH_2 (b) $CH_3CH_2NH_2$
(c) CH_3CONH_2 (d) CH_3COONH_4

10. Coordination number of Zn in ZnS is
(a) 4 (b) 6
(c) 2 (d) None of these

11. What is the order of the reaction which has a rate expression

$$\text{rate} = k[A]^{\frac{3}{2}}[B]^{-1}\ ?$$

(a) $\frac{3}{2}$ (b) $\frac{1}{2}$ (c) zero (d) None of these

12. 99% of a first order reaction was completed in 32 min. When will 99.9% of the reaction complete ?
(a) 50 min (b) 46 min
(c) 49 min (d) 48 min

13. Which one of the following solutions exhibits the maximum elevation in boiling point ?
(a) 0.1 m NaCl (b) 0.1 m $FeCl_3$
(c) 0.1m $CaCl_2$ (d) 0.1 m $BaCl_2$

14. The coefficients of I^-, IO_3^- and H^+ in the redox reaction,

$$I^- + IO_3^- + H^+ \longrightarrow I_2 + H_2O$$

in the balanced form respectively are
(a) 5, 1, 6 (b) 1, 5, 6
(c) 6, 1, 5 (d) 5, 6, 1

15. Lyophilic sols are more stable than lyophobic sols because
(a) the colloidal particles have negative charge
(b) the colloidal particles have positive charge
(c) the colloidal particles are solvated
(d) None of the above

16. Which of the following will give positive iodoform test ?
(a) HCHO (b) CH_3OCH_3
(c) $(CH_3)_2CO$ (d) CH_3CH_2Cl

17. Starch is the polymer of
(a) glucose (b) fructose
(c) Both (a) and (b) (d) None of these

18. The type of isomerism present in nitropentamine chromium (III) chloride is

(a) optical (b) coordination
(c) linkage (d) polymerisation

19. The standard reduction potentials for two half-cell reactions are given below.

$$Cd^{2+}(aq) + 2e^- \longrightarrow Cd(s); E° = -0.40 \text{ V}$$

$$Ag^+(aq) + e^- \longrightarrow Ag(s), E° = 0.80 \text{ V}$$

The standard free energy change for the reaction

$$2Ag^+(aq) + Cd(s) \longrightarrow 2Ag(s) + Cd^{2+}(aq)$$

is given by

(a) 115.8 kJ (b) – 115.8 kJ
(c) – 231.6 kJ (d) 231.6 kJ

20. Which of the following ions has the highest magnetic moment?

(a) Zn^{2+} (b) Ti^{3+} (c) Sc^{3+} (d) Mn^{2+}

21. Cane sugar charrs on treating with conc. H_2SO_4. It is because

(a) H_2SO_4 is oxidising agent
(b) H_2SO_4 is dehydrating agent
(c) H_2SO_4 is an acid
(d) H_2SO_4 is a catalyst

22. Which of the following sets of quantum numbers is impossible arrangement ?

(a) $n = 3, m = -2, s = +\frac{1}{2}$
(b) $n = 4, m = 3, s = +\frac{1}{2}$
(c) $n = 5, m = 2, s = -\frac{1}{2}$
(d) $n = 3, m = -3, s = -\frac{1}{2}$

23. The formation of oxide ion $O^{2-}(g)$ require first an exothermic and then an endothermic step as shown below :

$$O(g) + e^- \rightarrow O^-(g); \quad \Delta H° = -142 \text{ kJ mol}^{-1}$$

$$O^-(g) + e^- \rightarrow O^{2-}(g); \Delta H° = 844 \text{ kJ mol}^{-1}$$

This is because

(a) Oxygen is more electronegative
(b) O^- ion has comparatively larger size than oxygen atom
(c) O^- ion will tend to resist the addition of another electron
(d) Oxygen has high electron affinity

24. In the standardization of $Na_2S_2O_3$ using $K_2Cr_2O_7$ by iodometry, the equivalent weight of $K_2Cr_2O_7$ is

(a) molecular weight/2
(b) molecular weight/6
(c) molecular weight/3
(d) same as molecular weight

25. The coordination number of central metal atom in a complex is determined by

(a) the number of ligands around a metal ion bonded by sigma bonds
(b) the number of ligands around a metal ion bonded by π bonds
(c) the number of ligands around a metal ion bonded by sigma and π bonds both
(d) None of the above

26. Ethylene diamine is an example of

(a) monodentate ligand (b) bidentate ligand
(c) tridentate ligand (d) polydentate ligand

27. Deviation from the ideal gas behaviour of a gas can be expressed as

(a) $Z = \frac{p}{VRT}$ (b) $Z = \frac{pV}{nRT}$
(c) $Z = \frac{nRT}{pV}$ (d) $Z = \frac{VR}{pT}$

28. To a 4 L of 0.2 M solution of NaOH 2 L of 0.5 M NaOH are added. The molarity of resulting solution is

(a) 0.9 M (b) 0.3 M
(c) 1.8 M (d) 0.18 M

29. Reimer-Tiemann reaction involves

(a) carbonium ion intermediate
(b) carbene intermediate
(c) carbanion intermediate
(d) free radical intermediate

30. Aldol condensation will not observed in

(a) chloral
(b) phenylacetaldehyde
(c) nitroethanal
(d) hexanal

31. The entropy change for the fusion of 1 mol of a solid which melts at 27°C is (latent heat of fusion =600 cal mol^{-1})

(a) 2 cal K^{-1} (b) 22.2 cal K^{-1}
(c) 180 k cal K (d) 0.5 cal^{-1} K

32. The solubility product of $Al_2(SO_4)_3$ is given by the expression

(a) $[Al^{3+}][SO_4^{2-}]$ (b) $[Al^{3+}]^2[SO_4^{2-}]$
(c) $[Al^{3+}]^3[SO_4^{2-}]^2$ (d) $[Al^{3+}]^2[SO_4^{2-}]^3$

33. S_N1 reactions are favoured by

(a) non-polar solvents
(b) bulky groups on the carbon atom attached to the halogen atom
(c) small groups on carbon atom attached to the halogen atom
(d) None of the above

34. The order of relative bond strength of oxygen molecule, superoxide ion and peroxide ion is

(a) $O_2 < O_2^- < O_2^{2-}$ (b) $O_2 > O_2^- > O_2^{2-}$
(c) $O_2 < O_2^- > O_2^{2-}$ (d) $O_2 > O_2^- < O_2^{2-}$

35. 1.2 g of Mg (at. mass 24) will produce MgO equal to

(a) 0.05 mol (b) 40 g (c) 40 mg (d) 4 g

36. The enthalpy change for the transition of liquid water to steam is 40.8 kJ mol^{-1} at 373 K. Calculate the entropy of vaporisation of water.

(a) 109.4 JK^{-1} mol^{-1} (b) 10.94 JK^{-1} mol^{-1}
(c) 1.094 JK^{-1} mol^{-1} (d) 1009.4 JK^{-1} mol^{-1}

37. The function of $AlCl_3$ in the Friedel-Craft's reaction is

(a) to absorb water
(b) to absorb HCl
(c) to produce electrophile
(d) to produce nucleophile

38. Which of the following is paramagnetic ?

(a) O_2 (b) N_2 (c) O_2^{2-} (d) H_2

39. The number of stereoisomers for pent-3-en-2-ol is

(a) 2 (b) 4 (c) 3 (d) 5

40. When attraction between $A - B$ is more than that of $A - A$ and $B - B$, the solution will show

(a) positive deviation from Raoult's law
(b) negative deviation from Raoult's law
(c) no deviation from Raoult's law
(d) Cannot be predicted

41. The noble gas used for treatment of cancer is

(a) Helium (b) Argon
(c) Radon (d) Krypton

42. Among the following species, which has zero bond order ?

(a) H_2 (b) H_2^+
(c) He_2 (d) He_2^+

43. 0.16 g of dibasic acid requires 25 mL decimolar NaOH solution for complete neutralisation. The molecular mass of the acid is

(a) 256 (b) 64
(c) 32 (d) 128

44. The standard free energy change ($\Delta G°$) is related to equilibrium constant (K) as

(a) $\Delta G° = -2.303\ RT \ln K$
(b) $\Delta G° = 2.303\ RT \log K$
(c) $\Delta G° = RT \log K$
(d) $\Delta G° = -2.303\ RT \log K$

45. Degree of dissociation of an acid HCl is 95%. 0.192 g of the acid is present in 0.5 L of solution. The pH of the solution is

(a) 2 (b) 1
(c) 3 (d) 4

46. Molisch test is given by

(a) all carbohydrates (b) sucrose
(c) fructose (d) glucose

47. The number of amino acids in insulin is

(a) 21 (b) 574 (c) 51 (d) 5733

48. Low spin complexes are generally formed by elements of

(a) 3*d* series (b) 4*d* series
(c) 5*d* series (d) Both (b) and (c)

49. An optically active alkene with the M.F. C_6H_{12} which upon hydrogenation gives optically inactive alkane is

(a) 2-hexene (b) 3-methyl-2-pentene
(c) 2-methyl-2-pentene (d) 3-methyl-1-pentene

50. If uncertainty in position and velocity are equal, then uncertainty in momentum will be

(a) $\frac{1}{2}\sqrt{\frac{mh}{\pi}}$ (b) $\frac{1}{2}\sqrt{\frac{h}{\pi m}}$
(c) $\frac{h}{4\pi m}$ (d) $\frac{mh}{4\pi}$

Mathematics

1. The equation $3^{|3x+4|} = 9^{2x-2}, x > 0$ has the solution

(a) $\frac{7}{8}$ (b) $\frac{8}{7}$
(c) $\frac{-3}{4}$ (d) None of these

2. A parallelogram is constructed on the vectors $\mathbf{a} = 3\alpha - \beta, \mathbf{b} = \alpha + 3\beta$. If $|\alpha| = |\beta| = 2$ and the angle between α and β is $\frac{\pi}{3}$, then length of a diagonal of the parallelogram is

(a) $4\sqrt{3}$ (b) $4\sqrt{5}$
(c) $4\sqrt{7}$ (d) None of these

3. Every group of order 7 is

(a) not abelian (b) not cyclic
(c) cyclic (d) None of these

4. Let $U_n = 2 + 2^3 + 2^5 + ... + 2^{2n+1}$ and $V_n = 1 + 4 + 4^2 + ... + 4^{n-1}$. Then, $\lim_{n \to \infty} \frac{U_n}{V_n}$ is equal to

(a) 8 (b) 6 (c) $\frac{3}{4}$ (d) $\frac{3}{2}$

5. If $P(n): 1 + 3 + 5 + ... + (2n-1) = n^2$ is

(a) true for all $n \in N$ (b) true for $n > 1$
(c) true for no n (d) None of these

6. The minimum value of $x^2 + \frac{1}{1+x^2}$ is, at

(a) $x = 0$ (b) $x = 1$ (c) $x = 4$ (d) $x = 3$

7. The coefficient of x^{-9} in the expansion of $\left(\frac{x^2}{2} - \frac{2}{x}\right)^9$ is

(a) 512 (b) -512
(c) 521 (d) 251

8. The sum to n terms of the series $\frac{1}{2} + \frac{3}{4} + \frac{7}{8} + \frac{15}{16} + ...$ is

(a) $2^n - 1$ (b) $1 - 2^n$
(c) $n + 2^n - 1$ (d) $n - 1 + 2^{-n}$

9. The value of $\begin{vmatrix} a & a+b & a+2b \\ a+2b & a & a+b \\ a+b & a+2b & a \end{vmatrix}$ is equal to

(a) $9a^2 (a+b)$ (b) $9b^2 (a+b)$
(c) $a^2 (a+b)$ (d) $b^2 (a+b)$

10. The negation of the compound proposition $p \vee (\sim p \vee q)$ is

(a) $(p \wedge \sim q) \wedge \sim p$
(b) $(p \wedge \sim q) \vee \sim p$
(c) $(p \wedge \sim q) \vee \sim p$
(d) None of the above

11. In the group $[G = \{0, 1, 2, 3, 4\}, \times_5]$ the inverse of $(2 \times_5 2^{-1})$ is

(a) 0 (b) 1 (c) 2 (d) 3

12. The statement $P(n) = 9^n - 8^n$, when divided 8, always leaves the remainder

(a) 1 (b) 2 (c) 3 (d) 7

13. If $\tan^{-1} x + \tan^{-1} y + \tan^{-1} z = \pi$, then $x + y + z$ is

(a) xyz (b) 0 (c) 1 (d) $2xyz$

14. If $\left(\frac{1}{2}, \frac{1}{3}, n\right)$ are the direction cosines of a line, then the value of n is

(a) $\frac{\sqrt{23}}{6}$ (b) $\frac{23}{36}$ (c) $\frac{2}{3}$ (d) $\frac{3}{2}$

15. Derivative of x^x is

(a) $x^x(1 - \log x)$ (b) $1 + \log x$
(c) $x^x (1 + \log x)$ (d) $1 - \log x$

16. If $(3 - x) \equiv (2x - 5) \pmod 4$, then one of the values of x is

(a) 3 (b) 4 (c) 18 (d) 5

17. The product of the perpendicular from $(-1, 2)$ to the pair of lines $2x^2 - 5xy + 2y^2 + 3x - 3y + 1 = 0$ is

(a) $\frac{5}{12}$ (b) $\frac{12}{5}$ (c) $\frac{6}{5}$ (d) $\frac{5}{6}$

18. If $\mathbf{a} = -\hat{i} + 2\hat{j} - \hat{k}, \mathbf{b} = \hat{i} + \hat{j} - 3\hat{k}$ and $\mathbf{c} = -4\hat{i} - \hat{k}$, then $\mathbf{a} \times (\mathbf{b} \times \mathbf{c}) + (\mathbf{a} \cdot \mathbf{b})\, \mathbf{c}$

(a) $5\hat{i} + 5\hat{j} - 15\hat{k}$ (b) $\mathbf{0}$
(c) $12\hat{j} + 4\hat{k}$ (d) $-3\hat{i} + 6\hat{j} - 3\hat{k}$

19. The work done by the force $4\hat{i} - 3\hat{j} + 2\hat{k}$ in moving a particle along a straight line from the point $(3, 2, -1)$ to $(2, -1, 4)$ is

(a) 0 unit (b) 4 unit (c) 15 unit (d) 19 unit

20. The value of 'a' for which the function $f(x) = a \sin x + \left(\frac{1}{3}\right) \sin 3x$ has an extremum at $x = \frac{\pi}{3}$, is

(a) 1 (b) -1 (c) 2 (d) 0

21. If $|f(x)|$ is continuous at $x = a$, then $f(x)$ is

(a) continuous at $x = a$
(b) continuous at $x = -a$
(c) continuous at $x = \sqrt{a}$
(d) not be continuous at $x = a$

22. The general solution of $x(1 + y^2)^{1/2}\, dx + y(1 + x^2)^{1/2} dy = 0$ is

(a) $\sin^{-1} x + \sin^{-1} y = c$
(b) $x^2 + y^2 = (1 + x^2)^{1/2} + (1 + y^2)^{1/2} + c$
(c) $(1 + x^2)^{1/2} + (1 + y^2)^{1/2} = c$
(d) $\tan^{-1} x - \tan^{-1} y = c$

23. The area in square unit enclosed by the curve $x^2 y = 36$, the x-axis and the lines $x = 6$ and $x = 9$ is

(a) 2 sq unit (b) 1 sq unit
(c) 4 sq unit (d) 3 sq unit

24. $\int_0^{\frac{\pi}{2}} \frac{\sin^n \theta}{\sin^n \theta + \cos^n \theta}\, d\theta$ is equal to

(a) 1 (b) 0
(c) $\frac{\pi}{2}$ (d) $\frac{\pi}{4}$

25. Area of the triangle in the argand diagram formed by the complex numbers $z, iz, z + iz$, where $z = x + iy$ is

(a) $|z|$ (b) $|z|^2$
(c) $2|z|^2$ (d) $\frac{1}{2}|z|^2$

26. If $f(x) = \frac{\log(1 + ax) - \log(1 - bx)}{x}$ for $x \neq 0$ and $f(0) = k$ and $f(x)$ is continuous at $x = 0$, then k is equal to

(a) $a + b$ (b) $a - b$ (c) a (d) b

27. The least positive remainder when $123 \times 125 \times 127$ is divided by 124 is

(a) 4 (b) 120 (c) 121 (d) 130

28. By eliminating the arbitrary constants A and B from $y = Ax^2 + Bx$, we get the differential equation

(a) $\frac{d^3y}{dx^3} = 0$ (b) $x^2 \frac{d^2y}{dx^2} - 2x \frac{dy}{dx} + 2y = 0$
(c) $\frac{d^2y}{dx^2} = 0$ (d) $x^2 \frac{d^2y}{dx^2} + y = 0$

29. The equation $(\cos p - 1)\, x^2 + \cos px + \sin p = 0$ in the variable x, has real roots. Then, p can take any value in the interval

(a) $(0, \pi)$ (b) $(-\pi, 0)$
(c) $\left(-\frac{\pi}{2}, \frac{\pi}{2}\right)$ (d) $(-\pi, \pi)$

30. If $3 \sin^{-1}\left(\frac{2x}{1 + x^2}\right) - 4 \cos^{-1}\left(\frac{1 - x^2}{1 + x^2}\right) + 2 \tan^{-1}\left(\frac{2x}{1 - x^2}\right) = \frac{\pi}{3}$, then x is equal to

(a) $\frac{1}{\sqrt{3}}$ (b) $-\frac{1}{\sqrt{3}}$ (c) $\sqrt{3}$ (d) $-\frac{\sqrt{3}}{2}$

31. If $\sin(\pi \cos \theta) = \cos(\pi \sin \theta)$, then which of the following is correct?

(a) $\cos \theta = \frac{3}{2\sqrt{2}}$
(b) $\cos\left(\theta - \frac{\pi}{2}\right) = \frac{1}{2\sqrt{2}}$
(c) $\cos\left(\theta - \frac{\pi}{4}\right) = \frac{1}{2\sqrt{2}}$
(d) $\cos\left(\theta + \frac{\pi}{4}\right) = -\frac{1}{2\sqrt{2}}$

32. If $f(x) = \begin{cases} x + 1, & x \le 1 \\ 3 - ax^2, & x > 1 \end{cases}$ is continuous at $x = 1$, then the value of a is

(a) 1 (b) 2
(c) -3 (d) -2

33. The area bounded by $y = 1 + \frac{8}{x^2}$ and the ordinates $x = 2$ and $x = 4$ is

(a) 2 sq unit (b) 4 sq unit
(c) log 2 sq unit (d) log 4 sq unit

34. Everybody in a room shakes hands with everybody else. The total number of handshakes is 66. The number of persons in the room is

(a) 11 (b) 12 (c) 13 (d) 14

35. The condition that one root of the equation $ax^2 + bx + c = 0$ may be square of the other, is

(a) $a^2c + ac^2 + b^3 - 3abc = 0$
(b) $a^2c^2 + ac^2 + b^2 - 3abc = 0$
(c) $ac^2 + ac - b^3 - 3abc = 0$
(d) $a^2c + ac^2 - b^3 - 3abc = 0$

36. The value of $\sum_{r=1}^{\infty} [3 \cdot 2^{-r} - 2 \cdot 3^{1-r}]$ is

(a) 2 (b) 1/2 (c) 1 (d) 0

37. If $x(1 + y^2)\,dx + y(1 + x^2)\,dy = 0$ and $y(0) = 1$, then $x^2y^2 + x^2 + y^2$ is equal to

(a) 0 (b) 1 (c) 2 (d) 3

38. If $f(x) = \frac{x + \cos x}{x - \cos x}$, $f'\left(\frac{\pi}{2}\right)$ is equal to

(a) $\frac{2(x\sin x + \cos x)}{(x - \cos x)^2}$ (b) $\frac{2(x\cos x + \sin x)}{(x - \cos x)^2}$
(c) $\frac{-2(x\sin x + \cos x)}{(x + \cos x)^2}$ (d) None of these

39. If $A(-1, 2), B(5, 1), C(6, 5)$ are the vertices of a parallelogram $ABCD$. The equation to the diagonal through B is

(a) $x + y + 6 = 0$ (b) $x + y - 6 = 0$
(c) $x - y - 4 = 0$ (d) $x - 2y - 1 = 0$

40. $\int_0^1 x \sin \pi x \, dx$ is equal to

(a) 1 (b) 1/2 (c) π (d) $1/\pi$

41. If a, b, c are different integers such that $(a, b) = c$, which of the following statements is true ?

(a) $\frac{a}{c}$ (b) $\frac{c}{b}$
(c) $\frac{b}{a}$ (d) None of these

42. If a, b and c are negative and different real numbers, then $\Delta = \begin{vmatrix} a & b & c \\ b & c & a \\ c & a & b \end{vmatrix}$ is

(a) < 0 (b) 0 (c) > 0 (d) ≤ 0

43. The product of 'n' positive numbers is unity. Then, their sum is

(a) a positive integer (b) divisible by n
(c) equal to $n + \frac{1}{n}$ (d) never less than n

44. The function $\sin x(1 + \cos x), 0 \leq x \leq \frac{\pi}{2}$ has maximum value, when x is equal to

(a) 0 (b) $\frac{\pi}{2}$
(c) $\frac{\pi}{6}$ (d) None of these

45. Differential equations of the family of curves $y = a\cos \mu x + b\sin \mu x$, where a and b are arbitrary constants is given by

(a) $\frac{d^2y}{dx^2} + \mu y = 0$ (b) $\frac{d^2y}{dx^2} - \mu^2 y = 0$
(c) $\frac{d^2y}{dx^2} + \mu^2 y = 0$ (d) None of these

46. The area included between the curves $y^2 = 2x$ and $x^2 = 2y$ is

(a) $\frac{3}{4}$ sq unit (b) $\frac{4}{3}$ sq unit
(c) $\frac{1}{2}$ sq unit (d) $\frac{4}{3}$ sq unit

47. If $A \cdot \text{adj}(A) = O$, then $|A|$ is

(a) 0 (b) $\frac{1}{|\text{adj } A|}$ (c) 1 (d) -1

48. If $|\mathbf{a} + \mathbf{b}| = |\mathbf{a} - \mathbf{b}|$, then **a** and **b** are

(a) parallel
(b) perpendicular
(c) angle between **a** and **b** is 45°
(d) angle between **a** and **b** is 60°

49. $\lim_{x \to 0} \left(\frac{(2 + x)\sin(2 + x) - 2\sin 2}{x} \right)$ is equal to

(a) sin 2 (b) cos 2
(c) 1 (d) 2 cos 2+ sin2

50. $\int \frac{2dx}{(e^x + e^{-x})^2}$ is equal to

(a) $-\frac{e^{-x}}{(e^x + e^{-x})} + c$ (b) $-\frac{1}{(e^x + e^{-x})} + c$

(c) $\frac{1}{(e^x + 1)^2} + c$ (d) $\frac{1}{(e^x - e^{-x})^2} + c$

51. If the rate of change in the circumference of a circle of 0.3 cm/s, then the rate of change in the area of the circle when the radius is 5 cm, is

(a) 1.5 sq cm/s (b) 0.5 sq cm/s
(c) 5 sq cm/s (d) 3 sq cm/s

52. If ω is a cube root of unity, then $(1 + \omega - \omega^2)(1 - \omega + \omega^2)$ is

(a) 1 (b) 0 (c) 2 (d) 4

53. The complex number z which satisfy the equation $\left|\frac{i + z}{i - z}\right| = 1$ lies on

(a) a circle $x^2 + y^2 = 1$ (b) the x-axis
(c) the y-axis (d) the line $x + y = 1$

54. If $A + B + C = \pi$, then

$$\begin{vmatrix} \sin(A + B + C) & \sin B & \cos C \\ -\sin B & 0 & \tan A \\ \cos(A + B) & -\tan A & 0 \end{vmatrix}$$ is equal to

(a) $\sin A$ (b) $\sin A \cos B$
(c) 0 (d) None of these

55. The minimum value of $x \log x$ is equal to

(a) e (b) $1/e$ (c) $-1/e$ (d) $2/e$

56. If $f''(0) = k$, then

$\lim_{x \to 0} \frac{2f(x) - 3f(2x) + f(4x)}{x^2}$ is equal to

(a) k (b) $2k$ (c) $3k$ (d) $4k$

57. If $\int_{-1}^{4} f(x)\,dx = 4$ and $\int_{2}^{4} [3 - f(x)]\,dx = 7$, then the value of $\int_{-1}^{2} f(x)\,dx$ is

(a) -2 (b) 3 (c) 5 (d) 8

58. The equation of a curve passing through the origin and satisfying the differential equation $\frac{dy}{dx} = (x - y)^2$ is

(a) $e^{2x}(1 - x + y) = 1 + x - y$
(b) $e^{2x}(1 + x - y) = 1 - x + y$
(c) $e^{2x}(1 - x + y) + (1 + x - y) = 0$
(d) $e^{2x}(1 + x + y) = 1 - x + y$

59. If $(\sqrt{3} - i)^{50} = 2^{48}(x - iy)$, then $x^2 + y^2$ is equal to

(a) 2 (b) 4 (c) 8 (d) 16

60. If the coefficients of the rth term and the $(r + 1)$th term in the expansion of $(1 + x)^{20}$ are in the ratio 1 : 2, then r is equal to

(a) 6 (b) 7 (c) 8 (d) 9

61. The area of a parallelogram with diagonals as $\mathbf{a} = 3\hat{\mathbf{i}} + \hat{\mathbf{j}} - 2\hat{\mathbf{k}}$ and $\mathbf{b} = \hat{\mathbf{i}} - 3\hat{\mathbf{j}} + 4\hat{\mathbf{k}}$ is

(a) $10\sqrt{3}$ (b) $\frac{10}{\sqrt{3}}$

(c) $5\sqrt{3}$ (d) $\frac{5}{\sqrt{3}}$

62. The sum of infinite terms of the geometric progression $\frac{\sqrt{2} + 1}{\sqrt{2} - 1}, \frac{1}{2 - \sqrt{2}}, \frac{1}{2}, \ldots$ is

(a) $\sqrt{2}(\sqrt{2} + 1)^2$ (b) $(\sqrt{2} + 1)^2$
(c) $5\sqrt{2}$ (d) $3\sqrt{2} + \sqrt{5}$

63. The first three terms in the expansion of $(1 + ax)^n$ $(n \neq 0)$ are $1, 6x$ and $16x^2$. Then, the values of a and n are respectively

(a) 2 and 9 (b) 3 and 2
(c) $\frac{2}{3}$ and 9 (d) $\frac{3}{2}$ and 6

64. If $f(x) = \begin{cases} \frac{x^2 - 9}{x - 3}, & \text{if } x \neq 3 \\ 2x + k, & \text{otherwise} \end{cases}$ is continuous at $x = 3$, then k is equal to

(a) 3 (b) 0
(c) -6 (d) $\frac{1}{6}$

65. The area of the region $\{(x, y) : x^2 + y^2 \leq 1 \leq x + y\}$ is

(a) $\frac{\pi^2}{5}$ sq unit (b) $\frac{\pi^2}{2}$ sq unit

(c) $\left(\frac{\pi}{4} - \frac{1}{2}\right)$ sq unit (d) $\frac{\pi}{4}$ sq unit

66. $\left|\frac{x}{2}-1\right|<3$ implies that x lies in the interval

(a) (– 4, 8) (b) (– 3, 6)
(c) (– 4, 6) (d) (– 3, 8)

67. $\sin^{-1}\frac{1}{\sqrt{5}}+\cos^{-1}\frac{3}{\sqrt{10}}$ is equal to

(a) $\pi/6$ (b) $\pi/4$ (c) $\pi/3$ (d) $2\pi/3$

68. If $f(x)=\sec x-\cot x$, then $f'\left(\frac{\pi}{6}\right)$ is equal to

(a) – 3 (b) $\frac{14}{3}$ (c) $\frac{11}{3}$ (d) 5

69. If **a**, **b**, **c** are the position vectors of A, B, C respectively such that $3\mathbf{a}+4\mathbf{b}-7\mathbf{c}=\mathbf{0}$, then C divides AB in the ratio

(a) 4 : 3 (b) 4 :7
(c) 3 : 7 (d) 3 : 4

70. If $\begin{bmatrix} a & 2 & 3 \\ b & 5 & -1 \end{bmatrix}\times\begin{bmatrix} 1 & 2 \\ 3 & 4 \\ -1 & 1 \end{bmatrix}=\begin{bmatrix} 4 & 13 \\ 12 & 11 \end{bmatrix}$, (a, b) is

(a) (1, – 2) (b) (– 1, – 4)
(c) (1, 3) (d) (1, – 4)

English & General Aptitude

Directions (Q. 1–5) : *Read the following passage carefully and answer the questions given below it.*

After submitting his resignation, Albert came out and took the long narrow road leading to the railway station which was one of the busiest roads in the city. Sad and depressed and worried about looking for a new job. Albert looked around for a cigarette shop. He walked up to the end of the road but found no tobacconist. It was odd that such a busy thoroughfare with thousands of people passing through did not even have a single cigarette shop. He suddenly felt that it was no longer necessary for him to hunt for a job. He decided to open a tobacco shop himself. It was bound to be profitable, he felt.

1. After submitting his resignation, Albert came out worried about

(a) a shelter (b) cigarettes
(c) a job (d) the next available train

2. Albert was sad and depressed because

(a) he had no money for the train journey
(b) he had to walk on a long road
(c) he was not able to buy cigarettes
(d) he was worried about finding a job

3. There was no cigarette shop on that road because

(a) smoking was banned in that area
(b) just by chance nobody had opened one on that road
(c) it was a very narrow road
(d) cigarette-shop owners do not make any profit

4. Albert decided not to look for a new job because

(a) the thought of having to look for a job greatly distressed him
(b) he did not want to work at all
(c) there was no hope of finding a job
(d) he saw the possibility of self- employment

5. A cigarette shop on a busy road was bound to be profitable because

(a) cigarette shops are known to make a great deal of profit
(b) any shop on a busy street would attract a large number of customers
(c) cigarettes are inexpensive items and people buy them willingly
(d) a cigarette shop on a busy road would attract a large number of customers

Directions (Q. 6–8) *In the following passage, there are blanks each of which has been numbered. These numbers are printed below the passage and against each, some words are suggested,one of which fits the blank appropriately. Find the appropriate word for each blank.*

Scientific psychology recognises a ...6... truth that no two individuals are...7... in this world even through., equality is a fostered norm of civil society, the truth is that men...8...unequals in varying hues or degrees.

6. (a) clear (b) real (c) simple (d) direct

7. (a) alike (b) close (c) liked (d) equal

8. (a) will (b) are (c) is (d) was

Directions (Q. 9–10) : *In each of the following questions, choose the alternative which best expresses the meaning of the word given in capital letters.*

9. INDIGENOUS
(a) Normal
(b) Internal
(c) Natural
(d) Native

10. OPULENT
(a) Greedy
(b) Hungry
(c) Heavy
(d) Wealthy

Directions (Q. 11–12) : *In each of the following questions, choose the alternative which can be substituted for the given words/sentence.*

11. Science of printing
(a) Calligraphy
(b) Typography
(c) Topography
(d) Cryptography

12. Man whose wife has been unfaithful to him
(a) Dandy
(b) Bastard
(c) Concubine
(d) Cuckold

Directions (Q. 13–15) : *In each of the following questions, find out the part which has an error. If there is no error, your answer is (d).*

13. We grieve our loss and cry helplessly (a)/ while we should be fighting for our rights (b)/and die a noble death (c)/ No error. (d)

14. Work hard (a) / lest you will (b) / fail (c) / No error. (d)

15. My friend did not see me (a)/ for many years (b)/when I met him last week. (c)/No error. (d)

Directions (Q. 16–25) : *In the following questions, choose the option which shows common feature in the relationship given in each question.*

16. Indira : Nehru : Benazir
(a) They were Presidents
(b) They were Prime Ministers
(c) They were sports persons
(d) They were Indian politicians

17. Mohinder : Gavaskar : Azaharuddin
(a) They were Indian Athlets
(b) They were Indian foreign ministers
(c) They were cicket umpires
(d) They are former Indian cricketers

18. Folketing : Stortling : Knesset
(a) They are the name of currencies
(b) They are the name of rivers
(c) They are the name of Parliaments
(d) They are the name of cities

19. Car : Bike : Bus
(a) They are accelerator
(b) They are mode of transport
(c) They have wheels
(d) They are run by a person

20. Irna : PTI : Xin-Era
(a) They are newspapers
(b) They are computer manufacturing companies
(c) They are news agencies
(d) They are publishing houses

21. Pitcher : Dusra : Bunker
(a) They are parliament's name
(b) They are dwelling places of animals
(c) They are sports terms
(d) They are terms related to cricket

22. Mamb : Krait : Viper
(a) They are insects
(b) They are haunting spirits
(c) These are boot polishes
(d) These are snakes

23. Metre : Mile : Kilometre
(a) They are units of electricity
(b) They are units of measuring anything
(c) They are units of distance
(d) They are units of weight

24. Crocodile : Whale : Hippopotamus
(a) They are animals
(b) They are domestic animals
(c) They are land animals
(d) They are water animals

25. Knot : Watt : Fathom
(a) The terms are used by sailors
(b) The terms are used for installing electricity
(c) The terms are connected with rope
(d) They are units of measurement

26. At which place has the Central Government decided to set up a sector specific SEZ in biotechnology ?

(a) Bengaluru (b) Hyderabad
(c) New Delhi (d) Cuttack

27. What is 'NSG'?

(a) National Service of Government
(b) Nuclear Supplier Group
(c) Non-Nuclear Supplier Group
(d) Non-Secular Group

28. Which is the tallest structure in the world ?

(a) Taipei Towers (b) Effel Tower
(c) Burj Dubai (d) Patronos Tower

29. Name the Country with which India has agreed to deepen security cooperation, bilaterally and in the multilateral fora, and strengthen terrorism in all its form and maifestations.

(a) France (b) Canada
(c) Russia (d) Nepal

30. Nepal Electricity Authority (NEA) has signed a joint venture agreement with CTGC to construct 750 MW west seti hydropower project. CTGC is a/an

(a) India company (b) Chinese company
(c) Russian company (d) American company

Answers

Physics

1.	(b)	2.	(b)	3.	(b)	4.	(a)	5.	(c)	6.	(d)	7.	(b)	8.	(a)	9.	(d)	10.	(b)
11.	(c)	12.	(b)	13.	(c)	14.	(c)	15.	(c)	16.	(b)	17.	(c)	18.	(b)	19.	(d)	20.	(d)
21.	(a)	22.	(d)	23.	(a)	24.	(a)	25.	(c)	26.	(b)	27.	(c)	28.	(d)	29.	(d)	30.	(c)
31.	(b)	32.	(b)	33.	(a)	34.	(b)	35.	(b)	36.	(c)	37.	(a)	38.	(c)	39.	(d)	40.	(a)
41.	(c)	42.	(b)	43.	(a)	44.	(c)	45.	(c)	46.	(c)	47.	(d)	48.	(d)	49.	(b)	50.	(a)

Chemistry

1.	(b)	2.	(a)	3.	(b)	4.	(d)	5.	(a)	6.	(a)	7.	(d)	8.	(d)	9.	(c)	10.	(a)
11.	(b)	12.	(d)	13.	(b)	14.	(a)	15.	(c)	16.	(c)	17.	(a)	18.	(c)	19.	(c)	20.	(d)
21.	(b)	22.	(d)	23.	(c)	24.	(b)	25.	(a)	26.	(b)	27.	(b)	28.	(b)	29.	(b)	30.	(a)
31.	(a)	32.	(b)	33.	(b)	34.	(b)	35.	(a)	36.	(a)	37.	(c)	38.	(a)	39.	(b)	40.	(b)
41.	(b)	42.	(c)	43.	(c)	44.	(d)	45.	(a)	46.	(a)	47.	(c)	48.	(d)	49.	(d)	50.	(a)

Mathematics

1.	(d)	2.	(c)	3.	(c)	4.	(a)	5.	(a)	6.	(a)	7.	(b)	8.	(d)	9.	(b)	10.	(a)
11.	(b)	12.	(a)	13.	(a)	14.	(a)	15.	(c)	16.	(b)	17.	(b)	18.	(a)	19.	(c)	20.	(c)
21.	(d)	22.	(a)	23.	(d)	24.	(d)	25.	(d)	26.	(a)	27.	(c)	28.	(b)	29.	(a)	30.	(a)
31.	(c)	32.	(a)	33.	(b)	34.	(b)	35.	(a)	36.	(d)	37.	(b)	38.	(d)	39.	(b)	40.	(d)
41.	(b)	42.	(c)	43.	(a)	44.	(d)	45.	(c)	46.	(b)	47.	(a)	48.	(b)	49.	(d)	50.	(a)
51.	(a)	52.	(d)	53.	(b)	54.	(c)	55.	(c)	56.	(c)	57.	(c)	58.	(a)	59.	(d)	60.	(b)
61.	(c)	62.	(a)	63.	(c)	64.	(b)	65.	(c)	66.	(a)	67.	(b)	68.	(b)	69.	(a)	70.	(d)

English & General Aptitude

1.	(c)	2.	(d)	3.	(b)	4.	(d)	5.	(d)	6.	(c)	7.	(a)	8.	(b)	9.	(d)	10.	(d)
11.	(b)	12.	(d)	13.	(b)	14.	(b)	15.	(a)	16.	(b)	17.	(d)	18.	(c)	19.	(b)	20.	(c)
21.	(c)	22.	(d)	23.	(c)	24.	(d)	25.	(d)	26.	(b)	27.	(b)	28.	(c)	29.	(a)	30.	(b)

Answer *with* Solutions

Physics

1. ***(b)*** For a dipole of length $2l$

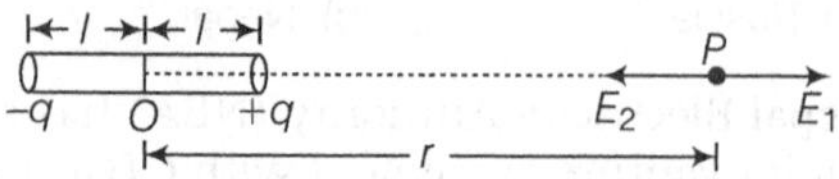

$$E = \frac{1}{4\pi\varepsilon_0 K} \cdot \frac{2pr}{(r^2 - l^2)^2}$$

When $(l << r)$, then l^2 may be neglected in comparison to r^2.

$$\therefore \quad E = \frac{1}{4\pi\varepsilon_0 K}\frac{2pr}{r^4} = \frac{1}{4\pi\varepsilon_0 K}\frac{2p}{r^3}\text{ N/C}$$

$$\Rightarrow \quad E \propto \frac{1}{r^3}$$

2. ***(b)*** Force due to magnetic field is

$$F = qvB\sin\theta$$

when $\quad \theta = 90°$

$$F = qvB \quad \text{...(i)}$$

Force due to electric field E is

$$F = qE \quad \text{...(ii)}$$

Equating Eqs. (i) and (ii), we get

$$|\vec{\mathbf{E}}| = v|\vec{\mathbf{B}}|$$

3. ***(b)*** The magnetic field is given by

$$B = \mu_0 ni$$

where $\quad \mu_0 = 4\pi \times 10^{-7}\text{ T mA}^{-1}$,

$$n = \frac{1000}{50 \times 10^{-2}},\ i = 5\text{ A}$$

$$\therefore \quad B = 4\pi \times 10^{-7} \times \frac{1000 \times 10^{-2}}{50 \times 10^{-2}} \times 5$$

$$B = 1.26 \times 10^{-2}\text{ T}$$

4. ***(a)*** From transformer ratio

$$\frac{V_s}{V_p} = \frac{N_s}{N_p}$$

$$\Rightarrow \quad V_s = \frac{V_p \times N_s}{N_p}$$

$$= \frac{220 \times 40000}{200}$$

$$= 44000\text{ V}$$

Potential difference per turn is

$$\frac{V_s}{N_s} = \frac{44000}{40000} = 1.1\text{ V}$$

5. ***(c)*** Slit width, $w = \dfrac{D\lambda}{d}$

Given, $\quad D = 80\text{cm} = 80 \times 10^{-2}\text{ m}$,

$$\lambda = 6000\text{Å} = 6000 \times 10^{-10}\text{ m},$$

$$d = \frac{5}{2}\text{mm} = \frac{5 \times 10^{-3}}{2}\text{ m}.$$

$$\therefore \quad w = \frac{80 \times 10^{-2} \times 6000 \times 10^{-10} \times 2}{5 \times 10^{-3}}$$

$$w = 0.192 \times 10^{-3}\text{m} = 0.192\text{ mm}$$

6. ***(d)*** According to Brewster's law, when unpolarised light is incident at polarising angle i_p on an interface separating a rarer medium from a denser medium of refractive index μ, such that

$$\mu = \tan i_p$$

then light reflected in the rarer medium is completely polarised.

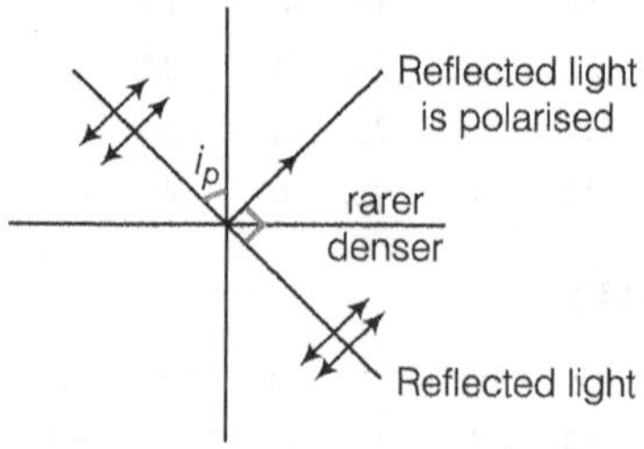

7. ***(b)*** Binding energy per nucleon increases with atomic number and is maximum for iron, after that it decreases.

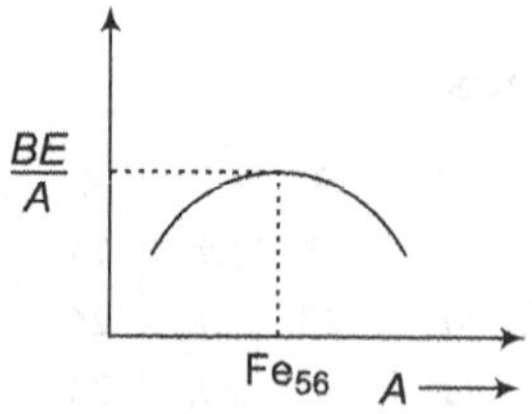

8. ***(a)*** From lens formula

$$\frac{1}{f} = ({}_a\mu_g - 1)\left(\frac{1}{R_1} - \frac{1}{R_2}\right)$$

$$= (1.5 - 1)\left(\frac{1}{R_1} - \frac{1}{R_2}\right) \quad \text{...(i)}$$

Also, $\quad {}_l\mu_g = \dfrac{\mu_g}{\mu_l} = \dfrac{1.5}{1.6}$

$$\therefore \quad \frac{1}{f'} = ({}_l\mu_g - 1)\left(\frac{1}{R_1} - \frac{1}{R_2}\right)$$

$$\frac{1}{f'} = \left(\frac{15}{16} - 1\right)\left(\frac{1}{R_1} - \frac{1}{R_2}\right) \quad \text{...(ii)}$$

Dividing Eq. (i) by Eq. (ii), we get

$$\frac{f}{f'} = \frac{\left(\frac{15}{16} - 1\right)}{(1.5 - 1)} = -\frac{1}{16 \times 5}$$

$$\Rightarrow \quad f = -160 \text{ cm}$$

9. *(d)* From law of conservation of momentum

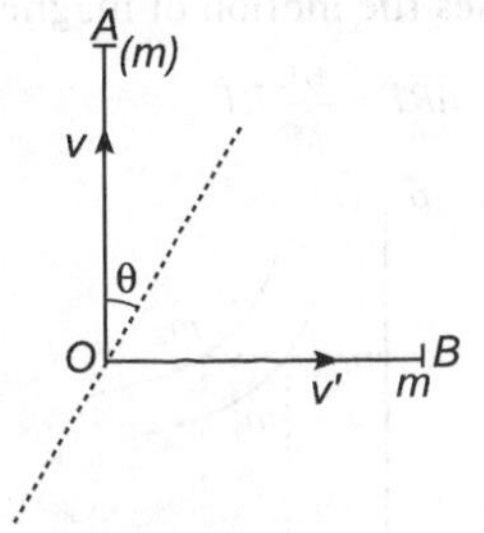

$$mv - mv' \cos\theta = 0 \quad \text{...(i)}$$

$$mv - mv' \sin\theta = 0 \quad \text{...(ii)}$$

From Eqs. (i) and (ii), we get

$$\theta = 45^\circ$$

$$v' = \frac{v}{\cos\theta} = v\sqrt{2}$$

10. *(b)* When two identical transverse or longitudinal progressive waves travel in a bounded medium with the same speed, but in opposite directions, then by their superposition a new type of wave is produced which appears stationary in the medium. Thus, is called stationary wave.

11. *(c)* From Bernoulli's theorem

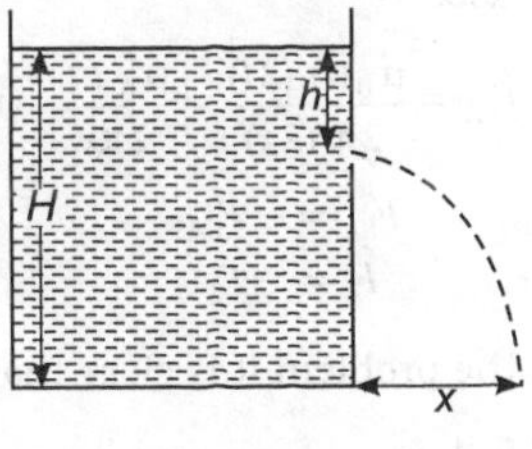

$$p + 0 + \rho g H = p + \frac{1}{2}\rho v^2 + \rho g (H - h)$$

$$\frac{1}{2}\rho v^2 = \rho g h$$

$$\Rightarrow \quad v = \sqrt{2gh}$$

12. *(b)* In N forces of equal magnitude works on a single point and their resultant is zero, then angle between any two forces is

$$\theta = \frac{360^\circ}{N} = \frac{360^\circ}{3} = 120^\circ$$

If these 3 vectors are represented by three sides of triangle, then they form equilateral triangle.

13. *(c)* $F = Bev$ and $B = \frac{\mu_0 i}{2\pi r}$

$$\therefore \quad F = \frac{\mu_0 i}{2\pi r} \times ev$$

$$= \frac{4\pi \times 10^{-7} \times 5}{2\pi \times 0.1} \times 1.6 \times 10^{-19} \times 5 \times 10^6$$

$$= 8 \times 10^{-18} \text{ N}$$

14. *(c)* $R' = \frac{9}{3} = 3\Omega$

$$R_{net} = 2\,\Omega$$

PD across AB

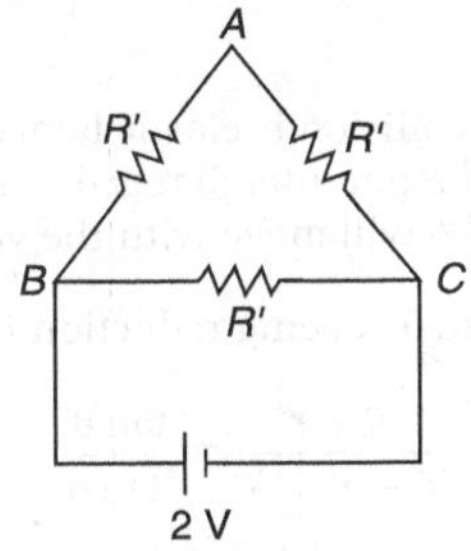

$$V' = i \times R' = \frac{V}{R_{net}} \times R = \frac{2}{2} \times 3 = 3\text{ V}$$

15. *(c)* $E_1 + E_2 = k \times 58$ ...(i)

$E_1 - E_2 = k \times 29$ ...(ii)

$$\therefore \quad \frac{E_1 + E_2}{E_1 - E_2} = 2$$

$$\Rightarrow \quad \frac{E_1}{E_2} = \frac{3}{1}$$

16. *(b)* When the lift is stationary $w = mg$

$$\Rightarrow \quad 49 = m \times 9.8$$

$$m = 5 \text{ kg}$$

When the lift is moving downward with an acceleration

$$R = m(9.8 - a) = 5(9.8 - 5) = 24 \text{ N}$$

17. *(c)* From Newton's law of cooling

$$\frac{\theta_1 - \theta_2}{t} = \left(\frac{\theta_1 + \theta_2}{2} - \theta_0\right)$$

Therefore,

$$\frac{60° - 40°}{7} \propto \left(\frac{60° + 40°}{2} - 10°\right) \quad ...(i)$$

$$\frac{40° - 28°}{t} \propto \left(\frac{40° + 28°}{2} - 10°\right) \quad ...(ii)$$

From Eqs. (i) and (ii), we get

$$t = 7 \text{ s}$$

18. *(b)* Einstein's equation is

$$V = \frac{h\nu}{e} - \frac{h\nu_0}{e}$$

Here, slope of line $= \frac{h}{e}$

$\Rightarrow h =$ Planck's constant $=$ slope $\times e$

19. *(d)* Width of central maximum

$$\beta = \frac{D\lambda}{d}, \; d \text{ being slit width.}$$

If the slit width is increased the central maximum becomes narrower and more light reaches the region.

20. *(d)* Since, collision is elastic hence, velocities of bobs *A* and *B* get interchanged. Thus, *A* will come to rest and *B* will move with the velocity of *A*.

21. *(a)* $I = k \tan\theta$, k being reduction factor and $I \propto \frac{1}{R}$

$$\Rightarrow \quad \frac{R + r}{(R + r) + R'} = \frac{\tan\theta'}{\tan\theta}$$

$$\Rightarrow \quad \frac{4 + 2}{6 + R'} = \frac{\tan 30°}{\tan 60°}$$

$$\Rightarrow \quad R' = 12\,\Omega$$

22. *(d)*

$$\beta = \frac{\alpha}{1 - \alpha} = \frac{0.9}{1 - 0.9} = 9$$

$$\Rightarrow \quad \frac{\Delta I_c}{\Delta I_b} = 9$$

$$\Rightarrow \quad I_c = 9 \times 2$$

$$\Rightarrow \quad \Delta I_c = 18\,\mu\text{A}$$

23. *(a)* Energy needed = Increment in surface energy

= (surface energy of *n* small drops)

− (surface energy of one big drop)

$= n\,4\pi r^2 T - 4\pi R^2 T = 4\pi T(nr^2 - R^2)$

24. *(a)*

$$p \propto T$$

$$\frac{p_1}{p_2} = \frac{T_1}{T_2}$$

$$\Rightarrow \quad \frac{p_2 - p_1}{p_1} = \frac{T_2 - T_1}{T_2}$$

$$\Rightarrow \quad \left[\frac{\Delta p}{p}\right]\% = \left[\frac{251 - 250}{250}\right] \times 100 = 0.4\%$$

25. *(c)* Net phase difference between current and voltage

$$= \frac{\pi}{6} - \left(-\frac{\pi}{6}\right) = \frac{\pi}{3} = 60°$$

Hence, current leads the voltage by 60°.

26. *(b)* According to Lenz's law acceleration of the magnet will be less than the acceleration due to gravity due to the fact that emf induced in the ring opposes the motion of magnet.

27. *(c)* $pV = nRT = \frac{m}{M} RT$

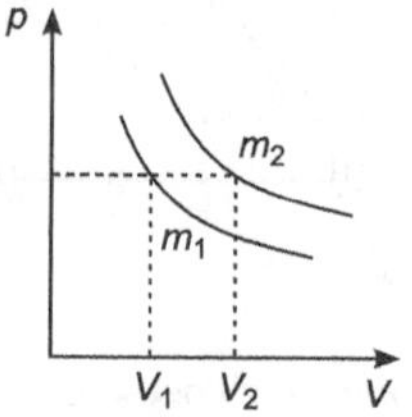

For 1st graph, $p = \frac{m_1}{M} \frac{RT}{V_1}$

For 2nd graph, $p = \frac{m_2}{M} \frac{RT}{V_2}$

Equating the two, we get

$$\frac{m_1}{m_2} = \frac{V_1}{V_2}$$

$$\Rightarrow \quad m \propto \frac{1}{V}$$

As $V_2 > V_1 \Rightarrow m_1 < m_2$

28. *(d)* $F_{BA} = \frac{\mu_0 I^2}{2\pi x}$ (attractive force)

and $F_{CA} = \frac{\mu_0 (I)(2I)}{2\pi (2x)} = \frac{\mu_0 I^2}{2\pi x}$ (repulsive force)

Hence, $F_{BA} = - F_{CA}$

$\Rightarrow \quad F_1 = - F_2$

29. *(d)* From the problem it is clear that $\angle e = \angle r_2 = 0$

From $A = r_1 + r_2$

$\Rightarrow \quad r_1 = A = 45°$

$$\therefore \quad \mu = \frac{\sin i}{\sin r_1} = \frac{\sin 60°}{\sin 45°} = \sqrt{\frac{3}{2}}$$

Also, from $i + e = A + \delta$

$\Rightarrow \quad 60 + 0 = 45 + \delta$

$\Rightarrow \quad \delta = 15°$

30. *(c)* $\frac{1}{f} = (\mu - 1)\left[\frac{1}{R_1} - \frac{1}{R_2}\right]$

$$= (1.6 - 1)\left[\frac{1}{60} - \frac{1}{\infty}\right] = \frac{1}{100}$$

$$f = 100 \text{ cm}$$

31. *(b)* Conservation of linear momentum gives

$$m_1 u_1 + m_2 u_2 = m_1 v_1 + m_2 v_2$$

$$\Rightarrow \quad m_2 v_2 = \frac{3u}{2} \quad \ldots\text{(i)}$$

Conservation of kinetic energy gives

$$\frac{1}{2} m_1 u_1^2 + \frac{1}{2} m_2 u_2^2 = \frac{1}{2} m_1 v_1^2 + \frac{1}{2} m_2 v_2^2$$

$$\Rightarrow \quad m_2 v_2^2 = \frac{15u^2}{8} \quad \ldots\text{(ii)}$$

Hence, on solving Eqs. (i) and (ii), we get

$$m_2 = 1.2 \text{ kg}$$

32. *(b)* Given,

$$10 = \frac{G \times 6 \times 10^{24} \times 1}{(6.4 \times 10^6)^2}$$

$$\Rightarrow \quad G = 6.82 \times 10^{-11} \text{ Nm}^2/\text{kg}^2$$

Therefore,

$$F_{\text{satellite}} = \frac{6.82 \times 10^{-11} \times 6 \times 10^{24} \times 200}{\left(\frac{3 \times 6.4 \times 10^6}{2}\right)^2} = 889 \text{ N}$$

33. *(a)* Electron concentration is greater than hole concentration in *n*-type semiconductor.

34. *(b)* *p-n* junction diode will be forward biased, if greater positive potential is given to *p*-side. So, diodes 2, 4, 5 are forward biased.

35. *(b)* $y_1 = a_1 \sin\left(\omega t - \frac{2\pi x}{\lambda}\right)$ and

$$y_2 = a_2 \cos\left(\omega t - \frac{2\pi x}{\lambda} + \phi\right)$$

$$= a_2 \sin\left(\omega t - \frac{2\pi x}{\lambda} + \phi + \frac{\pi}{2}\right)$$

So, phase difference $= \phi + \frac{\pi}{2}$

and path difference $\Delta = \frac{\lambda}{2\pi}\left(\phi + \frac{\pi}{2}\right)$

36. *(c)* $$I = \frac{E - e}{R} \quad \ldots\text{(i)}$$

and $$\eta \times 100 = \left(\frac{e}{E}\right) \times 100 \quad \ldots\text{(ii)}$$

$$\therefore \quad 30 = \frac{e}{50} \times 100 \quad \text{[from Eqs. (ii)]}$$

$$\Rightarrow \quad e = 15 \text{ V}$$

Thus, $$R = \frac{50 - 15}{12} = 2.9\ \Omega$$

37. *(a)* At point *P*,

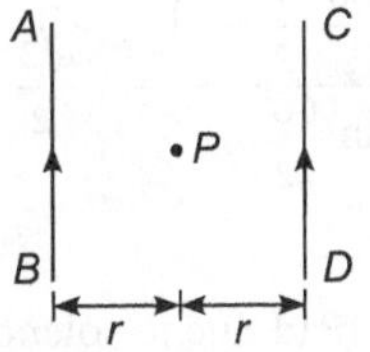

$$B_{\text{net}} = B_{AB} + B_{CD}$$

$$= \frac{\mu_0 i}{2r} + \frac{\mu_0 i}{2r} = \frac{\mu_0 i}{r}$$

38. *(c)* $S = \frac{I_g G}{I - I_g} = \frac{10^{-3} \times 20}{1 - 1 \times 10^{-3}} = 0.02\ \Omega$

39. *(d)* $\frac{V_s}{V_p} = \frac{I_s}{I_p}$

$$\Rightarrow \quad I_s = I_p \times \frac{V_s}{V_p} = 5 \times \frac{220}{22000} = 0.05 \text{ A}$$

40. *(a)* $\frac{I_{\max}}{I_{\min}} = \frac{(\sqrt{2I} + \sqrt{I})^2}{(\sqrt{2I} - \sqrt{I})^2} = \left(\frac{\sqrt{2}+1}{\sqrt{2}-1}\right)^2$

$$= \left[\frac{(\sqrt{2}+1)(\sqrt{2}+1)}{(\sqrt{2}-1)(\sqrt{2}+1)}\right]^2$$

$$= \left(\frac{2 + 1 + 2\sqrt{2}}{2-1}\right)^2$$

$$= \frac{9 + 8 + 12\sqrt{2}}{1} \approx 34$$

41. *(c)* From the question $F_1 = F_2$

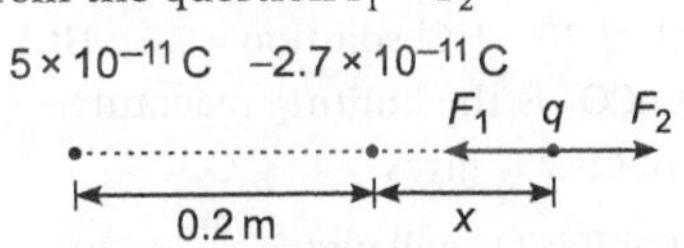

$$\Rightarrow \frac{1}{4\pi\varepsilon_0} \times \frac{5 \times 10^{-11} \times q}{(0.2 + x)^2} = \frac{1}{4\pi\varepsilon_0} \times \frac{2.7 \times 10^{-11} \times q}{x^2}$$

$$\Rightarrow \quad x = 0.556 \text{ m from IInd charge.}$$

42. *(b)* $\frac{4}{3}\pi r^3 \times n = \frac{4}{3}\pi R^3$

$$\Rightarrow \quad R = n^{1/3} r$$

$$\therefore \quad V_{\text{big}} = \frac{1}{4\pi\varepsilon_0} \times \frac{Q}{R} = \frac{1}{4\pi\varepsilon_0} \times \frac{nq}{n^{1/3} r} = n^{2/3} V$$

43. *(a)* The charge will flow from the outer surface of smaller one to bigger one only in case (a).

44. *(c)* $\mu = \dfrac{\sin\left(\dfrac{A+\delta_m}{2}\right)}{\sin\dfrac{A}{2}}$

$$\sqrt{3} = \frac{\sin\left(\dfrac{60^\circ+\delta_m}{2}\right)}{\sin\dfrac{60^\circ}{2}} \Rightarrow \frac{\sqrt{3}}{2} = \sin\left[30^\circ + \frac{\delta_m}{2}\right]$$

$\Rightarrow \delta_m = 60^\circ$

45. *(c)* Magnetic field due to solenoid is independent of diameter (Because $B = \mu_0 ni$).

46. *(c)* Resistance per unit length of wire

$$x = \frac{3}{100} = 3\times10^{-2}\ \Omega\ \text{cm}^{-1}$$

Equivalent length of wire $l' = \dfrac{R+3}{3\times10^{-2}}$,

R being additional resistance

$$\frac{E}{k} = \frac{R+3}{3\times10^{-2}}$$

$$\Rightarrow \frac{2}{1\times10^{-3}} = \frac{R+3}{3\times10^{-2}}$$

$$\Rightarrow R = 57\ \Omega$$

47. *(d)*

White	Brown	Red	Silver
9	1	10^2	10%

$\therefore \quad R = 91\times10^2 \pm 10\%\ \Omega$

ie, $\quad R = 9.1\ \text{k}\Omega$ with tolerance $\pm 10\%$

48. *(d)*

$$E = \Delta mc^2$$

$$E = \frac{0.3}{1000}\times(3\times10^8)^2 = 2.7\times10^{13}\ \text{J}$$

$$= \frac{2.7\times10^{13}}{3.6\times10^6} = 7.5\times10^6\ \text{kWh}$$

49. *(b)*

$$\lambda_{\text{photon}} = \frac{hc}{E}$$

$$\lambda_{\text{photon}} = \frac{h}{\sqrt{2mE}}$$

$$\Rightarrow \frac{\lambda_{\text{photon}}}{\lambda_{\text{electron}}} = c\sqrt{\frac{2m}{E}}$$

$$\frac{\lambda_{\text{photon}}}{\lambda_{\text{electron}}} \propto \frac{1}{\sqrt{E}}$$

50. *(a)* Applying Kirchhoff's law in the given figure.
At junction A

$$i + i_1 + i_2 = 1 \qquad \ldots\text{(i)}$$

For loop (1)

$$-60i + (15+5)\,i_1 = 0$$

$$\Rightarrow i_1 = 3i \qquad \ldots\text{(ii)}$$

For loop (2)

$$-(15+5)\,i_1 + 10i_2 = 0$$

$$\Rightarrow i_2 = 2i_1 = 6i \qquad \ldots\text{(iii)}$$

On solving Eqs. (i), (ii) and (iii), we get

$$i = 0.1\ \text{A}$$

Chemistry

1. *(b)* $CaCO_3 + 2HCl \longrightarrow CaCl_2 + CO_2 + H_2O$
100 g $\quad$ 73 g $\qquad\qquad$ 44 g

100 mL of 20% HCl solution = 20 g HCl

Here $CaCO_3$ is the limiting reactant.

100 g of $CaCO_3$ gives = 44 g CO_2

$\therefore$ 20 g of $CaCO_3$ will give $= \dfrac{44}{100}\times 20$

$= 8.80$ g of CO_2

2. *(a)* Mass of $H_2 = \dfrac{0.7}{22.4}\times 2 = \dfrac{14}{224}\ \text{g} = 0.0625\ \text{g}$

$\therefore$ 0.0625 g of hydrogen is displaced by x g of metal.

$\therefore$ 1 g of hydrogen is displaced by $\dfrac{x}{0.0625}$ g of metal.

$$\therefore \quad \frac{x}{0.0625} = 28$$

$\therefore$ Mass of metal $(x) = 28\times0.0625 = 1.75$ g

3. *(b)* In Balmer series electrons transfer from higher orbits to 2nd orbit.

$$n_2 = 3, 4, 5, 6, \ldots$$

$$n_1 = 2$$

4. *(d)* The electron affinity of Be is almost similar to that of Ne because they have stable (fully filled subshells) configuration.

5. *(a)* For a spontaneous process the value of ΔG must be negative.

6. *(a)* pH of an acid buffer is given by Henderson equation.

$$\text{pH} = -\log K_a + \log\frac{[\text{Salt}]}{[\text{Acid}]}$$

or

$$\text{pH} = \text{p}K_a + \log\frac{[\text{Salt}]}{[\text{Acid}]}$$

7. *(d)* In S_N1 reaction the racemization takes place due to inversion and retention of configuration.

8. *(d)* Phenol reacts with excess of bromine water to give 2, 4, 6-tribromophenol.

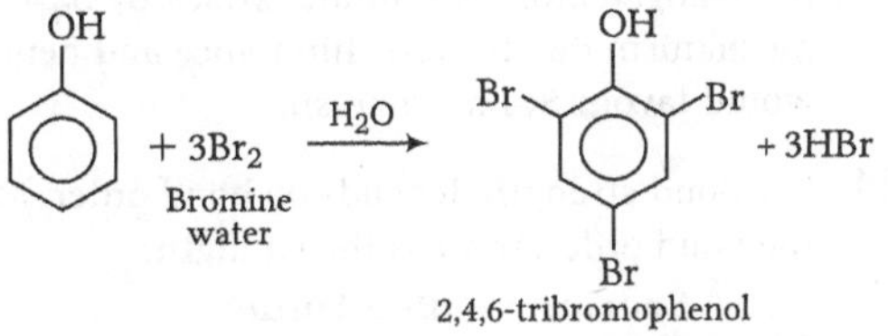

9. *(c)* $CH_3COOH + NH_3 \longrightarrow CH_3COONH_4$
Acetic acid — Ammonium acetate

$\xrightarrow{\Delta} CH_3CONH_2 + H_2O$
Acetamide

10. *(a)* In ZnS, S^- ions adopt fcc lattice in which Zn^{2+} ions are present at alternate tetrahedral sites. Hence, coordination number of S^{2-} is equal to coordination number of Zn^{2+} *i.e.* 4.

11. *(b)* $\because$ Rate $= k[A]^{3/2}[B]^{-1}$

$\therefore$ Order of reaction $= \frac{3}{2} - 1 = \frac{1}{2}$

12. *(d)* $k = \frac{2.303}{32} \log \frac{a}{a - 0.99a} = 0.1439 \text{ min}^{-1}$

$t = \frac{2.303}{0.1439} \log \frac{a}{a - 0.999a} = 48 \text{ min}$

13. *(b)* Greater the number of ions, greater will be elevation in boiling point. Hence, 0.1m $FeCl_3$ exhibits the maximum elevation in boiling point among these.

14. *(a)* The balanced equation is as

$$5I^- + IO_3^- + 6H^+ \longrightarrow 3I_2 + 3H_2O$$

15. *(c)* Lyophilic sols are more stable than lyophobic sols because the colloidal particles are solvated.

16. *(c)* The compounds which contain either $CH_3-\underset{|}{C}O$ or $CH_3-\underset{\underset{OH}{|}}{C}H^-$ group exhibit positive iodoform test. $(CH_3)_2CO$ due to presence of $CH_3-\underset{|}{C}=O$ group gives positive iodoform test

$$\begin{matrix} H_3C \\ H_3C \end{matrix}\!\!>\!C=O + NaOH + I_2 \longrightarrow CHI_3 + CH_3COONa + H_2O$$

17. *(a)* Starch is the main storage polysaccharide of plants. It is a polymer of α-D-glucose.

$$\underset{\text{Starch}}{(C_6H_{10}O_5)_n} \xrightarrow{H^+} \underset{\text{Maltose}}{C_{12}H_{22}O_{11}} \xrightarrow{H^+} \underset{\text{D-glucose}}{C_6H_{12}O_6}$$

18. *(c)* The nitro group can attach to metal through nitrogen as $(-NO_2)$ or through oxygen as nitro $(-ONO)$, hence nitropentamine chromium (III) chloride shows linkage isomerism.

19. *(c)* $2Ag^+(aq) + 2e^- \longrightarrow 2Ag(s); E° = 0.80$ V

$Cd(s) \longrightarrow Cd^{2+}(aq) + 2e^-; E° = 0.40$ V

$2Ag^+(aq) + Cd(s) \longrightarrow Cd^{2+}(aq) + 2Ag(s)$

$E° = 1.20$ V

$$\Delta G = -nFE°$$
$$= -2 \times 96500 \times 1.20$$
$$= -231.6 \text{ kJ}$$

20. *(d)* $\mu = \sqrt{n(n+2)}$

where, n = no. of unpaired electrons. Hence, the ion which has the maximum number of unpaired electron, will show the highest magnetic moment. Mn^{2+} has maximum 5 unpaired electrons, hence it shows the highest magnetic moment.

21. *(b)* Cane sugar reacts with conc. H_2SO_4 to give charrs. It is because H_2SO_4 acts as a dehydrating agent.

$$C_{12}H_{22}O_{11} + H_2SO_4(\text{conc.}) \longrightarrow 12C + 11H_2O + H_2SO_4$$

22. *(d)* For the value $n = 3$, the value of l are 0,1 and 2. Hence, m cannot have value 3.

23. *(c)* The formation of oxide ion $O^{2-}(g)$ require first an exothermic and then an endothermic step because the O^- ion will tend to resist the addition of another electron.

24. *(b)* Potassium dichromate acts like an oxidising agent in the presence of dilute H_2SO_4 as :

$$Cr_2O_7^{2-} + 14H^+ + 6e^- \longrightarrow 2Cr^{3+} + 7H_2O$$

$\therefore$ Equivalent mass of $K_2Cr_2O_7$

$$= \frac{\text{molecular mass of } K_2Cr_2O_7}{6}$$

25. *(a)* The number of atoms of the ligands that are directly bound to the central metal atom or ion by coordinate bond is known as the coordination number of the metal atom or ion.
Coordination number of metal = number of σ bonds formed by metal with ligands.

26. *(b)* Ethylene diamine is an example of bidentate ligand. Its structure is as :

$$\begin{array}{l} CH_2—NH_2 \\ | \\ CH_2—NH_2 \end{array}$$

27. *(b)* Deviation from ideal gas behaviour of a gas is expressed in terms of compressibility factor, Z

$$Z = \frac{pV}{nRT}$$

28. *(b)* Total moles of NaOH $= 4 \times 0.2 + 2 \times 0.5$

$$= 0.8 + 1.0 = 1.8$$

Total volume $= 4 + 2 = 6$ L

$$\therefore \text{Molarity} = \frac{1.8}{6} = 0.3\ \text{M}$$

29. *(b)* In the Reimer-Tiemann reaction carbene intermediate is formed

$$CHCl_3 + OH^- \longrightarrow H_2O + CCl_3^-$$

$$\longrightarrow Cl^- + \underset{\text{Dichlorocarbene}}{:CCl_2}$$

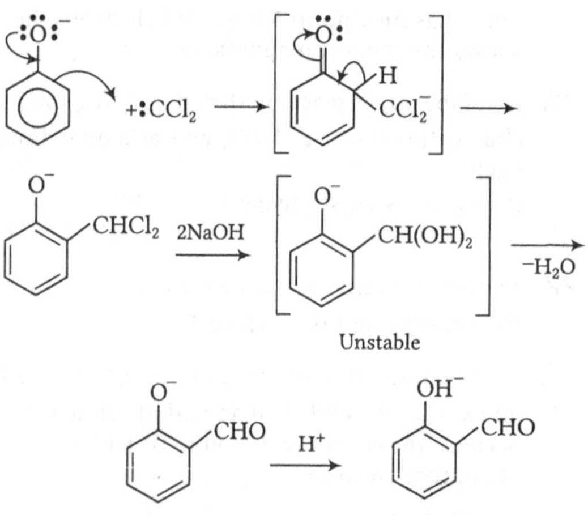

30. *(a)* Chloral (CCl_3CHO) does not contain any α-hydrogen atom, hence it will not give aldol condensation reaction.

31. *(a)* $\Rightarrow \Delta S_{fus} = \frac{\Delta H_{fus}}{T}$

Given, $T = 27 + 273$

$= 300$ K

$$\therefore \quad \Delta S_{fus} = \frac{600}{300} = 2\,\text{cal deg}^{-1}$$

32. *(d)* $Al_2(SO_4)_3 \rightleftharpoons 2Al^{3+} + 3SO_4^{2-}$

$$K_{sp} = [Al^{3+}]^2 [SO_4^{2-}]^3$$

33. *(b)* Bulky groups on the carbon atom attached to the halogen atom will hinder attack by S_N2 mechanism due to steric hindrance and hence, would favour S_N1 mechanism.

34. *(b)* Bond strength depends on bond order. More is the bond order, more is the strength.

Species	**Bond order**
O_2	2
O_2^-	1.5
O_2^{2-}	1.0

Hence, the order of bond strength is as :

$$O_2 > O_2^- > O_2^{2-}$$

35. *(a)* $\because$ 24 g of Mg produce MgO = 1 mol

$$\therefore \quad 1.2 \text{ g of Mg will produce MgO} = \frac{1 \times 1.2}{24} = 0.05 \text{ mol}$$

36. *(a)* $\Delta S_{vap} = \frac{\Delta H_{vap}}{T}$

$\because \Delta H_{vap} = 40.8$ kJ mol^{-1}, $T = 373$ K

$$\Delta S_{vap} = \frac{40.8 \times 10^3}{373} = 109.4\ \text{JK}^{-1}\ \text{mol}^{-1}$$

37. *(c)* The function of $AlCl_3$ in Friedel-Craft's reaction is to produce electrophile.

$$R—Cl + AlCl_3 \longrightarrow \underset{\text{Electrophile}}{R^+} + AlCl_4^-$$

38. *(a)* O_2 has two unpaired electrons in $\overset{*}{\pi}$ molecular orbitals, hence it is paramagnetic in nature.

39. *(b)* Pent-3-en-2-ol shows both optical as well as geometrical isomerism and hence, four stereoisomers are possible.

$$CH_3—\overset{*}{C}H(OH)—CH=CH—CH_3$$

40. *(b)* When A–B attractions are greater than A–A and B–B, then less vapours are formed and hence, solution shows negative deviation.

41. *(c)* Radon is radioactive with half-life period of 3.8 days and is used for the treatment of cancer.

42. *(c)* $He_2 : (\sigma 1s)^2 (\overset{*}{\sigma} 1s)^2$

$$\text{Bond order} = \frac{N_b - N_a}{2} = \frac{2-2}{2} = 0$$

43. *(d)* 0.16 g of dibasic acid is neutralised by = 25mL of 0.1M NaOH

$$H_2X + 2NaOH = Na_2X + H_2O$$

$$\text{Millimoles of acid} = \frac{1}{2} \times \text{millimoles of NaOH} = \frac{1}{2} \times 25 \times 0.1 = 1.25$$

$$\because \quad \text{Millimoles} = \frac{W \text{ (mg)}}{\text{mol. mass}}$$

$$\therefore \text{Mol. mass} = \frac{W \text{ (mg)}}{\text{millimoles}} = \frac{0.16 \times 1000}{1.25} = 128$$

44. *(d)* The relationship between the standard free energy ($\Delta G°$) and equilibrium constant (K) is as

$$\Delta G° = -2.303\, RT \log K$$

45. *(a)* $\text{Number of moles of HCl} = \frac{0.192}{36.5} = 0.0053$

$$\text{Volume} = 0.5\text{L}$$

$$\text{Molarity of HCl} = \frac{0.0053}{0.5} = 1.06 \times 10^{-2} \text{ M}$$

$$[H^+] \text{ in HCl solution} = 1.06 \times 10^{-2} \times \frac{95}{100} \approx 1 \times 10^{-2}$$

$$pH = -\log [H^+] = -\log (1 \times 10^{-2}) = 2$$

46. *(a)* Molisch test is given by all carbohydrates.
Molisch test : Carbohydrate solution + 10% alco. solution of α-naphthol + conc. H_2SO_4 along the side of test tube $\longrightarrow$ A reddish violet ring at the junction of two layers shows the presence of carbohydrate.

47. *(c)* Insulin is a peptide hormone. It is secreted from pancreas. It has 51 amino acids.

48. *(d)* As the crystal field splitting in case of complexes of metal ions in 4*d* and 5*d* transition series are much larger, the electrons of the metal ion are paired up first in lower energy set of orbitals before going to the higher energy set within the same *d*-subshell causing maximum pairing and thus, low spin (or spin paired) complexes are formed.

49. *(d)* $CH_2{=}CH{-}\overset{*}{C}H(CH_3){-}CH_2CH_3 \xrightarrow{H_2/Pt} CH_3CH_2{-}CH(CH_3){-}CH_2CH_3$

optically active 3-methyl-1-pentene → optically inactive

50. *(a)*

$$\because \quad \Delta x = \Delta v$$

$$\Delta x \times \Delta p \geq \frac{h}{4\pi}$$

$$\Delta x \times m \cdot \Delta v = \frac{h}{4\pi}$$

$$(\Delta v)^2 = \frac{h}{4\pi m} \quad (\because \Delta x = \Delta v)$$

$$\therefore \quad \Delta p = m \cdot \Delta v = m\sqrt{\frac{h}{4\pi m}} = \sqrt{\frac{mh}{4\pi}}$$

$$\Delta p = \frac{1}{2}\sqrt{\frac{mh}{\pi}}$$

Mathematics

1. *(d)* $3^{|3x+4|} = 9^{2x-2}, x > 0$

When $x > 0$, then

$$|3x + 4| = 3x + 4$$

$\therefore \quad 3^{3x+4} = 3^{2(2x-2)}$

$\Rightarrow \quad 3x + 4 = 4x - 4$

$\Rightarrow \quad x = 8$

2. *(c)* Given, $\vec{\mathbf{a}} = 3\vec{\alpha} - \vec{\beta}, \vec{\mathbf{b}} = \vec{\alpha} + 3\vec{\beta}$ and

$|\vec{\alpha}| = |\vec{\beta}| = 2$

$\therefore |\vec{\mathbf{a}} + \vec{\mathbf{b}}|^2 = |4\vec{\alpha} + 2\vec{\beta}|^2$

$= 16|\vec{\alpha}|^2 + 4|\vec{\beta}|^2 + 16|\vec{\alpha}||\vec{\beta}|\cos\frac{\pi}{3}$

$= 16(4) + 4(4) + 16(2 \times 2) \times \frac{1}{2}$

$= 64 + 16 + 32 = 112$

$\Rightarrow \quad |\vec{\mathbf{a}} + \vec{\mathbf{b}}| = 4\sqrt{7}$

3. *(c)* Since, every group of prime order is cyclic. Here, given number is prime. So, it is cyclic.

4. *(a)* Given, $U_n = 2 + 2^3 + 2^5 + 2^7 + \ldots + 2^{2n+1}$

and $V_n = 1 + 4 + 4^2 + \ldots + 4^{n-1}$

$\therefore \quad U_n = \frac{lr - a}{r - 1} = \frac{2^{2n+1} \cdot 4 - 2}{4 - 1} = \frac{2^{2n} \cdot 8 - 2}{3}$

and $\quad V_n = \frac{lr - a}{r - 1} = \frac{4^{n-1} \cdot 4 - 1}{4 - 1} = \frac{2^{2n} - 1}{3}$

Now, $\quad \frac{U_n}{V_n} = \frac{2^{2n} \cdot 8 - 2}{2^{2n} - 1} = \frac{8 - \frac{2}{2^{2n}}}{1 - \frac{1}{2^{2n}}}$

$\therefore \quad \lim_{n \to \infty} \frac{U_n}{V_n} = \lim_{n \to \infty} \frac{8 - \frac{2}{2^{2n}}}{1 - \frac{1}{2^{2n}}} = \frac{8 - 0}{1 - 0} = 8$

5. *(a)* Since, $P(1) : 1 = 1$ (true)

Let $\quad P(k) \equiv 1 + 3 + 5 + \ldots + (2k - 1) = k^2$

$\therefore P(k + 1) \equiv 1 + 3 + 5 + \ldots + (2k - 1) + 2k + 1$

$= k^2 + 2k + 1 = (k + 1)^2$

So, it holds for all n.

6. *(a)* Let $f(x) = x^2 + \frac{1}{1 + x^2}$

On differentiating w.r.t. x, we get

$$f'(x) = 2x - \frac{1}{(1 + x^2)^2} 2x$$

For a minimum, put $f'(x) = 0$

$\Rightarrow \quad x = 0$

So, the function has minimum value at $x = 0$.

7. *(b)* Let the coefficient of x^{-9} is in the $(r + 1)$th term in the expansion of $\left(\frac{x^2}{2} - \frac{2}{x}\right)^9$, then

$$T_{r+1} = {}^9C_r \left(\frac{x^2}{2}\right)^{9-r} \left(-\frac{2}{x}\right)^r$$

$$= {}^9C_r \frac{x^{18-2r}}{2^{9-r}} \cdot \frac{(-1)^r \cdot 2^r}{x^r}$$

$$= {}^9C_r \frac{x^{18-3r}}{2^{9-2r}} (-1)^r$$

For coefficient of x^{-9}, put

$x^{18-3r} = x^{-9}$

$\Rightarrow \quad 18 - 3r = -9 \Rightarrow 27 = 3r$

$\Rightarrow \quad r = 9$

$\therefore$ Coefficient of $x^{-9} = {}^9C_9 \frac{1}{2^{-9}} (-1)^9 = -2^9 = -512$

8. *(d)* $\frac{1}{2} + \frac{3}{4} + \frac{7}{8} + \frac{15}{16} + \ldots$upto n terms

$= \left(1 - \frac{1}{2}\right) + \left(1 - \frac{1}{4}\right) + \left(1 - \frac{1}{8}\right) \ldots$upto n terms

$= (1 + 1 + 1 + \ldots n \text{ terms})$

$- \left(\frac{1}{2} + \frac{1}{2^2} + \frac{1}{2^3} + \ldots n \text{ terms}\right)$

$= n - \frac{\frac{1}{2}\left(1 - \frac{1}{2^n}\right)}{1 - \frac{1}{2}} = n - 1 + \frac{1}{2^n} = n - 1 + 2^{-n}$

9. *(b)* $\begin{vmatrix} a & a+b & a+2b \\ a+2b & a & a+b \\ a+b & a+2b & a \end{vmatrix}$

$= \begin{vmatrix} 3(a+b) & 3(a+b) & 3(a+b) \\ a+2b & a & a+b \\ a+b & a+2b & a \end{vmatrix} (R_1 \to R_1 + R_2 + R_3)$

$= 3(a+b) \begin{vmatrix} 1 & 0 & 0 \\ a+2b & -2b & -b \\ a+2b & b & -b \end{vmatrix} \quad \begin{pmatrix} C_2 \to C_2 - C_1 \\ C_3 \to C_3 - C_1 \end{pmatrix}$

$= 3(a+b) \begin{vmatrix} -2b & -b \\ b & -b \end{vmatrix}$

$= 3b^2(a+b) \begin{vmatrix} 2 & 1 \\ -1 & 1 \end{vmatrix} = 3b^2(a+b)(2+1) = 9b^2(a+b)$

10. (a) $\sim[p \vee (\sim p \vee q)] \equiv \sim p \wedge \sim(\sim p \vee q)$

$$\equiv \sim p \wedge (\sim(\sim p) \wedge \sim q)$$
$$\equiv \sim p \wedge (p \wedge \sim q)$$

11. (b) Since, $2 \times_5 2^{-1}$ = Identity element 1

Therefore, $(2 \times_5 2^{-1})^{-1} = 1$

12. (a) Given, $P(n) = 9^n - 8^n$

$\therefore \quad P(1) = 9 - 8 = 1$

$\therefore \; P(1) - 1 = 0$, which is divisible by 8.

Now, $P(2) = 9^2 - 8^2 = 17$

When divided by 8, it leaves the remainder 1.

13. (a) $\tan^{-1} x + \tan^{-1} y + \tan^{-1} z = \pi$

$$\Rightarrow \quad \tan^{-1}\left\{\frac{x+y+z-xyz}{1-xy-yz-zx}\right\} = \pi$$

$\Rightarrow x + y + z - xyz = 0 \Rightarrow x + y + z = xyz$

14. (a) Since, $\left(\frac{1}{2}, \frac{1}{3}, n\right)$ are the direction cosines of a line, then $\left(\frac{1}{2}\right)^2 + \left(\frac{1}{3}\right)^2 + n^2 = 1 \Rightarrow \frac{1}{4} + \frac{1}{9} + n^2 = 1$

$$\Rightarrow n^2 = 1 - \frac{1}{4} - \frac{1}{9} = \frac{23}{36} \Rightarrow n = \frac{\sqrt{23}}{6}$$

15. (c) Let $y = x^x$

Taking log on both sides, we get

$$\log y = x \log x$$

On differentiating w.r.t. x, we get

$$\frac{1}{y}\frac{dy}{dx} = \frac{x}{x} + \log x$$

$$\Rightarrow \quad \frac{dy}{dx} = y(1 + \log x) = x^x (1 + \log x)$$

16. (b) $(3 - x) \equiv (2x - 5) \pmod 4$

$\Rightarrow \quad 3 - x - 2x + 5$ is divisible by 4

$\Rightarrow \quad 8 - 3x$ is divisible by 4

$\therefore \quad x = 4$

17. (b) $2x^2 - 5xy + 2y^2 + 3x - 3y + 1 = 0$

$\Rightarrow \quad (x - 2y + 1)(2x - y + 1) = 0$

$\therefore$ Two equations are $x - 2y + 1 = 0$ and $2x - y + 1 = 0$.

Length of perpendiculars from $(-1, 2)$ are

$$p_1 = \left|\frac{-1-4+1}{\sqrt{1+4}}\right| = \frac{4}{\sqrt{5}} \text{ and } p_2 = \left|\frac{-2-2+1}{\sqrt{5}}\right| = \frac{3}{\sqrt{5}}$$

$\therefore$ Product $= p_1 \cdot p_2 = \frac{12}{5}$

18. (a) $\mathbf{a} = -\hat{\mathbf{i}} + 2\hat{\mathbf{j}} - \hat{\mathbf{k}}$,

$\mathbf{b} = \hat{\mathbf{i}} + \hat{\mathbf{j}} - 3\hat{\mathbf{k}}$, $\mathbf{c} = -4\hat{\mathbf{i}} - \hat{\mathbf{k}}$

$\therefore \quad \mathbf{a} \times (\mathbf{b} \times \mathbf{c}) + (\mathbf{a} \cdot \mathbf{b})\mathbf{c} = (\mathbf{a} \cdot \mathbf{c})\mathbf{b}$

$= (4 + 1)(\hat{\mathbf{i}} + \hat{\mathbf{j}} - 3\hat{\mathbf{k}}) = 5\hat{\mathbf{i}} + 5\hat{\mathbf{j}} - 15\hat{\mathbf{k}}$

19. (c) Work done = (Force · displacement)

$= (4\hat{\mathbf{i}} - 3\hat{\mathbf{j}} + 2\hat{\mathbf{k}}) \cdot (-\hat{\mathbf{i}} - 3\hat{\mathbf{j}} + 5\hat{\mathbf{k}})$

$= (-4 + 9 + 10) = 15$ unit

20. (c) $f(x) = a \sin x + \frac{1}{3} \sin 3x$

On differentiating w.r.t. x, we get

$$f'(x) = a \cos x + \frac{3}{3} \cos 3x = 0, \text{ at } x = \frac{\pi}{3},$$

$$a \cos \frac{\pi}{3} + \cos \frac{\pi}{3} \cdot 3 = 0 \Rightarrow a \cdot \frac{1}{2} + (-1) = 0$$

$\therefore \quad a = 2$

21. (d) $|f(x)|$ is continuous at $x = a$, $f(x)$ is not be continuous at $x = a$.

$$\therefore \quad |f(x)| = \begin{cases} f(x), & x > a \\ -f(x), & x < a \end{cases}$$

22. (a) Given equation can be rewritten as

$$\frac{x}{(1+x^2)^{1/2}} dx + \frac{y}{(1+y^2)^{1/2}} dy = 0$$

On integrating, we get

$$\sqrt{1+x^2} + \sqrt{1+y^2} = c$$

23. (d) Required area $= \int_6^9 \frac{36}{x^2} dx = \left[-\frac{36}{x}\right]_6^9$

$$= -36\left[\frac{1}{9} - \frac{1}{6}\right] = 3 \text{ sq unit}$$

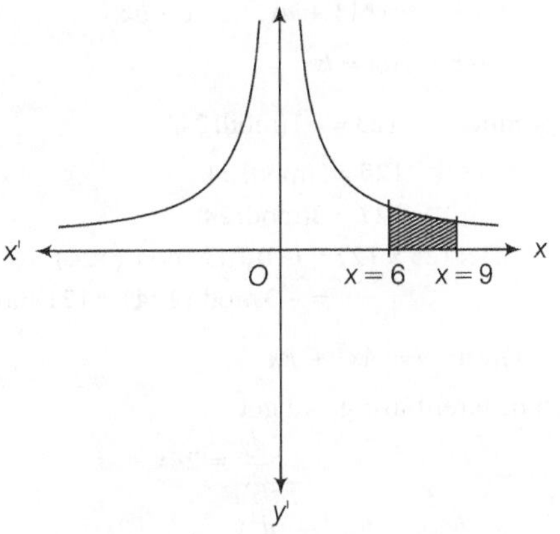

24. (d) Let $I = \int_0^{\pi/2} \frac{\sin^n \theta}{\sin^n \theta + \cos^n \theta} d\theta$ ...(i)

$$= \int_0^{\pi/2} \frac{\sin^n(\pi/2 - \theta)}{\sin^n(\pi/2 - \theta) + \cos^n(\pi/2 - \theta)} d\theta$$

$$= \int_0^{\pi/2} \frac{\cos^n \theta}{\cos^n \theta + \sin^n \theta} d\theta \quad \text{...(ii)}$$

On adding Eqs. (i) and (ii), we get

$$2I = \int_0^{\pi/2} \frac{\sin^n \theta + \cos^n \theta}{\cos^n \theta + \sin^n \theta} d\theta = \int_0^{\pi/2} d\theta = \frac{\pi}{2} \Rightarrow I = \frac{\pi}{4}$$

25. (d) Since, $iz = ze^{i\pi/2}$

This implies that iz is the vector obtained by rotating vector z in anticlock wise direction through 90°.

$\therefore \quad OA \perp AB$

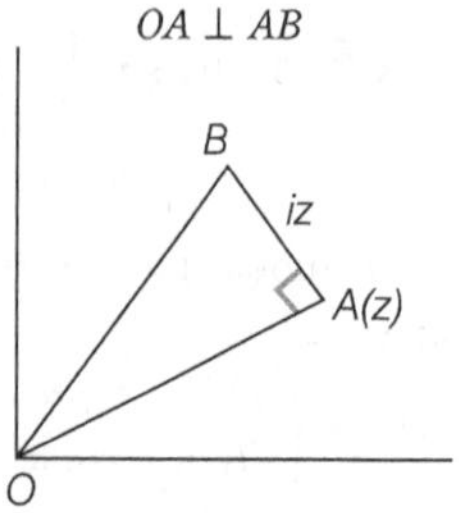

So, area of $\Delta OAB = \frac{1}{2} OA \times OB = \frac{1}{2}|z||iz| = \frac{1}{2}|z|^2$

26. (a) Given, $f(x) = \frac{\log(1 + ax) - \log(1 - bx)}{x}$

$f(x)$ is continuous at $x = k$ and $f(0) = k$.

$$\therefore \lim_{x \to 0} f(x) = \lim_{x \to 0} \frac{\log(1 + ax) - \log(1 - bx)}{x} \quad \left(\frac{0}{0} \text{ form}\right)$$

$$= \lim_{x \to 0} \left(\frac{1}{1 + ax} \cdot a + \frac{b}{1 - bx}\right) = a + b$$

$\therefore \quad a + b = f(0) = k$

27. (c) Since, $123 = -1 (\text{mod} 124)$

$125 = 1 (\text{mod} 124)$

$127 = 3 (\text{mod} 124)$

$\therefore \quad 123 \times 125 \times 127 = (-1)(1)(3) \text{ mod } (124)$

$= -3 \text{ mod } (124) = 121 \text{ mod} (124)$

28. (b) Given, $y = Ax^2 + Bx$ ...(i)

On differentiating, we get

$$\frac{dy}{dx} = 2Ax + B$$

$$\Rightarrow \quad \frac{d^2y}{dx^2} = 2A$$

$$\therefore \quad \frac{dy}{dx} - \frac{d^2y}{dx^2} x = B$$

From Eq. (i),

$$y = \frac{1}{2}\frac{d^2y}{dx^2} x^2 + x\left[\frac{dy}{dx} - \frac{d^2y}{dx^2} x\right]$$

$$\Rightarrow \quad y = -\frac{1}{2} x^2 \frac{d^2y}{dx^2} + x\frac{dy}{dx}$$

$$\therefore \quad x^2 \frac{d^2y}{dx^2} - 2x\frac{dy}{dx} + 2y = 0$$

29. (a) Since, the given equation has real roots.

$\therefore$ Discriminant,

$$D = \cos^2 p - 4(\cos p - 1)\sin p \geq 0$$

$$\Rightarrow \quad D = (\cos p - 2\sin p)^2 + 4\sin p(1 - \sin p) \geq 0$$

$$\Rightarrow \quad 0 < p < \pi$$

30. (a) Given,

$$3\sin^{-1}\left(\frac{2x}{1 + x^2}\right) - 4\cos^{-1}\left(\frac{1 - x^2}{1 + x^2}\right) + 2\tan^{-1}\left(\frac{2x}{1 - x^2}\right) = \frac{\pi}{3}$$

$$\Rightarrow 6\tan^{-1} x - 8\tan^{-1} x + 4\tan^{-1} x = \frac{\pi}{3}$$

$$\Rightarrow \quad 2\tan^{-1} x = \frac{\pi}{3}$$

$$\Rightarrow \quad \tan^{-1} x = \frac{\pi}{6}$$

$$\Rightarrow \quad x = \tan\frac{\pi}{6} = \frac{1}{\sqrt{3}}$$

31. (c) We have, $\sin(\pi\cos\theta) = \cos(\pi\sin\theta)$

$$\Rightarrow \quad \cos\left(\frac{\pi}{2} - \pi\cos\theta\right) = \cos(\pi\sin\theta)$$

$$\Rightarrow \quad \frac{\pi}{2} - \pi\cos\theta = 2n\pi \pm \pi\sin\theta$$

$$\Rightarrow \quad \frac{1}{2} - \cos\theta = 2n \pm \sin\theta$$

Taking '+' sign,

$$\Rightarrow \quad \frac{1}{2} - \cos\theta = 2n + \sin\theta$$

$$\Rightarrow \quad \frac{1}{\sqrt{2}}\cos\theta + \frac{1}{\sqrt{2}}\sin\theta = \frac{1}{2\sqrt{2}} - \sqrt{2}n$$

$$\Rightarrow \quad \cos\left(\theta - \frac{\pi}{4}\right) = \frac{1}{2\sqrt{2}} \quad (\text{for } n = 0)$$

32. (a) Since, $f(x)$ is continuous at $x = 1$.

$$\therefore \quad \lim_{h \to 0^+} f(1 + h) = f(1)$$

$$\Rightarrow \quad \lim_{h \to 0} 3 - a(1 + h)^2 = 2$$

$$\Rightarrow \quad 3 - a(1+0)^2 = 2$$
$$\Rightarrow \quad a = 1$$

33. *(b)* Required area $= \int_2^4 \left(1 + \frac{8}{x^2}\right) dx$

$$= \left[x - \frac{8}{x}\right]_2^4 = 4 - \frac{8}{4} - 2 + \frac{8}{2}$$
$$= 4 - 2 - 2 + 4$$
$$= 4 \text{ sq unit}$$

34. *(b)* Let the number of persons in the room be n, then

$$^nC_2 = 66$$
$$\Rightarrow \quad \frac{n(n-1)}{2} = 66$$
$$\Rightarrow \quad n^2 - n - 132 = 0$$
$$\Rightarrow \quad (n-12)(n+11) = 0$$
$$\Rightarrow \quad n = 12 \qquad (\because n \neq -11)$$

35. *(a)* Let one root of equation $ax^2 + bx + c = 0$ is α, then another root will be α^2.

$$\therefore \quad \alpha + \alpha^2 = -\frac{b}{a} \qquad \ldots\text{(i)}$$

and $$\alpha \cdot \alpha^2 = \frac{c}{a}$$

$$\Rightarrow \quad \alpha^3 = \frac{c}{a} \qquad \ldots\text{(ii)}$$

From Eq. (i),

$$(\alpha + \alpha^2)^3 = -\frac{b^3}{a^3}$$
$$\Rightarrow \quad \alpha^3 + \alpha^6 + 3\alpha^3(\alpha + \alpha^2) = -\frac{b^3}{a^3}$$
$$\Rightarrow \quad \frac{c}{a} + \frac{c^2}{a^2} + 3\frac{c}{a}\left(-\frac{b}{a}\right) = -\frac{b^3}{a^3}$$
$$\Rightarrow \quad a^2c + ac^2 - 3abc + b^3 = 0$$

36. *(d)* $\sum_{r=1}^{\infty} [3 \cdot 2^{-r} - 2 \cdot 3^{1-r}]$

$$= 3\sum_{r=1}^{\infty} \frac{1}{2^r} - 2\sum_{r=1}^{n} 3^{1-r}$$
$$= 3\left[\frac{1}{2} + \frac{1}{2^2} + \ldots \infty\right] - 2\left[1 + \frac{1}{3} + \frac{1}{3^2} + \ldots \infty\right]$$
$$= \frac{3}{2}\left[1 + \frac{1}{2} + \frac{1}{2^2} + \ldots\right] - 2\left[1 + \frac{1}{3} + \frac{1}{3^2} + \ldots \infty\right]$$
$$= \frac{3}{2}\left(\frac{1}{1 - \frac{1}{2}}\right) - 2\left(\frac{1}{1 - \frac{1}{3}}\right)$$
$$= \frac{3}{2} \times 2 - \frac{2 \times 3}{2} = 0$$

37. *(b)* Given that,

$$x(1+y^2)\,dx + y(1+x^2)\,dy = 0,\ y(0) = 1$$
$$\Rightarrow \quad \frac{x}{1+x^2}dx + \frac{y}{1+y^2}dy = 0$$

On integrating both sides, we get

$$\frac{1}{2}\log(1+x^2) + \frac{1}{2}\log(1+y^2) = c$$

Now,

$$y(0) = 1 \Rightarrow c = \frac{1}{2}\log(1+1) = \frac{1}{2}\log 2$$
$$\therefore \quad \sqrt{1+x^2}\sqrt{1+y^2} = \sqrt{2}$$
$$\Rightarrow \quad (1+x^2)(1+y^2) = 2$$
$$\Rightarrow \quad x^2 + y^2 + x^2y = 1$$

38. *(d)* Given, $f(x) = \frac{x + \cos x}{x - \cos x}$

On differentiating w.r.t. x, we get

$$f'(x) = \frac{\left[\begin{array}{l}(x - \cos x)(1 - \sin x) \\ \quad -(x + \cos x)(1 + \sin x)\end{array}\right]}{(x - \cos x)^2}$$
$$= \frac{-2x\sin x - 2\cos x}{(x - \cos x)^2}$$

39. *(b)* Diagonals of parallelogram bisects each other.

$\Rightarrow$ Diagonal BD passes through mid point of AC.

i.e. $$\left(\frac{5}{2}, \frac{7}{2}\right)$$

$\therefore$ Equation of diagonal passing through $(5, 1)$ and $\left(\frac{5}{2}, \frac{7}{2}\right)$ is

$$y - 1 = \frac{\frac{7}{2} - 1}{\frac{5}{2} - 5}(x - 5)$$
$$\Rightarrow \quad y - 1 = -1(x - 5)$$
$$\Rightarrow \quad y + x - 6 = 0$$

40. *(d)* $\int_0^1 x \sin \pi x\,dx$

$$= \left[-x\frac{\cos \pi x}{\pi}\right]_0^1 + \int_0^1 \frac{\cos \pi x}{\pi}dx$$
$$= \left[-x\frac{\cos \pi x}{\pi} + \frac{\sin \pi x}{\pi^2}\right]_0^1 = \left[\frac{-(-1)}{\pi}\right] = \frac{1}{\pi}$$

41. *(b)* Since, $(a, b) = c$

$\therefore$ c is the GCD of a and b.

$$\therefore \quad \frac{c}{b}$$

42. *(c)* Given, $\Delta = \begin{vmatrix} a & b & c \\ b & c & a \\ c & a & b \end{vmatrix}$

Applying, $C_1 \to C_1 + C_2 + C_3$ and taking common from C_1

$$\Delta = (a+b+c)\begin{vmatrix} 1 & b & c \\ 1 & c & a \\ 1 & a & b \end{vmatrix}$$

$$= -(a+b+c)[a^2+b^2+c^2-ab-bc-ca]$$

$$= -\frac{(a+b+c)}{2}[(a-b)^2+(b-c)^2+(c-a)^2]$$

> 0 as $a, b, c < 0$

43. *(a)* The product of 'n' positive integers is unity. Then their sum is a positive integer.

44. *(d)* Let $z = \sin x\,(1+\cos x)$

or $z = \sin x + \frac{1}{2}\sin 2x$

On differentiating w.r.t. x, we get

$$\frac{dz}{dx} = \cos x + \cos 2x$$

For maximum or minimum, put $\frac{dz}{dx} = 0$

$\Rightarrow \cos x + \cos 2x = 0$

$\Rightarrow 2\cos^2 x - 1 + \cos x = 0$

$\Rightarrow (\cos x + 1)(2\cos x - 1) = 0$

$\Rightarrow \cos x = -1, \frac{1}{2}$

$\Rightarrow x = \frac{\pi}{3}, \pi$

But $0 \le x \le \frac{\pi}{2}$, therefore we take only $x = \frac{\pi}{3}$.

$\therefore \frac{d^2z}{dx^2} = -\sin x - 2\sin 2x = -$ ve, for $x = \pi/3$

$\therefore$ It is maximum at $x = \pi/3$.

45. *(c)* Given,

$$y = a\cos \mu x + b\sin \mu x \quad \ldots\text{(i)}$$

On differentiating w.r.t. x, we get

$$\frac{dy}{dx} = -a\mu\sin\mu x + b\mu\cos\mu x$$

Again, differentiating, we get

$$\frac{d^2y}{dx^2} = -a\mu^2\cos\mu x - b^2\mu\sin\mu x$$

$$= -\mu^2 y \quad \text{[from Eq. (i)]}$$

$\Rightarrow \frac{d^2y}{dx^2} + \mu^2 y = 0$

46. *(b)* The point of intersection of the curves $y^2 = 2x$ and $x^2 = 2y$ are $O(0, 0)$, $A(2, 2)$.

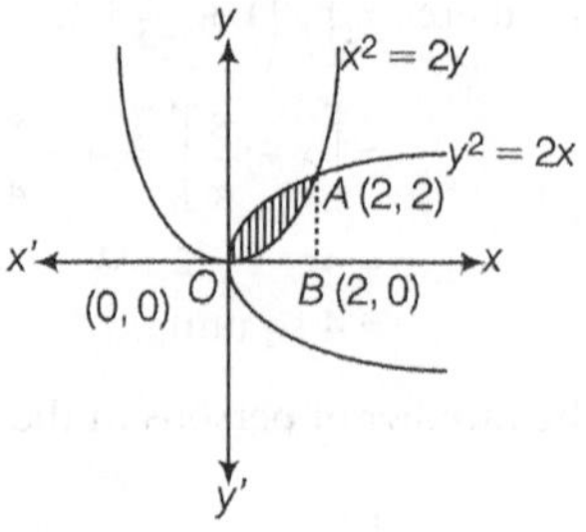

$\therefore$ Required area $= \int_0^2 \left(\sqrt{2x} - \frac{x^2}{2}\right) dx$

$$= \left[\sqrt{2}\frac{x^{3/2}}{3/2} - \frac{x^3}{6}\right]_0^2$$

$$= \left[\frac{2}{3}\cdot 2^2 - \frac{2^3}{6}\right] = \left[\frac{8}{3} - \frac{4}{3}\right]$$

$$= \frac{4}{3} \text{ sq unit}$$

47. *(a)* $A \cdot \text{adj}(A) = O \Rightarrow |A| I_n = O$

where n is order of A

$\Rightarrow |A| = 0$

48. *(b)* Given, $|\mathbf{a}+\mathbf{b}| = |\mathbf{a}-\mathbf{b}|$

$\Rightarrow |\mathbf{a}+\mathbf{b}|^2 = |\mathbf{a}-\mathbf{b}|^2$

$\Rightarrow |\mathbf{a}|^2 + |\mathbf{b}|^2 + 2\mathbf{a}\cdot\mathbf{b} = |\mathbf{a}|^2 + |\mathbf{b}|^2 - 2\mathbf{a}\cdot\mathbf{b}$

$\Rightarrow 4\mathbf{a}\cdot\mathbf{b} = 0$

Hence, $\mathbf{a}$ is perpendicular to $\mathbf{b}$.

49. *(d)* $\lim_{x\to 0} \frac{(2+x)\sin(2+x) - 2\sin 2}{x} \left(\frac{0}{0}\text{ form}\right)$

$$= \lim_{x\to 0} \frac{\sin(2+x) + (2+x)\cos(2+x)}{1}$$

$$= \sin 2 + 2\cos 2$$

50. *(a)* Let $I = \int \frac{2dx}{(e^x + e^{-x})^2} = \int \frac{2dx}{e^{2x} + e^{-2x} + 2}$

$$= \int \frac{2e^{2x}\,dx}{e^{4x} + 2e^{2x} + 1}$$

Put $e^{2x} = t$

$\Rightarrow 2e^{2x}\,dx = dt$

$\therefore I = \int \frac{dt}{t^2 + 2t + 1} = \int \frac{dt}{(t+1)^2} = -\frac{1}{t+1} + c$

$$= -\frac{1}{e^{2x}+1} = -\frac{e^{-x}}{e^x + e^{-x}} + c$$

51. *(a)* Circumference of circle, $C = 2\pi r$

$$\Rightarrow \quad \frac{dC}{dt} = 2\pi \frac{dr}{dt}$$

$$\Rightarrow \quad \frac{0.3}{2\pi} = \frac{dr}{dt} \qquad \left(\because \frac{dC}{dt} = 0.3 \text{ cm / s given}\right)$$

Now, $A = \pi r^2$

$$\frac{dA}{dt} = 2\pi r \frac{dr}{dt} \Rightarrow \frac{dA}{dt} = r \times 0.3$$

$$\Rightarrow \quad \left[\frac{dA}{dt}\right]_{r=5} = 5 \times 0.3 = 1.5 \text{ sq cm/s}$$

52. *(d)* If ω is a cube root of unity, then

$1 + \omega + \omega^2 = 0$ and $\omega^3 = 1$.

$$\therefore (1 + \omega - \omega^2)(1 - \omega + \omega^2) = (-\omega^2 - \omega^2)(-\omega - \omega)$$

$$= 2\omega^2 \cdot 2\omega = 4$$

53. *(b)* Let $z = x + iy$

$$\therefore \quad i + z = x + i(1 + y)$$

and $$i - z = -x + i(1 - y)$$

$$\because \quad \left|\frac{i + z}{i - z}\right| = 1 \Rightarrow |i + z| = |i - z|$$

$$\Rightarrow \quad x^2 + (1 + y)^2 = x^2 + (1 - y)^2$$

$$\Rightarrow x^2 + y^2 + 2y + 1 = x^2 + y^2 + 1 - 2y$$

$$\Rightarrow \quad 4y = 0$$

$$\Rightarrow \quad y = 0$$

i.e. x-axis.

54. *(c)* In a triangle ABC,

$$A + B + C = \pi,$$

$$\Rightarrow \sin(A + B + C) = 0, \cos(A + B) = -\cos C$$

$$\therefore \begin{vmatrix} \sin(A + B + C) & \sin B & \cos C \\ -\sin B & 0 & \tan A \\ \cos(A + B) & -\tan A & 0 \end{vmatrix}$$

$$= \begin{vmatrix} 0 & \sin B & \cos C \\ -\sin B & 0 & \tan A \\ -\cos C & -\tan A & 0 \end{vmatrix}$$

$$= 0$$

$\left(\because \text{It is the determinant of a skew symmetric matrix of odd order.}\right)$

55. *(c)* Let $y = x \log_e x$

On differentiating w.r.t. x, we get

$$\frac{dy}{dx} = x \cdot \frac{1}{x} + \log x = (1 + \log x)$$

Again, differentiating, we get

$$\frac{d^2y}{dx^2} = \frac{1}{x}$$

Put, $\frac{dy}{dx} = 0$ for maxima or minima.

$$\Rightarrow \quad 1 + \log x = 0$$

$$\Rightarrow \quad x = \frac{1}{e}$$

$$\therefore \quad \left(\frac{d^2y}{dx^2}\right)_{(x = 1/e)} = e$$

$$\therefore \quad y \text{ is minimum at } x = \frac{1}{e}$$

$$\therefore \quad y_{\min} = \frac{1}{e} \log_e\left(\frac{1}{e}\right) = -\frac{1}{e}$$

56. *(c)* $\because f''(0) = k$ and

$$\lim_{x \to 0} \frac{2f(x) - 3f(2x) + f(4x)}{x^2}$$

Using L'Hospital's rule,

$$= \lim_{x \to 0} \frac{2f'(x) - 6f'(2x) + 4f'(4x)}{2x}$$

Again, using L'Hospital's rule,

$$= \lim_{x \to 0} \frac{2f''(x) - 12f''(2x) + 16f''(4x)}{2}$$

$$= \frac{1}{2}[2f''(0) - 12f''(0) + 16f''(0)]$$

$$= \frac{1}{2}[2k - 12k + 16k] = 3k$$

57. *(c)* $\int_2^4 [3 - f(x)]\, dx = 7$

$$\Rightarrow \quad [3x]_2^4 - \int_2^4 f(x)\, dx = 7$$

$$\Rightarrow \quad 6 - \int_2^4 f(x)\, dx = 7$$

$$\Rightarrow \quad \int_2^4 f(x)\, dx = -1 \qquad \ldots\text{(i)}$$

and $$\int_{-1}^4 f(x)\, dx = 4 \qquad \ldots\text{(ii)}$$

$$\int_{-1}^4 f(x)\, dx = \int_{-1}^2 f(x)\, dx + \int_2^4 f(x)\, dx$$

$$\Rightarrow \quad 4 = \int_{-1}^2 f(x)\, dx - 1$$

$$\Rightarrow \quad \int_{-1}^2 f(x)\, dx = 5$$

58. *(a)* The given differential equation is

$$\frac{dy}{dx} = (x - y)^2$$

Put $$x - y = t$$

$$\Rightarrow \quad 1 - \frac{dy}{dx} = \frac{dt}{dx}$$

$$\Rightarrow \quad \frac{dy}{dx} = 1 - \frac{dt}{dx}$$

$\therefore \quad 1 - \frac{dt}{dx} = t^2$

$\Rightarrow \quad (1 - t^2) = \frac{dt}{dx}$

On integrating both sides, we get

$$\int \frac{1}{1-t^2}\, dt = \int 1\, dx$$

$\Rightarrow \quad \frac{1}{2} \log\left(\frac{1+t}{1-t}\right) = x + c$

$\Rightarrow \quad \left(\frac{1+t}{1-t}\right) = e^{2x + 2c}$

$\Rightarrow \quad \frac{1 + x - y}{1 - x + y} = Ae^{2x}$

$\because$ It passes through origin.

$\therefore \quad A = 1$

$\therefore$ Required curve is

$$(1 + x - y) = e^{2x}(1 - x + y)$$

59. *(d)* $\because \quad (\sqrt{3} - i) = 2\left(\cos\frac{\pi}{6} - i \sin\frac{\pi}{6}\right)$

$\therefore \quad (\sqrt{3} - i)^{50} = 2^{50}\left(\cos\frac{\pi}{6} - i \sin\frac{\pi}{6}\right)^{50}$

$= 2^{50}\left(\cos\frac{\pi}{3} - i \sin\frac{\pi}{3}\right)$

$= 2^{50}\left(\frac{1}{2} - i\frac{\sqrt{3}}{2}\right)$

But $\quad (\sqrt{3} - i)^{50} = 2^{48}(x - iy)$

$\therefore \quad x = 2$ and $y = 2\sqrt{3}$

$\therefore \quad x^2 + y^2 = 4 + 12 = 16$

60. *(b)* $T_r = T_{(r-1)+1} = {}^{20}C_{r-1} \cdot x^{r-1}$

$\therefore$ Coefficient of the rth term $= {}^{20}C_{r-1}$

and coefficient of the $(r + 1)$th term $= {}^{20}C_r$

$\therefore \quad \frac{{}^{20}C_{r-1}}{{}^{20}C_r} = \frac{1}{2}$

$\Rightarrow \quad \frac{r}{21 - r} = \frac{1}{2}$

$\Rightarrow \quad r = 7$

61. *(c)* Now, $\mathbf{a} \times \mathbf{b} = \begin{vmatrix} \hat{\mathbf{i}} & \hat{\mathbf{j}} & \hat{\mathbf{k}} \\ 3 & 1 & -2 \\ 1 & -3 & 4 \end{vmatrix}$

$= \hat{\mathbf{i}}(4 - 6) - \hat{\mathbf{j}}(12 + 2) + \hat{\mathbf{k}}(-9 - 1)$

$= -2\hat{\mathbf{i}} - 14\hat{\mathbf{j}} - 10\hat{\mathbf{k}}$

$\therefore$ Area of required parallelogram

$= \frac{1}{2}|\mathbf{a} \times \mathbf{b}| = \frac{1}{2}\sqrt{4 + 196 + 100} = \frac{10\sqrt{3}}{2} = 5\sqrt{3}$

62. *(a)* In the given series

$a = \frac{\sqrt{2}+1}{\sqrt{2}-1}$ and $r = \frac{1}{2+\sqrt{2}} = \frac{2-\sqrt{2}}{2}$

$\therefore \quad S_\infty = \frac{a}{1-r} = \frac{\frac{\sqrt{2}+1}{\sqrt{2}-1}}{1 - \frac{2-\sqrt{2}}{2}}$

$= \frac{\frac{\sqrt{2}+1}{\sqrt{2}-1}}{\frac{\sqrt{2}}{2}} = \frac{\sqrt{2}(\sqrt{2}+1)}{\sqrt{2}-1} \times \frac{\sqrt{2}+1}{\sqrt{2}+1}$

$= \sqrt{2}(\sqrt{2}+1)^2$

63. *(c)* $\because (1 + ax)^n = 1 + axn + \frac{n(n-1)}{2!}(ax)^2 + \ldots$

Given that,

$axn = 6x$

$\Rightarrow \quad an = 6 \quad \ldots\text{(i)}$

and $\quad \frac{a^2 n(n-1)}{2} = 16 \quad \ldots\text{(ii)}$

On solving Eqs. (i) and (ii), we get

$a = \frac{2}{3}$ and $n = 9$

64. *(b)* Since, $f(x)$ is continuous at $x = 3$, then

$\lim_{x \to 3} f(x) = f(3) \quad \ldots\text{(i)}$

$\therefore \lim_{x \to 3} \frac{x^2 - 9}{x - 3} = \lim_{x \to 3} x + 3 = 3 + 3 = 6$

and $\quad f(3) = 2 \times 3 + k = k + 6$

$\therefore$ From Eq. (i),

$6 = k + 6 \Rightarrow k = 0$

65. *(c)* Required area

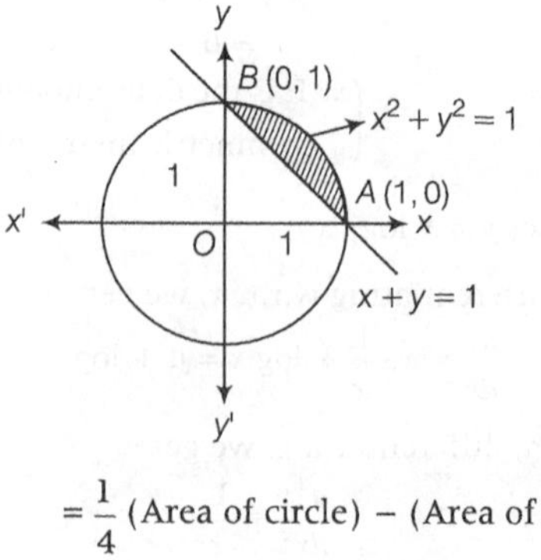

$= \frac{1}{4}$ (Area of circle) $-$ (Area of $\Delta\, AOB$)

$$= \frac{\pi (1)^2}{4} - \frac{1}{2} \times OA \times OB$$

$$= \left(\frac{\pi}{4} - \frac{1}{2}\right) \text{sq unit}$$

66. (*a*) $\left|\frac{x}{2} - 1\right| < 3 \Rightarrow -3 < \frac{x}{2} - 1 < 3$

$$\Rightarrow \quad -3 + 1 < \frac{x}{2} < 3 + 1$$

$$\Rightarrow \quad -2 < \frac{x}{2} < 4$$

$$\Rightarrow \quad -4 < x < 8$$

$$\therefore \quad x \in (-4, 8)$$

67. (*b*) $\sin^{-1} \frac{1}{\sqrt{5}} + \cos^{-1} \frac{3}{\sqrt{10}}$

Now, $\sin^{-1} \frac{1}{\sqrt{5}} = \tan^{-1} \left(\frac{1/\sqrt{5}}{\sqrt{1 - \frac{1}{5}}}\right) = \tan^{-1} \frac{1}{2}$

and $\cos^{-1} \frac{3}{\sqrt{10}} = \tan^{-1} \left(\frac{\sqrt{1 - \frac{9}{10}}}{\frac{3}{\sqrt{10}}}\right) = \tan^{-1} \frac{1}{3}$

$$\therefore \quad \tan^{-1} \frac{1}{2} + \tan^{-1} \frac{1}{3} = \tan^{-1} \left(\frac{\frac{1}{2} + \frac{1}{3}}{1 - \frac{1}{6}}\right) = \tan^{-1} (1) = \frac{\pi}{4}$$

68. (*b*) Given, $f(x) = \sec x - \cot x$

$$\therefore \quad f'(x) = \sec x \tan x + \text{cosec}^2 x$$

$$\Rightarrow \quad f'\left(\frac{\pi}{6}\right) = \sec \frac{\pi}{6} \tan \frac{\pi}{6} + \text{cosec}^2 \frac{\pi}{6}$$

$$= \frac{2}{\sqrt{3}} \cdot \frac{1}{\sqrt{3}} + 4 = \frac{14}{3}$$

69. (*a*) Let C divides AB in the ratio $m:1$. Then

$$\mathbf{c} = \frac{m\mathbf{b} + \mathbf{a}}{m + 1}$$

$$\Rightarrow \quad m\mathbf{c} + \mathbf{c} = m\mathbf{b} + \mathbf{a}$$

$$\Rightarrow \quad \mathbf{a} + m\mathbf{b} - m\mathbf{c} - \mathbf{c} = \mathbf{0}$$

$$\Rightarrow \quad \mathbf{a} + m\mathbf{b} + (-m - 1)\,\mathbf{c} = \mathbf{0} \quad \ldots\text{(i)}$$

Given, $3\mathbf{a} + 4\mathbf{b} - 7\mathbf{c} = \mathbf{0}$

$$\Rightarrow \quad \mathbf{a} + \frac{4}{3}\mathbf{b} - \frac{7}{3}\mathbf{c} = \mathbf{0} \quad \ldots\text{(ii)}$$

On comparing Eqs. (i) and (ii), we get

$$m = \frac{4}{3}$$

$\therefore$ Required ratio is 4 : 3

70. (*d*) $\begin{bmatrix} a & 2 & 3 \\ b & 5 & -1 \end{bmatrix} \begin{bmatrix} 1 & 2 \\ 3 & 4 \\ -1 & 1 \end{bmatrix}$

$$= \begin{bmatrix} a + 6 - 3 & 2a + 8 + 3 \\ b + 15 + 1 & 2b + 20 - 1 \end{bmatrix}$$

$$\Rightarrow \quad \begin{bmatrix} a + 3 & 2a + 11 \\ b + 16 & 2b + 19 \end{bmatrix} = \begin{bmatrix} 4 & 13 \\ 12 & 11 \end{bmatrix} \quad \text{(given)}$$

$$\Rightarrow \quad a + 3 = 4 \Rightarrow a = 1$$

and $b + 16 = 12 \Rightarrow b = 12 - 16 = -4$

$$\therefore \quad (a, b) = (1, -4)$$

English & General Aptitude

16. (*b*) Indira and Nehru were the Prime Ministers of India, while Benazir was the Prime Minister of Pakistan.

17. (*d*) All are former Indian cricketers.

18. (*c*) 'Folketing' is the parliament of 'Denmark'; 'Stortling' is the parliament of 'Norway', 'Knesset' is the parliament of Israel.

19. (*b*) They all are modes of transport.

20. (*c*) 'PTI' is an Indian news agency; 'Irna' is a news agency in Iran while 'Xin-Era' is a news agency in China.

21. (*c*) 'Pitcher' is a term used in Baseball; 'Dusra' is a term used in cricket and 'Bunker' is a term used in polo.

22. (*d*) These all are snakes.

23. (*c*) These all are units of distance.

24. (*d*) These all are water animals.

MANIPAL
Engineering Entrance Exam

Solved Paper 2017

Physics

1. In Young's experiment, using red light ($\lambda = 6600$ Å), 60 fringes are seen in the field of view. How many fringes will be seen by using violet light ($\lambda = 4400$ Å)?

(a) 10 (b) 20
(c) 45 (d) 90

2. Two identical air core capacitors are connected in series to a voltage source of 15 V. If one of the capacitors is filled with a medium of dielectric constant 4, the new potential across this capacitor is

(a) 5 V (b) 8 V
(c) 10 V (d) 12 V

3. A string of density 7.5 g cm^{-3} and area of cross-section 0.2 mm^2 is stretched under a tension of 20 N. When it is plucked at the mid-point, the speed of the transverse wave on the wire is

(a) 116 ms^{-1} (b) 40 ms^{-1}
(c) 200 ms^{-1} (d) 80 ms^{-1}

4. Three blocks of masses m_1, m_2 and m_3 are connected by massless string as shown kept on a frictionless table.

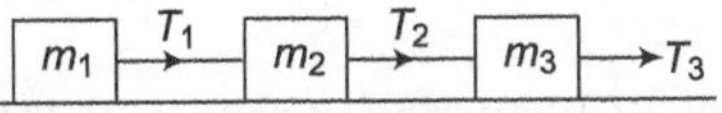

They are pulled with a force $T_3 = 40$ N.If $m_1 = 10$ kg, $m_2 = 6$ kg and $m_3 = 4$ kg, the tension T_2 will be

(a) 20 N (b) 40 N (c) 10 N (d) 32 N

5. Two identical cells whether connected in parallel or in series gives the same current when connected to an external resistance 1.5 Ω. Find the value of internal resistance of each cell.

(a) 1 Ω (b) 0.5 Ω
(c) zero (d) 1.5 Ω

6. Force between two identical charges placed at a distance of r in vacuum is F. Now a slab of dielectric of dielectric constant 4 is inserted between these two charges. If the thickness of the slab is $r/2$, then the force between the charges will become

(a) F (b) $\frac{3}{5}F$ (c) $\frac{4}{9}F$ (d) $\frac{F}{4}$

7. Water rises in a capillary tube to a height h. Choose false statement regarding capillary rise from the following.

(a) On the surface of Jupiter, height will be less than h
(b) In a lift moving up with constant acceleration height is less than h
(c) On the surface of moon the height is more than h
(d) In a lift moving down with constant acceleration height is less than h

8. The surface area of a black body is 5×10^{-4} m^2 and its temperature is 727°C. The energy radiated by it per minute is ($\sigma = 5.67 \times 10^{-8}$ J/m^2-s-K^4)

(a) 1.7×10^3 J (b) 2.5×10^2 J
(c) 8×10^3 J (d) 3×10^4 J

9. The inputs and outputs for different time intervals are given below the NAND gate

Time	Inputs *A*	Inputs *B*	Outputs *Y*
t_1 to t_2	0	1	*P*
t_2 to t_3	0	0	*Q*
t_3 to t_4	1	0	*R*
t_4 to t_5	1	1	*S*

The values taken by *P*, *Q*, *R*, *S* are respectively
(a) 1, 1, 1, 0
(b) 0, 1, 0, 1
(c) 0, 1, 0, 0
(d) 1, 0, 1, 1

10. An electric dipole is placed at an angle of 30° with an electric field of intensity 2×10^5 NC^{-1}. It experiences a torque equal to 4 Nm. Calculate the charge on the dipole if the dipole length is 2 cm.
(a) 8 mC
(b) 4 mC
(c) 8 μC
(d) 2 mC

11. In a gas, two waves of wavelengths 1 m and 1.01 m are superposed and produce 10 beats in 3 s. The velocity of sound in the medium is
(a) 300 m/s
(b) 336.7 m/s
(c) 360.2 m/s
(d) 270 m/s

12. If α and β are the current gain in the *CB* and *CE* configurations respectively of the transistor circuit, then $\frac{\beta - \alpha}{\alpha\beta} =$
(a) infinite
(b) 1
(c) 2
(d) 0.5

13. The apparent frequency of the whistle of an engine changes in the ratio 9 : 8 as the engine passes a stationary observer. If the velocity of the sound is 340 ms^{-1}, then the velocity of the engine is
(a) 40 ms^{-1}
(b) 20 ms^{-1}
(c) 340 ms^{-1}
(d) 180 ms^{-1}

14. The width of a single slit if the first minimum is observed at an angle 2° with a light of wavelength 6980 Å
(a) 0.2 mm
(b) 2×10^{-5} mm
(c) 2×10^5 mm
(d) 2 mm

15. The ratio of the resistance of conductor at temperature 15°C to its resistance at temperature 37.5°C is 4 : 5. The temperature coefficient of resistance the conductor is
(a) $\frac{1}{25}$ °C^{-1}
(b) $\frac{1}{50}$ °C^{-1}
(c) $\frac{1}{80}$ °C^{-1}
(d) $\frac{1}{75}$ °C^{-1}

16. The momentum of a body is increased by 25%. The kinetic energy is increased by about
(a) 25%
(b) 5%
(c) 56%
(d) 38%

17. The plane faces of two identical plano-convex lenses each having a focal length of 50 cm are placed against each other to form a usual biconvex lens. The distance from this lens combination at which an object must be placed to obtain a real, inverted image which has the same size as the object is
(a) 50 cm
(b) 25 cm
(c) 100 cm
(d) 40 cm

18. A short solenoid of length 4 cm, radius 2 cm and 100 turns is placed inside and on the axis of a long solenoid of length 80 cm and 1500 turns. A current of 3 A flows through the short solenoid. The mutual inductance of two solenoids is
(a) 2.96×10^{-4} H
(b) 5.3×10^{-5} H
(c) 3.52×10^{-3} H
(d) 8.3×10^{-5} H

19. The angle of a prism is 60° and its refractive index is $\sqrt{2}$. The angle of minimum deviation suffered by a ray of light in passing through it is
(a) about 20°
(b) 30°
(c) 60°
(d) 45°

20. A voltmeter has resistance of *G* ohms and range *V* volts. The value of resistance used in series to convert it into a voltmeter of range *nV* volts is
(a) nG
(b) $(n-1)G$
(c) $\frac{G}{n}$
(d) $\frac{G}{(n-1)}$

21. The half-life period of radium is 1600 yr. The fraction of a sample of radium that would remain after 6400 yr is
(a) $\frac{1}{4}$
(b) $\frac{1}{2}$
(c) $\frac{1}{8}$
(d) $\frac{1}{16}$

22. The temperatures of two bodies A and B are respectively 727°C and 327°C. The ratio $H_A : H_B$ of the rates of heat radiated by them is

(a) 727 : 327 (b) 5 : 3
(c) 25 : 9 (d) 625 : 81

23. A 2 kg copper block is heated to 500°C and then it is placed on a large block of ice at 0°C. If the specific heat capacity of copper is $400 \text{ J-kg}^{-1} \text{ °C}^{-1}$ and latent heat of fusion of water is $3.5 \times 10^5 \text{ J-kg}^{-1}$, the amount of ice that can melt is

(a) $\frac{7}{8}$ kg (b) $\frac{7}{5}$ kg (c) $\frac{8}{7}$ kg (d) $\frac{5}{7}$ kg

24. A man weighing 100 kg slides down a light rope with an acceleration of 1.8 ms^{-2}. If $g = 9.8 \text{ ms}^{-2}$, the tension of the rope is

(a) 180 N (b) 1160 N
(c) 800 N (d) weightlessness

25. A force of 20 N acts on a body of mass 2 kg initially at rest. Find the work done in 2 s.

(a) 400 J (b) 20 J (c) 10 J (d) 5 J

26. A charged particle of mass m and charge q is released from rest in a electric field of magnitude E. The kinetic energy of the particle after t second is

(a) $\frac{qEt}{2m}$ (b) $\frac{q^2E^2t}{2m}$
(c) $\frac{q^2E^2t^2}{m}$ (d) $\frac{q^2E^2t^2}{2m}$

27. The electric potential is + 100 V at a distance of 10 cm from a point charge q. Then, q is equal to

(a) $+ 1.1 \times 10^{-9}$ C (b) $+ 1.1 \times 10^{-3}$ C
(c) 3 C (d) 3×10^{-5} C

28. A cell of emf E is connected across a resistance R. The potential difference between the terminals of the cell is found to be V. The internal resistance of the cell must be

(a) $R\left[\frac{E}{V} - 1\right]$ (b) $R\left[\frac{E}{V} + 1\right]$
(c) ER (d) $(E - V)R$

29. Three 2 Ω resistors are connected to form a triangle. The resistance between any two corners is

(a) $\frac{4}{3}$ Ω (b) $\frac{3}{4}$ Ω
(c) 6 Ω (d) $\frac{2}{3}$ Ω

30. A thin rod of length $\frac{f}{3}$ is placed along the optic axis of a concave mirror of focal length f, such that its image which is real and elongated just touches the rod, the magnification is

(a) 1.5 (b) 2.5 (c) 3.5 (d) 4.5

31. Two coherent sources of intensity ratio 9 : 4 produce interference. What is the ratio between maxima and minima in the interference pattern produced ?

(a) 3 : 2 (b) 25 : 1
(c) 13 : 5 (d) 5 : 1

32. The angle of minimum deviation for a prism of angle A is $180 - 2A$. The refractive index is

(a) $\sin\frac{A}{2}$ (b) $\cos\frac{A}{2}$
(c) $\tan\frac{A}{2}$ (d) $\cot\frac{A}{2}$

33. An electron at rest is accelerated through a potential difference of 200 V. If the electron acquires a velocity of 8.4×10^8 cm/s, its $\frac{e}{m}$ ratio is

(a) 1.76×10^{11} C / kg (b) 2.5×10^9 C / kg
(c) 1.5×10^5 C / kg (d) 9.8×10^3 C / kg

34. The ratio of the wavelength of first line of Balmer series to the first line of Lyman series is

(a) 3 : 1 (b) 17 : 5
(c) 27 : 5 (d) 27 : 3

35. A p-n junction is designed to withstand current up to a maximum in 10 mA. A resistor $R = 200$ Ω is connected in series with it. When forward biased the diode has a potential drop of 0.5 V. The maximum voltage of the battery required to forward bias the diode is

(a) 2.5 V (b) 1.5 V (c) 2 V (d) 10 V

36. What is the Boolean equation for the figure ?

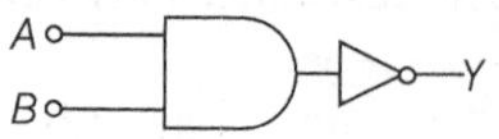

(a) $Y = A + B$
(b) $Y = \overline{A + B}$
(c) $Y = \overline{A} + \overline{B}$
(d) $Y = \overline{A \cdot B}$

37. A geo-stationary satellite is orbiting the earth at height of 6 R above the surface of the earth, R being the radius of earth. The time period of another satellite at a height of 2.5 R from the surface of the earth is

(a) 10 h
(b) $6/\sqrt{2}$ h
(c) 6 h
(d) $6\sqrt{2}$ h

38. For a colour of light the wavelength for air is 6000 Å and in water the wavelength is 4500 Å. Then the speed of light in water will be

(a) 5×10^{14} m/s
(b) 2.25×10^{8} m/s
(c) 4.0×10^{8} m/s
(d) zero

39. When an equilateral prism of refractive index $\sqrt{2}$ produces minimum deviation the angle of incidence at the first face must be

(a) 30°
(b) 42°
(c) 60°
(d) 75°

40. Two equal charges q are kept fixed at a and $+a$ along the x-axis. A particle of mass m and charge $\frac{q}{2}$ is brought to the origin and given a small displacement along the x-axis, then

(a) the particle executes oscillatory motion
(b) the particle remains stationary
(c) the particle executes, SHM along x-axis
(d) the particle executes SHM along y-axis

41. Two equal point charges, $Q = +\sqrt{2}\ \mu C$ are placed at each of the two opposite corners of a square and equal point charges q at each of the other two corners. The value of q, so that the resultant force on Q is zero is

(a) $+ 0.5\ \mu C$
(b) $- 0.5\ \mu C$
(c) $1\ \mu C$
(d) $- 1\ \mu C$

42. The linear momentum is increased by 50%, the kinetic energy will increase by

(a) 50%
(b) 100%
(c) 125%
(d) 25%

43. Two capacitors, one 4 pF and the other 6 pF, connected in parallel, are charged by a 100 V battery. The energy stored in the capacitors is

(a) 12×10^{-8} J
(b) 2.4×10^{-8} J
(c) 5.0×10^{-8} J
(d) 1.2×10^{-6} J

44. A voltmeter of resistance 998 Ω is connected across a cell of emf 2 V and internal resistance 2 Ω. The potential difference across the voltmeter is

(a) 1.99 V (b) 3.5 V (c) 5 V (d) 6 V

45. In the adjoining circuit, the resistances are given in ohm, X and Y are unknown resistances. The current through the 10 Ω resistance is 3A while that through the resistance X is 1A. No current passes through the galvanometer. The values of the unknown resistances X and Y are respectively

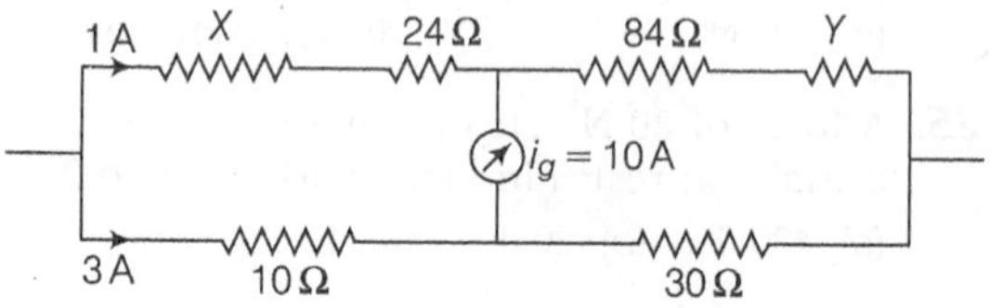

(a) 12 Ω and 12 Ω
(b) 12 Ω and 6 Ω
(c) 6 Ω and 12 Ω
(d) 6 Ω and 6 Ω

46. A sinusoidal voltage of peak value 300 V and an angular frequency $\omega = 400$ rad/s is applied to a series L-C-R circuit, in which $R = 3\ \Omega$, $L = 20$ mH and $C = 625\ \mu F$. The peak current in the circuit is

(a) $30\sqrt{2}$ A
(b) 60 A
(c) 100 A
(d) $60\sqrt{2}$ A

47. The photoelectric cut-off voltage in an experiment was found to be 1.5 V. The work function for the material used in the experiment was 4.2 eV. The maximum kinetic energy of the photoelectrons that emitted was

(a) 1.5 eV (b) 2.7 eV (c) 4.2 eV (d) 5.7 eV

48. First diffraction minima due to a single slit of width 10^{-4} cm is at $\theta = 30°$. Then wavelength of the light used is

(a) 4000 Å
(b) 5000 Å
(c) 6000 Å
(d) 6250 Å

49. When the inputs of a two input logic gate are 0 and 0, the output is 1. When the inputs are 1 and 0 the output is 0. The logic gate is of the type

(a) AND
(b) NAND
(c) NOR
(d) OR

50. A stationary police car sounds a siren with a frequency of 990 Hz. If the speed of sound is 330 m/s, an observer, driving towards the car with a speed of 33 m/s, will hear a frequency of

(a) 891 Hz (b) 900 Hz
(c) 1089 Hz (d) 1100 Hz

Chemistry

1. For f-orbital the values of m are

(a) $-2, -1, 0, +1, +2$
(b) $-3, -2, -1, 0, +1, +2, +3$
(c) $-1, 0, +1$
(d) $0, +1, +2, +3$

2. The mass of oxygen that would be required to produce enough CO which completely reduces 1.6 kg Fe_2O_3 (at. mass Fe = 56), is

(a) 240 g (b) 480 g (c) 720 g (d) 960 g

3. Which of the following is most acidic ?

(a) 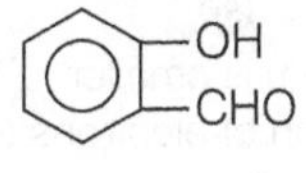

(b) 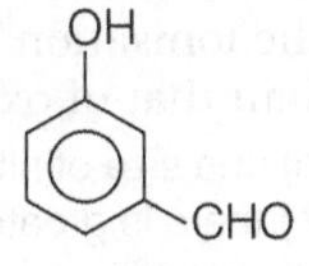

(c) 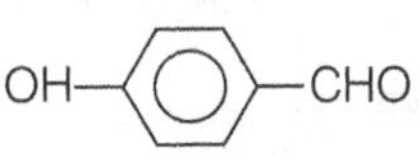

(d) 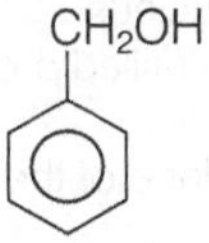

4. Purification of NaCl by passage of hydrogen chloride through brine is based on

(a) Common ion effect
(b) Distribution coeffficient
(c) Le-Chatelier's principle
(d) Distribution law

5. Wurtz reaction involves the reduction of alkyl halide with

(a) Zn/HCl (b) HI
(c) Zn/Cu Couple (d) Na in ether

6. Dry distillation of calcium benzoate gives

(a) benzaldehyde (b) acetophenone
(c) benzoic acid (d) benzophenone

7. Which one of the following reacts with Grignard reagent to form an addition product which can be hydrolysed to a carboxylic acid ?

(a) O_2 (b) CO_2
(c) SO_2 (d) None of these

8. Vinegar is dilute aqueous solution of

(a) ethanoic acid (b) benzoic acid
(c) citric acid (d) oxalic acid

9. If ionic radius of Cs^+ and Cl^- are 1.69 Å and 1.81 Å respectively, the edge length of unit cell is

(a) 4.04 Å (b) 3.50 Å
(c) 7.00 Å (d) None of these

10. What mass of calcium chloride in grams would be enough to produce 14.35 g of AgCl ? (Atomic mass Ca = 40, Ag = 108)

(a) 5.55 g (b) 8.295 g
(c) 11.19 g (d) 16.59 g

11. In hydrogen spectrum, the series of lines appearing in ultra violet region of electromagnetic spectrum are called

(a) Balmer lines (b) Lyman lines
(c) Pfund lines (d) Brackett lines

12. The set representing the correct order of first ionisation potential is

(a) K > Na > Li (b) Be > Mg > Ca
(c) B > C > N (d) Ge > Si > C

13. Fog is a colloidal solution of

(a) gas in gas (b) solid in gas
(c) liquid in gas (d) None of these

14. The oxidation number of N and Cl in $NOClO_4$ respectively are

(a) + 2 and + 7 (b) + 3 and + 7
(c) − 3 and + 5 (d) + 2 and − 7

15. Which of the following statements is correct regarding the slag obtained during the extraction of iron ?

(a) The slag is lighter and has lower melting point than the metal.
(b) The slag is lighter and has higher melting point than the metal.
(c) The slag is heavier and has lower melting point than metal.
(d) The slag is heavier and has higher melting point than the metal.

16. Which one of the following complexes is an outer orbital complex ?

(a) $[Fe(CN)_6]^{4-}$ (b) $[Mn(CN)_6]^{4-}$
(c) $[Co(NH_3)_6]^{3+}$ (d) $[Ni(NH_3)_6]^{2+}$

17. Which of the following is the most sweetest sugar ?

(a) Sucrose (b) Fructose
(c) Glucose (d) Lactose

18. What is the kinetic energy of 1 g of O_2 at 47°C ?

(a) 2.17×10^2 J (b) 2.24×10^2 J
(c) 1.24×10^2 J (d) None of these

19. The osmotic pressure of 0.2 molar solution of urea at 300 K is ($R = 0.082$ L atm $mol^{-1}K^{-1}$)

(a) 4.92 atm (b) 1 atm
(c) 0.25 atm (d) 27 atm

20. Reduction of glucose with $NaBH_4$ gives

(a) sorbitol (b) mannitol
(c) Both (a) and (b) (d) *n*-hexane

21. Decarboxylation of sodium propionate leads to the formation of

(a) methane (b) ethene
(c) propanone (d) ethane

22. Formic acid reacts with PCl_5 to form

(a) methyl chloride
(b) acetyl chloride
(c) formyl chloride
(d) carbon monoxide and hydrogen chloride

23. Which of the following conditions regarding the chemical process ensures its spontaneity at all temperatures ?

(a) $\Delta H < 0$; $\Delta S < 0$
(b) $\Delta H > 0$; $\Delta S < 0$
(c) $\Delta H < 0$; $\Delta S > 0$
(d) $\Delta H > 0$; $\Delta S > 0$

24. The equilibrium constant for a reaction is 10. $\Delta G°$ will be ($R = 8$ JK^{-1} mol^{-1}, $T = 300$ K)

(a) − 5527 kJ mol^{-1} (b) − 5.527 kJ mol^{-1}
(c) − 55.27 kJ mol^{-1} (d) + 5.527 kJ mol^{-1}

25. Which of the following is most basic ?

(a) $CH_3CH_2NH_2$ (b) CH_3CONH_2
(c) $C_6H_5CONCH_3$ (d) NH_2NH_2

26. The mass of $CaCO_3$ required to react with 25 mL of 0.75 M HCl is

(a) 0.94 g (b) 9.4 g
(c) 0.094 g (d) 0.49 g

27. For azimuthal quantum number $l = 3$, m can have

(a) only one value (b) 3 values
(c) 5 values (d) 7 values

28. The ionisation energy of nitrogen is larger than that of oxygen because

(a) the size of nitrogen atom is smaller
(b) there is greater attraction of electrons towards the nucleus
(c) half-filled *p*-orbitals of nitrogen have extra stability
(d) None of the above

29. The two-third life ($t_{2/3}$) of a first order reaction in which $k = 5.48 \times 10^{-14}$ per sec, is

(a) $\frac{2.303}{5.48 \times 10^{-14}} \log 3$ (b) $\frac{2.303}{5.48 \times 10^{-14}} \log 2$
(c) $\frac{2.303}{5.48 \times 10^{-14}} \log \frac{1}{3}$ (d) $\frac{2.303}{5.48 \times 10^{-14}} \log \frac{2}{3}$

30. The molecule (ion) having one unpaired electron is

(a) NO (b) CO (c) CN^- (d) O_2

31. Oxygen has an oxidation state of + 2 in

(a) H_2O_2 (b) H_2O
(c) F_2O (d) SO_2

32. Find the pH value of the mixture containing 50 cc M HCl and 30 cc M NaOH solution, assuming both to be completely ionised.

(a) 0.7051 (b) 0.6021
(c) 10.051 (d) 8.052

33. C_p / C_V for noble gases is

(a) 1.66 (b) 1.43 (c) 1.80 (d) 1.33

34. Carbylamine test is performed in alcholic KOH by heating a mixture of

(a) chloroform and silver powder
(b) trihalogenated methane and a primary amine
(c) an alkyl halide and a primary amine
(d) an alkyl cyanide and a primary amine

35. Relative acidity of the following is in the order

(a) $RCOOH > H_2CO_3 > C_6H_5OH > H_2O > ROH$
(b) $RCOOH > ROH > H_2CO_3 > C_6H_5OH > H_2O$
(c) $ROH > RCOOH > H_2CO_3 > C_6H_5OH > H_2O$
(d) $RCOOH > C_6H_5OH > ROH > H_2CO_3 > H_2O$

36. The compound which reacts fastest with Lucas reagent at room temperature is

(a) butan-1-ol (b) butan-2-ol
(c) 2-methyl propan-1-ol (d) 2-methyl propan-2-ol

37. $HCHO \xrightarrow{\text{Reduction}} A \xrightarrow{\text{P and } I_2} B \xrightarrow{\text{KCN}} C \xrightarrow{\text{Hydrolysis}} D$. '*D*' is

(a) acetic acid (b) ethylamine
(c) acetamide (d) None of these

38. The conjugate base of the acid H_2S is

(a) HS^- (b) S^{2-}
(c) S (d) None of these

39. Which one of the following is not coloured ?

(a) Cu^{2+} (b) Cu^+ (c) Ni^{2+} (d) Fe^{3+}

40. Glycine can be obtained from formaldehyde by

(a) Streker synthesis
(b) Williamson's synthesis
(c) Phthalimide synthesis
(d) Coupling reaction

41. The rate of S_N2 reaction is maximum when the solvent is

(a) CH_3OH (b) H_2O
(c) DMSO (d) benzene

42. The effective atomic number of central Cu (at. no. 29) metal in $[Cu(NH_3)_4]\,SO_4$ is

(a) 29 (b) 30
(c) 35 (d) 36

43. The reaction is spontaneous if the cell potential is

(a) positive (b) negative
(c) zero (d) infinite

44. In a face centred cubic lattice a unit cell is shared equally by how many unit cells ?

(a) 4 (b) 2
(c) 6 (d) 8

45. Which of the following will show geometrical isomerism ?

(a) 1-butene
(b) 1, 2-dibromobutene
(c) Propene
(d) *iso*-butylene

46. Which amino acids are essential building units of proteins ?

(a) α-amino acids
(b) β-amino acids
(c) γ-amino acids
(d) None of the above

47. $CH_2 = CH_2 + CH_2N_2 \longrightarrow \Delta$

Intermediate in the above reaction is

(a) $\overset{+}{C}H_2—\overset{-}{C}H_2$ (b) $CH_2CH_2\overset{+}{C}H_2\overset{-}{N}_2$
(c) $\overset{-}{C}H_2CH_2\overset{+}{C}H_2$ (d) $:CH_2$

48. The first compound of noble gas prepared by Bartlett was

(a) $XeOF_4$ (b) $Xe^+[PtF_6]^-$
(c) XeF_4 (d) XeF_6

49. $K_2Cr_2O_7$ on heating with aqueous caustic soda gives

(a) CrO_4^{2-} (b) $Cr(OH)_2$
(c) $Cr(OH)_3$ (d) $Cr_2O_7^{2-}$

50. The energy of second Bohr orbit of the hydrogen atom is -328 kJ mol^{-1}, hence the energy of fourth Bohr orbit would be

(a) -41 kJ mol^{-1} (b) -1312 kJ mol^{-1}
(c) -164 kJ mol^{-1} (d) -82 kJ mol^{-1}

Mathematics

1. If the roots $x^2 + ax + 9 = 0$ are complex, then

(a) $a < 6$ (b) $a < -6$
(c) $|a| < 6$ (d) $|a| > 6$

2. If the projection of **PQ** on OX, OY, OZ are respectively 12, 3 and 4, then the magnitude of **PQ** is

(a) 169 (b) 19 (c) 13 (d) 144

3. In the set Q^+ of all positive rational numbers, the operation * is defined by the formula $a * b = \frac{ab}{6}$. Then, the inverse of 9 with respect to * is

(a) 4 (b) 3 (c) $\frac{1}{9}$ (d) $\frac{1}{3}$

4. $\int_0^1 \frac{d}{dx}\left[\sin^{-1}\left(\frac{2x}{1+x^2}\right)\right] dx$ is equal to

(a) 0 (b) π
(c) $\frac{\pi}{2}$ (d) $\frac{\pi}{4}$

5. If $4\sin^{-1} x + \cos^{-1} x = \pi$, then x is equal to

(a) $\frac{1}{2}$ (b) 2 (c) 1 (d) $\frac{1}{3}$

6. $\sum_{n=1}^{\infty} \frac{(-1)^n}{n+1}$ is equal to

(a) e^{-1} (b) $\log 2 - 1$
(c) 1 (d) 0

7. If $\tan^{-1} x + \tan^{-1} y + \tan^{-1} z = \pi$, then the value of $x + y + z - xyz$ is

(a) 1 (b) 0
(c) -1 (d) $\frac{1}{2}$

8. $(p \wedge \sim q) \wedge (\sim p \vee q)$ is

(a) a contradiction (b) a tautology
(c) either (a) or (b) (d) neither (a) nor (b)

9. The coefficient of x in the expansion of $(1 + x + x^2 + x^3)^{-3}$ is

(a) 6 (b) 9
(c) 5 (d) -3

10. If $f(x + y) = f(x) \cdot f(y)$, $f(3) = 3$, $f'(0) = 11$, then $f'(3)$ is equal to

(a) $11 \cdot e^{33}$ (b) 33
(c) 11 (d) log 33

11. The line $\frac{x}{a} - \frac{y}{b} = 1$ cuts the x-axis at P. The equation of the line through P perpendicular to the given line is

(a) $x + y = ab$ (b) $x + y = a + b$
(c) $ax + by = a^2$ (d) $bx + ay = b^2$

12. $\cos^{-1} x + \cos^{-1} y = 2\pi$, then the value of $\sin^{-1} x + \sin^{-1} y$ is equal to

(a) 0 (b) π
(c) $-\pi$ (d) None of these

13. If $\log_3 2, \log_3(2^x - 5)$ and $\log_3\left(2^x - \frac{7}{2}\right)$ are in AP, the value of x is

(a) 2 (b) 3 (c) 0 (d) $\frac{1}{3}$

14. $\int_0^1 x(1-x)^{12}\, dx$ is equal to

(a) $\frac{1}{132}$
(b) $\frac{1}{156}$
(c) $\frac{1}{182}$
(d) None of the above

15. The middle term of $\left(\sqrt{x} - \frac{1}{\sqrt{x}}\right)^6$ is

(a) -20 (b) -1
(c) 1 (d) None of these

16. If every element of a group G is its own inverse, then G is

(a) finite (b) infinite
(c) not abelian (d) abelian

17. If $P(n)$ is a statement such that $P(3)$ is true. Assuming $P(k)$ is true $\Rightarrow P(k+1)$ is true for all $k \geq 3$, then $P(n)$ is true

(a) for all n (b) for $n \geq 3$
(c) for $n \geq 4$ (d) None of these

18. If $\Delta = \begin{vmatrix} 1 & a & a^2 \\ 1 & b & b^2 \\ 1 & c & c^2 \end{vmatrix} = k(a-b)(b-c)(c-a)$,

then k is equal to

(a) -2 (b) 1 (c) 2 (d) abc

19. The number of ways in which a team of 11 player can be selected from 22 players including 2 of them and excluding 4 of them is

(a) ${}^{16}C_{11}$ (b) ${}^{16}C_5$ (c) ${}^{16}C_9$ (d) ${}^{20}C_8$

20. If $\Delta_1 = \begin{vmatrix} x & a & b \\ b & x & a \\ a & b & x \end{vmatrix}$ and $\Delta_2 = \begin{vmatrix} x & b \\ a & x \end{vmatrix}$ are the given determinants, then

(a) $\Delta_1 = 3(\Delta_2)^2$ (b) $\left(\frac{d}{dx}\right)(\Delta_1) = 3\Delta_2$

(c) $\left(\frac{d}{dx}\right)(\Delta_1) = 3(\Delta_2)^2$ (d) $\Delta_1 = 3(\Delta_2)^{3/2}$

21. $\int_{-1/2}^{1/2} \frac{dx}{(1-x^2)^{1/2}}$ is equal to

(a) $\frac{\pi}{3}$ (b) $\frac{\pi}{4}$ (c) $\frac{\pi}{2}$ (d) 0

22. $\lim_{n \to \infty} \left[\frac{1}{n+1} + \frac{1}{n+2} + \dots + \frac{1}{6n}\right]$

(a) $\log 2$ (b) $\log(1+\sqrt{5})$

(c) $\log 6$ (d) 0

23. The degree of the differential equation

$$\left[5 + \left(\frac{dy}{dx}\right)^2\right]^{\frac{5}{3}} = x^5\left[\frac{d^2y}{dx^2}\right] \text{ is}$$

(a) 4 (b) 3 (c) 5 (d) 10

24. The difference of the focal distance of any point on the hyperbola is equal to its

(a) latusrectum

(b) eccentricity

(c) length of the transverse axis

(d) half the length of the transverse axis

25. If α, β, γ are the angles which a half ray makes with the positive direction of the axes, then $\sin^2\alpha + \sin^2\beta + \sin^2\gamma$ is equal to

(a) 1 (b) 2 (c) 0 (d) -1

26. If $f(x) = \begin{cases} 3x^2 + 12x - 1, & -1 \le x \le 2 \\ 37 - x, & 2 \le x \le 3 \end{cases}$, then

(a) $f(x)$ is decreasing on $[-1, 2]$

(b) $f'(2)$ does not exist

(c) $f(x)$ has the maximum value at $x = 2$

(d) None of the above

27. $\mathbf{a} = \frac{1}{7}(2\hat{\mathbf{i}} + 3\hat{\mathbf{j}} + 6\hat{\mathbf{k}}), \mathbf{b} = \frac{1}{7}(3\hat{\mathbf{i}} - \lambda\hat{\mathbf{j}} + 2\hat{\mathbf{k}})$. If $\mathbf{a}$ and $\mathbf{b}$ are mutually perpendicular, then value of λ is

(a) 2 (b) -1

(c) 6 (d) -6

28. The locus of the point of intersection of two perpendicular tangents to a circle is called

(a) point circle (b) circumcircle

(c) director circle (d) auxiliary circle

29. The area bounded by the curve $y = x^4 - 2x^3 + x^2 + 3$ with x-axis and ordinates corresponding to the minima of y is

(a) 1 sq unit (b) $\frac{91}{30}$ sq unit

(c) $\frac{30}{9}$ sq unit (d) 4 sq unit

30. $\int (e^{a \log x} + e^{x \log a})\, dx$ is equal to

(a) $\frac{x^{a+1}}{a+1} + c$

(b) $\frac{x^{a+1}}{a+1} + \frac{a^x}{\log a} + c$

(c) $x^{a+1} + a^x + c$

(d) $\frac{x^{a+1}}{a-1} + \frac{\log a}{a^x} + c$

31. $\lim_{n \to \infty} \left(\frac{1}{\sqrt{4n^2-1}} + \frac{1}{\sqrt{4n^2-2^2}} + \dots + \frac{1}{\sqrt{3n^2}}\right)$ is equal to

(a) 0 (b) 1

(c) $\frac{\pi}{3}$ (d) $\frac{\pi}{6}$

32. $\int \frac{dx}{x^2 + 4x + 13}$ is equal to

(a) $\log(x^2 + 4x + 130) + c$ (b) $\frac{1}{3}\tan^{-1}\left(\frac{x+2}{3}\right) + c$

(c) $\log(2x + 4) + c$ (d) $\frac{1}{x^2 + 4x + 13} + c$

33. Consider the hyperbola $\frac{x^2}{a^2}-\frac{y^2}{b^2}=1$ the area of the triangle formed by the asymptotes and the tangent drawn to it at $(a, 0)$ is

(a) $\frac{1}{2}ab$ (b) ab (c) $2ab$ (d) $4ab$

34. The differential equation of the family of lines passing through the origin is

(a) $x\frac{dy}{dx}+y=0$ (b) $x+\frac{dy}{dx}=0$
(c) $x\frac{dy}{dx}-y=0$ (d) $\frac{dy}{dx}=x$

35. If $P=\begin{bmatrix} i & 0 & -i \\ 0 & -i & i \\ -i & i & 0 \end{bmatrix}$ and $Q=\begin{bmatrix} -i & i \\ 0 & 0 \\ i & -i \end{bmatrix}$, then PQ is equal to

(a) $\begin{bmatrix} -2 & 2 \\ 1 & -1 \\ 1 & -1 \end{bmatrix}$ (b) $\begin{bmatrix} 2 & -2 \\ -1 & 1 \\ -1 & 1 \end{bmatrix}$

(c) $\begin{bmatrix} 2 & -2 \\ -1 & 1 \end{bmatrix}$ (d) $\begin{bmatrix} 1 & 0 & 0 \\ 0 & 1 & 0 \\ 0 & 0 & 1 \end{bmatrix}$

36. If $x=\sin^{-1}(3t-4t^3)$ and $y=\cos^{-1}(\sqrt{1-t^2})$, then $\frac{dy}{dx}$ is equal to

(a) 1/3 (b) 2/5
(c) 3/2 (d) 2/3

37. If $x^2+px+q=0$ is the quadratic equation whose roots are $a-2$ and $b-2$, where a and b are the roots of $x^2-3x+1=0$, then

(a) $p=1, q=5$ (b) $p=5, q=1$
(c) $p=1, q=1$ (d) $p=1, q=-1$

38. If the position vectors of A, B, C are respectively $\hat{\mathbf{i}}-2\hat{\mathbf{j}}+\hat{\mathbf{k}}$, $2\hat{\mathbf{i}}+3\hat{\mathbf{j}}$ and $2\hat{\mathbf{i}}-\hat{\mathbf{k}}$, then the projection of **AB** on **BC** is equal to

(a) $\frac{-14}{\sqrt{10}}$ (b) $\sqrt{5}$ (c) $\sqrt{7}$ (d) 2

39. The area between $y^2+4x-8=0$, the x-axis and the line $x=1$ is

(a) $\frac{4}{3}$ sq unit (b) $\frac{2}{3}$ sq unit
(c) $\frac{1}{3}$ sq unit (d) 1 sq unit

40. The solution of the equation $(2y-1)\,dx-(2x+3)\,dy=0$ is

(a) $\frac{2x-1}{2y+3}=c$ (b) $\frac{2x+3}{2y-1}=c$
(c) $\frac{2x-1}{2y-1}=c$ (d) $\frac{2y+1}{2x-3}=c$

41. If the nth term of the geometric progression, $5, -\frac{5}{2}, \frac{5}{4}, -\frac{5}{8}, \ldots$ is $\frac{5}{1024}$, then the value of n is

(a) 11 (b) 10 (c) 9 (d) 4

42. If $77x \equiv 88 \pmod 5$, then x is equal to

(a) 4 (b) 61 (c) 59 (d) 48

43. Negation of "Paris is in France and London is in England" is

(a) Paris is in England and London is in France.
(b) Paris is not in France or London is not in England.
(c) Paris is in England or London is in France.
(d) None of the above

44. If $C(2n, 3) : C(n, 2) = 44 : 3$, then n is equal to

(a) 6 (b) 7 (c) 5 (d) 8

45. If $f(x)=\frac{3x+\tan^2 x}{x}$ is continuous at $x=0$, then $f(0)$ is equal to

(a) 3 (b) 2 (c) 4 (d) 0

46. $0.2+0.22+0.222+\ldots$ to n terms is equal to

(a) $\left(\frac{2}{9}\right)-\left(\frac{2}{81}\right)(1-10^{-n})$
(b) $n-\left(\frac{1}{9}\right)(1-10^{-n})$
(c) $\left(\frac{2}{9}\right)\left[n-\left(\frac{1}{9}\right)(1-10^{-n})\right]$
(d) $\left(\frac{2}{9}\right)$

47. If ω is an imaginary cube root of 1, then $(1+\omega-\omega^2)^5+(1-\omega+\omega^2)^5$ is equal to

(a) 16 (b) 8 (c) 9 (d) 32

48. The equation of the circle circumscribing the triangle formed by the lines $x+y=6$, $2x+y=4$ and $x+2y=5$ is

(a) $x^2+y^2+17x+19y-50=0$
(b) $x^2+y^2-17x-19y-50=0$
(c) $x^2+y^2+17x-19y-50=0$
(d) $x^2+y^2-17x-19y+50=0$

49. $\lim_{x \to 0} \frac{\sin 3x - \sin x}{\sin x}$ is

(a) −2 (b) 2
(c) 0 (d) None of these

50. If the product of roots of the equation $mx^2 + 6x + (2m - 1) = 0$ is -1, then the value of m is

(a) 1 (b) −1 (c) $\frac{1}{3}$ (d) $\frac{-1}{3}$

51. $\int_0^{2a} \frac{f(x)}{f(x) + f(2a - x)} dx$ is equal to

(a) a (b) $-a$ (c) 1 (d) 0

52. $\int \frac{\sqrt{\tan x}}{\sin x \cos x} dx$ is equal to

(a) $2 \tan x + c$ (b) $\sqrt{\cot x} + c$
(c) $2\sqrt{\tan x} + c$ (d) $\tan^2 x + c$

53. The equation of the tangent to the curve $(1 + x^2) y = 2 - x$, where it crosses the x-axis, is

(a) $x + 5y = 2$ (b) $x - 5y = 2$
(c) $5x - y = 2$ (d) $5x + y - 2 = 0$

54. $\lim_{x \to 0} \frac{1 - \cos mx}{1 - \cos nx}$ is equal to

(a) $\frac{m}{n}$ (b) $\frac{m^2}{n^2}$ (c) 0 (d) $\frac{n^2}{m^2}$

55. If **a**, **b**, **c** are any three mutually perpendicular vectors of equal magnitude a, then $|\mathbf{a} + \mathbf{b} + \mathbf{c}|$

(a) a (b) $\sqrt{2}a$ (c) $\sqrt{3}a$ (d) $2a$

56. For what values of n is the graph K_n Eulerian ?

(a) Odd values of n
(b) Even values of n
(c) Both odd and even values of n
(d) None of the above

57. When 5^{20} is divided by 7, the remainder is

(a) 1 (b) 3 (c) 4 (d) 6

58. If $a * b$ denote the bigger among a and b and $a \cdot b = (a * b) + 3$, then $4 \cdot 7$ is equal to

(a) 4 (b) 31 (c) 10 (d) 8

59. In a tree on a vertices there is exactly one vertex with degree 2 and remaining vertices are of degree either 1 or 3. Then the number of pendant vertices is

(a) 8 (b) 5 (c) 4 (d) 6

60. $\frac{1}{\log_{25} 10} + \frac{1}{\log_4 10} + \frac{1}{\log_{\sqrt{2}} 10} + \frac{1}{\log_{\sqrt{5}} 10}$ is equal to

(a) 3/2 (b) 2 (c) 3 (d) 5/2

61. If the angle between two lines represented by $2x^2 + 5xy + 3y^2 + 7y + 4 = 0$ is $\tan^{-1} m$, then m is equal to

(a) 1/5 (b) 1 (c) 7/5 (d) 7

62. The foot of the perpendicular from $(-2, 3)$ to the line $2x - y - 3 = 0$

(a) $(-2, 3)$ (b) $(2, 1)$ (c) $(3, 2)$ (d) $(1, 2)$

63. $\int_{-\frac{\pi}{2}}^{\frac{\pi}{2}} |\sin x| dx$ is

(a) 2 (b) 0 (c) $\frac{\pi}{2}$ (d) 1

64. If $\begin{bmatrix} 2 + x & 3 & 4 \\ 1 & -1 & 2 \\ x & 1 & -5 \end{bmatrix}$ is a singular matrix, then x is

(a) $\frac{13}{25}$ (b) $-\frac{25}{13}$ (c) $\frac{5}{13}$ (d) $\frac{25}{13}$

65. In triangle ABC, $a(b^2 + c^2)\cos A + b(c^2 + a^2)\cos B + c(a^2 + b^2)\cos C$ is equal to

(a) abc (b) $2\,abc$ (c) $3abc$ (d) $4abc$

66. The function $y = a(1 - \cos x)$ is maximum when x is equal to

(a) π (b) $\frac{\pi}{2}$ (c) $-\frac{\pi}{2}$ (d) $-\frac{\pi}{6}$

67. The volume of the parallelopiped whose sides are given by $\mathbf{OA} = 2\hat{\mathbf{i}} - 3\hat{\mathbf{j}}$, $\mathbf{OB} = \hat{\mathbf{i}} + \hat{\mathbf{j}} - \hat{\mathbf{k}}$, $\mathbf{OC} = 3\hat{\mathbf{i}} - \hat{\mathbf{k}}$

(a) $\frac{4}{13}$ cu unit (b) 4 cu unit
(c) $\frac{2}{7}$ cu unit (d) None of these

68. Is it possible for wheel W_n $(n \geq 3)$ to be bipartite?

(a) No (b) Yes
(c) Do not say (d) None of these

69. The GCD of 364 and 462 is

(a) 3 (b) 11 (c) 14 (d) 7

70. The system $x + 4y - 2z = 3$, $3x + y + 5z = 7$, $2x + 3y + z = 5$ has

(a) infinite number of solutions
(b) unique solution
(c) trivial solution
(d) no solution

English & General Aptitude

Directions (Q. 1–4) : *Read the following passage carefully and answer the questions given below it.*

Most employees decide their own working hours, set production quotas, improving product and processes, are responsible for their own quality and for approval of leadership appointments. Everyone votes on major corporate decisions and on how to split the profits.

As confidence in its novel approach has grown, Semco has happily abolished a lot more of the conventions by which businesses are usually run. No secretaries, receptionists or personal assistants. Reserved parking spaces and dining rooms, dress codes and almost all rules have gone, including those for travel and expenses.

1. From the passage it is clear that the novel approach referred to is

(a) aristocratic (b) autocratic
(c) democratic (d) bureaucratic

2. The employees referred to are

(a) employees of a public sector undertaking
(b) employees of a private company
(c) employees of essential services
(d) government employees

3. In the given passage, 'Semco' is the name of

(a) a business establishment
(b) a brand of the product
(c) a novel approach to things and affairs
(d) the leader of secretaries, receptionists and personal assistants

4. The term 'leadership appointments' has been used in this passage to signify

(a) appointing officers-in-charge of various units/sections
(b) selecting political leaders
(c) selecting company directors
(d) choosing trade union leaders

Directions (Q. 5–8) : *In each of the following sentences, find out the part which has an error. If there is no error, your answer is (d).*

5. It cannot be forecasted (a)/how society will emerge (b)/a generation hence. (c)/No error (d).

6. The Prime Minister has said that India would not have spent so much on defence (a)/if some of the neighbouring countries (b)/adopted the policy of restricting defence expenditure (c)/No error (d).

7. I've been to a few of his lectures, (a)/but understood little of (b)/what he has said. (c)/No error (d).

8. More leisure, as well as an abundance of goods, (a)/are attainable (b)/through automation (c)/No error (d).

Directions (Q. 9–10) : *In each of the following questions, choose the alternative which best expresses the meaning of the word given in capital letters.*

9. COMPETENCE

(a) Efficiency
(b) Competition
(c) Ability
(d) Compensation

10. ADJUNCT

(a) Adaptation (b) Addition
(c) Decree (d) Oath

Directions (Q. 11–12) : *Choose the correct sequence of the parts P, Q, R and S to make a proper sentence.*

11. (1) There are

(P) laid on school games
(Q) who support the emphasis
(R) still those devotees of sports
(S) and for whom sports is
(6) a kind of religion.

The proper sequence should be

(a) PQRS (b) PSQR
(c) RPSQ (d) RQPS

12. (1) The fact that

(P) go to the police (Q) did not let him
(R) to speak the truth (S) he was a murderer
(6) about the theft.

The proper sequence should be

(a) PRSQ (b) QSPR
(c) RQSP (d) SQPR

Directions (Q. 13–15) : *In each of the following questions, choose the most suitable alternative to fill in the blank.*

13. True happiness consists .... giving pleasure to others.

(a) for (b) by
(c) in (d) of

14. The meeting was presided .... by the Prime Minister.

(a) on (b) over
(c) in (d) upon

15. He is definitely ......

(a) wrongly (b) in wrong
(c) in a wrong (d) in the wrong

Directions (Q. 16–25) : *In the following questions, choose the option which shows common feature in the relationship given in each question.*

16. Ganga : Narmada : Tapti

(a) They are name of rivers
(b) They are dance form of India
(c) They are the currency of different countries
(d) They are the parliaments name of different countries

17. Leap : Frisk : Trot

(a) They are youngone of animals
(b) They are Indian monuments
(c) They are movement of animals
(d) They are the name of famous zoological parks

18. Pen : Rubber : Pencil

(a) They are goods for all purpose
(b) They are stationery goods
(c) They are famous Indian sites
(d) They are sports terms

19. Sunday : Monday : Saturday

(a) They are name of the years
(b) They are name of the months
(c) They are name of the week days
(d) They are name of the rivers

20. Sale : Tale : Male

(a) They have 2 vowels
(b) They have 4 consonants
(c) The words have no vowels
(d) The words have no consonants

21. Peso : Won : Taka

(a) They are famous monuments
(b) They are name of the young ones of animals
(c) They are synonymous words
(d) None of the above

22. Jaipur : Bengaluru : Mumbai

(a) They are the cities in Rajasthan
(b) They are the famous business cities of India
(c) They are the three biggest villages of India
(d) They are the capitals of Indian states

23. Colombo : Kathmandu : Havana

(a) They are African cities
(b) They are European cties
(c) They are capitals of countries
(d) They are sports cities

24. Squeak : Hiss : Howl

(a) They are names of animals
(b) They are currencies
(c) They are biggest animals on earth
(d) They are sound produced by animals

25. Kathak : Bharatnatyam : Odissi

(a) They are the name of music instruments
(b) They are the classical dance forms of India
(c) They are the folk dance forms of India
(d) They are the names of Indian tribes

Directions (Q. 26 – 30) : In the following questions, choose the group of words that shows the same relationship as given at the top of every question.

26. The first regular session of United Nations Organisation was held in

(a) Jan, 1946 (b) Jun, 1946
(c) Jan, 1947 (d) Jun, 1947

27. Tbilisi is the capital of
(a) Finland
(b) Georgia
(c) Germany
(d) Hungary

28. Which defence production undertakings is engaged in the design, development and manufacturing of electronic equipments ?
(a) HAL (b) BEL
(c) BEML (d) BDL

29. Shanti Swaroop Bhatnagar Awards are given in the field of
(a) sports players
(b) scientists for their exceptional performance
(c) medical sciences
(d) music

30. Central Electronics Engineering Research Institute is located at
(a) Durgapur (b) Karaikudi
(c) Pilani (d) Roorkee

Answers

Physics

1.	(b)	2.	(c)	3.	(a)	4.	(d)	5.	(d)	6.	(d)	7.	(d)	8.	(a)	9.	(a)	10.	(d)
11.	(b)	12.	(b)	13.	(a)	14.	(b)	15.	(d)	16.	(c)	17.	(a)	18.	(a)	19.	(b)	20.	(b)
21.	(d)	22.	(d)	23.	(c)	24.	(c)	25.	(a)	26.	(d)	27.	(a)	28.	(a)	29.	(a)	30.	(a)
31.	(d)	32.	(d)	33.	(a)	34.	(a)	35.	(b)	36.	(d)	37.	(d)	38.	(b)	39.	(b)	40.	(c)
41.	(b)	42.	(c)	43.	(c)	44.	(a)	45.	(d)	46.	(b)	47.	(a)	48.	(b)	49.	(c)	50.	(c)

Chemistry

1.	(b)	2.	(b)	3.	(c)	4.	(a)	5.	(d)	6.	(d)	7.	(b)	8.	(a)	9.	(a)	10.	(a)
11.	(b)	12.	(b)	13.	(c)	14.	(b)	15.	(a)	16.	(d)	17.	(b)	18.	(c)	19.	(a)	20.	(a)
21.	(d)	22.	(d)	23.	(c)	24.	(c)	25.	(a)	26.	(a)	27.	(d)	28.	(c)	29.	(a)	30.	(a)
31.	(c)	32.	(b)	33.	(a)	34.	(b)	35.	(a)	36.	(d)	37.	(a)	38.	(a)	39.	(b)	40.	(a)
41.	(a)	42.	(c)	43.	(a)	44.	(c)	45.	(b)	46.	(a)	47.	(d)	48.	(b)	49.	(a)	50.	(d)

Mathematics

1.	(c)	2.	(c)	3.	(a)	4.	(c)	5.	(a)	6.	(b)	7.	(b)	8.	(a)	9.	(d)	10.	(a)
11.	(c)	12.	(c)	13.	(b)	14.	(c)	15.	(a)	16.	(d)	17.	(b)	18.	(b)	19.	(c)	20.	(b)
21.	(a)	22.	(c)	23.	(b)	24.	(c)	25.	(b)	26.	(b)	27.	(c)	28.	(c)	29.	(b)	30.	(b)
31.	(d)	32.	(b)	33.	(b)	34.	(c)	35.	(b)	36.	(a)	37.	(d)	38.	(a)	39.	(a)	40.	(b)
41.	(a)	42.	(a)	43.	(b)	44.	(a)	45.	(a)	46.	(c)	47.	(d)	48.	(d)	49.	(b)	50.	(c)
51.	(a)	52.	(c)	53.	(a)	54.	(b)	55.	(c)	56.	(a)	57.	(a)	58.	(c)	59.	(b)	60.	(d)
61.	(a)	62.	(b)	63.	(a)	64.	(b)	65.	(c)	66.	(a)	67.	(b)	68.	(a)	69.	(c)	70.	(d)

English & General Aptitude

1.	(c)	2.	(b)	3.	(d)	4.	(d)	5.	(a)	6.	(c)	7.	(c)	8.	(b)	9.	(c)	10.	(b)
11.	(d)	12.	(d)	13.	(c)	14.	(b)	15.	(d)	16.	(a)	17.	(c)	18.	(b)	19.	(c)	20.	(a)
21.	(d)	22.	(d)	23.	(c)	24.	(d)	25.	(b)	26.	(a)	27.	(b)	28.	(b)	29.	(b)	30.	(c)

Answer *with* Solutions

Physics

1. *(b)* Using red light ($\lambda = 6600$ Å) 60 fringes are seen.

Hence, range of field of view is

$$60 \times w = 60 \times \frac{D\lambda}{d}$$

Using light of wavelength λ', n fringes are seen, then

$$\Rightarrow \quad 60 \times \frac{D\lambda}{d} = n \times \frac{D\lambda'}{d}$$

$$60 \times \lambda = n \times \lambda'$$

$$\Rightarrow \quad n = 60 \times \frac{\lambda}{\lambda'} = 60 \times \frac{6600}{4400} = 90$$

2. *(c)* Let capacitance of each capacitor is C. Then equivalent capacitance in series is

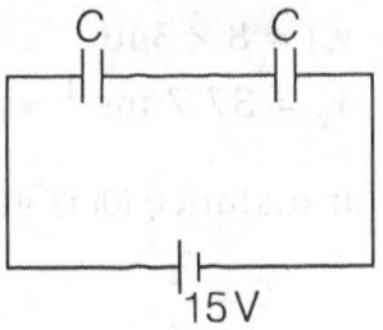

$$\frac{1}{C'} = \frac{1}{C} + \frac{1}{C} = \frac{2}{C}$$

$$\Rightarrow \quad C' = \frac{C}{2}$$

Charge $\quad Q = C'V = \frac{C}{2} \times 15 \quad \ldots\text{(i)}$

When filled with dielectric

$$C_1 = 4C, C_2 = C$$

$$\frac{1}{C'} = \frac{1}{4C} + \frac{1}{C} = \frac{5}{4C}$$

$$\Rightarrow \quad C' = \frac{4C}{5}$$

Since, charge is conserved

$Q = C'V' = \frac{4C}{5} V' \quad \ldots\text{(ii)}$

From Eqs. (i) and (ii), we get

$$\frac{C}{2} \times 15 = \frac{4C}{5} V'$$

$$\Rightarrow \quad V' = \frac{15 \times 5}{4 \times 2} = 9.4\ \text{V} \approx 10\ \text{V}$$

3. *(a)* The speed of transverse wave

$$v = \sqrt{\frac{T}{m}}$$

Given, $T = 20$ N, $\frac{M}{l} = \frac{d \times Al}{l} = d \times A$

$$\therefore \quad v = \sqrt{\frac{20 \times 10^{-3}}{7.5 \times 0.2 \times (10^{-3})^2}}$$

$$v \approx 116\ \text{ms}^{-1}$$

4. *(d)* Common acceleration

$$a = \frac{F}{m_1 + m_2 + m_3}$$

$$a = \frac{40}{10 + 6 + 4} = 2\ \text{m/s}^2$$

Equation of motion of m_3 is

$$T_3 - T_2 = m_3 a$$

$$40 - T_2 = 4 \times 2 \Rightarrow T_2 = 32\ \text{N}$$

5. *(d)* Let n cells be in series and m in parallel, then

$$\frac{nE}{R + nr} = \frac{E}{R + \frac{r}{m}}$$

$$\Rightarrow \quad n\left[R + \frac{r}{m}\right] = R + nr$$

$$\Rightarrow \quad nRm + nr = Rm + mnr$$

$$\Rightarrow \quad 6 + 2r = 3 + 4r$$

$$\Rightarrow \quad 2r = 3$$

$$\Rightarrow \quad R = 1.5\ \Omega$$

6. *(d)* From Coulomb's law the force (F) between two charges is

$$F = \frac{1}{4\pi\varepsilon_0 k} \frac{q^2}{r^2}$$

First case : $\quad k = 1$

$F = \frac{1}{4\pi\varepsilon_0} \cdot \frac{q^2}{r^2} \quad \ldots\text{(i)}$

Second case : $\quad k = 4$

$F' = \frac{1}{4\pi\varepsilon_0 \times 4} \cdot \frac{q^2}{r^2} \quad \ldots\text{(ii)}$

Dividing Eq. (i) by Eq. (ii), we have

$$\frac{F}{F'} = 4 \Rightarrow \quad F' = \frac{F}{4}$$

7. *(d)* The height (h) to which water rises in a capillary tube is given by

$h = \frac{2T\cos\theta}{r\rho g}$ where θ is angle of contact, r the radius, ρ the density and g acceleration due to gravity.

When lift moves down with constant acceleration, height is less than h, because effective value of acceleration due to gravity increases hence h decreases.

8. *(a)* From Stefan's law

$$E = \sigma T^4 A$$

Given,

$$T = 727°C = 727 + 273 = 1000\ K,$$
$$A = 5 \times 10^{-4}\ m^2$$

$\therefore$ $$\text{Energy} = 5.67 \times 10^{-8}\ (1000)^4\ (5 \times 10^{-4})\ 60$$

$\therefore$ $$E = 1.7 \times 10^3\ J$$

9. *(a)* NAND gate is obtained when the output of AND gate is made as the input of NOT gate Boolean expression for NAND gate is

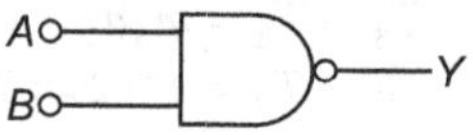

$$Y = \overline{A . B}$$

A	B	Y
0	0	1
1	0	1
0	1	1
1	1	0

10. *(d)* Torque on dipole is

$$\tau = pE \sin\theta$$
$$\tau = q \times 2l E \sin\theta$$

$\Rightarrow$ $$q = \frac{\tau}{2El \sin\theta}$$

Given, $\tau = 4$ Nm, $E = 2 \times 10^5\ NC^{-1}$,

$$l = 2 \times 10^{-2}\ m, \theta = 30°$$

$\therefore$ $$q = \frac{4}{2 \times 2 \times 10^5 \times 2 \times 10^{-2} \times \sin 30°}$$

or $q = 2$ mC

11. *(b)* The velocity (v) is given by

$$v = f\lambda$$

$\Rightarrow$ $$f = \frac{v}{\lambda}$$

Number of beats = difference in frequencies

$\therefore$ $$f_1 - f_2 = v\left(\frac{1}{\lambda_1} - \frac{1}{\lambda_2}\right)$$

$$\frac{10}{3} = v\left(\frac{1}{1} - \frac{1}{1.01}\right)$$

$$\frac{10}{3} = v\left(\frac{0.01}{1.01}\right)$$

$\Rightarrow$ $$v = \frac{10 \times 1.01}{3 \times 0.01} = 336.7\ m/s$$

12. *(b)* The relation between α and β is

$$\beta = \frac{\alpha}{1 - \alpha}$$

Putting this value in the given equation, we have

$$\frac{\beta - \alpha}{\alpha\beta} = \frac{\frac{\alpha}{1-\alpha} - \alpha}{\alpha \cdot \frac{\alpha}{1-\alpha}} = \frac{\alpha^2}{\alpha^2} = 1$$

13. *(a)* From Doppler's effect, perceived frequency

$$f' = f\left(\frac{v - v_o}{v - v_s}\right)$$

$$\frac{9}{8} = \frac{340}{340 - v_s}$$

$\Rightarrow$ $$9\ (340 - v_s) = 8 \times 340$$

$\therefore$ $$v_s = 37.7\ ms^{-1} \approx 40\ ms^{-1}$$

14. *(b)* The angular distance (θ) is given by

$$\theta = \frac{\lambda}{d}$$

Given, $\theta = 2° = \frac{\pi}{180} \times 2$, $\lambda = 6980$ Å

$$= 6980 \times 10^{-10}\ m$$

$\therefore$ $$d = \frac{\lambda}{\theta} = \frac{6980 \times 10^{-10} \times 180}{3.14 \times 2}$$

$\Rightarrow$ $$d \approx 2 \times 10^{-5}\ mm$$

15. *(d)* The resistance at temperature t is

$$R_t = R_0\ (1 + \alpha t)$$
$$R_t' = R_0\ (1 + \alpha t')$$

$\Rightarrow$ $$4 = 1 + 15\alpha \qquad \ldots(i)$$

$$5 = 1 + 37.5\alpha \qquad \ldots(ii)$$

$\therefore$ $$\frac{4}{5} = \frac{1 + 15\alpha}{1 + 37.5\alpha}$$

$\Rightarrow$ $$4\ (1 + 37.5\alpha) = 5\ (1 + 15\alpha)$$

$\Rightarrow$ $$75\alpha = 1 \Rightarrow \alpha = \frac{1}{75}\ °C^{-1}$$

16. *(c)* The relation between K and p is

$$p = \sqrt{2Km}$$

when increased by 25%, we have

$$p' = p + \frac{25}{100}p = \frac{5p}{4}$$

$$\Rightarrow \quad \frac{4p}{5p} = \sqrt{\frac{K_1}{K_2}}$$

$$\Rightarrow \quad \frac{K_1}{K_2} = \frac{16}{25}$$

Hence, percentage increase is

$$= \frac{25-16}{16} \times 100 = \frac{9}{16} \times 100 = 56.25\%$$

17. *(a)* The focal length of the combination is

$$\frac{1}{F} = \frac{1}{f_1} + \frac{1}{f_2}$$

Given, $f_1 = 50$ cm, $f_2 = 50$ cm

$$\therefore \quad \frac{1}{F} = \frac{1}{50} + \frac{1}{50} = \frac{2}{50}$$

$$\Rightarrow \quad F = \frac{50}{2} = 25 \text{ cm}$$

Object when placed at centre of curvature forms a real, inverted image of same size as object $= (2 \times 25 = 50 \text{ cm})$.

18. *(a)* $M_{21} = \dfrac{\mu_0 N_1 N_2 A_2}{l_1}$

$$\therefore \quad M_{21} = \frac{(4 \times 3.14 \times 10^{-7}) \times 1500 \times 100 \times \{3.14\,(2 \times 10^{-2})^2\}}{80 \times 10^{-4}}$$

$$M_{21} = 2.96 \times 10^{-4} \text{ H}$$

$$\Rightarrow \quad M_{12} = M_{21} = 2.96 \times 10^{-4} \text{ H}$$

19. *(b)* Refractive index, $\mu = \dfrac{\sin\left(\dfrac{A + \delta_m}{2}\right)}{\sin\left(\dfrac{A}{2}\right)}$

$$\Rightarrow \quad \sqrt{2} = \frac{\sin\left(\dfrac{60^\circ + \delta_m}{2}\right)}{\sin\left(\dfrac{60^\circ}{2}\right)}$$

$$\Rightarrow \quad \sqrt{2} \times \sin 30^\circ = \sin\left(\frac{60^\circ + \delta_m}{2}\right)$$

$$\Rightarrow \quad \sin 45^\circ = \sin\left(\frac{60^\circ + \delta_m}{2}\right)$$

$$\Rightarrow \quad \delta_m = 30^\circ$$

20. *(b)* Suppose resistance R is connected in series with voltmeter as shown. By Ohm's law

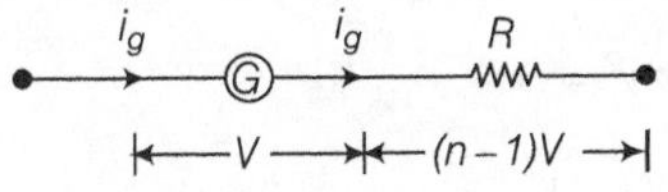

$$i_g R = (n-1)\,V$$

$$\Rightarrow \quad R = (n-1)\,G$$

where $\quad i_g = \dfrac{V}{G}$

21. *(d)* Fraction $= \dfrac{N}{N_0}$

$$= \left(\frac{1}{2}\right)^{\frac{6400}{1600}} = \left(\frac{1}{2}\right)^4 = \frac{1}{16}$$

22. *(d)* As we know $Q \propto T^4$

$$\Rightarrow \quad \frac{H_A}{H_B} = \left[\frac{273 + 727}{273 + 327}\right]^4 = \frac{625}{81}$$

23. *(c)* Heat emitted by copper = Heat gained by ice

$$mc\Delta\theta = m'L$$

$$\Rightarrow \quad m' = \frac{mc\Delta\theta}{L}$$

Given, $m = 2$ kg, $c = 400$ J-kg^{-1} C^{-1},

$\Delta\theta = 500$, $L = 3.5 \times 10^5$ J-kg^{-1}

$$\therefore \quad m' = \frac{2 \times 400 \times 500}{3.5 \times 10^5} = \frac{8}{7} \text{ kg}$$

24. *(c)* The resultant of tension and weight given acceleration to the man, hence

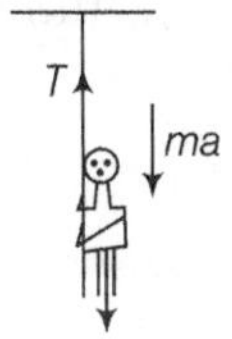

$$mg - T = ma$$

$$\Rightarrow \quad T = mg - ma = m\,(g - a)$$

Given, $\quad m = 100$ kg,

$g = 9.8$ m/s^2, $a = 1.8$ ms^{-2}.

$$\therefore \quad T = 100\,(9.8 - 1.8) = 800 \text{ N}$$

25. *(a)* $\quad F = \dfrac{dp}{dt} \Rightarrow p = F\,dt$

where p = momentum = mv

$$\therefore \quad mv = Fdt$$

$$\Rightarrow \quad v = \frac{Fdt}{m}$$

Given, $F = 20$ N, $dt = 2$ s, $m = 2$ kg

$$\therefore \quad v = \frac{20 \times 2}{2} = 20 \text{ m/s}^2$$

Work done = KE $= \dfrac{1}{2} \times 2 \times (20)^2 = 400$ J

26. *(d)* Force on a particle with charge q in electric field E is

$$F = qE = ma$$

$$\therefore \quad a = \frac{qE}{m}$$

here $\quad a = \dfrac{v}{t}$

$\therefore \quad \frac{v}{t} = \frac{qE}{m} \Rightarrow v = \frac{qEt}{m}$

Also, kinetic energy is due to velocity (v) is

$$K = \frac{1}{2}mv^2 = \frac{1}{2}m\left(\frac{qEt}{m}\right)^2 = \frac{q^2E^2t^2}{2m}$$

27. *(a)* $\quad V = \frac{1}{4\pi\varepsilon_0}\frac{q}{r}$

Given, $\frac{1}{4\pi\varepsilon_0} = 9 \times 10^9$, $V = +100$ volt,

$r = 10$ cm $= 0.10$ m.

$\therefore \quad 100 = (9 \times 10^9) \times \frac{q}{0.10}$

$\Rightarrow \quad q = \frac{100 \times 0.10}{9 \times 10^9} = 1.1 \times 10^{-9}$ C

28. *(a)* Internal resistance (r) is given by

$$r = \frac{E - V}{i}$$

Also, $\quad i = \frac{V}{R}$, we have

$$r = \frac{E - V}{V/R} \Rightarrow r = R\left(\frac{E}{V} - 1\right)$$

29. *(a)* In triangle formation, two resistors are in series and their sum is connected in parallel. Therefore,

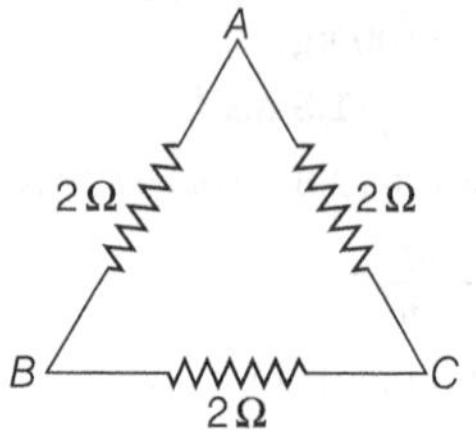

$R' = 2\Omega + 2\Omega = 4\,\Omega$

$\frac{1}{R''} = \frac{1}{4} + \frac{1}{2} = \frac{1+2}{4} = \frac{3}{4}\,\Omega$

$R'' = \frac{4}{3}\,\Omega$

30. *(a)* $u = PA = PC - AC = 2f - \frac{f}{3} = \frac{5f}{3}$

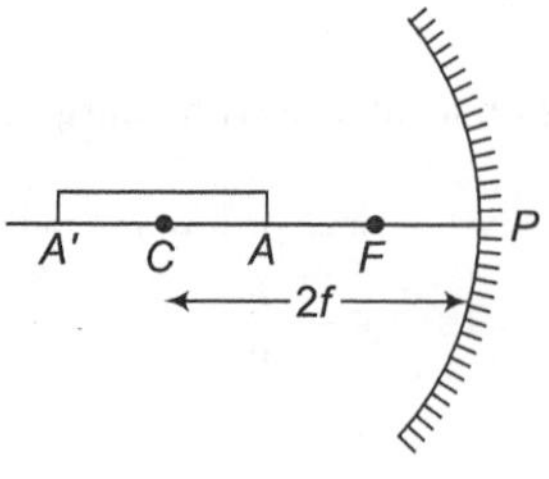

From mirror formula

$$\frac{1}{f} = \frac{1}{v} + \frac{1}{u}$$

$\Rightarrow \quad \frac{1}{v} = \frac{1}{f} - \frac{1}{u} = \frac{1}{f} - \frac{3}{5f} = \frac{2}{5f}$

Length of image

$$CA' = PA' - PC = \frac{5f}{2} - 2f = \frac{f}{2}$$

$\therefore$ Magnification $(m) = \frac{CA'}{CA} = \frac{f/2}{f/3} = 1.5$

31. *(d)* Intensity ($I = ka^2$), where a is amplitude and k is a constant.

$\therefore \quad I_1 = ka_1^2$

$\Rightarrow \quad \frac{I_1}{I_2} = \frac{a_1^2}{a_2^2} = \frac{9}{4}$

$\Rightarrow \quad \frac{a_1}{a_2} = \frac{3}{2}$

Maximum intensity

$$I_{max} = (a_1 + a_2)$$
$$= (3 + 2) = 5$$

Minimum intensity

$$I_{min} = (a_1 - a_2) = (3 - 2) = 1$$

$\therefore \quad \frac{I_{max}}{I_{min}} = \frac{5}{1}$

32. *(d)* Refractive index is given by

$$n = \frac{\sin\left(\frac{A + \delta_m}{2}\right)}{\sin\frac{A}{2}}$$

Given, $\quad \delta_m = 180° - 2A$

$\therefore \quad n = \frac{\sin\left(\frac{A + 180° - 2A}{2}\right)}{\sin\frac{A}{2}}$

$$n = \frac{\sin\left(\frac{180°}{2} - \frac{A}{2}\right)}{\sin\frac{A}{2}}$$

$\Rightarrow \quad n = \frac{\cos\frac{A}{2}}{\sin\frac{A}{2}} = \cot\frac{A}{2}.$

33. *(a)* $\frac{1}{2}mv^2 = eV$

$\Rightarrow \quad \frac{e}{m} = \frac{v^2}{2V}$

Given, $V = 200$ volt, $v = 8.4 \times 10^8$ cm/s

$$= 8.4 \times 10^6 \text{ m/s}.$$

$$\therefore \quad \frac{e}{m} = \frac{(8.4 \times 10^6)^2}{2 \times 200} = 1.76 \times 10^{11} \text{ C/kg}$$

34. *(a)* $\frac{1}{\lambda} = R\left(\frac{1}{n_1^2} - \frac{1}{n_2^2}\right)$

For Lyman series $n = 1$

For Balmer series $n = 2$

$$\therefore \quad \frac{1}{\lambda_L} = R\left(\frac{1}{1^2}\right), \frac{1}{\lambda_B} = R\left(\frac{1}{1^2} - \frac{1}{2^2}\right) \Rightarrow \frac{\lambda_B}{\lambda_L} = \frac{3}{1}$$

35. *(b)* When emf of cell is E, the potential drop is

$$V = E - ir$$

where i is current and r the internal resistance.

Given, $i = 10$ mA $= 10 \times 10^{-3}$ A, $R = 200\ \Omega$,

$$V = 0.5 \text{ volt}.$$

$$\Rightarrow \quad E = V + ir$$

$$E = 0.5 + 10 \times 10^{-3} \times 200$$

$$E = 1.5 \text{ V}$$

36. *(d)* In the following figure, output of AND gate is made as input of NOT gate, we get NAND gate

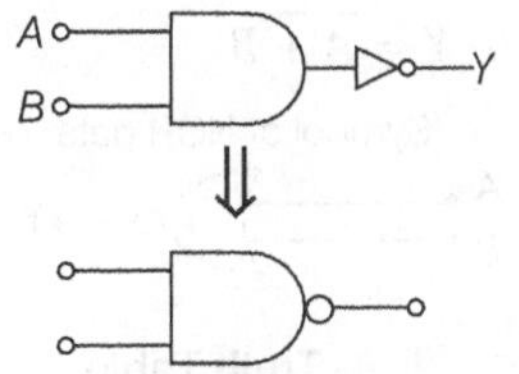

NAND gate

$$Y = \overline{A \cdot B}$$

37. *(d)* Distance of the satellite from the centre are $7R$ and $3.5\,R$ respectively.

$$\frac{T_2}{T_1} = \left(\frac{R_2}{R_1}\right)^{3/2}$$

$$\Rightarrow \quad T_2 = 24\left[\frac{3.5\,R}{7R}\right]^{3/2} = 6\sqrt{2} \text{ h}$$

38. *(b)* We know that $\quad v \propto \lambda$

$$\Rightarrow \quad \frac{v_1}{v_2} = \frac{\lambda_1}{\lambda_2}$$

$$\therefore \quad v_2 = \frac{v_1}{\lambda_1} \times \lambda_2$$

$$= 3 \times 10^8 \times \frac{4500}{6000} = 2.25 \times 10^8 \text{ m/s}$$

39. *(b)* The angle of minimum deviation is given by

$$\delta_m = (n-1)\,A$$

$$\delta_m = (\sqrt{2} - 1)\,60$$

Angle of incidence, $i = \frac{A + \delta_m}{2}$

$$i = \frac{60 + (\sqrt{2} - 1)\,60}{2}$$

$$i = \frac{60\sqrt{2}}{2} = 30\sqrt{2} = 42^\circ$$

40. *(c)* From Coulomb's law

$$F = \frac{1}{4\pi\varepsilon_0}\frac{q_1 q_2}{r^2}$$

$$F = \frac{1}{4\pi\varepsilon_0}\frac{q \times \frac{q}{2}}{(a + x)^2} - \frac{1}{4\pi\varepsilon_0} \cdot \frac{q \times \frac{q}{2}}{(a - x)^2}$$

$$= \frac{1}{4\pi\varepsilon_0}\frac{q^2}{2}\left[\frac{1}{(a + x)^2} - \frac{1}{(a - x)^2}\right]$$

$$= \frac{1}{4\pi\varepsilon_0} \cdot \frac{q^2}{2}\left[-\frac{4\,ax}{(a^2 - x^2)^2}\right]$$

When $x << a$, then

$$F = -\frac{2q^2}{4\pi\varepsilon_0 a^3}\,x$$

$$\Rightarrow \quad F \propto -x$$

Hence, SHM along x-axis.

41. *(b)* From Coulomb's law

$$F = \frac{1}{4\pi\varepsilon_0}\frac{Q^2}{2a^2}$$

F_1 and F_2 will be directed as shown, for this both q should be negative

$$F_1 = F_2 = \frac{1}{4\pi\varepsilon_0} \cdot \frac{Qq}{a^2}$$

$$F_{12} = \frac{1}{4\pi\varepsilon_0} \cdot \frac{\sqrt{2}\,Qq}{a^2}$$

For equilibrium of Q, we have

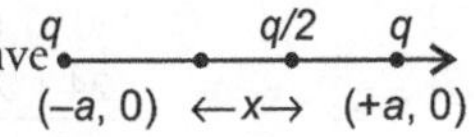

$$F = -F_{12}$$

$$\frac{Q^2}{2a^2} = -\frac{\sqrt{2}\,Qq}{a^2}$$

$$\Rightarrow \quad q = \frac{Q}{2\sqrt{2}}$$

Given $Q = +\sqrt{2}\,\mu\text{C}$

$$\therefore \quad q = -\frac{\sqrt{2}\,\mu\text{C}}{2\sqrt{2}} = -0.5\,\mu\text{C}$$

42. *(c)* Let $p_1 = p$, $p_2 = p_1 + 50\%$ of p_1

$$= p_1 + \frac{p_1}{2} = \frac{3p_1}{2}$$

$$E \propto p^2$$

$$\Rightarrow \quad \frac{E_2}{E_1} = \left(\frac{p_2}{p_1}\right)^2 = \left(\frac{3p_1/2}{p_1}\right)^2 = \frac{9}{4}$$

$$\Rightarrow \quad E_2 = 2.25E_1 = E_1 + 1.25\,E_1$$

$$\therefore \quad E_2 = E_1 + 125\%\text{of } E_1$$

ie, kinetic energy will increase by 125%

43. *(c)* The energy stored in capacitor is given by

$$E = \frac{1}{2}CV^2$$

Resultant capacitance

$$C' = C_1 + C_2 = 4 + 6 = 10\,\text{pF}$$

$$E = \frac{1}{2} \times 10 \times 10^{-12} \times (100)^2$$

$$(1\,\text{pF} = 10^{-12}\,\text{F})$$

$$= 5 \times 10^{-8}\,\text{J}$$

44. *(a)* From Ohm's law

$$i = \frac{E}{R + r} = \frac{2}{998 + 2} = 2 \times 10^{-3}\,\text{A}$$

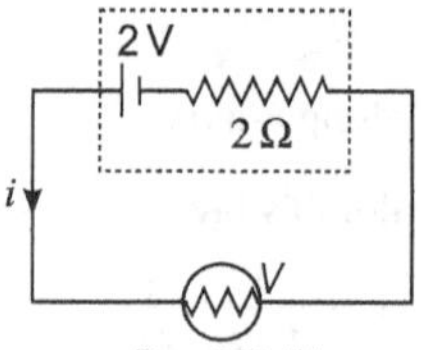

$R = 988\,W$

Potential difference across the voltmeter is

$$V = iR = (2 \times 10^{-3}) \times 998 = 1.996\,\text{V}$$

45. *(d)* When no current passes through the galvanometer, the bridge is balanced hence,

$$\frac{P}{Q} = \frac{R}{S}$$

$$\Rightarrow \quad \frac{24 + X}{10} = \frac{84 + Y}{30}$$

Also, applying Kirchhoff's law

$$\Sigma\, iR = \Sigma\, i'R' \qquad \text{(parallel)}$$

$$(X + 24) + (84) + Y = 3\,(10 + 30)$$

$$\Rightarrow \quad X + Y = 12 \qquad \ldots\text{(i)}$$

and $\quad 3X - Y = 12 \qquad \ldots\text{(ii)}$

From Eqs. (i) and (ii), we get

$X = 6$, $Y = 6$ (neglecting negative value)

46. *(b)* The impedance of the circuit is

$$Z = \sqrt{R^2 + (X_L - X_C)^2}$$

$$X_L = \omega L = 400 \times 20 \times 10^{-3} = 8\text{H}$$

$$X_C = \frac{1}{\omega C} = \frac{1}{400 \times 625 \times 10^{-6}} = 4\,\text{C}$$

$$Z = \sqrt{(3)^2 + (8 - 4)^2} = 5$$

$$i = \frac{E}{Z} = \frac{300}{5} = 60\,\text{A}$$

47. *(a)* If stopping potential is V_0, then maximum kinetic energy of photoelectrons is given by

$$E_k = eV$$

Given, $\quad V = 1.5\,\text{volt}$

$$E_k = 1.5\,\text{eV}$$

48. *(b)* $a \sin\theta = 1 \times \lambda$

$$\Rightarrow \quad \lambda = a \sin\theta = 10^{-4} \times \frac{1}{2} = \frac{10 \times 10^{-5}}{2}$$

$$= 5 \times 10^{-5}\,\text{cm} = 5000\,\text{Å}$$

49. *(c)* NOR gate is obtained when the ouput of OR gate is made as the input of NOT gate, Boolean expression for NOR gate is

$$Y = \overline{A + B}$$

Symbol of NOR gate

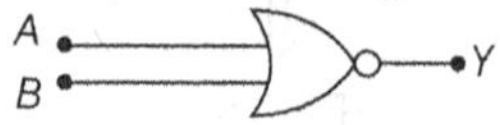

Truth Table

A	B	Y
0	0	1
1	0	0
0	1	0
1	1	0

50. *(c)* From Doppler's effect the perceived frequency f' is given by

$$f' = f\left(\frac{v + v_o}{v}\right)$$

Given, $f = 990$ Hz, $v = 300$ m/s, $v_o = 33$ m/s

$$\therefore \quad f' = 990\left(\frac{330 + 33}{330}\right)$$

$$= 990 \times \frac{363}{330} = 1089\,\text{Hz}$$

Chemistry

1. *(b)* For f orbital, $l = 3$

Hence, $m = -3, -2, -1, 0, +1, +2, +3$

2. *(b)* $3C + \frac{3}{2}O_2 \longrightarrow 3CO$

$Fe_2O_3 + 3CO \longrightarrow 2Fe + 3CO_2$

1 mole of $Fe_2O_3 \equiv$ 3 mole of $CO \equiv \frac{3}{2}$ mole of O_2

$\because$ 160 g of Fe_2O_3 require $O_2 = \frac{3}{2} \times 32 = 48$ g

$\therefore$ 1.60 kg of Fe_2O_3 will require $O_2 = 480$ g

3. *(c)* The phenoxide ion left from *o*-hydroxybenzaldehyde and *p*-hydroxybenzaldehyde are stabilized by $-I$ and $-R$ effect of the —CHO group. But in *o*-isomer due to chelation it is difficult to remove the H-atom. Hence, *p*-hydroxybenzaldehyde is the strongest acid.

4. *(a)* Purification of NaCl by passage of hydrogen chloride through brine is based on common ion effect.

5. *(d)* Wurtz reaction involves the reduction of alkyl halide with Na in ether.

$$RX + 2Na + XR \xrightarrow{\text{Dry ether}} R-R + 2NaX$$

6. *(d)* Dry distillation of calcium benzoate gives benzophenone.

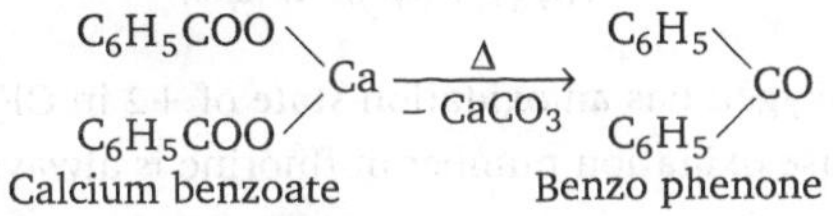

7. *(b)*

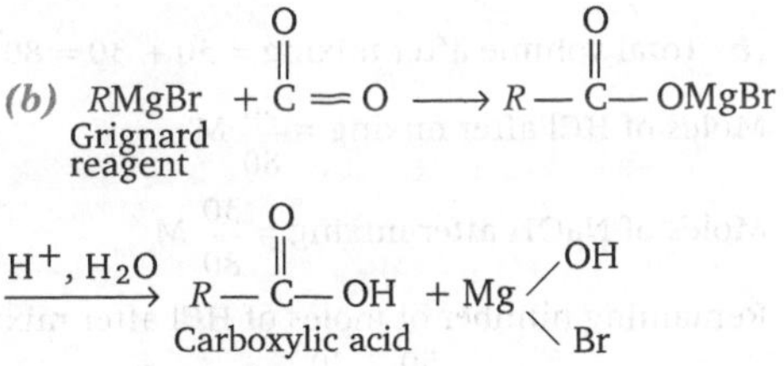

8. *(a)* The dilute aqueous solution (7 to 8%) of ethanoic acid (acetic acid) is called vinegar.

9. *(a)*

$$2(r^+ + r^-) = \sqrt{3}a$$

$$2(1.69 + 1.81) = \sqrt{3}a$$

$$2(3.5) = \sqrt{3}a$$

$$7.0 = \sqrt{3}a$$

$$a = \frac{7.0}{\sqrt{3}} = 4.04 \text{ Å}$$

10. *(a)* $CaCl_2 + 2AgNO_3 \longrightarrow Ca(NO_3)_2 + 2AgCl$

111g $\qquad\qquad$ 2×43.5 g

$CaCl_2$ required to produce 2×143.5 g of AgCl $= 111$ g

$CaCl_2$ required to produce 14.35 g of AgCl

$$= \frac{111 \times 14.35}{2 \times 143.5} = 5.55 \text{ g}$$

11. *(b)* In the hydrogen spectrum the Lyman series of lines appears in ultra violet region of electromagnetic spectrum.

12. *(b)* The first ionisation potential (IE_1) decreases from top to bottom in a group.

Hence, the order of ionisation potential is as :

$$Be > Mg > Ca$$

13. *(c)* Fog is a liquid in gas colloidal solution.

14. *(b)* $NOClO_4 \equiv (NO^+)\ (ClO_4^-)$

Let the O.N. of N in NO^+ is x.

$$x + (-2) = +1$$

$$x = 3$$

Let O.N. of Cl in ClO_4^- is y.

$$y + 4(-2) = -1$$

$$y - 8 = -1$$

$$y = +7$$

15. *(a)* Slag formed during the extraction of iron is always lighter than the molten metal. Its melting point should also be less than the molten metal.

16. *(d)*

Complex ion	Hybridisation on central atom
$[Fe(CN)_6]^{4-}$	d^2sp^3 (Inner)
$[Mn(CN)_6]^{4-}$	d^2sp^3 (Inner)
$[Co(NH_3)_6]^{3+}$	d^2sp^3 (Inner)
$[Ni(NH_3)_6]^{2+}$	sp^3d^2 (Outer)

17. *(b)*

Sugar :	Lactose	Glucose	Sucrose	Fructose
Relative sweetness	16	74	100	173

18. *(c)* $\Rightarrow$ $KE = \frac{3}{2}RT$ (for 1 mole)

Given, $T = 47 + 273 = 320$ K

$$KE = \frac{3}{2} \times 8.314 \times 320 = 3990.7 \text{ J mol}^{-1}$$

Molar mass of $O_2 = 32 \text{g mol}^{-1}$

Molar mass of $O_2 = 32 \text{g mol}^{-1}$

$$\text{KE (per gram of } O_2) = \frac{3990.7}{32} \text{ J g}^{-1}$$
$$= 124.7 \text{ J g}^{-1}$$
$$= 1.24 \times 10^2 \text{ J g}^{-1}$$

19. *(a)* $\pi = \frac{n}{V} RT = \frac{0.2 \times 0.082 \times 300}{1.0} = 4.92$ atm

20. *(a)* Reduction of glucose with $NaBH_4$ gives sorbitol.

$$\begin{array}{c} CHO \\ | \\ (CHOH)_4 \\ | \\ CH_2OH \end{array} + 2[H] \xrightarrow{NaBH_4} \begin{array}{c} CH_2OH \\ | \\ (CHOH)_4 \\ | \\ CH_2OH \end{array}$$

Glucose — Sorbitol

21. *(d)* Decarboxylation of sodium propionate leads to the formation of ethane.

$$CH_3CH_2COONa + NaOH \xrightarrow{CaO,\ 360\ K} \underset{\text{Ethane}}{CH_3—CH_3} + Na_2CO_3$$

22. *(d)* Formic acid reacts with PCl_5 to give carbon monoxide and hydrogen chloride.

$$HCOOH + PCl_5 \rightarrow \underset{\text{unstable}}{[HCOCl]} \rightarrow CO + HCl$$

23. *(c)* $\Delta H < 0$ and $\Delta S > 0$ are always favourable conditions for spontaneity at all temperatures.

24. *(b)* $\Delta G° = -RT \ln K$

$$\Delta G° = -2.303\, RT \log K$$
$$= -2.303 \times 8 \times 300 \times \log 10$$
$$= -2.303 \times 8 \times 300 \times 1$$
$$= -5527 \text{J mol}^{-1} = -5.527 \text{ kJ mol}^{-1}$$

25. *(a)* Due to $-R$ effect of the C=O group the electron density on the N-atom decreases, hence amides are weaker base than NH_3 and aliphatic amines. Further due to $-I$ effect of NH_2 group, the electron density on the other N-atom decreases. As a result, NH_2NH_2 is a much weaker base than NH_3 and aliphatic amines. Therefore, ethylamine is most basic among these.

26. *(a)* $CaCO_3 + 2HCl \longrightarrow CaCl_2 + CO_2 + H_2O$

$$25 \text{ mL of } 0.75 \text{ M HCl} = \frac{25}{1000} \times 0.75 = 0.01875 \text{ mol}$$

$$\text{Moles of } CaCO_3 \text{ required} = \frac{\text{moles of HCl}}{2}$$
$$= \frac{0.01875}{2} = 9.375 \times 10^{-3} \text{ mol}$$

Mass of $CaCO_3$ required
$$= 9.375 \times 10^{-3} \text{ mol} \times 100 \text{ g mol}^{-1}$$
$$= 0.9375 \text{ g} = 0.94 \text{ g}$$

27. *(d)* There are $(2l + 1)$ values of m for the each value of l.

$\because \quad l = 3$

$\therefore$ Total values of $m = (2 \times 3 + 1) = 7$

28. *(c)* The ionisation energy of nitrogen is larger than that of oxygen because half-filled p-orbitals of nitrogen have extra stability while oxygen has $2p^4$ electronic configuration of the outermost shell, hence its electrons can be easily removed.

29. *(a)* For two third of a reaction

$$[A]_0 = a, [A] = a - \frac{2}{3}a = \frac{a}{3}$$
$$t_{2/3} = \frac{2.303}{k} \log \frac{[A]_0}{[A]}$$
$$= \frac{2.303}{5.48 \times 10^{-14}} \log \frac{a}{a/3}$$
$$= \frac{2.303}{5.48 \times 10^{-14}} \log 3$$

30. *(a)* Nitric oxide (NO) has one unpaired electron and it is paramagnetic in nature. It has 15 electrons.

$$\text{NO}: KK\ \sigma(2s)^2\ \overset{*}{\sigma}(2s)^2\ \sigma(2p_z)^2$$
$$\pi(2p_x)^2\ \pi(2p_y)^2\ \overset{*}{\pi}(2p_x)^1$$

31. *(c)* Oxygen has an oxidation state of +2 in OF_2 because oxidation number of fluorine is always −1.

32. *(b)* Total volume after mixing = 50 + 30 = 80 cc

$$\text{Moles of HCl after mixing} = \frac{50}{80} \text{ M}$$

$$\text{Moles of NaOH after mixing} = \frac{30}{80} \text{ M}$$

Remaining number of moles of HCl after mixing
$$= \frac{50}{80} - \frac{30}{80} = 0.25 \text{ M}$$
$$[H^+] = 0.25 = 2.5 \times 10^{-1}$$
$$\text{pH} = -\log [2.5 \times 10^{-1}]$$
$$= 1 - 0.3979 = 0.6021$$

33. *(a)* Noble gases are monoatomic. The value of C_P / C_V for noble gases is 1.66.

34. *(b)* Carbylamine test is performed in alcoholic KOH by heating a mixture of trihalogenated methane and a primary amine.

$$\underset{\text{Primary amine}}{RNH_2} + \underset{\text{Chloroform}}{CHCl_3} + 3KOH(\text{alc.}) \xrightarrow{\Delta} \underset{\text{Alkyl isocyanide}}{RNC} + 3KCl + 3H_2O$$

35. *(a)* The order of acidity is as :

$$\xrightarrow[\text{Decreasing acid strength (Decreasing } K_a)]{RCOOH > H_2CO_3 > C_6H_5OH > H_2O > ROH}$$

36. *(d)* The order of reactivity of alcohols towards Lucas reagent is as

Tertiary > Secondary > Primary

Since 2-methyl propan-2-ol is a tertiary alcohol, hence it reacts fastest with Lucas reagent at room temperature.

37. *(a)* $\underset{\text{Methanal}}{HCHO} \xrightarrow{\text{Reduction}} \underset{\text{Methanol}}{CH_3OH} \xrightarrow{P + I_2} \underset{\text{Methyl iodide}}{CH_3I}$

$\xrightarrow{KCN} \underset{\text{Methyl cyanide}}{CH_3CN} \xrightarrow{\text{Hydrolysis}} \underset{\text{Acetic acid}}{CH_3COOH}$

38. *(a)* The conjugate base of the acid H_2S is HS^-.

$$\underset{\text{Acid}}{H_2S} \rightleftharpoons \underset{\text{Conjugate base}}{HS^-} + H^+$$

39. *(b)* Transition metal ion having electronic configuration $(n-1)\,d^{1-9}$ forms coloured ion.

Cu^{2+} $(Z = 29)$: [Ar] $3d^9$

Cu^+ $(Z = 29)$: [Ar] $3d^{10}$

Ni^{2+} $(Z = 28)$: [Ar] $3d^8$

Fe^{3+} $(Z = 26)$: [Ar] $3d^5$

Hence, Cu^+ ion is not coloured.

40. *(a)* Glycine can be obtained from formaldehyde by Streker synthesis.

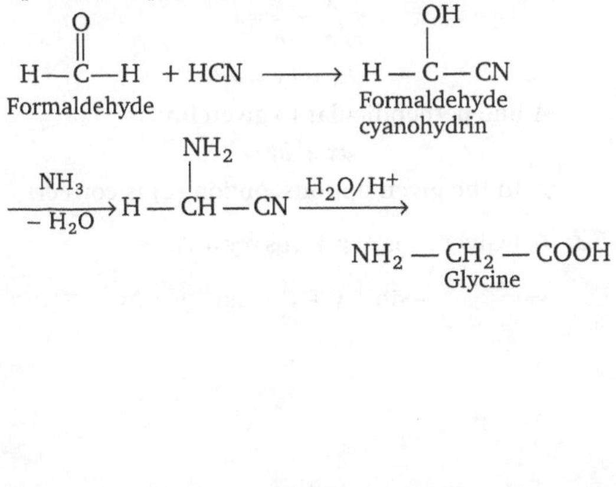

41. *(c)* Polar aprotic solvents such as DMSO increase the rate of S_N2 reactions.

42. *(c)* EAN of Cu in $[Cu(NH_3)_4]\,SO_4$ is as :

EAN = Z − (O.N.) + 2 (C.N.)

EAN = 29 − 2 + 4 × 2 = 35

43. *(a)* For a reaction to be spontaneous ΔG should be negative.

$$\Delta G = -nFE^\circ_{\text{cell}}$$

From this reaction, ΔG will be negative when E°_{cell} is positive.

44. *(c)* In a face centred cubic lattice, a unit cell is shared equally by six unit cells.

45. *(b)* 1, 2-dibromobutene has different substituents on each C-atom of the double bond, hence it shows geometrical isomerism.

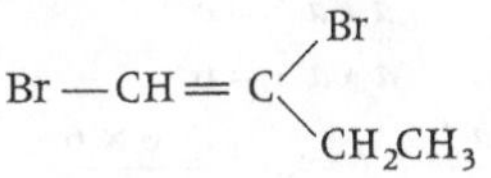

46. *(a)* α-amino acids are essential building units of proteins.

47. *(d)* $CH_2N_2 \xrightarrow[-N_2]{} \underset{\text{Carbene}}{:CH_2}$

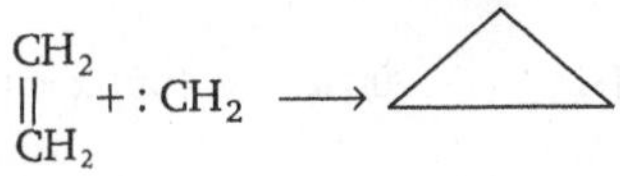

48. *(b)* The first ever compound of noble gas $Xe^+\,[PtF_6]^-$ was prepared by Neil Bartlett in 1962.

49. *(a)* $K_2Cr_2O_7$ on heating with aqueous alkalies gives chromates

$$K_2Cr_2O_7 + 2NaOH \longrightarrow K_2CrO_4 + Na_2CrO_4 + H_2O$$

50. *(d)* Energy of electron in *n*th orbit of hydrogen,

$$E_n = -\frac{E}{n^2}$$

where, E is a constant.

$$\Rightarrow \quad \frac{E_2}{E_4} = \frac{(n_4)^2}{(n_2)^2}$$

$$\frac{-328}{E_4} = \frac{(4)^2}{(2)^2}$$

$$\therefore \quad E_4 = \frac{-328 \times 4}{16} = -82 \text{ kJ mol}^{-1}$$

Mathematics

1. *(c)* Since, roots are complex.

$\therefore \quad a^2 - 36 < 0 \quad (\because \text{discriminant} < 0)$

$\Rightarrow \quad a^2 < 36$

$\Rightarrow \quad |a| < 6$

2. *(c)* Since, projection of **PQ** on OX, OY, OZ are 12, 3 and 4 respectively, then $\mathbf{PQ} = 12\hat{\mathbf{i}} + 3\hat{\mathbf{j}} + 4\hat{\mathbf{k}}$

$\therefore \quad |\mathbf{PQ}| = \sqrt{12^2 + 3^2 + 4^2} = \sqrt{169}$

$= 13$

3. *(a)* Let e be the identity.

$\therefore \quad a * e = a$

$\Rightarrow \quad \frac{ab}{6} = a$

$\Rightarrow \quad e = 6$

$\because \quad a * a^{-1} = e$

$\therefore \quad a * a^{-1} = 6$

$\Rightarrow \quad \frac{a\,a^{-1}}{6} = 6 \Rightarrow a^{-1} = \frac{6 \times 6}{a}$

$\therefore \quad 9^{-1} = \frac{6 \times 6}{9} = 4$

4. *(c)* Let $I = \int_0^1 \frac{d}{dx}\left[\sin^{-1}\left(\frac{2x}{1+x^2}\right)\right] dx$

$\therefore \quad I = \int_0^1 \frac{d}{dx}[\sin^{-1}(\sin 2\theta)]\, dx \quad (\text{put } x = \tan\theta)$

$= 2\int_0^1 \frac{d}{dx}[\tan^{-1} x]\, dx$

$= 2[\tan^{-1}(1) - \tan^{-1}(0)]$

$= 2 \cdot \frac{\pi}{4} = \frac{\pi}{2}$

5. *(a)* Given, $4\sin^{-1} x + \cos^{-1} x = \pi$

$\Rightarrow \quad 4\sin^{-1} x + \frac{\pi}{2} - \sin^{-1} x = \pi$

$\Rightarrow \quad 3\sin^{-1} x = \frac{\pi}{2}$

$\Rightarrow \quad x = \sin\left(\frac{\pi}{6}\right) = \frac{1}{2}$

6. *(b)* $\sum_{n=1}^{\infty} \frac{(-1)^n}{n+1}$

$= -\frac{1}{2} + \frac{1}{3} - \frac{1}{4} + \dots \infty$

$= 1 - \frac{1}{2} + \frac{1}{3} - \frac{1}{4} + \dots \infty - 1 = \log 2 - 1$

7. *(b)* Given, $\tan^{-1} x + \tan^{-1} y + \tan^{-1} z = \pi$

$\Rightarrow \quad \tan^{-1}\left(\frac{x + y + z - xyz}{1 - xy - yz - zx}\right) = \pi$

$\Rightarrow \quad x + y + z - xyz = 0$

8. *(a)*

p	q	$\sim p$	$\sim q$	$p \wedge \sim q$	$\sim p \vee q$	$(p \wedge \sim q) \wedge (\sim p \vee q)$
T	T	F	F	F	T	F
T	F	F	T	T	F	F
F	T	T	F	F	T	F
F	F	T	T	F	T	F

Clearly, $(p \wedge \sim q) \wedge (\sim p \vee q)$ is a contradiction.

9. *(d)* $(1 + x + x^2 + x^3)^{-3} = \left[\frac{(x^4 - 1)}{x - 1}\right]^{-3}$

$= (x^4 - 1)^{-3}(x - 1)^3$

$= (-1)^{-3}(1 - x^4)^{-3}(x - 1)^3$

$= (-1)(1 + 3x^4 + \dots)(x^3 - 3x^2 + 3x - 1)$

$\therefore$ Coefficient of x in $(1 + x + x^2 + x^3)^{-3}$ is -3.

10. *(a)* $f'(3) = \lim_{h \to 0} \frac{f(3 + h) - f(3)}{h}$

$= \lim_{h \to 0} \frac{f(3)\,f(h) - f(3)}{h} \quad [\because f(x + y) = f(x)\,f(y)]$

Now, $\quad f(x + 0) = f(x)\,f(0)$

$\Rightarrow \quad f(x)\,[f(0) - 1] = 0$

Either $f(x) = 0$ or $f(0) = 1$

$\therefore \quad f'(3) = f(3) \lim_{h \to 0} \frac{f(h) - f(0)}{h}$

$= f(3)\,f'(0) = f(3) \cdot 11$

$= 3 \times 11 = 33$

11. *(c)* The equation of given line is

$\frac{x}{a} - \frac{y}{b} = 1$

$\Rightarrow \quad bx - ay = ab$

A line perpendicular to given line is

$ax + by = \lambda$

$\therefore$ In the given options, option (c) is correct.

12. *(c)* Given, $\cos^{-1} x + \cos^{-1} y = 2\pi$

$\Rightarrow \quad \frac{\pi}{2} - \sin^{-1} x + \frac{\pi}{2} - \sin^{-1} y = 2\pi$

$\Rightarrow \quad \pi - 2\pi = \sin^{-1} x + \sin^{-1} y$

$\Rightarrow \quad \sin^{-1} x + \sin^{-1} y = -\pi$

13. *(b)* $\log_3 2, \log_3 (2^x - 5), \log_3 \left(2^x - \frac{7}{2}\right)$ are in AP.

$$\Rightarrow \log_3 (2^x - 5) = \frac{\log_3 \left(2^x - \frac{7}{2}\right) + \log_3 2}{2}$$

$$\Rightarrow (2^x - 5)^2 = 2 \cdot \left(2^x - \frac{7}{2}\right)$$

$$\Rightarrow (2^x)^2 + 25 - 10 \cdot 2^x = 2 \cdot 2^x - 7$$

$$\Rightarrow (2^x)^2 - 12 \cdot 2^x + 25 + 7 = 0$$

$$\Rightarrow (2^x)^2 - 12 \cdot 2^x + 32 = 0$$

$$\Rightarrow 2^x = 8 \text{ or } 2^x = 4$$

$$\Rightarrow x = 3 \text{ or } 2$$

$\Rightarrow x = 3$ ($\because x = 2$ does not satisfy the given series)

14. *(c)* $\int_0^1 x(1-x)^{12}\, dx$

$$= \left[\frac{x(1-x)^{13}}{-13} + \int_0^1 \frac{(1-x)^{13}}{13}\, dx\right]_0^1$$

$$= \left[\frac{x(1-x)^{13}}{-13}\right]_0^1 + \left[\frac{(1-x)^{14}}{-14 \times 13}\right]_0^1$$

$$= [0 - 0] + \left[0 - \frac{1}{-14 \times 13}\right] = \frac{1}{182}$$

15. *(a)* Middle term is $\left(\frac{n}{2} + 1\right)$th term for n even.

$\therefore$ Middle term $= \left(\frac{6}{2} + 1\right) = 4$ th term

$$\therefore \quad T_4 = {}^6C_3 (\sqrt{x})^3 \left(-\frac{1}{\sqrt{x}}\right)^{6-3}$$

$$= \frac{6!}{3!\,3!} (\sqrt{x})^3 \frac{(-1)^3}{(\sqrt{x})^3}$$

$$= \frac{-6 \times 5 \times 4}{6}$$

$$= -20$$

16. *(d)* Let a and b be any element of group G.

$\therefore \quad a = a^{-1}, b = b^{-1} \ \forall\ a, b \in G$

Now, $(ab)^{-1} = ab$ [$\because a, b \in G \Rightarrow ab \in G$]

$\Rightarrow \quad b^{-1}a^{-1} = ab \Rightarrow ba = ab \ \forall\ a, b \in G$

Hence, commutative law holds in group G.
Therefore, G is abelian group.

17. *(b)* Since, $P(3)$ is true.

Assume $P(k)$ is true $\Rightarrow P(k+1)$ is true means, if $P(3)$ is true $\Rightarrow P(4)$ is true $\Rightarrow P(5)$ is true and so on. So, statement is true for all $n \geq 3$.

18. *(b)* $\Delta = \begin{vmatrix} 1 & a & a^2 \\ 1 & b & b^2 \\ 1 & c & c^2 \end{vmatrix} = \begin{vmatrix} 1 & a & a^2 \\ 0 & b-a & b^2 - a^2 \\ 0 & c-a & c^2 - a^2 \end{vmatrix}$

(using $R_2 \to R_2 - R_1$ and $R_3 \to R_3 - R_1$)

$$= (b-a)(c-a) \begin{vmatrix} 1 & a & a^2 \\ 0 & 1 & b+a \\ 0 & 1 & c+a \end{vmatrix}$$

$$= (b-a)(c-a)(c+a-b-a)$$

$\Rightarrow (b-a)(c-a)(c-b)$

$= k(a-b)(b-c)(c-a)$ (given)

$\Rightarrow \quad k = 1$

19. *(c)* Number of required ways $= {}^{22-4-2}C_{11-2}$

$= {}^{16}C_9$

20. *(b)* Given, $\Delta_1 = \begin{vmatrix} x & a & b \\ b & x & a \\ a & b & x \end{vmatrix}, \Delta_2 = \begin{vmatrix} x & b \\ a & x \end{vmatrix}$

$$\therefore \frac{d}{dx}(\Delta_1) = \begin{vmatrix} 1 & 0 & 0 \\ b & x & a \\ a & b & x \end{vmatrix} + \begin{vmatrix} x & a & b \\ 0 & 1 & 0 \\ a & b & x \end{vmatrix} + \begin{vmatrix} x & a & b \\ b & x & a \\ 0 & 0 & 1 \end{vmatrix}$$

$$= \begin{vmatrix} x & a \\ b & x \end{vmatrix} + \begin{vmatrix} x & b \\ a & x \end{vmatrix} + \begin{vmatrix} x & a \\ b & x \end{vmatrix} = 3\Delta_2$$

$$\therefore \frac{d}{dx}(\Delta_1) = 3\Delta_2$$

21. *(a)* Let $I = \int_{-1/2}^{1/2} \frac{dx}{(1-x^2)^{1/2}}$

Again, let $f(x) = \frac{1}{(1-x^2)^{1/2}}$

$$f(-x) = \frac{1}{(1-x^2)^{1/2}} = f(x)$$

$$\therefore \quad I = 2\int_0^{1/2} \frac{dx}{(1-x^2)^{1/2}} = \left[2\sin^{-1}\frac{x}{1}\right]_0^{1/2}$$

$$= 2\sin^{-1}\frac{1}{2} = 2 \times \frac{\pi}{6} = \frac{\pi}{3}$$

22. *(c)* $\lim_{n \to \infty} \left[\frac{1}{n+1} + \frac{1}{n+2} + \ldots + \frac{1}{n+5n}\right]$

$$= \lim_{n \to \infty} \sum_{r=0}^{5n} \frac{1}{n+r} = \lim_{n \to \infty} \left[\frac{1}{n} \sum_{r=0}^{5n} \frac{n}{n+r}\right]$$

$$= \lim_{n \to \infty} \left[\frac{1}{n} \sum_{r=0}^{5n} \frac{1}{1 + (r/n)}\right]$$

$$= \int_0^5 \frac{1}{1+x}\, dx = [\log(1+x)]_0^5$$

$$= \log 6 - \log 1 = \log 6$$

23. *(b)* $\left[5+\left(\frac{dy}{dx}\right)^2\right]^{5/3} = x^5\left[\frac{d^2y}{dx^2}\right]$

$\Rightarrow \quad \left[5+\left(\frac{dy}{dx}\right)^2\right]^5 = x^{15}\left[\frac{d^2y}{dx^2}\right]^3$

$\therefore$ Degree = 3

24. *(c)* The difference of the focal distance at any point on the hyperbola is same as length of transverse axis.

25. *(b)* Since, a ray makes angles α, β, γ with the positive direction of the axes.

$\therefore \quad \cos^2\alpha + \cos^2\beta + \cos^2\gamma = 1$

$\Rightarrow \quad 1 - \sin^2\alpha + 1 - \sin^2\beta + 1 - \sin^2\gamma = 1$

$\Rightarrow \quad \sin^2\alpha + \sin^2\beta + \sin^2\gamma = 2$

26. *(b)* Given, $f(x) = \begin{cases} 3x^2 + 12x - 1, & -1 \le x \le 2 \\ 37 - x, & 2 \le x \le 3 \end{cases}$

On differentiating, we get

$f'(x) = \begin{cases} 6x + 12, & -1 \le x \le 2 \\ -1, & 2 \le x \le 3 \end{cases}$

$\Rightarrow \quad f'(x) > 0$ for $-1 \le x \le 2$

$\therefore f(x)$ is increasing in $[-1, 2]$

$\lim_{x \to 2^+} f'(x) = -1$ and $\lim_{x \to 2^-} f'(x) = 24$

$\therefore f'(2)$ does not exist.

27. *(c)* Given,

$\mathbf{a} = \frac{1}{7}(2\hat{\mathbf{i}} + 3\hat{\mathbf{j}} + 6\hat{\mathbf{k}}), \mathbf{b} = \frac{1}{7}(3\hat{\mathbf{i}} - \lambda\hat{\mathbf{j}} + 2\hat{\mathbf{k}})$

Since, the vectors **a** and **b** are perpendicular.

$\therefore \quad \mathbf{a} \cdot \mathbf{b} = 0$

$\Rightarrow \frac{1}{7}(2\hat{\mathbf{i}} + 3\hat{\mathbf{j}} + 6\hat{\mathbf{k}}) \cdot \frac{1}{7}(3\hat{\mathbf{i}} - \lambda\hat{\mathbf{j}} + 2\hat{\mathbf{k}}) = 0$

$\Rightarrow \quad \frac{1}{49}(6 - 3\lambda + 12) = 0$

$\Rightarrow \quad 3\lambda = 18$

$\Rightarrow \quad \lambda = 6$

28. *(c)* The locus of the point of intersection of two perpendicular tangents to a circle is called director circle.

29. *(b)* The equation of given curve is

$$y = x^4 - 2x^3 + x^2 + 3$$

On differentiating w.r.t. x, we get

$$\frac{dy}{dx} = 4x^3 - 6x^2 + 2x$$

Again, differentiating, we get

$$\frac{d^2y}{dx^2} = 12x^2 - 12x + 2$$

Put, $\frac{dy}{dx} = 0$ for maxima or minima.

$\Rightarrow \quad 2x(2x^2 - 3x + 1) = 0$

$\Rightarrow \quad x = 0, 1, \frac{1}{2}$

$\therefore \quad \left(\frac{d^2y}{dx^2}\right)_{(x=0)} = 2$

$\therefore$ Function is minimum at $x = 0$

and $\left(\frac{d^2y}{dx^2}\right)_{x=1} = 12 - 12 + 2 = 2$

$\therefore$ Function is minimum at $x = 1$

Also, $\left(\frac{d^2y}{dx^2}\right)_{\left(x=\frac{1}{2}\right)} = -1 < 0$

$\therefore$ Function is maximum at $x = \frac{1}{2}$

$\therefore$ Required area $= \int_0^1 (x^4 - 2x^3 + x^2 + 3)\,dx$

$= \left[\frac{x^5}{5} - \frac{2x^4}{4} + \frac{x^3}{3} + 3x\right]_0^1$

$= \frac{1}{5} - \frac{1}{2} + \frac{1}{3} + 3$

$= \frac{91}{30}$ sq unit

30. *(b)* $\int (e^{a\log x} + e^{x\log a})\,dx = \int (e^{\log x^a} + e^{\log a^x})\,dx$

$= \int (x^a + a^x)\,dx = \frac{x^{a+1}}{a+1} + \frac{a^x}{\log_e a} + c$

31. *(d)* $\lim_{n\to\infty}\left(\frac{1}{\sqrt{4n^2-1}} + \frac{1}{\sqrt{4n^2-2^2}} + \ldots + \frac{1}{\sqrt{3n^2}}\right)$

$= \lim_{n\to\infty}\frac{1}{n}\left[\frac{1}{\sqrt{4-\left(\frac{1}{n}\right)^2}} + \frac{1}{\sqrt{4-\left(\frac{2}{n}\right)^2}} + \ldots + \frac{1}{\sqrt{4-\left(\frac{n}{n}\right)^2}}\right]$

$= \lim_{h\to 0} h\left[\frac{1}{\sqrt{4-h^2}} + \frac{1}{\sqrt{4-(2h)^2}} + \ldots + \frac{1}{\sqrt{4-(nh)^2}}\right]$

$= \int_0^1 \frac{dx}{\sqrt{4-x^2}} = \left[\sin^{-1}\frac{x}{2}\right]_0^1$

$= \sin^{-1}\left(\frac{1}{2}\right) - \sin^{-1}(0) = \frac{\pi}{6}$

32. *(b)* Let $I = \int \frac{dx}{x^2 + 4x + 13}$

$$= \int \frac{dx}{x^2 + 4x + 4 + 3^2} = \int \frac{dx}{(x+2)^2 + 3^2}$$

$$= \frac{1}{3} \tan^{-1}\left(\frac{x+2}{3}\right) + c$$

33. *(b)* Equation of tangent at $(a, 0)$ is $x = a$ and the point of intersection of $x = a$ and the asymptotes will be obtained by solving $x = a$ and the equation of asymptotes $bx \pm ay = 0$

$\therefore \quad ab \pm ay = 0$

$\Rightarrow \quad y = \pm b$

Now, point of intersections are (a, b) and $(a, -b)$.

$\therefore$ Area of triangle $= \frac{1}{2}(ab \times 2) = ab$ sq unit

34. *(c)* The equation of line passing through the origin is given by

$$y = mx \quad \ldots \text{(i)}$$

On differentiating w.r.t. x, we get

$$\frac{dy}{dx} = m \Rightarrow \frac{dy}{dx} = \frac{y}{x} \quad \text{[from Eq. (i)]}$$

$$\Rightarrow \quad x\frac{dy}{dx} - y = 0$$

35. *(b)* $\therefore PQ = \begin{bmatrix} i & 0 & -i \\ 0 & -i & i \\ -i & i & 0 \end{bmatrix} \begin{bmatrix} -i & i \\ 0 & 0 \\ i & -i \end{bmatrix}$

$$= \begin{bmatrix} -i^2 - i^2 & i^2 + i^2 \\ 0 + 0 + i^2 & -i^2 \\ i^2 & -i^2 \end{bmatrix} = \begin{bmatrix} 2 & -2 \\ -1 & 1 \\ -1 & 1 \end{bmatrix}$$

36. *(a)* $x = \sin^{-1}(3t - 4t^3)$

and $\quad y = \cos^{-1}(\sqrt{1 - t^2})$

Let $\quad t = \sin\theta$

$\therefore \quad x = \sin^{-1}(3\sin\theta - 4\sin^3\theta)$

$\quad = \sin^{-1}(\sin 3\theta) = 3\sin^{-1} t$

$\therefore \quad \frac{dx}{dt} = \frac{3}{\sqrt{1-t^2}}$

and $\quad y = \cos^{-1}(\sqrt{1 - \sin^2\theta})$

$\quad = \cos^{-1}(\sqrt{\cos^2\theta}) = \sin^{-1} t$

$\therefore \quad \frac{dy}{dt} = \frac{1}{\sqrt{1-t^2}}$

$\therefore \quad \frac{dy}{dx} = \frac{dy/dt}{dx/dt} = \frac{1/\sqrt{1-t^2}}{3/\sqrt{1-t^2}} = \frac{1}{3}$

37. *(d)* Since, a and b are the roots of equation

$$x^2 - 3x + 1 = 0$$

$\therefore \quad a + b = 3$ and $ab = 1$

Now, the given roots are $a - 2$ and $b - 2$

$\therefore$ Sum of roots $= a - 2 + b - 2$

$\quad = 3 - 4 = -1$

and products of roots $= (a-2)(b-2)$

$\quad = ab - 2(a + b) + 4$

$\quad = 1 - 6 + 4 = -1$

$\therefore$ Required equation is $x^2 + x - 1 = 0$

Which is equivalent to $x^2 + px + q = 0$

$\therefore \quad p = 1$ and $q = -1$

38. *(a)* Since, $\mathbf{AB} = \hat{\mathbf{i}} + 5\hat{\mathbf{j}} - \hat{\mathbf{k}}$

and $\quad \mathbf{BC} = -3\hat{\mathbf{j}} - \hat{\mathbf{k}}$

$\therefore$ Projection of **AB** on $\mathbf{BC} = \frac{\mathbf{AB} \cdot \mathbf{BC}}{|\mathbf{BC}|}$

$$= \frac{(\hat{\mathbf{i}} + 5\hat{\mathbf{j}} - \hat{\mathbf{k}}) \cdot (-3\hat{\mathbf{j}} - \hat{\mathbf{k}})}{\sqrt{(-3)^2 + (-1)^2}} = \frac{-15+1}{\sqrt{10}} = -\frac{14}{\sqrt{10}}$$

39. *(a)* Given equation can be rewritten as

$$y^2 = -4(x - 2)$$

$\therefore$ Required area $= 2\int_1^2 y\,dx = \int_1^2 \sqrt{8 - 4x}\,dx$

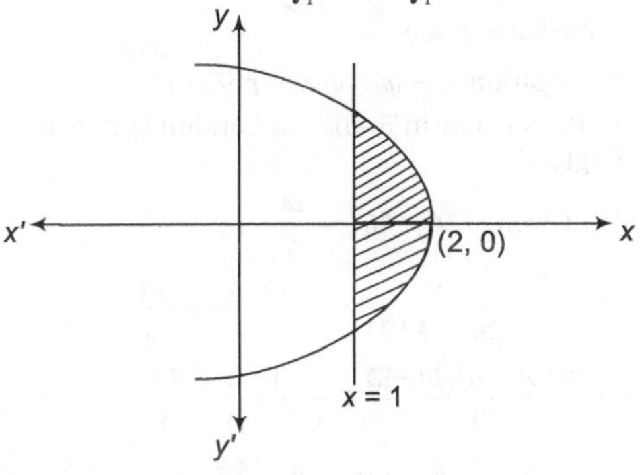

$$= \left[\frac{(8-4x)^{3/2}}{\frac{3}{2}(-4)}\right]_1^2 = -\frac{1}{6}[0 - 8]$$

$$= \frac{4}{3} \text{ sq unit}$$

40. *(b)* The given equation is

$$(2y - 1)\,dx - (2x + 3)\,dy = 0$$

or it can be rewritten as

$$\frac{dx}{(2x+3)} - \frac{dy}{(2y-1)} = 0$$

On integrating both sides, we get

$$\int \frac{1}{(2x+3)}\,dx - \int \frac{1}{(2y-1)}\,dy = 0$$

$$\Rightarrow \frac{1}{2}\log(2x+3) - \frac{1}{2}\log(2y-1) = \frac{1}{2}\log c$$

$$\Rightarrow \log\left(\frac{2x+3}{2y-1}\right) = \log c \Rightarrow \frac{2x+3}{2y-1} = c$$

41. *(a)* According to the question,

$$5\left(-\frac{1}{2}\right)^{n-1} = \frac{5}{1024}$$

$$\Rightarrow \left(-\frac{1}{2}\right)^{n-1} = \left(-\frac{1}{2}\right)^{10}$$

$$\Rightarrow n-1 = 10$$

$$\Rightarrow n = 11$$

42. *(a)* Give that, $77x \equiv 88 \pmod 5$

$\Rightarrow 77x - 88$ is divisible by 5.

$\therefore \quad x = 4$

[$\because 77 \times 4 - 88 = 308 - 88 = 220$ is divisible by 5]

43. *(b)* Let p : Paris is in France.

q : London is in England.

$\therefore$ We have, $p \wedge q$

Its negation is $\sim(p \wedge q) = \sim p \vee \sim q$

ie, Paris is not in France or London is not in England.

44. *(a)* Given, ${}^{2n}C_3 : {}^{n}C_2 = \frac{44}{3}$

$$\Rightarrow \frac{2n!}{(2n-3)!\,3!} \times \frac{2!(n-2)!}{n!} = \frac{44}{3}$$

$$\Rightarrow \frac{2n(2n-1)(2n-2)}{3} \times \frac{1}{n(n-1)} = \frac{44}{3}$$

$$\Rightarrow \frac{2}{3} \times 2(2n-1) = \frac{44}{3}$$

$$\Rightarrow 2n - 1 = 11 \Rightarrow 2n = 12 \Rightarrow n = 6$$

45. *(a)* Now, $\lim_{x \to 0} f(x) = \lim_{x \to 0} \frac{3x + \tan^2 x}{x}$

$$= \lim_{x \to 0} \frac{3 + 2\tan x \sec^2 x}{1}$$

$$= 3$$

Since, $f(x)$ is continuous at $x = 0$

$\therefore \quad f(0) = 3$

46. *(c)* $0.2 + 0.22 + 0.222 + \ldots n$ terms

$= 2(0.1 + 0.11 + 0.111 + \ldots n \text{ terms})$

$$= 2\left(\frac{1}{10} + \frac{11}{100} + \frac{111}{1000} + \ldots n \text{ terms}\right)$$

$$= \frac{2}{9}\left(\frac{9}{10} + \frac{99}{100} + \frac{999}{1000} + \ldots n \text{ terms}\right)$$

$$= \frac{2}{9}\left(1 - \frac{1}{10} + 1 - \frac{1}{100} + 1 - \frac{1}{1000} + \ldots n \text{ terms}\right)$$

$$= \frac{2}{9}\left[n - \left(\frac{1}{10} + \frac{1}{100} + \frac{1}{1000} + \ldots n\right)\right]$$

$$= \frac{2}{9}\left[n - \frac{1}{10}\,\frac{\left\{1 - \left(\frac{1}{10}\right)^n\right\}}{\left(1 - \frac{1}{10}\right)}\right]$$

$$= \frac{2}{9}\left[n - \frac{1}{10} \times \frac{10}{9} \cdot \left(\frac{10^n - 1}{10^n}\right)\right]$$

$$= \frac{2}{9}\left[n - \frac{1}{9}(1 - 10^{-n})\right]$$

47. *(d)* $(1 + \omega - \omega^2)^5 + (1 - \omega + \omega^2)^5$

$$= (-\omega^2 - \omega^2)^5 + (-\omega - \omega)^5$$

$$= -32(\omega^2)^5 + (-32)\omega^5$$

$$= -32(\omega + \omega^2) = 32$$

48. *(d)* Lines, $x + y = 6$, $2x + y = 4$ and $x + 2y = 5$ intersect at points $(-2, 8)$, $(7, -1)$ and $(1, 2)$. Now, all these points lie on

$$x^2 + y^2 - 17x - 19y + 50 = 0$$

49. *(b)* $\lim_{x \to 0} \frac{\sin 3x - \sin x}{\sin x}$

$$= \lim_{x \to 0} \frac{2\cos 2x \sin x}{\sin x} = 2$$

50. *(c)* Product of roots of equation is

$$\frac{2m-1}{m} = (-1) \text{ given}$$

$$\Rightarrow 2m - 1 = -m$$

$$\Rightarrow 3m = 1 \Rightarrow m = \frac{1}{3}$$

51. *(a)* Let $I = \int_0^{2a} \frac{f(x)}{f(x) + f(2a - x)}\,dx$ ...(i)

$$I = \int_0^{2a} \frac{f(2a-x)}{f(2a-x) + f(x)}\,dx \quad \text{...(ii)}$$

On adding Eqs. (i) and (ii), we get

$$2I = \int_0^{2a} \frac{f(2a-x) + f(x)}{f(2a-x) + f(x)}\,dx = \int_0^{2a} dx = 2a$$

$\Rightarrow I = a$

52. (*c*) Let $I = \int \frac{\sqrt{\tan x}}{\sin x \cos x} dx = \int \frac{\sqrt{\tan x}}{\tan x} \sec^2 x \, dx$

$$= \int \frac{1}{\sqrt{\tan x}} \cdot \sec^2 x \, dx = \frac{\sqrt{\tan x}}{1/2} = 2\sqrt{\tan x} + c$$

53. (*a*) The given equation of curve is $(1 + x^2) y = 2 - x$, it meets x-axis at $(2, 0)$

or it can be rewritten as $y = \frac{2 - x}{1 + x^2}$

On differentiating w.r.t. x, we get

$$\frac{dy}{dx} = \frac{(1 + x^2)(-1) - (2 - x)(2x)}{(1 + x^2)^2}$$

$$= \frac{-1 - x^2 - 4x + 2x^2}{(1 + x^2)^2}$$

$$\Rightarrow \left(\frac{dy}{dx}\right)_{(2, 0)} = \frac{-1 - 4 - 8 + 8}{25} = -\frac{1}{5}$$

$\therefore$ Required equation of tangent is

$$(y - 0) = -\frac{1}{5}(x - 2) \Rightarrow x + 5y = 2$$

54. (*b*) $\lim_{x \to 0} \left(\frac{1 - \cos mx}{1 - \cos nx}\right)$

$$= \lim_{x \to 0} \frac{2\sin^2 (mx/2)}{(mx/2)^2} \cdot \frac{(nx/2)^2}{2\sin^2 (nx/2)} \cdot \frac{m^2}{n^2} = \frac{m^2}{n^2}$$

55. (*c*) $|\mathbf{a} + \mathbf{b} + \mathbf{c}|^2 = |\mathbf{a}|^2 + |\mathbf{b}|^2 + |\mathbf{c}|^2 + 2(\mathbf{a} \cdot \mathbf{b} + \mathbf{b} \cdot \mathbf{c} + \mathbf{c} \cdot \mathbf{a})$

$$= a^2 + a^2 + a^2$$

($\because$ **a**, **b** and **c** are mutually perpendicular)

$$\Rightarrow \quad |\mathbf{a} + \mathbf{b} + \mathbf{c}| = \sqrt{3}a$$

56. (*a*) We know that K_n, the complete graph of n vertices is a connected graph in which degree of each vertex is $n - 1$. Since, a graph is Eulerian if and only if it is connected and degree of each vertex is even, we conclude that K_n is an Euler graph if and only if n is odd.

57. (*a*) Now, $5^2 \equiv 4 \pmod 7$

$$5^4 \equiv 16 \pmod 7 = 2 \pmod 7$$

$$(5^4)^5 \equiv 2^5 \pmod 7$$

i.e. $\quad 5^{20} \equiv 32 \pmod 7 = 4 \pmod 7$

$\therefore$ Remainder is 4.

58. (*c*) Since, $a \cdot b = (a * b) + 3$

$\therefore \quad 4 \cdot 7 = (4 * 7) + 3 = 7 + 3 = 10$

59. (*b*) Total degree $= 2(n - 1) = 2n - 2$

Let p be the number of pendant vertices. Then $(n - p - 1)$ are vertices with degree 3, one vertex of degree 2. Total degree $= 2 + p + 3(n - p - 1)$

$$\Rightarrow \quad 2n - 2 = -1 - 2p + 3n$$

$$\Rightarrow \quad p = \frac{n + 1}{2} = \frac{9 + 1}{2} \quad (\because n = 9 \text{ given})$$

$$= 5$$

60. (*d*) $\frac{1}{\log_{25} 10} + \frac{1}{\log_4 10} + \frac{1}{\log_{\sqrt{2}} 10} + \frac{1}{\log_{\sqrt{5}} 10}$

$$= \frac{\log 25}{\log 10} + \frac{\log 4}{\log 10} + \frac{\log \sqrt{2}}{\log 10} + \frac{\log \sqrt{5}}{\log 10}$$

$$= \frac{2\log 5 + \frac{1}{2}\log 5}{\log 10} + \frac{2\log 2 + \frac{1}{2}\log 2}{\log 10}$$

$$= \frac{5}{2}\left(\frac{\log 5 + \log 2}{\log 10}\right) = \frac{5 \log 10}{2 \log 10} = \frac{5}{2}$$

61. (*a*) Here, $a = 2, b = 3, h = \frac{5}{2}$

$$\therefore \quad \tan \theta = \frac{2\sqrt{\left(\frac{5}{2}\right)^2 - 6}}{2 + 3}$$

$$\Rightarrow \quad m = \frac{1}{5}$$

62. (*b*) Let AB be the line $2x - y - 3 = 0$ ...(i)

Its slope = 2

Slope of $PM = -\frac{1}{2}$

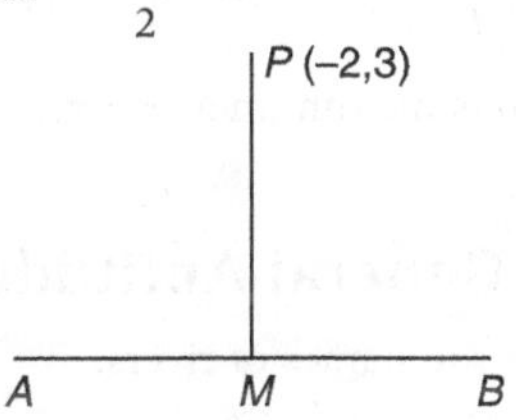

Equation of PM is

$$y - 3 = -\frac{1}{2}(x + 2)$$

$\Rightarrow \quad x + 2y - 4 = 0$ ...(ii)

On solving Eqs. (i) and (ii), we get

$$x = 2, y = 1$$

$\therefore$ Foot of perpendicular is (2, 1).

63. (*a*) $\int_{-\pi/2}^{\pi/2} |\sin x| \, dx = 2\int_0^{\pi/2} \sin x \, dx$

$$= 2[-\cos x]_0^{\pi/2} = 2$$

64. (b) $\because \begin{bmatrix} 2+x & 3 & 4 \\ 1 & -1 & 2 \\ x & 1 & -5 \end{bmatrix}$ is a singular matrix.

$$\therefore \quad \begin{vmatrix} 2+x & 3 & 4 \\ 1 & -1 & 2 \\ x & 1 & -5 \end{vmatrix} = 0$$

$$\Rightarrow (2+x)(5-2) - 3(-5-2x) + 4(1+x) = 0$$

$$\Rightarrow \quad 6 + 3x + 15 + 6x + 4 + 4x = 0$$

$$\Rightarrow \quad 13x + 25 = 0$$

$$\Rightarrow \quad x = -\frac{25}{13}$$

65. (c) $(b^2 + c^2)\, a \cos A + (c^2 + a^2)\, b \cos B + (a^2 + b^2)\, c \cos C$

$= ab\,(b \cos A + a \cos B) + bc\,(b \cos C + c \cos B) + ca\,(a \cos C + c \cos A)$

$= abc + abc + abc = 3\,abc$

66. (a) Given, $y = a\,(1 - \cos x)$

On differentiating w.r.t. x, we get

$$\frac{dy}{dx} = a \sin x$$

Again differentiating w.r.t. x, we get

$$\frac{d^2y}{dx^2} = a \cos x$$

Put $\frac{dy}{dx} = 0$ for maxima or minima

$$\Rightarrow \quad a \sin x = 0 \Rightarrow x = \pi$$

$$\therefore \quad \left(\frac{d^2y}{dx^2}\right)_{x=\pi} = a \cos \pi = -a$$

$\therefore$ Function is maximum at $x = \pi$.

67. (b) Volume of parallelopiped $= \begin{vmatrix} 2 & -3 & 0 \\ 1 & 1 & -1 \\ 3 & 0 & -1 \end{vmatrix}$

$= 2(-1) + 3(-1+3) = -2 + 6 = 4$ cu unit.

68. (a) It is not possible for wheel W_n $(n \geq 3)$ to be bipartite.

69. (c)

```
364)462(1
    364
    ---
     98)364(3
        294
        ---
         70)98(1
            70
            --
            28)70(2
               56
               --
               14)28(2
                  28
                  --
                   ×
```

$\therefore$ GCD of 364 and 462 is 14.

70. (d) Given system of equations are

$$x + 4y - 2z = 3$$
$$3x + y + 5z = 7$$
and $$2x + 3y + z = 5$$

$$\therefore \quad \Delta = \begin{vmatrix} 1 & 4 & -2 \\ 3 & 1 & 5 \\ 2 & 3 & 1 \end{vmatrix}$$

$$= 1(1-15) - 4(3-10) - 2(9-2)$$
$$= -14 + 28 - 14 = 0$$

and $$\Delta_2 = \begin{vmatrix} 1 & 3 & -2 \\ 3 & 7 & 5 \\ 2 & 5 & 1 \end{vmatrix}$$

$$= 1 \neq 0$$

$\therefore$ No solution will exist.

English & General Aptitude

16. (a) All are the names of rivers.

17. (c) 'Leap' is the movement of rabbit; 'Frisk' is the movement of 'Lamb' and 'Trot' is the movement of 'Donkey'.

18. (b) All are stationery goods.

19. (c) All are week days.

20. (a) They have two vowels 'a' and 'e'.

21. (d) None of the options is correct as they are the names of currencies.

22. (d) 'Jaipur' is the capital of 'Rajasthan'; 'Bengaluru' is the capital of 'Karnataka' and 'Mumbai' is the capital of 'Maharashtra'.

23. (c) 'Colombo' is the capital of Sri Lanka; 'Kathmandu' is the capital of 'Nepal' and 'Havana' is the capital of 'Cuba'.

24. (d) 'Squeak' is the sound produced by Mice; Hiss is the sound produced by Snake and Howl is the sound produced by Jackal.

25. (b) 'Kathak' is the classical dance of North India; 'Bharatnatyam' is the classical dance of Tamilnadu while 'Odissi' is the classical dance of Orissa.

SOLVED PAPER 2016

Manipal

Engineering Entrance Exam

Physics

1. Two particles of charges $+e$ and $+2e$ are at 16 cm away from each other. Where should another charge q be placed between them, so that the system remains in equilibrium?
   (a) 24 cm from $+e$
   (b) 12.23 cm from $+e$
   (c) 80 cm from $+e$
   (d) 6.63 cm from $+e$

2. How does the energy difference between the consecutive energy level vary as the quantum number n increases?
   (a) Increases
   (b) Decreases
   (c) Remains unchanged
   (d) First increases and then decreases

3. The ionisation potential of hydrogen atom is 13.6 eV. How much energy need to be supplied to ionise the hydrogen atom in the first excited state?
   (a) 13.6 eV
   (b) 27.2 eV
   (c) 3.4 eV
   (d) 6.8 eV

4. If 10% of the main current is to be passed through the moving coil galvanometer of resistance 99 Ω, then the required shunt resistance will be
   (a) 9.9 Ω
   (b) 10 Ω
   (c) 11 Ω
   (d) 9 Ω

5. A mixture of two gases is contained in a vessel. The gas 1 is monoatomic and gas 2 is diatomic and the ratio of their molecular masses $M_1/M_2 = 1/4$. What is the ratio of root mean square speeds of the molecules of two gases?
   (a) 2 (b) 4 (c) 8 (d) 16

6. A current carrying coil is bent sharply so as to convert it into a double loop both carrying current in the same direction. If B be the initial magnetic field at the centre, then what will be the final concentric magnetic field?
   (a) $2B$ (b) $4B$
   (c) $8B$ (d) Zero

7. White light reflected at normal incidence from a soap film has minima at 6500 Å and maxima at 7500 Å in the visible region with not minimum in between. If μ is $\frac{5}{3}$ for the, thin film, the thickness of film is
   (a) 7.40×10^{-7} cm
   (b) 9.75×10^{-5} mm
   (c) 9.70×10^{-7} cm
   (d) 9.75×10^{-4} mm

8. A long wire carries a steady current. It is bent into a circle of one turn and the magnetic field at the centre of the coil is B. It is then bent into a circular loop of n turns. The magnetic field at the centre of the coil will be
   (a) nB (b) n^2B (c) $2nB$ (d) $2n^2B$

9. Two springs are connected to a block of mass M placed on a frictionless surface as shown below. If both the springs have a spring constant k, then the frequency of oscillation of the block is

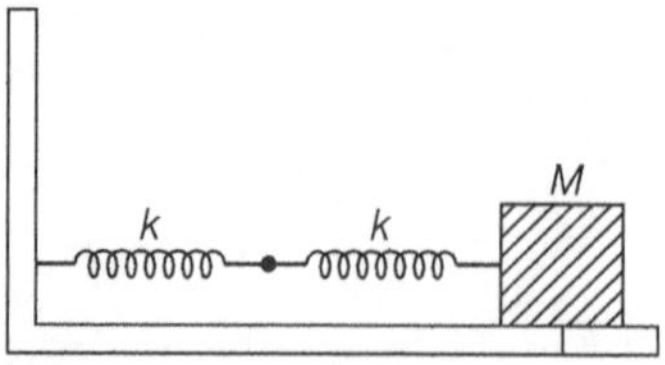

(a) $\frac{1}{2\pi}\sqrt{\frac{k}{M}}$ (b) $\frac{1}{2\pi}\sqrt{\frac{k}{2M}}$

(c) $\frac{1}{2\pi}\sqrt{\frac{2k}{M}}$ (d) $\frac{1}{2\pi}\sqrt{\frac{M}{k}}$

10. Four identical balls, A, B, C and D are placed on a smooth horizontal plane such that B, C and D are in one line and are in contact and A is kept along the same line but a distance away. Now, A is set in motion with speed v along the line and it strikes the ball B. After this collision, which is elastic.

(a) Balls A, B, C will be at rest and D will move with speed v

(b) A will be at rest, where B, C and D will move together with speed $v/3$

(c) A will rebound with speed $v/2$ while B, C and D will move together with speed $v/2$

(d) A, B, C and D will move together with speed $v/4$

11. Two coils X and Y are placed in a circuit such that a current changes by 3 A in coil X and magnetic flux changes of 1.2 Wb occurs in Y. The value of mutual inductance of the coils is

(a) 0.2 H (b) 0.4 H

(c) 0.6 H (d) 3.6 H

12. A projectile has the maximum range 500 m. If the projectile is thrown up an inclined plane of 30° with the same (magnitude) velocity, the distance covered by it along the inclined plane will be

(a) 250 m (b) 500 m

(c) 400 m (d) 1000 m

13. Nuclear reactor in which U-235 is used as fuel. Uses 2 kg of U-235 in 30 days. Then, power output of the reactor will be (given energy released per fission = 185 MeV)

(a) 43.5 MW (b) 58.5 MW

(c) 69.6 MW (d) 73.1 MW

14. Two capacitors of capacitances 3μF and 6 μF are charged to a potential of 12 V each. They are now connected to each other, with the positive plate of each joined to the negative plate of the other. The potential difference across each will be

(a) 4 V (b) 12 V

(c) zero (d) 3 V

15. The combination of gates shown in the figure below produces

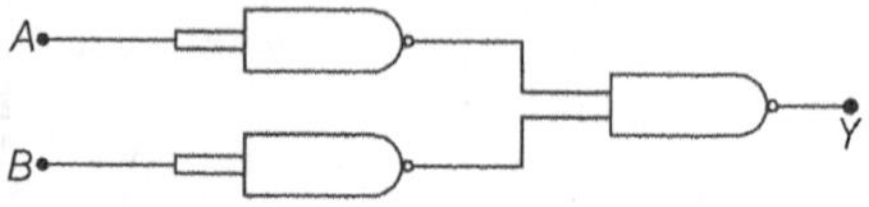

(a) NOR gate (b) OR gate

(c) AND gate (d) XOR gate

16. A thin dielectric rod of length l lies along X-axis. One end of the rod is placed at the origin and the other end of the rod is placed at $(l, 0)$. A total charge Q is distributed uniformly along the length of the rod. What is the potential at a point $(x, 0)$ when $x > l$?

(a) $\frac{Q}{4\pi\varepsilon_0}\cdot\log_e\left(\frac{x}{x-1}\right)$ (b) $\frac{Q}{4\pi\varepsilon_0 l}\cdot\log_e\left(\frac{x}{l}\right)$

(c) $\frac{Q}{4\pi\varepsilon_0 l}\cdot\log_e\left(\frac{x-l}{x}\right)$ (d) $\frac{Q}{4\pi\varepsilon_0 l}$

17. A ray of light is travelling through a medium of refractive index $\frac{1}{\sqrt{2}}$ with respect to air. When it is incident on the surface making an angle 45° with the surface, which of the following will take place?

(a) Angle of refraction will be 45°

(b) Angle of refraction will be 90°

(c) The ray will be internally reflected

(d) None of the above

18. If a wave travelling in positive x-direction with $A = 0.2$ m/s, velocity $= 360$ m/s and $\lambda = 60$ m, then correct expression for the wave is

(a) $y = 0.2 \sin\left[2\pi\left(6t + \frac{x}{60}\right)\right]$

(b) $y = 0.2 \sin\left[\pi\left(6t + \frac{x}{60}\right)\right]$

(c) $y = 0.2 \sin\left[2\pi\left(6t - \frac{x}{60}\right)\right]$

(d) $y = 0.2 \sin\left[\pi\left(6t - \frac{x}{60}\right)\right]$

19. In a potentiometer experiment for measuring the emf of a cell, the null point is at 240 cm when we have a 500 Ω resistor in series with the cell and galvanometer. If the series resistance reduced to half, then the null point will be at

(a) 480 cm (b) 240 cm
(c) 120 cm (d) 60 cm

20. Two coils each of self-inductance L are connected in parallel. If they are separated by a large distance, then what will be the self-inductance of combination?

(a) $\frac{L}{4}$ (b) $\frac{L}{2}$
(c) L (d) $2L$

21. Two bar magnets having same geometry with magnetic moments m and $2m$, are firstly placed in such a way that their similar poles one are on same side. Then, its time period of oscillation is T_1. Now, the polarity of the one of the magnet is reversed. Then, time period of oscillation is T_2, then

(a) $T_1 < T_2$ (b) $T_1 = T_2$
(c) $T_1 > T_2$ (d) $T_2 = \infty$

22. A solenoid of 2.5 m length and 2.0 cm diameter possesses 10 turns per cm. A current of 0.5 A is flowing through it. The magnetic induction at axis inside the solenoid is

(a) $2\pi \times 10^{-4}$ T (b) $2\pi \times 10^{-5}$ T
(c) $2\pi \times 10^{-6}$ T (d) $2\pi \times 10^{-7}$ T

23. In an *L-C-R* circuit which one of the following statements is correct?

(a) L and R oppose each other
(b) R value increase with frequency
(c) The inductive reactance increases with frequency
(d) The capacitive reactance increases with frequency

24. What will be the kinetic energy of an electron having de-Broglie wavelength 2Å?

(a) 37.5 eV (b) 75 eV
(c) 150 eV (d) 300 eV

25. In an *L-C-R* series circuit, the values of R, X_L and X_C are 120 Ω, 180Ω and 130Ω, what is the impedance of the circuit?

(a) 120 Ω (b) 130 Ω
(c) 180 Ω (d) 330 Ω

26. Particle which can be added to the nucleus of an atom without changing its chemical properties are

(a) neutrons
(b) electrons
(c) protons
(d) α-particle

27. Two sources give interference pattern which is observed on a screen, placed at a distance D from the sources. The fringe width is given as $2W$. If the distance of sources from the screen is doubled, then the fringe width will

(a) become $W/2$
(b) remains the same
(c) become W
(d) become $4W$

28. A particle moves along X-axis from $x = 0$ to $x = 5$ cm under the influence of force given by $F = (7 - 2x + 3x^2)$ N. The work done in the process is

(a) 70 J (b) 270 J
(c) 35 J (d) 135 J

29. A sphere of mass m moving with a constant velocity u hits another stationary sphere of the same mass. If e is the coefficient of restitution, then the ratio of velocities of the two spheres after collision $\left(\frac{V_1}{V_2}\right)$ will be

(a) $\frac{1-e}{1+e}$ (b) $\frac{1+e}{1-e}$
(c) $\frac{e+1}{1-e}$ (d) $\frac{e-1}{e+1}$

30. The molar specific heat of oxygen at constant pressure $C_p = 7.03$ cal/mol °C and $R = 8.31$ J/mol°C. The amount of heat taken by 5 moles of oxygen when heated at constant volume from 10°C to 20°C will be approximately.

(a) 25 cal (b) 50 cal
(c) 253 cal (d) 500 cal

31. The half-life of ^{215}At is 100 μs. The time taken for the radioactivity of a sample of this nucleus to decay to $\frac{1}{16}$th of its initial value is

(a) 6.3 μs (b) 40 μs
(c) 300 μs (d) 400 μs

32. A mass m is placed on a wedge (triangular block) of mass M. The wedge moves on a smooth horizontal surface what should be the force F applied on the wedge to the right, so that m remains stationary with respect to wedge. (Ignore any friction)

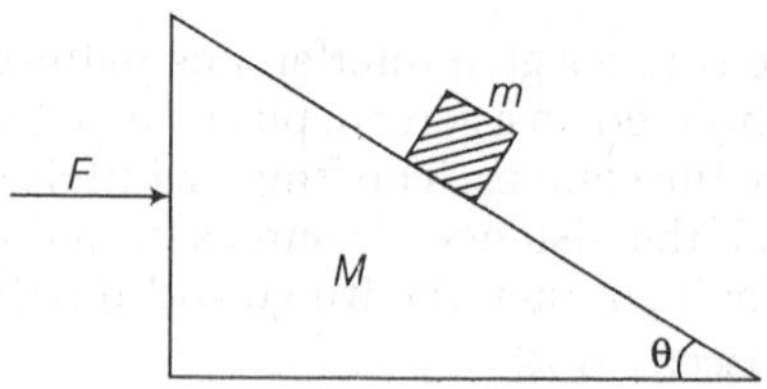

(a) $g \tan\theta$
(b) $mg\cos\theta$
(c) $(M+m)\, g\tan\theta$
(d) $(M+m)g\,\text{cosec}\,\theta$

33. A block of mass 1 kg slides down a curved track that is one quadrant of a circle of radius 1 m. Speed of the block at the bottom is 2 m/s. Work done by the frictional force on the block when it reaches at the bottom is

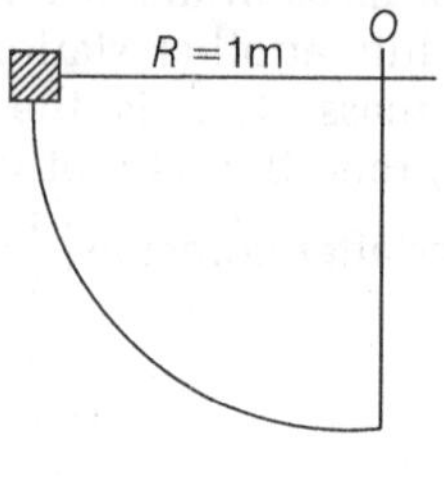

(a) 8 J (b) −8 J
(c) 4 J (d) 0 J

34. In the arrangement, the minimum value of tension in the string to prevent it from sliding down is

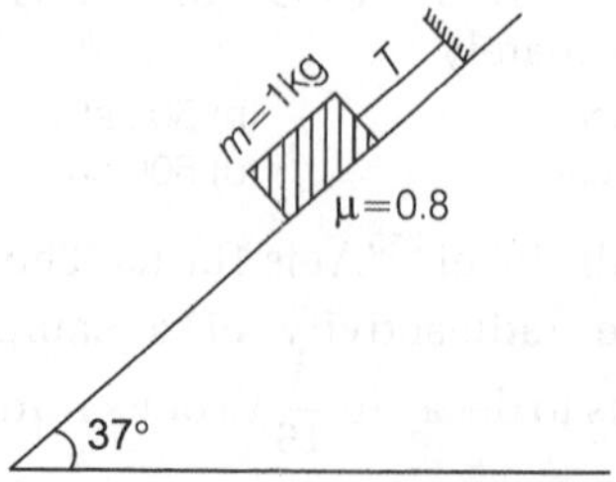

(a) 6 N (b) 8 N (c) 10.8 N (d) zero

35. In a planetary motion, the areal velocity of position vector of a planet depends on angular velocity (ω) and distance (r) of the planet from the sun. The correct relation for areal velocity $\left(\frac{dA}{dt}\right)$ is

(a) $\frac{dA}{dt} \propto \omega r$
(b) $\frac{dA}{dt} \propto \omega^2 r$
(c) $\frac{dA}{dt} \propto \omega r^2$
(d) $\frac{dA}{dt} \propto \sqrt{\omega r}$

36. Two blocks A and B are placed one over the other on a smooth horizontal surface. The maximum horizontal force that can be applied on the upper block B, so that A and B move without separation is 49N. The coefficient of friction between A and B is

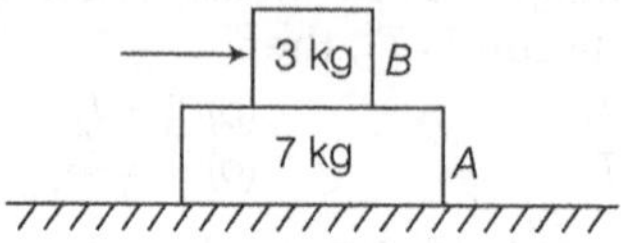

(a) 0.2 (b) 0.3
(c) 0.5 (d) 0.8

37. In the system shown in figure $M_1 > M_2$ and pulley and threads are ideal. System is held at rest by thread BC. Just after thread BC is burnt.

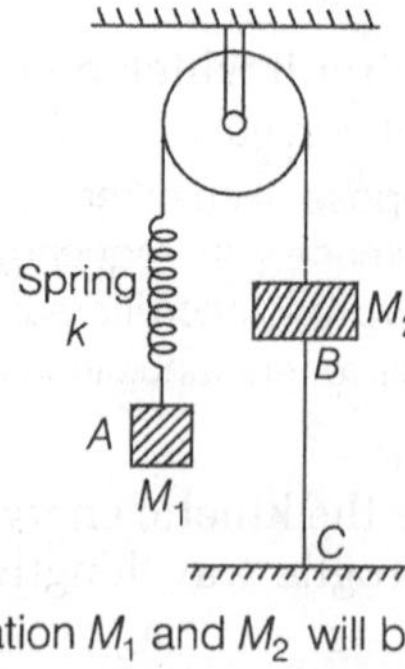

(a) Acceleration M_1 and M_2 will be upward
(b) Magnitude of acceleration of both masses will be $\frac{M_1 - M_2}{M_1 + M_2} g$
(c) Acceleration of M_1 and M_2 will be equal to zero
(d) Acceleration of M_1 will be equal to zero, which that of M_2 will be $\frac{M_1 - M_2}{M_2} g$ upward

38. A thin walled glass sphere of radius 2.5 cm is filled with water. An object (*O*) is placed at 7.5 cm from the surface of the sphere. Neglecting the effect of glass wall, at what distance the image (*I*) of the object, measured from the centre of sphere is formed?

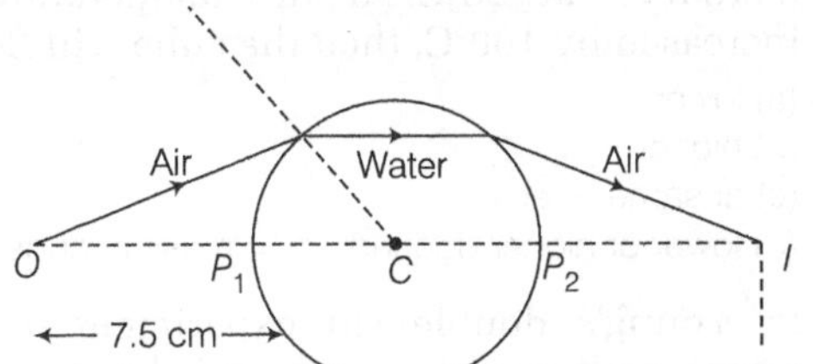

(a) 20 cm (b) 15 cm (c) 10 cm (d) 7.5 cm

39. An artificial satellite moving in a circular orbit around the earth has a total (kinetic + potential) energy E_0, its potential energy is

(a) $-E_0$ (b) $1.5E_0$ (c) $2E_0$ (d) E_0

40. Two junction diodes one of germanium (Ge) and other of silicon (Si) are connected as shown in figure to a battery of emf 12 V and a load resistance 10 k Ω. The germanium diode conducts at 0.3 V and silicon diode at 0.7 V. When a current flows in the circuit, then the potential of terminal *Y* will be

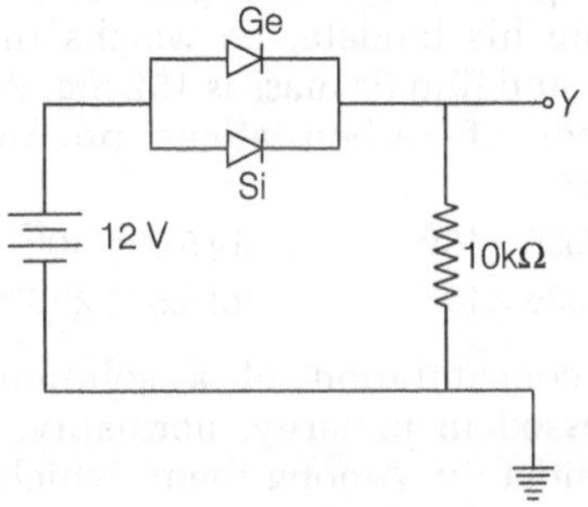

(a) 12 V (b) 11 V
(c) 11.3 V (d) 11.7 V

41. Fifty six tuning forks are arranged in a series such that each fork gives 6 beats per second with the preceding one. Assuming the frequency of the last fork to be double of the first fork, the frequency of the last fork should be

(a) 220 Hz (b) 330 Hz
(c) 440 Hz (d) 660 Hz

42 The wavelength of incident light falling on a photosensitive surface is changed from 2000 Å to 2100 Å. The corresponding change in stopping potential is

(a) 0.03 V (b) 0.3 V
(c) 3 V (d) 3.3 V

43. The path difference between two waves

$$y_1 = a_1 \sin\left(\omega t - \frac{2\pi x}{\lambda}\right) \text{ and}$$

$$y_2 = a_2 \cos\left(\omega t - \frac{2\pi x}{\lambda} + \phi\right) \text{ is}$$

(a) $\frac{\lambda\phi}{2\pi}$ (b) $\frac{\lambda}{2\pi}\left(\phi + \frac{\pi}{2}\right)$
(c) $\frac{2\pi}{\lambda}\left(\phi - \frac{\pi}{2}\right)$ (d) $\frac{2\pi\phi}{\lambda}$

44. In photoelectric effect, the photo current

(a) increases with increase of frequency of incident photon
(b) decreases with increase of frequency of incident photon
(c) does not depend on the frequency of photon but depends only on intensity of incident light
(d) depends both on intensity and frequency of incident beam

45. A copper ball of radius *r* is moving with a uniform velocity *v* in the mustard oil and the dragging force acting on the ball is *F*. The dragging force on the copper ball of radius 2*r* with uniform velocity 2*v* in the mustard oil is

(a) *F* (b) 2*F*
(c) 4*F* (d) 8*F*

46. In a photoemissive cell with exciting wave length λ, the fastest electron has a speed *v*. If the exciting wavelength is changed to 3λ/4, then the speed of the fastest emitted electron will be

(a) $v\left(\frac{3}{4}\right)^{\frac{1}{2}}$

(b) $v\left(\frac{4}{3}\right)^{\frac{1}{2}}$

(c) less than $v\left(\frac{4}{3}\right)^{\frac{1}{2}}$

(d) greater than $v\left(\frac{4}{3}\right)^{\frac{1}{2}}$

47. A block of wood of side 40 cm floats in water in such a way that its lower face is 5 cm below the free surface of water. What is the weight of the block?

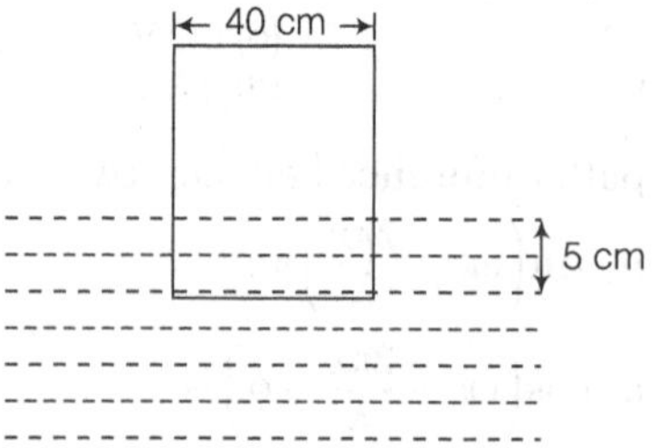

(a) 64 kg
(b) 16 kg
(c) 8 kg
(d) cannot be determined as density of wood is not given

48. A stone is projected from the ground with velocity 50 m/s at an angle 30° with horizontal. It crosses a wall after 3 s. How far beyond the wall the stone will strike the ground? ($g = 10\ m/s^2$)

(a) 90.2 m (b) 12.5 m
(c) 86.5 m (d) $125\sqrt{3}$ m

49. An iron cube floats in a vessel containing mercury at 20°C. If the temperature is increased by 100°C, then the cube will float

(a) lower
(b) higher
(c) at same level
(d) lower or higher depending on mass of cube

50. In Young's double slit experiment, when wavelength used is 600 Å and the screen is 40 cm from the slits, then the fringes are 0.012 cm apart. What is the distance between the slits?

(a) 0.024 cm
(b) 2.4 cm
(c) 0.24 cm
(d) 0.2 cm

Chemistry

1. A gaseous compounds having carbon, hydrogen and oxygen was burnt in the presence of oxygen. After combustion, 1 volume of the gaseous compound produces 2 volumes of CO_2 and 2 volumes of steam. Determines the molecular formula of the compound, it all the volumes were measured under the same conditions of pressure and temperature.

(a) $C_2H_4O_2$ (b) $C_3H_6O_3$
(c) $C_4H_8O_2$ (d) $C_5H_{10}O_3$

2. A silicon chip used in an integrated circuit of computer has a mass of 5.68 mg. The number of Si atoms present in this chip are

(a) 1.44×10^{20} atoms (b) 1.8×10^{20} atoms
(c) 2.1×10^{20} atoms (d) None of these

3. Cost of sugar ($C_{12}H_{22}O_{11}$) is ₹ 50 per kg. Calculate its cost per mol.

(a) ₹ 17.1 per mol
(b) ₹ 27.1 per mol
(c) ₹ 14.1 per mol
(d) None of the above

4. A man writes his biodata with carbon pencil on the plane paper having mass 150 mg. After writing his biodata, he weighs the written paper and find its mass is 152 mg. What is the number of carbon atoms present in the paper?

(a) 1.0036×10^{20} (b) 5.02×10^{20}
(c) 1.0036×10^{23} (d) 0.502×10^{20}

5. The concentration of a solution can be expressed in molarity, normality, formality and molality. Among them, which mode of expression is the most accurate for the all conditions?

(a) Molarity (b) Formality
(c) Normality (d) Molality

6. Calculate the energy of photons having wavelength, 5×10^{-7}m falls on a metal surface of work function, 3.4×10^{-19} J.

(a) 3.97×10^{-19} J
(b) 3.55×10^{-19} J
(c) 2.97×10^{-19} J
(d) 2.57×10^{-19} J

7. Select the correct options according to Hund's rule and Pauli exclusion principle.

I. [⇅ | |]   II. [⇅ | ⇅ |]

III. [⇅ | ↑ | ↑]   IV. [⇅ | ⇅ | ↑]

Codes
(a) Both I and II
(b) Both II and III
(c) Both I and III
(d) Both III and IV

8. The electronic configurations of chromium (at. no. 24) and copper (at. no. 29) are
(a) $[Ar]4s^1 3d^5$ and $[Ar]\ 4s^1 3d^{10}$
(b) $[Ar]\ 4s^2 3d^4$ and $[Ar]\ 4s^1 3d^{10}$
(c) $[Ar]\ 4s^1 3d^5$ and $[Ar]\ 4s^2 3d^9$
(d) $[Ar]\ 4s^2 3d^4$ and $[Ar]\ 4s^2 3d^9$

9. The Pauli exclusion principle applies to the following option.
(a) H (b) H^+
(c) H^- (d) He^+

10. Metals are usually used as catalysts belong to
(a) *s*-block (b) *p*-block
(c) *d*-block (d) *f*-block

11. Which of the following has the highest electron affinity?
(a) K (b) O^- (c) F^- (d) O

12. Which of the following compounds has lowest boiling point?
(a) KCl (b) CuCl
(c) $CuCl_2$ (d) CsCl

13. Which of the following compounds has zero dipole moment?
(a) CH_2Cl_2 (b) NH_3 (c) CH_4 (d) PH_3

14. Which of the following compounds show hydrogen bonding?
(a) HCl (b) C_2H_6
(c) RCH_2CHO (d) RCH_2NHCH_3

15. In van der Waals' equation of gaseous state, the constant '*b*' is a measure of
(a) volume occupied by the molecules
(b) intermolecular repulsions
(c) intermolecular attraction
(d) pressure correction

16. Which of the following is incorrect statement for liquid?
(a) Surface tension of a liquid decreases with increase in temperature
(b) Viscosity of a liquid decreases with decrease in temperature
(c) Vapour pressure of a liquid increases with increase in temperature
(d) Droplets of a liquid on flat surface are slightly flattened due to gravity

17. Thermodynamically, in which form of carbon is the most stable?
(a) Coal (b) Fullerenes
(c) Diamond (d) Graphite

18. Enthalpy of formation of HF and HCl are – 161 kJ and – 92 kJ, respectively. Select the incorrect statement on the basis of the above fact.
(a) HF is more stable than HCl
(b) HCl is more stable than HF
(c) Affinity of fluorine to hydrogen is greater than affinity of chlorine to hydrogen
(d) HF and HCl are exothermic compound

19. A reaction is spontaneous at low temperature, but non-spontaneous at high temperature. Which of the following condition is true for the chemical reaction?
(a) $\Delta H < 0, \Delta S < 0$ (b) $\Delta H > 0, \Delta S = 0$
(c) $\Delta H < 0, \Delta S > 0$ (d) $\Delta H > 0, \Delta S > 0$

20. The chemical reaction of a gas phase is given below.

$$2A(g) + B(g) \rightleftharpoons C(g) + D(g)$$

Which one of the following changes will affect the value of K_C?
(a) Addition of inert gas
(b) Increasing in temperature
(c) Addition of reactants
(d) Addition of catalyst

21. For the chemical equilibrium,

$$2NO_2(g) \rightleftharpoons N_2O_4(g); +14.6\ \text{kcal}$$

increase in temperature
(a) favours the formation of N_2O_4
(b) favours the decomposition of N_2O_4
(c) does not affect equilibrium
(d) stop the reaction

22. The property of the alkaline earth metals which increases with their atomic number?

(a) Ionisation energy
(b) Electronegativity
(c) Solubility of their hydroxides in water
(d) Solubility of their sulphates in water

23. Which of the following element has maximum photoelectric effect?

(a) Li (b) Na
(c) K (d) Cs

24. Among boron halides, which is the strongest Lewis is acid?

(a) BF_3 (b) BI_3
(c) BCl_3 (d) BBr_3

25. The compound gives blood red colouration when its Lassaigne's extract is treated with alkali and ferric chloride?

(a) Benzamide (b) Phenyl hydrozine
(c) Diphenyl sulphide (d) Thiourea

26. The IUPAC name for the hydrocarbon represented by

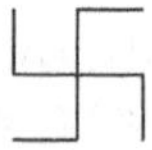

(a) 3, 3-diethylpentane (b) 2-ethylpentane
(c) tetraethyl carbon (d) *neo*-nonane

27. Keto-enol tautomerism is not observed in

(a) $CH_3COCH_2COCH_3$ (b) $C_6H_5COCH_2COCH_3$
(c) $C_6H_5COCH{=}CH_2$ (d) $C_6H_5COC_6H_5$

28. Which of the following statements are true with respect to electronic displacement in a covalent bond?

I. Inductive effect operates through π-bond.
II. Resonance effect operates through σ-bond.
III. Inductive effect operates through σ-bond.
IV. Resonance effect operates through π-bond.
V. Resonance and inductive effects operate through σ-bond.

(a) III and IV (b) I and II
(c) II and IV (d) I and V

29. The most susceptible to nucleophilic attack at the carbonyl group among the given compound is

(a) CH_3COOCH_3 (b) CH_3CONH_2
(c) CH_3COOCH_2 (d) CH_3COCl

30. Among the following species, the most stable carbonium ion is

(a) $C_6H_5\overset{+}{C}HC_6H_5$ (b) $C_6H_5\overset{+}{C}H_2$
(c) $CH_3\overset{+}{C}H_2$ (d) $C_6H_5CH_2\overset{+}{C}H_2$

31. For the following,

(i) I^- (ii) Cl^- (iii) Br^-

the increasing order of nucleophilicity would be

(a) $Br^- < Cl^- < I^-$ (b) $I^- < Cl^- < Br^-$
(c) $Cl^- < Br^- < I^-$ (d) $I^- < Br^- < Cl^-$

32. Which will undergo fastest S_N2 substitution reaction when treated with NaOH?

(a) $\begin{array}{c} H \\ | \\ H-C-CH_2-CH_2-CH_3 \\ | \\ Br \end{array}$

(b) $\begin{array}{c} CH_3 \\ | \\ H-C-Br \\ | \\ C_2H_5 \end{array}$

(c) $\begin{array}{c} CH_3 \\ | \\ CH_3-C-Br \\ | \\ CH_3 \end{array}$

(d) $\begin{array}{c} CH_3 \\ | \\ H_5C_2-C-Br \\ | \\ H \end{array}$

33. Which one of the following is aromatic?

(a) Cycloheptatrienyl cation
(b) Cycloheptrariene
(c) Cyclooctateraene
(d) Cyclopentadienyl cation

34. The following compound which cannot be prepared by Wurtz reaction?

(a) Methane (b) Ethane (c) Butane (d) Propane

35. Anti-Markownikoff's rule addition reaction is not observed in the following compound.

(a) propene (b) 1-butene
(c) but-2-ene (d) pent-2-ene

36. The most reactive towards electrophilic reagent is

(a) 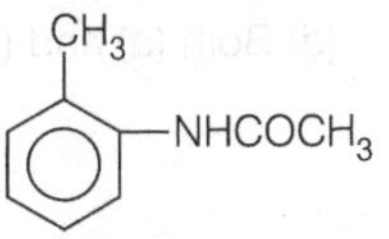 (b) 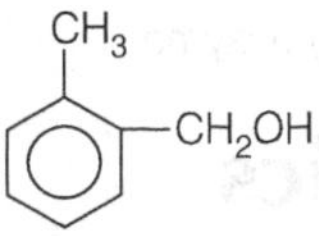

(c) 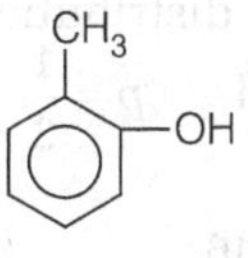 (d) 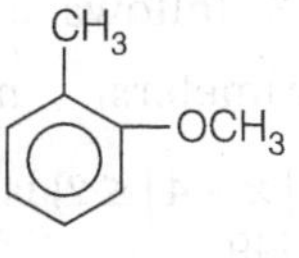

37. The maximum heat of hydrogenation among the following is

(a)  (b) 

(c)  (d) 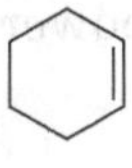

38. The process of 'eutrophication' is due to

(a) increase in concentration of radioactive substances in water
(b) the reduction in concentration of the dissolved oxygen in water due to phosphate pollution in water
(c) increase in concentration of fluoride ion in water
(d) increase in concentration of insecticide in water

39. Match the Column I and Column II and pick the correct matching from the codes given below.

Column I	Column II
A. Peroxy acetyl nitrate	1. Waste incineration
B. Polycyclic aromatic	2. Global warming
C. Dioxins	3. Photochemical smog
D. Indigo	4. Carcinogens
E. IR active molecules	5. Vat dye

	A	B	C	D	E
(a)	3	4	1	5	2
(b)	1	2	3	4	5
(c)	3	5	1	2	4
(d)	5	3	1	2	4

40. The flame colours of metal ions are due to

(a) metal excess defect
(b) metal deficiency defect
(c) Schottky defect
(d) Frenkel defect

41. The gas mixture which is used by the divers inside the sea is

(a) $O_2 + N_2$ (b) $O_2 + Ar$
(c) $O_2 + Xe$ (d) $O_2 + He$

42. van't Hoff factor, when benzoic acid is dissolved in benzene, will be

(a) 2 (b) 1 (c) 0.5 (d) 1.5

43. 1 C electricity deposits

(a) half of electrochemical equivalent of Ag
(b) electrochemical equivalent of Ag
(c) 96500 g of Ag
(d) 10.8 g of Ag

44. Which of the following does not influence the rate of reaction?

(a) Molecularity of the reaction
(b) Temperature of the reaction
(c) Concentration of the reactants
(d) Nature of the reactants

45. During the adsorption of krypton on activated charcoal at low temperature.

(a) $\Delta H < 0$ and $\Delta S > 0$
(b) $\Delta H > 0$ and $\Delta S > 0$
(c) $\Delta H < 0$ and $\Delta S < 0$
(d) $\Delta H > 0$ and $\Delta S < 0$

46. In which of the following the pyrophosphate bond of ATP is broken down and linkage takes place between two molecules?

(a) I some rases (b) Lyases
(c) Transferases (d) Ligases

47. Which of the following option best describes the 'hydrogen economy?.'

(a) Using dihydrogen in an efficient manner
(b) Dihydrogen is carried and stored in the form of liquid or gas
(c) Addition of dihydrogen to the reactant
(d) Both (a) and (b)

48. Electrolytic reduction of alumina to aluminium by Hall-Heroult process is carried out

(a) in the presence of NaCl
(b) in the presence of Hnorite
(c) in the presence of cryolite which forms a melt with lower melting temperature
(d) in the presence of cryolite which forms a melt with higher melting temperature

49. Which of the following anions is present in the chain structure of silicates?

(a) $(Si_2O_5^{2-})_n$ (b) $(Si_2O_3^{2-})_n$ (c) $Si_2O_4^{4-}$ (d) $Si_2O_7^{6-}$

50. Myosin(s) is/are

(a) fibrous proteins (b) globular proteins
(c) enzyme (d) Both (a) and (b)

Mathematics

1. If $2a + 3b + 6c = 0$, then atleast one root of the equation $ax^2 + bx + c = 0$ lies in the interval

(a) (0, 1) (b) (1, 2)
(c) (2, 3) (d) None of these

2. If **a** and **b** be two perpendicular unit vectors such that $\mathbf{x} = \mathbf{b} - (\mathbf{a} \times \mathbf{x})$, then $|\mathbf{x}|$ is equal to

(a) 1 (b) $\sqrt{2}$
(c) $\frac{1}{\sqrt{2}}$ (d) $\sqrt{3}$

3. If $aN = \{an : n \in N\}$ and $bN \cap cN = dN$, where $a, b, c \in N$ and b, c are coprime, then

(a) $b = cd$ (b) $c = bd$
(c) $d = bc$ (d) None of these

4. If $f(x) = \int_0^1 e^{|t - x|} dt \quad (0 \le x \le 1)$, then the maximum value of $f(x)$ is

(a) $\sqrt{e} - 1$ (b) $2(e - 1)$
(c) $(e - 1)$ (d) $2(\sqrt{e} - 1)$

5. The mean deviation from the mean for the set of observations $-1, 0, 4$ is

(a) less than 3
(b) less than 1
(c) greater than 2.5
(d) greater than 4.9

6. If $f(x)$ is a non-negative continuous function for all $x \ge 1$ such that $f'(x) \le pf(x)$, where $p > 0$ and $f(1) = 0$, then $[f(\sqrt{e}) + f(\sqrt{\pi})]$ is equal to

(a) 0 (b) negative
(c) positive (d) None of these

7. If $t_n = \sum_{r=0}^{n} \frac{1}{({}^nC_r)^k}$ and $S_n = \sum_{r=0}^{n} \frac{r}{({}^nC_r)^k}$, where $k \in Z^+$, then $\cos^{-1}\left(\frac{S_n}{nt_n}\right)$ is

(a) $\frac{\pi}{6}$ (b) $\frac{\pi}{3}$
(c) $\frac{\pi}{4}$ (d) $\frac{\pi}{2}$

8. If X follows a binomial distribution with parameters $n = 8$ and $p = \frac{1}{2}$, then $P(|x - 4| \le 2)$ is equal to

(a) $\frac{119}{128}$ (b) $\frac{116}{128}$
(c) $\frac{29}{128}$ (d) None of these

9. Sum of the last 30 coefficients in the expansion of $(1 + x)^{59}$, when expanded in ascending power of x is

(a) 2^{59} (b) 2^{58}
(c) 2^{30} (d) 2^{29}

10. If $\theta = \sin^{-1} x + \cos^{-1} x - \tan^{-1} x$, $1 \le x < \infty$, then the smallest interval in which θ lies is

(a) $\frac{\pi}{2} \le \theta \le \frac{3\pi}{4}$ (b) $0 \le \theta \le \frac{\pi}{4}$
(c) $-\frac{\pi}{4} \le \theta \le 0$ (d) $\frac{\pi}{4} \le \theta \le \frac{\pi}{2}$

11. If $2\tan^{-1}(\cos x) = \tan^{-1}(2\operatorname{cosec} x)$, then the value of x is

(a) $\frac{3\pi}{4}$ (b) $\frac{\pi}{4}$
(c) $\frac{\pi}{3}$ (d) None of these

12. If $\cos^{-1} p + \cos^{-1} q + \cos^{-1} r = 3\pi$, then $p^2 + q^2 + r^2 + 2pqr$ is equal to

(a) 3 (b) 1 (c) 2 (d) -1

13. The points on the curve $xy^2 = 1$ which are nearest to the origin, are

(a) $\left[\left(\frac{1}{2}\right)^{1/3}, \pm\left(\frac{1}{2}\right)^{-1/6}\right]$ (b) $\left[\left(\frac{1}{2}\right)^{1/3}, 2^{-1/6}\right]$
(c) $\left[2^{1/3}, \pm\left(\frac{1}{2}\right)^{-1/6}\right]$ (d) None of these

14. The statement $p \to (q \to p)$ is equivalent to

(a) $p \to (p \leftrightarrow q)$ (b) $p \to (p \to q)$
(c) $p \to (p \vee q)$ (d) $p \to (p \wedge q)$

15. Total number of polynomials of the form $x^3 + ax^2 + bx + c$ that are divisible by $x^2 + 1$, where $a, b, c \in \{1, 2, 3, \ldots, 10\}$ is equal to

(a) 90 (b) 45 (c) 5 (d) 10

16. Total number of regions in which 'n' coplanar lines can divide the plane, it is known that no two lines are parallel and no three of them are concurrent, is equal to

(a) $\frac{1}{2}(n^2 + n + 2)$ (b) $\frac{1}{2}(n + 3n^2)$

(c) $\frac{1}{2}(3n + n^2)$ (d) $(n^2 - n + 2)$

17. The value of $\lim_{n\to\infty}\left[\sqrt[3]{n^2 - n^3} + n\right]$ is

(a) $\frac{1}{3}$ (b) $-\frac{1}{3}$ (c) $\frac{2}{3}$ (d) $-\frac{2}{3}$

18. The coefficient of the term independent of x in the expansion of $\left(\frac{x+1}{x^{2/3} - x^{1/3} + 1} - \frac{x-1}{x - x^{1/2}}\right)^{10}$ is

(a) 210 (b) 105

(c) 70 (d) 112

19. If $\begin{bmatrix} \alpha & \beta \\ \gamma & -\alpha \end{bmatrix}$ is to be the square root of two rowed unit matrix, then α, β and γ should satisfy the relation

(a) $1 + \alpha^2 + \beta\gamma = 0$ (b) $1 - \alpha^2 - \beta\gamma = 0$

(c) $1 - \alpha^2 + \beta\gamma = 0$ (d) $\alpha^2 + \beta\gamma - 1 = 0$

20. The differential equation of the family of curves $y = Ae^{3x} + Be^{5x}$, where A and B are arbitrary constants, is

(a) $\frac{d^2y}{dx^2} + 8\frac{dy}{dx} + 15y = 0$

(b) $\frac{d^2y}{dx^2} - \frac{dy}{dx} + y = 0$

(c) $\frac{d^2y}{dx^2} - 8\frac{dy}{dx} + 15y = 0$

(d) None of the above

21. The derivative of $\sin^{-1}(2x\sqrt{1 - x^2})$ with respect to $\sin^{-1}(3x - 4x^3)$ is

(a) $\frac{2}{3}$ (b) $\frac{3}{2}$

(c) $\frac{1}{2}$ (d) 1

22. $\int \frac{(2x^{12} + 5x^9)}{(1 + x^3 + x^5)^3} dx$ equals

(a) $\frac{x^{10}}{2(x^5 + x^3 + 1)^2} + C$

(b) $\frac{x^2 + 2x}{(x^5 + x^3 + 1)} + C$

(c) $\log(x^5 + x^3 + 1 + \sqrt{(2x^{12} + 5x^9)} + C)]$

(d) None of the above

23. $\begin{vmatrix} \sin\theta & \cos\theta & \sin 2\theta \\ \sin\left(\theta + \frac{2\pi}{3}\right) & \cos\left(\theta + \frac{2\pi}{3}\right) & \sin\left(2\theta + \frac{4\pi}{3}\right) \\ \sin\left(\theta - \frac{2\pi}{3}\right) & \cos\left(\theta - \frac{2\pi}{3}\right) & \sin\left(2\theta - \frac{4\pi}{3}\right) \end{vmatrix}$ equals

(a) $3\sin\theta$ (b) $\sin^3\theta$

(c) 0 (d) None of these

24. Area bounded by the curves $y = x^2$ and $y = 2 - x^2$ is

(a) $\frac{8}{3}$ sq units (b) $\frac{3}{8}$ sq unit

(c) $\frac{3}{2}$ sq units (d) None of these

25. $\lim_{n\to\infty}\left(\frac{1^2}{1 - n^3} + \frac{2^2}{1 - n^3} + \ldots + \frac{n^2}{1 - n^3}\right)$ is equal to

(a) $\frac{1}{3}$ (b) $-\frac{1}{3}$ (c) $\frac{1}{6}$ (d) $-\frac{1}{6}$

26. The value of $\int \frac{\sin x + \cos x}{3 + \sin 2x} dx$ is

(a) $\frac{1}{4}\log\left(\frac{2 - \sin x + \cos x}{2 + \sin x - \cos x}\right) + C$

(b) $\frac{1}{2}\log\left(\frac{2 + \sin x}{2 - \sin x}\right) + C$

(c) $\frac{1}{4}\log\left(\frac{1 + \sin x}{1 - \sin x}\right) + C$

(d) None of the above

27. The equation of the circle passing through the point $(2a, 0)$ and whose radical axis is $x = \frac{a}{2}$ with respect to the circle $x^2 + y^2 = a^2$, will be

(a) $x^2 + y^2 - 2ax = 0$ (b) $x^2 + y^2 + 2ax = 0$

(c) $x^2 + y^2 + 2ay = 0$ (d) $x^2 + y^2 - 2ay = 0$

28. The coordinates of the point on the parabola $y = x^2 + 7x + 2$, which is nearest to the straight line $y = 3x - 3$, are

(a) $(-2, -8)$ (b) $(1, 10)$
(c) $(2, 20)$ (d) $(-1, -4)$

29. The area bounded by the parabolas $y^2 = 4a(x + a)$ and $y^2 = -4a(x - a)$ is

(a) $\frac{16}{3}a^2$ (b) $\frac{8}{3}a^2$
(c) $\frac{4}{3}a^2$ (d) None of these

30. The solution of the differential equation $\frac{dy}{dx} = \sin(x + y)\tan(x + y) - 1$ is

(a) $\operatorname{cosec}(x + y) + \tan(x + y) = x + C$
(b) $x + \operatorname{cosec}(x + y) = C$
(c) $x + \tan(x + y) = C$
(d) $x + \sec(x + y) = C$

31. Fifteen coupons are numbered 1, 2, ..., 15, respectively. Seven coupons are selected at random one at a time with replacement. The probability that the largest number appearing on a selected coupon is 9, is

(a) $\left(\frac{9}{10}\right)^6$ (b) $\left(\frac{8}{15}\right)^7$
(c) $\left(\frac{3}{5}\right)^7$ (d) None of these

32. Let $h(x) = \min(x, x^2)$, for every real numbers x. Then,

(a) h is not continuous for all x
(b) h is differentiable for all x
(c) $h'(x) \neq 1$, for all $x > 1$
(d) h is not differentiable at two values of x

33. If the function

$$f(x) = \begin{cases} (\cos x)^{1/x}, & x \neq 0 \\ k, & x = 0 \end{cases}$$ is continuous

at $x = 0$, then the value of k is

(a) 8 (b) 1
(c) -1 (d) None of these

34. If $f(x) = \frac{e^{1/x} - 1}{e^{1/x} + 1}$, $x \neq 0$ and $f(0) = 0$, then $f(x)$ is

(a) left continuous at 0 (b) right continuous at 0
(c) discontinuous at 0 (d) continuous at 0

35. The value of $\sum_{p=1}^{6} 2\left(\sin\frac{2p\pi}{7} - i\cos\frac{2p\pi}{7}\right)$ is

(a) $2i$ (b) $-2i$
(c) 2 (d) 1

36. A letter lock contains 5 rings each marked with four different letters. The number of all possible unsuccessful attempts to open the lock is

(a) 625 (b) 1024
(c) 624 (d) 1023

37. The domain of definition of the function $f(x) = \frac{1}{\sqrt{|[|x|] - 1]| - 5}}$ is

(a) $(-\infty, -7] \cup [7, \infty)$
(b) $(-\infty, \infty)$
(c) $(7, \infty)$
(d) $(-7, \infty)$

38. If $y^{1/m} + y^{-1/m} = 2x$, then $(x^2 - 1)y_2 + xy_1$ is equal to

(a) $m^2 y$ (b) $-m^2 y$
(c) $\pm m^2 y$ (d) None of these

39. The condition that the straight line $cx - by + b^2 = 0$ may touch the circle $x^2 + y^2 = ax + by$ is

(a) $abc = 1$ (b) $a = c$
(c) $b = ac$ (d) None of these

40. If $0 < \alpha, \beta, \gamma < \frac{\pi}{2}$ such that $\alpha + \beta + \gamma = \frac{\pi}{2}$ and $\cot\alpha, \cot\beta, \cot\gamma$ are in AP, then the value of $\cot\alpha\cot\gamma$ is

(a) 1 (b) 3
(c) $\cot^2\beta$ (d) None of these

41. In a ΔABC, if $b = 2$, $\angle B = 30°$, then the area of the circumcircle of ΔABC (in sq units) is

(a) π (b) 2π
(c) 4π (d) 6π

42. The sum of the series

$$\frac{3}{4 \cdot 8} - \frac{3 \cdot 5}{4 \cdot 8 \cdot 12} + \frac{3 \cdot 5 \cdot 7}{4 \cdot 8 \cdot 12 \cdot 16} - \ldots \text{ is}$$

(a) $\sqrt{\frac{3}{2}} - \frac{3}{4}$ (b) $\sqrt{\frac{2}{3}} - \frac{3}{4}$
(c) $\sqrt{\frac{3}{2}} - \frac{1}{4}$ (d) $\sqrt{\frac{2}{3}} - \frac{1}{4}$

43. The value of $\int_0^{\pi} |\sin^3 \theta|\, d\theta$ is

(a) 0 (b) π
(c) $\frac{4}{3}$ (d) $\frac{3}{8}$

44. The value of $\int \frac{\cos x + x \sin x}{x^2 + x \cos x}\, dx$ is

(a) $\log \left| \frac{\sin x}{1 + \cos x} \right| + C$

(b) $\log \left| \frac{\sin x}{x + \cos x} \right| + C$

(c) $\log \left| \frac{2 \sin x}{x + \cos x} \right| + C$

(d) $\log \left| \frac{x}{x + \cos x} \right| + C$

45. Tangent to the ellipse $\frac{x^2}{32} + \frac{y^2}{18} = 1$ having slope $\frac{-3}{4}$ meet the coordinate axis at A and B. Then, the area of ΔAOB, where O is the origin, is

(a) 12 sq units (b) 8 sq units
(c) 24 sq units (d) 32 sq units

46. The value of $\lim_{x \to \pi/6} \frac{2\sin^2 x + \sin x - 1}{2\sin^2 x - 3\sin x + 1}$ is

(a) 3 (b) -3 (c) 6 (d) 0

47. If $\sin A + \cos B = a$ and $\sin B + \cos A = b$, then $\sin (A + B)$ is equal to

(a) $\frac{a^2 + b^2}{2}$ (b) $\frac{a^2 - b^2 + 2}{2}$
(c) $\frac{a^2 + b^2 - 2}{2}$ (d) None of these

48. In a ΔABC, if $\sin A \sin B = \frac{ab}{c^2}$, then the triangle is

(a) equilateral (b) isosceles
(c) right angled (d) obtuse angled

49. The direction cosines l, m, n of two lines are connected by the relations $l + m + n = 0$ $lm = 0$, then the angle between them is

(a) $\frac{\pi}{3}$ (b) $\frac{\pi}{4}$
(c) $\frac{\pi}{2}$ (d) 0

50. The foci of the conic section

$25x^2 + 16y^2 - 150x = 175$ are

(a) $(0, \pm 3)$ (b) $(0, \pm 2)$
(c) $(3, \pm 3)$ (d) $(0, \pm 1)$

51. If $x - 2y = 4$, then the minimum value of xy is

(a) -2 (b) 0 (c) 1 (d) -3

52. If **b** and **c** are any two non-collinear unit vectors and **a** is any vector, then

$(\mathbf{a} \cdot \mathbf{b})\, \mathbf{b} + (\mathbf{a} \cdot \mathbf{c}) + \frac{\mathbf{a} \cdot \mathbf{b} \times \mathbf{c}}{|\mathbf{b} \times \mathbf{c}|} \cdot (\mathbf{b} \times \mathbf{c})$ is equal to

(a) 0 (b) **a** (c) **b** (d) **c**

53. If θ is the semi-vertical angle of a cone of maximum volume and given slant height, then $\tan \theta$ is given by

(a) 2 (b) 1 (c) $\sqrt{2}$ (d) $\sqrt{3}$

54. The period of the function

$f(x) = \frac{\sin 8x \cos x - \sin 6x \cos 3x}{\cos 2x \cos x - \sin 3x \sin 4x}$ is

(a) π (b) 2π
(c) $\frac{\pi}{2}$ (d) None of these

55. The equation of the plane through the points $(2, 3, 1)$ and $(4, -5, 3)$ parallel to X-axis is

(a) $x + 4z = 7$ (b) $y - 4z = 7$
(c) $y + 4z = -7$ (d) $y + 4z = 7$

56. If Q is the image of the point $P(2, 3, 4)$ under the reflection in the plane $x - 2y + 5z = 6$, then the equation of the line PQ is

(a) $\frac{x-2}{-1} = \frac{y-3}{2} = \frac{z-4}{5}$

(b) $\frac{x-2}{1} = \frac{y-3}{-2} = \frac{z-4}{5}$

(c) $\frac{x-2}{-1} = \frac{y-3}{-2} = \frac{z-4}{5}$

(d) $\frac{x-2}{1} = \frac{y-3}{2} = \frac{z-4}{5}$

57. The equation of the plane passing through the line of intersection of the planes $x + y + z = 6$ and $2x + 3y + 4z + 5 = 0$ and perpendicular to the plane $4x + 5y - 3z = 8$ is

(a) $x + 7y + 13z - 96 = 0$ (b) $x + 7y + 13z + 96 = 0$
(c) $x + 7y - 13z - 96 = 0$ (d) $x - 7y + 13z + 96 = 0$

58. $1+\frac{2^3}{2!}+\frac{3^3}{3!}+\frac{4^3}{4!}+\ldots\infty$ equals

(a) 5e (b) 4e (c) 3e (d) 2e

59. The number of values of x, where $f(x)=\cos x+\cos\sqrt{2}x$ attains its maximum is

(a) 1 (b) 0
(c) 2 (d) infinite

60. A vector perpendicular to the plane containing the points $A(1,-1,2)$, $B(2,0,-1)$, $C(0,2,1)$ is

(a) $4\hat{i}+8\hat{j}-4\hat{k}$ (b) $8\hat{i}+4\hat{j}+4\hat{k}$
(c) $3\hat{i}+\hat{j}+2\hat{k}$ (d) $\hat{i}+\hat{j}-\hat{k}$

61. If $P=\begin{bmatrix}\sqrt{3}/2 & 1/2\\ -1/2 & \sqrt{3}/2\end{bmatrix}, A=\begin{bmatrix}1 & 1\\ 0 & 1\end{bmatrix}$ and $Q=PAP'$, then $P'Q^{2005}P$ is

(a) $\begin{bmatrix}1 & 1\\ 2005 & 1\end{bmatrix}$ (b) $\begin{bmatrix}1 & 2005\\ 0 & 1\end{bmatrix}$

(c) $\begin{bmatrix}1 & 0\\ 0 & 1\end{bmatrix}$ (d) $\begin{bmatrix}1 & 2005\\ 2005 & 1\end{bmatrix}$

62. $\int\cos^{-3/7}x\sin^{-11/7}x\,dx$ is equal to

(a) $\log|\sin^{4/7}x|+C$

(b) $\frac{4}{7}\tan^{4/7}x+C$

(c) $\frac{-7}{4}\tan^{-4/7}x+C$

(d) $\log|\cos^{3/7}x|+C$

63. All the chords of the hyperbola $3x^2-y^2-2x+4y=0$, subtending a right angle at the origin pass through the fixed point

(a) (1, –2) (b) (–1, 2)
(c) (1, 2) (d) None of these

64. The vector equation of the plane through the point $\hat{\mathbf{i}}+2\hat{\mathbf{j}}-\hat{\mathbf{k}}$ and perpendicular to the line of intersection of the plane $\mathbf{r}\cdot(3\hat{\mathbf{i}}-\hat{\mathbf{j}}+\hat{\mathbf{k}})=1$ and $\mathbf{r}\cdot(\hat{\mathbf{i}}-4\hat{\mathbf{j}}+2\hat{\mathbf{k}})=2$ is

(a) $\mathbf{r}\cdot(2\hat{i}+7\hat{j}-13\hat{k})=1$
(b) $\mathbf{r}\cdot(2\hat{i}-7\hat{j}-13\hat{k})=1$
(c) $\mathbf{r}\cdot(2\hat{i}+7\hat{j}+13\hat{k})=0$
(d) None of the above

65. The degree of the differential equation $x=1+\left(\frac{dy}{dx}\right)+\frac{1}{2!}\left(\frac{dy}{dx}\right)^2+\frac{1}{3!}\left(\frac{dy}{dx}\right)^3+\ldots$

(a) 3 (b) 2
(c) 1 (d) Not defined

66. The locus of the mid-points of the chords of the circle $x^2+y^2=16$ which are tangent to the hyperbola $9x^2-16y^2=144$, is

(a) $(x^2+y^2)^2=16x^2+9y^2$
(b) $(x^2-y^2)^2=16x^2-9y^2$
(c) $(x^2+y^2)^2=16x^2-9y^2$
(d) None of the above

67. The number of positive integers satisfying the inequality ${}^{n+1}C_{n-2}-{}^{n+1}C_{n-1}\le 50$ is

(a) 9 (b) 8 (c) 7 (d) 6

68. The intercepts on the straight line $y=mx$ by the line $y=2$ and $y=6$ is less than 5, then m belongs to

(a) $\left]-\frac{4}{3},\frac{4}{3}\right[$

(b) $\left]\frac{4}{3},\frac{3}{8}\right[$

(c) $\left]-\infty,-\frac{4}{3}\right[\cup\left]\frac{4}{3},\infty\right[$

(d) $\left]\frac{4}{3},\infty\right[$

69. A line passing through origin and is perpendicular to two given lines $2x+y+6=0$ and $4x+2y-9=0$. The ratio in which the origin divides this line, is

(a) 1 : 2 (b) 2 : 1 (c) 4 : 2 (d) 4 : 3

70. Let A and B be two events such that $P(\overline{A\cup B})=\frac{1}{6}$, $P(A\cap B)=\frac{1}{4}$ and $P(\overline{A})=\frac{1}{4}$, where $\overline{A}$ stands for complement of event A. Then, the events A and B are

(a) mutually exclusive and independent
(b) independent but not equally likely
(c) equally likely but not independent
(d) equally likely and mutually exclusive

English & General Aptitude

Directions (Q. Nos. 1-5) *Rearrange the following five sentences (A), (B), (C), (D) and (E) in proper sequence to form a meaningful paragraph and then answer the questions given below them.*

(A) The history of mankind is full of such fightings between communities, nations and people.

(B) From the primitive weapons of warfare, man has advanced to the modern nuclear weapons.

(C) Ever since the dawn of civilisation, man has been fighting with man.

(D) A modern war is scientific in character, but the effect is the same, wiping human existence out of this Earth.

(E) The only difference now seems to be in the efficiency of the instruments used for killing each other.

1. Which of the following should be the FIRST sentence?

(a) A (b) B
(c) C (d) D
(e) E

2. Which of the following should be the SECOND sentence?

(a) A (b) B
(c) C (d) D
(e) E

3. Which of the following should be the THIRD sentence?

(a) A (b) B
(c) C (d) D
(e) E

4. Which of the following should be the FOURTH sentence?

(a) A (b) B (c) C (d) D
(e) E

5. Which of the following should be the FIFTH (LAST) sentence?

(a) A (b) B
(c) C (d) D
(e) E

Directions (Q. Nos. 6-10) *In these questions, look at the underlined part of the given sentence. Below the sentence are given three possible substitutions for the underlined part. If any of substitutions; (a), (b) or (c) is better than the underlined part, choose that substitution as your response. If none of the substitution improves the sentence, choose (d) as your response. Thus, a 'No improvement' response will be signified by the letter (d).*

6. It is no good to cry over spilt milk.

(a) It is no good crying
(b) It is of no good to cry
(c) It is of no good crying
(d) No improvement

7. He has been working off and on for several years to compile a dictionary.

(a) on or off
(b) on and off
(c) regularly
(d) No improvement

8. Rohit assured Sunita that he would look at her work while she was on leave.

(a) would overlook
(b) would look after
(c) will look
(d) No improvement

9. Newton wanted to know why did the apple fall to the ground.

(a) know that why did the apple fall
(b) know why the apple fell
(c) know that why the apple fell
(d) No improvement

10. He was extremely unhappy because of the inordinately delay.

(a) the inordinate delaying
(b) the inordinate delay
(c) the inordinately delaying
(d) No improvement

Directions (Q. Nos. 11-12) *Each of the following items consists of a sentence followed by four words or group of words. Select the opposite of the word occurring in the sentence as per the context.*

11. His family <u>severed</u> ties with him for marrying inter-caste.

(a) joined (b) included
(c) detached (d) disrupted

12. Domestic violence is a very <u>inhuman</u> act.

(a) indifferent
(b) compassionate
(c) terrible
(d) ferocious

Directions (Q. Nos. 13-14) *Four words are given in each question, out of which only one word is correctly spelt. find the correctly spelt word and mark your answer.*

13. (a) arcitecture
(b) architectere
(c) architecetur
(d) architecture

14. (a) epilogue
(b) eppilogue
(c) epiloge
(d) epiloug

Fill in the blanks with correct prepositions.

15. The poor have to work .... morning to evening.

(a) in (b) to
(c) from (d) before

16. John, in the morning, started walking towards North and then turn towards opposite side of the Sun. He then turn left again and stops. Which direction is he facing now?

(a) North
(b) West
(c) South
(d) East

17. If $64 + 7 = 460$ and $25 + 8 = 212$, then $43 + 8 = ?$

(a) 360 (b) 376
(c) 332 (d) 356

18. What is the mirror image of the following figure?

Question Figure

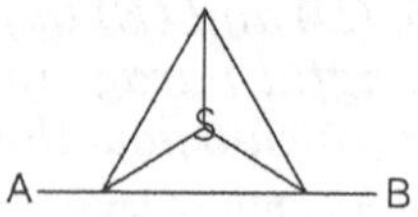

Answer Figures

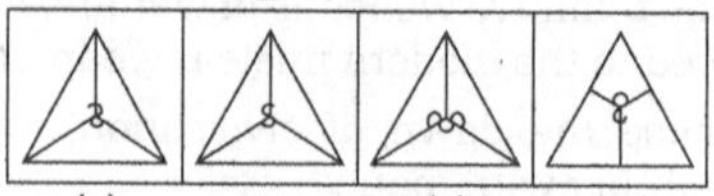

(a) (b) (c) (d)

19. If '+' mean '×', '−' means '÷', '×' means '−' and '÷' means '+', then find the value of the following equation. $6 + 64 - 8 \div 45 \times 8$

(a) 85 (b) 76
(c) 87 (d) 75

20. Which of the following has the given figure embedded in it?

Question Figure

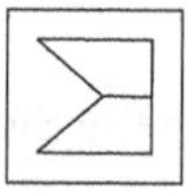

Answer Figures

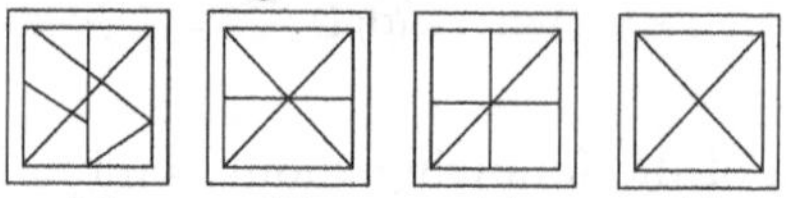

(a) (b) (c) (d)

21. 1, 4, 2, 3, 2, ?

(a) 2 (b) 5
(c) 3 (d) 4

22.

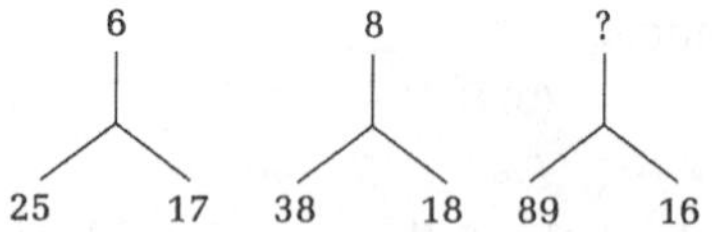

(a) 13 (b) 15 (c) 17 (d) 19

23. Two persons ride towards each other from two places 55 km apart, one riding at 12 km/h and the other at 10 km/h. When will they be 11 km apart?

(a) 2 h and 30 min (b) 1 h and 30 min
(c) 2 h (d) 2 h and 45 min

24. If $x - \frac{1}{x} = 2$, then what is the value of $x^2 + \frac{1}{x^2}$?

(a) 4 (b) 5 (c) 3 (d) 6

25. The population of a state is 20000. It increases by 20% during the first year and 30% during the second year. The population after two years will be

(a) 32000
(b) 40000
(c) 31200
(d) 30000

26. A reduction of 20% in the price of rice enable a buyer to buy 5 kg more for ₹ 1200. The reduced price per kg of rice will be

(a) 36 (b) 45
(c) 48 (d) 60

27. A man spends 60% of his income on different expenditures. His income is increased by 20% and his expenditure also increased by 10%. Find the percentage decrease in his saving.

(a) 10%
(b) 15%
(c) 20%
(d) 25%

28. Find the amount which Shyam will get on ₹ 4096, if he gave it for 18 months at $12\frac{1}{2}\%$ per annum, interest being compounded half yearly.

(a) ₹ 5813
(b) ₹ 4515
(c) ₹ 4913
(d) ₹ 5713

29. *A* works twice as fast as *B*. If *B* can complete a piece of work independently in 12 days, then what will be the number of days taken by *A* and *B* together to finish the work?

(a) 4
(b) 6
(c) 8
(d) 18

30. A well of diameter 3 m is dug 14 m deep. The Earth taken out of it has been spread evenly all around it in the shape of a circular ring of width 4 m to form an embankment. Find the height of the embankment.

(a) 4.25 m
(b) 2.250 m
(c) 1.125 m
(d) 1.750 m

Answers

Physics

1. (d)	**2.** (b)	**3.** (c)	**4.** (c)	**5.** (a)	**6.** (b)	**7.** (a)	**8.** (b)	**9.** (b)	**10.** (a)
11. (b)	**12.** (b)	**13.** (b)	**14.** (b)	**15.** (b)	**16.** (c)	**17.** (b)	**18.** (c)	**19.** (b)	**20.** (b)
21. (a)	**22.** (a)	**23.** (c)	**24.** (a)	**25.** (b)	**26.** (a)	**27.** (d)	**28.** (d)	**29.** (d)	**30.** (c)
31. (d)	**32.** (c)	**33.** (b)	**34.** (d)	**35.** (c)	**36.** (c)	**37.** (d)	**38.** (c)	**39.** (c)	**40.** (d)
41. (d)	**42.** (b)	**43.** (b)	**44.** (c)	**45.** (c)	**46.** (d)	**47.** (c)	**48.** (c)	**49.** (a)	**50.** (d)

Chemistry

1. (a)	**2.** (d)	**3.** (a)	**4.** (a)	**5.** (d)	**6.** (a)	**7.** (d)	**8.** (a)	**9.** (c)	**10.** (c)
11. (d)	**12.** (c)	**13.** (c)	**14.** (d)	**15.** (a)	**16.** (b)	**17.** (d)	**18.** (b)	**19.** (a)	**20.** (b)
21. (b)	**22.** (c)	**23.** (d)	**24.** (b)	**25.** (d)	**26.** (a)	**27.** (d)	**28.** (a)	**29.** (d)	**30.** (a)
31. (d)	**32.** (a)	**33.** (a)	**34.** (a)	**35.** (c)	**36.** (c)	**37.** (b)	**38.** (b)	**39.** (a)	**40.** (a)
41. (d)	**42.** (c)	**43.** (b)	**44.** (a)	**45.** (c)	**46.** (d)	**47.** (d)	**48.** (c)	**49.** (b)	**50.** (a)

Mathematics

1. (a)	**2.** (c)	**3.** (c)	**4.** (c)	**5.** (a)	**6.** (a)	**7.** (b)	**8.** (a)	**9.** (b)	**10.** (b)
11. (b)	**12.** (b)	**13.** (a)	**14.** (c)	**15.** (d)	**16.** (a)	**17.** (a)	**18.** (a)	**19.** (b)	**20.** (c)
21. (a)	**22.** (a)	**23.** (c)	**24.** (a)	**25.** (b)	**26.** (a)	**27.** (a)	**28.** (a)	**29.** (a)	**30.** (b)
31. (d)	**32.** (d)	**33.** (b)	**34.** (c)	**35.** (a)	**36.** (d)	**37.** (a)	**38.** (a)	**39.** (b)	**40.** (b)
41. (c)	**42.** (b)	**43.** (c)	**44.** (d)	**45.** (c)	**46.** (b)	**47.** (c)	**48.** (c)	**49.** (a)	**50.** (c)
51. (a)	**52.** (b)	**53.** (c)	**54.** (c)	**55.** (d)	**56.** (b)	**57.** (b)	**58.** (a)	**59.** (a)	**60.** (b)
61. (b)	**62.** (c)	**63.** (a)	**64.** (b)	**65.** (c)	**66.** (c)	**67.** (b)	**68.** (c)	**69.** (d)	**70.** (b)

General English & Aptitude

1. (c)	**2.** (a)	**3.** (b)	**4.** (d)	**5.** (e)	**6.** (a)	**7.** (d)	**8.** (b)	**9.** (b)	**10.** (b)
11. (a)	**12.** (b)	**13.** (d)	**14.** (a)	**15.** (c)	**16.** (c)	**17.** (d)	**18.** (a)	**19.** (a)	**20.** (b)
21. (d)	**22.** (b)	**23.** (c)	**24.** (d)	**25.** (c)	**26.** (c)	**27.** (a)	**28.** (c)	**29.** (a)	**30.** (d)

Explanations

Physics

1. 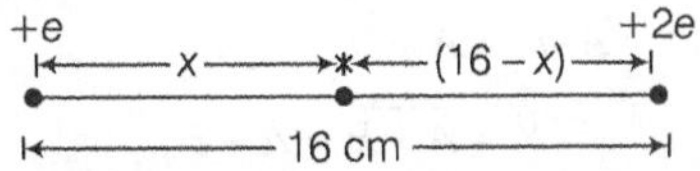

For the system to be in equilibrium, we have force on charge q due to $+e$ = force on charge q due to $+2e$

i.e. $$\frac{1}{4\pi\varepsilon_0}\cdot\frac{e.q}{x^2}=\frac{1}{4\pi\varepsilon_0}\cdot\frac{2e\cdot q}{(16-x)^2}$$

$$\Rightarrow \quad 2x^2=(16-x)^2$$

$$\Rightarrow \quad 2x^2=256+x^2-32x$$

$$[\because (a-b)^2=a^2+b^2-2ab]$$

$$\Rightarrow x^2+32x-256=0 \qquad \ldots(i)$$

On solving Eq. (i), we get

$$x=6.63\text{ cm}$$

2. We know that, $E_n \propto \frac{1}{n^2}$.

As n increases $(E_n - E_{n+1})$ decreases.

3. Ionisation energy of nth state $= -E_n$

For the first excited state, $n = 2$

and energy in that state is, $\frac{13.6}{2^2}\text{eV} = \frac{13.6}{4}\text{eV}$

$= 3.4$ eV

4. Here, $I_g = \frac{I}{10}$, $G = 99\ \Omega$

$\because$ The shunt, $$S=\frac{I_g}{I-I_g}\cdot G$$

$\therefore$ $$S=\frac{I/10}{I-\frac{I}{10}}\times 99$$

$\Rightarrow$ $$S=11\Omega$$

5. From kinetic theory of gases

$$p=\frac{1}{3}\frac{M}{V}\cdot e^2$$

Since, p and V are constant (same)

$\therefore$ $$M_1e_1^2=M_2\,e_2^2$$

$\Rightarrow$ $$\frac{e_1^2}{e_2^2}=\frac{M_2}{M_1}$$

$\Rightarrow$ $$\frac{e_1}{e_2}=\sqrt{\frac{M_2}{M_1}}=\sqrt{4/1}=2 \qquad \left[\text{given, } \frac{M_1}{M_2}=\frac{1}{4}\right]$$

6. Radius of the double loop, $r=\frac{R}{2}$

where, R is the radius of single loop B (centre)

$$=\frac{\mu_0}{4\pi}\cdot\frac{2\pi I}{R}.$$

So, magnetic field due to each coil is double of that due to single coil.

and total magnetic field

$$=2B+2B=4B$$

7. For minima to occur,

$$2\mu t=2n\frac{\lambda_1}{2} \qquad \ldots(i)$$

and for maxima

$$2\mu t=(2n-1)\frac{\lambda_2}{2} \qquad \ldots(ii)$$

From Eqs. (i) and (ii), we get

$$(2n-1)\lambda_2=2n\lambda_1$$

or $$(2n-1)7500=(2n)6500$$

$$[\text{given, } \lambda_2=7500\text{ Å}, \lambda_1=6500\text{ Å}]$$

or $$1500n-7500=13000n$$

or $$2000n=7500$$

or $$n=3.75$$

Substitution the value of (n) in Eq. (i), we get

$$2\times\frac{5}{3}\times t=2\times 3.75\times\frac{6500}{2}\times 10^{-10} \qquad \left[\because \mu=\frac{5}{3}\right]$$

$\Rightarrow$ $$t=7.40\times 10^{-7}\text{ cm}$$

8. $\therefore$ Current, $I=2\pi R=2\pi n\cdot r$ $\qquad \left[\because r=\frac{R}{n}\right]$

The magnetic field at the centre of the coil,

$$B=\frac{\mu_0}{4\pi}\cdot\frac{2\pi I}{R} \qquad \ldots(i)$$

and $$B'=\left(\frac{\mu_0}{4\pi}\cdot\frac{2\pi I}{r}\right)\cdot n \qquad \ldots(ii)$$

From Eqs. (i) and (ii), we get

$$B'=n^2B$$

9. The frequency of oscillation of a vibrating spring-block, system is given by

$$\nu=\frac{1}{2\pi}\sqrt{\frac{k_{\text{eff}}}{M}} \qquad \ldots(i)$$

Since, the springs are attached in series, each having spring constant k, so we have

$$\frac{1}{k_{\text{eff}}}=\frac{1}{k}+\frac{1}{k}=\frac{2}{k}$$

$\Rightarrow$ $$k_{\text{eff}}=\frac{k}{2}.$$

From Eq. (i), we get $\nu=\frac{1}{2\pi}\sqrt{\frac{k_{\text{eff}}}{M}}=\frac{1}{2\pi}\sqrt{\frac{k}{2M}}.$

10. It is a case of elastic collision, whole of energy of A is transferred to D.

11. Given, $d\phi = Wb, M = ?$

$$dI = 3\text{A}$$

As, $$\phi = MI$$

$\therefore$ $$d\phi = M\,dI$$

$\therefore$ $$M = \frac{d\phi}{dI} = \frac{1.2}{3} = 0.4\text{ H}$$

12. Since, range is maximum, therefore $\theta = 45°$.

Hence, using $R = u^2 \sin 2\theta/g$.

We find $500 = \frac{u^2}{g}$, i.e. $u = [500g]^{\frac{1}{2}}$.

Distance covered along the inclined plane can be obtained using relation

$$v^2 - v_0^2 = 2ax$$

$\Rightarrow$ $$0 - u^2 = 2 \times (-g\sin 30°) \times x$$

$[\because a = -g\sin 30°]$

$\therefore$ $$x = \frac{u^2}{g} = 500\text{ m}$$

13. Number of atoms in 2 kg of uranium

$$= \frac{6.02 \times 10^{23}}{235} \times 2000$$

$$= 5.12 \times 10^{24}$$

Therefore, energy obtained from these atoms

$$= 5.12 \times 10^{24} \times 185\text{ MeV}$$

$$= 5.12 \times 10^{24} \times 185 \times 10^6\text{ eV}$$

Energy obtained per second

$$= \frac{5.12 \times 10^{24} \times 185 \times 10^6 \times 16 \times 10^{-19}}{30 \times 24 \times 60 \times 60}$$

Solving the above expression

$$= 58.47 \times 10^6\text{ W}$$

$$\cong 58.5\text{ MW}$$

14. Given, $C_1 = 3\mu\text{F}, C_2 = 6\mu\text{F}, V_1 = V_2 = 12\text{ V}$

$$V = ?$$

The common potential across capacitors is given by

$$V = \frac{C_1V_1 + C_2V_2}{C_1 + C_2}$$

$$V = \frac{3 \times 10^{-6} \times 12 + 6 \times 10^{-6} \times 12}{(3+6) \times 10^{-6}} = 12\text{V}$$

15.

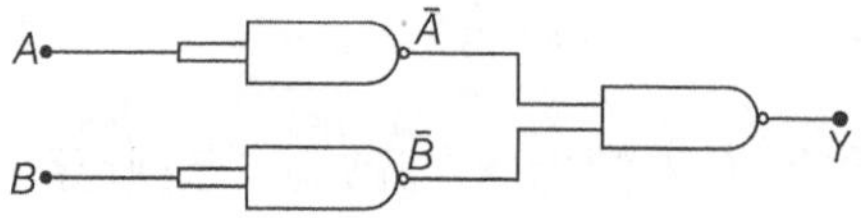

Here, the output of these gate

$$Y = \overline{\overline{A} \cdot \overline{B}}$$

But according to Boolean algebra,

$$\overline{A} \cdot \overline{B} = \overline{A + B}$$

$\therefore$ $$Y = \overline{\overline{A + B}} = A + B$$

or $$Y = A + B$$

It means the whole circuit of gates behaves as OR gate.

16.

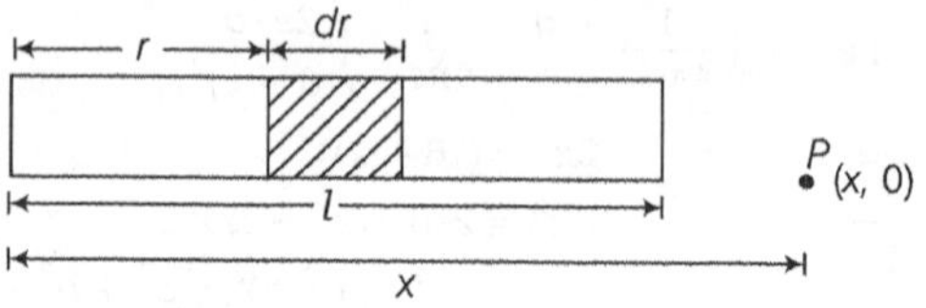

The potential due to a small element of length dr having charge dQ of the dielectric rod is given by

$$dV = \frac{dQ}{4\pi\varepsilon_0 r}$$

or $$dV = \frac{\lambda\,dr}{4\pi\varepsilon_0 r} \qquad \left[\because \lambda = \frac{dQ}{dr},\ \lambda\,dr = dQ\right]$$

or $$V = \frac{\lambda}{4\pi\varepsilon_0}\int_x^{x-l}\frac{dr}{r}$$

$$= \frac{\lambda}{4\pi\varepsilon_0}[\log_e r]_x^{x-l}$$

$$= \frac{\lambda}{4\pi\varepsilon_0} \cdot \log\left(\frac{x-l}{x}\right)$$

$$V = \frac{Q}{4\pi\varepsilon_0 l} \cdot \log e\left(\frac{x-l}{x}\right)$$

When $\lambda = \frac{Q}{l}$ (for whole length of rod)

17. Given, $\mu = \frac{1}{\sqrt{2}}$ w.r.t. air $i = 45°, r = ?$

Also, $$\mu = \frac{\sin i}{\sin r}$$

$\Rightarrow$ $$\frac{1}{\sqrt{2}} = \frac{\sin 45°}{\sin r} \Rightarrow \frac{1}{\sqrt{2}} = \frac{1/\sqrt{2}}{\sin r}$$

$\Rightarrow$ $$\sin r = 1$$

$\Rightarrow$ $$r = 90°$$

Note that the ray is incident from denser medium and is refracted into a rarer medium air.

18. General equation for a plane progressive wave travelling along positive x-direction is given by

$$y = A\sin(\omega t - kx)$$

$$= A\sin\left[2\pi\left(vt - \frac{x}{\lambda}\right)\right] \qquad \ldots\text{(i)}$$

Here, it is given

$$v = 360\text{ m/s}, \lambda = 60\text{ m}$$

We know that, frequency $\nu = \frac{v}{\lambda} = \frac{360}{60} = 6\text{ Hz}$

Substituting $A = 0.2\text{ m/s}, \nu = 6\text{ Hz}$

$\lambda = 60$ m in Eq. (i), we get

$$y = 0.2 \sin\left[2\pi\left(6t - \frac{x}{60}\right)\right]$$

19. The resistance in series with galvanometer does not affect the null point.
So, it will obtained at 240 cm.

20. In parallel combination, the resultant inductance L_P,

$$\frac{1}{L_P} = \frac{1}{L_1} + \frac{1}{L_2}$$

$$= \frac{1}{L} + \frac{1}{L} = \frac{2}{L}$$

$$\Rightarrow \quad L_P = \frac{L}{2}$$

21. When similar poles are on same side time period of oscillation T_1 is given by

$$T_1 = 2\pi\sqrt{\frac{I_1 + I_2}{(m_1 + m_2)B_H}} \quad \text{[given]}$$

$$m_1 = 2m, m_2 = m$$

$$T_1 = 2\pi\sqrt{\frac{I_1 + I_2}{(3m)B_H}} \quad \ldots\text{(i)}$$

When the polarity of magnet is reversed, time period of oscillation T_2 is given by

$$T_2 = 2\pi\sqrt{\frac{I_1 + I_2}{(m_1 - m_2)B_H}}$$

$$T_2 = 2\pi\sqrt{\frac{I_1 + I_2}{mB_H}} \quad \ldots\text{(ii)}$$

On comparing Eqs. (i) and (ii), we get

$$T_2 > T_1$$

or $\quad T_1 < T_2$

22. The magnetic induction along the axis of solenoid

$$B = \mu_0 nI$$

where, n = number of turns per unit length

and $\quad I$ = current

$$\because \quad \frac{\mu_0}{4\pi} = 10^{-7}\text{ T/A}$$

$$\Rightarrow \quad \mu_0 = 4\pi \times 10^{-7}\text{ T/A}$$

Given, $\quad n = 10\text{ cm}^{-1} = 1000\text{ m}^{-1}$

$$I = 0.5\text{ A}$$

$$\therefore \quad B = 4\pi \times 10^{-7} \times 1000 \times 0.5$$

$$= 2\pi \times 10^{-4}\text{ T}$$

23. In *L-C-R* circuit, inductive reactance is given by

$$X_L = \omega L$$

$$= 2\pi fL \quad [\because \omega = 2\pi f]$$

$$\therefore \quad X_L \propto f$$

Thus, inductive reactance increases with increase frequency.

24. The de-Broglie wavelength associated with electron

$$\lambda_e = \frac{1.227}{\sqrt{V}}\text{nm}$$

$$V = \frac{(12.27)^2}{\lambda_e^2} \times 10^{-2}$$

$$= \frac{150}{\lambda_e^2} \times 10^{-2}$$

$$\Rightarrow \quad V = \frac{150}{(0.2)^2}$$

$$= \frac{150 \times 10^{-2}}{4 \times 10^{-2}}$$

$$\therefore \quad E = \text{V eV} = 37.5\text{ eV}$$

25. Given, $R = 120\,\Omega$

$$X_L = 180\,\Omega, \; X_C = 130\,\Omega$$

The impedance of *L-C-R* circuit

$$Z = \sqrt{R^2 + (X_L - X_C)^2}$$

$$Z = \sqrt{(120)^2 + (180 - 130)^2}$$

$$= 130\,\Omega$$

26. Neutrons, as they are neutral particles.

27. Given, $W_1 = 2W$,

Fringe width, $\quad W = \frac{\lambda D}{d}$

$$\Rightarrow \quad 2W = \frac{\lambda D}{d}$$

When, $\quad D' = 2D$

$$W_2 = \frac{\lambda D'}{d}$$

$$\Rightarrow \quad W_2 = \frac{\lambda 2D}{d}$$

$$= 2 \times 2W = 4W$$

28. The given force varies with distance.
Therefore, work done,

$$W = \int_0^{x=5} F \cdot dx$$
$$= \int_0^5 (7 - 2x + 3x^2)dx$$
$$= \left[7x - \frac{2x^2}{2} + \frac{3x^3}{3}\right]_0^5$$
$$= [7 \times 5 - (5)^2 + (5)^3 - 0]$$
$$= 135 \text{ J}$$

29. Conservation of momentum (gives)

$$mu + 0 = mv_1 + mv_2$$
$$\Rightarrow \quad v_1 + v_2 = u \quad \ldots\text{(i)}$$

Newton's experimental formula,

$$v_1 - v_2 = -e[u_1 - u_2] \text{ gives}$$
$$v_1 - v_2 = -eu \quad \ldots\text{(ii)}$$

From Eqs. (i) and (ii), we get

$$v_1 = \frac{(1-e)u}{2}$$
$$v_2 = \frac{(1+e)u}{2}$$
$$\therefore \quad \frac{v_1}{v_2} = \frac{1-e}{1+e}$$

30. Given, $C_p = 7.03$ cal/mol °C, $R = 8.315$/mol°C,
$n = 5$ moles, $T_2 = 20°C$, $T_1 = 10°C$,
$\Delta Q = ?$

$\therefore$ Heat, $\Delta Q = nC_V \Delta T$

$$= n(C_p - R)\Delta T \quad [\because C_p - C_V = R]$$
$$= 5\left[7.03 - \frac{8.31}{4.2}\right] \times (20 - 10)$$
$$\simeq 253 \text{ cal}$$

31. Activity, $\left(\frac{R}{R_0}\right) = \left(\frac{1}{2}\right)^n$

$$\Rightarrow \quad \frac{1}{16} = \left(\frac{1}{2}\right)^n$$
$$\Rightarrow \quad \left(\frac{1}{2}\right)^4 = \left(\frac{1}{2}\right)^n$$
$$\Rightarrow \quad n = 4$$
$$\Rightarrow \quad \frac{t}{T} = 4 \quad \left[\because n = \frac{t}{T}\right]$$
$$\Rightarrow \quad t = 4T$$
$$= 4 \times 100 \text{ µs}$$
$$\therefore \quad t = 400 \text{ µs}$$

32. From figure, $mg\sin\theta = ma\cos\theta$

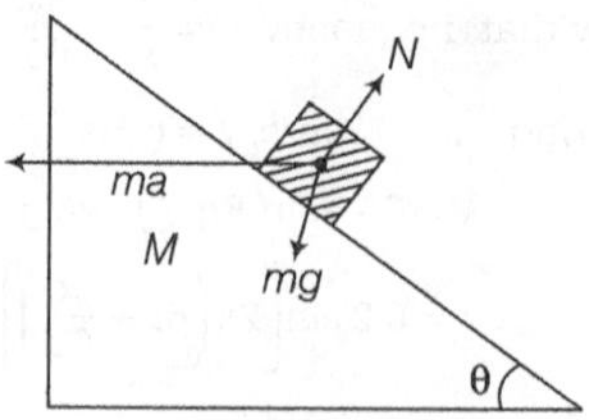

$$\Rightarrow \quad a = g\tan\theta$$

$\therefore$ Force,
$$F = (M + m)a$$
$$= (M + m)g\tan\theta$$

33. Given, $m = 1$ kg,

$$v = 2\text{m/s}$$
$$R = 1\text{m}$$
$$W = ?$$

From energy conservation,

$$mgh + W = \frac{1}{2}mv^2$$
$$W = \frac{1}{2}mv^2 - mgh$$
$$= \frac{1}{2} \times 1(2)^2 - 1 \times 10 = -8\text{J}$$

34. Limiting friction along the plane upward

$$= \mu_s \, mg\cos\theta$$
$$= 0.8 \times 1 \times 10 \times \cos 37°$$
$$= 0.8 \times 10 \times \frac{4}{5}$$
$$= 6.4 \text{ N}$$

Force responsible for downward motion

$$= mg\sin\theta$$
$$= 1 \times 10 \times \sin 37°$$
$$= 1 \times 10 \times \frac{3}{5}$$
$$= 6\text{N}$$

As, downward force < limiting friction.
So, frictional force = downward force.
Hence, tension in string is zero.

35. Areal velocity, $\frac{dA}{dt} = \frac{L}{2m}$

where, L = angular momentum

$$\Rightarrow \quad \frac{dA}{dt} = \frac{mr^2\omega}{2m} \quad [\because L = mr^2\omega]$$
$$\propto r^2\omega$$

36. Given, $F = 49\text{N}$, $M_A = 7\text{kg}$, $M_B = 3\text{kg}$, $\mu = ?$

Equation of motion of two blocks-system

$$F_{max} = \mu(M_A + M_B)\cdot g$$

$$\mu = \frac{F}{(M_A + M_B)\cdot g}$$

$$= \frac{49}{(7+3)\cdot 9\cdot 8}$$

$$= \frac{49}{10 \times 9.8}$$

$$= 0.5$$

37. As $M_1 > M_2$, so tension in thread connecting the block B and support C is $(M_1 - M_2)g$. When this thread is burnt, this tension disappears. In this case tension in spring is $M_1 g$ and it is elongated. So, tension in string connecting A and B is $M_1\, g$. Hence, resultant force on A remains zero because tension in string is balanced by tension in spring. The block B has a net upward force $(M_1 - M_2)\, g$, so initial acceleration of block B is

$$\frac{(M_1 - M_2)g}{M_2}$$

Thus, initial acceleration of M_1 is zero and that of M_2 is $\frac{(M_1 - M_2)g}{M_2}$ upward.

38. For refraction at first surface,

$$u = -7.5\text{ cm},\ R_1 = 2.5\text{ cm}$$

$$\mu_1 = 1,\ \mu_2 = \frac{4}{3}$$

$$\therefore \quad \frac{\mu_2}{v'} - \frac{\mu_1}{u} = \frac{\mu_2 - \mu_1}{R_1}$$

$$\frac{4/3}{v'} - \frac{1}{(-7.5)} = \frac{4/3 - 1}{2.5}$$

$$\Rightarrow \quad \frac{4}{3v'} = \frac{1}{7.5} - \frac{1}{7.5} = 0$$

$$\Rightarrow \quad v' = \infty$$

It means the ray is parallel within the sphere.

For refraction at second surface,

$$u = -\infty,\ \mu_1 = \frac{4}{3}$$

$$\mu_2 = 1,\ R_2 = -2.5\text{ cm}$$

$$\therefore \quad \frac{\mu_2}{v'} - \frac{\mu_1}{u} = \frac{\mu_2 - \mu_1}{R_2} \text{ gives}$$

$$\Rightarrow \quad \frac{1}{v'} - \frac{4/3}{(-\infty)} = \frac{1-(4/3)}{(-2.5)}$$

$$\Rightarrow \quad v = +7.5\text{ cm}$$

i.e. Image is formed at distance 7.5 cm to the right of P_2. Hence, distance of final image P from centre $C = 7.5 + 2.5 = 10$ cm.

39. Total energy, $E_0 = -\dfrac{GMm}{2r}$ ...(i)

and potential energy, $V = -\dfrac{GMm}{r}$ ...(ii)

From Eqs. (i) and (ii), we get

$$= 2E_0$$

40. Ge conducts at 0.3 V and Si Conducts at 0.7 V. Both Ge and Si are connected in parallel. When current begins to flow, then the potential difference remains at 0.3 V, so no current flows through Si-diode.

$\therefore$ Potential difference across resistive load $= 12 - 0.3$

$$= 11.7\text{ V}$$

$\therefore$ Potential of $Y = 11.7$ V

41. From question,

$$n_1 + 55 \times 6 = n_2$$

$$\Rightarrow \quad n + 55 \times 6 = 2n$$

$$\Rightarrow \quad n = 330\text{ Hz}$$

$$\therefore \quad n_2 = 2n = 2 \times 330 = 660\text{ Hz}$$

42. Given, $\lambda_1 = 2000\text{ Å} = 2000 \times 10^{-10}$ m

$$= 2 \times 10^{-7}\text{ m}$$

$$\lambda_2 = 2100\text{ Å} = 2.1 \times 10^{-7}\text{ m}$$

$$\therefore \quad \frac{hc}{\lambda_1} = W + eV_0 \quad \text{...(i)}$$

$$\text{and} \quad \frac{hc}{\lambda_2} = W + eV_0' \quad \text{...(ii)}$$

Subtracting Eq. (ii) from Eq. (i), we get

$$hc\left(\frac{1}{\lambda_1} - \frac{1}{\lambda_2}\right) = e(V_0 - V_0')$$

Change in stopping potential

$$\Delta V = V_0 - V_0'$$

$$= \frac{hc}{e}\left(\frac{1}{\lambda_1} - \frac{1}{\lambda_2}\right)$$

$$= \frac{6.6 \times 10^{-34} \times 3 \times 10^8}{1.6 \times 10^{-19}}\left(\frac{1}{2 \times 10^{-7}} - \frac{1}{2.1 \times 10^{-7}}\right)$$

$$= \frac{6.6 \times 3}{1.6}\left(\frac{1}{2} - \frac{1}{2.1}\right)$$

$$= \frac{6.6 \times 3 \times 0.1}{1.6 \times 2 \times 2.1} = 0.3\text{ V}$$

43. Given, $y_1 = a_1 \sin\left(\omega t - \frac{2\pi x}{\lambda}\right)$ and

$$y_2 = a_2 \cos\left(\omega t - \frac{2\pi x}{\lambda} + \phi\right)$$

$$= a_2 \sin\left(\frac{\pi}{2} + \omega t - \frac{2\pi x}{\lambda} + \phi\right)$$

$\therefore$ Phase of first wave, $\phi_1 = \left(\omega t - \frac{2\pi x}{\lambda}\right)$

and phase of second wave, $\phi_2 = \frac{\pi}{2} + \omega t - \frac{2\pi x}{\lambda} + \phi$

$\therefore$ Phase difference, $\Delta\phi = \phi_2 - \phi_1 = \frac{\pi}{2} + \phi$

$\therefore$ Path difference $= \frac{\lambda}{2\pi} \times$ phase difference

$$= \frac{\lambda}{2\pi}\left(\frac{\pi}{2} + \phi\right)$$

44. When intensity of incident photon increases number of electrons emitted from the surface increases. Thus current increases. It is independent of frequency of photon.

45. From Stokes' law, $F = 6\pi\eta rv$

wehre, η = coefficient of viscosity

and $F' = 6\pi\eta(2r)(2v) = 6\pi\eta rv \times 4$

$\because \quad \frac{F'}{F} = 4$

$$F' = 4F$$

46. From Einstein's photoelectric equation,

$$\frac{1}{2}mv^2 = \frac{hc}{\lambda} - W \quad \ldots(i)$$

and $$\frac{1}{2}mv'^2 = \frac{hc}{3\lambda/4} - W \quad \ldots(ii)$$

On dividing Eq. (ii) by Eq. (i), we get

$$\left(\frac{v'}{v}\right)^2 = \frac{\frac{4hc}{3\lambda} - W}{\frac{hc}{\lambda} - W}$$

$\because \quad W > 0$

$\therefore \quad v' > v\left(\frac{4}{3}\right)^{\frac{1}{2}}$

47. Weight of wooden block = Weight of water displaced

$$Mg = V_{in}\sigma_w g$$

$$= (40 \times 40 \times 5) \times 10^{-6} \times 10^3 \text{ g}$$

$\Rightarrow \quad M = 8 \text{ kg}$

48. Time, $T = \frac{2u\sin\theta}{g}$

$$T = \frac{2 \times 50 \times \sin 30°}{10}$$

$$= \frac{2 \times 50 \times \frac{1}{2}}{10} = 5\text{s}$$

Time spent after crossing the wall, $t_2 = 5 - 3 = 2\text{s}$

$\therefore$ Distance, $x_2 = u\cos\theta \times 1_2$

$$= 50 \times \cos 30° \times 2$$

$$= 50 \times \frac{\sqrt{3}}{2} \times 2$$

$$= 50\sqrt{3}$$

$$= 50 \times 1.73$$

$$= 86.5 \text{ m}$$

49. Lower (conceptual question).

When temperature is increased, then the density will decrease, so cube will float lower.

50. Given, $D = 40 \text{ cm} = 40 \times 10^{-2}$ m

$$\lambda = 600 \text{ Å} = 6000 \times 10^{-10} \text{ m}$$

$$W = 0.012 \text{ cm} = 0.012 \times 10^{-2} \text{ m}$$

$\therefore$ Fringe width,

$$W = \frac{D\lambda}{d}$$

$\Rightarrow \quad d = \frac{D\lambda}{W}$

where, d = distance between the slits

$$d = \frac{40 \times 10^{-2} \times 6000 \times 10^{-10}}{0.012 \times 10^{-2}}$$

$$= 2 \times 10^{-3} \text{m} = 2 \text{ mm}$$

$$= 0.2 \text{ cm}$$

Chemistry

1. Suppose the chemical formula of the compound be $C_xH_yO_z$ chemical reaction related to combustion of the compound.

$C_xH_yO_z$ +	O_2 ⟶	CO_2 +	H_2O	(Steam)
1 vol.	2 vol.	2 vol.	2 vol.	Given
1. mol	2 mol	2 mol	2 mol.	(By Avogadro hypothesis)
	or	or	or	
	4 atoms	2 atom C and 4 atom O	4 atom H and 2 atom O	

Since, 2 molecules of CO_2 have 2 atoms of carbon and these carbon must have come from the compound.
Hence, the value of x is 2. Again, 2 molecules of steam contain 4 atoms of hydrogen and they have come from the compound. Hence, the value of y will be 4. Again, total oxygen atoms in the product side = 6 But 4 oxygen atoms must have come from the oxygen molecules. Hence 2 oxygen atoms definitely come from the compound. Hence, the value of $z = 2$. So, the molecular formula of the compound is $C_2H_4O_2$.

2. Molar mass of silicon = 28 g = 28000 mg
∵ 28000 mg of silicon contain 6.022×10^{23} atoms

$$\therefore 5.68 \text{ mg of silicon contain } \frac{6.022 \times 10^{23}}{28000} \times 5.68$$
$$= 0.00122 \times 10^{23}$$
$$= 1.22 \times 10^{20} \text{ atoms}$$

3. Molar mass of sugar $= 12 \times 12 + 1 \times 22 + 11 \times 16$
$$= 144 + 22 + 176$$
$$= 342 \text{ g}$$
∵ Cost of 1000 g of sugar is ₹ 50.
$$\therefore \text{Cost of 342 g of sugar} = \frac{50}{1000} \times 342$$
$$= ₹\ 17.1$$

4. Mass of C-atoms = 152 − 150 = 2 mg
Molar mass of C-atoms = 12 g = 12000 mg
∵ 12000 mg of carbon contains 6.022×10^{23} atoms carbon
$$\therefore 2 \text{ mg of carbon contains} = \frac{6.022 \times 10^{23}}{12000} \times 2$$
$$= 0.0010036 \times 10^{23}$$
$$= 1.0036 \times 10^{20} \text{ atoms}$$

5. Since, molarity normality and formality are calculated against the volume of the solution and the volume of the solution changes with change in temperature. Therefore, these qualities do not remain constant with temperature.
$$\text{Molality } (m) = \frac{\text{Moles of solute}}{\text{Mass of solvent (in kg)}}$$
The molality of a solution remains independent of temperature, because it involves only mass, which is independent of temperature.

6. Energy of photons,
$$E = h\nu = \frac{hc}{\lambda}$$
$$= \frac{(6.626 \times 10^{-34} \text{ Js}) \times (3 \times 10^8 \text{ ms}^{-1})}{5 \times 10^{-7}}$$
$$= 3.975 \times 10^{-19} \text{ J}$$

7. According to Pauli exclusion principle, no two electron in an atom can have the same set of four quantum numbers.
Hund's rule states that the pairing of electrons in the orbital of a particular sub-shell (p, d or f) does not take place until all the orbitals of the subshell are singly filled.
Option I and II do not follow Hund's rule and Paulis exclusion principle.

8. Chromium and copper have five and ten electrons in 3d-orbitals rather than four and nine because fully-filled and half-filled orbitals have lower energy and thus have extra stability. Therefore, chromium and copper have d^5 and d^{10}-configurations.

9. Since, He^+ and H have one electron each, while H^+ has no electron, hence Pauli exclusion principle does not apply to them. However, H^- has two electrons, hence this principle applies on it.

10. d-block metals, e.g., Ni, Pt, Fe, etc., are used as catalysts.

11. In general, electron affinity increases in a period from left to right. This is due to decrease in size and increase in nuclear charge as the atomic number increases in a period. Potassium is a metallic element which is electron donor. O^- and F^- ions gain stability by taking one electron.
Only O (oxygen atoms) has tendency to gain electron to attain stability.

12. According to Fajan's rule, as the size of cation decreases, it polarising power increases. Hence, K^+ and Cs^+ having bigger cations contain less polarising power. Cu^{2+} polarise Cl^- ions more than Cu^+ due to its smaller size. Therefore, $CuCl_2$ has more covalent character and hence, its boiling point is less.

13. CH_4 has regular tetrahedral shape, so its dipole moment is zero.

14. RCH_2NHCH_3 shows the hydrogen bonding, because H-atom is directly attached to N-atom.

15. van der Waals' equation for 1 mole of a real gas is

$$\left(p + \frac{a}{V^2}\right)(V - b) = RT$$

where, b is volume correction. It arises due to effective size of molecules.

16. Viscosity of a liquid increases with decrease in temperature as kinetic energy of the molecules decreases and hence, intermolecular attraction increases which in turn increases the viscosity of a liquid.

17. Thermodynamically, graphite is the most stable form of carbon.

18. HF is more stable than HCl because more energy is produced in the formation of HF. It means HF has less energy than HCl and hence is more stable. So, we need more energy to break H—F bond. Hence, HF is more stable.

19. $\Delta H < 0, \Delta S < 0$

$$\Delta G = \Delta H - T\Delta S$$

When, $\Delta H < 0$ and $\Delta S < 0$, then ΔG will be negative only at low temperature.

20. K_C is a function of temperature

21. The given reaction is exothermic and according to Le-Chatelier principle, high temperature shifts the equilibrium in backward direction, i.e. high temperature favours the decomposition of N_2O_4.

22. Solubility of alkaline earth metal hydroxide in water increases with increase in atomic number.

23. The phenomenon of emission of electrons from the surface of metal by visible light is called photoelectric effect.

All alkali metals (except lithium) exhibits photoelectric effect. This ability to exhibit photoelectric effect is due to low value of their ionisation energies. Among alkali metals, Cs having least ionisation energy, exhibits maximum photoelectric effect.

24. Generally, as the electronegativity increases, acidic character increases, but in case of boron halides. BI_3 is the most acidic, i.e. strongest Lewis acid. This is because there is back donation of electrons from F-atom to boron in case of BF_3. This makes BF_3 less electron deficient. The tendency of back donation decreases as the size of halogen atom increases due to large energy difference. Hence, BI_3 is the most electron deficient and thus, the strongest Lewis acid.

25. Thiourea contains both S and N and hence gives blood red colouration when its Lassaigne's extract is treated with alkali and $FeCl_3$.

$$3NaCNS + FeCl_3 \longrightarrow \underset{\substack{\text{Ferric} \\ \text{sulphocyanide} \\ \text{(Blood red)}}}{Fe(CNS)_3} + 3NaCl$$

26.

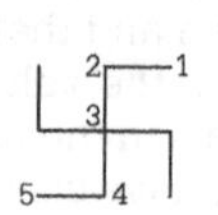

3, 3-diethylpentane

27. $C_6H_5COC_6H_5$ does not exhibit tautomerism as it contains no α-H atom.

28. Inductive effect is a permanent effect operates on σ-electrons, while resonance involves delocalisation of π-electrons.

29. Cl^- is the weakest base, so good leaving group.

30. $C_6H_5\overset{+}{C}HC_6H_5$ is the most stable, since the positive charges can be delocalised over both of phenyl rings.

31. Halogens are the most reactive elements. High reactivity is due to their low dissociation energy. Further, among halogen, reactivity decreases as we move down the group. Hence, the correct order of nucleophilicity would be

$$I^- < Br^- < Cl^-$$

32. Primary alkyl halides undergo bimolecular nucleophilic substitution most readily.

33. To be aromatic the following conditions must be fulfilled.

(i) Molecules should have $(4\pi + 2)\pi$-electrons in conjugation.

(ii) Molecule should be planar.

Cycloheptatriene has $(4n + 2)\pi$-electrons, but not in conjugation.

34. Compounds containing two or more carbon atoms are prepared by Wurtz reaction. Thus, methane (CH_4) cannot be prepared by Wurtz reaction.

35. Anti-Markownikoff's rule addition reaction is not observed in case of but-2-ene. The reason is that symmetrical alkenes do not exhibit anti-Markownikoff's addition.

36. Due to greater electron releasing effect,

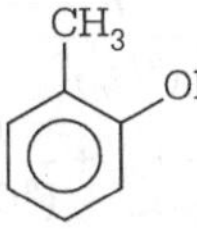

OH is the most reactive towards electrophilic reagent.

37. As, the conjugation increases, heat hydrogenation decreases. Thus, alkene 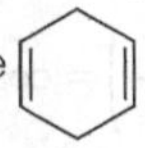 with two isolated double bonds has the highest heat of hydrogenation.

38. The enrichment of water by plant nutrients is known as eutrophication. It is due to the reduction in concentration of dissolved oxygen in water due to phosphate pollution.

39. Peroxy acetyl nitrate is used in photochemical smog. Polycyclic aromatic hydrocarbons are used as carcinogens. Dioxins are used to burn the waste into ashes. IR active molecules are used in global warming. Indigo is called as vat dye because the dying with indigotin was carried out in wooden vats in the form of water soluble indigotin-white.

40. Flame colours of metal ions are due to metal excess defect.

41. Helium-oxygen mixture is used by deep sea divers because of its very low solubility in blood.

42. When benzoic acid dissolve in benzene, it will be dimerise due to the intermolecular hydrogen bonding.

$$2C_6H_5COOH \xrightarrow{\text{Benzene}} (C_6H_5COOH)_2$$

van't Hoff factor =

$$= \frac{\text{Number of particles after association or dissociation}}{\text{Number of particles before associations or dissociations}}$$

$$= \frac{1}{2} = 0.5$$

43. From Faraday's first law,

Mass deposited, $W = Z \cdot Q$

when $Q = 1\,C$, then $W = Z$ = electrochemical equivalent.

44. Nature and concentration of the reactants its and temperature of the reaction influence the rate of reaction. But molecularity does not affect the rate of reaction as it includes the number of atoms, ions or molecules that must collide with one another to result into a chemical reaction.

45. Adsorption is an exothermic process thus, ΔH is negative (i.e. $\Delta H < 0$). Moreover, adsorption results in more ordered arrangements of molecules, thus entropy decreases (i.e. $\Delta S < 0$).

Hence, low temperature favours the reaction.

46. Ligases catalyse reactions in which the pyrophosphate bond of ATP is broken down and linkage takes place between two molecules.

47. It is a method of using dihydrogen in an efficient manner and carried or stored in the form of liquid or gas. dihydrogen releases more energy than petrol and in more eco-friendly and also can be used in fuel cells.

48. In the metallurgy of alumina, purified Al_2O_3 is mixed with Na_3AlF_6 or CaF_2 which lowers the melting point of the mix and brings conductivity.

$$2Al_2O_3 + 3C \longrightarrow 4Al + 3CO_2$$

49. In $(Si_2O_3^{2-})_n$ two oxygen atoms per SiO_4 tetrahedron are shared giving polymeric anions chain. e.g. Spodumene, $LiAl(SiO_3)_2$ and diposide, $CaMg\,(SiO_3)_2$.

50. Myosins are fibrous proteins present in muscles.

Mathematics

1. Consider the function

$$f(x) = \frac{ax^3}{3} + \frac{bx^2}{2} + cx + d$$

We have, $f(0) = d$

and $f(1) = \frac{a}{3} + \frac{b}{2} + c + d$

$$= \frac{2a + 3b + 6c}{6} + d$$

$$= \frac{0}{6} + d \quad [\because 2a + 3b + 6c = 0]$$

Therefore, 0 and 1 are roots of the polynomial $f(x)$. Hence, according to Rolle's theorem, there exists atleast one root of the polynomial $f'(x) = ax^2 + bx + c$ lying between 0 and 1.

2. We have,

$$\mathbf{a} \cdot \mathbf{x} = 0$$

Also, $\mathbf{x} = \mathbf{b} + (\mathbf{x} \times \mathbf{a})$

$\Rightarrow \mathbf{x} \times \mathbf{a} = \mathbf{b} \times \mathbf{a} + (\mathbf{x} \times \mathbf{a}) \times \mathbf{a}$

$\Rightarrow \mathbf{x} \times \mathbf{a} = \mathbf{b} \times \mathbf{a} + (\mathbf{x} \cdot \mathbf{a})\mathbf{a} - (\mathbf{a} \cdot \mathbf{a})\mathbf{x}$

$\Rightarrow \mathbf{x} - \mathbf{b} = \mathbf{b} \times \mathbf{a} - \mathbf{x}$

$|2\mathbf{x}| = |\mathbf{b} + \mathbf{b} \times \mathbf{a}|$

$\Rightarrow |\mathbf{x}| = \frac{1}{\sqrt{2}}$

3. We have, $b\mathbb{N} = \{bn : n \in \mathbb{N}\}$, $c\mathbb{N} = \{cn : n \in \mathbb{N}\}$

and $d\mathbb{N} = \{dn : n \in \mathbb{N}\}$

Also, $bc \in b\mathbb{N} \cap c\mathbb{N} = d\mathbb{N}$ [given]

$\therefore$ $bc \in b\mathbb{N} \cap c\mathbb{N}$ or $bc \in d\mathbb{N}$

$\therefore$ $bc = d$

[as b and c are coprime]

4. Given, $f(x) = \int_0^x e^{|t-x|}\, dt + \int_x^1 e^{|t-x|}\, dt$

$$= e^x + e^{1-x} - 2$$

$\therefore$ $f'(x) = e^x - e^{1-x}$

Clearly, $x = \frac{1}{2}$ is the point of minima of $f(x)$.

Also, $f'(0) = f(1) = e - 1$

5. Meen $(\bar{x}) = \frac{-1 + 0 + 4}{3}$

$\therefore$ Mean deviation $= \frac{1}{3}[|-1-1| + |0-1| + |4-1|]$

$$= 2$$

6. $f'(x) - pf(x) \le 0$

$\Rightarrow \frac{d}{dx} e^{-px} f(x) \le 0$

Let $g(x) = e^{-px} f(x)$

$g'(x) \le 0, \forall\, x \ge 1$

$\Rightarrow g(x) \le g(1), \forall\, x \ge 1$

$\Rightarrow g(x) \le 0$

$\Rightarrow f(x) \le 0$

But given, $f(x) \ge 0$

$\therefore$ $f(x) = 0, \forall x \ge 1$

7. $S_n = \frac{0}{({}^nC_0)^k} + \frac{1}{({}^nC_1)^k} + \frac{2}{({}^nC_2)^k} + \ldots + \frac{n}{({}^nC_n)^k}$

$\Rightarrow S_n = \frac{n}{({}^nC_0)^k} + \frac{n-1}{({}^nC_1)^k} + \frac{n-2}{({}^nC_2)^k} + \ldots + \frac{0}{({}^nC_0)^k}$

$\Rightarrow 2S_n = nt_n$

$\Rightarrow \frac{S_n}{nt_n} = \frac{1}{2}$

Now, $\cos^{-1}\left(\frac{S_n}{nt_n}\right) = \cos^{-1}\left(\frac{1}{2}\right) = \frac{\pi}{3}$

8. We have,

$P(|x-4| \le 2) = P(-2 \le x - 4 \le 2) = P(2 \le x \le 6)$

$= 1 - [P(x=0) + P(x=1) + P(x=7) + P(x=8)]$

$$= 1 - \left[{}^8C_0\left(\frac{1}{2}\right)^8 + {}^8C_1\left(\frac{1}{2}\right)^8 + {}^8C_7\left(\frac{1}{2}\right)^8 + {}^8C_8\left(\frac{1}{2}\right)^8\right]$$

$$= 1 - \left(\frac{1}{2}\right)^8 (1 + 8 + 8 + 1) = 1 - \frac{18}{2^8} = \frac{119}{128}$$

9. Since, total number of terms $= 59 + 1 = 60$

$\therefore$ Required sum $= \frac{2^{59}}{2} = 2^{58}$

10. $\theta = \sin^{-1} x + \cos^{-1} x - \tan^{-1} x = \frac{\pi}{2} - \tan^{-1} x$

$$\left[\because \sin^{-1} x + \cos^{-1} x = \frac{\pi}{2}\right]$$

$\Rightarrow \theta = \cot^{-1} x$

Since, $1 \le x < \infty$, therefore $0 \le \theta \le \frac{\pi}{4}$.

11. Given, $2\tan^{-1}(\cos x) = \tan^{-1}(2 \operatorname{cosec} x)$

$\therefore \tan^{-1}\left(\frac{2\cos x}{1 - \cos^2 x}\right) = \tan^{-1}(2 \operatorname{cosec} x)$

$\Rightarrow \frac{2\cos x}{1 - \cos^2 x} = 2 \operatorname{cosec} x$

$\Rightarrow \sin x = \cos x \Rightarrow x = \frac{\pi}{4}$

12. $\cos^{-1} p + \cos^{-1} q + \cos^{-1} r = 3\pi$

We know that, if $y = \cos^{-1} x$, then $-1 \le x \le 1$ and $0 \le y \le \pi$.

Hence, the given equation will hold only when each is π.

$\therefore \quad p = q = r = \cos \pi = -1$

$\therefore \quad p^2 + q^2 + r^2 + 2pqr$

$= (-1)^2 + (-1)^2 + (-1)^2 + 2(-1)(-1)(-1)$

$= 1 + 1 + 1 - 2 = 3 - 2 = 1$

13. Let $P(x, y)$ be such a point, then

$$OP^2 = x^2 + y^2$$

Let $\quad s = OP^2 = x^2 + \frac{1}{x}$

$\Rightarrow \quad \frac{ds}{dx} = 2x - \frac{1}{x^2}$

For maximum or minimum, put $\frac{ds}{dx} = 0$

$\Rightarrow \quad 2x - \frac{1}{x^2} = 0 \Rightarrow x^3 = \frac{1}{2}$

$\therefore \quad x = \left(\frac{1}{2}\right)^{\frac{1}{3}}$

Also, $\quad \frac{d^2s}{dx^2} = 2 + \frac{2}{x^3}$

Now, $\left(\frac{d^2s}{dx^2}\right)_{x = \left(\frac{1}{2}\right)^{1/3}} = 2 + \frac{2}{1/2} > 0$

$\therefore$ s is minimum at $x = \left(\frac{1}{2}\right)^{1/3}$.

Thus, for nearest point $x = \left(\frac{1}{2}\right)^{1/3}$,

$$y = \pm\left(\frac{1}{2}\right)^{-1/6}$$

14.

q	p	$q \to p$	$p \to (q \to p)$	$p \lor q$	$p \to (p \lor q)$
T	T	T	T	T	T
T	F	F	T	T	T
F	T	T	T	T	T
F	F	T	T	F	T

$\therefore$ Statement $p \to (q \to p)$ is equivalent to $p \to (p \lor q)$.

15. $(x^2 + 1) = (x - 2)(x + 2)$

Let $f(x) = x^3 + ax^2 + bx + c$; $a, b, c \in \{1, 2, \dots, 10\}$

$f(x)$ is divisible by $x^2 + 1$ iff $f(i) = 0$, $f(-i) = 0$

$\therefore \quad i^3 + ai^2 + bi + c = 0$

$\Rightarrow \quad -i - a + bi + c = 0$

$\Rightarrow \quad a = c, b = 1$

and $\quad -i^3 + ai^2 - bi + c = 0$

$\Rightarrow \quad i - a - bi + c = 0$

$\Rightarrow \quad a = c, b = 1$

$\therefore$ Total number of selection of triplets

$(a, b, c) = (1, 1, 1), (2, 1, 2), \dots, (10, 1, 10)$

Hence, number of such polynomials = 10.

16. Let number of regions for n lines be $R(n)$.

Clearly, $R(1) = 2$, $R(2) = 4$, ...

$R(n) = R(n-1) + n$ i.e. $R(n) - R(n-1) = n$

Putting $n = 2, 3, \dots, n$ and then adding, we get

$R(n) - R(1) = 2 + 3 + \dots + n$

$= 2 + 2 + 3 + 4 + \dots + n = 1 + \frac{n(n+1)}{2}$

$= \frac{n^2 + n + 2}{2}$

17. $\lim_{n \to \infty}\left[\sqrt[3]{n^2 - n^3} + n\right] = \lim_{n \to \infty} n\left[\left(-1 + \frac{1}{n}\right)^{1/3} + 1^{1/3}\right]$

$= \lim_{n \to \infty} n \cdot \frac{\left(\frac{1}{n} - 1\right) + 1}{\left(\frac{1}{n} - 1\right)^{2/3} + 1 - \left(\frac{1}{n} - 1\right)^{1/3}}$

$\left[\because a + b = \frac{a^3 + b^3}{a^2 - ab + b^2}\right]$

$= \lim_{n \to \infty} \frac{1}{\left(\frac{1}{n} - 1\right)^{2/3} + 1 - \left(\frac{1}{n} - 1\right)^{1/3}}$

$= \frac{1}{1 + 1 + 1} = \frac{1}{3}$

18. $\frac{x+1}{x^{2/3} - x^{1/3} + 1} - \frac{x - 1}{x - x^{1/2}}$

$= \frac{(x^{1/3})^3 + 1^3}{x^{2/3} - x^{1/3} + 1} - \frac{x - 1}{x^{1/2}(x^{1/2} - 1)}$

$= \frac{(x^{1/3} + 1)(x^{2/3} - x^{1/3} + 1)}{x^{2/3} - x^{1/3} + 1} - \frac{x^{1/2} + 1}{x^{1/2}}$

$= x^{1/3} + 1 - 1 - x^{-1/2}$

$= x^{1/3} - x^{-1/2}$

$$\therefore \left(\frac{x+1}{x^{2/3}-x^{1/3}+1}-\frac{x-1}{x-x^{1/2}}\right)^{10}=(x^{1/3}-x^{-1/2})^{10}$$

T_{r+1} for $(x^{1/3}-x^{-1/2})$ is

$^{10}C_r\,(x^{1/3})^{10-r}\,(-1)^r\,(x^{-1/2})^r$.

For term independent of x,

$$\frac{10-r}{3}-\frac{r}{2}=0$$

$$\Rightarrow \quad 20-2r-3r=0$$

$$\Rightarrow \quad r=4$$

Hence, required coefficient $= {}^{10}C_4\,(-1)^4=210$.

19. Since, $\begin{bmatrix}\alpha & \beta\\ \gamma & -\alpha\end{bmatrix}$ is a square root of I_2 i.e. two rowed unit matrix.

$$\therefore \quad \begin{bmatrix}\alpha & \beta\\ \gamma & -\alpha\end{bmatrix}^2=\begin{bmatrix}1 & 0\\ 0 & 1\end{bmatrix}$$

$$\Rightarrow \quad \begin{bmatrix}1 & 0\\ 0 & 1\end{bmatrix}=\begin{bmatrix}\alpha & \beta\\ \gamma & -\alpha\end{bmatrix}\begin{bmatrix}\alpha & \beta\\ \gamma & -\alpha\end{bmatrix}$$

$$\Rightarrow \quad \begin{bmatrix}1 & 0\\ 0 & 1\end{bmatrix}=\begin{bmatrix}\alpha^2+\beta\gamma & \alpha\beta-\alpha\beta\\ \gamma\alpha-\gamma\alpha & \gamma\beta+\alpha^2\end{bmatrix}$$

$$\Rightarrow \quad \begin{bmatrix}1 & 0\\ 0 & 1\end{bmatrix}=\begin{bmatrix}\alpha^2+\beta\gamma & 0\\ 0 & \alpha^2+\beta\gamma\end{bmatrix}$$

$$\therefore \quad \alpha^2+\beta\gamma=1$$

$$\Rightarrow 1-\alpha^2-\beta\gamma=0$$

20. Given, $y=Ae^{3x}+Be^{5x}$ ...(i)

$$\Rightarrow \quad \frac{dy}{dx}=3Ae^{3x}+5Be^{5x} \quad \text{...(ii)}$$

and $\quad \dfrac{d^2y}{dx^2}=9Ae^{3x}+25Be^{5x}$ ...(iii)

From Eqs. (i), (ii) and (iii), we have

$$\frac{d^2y}{dx^2}-8\frac{dy}{dx}+15y=0$$

which is the required differential equation.

21. Let $y=\sin^{-1}(2x\sqrt{1-x^2})$

and $\quad z=\sin^{-1}(3x-4x^3)$

$\Rightarrow \quad y=2\sin^{-1}x$ and

$\quad z=3\sin^{-1}x$

$$\Rightarrow \quad y=\frac{2}{3}z$$

$$\therefore \quad \frac{dy}{dz}=\frac{2}{3}$$

22. Let

$$I=\int\frac{2x^{12}+5x^9}{(1+x^3+x^5)^3}\,dx=\int\frac{5x^{-6}+2x^{-3}}{(1+x^{-2}+x^{-5})^3}\,dx$$

Put $1+x^{-2}+x^{-5}=t$, $(-2x^{-3}-5x^{-6})\,dx=dt$

$$\therefore \quad I=-\int\frac{dt}{t^3}=\frac{1}{2t^2}=\frac{1}{2(1+x^{-2}+x^{-5})^2}$$

$$=\frac{x^{10}}{2(x^5+x^3+1)^2}+C$$

23. Let

$$\Delta=\begin{vmatrix}\sin\theta & \cos\theta & \sin 2\theta\\ \sin\left(\theta+\frac{2\pi}{3}\right) & \cos\left(\theta+\frac{2\pi}{3}\right) & \sin\left(2\theta+\frac{4\pi}{3}\right)\\ \sin\left(\theta-\frac{2\pi}{3}\right) & \cos\left(\theta-\frac{2\pi}{3}\right) & \sin\left(2\theta-\frac{4\pi}{3}\right)\end{vmatrix}$$

$$=\begin{vmatrix}\sin\theta & \cos\theta & \sin 2\theta\\ 2\sin\theta\cos\frac{2\pi}{3} & 2\cos\theta\cos\frac{2\pi}{3} & 2\sin 2\theta\cos\frac{4\pi}{3}\\ \sin\left(\theta-\frac{2\pi}{3}\right) & \cos\left(\theta-\frac{2\pi}{3}\right) & \sin\left(2\theta-\frac{4\pi}{3}\right)\end{vmatrix}$$

Applying $R_2\to R_2+R_3$,

$$=-\begin{vmatrix}\sin\theta & \cos\theta & \sin 2\theta\\ \sin\theta & \cos\theta & \sin 2\theta\\ \sin\left(\theta-\frac{2\pi}{3}\right) & \cos\left(\theta-\frac{2\pi}{3}\right) & \sin\left(2\theta-\frac{4\pi}{3}\right)\end{vmatrix}=0$$

as $\quad R_1=R_2$

24. Required area = 2 Area of curve $OABO$

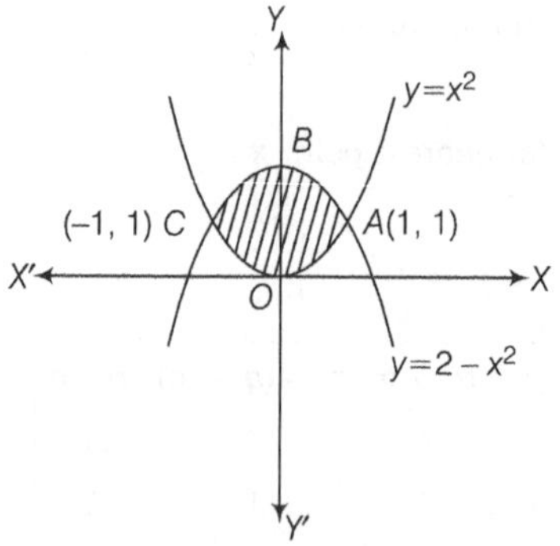

$$=2\int_0^1[(2-x^2)-(x^2)]\,dx=2\int_0^1(2-2x^2)\,dx$$

$$=4\left[x-\frac{x^3}{3}\right]_0^1=\frac{8}{3}\text{ sq units}$$

25. $\lim_{n\to\infty} \frac{1}{1-n^3} \sum_{r=1}^{n} r^2$

$$= \lim_{n\to\infty} \frac{1}{n^3\left(\frac{1}{n^3}-1\right)} \cdot \frac{n(n+1)(2n+1)}{6}$$

$$= \lim_{n\to\infty} \frac{n^3\left(1+\frac{1}{n}\right)\left(2+\frac{1}{n}\right)}{n^3\left(\frac{1}{n^3}-1\right)6} = -\frac{1}{3}$$

26. Let $I = \int \frac{\sin x + \cos x}{3+\sin 2x}\, dx$

$$= \int \frac{\sin x + \cos x}{4+\sin 2x - 1}\, dx$$

$$\Rightarrow I = \int \frac{\sin x + \cos x}{4-(\sin x - \cos x)^2}\, dx$$

Put $\sin x - \cos x = t$

$\Rightarrow$ $(\cos x + \sin x)dx = dt$

$\therefore$ $I = \int \frac{dt}{4-t^2} = \frac{1}{4}\log \frac{2-t}{2+t} + C$

$$= \frac{1}{4}\log\left(\frac{2-\sin x+\cos x}{2+\sin x - \cos x}\right) + C$$

27. Let the equation of circle be

$$x^2+y^2+2gx+2fy+c=0.$$

Since, it passes through the point $(2a, 0)$, so

$$4a^2+4ag+c=0 \quad \ldots\text{(i)}$$

Also, its radical axis with $x^2+y^2=a^2$ is

$$2gx+2fy+c+a^2=0$$

But the radical axis is $x=\frac{a}{2}$, so we get

$$\frac{2g}{1} = \frac{c+a}{-a/2} \text{ and } f=0$$

or $\quad ag+c+a=0 \quad \ldots\text{(ii)}$

From Eqs. (i) and (ii), we get g and c

Hence, the equation is $x^2+y^2-2ax=0$

28. Any point on the parabola is (x, x^2+7x+2).

Its distance from the line $y=3x-3$ is given by

$$p = \left|\frac{3x-(x^2+7x+2)-3}{\sqrt{9+1}}\right|$$

$$= \left|\frac{x^2+4x+5}{\sqrt{10}}\right|$$

$$= \frac{x^2+4x+5}{\sqrt{10}}$$

[as $x^2+4x+5>0, \forall x \in R$]

$\therefore$ $\frac{dp}{dx}=0$

$\Rightarrow$ $x=-2$

So, the required point is $(-2, -8)$.

29. Required area $= 4\int_0^a \sqrt{4a(a-x)}\, dx$

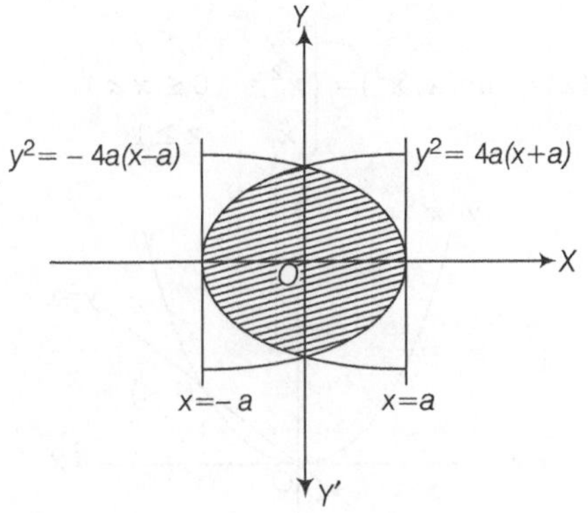

$$= 4\cdot 2\sqrt{a}\cdot\left[\frac{-2(a-x)^{3/2}}{3}\right]_0^a$$

$$= \frac{16}{3}a^2$$

30. We have,

$$\frac{dy}{dx} = \sin(x+y)\tan(x+y)-1 \quad \ldots\text{(i)}$$

Put $x+y=v$, we get

$$1+\frac{dy}{dx} = \frac{dv}{dx}$$

$$\Rightarrow \frac{dy}{dx} = \frac{dv}{dx}-1$$

Eq. (i) reducing to

$$\frac{dv}{dx} = \sin v \tan v$$

$$\Rightarrow \frac{dv}{\sin v \tan v} = dx$$

$$\Rightarrow \int \operatorname{cosec} v \cot v\, dv = \int dx$$

$$\Rightarrow -\operatorname{cosec} v \cot v = x + C$$

$$\Rightarrow -\operatorname{cosec}(x+y) = x+C$$

$$\Rightarrow x+\operatorname{cosec}(x+y) = C$$

31. Every time the selected coupon must be from 1 to 9. The probability of such a selection at any time is $\frac{9}{15} = \frac{3}{5}$ and this can happen 7 times. i.e. $\left(\frac{9}{15}\right)^7$.

Put of these $\left(\frac{9}{15}\right)^7, \left(\frac{8}{15}\right)^7$ cases do not contain the number 9.

Thus, the required probability $= \left(\frac{9}{15}\right)^7 - \left(\frac{8}{15}\right)^7$

32. $h(x) = \min(x, x^2) = \begin{cases} x, & x < 0 \\ x^2, & 0 \le x < 1 \\ x, & x \ge 1 \end{cases}$

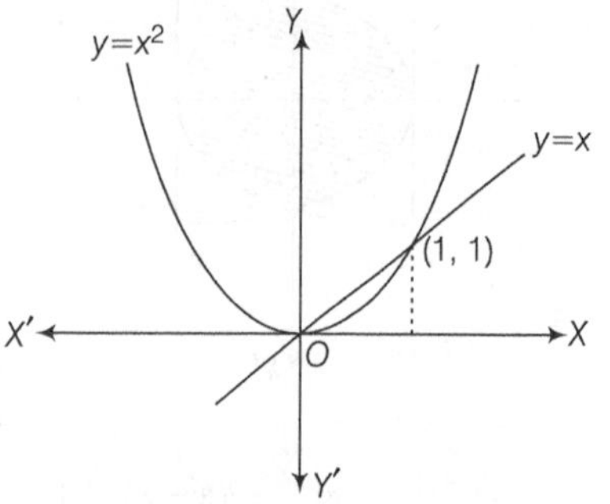

It is evident from the above graph that the given function is continuous for all x.

Since, there are sharp edges at $x = 0$ and $x = 1$, the function is not differentiable at these points.

Also, at $x \ge 1$, the function represents a straight line having slope 1, therefore $h'(x) = 1, \forall x \ge 1$.

33. For $f(x)$ to be continuous at $x = 0$, we must have

$$\lim_{x \to 0} f(x) = f(0)$$

$$\Rightarrow \lim_{x \to 0} (\cos x)^{1/x} = k$$

$$\Rightarrow \lim_{x \to 0} \{1 + (\cos x - 1)\}^{1/x} = k$$

$$\Rightarrow e^{\lim_{x \to 0} \frac{\cos x - 1}{x}} = k$$

$$\Rightarrow e^{-\lim_{x \to 0} \frac{2\sin^2 x/2}{x}} = k$$

$$\Rightarrow e^0 = k$$

$$\therefore k = 1$$

34. $\because f'(0^-) = \lim_{h \to 0} f(0 - h)$

$$\Rightarrow f'(0^-) = \lim_{h \to 0} \frac{e^{-1/h} - 1}{e^{1/-h} + 1}$$

$$\Rightarrow f'(0^-) = \frac{0 - 1}{0 + 1}$$

$$\therefore f'(0^-) = -1$$

and $f'(0^+) = \lim_{h \to 0} f(0 + h) = \lim_{h \to 0} \frac{e^{1/h} - 1}{e^{1/h} + 1}$

$$\Rightarrow f'(0^+) = \frac{1 - 0}{1 + 0} = 1$$

35. Let $E = -2i \sum_{p=1}^{6} \left(\cos \frac{2p\pi}{7} + i \sin \frac{2p\pi}{7}\right)$

$$\Rightarrow E = -2i \sum_{p=1}^{6} e^{i\left(\frac{2p\pi}{7}\right)}$$

Let $\alpha = e^{i\left(\frac{2\pi}{7}\right)}$ ...(i)

$$\therefore E = -2i(\alpha + \alpha^2 + \dots + \alpha^6)$$

$$\Rightarrow E = -2i[(1 + \alpha + \alpha^2 + \dots + \alpha^6) - 1]$$

$$\Rightarrow E = -2i\left[\left(\frac{1 - \alpha^7}{1 - \alpha}\right) - 1\right]$$

Since, $\alpha^7 = \left(e^{i\left(\frac{2\pi}{7}\right)}\right)^7 = e^{i2\pi} = 1$

$$\Rightarrow E = -2i\left\{\frac{1 - \left(e^{i\frac{2\pi}{7}}\right)^7}{1 - e^{i\left(\frac{2\pi}{7}\right)}} - 1\right\}$$

$$\Rightarrow E = -2i\left\{\frac{1 - 1}{1 - e^{i\left(\frac{2\pi}{7}\right)}} - 1\right\}$$

$$\Rightarrow E = -2i(0 - 1)$$

$$\therefore E = 2i$$

36. Number of options on 1st ring = 4

Number of options on 2nd ring = 4

Number of options on 3rd ring = 4

Number of options on 4rd ring = 4

Number of options on 5rd ring = 4

$\therefore$ Total number of ways of all the options

$= 4 \times 4 \times 4 \times 4 \times 4 = 1024$

$\therefore$ Number of all possible unsuccesful attempts to open the lock is $= 1024 - 1 = 1023$

37. Domain of f is defined, if

$$|[x| - 1]| - 5 > 0$$

$$\Rightarrow |[|x| - 1]| > 5$$

$\Rightarrow$ either $[|x| - 1] < -5$ or $[|x| - 1] > 5$

***Case* I**

$$[|x| - 1] < -5$$

$\Rightarrow$ $|x| - 1 < -5$

$\Rightarrow$ $|x| < -4$, which is not possible.

***Case* II**

$$[|x| - 1] > 5$$

$\Rightarrow$ $|x| - 1 \geq 6$

$\Rightarrow$ $|x| \geq 7$

$\therefore$ $x \in (-\infty, -7] \cup [7, \infty)$

38. We have, $y^{1/m} + y^{-1/m} = 2x$

or $y^{1/m} + \dfrac{1}{y^{1/m}} = 2x$

or $(y^{1/m})^2 - 2xy^{1/m} + 1 = 0$

$\therefore y^{1/m} = x \pm \sqrt{(x^2 - 1)}$

or $y = \left[x \pm \sqrt{x^2 - 1}\right]^m$

$\therefore$ $y_1 = m[x \pm \sqrt{(x^2 - 1)}]^{m-1}\left[1 \pm \dfrac{2x}{2\sqrt{x^2 - 1}}\right]$

$\Rightarrow$ $y_1 = \dfrac{\pm m\, [x \pm \sqrt{(x^2 - 1)}]^m}{\sqrt{(x^2 - 1)}} = \dfrac{\pm my}{\sqrt{(x^2 - 1)}}$

$\Rightarrow$ $(x^2 - 1)y_1^2 = m^2y^2$

$\Rightarrow$ $(x^2 - 1)\, 2y_1y_2 + 2xy_1^2 = 2m^2yy_1$

$\Rightarrow$ $(x^2 - 1)\, y_2 + xy_1 = m^2y$

39. We have, $cx - by + b^2 = 0$ is a tangent of circle

$x^2 + y^2 - ax - by = 0.$

$\therefore$ Centre $= \left(\dfrac{a}{2}, \dfrac{b}{2}\right)$, Radius $(r) = \sqrt{\dfrac{a^2 + b^2}{4}}$

Distance from centre to the line $cx - by + b^2 = 0$,

$\Rightarrow$ $r = \left|\dfrac{\dfrac{ac}{2} - \dfrac{b^2}{2} + b^2}{\sqrt{b^2 + c^2}}\right|$

$\Rightarrow$ $\sqrt{\dfrac{a^2 + b^2}{4}} = \dfrac{\dfrac{ac + b^2}{2}}{\sqrt{b^2 + c^2}}$

$\Rightarrow$ $(b^2 + c^2)(a^2 + b^2) = (ac + b^2)^2$

$\Rightarrow$ $b^4 + b^2(a^2 + c^2) + a^2c^2 = b^4 + 2b^2ac + a^2c^2$

$\Rightarrow$ $a^2 + c^2 = 2ac$

$\Rightarrow$ $(a - c)^2 = 0$

$\Rightarrow$ $a = c$

40. Given, $\alpha + \beta + \gamma = \dfrac{\pi}{2}$

$\Rightarrow \cot\alpha + \cot\beta + \cot\gamma = \cot\alpha \cot\beta \cot\gamma$

$\therefore$ $\cot\alpha, \cot\beta, \cot\gamma$ are in AP.

$\Rightarrow$ $3\cot\beta = \cot\alpha \cot\beta \cot\gamma$

$\Rightarrow$ $\cot\alpha \ \cot\gamma = 3$

41. Since, $R = \dfrac{b}{2\sin B} = \dfrac{2}{2\sin 30^\circ} = \dfrac{2}{1}$

$\therefore$ Area of circumcircle $= \pi R^2 = \pi\,(2)^2$

$= 4\pi$ sq units

42. $\dfrac{3}{4\cdot 8} - \dfrac{3\cdot 5}{4\cdot 8\cdot 12} + \dfrac{3\cdot 5\cdot 7}{4\cdot 8\cdot 12\cdot 16} - \ldots + \dfrac{3}{4} - \dfrac{3}{4}$

$= 1 - \dfrac{1}{4} + \dfrac{1\cdot 3}{2\cdot 4\cdot 4} - \dfrac{1\cdot 3\cdot 5}{4\cdot 4\cdot 2\cdot 4\cdot 3} + \ldots - \dfrac{3}{4}$

$= \left[1 + \dfrac{1}{1!}\left(-\dfrac{1}{4}\right) + \dfrac{1(1+2)}{2!}\left(-\dfrac{1}{4}\right)^2 + \dfrac{1\,(1+2)\,(1+4)}{3!}\left(-\dfrac{1}{4}\right)^3 + \ldots\right] - \dfrac{3}{4}$

$= \left(1 - \dfrac{1}{4}\right)^{-1/2} - \dfrac{3}{4} = \sqrt{\dfrac{2}{3}} - \dfrac{3}{4}$

43. Let $I = \int_0^\pi \sin^3\theta\, d\theta$

$[\because \sin\theta > 0 \text{ for } 0 < \theta < \pi]$

$= \int_0^\pi \sin\theta\,(1 - \cos^2\theta)\, d\theta$

Put $\cos\theta = t$

$\Rightarrow$ $-\sin\theta\, d\theta = dt$

$\therefore$ $I = \int_{-1}^{1}(1 - t^2)dt = \left[t - \dfrac{t^3}{3}\right]_{-1}^{1}$

$= 2 - \dfrac{2}{3} = \dfrac{4}{3}$

44. $\displaystyle\int \frac{\cos x + x\sin x}{x^2 + x\cos x}\, dx$

$= \displaystyle\int \frac{x + \cos x - x + x\sin x}{x^2 + x\cos x}\, dx$

$= \displaystyle\int \frac{dx}{x} - \int \frac{x(1 - \sin x)}{x(x + \cos x)}\, dx$

$= \log x - \log(x + \cos x) + C$

$= \log\left|\dfrac{x}{x + \cos x}\right| + C$

45. Equation of tangent with the slope $\dfrac{-3}{4}$ is

$y = \dfrac{-3}{4}x + C.$

According to condition of tangency,

$$C = \sqrt{32 \times \left(\frac{-3}{4}\right)^2 + 18} = \sqrt{18 + 18} = 6$$

$\therefore \quad y = \frac{-3}{4}x + 6$

$\Rightarrow 4y + 3x = 24$

Since, it meet the coordinate axes in A and B.

$\therefore \; A \equiv (8, 0)$ and $B \equiv (0, 6)$

$\therefore$ Required area $= \frac{1}{2} \times 8 \times 6 = 24$ sq units

46. $\lim_{x \to \frac{\pi}{6}} \frac{2\sin^2 x + \sin x - 1}{2\sin^2 x - 3\sin x + 1}$

$= \lim_{x \to \frac{\pi}{6}} \frac{4\sin x \cos x + \cos x}{4\sin x \cos x - 3\cos x}$ [by L' Hospital's rule]

$= \lim_{x \to \frac{\pi}{6}} \frac{\cos x(4\sin x + 1)}{\cos x(4\sin x - 3)}$

$= \frac{4\sin\frac{\pi}{6} + 1}{4\sin\frac{\pi}{6} - 3} = -3$

47. On squaring and adding the given equations, we get

$\sin^2 A + \cos^2 B + 2\sin A \cos B + \sin^2 B + \cos^2 A + 2\sin B \cos A = a^2 + b^2$

$\Rightarrow \quad 2\sin(A + B) + 2 = a^2 + b^2$

$\Rightarrow \quad \sin(A + B) = \frac{a^2 + b^2 - 2}{2}$

48. Given, $\sin A \sin B = \frac{ab}{c^2}$

$\Rightarrow c^2 = \frac{ab}{\sin A \sin B} = \left(\frac{a}{\sin A}\right)\left(\frac{b}{\sin B}\right)$

$\Rightarrow c^2 = \left(\frac{c}{\sin C}\right)^2 \quad \left[\because \frac{a}{\sin A} = \frac{b}{\sin B} = \frac{c}{\sin C}\right]$

$\Rightarrow \quad \sin^2 C = 1$

$\Rightarrow \quad C = 90°$

Hence, ΔABC is a right angled triangle.

49. Given, $l + m + n = 0$ ...(i)

and $lm = 0$

Either $m = 0$ or $l = 0$

If $l = 0$, then put in Eq. (i), we get $m = -n$

$\therefore$ Direction ratios are $0, -n, n$ i.e. $0, -1, 1$

If $m = 0$, then put in Eq. (i), we get $l = -n$

$\therefore$ Direction ratios are $-n, 0, n$ i.e. $-1, 0, 1$.

$\therefore \; \cos\theta = \frac{0 \times (-1) + (-1) \times 0 + 1 \times 1}{\sqrt{(0)^2 + (-1)^2 + (1)^2}\sqrt{(-1)^2 + (0)^2 + (1)^2}} = \frac{1}{2}$

$\Rightarrow \quad \theta = \frac{\pi}{3}$

50. Given equation can be written as

$$\frac{(x-3)^2}{16} + \frac{y^2}{25} = 1$$

Here, $a^2 = 16$ and $b^2 = 25$

$\therefore \quad e = \sqrt{1 - \frac{16}{25}} = \frac{3}{5}$

Hence, the foci of conic section are $(0, \pm be)$ i.e. $(3, \pm 3)$.

51. Given, $x - 2y = 4$

Let $A = xy$

$\Rightarrow \quad A = 2y^2 + 4y$

$\Rightarrow \quad \frac{dA}{dy} = 4 + 4y$

For extremum value, $\frac{dA}{dy} = 0 \Rightarrow y = -1$

Now, $\frac{d^2A}{dy^2} = 4 > 0$, (minimum)

At $y = -1, x = 4 + 2(-1) = 2$

$\therefore \; A = xy = 2(-1) = -2$

$\therefore$ Hence, the minimum value of xy is -2.

52. Let $\mathbf{b} = \hat{\mathbf{j}}$ and $\mathbf{c} = \hat{\mathbf{j}}$

$\therefore \quad |\mathbf{b} \times \mathbf{c}| = |\hat{\mathbf{k}}| = 1$

Again, let $\mathbf{a} = a_1\hat{\mathbf{i}} + a_2\hat{\mathbf{j}} + a_3\hat{\mathbf{j}}$

Now, $\mathbf{a} \cdot \mathbf{b} = \mathbf{a} \cdot \hat{\mathbf{i}} = \mathbf{a}_1, \mathbf{b} \cdot \bar{\mathbf{c}} = \mathbf{a} \cdot \hat{\mathbf{k}} = \mathbf{a}_2$

and $\mathbf{a} \cdot \frac{\mathbf{b} \cdot \mathbf{c}}{|\mathbf{b} \times \mathbf{c}|} = \mathbf{a} \cdot \hat{\mathbf{k}} = \mathbf{a}_3$

$\therefore \; (\mathbf{a} \cdot \mathbf{b})\,\mathbf{b} + (\mathbf{a} \cdot \mathbf{c})\,\mathbf{c} + \frac{\mathbf{a} \cdot (\mathbf{b} \times \mathbf{c})}{|\bar{\mathbf{b}} \times \mathbf{c}|} \cdot (\mathbf{b} \times \mathbf{c})$

$= a_1\mathbf{b} + a_2\mathbf{c} + a_3(\mathbf{b} \times \mathbf{c}) = a_1\hat{\mathbf{i}} + a_2\hat{\mathbf{j}} + a_3\hat{\mathbf{k}} = \mathbf{a}$

53. Volume of cone, $V = \frac{\pi}{3}r^2 h$

$\Rightarrow \quad V = \frac{\pi}{3}r^2\sqrt{l^2 - r^2}$

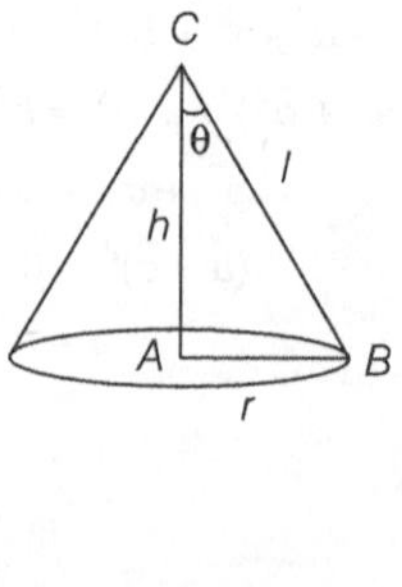

On differentiating w.r.t. r, we get

$$\frac{dV}{dr} = \frac{\pi}{3}\left[2r\sqrt{l^2 - r^2} + \frac{r^2}{2\sqrt{l^2 - r^2}}(-2r)\right]$$

Put $\frac{dV}{dr} = 0$, we get

$$r = \pm l\sqrt{\frac{2}{3}}$$

At $r = l\sqrt{\frac{2}{3}}, \frac{d^2V}{dr^2} < 0,$ (maximum)

$$\therefore \quad h = \sqrt{l^2 - \frac{2}{3}l^2} = \frac{l}{\sqrt{3}}$$

In ΔABC, $\tan\theta = \frac{r}{h} = \frac{l\sqrt{\frac{2}{3}}}{l/\sqrt{3}} = \sqrt{2}$

54. $f(x) = \frac{2\sin 8x \cos x - 2\sin 6x \cos 3x}{2\cos 2x \cos x - 2\sin 3x \sin 4x}$

$$= \frac{(\sin 9x + \sin 7x) - (\sin 9x + \sin 3x)}{(\cos 3x + \cos x) + (\cos 7x - \cos x)}$$

$$= \frac{\sin 7x - \sin 3x}{\cos 7x + \cos 3x} = \frac{2\cos 5x \sin 2x}{2\cos 2x \cos 5x} = \tan 2x$$

$\therefore$ Period of $f(x) = \frac{\pi}{2}$

55. We know that the equation of a plane parallel to X-axis is

$$by + cz + d = 0.$$

Since, it passes through the points $(2, 3, 1)$ and $(4, -5, 3)$.

$\therefore \quad 3b + c + d = 0$

and $\quad -5b + 3c + d = 0$

$\Rightarrow \quad \frac{b}{1-3} = \frac{c}{-8} = \frac{d}{14}$

$\Rightarrow \quad \frac{b}{-2} = \frac{c}{-8} = \frac{d}{14}$

$\therefore$ Equation of plane is $-2y - 8z + 14 = 0$

i.e. $\quad y + 4z = 7.$

56. Since, the point Q is image of P, therefore PQ is perpendicular to the plane $x - 2y + 5z = 6$

$\therefore$ Required equation of the line is

$$\frac{x-2}{1} = \frac{y-3}{-2} = \frac{z-4}{5}.$$

57. The intersection of two planes is

$$(x + y + z - 6) + \lambda(2x + 3y + 4z + 5) = 0$$

$$\Rightarrow (1 + 2\lambda)x + (1 + 3\lambda)y + (1 + 4\lambda)z + (-6 + 5\lambda) = 0 \quad \text{...(i)}$$

Since, this plane is perpendicular to the plane $4x + 5y - 3z - 8 = 0.$

$\therefore \quad (1 + 2\lambda)4 + (1 + 3\lambda)5 + (1 + 4\lambda)(-3) = 0$

$\Rightarrow \quad \lambda = \frac{-6}{11}$

On putting the value of λ in Eq. (i), we get

$$\left(-\frac{1}{11}\right)x + \left(-\frac{7}{11}\right)y + \left(-\frac{13}{11}\right)z + \left(-\frac{96}{11}\right) = 0$$

$\Rightarrow \quad x + 7y + 13z + 96 = 0$

58. Here, $T_n = \frac{n^3}{n!} = \frac{n^2 - 1}{(n-1)!} + \frac{1}{(n-1)!}$

$$= \frac{n+1}{(n-2)!} + \frac{1}{(n-1)!}$$

$$= \frac{1}{(n-3)!} + \frac{3}{(n-2)!} + \frac{1}{(n-1)!}$$

$$\therefore \Sigma T_n = \Sigma\left(\frac{1}{(n-3)!} + \frac{3}{(n-2)!} + \frac{1}{(n-1)!}\right)$$

$$= e + 3e + e = 5e$$

59. Given, $f(x) = \cos x + \cos\sqrt{2}x$

$\Rightarrow \quad f'(x) = -\sin x - \sqrt{2}\sin\sqrt{2}x$

$\Rightarrow \quad f''(x) = -\cos x - 2\cos\sqrt{2}x$

For extremum value, put $f'(x) = 0$

$\Rightarrow \quad -\sin x - \sqrt{2}\sin\sqrt{2}x = 0$

$\Rightarrow \quad x = 0$

At $x = 0$, $f''(x) < 0$ (maximum)

Hence, $f(x)$ is maximum only once.

60. We know that a vector perpendicular to the plane containing the point **A,B** and **C** is given by $\mathbf{A} \times \mathbf{B} + \mathbf{B} \times \mathbf{C} + \mathbf{C} \times \mathbf{A}$.

Given, $\mathbf{A} = \hat{\mathbf{i}} - \hat{\mathbf{j}} + 2\hat{\mathbf{k}}, \quad = 2\hat{\mathbf{i}} + 0\hat{\mathbf{j}} - \hat{\mathbf{k}}$

and $\quad \mathbf{C} = 0\hat{\mathbf{i}} + 2\hat{\mathbf{j}} + \hat{\mathbf{k}}$

Now, $\mathbf{A} \times \mathbf{B} = \begin{vmatrix} \hat{\mathbf{i}} & \hat{\mathbf{j}} & \hat{\mathbf{k}} \\ 1 & -1 & 2 \\ 2 & 0 & -1 \end{vmatrix} = \hat{\mathbf{i}} + 5\hat{\mathbf{j}} + 2\hat{\mathbf{k}}$

$\mathbf{B} \times \mathbf{C} = \begin{vmatrix} \hat{\mathbf{i}} & \hat{\mathbf{j}} & \hat{\mathbf{k}} \\ 2 & 0 & -1 \\ 0 & 2 & 1 \end{vmatrix} = 2\hat{\mathbf{i}} - 2\hat{\mathbf{j}} + 4\hat{\mathbf{k}}$

$\mathbf{C} \times \mathbf{A} = \begin{vmatrix} \hat{\mathbf{i}} & \hat{\mathbf{j}} & \hat{\mathbf{k}} \\ 0 & 2 & 1 \\ 1 & -1 & 2 \end{vmatrix} = 5\hat{\mathbf{i}} + \hat{\mathbf{j}} - 2\hat{\mathbf{k}}$

Thus, $\mathbf{A} \times \mathbf{B} + \mathbf{B} \times \mathbf{C} + \mathbf{C} \times \mathbf{A}$

$$= (\hat{\mathbf{i}} + 5\hat{\mathbf{j}} + 2\hat{\mathbf{k}}) + (2\hat{\mathbf{i}} - 2\hat{\mathbf{j}} + 4\hat{\mathbf{k}}) + (5\hat{\mathbf{i}} + \hat{\mathbf{j}} - 2\hat{\mathbf{k}})$$

$$= 8\hat{\mathbf{i}} + 4\hat{\mathbf{j}} + 4\hat{\mathbf{k}}$$

61. We have,

$$P'P = \begin{bmatrix} \sqrt{3}/2 & 1/2 \\ -1/2 & \sqrt{3}/2 \end{bmatrix}\begin{bmatrix} \sqrt{3}/2 & -1/2 \\ 1/2 & \sqrt{3}/2 \end{bmatrix} = \begin{bmatrix} 0 & 0 \\ 0 & 1 \end{bmatrix}$$

$\Rightarrow \quad P'P = I$ or $P' = P^{-1}$

As, $\quad Q = PAP'$

$\therefore P'Q^{2005}P = P'[(PAP')(PAP') \ldots 2005 \text{ times}]\, P$

$$= \underbrace{(P'P)A(P'P)A(P'P)\ldots(P'P)A(P'P)}_{2005 \text{ times}}$$

$$= IA^{2005} = A^{2005}$$

Now, $A = \begin{bmatrix} 1 & 1 \\ 0 & 1 \end{bmatrix}, A^2 = \begin{bmatrix} 1 & 1 \\ 0 & 1 \end{bmatrix}\begin{bmatrix} 1 & 1 \\ 0 & 1 \end{bmatrix} = \begin{bmatrix} 1 & 2 \\ 0 & 1 \end{bmatrix}$

$$A^3 = \begin{bmatrix} 1 & 2 \\ 0 & 1 \end{bmatrix}\begin{bmatrix} 1 & 1 \\ 0 & 1 \end{bmatrix} = \begin{bmatrix} 1 & 3 \\ 0 & 1 \end{bmatrix} \ldots A^{2005}$$

$$= \begin{bmatrix} 1 & 2005 \\ 0 & 1 \end{bmatrix}$$

$$\therefore \quad P'Q^{2005}P = \begin{bmatrix} 1 & 2005 \\ 0 & 1 \end{bmatrix}$$

62. $m + n = -\dfrac{3}{7} + \left(\dfrac{-11}{7}\right) = -2$ (–ve integer)

Let $I = \int \cos^{-3/7} x \,(\sin^{(-2+\frac{3}{7})} x)\, dx$

$$= \int \cos^{-3/7} x \sin^{-2} x \sin^{3/7} x dx$$

$$= \int \frac{\text{cosec}^2 x}{\left(\dfrac{\cos^{3/7} x}{\sin^{3/7} x}\right)} dx = \int \frac{\text{cosec}^2 x\, dx}{\cot^{3/7} x}$$

Put $\cos x = t \Rightarrow -\text{cosec}^2 x dx = dt$

$$\therefore \quad I = -\int \frac{dt}{t^{3/7}} = -\frac{t^{\frac{3}{7}+1}}{-\frac{3}{7}+1} + C$$

$$= -\frac{7}{4} t^{4/7} + c = -\frac{7}{4}\cot^{4/7} x + C$$

$$= -\frac{7}{4}\tan^{-4/7} x + C$$

63. Let $ax + by = 1$ be the chord ...(i)

Making the equation of hyperbola homogeneous using Eq. (i), we get

$$3x^2 - y^2 + (-2x + 4y)(ax + by) = 0$$

or $\quad (3 - 2a)x^2 + (-1 + 4b)y^2 + (-2b + 4a)xy = 0$

Since, the angle subtended at the origin is a right angle.

$\therefore$ Coefficient of x^2 + Coefficient of $y^2 = 0$

$\Rightarrow \quad (3 - 2a) + (-1 + 4b) = 0$

$\Rightarrow \quad a = 2b + 1$

$\therefore$ Chords are $(2b + 1)x + by - 1 = 0$

or $\quad b(2 + y) + (x - 1) = 0,$

which clearly pass through the fixed point $(1, -2)$.

64. The line of intersection of the planes

$\mathbf{r} \cdot (3\hat{\mathbf{i}} - \hat{\mathbf{j}} + \hat{\mathbf{k}}) = 1$ and $\mathbf{r} \cdot (\hat{\mathbf{i}} + 4\hat{\mathbf{j}} - 2\hat{\mathbf{k}}) = 2$

is common to both the planes.

Therefore, it is perpendicular to normals to the two planes i.e. $\mathbf{n}_1 = 3\hat{\mathbf{i}} - \hat{\mathbf{j}} + \hat{\mathbf{k}}$ and $\mathbf{n}_2 = \hat{\mathbf{i}} + 4\hat{\mathbf{j}} - 2\hat{\mathbf{k}}$.
Hence, it is parallel to the vector

$$\mathbf{n}_1 \times \mathbf{n}_2 = -2\hat{\mathbf{i}} + 7\hat{\mathbf{j}} + 13\hat{\mathbf{k}}.$$

Thus, we have to find the equation of the plane passing through $\mathbf{a} = \hat{\mathbf{i}} + 2\hat{\mathbf{j}} - \hat{\mathbf{k}}$ and normal to the vector $\mathbf{n} = \mathbf{n}_1 \times \mathbf{n}_2$.
The equation of the required plane is

$$(\mathbf{r} - \mathbf{a}) \cdot \mathbf{n} \text{ or } \mathbf{r} \cdot \mathbf{n} = \mathbf{a} \cdot \mathbf{n}$$

or $\mathbf{r} \cdot (-2\hat{\mathbf{i}} + 7\hat{\mathbf{j}} + 13\hat{\mathbf{k}})$

$$= (\hat{\mathbf{i}} + 2\hat{\mathbf{i}} + \hat{\mathbf{k}}) \cdot (-2\hat{\mathbf{i}} + 7\hat{\mathbf{j}} + 13\hat{\mathbf{k}})$$

or $\mathbf{r} \cdot (2\hat{\mathbf{i}} - 7\hat{\mathbf{j}} - 13\hat{\mathbf{k}}) = 1$

65. Given, $x = 1 + \dfrac{dy}{dx} + \dfrac{1}{2!}\left(\dfrac{dy}{dx}\right)^2 + \dfrac{1}{3!}\left(\dfrac{dy}{dx}\right)^3 + \ldots$

or $\quad x = e^{\frac{dy}{dx}}$

$\Rightarrow \quad \dfrac{dy}{dx} = \log_e x$

Hence, the degree of the above differential equation is 1.

66. Let $P(x_1, y_1)$ be the mid-point of a chord of the circle.

The equation of the chord is $T = S_1$

i.e. $\quad xx_1 + yy_1 - 16 = x_1^2 + y_1^2 - 16$

or $\quad y = -\dfrac{x_1}{y_1}x + \dfrac{x_1^2 + y_1^2}{y_1}$

Since, it touches the hyperbola $\dfrac{x^2}{4^2} - \dfrac{y^2}{3^2} = 1.$

$\therefore \left(\dfrac{x_1^2 + y_1^2}{y_1}\right) = 4^2 \cdot \dfrac{x_1^2}{y_1^2} - 3^2 \quad [\because c^2 = a^2m^2 - b^2]$

$\therefore$ Locus of (x_1, y_1) is

$$\frac{(x^2 + y^2)^2}{y^2} = \frac{16x^2 - 9y^2}{y^2}$$

or $\quad (x^2 + y^2)^2 = 16x^2 - 9y^2$

67. Given, ${}^{n+1}C_{n-2} - {}^{n+1}C_{n-1} \le 50$

$$\Rightarrow \quad \frac{(n-1)!}{3!(n-2)!} - \frac{(n+1)!}{2!(n-2)!} \le 50$$

$$\Rightarrow \quad \frac{(n+1)!}{3!}\left[\frac{1}{(n-2)!} - \frac{3}{(n-1)!}\right] \le 50$$

$$\Rightarrow \quad (n+1)!\left(\frac{n-1-3}{(n-1)!}\right) \le 300$$

$$\Rightarrow \quad (n+1)n(n-4) \le 300$$

For $n = 8$, it satisfy to the above inequality.

But $n = 1$ it does not satisfy the above inequality.

68. Given lines are $y = mx, y = 2$ and $y = 6$.

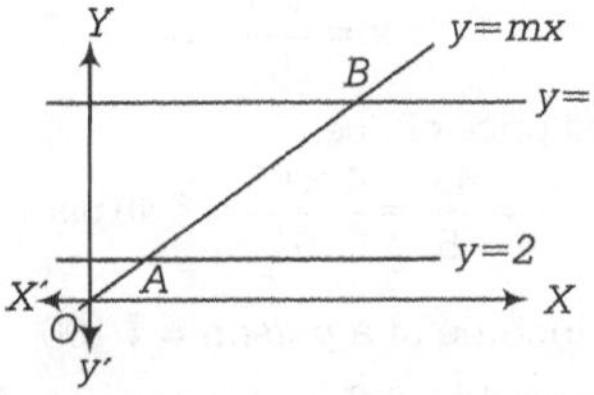

Here, coordinates of points A and B are $\left(\frac{2}{m}, 2\right)$, $\left(\frac{6}{m}, 6\right)$, respectively.

$$\therefore \quad AB = \sqrt{\left(\frac{2}{m} - \frac{6}{m}\right)^2 + (2-6)^2} < 5 \qquad \text{[given]}$$

$$\Rightarrow \quad \left(\frac{2}{m} - \frac{6}{m}\right)^2 + (4)^2 < 25$$

$$\Rightarrow \quad \left(\frac{2}{m} - \frac{6}{m}\right)^2 < 9$$

$$\Rightarrow \quad -3 < \frac{2}{m} - \frac{6}{m} < 3 \Rightarrow -\frac{4}{3} > m > \frac{4}{3}$$

$$\therefore \quad m \in \left]-\infty, -\frac{4}{3}\right[\cup \left]\frac{4}{3}, \infty\right[$$

69. Equation of line perpendicular to $2x + y + 6 = 0$ and passes through origin is $x - 2y = 0$

Now, the point of intersection of $2x + y + 6 = 0$ and $x - 2y = 0$ is $\left(-\frac{12}{5}, -\frac{6}{5}\right)$.

Similarly, the point of intersection of $x - 2y = 0$ and $4x + 2y - 9 = 0$ is $\left(\frac{9}{5}, \frac{9}{10}\right)$.

Let the origin divide the line $x - 2y = 0$ in the ratio $\lambda : 1$

$$\therefore \quad x = \frac{\frac{9}{5}\lambda - \frac{12}{5}}{\lambda + 1} = 0$$

$$\Rightarrow \quad \frac{9}{5}\lambda = \frac{12}{5}$$

$$\Rightarrow \quad \lambda = \frac{12}{9} = \frac{4}{3}$$

70. Given, $P(\overline{A \cup B}) = \frac{1}{6}$

$$\Rightarrow \quad 1 - P(A \cup B) = \frac{1}{6}$$

$$\Rightarrow \quad 1 - P(A) - P(B) + P(A \cap B) = \frac{1}{6}$$

$$\Rightarrow \quad P(\overline{A}) - P(B) + P(A \cap B) = \frac{1}{6}$$

$$\Rightarrow \quad \frac{1}{4} - P(B) + \frac{1}{4} = \frac{1}{6}$$

$$\Rightarrow \quad P(B) = \frac{1}{3}$$

and $\quad P(A) = 1 - P(\overline{A}) = 1 - \frac{1}{4} = \frac{3}{4}$

Since, $P(A \cap B) = P(A) \cdot P(B)$, so the events A and B are independent events but not equally likely.

General Aptitude

16. Walking diagram of John is as follows

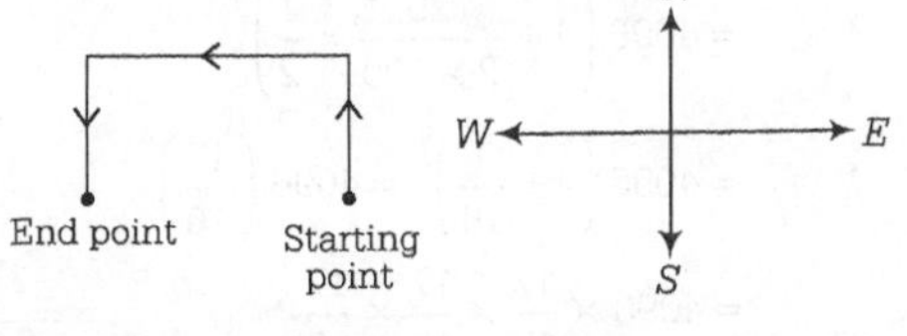

It is clear from the above diagram that John is in facing South direction now.

17. As, $64 + 7 = 460$

$$\Rightarrow \quad 64 \times 7 + 12 = 460$$

and $\quad 25 + 8 = 212$

$$\Rightarrow \quad 25 \times 8 + 12 = 212$$

Same as, $\quad 43 + 8 = ?$

$$\Rightarrow \quad 43 \times 8 + 12 = \boxed{356}$$

$$\therefore \quad ? = 356$$

19. Given equation

$$6 + 64 - 8 \div 45 \times 8$$

Now, changing the sign as per the question,

$$6 \times 64 \div 8 + 45 - 8$$

$$= \frac{6 \times 64}{8} + 45 - 8$$

$$= 6 \times 8 + 45 - 8$$

$$= 48 + 45 - 8$$

$$= 93 - 8 = 85$$

21.

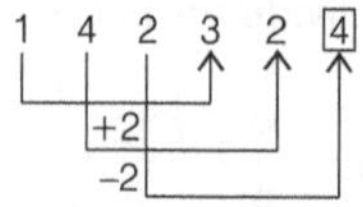

22. As,

$$25 + 17 = 42 \Rightarrow \frac{42}{7} = 6$$

and $$38 + 18 = 56 \Rightarrow \frac{56}{7} = 8$$

Similarly, $$89 + 16 = 105$$

$$\Rightarrow \frac{105}{7} = \boxed{15}$$

23. Relative speed of both the persons

$$= 12 + 10 = 22 \text{ km/h}$$

Now, distance between both of them

$$= 55 - 11 = 44 \text{ km}$$

$\therefore$ Required time, when distance is 11 km between both of them $= \frac{\text{Total distance}}{\text{Relative speed}}$

$$= \frac{44}{22} = 2 \text{ h}$$

Hence, they will be apart 11 km after 2 h.

24. Given, $x - \frac{1}{x} = 2$

On squaring both the sides, we get

$$x^2 + \frac{1}{x^2} - 2 = 4$$

$$\Rightarrow x^2 + \frac{1}{x^2} = 4 + 2$$

$$\Rightarrow x^2 + \frac{1}{x^2} = 6$$

25. Population after 2 yr

$$= 20000\left(1 + \frac{20}{100}\right)\left(1 + \frac{30}{100}\right)$$

$$= 20000 \times \frac{120}{100} \times \frac{130}{100}$$

$$= 2 \times 120 \times 130 = 31200$$

26. Let initial price of rice = ₹ x per kg

Now, a reduction of 20% in price,

Reduced price of rice

$$= x \times \frac{80}{100} = ₹\ \frac{4x}{5} \text{ per kg}$$

Now, according to the question,

$$\frac{1200}{\frac{4x}{5}} - \frac{1200}{x} = 5$$

$$\Rightarrow \frac{6000}{4x} - \frac{1200}{x} = 5$$

$$\Rightarrow \frac{1500}{x} - \frac{1200}{x} = 5 \Rightarrow \frac{300}{x} = 5$$

$$\Rightarrow x = \frac{300}{5} = 60$$

$\therefore$ Reduced price of rice

$$= \frac{4x}{5} = \frac{4 \times 60}{5} = ₹\ 48 \text{ per kg}$$

27. Let total income of a person = ₹ 100

$\therefore$ Then, spend on different expenses = ₹ 60

$\therefore$ His savings = 100 − 60 = ₹ 40

Now, income after increment of 20%

$$= \frac{120}{100} \times 100 = ₹\ 120$$

and expences after increment of 10%

$$= 120 \times (60 + 10)\%$$

$$= 120 \times \frac{70}{100} = ₹\ 84$$

$\therefore$ New savings = 120 − 84 = ₹ 36

Now, percentage decrease

$$= \left(\frac{40 - 36}{40}\right) \times 100\%$$

$$= \frac{4}{40} \times 100\% = \frac{100}{10}\% = 10\%$$

28. Here, $P =$ ₹ 4096, $R = 12\frac{1}{2}\% = \frac{25}{2}\%$,

and $T = 18 \text{ months} = \frac{18}{12} \text{ yr} = \frac{3}{2} \text{ yr}$

Now, money received to Shyam

$$= 4096\left(1 + \frac{25}{2 \times 100} \times \frac{1}{2}\right)^{\frac{3}{2} \times 2}$$

$$= 4096\left(1 + \frac{1}{16}\right)^3 = 4096\left(\frac{17}{16}\right)^3$$

$$= 4096 \times \frac{17}{16} \times \frac{17}{16} \times \frac{17}{16}$$

$$= 17 \times 17 \times 17 = ₹\ 4913$$

29. Time taken to complete the work by $B = 12$ days

Since, A work twice as fast as B.

$\therefore$ Time taken to complete the work by $A = 6$ days

Now, one day work of $B = \frac{1}{12}$

and one day work of $A = \frac{1}{6}$

$$\therefore \text{One day work of } (A + B) = \frac{1}{6} + \frac{1}{12}$$

$$= \frac{2+1}{12} = \frac{3}{12} = \frac{1}{4}$$

Hence, time taken by A and B together to finish the work $= 4$ days.

30. Let the height of embankment $= h$ m

Given, radius of well $= \frac{3}{2} = 1.5$ m

$$\therefore \frac{4}{3} \times \frac{22}{7} \times [(4)^2 - (1.5)^2] \times h = \frac{22}{7} \times (1.5)^2 \times 14$$

$$\Rightarrow \frac{4}{3} \times (16 - 2.25) \times h = 2.25 \times 14$$

$$\Rightarrow \frac{4}{3} \times 13.75 \times h = 2.25 \times 14$$

$$\Rightarrow h = \frac{2.25 \times 14 \times 3}{4 \times 13.75} = \frac{225 \times 7 \times 3}{2 \times 1375} = \frac{189}{110}$$

$$= 1.718 \approx 1.750 \text{ m}$$

Hence, the height of embankment is 1.750 m.

Solved Paper 2015

Manipal

Engineering Entrance Exam

Physics

1. Two point charges $2q$ and $8q$ are placed at a distance r apart. Where should a third charge $-q$ be placed between them, so that the electrical potential energy of the system is minimum?

(a) At a distance of $r/3$ from $2q$
(b) At a distance of $2r/3$ from $2q$
(c) At a distance of $r/16$ from $2q$
(d) None of the above

2. If levels 1 and 2 are separated by an energy $E_2 - E_1$, such that the corresponding transition frequency falls in the middle of the visible range, calculate the ratio of the populations of two levels in the thermal equilibrium at room temperature.

(a) 1.1577×10^{-38} (b) 2.9×10^{-35}
(c) 2.168×10^{-36} (d) 1.96×10^{-20}

3. The energy of a hydrogen atom in its ground state is -13.6 eV. The energy of the level corresponding to the quantum number $n = 5$ is

(a) -5.40 eV (b) -0.54 eV
(c) -8.5 eV (d) -2.72 eV

4. We have a galvanometer of resistance 25 Ω. It is shunted by 2.5 Ω wire. The part of the total current that flows through the galvanometer is given as

(a) $\frac{i}{i_0} = \frac{4}{11}$ (b) $\frac{i}{i_0} = \frac{3}{11}$ (c) $\frac{i}{i_0} = \frac{2}{10}$ (d) $\frac{i}{i_0} = \frac{1}{11}$

5. Two moles of helium are mixed with n moles of hydrogen. The root mean square (rms) speed of gas molecules in the mixture is $\sqrt{2}$ times the speed of sound in the mixture. Then, the value of n is

(a) 1 (b) 3 (c) 2 (d) 3/2

6. A coil having n turns and resistance $R\Omega$ is connected with a galvanometer of resistance $4R\Omega$. This combination is moved in time t second from a magnetic field w_1 weber to w_2 weber. The induced current in the circuit is

(a) $\frac{-(w_2 - w_1)}{5Rt}$
(b) $\frac{-n(w_2 - w_1)}{5Rt}$
(c) $\frac{-(w_2 - w_1)}{Rnt}$
(d) $\frac{-n(w_2 - w_1)}{Rt}$

7. A thin film of soap solution ($\mu_s = 1.4$) lies on the top of a glass plate ($\mu_g = 1.5$). When incident light is almost normal to the plate, two adjacent reflection maxima are observed at two wavelengths 420 nm and 630 nm. The minimum thickness of the soap solution is

(a) 420 nm (b) 450 nm
(c) 630 nm (d) 1260 nm

8. A coil in the shape of an equilateral triangle of side l is suspended between the pole pieces of a permanent magnet such that **B** is in the plane of the coil. If due to current i in the triangle, a torque τ acts on it. The side l of the triangle is

(a) $\frac{2}{\sqrt{3}}\left(\frac{\tau}{Bi}\right)$
(b) $2\left(\frac{\tau}{\sqrt{3}\,Bi}\right)^{1/2}$
(c) $\frac{2}{\sqrt{3}}\left(\frac{\tau}{Bi}\right)^{1/2}$
(d) $\frac{1}{\sqrt{3}}\frac{\tau}{Bi}$

9. The two blocks of masses m_1 and m_2 are kept on a smooth horizontal table as shown in the figure. Block of mass m_1 but not m_2 is fastened to the spring. If now both the blocks are pushed to the left, so that the spring is compressed at a distance d. The amplitude of oscillation of block of mass m_1 after the system released, is

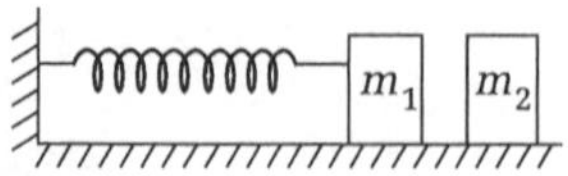

(a) $d\sqrt{\dfrac{m_1}{m_1+m_2}}$ (b) $d\sqrt{\dfrac{m_2}{m_1+m_2}}$

(c) $d\sqrt{\dfrac{2m_2}{m_1+m_2}}$ (d) $d\sqrt{\dfrac{2m_1}{m_1+m_2}}$

10. A juggler keeps on moving four balls in air throwing the balls after regular intervals. When one ball leaves his hand (speed $= 20\ \text{ms}^{-1}$), the position of other balls (height in metre) will be (take $g = 10\ \text{ms}^{-2}$)

(a) 10, 20, 10
(b) 15, 20, 15
(c) 5, 15, 20
(d) 5, 10, 20

11. Two coils have mutual inductance 0.005 H. The current changes in the first coil according to equation $I = I_0 \sin \omega t$, where $I_0 = 10$ A and $\omega = 100\ \pi$ rad/s. The maximum value of emf in the second coil is

(a) $12\ \pi$ (b) 8π
(c) 5π (d) 2π

12. A ball is projected from the point O with velocity 20 m/s at an angle of 60° with horizontal as shown in the figure. At highest point of its trajectory, it strikes a smooth plane of inclination 30° at point A. The collision is perfectly inelastic. The maximum height from the ground attained by the ball is

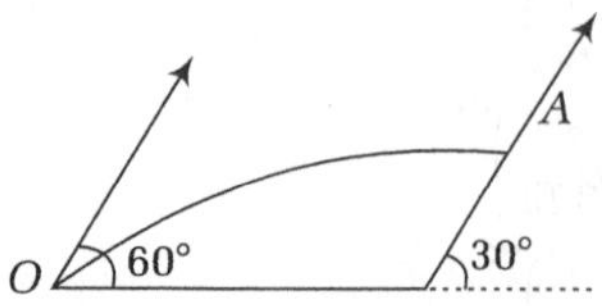

(a) 18.75 m (b) 15 m
(c) 22.5 m (d) 20.25 m

13. In a nuclear reactor, ^{235}U undergoes fission liberating 200 MeV of energy. The reactor has a 10% efficiency and produces 1000 MW power. If the reactor is to function for 10 yr, then find the total mass of uranium required.

(a) 36.5×10^3 kg (b) 36×10^2 kg
(c) 39.5×10^3 kg (d) 38.2×10^3 kg

14. A capacitor of capacitance 10 μF is charged to potential 50 V with a battery. The battery is now disconnected and an additional charge 200 μC is given to the positive plate of the capacitor. The potential difference across the capacitor will be

(a) 50 V (b) 80 V
(c) 100 V (d) 60 V

15. The following configuration of gate is equivalent to

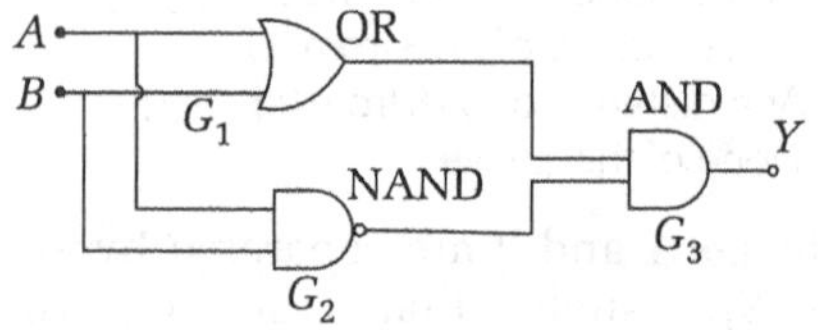

(a) NAND (b) XOR
(c) OR (d) None of these

16. Under what conditions current passing through the resistance R can be increased by short circuiting the battery of emf E_2? The internal resistances of the two batteries are r_1 and r_2 respectively.

(a) $E_2 r_1 > E_1 (R + r_2)$ (b) $E_1 r_2 < E_2 (r_1 + R)$
(c) $E_2 r_2 < E (R + r_2)$ (d) $E_1 r_1 > E_2 (r_1 + R)$

17. A rectangular glass slab $ABCD$ of refractive index n_1 is immersed in water of refractive index n_2 $(n_1 > n_2)$. A ray of light is incident at the surface AB of the slab as shown. The maximum value of the angle of incidence α_{max} such that the ray comes out only from the another surface CD is given by

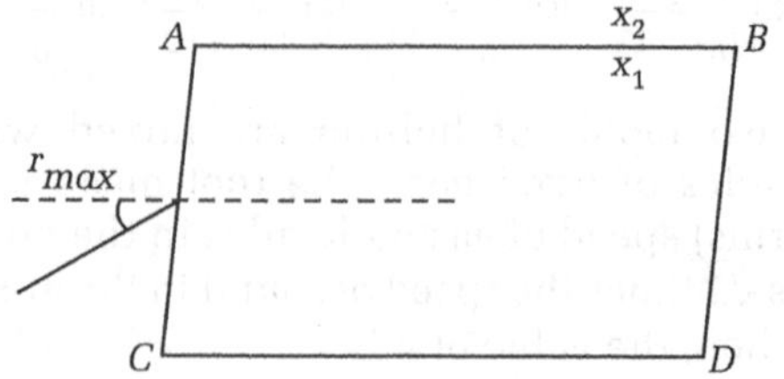

(a) $\sin^{-1}\left[\frac{n_1}{n_2}\cos\left(\sin^{-1}\left(\frac{n_2}{n_1}\right)\right)\right]$

(b) $\sin^{-1}\left[n_1\cos\left(\sin^{-1}\left(\frac{1}{2}\right)\right)\right]$

(c) $\sin^{-1}\left(\frac{n_1}{n_2}\right)$

(d) $\sin^{-1}\left(\frac{n_2}{n_1}\right)$

18. A sinusoidal wave travelling in the positive direction on stretched string has amplitude 20 cm, wavelength 1 m and wave velocity 5 m/s. At $x = 0$ and $t = 0$, it is given that $y = 0$ and $\frac{dy}{dt} < 0$. Find the wave function $y(x,t)$.

(a) $y(x,t) = (0.02\text{ m})\sin[(2\pi\text{m}^{-1})x + (10\pi\text{s}^{-1})t]\text{m}$
(b) $y(x,t) = (0.02\text{ m})\cos[(10\pi\text{s}^{-1})t + (2\pi\text{m}^{-1})x]\text{m}$
(c) $y(x,t) = (0.02\text{ m})\sin[(2\pi\text{m}^{-1})x - (10\pi\text{s}^{-1})t]\text{m}$
(d) $y(x,t) = (0.02\text{ m})\sin[(\pi\text{m}^{-1})x + (5\pi\text{s}^{-1})t]\text{m}$

19. The length of a potentiometer wire is l. A cell of emf E is balanced at length $l/3$ from the positive end of the wire. If the length of the wire is increased by $l/2$. Then, at what distance will the same cell give a balance point?

(a) $\frac{2l}{3}$ (b) $\frac{l}{2}$ (c) $\frac{l}{6}$ (d) $\frac{4l}{3}$

20. Find the inductance of a unit length of two parallel wires, each of radius a, whose centres are at distance d apart and carry equal currents in opposite directions. Neglect the flux within the wire.

(a) $\frac{\mu_0}{2\pi}\ln\left(\frac{d-a}{a}\right)$ (b) $\frac{\mu_0}{\pi}\ln\left(\frac{d-a}{a}\right)$
(c) $\frac{3\mu_0}{\pi}\ln\left(\frac{d-a}{a}\right)$ (d) $\frac{\mu_0}{3\pi}\ln\left(\frac{d-a}{a}\right)$

21. Two wires of same lengths are shaped into a square and a circle. If they carry same current, then the ratio of their magnetic moments is

(a) $2:\pi$ (b) $\pi:2$ (c) $\pi:4$ (d) $4:\pi$

22. A circular coil of 100 turns has an effective radius of 0.05 m and carries a current of 0.1 A. How much work is required to turn it in an external magnetic field of 1.5 Wb/m^2 through 180° about its axis perpendicular to the magnetic field? The plane of the coil is initially perpendicular to the magnetic field.

(a) 0.456 J (b) 2.65 J (c) 0.2355 J (d) 3.95 J

23. The time constant of L-R circuit is

(a) LR (b) $\frac{L}{R}$
(c) $\frac{R}{L}$ (d) $\frac{1}{LR}$

24. In wave mechanics, the angular momentum of an electron is given by

(a) $\sqrt{l\frac{h}{2\pi}}$ (b) $\sqrt{l(l+1)}\frac{h}{2\pi}$
(c) $\sqrt{\frac{h2\pi}{l(l+1)}}$ (d) $\sqrt{\frac{h2\pi}{l}}$

25. The alternating voltage and current in an electric circuit are respectively given by

$E = 100\sin 100\pi t$, $I = 5\sin 100\pi t$

The reactance of the circuit will be

(a) 1 Ω (b) 0.05 Ω
(c) 20 Ω (d) zero

26. The biological damage caused by 1 gray α-radiation is compared with 1 gray γ-radiation in the same type of human tissue. The damage caused by the γ-radiation is

(a) more serious as compared to the damage caused by the α-radiation.
(b) less serious as compared to the damage caused by the α-radiation
(c) equally serious as the damage caused by the α-radiation
(d) incomparable with the damage caused by the α-radiation, because γ-radiation are not particles.

27. Two identical coherent sources placed on a diameter of a circle of radius R at separation $x\,(<< R)$ symmetrically about the centre of the circle. The sources emit identical wavelength λ each. The number of points on the circle with maximum intensity is $(x = 5\lambda)$

(a) 20 (b) 22
(c) 24 (d) 26

28. Two point masses 1 and 2 move with uniform velocities $\mathbf{v}_1$ and $\mathbf{v}_2$ respectively. Their initial position vectors are $\mathbf{r}_1$ and $\mathbf{r}_2$, respectively. Which of the following should be satisfied for the collision of the point masses?

(a) $\frac{\mathbf{r}_1 - \mathbf{r}_2}{|\mathbf{r}_2 - \mathbf{r}_1|} = \frac{\mathbf{v}_2 - \mathbf{v}_1}{|\mathbf{v}_2 - \mathbf{v}_1|}$ (b) $\frac{\mathbf{r}_2 - \mathbf{r}_1}{|\mathbf{r}_1 - \mathbf{r}_1|} = \frac{\mathbf{v}_2 - \mathbf{v}_1}{|\mathbf{v}_2 - \mathbf{v}_1|}$
(c) $\frac{\mathbf{r}_2 - \mathbf{r}_1}{|\mathbf{r}_2 + \mathbf{r}_1|} = \frac{\mathbf{v}_2 - \mathbf{v}_1}{|\mathbf{v}_2 + \mathbf{v}_1|}$ (d) $\frac{\mathbf{r}_2 + \mathbf{r}_1}{|\mathbf{r}_2 + \mathbf{r}_1|} = \frac{\mathbf{v}_2 - \mathbf{v}_1}{|\mathbf{v}_2 + \mathbf{v}_1|}$

29. A neutron moving with a speed v makes a head on collision with a hydrogen atom in ground state kept at rest. The minimum kinetic energy of neutron for which inelastic collision will take place is

(a) 10.2 eV (b) 20.4 eV
(c) 12.1 eV (d) 16.8 eV

30. The specific heat at constant volume for the monoatomic argon is 0.075 kcal/kg-K, whereas its gram molecular specific heat is $C_V = 2.98$ cal/mol K. The mass of the carbon atom is

(a) 6.60×10^{-23}g (b) 3.30×10^{-23}g
(c) 2.20×10^{-23}g (d) 13.20×10^{-23}g

31. Order of magnitude of density of uranium nuclus is $(m_p = 1.6 \times 10^{-27}\text{kg})$

(a) 10^{20} kg/m^3 (b) 10^{17} kg/m^3
(c) 10^{14} kg/m^3 (d) 10^{11} kg/m^3

32. A block of mass m is lying on the edge having inclination angle $\alpha = \tan^{-1}\left(\frac{1}{5}\right)$. Wedge is moving with a constant acceleration, $a = 2\text{ms}^{-2}$. The minimum value of coefficient of friction μ, so that m remains stationary with respect to wedge is

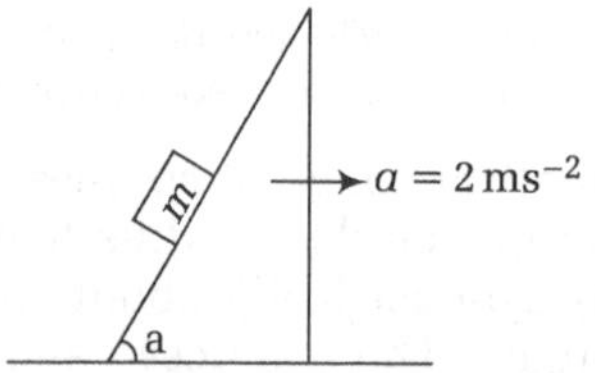

(a) $\frac{2}{a}$ (b) $\frac{5}{12}$ (c) $\frac{1}{5}$ (d) $\frac{2}{5}$

33. A small particle of mass m is released from rest from point A inside a smooth hemispherical bowl as shown in the figure. The ratio (x) of magnitude of centripetal force and normal reaction on the particle at any point B varies with θ as

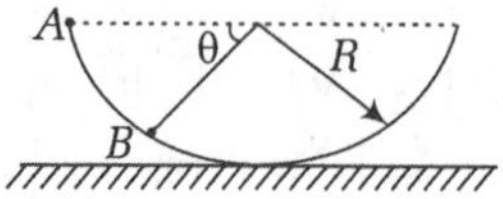

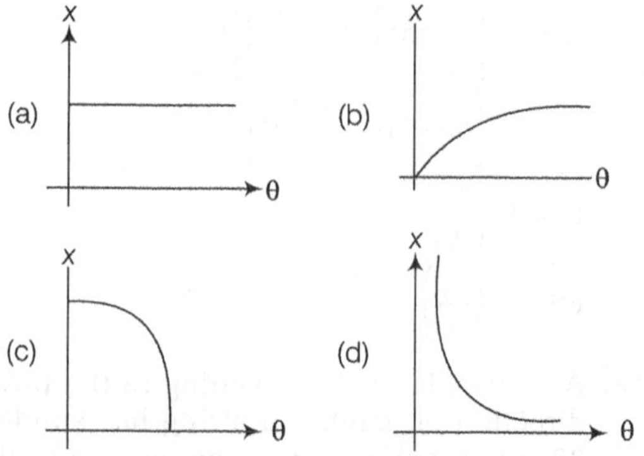

34. A solid cylinder is rolling down on an inclined plane of angle θ. The coefficient of static friction between the plane and the cylinder is μ_s. The condition for the cylinder not to slip is

(a) $\tan\theta \geq 3\mu_s$ (b) $\tan\theta > 3\mu_s$
(c) $\tan\theta \leq 3\mu_s$ (d) $\tan\theta < 3\mu_s$

35. For a given density of a planet, the orbital speed of satellite near the surface of the planet of radius R is proportional to

(a) $R^{1/2}$ (b) $R^{3/2}$
(c) $R^{-1/2}$ (d) R^0

36. A large slab of mass 5 kg lies on a smooth horizontal surface, with a block of mass 4 kg lying on the top of it, the coefficient of friction between the block and the slab is 0.25. If the block is pulled horizontally by a force of $F = 6$ N, the work done by the force of friction on the slab between the instants $t = 2$ s and $t = 3$ s is

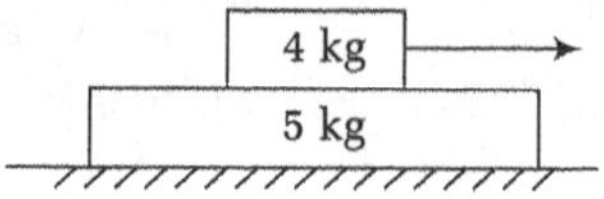

(a) 2.4 J (b) 5.55 J
(c) 4.44 J (d) 10 J

37. Two unequal masses are connected on two sides of a light string passing over a light and smooth pulley as shown in the figure. The system is released from the rest. The larger mass is stoped for a moment, 1s after the system is set into motion. The time elapsed before the string is tight again is

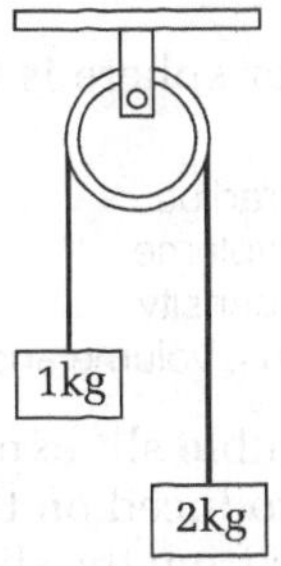

(a) 1/4 s (b) 1/2 s (c) 2/3 s (d) 1/3 s

38. Figure shows an irregular block of material of refractive index $\sqrt{2}$. A ray of light strikes the face AB as shown in figure. After refraction, it is incident on a spherical surface CD of radius of curvature 0.4 m and enters a medium of refractive index 1.514 to meet PQ at E. Find the distance OE upto two places of decimal.

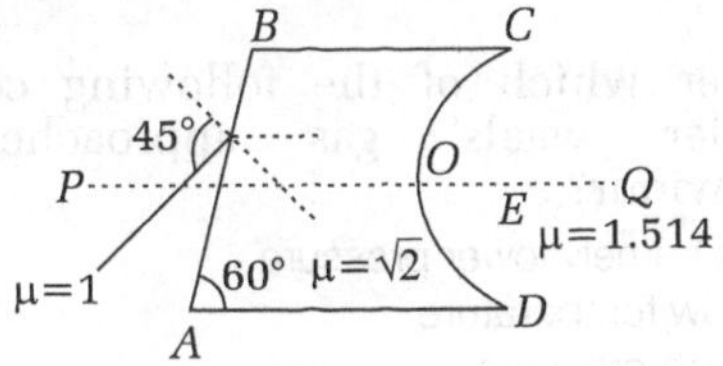

(a) 7 m (b) 7.29 m (c) 6.06 m (d) 8.55 m

39. The ratio of the energy required to raise a satellite upto a height h above the earth to the kinetic energy of the satellite into the orbit is

(a) $h : R$ (b) $R : 2h$ (c) $2h : R$ (d) $R : h$

40. In the circuit diagram, the current through the Zener diode is

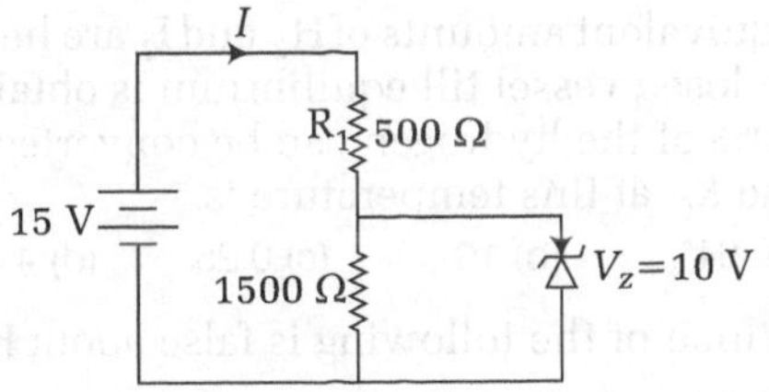

(a) 10 mA (b) 3.33 mA (c) 6.67 mA (d) 0 mA

41. The frequency of sonometer wire is f, but when the weights producing the tensions are completely immersed in water, the frequency becomes $f/2$ and on immersing the weights in a certain liquid, the frequency becomes $f/3$. The specific gravity of the liquid is

(a) 4/3 (b) 16/9
(c) 15/12 (d) 32/27

42. Find the frequency of light which ejects electron from a metal surface fully stopped by a retarding potential of 3V. The photoelectric effect begins in this metal at a frequency of 6×10^{15}Hz.

(a) 1.324×10^{15}Hz
(b) 2.295×10^{16}Hz
(c) 3.678×10^{18}Hz
(d) 2.7×10^{14}Hz

43. Equations of a stationary and a travelling waves are as follows, $y_1 = a \sin kx \cos \omega t$ and $y_2 = a \sin(\omega t - kx)$. The phase difference between two points $x_1 = \frac{\pi}{3k}$ and $x_2 = \frac{3\pi}{2k}$ are ϕ_1 and ϕ_2 respectively for two waves. The ratio $\frac{\phi_1}{\phi_2}$ is

(a) 1 (b) 5/6
(c) 3/4 (d) 6/7

44. Out of a photon and an electron, the equation $E = Pc$, is valid for

(a) both (b) neither
(c) photon only (d) electron only

45. A small block of wood of specific gravity 0.5 is submerged at a depth of 1.2 m in a vessel filled with the water. The vessel is accelerated upwards with an acceleration $a_0 = g/2$. Time taken by the block to reach the surface, if it is released with zero initial velocity is

(a) 0.6 s (b) 0.4 s
(c) 1.2 s (d) 1 s

46. An electron beam accelerated from rest through a potential difference of 5000 V in vaccum is allowed to impinge on a surface normally. The incident current is 50 μA and if the electron comes to rest on striking the surface, the force on it is

(a) 1.1924×10^{-8}N (b) 2.1×10^{-8}N
(c) 1.6×10^{-8}N (d) 1.6×10^{-6}N

47. A uniform rod of length 2 m, specific gravity 0.5 and mass 2 kg is hinged at one end to the bottom of a tank of water (specific gravity = 1.0) filled upto a height of 1 m as shown in the figure. Taking the case $\theta \neq 0°$, the force exerted by the hinge on the rod is

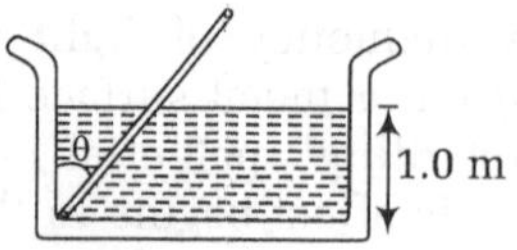

(a) 10.2 N upwards
(b) 4.2 N downwards
(c) 8.3 N downwards
(d) 6.2 N upwards

48. A projectile is thrown in upward direction making an angle of 60° with the horizontal direction with a velocity of 150 ms^{-1}. Then, the time after which its inclination with horizontal is 45°, is

(a) $15(\sqrt{3}-1)$ s
(b) $15(\sqrt{3}+1)$ s
(c) $7.5(\sqrt{3}-1)$ s
(d) $7.5(\sqrt{3}+1)$ s

49. When a copper sphere is heated, percentage change is

(a) maximum in radius
(b) maximum in volume
(c) maximum in density
(d) equal in radius, volume and density

50. In Young's double slit experiment, fringes of width β are produced on the screen kept at a distance of 1m from the slit. When the screen is moved away by 5×10^{-2}m, fringe width changes by 3×10^{-5}m. The separation between the slits is 1×10^{-3}m. The wavelength of light used is......nm.

(a) 500
(b) 600
(c) 700
(d) 400

Chemistry

1. How many moles of $Al_2(SO_4)_3$ would be in 50 g of the substance?

(a) 0.0843 mol
(b) 0.952 mol
(c) 0.481 mol
(d) 0.140 mol

2. The ionisation energy of hydrogen atom is 13.6 eV. What will be the ionisation energy of He^+?

(a) 13.6 eV
(b) 54.4 eV
(c) 122.4 eV
(d) Zero

3. In which of the following arrangement the order is not according to the property indicated against it?

(a) Li < Na < K < Rb (increasing metallic radius)
(b) I < Br < F < Cl (increasing electron gain enthalpy, with negative sign)
(c) B < C < N < O (increasing first ionisation enthalpy)
(d) $Al^{3+} < Mg^{2+} < Na^+ < F^-$ (increasing ionic size)

4. The H–O–H bond angle in H_2O is 104.5°. This fact can be explained with the help of

(a) valence shell electron pair repulsion theory (VSEPR)
(b) molecular orbital theory
(c) presence of hydrogen bond
(d) electronegativity difference between hydrogen and oxygen atoms

5. Under which of the following condition, vander waals' gas approaches ideal behaviour?

(a) Extremely lower pressure
(b) Low temperature
(c) High pressure
(d) Low product of pV

6. The enthalpy of combustion of carbon and carbon monoxide are – 393.5 and – 283 kJ/mol respectively. The enthalpy of formation of carbon monoxide per mole is

(a) 110.5 kJ
(b) 676.5 kJ
(c) – 676.5 kJ
(d) – 110.5 kJ

7. Equivalent amounts of H_2 and I_2 are heated in a closed vessel till equilibrium is obtained. If 80% of the hydrogen can be converted to HI, the K_C at this temperature is

(a) 64
(b) 16
(c) 0.25
(d) 4

8. Which of the following is false about H_2O_2?

(a) Act as both oxidising and reducing agent
(b) Two OH bond lie in the same plane
(c) Pale blue liquid
(d) Can be oxidised by ozone

9. KO_2 is used in space and submarines because it

(a) absorb CO_2 and increase O_2 concentration
(b) absorb moisture
(c) absorb CO_2
(d) produce ozone

10. The relative Lewis acid character of boron trihalides is the order

(a) $BI_3 > BBr_3 > BF_3 > BCl_3$
(b) $BI_3 > BBr_3 > BCl_3 > BF_3$
(c) $BF_3 > BCl_3 > BBr_3 > BI_3$
(d) $BCl_3 > BF_3 > BI_3 > BBr_3$

11. The IUPAC name of

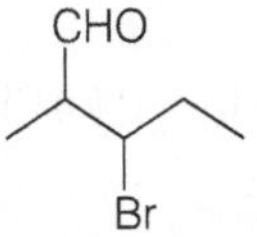

is

(a) 2-methyl-3-bromo hexanal
(b) 3-bromo-2-methyl butanal
(c) 2-bromo-3-bromo butanal
(d) 3-bromo-2-methyl pentanal

12. Which of the following would react most readily with nucleophiles?

(a) 

(b) 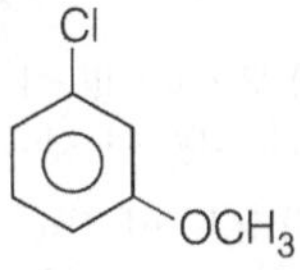

(c) 

(d) 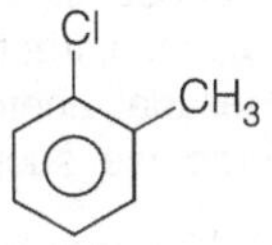

13. Ethyl acetoacetate shows, Which type of isomerism?

(a) Chain (b) Optical
(c) Metamerism (d) Tautomerism

14. $C_7H_8 \xrightarrow{3Cl_2,\text{ heat}} A \xrightarrow{Fe/Br_2} B \xrightarrow{Zn/HCl} C$

Here the compound C is

(a) 3-bromo -2,4,5,6-trichloro tolucne
(b) *o*-bromo toluene
(c) *p*-bromo toluene
(d) *m*-bromo toluene

15. An important product in the ozone deplation by chlorofluoro carbons is

(a) Cl_2 (b) OCl
(c) OF_2 (d) O_2F_2

16. The crystalline structure of NaCl is

(a) hexagonal close packing
(b) face centred cubic
(c) square planar
(d) body centred cubic

17. 40% by weight solution will contain how much mass of the solute in 1L solution, density of the solution is 1.2 g/mL?

(a) 480 g (b) 48 g
(c) 38 g (d) 380 g

18. The standard reduction potential $E°$ for half reactions are,

$$Zn \longrightarrow Zn^{2+} + 2e^- \quad ; E° = +0.76V$$

$$Fe^{2+} + 2e^- \longrightarrow Fe \quad ; E° = +0.41$$

The emf of the cell reaction;

$$Fe^{2+} + Zn \longrightarrow Zn^{2+} + Fe \text{ is}$$

(a) − 0.35 V (b) + 0. 35 V
(c) + 1.17 V (d) − 1.17 V

19. $N_2 + 3H_2 \rightleftharpoons 2NH_3 ; + 22\text{ kcal}$

The activation energy for the forward reaction is 50 kcal. What is the activation energy for the backword reaction?

(a) − 72 kcal (b) − 28 kcal
(c) + 28 kcal (d) + 72 kcal

20. The coagulating power of an electrolyte for arsenious sulphide decrease in order

(a) $Na^+ > Al^{3+} > Ba^{2+}$ (b) $PO_4^{3-} > SO_4^{2-} > Cl^-$
(c) $Cl^- > SO_4^{2-} > Cl^-$ (d) $Al^{3+} > Ba^{2+} > Na^+$

21. ${}^{228}_{88}X - 3\alpha, -\beta \longrightarrow Y$. The element Y is

(a) ${}^{216}_{82}Pb$ (b) ${}^{217}_{82}Pb$ (c) ${}^{218}_{83}Bi$ (d) ${}^{216}_{83}Bi$

22. Hydrolysis of NCl_3 gives NH_3 and X. Which of the following is X?

(a) $HClO_4$ (b) $HClO_3$
(c) $HOCl$ (d) $HClO_2$

23. In the reaction,

$$4M + 8CN^- + 2H_2O + O_2 \longrightarrow 4[M(CN)_2] + 4OH^-$$

identify the metal M.

(a) Copper (b) Iron
(c) Silver (d) Zinc

24. In which of the following complex ion, the central matal ion is in a state of sp^3d^2 hybridisation?

(a) $[CoF_6]^{3-}$ (b) $[Co(NH_3)_6]^{3+}$
(c) $[Fe(CN)_6]^{3-}$ (d) $[Cr(NH_3)_6]^{3+}$

25. In the reaction sequence,

$$C_2H_5Cl + KCN \xrightarrow{C_2H_5OH} X \xrightarrow[\Delta]{H_3O^{\oplus}} Y,$$

what is the molecular formula of Y?

(a) $C_3H_6O_2$ (b) C_3H_5N
(c) $C_2H_4O_2$ (d) C_2H_6O

26. 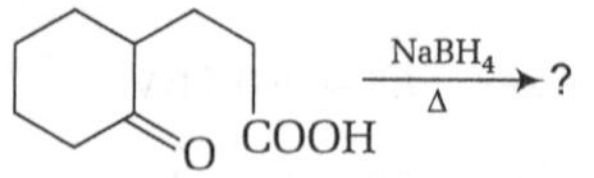

(a) 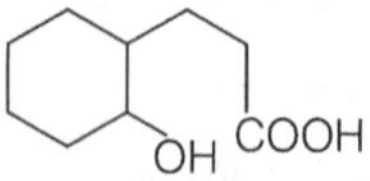

(b) 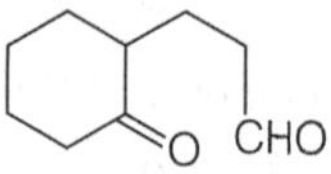

(c) 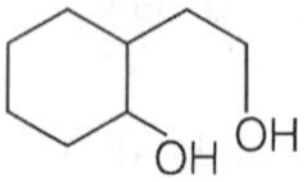

(d) 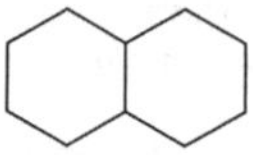

27. The product P in the reaction is

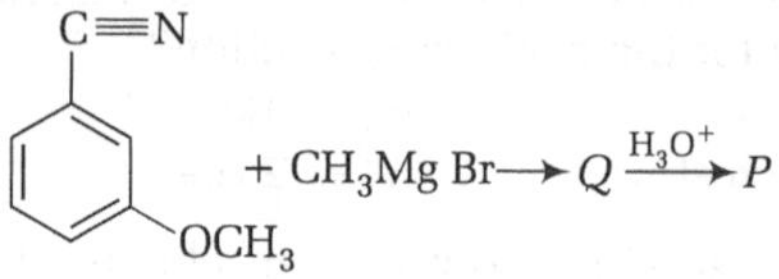

(a) 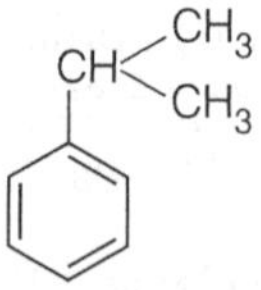

(b) 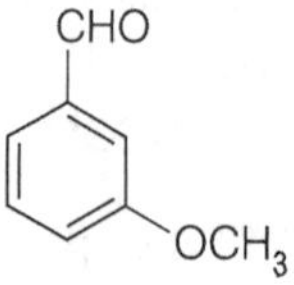

(c) 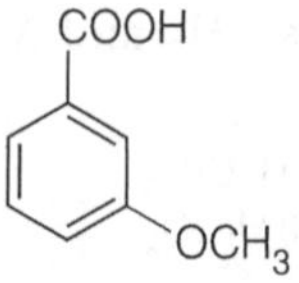

(d) 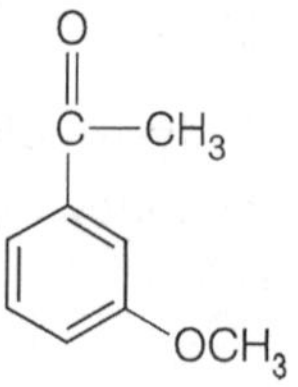

28. An organic compound of molecular formula C_3H_6O did not give a silver mirror with Tollen's reagent, but gave an oxime with hydroxylamine, it may be

(a) CH_3COCH_3
(b) C_2H_5CHO
(c) $CH_2{=}CH—CH_2—OH$
(d) $CH_3—O—CH{=}CH_2$

29. Given the following sequence of reactions,

$$CH_3CH_2I \xrightarrow{NaCN} A \xrightarrow[\text{Partial hydrolysis}]{OH^-} B \xrightarrow{Br_2/NaOH} C.$$ The major product C is

(a) $CH_3CH_2NH_2$ (b) $CH_3CH_2—\overset{O}{\overset{\|}{C}}—NHBr$
(c) $CH_3CH_2COONH_4$ (d) $CH_3{\cdot}CH_2\overset{O}{\overset{\|}{C}}—NBr_2$

30. Coupling of diazonium salts of the following takes place in the order

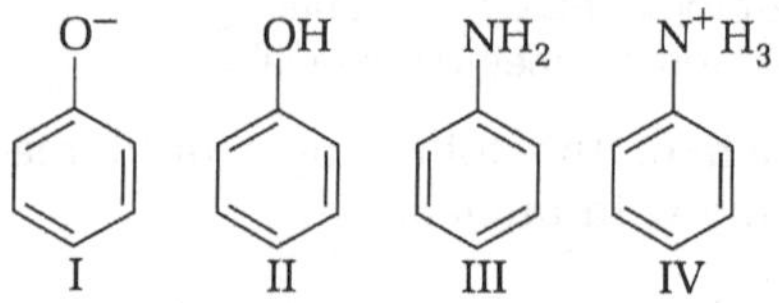

(a) IV < II < III < I (b) IV < III < II < I
(c) II < IV < I < III (d) I < II < III < IV

31. Which of the following pairs give positive Tollen's test?

(a) Glucose, sucrose
(b) Glucose, fractose
(c) Hexanal, acetophenone
(d) Fractose, sucrose

32. Which one is chain growth polymer?

(a) Teflon (b) Nylon-6 (c) Nylon-66 (d) Bakelite

33. Sodium alkyl benzene sulphonate is used as

(a) soap (b) fertilizer
(c) detergent (d) pesticide

34. Brown ring is made for

(a) NO_3^- (b) Cl^- (c) I^- (d) Br^-

35. The purest zinc is made by

(a) electrolytic refining (b) zone refining
(c) The van-Arkel method (d) Mond process

36. Which of the following is the strongest oxidising agent?

(a) HOCl (b) $HClO_2$ (c) $HClO_3$ (d) $HClO_4$

37. In aerosol, the dispersion medium is

(a) solid (b) liquid
(c) gas (d) Any of these

38. For the two gaseous reactions, following data

$A \longrightarrow B; k_1 = 10^{10}\, e^{-20,000/T}$

$C \longrightarrow D; k_2 = 10^{12}\, e^{-24,606/T}$

the temperature at which k_1 becomes equal to k_2 is

(a) 400 K (b) 1000 K
(c) 800 K (d) 1500 K
(e) 500 K

39. How long (in hours) must a current of 5.0 A be maintained to electroplate 60g of calcium from molten $CaCl_2$?

(a) 27 h (b) 8.3 h
(c) 11 h (d) 16 h

40. Calculate the molal depression constant of a solvent which has freezing point 16.6° C and latent heat of fusion 180.75 J/g.

(a) 2.68 (b) 3.86
(c) 4.68 (d) 2.86

41. In CsCl type structure, the coordination number of Cs^+ and Cl^- respectively are

(a) 6, 6 (b) 6,8
(c) 8,8 (d) 8,6

42. 9.2 g N_2O_4 is heated in 1L vessel till equilibrium state is established.

$$N_2O_4\,(g) \rightleftharpoons 2NO_2\,(g)$$

In equilibrium state, 50% N_2O_4 was dissociated, equilibrium constant will be (molecular wt. of $N_2O_4 = 92$)

(a) 0.1 (b) 0.4
(c) 0.3 (d) 0.2

43. Enthalpy is equal to

(a) $T^2 \left[\frac{\delta (G/T)}{\delta T}\right]_p$

(b) $-T^2 \left[\frac{\delta (G/T)}{\delta T}\right]_p$

(c) $T^2 \left[\frac{\delta (G/T)}{\delta T}\right]_V$

(d) $-T^2 \left[\frac{\delta (G/T)}{\delta T}\right]_V$

44. The isoelectronic pair is

(a) Cl_2O, ICl_2^- (b) Cl_2^-, ClO_2
(c) IF_2^+, I_3^- (d) ClO_2^-, ClF_2^+

45. Which of the following has largest ionic radius?

(a) Cs^+ (b) Li^+
(c) Na^+ (d) K^+

46. The ground state term symbol for an electronic state is governed by

(a) Heisenberg principle (b) Hund's rule
(c) Aufbau principle (d) Pauli exclusion principle

47. 2.76 g of silver carbonate on being strongly heated yeild a residue weighing

(a) 2.16 g (b) 2.48 g
(c) 2.64 g (d) 2.32 g

48. The following reaction is an example of ...... reaction.

$$C_2H_4Br_2 \xrightarrow{\text{Alc} \cdot \text{KOH}} C_2H_2$$

(a) addition
(b) dehydrobromination
(c) substitution
(d) bromination

49. I_2 dissolve in KI solution due to formation of

(a) KI_2 and I^-
(b) K^+; I^- and I_2
(c) I_3^-
(d) None of the above

50. 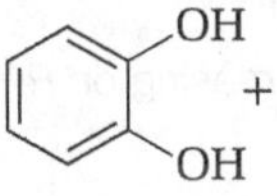

$+ CH_2I_2 + NaOH \longrightarrow$

The product is

(a) 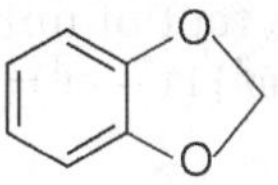

(b) 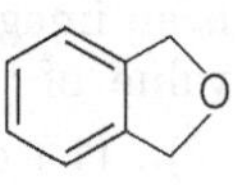

(c) 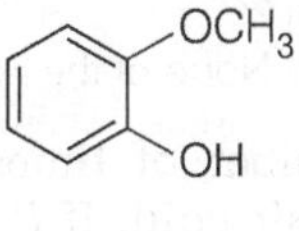

(d) 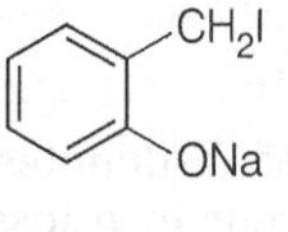

Mathematics

1. The approximate value of $(1.0002)^{3000}$ is

(a) 1.2 (b) 1.4
(c) 1.6 (d) 1.8

2. $(\mathbf{a} \cdot \hat{\mathbf{i}})(\mathbf{a} \times \hat{\mathbf{i}}) + (\mathbf{a} \cdot \hat{\mathbf{j}})(\mathbf{a} \times \hat{\mathbf{j}}) + (\mathbf{a} \cdot \hat{\mathbf{k}})(\mathbf{a} \times \hat{\mathbf{k}})$ is equal to

(a) $3\mathbf{a}$ (b) $\mathbf{a}$
(c) $\mathbf{0}$ (d) None of these

3. If $A = \{(x, y) : x^2 + y^2 = 25\}$ and $B = \{(x, y) : x^2 + y^2 = 144\}$; then $A \cap B$ contains

(a) one point (b) three points
(c) two points (d) four points

4. The value of c prescribed by Lagrange's mean value theorem, when $f(x) = \sqrt{x^2 - 4}$, $a = 2$ and $b = 3$, is

(a) 2.5 (b) $\sqrt{5}$
(c) $\sqrt{3}$ (d) $\sqrt{3} + 1$

5. The mean deviation from the mean of the series $a, a + d, a + 2d, \dots, a + 2nd$, is

(a) $n(n + 1)d$ (b) $\frac{n(n + 1)d}{2n + 1}$
(c) $\frac{n(n + 1)d}{2n}$ (d) $\frac{n(n - 1)d}{2n + 1}$

6. If $f(x) = xe^{x(1 - x)}$, then $f(x)$ is

(a) increasing on $\left[-\frac{1}{2}, 1\right]$ (b) decreasing on R
(c) increasing on R (d) decreasing on $\left[-\frac{1}{2}, 1\right]$

7. If ω is an imaginary cube root of unity, then the value of $(1 + \omega)(1 + \omega^2)(1 + \omega^3)(1 + \omega^4)(1 + \omega^5)\dots(1 + \omega^{3n})$ is

(a) 2^{3n} (b) 2^{2n}
(c) 2^n (d) None of these

8. Let X denotes the number of times heads occur in n tosses of a fair coin. If $P(X = 4)$, $P(X = 5)$ and $P(X = 6)$ are in AP, then the value of n is

(a) 7, 14 (b) 10,14
(c) 12, 7 (d) 14, 12

9. If there is a term containing x^{2r} in $\left(x + \frac{1}{x^2}\right)^{n-3}$, then

(a) $n - 2r$ is a positive integral multiple of 3.
(b) $n - 2r$ is even
(c) $n - 2r$ is odd
(d) None of the above

10. The value of $\sin^{-1}\left\{\cot\left(\sin^{-1}\sqrt{\frac{2 - \sqrt{3}}{4}} + \cos^{-1}\frac{\sqrt{12}}{4} + \sec^{-1}\sqrt{2}\right)\right\}$ is

(a) $\frac{\pi}{4}$ (b) $\frac{\pi}{6}$ (c) 0 (d) $\frac{\pi}{2}$

11. If $\cos^{-1}\frac{x}{2} + \cos^{-1}\frac{y}{3} = \theta$, then $9x^2 - 12xy\cos\theta + 4y^2$ is equal to

(a) 36 (b) $-36\sin^2\theta$
(c) $36\sin^2\theta$ (d) $36\cos^2\theta$

12. If $\tan(\sec^{-1} x) = \sin\left(\cos^{-1}\frac{1}{\sqrt{5}}\right)$, then x is equal to

(a) $\pm\frac{3}{\sqrt{5}}$ (b) $\pm\frac{\sqrt{5}}{3}$
(c) $\pm\sqrt{\frac{3}{5}}$ (d) None of these

13. The equation of the tangent to the curve $y = (2x - 1)e^{2(1 - x)}$ at the point of its maximum, is

(a) $y - 1 = 0$ (b) $x - 1 = 0$
(c) $x + y - 1 = 0$ (d) $x - y + 1 = 0$

14. The proposition $(p \to \sim p) \wedge (\sim p \to p)$ is

(a) a tautology
(b) a contradiction
(c) neither tautology nor contradiction
(d) both tautology and contradiction

15. The number of ways in which four letters can be selected from the word 'DEGREE', is

(a) 7 (b) 6
(c) $\frac{6!}{3!}$ (d) None of these

16. If $PQRS$ is a convex quadrilateral with 3, 4, 5 and 6 points marked on sides PQ, QR, RS and PS respectively. Then, the number of triangles with vertices on different sides is

(a) 220 (b) 270
(c) 282 (d) 342

17. $\lim_{x \to -1} \left(\dfrac{x^4 + x^2 + x + 1}{x^2 - x + 1} \right)^{\frac{1 - \cos(x+1)}{(x+1)^2}}$ is equal to

(a) 1 (b) $\sqrt{\dfrac{2}{3}}$
(c) $\sqrt{\dfrac{3}{2}}$ (d) $e^{1/2}$

18. The term independent of x in the expansion of $\left(x - \dfrac{1}{x}\right)^4 \left(x + \dfrac{1}{x}\right)^3$, is

(a) -3 (b) 0
(c) 1 (d) 3

19. If A is a square matrix such that $A^2 = A$ and $(1 + A)^n = I + \lambda A$, then λ is equal to

(a) $2n - 1$ (b) $2^n - 1$
(c) $2n + 1$ (d) None of these

20. The functions $u = e^x \sin x$ and $v = e^x \cos x$ satisfy the equation

(a) $v\dfrac{du}{dx} - u\dfrac{dv}{dx} = u^2 + v^2$ (b) $\dfrac{d^2u}{dx^2} = 2v$
(c) $\dfrac{d^2v}{dx^2} = -2u$ (d) All of these

21. The derivative of $\tan^{-1}\left(\dfrac{\sqrt{1 + x^2} - 1}{x}\right)$ with respect to $\tan^{-1}\left(\dfrac{2x\sqrt{1 - x^2}}{1 - 2x^2}\right)$ at $x = 0$, is

(a) $\dfrac{1}{8}$ (b) $\dfrac{1}{4}$ (c) $\dfrac{1}{2}$ (d) 1

22. $\displaystyle\int \frac{1 - x^2}{(1 + x^2)\sqrt{1 + x^4}} dx$ is equal to

(a) $\sqrt{2} \sin^{-1}\left\{\dfrac{\sqrt{2}x}{x^2 + 1}\right\} + C$ (b) $\dfrac{1}{\sqrt{2}} \sin^{-1}\left\{\dfrac{\sqrt{2}x}{x^2 + 1}\right\} + C$
(c) $\dfrac{1}{2} \sin^{-1}\left\{\dfrac{\sqrt{2}x}{x^2 + 1}\right\} + C$ (d) None of these

23. In a ΔABC, if $\begin{vmatrix} 1 & a & b \\ 1 & c & a \\ 1 & b & c \end{vmatrix} = 0$, then $\sin^2 A + \sin^2 B + \sin^2 C$ is equal to

(a) $\dfrac{3\sqrt{3}}{2}$ (b) $\dfrac{9}{4}$
(c) $\dfrac{5}{4}$ (d) 2

24. The area of the region enclosed by the curves $y = x$, $x = e$, $y = \dfrac{1}{x}$ and the positive X-axis, is

(a) $\dfrac{1}{2}$ sq unit (b) 1 sq unit
(c) $\dfrac{3}{2}$ sq units (d) $\dfrac{5}{2}$ sq units

25. The value of $\lim_{x \to \infty} \left\{ \dfrac{a_1^{1/x} + a_2^{1/x} + \ldots + a_n^{1/x}}{n} \right\}^{nx}$, is

(a) $a_1 + a_2 + \ldots + a_n$ (b) $e^{a_1 + a_2 + \ldots + a_n}$
(c) $\dfrac{a_1 + a_2 + \ldots + a_n}{n}$ (d) $a_1 a_2 \ldots a_n$

26. $\displaystyle\int \frac{\sec^2 x}{(\sec x + \tan x)^{9/2}} dx$ equals

(a) $-\dfrac{1}{(\sec x + \tan x)^{11/2}}\left\{\dfrac{1}{11} - \dfrac{1}{7}(\sec x + \tan x)^2\right\} + K$

(b) $\dfrac{1}{(\sec x + \tan x)^{11/2}}\left\{\dfrac{1}{11} - \dfrac{1}{7}(\sec x + \tan x)^2\right\} + K$

(c) $-\dfrac{1}{(\sec x + \tan x)^{11/2}}\left\{\dfrac{1}{11} + \dfrac{1}{7}(\sec x + \tan x)\right\} + K$

(d) None of the above

27. The centre of the circle passing through $(0, 0)$ and $(1, 0)$ and touching the circle $x^2 + y^2 = 9$, is

(a) $\left(\dfrac{3}{2}, \dfrac{1}{2}\right)$ (b) $\left(\dfrac{1}{2}, \dfrac{3}{2}\right)$
(c) $\left(\dfrac{1}{2}, \dfrac{1}{2}\right)$ (d) $\left(\dfrac{1}{2}, \pm\sqrt{2}\right)$

28. If the line $x - 1 = 0$ is the directrix of the parabola $y^2 - kx + 8 = 0$, then one of the value of k is

(a) $\dfrac{1}{8}$ (b) 8 (c) 4 (d) $\dfrac{1}{4}$

29. The area of the region bounded by the parabola $(y-2)^2=(x-1)$, the tangent to the parabola at the point (2, 3) and the X-axis, is
(a) 3 (b) 6
(c) 9 (d) 12

30. Solution of the differential equation
$$x=1+xy\frac{dy}{dx}+\frac{(xy)^2}{2!}\left(\frac{dy}{dx}\right)^2+\frac{(xy)^3}{3!}\left(\frac{dy}{dx}\right)^3$$
+... is
(a) $y=\log_e(x)+C$ (b) $y=(\log_e x)^2+C$
(c) $y=\pm\sqrt{(\log_e x)^2+2C}$ (d) $xy=x^y+C$

31. Numbers 1, 2, 3, ..., 100 are written down on each of the cards A, B and C. One number is selected at random from each of the cards. The probability that the numbers so selected can be the measures (in cm) of three sides of a right angled triangle, is
(a) $\frac{4}{100^3}$ (b) $\frac{3}{50^3}$
(c) $\frac{3!}{100^3}$ (d) None of these

32. If $f(x)=\begin{cases}\dfrac{\sin(\cos x)-\cos x}{(\pi-2x)^3}, & x\neq\dfrac{\pi}{2}\\ k, & x=\dfrac{\pi}{2}\end{cases}$

is a continuous at $x=\frac{\pi}{2}$, then k is equal to
(a) 0 (b) $-\frac{1}{6}$ (c) $-\frac{1}{24}$ (d) $-\frac{1}{48}$

33. If [] denotes the greatest integer function, then $f(x)=[x]+\left[x+\frac{1}{2}\right]$
(a) is continuous at $x=\frac{1}{2}$
(b) is discontinuous at $x=\frac{1}{2}$
(c) $\lim_{x\to(1/2)^+}f(x)=2$
(d) $\lim_{x\to(1/2)^-}f(x)=1$

34. If $f(x)=\min\{1,x^2,x^3\}$, then
(a) $f(x)$ is not everywhere continuous
(b) $f(x)$ is continuous and differentiable everywhere
(c) $f(x)$ is not differentiable at two points
(d) $f(x)$ is not differentiable at one point

35. If $x_n=\cos\frac{\pi}{3^n}+i\sin\frac{\pi}{3^n}$, then $x_1\cdot x_2\cdot x_3\ldots$ is equal to
(a) 1 (b) -1
(c) i (d) $-i$

36. The total number of natural numbers of 6 digits that can be made with digits 1, 2, 3, 4, if all digits are to appear in the same number at least once, is
(a) 1560 (b) 840
(c) 1080 (d) 480

37. The number of solutions of the equation $9x^2-18|x|+5=0$ belonging to the domain of definition of $\log_e\{(x+1)(x+2)\}$, is
(a) 1 (b) 2
(c) 3 (d) 4

38. If $2^x+2^y=2^{x+y}$, then $\frac{dy}{dx}$ is equal to
(a) $\frac{2^x+2^y}{2^x-2^y}$ (b) $\frac{2^x+2^y}{1+2^{x+y}}$
(c) $2^{x-y}\left(\frac{2^y-1}{1-2^x}\right)$ (d) $\frac{2^{x+y}-2^x}{2^y}$

39. If $3x+y=0$ is a tangent to the circle with centre at the point $(2,-1)$, then the equation of the other tangent to the circle from the origin, is
(a) $x-3y=0$ (b) $x+3y=0$
(c) $3x-y=0$ (d) $2x+y=0$

40. If $\sin A+\cos A=m$ and $\sin^3 A+\cos^3 A=n$, then
(a) $m^3-3m+n=0$ (b) $n^3-3n+2m=0$
(c) $m^3-3m+2n=0$ (d) $m^3+3m+2n=0$

41. In ΔABC, if $a=4, b=3$, $\angle A=60°$. Then, c is the root of the equation
(a) $c^2-3c-7=0$ (b) $c^2+3c+7=0$
(c) $c^2-3c+7=0$ (d) $c^2+3c-7=0$

42. The sum to n terms of the series
$$\frac{3}{1^2}+\frac{5}{1^2+2^2}+\frac{7}{1^2+2^2+3^2}+\ldots,\text{ is}$$
(a) $\frac{6n}{n+1}$ (b) $\frac{9n}{n+1}$
(c) $\frac{12n}{n+1}$ (d) $\frac{3n}{n+1}$

43. The value of $\int_{\sqrt{\ln 2}}^{\sqrt{\ln 3}} \frac{x \sin x^2}{\sin x^2 + \sin(\ln 6 - x^2)} dx$ is

(a) $\frac{1}{4}\ln\frac{3}{2}$ (b) $\frac{1}{2}\ln\frac{3}{2}$
(c) $\ln\frac{3}{2}$ (d) $\frac{1}{6}\ln\frac{3}{2}$

44. $\int_1^4 \log_e [x]\, dx$ equals

(a) $\log_e 2$
(b) $\log_e 3$
(c) $\log_e 6$
(d) None of the above

45. The radius of the circle passing through the foci of the ellipse $9x^2 + 16y^2 = 144$ and having its centre at (0, 3), is

(a) 4 (b) 3 (c) $\sqrt{12}$ (d) $\frac{7}{2}$

46. If $f(x) = \frac{\sin(e^{x-2} - 1)}{\log(x-1)}$, then $\lim_{x \to 2} f(x)$ is given by

(a) -2 (b) -1
(c) 0 (d) 1

47. If $\tan A$ and $\tan B$ are the roots of the equation $x^2 - ax + b = 0$, then the value of $\sin^2(A + B)$ is

(a) $\frac{a^2}{a^2 + (1-b)^2}$ (b) $\frac{a^2}{a^2 + b^2}$
(c) $\frac{a^2}{(a+b)^2}$ (d) $\frac{a^2}{b^2 + (1-a)^2}$

48. In a ΔABC, if $a^2 + b^2 + c^2 = ac + \sqrt{3}ab$, then the triangle is

(a) equilateral
(b) right angled and isosceles
(c) right angled and not isosceles
(d) None of the above

49. Let **a**, **b** and **c** be non-zero vectors such that no two are collinear and $(\mathbf{a} \times \mathbf{b}) \times \mathbf{c} = \frac{1}{3}|\mathbf{b}||\mathbf{c}|\mathbf{a}$. If θ is the acute angle between the vectors **b** and **c**, then $\sin\theta$ equals

(a) $\frac{2\sqrt{2}}{3}$ (b) $\frac{\sqrt{2}}{3}$
(c) $\frac{2}{3}$ (d) $\frac{1}{3}$

50. The distance of the point (1, 1) from the line $2x - 3y - 4 = 0$ in the direction of the line $x + y = 1$, is

(a) $\sqrt{2}$ (b) $5\sqrt{2}$
(c) $\frac{1}{\sqrt{2}}$ (d) $\frac{1}{2}$

51. The two tangents to the curve $ax^2 + 2hxy + by^2 = 1$, $a > 0$ at the points, where it crosses X-axis, are

(a) parallel (b) perpendicular
(c) inclined at an angle $\frac{\pi}{4}$ (d) None of these

52. The greatest value of the function $f(x) = xe^{-x}$ in $[0, \infty)$, is

(a) 0 (b) $\frac{1}{e}$
(c) $-e$ (d) e

53. If an isosceles triangle of vertical angle 2θ is inscribed in a circle of radius a. Then, area of the triangle is maximum, when θ is equal to

(a) $\frac{\pi}{6}$ (b) $\frac{\pi}{4}$
(c) $\frac{\pi}{3}$ (d) $\frac{\pi}{2}$

54. The range of the function $f(x) = \log_5(25 - x^2)$ is

(a) [0, 5] (b) [0, 2)
(c) (0, 2) (d) None of these

55. The image of the point P (1, 3, 4) in the plane $2x - y + z + 3 = 0$, is

(a) (3, 5, − 2) (b) (− 3, 5, 2)
(c) (3, − 5, 2) (d) (3, 5, 2)

56. Equation of the plane that contains the lines $\mathbf{r} = (\hat{i} + \hat{j}) + \lambda(\hat{i} + 2\hat{j} - \hat{k})$ and $\mathbf{r} = (\hat{i} + \hat{j}) + \mu(-\hat{i} + \hat{j} - 2\hat{k})$, is

(a) $\mathbf{r} \cdot (2\hat{i} + \hat{j} - 3\hat{k}) = -4$ (b) $\mathbf{r} \times (-\hat{i} + \hat{j} + \hat{k}) = 0$
(c) $\mathbf{r} \cdot (-\hat{i} + \hat{j} + \hat{k}) = 0$ (d) None of these

57. The distance of the point (3, 8, 2) from the line $\frac{x-1}{2} = \frac{y-3}{4} = \frac{z-2}{3}$ measured parallel to the plane $3x + 2y - 2z + 15 = 0$, is

(a) 2 (b) 3
(c) 6 (d) 7

58. Let u_1 and u_2 be two urns such that u_1 contains 3 white, 2 red balls and u_2 contains only 1 white ball. A fair coin is tossed. If head appears, then 1 ball is drawn at random from urn u_1 and put into u_2. However, if tail appears, then 2 balls are drawn at random from u_1 and put into u_2. Now, 1 ball is drawn at random from u_2. Then, probability of the drawn ball from u_2 being white is

(a) $\frac{13}{30}$ (b) $\frac{23}{30}$ (c) $\frac{19}{30}$ (d) $\frac{11}{30}$

59. Let $D_k = \begin{vmatrix} a & 2^k & 2^{16}-1 \\ b & 3(4^k) & 2(4^{16}-1) \\ c & 7(8^k) & 4(8^{16}-1) \end{vmatrix}$, then the value of $\sum_{k=1}^{16} D_k$ is

(a) 0 (b) $a+b+c$
(c) $ab+bc+ca$ (d) None of these

60. The eccentricity of the conic $x^2 - 4x + 4y^2 = 12$ is

(a) $\frac{\sqrt{3}}{2}$ (b) $\frac{2}{\sqrt{3}}$
(c) $\sqrt{3}$ (d) None of these

61. The value of $\int \frac{1}{\sin\left(x-\frac{\pi}{3}\right)\cos x} dx$ is

(a) $2\log\left|\sin x + \sin\left(x-\frac{\pi}{3}\right)\right| + C$
(b) $2\log\left|\sin x \cdot \sin\left(x-\frac{\pi}{3}\right)\right| + C$
(c) $2\log\left|\sin x - \sin\left(x-\frac{\pi}{3}\right)\right| + C$
(d) None of the above

62. Two common tangents to the circle $x^2 + y^2 = 2a^2$ and parabola $y^2 = 8ax$ are

(a) $x = \pm(y+2a)$ (b) $y = \pm(x+2a)$
(c) $x = \pm(y+a)$ (d) $y = \pm(x+a)$

63. The equation of the plane through the intersection of the planes $x+y+z=1$ and $2x+3y-z+4=0$ and parallel to X-axis, is

(a) $y-3z+6=0$
(b) $3y-z+6=0$
(c) $y+3z+6=0$
(d) $3y-2z+6=0$

64. Let $\mathbf{a} = 2\hat{\mathbf{i}} + \hat{\mathbf{j}} - 2\hat{\mathbf{k}}$ and $\mathbf{b} = \hat{\mathbf{i}} + \hat{\mathbf{j}}$. If $\mathbf{c}$ is a vector such that $\mathbf{a}\cdot\mathbf{c} = |\mathbf{c}|, |\mathbf{c}-\mathbf{a}| = 2\sqrt{2}$ and the angle between $\mathbf{a}\times\mathbf{b}$ and $\mathbf{c}$ is 30°. Then, $|(\mathbf{a}\times\mathbf{b})\times\mathbf{c}|$ is equal to

(a) $\frac{2}{3}$ (b) $\frac{3}{2}$
(c) 2 (d) 3

65. Order and degree of a differential equation

$$\frac{d^2y}{dx^2} = \left\{y + \left(\frac{dy}{dx}\right)^2\right\}^{1/4} \text{ are}$$

(a) 4 and 2 (b) 1 and 2
(c) 1 and 4 (d) 2 and 4

66. A cone whose height is always equal to its diameter, is increasing in volume at the rate of 40 cm^3/s. At what rate is the radius increasing when its circular base area is 1 m^2 ?

(a) 1 mm/s (b) 0.001 cm/s
(c) 2 mm/s (d) 0.002 cm/s

67. The sum to infinity of the series

$$1 + \frac{2}{3} + \frac{6}{3^2} + \frac{10}{3^3} + \frac{14}{3^4} + \ldots \text{ is}$$

(a) 2 (b) 3
(c) 4 (d) 6

68. The equations of the straight lines passing through the point (4, 3) and making intercepts on the coordinate axes whose sum is – 1, is

(a) $\frac{x}{2} - \frac{y}{3} = 1$ and $\frac{x}{-2} + \frac{y}{1} = 1$
(b) $\frac{x}{2} - \frac{y}{3} = -1$ and $\frac{x}{-2} + \frac{y}{1} = -1$
(c) $\frac{x}{2} + \frac{y}{3} = 1$ and $\frac{x}{2} + \frac{y}{1} = 1$
(d) None of the above

69. One possible condition for the three points $(a, b), (b, a)$ and $(a^2, -b^2)$ to be collinear, is

(a) $a-b=2$ (b) $a+b=2$
(c) $a=1+b$ (d) $a=1-b$

70. The number of positive integral solutions of $x^2+9 < (x+3)^2 < 8x+25$, is

(a) 2 (b) 3
(c) 4 (d) 5

General English & Aptitude

1. **John M Keynes was advocate of which of the following suggestions?**

(a) Pending money recklessly during recessions is suicidal
(b) Exorbitant spending during recessions is likely to boost economy
(c) Aggressive deficit spending is likely to be fatal for economic meltdown
(d) Government stimulus to economy may not help because of red-tapism
(e) None of the above

2. **Which of the following is true about Keynes philosophy?**

(a) Actual spending money during meltdown is more important than where and on what it is spent
(b) Government should be selective in approach for spending money during recession
(c) Filling old bottles with banknotes and burying them is an atrocious proposal
(d) Creating jobs and prosperity during recessions is almost an impracticable proposal
(e) None of the above

3. **The author of the passage calls Barack Obama and his team as 'Keynesians' because**

(a) Barack Obama has been reluctant to follow Keynes' philosophy
(b) his team is advising Barack to refrain from Keynes' philosophy
(c) Barack Obama and his team have decided to fill old bottles with banknotes
(d) building houses has been under the active consideration of Barack Obama and his team
(e) None of the above

4. **What, according to Keynes, is the 'aggregate demand'?**

(a) Goods and Services Sector
(b) Stimulation of a short-term activity
(c) Attempting to rev up the sluggish economy
(d) Pumping one trillion dollars into economy
(e) None of the above

5. **Highways, bridges, ethanol plants etc., are considered by the author as**

(a) reasonably appropriate propositions to spend money on
(b) measures that affect the environment adversely
(c) imprudent proposals to waste money on
(d) tax saving schemes bestowed on builders
(e) None of the above

6. **Obama's upcoming American Recovery and Reinvestment Plan focuses on which of the following?**

A. **Recovery of all debts from the debtors in a phased manner.**
B. **Pumping money very liberally in projects that are mandatory.**
C. **Investing money recklessly in any project regardless or its utility.**

(a) Only A
(b) Only B
(c) Only C
(d) B and C
(e) All of the above

Directions (Q. Nos. 7 and 8) *Choose the word which is most opposite in meaning of the word printed in **bold** as used in the passage.*

7. **Moribund**

(a) Declining
(b) Waning
(c) Thriving
(d) Pessimistic
(e) Glorifying

8. **Beleaguered**

(a) Carefree
(b) Harassed
(c) Stressful
(d) Uneventful
(e) Evaporating

Directions (Q.Nos. 9 and 10) *Choose the word which is most nearly the same in meaning as the word given in **bold** as used in the passage.*

9. **Apocalyptic**

(a) Unwelcome
(b) Disastrous
(c) Risk-free
(d) Joyous
(e) Ceremonious

10. **Resuscitate**

(a) Meltdown
(b) Devastate
(c) Mislead
(d) Save
(e) Deactivate

Directions (Q.Nos. 11-15) *Rearrange the following seven sentences* (A), (B), (C), (D), (E), (F) *and* (G) *in the proper sequence to form a meaningful paragraph, then answer the questions given below them.*

A. People thoroughly dedicated to social service but not fulfilling the eligibility requirements would not be able to contest elections.

B. Those who fulfil the stipulated criteria of age and formal education may not be necessarily devoted to social service.

C. This system has both advantages and disadvantages.

D. Therefore, imposing such eligibility requirements is likely to be counter-productive.

E. In certain democratic countries, elections can be contested by anybody.

F. People would be deprived of the probable benefit accrued from services of such people.

G. There are no eligibility requirements of formal education and upper age limit stipulated in their Constitution.

11. Which sentence should be the **FOURTH** in the paragraph?

(a) A (b) B
(c) C (d) D
(e) E

12. Which sentence should be the **LAST** in the paragraph?

(a) A (b) B
(c) C (d) D
(e) E

13. Which sentence should be the **FIRST** in the paragraph?

(a) G (b) F
(c) E (d) D
(e) C

14. Which sentence should be the **SECOND** in the paragraph?

(a) G (b) F
(c) E (d) D
(e) C

15. Which sentence should be the **THIRD** in the paragraph?

(a) A (b) B
(c) C (d) D
(e) E

Directions (Q. Nos. 16-20) *In each of the following sentences there are two blank spaces. Below each five pairs of words have been denoted by* (a), (b), (c), (d) *and* (e). *Find out which pair of words can be filled up in the blanks in the sentence in the same sequence to make the sentence meaningfully complete.*

16. As public resources are .... there is a need for .... of the private sector in development.

(a) exhaustive, participation
(b) relative, solicitation
(c) necessary, regulation
(d) dwindling, coveting
(e) limited, involvement

17. The Government is likely to .... a new rural infrastructure .... by the end of March.

(a) formulate, development
(b) frame, policy
(c) sanction, funds
(d) inspect, clearance
(e) conceive, inflow

18. State owned banks may lose .... Government business if they .... to fulfil their developmental role.

(a) valuable, compete (b) approved, omit
(c) lucrative, fail (d) significant, neglected
(e) profitable, succeed

19. The company was .... with a .... to improve power distribution in rural areas.

(a) initiated, backing (b) established, aim
(c) promote, priority (d) started, mandate
(e) acquired, belief

20. The Finance Minister is .... that the country will .... a 9% growth rate.

(a) assured, check (b) convinced, ensure
(c) confident, post (d) proposed, achieve
(e) challenged, sustain

21. In an examination, Raju got more marks than Mukesh but not as many as Priya. Priya got more marks than Gaurav and Kavita. Gaurav got less marks than Mukesh but his marks are not the lowest in the group. Who is second in the descending order of marks?

(a) Priya (b) Kavita
(c) Raju (d) Gaurav

Directions (Q. Nos. 22 and 23) Select the missing number from the given responses.

22.

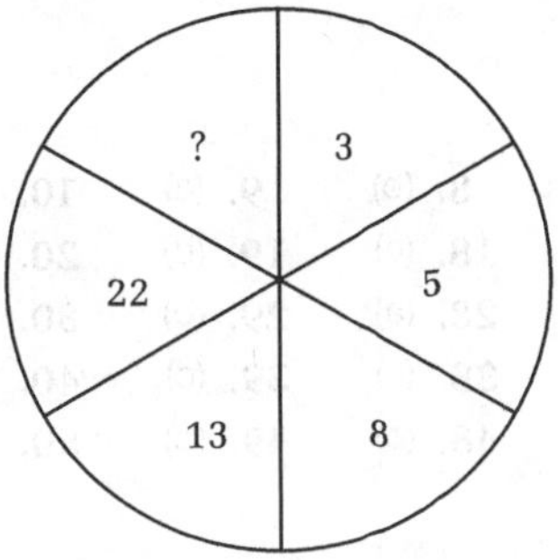

(a) 1 (b) 26 (c) 39 (d) 45

23.

64	36	2
81	25	4
144	16	?

(a) 6 (b) 8 (c) 3 (d) 16

24. Ashish walked 50 m towards East and took a right turn and walked 40 m. He again took a right turn and walked 50 m. How for is he from the starting point?

(a) 10 m (b) 25 m
(c) 30 m (d) 40 m

25. Which of the answer figure is exactly the mirror image of the given figure, when the mirror is held on the line *AB*?

(a) (b) (c) (d)

26. The least number which when divided by 5, 6, 7 and 8 leaves a remainder 3 but when divided by 9 leaves no remainder, is

(a) 1677
(b) 1683
(c) 2523
(d) 3363

27. If $\frac{x+y}{x-y}=\frac{5}{2}$, then value of $\frac{x}{y}$ is

(a) $\frac{3}{8}$ (b) $\frac{8}{3}$
(c) $\frac{5}{3}$ (d) $\frac{3}{5}$

28. Ram bought a refrigerator with 5% discount on the labelled price. Had he bought it with 20% discount, he would have saved ₹ 1500. At what price did he buy the refrigerator?

(a) ₹ 10000
(b) ₹ 9000
(c) ₹ 8000
(d) ₹ 9500

29. The ratio of the incomes of *A* and *B* is 5 : 4 and the ratio of their expenditures is 3 : 2. If at the end of the year, each saves ₹1600, then the income of *A* is

(a) ₹ 3400
(b) ₹ 3600
(c) ₹ 4000
(d) ₹ 4400

30. The difference between the compound interest and simple interest on a certain sum of money for 3yr at $6\frac{2}{3}$% per annum is ₹ 184, then what is the sum?

(a) ₹ 13500
(b) ₹ 12500
(c) ₹ 11500
(d) ₹ 10500

Answers

Physics

1. (a)	2. (a)	3. (b)	4. (d)	5. (c)	6. (b)	7. (b)	8. (b)	9. (a)	10. (b)
11. (c)	12. (a)	13. (d)	14. (d)	15. (b)	16. (b)	17. (a)	18. (c)	19. (b)	20. (b)
21. (c)	22. (c)	23. (b)	24. (b)	25. (c)	26. (b)	27. (a)	28. (a)	29. (b)	30. (a)
31. (b)	32. (b)	33. (a)	34. (c)	35. (d)	36. (b)	37. (d)	38. (c)	39. (c)	40. (b)
41. (d)	42. (a)	43. (d)	44. (c)	45. (b)	46. (a)	47. (c)	48. (c)	49. (b)	50. (b)

Chemistry

1. (d)	2. (b)	3. (c)	4. (a)	5. (a)	6. (d)	7. (a)	8. (b)	9. (a)	10. (b)
11. (d)	12. (c)	13. (d)	14. (d)	15. (b)	16. (b)	17. (a)	18. (a)	19. (d)	20. (d)
21. (d)	22. (c)	23. (c)	24. (a)	25. (a)	26. (a)	27. (d)	28. (a)	29. (a)	30. (a)
31. (b)	32. (a)	33. (c)	34. (a)	35. (b)	36. (d)	37. (c)	38. (b)	39. (d)	40. (b)
41. (c)	42. (d)	43. (b)	44. (a)	45. (a)	46. (b)	47. (a)	48. (b)	49. (c)	50. (a)

Mathematics

1. (c)	2. (c)	3. (d)	4. (b)	5. (b)	6. (a)	7. (c)	8. (a)	9. (a)	10. (c)
11. (c)	12. (a)	13. (a)	14. (b)	15. (a)	16. (d)	17. (b)	18. (b)	19. (b)	20. (d)
21. (b)	22. (b)	23. (b)	24. (c)	25. (d)	26. (c)	27. (d)	28. (c)	29. (c)	30. (c)
31. (d)	32. (d)	33. (b)	34. (d)	35. (c)	36. (a)	37. (c)	38. (c)	39. (a)	40. (c)
41. (a)	42. (a)	43. (a)	44. (c)	45. (a)	46. (d)	47. (a)	48. (c)	49. (a)	50. (a)
51. (a)	52. (b)	53. (a)	54. (d)	55. (b)	56. (c)	57. (d)	58. (b)	59. (a)	60. (a)
61. (d)	62. (b)	63. (a)	64. (b)	65. (d)	66. (d)	67. (b)	68. (a)	69. (c)	70. (d)

General English & Aptitude

1. (b)	2. (a)	3. (e)	4. (a)	5. (c)	6. (a)	7. (c)	8. (a)	9. (b)	10. (d)
11. (a)	12. (d)	13. (c)	14. (a)	15. (c)	16. (e)	17. (b)	18. (c)	19. (d)	20. (c)
21. (c)	22. (c)	23. (b)	24. (d)	25. (a)	26. (b)	27. (b)	28. (a)	29. (c)	30. (a)

Solutions

Physics

1. Let the third charge $-q$ is placed at a distance of x from the charge $2q$. Then, the potential energy of the system is

$$U = k\left[\frac{2q\,(8q)}{r} - \frac{(2q)\,(q)}{x} - \frac{(8q)\,(q)}{r-x}\right]$$

Here, $k = \dfrac{1}{4\pi\varepsilon_0}$

U is minimum, where $\dfrac{2}{x} + \dfrac{8}{r-x}$ is maximum

Let $\dfrac{2}{x} + \dfrac{8}{r-x} = y$

For y to be maximum, $\dfrac{dy}{dx} = 0$

$$-\frac{2}{x^2} + \frac{8}{(r-x)^2} = 0$$

$$\Rightarrow \quad \frac{x}{r-x} = \sqrt{\frac{2}{8}} = \frac{1}{2} \Rightarrow x = \frac{r}{3}$$

But at $x = \dfrac{r}{3}$, $\dfrac{d^2y}{dx^2}$ = positive i.e. at $x = \dfrac{r}{3}$, y is minimum or U is maximum. So, there cannot be any point between them where the potential energy is minimum.

2. At thermal equilibrium, the ratio $\dfrac{N_2}{N_1}$ is given as

$$\frac{N_2}{N_1} = \exp\left(-\frac{E_2 - E_1}{kT}\right)$$

The middle of the visible range is taken at $\lambda = 550$ nm

$$\Rightarrow \quad E_2 - E_1 = \frac{hc}{\lambda} = 3.16 \times 10^{-19}\ \text{J}$$

$$\Rightarrow \quad \frac{N_2}{N_1} = \exp\left(\frac{-3.16 \times 10^{-19}\ \text{J}}{(1.38 \times 10^{-23} 1/k) \cdot (300\ k)}\right)$$

$$\Rightarrow \quad \frac{N_2}{N_1} = 1.1577 \times 10^{-38}$$

3. The energy of the hydrogen atom in nth state is givne by

$$E_n = \frac{-13.6}{(n^2)}\ \text{eV}$$

Here, $n = 5$

$$\therefore \quad E = \frac{-13.6}{(5)^2}\ \text{eV} = -0.54\ \text{eV}$$

4. When a galvanometer is connected to the shunt resistance, then we have potential drop across the galvanometer = potential drop across the shunt

i.e. $iG = (i - i_0)\,S$ ...(i)

Here, i = current through the galvanometer
G = resistance of the galvanometer
i_0 = current through the shunt resistance
S = shunt resistance

From Eq. (i), we have the part of the total current that flows through the galvanometer is

$$\frac{i}{i_0} = \frac{S}{G+S}$$

Substituting, $S = 2.5\ \Omega$, $G = 25\ \Omega$, we get

$$\frac{i}{i_0} = \frac{1}{11}$$

5. $\because$ $v_{\text{rms}} = \sqrt{\dfrac{3RT}{M}}$ and $v_{\text{sound}} = \sqrt{\dfrac{\gamma RT}{M}}$,

$$v_{\text{rms}} = 2\,v_{\text{sound}}$$

i.e. $\gamma = \dfrac{3}{2}$ = ratio of $\dfrac{C_p}{C_V}$ for the mixture

$$C_V = \frac{n_1 C_{V_1} + n_2 C_{V_2}}{n_1 + n_2}$$

and $$C_p = \frac{n_1 C_{p_1} + n_2 C_{p_2}}{n_1 + n_2}$$

$$\therefore \quad \gamma = \frac{C_P}{C_V} = \frac{n_1 C_{p_1} + n_2 C_{p_2}}{n_1 C_{V_1} + n_2 C_{V_2}}$$

$$\therefore \quad \frac{3}{2} = \frac{2\left(\frac{5}{2}R\right) + n\left(\frac{7}{2}R\right)}{2\left(\frac{3}{2}R\right) + n\left(\frac{5}{2}R\right)}$$

$$\Rightarrow \quad \frac{3}{2} = \frac{10 + 7n}{6 + 5n}$$

$$\Rightarrow \quad n = 2$$

6. Current is given by

$$I = \frac{E}{R'},$$

Substituting, $E = -\dfrac{nd\phi}{dt}$, we get

$$I = -\frac{n}{R'} \cdot \frac{d\phi}{dt} \quad \text{...(i)}$$

Given that w_1 and w_2 are the values of the flux associated with one turn of the coil.

Therefore, Eq. (i) becomes,

$$I = -\frac{n}{R'}\left[\frac{w_2 - w_1}{t_2 - t_1}\right] \quad \ldots\text{(ii)}$$

Total resistance of the combination is $R' = R + 4R = 5R$, substituting the values of R' and $(t_2 - t_1) = t$ in Eq. (ii),we get

$$I = -\frac{n}{5R}\left(\frac{w_2 - w_1}{t}\right)$$

$$\Rightarrow \quad I = \frac{-n(w_2 - w_1)}{5Rt}$$

7. For the reflection at the air-soap solution interface, the phase difference is π.

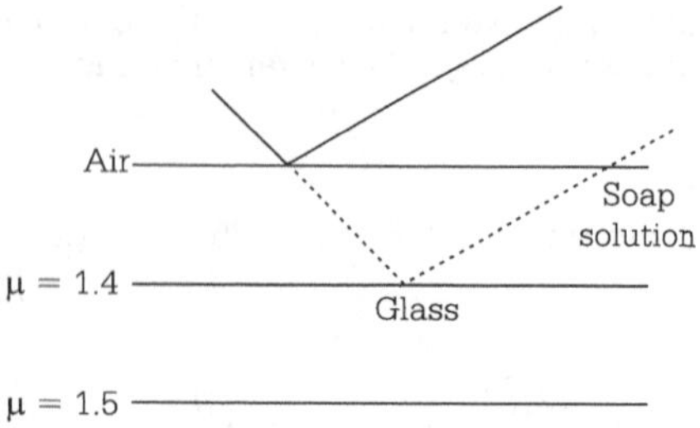

For reflection at the interface of soap solution to the glass also there will be a phase difference of π.

$\therefore$ The condition for maximum intensity $= 2\mu t = n\lambda$ ...(i)

For n, $\quad n\lambda_1 = (n-1)\lambda_2$

$$n(420) = (n-1)\,630$$

$$n(630 - 420) = 630$$

$$\therefore \quad n(210) = 630$$

$$\Rightarrow \quad n = 3$$

This is the maximum order where they coincide,

From Eq. (i),

$$\therefore \quad 2 \times 1.4 \times t = 3 \times 420$$

$$\Rightarrow \quad t = \frac{2 \times 420}{2 \times 1.40} = 450 \text{ nm}$$

8. The current flowing clockwise in an equilateral triangle has a magnetic field in the direction of $\hat{\mathbf{k}}$.

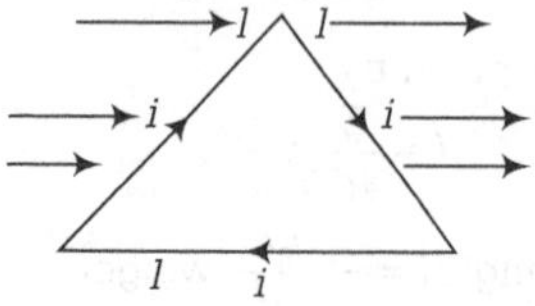

$$\tau = BiNA \sin\theta$$

$$\tau = BiNA \sin 90^\circ$$

$$\tau = Bi \times \frac{\sqrt{3}}{4} l^2 \times 1$$

[$\because$ Area of equilateral triangle $= \frac{\sqrt{3}}{4} l^2$ and $N = 1$, (assuming)]

$$\Rightarrow \quad l^2 = \frac{4\tau}{\sqrt{3}\,Bi} \Rightarrow l = 2\left[\frac{\tau}{Bi\sqrt{3}}\right]^{1/2}$$

9. Block of mass m_2 shoots off carrying some kinetic energy away from the system. To find its speed, potential energy of spring = maximum kinetic energy of blocks.

$$\frac{kd^2}{2} = (m_1 + m_2)\frac{v^2}{2}$$

[k = force constant of spring]

$$v^2 = \frac{kd^2}{m_1 + m_2} \text{ with } m_1 \text{ along on the spring.}$$

Maximum potential energy
= Maximum kinetic energy of m_2

$$\Rightarrow \quad \frac{1}{2}kA^2 = \frac{1}{2}m_1 v^2 \Rightarrow kA^2 = \frac{km_1 d^2}{m_1 + m_2}$$

$$\Rightarrow \quad A = d\sqrt{\frac{m_1}{m_1 + m_2}}$$

10. Time taken by the small ball to return to the hands of the juggler is $\frac{2v}{g} = \frac{2 \times 20}{10} = 4$ s. So, he is throwing the balls after 1 s each. Let at some instant, he throws the ball number 4. Before 1 s of throwing it, he throws ball 3. So, the height of ball 3 is

$$h = ut - \frac{1}{2}gt^2$$

$$h_3 = 20 \times 1 - \frac{1}{2}(10)(1)^2 = 15 \text{ m}$$

Before 2 s, he throws ball 2. So, the height of ball 2 is

$$h_2 = 20 \times 2 - \frac{1}{2} \times 10\,(2)^2 = 20 \text{ m}$$

Before 3 s, he throws ball 1. So, the height of ball 1 is

$$h_1 = 20 \times 3 - \frac{1}{2} \times 10\,(3)^2$$

$$\Rightarrow \quad h_1 = 15 \text{ m}$$

11. The instantaneous current in the AC circuit is given by $I = I_0 \sin\omega t$

$$\frac{dI}{dt} = \frac{d}{dt}(I_0 \sin\omega t)$$

$$\frac{dI}{dt} = (I_0 \cos\omega t)\,\omega$$

Substituting $I_0 = 10$ A, $\omega = 100\,\pi$

$\cos(100\,\pi) = 1$, we get

$$\frac{dI}{dt} = 1000\,\pi$$

Hence, induced emf is given by

$$E = M\frac{dI}{dt}$$

On putting $M = 0.005$ H, $\frac{dI}{dt} = 1000\,\pi$, we get

$$E = (5\pi)\text{ V}$$

12. Speed of ball before collision is

$$v = u\cos 60^\circ = 20 \times \frac{1}{2} = 10\text{ m/s}$$

Since, collision is perfectly inelastic ($e = 0$), so the ball will not bounce. It will move along the plane with velocity

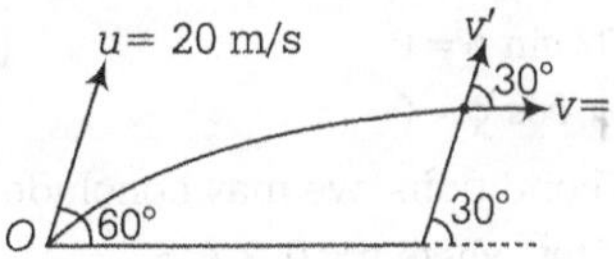

$$v' = v\cos 30^\circ = \frac{10\sqrt{3}}{2} = 5\sqrt{3}\text{ m/s}$$

∴ Maximum height attained by the ball

$$H = \frac{u^2\sin^2 60^\circ}{2g} + \frac{v'^2}{2g}$$

$$H = \frac{(20)^2\left(\frac{\sqrt{3}}{2}\right)^2}{20} + \frac{(5\sqrt{3})^2}{20}$$

$$= \frac{300}{20} + \frac{75}{20} = 18.75\text{ m}$$

13. Total energy produced by the reactor in time $t = 10$ yr.

$$E = 1000 \times 10^6 \times 10 \times 3.15 \times 10^7\text{ J}$$

$$= 3.15 \times 10^{17}\text{ J}$$

$\because$ Efficiency $= \frac{\text{output energy}}{\text{input energy}}$

Input energy caused by fission $= \frac{\text{output energy}}{\text{efficiency}}$

$$= \frac{3.15 \times 10^{17}}{(10/100)} = 3.15 \times 10^{18}\text{ J}$$

Energy produced by one fission of $^{235}U = 200$ MeV

$$= 200 \times 1.6 \times 10^{-13}\text{ J}$$

$$= 3.2 \times 10^{-11}\text{ J}$$

Therefore, number of fissions required

$$= \frac{\text{total energy}}{\text{energy per fission}}$$

$$= \frac{3.15 \times 10^{18}}{3.2 \times 10^{-11}} = 9.8 \times 10^{28}$$

Hence, mass of uranium required is given by

$$m = \frac{N}{N_a} \times 235 = \frac{9.8 \times 10^{28}}{6.02 \times 10^{26}}$$

$$= 38.2 \times 10^3\text{ kg}$$

14. Charge acquired by the plates of the capacitor $q_0 = CV = (10\,\mu\text{F}) \times (50\text{V}) = 500\,\mu\text{C}$

Now, let the charge distribution is as follows.

Total charge on positive plate has now become 700 μC while that in negative plate is still − 500 μC,

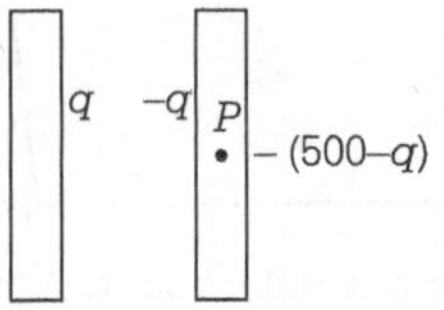

Here, charges are in μC.

Net electric field at point P is zero.

$$\therefore \quad \frac{(700 - q)}{2A\varepsilon_0} + \frac{q}{2A\varepsilon_0} + \left(\frac{500 - q}{2A\varepsilon_0}\right) = \frac{q}{2A\varepsilon_0}$$

$$\Rightarrow \quad q = 600\,\mu\text{C}$$

∴ Potential difference between the plates is

$$\Delta V = \frac{q}{C} = \frac{600\,\mu\text{C}}{10\,\mu\text{F}} = 60\text{ V}$$

15.

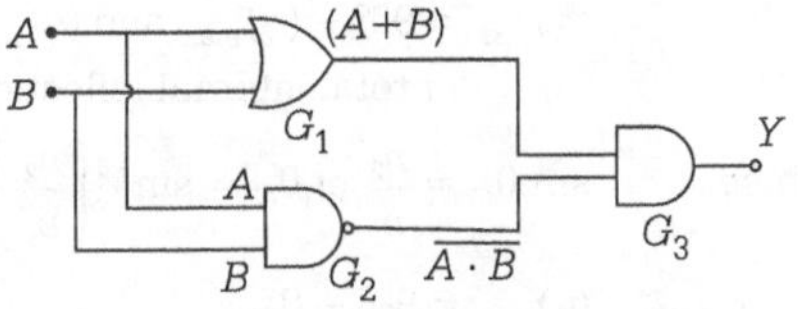

$$Y = (A + B)\cdot\overline{AB}$$

The given output equation can also be written as

$Y = (A + B)\cdot(\overline{A} + \overline{B})$ (Demorgan's theorem)

$$\Rightarrow \quad Y = A\overline{A} + A\overline{B} + B\overline{A} + B\overline{B}$$

$$\Rightarrow \quad Y = 0 + A\overline{B} + \overline{A}B + 0$$

$$\Rightarrow \quad Y = \overline{A}B + A\overline{B}$$

This is the expression for XOR gate.

16. The current through the circuit before the battery of emf E_2 is short circuited, is

$$i_1 = \frac{E_1 + E_2}{r_1 + r_2 + R} \quad \ldots\text{(i)}$$

After short circuiting the battery of emf E_2, current through the resistance R would be

$$i_2 = \frac{E_1}{R + r_1} \quad \ldots\text{(ii)}$$

Since, from Eqs. (i) and (ii), we get

$$i_2 > i_1$$

$$\therefore \quad \frac{E_1}{R + r_1} > \frac{E_1 + E_2}{R + r_1 + r_2}$$

This gives $\quad E_1 r_2 > E_2 (R + r_1)$

17.

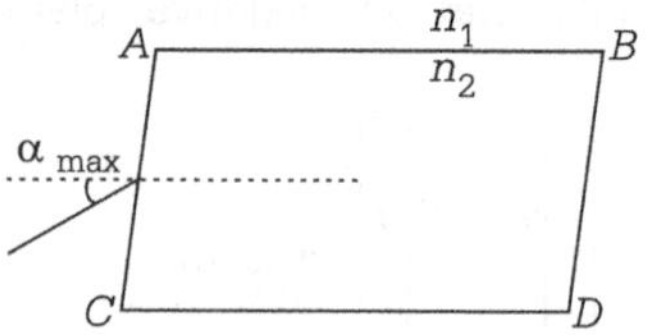

Rays comes out only from CD, means rays after refraction from AB get totally internally reflected at AD.

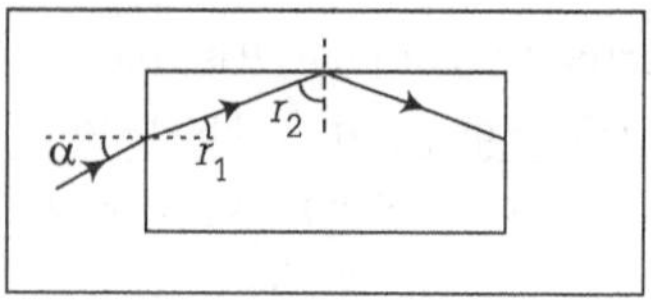

From the figure,

$$r_1 + r_2 = 90$$

$$r_1 = 90 - r_2$$

$$(r_1)_{max} = 90° - (r_2)_{max} \text{ and } (r_2)_{min} = \theta_C$$

(for total internal reflection at AD)

where, $\quad \sin\theta_C = \frac{n_2}{n_1}$ or $\theta_C = \sin^{-1}\left(\frac{n_2}{n_1}\right)$

$$\therefore \quad (r_1)_{max} = 90° - \theta_C$$

Now, applying Snell's law at face AB, we wet

$$\frac{n_1}{n_2} = \frac{\sin\alpha_{max}}{\sin(r_1)_{max}} = \frac{\sin\alpha_{max}}{\sin(90 - \theta_C)} = \frac{\sin\alpha_{max}}{\cos\theta_C}$$

$$\sin\alpha_{max} = \frac{n_1}{n_2}\cos\theta_C$$

$$\therefore \quad \alpha_{max} = \sin^{-1}\left[\frac{n_1}{n_2}\cos\theta_C\right]$$

$$= \sin^{-1}\left[\frac{n_1}{n_2}\cos\left(\sin^{-1}\left(\frac{n_2}{n_1}\right)\right)\right]$$

18. We start a general form for a rightward moving wave,

$$y(x,t) = A\sin(kx - \omega t + \phi) \quad \ldots\text{(i)}$$

The given amplitude is $A = 2\text{ cm} = 0.02\text{ m}$

The wavelength is given as

$$\lambda = 1\text{ m}$$

Wave number $= k = 2\pi/\lambda = 2\pi\,\text{m}^{-1}$

Angular frequency,

$$\omega = vk = 10\,\pi\text{ rad/s}$$

From Eq. (i),

$$y(x, t) = (0.02)\sin[2\pi(x - 5t) + \phi]$$

$\because \quad$ For $x = 0, t = 0$

$$y = 0 \text{ and } \frac{\partial y}{\partial t} < 0$$

i.e. $\quad 0.02\sin\phi = 0 \quad$ (as $y = 0$)

and $\quad -0.2\pi\cos\phi < 0$

From these conditions, we may conclude that

$\phi = 2n\pi$, where $n = 0, 2, 4, 6, \ldots$

Therefore,

$$y(x, t) = (0.02\text{ m})\sin[(2\pi\text{m}^{-1})x - (10\pi\text{s}^{-1})t]\,m$$

19. Let x be the desired length.

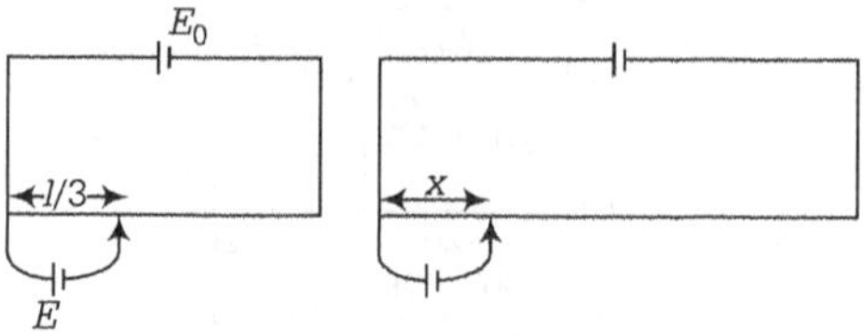

Potential gradient in the first case $= \frac{E_0}{l}$

$$E = \left(\frac{l}{3}\right)\cdot\left(\frac{E_0}{l}\right) = \frac{E_0}{3} \quad \ldots\text{(i)}$$

Potential gradient in the second case.

$$= \frac{E_0}{3l/2} = \frac{2E_0}{3l}$$

$$\therefore \quad E = (x)\frac{2E_0}{3l} \quad \ldots\text{(ii)}$$

From Eqs. (i) and (ii), we get

$$\frac{E_0}{3} = \left(\frac{2E_0}{3l}\right)x$$

$$\Rightarrow \quad x = \frac{l}{2}$$

20. Since, the wires are infinite, so the system of these two wires can be considered as closed rectangle of infinite length and breadth equal to d.

Flux through strip

$$\phi = \int_a^{d-a} \frac{\mu_0 I}{2\pi r}(l dr) = \frac{\mu_0 Il}{2\pi} \ln\left(\frac{d-a}{a}\right)$$

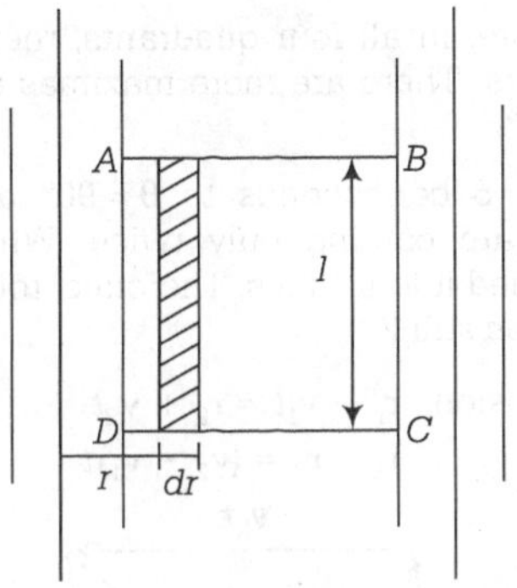

The other wire produces the same result, so the total flux through the rectangle $ABCD$ is

$$\phi_{\text{total}} = \frac{\mu_0 Il}{\pi} \ln\left(\frac{d-a}{a}\right)$$

The total inductance of length l

$$L = \frac{\phi_{\text{total}}}{I} = \frac{\mu_0 l}{\pi} \ln\left(\frac{d-a}{a}\right)$$

$$\text{Inductance per unit length} = \frac{L}{l} = \frac{\mu_0}{\pi} \ln\left(\frac{d-a}{a}\right)$$

21. Magnetic moment of a loop is given by $p_m = IA$

where, I = current through the loop

A = area of loop

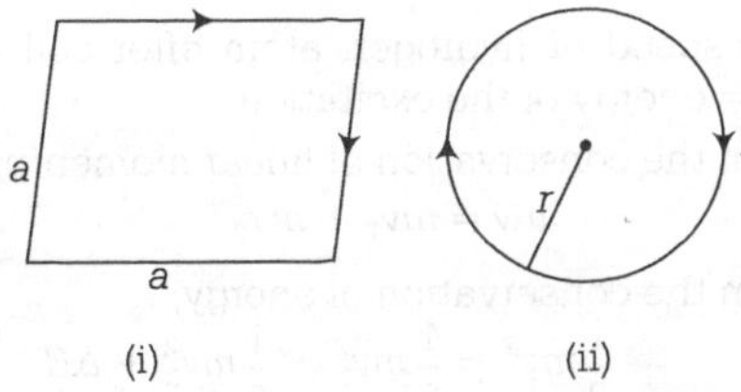

Perimeter of square should be equal to the circumference of circle, since wires have same length.

Therefore, $4a = 2\pi r \Rightarrow a = \dfrac{\pi r}{2}$ ...(i)

In Fig. (i), magnetic moment of square loop is given by

$$p_{m_1} = I(A_1) \quad \Rightarrow$$

$$\Rightarrow \quad p_{m_1} = \frac{I\pi^2 r^2}{4} \quad \text{[from Eq. (i)] ...(ii)}$$

In Fig. (ii), magnetic moment of circular loop is given by

$$p_{m_2} = I(A_2)$$

$$p_{m_2} = I(\pi r^2) \quad \text{...(iii)}$$

From Eqs. (ii) and (iii), we get

$$\Rightarrow \quad \frac{p_{m_1}}{p_{m_2}} = \frac{\pi}{4}$$

22. The potential energy of circular loop in the magnetic field $= -MB\cos\theta$, where θ is the angle between normal to plane of coil and magnetic field B.

Initial potential energy,

$$U_i = -NiAB\cos 0° = -NiAB$$

When coil is turned through 180°, therefore final potential energy,

$$U_f = -NiAB\cos 180° = NiAB$$

∴ Required work, W = gain in potential energy

$$= U_f - U_i = NiAB - (-NiAB) = 2NiAB$$

Here, $N = 100$, $i = 0.1$ A, $B = 1.5\ \text{Wb/m}^2$ and radius $r = 0.05$ m

∴ $A = \pi r^2 = 3.14 \times (0.05)^2 = 7.85 \times 10^{-3}\ \text{m}^2$

∴ Work, $W = 2 \times 100 \times 0.1 \times 7.85 \times 10^{-3} \times 1.5$

⇒ $W = 0.2355$ J

23. The instantaneous current in L-R circuit during growth is given by

$$I = I_0(1 - e^{-\frac{R}{L}t})$$

Since, all the exponential terms are dimensionless, therefore, it follows that factor $\frac{R}{L}t$ should be dimensionless i.e.

$$\frac{R}{L}t = [M^0L^0T^0]$$

$$\Rightarrow \quad \frac{R}{L} = \frac{[M^0L^0T^0]}{[T]} = [T^{-1}]$$

$$\Rightarrow \quad \left[\frac{L}{R}\right] = [T]$$

Hence, the time constant of L-R circuit is $\left[\frac{L}{R}\right]$.

24. In wave mechanics, the magnitude of angular momentum of an electron is

$$L = \sqrt{l(l+1)}\frac{h}{2\pi}$$

where, l can have the values 0, 1, 2......$(n+1)$ n is principal quantum number.

25. The general equations of alternating voltage and current in an AC circuit are as follows

$$E = E_0 \sin \omega t \quad \ldots(i)$$

$$I = I_0 \sin \omega t \quad \ldots(ii)$$

Given that,

$$E = 100 \sin 100\pi t \quad \ldots(iii)$$

$$I = 5 \sin 100\, \pi t \quad \ldots(iv)$$

Comparing Eqs. (ii) and and (iv), we get,

Peak value of current, $I_0 = 5\,A$

Comparing Eqs. (i) and (iii), we get

Peak value of voltage, $E_0 = 100$ V

Now, r.m.s. value of voltage is given by

$$E_{rms} = \frac{E_0}{\sqrt{2}} = \frac{100}{\sqrt{2}}\ V$$

Similarly, r.m.s. value of current is given by

$$I_{rms} = \frac{I_0}{\sqrt{2}} = \frac{5}{\sqrt{2}}\ A$$

Hence, reactance of the circuit is given by

$$Z = \frac{E_{rms}}{I_{rms}} = \frac{\frac{100}{\sqrt{2}}}{\frac{5}{\sqrt{2}}} = 20\,\Omega$$

26. The damage caused by the γ-radiation is less serious as compared to damage caused by the α-radiation. This is because α-radiation is more dangerous than beta or gamma if ingested because its power to ionize or to disrupt, atoms is 20 times more than β and γ. But if the source is outside the body or at a distance gamma radiation is much dangerous because it could penertrate thick walls of skin.

27. Path difference at P is

$$\Delta x = 2\left(\frac{x}{2} \cos \theta\right) = x \cos \theta$$

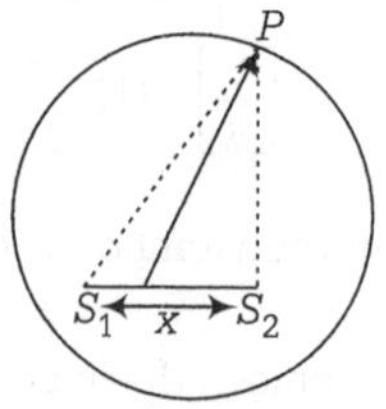

For intensity to be maximum

$$\Delta x = n\lambda \quad (n = 0, 1, \ldots\ldots)$$

$$\Rightarrow \quad x \cos \theta = n\lambda$$

$$\cos \theta = \frac{n\lambda}{x} \text{ or } \cos \theta \ngtr 1$$

$$\therefore \quad \frac{n\lambda}{x} \ngtr 1$$

$$\therefore \quad n \ngtr \frac{x}{\lambda}$$

Substituting $x = 5\lambda$

$$n \ngtr 5 \text{ or } n = 1, 2, 3, 4, 5, \ldots\ldots$$

Therefore, in all four quadrants, there can be 20 maximas. There are more maximas at $\theta = 0°$ and $\theta = 180°$.

But $n = 5$ corresponds to $\theta = 90°$ and $\theta = 270°$, which are coming only twice. While we have multiplied it four times. Therefore, total number of maximas still 20.

28. For collision, $\mathbf{r}_1 + \mathbf{v}_1 t = \mathbf{r}_2 + \mathbf{v}_2 t$

$$\Rightarrow \quad \mathbf{r}_1 - \mathbf{r}_2 = (\mathbf{v}_2 - \mathbf{v}_1)t$$

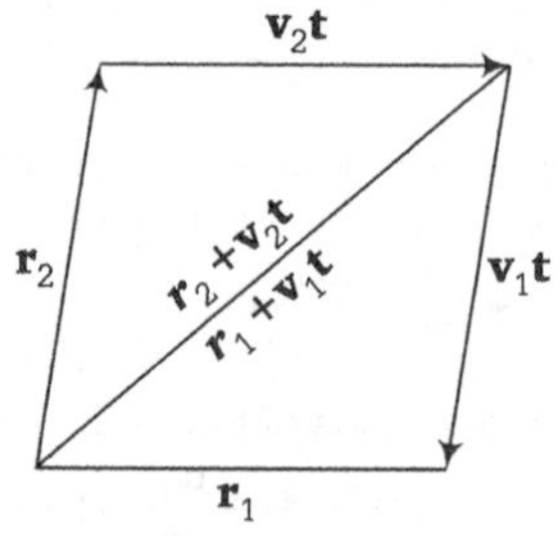

Equating unit vectors, we get

$$\frac{\mathbf{r}_1 - \mathbf{r}_2}{|\mathbf{r}_2 - \mathbf{r}_1|} = \frac{\mathbf{v}_1 - \mathbf{v}_2}{|\mathbf{v}_2 - \mathbf{v}_1|}$$

29. Let v = speed of neutron before collision

v_1 = speed of neutron after collision

v_2 = speed of hydrogen atom after collision and ΔE = energy of the excitation

From the conservation of linear momentum,

$$mv = mv_1 + mv_2 \quad \ldots(i)$$

From the conservation of energy,

$$\frac{1}{2}mv^2 = \frac{1}{2}mv_1^2 + \frac{1}{2}mv_2^2 + \Delta E \quad \ldots(ii)$$

from Eq. (i),

$$v^2 = v_1^2 + v_2^2 + 2v_1 v_2$$

From Eq. (ii),

$$v^2 = v_1^2 + v_2^2 + \frac{2\Delta E}{m}$$

$$\therefore \quad 2v_1 v_2 = \frac{2\Delta E}{m}$$

$$\therefore \quad (v_1 - v_2)^2 = (v_1 + v_2)^2 - 4v_1 v_2$$

$$= v^2 - \frac{4\Delta E}{m}$$

As, $v_1 - v_2$ must be real

$$\Rightarrow \quad v^2 - \frac{4\Delta E}{m} \geq 0$$

or $$\frac{1}{2}mv^2 \geq 2\Delta E$$

The maximum energy that can be absorbed by hydrogen atom in ground state to go into excited state is 10.2 eV. Therefore, the minimum kinetic energy required is

$$\frac{1}{2}mv_{min}^2 = 2 \times 10.2 \text{ eV} = 20.4 \text{ eV}$$

30. Molar specific heat
= Molecular weight × gram specific heat.

$$C_V = M \times c_V$$

$$2.98 \text{ cal/mol-K} = M \times 0.075 \text{ kcal/kg-K}$$

$$= \frac{M \times 0.075 \times 10^3}{10^3} \text{ cal/g-K}$$

∴ Molecular weight of argon,

$$M = \frac{2.98}{0.075} = 39.7 \text{ g}$$

i.e. mass of 6.023×10^{23} atom = 39.7 g

Therefore, mass of single atom $= \dfrac{39.7}{6.023 \times 10^{23}}$

$$= 6.60 \times 10^{-23} \text{ g}$$

31. Radius of a nucleus is given by

$$R = R_0 A^{1/3} \text{ (where, } R_0 = 1.25 \times 10^{-15} \text{ m)}$$

$$= 1.25 \times 10^{-15} A^{1/3} \text{ m}$$

Here, A is the mass number and mass of the uranium nucleus will be

$$m \approx Am_p,\ m_p = \text{mass of proton}$$

$$= A(1.67 \times 10^{-27} \text{ kg})$$

$$\therefore \quad \text{Density} = \frac{\text{Mass}}{\text{Volume}} = \frac{m}{\frac{4}{3}\pi R^3}$$

$$= \frac{A(1.67 \times 10^{-27})}{A \times \frac{4}{3} \times \frac{22}{7} \times (1.25 \times 10^{-15})^3}$$

$$\Rightarrow \quad \rho = 2 \times 10^{17} \text{ kg/m}^3$$

32. FBD of m in frame of wedge

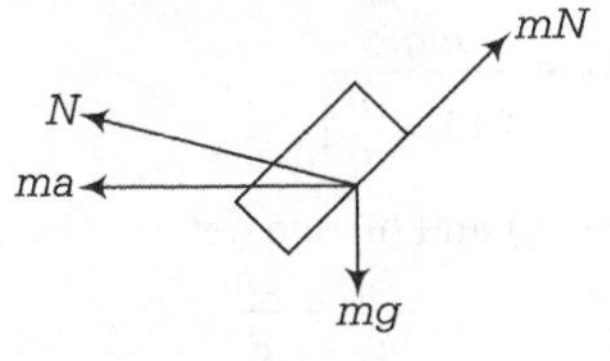

$$N = mg \cos(\alpha) - ma \sin(\alpha)$$

Now, $f = \mu N = ma \cos\alpha + mg \sin\alpha$

$$\Rightarrow \quad \mu = \frac{a \cos\alpha + g \sin\alpha}{g \cos\alpha - a \sin\alpha}$$

$$\Rightarrow \quad \mu = \frac{a + g \tan\alpha}{g - a \tan\alpha} = \frac{5}{12} \Rightarrow \mu = \frac{5}{12}$$

33. ∵ $h = R \sin\theta$

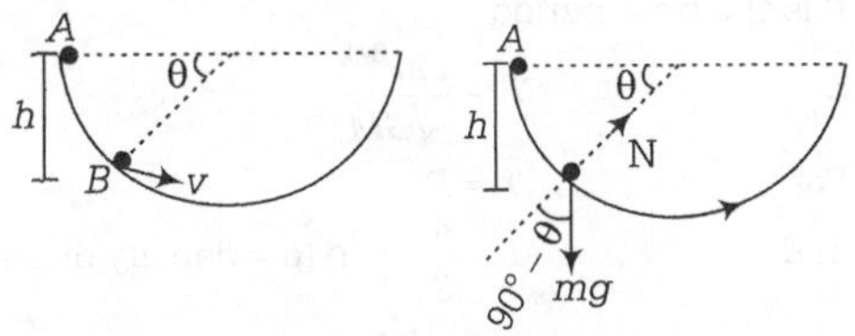

∴ Speed of the particle at point B,

$$v^2 = 2gh = 2Rg \sin\theta$$

Centripetal force, $F = \dfrac{mv^2}{R} = 2mg \sin\theta$ ...(i)

Equation of motion gives

$$N - mg \sin\theta = \frac{mv^2}{R} = 2\, mg \sin\theta$$

34. ∵ Linear acceleration for rolling,

$$a = \frac{g \sin\theta}{\sqrt{1 + k^2/R^2}}$$

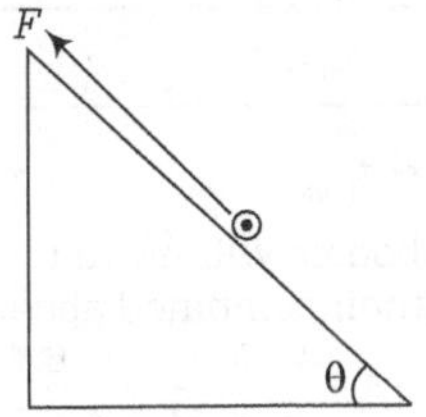

For cylinder, $\dfrac{k^2}{R^2} = \dfrac{1}{2}$

$$\therefore \quad a_{\text{cylinder}} = \frac{2}{3} g \sin\theta$$

For rotation, the torque

$$FR = I\alpha \cdot (MR^2\alpha)/2$$

(where, F = force of friction)

But $R\alpha = a \quad \therefore \quad F = \dfrac{M}{2}a$

$$F = \frac{M}{2} \cdot \frac{2}{3} g \sin\theta = \frac{M}{3} g \sin\theta$$

$$\therefore \quad \mu_s = \frac{F}{N}, \text{ where } N \text{ is normal reaction,}$$

$$\mu_s = \frac{\frac{M}{3} g \sin\theta}{Mg \cos\theta} = \frac{\tan\theta}{3}$$

∴ For rolling without slipping of a roller down, the inclined plane, $\tan\theta \leq 3\mu_s$.

35. Applying Newton's second law to a circular orbit, we have

$$\frac{mv^2}{r} = \frac{4\pi^2 rm}{T^2} = \frac{GMm}{r^2}$$

where, m is the mass of satellite, and v is the orbital speed

T is the time period

$$\therefore \quad T = \frac{2\pi r^{3/2}}{\sqrt{GM}}$$

For $\quad r \approx R$

and $\quad M = \frac{4}{3}\pi R^3 \rho$ (ρ = density of planet)

$$\because \quad T = \sqrt{\frac{3\pi}{\rho G}}$$

i.e. T is independent of R

36. Maximum frictional force between the slab and the block

$$f_{max} = \mu N = \mu mg = \frac{1}{4} \times 4 \times 10 = 10 \text{ N}$$

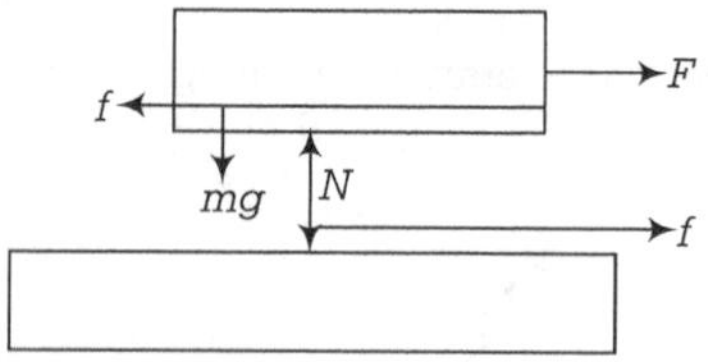

Evidently, $f \propto f_{max}$

So, the two bodies will move together as a single unit. If a be their combined acceleration, then

$$a = \frac{F}{m+M} = \frac{6}{4+5} = \frac{2}{3} \text{ ms}^{-2}$$

Therefore, frictional force acting can be obtained as

$$f = Ma = \frac{2}{3} \times 5 = \frac{10}{3} \text{N}$$

Using $\quad S = \frac{1}{2}at^2$

$$S(2) = \frac{1}{2} \times \frac{2}{3}(2)^2 = \frac{4}{3}$$

and $\quad S(3) = \frac{1}{2} \times \frac{2}{3} \times (3)^2 = 3$

Therefore, work done by friction = $F[S(3) - S(2)]$

$$= \frac{10}{3}\left[3 - \frac{4}{3}\right] = \frac{50}{9} = 5.55 \text{ J}$$

37. Net pulling force = $2g - 1g = 10$ N

Mass being pulled = 2 + 1 = 3 kg

$\therefore$ Acceleration of the system is $a = \frac{10}{3}$ m/s^2

$\therefore$ Velocity of both the blocks at $t = 1$ s will be

$$v_0 = at = \left(\frac{10}{3}\right)(1) = \frac{10}{3} \text{ m/s}$$

Now at this moment, velocity of 2 kg becomes zero, while that of 1 kg block is 10 m/s upwards. Hence, string becomes tight again when displacement of 1 kg block = displacement of 2 kg block.

$$\Rightarrow \quad v_0 t - \frac{1}{2}gt^2 = \frac{1}{2}gt^2$$

$$\Rightarrow \quad t = \frac{v_0}{g} = \frac{\frac{10}{3}}{10} = \frac{1}{3} \text{ s}$$

38. *(c)* From Snell's law,

$$\mu_1 \sin i = \mu_2 \sin r$$

$$\Rightarrow \quad \sin r = \frac{\mu_1}{\mu_2}\sin i = \frac{1}{\sqrt{2}}\sin 45°$$

$$\Rightarrow \quad \frac{1}{2} = \sin r \Rightarrow r = 30°$$

This means that the ray becomes parallel to side AD inside the slab.

This implies for the second face CD, $u = \infty$.

Given $\quad R = 0.4$ m

$$\frac{\mu_3}{v} - \frac{\mu_2}{v} = \frac{\mu_3 - \mu_2}{R}$$

$$\Rightarrow \quad \frac{1.514}{v} - \frac{\sqrt{2}}{\infty} = \frac{1.514 - \sqrt{2}}{R}$$

$$v = \frac{1.514 \times 0.4}{1.514 - 1.414} = \frac{1.514 \times 0.4}{0.1}$$

$$v = 6.056 = 6.06 \text{ m (upto 2 decimal places)}$$

39. Energy required to raise a satellite upto a height h,

$$E_1 = \Delta U = \frac{mgh}{1 + \frac{h}{R}} \quad \text{...(i)}$$

E_2 = energy required to put in orbit

$$= \frac{1}{2}mv_0^2$$

$$= \frac{1}{2}m\left(\frac{GM}{r}\right) \text{ as } v_0 = \text{orbital speed}$$

$$= \frac{1}{2}m\left(\frac{GM}{R+h}\right) = \frac{1}{2}m\left(\frac{GM}{R^2}\right)\frac{R}{1 + \frac{h}{R}}$$

$$\Rightarrow \quad E_2 = \frac{mgR}{2\left(1 + \frac{h}{R}\right)} \quad \text{...(ii)}$$

From Eqs. (i) and (ii), we get

$$\frac{E_1}{E_2} = \frac{2h}{R}$$

40.

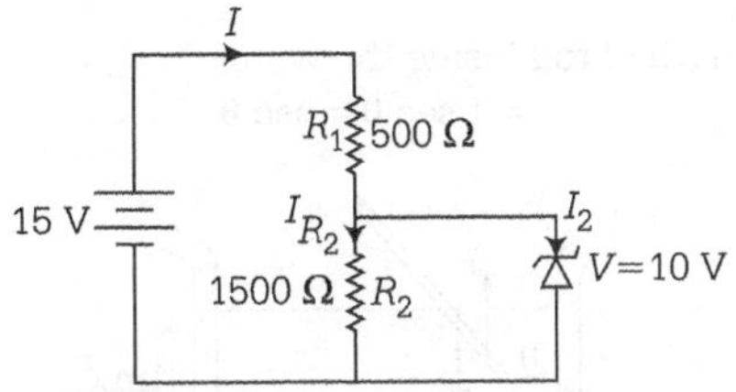

The voltage drop across R_2 is

$$V_{R_2} = V_z = 10\text{ V}$$

The current through R_2 is

$$I_{R_2} = \frac{V_{R_2}}{R_2} = \frac{10}{1500} = 0.667 \times 10^{-2}\text{ A}$$

$$= 6.67 \times 10^{-3}\text{ A} = 6.67\text{ mA}$$

The voltage drop across R_1 is

$$V_{R_1} = 15\text{V} - V_{R_2} = 15\text{V} - 10\text{ V} = 5\text{ V}$$

The current through R_1 is

$$I_{R_1} = \frac{V_{R_1}}{R_1} = \frac{5\text{V}}{500\,\Omega} = 10^{-2}\text{ A} = 10 \times 10^{-3} = 10\text{ mA}$$

The current through Zener diode is

$$I_z = I_{R_1} - I_{R_2} = (10 - 6.67)\text{ mA} = 3.33\text{ mA}$$

41. $\because f \propto \sqrt{T}$

$$\frac{f_{\text{air}}}{f_{\text{water}}} = \sqrt{\frac{W_{\text{air}}}{W_{\text{water}}}} = \sqrt{\frac{V\rho g}{V\rho g - V\rho_w g}}$$

$$= \frac{f}{\frac{f}{2}} = \sqrt{\frac{\rho}{\rho - \rho_w}}$$

$$\Rightarrow \quad 2 = \sqrt{\frac{\rho}{\rho - \rho_w}}$$

$$\Rightarrow \quad 4\rho - 4\rho_w = \rho$$

$$\Rightarrow \quad \rho = \frac{4}{3}\rho_w$$

Similarly in second case,

$$\frac{f}{\frac{f}{3}} = \sqrt{\frac{\rho}{\rho - \rho_L}}$$

$$\Rightarrow \quad 3 = \sqrt{\frac{\frac{4}{3}\rho_w}{4 - \frac{3\rho_L}{\rho_w}}} = \sqrt{\frac{4}{4 - \frac{3\rho_L}{\rho_w}}}$$

Here, $\frac{\rho_L}{\rho_w}$ = specific gravity (say S)

$$\Rightarrow \quad 9 = \frac{4}{4 - 3S}$$

$$\Rightarrow \quad 36 - 27S = 4 \Rightarrow S = \frac{32}{27}$$

42. According to Einstein's photoelectric equation,

$$E = h\nu - w$$

If V_s is retarding or stopping potential and ν_0 is the threshold frequency, then the above equation becomes

$$eV_s = h\nu - h\nu_0$$

$$h\nu = eV_s + h\nu_0$$

$$\nu = \frac{eV_s}{h} + \nu_0$$

Hence, $e = 1.6 \times 10^{-19}\text{ C}$

$$V_s = 3\text{ V}, \nu_0 = 6 \times 10^{-14}\text{ Hz}$$

Therefore, required frequency,

$$\nu = \frac{1.6 \times 10^{-19} \times 3}{6.63 \times 10^{-34}} + 6 \times 10^{14}$$

$$= \frac{4.8}{6.63} \times 10^{15} + 6 \times 10^{14}$$

$$= 0.723 \times 10^{15} + 6 \times 10^{14}$$

$$= 7.23 \times 10^{14} + 6 \times 10^{14}$$

$$= 13.23 \times 10^{14}\text{ Hz}$$

$$= 1.323 \times 10^{15}\text{ Hz}$$

43. At $x_1 = \frac{\pi}{3k}$ and $x_2 = \frac{3\pi}{2k}$

$\sin kx_1$ or $\sin kx_2$ is not zero.

Therefore, neither of x_1 nor x_2 is a node.

$$\Delta x = x_2 - x_1 = \left(\frac{3}{2} - \frac{1}{3}\right)\frac{\pi}{k} = \frac{7\pi}{6k}$$

Since, $\frac{2\pi}{k} > \Delta x > \frac{\pi}{k}$

$$\lambda > \Delta x > \frac{\lambda}{2} \qquad \left\{k = \frac{2\pi}{\lambda}\right\}$$

Therefore, $\phi_1 = \pi$

and $\phi_2 = k$

$$\Delta x = \frac{7\pi}{6}$$

Therefore, $\frac{\phi_1}{\phi_2} = \frac{6}{7}$

44. Energy is given by

$$E = \frac{m_0 c^2}{\sqrt{1 - \frac{v^2}{c^2}}}$$

$$\Rightarrow \quad E^2 = \frac{m_0^2 c^6}{c^2 - v^2}$$

Momentum P is given by

$$P = \frac{m_0 v}{\sqrt{1 - \left(\frac{v^2}{c^2}\right)}}$$

$$P^2c^2 = \frac{m_0^2c^4v^2}{c^2 - v^2}$$

$\therefore \quad E^2 - P^2c^2 = m_0^2c^4$

$\Rightarrow \quad E^2 = P^2c^2 + m_0^2c^4$

For photon, rest mass

$m_0 = 0, \quad$ so $E = Pc$

For electron, $m_0 \neq 0, \quad$ so $E \neq Pc$

45. Let m be the mass of block and a its acceleration in upward direction. Then,

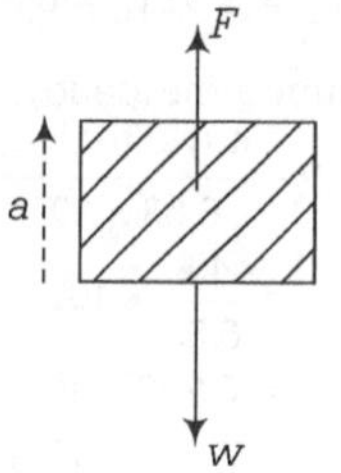

$$a = \frac{\text{upward thrust} - \text{weight}}{m} = \frac{F - w}{m}$$

$$\Rightarrow \quad a = \frac{\left(\frac{m}{0.5 \times 10^3}\right) \times (10^3) \times (g + a_0) - mg}{m}$$

$$\Rightarrow \quad a = 2\left(g + \frac{g}{2}\right) - g = 2g$$

Acceleration of block relative to water

$$a_r = a - a_0 = 2g - \frac{g}{2} = \frac{3g}{2}$$

$$\therefore \quad h = \left(\frac{1}{2}\right) a_r t^2$$

$$\Rightarrow \quad t = \sqrt{\frac{2h}{a_r}} = \sqrt{\frac{2h}{\frac{3}{2}g}} = 2\sqrt{\frac{h}{3g}} = 2\sqrt{\frac{1.2}{3 \times 10}}\, m$$

$$\Rightarrow \quad t = 0.4 \text{ s}$$

46. Energy $= \frac{1}{2}mv^2 = 5000$ eV

$= 5000 \times 1.6 \times 10^{-19}$ J

$mv_2 = \sqrt{2 \times 5000 \times (1.6 \times 10^{-19})}$ m

$= 4 \times 10^{-8}\sqrt{m}$

Number of electrons striking per second is

$$n = \frac{q}{e} = \frac{It}{e} = \frac{50 \times 10^{-6} \times 1}{1.6 \times 10^{-19}}$$

$$\Rightarrow \quad n = 31.25 \times 10^{13}$$

Force = Change of momentum per second

$= n\,(mv) = 31.25 \times 10^{13} \times 4 \times 10^{-8}\sqrt{m}$

$= 125 \times 10^5\sqrt{9.1 \times 10^{-31}} = 1.1924 \times 10^{-8}$ N

47. Length of rod inside the water

$= 1 \sec\theta = \sec\theta$

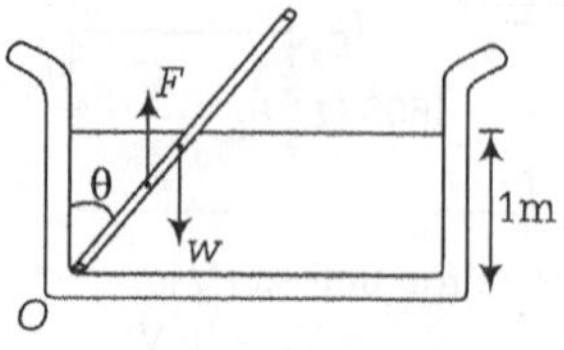

Upthrust, $\quad F = \left(\frac{2}{2}\right)(\sec\theta)\left(\frac{1}{500}\right)(1000)(10)$

$\Rightarrow \quad F = 20 \sec\theta$

Weight of rod, $w = 20 \times 10 = 20$ N

For rotational equilibrium of rod, net torque about O should be zero.

$\therefore F\left(\frac{\sec\theta}{2}\right)(\sin\theta) = w = (1 \sin\theta)$

$\Rightarrow \quad \frac{20}{2}\sec^2\theta = 20$

$\Rightarrow \quad \theta = 45°$

$\Rightarrow \quad F = 20 \sec 45°$

$F = 20\sqrt{2}$ N

For vertical equilibrium of rod, force exerted by the hinge on the rod will be $(20\sqrt{2} - 20)$ N downwards i.e. 8.3 N downwards

48. At the two points of the trajectory during projectile motion, the horizontal component of velocity is same. Then,

$$150 \times \frac{1}{2} = v \times \frac{1}{\sqrt{2}} \quad \Rightarrow \quad v = \frac{150}{\sqrt{2}} \text{ ms}^{-1}$$

Initially, $\quad u_y = u \sin 60 = \frac{150\sqrt{3}}{2}$ ms^{-1}

Finally, $\quad v_y = v \sin 45° = \frac{150}{\sqrt{2}} \times \frac{1}{\sqrt{2}} = \frac{150}{2}$ ms^{-1}

But $\quad v_y = u_y + a_y t$

$$\frac{150}{2} = \frac{150\sqrt{3}}{2} - 10t$$

$$\Rightarrow \quad 10t = \frac{150(\sqrt{3} - 1)}{2}$$

$$\Rightarrow \quad t = 7.5(\sqrt{3} - 1) \text{ s}$$

49. Let R be the radius of sphere, V its volume and ρ its density. Then,

$$\Delta R = R\,\alpha\,\Delta\theta$$

and percentage change in radius

$$\frac{\Delta R}{R} \times 100 = 100\,\alpha\,\Delta\theta \quad \ldots\text{(i)}$$

Now, $\Delta V = \gamma v \Delta\theta = 3\alpha v \Delta\theta$ $\{\because \gamma = 3\alpha\}$

$\therefore$ Percentage increase in volume

$$= \frac{\Delta V}{V} \times 100 = 300\ \alpha\Delta\theta \quad \ldots\text{(ii)}$$

Again, $\rho' = \dfrac{\rho}{1+\gamma\Delta\theta} = \dfrac{\rho}{1+3\alpha\Delta\theta}$

$$\therefore \quad \Delta\rho = \rho - \rho'$$

$$= \rho\left[1 - \frac{1}{1+3\ \alpha\Delta\theta}\right] = \frac{(3\alpha\Delta\theta)\ \rho}{1+3\alpha\Delta\theta}$$

Percentage change in density

$$= \frac{\Delta\rho}{\rho} \times 100 = \frac{300\ \alpha\Delta\theta}{1+3\alpha\Delta\theta} \quad \ldots\text{(iii)}$$

From Eqs. (i), (ii) and (iii), we see that the percentage change is maximum in volume.

50. Since, we know that in Young's double slit experiment.

$$\beta = \frac{D\lambda}{d} \quad \ldots\text{(i)}$$

$$\beta' = \frac{D'\lambda}{d} \quad \ldots\text{(ii)}$$

Subtract Eq. (ii) from Eq. (i), we get

$$\beta - \beta' = \frac{\lambda\,(D - D')}{d}$$

Substituting the given values, we get

$$3 \times 10^{-5} = \frac{\lambda \times 5 \times 10^{-2}}{1 \times 10^{-3}}$$

$$\Rightarrow \quad \lambda = \frac{3 \times 10^{-8}}{5 \times 10^{-2}} = 0.6 \times 10^{-6}\ \text{m}$$

$$\Rightarrow \quad \lambda = 600 \times 10^{-9}\ \text{m}$$

$$\Rightarrow \quad \lambda = 600\ \text{nm}$$

Chemistry

1. Number of moles $= \dfrac{\text{Mass}}{\text{Molecular mass}}$

Given, mass of $Al_2(SO_4)_3 = 50$ g

Molecular mass of

$Al_2(SO_4)_3 = 2 \times 27 + 3(32 + 4 \times 16) = 342$ g

$\therefore$ moles of $Al_2(SO_4)_3 = \dfrac{50}{342} = 0.14$ mol

2. Ionisation energy of $He^+ = 13.6 \times Z^2$eV

$= 13.6 \times (2)^2$ eV

$= 13.6 \times 4$ eV $= 54.4$ eV

3. (a) Metallic radii increase in a group from top to bottom. Thus, Li < Na < K < Rb is true.

(b) Electron gain enthalpy of Cl > F and afterwards it decreases along a group.

(c) Ionisation enthalpy increase along a period left to right but due to presence of half-filled orbital in N, ionisation enthalpy of N > O. Thus B < C < N < O is incorrect.

(d) For isoelectronic species,

$$\text{ionic size} \propto \frac{1}{\text{atomic number}}$$

Thus, $Al^{3+} < Mg^{2+} < Na^+ < F^-$ is true.

4. The $H - \ddot{O} - H$ bond angle in H_2O is 104.5° due to presence of two lone pair of electrons. This fact can be explained with the help of valence shell electron pair repulsion theory (VSEPR).

5. Van der Waals' gas approaches ideal behaviour at extremely low pressure and high temperature.

6. I : $C(s) + O_2(g) \longrightarrow CO_2$; $\Delta H = -393.5$ kJ/mol

II : $CO(g) + \frac{1}{2}O_2(g) \longrightarrow CO_2(g)$;

$\Delta H = -283.0$ kJ/ mol

On substracting II from I

III : $C(s) + \frac{1}{2}O_2(g) \longrightarrow CO(g)$;

$\Delta H = -110.5$ kJ/mol

This equation III also represents formation of 1 mole of CO and thus enthalpy of formation of $CO(g)$ is -110.5 kJ / mol

7.

	H_2 +	I_2 $\rightleftharpoons$	2HI
At initial	1	1	0
At equili	1 − 0.8	1 − 0.8	2 × 0.8
	= 0.2	= 0.2	= 1.6

$$\because \quad K_C = \frac{[HI]^2}{[H_2][I_2]}$$

$$\therefore \quad K_C = \frac{1.6 \times 1.6}{0.2 \times 0.2}$$

$$\Rightarrow \quad K_C = 64$$

8. H_2O_2 is a pale blue liquid, it can be oxidised by ozone. H_2O_2 acts as both oxidising and reducing agent. The value of dipole moment of H_2O_2 is 2.1 D. Which suggest it cannot be planar infact it has open book like structure.

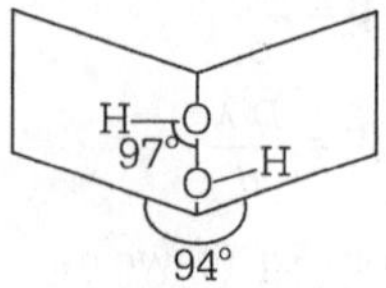

Structure of H_2O_2

Two O—H lie in different plane

9. KO_2 absorb CO_2 and increase O_2 concentration. So, it is used in space and submarines.

10. $BI_3 > BBr_3 > BCl_3 > BF_3$

This order can be easily explained on the basis of the tendaney of the halogen atom to back donate its lone pair of electrones to empty p–orbital of the boron atom through $p\pi$ - $p\pi$ bonding.

11.

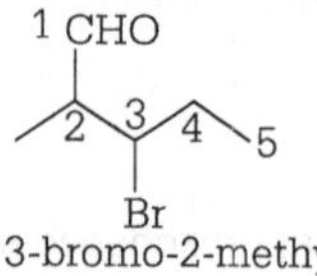

3-bromo-2-methylpentanal

12. Nucleophile always attacks on electron deficient site. Presence of electron withdrawing group such as NO_2 , CHO etc., decreases the electron density on benzene nucleous, hence such group activates the ring towards nucleophilic attack. While presence of electron releasing groups such as *R* or O*R* increase the electron density, thus deactivate the nucleus toward nucleophilie attack. NO_2 group activate the ring more than Cl toward nucleophilic attack hence react readily with nucleophile.

13. Ethyl acetoacetate shows tautomerism.

$$CH_3—\overset{\overset{O}{\|}}{C}—CH_2—COOC_2H_5$$

Keto form

$$\longleftrightarrow CH_3—\overset{\overset{OH}{|}}{C}= CH—COOC_2H_5.$$

Enol form

14.

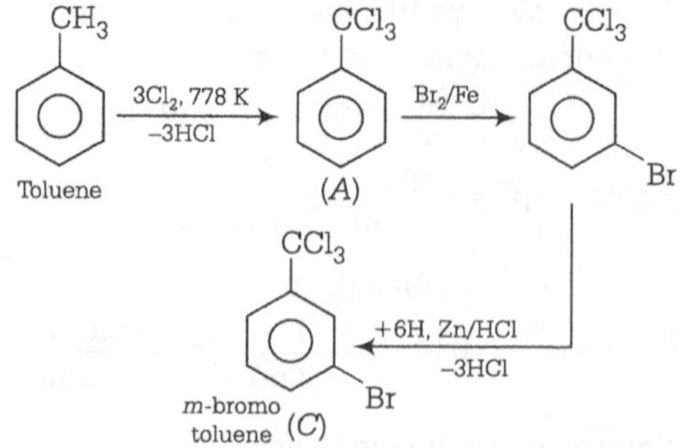

15. We know that,

$$O_3 + CCl_2F_2 \longrightarrow 2OCl + OF_2$$

Thus, in this reaction OCl is produced.

16. Sodium chloride (NaCl) has face centred cubic structure. It contains 4 Na^+ and 4 Cl^- ions. Each Na^+ ino is surrounded by 8 Cl^- ions and each Cl^- ion is also surrounded by 8 Na^+ ions.

17. $$\text{Molarity} = \frac{\text{\% by weight of solute} \times \text{density of solution} \times 10 \text{ (in litre)}}{\text{molecular weight of the solute}}$$

$$\text{Molarity} = \frac{40 \times 1.2 \times 10}{M \times 1000}$$

Again, since

$$\text{Molarity} = \frac{\text{weight of the solute}/M}{\text{volume of the solution (in litre)}}$$

Therefore, weight of solute = 480 g

18. $E_{\text{cell}} = E°_{\text{(cathode)}} - E°_{\text{(anode)}}$

$= 0.41 - 0.76 = -0.35$ V

19. $N_2(g) + 3H_2(g) \rightleftharpoons 2NH_3(g); + 22$ kcal

$\because$ The activation energy for the forward reaction $= 50$ kcal

$\therefore$ The activation energy for the backward reaction $= 50 + 72 = +72$ kcal

20. Coagulating power of an electrolyte for arsenious sulphide decreases as $Al^{3+} > Ba^{2+} > Na^+$

21. $^{228}_{88}X \xrightarrow{-3\alpha} {}^{216}_{82}Pb \xrightarrow{-\beta} {}^{216}_{83}Bi\ (Y)$

22. Hydrolysis of NCl_3 gives NH_3 or NH_4OH and HClO as

$$NCl_3 + 4H_2O \longrightarrow NH_4OH + 3HOCl$$

23. $4Ag + 8CN^- + 2H_2O + O_2 \longrightarrow 4[Ag(CN)_2]^- + 4OH^-$

This process is called cynide process. It is used in the extraction of silver from argentite (Ag_2S).

24. (a) $[Co\ F_6]^{3-}$ ion

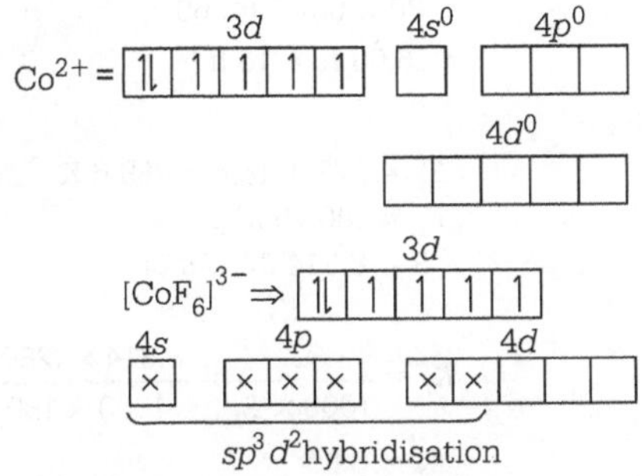

F^- is a weak ligand. It cannot pair up d-electrons and forms outer orbital octahedral complex.

NH_3 and CN^- are strong ligands. So, they pair up d-electrons and form their inner orbital complex.

(b) $[Co(NH_3)_6]^{3+}$ ion

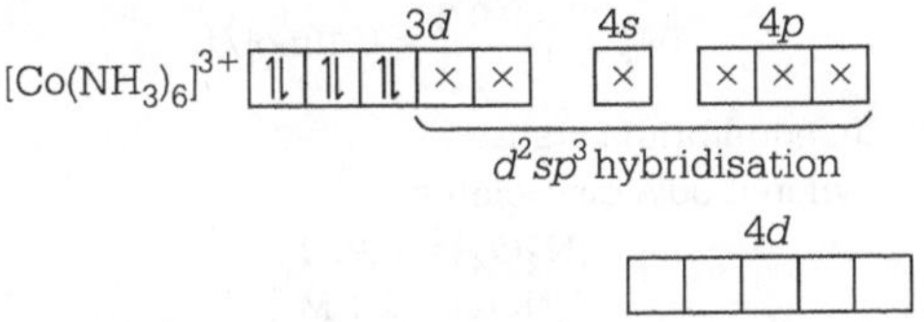

$[Fe(CN)_6]^{3-}$ ion

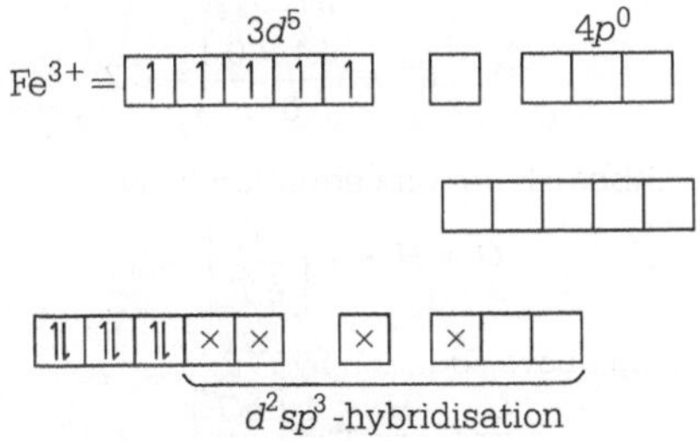

(d) $[Cr(NH_3)_6]^{3+}$ ion

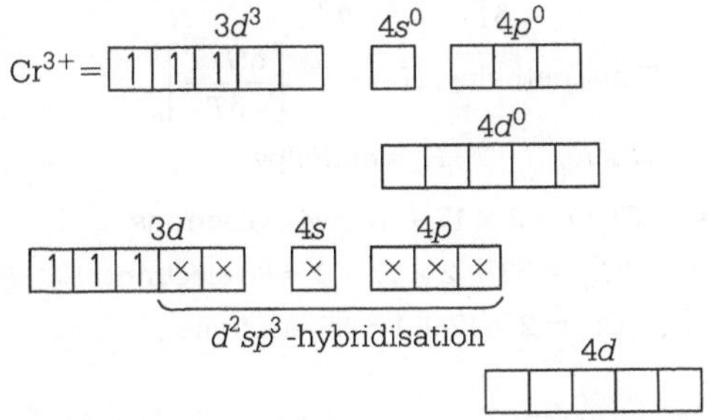

25. $C_2H_5Cl + KCN \xrightarrow{C_2H_5OH} \underset{(X)}{C_2H_5CN} + KCl$

$C_2H_5CN \xrightarrow[\Delta]{H_3O^+,\ 2H_2O} C_2H_5COOH + NH_3$

or

$(C_3H_6O_2)$

(y)

26. $NaBH_4$ reduces aldehyde into 1° alcohol,

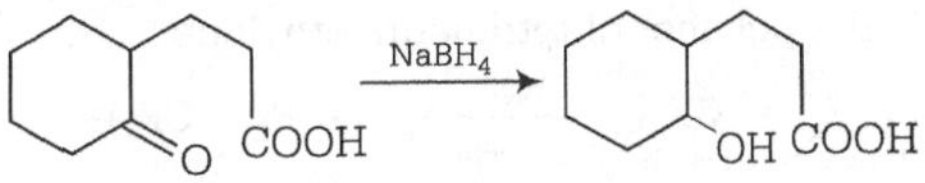

27.

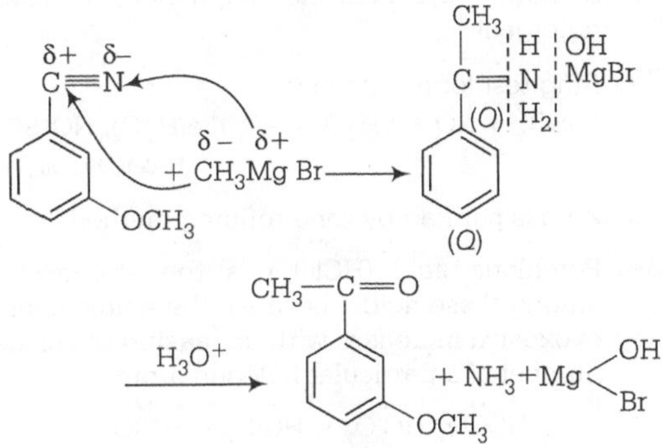

3-methoxy acetophenone

(P)

28. (i) Organic compound give an oxime with hydroxyl amine, therefore, it must be an aldehyde or ketone.

(ii) Organic compound did not give silver mirror with Tollen's reagent, therefore it cannot be an aldehyde. Therefore, compound is ketone and its molecular formula with be CH_3COCH_3.

29. $CH_3CH_2I \xrightarrow{NaCN} \underset{(A)}{CH_3CH_2CN} \xrightarrow{OH^-}$

$\underset{(B)}{CH_3CH_2CONH_2} \xrightarrow{Br_2/NaOH} \underset{(C)}{CH_3CH_2NH_2}$

30. Coupling of diazonium salts takes place in the following order as

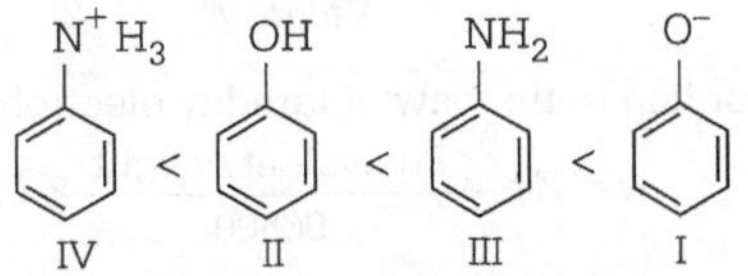

31. Aldehyde and α-hydroxy ketones give positive Tollen's test. Glucose has an aldehyde group and fructose is an α-hydroxy ketone.

32. Chain growth polymers are formed by the chain growth polymerisation or chain polymerisation. This polymerisation process involves in a series of reaction each of which consumes a relative particles and produces another similar particle resulting a chain reaction. Teflon is a chain growth polymer. It is the polymer of tetraflouro ethylene.

$$\underset{\text{Tetrofluoro ethylene}}{n(CF_2 = CF_2)} \xrightarrow[\text{Pressure}]{\text{High}} \underset{\text{Teflon}}{-(CF_2-CF_2)_n-}$$

33. Sodium alkyl benzene sulphonate is used as detergent.

34. Ring test for nitrate ions is

$$FeSO_4 + NO + 5H_2O \longrightarrow \underset{\text{Dark brown ring}}{[Fe(H_2O)_5NO]SO_4}$$

35. Zinc is purified by zone-refining method.

36. Perchloric acid ($HClO_4$) is the strongest acid among these acidic because the acidic character of oxoacid increases with increasing the oxidation number of a particular halogen atom.

$$\overset{+1}{HOCl} < \overset{+2}{HClO_2} < \overset{+5}{HClO_3} < \overset{+7}{HClO_4}$$

37. Aerosol is colloidal system of solid in gas, e.g, smoke. So, dispersion medium in aerosol is gas.

38. Given, $k_1 = 10^{10}\, e^{-20,000/T}$

$$k_2 = 10^{12}\, e^{-24,606/T}$$

$$k_1 = k_2 \Rightarrow 10^{10}\, e^{-20,000/T} = 10^{12}\, e^{-24,606/T}$$

$$e^{-\frac{20,000}{T} + \frac{24606}{T}} = 10^2$$

$$e^{\frac{4,606}{T}} = 10^2$$

On taking log both sides,

$$\frac{4.606}{2.303\,T} = \log 10^2$$

$$2 \log 10 \times T = \frac{4606}{2.303}$$

$$T = \frac{4606}{2.303 \times 2} = \frac{4606}{4.606} = 1000 \text{ K}$$

39. $w = 60\text{g}, i = 5\text{A}$

$$\text{Equivalent of Ca} = \frac{\text{atomic weight}}{\text{Valency}} = \frac{40}{2} = 20$$

According to first law of faraday electrolysis

$$\text{w} = Zit = \frac{\text{Equivalent weight}}{96500} \times i \times T$$

$$60 = \frac{20}{96500} \times 5 \times T$$

$$T = \frac{96500 \times 60}{20 \times 5}\, s$$

$$T = \frac{96500 \times 60}{20 \times 5 \times 60 \times 60}\, \text{h}$$

$$= 16.08 \text{ h.} \approx 16 \text{ h.}$$

40. Given,

$$T_f = 273 + 16.6 = 289.6 \text{ K}$$

$$L_f = 180.75 \text{ J/g}$$

$$R = 8.314 \text{ J/K/mol}$$

$$K_f = ?,$$

$$\therefore \quad K_f = \frac{RT_f^2}{1000 \times L_f} = \frac{8.314 \times (289.6)^2}{1000 \times 180.75}$$

$$\therefore \quad K_f = 3.86$$

41. The coordination number is 8:8 in $Cs^+ : Cl^-$.

The coordination number is 6:6 in $Na^+ : Cl^-$.

42. $N_2O_4\,(g) \rightleftharpoons 2NO_2\,(g)$

Molar concentration of

$$[N_2O_4] = \frac{9.2}{92} = 0.1 \text{ mol/L}$$

In equilibrium state,
when it 50% dissociates

$$[N_2O_4] = 0.05 \text{ M}$$

$$[NO_2] = 0.1 \text{ M}$$

Now, from $K_C = \dfrac{[NO_2]^2}{[N_2O_4]}$

$$K_C = \frac{0.1 \times 0.1}{0.05} = 0.2$$

43. The Gibbs-Helmholtz equation is as

$$G = H + T\left[\frac{\delta G}{\delta T}\right]_p$$

Dividing above equation by T^2

$$\frac{G}{T^2} = \frac{H}{T^2} + \frac{1}{T}\left[\frac{\delta G}{\delta T}\right]_p$$

This on rearrangement becomes.

$$\left[\frac{\delta\,(G/T)}{\delta T}\right] = -\frac{H}{T^2}$$

Thus, enthalpy, $H = -T^2\left[\dfrac{\delta G/T}{\delta T}\right]_p$

Where, H = enthalpy

44. $Cl_2O = 2 \times 17 + 16 = 42$ electrons

$ICl_2^- = 53 + 2 \times 17 + 1 = 88$ electrons

$Cl_2^- = 2 \times 17 + 1 = 35$ electrons

$IF_2^+ = 53 + 2 \times 9 - 1 = 70$ electrons

$ClO_2 = 17 + 2 \times 8 = 33$ electrons

$ClO_2 = 17 + 2 \times 8 + 1 = 33$ electrons

$ClO_2^- = 17 + 2 \times 8 + 1 = 34$ electrons

$ClF_2^+ = 17 + 2 \times 9 - 1 = 34$ electrons

ClO_2^- and ClF_2^+ contain 34 electrons each.

Hence they are isoelectronic.

45. Atomic and ionic radii increase from top to bottom in a group due to the inclusion of another shell at every step. Hence, Cs^+ ion will be the largest among given IA group ions, i.e.

Na^+, Li^+ and K^+.

46. In the ground state of an atom, the number of states is limited by Hund's rule. These are $\frac{Ln}{Lr.(n-r)!}$ ways in which electron in orbital may be arranged which do not violate Paulis exclusion principle.

Where, n = number of maximum electrons that can be filled in an orbital

and r = number of electrons present in orbital.

But the valid ground state term is calculated by Hund's rule of maximum multiplicity. As Hund's rule gives the most stable electronic configuration of electrons.

47. $\underset{2 \times 276\ g}{2\,Ag_2CO_3} \xrightarrow[\Delta]{} \underset{4 \times 108 g}{4Ag} + 2CO_2 + O_2$

$\therefore 2 \times 276$ g of Ag_2CO_3 gives $= 4 \times 108$ g Ag

$\therefore$ 1 g of Ag_2CO_3 will give $= \frac{4 \times 108}{2 \times 276}$

$\therefore$ 2.76 g of Ag_2CO_3 will give $= \frac{4 \times 108 \times 2.76}{2 \times 276}$

$= 2.16$ g

48. $\begin{matrix} CH_2Br \\ | \\ CH_2Br \end{matrix} + 2\ KOH \xrightarrow[\text{Alc.}]{\Delta} \underset{\text{Acetylene}}{\begin{matrix} CH \\ ||| \\ CH \end{matrix}} + 2KBr + 2H_2O$

This is dehydrohalogenation reaction.

49. I_2 form complex ion I_3^- in KI solution due to which it dissolve in it.

50. Catechol is most acidic out of all dihydric phenols.

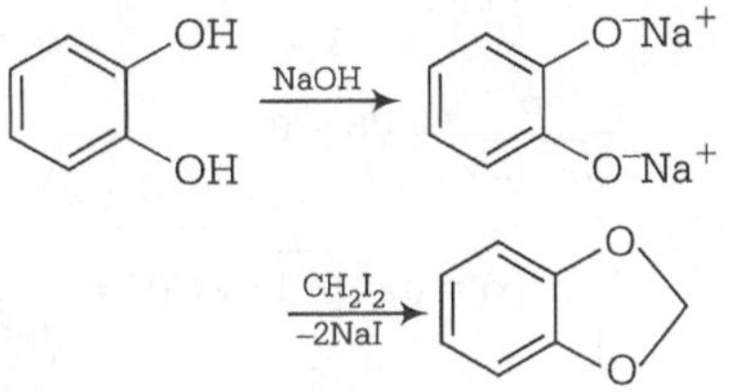

The reaction is Williamson's synthesis type reaction.

Mathematics

1. Let $y = x^{3000}$, $x = 1$ and $x + \Delta x = 1.0002$

Then, $\Delta x = 1.0002 - 1 = 0.0002$

Also, $y = 1$, when $x = 1$

Now, $y = x^{3000}$

$\Rightarrow \frac{dy}{dx} = 3000\,x^{2999}$

$\Rightarrow \left[\frac{dy}{dx}\right]_{x=1} = 3000$

$\because \Delta y = \frac{dy}{dx}\Delta x$

$\therefore \Delta y = 3000 \times 0.0002 = \frac{6}{10} = 0.6$

Hence, $(1.0002)^{3000} = y + \Delta y = 1 + 0.6 = 1.6$

2. Let $\mathbf{a} = a_1\hat{\mathbf{i}} + a_2\hat{\mathbf{j}} + a_3\hat{\mathbf{k}}$.

Then, $\mathbf{a}\cdot\hat{\mathbf{i}} = a_1, \mathbf{a}\cdot\hat{\mathbf{j}} = a_2, \mathbf{a}\cdot\hat{\mathbf{k}} = a_3$

$\mathbf{a} \times \hat{\mathbf{i}} = -a_2\hat{\mathbf{k}} + a_3\hat{\mathbf{j}}, \mathbf{a} \times \hat{\mathbf{j}} = a_1\hat{\mathbf{k}} - a_3\hat{\mathbf{i}}$

and $\mathbf{a} \times \hat{\mathbf{k}} = -a_1\hat{\mathbf{j}} + a_2\hat{\mathbf{i}}$

$\therefore (\mathbf{a}\cdot\hat{\mathbf{i}})(\mathbf{a}\times\hat{\mathbf{i}}) + (\mathbf{a}\cdot\hat{\mathbf{j}})(\mathbf{a}\times\hat{\mathbf{j}}) + (\mathbf{a}\cdot\hat{\mathbf{k}})(\mathbf{a}\times\hat{\mathbf{k}})$

$= a_1(-a_2\hat{\mathbf{k}} + a_3\hat{\mathbf{j}}) + a_2(a_1\hat{\mathbf{k}} - a_3\hat{\mathbf{i}}) + a_3(-a_1\hat{\mathbf{j}} + a_2\hat{\mathbf{i}})$

$= \mathbf{0}$

3. Clearly, A is the set of all points on the circle $x^2 + y^2 = 25$ and B is the set of all points on the ellipse $x^2 + 9y^2 = 144$. These two intersect at four points P, Q, R and S.

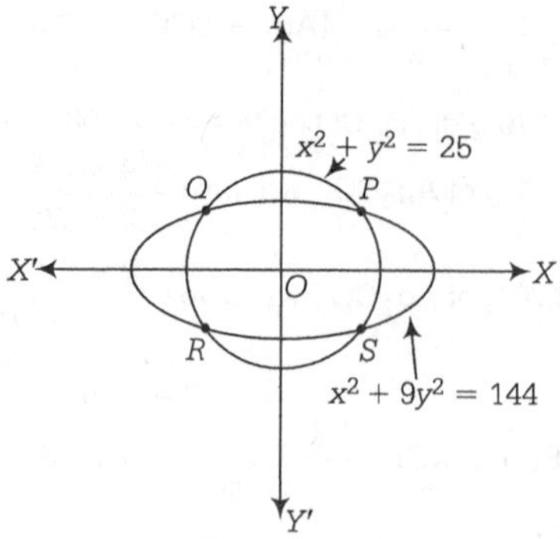

Hence, $A \cap B$ contains four points.

4. Clearly, $f(x) = \sqrt{x^2 - 4}$ is continuous on [2, 3] and differentiable on (2, 3).

So, by Lagrange's mean value theorem, there exists $c \in (2, 3)$ such that

$$f'(c) = \frac{f(3) - f(2)}{3 - 2}$$

$$\Rightarrow \quad \frac{c}{\sqrt{c^2 - 4}} = \sqrt{5} - 0$$

$$\left[\because f(x) = \sqrt{x^2 - 4} \Rightarrow f'(x) = \frac{x}{\sqrt{x^2 - 4}}\right]$$

$$\Rightarrow \quad c^2 = 5(c^2 - 4)$$

$$\Rightarrow \quad 4c^2 = 20$$

$$\Rightarrow \quad c = \sqrt{5}$$

5. The mean of the series $a, a + d, a + 2d, \ldots a + 2nd$

is $\bar{x} = \frac{1}{2n + 1}[a + a + d + a + 2d + \ldots + a + 2nd]$

$$\Rightarrow \bar{x} = \frac{1}{2n + 1}\left\{\frac{2n + 1}{2}(a + a + 2nd)\right\} = a + nd$$

$\therefore$ Mean deviation from mean,

$$\text{MD} = \frac{1}{2n + 1}\sum_{r = 0}^{2n}|(a + rd) - (a + nd)|$$

$$\Rightarrow \quad \text{MD} = \frac{1}{2n + 1}\sum_{r = 0}^{2n}|r - n|\, d$$

$$\Rightarrow \quad \text{MD} = \frac{1}{2n + 1}\{2d(1 + 2 + \ldots + n)\} = \frac{n(n + 1)\, d}{2n + 1}$$

6. We have,

$$f(x) = xe^{x(1 - x)}$$

On differentiating both sides w.r.t. x, we get

$$f'(x) = e^{x(1 - x)} + x(1 - 2x)\, e^{x(1 - x)}$$

$$\Rightarrow \quad f'(x) = (1 + x - 2x^2)\, e^{x(1 - x)}$$

$$\Rightarrow \quad f'(x) = -(x - 1)(2x + 1)\, e^{x(1 - x)}$$

Since, $e^{x(1 - x)} > 0$ for all x. Therefore, signs of $f'(x)$ for different values of x are as shown below:

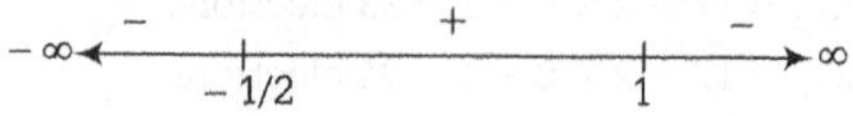

Clearly, $f(x)$ is increasing on $\left[-\frac{1}{2}, 1\right]$ and decreasing on $\left(-\infty, -\frac{1}{2}\right] \cup [1, \infty)$.

7. We have,

$$(1 + \omega)(1 + \omega^2)(1 + \omega^3)(1 + \omega^4)\ldots(1 + \omega^{3n})$$

$$= [\{(1 + \omega)(1 + \omega^2)\}\{(1 + \omega^4)(1 + \omega^5)\}\ldots \{(1 + \omega^{3n - 2})(1 + \omega^{3n - 1})\} \times (1 + \omega^3)(1 + \omega^6)(1 + \omega^9)\ldots(1 + \omega^{3n})]$$

$$= 2^n[\{(-\omega^2)(-\omega)\}\{(-\omega^2)(-\omega)]\ldots\{(-\omega^2)(-\omega)\}]$$

$$\Rightarrow [(1 + w)(1 + w^2)\{(1 + w)(1 + w^2)\}\{(1 + w)(1 + w^2)\} \ldots \{(1 + w)(1 + w^2)\}] \times (1 + 1)(1 + 1) \ldots (1 + 1)$$

$$= 2^n$$

8. Clearly, X is a binomial variate with parametres n and $p = \frac{1}{2}$ such that

$$P(X = r) = {}^nC_r p^r q^{n - r} = {}^nC_r\left(\frac{1}{2}\right)^r\left(\frac{1}{2}\right)^{n - r}$$

$$= {}^nC_r\left(\frac{1}{2}\right)^n$$

Now, $P(X = 4)$, $P(X = 5)$ and $P(X = 6)$ are in AP.

$$\Rightarrow \quad 2P(X = 5) = P(X = 4) + P(X = 6)$$

$$\Rightarrow \quad 2 \cdot {}^nC_5\left(\frac{1}{2}\right)^n = {}^nC_4\left(\frac{1}{2}\right)^n + {}^nC_6\left(\frac{1}{2}\right)^n$$

$$\Rightarrow \quad 2 \cdot {}^nC_5 = {}^nC_4 + {}^nC_6$$

$$\Rightarrow \quad 2 \cdot \frac{n!}{(n - 5)!\, 5!} = \frac{n!}{(n - 4)!\, 4!} + \frac{n!}{(n - 6)!\, 6!}$$

$$\Rightarrow \quad \frac{2}{5(n - 5)} = \frac{1}{(n - 4)(n - 5)} + \frac{1}{6 \times 5}$$

$$\Rightarrow \quad n^2 - 21n + 98 = 0$$

$$\Rightarrow \quad (n - 7)(n - 14) = 0$$

$$\therefore \quad n = 7 \text{ or } 14$$

9. Suppose $(s + 1)$th term contains x^{2r}.

Then, we have

$$T_{s + 1} = {}^{n - 3}C_s x^{n - 3 - s}\left(\frac{1}{x^2}\right)^s$$

$$= {}^{n - 3}C_s x^{n - 3 - 3s}$$

This will contain x^{2r}, if

$$n-3-3s=2r$$

$$\Rightarrow \quad s=\frac{n-3-2r}{3}$$

$$\Rightarrow \quad s=\frac{n-2r}{3}-1$$

$$\Rightarrow \quad s+1=\frac{n-2r}{3}$$

$$\Rightarrow \quad n-2r=3(s+1)$$

Hence, $(n-2r)$ is a positive integral multiple of 3.

10. We have, $\sin^{-1}\left\{\cot\left(\sin^{-1}\sqrt{\frac{2-\sqrt{3}}{4}}\right.\right.$

$$\left.\left.+\cos^{-1}\frac{\sqrt{12}}{4}+\sec^{-1}\sqrt{2}\right)\right\}$$

$$=\sin^{-1}\left\{\cot\left(\sin^{-1}\frac{\sqrt{3}-1}{2\sqrt{2}}+\cos^{-1}\frac{\sqrt{3}}{2}+\sec^{-1}\sqrt{2}\right)\right\}$$

$$=\sin^{-1}\left\{\cot\left(\frac{\pi}{12}+\frac{\pi}{6}+\frac{\pi}{4}\right)\right\}$$

$$=\sin^{-1}\left(\cot\frac{\pi}{2}\right)$$

$$=\sin^{-1}0=0$$

11. We have,

$$\cos^{-1}\frac{x}{2}+\cos^{-1}\frac{y}{3}=\theta$$

$$\Rightarrow \quad \cos^{-1}\left\{\frac{xy}{6}-\sqrt{1-\frac{x^2}{4}}\cdot\sqrt{1-\frac{y^2}{9}}\right\}=\theta$$

$$\Rightarrow \quad xy-\sqrt{4-x^2}\cdot\sqrt{9-y^2}=6\cos\theta$$

$$\Rightarrow \quad (xy-6\cos\theta)^2=(4-x^2)(9-y^2)$$

$$\Rightarrow \quad -12xy\cos\theta+36\cos^2\theta=36-4y^2-9x^2$$

$$\Rightarrow \quad 9x^2+4y^2-12xy\cos\theta=36\sin^2\theta$$

12. We have,

$$\tan(\sec^{-1}x)=\sin\left(\cos^{-1}\frac{1}{\sqrt{5}}\right)$$

$$\Rightarrow \quad \sqrt{\sec^2(\sec^{-1}x)-1}=\sqrt{1-\cos^2\left(\cos^{-1}\frac{1}{\sqrt{5}}\right)}$$

$$\Rightarrow \quad \sqrt{x^2-1}=\sqrt{1-\frac{1}{5}}$$

$$\Rightarrow \quad x^2-1=\frac{4}{5}$$

$$\Rightarrow \quad x^2=\frac{9}{5}$$

$$\Rightarrow \quad x=\pm\frac{3}{\sqrt{5}}$$

13. We have,

$$y=(2x-1)e^{2(1-x)}$$

On differentiating both sides w.r.t. x, we get

$$\frac{dy}{dx}=2e^{2(1-x)}-2(2x-1)e^{2(1-x)}$$

$$\Rightarrow \quad \frac{dy}{dx}=2e^{2(1-x)}(2-2x)=4e^{2(1-x)}(1-x)$$

At points of maximum, we must have

$$\frac{dy}{dx}=0\Rightarrow x=1$$

Now, $$\frac{d^2y}{dx^2}=-8e^{2(1-x)}(1-x)-4e^{2(1-x)}$$

$$\Rightarrow \quad \left[\frac{d^2y}{dx^2}\right]_{x=1}=-4<0$$

So, y is maximum at $x=1$. Clearly, $y=1$ for $x=1$.

Thus, the point of maximum is (1, 1).

The equation of the tangent at (1, 1) is

$$y-1=0(x-1)$$

$$\Rightarrow \quad y=1$$

14. The truth table of $(p\rightarrow\sim p)\wedge(\sim p\rightarrow p)$ is as follows:

p	$\sim p$	$p\rightarrow\sim p$	$\sim p\rightarrow p$	$(P\rightarrow\sim P)\wedge(\sim p\rightarrow p)$
T	F	F	T	F
F	T	T	F	F

Clearly, last column of the above truth table contains F only. So, the given statement is a contradiction.

15. There are 6 letters in the word 'DEGREE', namely 3E's and D, G, R. Four letters out of six can be selected in the following ways:

(i) 3 like letters and 1 different, *viz.* EEE + D, G, R

(ii) 2 like letters and 2 different, *viz.* EEE + any two of D, G, R

(iii) All different letters, *viz.* EDGR

So, the total number of ways

$$={}^3C_1+{}^3C_2+1=3+3+1=7$$

16. Total number of triangles = Number of triangles with vertices on sides ($PQ, QR, RS + QR, RS, PS + RS, PS, PQ + PS, PQ, QR$)

$$= {}^3C_1 \times {}^4C_1 \times {}^5C_1 + {}^4C_1 \times {}^5C_1 \times {}^6C_1 + {}^5C_1 \times {}^6C_1 \times {}^3C_1 + {}^6C_1 \times {}^3C_1 \times {}^4C_1$$

$= 60 + 120 + 90 + 72 = 342$

17. We observe that

$$\frac{x^4 + x^2 + x + 1}{x^2 - x + 1} \to \frac{2}{3} \text{ as } x \to -1$$

But $\dfrac{1 - \cos(x + 1)}{(x + 1)^2} \to \dfrac{0}{0}$ as $x \to -1$

$$\therefore \lim_{x \to -1} \left(\frac{x^4 + x^2 + x + 1}{x^2 - x + 1}\right)^{\frac{1 - \cos(x+1)}{(x+1)^2}}$$

$$= \left(\frac{2}{3}\right)^{\lim_{x \to -1} \frac{1 - \cos(x+1)}{(x+1)^2}} = \left(\frac{2}{3}\right)^{\lim_{x \to -1} \frac{2 \sin^2\left(\frac{x+1}{2}\right)}{4\left(\frac{x+1}{2}\right)^2}} = \left(\frac{2}{3}\right)^{1/2}$$

18. We have,

$$\left(x - \frac{1}{x}\right)^4 \left(x + \frac{1}{x}\right)^3$$

$$= \left({}^4C_0 x^4 - {}^4C_1 x^2 + {}^4C_2 - {}^4C_3 \frac{1}{x^2} + {}^4C_4 \frac{1}{x^4}\right) \times \left({}^3C_0 x^3 + {}^3C_1 x + {}^3C_2 \frac{1}{x} + {}^3C_3 \frac{1}{x^3}\right)$$

Clearly, there is no term free from x on RHS. Therefore, the term independent of x on LHS is zero.

19. We have,

$$A^2 = A$$

$\therefore (I + A)^2 = (I + A)(I + A) = I + 2A + A^2 = I + 3A$

and $(I + A)^3 = (I + A)^2 (I + A)$

$= (I + 3A)(I + A)$ $[\because (I + A)^2 = I + 3A]$

$= I + 4A + 3A^2$

$= I + 7A$ $[\because A^2 = A]$

Thus, we have

$(I + A)^2 = I + 3A$ and $(I + A)^3 = I + 7A$

$\Rightarrow (I + A)^2 = I + (2^2 - 1)A$

and $(I + A)^3 = I + (2^3 - 1)A$

Hence, $(I + A)^n = I + (2^n - 1)A$

$\therefore (I + A)^n = I + \lambda A$

$\Rightarrow \lambda = 2^n - 1$

20. We have, $u = e^x \sin x$ and $v = e^x \cos x$

On differentiating both sides w.r.t. x, we get

$$\frac{du}{dx} = e^x \sin x + e^x \cos x = u + v$$

and $\dfrac{dv}{dx} = e^x \cos x - e^x \sin x = v - u$

$$\therefore v\frac{du}{dx} - u\frac{dv}{dx} = v(u + v) - u(v - u) = u^2 + v^2$$

Now, $\dfrac{d^2u}{dx^2} = \dfrac{du}{dx} + \dfrac{dv}{dx} = u + v + v - u = 2v$

and $\dfrac{d^2v}{dx^2} = \dfrac{dv}{dx} - \dfrac{du}{dx}$

$= (v - u) - (v + u) = -2u$

21. Let $y = \tan^{-1}\left(\dfrac{\sqrt{1 + x^2} - 1}{x}\right)$

Putting $x = \tan\theta$, we get

$$y = \tan^{-1}\left(\frac{\sec\theta - 1}{\tan\theta}\right) = \tan^{-1}\left(\tan\frac{\theta}{2}\right)$$

$$= \frac{1}{2}\tan^{-1} x$$

On differentiating both sides w.r.t. x, we get

$$\frac{dy}{dx} = \frac{1}{2(1 + x^2)}$$

and $z = \tan^{-1}\left(\dfrac{2x\sqrt{1 - x^2}}{1 - 2x^2}\right)$

Putting $x = \sin\theta$, we get

$$z = \tan^{-1}\left(\frac{2\sin\theta\cos\theta}{\cos 2\theta}\right) = \tan^{-1}(\tan 2\theta)$$

$\Rightarrow z = 2\theta = 2\sin^{-1} x$

On differentiating both sides w.r.t. x, we get

$$\frac{dz}{dx} = \frac{2}{\sqrt{1 - x^2}}$$

Thus, $\dfrac{dy}{dz} = \dfrac{dy/dx}{dz/dx} = \dfrac{1}{4(1 + x^2)}\sqrt{1 - x^2}$

$$\therefore \left[\frac{dy}{dz}\right]_{x=0} = \frac{1}{4}$$

22. Let $I = \displaystyle\int \frac{1 - x^2}{(1 + x^2)\sqrt{1 + x^4}}\, dx$

$$= \int \frac{\frac{1}{x^2} - 1}{\left(x + \frac{1}{x}\right)\sqrt{x^2 + \frac{1}{x^2}}}\, dx$$

$$= -\int \frac{1}{\left(x+\frac{1}{x}\right)\sqrt{\left(x+\frac{1}{x}\right)^2-(\sqrt{2})^2}}\, d\left(x+\frac{1}{x}\right)$$

$$= \frac{1}{\sqrt{2}} \operatorname{cosec}^{-1}\left(\frac{x+\frac{1}{x}}{\sqrt{2}}\right) + C$$

$$= \frac{1}{\sqrt{2}} \sin^{-1}\left(\frac{\sqrt{2}x}{x^2+1}\right) + C$$

23. We have,

$$\begin{vmatrix} 1 & a & b \\ 1 & c & a \\ 1 & b & c \end{vmatrix} = 0$$

$$\Rightarrow \begin{vmatrix} 1 & a & b \\ 0 & c-a & a-b \\ 0 & b-a & c-b \end{vmatrix} = 0$$

[applying $R_2 \to R_2 - R_1$ and $R_3 \to R_3 - R_1$]

$$\Rightarrow (c-a)(c-b) + (a-b)^2 = 0$$

$$\Rightarrow a^2 + b^2 + c^2 - ab - bc - ca = 0$$

$$\Rightarrow 2a^2 + 2b^2 + 2c^2 - 2ab - 2bc - 2ca = 0$$

$$\Rightarrow (a-b)^2 + (b-c)^2 + (c-a)^2 = 0$$

$$\Rightarrow a = b = c$$

$\therefore$ ΔABC is an equilateral triangle.

$$\Rightarrow A = B = C = \frac{\pi}{3}$$

$$\therefore \sin^2 A + \sin^2 B + \sin^2 C = 3\sin^2\frac{\pi}{3} = \frac{9}{4}$$

24. Required area $= \int_0^1 x\, dx + \int_1^e \frac{1}{x} dx$

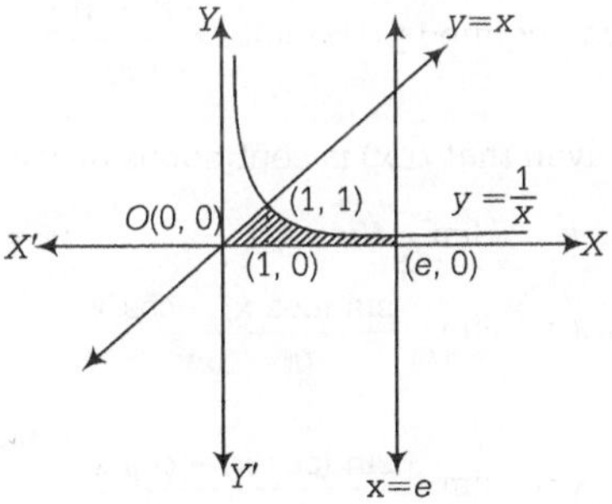

$$= \left[\frac{x^2}{2}\right]_0^1 + [\log x]_1^e = \frac{1}{2} + 1 = \frac{3}{2}$$

25. Let $x = \frac{1}{y}$. Then,

$$\lim_{x \to \infty} \left\{\frac{a_1^{1/x} + a_2^{1/x} + \ldots + a_n^{1/x}}{n}\right\}^{nx}$$

$$= \lim_{y \to 0} \left\{\frac{a_1^y + a_2^y + \ldots + a_n^y}{n}\right\}^{n/y}$$

$$= \lim_{y \to 0} \left\{1 + \frac{a_1^y + a_2^y + \ldots + a_n^y - n}{n}\right\}^{n/y}$$

$$= e^{\lim_{y \to 0}\left\{\frac{a_1^y - 1}{y} + \frac{a_2^y - 1}{y} + \ldots + \frac{a_n^y - 1}{y}\right\}}$$

$$= e^{\log a_1 + \log a_2 + \ldots + \log a_n}$$

$$= e^{\log (a_1 a_2 \ldots a_n)}$$

$$= a_1 a_2 a_3 \ldots a_n$$

26. Let $I = \int \frac{\sec^2 x}{(\sec x + \tan x)^{9/2}} dx$

Put $\sec x + \tan x = t$

Then, $\sec x - \tan x = \frac{1}{t}$

and $\sec x (\sec x + \tan x)\, dx = dt$

$$\Rightarrow \sec x\, dx = \frac{1}{t} dt$$

Also, $\sec x = \frac{1}{2}\left(t + \frac{1}{t}\right)$

$$\therefore I = \frac{1}{2}\int \frac{\frac{1}{t}\left(t+\frac{1}{t}\right)}{t^{9/2}} dt$$

$$\Rightarrow I = \frac{1}{2}\int \frac{1}{t^{9/2}} + \frac{1}{t^{13/2}} dt$$

$$\Rightarrow I = -\frac{1}{7t^{7/2}} - \frac{1}{11t^{11/2}} + K$$

$$\Rightarrow I = -\frac{1}{t^{11/2}}\left\{\frac{t^2}{7} + \frac{1}{11}\right\} + K$$

$$\Rightarrow I = -\frac{1}{(\sec x + \tan x)^{11/2}} \left\{\frac{1}{11} + \frac{1}{7}(\sec x + \tan x)^2\right\} + K$$

27. Let the equation of the circle be

$$x^2 + y^2 + 2gx + 2fy + c = 0$$

which passes through (0, 0) and (1, 0).

$$\therefore c = 0 \text{ and } 1 + 2g + c = 0$$

$$\Rightarrow c = 0, g = -\frac{1}{2}$$

The circle $x^2 + y^2 + 2gx + 2fy + c = 0$ touches the circle $x^2 + y^2 = 9$.

$\therefore \quad \sqrt{g^2 + f^2} = 3 \pm \sqrt{g^2 + f^2 - c}$ $\quad [\because C_1C_2 = r_1 \pm r_2]$

$\Rightarrow \quad \sqrt{\frac{1}{4} + f^2} = 3 \pm \sqrt{\frac{1}{4} + f^2}$

$\Rightarrow \quad 3 = 2\sqrt{\frac{1}{4} + f^2}$

$\Rightarrow \quad f^2 = \frac{9}{4} - \frac{1}{4} = 2$

$\Rightarrow \quad f = \pm\sqrt{2}$

Hence, the coordinates of the centre are $\left(\frac{1}{2}, \pm\sqrt{2}\right)$.

28. Given equation of parabola is

$$y^2 - kx + 8 = 0$$

$\Rightarrow \quad y^2 = k\left(x - \frac{8}{k}\right)$

$\Rightarrow \quad (y - 0)^2 = k\left(x - \frac{8}{k}\right)$

The equation of the directrix of this parabola is

$$x - \frac{8}{k} = -\frac{k}{4} \quad [\because x = -a]$$

$\Rightarrow \quad x = \frac{8}{k} - \frac{k}{4}$

But the equation of the directrix is given as $x - 1 = 0$.

$\therefore \quad \frac{8}{k} - \frac{k}{4} = 1$

$\Rightarrow \quad k^2 + 4k - 32 = 0$

$\Rightarrow \quad (k - 4)(k + 8) = 0$

$\therefore \quad k = -8, 4$

29. Given equation of parabola is

$$y^2 - 4y - x + 5 = 0$$

Then, the equation of tangent at $(2, 3)$ is

$$3y - 2(y + 3) - \frac{(x + 2)}{2} + 5 = 0$$

$\Rightarrow \quad 2y - x - 4 = 0$

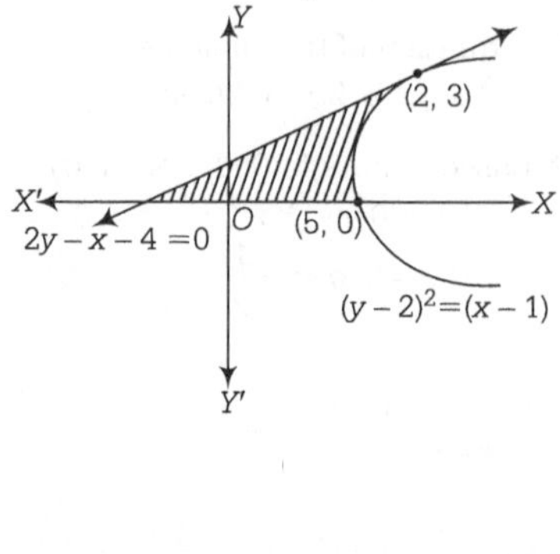

$\therefore$ Required area is given by

$$A = \int_0^3 (x_2 - x_1)\, dy$$

$$= \int_0^3 [\{(y - 2)^2 + 1\} - \{2y - 4\}]\, dy$$

$$= \int_0^3 (y^2 - 6y + 9)\, dy = \int_0^3 (3 - y)^2 dy$$

$$= -\left[\frac{(3 - y)^3}{3}\right]_0^3$$

$\therefore \quad A = 9$

30. We have,

$$x = e^{xy\frac{dy}{dx}}$$

$\Rightarrow \quad \log x = xy\frac{dy}{dx}$

$\Rightarrow \quad ydy = \frac{\log x}{x} dx$

$\Rightarrow \quad ydy = \log x \, d(\log x)$

On integrating both sides, we get

$$\frac{y^2}{2} = \frac{(\log_e x)^2}{2} + C$$

$\Rightarrow \quad y^2 = (\log_e x)^2 + 2C$

$\Rightarrow \quad y = \pm\sqrt{(\log_e x)^2 + 2C}$

31. Number of ways of selecting three numbers one on each card $= 100 \times 100 \times 100 = 100^3$

We know that $(2n + 1), (2n^2 + 2n)$ and $(2n^2 + 2n + 1)$ are Pythagorean triplets.

$\therefore$ For $n = 1, 2, 3, 4, 5, 6$, we get the lengths of three sides of a right angled triangle such that its hypotenuse is less than or equal to 100 cm.

Now, one Pythagorean triplet (e.g. 3, 4, 5; 5, 12, 13; etc.) can be chosen in 3! ways.

$\therefore$ Number of ways of selecting 6 Pythagorean triplets $= 6 \times 3!$

Hence, required probability $= \frac{6 \times 3!}{100^3}$

32. It is given that $f(x)$ is continuous at $x = \frac{\pi}{2}$.

$\therefore \quad k = \lim_{x \to \pi/2} f(x)$

$\Rightarrow \quad k = \lim_{x \to \pi/2} \frac{\sin(\cos x) - \cos x}{(\pi - 2x)^3}$

$\Rightarrow \quad k = \lim_{x \to \pi/2} \frac{\sin(\cos x) - \cos x}{\cos^3 x} \times \frac{\sin^3\left(\frac{\pi}{2} - x\right)}{8\left(\frac{\pi}{2} - x\right)^3}$

$\therefore \quad k = -\frac{1}{6} \times \frac{1}{8} = -\frac{1}{48} \quad \left[\because \lim_{x \to 0} \frac{\sin x - x}{x^3} = -\frac{1}{6}\right]$

33. We have,

$$f(x) = [x] + \left[x + \frac{1}{2}\right] = \begin{cases} 0, & \text{if } 0 < x < \frac{1}{2} \\ 1, & \text{if } x = \frac{1}{2} \\ 1, & \text{if } \frac{1}{2} < x < 1 \end{cases}$$

Clearly, $f(x)$ is discontinuous at $x = \frac{1}{2}$.

Also, $\lim_{x \to 1/2^-} f(x) = 0$ and $\lim_{x \to 1/2^+} f(x) = 1$

34. It is evident from the graph of $f(x)$ that

$$f(x) = \begin{cases} 1, & x \geq 1 \\ x^3, & x < 1 \end{cases}$$

Clearly, $f(x)$ is everywhere continuous but it is not differentiable at $x = 1$.

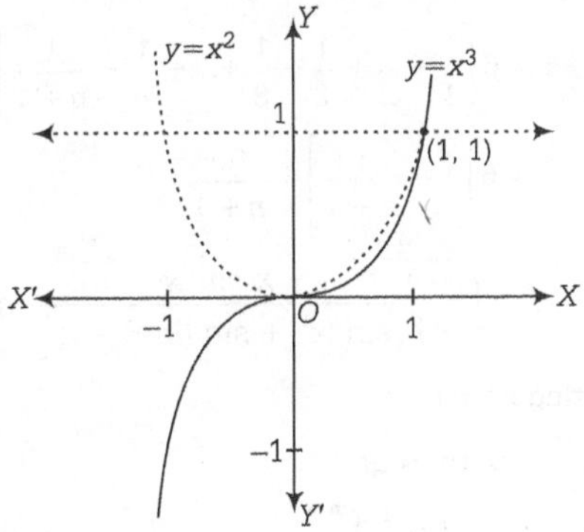

35. We have,

$$x_n = \cos\left(\frac{\pi}{3^n}\right) + i \sin\left(\frac{\pi}{3^n}\right)$$

$\therefore x_1 \cdot x_2 \cdot x_3 \ldots$

$$= \cos\left(\frac{\pi}{3} + \frac{\pi}{3^2} + \ldots\right) + i \sin\left(\frac{\pi}{3} + \frac{\pi}{3^2} + \ldots\right)$$

$$= \cos\left(\frac{\pi/3}{1 - \frac{1}{3}}\right) + i \sin\left(\frac{\pi/3}{1 - \frac{1}{3}}\right)$$

$$= \cos\frac{\pi}{2} + i \sin\frac{\pi}{2} = 0 + i = i$$

36. There can be 2 types of numbers.

(i) Any one of the digits 1, 2, 3, 4 repeats thrice and the remaining digits only once i.e., of the type 1, 2, 3, 4, 4, 4.

(ii) Any two of the digits 1, 2, 3, 4 repeat twice and the remaining two only once i.e., of the type 1, 2, 3, 3, 4, 4.

Number of numbers of the type 1 2 3 4 4 4

$$= \frac{6!}{3!} \times {}^4C_1 = 480$$

Number of numbers of the type 1 2 3 3 4 4

$$= \frac{6!}{2!2!} \times {}^4C_2 = 1080$$

So, the required number = 480 + 1080 = 1560

37. We have,

$$9x^2 - 18|x| + 5 = 0$$

$$\Rightarrow \quad 9|x|^2 - 18|x| + 5 = 0$$

$$\Rightarrow \quad (3|x| - 1)(3|x| - 5) = 0$$

$$\Rightarrow \quad |x| = \frac{1}{3} \text{ or } |x| = \frac{5}{3}$$

$$\Rightarrow \quad x = \pm\frac{1}{3} \text{ or } x = \pm\frac{5}{3}$$

Now, $\log_e\{(x + 1)(x + 2)\}$ is defined, if

$$(x + 1)(x + 2) > 0$$

$$\Rightarrow \quad x < -2 \text{ or } x > -1$$

Clearly, $x = -\frac{5}{3}$ does not satisfy this condition. But all other values of x satisfy the above conditions. Hence, the number of solutions lying in the domain of definition of the given function is 3.

38. We have,

$$2^x + 2^y = 2^{x+y}$$

On differentiating both sides w.r.t. x, we get

$$2^x \log 2 + 2^y \log 2 \frac{dy}{dx} = 2^{x+y} \log 2\left(1 + \frac{dy}{dx}\right)$$

$$\Rightarrow \quad (2^y - 2^{x+y})\frac{dy}{dx} = 2^{x+y} - 2^x$$

$$\Rightarrow \quad \frac{dy}{dx} = \frac{2^x(2^y - 1)}{2^y(1 - 2^x)} = 2^{x-y}\left(\frac{2^y - 1}{1 - 2^x}\right)$$

39. Since, $3x + y = 0$ is a tangent to the circle with centre at $(2, -1)$.

$\therefore$ Radius = Length of the perpendicular from $(2, -1)$ on $3x + y = 0$

$$\Rightarrow \quad \text{Radius} = \frac{6 - 1}{\sqrt{9 + 1}} = \frac{5}{\sqrt{10}} = \sqrt{\frac{5}{2}}$$

So, the equation of the circle is

$$(x-2)^2+(y+1)^2=\frac{5}{2}$$

$$\Rightarrow \quad x^2+y^2-4x+2y+\frac{5}{2}=0$$

The combined equation of the tangents drawn from the origin to this circle is

$$SS_1=T^2$$

where, $S=x^2+y^2-4x+2y+\frac{5}{2}$,

$$S_1=0^2+0^2-4\times 0+2\times 0+\frac{5}{2}$$

$$=\frac{5}{2}$$

and $T=x(0)+y(0)-2(x+0)+(y+0)+\frac{5}{2}$

$$=-2x+y+\frac{5}{2}$$

$$\therefore \left(x^2+y^2-4x+2y+\frac{5}{2}\right)\left(\frac{5}{2}\right)=\left(-2x+y+\frac{5}{2}\right)^2$$

$$\Rightarrow \quad 3x^2-8xy-3y^2=0$$

$$\Rightarrow \quad 3x+y=0 \text{ and } x-3y=0$$

40. We have,

$\sin A+\cos A=m$ and $\sin^3 A+\cos^3 A=n$

Now, $\quad \sin A+\cos A=m$

$$\Rightarrow \quad (\sin A+\cos A)^3=m^3$$

$$\Rightarrow \quad \sin^3 A+\cos^3 A+3\sin A\cos A(\sin A+\cos A)=m^3$$

$$\Rightarrow \quad n+3\sin A\cos A\, m=m^3 \quad \text{...(i)}$$

Again, $\quad \sin A+\cos A=m$

$$\Rightarrow \sin^2 A+\cos^2 A+2\sin A\cos A=m^2$$

$$\Rightarrow \quad \sin A\cos A=\frac{m^2-1}{2} \quad \text{...(ii)}$$

From Eqs. (i) and (ii), we get

$$n+3m\frac{(m^2-1)}{2}=m^3$$

$$\Rightarrow \quad 2n+3m^3-3m=2m^3$$

$$\Rightarrow \quad m^3-3m+2n=0$$

41. We have, $a=4$, $b=3$ and $\angle A=60^\circ$

$$\because \quad \cos A=\frac{b^2+c^2-a^2}{2bc}$$

$$\therefore \quad \cos 60^\circ=\frac{9+c^2-16}{6c}$$

$$\Rightarrow \quad \frac{1}{2}\times 6c=c^2-7$$

$$\Rightarrow \quad c^2-3c-7=0$$

42. Let T_r be the rth term of the given series. Then,

$$T_r=\frac{2r+1}{1^2+2^2+\ldots+r^2}=\frac{2r+1}{\frac{r(r+1)(2r+1)}{6}}$$

$$=\frac{6}{r(r+1)}$$

$$\Rightarrow \quad T_r=6\left(\frac{1}{r}-\frac{1}{r+1}\right)$$

So, the required sum is given by

$$\sum_{r=1}^{n} T_r=6\sum_{r=1}^{n}\left(\frac{1}{r}-\frac{1}{r+1}\right)$$

$$=6\left[\frac{1}{1}-\frac{1}{2}+\frac{1}{2}-\frac{1}{3}+\ldots+\frac{1}{n}-\frac{1}{n+1}\right]$$

$$=6\left[1-\frac{1}{n+1}\right]=\frac{6n}{n+1}$$

43. Let $\quad I=\int_{\sqrt{\ln 2}}^{\sqrt{\ln 3}}\frac{x\sin x^2}{\sin x^2+\sin(\ln 6-x^2)}dx$

Putting $x^2=t$

$$\Rightarrow \quad 2x\,dx=dt$$

$$\therefore \quad I=\frac{1}{2}\int_{\ln 2}^{\ln 3}\frac{\sin t}{\sin t+\sin(\ln 6-t)}dt \quad \text{...(i)}$$

Using $\int_a^b f(x)\,dx=\int_a^b f(a+b-x)\,dx$, we get

$$I=\frac{1}{2}\int_{\ln 2}^{\ln 3}\frac{\sin(\ln 6-t)}{\sin(\ln 6-t)+\sin t}dt \quad \text{...(ii)}$$

On adding Eqs.(i) and (ii), we get

$$2I=\int_{\ln 2}^{\ln 3}1\,dt=[t]_{\ln 2}^{\ln 3}$$

$$\Rightarrow \quad 2I=\frac{1}{2}(\ln 3-\ln 2)=\frac{1}{2}\ln\left(\frac{3}{2}\right)$$

$$\Rightarrow \quad I=\frac{1}{4}\ln\left(\frac{3}{2}\right)$$

44. Let $I=\int_1^4 \log_e[x]\,dx$

$$=\int_1^2\log_e 1\,dx+\int_2^3\log_e 2dx+\int_3^4\log_e 3\,dx$$

$$=\log_e 2+\log_e 3=\log_e 6$$

45. We have,

$$\frac{x^2}{16} + \frac{y^2}{9} = 1$$

The eccentricity e of the ellipse is given by

$$e = \sqrt{1 - \frac{9}{16}} = \frac{\sqrt{7}}{4}$$

So, the coordinates of the foci are $(\pm \sqrt{7}, 0)$.

$\therefore$ Radius of the circle

$$= \sqrt{(\sqrt{7} - 0)^2 + (0 - 3)^2}$$

$$= \sqrt{7 + 9.} = \sqrt{16} = 4$$

46. We have,

$$\lim_{x \to 2} f(x) = \lim_{x \to 2} \frac{\sin (e^{x-2} - 1)}{\log (x - 1)}$$

$$= \lim_{x \to 2} \left\{ \frac{\sin (e^{x-2} - 1)}{e^{x-2} - 1} \cdot \frac{e^{x-2} - 1}{x - 2} \cdot \frac{x - 2}{\log [1 + (x - 2)]} \right\}$$

$$= 1 \times 1 \times 1 = 1$$

47. Since, $\tan A$ and $\tan B$ are the roots of the equation $x^2 - ax + b = 0$. Therefore,

$$\tan A + \tan B = a \text{ and } \tan A \tan B = b$$

Now, $\tan (A + B) = \dfrac{\tan A + \tan B}{1 - \tan A \tan B} = \dfrac{a}{1 - b}$

Now, $\sin^2 (A + B) = \dfrac{1}{2} [1 - \cos 2 (A + B)]$

$$\Rightarrow \sin^2 (A + B) = \frac{1}{2} \left\{ 1 - \frac{1 - \tan^2 (A + B)}{1 + \tan^2 (A + B)} \right\}$$

$$\Rightarrow \sin^2 (A + B) = \frac{1}{2} \left\{ 1 - \frac{1 - \dfrac{a^2}{(1 - b)^2}}{1 + \dfrac{a^2}{(1 - b)^2}} \right\}$$

$$\Rightarrow \sin^2 (A + B) = \frac{1}{2} \left\{ \frac{2a^2}{a^2 + (1 - b)^2} \right\}$$

$$\Rightarrow \sin^2 (A + B) = \frac{a^2}{a^2 + (1 - b)^2}$$

48. We have, $a^2 + b^2 + c^2 = ac + \sqrt{3}\, ab$

$$\Rightarrow \frac{a^2}{4} + c^2 - ac + \frac{3a^2}{4} + b^2 - \sqrt{3}ab = 0$$

$$\Rightarrow \left(\frac{a}{2} - c\right)^2 + \left(\frac{\sqrt{3}a}{2} - b\right)^2 = 0$$

$$\Rightarrow \frac{a}{2} - c = 0 \text{ and } \frac{\sqrt{3}}{2} a - b = 0$$

$$\Rightarrow a = 2c \text{ and } \sqrt{3}a = 2b$$

$$\Rightarrow a = \frac{2b}{\sqrt{3}} = 2c = \lambda \quad \text{[say]}$$

$$\Rightarrow a = \lambda,\ b = \frac{\sqrt{3}}{2}\lambda \text{ and } c = \frac{\lambda}{2}$$

Now, $b^2 + c^2 = \left(\dfrac{\sqrt{3}\lambda}{2}\right)^2 + \left(\dfrac{\lambda}{2}\right)^2$

$$= \frac{3\lambda^2}{a} + \frac{\lambda^2}{4} = \lambda^2$$

$\therefore \quad b^2 + c^2 = a^2$

Hence, the triangle is right angled but not isosceles.

49. We have,

$$(\mathbf{a} \times \mathbf{b}) \times \mathbf{c} = \frac{1}{3} |\mathbf{b}| |\mathbf{c}| \mathbf{a}$$

$$\Rightarrow (\mathbf{a} \cdot \mathbf{c})\, \mathbf{b} - (\mathbf{b} \cdot \mathbf{c})\, \mathbf{a} = \frac{1}{3} |\mathbf{b}| |\mathbf{c}| \mathbf{a}$$

$$\Rightarrow (\mathbf{a} \cdot \mathbf{c})\, \mathbf{b} - \left\{ (\mathbf{b} \cdot \mathbf{c}) + \frac{1}{3} |\mathbf{b}| |\mathbf{c}| \right\} \mathbf{a} = \mathbf{0}$$

$$\Rightarrow \mathbf{a} \cdot \mathbf{c} = 0 \text{ and } \mathbf{b} \cdot \mathbf{c} + \frac{1}{3} |\mathbf{b}| |\mathbf{c}| = 0$$

[$\because$ **a**, **b** are non-collinear]

$$\Rightarrow |\mathbf{b}| |\mathbf{c}| \cos \theta + \frac{1}{3} |\mathbf{b}| |\mathbf{c}| = 0$$

$$\Rightarrow \cos \theta = -\frac{1}{3}$$

$$\Rightarrow \sin \theta = \sqrt{1 - \cos^2 \theta} = \sqrt{1 - \left(-\frac{1}{3}\right)^2}$$

$$= \sqrt{1 - \frac{1}{9}} = \sqrt{\frac{8}{9}} = \frac{2\sqrt{2}}{3}$$

50. The equation of a line through $P(1, 1)$ and parallel to $x + y = 1$, is

$$\frac{x - 1}{\cos \frac{3\pi}{4}} = \frac{y - 1}{\sin \frac{3\pi}{4}}$$

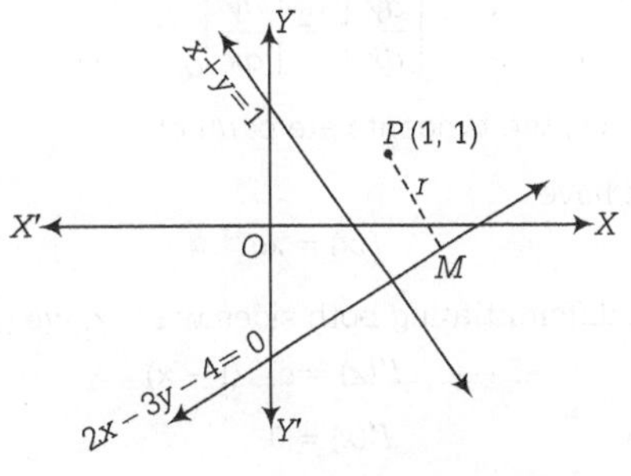

Let $PM = r$.
Then, the coordinates of M given by

$$\frac{x-1}{\cos\frac{3\pi}{4}} = \frac{y-1}{\sin\frac{3\pi}{4}} = r \text{ is } \left(1 - \frac{r}{\sqrt{2}}, 1 + \frac{r}{\sqrt{2}}\right).$$

$\because$ M lies on $2x - 3y - 4 = 0$.

$$\therefore \quad 2\left(1 - \frac{r}{\sqrt{2}}\right) - 3\left(1 + \frac{r}{\sqrt{2}}\right) - 4 = 0$$

$$\Rightarrow \quad 2 - \frac{2r}{\sqrt{2}} - 3 - \frac{3r}{\sqrt{2}} - 4 = 0$$

$$\Rightarrow \quad -5 - \frac{5r}{\sqrt{2}} = 0$$

$$\Rightarrow \quad r = -\sqrt{2}$$

Hence, $PM = \sqrt{2}$

51. At the point where the given curve crosses X-axis, we have

$$y = 0$$

$$\Rightarrow \quad ax^2 = 1$$

$$[\text{putting } y = 0 \text{ in } ax^2 + 2hyx + by^2 = 1]$$

$$\Rightarrow \quad x = \pm\frac{1}{\sqrt{a}}$$

Thus, the given curve cuts X-axis at $P\left(\frac{1}{\sqrt{a}}, 0\right)$ and $Q\left(-\frac{1}{\sqrt{a}}, 0\right)$.

Now, $ax^2 + 2hxy + by^2 = 1$

On differentiating both sides w.r.t. x, we get

$$2ax + 2h\left(x\frac{dy}{dx} + y\right) + 2by\frac{dy}{dx} = 0$$

$$\Rightarrow \quad \frac{dy}{dx} = -\frac{ax + hy}{hx + by}$$

$$\Rightarrow \quad \left[\frac{dy}{dx}\right]_P = -\frac{a}{h} \text{ and } \left[\frac{dy}{dx}\right]_Q = -\frac{a}{h}$$

$$\Rightarrow \quad \left[\frac{dy}{dx}\right]_P = \left[\frac{dy}{dx}\right]_Q$$

Hence, the tangents are parallel.

52. We have,

$$f(x) = xe^{-x}$$

On differentiating both sides w.r.t. x, we get

$$f'(x) = e^{-x}(1 - x)$$

Put $\quad f'(x) = 0$

$$\Rightarrow \quad x = 1$$

Now, $\quad f(0) = 0$

and $\quad \lim_{x \to \infty} f(x) = \lim_{x \to \infty} xe^{-x}$

$$= \lim_{x \to \infty} \frac{x}{e^x} = \lim_{x \to \infty} \frac{1}{e^x} = 0$$

Hence, the greatest value of $f(x)$ is $\frac{1}{e}$.

53. We have,

$AD = a + a\cos 2\theta$ and $BC = 2BD = 2a\sin 2\theta$

$\therefore$ Area of ΔABC is given by

$$\Delta = \frac{1}{2}BC \times AD = \frac{2}{2}a^2(\sin 2\theta + \sin 2\theta\cos 2\theta)$$

$$\Rightarrow \quad \Delta = a^2\sin 2\theta + \frac{1}{2}a^2\sin 4\theta$$

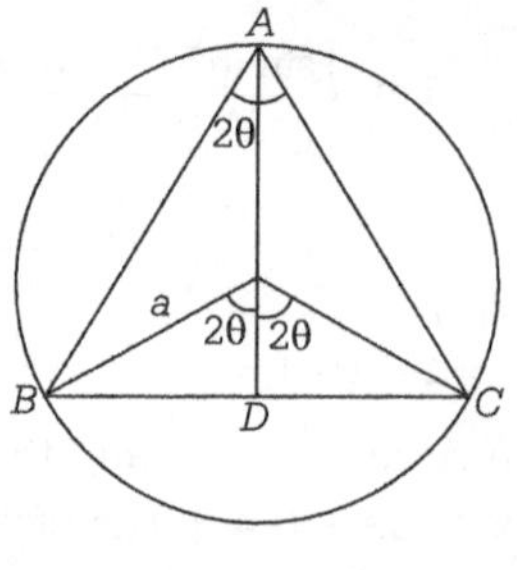

On differentiating both sides w.r.t. θ, we get

$$\frac{d\Delta}{d\theta} = 2a^2\cos 2\theta + 2a^2\cos 4\theta$$

For maximum or minimum, put $\frac{d\Delta}{d\theta} = 0$.

$$\Rightarrow \quad \cos 2\theta = -\cos 4\theta$$

$$\Rightarrow \quad 2\theta = \pi - 4\theta$$

$$\Rightarrow \quad \theta = \frac{\pi}{6}$$

Now, $\frac{d^2\Delta}{d\theta^2} = -2a^2\sin 2\theta \times 2 - 2a^2\sin 4\theta \times 4$

$$= -4a^2\sin 2\theta - 8a^2\sin 4\theta$$

At $\quad \theta = \frac{\pi}{6}, \frac{d^2\Delta}{d\theta^2} = -4a^2\sin\frac{\pi}{3} - 8a^2\sin\frac{2\pi}{3}$

$$= -4a^2 \times \frac{\sqrt{3}}{2} - 8a^2 \times \frac{\sqrt{3}}{2}$$

$$= -6a^2\sqrt{3} < 0$$

Hence, Δ is maximum for $\theta = \frac{\pi}{6}$.

54. Clearly, $f(x)$ is defined, if

$$25 - x^2 > 0$$

$$\Rightarrow \quad -5 < x < 5$$

Let $y = \log_5(25 - x^2)$, then

$$5^y = 25 - x^2$$

$$\Rightarrow \quad x^2 = 25 - 5^y$$

$$\Rightarrow \quad x = \pm\sqrt{25 - 5^y}$$

For x to be real, we must have

$$25 - 5^y \geq 0$$

$$\Rightarrow \quad 5^y \leq 25$$

$$\Rightarrow \quad y \leq 2$$

Also, $y = f(x) \to -\infty$ as $x \to \pm 5$.

Hence, range $(f) = (-\infty, 2]$

55. We know that, the image of the point (x_1, y_1, z_1) in the plane $ax + by + cz + d = 0$ is given by

$$\frac{x - x_1}{a} = \frac{y - y_1}{b} = \frac{z - z_1}{c} = \frac{-2(ax_1 + by_1 + cz_1 + d)}{a^2 + b^2 + c^2}$$

So, the image of the point $P(1, 3, 4)$ in the plane $2x - y + z + 3 = 0$ is given by

$$\frac{x - 1}{2} = \frac{y - 3}{-1} = \frac{z - 4}{1} = \frac{-2(2 - 3 + 4 + 3)}{4 + 1 + 1}$$

$$\Rightarrow \quad \frac{x - 1}{2} = \frac{y - 3}{-1} = \frac{z - 4}{1} = -2$$

$$\Rightarrow \quad x = -3, y = 5, z = 2$$

Hence, the required point is $(-3, 5, 2)$.

56. The lines are parallel to the vectors $\mathbf{b}_1 = \hat{\mathbf{i}} + 2\hat{\mathbf{j}} - \hat{\mathbf{k}}$ and $\mathbf{b}_2 = -\hat{\mathbf{i}} + \hat{\mathbf{j}} - 2\hat{\mathbf{k}}$.

$\therefore$ The plane normal to the vector $\mathbf{n}$ is

$$\mathbf{b}_1 \times \mathbf{b}_2 = \begin{vmatrix} \hat{\mathbf{i}} & \hat{\mathbf{j}} & \hat{\mathbf{k}} \\ 1 & 2 & -1 \\ -1 & 1 & -2 \end{vmatrix} = -3\hat{\mathbf{i}} + 3\hat{\mathbf{j}} + 3\hat{\mathbf{k}}$$

The required plane passes through $(\hat{\mathbf{i}} + \hat{\mathbf{j}})$ and is normal to the vector $\mathbf{n}$.

$\therefore$ Its equation is

$$\mathbf{r} \cdot \mathbf{n} = \mathbf{a} \cdot \mathbf{n}$$

$$\Rightarrow \quad \mathbf{r} \cdot (-3\hat{\mathbf{i}} + 3\hat{\mathbf{j}} + 3\hat{\mathbf{k}}) = (\hat{\mathbf{i}} + \hat{\mathbf{j}}) \cdot (-3\hat{\mathbf{i}} + 3\hat{\mathbf{j}} + 3\hat{\mathbf{k}})$$

$$\Rightarrow \quad \mathbf{r} \cdot (-3\hat{\mathbf{i}} + 3\hat{\mathbf{j}} + 3\hat{\mathbf{k}}) = -3 + 3$$

$$\Rightarrow \quad \mathbf{r} \cdot (-\hat{\mathbf{i}} + \hat{\mathbf{j}} + \hat{\mathbf{k}}) = 0$$

57. Let $P(3, 8, 2)$ be the given point.

Let the equation of a line through $P(3, 8, 2)$ and parallel to the plane $3x + 2y - 2z + 15 = 0$ be

$$\frac{x - 3}{a} = \frac{y - 8}{b} = \frac{z - 2}{c} \quad \ldots(i)$$

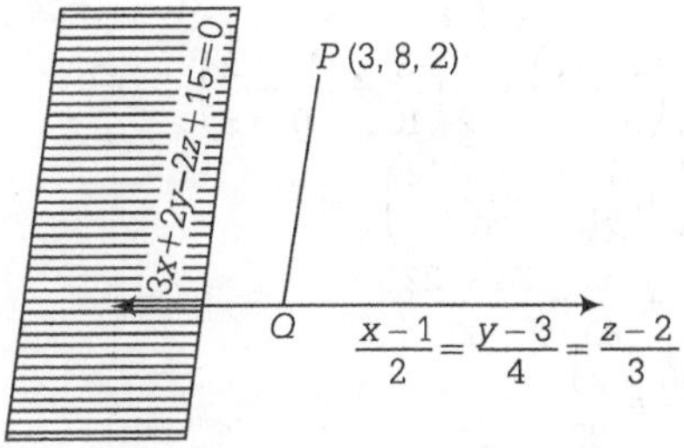

Then, normal to the plane is perpendicular to the parallel line.

Then, $\quad 3a + 2b - 2c = 0 \quad \ldots(ii)$

Line (i) intersects the line $\frac{x - 1}{2} = \frac{y - 3}{4} = \frac{z - 2}{3}$ at Q.

$$\therefore \quad \begin{vmatrix} 3 - 1 & 8 - 3 & 2 - 2 \\ a & b & c \\ 2 & 4 & 3 \end{vmatrix} = 0$$

$$\Rightarrow \quad \begin{vmatrix} 2 & 5 & 0 \\ a & b & c \\ 2 & 4 & 3 \end{vmatrix} = 0$$

$$\Rightarrow \quad 2(3b - 4c) - 5(3a - 2c) = 0$$

$$\Rightarrow \quad 15a - 6b - 2c = 0 \quad \ldots(iii)$$

On solving Eqs. (ii) and (iii) by cross-multiplication, we get

$$\frac{a}{-4 - 12} = \frac{b}{-30 + 6} = \frac{c}{-18 - 30}$$

$$\Rightarrow \quad \frac{a}{-16} = \frac{b}{-24} = \frac{c}{-48} \Rightarrow \frac{a}{2} = \frac{b}{3} = \frac{c}{6}$$

Substituting a, b and c in Eq. (i), we get

$$\frac{x - 3}{2} = \frac{y - 8}{3} = \frac{z - 2}{6} = \lambda \quad \text{[say]}$$

Let the coordinates of Q be $(2\lambda + 3, 3\lambda + 8, 6\lambda + 2)$. It lies on the line $\frac{x - 1}{2} = \frac{y - 3}{4} = \frac{z - 2}{3}$

$$\therefore \quad \lambda + 1 = \frac{3\lambda + 5}{4} = 2\lambda$$

$$\Rightarrow \quad \lambda = 1$$

$\therefore$ Coordinate of $Q = (2 + 3, 3 + 8, 6 + 2)$ i.e. $Q = (5, 11, 8)$.

Hence, $PQ = \sqrt{(5 - 3)^2 + (11 - 8)^2 + (8 - 2)^2}$

$$= \sqrt{4 + 9 + 36} = 7$$

58. By total probability theorem,

$$\text{Required probability} = \frac{1}{2} \times \left(\frac{3}{5} \times 1 + \frac{2}{5} \times \frac{1}{2}\right) + \frac{1}{2}\left(\frac{{}^3C_2}{{}^5C_2} \times 1 + \frac{{}^2C_2}{{}^5C_2} \times \frac{1}{3} + \frac{{}^3C_1 \times {}^2C_1}{{}^5C_2} \times \frac{2}{3}\right)$$

$$= \frac{1}{2}\left(\frac{3}{5} + \frac{1}{5}\right) + \frac{1}{2}\left(\frac{3}{10} + \frac{1}{30} + \frac{2}{5}\right)$$

$$= \frac{2}{5} + \frac{1}{2}\left(\frac{9 + 1 + 12}{30}\right)$$

$$= \frac{2}{3} + \frac{22}{60} = \frac{24 + 22}{60}$$

$$= \frac{46}{60} = \frac{23}{30}$$

59. We have,

$$\sum_{k=1}^{16} D_k = \begin{vmatrix} a & \left(\sum_{k=1}^{16} 2^k\right) & 2^{16} - 1 \\ b & 3\left(\sum_{k=1}^{16} 4^4\right) & 2\,(4^{16} - 1) \\ c & 7\left(\sum_{k=1}^{16} 8^k\right) & 4\,(8^{16} - 1) \end{vmatrix}$$

$$\Rightarrow \sum_{k=1}^{16} D_k = \begin{vmatrix} a & 2\left(\frac{2^{16} - 1}{2 - 1}\right) & 2^{16} - 1 \\ b & 3 \times 4\left(\frac{4^{16} - 1}{4 - 1}\right) & 2\,(4^{16} - 1) \\ c & 7 \times 8\left(\frac{8^{16} - 1}{8 - 1}\right) & 4\,(8^{16} - 1) \end{vmatrix}$$

$$\Rightarrow \sum_{k=1}^{16} D_k = \begin{vmatrix} a & 2\,(2^{16} - 1) & (2^{16} - 1) \\ b & 4\,(4^{16} - 1) & 2\,(4^{16} - 1) \\ c & 8\,(8^{16} - 1) & 4\,(8^{16} - 1) \end{vmatrix}$$

$$\Rightarrow \sum_{k=1}^{16} D_k = 2\begin{vmatrix} a & (2^{16} - 1) & (2^{16} - 1) \\ b & 2\,(4^{16} - 1) & 2\,(4^{16} - 1) \\ c & 4\,(8^{16} - 1) & 4\,(8^{16} - 1) \end{vmatrix}$$

[taking 2 common from C_2]

$$\Rightarrow \sum_{k=1}^{16} D_k = 2 \times 0 = 0$$

60. We have,

$$x^2 - 4x + 4y^2 = 12$$

$$\Rightarrow (x - 2)^2 + 4\,(y - 0)^2 = 8$$

$$\Rightarrow \frac{(x - 2)^2}{(2\sqrt{2})^2} + \frac{(y - 0)^2}{(\sqrt{2})} = 1$$

which is an ellipse whose major and minor axes are $2a = 2\sqrt{2}$ and $2b = \sqrt{2}$ repectively.

$\therefore$ Its eccentricity e is given by

$$e = \sqrt{1 - \frac{b^2}{a^2}}$$

$$= \sqrt{1 - \frac{2}{8}} = \frac{\sqrt{3}}{2}$$

61. Let $I = \int \frac{1}{\sin\left(x - \frac{\pi}{3}\right)\cos x}\,dx$

$$= \frac{1}{\cos\frac{\pi}{3}} \int \frac{\cos\left\{x - \left(x - \frac{\pi}{3}\right)\right\}}{\sin\left(x - \frac{\pi}{3}\right)\cos x}\,dx$$

$$= 2\int \frac{\cos x \cos\left(x - \frac{\pi}{3}\right) + \sin x \sin\left(x - \frac{\pi}{3}\right)}{\sin\left(x - \frac{\pi}{3}\right)\cos x}\,dx$$

$$= 2\int \left\{\cot\left(x - \frac{\pi}{3}\right) + \tan x\right\} dx$$

$$= 2\left\{\log\left|\sin\left(x - \frac{\pi}{3}\right)\right| + \log|\sec x|\right\} + C$$

$$= 2\log\left|\sin\left(x - \frac{\pi}{3}\right)\sec x\right| + C$$

62. The equation of any tangent to $y^2 = 8ax$ is

$$y = mx + \frac{2a}{m} \quad \text{...(i)}$$

If it touches $x^2 + y^2 = 2a^2$, then

$$\left(\frac{2a}{m}\right)^2 = 2a^2(1 + m^2)$$

$$[\because c^2 = a^2(1 + m^2)]$$

$$\Rightarrow 2 = m^2\,(m^2 + 1)$$

$$\Rightarrow m^4 + m^2 - 2 = 0$$

$$\Rightarrow (m^2 + 2)\,(m^2 - 1) = 0$$

$$\Rightarrow m^2 - 1 = 0$$

$$\Rightarrow m = \pm 1$$

Putting the values of m in Eq. (i), we get $y = \pm(x + 2a)$ as the equations of common tangents.

63. The equation of the plane through the intersection of the planes $x + y + z = 1$ and $2x + 3y - z + 4 = 0$ is

$$(x + y + z - 1) + \lambda (2x + 3y - z + 4) = 0$$

or $(2\lambda + 1) x + (3\lambda + 1) y + (1 - \lambda) z + 4\lambda - 1 = 0$ ...(i)

It is parallel to X-axis.

i.e. $\frac{x}{1} = \frac{y}{0} = \frac{z}{0}$

$\therefore \; 1 (2\lambda + 1) + 0 \times (3\lambda + 1) + 0 (1 - \lambda) = 0$

$\Rightarrow \quad \lambda = -\frac{1}{2}$

On substituting $\lambda = -\frac{1}{2}$ in Eq. (i), we get

$$\left(2 \times -\frac{1}{2} + 1\right)x + \left(3 \times \frac{-1}{2} + 1\right)y + \left(1 + \frac{1}{2}\right)z + 4 \times \left(-\frac{1}{2}\right) - 1 = 0$$

$\Rightarrow \quad \frac{-1}{2}y + \frac{3}{2}z - 3 = 0$

$\Rightarrow \quad y - 3z + 6 = 0$

which is the equation of the required plane.

64. We have, $\mathbf{a} \times \mathbf{b} = \begin{vmatrix} \hat{\mathbf{i}} & \hat{\mathbf{j}} & \hat{\mathbf{k}} \\ 2 & 1 & -2 \\ 1 & 1 & 0 \end{vmatrix} = 2\hat{\mathbf{i}} - 2\hat{\mathbf{j}} + \hat{\mathbf{k}}$

and $|\mathbf{a} \times \mathbf{b}| = \sqrt{2^2 + 2^2 + 1^2} = \sqrt{4 + 4 + 1} = 3$

$\therefore \quad |(\mathbf{a} \times \mathbf{b}) \times \mathbf{c}| = |\mathbf{a} \times \mathbf{b}|\,|\mathbf{c}| \sin 30^\circ = \frac{3}{2}|\mathbf{c}|$

Given, $|\mathbf{c} - \mathbf{a}| = 2\sqrt{2}$

$\Rightarrow \quad |\mathbf{c} - \mathbf{a}|^2 = 8$

$\Rightarrow \quad |\mathbf{c}|^2 + |\mathbf{a}|^2 - 2(\mathbf{a} \cdot \mathbf{c}) = 8$

$\Rightarrow \quad |\mathbf{c}|^2 + 9 - 2|\mathbf{c}| = 8$

$\Rightarrow \quad |\mathbf{c}|^2 - 2|\mathbf{c}| + 1 = 0$

$\Rightarrow \quad |\mathbf{c}| = 1$

Hence, $|(\mathbf{a} \times \mathbf{b}) \times \mathbf{c}| = \frac{3}{2}$

65. The given differential equation can be written as

$$\left(\frac{d^2y}{dx^2}\right)^4 = \left\{y + \left(\frac{dy}{dx}\right)^2\right\}$$

Clearly, its order and degree are 2 and 4 respectively.

66. Let h be the height, r be the radius of the base and V be the volume of the cone at time t. Then,

$$V = \frac{1}{3}\pi r^2 h$$

$\Rightarrow \quad V = \frac{2}{3}\pi r^3 \quad [\because h = 2r, \text{ given}]$

On differentiating both sides w.r.t. t, we get

$$\frac{dV}{dt} = 2\pi r^2 \frac{dr}{dt}$$

$\Rightarrow \quad 40 = 2 (10)^4 \frac{dr}{dt}$

$[\because \pi r^2 = 1 \text{ m}^2 = 10^4 \text{ cm}^2 \text{ and } \frac{dV}{dt} = 40 \text{ cm}^3/\text{s, given}]$

$\Rightarrow \quad \frac{dr}{dt} = \frac{2}{1000} \text{ cm/s} = 0.002 \text{ cm/s}$

67. Let S be the sum of the given series.

i.e. $S = 1 + \frac{2}{3} + \frac{6}{3^2} + \frac{10}{3^3} + \ldots$

$\Rightarrow \quad (S - 1) = \frac{2}{3} + \frac{6}{3^2} + \frac{10}{3^3} + \ldots$ ...(i)

$\Rightarrow \quad (S - 1) \times \frac{1}{3} = \frac{2}{3^2} + \frac{6}{3^3} + \frac{10}{3^4} + \ldots$ ...(ii)

On subtracting Eq. (ii) from Eq. (i), we get

$$\frac{2}{3}(S - 1) = \frac{2}{3} + \frac{4}{3^2} + \frac{4}{3^3} + \frac{4}{3^4} + \ldots$$

$\Rightarrow \quad \frac{2}{3}(S - 1) = \frac{2}{3} + \frac{\frac{4}{3^2}}{1 - \frac{1}{3}}$

$\Rightarrow \quad \frac{2}{3}(S - 1) = \frac{2}{3} + \frac{2}{3}$

$\Rightarrow \quad S - 1 = 2$

$\Rightarrow \quad S = 3$

68. Let the equation of the line be

$$\frac{x}{a} + \frac{y}{b} = 1 \quad \ldots(i)$$

which passes through (4, 3).

$\therefore \quad \frac{4}{a} + \frac{3}{b} = 1$ ...(ii)

It is given that $a + b = -1$ ...(iii)

On solving Eqs. (i) and (ii), we get

$$a = -2, b = 1 \text{ or } a = 2, b = -3$$

On substituting the values of a and b in Eq. (i), we get

$$\frac{x}{-2} + \frac{y}{1} = 1 \text{ and } \frac{x}{2} + \frac{y}{-3} = 1$$

69. Given points will be collinear, if

$$\begin{vmatrix} a & b & 1 \\ b & a & 1 \\ a^2 & -b^2 & 1 \end{vmatrix} = 0$$

$$\Rightarrow \begin{vmatrix} a & b & 1 \\ b-a & a-b & 0 \\ a^2-a & -b^2-b & 0 \end{vmatrix} = 0$$

[applying $R_2 \to R_2 - R_1$, $R_3 \to R_3 - R_1$]

$$\Rightarrow (a-b)\begin{vmatrix} a & b & 1 \\ -1 & 1 & 0 \\ a^2-a & -b^2-b & 0 \end{vmatrix} = 0$$

$$\Rightarrow (a-b)(b^2+b-a^2+a) = 0$$

$$\Rightarrow (a-b)\{(a+b)-(a^2-b^2)\} = 0$$

$$\Rightarrow (a-b)(a+b)(1-a+b) = 0$$

$$\Rightarrow a = b \text{ or } a+b=0 \text{ or } a = 1+b$$

70. We have,

$$x^2 + 9 < (x+3)^2 < 8x + 25$$

$$\Rightarrow x^2 + 9 < x^2 + 6x + 9 < 8x + 25$$

$$\Rightarrow x^2 + 9 < x^2 + 6x + 9$$

and $$x^2 + 6x + 9 < 8x + 25$$

$$\Rightarrow 6x > 0 \text{ and } x^2 - 2x - 16 < 0$$

$$\Rightarrow x > 0 \text{ and } 1 - \sqrt{17} < x < 1 + \sqrt{17}$$

$$\Rightarrow 0 < x < 1 + \sqrt{17}$$

$$\therefore x = 1, 2, 3, 4, 5$$

Solved Paper 2014

Manipal

Engineering Entrance Exam

Physics

1. The equation $\left(p + \frac{a}{V^2}\right)(V - b) = $ constant. The unit of a is

(a) Dyne × cm^5 (b) Dyne × cm^4
(c) Dyne/cm^3 (d) Dyne/cm^2

2. If L, C and R denote the inductance, capacitance and resistance respectively, the dimensional formula for C^2LR is

(a) $[ML^{-2}T^{-1}I^0]$ (b) $[M^0L^0T^3I^0]$
(c) $[M^{-1}L^{-2}T^6I^2]$ (d) $[M^0L^0T^2I^0]$

3. Number of particles is given by $n = -D\frac{n_2 - n_1}{x_2 - x_1}$ crossing a unit area perpendicular to X-axis is unit time, where n_1 and n_2 are number of particles per unit volume for the value of x meant to x_2 and x_1. Find dimensions of D called as diffusion constant.

(a) $[M^0LT^2]$ (b) $[M^0L^2T^{-4}]$ (c) $[M^0LT^{-3}]$ (d) $[M^0L^2T^{-1}]$

4. A body starts from the origin and moves along the x-axis such that velocity at any instant is given by $(4t^3 - 2t)$, where t is in second and velocity is in m/s. What is the acceleration of the particle, when it is $2m$ from the origin?

(a) 28 m/s^2 (b) 22 m/s^2
(c) 12 m/s^2 (d) 10 m/s^2

5. The acceleration-time graph of a body is shown below

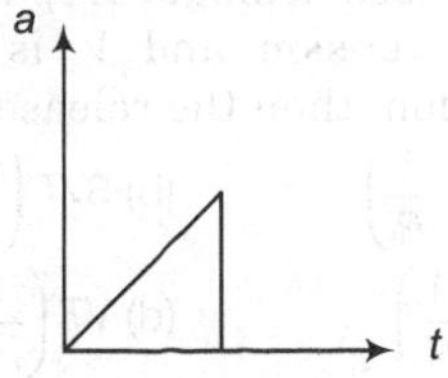

The most probable velocity-time graph of the body is

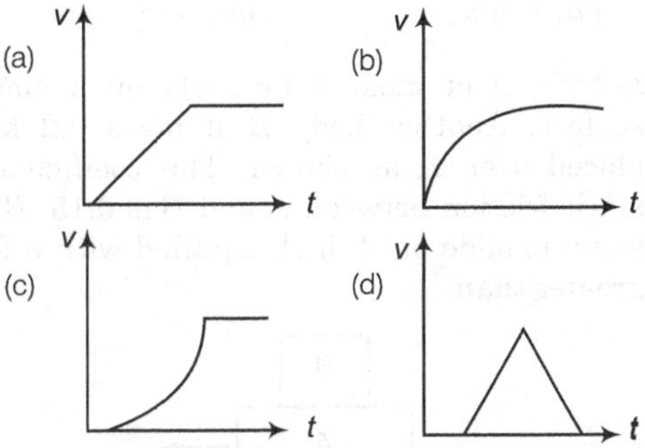

6. Two seconds after projection, a projectile is travelling in a direction inclined at 30° to the horizontal. After one more sec, it is travelling horizontally, the magnitude and direction of its velocity are

(a) $2\sqrt{20}$ m/s, 60° (b) $20\sqrt{3}$ m/s, 60°
(c) $6\sqrt{40}$ m/s, 30° (d) $40\sqrt{6}$ m/s, 30°

7. The kinetic energy k of a particle moving along a circle of radius R depends on the distance covered. It is given as KE $= as^2$, where a is a constant. The force acting on the particle is

(a) $2a\frac{s^2}{R}$ (b) $2as\left(1 + \frac{s^2}{R^2}\right)^{1/2}$ (c) $2as$ (d) $2a\frac{R^2}{s}$

8. Three blocks of masses 2 kg, 3 kg and 5 kg are connected to each other with light string and are then placed on a frictionless surface as shown in the figure. The system is pulled by a force $F = 10$ N, then tension $T_1 =$

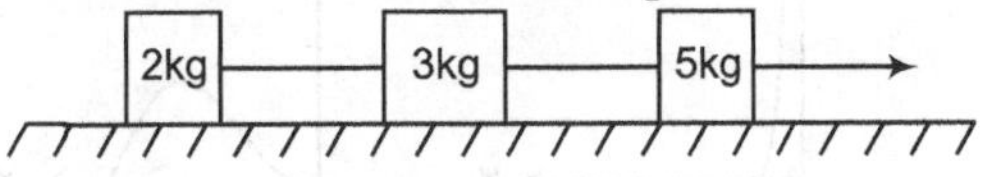

(a) 1 N (b) 5 N (c) 8 N (d) 10 N

9. The acceleration of block B in the figure will be

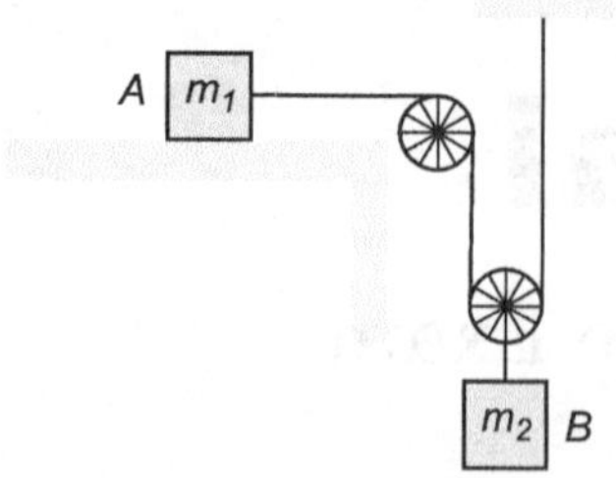

(a) $\dfrac{m_2 g}{(4m_1 + m_2)}$ (b) $\dfrac{2m_2 g}{(4m_1 + m_2)}$
(c) $\dfrac{2m_1 g}{(m_1 + 4m_2)}$ (d) $\dfrac{2m_1 g}{(m_1 + m_2)}$

10. A body A of mass 1 kg rests on a smooth surface. Another body B of mass 0.2 kg is placed over A as shown. The coefficient of static friction between A and B is 0.15. B will begin to slide on A, if A is pulled with a force greater than

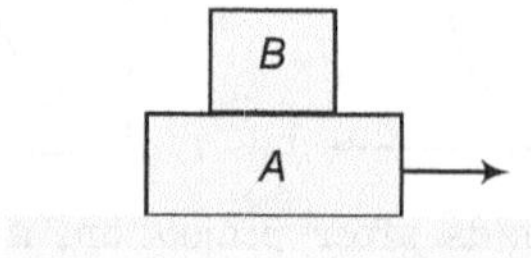

(a) 1.764 N (b) 0.1764 N
(c) 0.3 N (d) It will not slide for any F

11. A particle which is constrained to move along the x-axis, is subjected to a force in the same direction which varies with the distance x of the particle from the origin is $F(x) = -kx + ax^3$. Here, k and a are positive constants. For $x \geq 0$, the functional from of the potential energy $U_{(x)}$ of the particle is

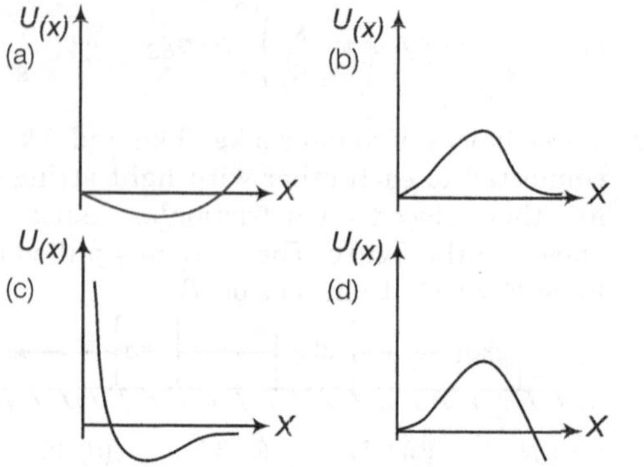

12. Four thin rods of same mass M and same length l, form a square as shown in figure. Moment of inertia of this system about an axis through centre O and perpendicular to its plane is

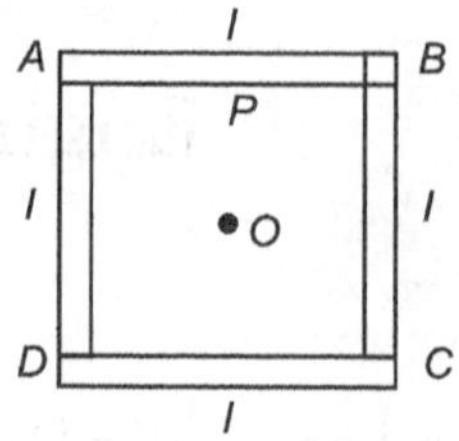

(a) $\frac{4}{3}Ml^2$ (b) $\frac{Ml^2}{3}$ (c) $\frac{Ml^2}{6}$ (d) $\frac{2}{3}Ml^2$

13. A wheel is rotating with an angular speed of 20 rad/s. It is stopped to rest by applying a constant torque in 4s. If the moment of inertia of the wheel about its axis is 0.20 kg/m^2, then the work done by the torque in two seconds will be

(a) 10 J (b) 20 J
(c) 30 J (d) 40 J

14. The strain stress curves of three wires of different materials are shown in the figure. P, Q and R are the elastic limits of the wires. The figure shows that

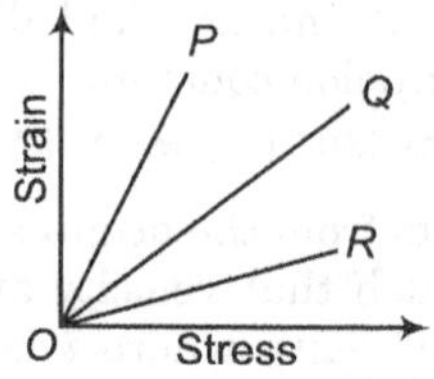

(a) elasticity of wire P is maximum
(b) elasticity of wire Q is maximum
(c) tensile strength of R is maximum
(d) None of the above is true

15. Several spherical drops of a liquid of radius r coalesce to form a single drop to radius R. If T is surface tension and V is volume under consideration, then the release of energy is

(a) $3VT\left(\frac{1}{r} + \frac{1}{R}\right)$ (b) $3VT\left(\frac{1}{r} - \frac{1}{R}\right)$
(c) $VT\left(\frac{1}{r} - \frac{1}{R}\right)$ (d) $VT\left(\frac{1}{r^2} + \frac{1}{R^2}\right)$

16. Two substances of densities ρ_1 and ρ_2 are mixed in equal volume and the relative density of mixture is 4. When they are mixed in equal masses, the relative density of the mixture is 3. The values of ρ_1 and ρ_2 are

(a) $\rho_1 = 6$ and $\rho_2 = 2$ (b) $\rho_1 = 3$ and $\rho_2 = 5$
(c) $\rho_1 = 12$ and $\rho_2 = 4$ (d) None of these

17. The pressure p, volume V and temperature T of a gas in the jar A and the other gas in the jar B at pressure $2p$, volume $V/4$ and temperature $2T$, then the ratio of the number of molecules in the jar A and B will be

(a) 1 : 1 (b) 1 : 2 (c) 2 : 1 (d) 4 : 1

18. Two identical containers each of volume V_0 are joined by a small pipe. The containers contain identical gases at temperature T_0 and pressure P_0. One container is heated to temperature $2T_0$ while maintaining the other at the same temperature. The common pressure of the gas is P and n is the number of moles of gas in container at temperature $2T_0$

(a) $p = 2p_0$ (b) $p = \frac{4}{3}p_0$
(c) $n = \frac{2}{3}\frac{p_0V_0}{RT_0}$ (d) $n = \frac{3}{2}\frac{p_0V_0}{RT_0}$

19. An experiment is carried on a fixed amount of gas at different temperature and at high pressure such that it deviates from the ideal gas behaviour. The variation of $\frac{pV}{RT}$ with p is shown in the diagram. The correct variation will correspond to

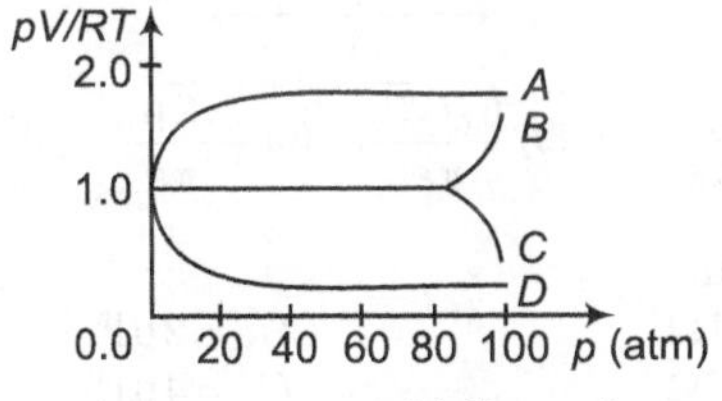

(a) Curve A (b) Curve B
(c) Curve C (d) Curve D

20. Liquid is filled in a flask upto a certain point. When, the flask is heated, the level of the liquid

(a) immediately starts increasing
(b) initially falls and then rises
(c) rises abruptly
(d) falls abruptly

21. Two rods (one semi-circular and other straight) of same material and of same cross-sectional area are joined as shown in the figure. The points A and B are maintained at different temperature. The ratio of the heat transferred through a cross-section of a semi-circular rod to the heat transferred through a cross-section of the straight rod in a given time is

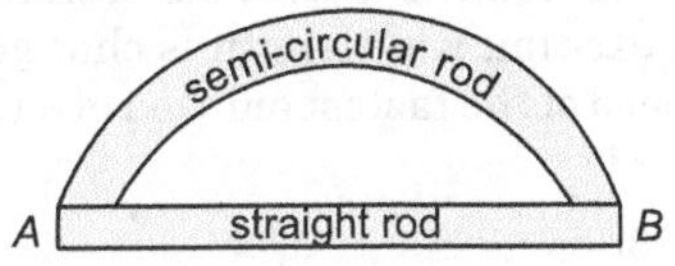

(a) $2 : \pi$ (b) 1:2
(c) $\pi : 2$ (d) 3:2

22. Two particles executes SHM of same amplitude and frequency along the same straight line. They pass one another when going in opposite directions. Each time their displacement is half of their amplitude. The phase difference between them is

(a) 30° (b) 60°
(c) 90° (d) 120°

23. In the following circuit reading of voltmeter V is

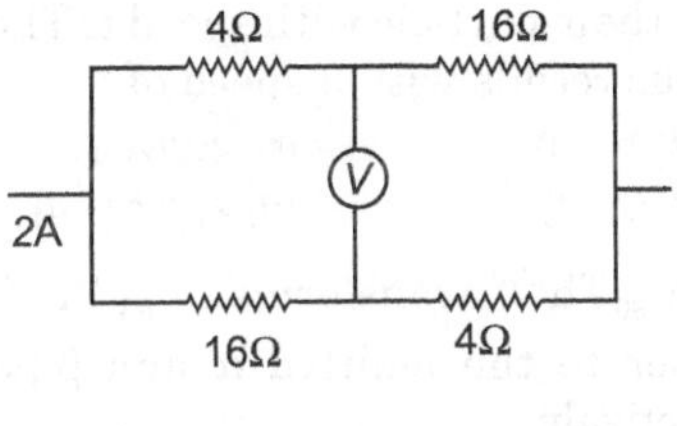

(a) 12 V (b) 8 V
(c) 20 V (d) 16 V

24. The amount of work done in increasing the voltage across the plates of a capacitor from 5V to 10V is W. The work done in increasing it from 10 V to 5 V will be

(a) 0.6 W
(b) 1 W
(c) 1.25 W
(d) 1.67 W

25. The de-Broglie wavelength of a neutron at 27°C is λ. What will be its wavelength at 927°C ?

(a) $\lambda/2$ (b) $\lambda/3$
(c) $\lambda/4$ (d) $\lambda/9$

26. The minimum intensity of light to be detected by human eye is 10^{-10} W/m^2. The number of photons of wavelength 5.6×10^{-7} m entering the eye, with pupil area 10^{-6} m^2, per second for vision will be nearly

(a) 100 (b) 200
(c) 300 (d) 400

27. In a photoemissive cell with exciting wavelength λ, the fastest electron has speed v. If the exciting wavelength is changed by $3\lambda/4$, the speed of the fastest emitted electron will be

(a) $v(3/4)^{1/2}$
(b) $v(4/3)^{1/2}$
(c) less than $v(4/3)^{1/2}$
(d) greater than $v(4/3)^{1/2}$

28. The ratio of the wavelength for $2 \rightarrow 1$ transition in Li^+, He^+ and H is

(a) 1 : 2 : 3 (b) 1 : 4 : 9
(c) 4 : 9 : 36 (d) 3 : 2 : 1

29. The fraction of atoms of radioactive element that decays in 6 days is 7/8. The fraction that decays in 10 days will be

(a) 77/80 (b) 71/80 (c) 31/32 (d) 15/16

30. Nucleus of mass number A, originally at rest, emits the α-particle with speed v. The daughter nucleus recoils with a speed of

(a) $2v/(A+4)$ (b) $4v/(A+4)$
(c) $4v/(A-4)$ (d) $2v/(A-4)$

31. When ${}_{90}Th^{228}$ transforms to ${}_{83}Bi^{212}$, then the number to the emitted α and β-particles is, respectively

(a) $8\alpha, 7\beta$ (b) $4\alpha, 7\beta$
(c) $4\alpha, 4\beta$ (d) $4\alpha, 1\beta$

32. Two Cu^{64} nuclei touch each other. The electrostatics repulsive energy of the system will be

(a) 0.788 MeV (b) 7.88 MeV
(c) 126.15 MeV (d) 788 MeV

33. A short bar magnet is placed with its south pole towards geographical north. The neutral points are situated at a distance of 20 cm from the centre of the magnet. If $B_H = 0.3 \times 10^{-4}$ Wb/m^2 then the magnetic moment of the magnet is

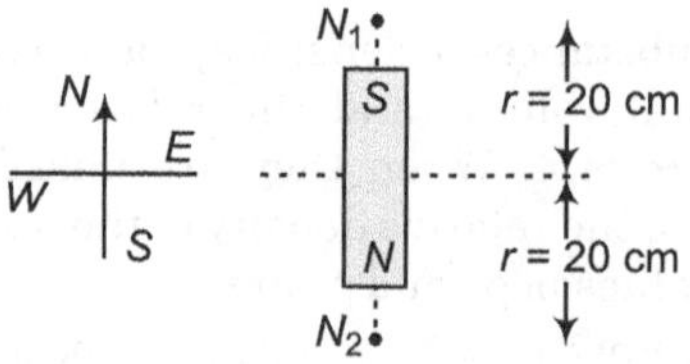

(a) 9000ab – amp × cm^2 (b) 900ab – amp × cm^2
(c) 1200ab – amp × cm^2 (d) 225ab – amp × cm^2

34. A magnet makes 40 oscillations per minute at a place having magnetic field intensity $B_H = 0.1 \times 10^{-5}$. At another place, it takes 2.5 s to complete one vibration. The value of the earth's horizontal field at the place

(a) 0.25×10^{-6} T (b) 0.36×10^{-6} T
(c) 0.66×10^{-8} T (d) 1.2×10^{-6} T

35. The coereivity of a small bar magnet is 4×10^3 A/m. It is inserted inside a solenoid of 500 turns and length 1 m to demagnetise it. The amount of current to be passed through the solenoid will be

(a) 2.5 A (b) 5 A (c) 8 A (d) 10 A

36. A wire in the form of a square of side a carries a current i. Then, the magnetic induction at the centre of the square wire is (magnetic permeability of free space $= \mu_0$).

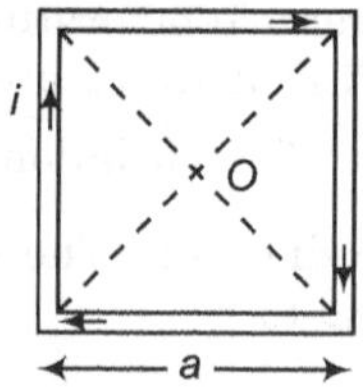

(a) $\frac{\mu_0 i}{2\pi a}$ (b) $\frac{\mu_0 i\sqrt{2}}{\pi a}$ (c) $\frac{2\sqrt{2}\mu_0 i}{\pi a}$ (d) $\frac{\mu_0 i}{\sqrt{2}\pi a}$

37. Given,

$R_1 = 1\Omega$ $C_1 = 2\mu F$
$R_2 = 2\Omega$ $C_2 = 4\mu F$

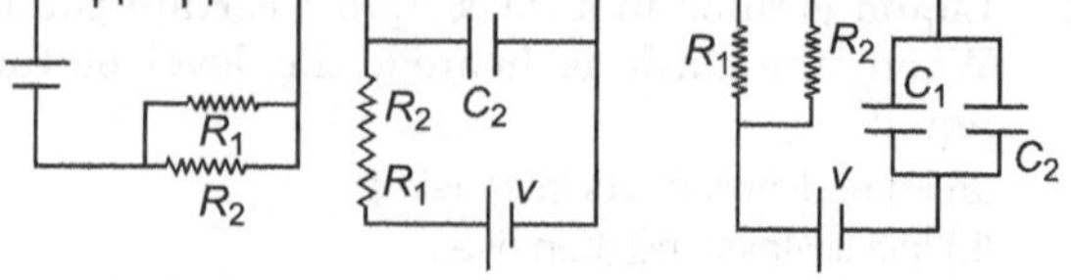

(I) (II) (III)

The time constant (in μs) for the circuits I, II, III are respectively

(a) 18, 8/9, 4 (b) 18, 4, 8/9
(c) 4, 8/9, 18 (d) 8/9, 18, 4

38. Two blocks A and B of masses $2m$ and m, respectively, are connected by a massless and inextensible string. The whole system is suspended by a massless spring as shown in the figure. The magnitudes of acceleration of A and B, immediately after the string is cut, are respectively

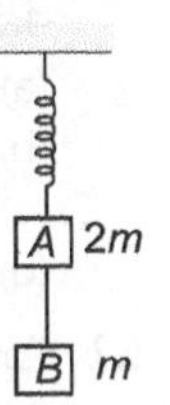

(a) $g, g/2$ (b) $g/2, g$
(c) g, g (d) $g/2, g/2$

39. A point object is placed at a distance of 20 cm from a thin plano-convex lens of focal length 15 cm, if the plane surface is silvered. The image will form at

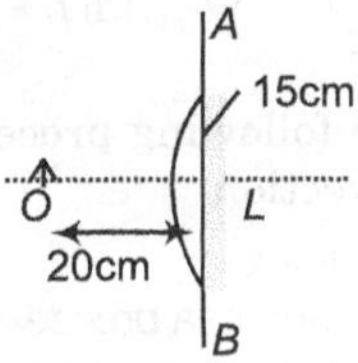

(a) 60 cm left of AB
(b) 30 cm left of AB
(c) 12 cm left of AB
(d) 60 cm right of AB

40. A biconvex lens of focal length f forms a circular image of sun of radius r in focal plane. Then,

(a) $\pi r^2 \propto f$
(b) $\pi r^2 \propto f^2$
(c) if lower half part is covered by black sheet, then area of the image is equal to $\pi r^2/2$
(d) if f is doubled, intensity will increase

41. Given a sample of radium-226 having half-life of 4 days. Find, the probability, a nucleus disintegrates after 2 half lives.

(a) 1 (b) 1/2
(c) 1.5 (d) 3/4

42. Graph of position of image *versus* position of point object from a convex lens is shown. Then, focal length of the lens is

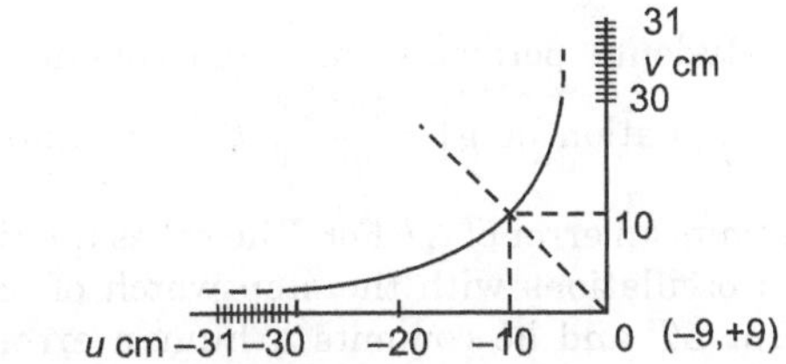

(a) 0.50 ± 0.05 cm (b) 0.50 ± 0.10 cm
(c) 5.00 ± 0.05 cm (d) 5.00 ± 0.10 cm

43. A massless rod is suspended by two identical strings AB and CD of equal length. A block of mass m is suspended from point O such that BO is equal to 'x'. Further, it is observed that the frequency of 1st harmonic (fundamental frequency) in AB is equal to 2nd harmonic frequency in CD. Then, length of BO is

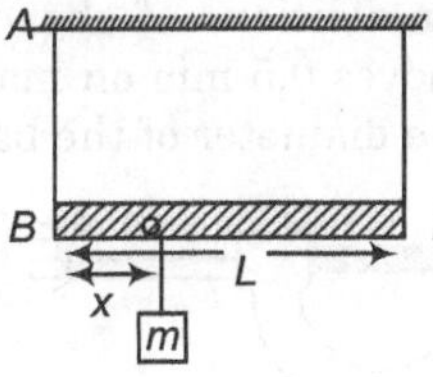

(a) $L/5$ (b) $4L/5$
(c) $3L/4$ (d) $L/4$

44. A system of binary stars of masses m_A and m_B are moving in circular orbits of radii r_A and r_B respectively. If T_A and T_B are the time periods of masses m_A and m_B respectively, then

(a) $\frac{T_A}{T_B} = \left(\frac{r_A}{r_B}\right)^{3/2}$
(b) $T_A > T_B$ (if $r_A > r_B$)
(c) $T_A > T_B$ (if $m_A > m_B$)
(d) $T_A = T_B$

45. A solid sphere of mass M, radius R and having moment of inertia about an axis passing through the centre of mass as I, is recast into a disc of thickness t, whose moment of inertia about an axis passing through its edge and perpendicular to its plane remains I. Then, radius of the disc will be

(a) $\frac{2R}{\sqrt{15}}$ (b) $R\sqrt{\frac{2}{15}}$
(c) $\frac{4R}{\sqrt{15}}$ (d) $\frac{R}{4}$

46. A student performs an experiment for determination of $g\left(=\dfrac{4\pi^2 l}{T^2}\right)$, $l \approx 1$ m, and he commits an error of Δl. For T he takes the time of n oscillations with the stop watch of least count ΔT and he commits a human error of 0.1 s. For which of the following data, the measurement of g will be most accurate?

	Δl	ΔT	n	Amplitude of oscillation
(a)	5 mm	0.2 s	10	5 mm
(b)	5 mm	0.2 s	20	5 mm
(c)	5 mm	0.1 s	20	1 mm
(d)	1 mm	0.1 s	50	1 mm

47. The circular divisions of shown screw gauge are 50. It moves 0.5 mm on main scale in one rotation. The diameter of the ball is

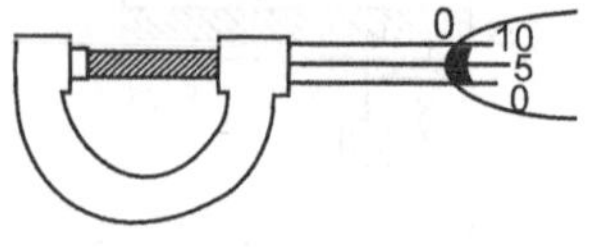

(a) 2.25 mm (b) 2.20 mm
(c) 1.20 mm (d) 1.25 mm

48. Consider the cylindrical element as shown in the figure. Current flowing the through element is I and resistivity of material of the cylinder is ρ. Choose the correct option out the following.

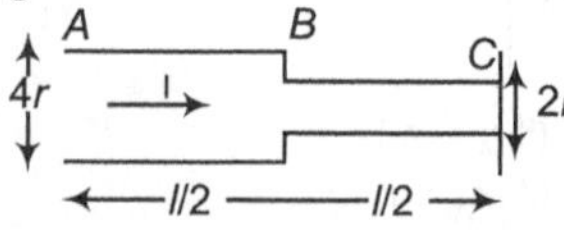

(a) Power loss in first half is four times the power loss in second half
(b) Voltage drop in first half is twice of voltage drop in second half
(c) Current density in both halves are equal
(d) Electric field in both halves is equal

49. Ideal gas is contained in a thermally insulated and rigid container and it is heated through a resistance 100 Ω by passing a current of 1 A for five minutes, then change in internal energy of the gas is

(a) 0 kJ (b) 30 kJ
(c) 10 kJ (d) 20 kJ

50. Two litre of water at initial temperature of 27°C is heated by a heater of power 1 kW. If the lid of kettle is opened, then heat is lost at the constant rate of 160 J/s. Find, the time required to raise the temperature of water to 77°C with the lid open (specific heat of water 4.2 kJ/kg)

(a) 5 min 40 s
(b) 14 min 20 s
(c) 8 min 20 s
(d) 16 min 10 s

51. An open organ pipe resonated with frequency f_1 and 2nd harmonic. Now, one end is closed and the frequency is slowly increased, then it resonates with frequency f_2 and nth harmonic then,

(a) $n = 3, f_2 = \frac{3}{4} f_1$ (b) $n = 5, f_2 = \frac{3}{4} f_1$
(c) $n = 3, f_2 = \frac{5}{4} f_1$ (d) $n = 5, f_2 = \frac{5}{4} f_1$

52. Which of the following process does not occur through convection

(a) boiling of water
(b) land breeze and sea breeze
(c) circulation of air around furnace
(d) heating of glass bulb through filament

53. T_1 is the time period of simple pendulum. The point of suspension moves vertically upwards according to $y = kt^2$, where $k = 1\,\text{m/s}^2$. New time period is T_2, then $\dfrac{T_1^2}{T_2^2} = ?\ (g = 10\,\text{m/s}^2)$

(a) 4/5 (b) 6/5
(c) 5/6 (d) 1

54. In the figure shown, a cubical block is held stationary against a rough wall by applying force 'F', then incorrect statement among the following is

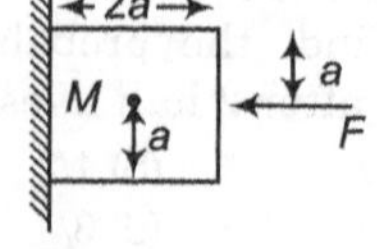

(a) frictional force $f = mg$
(b) $F = N$, N is normal reaction
(c) F does not apply any torque
(d) N does not apply any torque

55. Three infinitely charged sheets are kept parallel to x-y plane having charge densities as shown. Then, the value of electric field at 'P' is

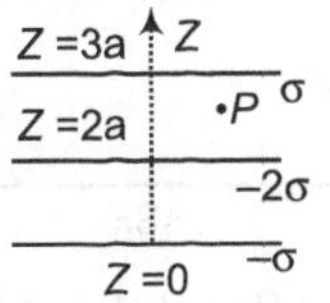

(a) $\frac{-4\sigma}{\varepsilon_0}\hat{k}$ (b) $\frac{4\sigma}{\varepsilon_0}\hat{k}$
(c) $\frac{-2\sigma}{\varepsilon_0}\hat{k}$ (d) $\frac{2\sigma}{\varepsilon_0}\hat{k}$

56. A circular disc of radius $\frac{R}{3}$ is cut from a circular disc of radius R and mass $9M$ as shown. Then, moment of inertia of remaining disc about 'O' perpendicular to the plane of the disc is

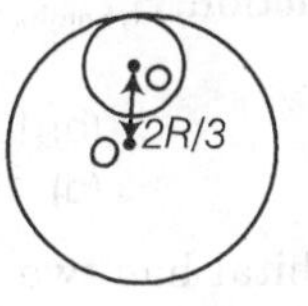

(a) $4\,MR^2$ (b) $9\,MR^2$
(c) $\frac{37}{9}\,MR^2$ (d) $\frac{40}{9}\,MR^2$

57. A cylindrical conducting rod is kept with its axis along positive z-axis, where a uniform magnetic field exists parallel to z-axis. The current induced in the cylinder is
(a) zero
(b) clockwise as seen from $+z$-axis
(c) anti-clockwise as seen from $+z$-axis
(d) opposite to the direction of magnetic field

58. Ratio of area of hole to beaker is 0.1. Height of liquid in beaker is 3m, and hole is at the height of 52.5 cm from the bottom of beaker, find the square of the velocity of liquid coming out from the hole
(a) 50 $(\text{m/s})^2$ (b) 50.5 $(\text{m/s})^2$
(c) 51 $(\text{m/s})^2$ (d) 42 $(\text{m/s})^2$

59. Temperature of a gas is 20°C and pressure is changed from 1.01×10^5 Pa to 1.165×10^5 Pa. If volume is decreased isothermally by 10%. Bulk modulus of gas is
(a) 1.55×10^5 (b) 0.155×10^5
(c) 1.4×10^5 (d) 1.01×10^5

60. The atomic number (Z) of an element whose k_a wavelength is λ is 11. The atomic number of an element whose k_a, wavelength is 4λ is equal to
(a) 6 (b) 11 (c) 44 (d) 4

Chemistry

1. Which of the following compounds will not react with Mg metal in the presence of ether to give a Grignard reagent?
(a) $CH_2{=}CHBr$
(b) CH_3CH_2I
(c) $CH_2OH\,CH_2Br$
(d) $CH_3OCH_2CH_2I$

2. How many enols (including stereo isomers) exist for 3-hexanone?
(a) Two (b) Three
(c) Four (d) Six

3. Which of the following processes does not yield a primary amine?
(a) $CH_3COCH_3 + NH_3 + H_2 \xrightarrow{Ni}$
(b) $CH_3CH_2CN + H_2 \xrightarrow[140°C]{Ni}$
(c) $CH_3CHO + H_2NOH \xrightarrow{Na,\ C_2H_5OH}$
(d) $CH_3CHO + N_2CH_2 \xrightarrow{NaBH_4}$

4. An aqueous solution of phosphoric acid, H_3PO_4 being titrated has molarity equal to 0.25 M. Which of the following could be normality of this solution?
(a) 0.25 N (b) 0.50 N
(c) 0.75 N (d) All of these

5. For the reaction represented by the equation,

$$CX_4 + 2O_2 \longrightarrow CO_2 + 2X_2O$$

9.0 g of CX_4 completely reacts with 1.74 g of oxygen. The approximate molar mass of X will be
(a) 20 (b) 40
(c) 60 (d) 80

6. Which of the following species contain an element in an oxidation state that is not a whole number?

(a) VO_4^{3-} (b) Mn_2O_3 (c) $S_4O_6^{2-}$ (d) Cl_2O_7

7. Gold is extracted by hydrometallurgical process, based on its property

(a) of being electropositive
(b) of being less reactive
(c) to form complexes which are water soluble
(d) to form salts which are water soluble

8. Due to lanthanide contraction

(a) Fe, Co, Ni have equal size
(b) Zr and Hf have equal size
(c) all *f*-block ions have equal size
(d) all isoelectronic ions have equal size

9. The element which forms ions in dimeric state is

(a) mercury (b) nickel (c) iron (d) cadmium

10. The high oxidising power of fluorine in due to its

(a) high electron affinity
(b) high heat of dissociation and low heat of hydration
(c) low heat of dissociation and high heat of hydration
(d) high heat of dissociation and high heat of hydration

11. Which one of the following is the weakest acid?

(a) HF (b) HCl (c) HBr (d) HI

12. Find the pressure of a sample of CCl_4 if 1.0 mole occupies 35L at 77°C. Assume CCl_4 obeys the van der Waals equation of state. ($a = 20\ L^2 mol^{-2}$ atm and $b = 0.14\ L\,mol^{-1}$)

(a) 0.56 atm (b) 0.572 atm
(c) 0.82 atm (d) 0.807 atm

13. KBr crystallises in NaCl type of crystal lattice and its density is 2.75 g/cm^3. Number of unit cells of KBr present in a 1.00 mm^3 grain of KBr are

(a) $3\cdot6\times10^{18}$ (b) $3\cdot6\times10^{16}$
(c) $2\cdot34\times10^{23}$ (d) $6\cdot02\times10^{24}$

14. A body centre cubic lattice is made up of two different types of atoms *A* and *B*. Atom *A* occupies the body centre and *B* occupying the corner positions. One of the corners is left unoccupied per unit cell. Empirical formula of such a solid is

(a) AB (b) A_2B_3 (c) A_5B_7 (d) A_8B_7

15. A substance is carried out through the following transformations

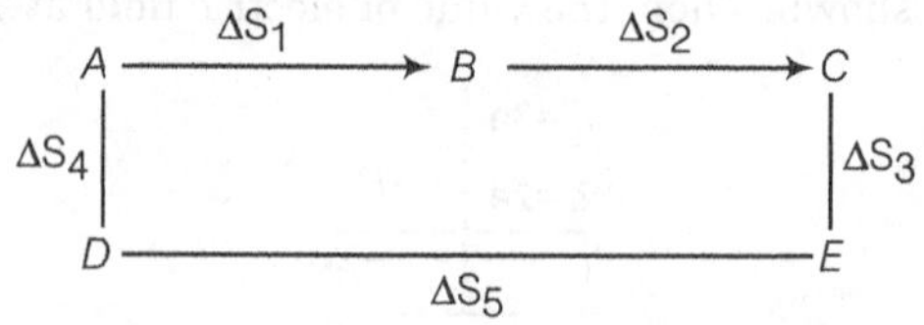

The equation $\Delta S_1 + \Delta S_2 = \Delta S_3 + \Delta S_4 + \Delta S_5$

(a) is true only if the steps are carried out reversibly
(b) is always true because entropy is a state function
(c) may be true but need more informations on the processes
(d) is incorrect

16. At 298 K the entropy of rhombic sulphur is 32.04 J/mol K and that of monoclinic sulphur is 32.68 J/mol K. The heat of their combustion are respectively–298246 and −297948 J mol^{-1}. ΔG for the reaction; $S_{rhombic} \longrightarrow S_{monoclinic}$ will be

(a) 107. 28 J (b) 10. 728 J
(c) 107. 28 kJ (d) 10728 J

17. An atomic orbital has two angular nodes and one radial node. It is a

(a) 2*p* orbital (b) 3*p* orbital
(c) 3*d* orbital (d) 4*d* orbital

18. Which of the following sets of quantum numbers could represent the last electron added to complete the electron configuration for a ground state atom of Br($Z = 35$) according to the Aufbau principle, $n\,;l\,;m_l\,;m_s$

(a) $4\,;0\,;0\,;-\frac{1}{2}$ (b) $4\,;1\,;1\,;-\frac{1}{2}$
(c) $3\,;1\,;1\,;-\frac{1}{2}$ (d) $4\,;1\,;2\,;+\frac{1}{2}$

19. Order the species CF^+, CF and CF^- according to increasing C—F bond length

(a) $CF^+ < CF < CF^-$ (b) $CF^+ < CF^- < CF$
(c) $CF^- < CF < CF^+$ (d) $CF < CF^- < CF^+$

20. When the substances CCl_4, CH_4 and CF_4 are arranged is order of increasing boiling point, what is the correct order?

(a) CCl_4, CF_4, CH_4
(b) CH_4, CF_4, CCl_4
(c) CF_4, CCl_4, CH_4
(d) CF_4, CH_4, CCl_4

21. The nitrite ion in represented by

(I) $[O-N=O]^-$

(II) $[N=O-O]^-$

Which of the structures represents possible resonance forms of this ion?

(a) Only I (b) Only II
(c) Both I and II (d) Neither I and II

22. Equal moles of $NO_2(g)$ and $NO(g)$ are to be placed in a container to produce N_2O according to the reaction,

$$NO_2 + NO \rightleftharpoons N_2O + O_2;\ K_C = 0.914$$

How many moles of NO_2 and NO be placed in the 5.0 L container to have an equilibrium concentration of N_2O to be 0.05 M?

(a) 0.5115 (b) 0.1023 (c) 0.0526 (d) 0.2046

23. The rate of a chemical reaction involving liquids only will not be influenced by the

(a) temperature of the system
(b) number of reacting particles per mL
(c) stability of the bonds
(d) pressure of the system

24. Consider the following reaction in aqueous solution.

$$5Br^-(aq) + BrO_3^-(aq) + 6H^+(aq) \longrightarrow 3Br_2(aq) + 3H_2O(l)$$

If the rate of appearance of Br_2 at a particular moment during the reaction is 0.025 Ms^{-1}, what is the rate of disappearance (in Ms^{-1}) of Br^- at that moment?

(a) 0.025 ms^{-1} (b) 0.042 ms^{-1}
(c) 0.075 ms^{-1} (d) 0.125 ms^{-1}

25. Shown below are four possible synthesis of propionaldehyde. Which one would not work?

(a) 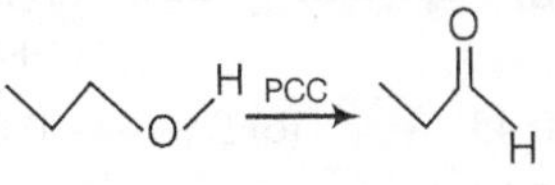

(b) 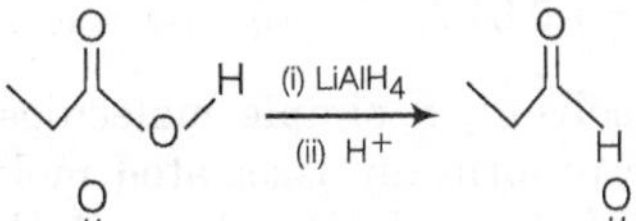

(c) 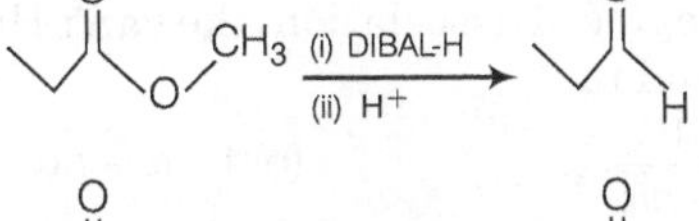

(d) 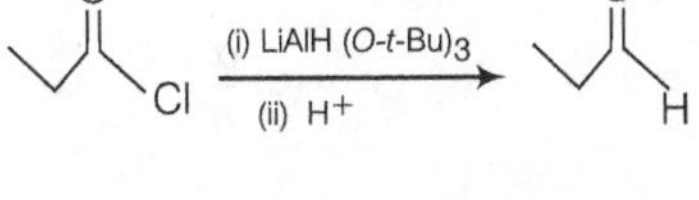

26. The compound $C_6H_5CH=N-N=CHC_6H_5$ is produced by the reaction of an axcess of benzaldehyde with which compound?

(a) NH_3 (b) NH_2NH_2
(c) Hydroxyl amine (d) Phenyl hydrazine

27. Which of the following statements is not true for acetaldehyde?

(a) It will undergo an aldol condensation reaction.
(b) It will undergo a Cannizzaro reaction.
(c) It will undergo a haloform reaction
(d) Enolisation is catalysed by acid or base

28. Which of the following compounds does not undergo mutarotation?

(a) Glucose (b) Sucrose
(c) Ribose (d) Fructose

29. If two isomers have been classified correctly as anomers, they may be also called

(a) conformers (b) enantiomers
(c) tautomers (d) diastereomers

30. Which two of the following aldohexoses give the same osazone derivative?

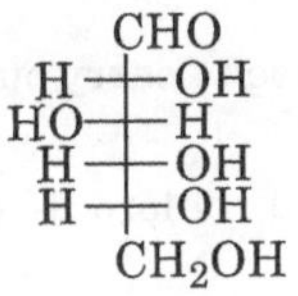
I.

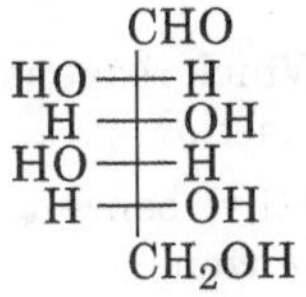
II.

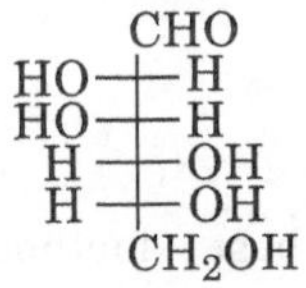
III.

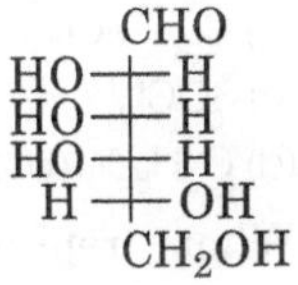
IV.

(a) I and IV (b) I and III
(c) II and III (d) III and IV

31. Which of the following describes the overall three dimensional folding of a polypeptide?

(a) Primary structure (b) Secondary structure
(c) Tertiary structure (d) Quaternary structure

32. What is incorrect regarding natural rubber?

(a) It is *cis*-polyisoprene
(b) It is amorphous in nature
(c) It has syndiotactic stereochemistry
(d) Molecules of rubber have coiled conformation

33. Which one of the following is the structure of polyacrylonitrile?

(a) 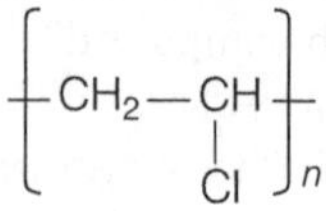 (b) 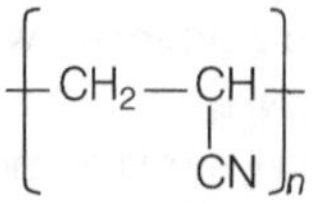

(c) 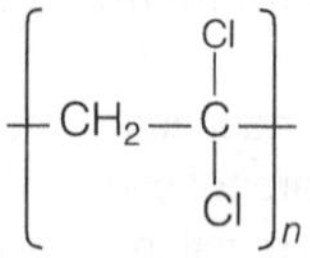 (d) 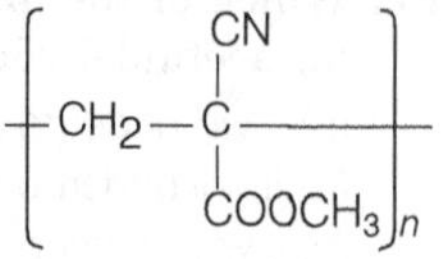

34. Which of the aromatic compounds react fastest with methoxide ion?

(a) 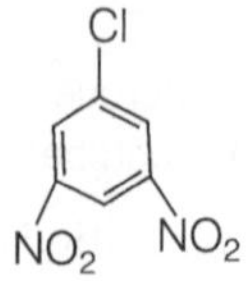 (b) 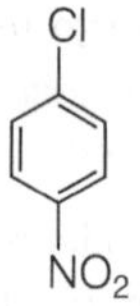

(c) 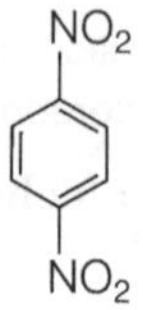 (d) 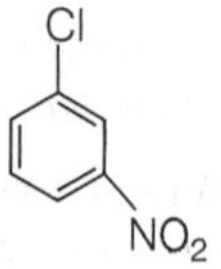

35. Which reagents would you use to carry out the reaction

Ethyl benzene ⟶ 2 and 4- chloro 1- ethyl benzene?

(a) Cl_2, light and heat
(b) Cl_2, $FeCl_3$
(c) $SOCl_2$
(d) C_2H_5Cl, $AlCl_3$

36. What represents the best method for producing salicylic acid?

(a) Benzoic acid, strong base, high temperature
(b) Toluene, strong base, water, high temperature
(c) Phenol, base, carbondioxide then acid
(d) Bromobenzene, carbonic acid, high temperature

37. Which pair is usually required for a nucleophilic aromatic substitution reaction?

(a) An $-NO_2$ substitution and a strong electrophile
(b) A ring bearing a strong activating group and a strong acid
(c) An aryl halide with an $-NO_2$ and a strong nucleophile
(d) An unsubstituted benzene ring and a strong electrophile

38. What reaction does hydroquinone undergo most readily?

(a) S_N2 substitution (b) Hydrolysis
(c) Oxidation (d) Reduction

39. In physical adsorption, the adsorbent and adsorbate are held together by the

(a) ionic forces (b) covalent forces
(c) van der Waals' forces (d) H-bonding

40. Potassium stearate is obtained by the saponification of an oil or fat. It has formula $CH_3-(CH_2)_{16}-COOK$. The molecule has a lyophobic terminal CH_3 and a lyophilic terminal $-COOK$. Potassium stearate is an example of

(a) lyophilic colloid
(b) lyophobic colloid
(c) macromolecular colloid
(d) micelle

41. Suppose that gold is being plated on to another metal in an electrolytic cell. The half-cell reaction producing the Au(s) is $AuCl_4^- \rightarrow Au(s) + 4Cl^- - 3e^-$

If a 0.30 A current runs for 15.00 min, what mass of Au(s) will be plated, assuming all the electrons are used in the reduction of $AuCl_4^-$? The Faraday constant is 96485 C/mol and molar mass of Au is 197.

(a) 0.184 g Au (b) 0.551 g Au
(c) 1.84 g Au (d) 0.613 g Au

42. What is the cell potential (standard emf, E^0) for the reaction below?

$[E^0(Fe^{2+}(aq)Fe) = 0.44V$ and $E^0(O_2(g)/H_2O/OH^-) = +0.4V]$

$$2Fe(s) + O_2(g) + 2H_2O(l) \rightleftharpoons 2Fe^{2+}(aq) + 4OH^-(aq)$$

(a) $E^0_{cell} = -0.48\,V$ (b) $E^0_{cell} = -0.04\,V$
(c) $E^0_{cell} = +0.84\,V$ (d) $E^0_{cell} = +1.28\,V$

43. If in a solvent, n simple molecules of solute combine to form an associated molecule, α is the degree of association, the van't Hoff's factor is equal to

(a) $\frac{1}{1-n\alpha}$ (b) $1-\alpha+n\alpha$
(c) $1-\alpha+\frac{\alpha}{n}$ (d) $\frac{\alpha}{n}-1+\alpha$

44. If the osmotic pressure of a 0.010 M aqueous solution of sucrose at 27°C is 0.25 atm, then the osmotic pressure of a 0.010 M aqueous solution of NaCl at 27°C is
(a) 0.062 atm (b) 0.12 atm
(c) 0.25 atm (d) 0.50 atm

45. A certain radioactive isotope ${}_{Z}^{A}X$ (half-life = 10 days) decays to give ${}_{Z-2}^{A-4}X$. If 1.0 g of X is kept in a sealed vessel find the volume of He accumulated at STP in 20 days. Molar mass of ${}_{Z}^{A}X = 253$ u
(a) 22.40 mL (b) 33.40 mL
(c) 55.40 mL (d) 66.40 mL

46. Raw milk sours in about 4 h at 27°C, but in about 48 h in a refrigerator at 17°C. What is the activation energy for souring of milk?
(a) 78.3 kJ mol^{-1}
(b) 46.21 kJ mol^{-1}
(c) 23.5 kJ mol^{-1}
(d) 80.8 kJ mol^{-1}

47. If 200 mL of a 0.031 molar solution of H_2SO_4 are added to 84 mL of a 0.150 M KOH solution, what is the pH of the resulting solution?
(a) 12.4 (b) 1.7
(c) 2.2 (d) 10.9

48. $(SiH_3)_3N$ is
(a) planar (b) pyramidal
(c) octahedral (d) tetrahedral

49. Solid N_2O_5 is
(a) ionic (b) covalent
(c) coordinate covalent (d) metallic

50. Which of the following phosphorus is the most reactive?
(a) Red phosphorus
(b) White phosphorus
(c) Scarlet phosphorus
(d) Violet phosphorus

51. The activity of alkaline earth metals as reducing agent
(a) decreases from Be to Ba
(b) increases from Be to Ba
(c) increases from Be to Ca and decreases from Ca to Ba
(d) decreases from Be to Ca and increases from Ca to Ba

52. Baking powder is more commonly used to make cakes or bread "rise". Filler in baking powder is
(a) $NaHCO_3$ (b) $Ca(H_2PO_4)_2$
(c) starch (d) $NaAl(SO_4)_2$

53. Strongest conjugate base is
(a) ClO^- (b) ClO_2^- (c) ClO_3^- (d) ClO_4^-

54. Out of the following redox reactions

I. $NH_4NO_3 \xrightarrow{\Delta} N_2O + 2H_2O$

II. $NH_4NO_2 \xrightarrow{\Delta} N_2 + 2H_2O$

III. $PCl_5 \xrightarrow{\Delta} PCl_3 + Cl_2$

disproportionation is not shown in
(a) I and II (b) II and III
(c) I and III (d) I, II and III

55. The complex $[Fe(H_2O)_5NO]^{2+}$ is formed in the ring test for nitrate ion when freshly prepared $FeSO_4$ solution is added to aqueous solution of NO_3^- followed by addition of Conc. H_2SO_4. This complex is formed by charge transfer in which
(a) Fe^{2+} changes to Fe^{3+} and NO^+ changes to NO
(b) Fe^{2+} changes to Fe^{3+} and NO changes to NO^+
(c) Fe^{2+} changes to Fe^+ and NO changes to NO^+
(d) No charge transfer takes place

56. Which pair is different from the others?
(a) Li — Mg (b) B — Al
(c) Na — K (d) Ca — Mg

57. The dominant factor in determining the IE of the elements on moving down the groups is its
(a) atomic radius (b) effective nuclear charge
(c) Both (a) and (b) (d) None of the above

58. Valence electrons in the element A are 3 and that in element B are 6. Most probable compound formed from A and B is
(a) A_2B (b) AB_2
(c) A_6B_3 (d) A_2B_3

59. Which of the following in the strongest base?

(a) 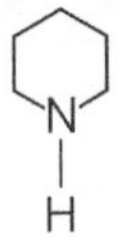

(b) 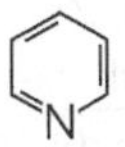

(c) 

(d) 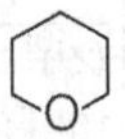

60. Which of the following best define the term hyperconjugation?

(a) Transfer of H from a σ_{C-H} bond to the π system

(b) Participation of an H-atom on terminal carbon in the electron sharing

(c) Delocalisation of electron density from a properly oriented σ-bond into π-system

(d) Interaction between a properly aligned lone electron pair with the empty π-system

Mathematics

1. If $a = \cos\frac{2\pi}{7} + i\sin\frac{2\pi}{7}$, then the quadratic equation whose roots are $\alpha = a + a^2 + a^4$ and $\beta = a^3 + a^5 + a^6$, is

(a) $x^2 - x + 2 = 0$ (b) $x^2 + 2x + 2 = 0$

(c) $x^2 + x + 2 = 0$ (d) $x^2 + x - 2 = 0$

2. If $1, \omega, \omega^2, \ldots \omega^{n-1}$ are n, n^{th} roots of unity, then the value of $(9-\omega)\cdot(9-\omega^2)\cdot(9-\omega^3)\ldots(9-\omega^{n-1})$ will be

(a) $\frac{9^n + 1}{8}$ (b) $9^n - 1$

(c) $\frac{9^n - 1}{8}$ (d) $9^n + 1$

3. Middle term in the expansion of $\left(x^2 + \frac{1}{x^2} + 2\right)^n$ is

(a) $\frac{n!}{(n!)^2}$ (b) $\frac{(2n!)}{(n!)^2}$ (c) $\frac{(2n-1)!}{n!}$ (d) $\frac{(2n)!}{n!}$

4. The common tangent of the parabolas $y^2 = 4x$ and $x^2 = -8y$ is

(a) $y = x + 2$ (b) $y = x - 2$

(c) $y = 2x + 3$ (d) None of these

5. 10 different toys are to be distributed among 10 children. Total number of ways of distributing these toys, so that exactly two children do not get any toy, is

(a) $\frac{10!}{2!\cdot 3!\cdot 7!}$

(b) $\frac{10!}{(2!)^4 \times 6!}$

(c) $(10!)^2\left[\frac{1}{(2!)^4 \times 6!} + \frac{1}{2! \times 3!}\right]$

(d) $\frac{10! \times 10!}{(2!)^2 \times 6!}\left[\frac{25}{84}\right]$

6. The variance of first n natural numbers is

(a) $\frac{n(n+1)}{12}$ (b) $\frac{(n+1)(n+5)}{12}$

(c) $\frac{(n+1)(n-5)}{12}$ (d) $\frac{n(n-5)}{12}$

7. A circle of radius 2 is touching both the axes and a circle with centre (6, 5). The distance between their centres is

(a) 8 units (b) 5 units

(c) 7 units (d) None of these

8. $S = \{1, 2, 3, \ldots 20\}$ is to be partitioned into four sets A, B, C and D of equal size. The number of ways it can be done, is

(a) $\frac{20!}{4! \times 5!}$ (b) $\frac{20!}{4^5}$

(c) $\frac{20!}{(5!)^4}$ (d) $\frac{20!}{(4!)^5}$

9. Out of 50 tickets numbered $00, 01, 02, \ldots, 49$, one ticket is drawn randomly, the probability of the ticket having the product of its digits 7, given that the sum of the digits is 8, is

(a) $\frac{1}{14}$ (b) $\frac{3}{14}$

(c) $\frac{1}{5}$ (d) None of these

10. If the term free from x in the expansion of $\left(\sqrt{x} - \frac{k}{x^2}\right)^{10}$ is 405, then the value of k is

(a) ± 1 (b) ± 3

(c) ± 4 (d) ± 2

11. If $\tan\frac{\alpha}{2}$, and $\tan\frac{\beta}{2}$ are the roots of $8x^2 - 26x + 15 = 0$, then $\cos(\alpha + \beta)$ is equal to

(a) $\frac{627}{725}$ (b) $-\frac{627}{725}$

(c) -1 (d) None of these

12. If $a\cos^3\alpha + 3a\cos\alpha\sin^2\alpha = m$ and $a\sin^3\alpha + 3a\cos^2\alpha\sin\alpha = n$, then $(m+n)^{2/3} + (m-n)^{2/3}$ is equal to

(a) $2a^3$ (b) $2a^{1/3}$
(c) $2a^{2/3}$ (d) $2a^3$

13. In a triangle, the lengths of two larger sides are 10 cm and 9 cm. If the angles of the triangle are in AP, then the length of the third side is

(a) $\sqrt{5}-\sqrt{6}$ (b) $\sqrt{5}+\sqrt{6}$
(c) $\sqrt{5}\pm\sqrt{6}$ (d) $5\pm\sqrt{6}$

14. The distance of the point (3, 5) from $2x+3y-14=0$ measured parallel to $x-2y=1$ is

(a) $\frac{7}{\sqrt{5}}$ (b) $\frac{7}{\sqrt{13}}$
(c) $\sqrt{5}$ (d) $\sqrt{13}$

15. If the line $\frac{x}{a}+\frac{y}{b}=1$ moves such that $\frac{1}{a^2}+\frac{1}{b^2}=\frac{1}{c^2}$, then the locus of the foot of the perpendicular from the origin to the line is

(a) straight line (b) circle
(c) parabola (d) ellipse

16. A rhombus is inscribed in the region common to the two circles $x^2+y^2-4x-12=0$ and $x^2+y^2+4x-12=0$ with two of its vertices on the line joining the centres of the circles. The area of rhombus is

(a) $8\sqrt{3}$ (b) $4\sqrt{2}$
(c) $16\sqrt{3}$ (d) $16\sqrt{2}$

17. If the line $y-\sqrt{3}x+3=0$ cuts the parabola $y^2=x+2$ at A and B, then $PA\cdot PB$ is equal to where [where $P=(\sqrt{3},0)$].

(a) $\frac{4(2-\sqrt{3})}{3}$ (b) $\frac{4(\sqrt{3}+2)}{3}$
(c) $\frac{4\sqrt{3}}{3}$ (d) $\frac{2(\sqrt{3}+2)}{3}$

18. AB is a chord of the parabola $y^2=4ax$ with vertex A, BC is drawn perpendicular to AB meeting the axis at C. The projection of BC on the axis of the parabola is

(a) a (b) $2a$
(c) $4a$ (d) $8a$

19. The radius of the circle passing through the foci of the ellipse $\frac{x^2}{4}+\frac{4y^2}{7}=1$ and having its centre at $\left(\frac{1}{2},2\right)$ is

(a) $\sqrt{5}$ (b) 3
(c) $\sqrt{12}$ (d) $\frac{7}{2}$

20. The area of the quadrilateral formed by the tangents at the end point of latusrectum to the ellipse $\frac{x^2}{9}+\frac{y^2}{5}=1$ is

(a) $\frac{27}{4}$ (b) 9 (c) $\frac{27}{2}$ (d) 27

21. If the chords of the hyperbola $x^2-y^2=a^2$ touch the parabola $y^2=4ax$. Then, the locus of the middle points of these chords is

(a) $y^2=(x-a)x^3$ (b) $y^2(x-a)=x^3$
(c) $x^2(x-a)=x^3$ (d) None of these

22. $\lim_{x\to 0}\frac{e^{\sin x}-1}{x}$ is equal to

(a) 0 (b) e
(c) 1 (d) Does not exist

23. If p and q are two statements, then $(p\wedge q)\vee(q\Leftrightarrow p)$ is

(a) tautology
(b) contradiction
(c) neither tautology nor contradiction
(d) either tautology or contradiction

24. The value of the determinant $\begin{vmatrix}(a^x+a^{-x})^2 & (a^x-a^{-x})^2 & 1\\(b^x+b^{-x})^2 & (b^x-b^{-x})^2 & 1\\(c^x+c^{-x})^2 & (c^x-c^{-x})^2 & 1\end{vmatrix}$ is

(a) 0 (b) $2abc$
(c) $a^2b^2c^2$ (d) None of these

25. If $\Delta_r=\begin{vmatrix}1 & n & n\\2r & n^2+n+1 & n^2+n\\2r-1 & n^2 & n^2+n+1\end{vmatrix}$ and $\sum_{r=1}^{n}\Delta_r=56$, then n is equal to

(a) 4 (b) 6 (c) 7 (d) 8

26. If a, b, c are cube roots of unity, then $\begin{vmatrix} e^a & e^{2a} & e^{3a}-1 \\ e^b & e^{2b} & e^{3b}-1 \\ e^c & e^{2c} & e^{3c}-1 \end{vmatrix}$ is equal to

(a) 0 (b) e (c) e^2 (d) e^3

27. If $fog = |\sin x|$ and $gof = \sin^2 \sqrt{x}$, then $f(x)$ and $g(x)$ are

(a) $f(x) = \sqrt{\sin x}$, $g(x) = x^2$

(b) $f(x) = |x|$, $g(x) = \sin x$

(c) $f(x) = \sqrt{x}$, $g(x) = \sin^2 x$

(d) $f(x) = \sin \sqrt{x}$, $g(x) = x^2$

28. Let $f : R \to R$ be defined by $f(x) = \dfrac{x}{\sqrt{1+x^2}}$, then $(fofof)(x)$ is

(a) $\dfrac{x}{\sqrt{1+x^2}}$ (b) $\dfrac{x}{\sqrt{1+3x^2}}$

(c) x (d) 1

29. If $f(x) = \dfrac{a^x + a^{-x}}{2}$ and $f(x+y) + f(x-y) = kf(x)\,f(y)$, then k is equal to

(a) 2 (b) 4

(c) –2 (d) None of these

30. If $\tan^{-1}\left(\dfrac{a}{x}\right) + \tan^{-1}\left(\dfrac{b}{x}\right) = \dfrac{\pi}{2}$, then x is equal to

(a) $\sqrt{ab}$ (b) $\sqrt{2\,ab}$ (c) $2ab$ (d) ab

31. $\cos^{-1}\{\cos 2 \cot^{-1}(\sqrt{2}-1)\}$ is equal to

(a) $\sqrt{2}-1$ (b) $\dfrac{\pi}{4}$ (c) $\dfrac{3\pi}{4}$ (d) 0

32. $\sum_{m=1}^{n} \tan^{-1}\left(\dfrac{2m}{m^4+m^2+2}\right)$ is equal to

(a) $\tan^{-1}\left(\dfrac{n^2+n}{n^2+n+2}\right)$ (b) $\tan^{-1}\left(\dfrac{n^2-n}{n^2-n+2}\right)$

(c) $\tan^{-1}\left(\dfrac{n^2+n+2}{n^2+n}\right)$ (d) None of these

33. $\lim_{x \to \frac{\pi}{2}} \dfrac{\left(1-\tan\dfrac{x}{2}\right)(1-\sin x)}{\left(1+\tan\dfrac{x}{2}\right)(\pi-2x)^3}$ is equal to

(a) $\dfrac{1}{8}$ (b) 0 (c) $\dfrac{1}{32}$ (d) ∞

34. $f(x) = \begin{cases} \dfrac{\sin^3(\sqrt{3})\cdot\log(1+3x)}{(\tan^{-1}\sqrt{x})^2\,(e^{5\sqrt{x}}-1)\,x}, & x \neq 0 \\ a, & x = 0 \end{cases}$ is continuous in [0, 1], if 'a' equals to

(a) 0 (b) $\dfrac{3}{5}$

(c) 2 (d) $\dfrac{5}{3}$

35. If $y = |\sin x|^{|x|}$, then the value of $\dfrac{dy}{dx}$ at $x = \dfrac{-\pi}{6}$ is

(a) $\dfrac{2^{-\frac{\pi}{6}}}{6}[6\log 2 - \sqrt{3}\pi]$

(b) $2^{\frac{\pi}{6}}[6\log 2 + \sqrt{3}\pi]$

(c) $\dfrac{2^{-\frac{\pi}{6}}}{6}[6\log 2 + \sqrt{3}\pi]$

(d) None of the above

36. The function x^x is increasing, when

(a) $x > \dfrac{1}{e}$ (b) $x < \dfrac{1}{e}$

(c) $x < 0$ (d) for all x

37. The length of the longest interval, in which $f(x) = 3\sin x - 4\sin^3 x$ is increasing, is

(a) $\dfrac{\pi}{3}$ (b) $\dfrac{\pi}{2}$ (c) $\dfrac{3\pi}{2}$ (d) π

38. The triangle formed by the tangent to the curve $f(x) = x^2 + bx - b$ at the point (1, 1) and the coordinate axes lies in the first quadrant. If its area is 2, then the value of b is

(a) –1 (b) 3

(c) –3 (d) 1

39. The equation of the tangent to the curve $\left(\dfrac{x}{a}\right)^n + \left(\dfrac{y}{b}\right)^n = 2$ at (a, b) is

(a) $\dfrac{x}{a} + \dfrac{y}{b} = 2$ (b) $\dfrac{x}{a} + \dfrac{y}{b} = \dfrac{1}{2}$

(c) $\dfrac{x}{b} - \dfrac{y}{a} = 2$ (d) $ax + by = 2$

40. The minimum radius vector of the curve $\dfrac{a^2}{x^2} + \dfrac{b^2}{y^2} = 1$ is of length

(a) $a - b$ (b) $a + b$

(c) $2a + b$ (d) None of these

41. The largest term in the sequence $a_n = \dfrac{n^2}{n^3 + 200}$ is given by

(a) $\dfrac{529}{49}$ (b) $\dfrac{8}{89}$

(c) $\dfrac{49}{543}$ (d) None of these

42. $f(x) = \begin{cases} |x^3 + x^2 + 3x + \sin x| \cdot \left(3 + \sin \dfrac{1}{x}\right), & x \neq 0 \\ 0, & x = 0 \end{cases}$

The number of points, where $f(x)$ attains its minimum value, is

(a) 1 (b) 2

(c) 3 (d) infinitely many

43. $\displaystyle\int \frac{x^2 - 1}{(x^4 + 3x^2 + 1)\tan^{-1}\left(x + \frac{1}{x}\right)} dx$ is equal to

(a) $\tan^{-1}\left(x + \dfrac{1}{x}\right) + C$

(b) $\cot^{-1}\left(x + \dfrac{1}{x}\right) + C$

(c) $\log\left(x + \dfrac{1}{x}\right) + C$

(d) $\log\left[\tan^{-1}\left(x + \dfrac{1}{x}\right)\right] + C$

44. $\displaystyle\int \frac{\cos x}{\sin^2 x \cdot (\sin x + \cos x)} dx$ is equal to

(a) $\log\left|\dfrac{1 + \tan x}{\tan x}\right| - \cot x + C$

(b) $\log\left|\dfrac{\tan x}{1 + \tan x}\right| + C$

(c) $\log\left|\dfrac{\tan x}{1 + \tan x}\right| - \tan x + C$

(d) $\log\left|\dfrac{\tan x}{1 + \tan x}\right| + \cot x + C$

45. $\displaystyle\int \frac{dx}{\sin x - \cos x + \sqrt{2}}$ is equal to

(a) $-\dfrac{1}{\sqrt{2}}\tan\left(\dfrac{x}{2} + \dfrac{\pi}{8}\right) + C$

(b) $\dfrac{1}{\sqrt{2}}\tan\left(\dfrac{x}{2} + \dfrac{\pi}{8}\right) + C$

(c) $\dfrac{1}{\sqrt{2}}\cot\left(\dfrac{x}{2} + \dfrac{\pi}{8}\right) + C$

(d) $-\dfrac{1}{\sqrt{2}}\cot\left(\dfrac{x}{2} + \dfrac{\pi}{8}\right) + C$

46. $\displaystyle\int_0^{\frac{3\pi}{2}} \sin\left[\frac{2x}{\pi}\right] \cdot dx$, where $[\cdot]$ denotes the greatest integer function, is equal to

(a) $\dfrac{\pi}{2}(\sin 1 + \cos 1)$ (b) $\dfrac{\pi}{2}(\sin 1 + \sin 2)$

(c) $\dfrac{\pi}{2}(\sin 1 - \cos 1)$ (d) $\dfrac{\pi}{2}(\sin \pi + \sin 2)$

47. The value of $\displaystyle\int_{1/n}^{(an-1)/n} \frac{\sqrt{x}}{\sqrt{a - x} + \sqrt{x}} dx$ is equal to

(a) $\dfrac{a}{2}$ (b) $\dfrac{n \cdot a + 2}{2n}$

(c) $\dfrac{n \cdot a - 2}{2n}$ (d) $\dfrac{n \cdot a}{2}$

48. The area bounded by $y = x|\sin x|$ and X-axis between $x = 0$ and $x = 2\pi$ is

(a) 2π sq units (b) 3π sq units

(c) π sq units (d) 4π sq units

49. The area enclosed by the curves $|y + x| \leq 1$, $|y - x| \leq 1$ and $2x^2 + 2y^y = 1$ is

(a) $\left(2 + \dfrac{\pi}{2}\right)$sq units (b) $\left(2 - \dfrac{\pi}{2}\right)$sq units

(c) $\left(3 + \dfrac{\pi}{2}\right)$sq units (d) $\left(3 - \dfrac{\pi}{4}\right)$sq units

50. The real value of m for which the substitution $y = u^m$ will transform the differential equation $2x^4 \cdot y\dfrac{dy}{dx} + y^4 = 4x^6$ into a homogeneous equation, is

(a) $m = 0$ (b) $m = 1$ (c) $m = \dfrac{3}{2}$ (d) $m = \dfrac{2}{3}$

51. The equation of the curve for which the square of the ordinate is twice the rectangle contained by the abscissa and the intercept of the normal on the X-axis and passing through (2, 1), is

(a) $x^2 + y^2 - x = 0$

(b) $4x^2 + 2y^2 - 9y = 0$

(c) $2x^2 + 4y^2 - 9x = 0$

(d) $4x^2 + 2y^2 - 9x = 0$

52. If $|\mathbf{a}| = 1, |\mathbf{b}| = 4, \mathbf{a} \cdot \mathbf{b} = 2$ and $\mathbf{c} = 2\mathbf{a} \times \mathbf{b} - 3\mathbf{b}$, then the angle between $\mathbf{b}$ and $\mathbf{c}$ is

(a) $\dfrac{\pi}{6}$ (b) $\dfrac{5\pi}{6}$

(c) $\dfrac{\pi}{3}$ (d) $\dfrac{2\pi}{3}$

53. If a, b, c are p^{th} and q^{th} and r^{th} terms of a GP, then the vectors $\log a\,\hat{\mathbf{i}} + \log b\,\hat{\mathbf{j}} + \log c\,\hat{\mathbf{k}}$ and $(q-r)\,\hat{\mathbf{i}} + (r-p)\,\hat{\mathbf{j}} + (p-q)\,\hat{\mathbf{k}}$ are

(a) equal (b) parallel
(c) perpendicular (d) None of these

54. The line joining the points $(1, 1, 2)$ and $(3, -2, 1)$ meets the plane $3x + 2y + z = 6$ at the point

(a) (1, 1, 2) (b) (3, −2, 1)
(c) (2, − 3, 1) (d) (3, 2, 1)

55. The distance of the point $(1, -5, 9)$, from the plane $\mathbf{r} \cdot (\hat{\mathbf{i}} - \hat{\mathbf{j}} + \hat{\mathbf{k}}) = 5$ measured along the line $\mathbf{r} = \hat{\mathbf{i}} + \hat{\mathbf{j}} + \hat{\mathbf{k}}$ is

(a) $3\sqrt{5}$ (b) $10\sqrt{3}$ (c) $5\sqrt{3}$ (d) $3\sqrt{10}$

56. The image of the point with position vector $\hat{\mathbf{i}} + 3\hat{\mathbf{k}}$ in the plane $\mathbf{r} \cdot (\hat{\mathbf{i}} + \hat{\mathbf{j}} + \hat{\mathbf{k}}) = 1$ is

(a) $\hat{\mathbf{i}} + 2\hat{\mathbf{j}} + \hat{\mathbf{k}}$ (b) $\hat{\mathbf{i}} - 2\hat{\mathbf{j}} + \hat{\mathbf{k}}$
(c) $-\hat{\mathbf{i}} - 2\hat{\mathbf{j}} + \hat{\mathbf{k}}$ (d) $\hat{\mathbf{i}} + 2\hat{\mathbf{j}} - \hat{\mathbf{k}}$

57. In a trial, the probability of success is twice the probability of failure. In six trial, the probability of atleast four successes will be

(a) $\frac{496}{729}$ (b) $\frac{400}{729}$ (c) $\frac{500}{729}$ (d) $\frac{600}{729}$

58. If the integers m and n are chosen at randon between 1 and 100, then the probability that a number of the form $7^m + 7^n$ is divisible by 5, equals

(a) $\frac{1}{4}$ (b) $\frac{1}{7}$ (c) $\frac{1}{8}$ (d) $\frac{1}{49}$

59. If $\sin^{-1} a + \sin^{-1} b + \sin^{-1} c = \pi$, then the value of $a \cdot \sqrt{(1-a^2)} + b \cdot \sqrt{(1-b^2)} + c \cdot \sqrt{(1-c^2)}$ will be

(a) $2abc$ (b) abc (c) $\frac{1}{2}abc$ (d) $\frac{1}{3}abc$

60. The value of λ and μ for which the system of equations $x + y + z = 6$, $x + 2y + 3z = 10$ and $x + 2y + \lambda z = \mu$ have no solution, are

(a) $\lambda = 3, \mu \neq 10$ (b) $\lambda \neq 3, \mu = 10$
(c) $\lambda \neq 3, \mu \neq 10$ (d) None of these

61. The range of $f(x) = \sec\left(\frac{\pi}{4}\cos^2 x\right), -\infty < x < \infty$ is

(a) $[1, \sqrt{2}]$ (b) $[1, \infty]$
(c) $[-\sqrt{2}, -1] \cup [1, \sqrt{2}]$ (d) $[-\infty, -1] \cup [1, \infty]$

62. The value of the $\lim_{x \to 0}\left(\frac{a^x + b^x + c^x}{3}\right)^{2/x}$, $(a, b, c > 0)$ is

(a) $(abc)^3$ (b) abc
(c) $(abc)^{1/3}$ (d) None of these

63. The function $f(x) = [x] \cos\left[\frac{2x-1}{2}\right]\pi$, where $[\cdot]$ denotes the greatest integer funtion, is discontinuous at

(a) all x
(b) no x
(c) all integral points
(d) x which is not an integer

64. The function $f(x) = (x^2 - 1)\,|x^2 - 3x + 2| + \cos|x|$ is non-differentiable at

(a) −1 (b) 0
(c) 1 (d) 2

65. If $f(x + y) = f(x) \cdot f(y)$ for all x and y and $f(15) = 2$, $f'(0) = 3$, then $f'(5)$ will be

(a) 2 (b) 4
(c) 6 (d) 8

66. If $x \sin(a + y) + \sin a \cos(a + y) = 0$, then $\frac{dy}{dx}$ is equal to

(a) $\frac{\sin^2(a+y)}{\sin a}$ (b) $\frac{\cos^2(a+y)}{\cos a}$
(c) $\frac{\sin^2(a+y)}{\cos a}$ (d) $\frac{\cos^2(a+y)}{\sin a}$

67. The approximate value of $f(5.001)$, where $f(x) = x^3 - 7x^2 + 15$, is

(a) − 34.995 (b) − 33.995
(c) − 33.335 (d) − 35.993

68. What are the values of c for which Rolle's theorem for the function $f(x) = x^3 - 3x^2 + 2x$ in the interval [0, 2] is verified?

(a) $c = \pm 1$ (b) $c = 1 \pm \frac{1}{\sqrt{3}}$
(c) $c = \pm 2$ (d) None of these

69. The equation of the plane through the intersection of the planes $3x - y + 2z - 4 = 0$ and $x + y + z - 2 = 0$ and the point (2, 2, 1) is

(a) $7x + 5y + 4z + 8 = 0$ (b) $7x + 5y + 4z - 8 = 0$
(c) $7x - 5y + 4z - 8 = 0$ (d) None of these

70. $\lim_{x \to 0} (1^{\text{cosec}^2 x} + 2^{\text{cosec}^2 x} + \ldots + n^{\text{cosec}^2 x})^{\sin^2 x}$ is equal to

(a) 1 (b) $\frac{1}{n}$

(c) n (d) 0

71. $\int_0^{\pi/2} \frac{\sin x - \cos x}{1 + \sin x \cdot \cos x} dx$ is equal to

(a) 0 (b) $\frac{\pi}{4}$

(c) $\frac{\pi}{2}$ (d) π

72. The area of the region bounded by $1 - y^2 = |x|$ and $|x| + |y| = 1$ is

(a) 1/3 sq unit

(b) 2/3 sq unit

(c) 4/3 sq units

(d) 1 sq unit

73. The equations of the lines passing through the point (1, 0) and at a distance $\frac{\sqrt{3}}{2}$ from the origin are

(a) $\sqrt{3}x + y - \sqrt{3} = 0, \sqrt{3}x - y - \sqrt{3} = 0$

(b) $\sqrt{3}x + y + \sqrt{3} = 0, \sqrt{3}x - y + \sqrt{3} = 0$

(c) $x + \sqrt{3}y - \sqrt{3} = 0, x - \sqrt{3}y - \sqrt{3} = 0$

(d) None of the above

74. The equation of circle which passes through the origin and cuts off intercepts 5 and 6 from the positive parts of the axes respectively, is $\left(x - \frac{5}{2}\right)^2 + (y - 3)^2 = \lambda$, where λ is

(a) $\frac{61}{4}$ (b) $\frac{6}{4}$

(c) $\frac{1}{4}$ (d) 0

75. A perpendicular is drawn from the point $P(2, 4, -1)$ to the line $\frac{x+5}{1} = \frac{y+3}{4} = \frac{z-6}{-9}$. The equation of the perpendicular from P to the given line is

(a) $\frac{x-2}{6} = \frac{y-4}{3} = \frac{z+1}{2}$ (b) $\frac{x+2}{6} = \frac{y-4}{3} = \frac{z+1}{2}$

(c) $\frac{x+2}{-6} = \frac{y-4}{3} = \frac{z+1}{2}$ (d) $\frac{x+2}{6} = \frac{y+4}{3} = \frac{z+1}{2}$

76. If $\cos^2 A + \cos^2 C = \sin^2 B$, then ΔABC is

(a) equilateral

(b) right angled

(c) isosceles

(d) None of the above

77. If $\begin{vmatrix} x+y+2z & x & y \\ z & y+z+2x & y \\ z & x & z+x+2y \end{vmatrix} = k(x+y+z)^3$, then the value of k is

(a) 1 (b) 2

(c) 4 (d) 8

78. The minimum value of $\frac{x}{\log x}$ is

(a) e (b) $\frac{1}{e}$

(c) e^2 (d) e^3

79. One mapping (function) is selected at random from all the mappings of the set $A = \{1, 2, 3, \ldots, n\}$ into itself. The probability that the mapping selected is one-one, is

(a) $\frac{n!}{n^{n-1}}$

(b) $\frac{n!}{n^n}$

(c) $\frac{n!}{2n^n}$

(d) None of the above

80. In how many ways, can a student choose a program of 5 courses, if 9 courses are available and 2 specific courses are compulsory for every student?

(a) 34 (b) 36 (c) 35 (d) 37

English

Directions (Q.Nos. 1-5) *Out of the four alternatives, choose the one which best expresses the meaning of the given word.*

1. Remorse
(a) Warmth (b) Delight
(c) Repentance (d) Satisfaction

2. Stigma
(a) Gedit (b) Pride
(c) Shame (d) Clarity

3. Blemish
(a) Adornment (b) Blank
(c) Embellishment (d) Flaw

4. Deteriorate
(a) Boost (b) Improve (c) Build (d) Descend

5. Knave
(a) Fraud (b) Angel
(c) Ugly (d) Perfect

Directions (Q.Nos. 6-10) *Choose the word opposite in meaning to the given word and mark it in the answer sheet.*

6. Enormous
(a) Soft (b) Average
(c) Tiny (d) Weak

7. Commissioned
(a) Started (b) Closed
(c) Finished (d) Terminated

8. Artificial
(a) Red (b) Natural
(c) Truthful (d) Solid

9. Exodus
(a) Influx (b) Home-coming
(c) Return (d) Restoration

10. Mortal
(a) Divine (b) Immortal
(c) Spiritual (d) Eternal

Directions (Q.Nos. 11-15) *Four alternatives are given in each group, one word is correctly spelt. Find the correctly spelt word.*

11. (a) Addvertise (b) Surender
(c) Admiror (d) Pitiable

12. (a) Humilliation (b) Veterinary
(c) Appology (d) Resturant

13. (a) Accute (b) Forstry
(c) Personel (d) Omnious

14. (a) Diference (b) Homogenious
(c) Greatful (d) Possession

15. (a) Committee (b) Commentery
(c) Involuntry (d) Anxiety

Directions (Q.Nos. 16-20) *Out of the four alternatives, choose the one which can be substituted for the given words/sentences.*

16. To examine ones to own thoughts and feelings
(a) Introspection (b) Retrospection
(c) Euthanasia (d) Circumspection

17. The custom of having more than one husband at a time
(a) Polygamy (b) Polyandry
(c) Debauchery (d) Bigamy

18. Murder of a brother
(a) Patricide (b) Regicide
(c) Homicide (d) Fratricide

19. Study of mankind
(a) Pathology (b) Physiology
(c) Philology (d) Anthropology

20. Through which life cannot pass
(a) Dull (b) Dark
(c) Obscure (d) Opaque

Directions (Q.Nos. 21-30) *Some of the words have been left out. First read the passage over and try to understand what it is about. Then, fill in the blanks with the help of the alternatives given.*

Something has happened in the last twenty years that surely must (21) anything that has happened before. Some historians are already saying that thrust (22) space represents a vital turning point in history. Moon flights are considered (23) less than steps in human evolution (24) to the time when life on earth emerged from the sea and established itself on land. Of course, not everyone (25) enraptured by space. Critics have often said that space flight has been an (26) use of resources that should have (27) to feeding, clothing and housing people. There is, however, no proof that if we had **(28)** been working on space, we would have done anything of great human value. In fact, research and exploration have a **(29)** spin-offs, quite apart from the fact that they demonstrate that **(30)** is alive and insatiably curious.

21. (a) terminate (b) transcend
(c) precede (d) recede

22. (a) on (b) upon
(c) in (d) into

23. (a) nothing (b) certainly
(c) sufficiently (d) probably

24. (a) exceeding (b) contrasting
(c) comparable (d) matching

25. (a) was (b) has been
(c) had been (d) being

26. (a) economical (b) extravagant
(c) appropriate (d) benevolent

27. (a) devoted (b) allotted
(c) reserved (d) gone

28. (a) not (b) occasionally
(c) seldom (d) possibly

29. (a) renowned (b) renounced
(c) remarkable (d) relevant

30. (a) one (b) man
(c) human (d) individual

Directions (Q.Nos. 31-35) *Some of the sentences have errors and some are correct. Find out which part of a sentence has a error and corresponding to the appropriate letter* (a), (b) *and* (c) *If the sentence free from error, then mark* (d).

31. You will not(a)/get well until(b)/you take medicine regularly.(c)/No error(d)

32. You must either(a)/ inform the police else be(b)/ prepared to suffer any loss.(c)/ No error(d)

33. Both the rich along with the poor(a)/ are responsible for a great many vices(b)/with which our society as well as country is in flicted.(c)/ No error(d)

34. Have you ever(a)/ met such a boy(b)/ who has not travelled by train?(c)/ No error(d)

35. The robber had hardly(a)/ put the ornaments in his bag(b)/ than the house wife woke up.(c)/ No error(d)

Directions (Q.Nos. 36-40) *Sentences are given with blanks to be filled with an appropriate word. Four alternatives are suggested for each question. Choose the correct alternative.*

36. I wish my bus .... so late, then I could have reached home before breakfast
(a) had not arrived (b) does not arrive
(c) has not arrived (d) arrived

37. .... you examine it carefully you will notice some flaws.
(a) Despots (b) If
(c) Unless (d) Until

38. The contractor did not keep his promise that the work .... be finished before the end of the week.
(a) shall (b) could (c) would (d) may

39. If you come across my umbrella anywhere bring it to me, ...?
(a) is not it (b) do not you
(c) will you (d) let you

40. Miss Pillai teaches very well ....?
(a) did not she
(b) does not she
(c) was not it
(d) would not she

Answers

Physics

1. (b)	**2.** (b)	**3.** (d)	**4.** (b)	**5.** (c)	**6.** (b)	**7.** (b)	**8.** (c)	**9.** (a)	**10.** (a)
11. (d)	**12.** (a)	**13.** (c)	**14.** (c)	**15.** (b)	**16.** (a)	**17.** (d)	**18.** (b,c)	**19.** (b)	**20.** (b)
21. (a)	**22.** (d)	**23.** (a)	**24.** (d)	**25.** (a)	**26.** (c)	**27.** (d)	**28.** (c)	**29.** (c)	**30.** (c)
31. (d)	**32.** (c)	**33.** (c)	**34.** (b)	**35.** (c)	**36.** (c)	**37.** (d)	**38.** (b)	**39.** (c)	**40.** (b)
41. (b)	**42.** (d)	**43.** (a)	**44.** (d)	**45.** (a)	**46.** (d)	**47.** (c)	**48.** (b)	**49.** (b)	**50.** (c)
51. (d)	**52.** (d)	**53.** (b)	**54.** (d)	**55.** (c)	**56.** (a)	**57.** (a)	**58.** (a)	**59.** (a)	**60.** (a)

Chemistry

1. (c)	**2.** (c)	**3.** (d)	**4.** (d)	**5.** (d)	**6.** (c)	**7.** (c)	**8.** (b)	**9.** (a)	**10.** (c)
11. (a)	**12.** (d)	**13.** (a)	**14.** (d)	**15.** (b)	**16.** (a)	**17.** (d)	**18.** (b)	**19.** (a)	**20.** (b)
21. (a)	**22.** (a)	**23.** (d)	**24.** (b)	**25.** (b)	**26.** (b)	**27.** (b)	**28.** (b)	**29.** (d)	**30.** (b)
31. (c)	**32.** (c)	**33.** (b)	**34.** (b)	**35.** (b)	**36.** (c)	**37.** (c)	**38.** (c)	**39.** (c)	**40.** (d)
41. (a)	**42.** (c)	**43.** (c)	**44.** (d)	**45.** (d)	**46.** (a)	**47.** (d)	**48.** (a)	**49.** (a)	**50.** (b)
51. (b)	**52.** (c)	**53.** (a)	**54.** (d)	**55.** (c)	**56.** (a)	**57.** (a)	**58.** (d)	**59.** (a)	**60.** (c)

Mathematics

1. (c)	**2.** (c)	**3.** (b)	**4.** (d)	**5.** (d)	**6.** (c)	**7.** (b)	**8.** (c)	**9.** (c)	**10.** (b)
11. (b)	**12.** (c)	**13.** (d)	**14.** (c)	**15.** (b)	**16.** (a)	**17.** (b)	**18.** (c)	**19.** (a)	**20.** (d)
21. (b)	**22.** (c)	**23.** (c)	**24.** (a)	**25.** (c)	**26.** (a)	**27.** (c)	**28.** (b)	**29.** (a)	**30.** (a)
31. (c)	**32.** (a)	**33.** (c)	**34.** (b)	**35.** (a)	**36.** (a)	**37.** (a)	**38.** (c)	**39.** (a)	**40.** (b)
41. (c)	**42.** (a)	**43.** (d)	**44.** (a)	**45.** (d)	**46.** (b)	**47.** (c)	**48.** (d)	**49.** (b)	**50.** (c)
51. (d)	**52.** (b)	**53.** (c)	**54.** (b)	**55.** (b)	**56.** (b)	**57.** (a)	**58.** (a)	**59.** (a)	**60.** (a)
61. (a)	**62.** (d)	**63.** (c)	**64.** (d)	**65.** (c)	**66.** (a)	**67.** (a)	**68.** (b)	**69.** (c)	**70.** (c)
71. (a)	**72.** (b)	**73.** (a)	**74.** (a)	**75.** (a)	**76.** (b)	**77.** (b)	**78.** (a)	**79.** (b)	**80.** (c)

English

1. (c)	**2.** (c)	**3.** (d)	**4.** (d)	**5.** (a)	**6.** (c)	**7.** (d)	**8.** (b)	**9.** (a)	**10.** (b)
11. (d)	**12.** (b)	**13.** (d)	**14.** (d)	**15.** (d)	**16.** (a)	**17.** (b)	**18.** (d)	**19.** (d)	**20.** (d)
21. (b)	**22.** (d)	**23.** (a)	**24.** (c)	**25.** (b)	**26.** (b)	**27.** (d)	**28.** (a)	**29.** (c)	**30.** (b)
31. (b)	**32.** (b)	**33.** (a)	**34.** (c)	**35.** (c)	**36.** (a)	**37.** (b)	**38.** (c)	**39.** (c)	**40.** (b)

Hints & Solutions

Physics

1. According to the principle of dimensional homogenity $[p] = \left[\dfrac{a}{V^2}\right]$

$\Rightarrow \quad [a] = [p][V^2] = [ML^{-1}T^{-2}][L^6] = [ML^5T^{-2}]$

Hence, the units of $a = \text{gm} \times \text{cm}^5 \times \text{s}^{-2}$
$= \text{Dyne} \times \text{cm}^4$

2. Given, $[C^2LR] = \left[C^2L^2\dfrac{R}{L}\right] = \left[(LC)^2\left(\dfrac{R}{L}\right)\right]$

and we know that frequency of LC circuits $f = \dfrac{1}{2\pi}\dfrac{1}{\sqrt{LC}}$. Here the dimension of LC is equal to $[T^2]$.

$\left[\dfrac{L}{R}\right]$ gives the time constant of L - R circuit, so that the dimension of $\dfrac{L}{R}$ is equal to [T].

Hence the required dimensions

$\left[(LC)^2\left(\dfrac{R}{L}\right)\right] = [T^2]^2[T^{-1}] = [T^3]$.

3. Given, number of particle passing from unit area in unit time $= n$

$= \dfrac{\text{Number of particle}}{A \times t} = \dfrac{[M^0L^0T^0]}{[L^2][T]} = [L^{-2}T^{-1}]$

$[n_1] = [n_2] =$ Number of particle in unit volume $= [L^{-3}]$

Now, from the given formula $[D] = \dfrac{[n][x_2 - x_1]}{[n_2 - n_1]}$

$= \dfrac{[L^{-2}T^{-1}][L]}{[L^{-3}]}$

$= [L^2T^{-1}]$

4. Given that $v = 4t^3 - 2t$

$x = \int v\,dt,\ x = t^4 - t^2 + C$, at $t = 0,\ x = 0$

$\Rightarrow \quad C = 0$

When particle is $2m$ away from the origin, then

$2 = t^4 - t^2 \Rightarrow t^4 - t^2 - 2 = 0$

$\Rightarrow (t^2 - 2)(t^2 + 1) = 0$

$\Rightarrow \quad t = \sqrt{2}$ s

$a = \dfrac{dv}{dt} = \dfrac{d}{dt}(4t^3 - 2t)$

$a = 12t^2 - 2$

For $\quad t = \sqrt{2}$ sec

$a = 12 \times (\sqrt{2})^2 - 2 = 22 \text{ m/s}^2$

5. From given a - t graph acceleration is increasing at constant rate

$\therefore \dfrac{da}{dt} = k$ (constant) $\Rightarrow a = kt$ (by integration)

$\Rightarrow \quad \dfrac{dv}{dt} = kt \Rightarrow dv = kt\,dt$

$\Rightarrow \quad \int dv = k\int t\,dt \Rightarrow v = \dfrac{kt^2}{2}$

i.e., v is dependent on time parabolically and parabola is symmetric about v-axis and suddenly acceleration becomes zero. *i.e.*, velocity becomes constant. Hence the most stable graph is (c).

6. Let in 2 s body reaches upto point A and after one more sec upto point B.

Total time of ascent for a body is given 3 s

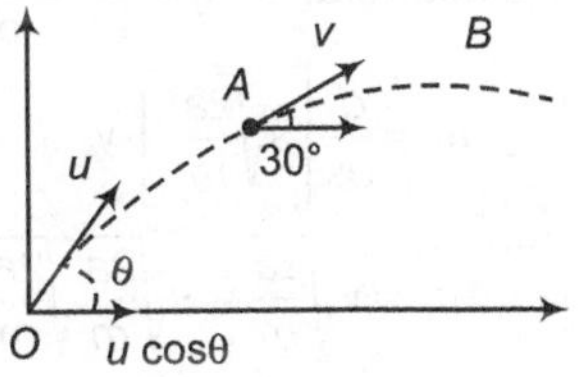

i.e., $\quad t = \dfrac{u\sin\theta}{g} = 3$

$\therefore \quad u\sin\theta = 10 \times 3$

$u\sin\theta = 30 \quad \ldots\text{(i)}$

Horizontal component of velocity remains always constant

$u\cos\theta = v\cos 30° \quad \ldots\text{(ii)}$

For vertical upward motion between point O and A

$v\sin 30° = u\sin\theta - g \times 2 \quad$ [Using $v = u - gt$]

$v \sin 30° = 30 - 20$ [As $u \sin\theta = 30$]

$\therefore \quad v = 20$ m/s

Substituting this value v, we get in Eq. (ii)

$u\cos\theta = 20\cos 30° = 10\sqrt{3}$ ...(iii)

From Eqs. (i) and (iii)

$$u = 20\sqrt{3} \text{ and } \theta = 60°$$

7. In non-uniform circular motion two forces will work on a particle F_c and F_t.

So, the net force $F_{Net} = \sqrt{F_c^2 + F_t^2}$ ...(i)

Centripetal force $F_c = \dfrac{mv^2}{R} = \dfrac{2as^2}{R}$ ...(ii)

[given $\frac{1}{2}mv^2 = as^2$ given]

Again from $\frac{1}{2}mv^2 = as^2$

$\Rightarrow \quad v^2 = \dfrac{2as^2}{m}$

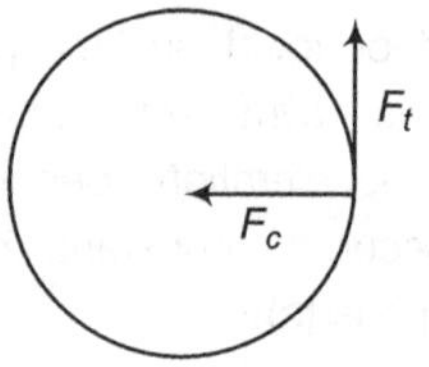

$\Rightarrow \quad v = s\sqrt{\dfrac{2a}{m}}$

Tangential acceleration $a_t = \dfrac{dv}{dt} = \dfrac{dv}{ds}\cdot\dfrac{ds}{dt}$

$\Rightarrow \quad a_t = \dfrac{d}{ds}\left[s\sqrt{\dfrac{2a}{m}}\right]\cdot v$

$a_t = v\sqrt{\dfrac{2a}{m}} = s\sqrt{\dfrac{2a}{m}}\sqrt{\dfrac{2a}{m}} = \dfrac{2as}{m}$

and $\quad F_t = ma_t = 2as$ ...(iii)

Now, on substituting value of F_c and F_t in Eq.(i), we get

$$\therefore F_{Net} = \sqrt{\left(\frac{2as^2}{R}\right)^2 + (2as)^2} = 2as\left[1 + \frac{s^2}{R^2}\right]^{1/2}$$

8. We have $T_1 = \dfrac{(m_2 + m_3)F}{m_1 + m_2 + m_3}$

Now $T_1 = \dfrac{(3+5)10}{2+3+5} = 8$ N

9. When the block m_2 moves downward with acceleration a, the acceleration of mass m_1 will be $2a$, because it covers double distance in the same time in comparison to m_2.

Let T is the tension in the string.

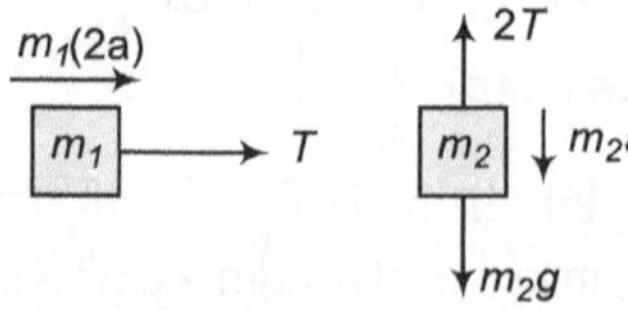

From the free body diagram of A and B we get

$T = m_1 2a$ ...(i)

$m_2g - 2T = m_2a$ ...(ii)

From Eqs. (i) and (ii), we get

$$a = \frac{m_2g}{(4m_1 + m_2)}$$

10. B will begin to slide on A, if pseudo force is more than limiting friction $F > F_1 \Rightarrow m\left(\dfrac{F}{m+M}\right) > \mu_s R$

$\Rightarrow \quad m\left(\dfrac{F}{m+M}\right) > 0.15mg$

$\therefore \quad F > 1.764$ N

11. We know $F = -\dfrac{dU}{dx}$

$\Rightarrow dU = -F\cdot dx$

$\Rightarrow U = -\int_0^x(-kx + ax^3)dx \Rightarrow U = \dfrac{kx^2}{2} - \dfrac{ax^4}{4}$

$\therefore$ We get $U = 0$ at $x = 0$ and $x = \sqrt{\dfrac{2k}{a}}$

Also we get U = negative for $x > \sqrt{\dfrac{2k}{a}}$

From the given function we can see that $F = 0$ at $x = 0$ *i.e.*, slope of U-x graph is zero at $x = 0$.

12. Moment of inertia of rod AB about point $P = \dfrac{1}{12}Ml^2$

MI of rod AB about point $O = \dfrac{Ml^2}{12} + M\left(\dfrac{l}{2}\right)^2 = \dfrac{1}{3}Ml^2$ [from the theorem of parallel axis]

and the system consists of 4 rods of similar type so by the symmetry $I_{system} = \dfrac{4}{3}Ml^2$

13. $\omega_1 = 20 \text{rad/s}, \quad \omega = 0, \quad t = 4\,\text{s}$. So angular retardation $\alpha = \dfrac{\omega_1 - \omega_2}{t} = \dfrac{20}{4} = 5\,\text{rad/s}^2$

Now, angular speed after 2s $\omega_2 = \omega_1 - \alpha t = 20 - 5 \times 2 = 10\,\text{rad/s}$

Work done by torque in 2 s = loss in kinetic energy $= \dfrac{1}{2} I(\omega_1^2 - \omega_1^2) = \dfrac{1}{2}(0.20)((20)^2 - (10)^2)$

$$= \frac{1}{2}(0.20)(400 - 100)$$

$$= \frac{1}{2}(0.20) \times (300) = \frac{1}{2} \times 60 = 30$$

14. On the graph stress is represented on x-axis and strain y-axis

So from the graph $y = \cot\theta = \dfrac{1}{\tan\theta} \propto \dfrac{1}{\theta}$

[where, θ is the angle from stress axis]

$\therefore \quad Y_P < Y_Q < Y_R \quad$ [As $\theta_P < \theta_Q < \theta_R$]

We can say that elasticity of wire P is minimum and R is maximum.

15. Energy released

$$= 4\pi T R^3\left(\frac{1}{r} - \frac{1}{R}\right) = 3\left(\frac{4}{3}\pi R^3\right)T\left(\frac{1}{r} - \frac{1}{R}\right)$$

$$= 3VT\left(\frac{1}{r} - \frac{1}{R}\right)$$

16. When substances are mixed in equal volume then density $= \dfrac{\rho_1 + \rho_2}{2} = 4$

$$\Rightarrow \quad \frac{\rho_1 + \rho_2}{2} = 4 \Rightarrow \rho_1 + \rho_2 = 8 \quad \text{...(i)}$$

When substances are mixed in equal masses then density $= \dfrac{2\rho_1 \rho_2}{\rho_1 + \rho_2} = 3$

$$\Rightarrow \quad 2\rho_1\rho_2 = 3(\rho_1 + \rho_2) \quad \text{...(ii)}$$

On solving (i) and (ii) we get

$\rho_1 = 6$ and $\rho_2 = 2$.

17. We have, ideal gas equation $pV = \mu RT = \left(\dfrac{N}{N_A}\right)RT$

where, N = number of molecule, N_A = Avogadro number

$$\therefore \quad \frac{N_1}{N_2} = \left(\frac{p_1}{p_2}\right)\left(\frac{V_1}{V_2}\right)\left(\frac{T_2}{T_1}\right) = \left(\frac{p}{2p}\right)\left(\frac{V}{V/4}\right)\left(\frac{2T}{T}\right) = \frac{4}{1}$$

18. Initially for container A, $p_0V_0 = n_0RT_0$

For container B, $p_0V_0 = n_0RT_0$

$$\therefore \quad n_0 = \frac{p_0V_0}{RT_0}$$

Total number of moles $= n_0 + n_0 = 2n_0$

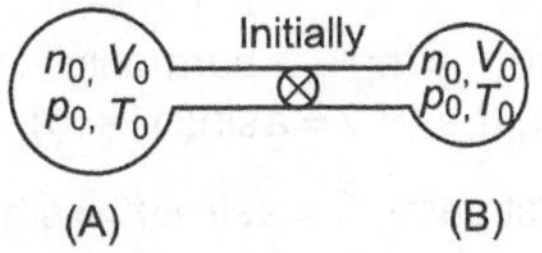

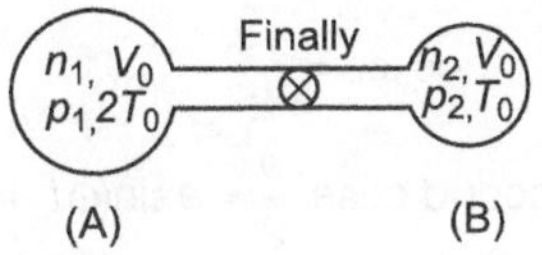

Since, even on heating the total number of moles is conserved

Hence, $\quad n_1 + n_2 = 2n_0 \quad$ ...(i)

If p be the common pressure, then

For container A, $pV_0 = n_1R\,2T_0 \quad \therefore \; n_1 = \dfrac{pV_0}{2RT_0}$

For container A, $pV_0 = n_2RT_0 \quad \therefore \; n_2 = \dfrac{pV_0}{RT_0}$

Substituting the value of n_0, n_1 and n_2 in Eq. (i) we get

$$\frac{pV_0}{2RT_0} + \frac{pV_0}{RT_0} = \frac{2 \cdot p_0V_0}{RT_0} \Rightarrow p = \frac{4}{3}p_0$$

Number of moles in container A (at temperature $2T_0$)

$$= n_1 = \frac{PV_0}{2RT_0} = \left(\frac{4}{3}p_0\right)\frac{V_0}{2RT_0}$$

$$= \frac{2}{3}\frac{p_0V_0}{RT_0} \quad \left[\text{As } p = \frac{4}{3}p_0\right]$$

19. At lower pressure, we can assume that given gas behaves as ideal gas so $\dfrac{pV}{RT}$ = constant but when pressure increase, the decrease in volume will not take place in same proportion so $\dfrac{pV}{RT}$ will increases.

20. Since both the liquid and the flask undergoes volume expansion and the flask expands first therefore the level of the liquid initially falls and then rises.

21. $\frac{dQ}{dt} = \frac{KA\Delta\theta}{l}$, for both rods K, A and $\Delta\theta$ are same

$\therefore \quad \frac{dQ}{dt} \propto \frac{1}{l}$

so $\frac{(dQ/dt)_{\text{semi-circular}}}{(dQ/dt)_{\text{straight}}} = \frac{l_{\text{straight}}}{l_{\text{semi-circular}}} = \frac{2r}{\pi r} = \frac{2}{\pi}$

22. Let two simple harmonic motions are $y = a\sin\omega t$ and $y = a\sin(\omega t + \phi)$

In the first case, $\frac{a}{2} = a\sin\omega t \Rightarrow \sin\omega t = \frac{1}{2}$

$\therefore \quad \cos\omega t = \frac{\sqrt{3}}{2}$

In the second case, $\frac{a}{2} = a\sin(\omega t + \phi)$

$\Rightarrow \frac{1}{2} = [\sin\omega t \cdot \cos\phi + \cos\omega t \sin\phi]$

$\Rightarrow \frac{1}{2} = \left[\frac{1}{2}\cos\phi + \frac{\sqrt{3}}{2}\sin\phi\right]$

$\Rightarrow \quad 1 - \cos\phi = \sqrt{3}\sin\phi$

$\Rightarrow \quad (1-\cos\phi)^2 = 3\sin^2\phi$

$\Rightarrow \quad (1-\cos\phi)^2 = 3(1-\cos^2\phi)$

On solving we get

$\cos\phi = +1 \text{ or } \cos\phi = \frac{-1}{2}$

i.e., $\quad \phi = 0 \text{ or } \phi = 120°$

23. PD between X and Y is $V_{XY} = V_X - V_Y = 1 \times 4$

and PD between X and Z is $V_{XZ} = V_X - V_Z = 1 \times 16 = 16\text{ V}$

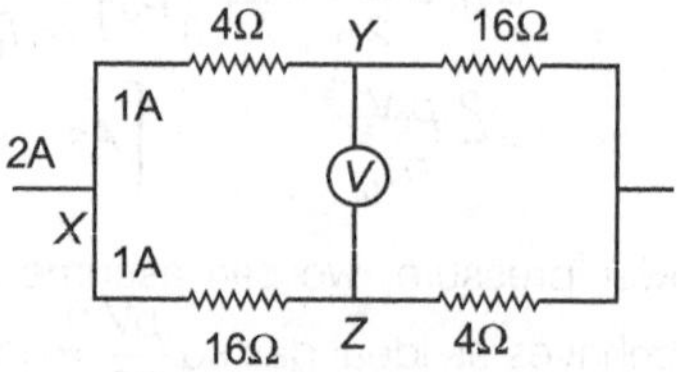

Now, we get potential difference between Y and Z *i.e.*, reading of voltmeter is $V_Y - V_Z = 12\text{ V}$

24. As we know that work done

$= U_{\text{final}} - U_{\text{initial}} = \frac{1}{2}C(V_2^2 - V_1^2)$

When potential difference increases from 5V to 10 V then $W = \frac{1}{2}C(10^2 - 5^2)$ ...(i)

When potential difference increases form 10 V to 15 V, then $W' = \frac{1}{2}C(15^2 - 10^2)$ ...(ii)

On solving Eqs. (i) and (ii) we get

$W' = 1.67\,W$

25. $\lambda_{\text{neutron}} \propto \frac{1}{\sqrt{T}} \Rightarrow \frac{\lambda_1}{\lambda_2} = \sqrt{\frac{T_2}{T_1}}$

$\Rightarrow \frac{\lambda_1}{\lambda_2} = \sqrt{\frac{(273+927)}{(273+27)}} = \sqrt{\frac{1200}{300}} = 2 \Rightarrow \lambda_2 = \frac{\lambda}{2}$

26. On using $I = \frac{P}{A}$; where P = radiation power

$\Rightarrow \quad P = I \times A \Rightarrow \frac{nhc}{t\lambda} = IA \Rightarrow \frac{n}{t} = \frac{IA\lambda}{hc}$

Hence, number of photons entering per sec the eye $\left(\frac{n}{t}\right) = \frac{10^{-10} \times 10^{-6} \times 5.6 \times 10^{-7}}{6.6 \times 10^{-34} \times 3 \times 10^{8}} = 300$

27. From $E = W_0 + \frac{1}{2}mv^2_{\text{max}} \Rightarrow v_{\text{max}} = \sqrt{\frac{2E}{m} - \frac{2W_0}{m}}$

(where $E = \frac{hc}{\lambda}$)

If wavelength of incident light charges from λ to $\frac{3\lambda}{4}$ (decreases).

Let energy of incident light charges from E to E' and speed of fastest electron changes from v to v' then

$v = \sqrt{\frac{2E}{m} - \frac{2W_0}{m}}$ ...(i)

and $\quad v' = \sqrt{\frac{2E'}{m} - \frac{2W_0}{m}}$ ...(ii)

As $\quad E \propto \frac{1}{\lambda} \Rightarrow E' = \frac{4}{3}E$

Hence, $\quad v' = \sqrt{\frac{2\left(\frac{4}{3}E\right)}{m} - \frac{2W_0}{m}}$

$\Rightarrow \quad v' = \left(\frac{4}{3}\right)^{1/2}\sqrt{\frac{2E}{m} - \frac{2W_0}{m\left(\frac{4}{3}\right)^{1/2}}}$

$\Rightarrow \quad v' = \left(\frac{4}{3}\right)^{1/2} X = \sqrt{\frac{2E}{m} - \frac{2W_0}{m\left(\frac{4}{3}\right)^{1/2}}} > v$

So $\quad v' > \left(\frac{4}{3}\right)^{1/2} v.$

28. Using $\frac{1}{\lambda}=RZ^2\left(\frac{1}{n_1^2}-\frac{1}{n_2^2}\right) \Rightarrow \lambda \propto \frac{1}{Z^2} \Rightarrow$

$\lambda_{Li} : \lambda_{He} : \lambda_H = \frac{1}{9} : \frac{1}{4} : \frac{1}{1} = 4:9:36$

29. By using $N = N_0\left(\frac{1}{2}\right)^{1/T_{1/2}}$

$$\Rightarrow \quad t = \frac{T_{12}\log_e\left(\frac{N_0}{N}\right)}{\log_e(2)}$$

$$\Rightarrow \quad t \propto \log_e \frac{N_0}{N}$$

$$\Rightarrow \quad \frac{t_1}{t_2} = \frac{\log_e\left(\frac{N_0}{N}\right)_1}{\log_e\left(\frac{N_0}{N}\right)_2}$$

Hence, $\frac{6}{10} = \frac{\log_e(8/1)}{\log_e(N_0/N)}$

$$\Rightarrow \quad \log_e \frac{N_0}{N} = \frac{10}{6}\log_e(8) = \log_e 32$$

$$\Rightarrow \quad \frac{N_0}{N} = 32$$

So, fraction that decays $= 1 - \frac{1}{32} = \frac{31}{32}$

30. According to conservation of momentum $4v = (A-4)v' \Rightarrow v' = \frac{4v}{A-4}$

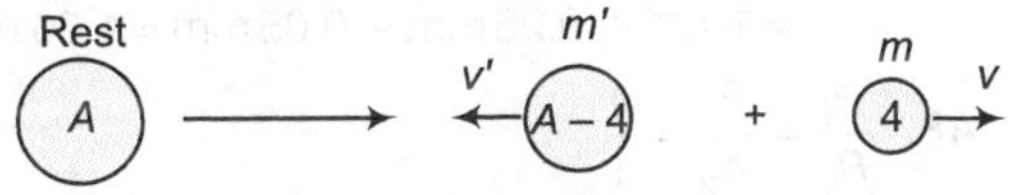

31. ${}_{Z=90}\text{Th}^{A=228} \longrightarrow {}_{Z'=83}\text{Bi}^{A'=212}$

Number of α-particles emitted

$$n_\alpha = \frac{A-A'}{4} = \frac{228-212}{4} = 4$$

Number of β-particles emitted

$$n_\beta = 2n_\alpha - Z + Z' = 2\times 4 - 90 + 83 = 1.$$

32. Radius of each nucleus $R = R_0(A)^{1/3} = 1.2(64)^{1/3}$ $= 4.8\,\text{m}$

Distance between two nuclei $(r) = 2R$

So, potential energy $U = \frac{k\cdot q^2}{r}$

$$= \frac{9\times 10^9 \times (1.6\times 10^{-19}\times 29)^2}{2\times 4.8\times 10^{-15}\times 1.6\times 10^{-19}} = 126.15\,\text{MeV}$$

33. At neutral point magnetic field due to magnet = Horizontal component of the earth's magnetic field.

$$\Rightarrow \frac{\mu_0}{4\pi}\cdot\frac{2M}{r^3} = B_H \Rightarrow \frac{10^{-7}\times 2\times M\times 1}{(0.2)^3} = 0.3\times 15^4$$

$$\Rightarrow M = 1.2\,\text{amp}\times\text{m}^2 = 1200\,ab-\text{amp}\times\text{cm}^2.$$

34. By using $T = 2\pi\sqrt{\frac{1}{MB_H}} \Rightarrow \frac{T_1}{T_2} = \sqrt{\frac{(B_H)_2}{(B_H)_1}}$

$$\Rightarrow \quad \frac{60/40}{2.5} = \sqrt{\frac{(B_H)_2}{0.1\times 10^{-5}}}$$

$$\Rightarrow \quad (B_H)_2 = 0.36\times 10^{-6}\,\text{T}$$

35. $H = ni \Rightarrow i = \frac{H}{n} = \frac{4\times 10^3}{500} = 8\,\text{A}$

36. Magnetic field due to one side of the square at centre O.

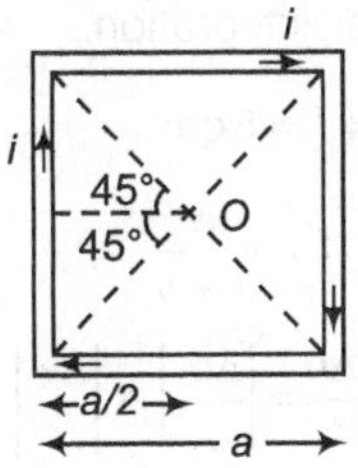

$$B_1 = \frac{\mu_0}{4\pi}\cdot\frac{2i\sin 45°}{a/2}$$

$$\Rightarrow \quad B_1 = \frac{\mu_0}{4\pi}\cdot\frac{2\sqrt{2}\,i}{a}$$

Hence, magnetic field at centre due to all sides

$$B_{net} = 4B_1 = \frac{\mu_0(2\sqrt{2}\,i)}{\pi a}$$

37. $\tau_1 = \frac{8}{9}\mu s$

$\tau_2 = 18\mu s$

$\tau_3 = 4\mu s$

38. $a_A = \frac{g}{2}$

$a_B = g$

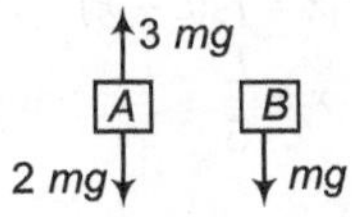

39. $-\frac{1}{F} = \frac{2}{f_l} + \frac{1}{\infty} \Rightarrow F = -\frac{15}{2}$

$$-\frac{2}{15} = \frac{1}{v} - \frac{1}{20}$$

$\Rightarrow v = -12$ cm *i.e.*, 12 cm left of *AB*.

40. $r = f \tan\alpha$

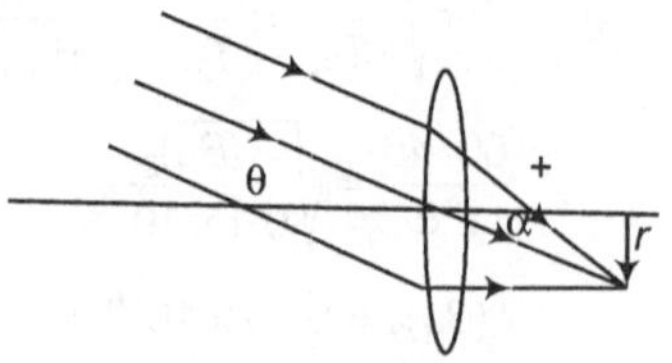

Hence, $\pi r^2 \propto f^2$

41. Disintegration of each nuclei is independent of any factor. Hence, each nuclei has same change of disintegration.

42. $\frac{1}{f} = \frac{1}{v} - \frac{1}{u} \Rightarrow f = 5\,\text{cm}$

$$f = \frac{uv}{u+v}$$

$$\frac{\Delta f}{f} = \left|\frac{\Delta u}{u}\right| + \left|\frac{\Delta v}{v}\right| + \frac{|\Delta u| + |\Delta v|}{|u| + |v|}$$

$$\Delta f = 0.15$$

The most appropriate answer is 5.00 ± 0.10 cm.

43. $\frac{1}{2l}\sqrt{\frac{T_1}{\mu}} = \frac{1}{l}\sqrt{\frac{T_2}{\mu}}$

$$T_2 = \frac{T_1}{4}$$

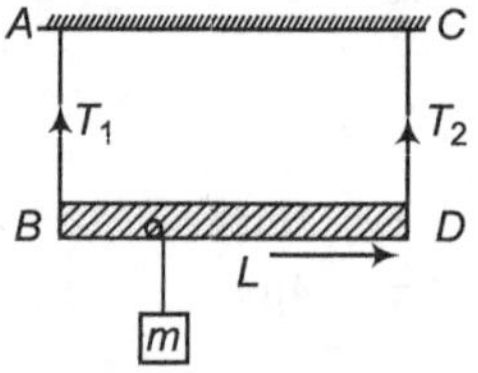

For rotational equilibrium,

$$T_1 x = T_2(L - x)$$

$\Rightarrow$ $x = \frac{L}{5}$

44. $\frac{Gm_A m_B}{(r_A + r_B)^2} = \frac{m_A r_A 4\pi^2}{T_A^2} = \frac{m_B r_B 4\pi^2}{T_B^2}$

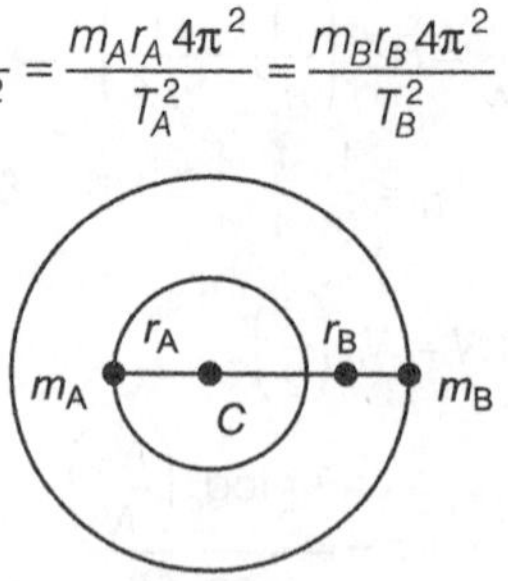

$\Rightarrow$ $m_A r_A = m_B r_B$

$\therefore$ $T_A = T_B$

45. $\frac{2}{5}MR^2 = \frac{3}{2}Mr^2$

$$r = \frac{2R}{\sqrt{15}}$$

46. Option (d) is correct.

47. Zero error $= 5 \times \frac{0.5}{50} = 0.05\,\text{mm}$

Actual measurement

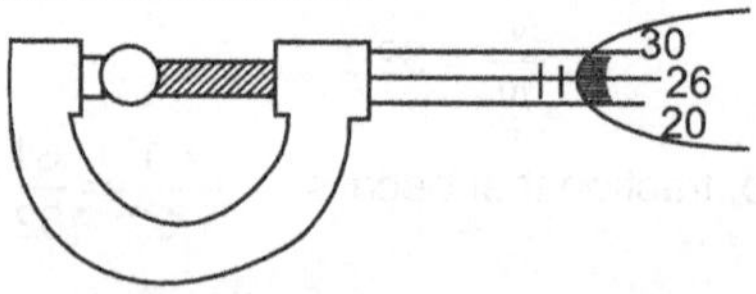

$$= 2 \times 0.5\,\text{mm} + 25 \times \frac{0.5}{50} - 0.05\,\text{mm}$$

$$= 1\,\text{mm} + 0.25\,\text{mm} - 0.05\,\text{mm} = 1.20\,\text{mm}$$

48. $\frac{R_1}{R_2} = \frac{A_1}{A_2} = \frac{4}{1}$

$$\frac{P_1}{P_2} = \frac{I^2 R_1}{I^2 R_2} = \frac{4}{1} \Rightarrow \frac{V_1}{V_2} = \frac{IR_1}{IR_2} = \frac{4}{1}$$

$$\frac{J_1}{J_2} = \frac{1}{4}$$

49. $\Delta W = 0 \quad \therefore \quad \Delta Q = \Delta U$

$$\Delta Q = \Delta U = I^2 R \Delta t = 1^2(100)(5 \times 60) = 30\,\text{kJ}$$

50. Rate of heat gain = 1000 − 160 = 840 J/s

$\therefore$ Required time $= \frac{2 \times 4.2 \times 10^3 \times (77 - 27)}{840}$

$= 500$ s $= 8$ min 20 s

51. $f_1 = \frac{1}{l}\sqrt{\frac{B}{\rho}}$

$$f_2 = \frac{n}{4l}\sqrt{\frac{B}{\rho}} \Rightarrow \frac{f_1}{f_2} = \frac{4}{n} \Rightarrow f_2 = \frac{n}{4} f_1$$

52. Heating of glass bulb is by radiation.

53. Acceleration of the point of suspension

$$a = \frac{d^2y}{dt^2} = 2k = 2\ \text{m/s}^2$$

$$T = 2\pi\sqrt{\frac{L}{g_{\text{eff}}}} \Rightarrow T_1 = 2\pi\sqrt{\frac{L}{10}} \text{ and } T_2 = 2\pi\sqrt{\frac{L}{12}}$$

$$\frac{T_1^2}{T_2^2} = \frac{6}{5}$$

54. For equilibrium, $f = Mg$

$$F = N$$

For rotational equilibrium normal will shift downward.

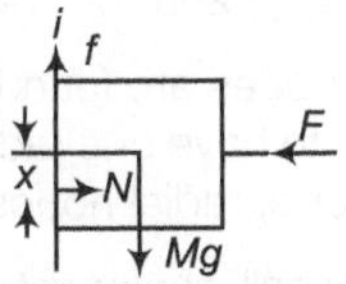

Hence torque due to friction about centre of mass = Torque due to normal reaction about centre of mass.

55. $\mathbf{E}_P = \frac{\sigma}{2\varepsilon_0}(-\hat{\mathbf{k}}) + \frac{(-2\sigma)}{2\varepsilon_0}(\hat{\mathbf{k}}) + \frac{(-\sigma)}{2\varepsilon_0}(\hat{\mathbf{k}})$

$$= \frac{-2\sigma}{\varepsilon_0}\hat{\mathbf{k}}$$

56. $I_0 = \frac{9MR^2}{2} - \left[\frac{M(R/3)^2}{2} + M\left(\frac{2R}{3}\right)^2\right] = 4MR^2$

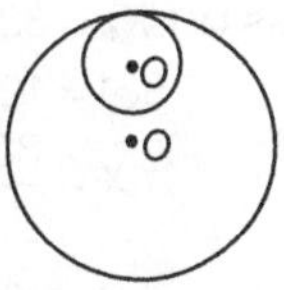

57. Since, B is constant,

$$\therefore \quad \frac{d\phi}{dt} = 0$$

$$\therefore \quad i = 0$$

58. $u^2 = \frac{2gh}{\left[1 - \left(\frac{A_0}{A}\right)^2\right]} = 50\ (\text{m/s})^2$

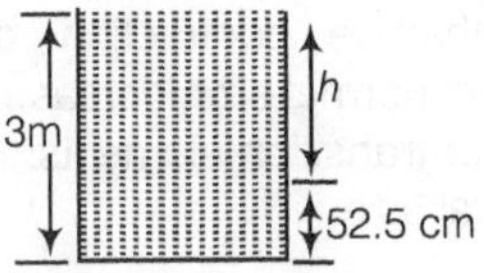

59. $B = -\frac{\Delta p}{\left(\frac{\Delta V}{V}\right)} = -\frac{(1.165 - 1.01) \times 10^5}{0.1}$

$$= 1.55 \times 10^5$$

60. $(Z-1)^2\lambda = \text{constant}$

$$\therefore \quad (10^2)\lambda = 4\lambda(Z-1)^2$$

$$\Rightarrow \quad Z = 6$$

Chemistry

1. Compounds containing acidic hydrogen cannot be used in Grignard's synthesis and the compound CH_2OHCH_2Br contains acidic hydrogen.
2. Four enols, two pairs of geometrical isomers.

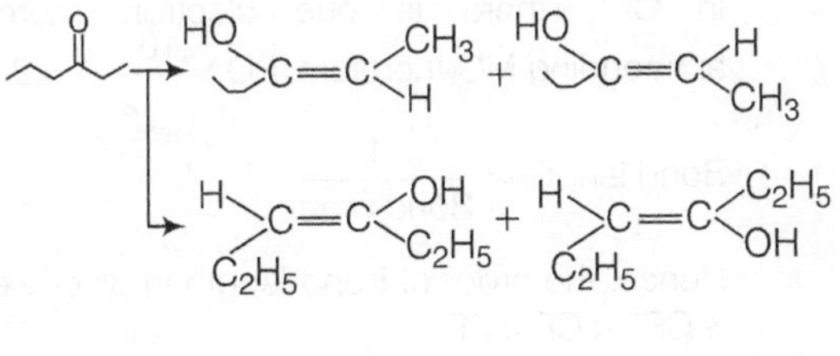

3. 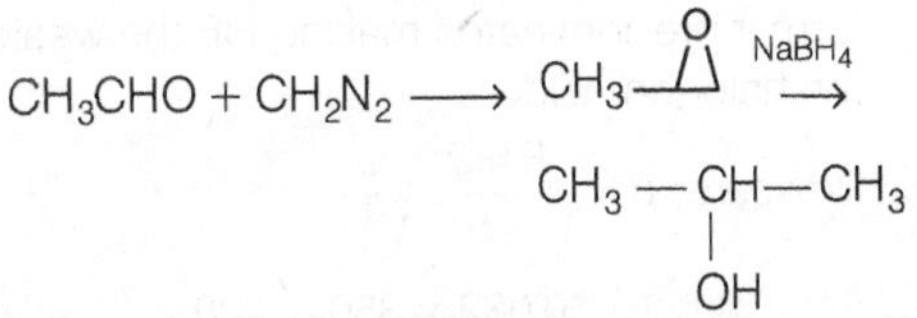

4. As phosphoric acid is a tribasic acid, its *n*-factor can be 1, 2 or 3. Hence, its normality can be M, *2M* or *3M*, Where, *M* is molarity of solution. Therefore, $N = 0.25$, 0.50 or 0.75.
5. $CX_4 + 2O_2 \longrightarrow CO_2 + 2X_2O$

1 mol   2 mol = 2 × 32 g.
9.0 g   1.74 g.

$\because$ 1.74 g O_2 reacts with $CX_4 = 9.0$ g

$\therefore$ 64 g O_2 will react with $CX_4 = \frac{9.0 \times 64}{1.74} = 331$

Molecular mass of $CX_4 = 331$

or $331 = 12 + 4XX$

$$X = \frac{331 - 12}{4} = 79.8 \approx 80$$

6. In $S_4O_6^{2-}$, let 'x' be the oxidation number of S.

Thus, the element X may be bromine.

$$4x - 12 = -2$$

$$x = \frac{10}{4} = 2.5$$

7. $4Au(s) + 8CN^- + O_2(g) + 2H_2O \longrightarrow \underset{\text{soluble complex}}{4[Au(CN)_2]^-} + 4OH^-$

8. The lanthanide contraction cancels almost exactly the normal size increase on descending a group of transition elements. Thus, Zr and Hf have equal size.

9. Mercurous ion is diamagnetic, this indicates that there is no unpaired electron, hence, mercurous ion is not Hg^+ but Hg_2^{2+}.

10. The high oxidising power of fluorine is due to its low heat of dissociation and high heat of hydration. Due to its smallest size, there is considerable repulsion in the non-bonding electrons which weakens the F—F bond. Thus, bond dissociation energy of F_2 is low.

11. HF has the highest bond dissociation enthalpy amongst all the halogen acids. The hydration enthalpy of the fluoride ion is also highest. The former predominates making HF the weakest of the halogen acids.

12. Pressure, $p = \frac{RT}{V - b} - \frac{a}{V^2}$

$$= \frac{0.082 \times 350}{35 - 0.14} - \frac{20}{(35)^2}$$

$$= 0.807 \text{ atm}$$

13. In a unit cell of KBr, there are four formula units of KBr therefore,

Density $(\rho) = \frac{4 \times 119}{6.023 \times 10^{23} \times a^3} = 2.75$

$$a^3 = 2.873 \times 10^{-22} \text{ cm}^3$$

$$= 2.778 \times 10^{-19} \text{mm}^3$$

Number of unit cells per mm^3

$$= \frac{10^{19}}{2.873} = 3.5 \times 10^{18}$$

14. There is one A per unit cell

Number of B per unit cell $= \frac{1}{8} \times 7 = \frac{7}{8}$

Empirical formula $= A_1 B_{7/8}$

$= A_8 B_7$

15. State functions do not depend upon path followed.

16. $S_{\text{rhombic}} \longrightarrow S_{\text{monoclinic}}$

$\Delta H = -297948 - (-298246) = 298$ J

$\Delta S = 32.68 - 32.04 = 0.64$

$\Delta G = \Delta H - T\Delta S = 298 - 298 \times 0.64 = 107.28$ J

17. Two angular nodes are found in d-orbitals. Also for a d-orbital to have one radial node it must be $4d$. [$\because$ Number of radial nodes $= n - l - 1$]

18. The valence shell configuration is $4s^2\ 4p^5$ and for the last electron $n = 4, l = 1$

$$m = -1, 0 \text{ or } +1, s = +\frac{1}{2} \text{ or } -\frac{1}{2}$$

Hence, $4, 1, 1, -\frac{1}{2}$

19. The molecular orbital electronic configuration of CF($15\bar{e}$) is $\sigma 1s^2, \overset{*}{\sigma} 1s^2\ \sigma 2s^2\ \overset{*}{\sigma} 2s^2\ \sigma 2p_z^2$

$\pi 2p_x^2 \approx \pi 2p_y^2, \overset{*}{\pi} 2p_x^1$

Bond order (BO) $= \frac{10 - 5}{2} = 2.5$

In CF^+, there is one electron less in antibonding MO therefore, BO $= \frac{10 - 4}{2} = 3$

In CF^-, there is one electron more in antibonding MO therefore, BO $= \frac{10 - 6}{2} = 2$

Bond length $\propto \frac{1}{\text{Bond order}}$

Hence, the order of bond length in given series is $CF^+ < CF < CF^-$.

20. All are non-polar, molar mass follows the boiling points.

21. In structure II, position of N is changed which is not allowed in resonance.

22. Equilibrium concentration of $N_2O = O_2 = 0.05M$

Let the equilibrium concentration of NO_2 and NO be x M

$$K_c = 0.917 = \frac{(0.05)^2}{x^2} \Rightarrow x = 0.0522 \text{ M}$$

Initial concentrations of NO_2 and NO were $0.05 + x = 0.1022$ M each.

Hence, moles of each gas NO_2 and NO taken initially $= 5 \times 0.1022 = 0.511$

23. Pressure of the system will not influenced the rate of a chemical reaction involving liquids.

24. The two rates are related as– $-\frac{1}{5}\frac{d[Br^-]}{dt} = -\frac{1}{3}\frac{d[Br_2]}{dt}$

$$-\frac{d[Br^-]}{dt} = \frac{5}{3}\frac{d[Br_2]}{dt} = \frac{5}{3} \times 0.025 = 0.042 \text{ Ms}^{-1}$$

25. $LiAlH_4$ will reduce aldehyde further to alcohol.

26. Condensation of NH_2NH_2 at both nitrogens with C_6H_5CHO gives the desired product.

27. Acetaldehyde contains α-H, so it undergoes aldol condensation rather than Cannizzaro reaction.

28. Sucrose does not have free hemiacetal group, so does not undergo mutarotation.

29. Anomers differs in configuration only at C-1 carbons, they are simultaneously diastereomers.

30. 1 and 3 are C-2 epimers. C-2 epimers give same osazone on treatment with phenyl hydrazine.

31. The tertiary structure of proteins refers to three dimensional folding of polymer chain.

32. Rubber has isotactic stereochemistry, not syndiotactic.

33. $n\text{CH}_2{=}\text{CH(CN)}$ (Acrylonitrile) $\xrightarrow{\text{Polymerisation}}$ $\left(\text{CH}_2{-}\text{CH(CN)}{-}\right)_n$ (Polyacrylonitrile)

34. A nitro from *para* positions activate more for S_NAr reaction due to electron withdrawing resonance effect than two nitro from *meta* positions by inductive effect.

35. Ethyl in *o/p* directing group. Direct chlorination of ethyl benzene with $Cl_2/FeCl_3$ would give mixture of 2 and 4-chloroethyl benzene.

36.

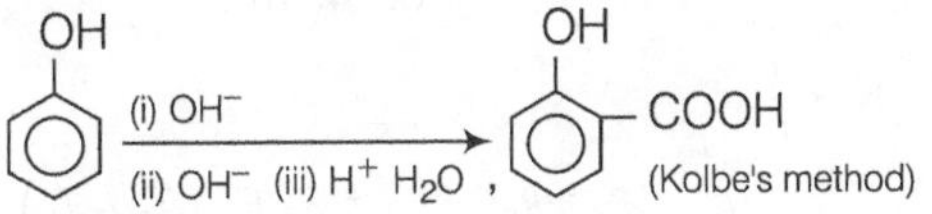

37. An aryl halide with $-NO_2$ substituents are highly activated for aromatic nucleophilic substitution reaction.

38.

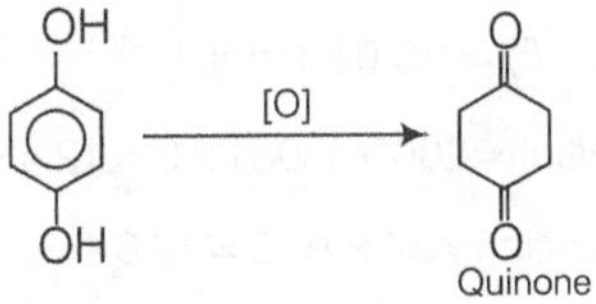

Quinone

39. In physical adsorption, the adsorbent and adsorbate are held together by the van der Waals' forces.

40. Potassium stearate is an example of micelle.

41. Mass of Au deposited = Number of Faraday passed × Eq. mass

$$= \frac{0.30 \times 15 \times 60}{96485} \times \frac{197}{3} = 0.184 \text{ g}$$

42. $E^0_{cell} = E^0_{cathode} - E^0_{anode}$

$$= 0.40 - (-0.44) = 0.84 \text{ V}$$

43. $\underset{(1-\alpha)}{nA} \rightleftharpoons \underset{\frac{\alpha}{n}}{An}$

$$i = 1 - \alpha + \frac{\alpha}{n}$$

44. Osmotic pressure $= iCRT$, For both the solutions, all other conditions are similar except '*i*' which is 1.0 for sucrose while 2.0 for NaCl.

Therefore, osmotic pressure of NaCl

$$= 2 \times 0.25$$

$$= 0.50 \text{atm}$$

45. Mass of X decay = 0.75 g

Moles of X decayed $= \frac{0.75}{253}$

$= $ moles of He formed

$V(\text{mL})$ of He at STP $= \frac{0.75 \times 22400}{253}$

$= 66.40$ mL

46. $K \propto \frac{1}{t}$

$$\log \frac{k_2}{k_1} = \log \frac{t_1}{t_2} = \frac{E_a}{R}\left(\frac{T_2 - T_1}{T_1 \cdot T_2}\right)$$

$$\log \frac{48}{4} = \frac{E_a}{8.314}\left(\frac{10}{300 \times 290}\right)$$

$$E_a = 78.0 \text{ kJ mol}^{-1}$$

47. mmol of H^+ (initial) $= 200 \times 0.031 \times 2 = 12.4$

mmol of OH^- (initial) $= 84 \times 0.15 = 12.6$

mmol of OH^- (left) after neutralisation $= 0.2$

$$[OH^-]_{final} = \frac{0.2}{284} = 7 \times 10^{-4} M$$

pOH = 3.15 and pH = 10.85

48. In $(SiH_3)_3N$, lone pair of the nitrogen is transferred to the empty *d*-orbitals of silicon ($p\pi - d\pi$ overlap) thereby causing planarity of unit.

49. Solid N_2O_5 is ionic. X-ray diffraction shows that solid N_2O_5 exists as $NO_2^+NO_3^-$ (nitronium nitrate).

50. White phosphorus is highly reactive, bursting into flames when exposed to air, and is thus stored under water.

51. The activity of alkaline earth metals as reducing agent increases from Be to Ba.

52. An improved combination of baking powder contains about 40% starch, 30% $NaHCO_3$, 20% $NaAl(SO_4)_2$ and 10% $Ca(H_2PO_4)_2$. Here starch acts as filler.

53. If acid is weak, its conjugate base is strong. Greater the oxidation number of Cl, stronger the acid and thus, weaker the conjugate base.

$$\underset{+1}{HClO} \quad \underset{+3}{HClO_2} \quad \underset{+5}{HClO_3} \quad \underset{+7}{HClO_4}$$
$$\longrightarrow$$

Oxidation number of Cl and acid strength increases

$$ClO^- \quad ClO_2^- \quad ClO_3^- \quad ClO_4^-$$
$$\longleftarrow$$

Conjugate base strength increases.

Hence, ClO^- is the strongest base.

54. I. $\underset{-3}{NH_4^+} \underset{+5}{NO_3^-} \longrightarrow \underset{+1}{N_2O} + 2H_2O$

It is a redox reaction but not a disproportionation reaction. Since, different N-species are involved.

II. $\underset{-3}{NH_4^+} \underset{+3}{NO_2^-} \longrightarrow \underset{0}{N_2}$

same as (I)

III. $\underset{+5\ -1}{PCl_5} \longrightarrow \underset{+3\ -1}{PCl_3} + \underset{0}{Cl_2}$

Thus, it is also not a disproportionation reaction.

55. $Fe^{2+} + e^- \longrightarrow Fe^+$

$NO \longrightarrow NO^+ + e^-$

Based on magnetic properties of the complex, it is found that iron has three unpaired electrons thus it exists as Fe^+ formed by reduction of Fe^{2+} by NO which is oxidised to NO^+.

56. Li – Mg → diagonal relationship.

B — Al, Na — K and Ca — Mg → belongs to same group.

57. Along a group Z (effective nuclear charge) is almost constant. Thus, IE depends mainly on r_n (atomic radius).

58. Valence electron in $A = 3$

Valence electron in $B = 6$

Thus, *A* is electropositive and *B* is electronegative. *A* can lose three electrons and *B* can gain two electrons to attain stable configuration.

Thus, A exists as A^{3+} and B as B^{2-} thus, compound is A_2B_3.

59. In 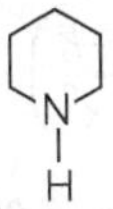 lone pair of nitrogen is not involved in resonance as well as it has less s-character than the nitrogen of pyridine, hence the former is better electron donor.

60. Hyperconjugation involve delocalisation of σ-electrons with adjacent pi-system.

Mathematics

1. Given, $a = \cos\frac{2\pi}{7} + i.\sin\frac{2\pi}{7}$

$\therefore \quad a^7 = \cos 2\pi + i \sin 2\pi$

$[\because e^{i\theta} = \cos\theta + i\sin\theta]$

$= 1$

Also, $\alpha = a + a^2 + a^4, \beta = a^3 + a^5 + a^6$

then the sum of roots,

$$S = \alpha + \beta = a + a^2 + a^3 + a^4 + a^5 + a^6$$

$$\Rightarrow \quad S = \frac{a(1-a^6)}{1-a} = \frac{a-a^7}{1-a}$$

$$= \frac{a-1}{1-a} = -1 \qquad [\because a^7 = 1]$$

Product of the roots,

$P = \alpha\beta = (a + a^2 + a^4)(a^3 + a^5 + a^6)$

$= a^4 + a^5 + 1 + a^6 + 1 + a^2 + 1 + a + a^3 \; [\because a^7 = 1]$

$= 3 + (a + a^2 + a^3 + a^4 + a^5 + a^6) = 3 - 1 = 2$

Hence, the required quadratic equation is

$$x^2 + x + 2 = 0$$

2. Let $x = (1)^{1/n}$

$\Rightarrow \quad x^n - 1 = 0$

or $\quad x^n - 1 = (x-1)(x-\omega)(x-\omega^2)\dots(x-\omega^{n-1})$

$$\Rightarrow \frac{x^n - 1}{x - 1} = (x-\omega)(x-\omega^2)(x-\omega^3)\dots(x-\omega^{n-1})$$

Putting $x = 9$ on both sides, we get

$$(9-\omega)(9-\omega^2)(9-\omega^3)\dots(9-\omega^{n-1}) = \frac{9^n - 1}{8}$$

3. The given expression is

$$\left(x^2 + \frac{1}{x^2} + 2\right)^n = \left[\left(x + \frac{1}{x}\right)^2\right]^n$$

$$= \left(x + \frac{1}{x}\right)^{2n}$$

The binomial contains $(2n+1)$ terms in its expansion.

$\therefore$ The middle term is T_{n+1}.

$$T_{n+1} = {}^{2n}C_n \, (x)^{2n-n} \left(\frac{1}{x}\right)^n.$$

$$= {}^{2n}C_n$$

$$= \frac{(2n)!}{(2n-n)!\,n!} = \frac{(2n)!}{(n!)^2}$$

4. The equation of the tangent to parabola $y^2 = 4x$ is $y = mx + \frac{1}{m}$

This is also the tangent to parabola $x^2 = -8y$

$$\therefore \quad x^2 = -8\left(mx + \frac{1}{m}\right)$$

$\Rightarrow mx^2 + 8m^2x + 8 = 0$

It has equal roots, if

$$64m^4 = 32m \qquad [\because b^2 = 4ac]$$

$$\Rightarrow \quad m^3 = \frac{1}{2} \Rightarrow m = \frac{1}{\sqrt[3]{2}}$$

Hence, equation of the tangent is $y = \frac{1}{\sqrt[3]{2}}x + \sqrt[3]{2}$.

5. There may be two cases.

***Case* I.** Two children get none and one get three and others get one each.

Then, total number. of ways $= \frac{10!}{2! \times 3! \times 7!} \times 10!$

***Case* II.** Two get none and two get 2 each and the others get one each.

Then, total number of ways $= \frac{10!}{(2!)^4 \times 6!} \times 10!$

Hence, total number of ways

$$= \frac{(10!)^2}{2! \times 3! \times 7!} + \frac{(10!)^2}{(2!)^4 \times 6!} = \frac{(10!)^2 \times 25}{(2!)^2 \times 6! \times 84}$$

6. Variance of first n natural numbers is given by

$$\text{Variance} = \frac{\Sigma n^2}{n} - \left(\frac{\Sigma n}{n}\right)^2$$

$$= \frac{n(n+1)(2n-1)}{6n} - \left(\frac{n(n+1)}{2n}\right)^2$$

$$= (n+1)\left[\frac{2n-1}{6} - \frac{(n+1)}{4}\right]$$

$$= \frac{(n+1)}{12}[4n-2-3n-3]$$

$$= \frac{(n+1)}{12} \times (n-5)$$

7. Radius of smaller circle is $r = 2$ cm.

Since, it is touching both the axes, so its centre is (2,2).

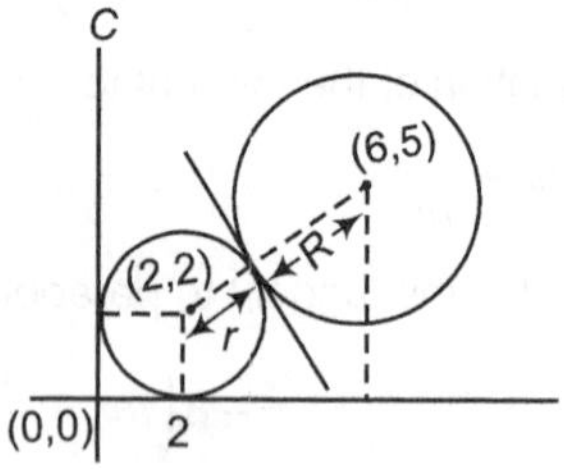

The distance between their centres

$$= r + R = \sqrt{(5-2)^2 + (6-2)^2}$$

$$= \sqrt{3^2 + 4^2} = \sqrt{25} = 5 \text{ units}$$

8. Set $S = \{1, 2, 3, \ldots, 20\}$ is to be partitioned into four sets of equal size *i.e.*, having 5-5 numbers.

The number of ways in which it can be done

$$= {}^{20}C_5 \times {}^{15}C_5 \times {}^{10}C_5 \times {}^{5}C_5$$

$$= \frac{20!}{15! \times 5!} \times \frac{15!}{10! \times 5!} \times \frac{10!}{5! \times 5!} \times 1 = \frac{(20!)}{(5!)^4}$$

9. Total number of cases $= {}^{50}C_1 = 50$

Let A be the event of selecting ticket with sum of digits '8'.

Favourable cases to A are $\{08, 17, 26, 35, 44\}$.

Let B be the event of selecting ticket with product of its digits '7'

Favourable cases to B is only $\{17\}$.

$$\text{Now, } P(B/A) = \frac{P(A \cap B)}{P(A)}$$

$$= \frac{1/50}{5/50} = \frac{1}{5}$$

10. General term in the expansion of $\left(\sqrt{x} - \frac{k}{x^2}\right)^{10}$ is

$$T_{r+1} = {}^{10}C_r (\sqrt{x})^{10-r} \cdot \left(\frac{-k}{x^2}\right)^r$$

$$= {}^{10}C_r x^{\frac{10-r}{2}} \cdot (-k)^r \cdot x^{-2r}$$

$$= {}^{10}C_r (-k)^r x^{\left(\frac{10-5r}{2}\right)}$$

$\because$ The term is free from x

$\therefore$ Put $\frac{10-5r}{2} = 0$

$\Rightarrow$ $r = 2$

Now, ${}^{10}C_2(-k)^2 = 405$

$\Rightarrow$ $\frac{10 \times 9}{1 \times 2} \cdot k^2 = 405$

$\Rightarrow$ $k^2 = \frac{405}{45} = 9$

$\Rightarrow$ $k = \pm 3$

11. The given equation is $8x^2 - 26x + 15 = 0$.

$\therefore$ The sum of roots,

$\tan\frac{\alpha}{2} + \tan\frac{\beta}{2} = \frac{26}{8} = \frac{13}{4}$ and product of roots,

$\tan\frac{\alpha}{2} \cdot \tan\frac{\beta}{2} = \frac{15}{8}$

$$\therefore \tan\left(\frac{\alpha+\beta}{2}\right) = \frac{\tan\frac{\alpha}{2} + \tan\frac{\beta}{2}}{1 - \tan\frac{\alpha}{2} \cdot \tan\frac{\beta}{2}}$$

$$= \frac{\frac{13}{4}}{1 - \frac{15}{8}} = -\frac{26}{7}$$

$$\text{Now, } \cos(\alpha+\beta) = \frac{1 - \tan^2\left(\frac{\alpha+\beta}{2}\right)}{1 + \tan^2\left(\frac{\alpha+\beta}{2}\right)}$$

$$\left[\because \cos 2\theta = \frac{1 - \tan^2\theta}{1 + \tan^2\theta}\right]$$

$$= \frac{1 - \left(-\frac{26}{7}\right)^2}{1 + \left(-\frac{26}{7}\right)^2} = \frac{49 - 676}{49 + 676} = -\frac{627}{725}$$

12. Given relations are

$a\cos^3\alpha + 3a\cos\alpha\sin^2\alpha = m$ ...(i)

$a\sin^3\alpha + 3a\cos^2\alpha\sin\alpha = n$ ...(ii)

On adding Eq. (i) and (ii) we get

$(m+n) = a\cos^3\alpha + 3a\cos\alpha\sin^2\alpha + 3a\cos^2\alpha\sin\alpha + a\sin^3\alpha$

$= a(\cos\alpha + \sin\alpha)^3$

Similarly on subtracting Eq. (ii) from Eq. (i), we get

$m - n = a(\cos\alpha - \sin\alpha)^3$

Now, $(m+n)^{2/3} + (m-n)^{2/3}$

$= a^{2/3}\{(\cos\alpha + \sin\alpha)^2 + (\cos\alpha - \sin\alpha)^2\}$

$= a^{2/3}\{2(\cos^2\alpha + \sin^2\alpha)\} = 2a^{2/3}$

13. Let in the figure, $\angle A > \angle B > \angle C$

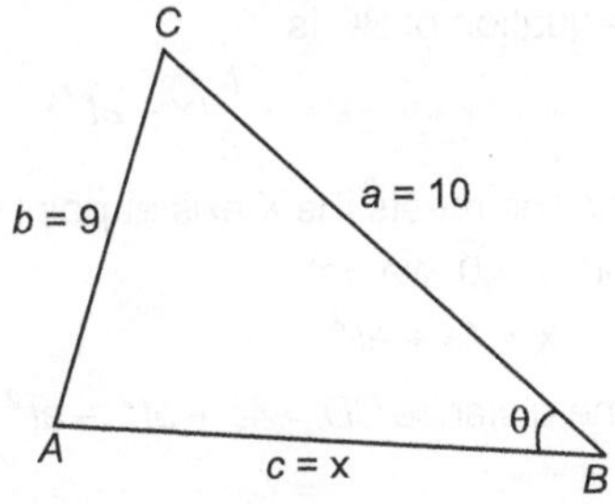

$\because$ Angles are in A.P.

$\therefore \quad \angle C = \theta - d, \angle B = \theta$

and $\quad \angle A = \theta + d$

As $\quad \angle A + \angle B + \angle C = 180°$

$\Rightarrow \theta + d + \theta + \theta - d = 180°$

$\Rightarrow \quad 3\theta = 180°$

$\Rightarrow \quad \theta = 60°$

Now in ΔABC,

$$\cos B = \frac{a^2 + c^2 - b^2}{2ac}$$

$$\cos 60° = \frac{(10)^2 + x^2 - 9^2}{2 \times 10 \times x}$$

$$\Rightarrow \frac{1}{2} = \frac{100 + x^2 - 9^2}{20x}$$

$\Rightarrow x^2 - 10x + 19 = 0$

$$\Rightarrow x = \frac{10 \pm \sqrt{100 - 76}}{2}$$

$$= \frac{10 \pm \sqrt{24}}{2} = 5 \pm \sqrt{6}$$

14. Let the equation of the line parallel to $x - 2y = 1$ is $x - 2y + \lambda = 0$

Since, it passes through (3, 5)

$\therefore \quad 3 - 10 + \lambda = 0 \Rightarrow \lambda = 7$

$\therefore$ The line is $x - 2y + 7 = 0$.

The point of intersection of $x - 2y + 7 = 0$ and $2x + 3y - 14 = 0$ is (1, 4).

$\therefore$ The distance between (3, 5) and (1, 4)

$= \sqrt{(3-1)^2 + (5-4)^2} = \sqrt{4+1} = \sqrt{5}$

15. Equation of line is $\frac{x}{a} + \frac{y}{b} = 1$ ...(i)

Let the foot of the perpendicular drawn from the origin to the line be $P(x_1, y_1)$.

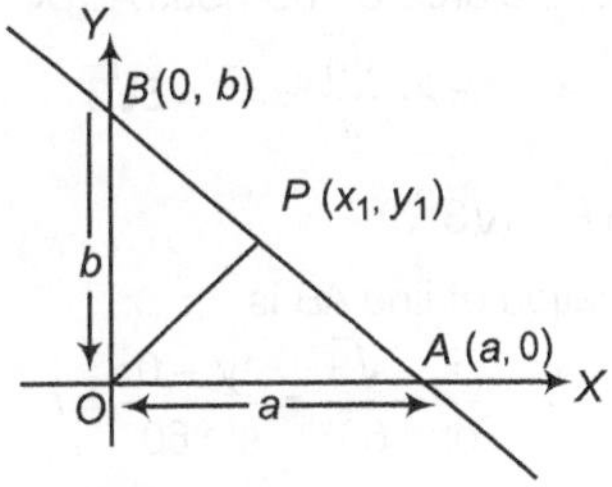

Since, $OP \perp AB$

$\therefore$ slope of $OP \times$ slope of $AB = -1$

$\Rightarrow \quad \frac{y_1}{x_1} \times \frac{b}{-a} = -1 \Rightarrow by_1 = ax_1$ ...(ii)

Since, P lies on the line AB, so

$$\frac{x_1}{a} + \frac{y_1}{b} = 1$$

$\Rightarrow \quad bx_1 + ay_1 = ab$ ...(iii)

From (ii) and (iii), we get

$$x_1 = \frac{ab^2}{a^2 + b^2} \text{ and } y_1 = \frac{a^2 b}{a^2 + b^2}$$

Now, $x_1^2 + y_1^2 = \left(\frac{ab^2}{a^2+b^2}\right)^2 + \left(\frac{a^2 b}{a^2+b^2}\right)^2$

$$= \frac{a^2b^4}{(a^2+b^2)^2} + \frac{a^4b^2}{(a^2+b^2)^2} = \frac{a^2b^2}{(a^2+b^2)^2}(a^2+b^2)$$

$$= \frac{a^2b^2}{a^2+b^2} = \frac{1}{\frac{1}{a^2} + \frac{1}{b^2}}$$

$$\Rightarrow x_1^2 + y_1^2 = \frac{1}{c^2} \qquad \left(\because \frac{1}{a^2} + \frac{1}{b^2} = \frac{1}{c^2}\right)$$

Thus, the locus of $P(x_1, y_1)$ is $x^2 + y^2 = c^2$, which is a circle.

16. Circles with centres $A(-2, 0)$ and $B(+2, 0)$, each of radius 4 has Y-axis as their common chord.

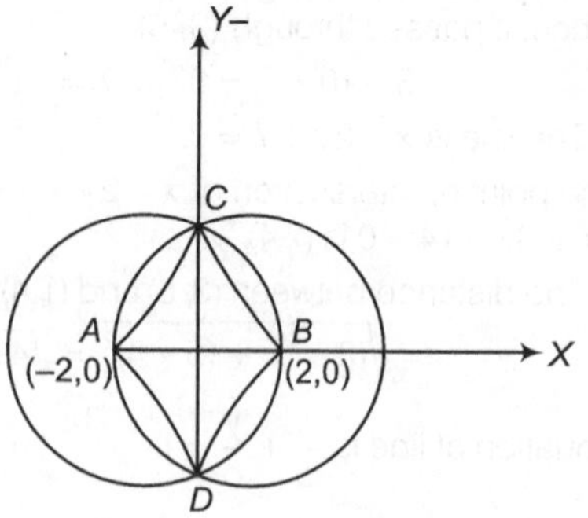

$\Rightarrow \Delta ABC$ is an equilateral triangle.

Hence, the area of rhombus $ADBC$ is

$$= 2 \cdot \frac{\sqrt{3}}{4} \times 4^2 = 8\sqrt{3}$$

17. Given $P = (\sqrt{3}, 0)$

$\therefore$ Equation of line AB is

$$\frac{x - \sqrt{3}}{\cos 60^\circ} = \frac{y - 0}{\sin 60^\circ} = r \quad \text{(say)}$$

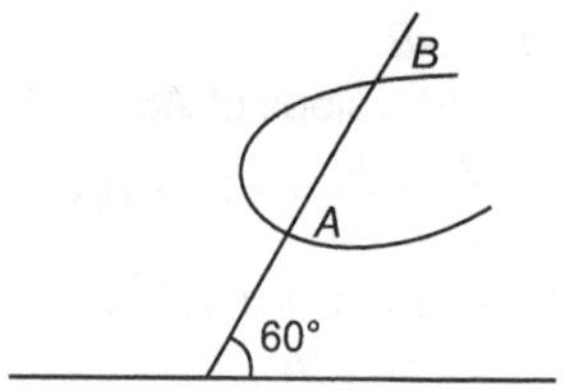

$$\Rightarrow \quad x = \sqrt{3} + \frac{r}{2}, \; y = \frac{r\sqrt{3}}{2}$$

$\therefore$ point $\left(\sqrt{3} + \frac{r}{2}, \frac{r\sqrt{3}}{2}\right)$ lies on $y^2 = x + 2$

$$\therefore \quad \frac{3r^2}{4} = \sqrt{3} + \frac{r}{2} + 2$$

$$\Rightarrow \quad \frac{3r^2}{4} - \frac{r}{2} - (2 + \sqrt{3}) = 0$$

Let the roots be r_1 and r_2, then the product

$$r_1 \times r_2 = PA \cdot PB = \left| \frac{-(2 + \sqrt{3})}{\frac{3}{4}} \right|$$

$$= \frac{4(2 + \sqrt{3})}{3}$$

18. Given equation of parabola is $y^2 = 4ax$.

Let the co-ordinates of B are $(at^2, 2at)$, then the slope of $AB = \frac{2}{t}$

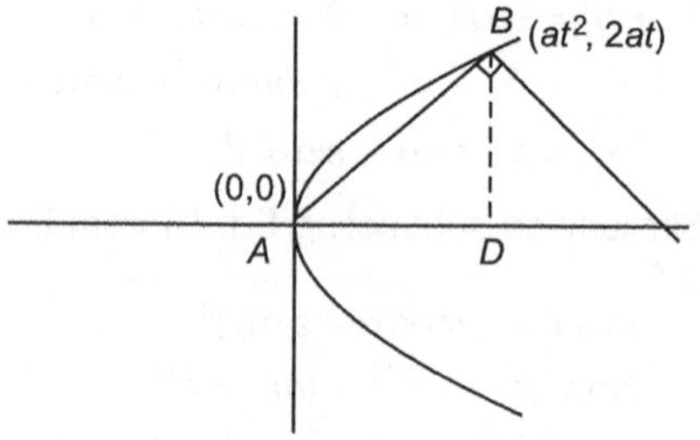

Since, $BC \perp AB$

$\therefore$ Slope of $BC = \frac{-t}{2}$

The equation of BC is

$$y - 2 = -\frac{t}{2}(x - at^2)$$

$\because$ This line meets the X-axis at point C.

So, put $y = 0$, we get

$$\Rightarrow \quad x = 4a + at^2$$

So, the distance $CD = 4a + at^2 - at^2$

$$= 4a$$

19. Given equation of ellipse is $\frac{x^2}{4} + \frac{y^2}{\frac{7}{4}} = 1$

Here, $\quad a^2 = 4$

and $\quad b^2 = \frac{7}{4}$

$\because \quad b^2 = a^2(1 - e^2)$

$\therefore \quad \frac{7}{4} = 4(1 - e^2)$

$\Rightarrow \quad e^2 = 1 - \frac{7}{16} = \frac{9}{16} \Rightarrow e = \frac{3}{4}$,

Thus, the foci are $\left(\pm\frac{3}{2}, 0\right)$.

The radius of the required circle

$$= \sqrt{\left(\frac{3}{2} - \frac{1}{2}\right)^2 + (0 - 2)^2}$$

$$= \sqrt{5}$$

20. The quadrilateral formed by the tangents at the end points of the latusrectum is a rhombus. It is symmetrical about the axes.

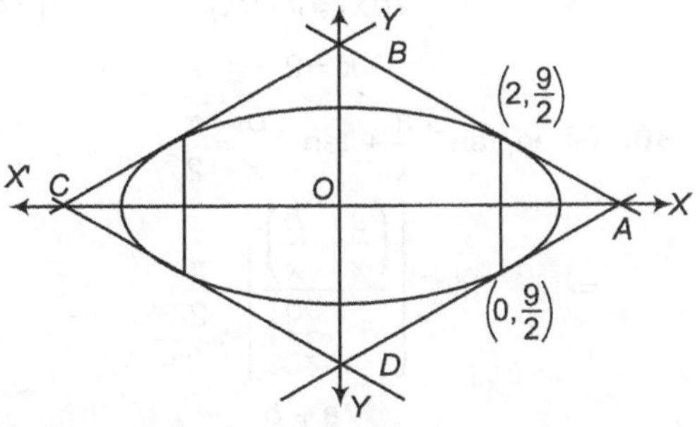

So, total area is four times the area of the right triangle formed by the tangents and the axes in the first quadrant.

Now, $ae = \sqrt{a^2 - b^2}$

$\Rightarrow \quad ae = 2$

$\therefore$ The co-ordinates of one end point of latusrectum are $\left(2, \frac{5}{3}\right)$.

The equation of tangent at that point is

$\frac{x}{\frac{9}{2}} + \frac{y}{3} = 1$, this meets the co-ordinate *axes* at

$A\left(0, \frac{9}{2}\right)$ and B (3, 0).

Area of $\Delta AOB = \frac{1}{2} \times \frac{9}{2} \times 3 = \frac{27}{4}$

Hence, the area of rhombus $ABCD$

$$= 4 \times \frac{27}{4} = 27 \text{ sq units}$$

21. Equation of chord of hyperbola $x^2 - y^2 = a^2$ with mid-point as (h, k) is given by

$$xh - yk = h^2 - k^2$$

$$\Rightarrow \quad y = \frac{h}{k}x - \frac{(h^2 - k^2)}{k}$$

This will touch the parabola $y^2 = 4ax$, if

$$-\left(\frac{h^2 - k^2}{k}\right) = \frac{a}{\frac{h}{k}}$$

$$\Rightarrow \quad ak^2 = h^3 + k^2 h$$

$\therefore$ Locus of the mid-point is $x^3 = y^2 (x - a)$

22. $\lim_{x \to 0} \frac{e^{\sin x} - 1}{x}$

$$= \lim_{x \to 0} \left[\frac{(e^{\sin x} - 1)}{\sin x} \times \frac{\sin x}{x}\right]$$

$$= \lim_{\sin x \to 0} \frac{e^{\sin x} - 1}{\sin x} \times \lim_{x \to 0} \frac{\sin x}{x}$$

$$= 1 \times 1 = 1$$

23. By truth table

p	q	$(p \wedge q)$	$\sim(p \wedge q)$	$q \Leftrightarrow p$	$\sim(q \Leftrightarrow p)$	$\sim(p \wedge q) \vee \sim(q \Leftrightarrow p)$
T	T	T	F	T	F	F
T	F	F	T	F	T	T
F	T	F	T	F	T	T
F	F	F	T	T	F	T

It is clear that the given statement is neither tautology nor contradiction.

24. Given determinant is

$$\Delta = \begin{vmatrix} (a^x + a^{-x})^2 & (a^x - a^{-x})^2 & 1 \\ (b^x + b^{-x})^2 & (b^x - b^{-x})^2 & 1 \\ (c^x + c^{-x})^2 & (c^x - c^{-x})^2 & 1 \end{vmatrix}$$

Applying $C_1 \to C_1 - C_2$, we get

$$\Delta = \begin{vmatrix} 4 & (a^x - a^{-x})^2 & 1 \\ 4 & (b^x - b^{-x})^2 & 1 \\ 4 & (c^x - c^{-x})^2 & 1 \end{vmatrix} = 4 \begin{vmatrix} 1 & (a^x - a^{-x})^2 & 1 \\ 1 & (b^x - b^{-x})^2 & 1 \\ 1 & (c^x - c^{-x})^2 & 1 \end{vmatrix}$$

$= 0$ ($\because$ two columns are identical)

25. Putting $r = 1, 2, 3, \ldots, n$ and using the formula

$$\Sigma 1 = n$$

and $\Sigma r = \frac{n(n+1)}{2}$

$$\Sigma(2r - 1) = 1 + 3 + 5 + \ldots = n^2$$

$$\therefore \sum_{r=1}^{n} \Delta_r = \begin{vmatrix} n & n & n \\ n(n+1) & n^2 + n + 1 & n^2 + n \\ n^2 & n^2 & n^2 + n + 1 \end{vmatrix} = 56$$

Applying $C_1 \to C_1 - C_3, C_2 \to C_2 - C_3$, we get

$$\begin{vmatrix} 0 & 0 & n \\ 0 & 1 & n^2 + n \\ -n-1 & -n-1 & n^2 + n + 1 \end{vmatrix} = 56$$

$$\Rightarrow \quad n(n+1) = 56$$

$$\Rightarrow \quad n^2 + n - 56 = 0$$

$\Rightarrow \quad (n+8)(n-7)=0$

$\Rightarrow \quad n=7, (n \neq -8)$

26. $\Delta = \begin{vmatrix} e^a & e^{2a} & e^{3a} \\ e^b & e^{2b} & e^{3b} \\ e^c & e^{2c} & e^{3c} \end{vmatrix} - \begin{vmatrix} e^a & e^{2a} & 1 \\ e^b & e^{2b} & 1 \\ e^c & e^{2c} & 1 \end{vmatrix}$

$= e^a \cdot e^b \cdot e^c \begin{vmatrix} 1 & e^a & e^{2a} \\ 1 & e^b & e^{2b} \\ 1 & e^c & e^{2c} \end{vmatrix} + \begin{vmatrix} e^a & 1 & e^{2a} \\ e^b & 1 & e^{2b} \\ e^c & 1 & e^{2c} \end{vmatrix}$

$= e^{(a+b+c)} \begin{vmatrix} 1 & e^a & e^{2a} \\ 1 & e^b & e^{2b} \\ 1 & e^c & e^{2c} \end{vmatrix} - \begin{vmatrix} 1 & e^a & e^{2a} \\ 1 & e^b & e^{2b} \\ 1 & e^c & e^{2c} \end{vmatrix}$

$= (e^{a+b+c} - 1) \begin{vmatrix} 1 & e^a & e^{2a} \\ 1 & e^b & e^{2b} \\ 1 & e^c & e^{2c} \end{vmatrix}$

$= 0 \quad (\because a+b+c=0)$

27. $fog = f\{g(x)\} = |\sin x| = \sqrt{\sin^2 x}$

Also, $gof = g\{f(x)\} = \sin^2 \sqrt{x}$

Obviously, $\sqrt{\sin^2 x} = \sqrt{g(x)}$

and $\sin^2 \sqrt{x} = \sin^2\{f(x)\}$

i.e., $g(x) = \sin^2 x$

and $f(x) = \sqrt{x}$

28. Given, $f(x) = \dfrac{x}{\sqrt{1+x^2}}$

$(fof)x = \{f(f(x))\}$

$= f\left(\dfrac{x}{\sqrt{1+x^2}}\right) = \dfrac{\frac{x}{\sqrt{1+x^2}}}{\sqrt{1+\frac{x^2}{1+x^2}}}$

$\Rightarrow \quad (fof)x = \dfrac{x}{\sqrt{1+2x^2}}$

Now, $(fofof)x = f\{(fof)x\}x = f\left(\dfrac{x}{\sqrt{1+2x^2}}\right)$

$= \dfrac{\frac{x}{\sqrt{1+2x^2}}}{\sqrt{1+\frac{x^2}{1+2x^2}}} = \dfrac{x}{\sqrt{1+3x^2}}$

29. On putting $y=0$, in the given relation, we get

$f(x) + f(x) = k \cdot f(x) \cdot f(0)$

$\Rightarrow \quad 2f(x) = k \cdot f(x) \quad [\because f(0)=1]$

$\therefore \quad k = 2$

30. Given, $\tan^{-1}\dfrac{a}{x} + \tan^{-1}\dfrac{b}{x} = \dfrac{\pi}{2}$

$\Rightarrow \quad \tan^{-1}\left[\dfrac{\left(\frac{a}{x}+\frac{b}{x}\right)}{1-\frac{ab}{x^2}}\right] = \dfrac{\pi}{2}$

$\Rightarrow \quad \dfrac{\frac{a+b}{x}}{\frac{x^2-ab}{x^2}} = \tan\dfrac{\pi}{2}$

$\Rightarrow \quad x^2 = ab$

$\Rightarrow \quad x = \sqrt{ab}$

31. $\cos^{-1}[\cos\{2 \cdot \cot^{-1}(\sqrt{2}-1)\}]$

$= \cos^{-1}[\cos\{2 \times 67.5\}]$

$= \cos^{-1}[\cos 135^\circ]$

$= 135^\circ$

$= \dfrac{3\pi}{4}$

32. We have,

$\sum_{m=1}^{n} \tan^{-1}\left(\dfrac{2m}{m^4+m^2+2}\right)$

$= \sum_{m=1}^{n} \tan^{-1}\left[\dfrac{2m}{1+(m^2+m+1)(m^2-m+1)}\right]$

$= \sum_{m=1}^{n} \tan^{-1}\left[\dfrac{(m^2+m+1)-(m^2-m+1)}{1+(m^2+m+1)(m^2-m+1)}\right]$

$= \sum_{m=1}^{n} \{\tan^{-1}[m^2+m+1] - \tan^{-1}[m^2-m+1]\}$

$= \tan^{-1}3 - \tan^{-1}1 + \tan^{-1}7 - \tan^{-1}3$

$+ (\tan^{-1}13 - \tan^{-1}7) + \ldots + \tan^{-1}(n^2+n+1)$

$- \tan^{-1}(n^2-n+1)$

$= \tan^{-1}\left(\dfrac{n^2+n+1-1}{1+(n^2+n+1)\cdot 1}\right) = \tan^{-1}\left(\dfrac{n^2+n}{2+n^2+n}\right)$

33. Put $x = \frac{\pi}{2} - h$ as $x \to \frac{\pi}{2}, h \to 0$

$$\therefore \text{ Given, limit} = \lim_{h \to 0} \frac{1 - \tan\left(\frac{\pi}{4} + \frac{h}{2}\right)(1 - \cos h)}{1 + \tan\left(\frac{\pi}{4} + \frac{h}{2}\right)(2h)^3}$$

$$= \lim_{h \to 0} \tan\left(\frac{h}{2}\right) \lim_{h \to 0} \frac{2 \cdot \sin^2\left(\frac{h}{2}\right)}{8h^3}$$

$$= \lim_{h \to 0} \frac{1}{4}\left(\frac{\tan\frac{h}{2}}{2 \times \frac{h}{2}}\right) \times \lim_{h \to 0}\left(\frac{\sin\frac{h}{2}}{\frac{h}{2}}\right)^2 \times \frac{1}{4}$$

$$= \frac{1}{32}$$

34. We have,

$$f(x) = \begin{cases} \dfrac{\sin^3(\sqrt{x}) \cdot \log(1 + 3x)}{x \cdot (\tan^{-1}\sqrt{x})^2 \cdot (e^{5\sqrt{x}} - 1)}, & x \neq 0 \\ a, & x = 0 \end{cases}$$

For continuity in [0, 1], $f(0) = f(0^+)$, otherwise it is discontinuous.

$$\therefore \lim_{x \to 0^+} \frac{\sin^3(\sqrt{x}) \cdot \log(1 + 3x)}{x \cdot (\tan^{-1}\sqrt{x})^2 \cdot (e^{5\sqrt{x}} - 1)}$$

$$= \lim_{x \to 0^+}\left[\frac{3}{5} \cdot \frac{\sin^3\sqrt{x}}{(\sqrt{x})^3} \cdot \frac{(\sqrt{x})^2}{(\tan^{-1}\sqrt{x})^2} \times \frac{\log(1 + 3x)}{3x} \cdot \frac{5\sqrt{x}}{e^{5\sqrt{x}} - 1}\right]$$

$$= \frac{3}{5} \lim_{x \to 0^+} \frac{\sin^3\sqrt{x}}{(\sqrt{x})^3} \cdot \frac{(\sqrt{x})^2}{\tan^{-1}\sqrt{x}} \times \frac{\log(1 + 3x)}{3x} \cdot \frac{5\sqrt{x}}{e^{5\sqrt{x}} - 1} = a$$

$$\Rightarrow \quad a = \frac{3}{5}$$

35. Given $y = |\sin x|^{|x|}$

In the neighbourhood of $-\frac{\pi}{6}$, $|x|$ and $|\sin x|$ both are negative

i.e., $y = (-\sin x)^{-x}$

Taking log on both sides, we get

$\log y = -x \log(-\sin x)$

$$\Rightarrow \frac{1}{y}\frac{dy}{dx} = (-x)\frac{1}{-\sin x}(-\cos x) + \log(-\sin x)(-1)$$

$$= -x \cdot \cot x - \log(-\sin x)$$

$$= -[x \cot x + \log(-\sin x)]$$

$$\Rightarrow \quad \frac{dy}{dx} = -y[x \cot x + \log(-\sin x)]$$

$$\therefore \left(\frac{dy}{dx}\right)_{x = -\frac{\pi}{6}} = (2)^{-\frac{\pi}{6}} \frac{[6 \log 2 - \sqrt{3}\pi]}{6}$$

36. Let $y = x^x$

$$\Rightarrow \frac{dy}{dx} = x^x(1 + \log x)$$

For increasing function, $\frac{dy}{dx} > 0$

$\Rightarrow \quad x^x\,(1 + \log x) > 0$

$\Rightarrow \quad 1 + \log x > 0$

$\Rightarrow \quad \log_e x > \log_e \frac{1}{e}$

$\Rightarrow \quad x > \frac{1}{e}$

$\therefore$ The function is increasing when $x > \frac{1}{e}$

37. Let $f(x) = 3\sin x - 4\sin^3 x = \sin 3x$

Since, $\sin x$ is increasing in the interval $\left[-\frac{\pi}{2}, \frac{\pi}{2}\right]$

$$\therefore \quad -\frac{\pi}{2} \leq 3x \leq \frac{\pi}{2} \Rightarrow -\frac{\pi}{6} \leq x \leq \frac{\pi}{6}$$

Thus, the length of interval $= \left[\frac{\pi}{6} - \left(-\frac{\pi}{6}\right)\right] = \frac{\pi}{3}$

38. Given curve is $y = f(x) = x^2 + bx - b$

On differentiating w.r.t.x, we get

$$\frac{dy}{dx} = 2x + b$$

The equation of the tagent at (1, 1) is

$$y - 1 = \left(\frac{dy}{dx}\right)_{(1,1)} (x - 1)$$

$\Rightarrow \quad y - 1 = (b + 2)(x - 1)$

$\Rightarrow \quad (2 + b)x - y = 1 + b$

$$\Rightarrow \quad \frac{x}{(1 + b)/(2 + b)} - \frac{y}{(1 + b)} = 1$$

So, $OA = \frac{1 + b}{2 + b}$ and $OB = -(1 + b)$

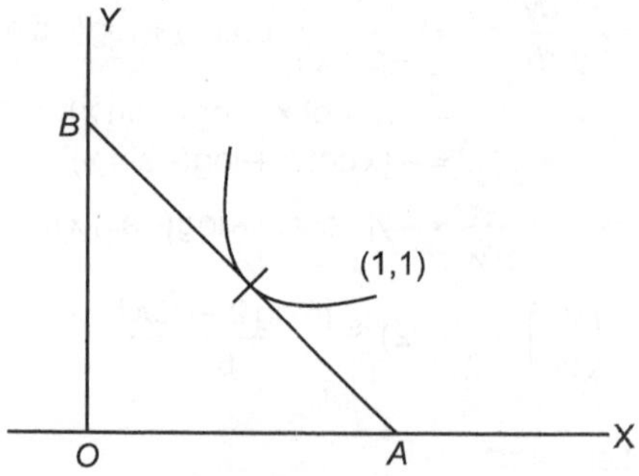

Now, area of ΔAOB

$$= \frac{1}{2}\frac{(1+b)}{(2+b)}[-(1+b)] = 2 \quad \text{(given)}$$

$$\Rightarrow \quad 4(2+b) + (1+b)^2 = 0$$

$$\Rightarrow \quad 8 + 4b + 1 + b^2 + 2b = 0$$

$$\Rightarrow \quad b^2 + 6b + 9 = 0$$

$$\Rightarrow \quad (b+3)^2 = 0 \Rightarrow b = -3$$

39. The given curve is $\left(\frac{x}{a}\right)^n + \left(\frac{y}{b}\right)^n = 2$

$$\Rightarrow n\left(\frac{x}{a}\right)^{n-1} \cdot \frac{1}{a} + n\left(\frac{y}{b}\right)^{n-1} \cdot \frac{1}{b}\frac{dy}{dx} = 0$$

$$\Rightarrow \quad \frac{dy}{dx} = -\frac{b}{a}\frac{\left(\frac{x}{a}\right)^{n-1}}{\left(\frac{y}{b}\right)^{n-1}}$$

At (a, b), $\frac{dy}{dx} = -\frac{b}{a}\frac{(1)^{n-1}}{(1)^{n-1}} = -\frac{b}{a}$

$\therefore$ The tangent at (a, b) is

$$\Rightarrow \quad y - b = -\frac{b}{a}(x - a)$$

$$\Rightarrow \quad ay - ab = -bx + ab$$

$$\Rightarrow \quad bx + ay = 2ab$$

$$\Rightarrow \quad \frac{x}{a} + \frac{y}{b} = 2$$

40. Given curve is $\frac{a^2}{x^2} + \frac{b^2}{y^2} = 1$

Let radius vector is 'r'

$$\therefore \quad r^2 = x^2 + y^2$$

$$\Rightarrow \quad r^2 = \frac{a^2y^2}{y^2 - b^2} + y^2 \qquad \left(\because \frac{a^2}{x^2} + \frac{b^2}{y^2} = 1\right)$$

For minumum value of r,

$$\frac{d(r^2)}{dy} = 0$$

$$\Rightarrow \quad \frac{-2y\,b^2a^2}{(y^2 - b^2)^2} + 2y = 0$$

$$\Rightarrow \quad y^2 = b(a+b)$$

$$\therefore \quad x^2 = a(a+b)$$

$$\Rightarrow \quad r^2 = (a+b)^2$$

$$\Rightarrow \quad r = a + b$$

41. Consider the function $f(x) = \frac{x^2}{x^3 + 200}$

$$f'(x) = x\frac{(400 - x^3)}{(x^3 + 200)^2} = 0$$

$$\Rightarrow \quad x^3 = 400$$

$$\Rightarrow \quad x = (400)^{1/3}$$

When $x < 0$, $f'(x) > 0$

$x > 0$, $f'(x) < 0$

$\therefore$ $f(x)$ has maximum at $x = (400)^{1/3}$

Since, $7 < (400)^{1/3} < 8$, either a_7 or a_8 is the greatest term of the sequence.

$$\because \quad a_7 = \frac{49}{543},\ a_8 = \frac{8}{89}$$

and $\quad \frac{49}{543} > \frac{8}{89}$

$a_7 = \frac{49}{543}$ is the greatest.

42. $f(x) = \begin{cases} |x^3 + x^2 + 3x + \sin x|\left(3 + \sin\frac{1}{x}\right) & x \neq 0 \\ 0, & x = 0 \end{cases}$

Let $\quad g(x) = x^3 + x^2 + 3x + \sin x$

$$g'(x) = 3x^2 + 2x + 3 + \cos x$$

$$= 3\left(x^2 + \frac{2x}{3} + 1\right) + \cos x$$

$$\Rightarrow g'(x) = 3\left\{\left(x + \frac{1}{3}\right)^2 + \frac{8}{9}\right\} + \cos x > 0$$

and $2 < 3 + \sin\left(\frac{1}{x}\right) < 4$

Hence, minimum value of $f(x)$ is 0 at $x = 0$

Hence, the number of points = 1

43. Let, $I = \int \frac{x^2 - 1}{(x^4 + 3x^2 + 1)\tan^{-1}\left(x + \frac{1}{x}\right)} dx$

On dividing numerator and denominator by x^2, we get

$$I = \int \frac{1 - \frac{1}{x^2}}{\left(x^2 + \frac{1}{x^2} + 3\right)\tan^{-1}\left(x + \frac{1}{x}\right)} dx$$

Put $x + \frac{1}{x} = t$

$\Rightarrow \left(1 - \frac{1}{x^2}\right)dx = dt$ and

$x^2 + \frac{1}{x^2} = t^2 - 2$

$$\therefore \quad I = \int \frac{dt}{(t^2 - 2 + 3)\tan^{-1} t}$$
$$= \int \frac{dt}{(1 + t^2)\tan^{-1} t}$$
$$= \log(\tan^{-1} t) + C$$
$$= \log\left[\tan^{-1}\left(x + \frac{1}{x}\right)\right] + C$$

44. Let, $I = \int \frac{\cos x}{\sin^2 x(\sin x + \cos x)} dx$

On dividing numerator and denominator by $\cos^3 x$, we get

$$I = \int \frac{\sec^2 x}{\tan^2 x\,(1 + \tan x)} dx$$

Put $\tan x = t \Rightarrow \sec^2 x dx = dt$

$$\therefore \quad I = \int \frac{dt}{t^2(1 + t)}$$
$$= \int \left[-\frac{1}{t} + \frac{1}{t^2} + \frac{1}{1 + t}\right] dt$$

(using partial fraction)

$$\Rightarrow \quad I = -\int \frac{1}{t} dt + \int t^{-2} dt + \int \frac{1}{1 + t} dt$$
$$= -\log|t| - \frac{1}{t} + \log|1 + t| + C$$
$$= -\frac{1}{\tan x} + \log\left|\frac{1 + \tan x}{\tan x}\right| + C$$
$$= -\cot x + \log\left|\frac{1 + \tan x}{\tan x}\right| + C$$

45. Let, $I = \int \frac{dx}{\sin x - \cos x + \sqrt{2}}$

$$\Rightarrow I = \int \frac{dx}{\sqrt{2}\left(\frac{1}{\sqrt{2}}\sin x - \frac{1}{\sqrt{2}}\cos x\right) + \sqrt{2}}$$
$$= \frac{1}{\sqrt{2}} \int \frac{dx}{1 - \cos\left(x + \frac{\pi}{4}\right)} = \frac{1}{\sqrt{2}} \int \frac{dx}{2\sin^2\left(\frac{x}{2} + \frac{\pi}{8}\right)}$$
$$= \frac{1}{2\sqrt{2}} \int \frac{dx}{\sin^2\left(\frac{x}{2} + \frac{\pi}{8}\right)} = \frac{1}{2\sqrt{2}} \int \operatorname{cosec}^2\left(\frac{x}{2} + \frac{\pi}{8}\right) dx$$
$$= \frac{1}{2\sqrt{2}}\left[\frac{-\cot\left(\frac{x}{2} + \frac{\pi}{8}\right)}{\frac{1}{2}}\right] + C$$
$$= -\frac{1}{\sqrt{2}}\cot\left(\frac{x}{2} + \frac{\pi}{8}\right) + C$$

46. $\int_0^{\frac{3\pi}{2}} \sin\left(\frac{2x}{\pi}\right) dx = \int_0^{\frac{\pi}{2}} \sin\left(\frac{2x}{\pi}\right) dx + \int_{\frac{\pi}{2}}^{\pi} \sin\left(\frac{2x}{\pi}\right) dx + \int_{\pi}^{\frac{3\pi}{2}} \sin\left(\frac{2x}{\pi}\right) dx$

$$= 0 + \sin 1 \int_{\frac{\pi}{2}}^{\pi} dx + \sin 2 \int_{\pi}^{\frac{3\pi}{2}} dx$$
$$= \frac{\pi}{2}(\sin 1 + \sin 2)$$

47. Let, $I = \int_{\frac{1}{n}}^{(an - 1)/n} \frac{\sqrt{x}}{\sqrt{a - x} + \sqrt{x}} dx$

$$= \int_{\frac{1}{n}}^{a - \frac{1}{n}} \frac{\sqrt{x}}{\sqrt{a - x} + \sqrt{x}} dx \quad \text{...(i)}$$
$$= \int_{\frac{1}{n}}^{a - \frac{1}{n}} \frac{\left(\sqrt{\frac{1}{n} + a - \frac{1}{n} - x}\right)}{\sqrt{a - \left(\frac{1}{n} + a - \frac{1}{n} - x\right)} + \sqrt{\frac{1}{n} + a - \frac{1}{n} - x}} dx$$
$$\Rightarrow \quad I = \int_{\frac{1}{n}}^{a - \frac{1}{n}} \frac{\sqrt{a - x}}{\sqrt{x} + \sqrt{a - x}} dx \quad \text{...(ii)}$$

On adding Eqs. (i) and (ii), we get

$$2I = \int_{\frac{1}{n}}^{a - \frac{1}{n}} 1\,dx = [x]_{\frac{1}{n}}^{a - \frac{1}{n}}$$
$$= a - \frac{1}{n} - \frac{1}{n} = \frac{na - 2}{n} \Rightarrow I = \frac{na - 2}{2n}$$

48. $y = x|\sin x|$

$$\Rightarrow y = \begin{cases} x\sin x, & 0 \le x \le \pi \\ -x\sin x, & \pi < x \le 2\pi \end{cases}$$

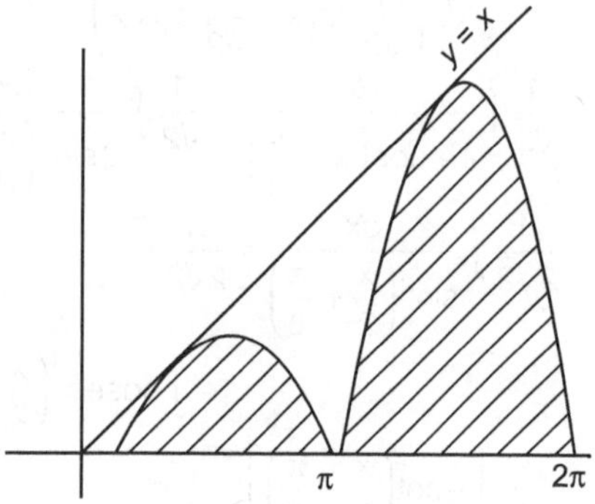

$\therefore$ Required Area $= \int_0^{\pi} x\sin x + \int_{\pi}^{2\pi} (-x\sin x)\,dx$

$$= [x(-\cos x)]_0^{\pi} - \int_0^{\pi}(-\cos x)\,dx$$

$$- [\{x(-\cos x)\}]_{\pi}^{2\pi} + \int_{\pi}^{2\pi}(-\cos x)\,dx$$

$$= [\pi + \sin x]_0^{\pi} + [(2\pi + \pi) - \sin x]_{\pi}^{2\pi}$$

$$= 4\pi \text{ sq units}$$

49. The curves $y + x \le 1$ and $|y - x| \le 1$ form a square $ABCD$ and $2x^2 + 2y^2 = 1$ is a circle of radius $\frac{1}{\sqrt{2}}$ and centre (0, 0).

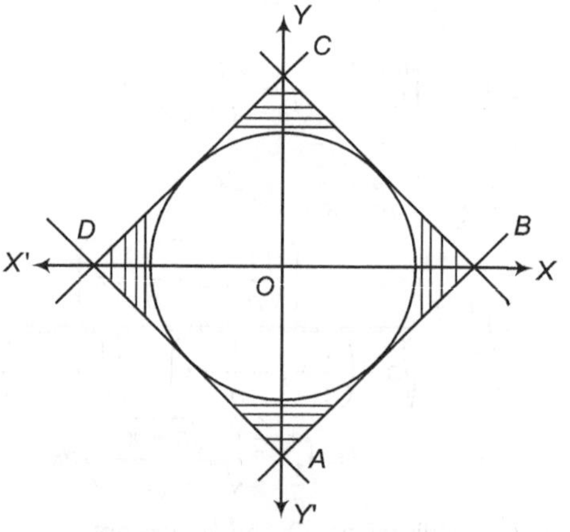

$\therefore$ Required Area

= Area of square – Area of circle

$$= \left(2 - \frac{\pi}{2}\right) \text{ sq units}$$

50. $y = u^m \Rightarrow \frac{dy}{dx} = m \cdot u^{m-1} \cdot \frac{du}{dx}$

Hence, $2x^4 \cdot mu^{m-1}u^m \cdot \frac{du}{dx} + u^{4m} = 4x^6$

$$\Rightarrow \quad \frac{du}{dx} = \frac{4x^6 - u^{4m}}{2mx^4 \cdot u^{2m-1}}$$

$$\therefore \quad 4m = 6 \Rightarrow m = \frac{3}{2}$$

51. Equation of the normal at (x_1, y_1) is

$$y - y_1 = \frac{dx}{dy}(x - x_1)$$

Put $\quad y = 0$, then

$$x = x_1 + y_1\frac{dy}{dx}$$

$$y_1^2 = 2x_1x$$

$$= 2x_1\left(x_1 + y_1\frac{dy}{dx}\right)$$

$$\Rightarrow \quad y_1^2 = 2x_1^2 + 2x_1y_1\frac{dy}{dx}$$

$$\Rightarrow \quad \frac{dy}{dx} = \frac{y_1^2 - 2x_1^2}{2x_1y_1} = \frac{\left(\frac{y_1}{x_1}\right)^2 - 2}{2\left(\frac{y_1}{x_1}\right)}$$

$$\Rightarrow \quad \frac{dy}{dx} = \frac{\left(\frac{y}{x}\right)^2 - 2}{2y/x}$$

Put $y = vx$

$$\Rightarrow \quad \frac{dy}{dx} = v + x\frac{dv}{dx},$$

We have, $v + \frac{xdv}{dx} = \frac{v^2 - 2}{2v}$

$$\Rightarrow \quad x \cdot \frac{dv}{dx} = -\frac{(2 + v^2)}{2v}$$

$$\Rightarrow \quad \frac{2v \cdot dv}{2 + v^2} + \frac{dx}{x} = 0$$

On integrating both sides, we get

$$\log(2 + v^2) + \log|x| = \log C$$

$$\Rightarrow \quad \log(2 + v^2)|x| = \log C$$

$$\Rightarrow \quad |x|\left(2 + \frac{y^2}{x^2}\right) = C$$

$\because$ It passes through (2, 1) then, $2\left(2+\frac{1}{4}\right)=C$

$\Rightarrow \quad C=\frac{9}{2}$

$\therefore \quad \frac{|x|(2x^2+y^2)}{x^2}=\frac{9}{2}$

$\Rightarrow \quad 2x^2+y^2=\frac{9}{2}|x|$

$\Rightarrow \quad 4x^2+2y^2-9|x|=0$

52. $|\mathbf{a}\times\mathbf{b}|^2=|\mathbf{a}|^2|\mathbf{b}|^2-2(\mathbf{a}\cdot\mathbf{b})^2=16-4=12$ and

$|\mathbf{c}|^2=(2\,\mathbf{a}\times\mathbf{b}-3\mathbf{b})^2=(2\,\mathbf{a}\times\mathbf{b}-3\mathbf{b})$

$=4|\mathbf{a}\times\mathbf{b}|^2+9|\mathbf{b}|^2=4.12+9.16$

$=192\Rightarrow|\mathbf{c}|=8\sqrt{3}$

Now, $\mathbf{b}\cdot\mathbf{c}=\mathbf{b}(2\,\mathbf{a}\times\mathbf{b}-3\mathbf{b})=-3|\mathbf{b}|^2=-48$

$\therefore \quad \cos\theta=\frac{\mathbf{b}\cdot\mathbf{c}}{|\mathbf{b}||\mathbf{c}|}=-\frac{48}{4\cdot 8\sqrt{3}}=-\frac{\sqrt{3}}{2}$

$\Rightarrow \quad \theta=\frac{5\pi}{6}$

53. Let the first term and common ratio of a GP be α and β, then

$a=\alpha\cdot\beta^{p-1}, b=\alpha\cdot\beta^{q-1}$ and $c=\alpha\cdot\beta^{r-1}$

$\therefore \quad \log a=\log\alpha+(p-1)\log\beta,$

$\log b=\log\alpha+(q-1)\log\beta$ and

$\log c=\log\alpha+(r-1)\log\beta$

The dot product of the given two vectors is

$(q-r)\log a+(r-p)\log b+(p-q)\log c$

$\Rightarrow (q-r)[\log\alpha+(p-1)\log\beta]+(r-p)$

$[\log\alpha+(q-1)\log\beta]+(p-q)[\log\alpha+(r-1)\log\beta]$

$\Rightarrow \log\alpha\,[q-r+r-p+p-q]+$

$\log\beta\,[(p-1)(q-r)+(r-p)(q-1)$

$+(r-1)(p-q)]$

$=0+0=0$

$\Rightarrow$ The two vectors are perpedicular.

54. The straight line joining the points (1, 1, 2) and (3, −2, 1) is $\frac{x-1}{2}=\frac{y-1}{-3}=\frac{z-2}{-1}=r$ (say)

$\therefore$ point is $(2r+1, 1-3r, 2-r)$ which lies on

$3x+2y+z=6$

$\therefore \quad 3(2r+1)+2(1-3r)+2-r=6$

$\Rightarrow \quad r=1$

So, required point is (3, −2, 1).

55. Let Q be the point where the line through $P(1,-5,9)$ and parallel to $\mathbf{r}=\hat{\mathbf{i}}+\hat{\mathbf{j}}+\hat{\mathbf{k}}$ meets the plane $\mathbf{r}(\hat{\mathbf{i}}-\hat{\mathbf{j}}+\hat{\mathbf{k}})=5$.

Equation of PQ, is

$$\frac{x-1}{1}=\frac{y+5}{1}=\frac{z-9}{1}=\lambda$$

$\therefore \quad x=\lambda+1,$

$y=\lambda-5$

and $\quad z=\lambda+9$

lies on the plane $x-y+z=5$

$\Rightarrow \quad \lambda+1-\lambda+5+\lambda+9=5$

$\Rightarrow \quad \lambda=-10$

$\therefore$ Co-ordinates of Q are Q (−9, −15, −1)

$\Rightarrow \quad PQ=\sqrt{10^2+10^2+10^2}=10\sqrt{3}$

56. Let Q be the image of the point $P(\hat{\mathbf{i}}+3\hat{\mathbf{k}})$ in the plane $\mathbf{r}\cdot(\hat{\mathbf{i}}+\hat{\mathbf{j}}+\hat{\mathbf{k}})=1$. Then PQ is normal to the plane. Since PQ passes through P and is normal to the given plane, therefore equation of PQ is

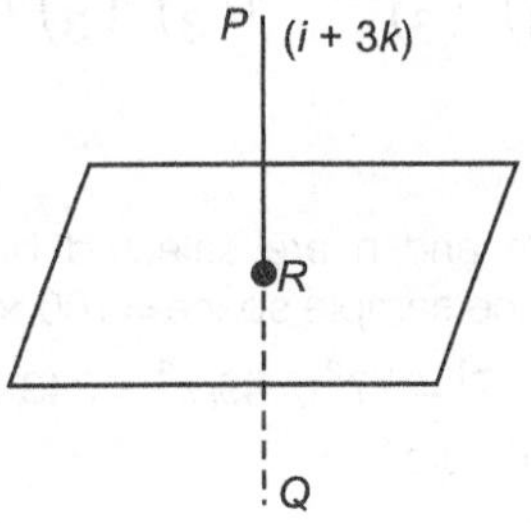

$$\mathbf{r}=(\hat{\mathbf{i}}+3\hat{\mathbf{k}})+\lambda(\hat{\mathbf{i}}+\hat{\mathbf{j}}+\hat{\mathbf{k}})$$

Since, Q lies on the line PQ, so, let the position vector of Q be $(\hat{\mathbf{i}}+3\hat{\mathbf{k}})+\lambda(\hat{\mathbf{i}}+\hat{\mathbf{j}}+\hat{\mathbf{k}})$

$\Rightarrow (1+\lambda)\hat{\mathbf{i}}+\lambda\hat{\mathbf{j}}+(3+\lambda)\hat{\mathbf{k}}$

Since R is the mid point of PQ, therefore position vector of R is

$$\frac{(1+\lambda)\mathbf{i}+\lambda\mathbf{j}+(3+\lambda)\hat{\mathbf{k}}+\hat{\mathbf{i}}+3\hat{\mathbf{k}}}{2}$$

or $\left(\frac{\lambda+2}{2}\right)\mathbf{i}+\left(\frac{\lambda}{2}\right)\hat{\mathbf{j}}+\left(3+\frac{\lambda}{2}\right)\hat{\mathbf{k}}$

or $\left(\frac{\lambda}{2}+1\right)\hat{\mathbf{i}}+\left(\frac{\lambda}{2}\right)\hat{\mathbf{j}}+\left(3+\frac{\lambda}{2}\right)\hat{\mathbf{k}}$

Since, R less on the plane $\mathbf{r}\cdot(\hat{\mathbf{i}}+\hat{\mathbf{j}}+\hat{\mathbf{k}})=1$

Therefore,

$$\left[\left(\frac{\lambda}{2}+1\right)\hat{i}+\left(\frac{\lambda}{2}\right)\hat{j}+\left(3+\frac{\lambda}{2}\right)\hat{k}\right]\cdot(\hat{i}+\hat{j}+\hat{k})=1$$

$$\Rightarrow \quad \left(\frac{\lambda}{2}+1+\frac{\lambda}{2}+3+\frac{\lambda}{2}\right)=1$$

$$\Rightarrow \quad \lambda=-2$$

So, the postion vector of Q is

$$(\hat{i}+3\hat{k})-2(\hat{i}+\hat{j}+\hat{k})=-\hat{i}-2\hat{j}+\hat{k}$$

57. Let the probability of success and failure be p and q, respectively

Then, $p=2q$

and $p+q=1$

$\Rightarrow \quad 3q-1$

$\Rightarrow \quad q=\frac{1}{3}$

$\therefore \quad p=\frac{2}{3}$

Required probability

$$={}^6C_4\left(\frac{2}{3}\right)^4\left(\frac{1}{3}\right)^2+{}^6C_5\left(\frac{2}{3}\right)^5\left(\frac{1}{3}\right)+{}^6C_6\left(\frac{2}{3}\right)^6$$

$$=\frac{496}{729}$$

58. Since, m and n are selected between 1 and 100, hence sample space $=100\times100$

Also, $7^1=1, 7^2=49, 7^3=343, 7^4=2401$

$7^5=16807$ etc.

Hence, 1, 3, 7 and 9 will be the last digits in the power of 7. Hence, for favourable cases.

nm

↓

1, 2 1, 2 1, 3 ... 1, 100

2, 1 2, 2 2, 3 ... 2, 100

100, 1 100, 2 100, 3 ... 100, 100

for $m=1$, $n=3, 7, 11, \ldots 97$

$\therefore$ favourable cases $=25$

for $m=2$, $n=4, 8, 12, 10 \ldots 100$

$\therefore$ favourable cases $=25$

Similarly for every m, favourable n are 25.

$\therefore$ Total favourable cases $=100\times25$

Hence, required probability $=\frac{100\times25}{100\times100}=\frac{1}{4}$

59. Let $\sin^{-1}a=A$

$$\sin^{-1}b=B$$

$$\sin^{-1}C=C$$

$\therefore \quad \sin A=a, \quad \sin B=b, \quad \sin C=C$

and $A+B+C=\pi$, then

$$\sin 2A+\sin 2B+\sin 2C = 4\sin A\sin B\sin C \quad \ldots(i)$$

$$\Rightarrow \sin A\cos A+\sin B\cos B+\sin C\cos C = 2\sin A\sin B\sin C$$

$$\Rightarrow \sin A\sqrt{1-\sin^2 A}+\sin B\sqrt{1-\sin^2 B}+\sin C\sqrt{1-\sin^2 C}=2\sin A\sin B\sin C \quad \ldots(ii)$$

$$\Rightarrow a\sqrt{1-a^2}+b\sqrt{1-b^2}+c\sqrt{1-c^2}=2abc$$

while $\sin^{-1}a+\sin^{-1}b+\sin^{-1}c=\pi$

60. Given system of equation is

$$x+y+z=6$$

$$x+2y+3z=10$$

$$x+2y+\lambda z=\mu$$

If given system of equation has no solution, then $D=0$ and atleast one of the determinants $D_1, D_2, D_3\neq0$

Here $D=0$

$$\Rightarrow \begin{vmatrix}1&1&1\\1&2&3\\1&2&\lambda\end{vmatrix}=0$$

$$\Rightarrow 1(2\lambda-6)-1(\lambda-3)+1(2-2)=0$$

$$\Rightarrow 2\lambda-6-\lambda+3+0=0$$

$$\Rightarrow \lambda-3=0$$

$$\Rightarrow \lambda=3$$

Also, $D_1=\begin{vmatrix}6&1&1\\10&2&3\\\mu&2&3\end{vmatrix}\neq0$

$$\Rightarrow 6(6-6)-1(30-3\mu)+1(20-2\mu)\neq0$$

$$\Rightarrow 0-30+3\mu+20-2\mu\neq0$$

$$\Rightarrow \mu-10\neq0$$

$$\Rightarrow \mu\neq10$$

61. Given, $f(x)=sec\left(\frac{\pi}{4}\cos^2 x\right)$

We know that, $0\leq\cos^2 x\leq1$

$\therefore \quad f(x)=sec\left(\frac{\pi}{4}\times1\right)=\sqrt{2}$

$f(x)=sec\left(\frac{\pi}{4}\times0\right)=1 \quad \therefore$ Range $\in[1, \sqrt{2}]$

62. Let $y = \lim_{x \to 10} \left(\frac{a^x + b^x + c^x}{3} \right)^{2/x}$

$\Rightarrow \log y = \lim_{x \to 10} \frac{2}{x} \log \left(\frac{a^x + b^x + c^x}{3} \right)$

$= \lim_{x \to 10} \frac{\log (a^x + b^x + c^x) - \log 3}{x}$

$\Rightarrow \log y = \log (abc)^{2/3}$ (by L-Hospital's rule)

$\Rightarrow \quad y = (abc)^{2/3}$

63. We have $f(x) = [x] \cos \left[\frac{2x-1}{2} \right] \pi$

Since, $g(x) = [x]$ is always discontinuous at all integral values of points. Hence, $f(x)$ is discontinuous for all integral points.

64. we know that funtion $|x|$ is not differentiable at $x = 0$

$\therefore \quad |x^2 - 3x + 2| = |(x-1)(x-2)|$

Hence, it is not differentiable at $x = 1$ and 2

Now, $f(x) = (x^2 - 1)|x^2 - 3x + 2| + \cos |x|$ is not differentiable at $x = 2$.

For $1 < x < 2$, $f(x) = -(x^2 - 1)(x^2 - 3x + 2) + \cos x$

For $2 < x < 3$, $f(x) = (x^2 - 1)(x^2 - 3x + 2) + \cos x$

$L f'(x) = -(x^2 - 1)(2x - 3) - 2x(x^2 - 3x + 2) - \sin x$

$Lf'(2) = -3 \sin 2$

$Rf'(x) = (x^2 - 1)(2x - 3) + 2x(x^2 - 3x + 2) - \sin x$

$Rf'(2) = (4 - 1)(4 - 3) + 0 - \sin 2 = 3 - \sin 2$

Hence, $Lf'(2) \neq Rf'(2)$.

So, $f(x)$ is not differentable at $x = 2$.

65. Let $x = 5$, $y = 0 \Rightarrow f(5 + 0) = f(5) \cdot f(0)$

$\Rightarrow \quad f(5) = f(5) f(0) \Rightarrow f(0) = 1$

$\therefore \quad f'(5) \lim_{h \to 0} = \frac{f(5+h) - f(5)}{h}$

$= \lim_{h \to 0} \frac{f(5) f(h) - f(5)}{h}$

$= \lim_{h \to 0} 2 \left(\frac{f(h) - 1}{h} \right) \quad [\because f(5) = 2]$

$= 2 \lim_{h \to 0} \left(\frac{f(h) - f(0)}{h} \right) = 2 \times f(0)$

$= 2 \times 3 = 6$

66. Here, $x \sin (a + y) + \sin a \cos (a + y) = 0 \quad \ldots$(i)

On differentiating w.r.t. x, we get

$\frac{d}{dx} [x \sin (a + y)] + \frac{d}{dx} [\sin a \cos (a + y)] = 0$

$\Rightarrow \left[x \frac{d}{dx} \{\sin (a + y)\} + \sin (a + y) \frac{d}{dx} (x) \right] + \sin a \frac{d}{dx} \cos (a + y) = 0$

(using product rule and chain rule)

$\Rightarrow \left[x \cos (a + y) \frac{d}{dx} (a + y) + \sin (a + y) \right] + \sin a \left[-\sin (a + y) \frac{d}{dx} (a + y) \right] = 0$

$\Rightarrow \left[x \cos (a + y) \left(0 + \frac{dy}{dx} \right) + \sin (a + y) \right] - \sin a \sin (a + y) \left(0 + \frac{dy}{dx} \right) = 0$

$\Rightarrow \quad x \cos (a + y) \frac{dy}{dx} + \sin (a + y) - \sin a \sin (a + y) \frac{dy}{dx} = 0$

$\Rightarrow \quad \frac{dy}{dx} [x \cos (a + y) - \sin a \sin (a + y)] = -\sin(a + y)$

$\left[\text{from Eq. (i), putting } x = -\sin a \frac{\cos (a + y)}{\sin (a + y)} \right]$

$\Rightarrow \quad \frac{dy}{dx} \left[-\sin a \frac{\cos^2 (a + y)}{\sin (a + y)} - \sin a \sin (a + y) \right] = -\sin (a + y)$

$\Rightarrow -\frac{dy}{dx} \left[\frac{\sin a \cos^2 (a + y) + \sin a \sin^2 (a + y)}{\sin (a + y)} \right] = -\sin (a + y)$

$\Rightarrow \frac{dy}{dx} = \sin (a + y) \left[\frac{\sin (a + y)}{\sin a \{\cos^2 (a + y) + \sin^2 (a + y)\}} \right]$

$\Rightarrow \quad \frac{dy}{dx} = \frac{\sin^2 (a + y)}{\sin a} \quad (\because \sin^2 \theta + \cos^2 \theta = 1)$

67. Firstly, break the number 5.001 as $x = 5$ and $\Delta x = 0.001$ and use the relation $f(x + \Delta x) \simeq f(x) + \Delta x f'(x)$.

Consider, $f(x) = x^3 - 7x^2 + 15$

$\Rightarrow \quad f'(x) = 3x^2 - 14x$

Let $\quad x = 5$

and $\quad \Delta x = 0.001$

Also, $f(x + \Delta x) \simeq f(x) + \Delta x\, f'(x)$

Therefore,

$f(x + \Delta x) \simeq (x^3 - 7x^2 + 15) + \Delta x\,(3x^2 - 14x)$

$\Rightarrow \quad f(5.001) \simeq (5^3 - 7 \times 5^2 + 15) + (3 \times 5^2 - 14 \times 5)(0.001)$

(as $x = 5$, $\Delta x = 0.001$)

$= 125 - 175 + 15 + (75 - 70)(0.001)$

$= -35 + (5)(0.001)$

$= -35 + 0.005 = -34.995$

68. Here, we observe that

(a) $f(x)$ is a polynomial, so it is continuous in the interval [0, 2].

(b) $f'(x) = 3x^2 - 6x + 2$ exists for all $x \in (0, 2)$.

So, $f(x)$ is differentiable for all $x \in (0, 2)$ and

(c) $f(0) = 0$, $f(2) = 2^3 - 3(2)^2 + 2(2) = 0$

$\Rightarrow f(0) = f(2)$

Thus, all the three conditions of Rolle's theorem are satisfied.

So, there must exist $c \in [0, 2]$ such that $f'(c) = 0$

$\Rightarrow \quad f'(c) = 3c^2 - 6c + 2 = 0$

$\Rightarrow \quad c = 1 \pm \dfrac{1}{\sqrt{3}} \in [0, 2]$

69. The equation of any plane through the intersection of the planes,

$3x - y + 2z - 4 = 0$ and $x + y + z - 2 = 0$, is

$(3x - y + 2z - 4) + \lambda\,(x + y + z - 2) = 0 \quad \ldots(\text{i})$

The plane passes through the point (2, 2, 1). Therefore, this point will satisfy Eq. (i).

$\therefore \quad (3 \times 2 - 2 + 2 \times 1 - 4) + \lambda\,(2 + 2 + 1 - 2) = 0$

$\Rightarrow \quad (6 - 4) + 3\lambda = 0$

$\Rightarrow \quad 2 + 3\lambda = 0$

$\Rightarrow \quad \lambda = \dfrac{-2}{3}$

On substituting this value of λ in Eq. (i), we get the required plane as

$$(3x - y + 2z - 4) - \frac{2}{3}(x + y + z - 2) = 0$$

$\Rightarrow \quad 9x - 3y + 6z - 12 - 2x - 2y - 2z + 4 = 0$

$\Rightarrow \quad 7x - 5y + 4z - 8 = 0$

This is the required equation of the plane.

70. Let $y = \operatorname{cosec}^2 x$,

Required limit $= \lim\limits_{y \to \infty} (1^y + 2^y + \ldots n^y)^{1/y}$ $\quad (0/\infty$ form)

$= \lim\limits_{y \to \infty} (n^y)^{1/y} \left[\left(\frac{1}{n}\right)^y + \left(\frac{2}{n}\right)^y + \ldots + \left(\frac{n-1}{n}\right)^y + 1\right]^{1/y}$

$= \lim\limits_{y \to \infty} n \left[\left(\frac{1}{n}\right)^y + \left(\frac{2}{n}\right)^y + \ldots + \left(\frac{n-1}{n}\right)^y + 1\right]^{1/y}$

$= n \cdot 1^0 = n \cdot 1 = n$

71. Let $I = \displaystyle\int_0^{\pi/2} \frac{\sin x - \cos x}{1 + \sin x \cos x}\, dx \quad \ldots(\text{i})$

$\Rightarrow I = \displaystyle\int_0^{\pi/2} \frac{\sin\left(\frac{\pi}{2} - x\right) - \cos\left(\frac{\pi}{2} - x\right)}{1 + \sin\left(\frac{\pi}{2} - x\right)\cos\left(\frac{\pi}{2} - x\right)}\, dx$

$\left[\because \int_0^a f(x)dx = \int_0^a f(a - x)dx\right]$

$= \displaystyle\int_0^{\pi/2} \frac{\cos x - \sin x}{1 + \cos x \sin x}\, dx \quad \ldots(\text{ii})$

$\left[\because \sin\left(\frac{\pi}{2} - x\right) = \cos x \text{ and } \cos\left(\frac{\pi}{2} - x\right) = \sin x\right]$

On adding Eqs. (i) and (ii), we get

$2I = \displaystyle\int_0^{\pi/2} \frac{0}{1 + \sin x \cos x}\, dx = 0 \Rightarrow I = 0$

72. Since, $|x| + |y| = 1$

$$\Rightarrow \begin{cases} x + y = 1, & x > 0, \; y > 0 \\ x - y = 1, & x > 0, \; y < 0 \\ -x + y = 1, & x < 0, \; y > 0 \\ -x - y = 1, & x < 0, \; y < 0 \end{cases}$$

and $\quad 1 - y^2 = |x|$

$$\Rightarrow \begin{cases} 1-y^2 = x, & x \geq 0 \\ 1-y^2 = -x, & x < 0 \end{cases}$$

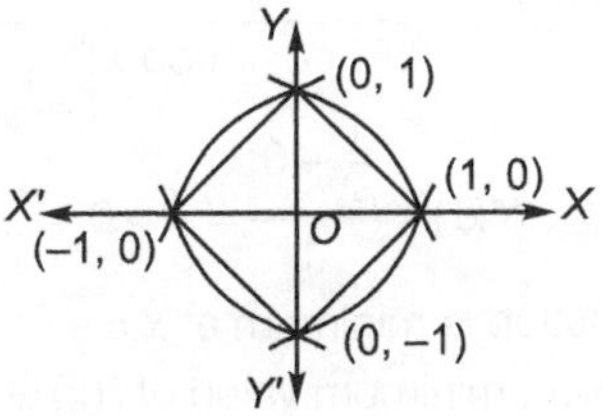

∴ Required area

$$= \left|2\int_0^1 \sqrt{(1-x)}\,dx\right| + \left|2\int_{-1}^0 \sqrt{(x+1)}\,dx\right| - 4\left(\frac{1}{2}\cdot 1\cdot 1\right)$$

$$= \frac{2}{3} \text{ sq unit}$$

73. Let slope of a line be m.

Now, the equation of a line passing through (1, 0) is

$$y - 0 = m(x-1)$$

$$\Rightarrow mx - y - m = 0$$

Distance from origin $= \frac{\sqrt{3}}{2}$

$$\Rightarrow \frac{|-m|}{\sqrt{1+m^2}} = \frac{\sqrt{3}}{2}$$

$$\Rightarrow 4m^2 = 3(1+m^2)$$

$$\Rightarrow m^2 = 3$$

$$\Rightarrow m = \pm\sqrt{3}$$

∴ Equations of lines are

$$\sqrt{3}x - y - \sqrt{3} = 0$$

and $$-\sqrt{3}x - y + \sqrt{3} = 0$$

$$\Rightarrow \sqrt{3}x - y - \sqrt{3} = 0$$

and $$\sqrt{3}x + y - \sqrt{3} = 0$$

74. From figure, we have

$$OP = 5, \quad OQ = 6$$

and $$OM = \frac{5}{2}, \quad CM = 3$$

∴ In $\Delta OMC, OC^2 = OM^2 + MC^2$

$$\Rightarrow OC^2 = \left(\frac{5}{2}\right)^2 + (3)^2$$

$$\Rightarrow OC = \frac{\sqrt{61}}{2}$$

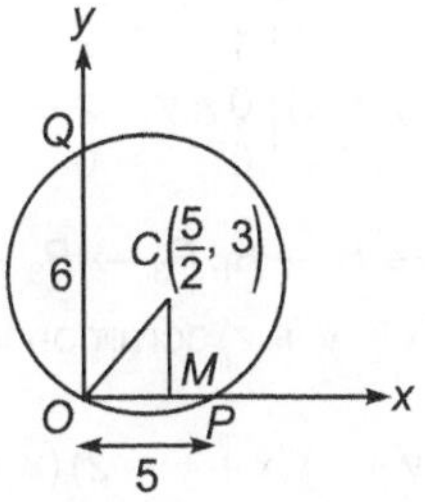

Thus, the required circle has its centre $\left(\frac{5}{2}, 3\right)$ and radius $\frac{\sqrt{61}}{2}$.

So, its equation is $\left(x - \frac{5}{2}\right)^2 + (y-3)^2 = \left(\frac{61}{4}\right)$.

Hence, $\lambda = \frac{61}{4}$

75. Obtain the foot of the perpendicular $N(-4, 1, -3)$.

Find the equations of the line passing through $P(2, 4, -1)$ and $N(-4, 1, -3)$

$$\frac{x-2}{6} = \frac{y-4}{3} = \frac{z+1}{2}$$

76. Given, $\cos^2 A + \cos^2 C = \sin^2 B$

Obviously it is not an equilateral triangle because $A = B = C = 60°$ does not satisfy the given condition. But $B = 90°$, then $\sin^2 B = 1$ and

$$\cos^2 A + \cos^2 C = \cos^2 A + \cos^2\left(\frac{\pi}{2} - A\right)$$

$$= \cos^2 A + \sin^2 A = 1$$

Hence, this satisfies the condition, so it is a right angled triangle but not necessarily isosceles triangle.

77. $$\Delta = \begin{vmatrix} x+y+2z & x & y \\ z & y+z+2x & y \\ z & x & z+x+2y \end{vmatrix}$$

$$= \begin{vmatrix} 2(x+y+z) & x & y \\ 2(x+y+z) & y+z+2x & y \\ 2(x+y+z) & x & z+x+2y \end{vmatrix}$$

(using $C_1 \to C_1 + C_2 + C_3$)

Take out $2(x+y+z)$ common from C_1, we get

$$\Delta = 2(x+y+z)\begin{vmatrix} 1 & x & y \\ 1 & y+z+2x & y \\ 1 & x & z+x+2y \end{vmatrix}$$

$$= 2(x+y+z)\begin{vmatrix} 1 & x & y \\ 0 & y+z+x & 0 \\ 0 & 0 & z+x+y \end{vmatrix}$$

(using $R_2 \to R_2 - R_1, R_3 \to R_3 - R_1$)

Take out $(x+y+z)$ common from R_2 and R_3, we get

$$\Delta = 2(x+y+z)(x+y+z)(x+y+z) \times \begin{vmatrix} 1 & x & y \\ 0 & 1 & 0 \\ 0 & 0 & 1 \end{vmatrix}$$

Expanding along R_3,we get

$$\Delta = 2(x+y+z)^3\,[(1)(1-0)]$$

$$= 2(x+y+z)^3 = k\,(x+y+z)^3 \text{ (given)}$$

$\therefore \quad k = 2$

78. Let $f(x) = \dfrac{x}{\log x}$

$$f'(x) = \frac{\log x - 1}{(\log x)^2}$$

Pur $f'(x) = 0$, for maxima or minima, we get

$$\log x - 1 = 0$$

$$\Rightarrow \quad x = e$$

Now, $f''(x) = \dfrac{(\log x)^2 \cdot \dfrac{1}{x} - (\log x - 1)\cdot \dfrac{2\log x}{x}}{(\log x)^4}$

$$\Rightarrow \quad f''(e) = \frac{\frac{1}{e} - 0}{1} = \frac{1}{e} > 0$$

$\therefore$ Function is minimum at $x = e$.

Hence, minimum value of $f(x)$ at $x = e$ is

$$f(e) = e.$$

79. Total number of mappings from a set A having elements into itself is n^n.

And the total number of one to one mapping is $n!$.

$\therefore$ Required probability $= \dfrac{n!}{n^n}$

80. Total number of available courses = 9

Out of these 5 courses have to be chosen. But it is given that 2 courses are compulsory for every student *i.e.*, you have to choose only 3 courses instead of 5, out of 7 instead of 9.

It can be done in ${}^7C_3$ ways $= \dfrac{7 \times 6 \times 5}{6} = 35$ ways.

Manipal

Engineering Entrance Exam

Solved Paper 2013

Physics

1. The angular velocities of three bodies in simple harmonic motion are $\omega_1, \omega_2, \omega_3$ with their respective amplitudes as A_1, A_2, A_3. If all the three bodies have same mass and velocity, then

(a) $A_1^2\omega_1^2 = A_2^2\omega_2^2 = A_3^2\omega^2$ (b) $A_1^2\omega_1 = A_2^2\omega_2 = A_3^2\omega_3$

(c) $A_1\omega_1^2 = A_2\omega_2^2, A_3\omega_3^2$ (d) $A_1\omega_1 = A_2\omega_2 = A_3\omega_3$

2. If value of surface tension of a liquid is 70 dyne/cm, then its value in N/m will be

(a) 7×10^3 N/m (b) 7×10^2 N/m

(c) 7×10^{-2} N/m (d) 70 N/m

3. A ballon contains 1500 m^3 of a helium at 27°C and 4 atmospheric pressure. The volume of helium at −3°C temperature and 2 atmo spheric pressure will be

(a) 2700 m^3 (b) 1900 m^3

(c) 1700 m^3 (d) 1500 m^3

4. A car is moving along a straight horizontal road with a speed v_0. If the coefficient of friction between the tyres and the road is μ, the shortest distance in which the car be stopped is

(a) $\frac{v_0^2}{\mu}$ (b) $\left(\frac{v_0}{\mu g}\right)^2$

(c) $\frac{v_0^2}{\mu g}$ (d) $\frac{v_0^2}{2\mu g}$

5. A car travelling on a straight track moves with uniform velocity of v_1 for some time and with unifrom velocity v_2 for the next equal time the average velocity on the car is

(a) $\frac{v_1 v_2}{2}$ (b) $\frac{v_1 v_2}{4}$

(c) $\frac{v_1 + v_2}{2}$ (d) $\frac{v_1 - v_2}{2}$

6. Threshold wavelength of a metal is 4000A°. If light of wavelength 3000Å irradiates the surface, the maximum kinetic energy of photoelectron is

(a) 1.7 eV (b) 1.6 eV (c) 1.5 eV (d) 1.0 eV

7. Simple pendulum of length l has a maximum angular displacement θ. The maximum kinetic energy of the bob is

(a) $mgl\,(1-\cos\theta)$ (b) 0.5 mgl

(c) mgl (d) 2 mgl

8. Radius of orbit of satellite of earth is R. Its kinetic energy is proportional to

(a) $\frac{1}{R}$ (b) $\frac{1}{\sqrt{R}}$

(c) R (d) $\frac{1}{R^{3/2}}$

9. The radius R of the soap bubble is doubled under isothermal condition. If T be the surface tension of soap bubble, the work done in doing so is given by

(a) $32\,\pi R^2 T$ (b) $24\pi R^2 T$

(c) $8\pi R^2 T$ (d) $4\pi R^2 T$

10. A body of specific heat 0.2 kcal/kg°C is heated through 100°C. The percentage increase in its mass is

(a) 9% (b) 9.3×10^{-11}%

(c) 10% (d) None of these

11. Two similar coils are kept mutually perpendicular such that their centres coincide. At the centre, find the ratio of the magnetic field due to one coil and the resultant magnetic field through both coils, if the same current is flown

(a) $1:\sqrt{2}$ (b) $1:2$
(c) $1:3$ (d) $\sqrt{3}:1$

12. A prism of refractive index $\sqrt{2}$ has refracting angle of 60°. At what angle a ray must be incident one it so that it suffers a minimum deviation?

(a) 45° (b) 60° (c) 90° (d) 180°

13. A elevator car whose floor to distance is 2.7 m starts ascending with a constant acceleration of 1.2 m/s^2, 2s after a bolt is begin to fall from the ceiling of the car. The free fall time of the bolt is ($g = 9.8$ m/s^e)

(a) $\sqrt{\frac{2.7}{9.8}}$ s (b) $\sqrt{\frac{5.4}{9.8}}$ s
(c) $\sqrt{\frac{5.4}{8.6}}$ s (d) $\sqrt{\frac{5.4}{11}}$ s

14. A wet open umbrella is held vertical and it whirld about the handle at a uniform rate of 21 revolutions in 44s. If the rim of the umbrella is circle of 1 m in diameter and the height of the rim above the flour is 4.9 m, the locus of the drop is a circle of radius

(a) $\sqrt{2.5}$ m (b) 1 m
(c) 3 m (d) 1.5 m

15. The moment of inertia of a body about a given axis is 1.2 kg-m^2. To produce a rotational kinetic energy of 1500 J an angular acceleration of 25 rad/s^2 must be applied for

(a) 8.5 s (b) 5 s
(c) 2 s (d) 1 s

16. A running man has half the kinetic energy of that of a boy of half of his mass. The man speed up 1 m/s, so as the have same kinetic energy as that of a boy. The original speed of the man is

(a) $(\sqrt{2}-1)$ m/s
(b) $\sqrt{2}$ m/s
(c) $\frac{1}{(\sqrt{2}-1)\text{ m/s}}$
(d) $\frac{1}{\sqrt{2}}$ m/s

17. An earth satelites S has orbit radius which is 4 times that of communication satellite C. The period of revolution of S will be

(a) 32 days (b) 18 days
(c) 8 days (d) 9 days

18. n small balls, each of mass m impinge elastically each second on a surface with a velocity u, then the force experienced by the surface in one second, will be

(a) 4 mnu (b) 2 mnu
(c) 1.5 mnu (d) 0.8 mnu

19. A circular disc is rotating with angular velocity ω. If a man standing at the edge of the disc walks towards its centre then the angular velocity of the disc will

(a) decraese (b) increase
(c) be halved (d) not change

20. The moment of inertia of a disc of mass m and radius R about an axis, which is tangential to the circumference of the disc and parallel to its diameter is

(a) $\frac{3}{2}mR^2$ (b) $\frac{2}{3}mR^2$
(c) $\frac{5}{4}mR^2$ (d) $\frac{4}{5}mR^2$

21. A body weights 500 N on the surface of the earth. How much would it weigh half way below the surface of the earth ?

(a) 1000 N (b) 500 N
(c) 250 N (d) 125 N

22. Escape velocity at surface of earth is 11.2 Km/s. Escape velocity from a planet whose mass is the same as that of earth and radius 1/4 that of earth, is

(a) 2.8 km/s (b) 15.6 km/s
(c) 22.4 km/s (d) 44.8 km/s

23. The Bulk Modulus for an incompressible liquid is

(a) zero (b) unity
(c) infinity (d) between 0 and 1

24. In a capillary tube, water rises to 3 mm the height of water that will rise in another capillary tube having one-third radius of the first is

(a) 1 mm (b) 3 mm (c) 6 mm (d) 9 mm

25. An object is placed at a distance 20 cm from the pole of a convex mirror of focal length 20 cm. The image is produced

(a) 13.3 cm (b) 20 cm (c) 25 cm (d) 10 cm

26. Angular momentum is conserved

(a) always
(b) never
(c) when external force is absent
(d) when external torque is absent

27. The plano-convex lens of focal length 20 cm and 30 cm are placed together to form a double convex lens, the final focal length will be

(a) 12 cm (b) 60 cm (c) 20 cm (d) 30 cm

28. Due to a force of $(6\hat{i} + 2\hat{j})$ N the displacement of a body is $(3\hat{i} - \hat{j})$ m, then the work done is

(a) 16 j (b) 12 j (c) 8 j (d) zero

29. For a body moving with relativistic speed if the velocity is doubled, then

(a) its linear momentum is doubled
(b) its linear momentum will be less than double
(c) its linear momentum will be more than double
(d) its linear momentum remains unchanged

30. Position of a body with acceleration a is given by $x = ka^m t^n$, here t is time, find the dimension of m and n

(a) $m = 1, n = 1$ (b) $m = 1, n = 2$
(c) $m = 2, n = 1$ (d) $m = 2, n = 2$

31. The correct relation between α and β in a transistor is

(a) $\beta = \frac{\alpha}{1-\alpha}$ (b) $\beta = \frac{\alpha}{1+\alpha}$
(c) $\beta = \frac{1+\alpha}{\alpha}$ (d) $\beta = 1 - \alpha$

32. A man crosses a 320 m wide river perpendicular to the current in 4 min. If in still water he can swim with a speed 5/3 times that of the current, then the speed of the current, in mm^{-1} in is

(a) 30 (b) 40 (c) 50 (d) 60

33. Two spheres of equal masses, one of which is a thin spherical shell and the other a solid, have the same moment of inertia about their respective diameters. The ratio of their radii will be

(a) 5 : 7 (b) 3 : 5
(c) $\sqrt{3} : \sqrt{5}$ (d) $\sqrt{3} : \sqrt{7}$

34. The speed with which the earth have to rotate on its axis so that a person on the equator would weight $(3/5)^{th}$ as much as present. [Radius of earth = 6400 km]

(a) 4.83×10^{-3} rads^{-1}
(b) 5.41×10^{-3} rads^{-1}
(c) 7.82×10^{-4} rads^{-1}
(d) 8.88×10^{-14} rads^{-1}

35. The rate of flow of glycerine of density 1.25×10^3 kgm^{-3} through the conical section of a pipe if the radii of its ends are 0.1 m and 0.04 m and the pressure drop across its length 10 Nm^{-2} is

(a) 6.93×10^{-4} m^{3}s^{-1}
(b) 7.8×10^{-4} m^{3}s^{-1}
(c) 10.4×10^{-5} m^{3}s^{-1}
(d) 14.5×10^{-5} m^{3}s^{-1}

36. The temperature of the black body increases from T to $2T$. The factor by which the rate of emission will increases is

(a) 4 (b) 2 (c) 16 (d) 8

37. A police jeep is chasing with velocity of 45 km/h a thief in another jeep moving with velocity 153 km/h. Police fires a bullet with muzzle velocity of 180 m/s. The velocity it will strike the car of the thief is

(a) 150 m/s (b) 27 m/s
(c) 450 m/s (d) 250 m/s

38. If the energy of a hydrogen atom in n^{th} orbit is E_n then energy in the nth orbit of a singly ionized helium atom will be

(a) $4E_n$ (b) $E_n/4$
(c) $2E_n$ (d) $E_n/2$

39. If the work function of a potential is 6.875 eV, its threshold wavelength will be (Take = $c = 3 \times 10^8$ m/s)

(a) 3600 Å (b) 2400 Å
(c) 1800 Å (d) 1200 Å

40. Which of the following is unipoler transistor ?

(a) *p-n-p* transistor
(b) *n-p-n* transistor
(c) field effect transistor
(d) point confact transistor

41. If the length of a pendulum is made 9 times and mass of bob is made 4 times, then the value of time period becomes

(a) $3T$ (b) $3/2T$
(c) $4T$ (d) $2T$

42. Two weights w_1 and w_2 are suspended to be the two strings on a frictionless pulley. When the pulled up with an acceleration g then the tension in the string is

(a) $\frac{4w_1w_2}{w_1+w_2}$ (b) $\frac{w_1w_2}{w_1+w_2}$
(c) $\frac{2w_1w_2}{w_1+w_2}$ (d) $\frac{w_1+w_2}{2}$

43. An observer loop at a tree of height 15 m with a telescope of magnifying power 10. To him the tree appears

(a) 10 times taller (b) 15 times taller
(c) 10 times nearer (d) 15 times nearer

44. An inductor is connected to a battary through a switch induced emf is a when the switch is pressed and e_2 when the switch is opened. Then

(a) $e_1 = e_2$ (b) $e_1 > e_2$
(c) $e_1 < e_2$ (d) $e_1 >/< e_2$

45. The speed of a wave on a string is 150 m/s when the tension is 120 N. The percentage increase in the tension in order to raise the wave speed by 20% is

(a) 44% (b) 40% (c) 20% (d) 10%

46. The energy supplied to Kolkata by the state electricity board during an average November week day was 40 GWh. If this energy could be obtained by the conservation of matter, how much mass would have to be annihilated?

(a) 1.6 g (b) 2.2 g (c) 4.0 g (d) 1.6 kg

47. A simple pendulum has time period T_1. The point of suspension is now moved upward according to the relation

$y = kt^2$ $(k = 1\text{ms}^2)$ where y is the vertical displacement. The time period now become T_2.

The ratio of $\frac{T_1^2}{T_2^2}$ is $g = 10 \text{ m/s}^2$

(a) 6/5 (b) 5/6
(c) 1 (d) 4/5

48. An electron is moving round the nucleus of a hydrogen atom in a circular orbit of radius r. The coulomb force **F** between the two is

(a) $k\frac{e^2}{r^3}r$ (b) $-k\frac{e^2}{r^3}r$ (c) $k\frac{e^2}{r^2}r$ (d) $-k\frac{e^2}{r^2}r$

49. A charge q is located at the centre of a cube. The electric flux through any face is

(a) $\frac{\pi q}{6(4\pi\varepsilon_0)}$ (b) $\frac{q}{6(4\pi\varepsilon_0)}$
(c) $\frac{2\pi q}{6(4\pi\varepsilon_0)}$ (d) $\frac{4\pi q}{6(4\pi\varepsilon_0)}$

50. In the Boolean Algebra $\mathbf{A \cdot B}$ is same

(a) **A + B** (b) A · B (c) **A − B** (d) A + B

51. When a force F_1 acts on a particle, frequency is 6Hz and when a force F_2 acts, frequency is 8 Hz. what is the frequency when both the forces act simultaneously in same direction?

(a) 12 Hz (b) 25 Hz (c) 10 Hz (d) 5 Hz

52. For a particle executing simple harmonic motion, the kinetic energy k is given by, $k = k_0 \cos^2 \omega t$. The maximum value of potential energy is

(a) k_0 (b) zero
(c) $k_0/2$ (d) not obtainable

53. A train of 150 m length is going towards north direction at a speed of 10 m/s. A parrot flies at a speed of 5 m/s towards south direction parallel to the railway track, the time taken by the parrot to cross the train is equal to

(a) 12 s (b) 8 s (c) 15 s (d) 10 s

54. Ice starts freezing in a lake with water at 0°C when the atmospheric temperature is – 10 °C. If the time taken for 1 cm of ice to be formed is 12 min, the time taken for the thickness of the ice to change from 1 to 2cm will be

(a) 12 min
(b) less than 12 min
(c) more than 12 min but less than 24 min
(d) more than 24 min

55. The wavelength of the k_a line for an element of atomic number 43 is λ. Then the wavelength of the K_α line for an element of atomic number 29 is

(a) $\left(\frac{43}{29}\right)\lambda$ (b) $\left(\frac{42}{28}\right)\lambda$ (c) $\left(\frac{9}{4}\right)\lambda$ (d) $\left(\frac{4}{9}\right)\lambda$

56. The ratio of the speed of an object to the speed of its real image of magnification m of a convex mirror is

(a) $-\frac{1}{m^2}$ (b) m^2

(c) $-m$ (d) $\frac{1}{m}$

57. The maximum current that flow in the fuse wire before it blows out, varies with the radius r as

(a) $r^{3/2}$ (b) r

(c) $r^{2/3}$ (d) $r^{1/2}$

58. A diatomic gas is heated at certain pressure. What fraction of the heat energy is used to increase the internal energy?

(a) 3/5 (b) 3/7 (c) 5/7 (d) 5/9

59. In interference pattern, the energy is

(a) created at the minimum
(b) destroyed at the minimum
(c) conserved but redistributed
(d) All of the above

60. A red flower kept in green light will appear

(a) red (b) yellow
(c) black (d) white

Chemistry

1. 0.01 M solution of KCl and $CaCl_2$ are prepared in water. The freezing point of KCl is found to be $-2°C$. What is the freezing point of $CaCl_2$ to be completely ionised?

(a) − 3°C (b) + 3°C
(c) − 2°C (d) − 4°C

2. 1 mol He and 3 mol N_2 exert a pressure of 16 atm. Due to a hole in the vessel in which mixture in placed, mixture leaks out. What is the composition of mixture effusing out initially?

(a) 0.22 (b) 044 (c) 0.66 (d) 0.88

3. What are the four quantum numbers of 19th electron in Sc ($Z = 21$)?

(a) $n = 4, l = 0, m = 0, m_s = +\frac{1}{2}$

(b) $n = 3, l = 1, m = -1, m_s = +\frac{1}{2}$

(c) $n = 4, l = 1, m = +1, m_s = -\frac{1}{2}$

(d) $n = 3, l = 2, m = +2, m_s = -\frac{1}{2}$

4. (A) +tap water → white turbidity soluble in *aq.* NH_3

$$(A) \xrightarrow{\Delta} \text{residue}\,(B) + NO_2 + O_2$$

Aqueous (A) gives brown ring on adding $FeSO_4$ and Conc. H_2SO_4. Identify (A).

(a) $NaNO_3$ (b) $AgNO_3$
(c) $NaNO_2$ (d) $AgNO_2$

5. Each of the following compounds has been dissolved in water to make its 0.001 M solution. Rank them in order of their increasing conductivity in solution (assume 100% ionisation in each case)

(a) $c < d < a < b$
(b) $c < d < b < a$
(c) $a < b < c < d$
(d) $d < c < a < b$

6. Metal carbonyls can have formula $M(CO)_x$, where x = number of CO units coordinated to metal M. What is the formula of the carbonyl of Fe (26) if EAN of Fe in metal carbonyl is 36?

(a) 2 (b) 3
(c) 4 (d) 5

7. Following method of extracting Zn is based on thermodynamics

$$A: 2ZnS + 3O_2 \longrightarrow 2ZnO + 2SO_2$$

$$B: ZnO + C \longrightarrow Zn + CO$$

If ΔG_f° (standard free energies of formation, in $kJmol^{-1}$) of ZnS = − 205.4,

ZnO = −318.0, SO_2 = −300.4 and of CO = − 137.3

Free energy changes of the above reaction A and B (respectively) will be

(a) −826.4 kJ, +180.9 kJ
(b) +826.4 kJ, −180.9 kJ
(c) −826.4 kJ, −180.9 kJ
(d) +826.4 kJ, +180.9 kJ

8. The final step for the extraction of copper form copper pyrites in Bessemer convertor involves the reaction.

(a) $Cu_2S + 2FeO \longrightarrow 2Cu + 2Fe + SO_2$
(b) $Cu_2S + 2CuO \longrightarrow 6Cu + SO_2$
(c) $4Cu_2O + FeS \longrightarrow 8Cu + 2FeSO_4$
(d) $2Cu_2O + FeS \longrightarrow 4Cu + Fe + SO_2$

9. If $CO_2(g)$ under pressure is passed into Na_2CrO_4 (aq), yellow colour solution changes to

(a) blue
(b) green
(c) red
(d) orange-red

10. Select incorrect statements.

(a) Ionisation energies of 5*d*- elements are greater than those of 3*d* and 4*d* elements.
(b) Cu (I) is diamagnetic while Cu (II) is paramagnetic.
(c) $[Ti(H_2O)_6]^{3+}$ is colored while $[Sc(H_2O)_6]^{+}$ is colourless.
(d) Transition elements cannot form complexes.

11. ClO_2 is an/a

(a) anhydride of $HClO_2$
(b) anhydride of $HClO_3$
(c) mixed anhydride of $HClO_2$ and $HClO_3$
(d) mixed anhydride of $HClO_3$ and $HClO_4$

12. Based on the following reaction identify alkene *A*

$$A \xrightarrow{O_3/(H_2O)/Zn} 2HCHO + \underset{\displaystyle CHO}{\underset{|}{CHO}}$$

(a) $CH_3-\overset{\overset{\displaystyle CH_2}{\|}}{C}-CH_3$
(b) $CH_2=CH-CH=CH_2$
(c) $CH_3-CH=CH-CH_3$
(d) $CH_3-CH_2-CH=CH_2$

13. Which one of the following statments about NO is wrong?

(a) NO is an odd electron molecule
(b) It is a free radical and highly reactive
(c) It readily forms complexes with transition metal ions
(d) It can be prepared by heating NH_4NO_3 at 250°C

14. KO_2 is used in life supports in space crafts, submarines and emergency breathing apparatus since it

(a) absorbs CO_2
(b) release O_2
(c) releases CO_2
(d) absorbs CO_2 and releases O_2

15. Calculate second electron affinity of oxygen for the process,

$$O^-(g) + e^-(g) \rightarrow O^{2-}(g)$$

by using the following data

(i) Heat of sublimation of

$$Mg\ (s) = +147.7 \text{ kJ mol}^{-1}$$

(ii) Ionisation energy of Mg(g) to form

$$Mg^{2+}(g) = +2189.0 \text{ kJ mol}^{-1}$$

(iii) Bond dissociation energy for

$$O_2 = +498.4 \text{ kJ mol}^{-1}$$

(iv) First electron affinity of

$$O(g) = -141.0 \text{ kJ mol}^{-1}$$

(v) Heat formation of

$$MgO = -601.7 \text{ kJ mol} - 1$$

(vi) Lattice energy of

$$MgO = -3791.0 \text{ kJ mol}^{-1}$$

(a) 235.6 kJ mol^{-1}
(b) 468.7 kJ mol^{-1}
(c) 544.4 kJ mol^{-1}
(d) 744.4 kJ mol^{-1}

16. 1 g of fuming H_2SO_4 (oleum : It is a mixture of concentrated H_2SO_4 saturated with SO_3 and having formula $H_2S_2O_7$) is diluted with H_2O. This solution is completely neutralised by 26.7mL of 0.8 N NaOH. Find the percentage of free SO_3 in the oleum.

(a) 20.73% (b) 43.80%
(c) 79.27% (d) 60.74%

17. ${}^{14}_{7}N + {}^{1}_{0}n \rightarrow {}^{14}_{6}C + {}^{1}_{1}H$ is written as

(a) ${}^{14}_{7}N(n, e){}^{1}_{1}H$
(b) ${}^{14}_{7}N\ (p, n)\ {}^{14}_{6}C$
(c) ${}^{14}_{7}N\ (n, p)\ {}^{14}_{6}C$
(d) ${}^{14}_{6}C\ (p, n)\ {}^{14}_{7}N$

18. Photographic paper is developed with alkaline hydroquinone

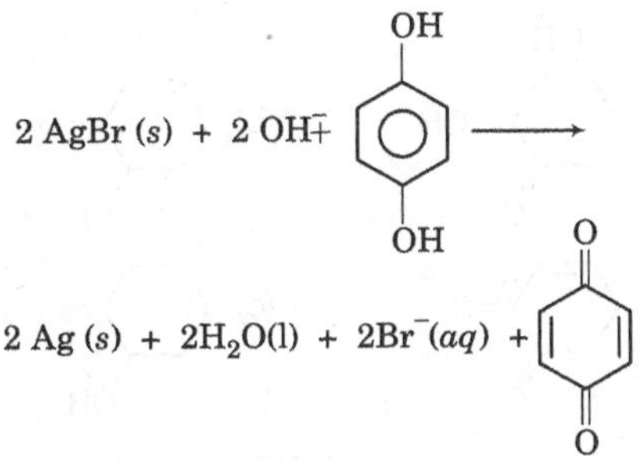

Select correct statement.

(a) Hydroquinone is the oxidant
(b) Ag^+ is the oxidant
(c) Br^- is the oxidant
(d) Ag^+ is the reductant

19. Which has maximum ionisation potential?

(a) N
(b) O
(c) O^+
(d) Na

20. Electronic configuration of Gd (64) is written as

(a) $[Xe]\ 4f^7\ 5d^1 6s^2$
(b) $[Xe]\ 4f^8\ 6s^2$
(c) $[Xe]\ 4f^9\ 6s^1$
(d) $[Xe]\ 4f^{10}$

21. Screening effect is not observed in

(a) He^+
(b) Li^{2+}
(c) Be^{3+}
(d) all cases

22. Which one of the following isomers of $PBrCl_3$ have no dipole moment?

(I) 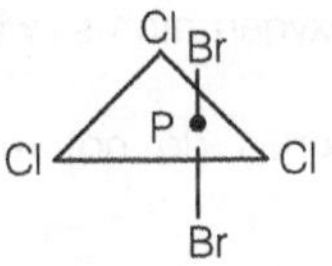

(II) 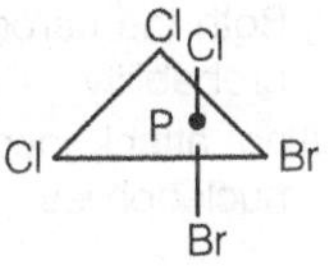

(III) 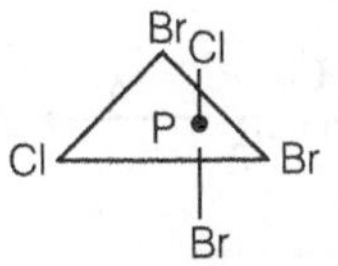

(a) Only I
(b) I and II
(c) II and III
(d) Only III

23. The type of hybridisation of P atom in PCl_5, PCl_4^+ and PCl_6^- is (respectively)

(a) sp^3, sp^3d sp^3d^2
(b) sp^3d, sp^3, sp^2d^2
(c) sp^3, sp^3d^2, sp^3
(d) sp^3d^2, sp^3, sp^3d

24. Which has a maximum repulsive interaction?

(a) *bp*–*bp* (b) *lp*–*lp* (c) *lp*–*bp* (d) Equal

25. Calculate the electronegativity of chlorine from bond energy of Cl—F bond (61 kcal mol^{-1}) F—F (38 kcal mol^{-1}) and Cl—Cl bond (58 kcal mol^{-1}) and electronegativity of fluorine 4.0 eV

(a) 1.42 eV (b) 1.89 eV (c) 2.67 eV (d) 3.22 eV

26. A sample of ammonium phosphate $(NH_4)_3PO_4$ contains 3.18 moles of H-atom. The number of moles of O-atom in the sample is

(a) 0.265 (b) 0.795 (c) 1.06 (d) 3.18

27. In the following compound keto group is at position

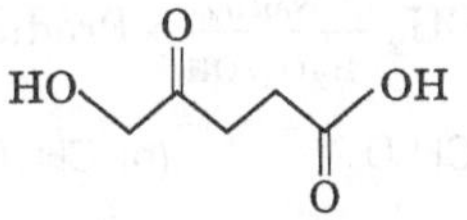

(a) 1 (b) 2 (c) 3 (d) 4

28. 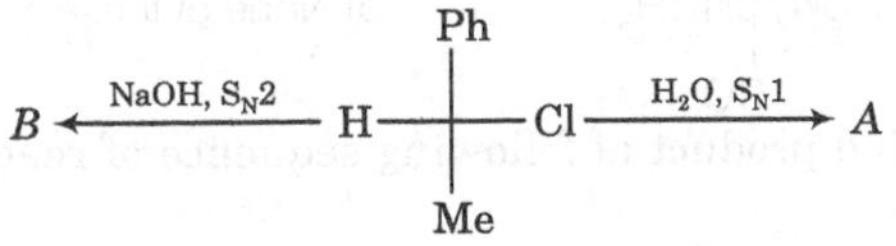

I. Formation of *A* has proceeded with racemisation.

II. Formation of *B* has proceeded with inversion.

Select the correct statement.

(a) I and II
(b) Only I
(c) Only III
(d) None of these

29. 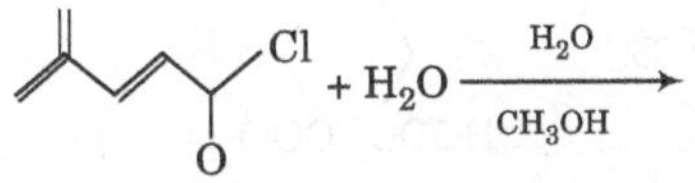

In this reaction, we get types of substituted alcohols (stereoisomers not considered)

(a) one (b) two (c) three (d) four

30. Electrophile $NO_2^{\oplus}$ attacks the following

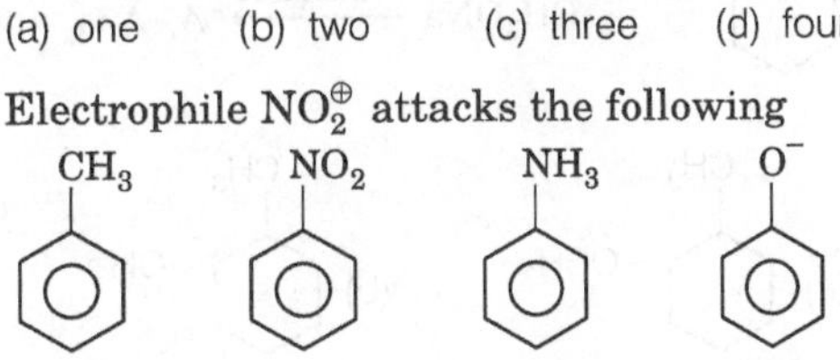

I II III IV

In which cases, $\overset{\oplus}{NO_2}$ will be at *meta* position?

(a) II and IV
(b) I, II and III
(c) II and III
(d) Only I

31. The reaction of 1- bromo-3- chlorocyclobutane with metallic sodium in dioxane gives

(a) 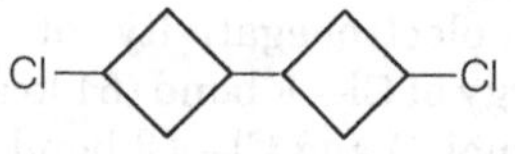

(b) 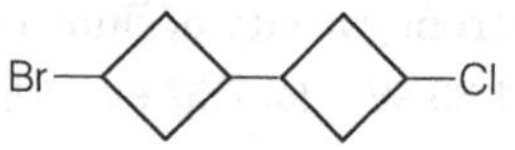

(c) 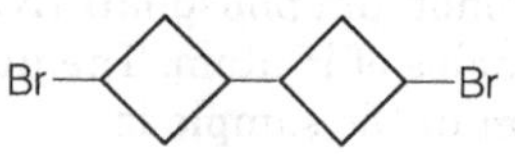

(d) 

32. $CH_3CH{=}CH_2 \xrightarrow[H_2O_2/OH^-]{NaBD_4}$ Product X, X is

(a) $CH_3CH(OH)CH_2D$ (b) $CH_3CH(D)CH_2OH$

(c) $CH_3CH(CO)CH_3$ (d) None of these

33. End product of following sequence of reaction is

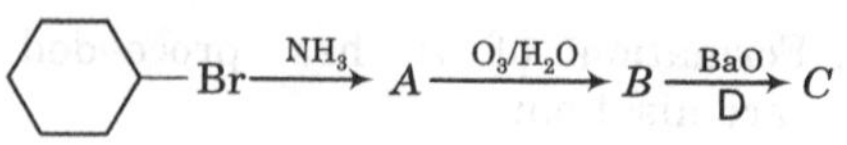

(a) 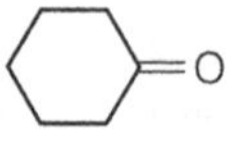 (b) 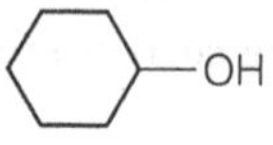

(c) 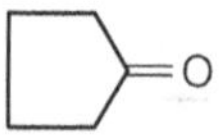 (d) 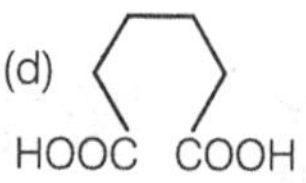

34. 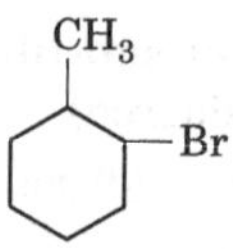 + $CH_3ONa \xrightarrow{CH_3OH}$ A, A is

(a) 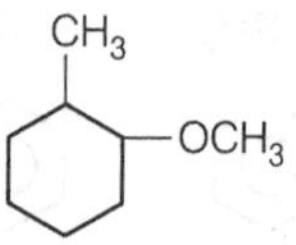 (b) 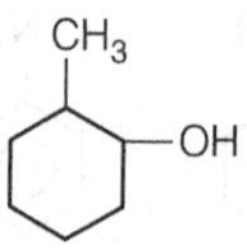

(c) 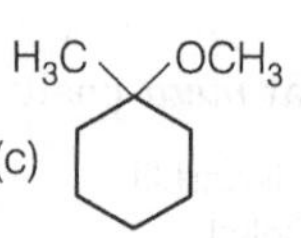 (d) 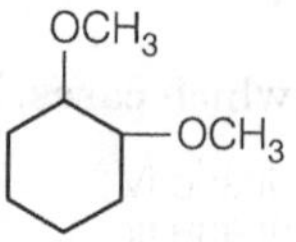

35. 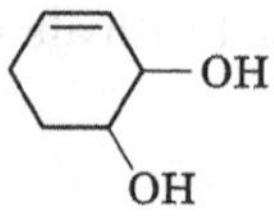 $\xrightarrow{MnO_2}$ A, A is

(a) 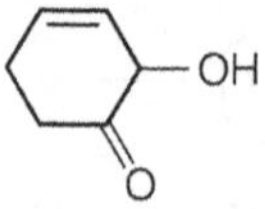 (b) 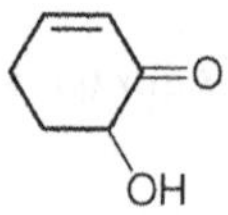

(c) 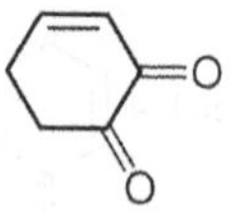 (d) 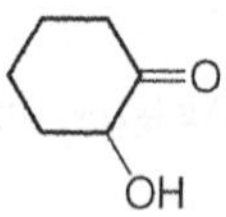

36. 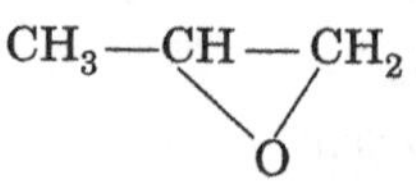 $\xrightarrow[H^+]{LiAlH_4}$ A, A can be

(a) $CH_3CH(OH)CH_3$ (b) $CH_3CH_2CH_2OH$

(c) Both (a) and (b) (d) None of these

37. $2CH_3\overset{O}{\overset{\|}{C}}OC_2H_5 \xrightarrow{C_2H_5ONa}$ A, A is formed by Claisen condensation. Which is/are true about A?

(a) A forms oxime
(b) A shows tautomerism
(c) A shows iodoform test
(d) All of the above are true

38. When a nucelophile encounters a ketone the site of attack is

(a) the carbon atom of the carbonyl
(b) the oxygen atom of the carbonyl
(c) Both the carbon and oxygen atoms with equal probability
(d) no attack occurs- ketones do not react with nucleophiles

39. 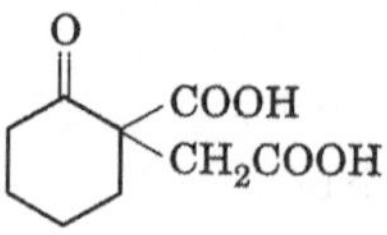 $\xrightarrow{\Delta}$ A, A is

(a) 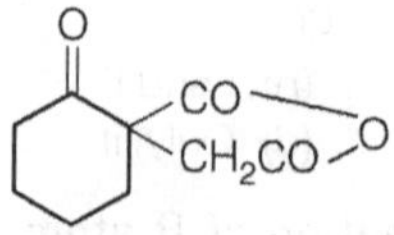 (b) 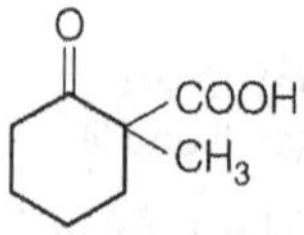

(c) 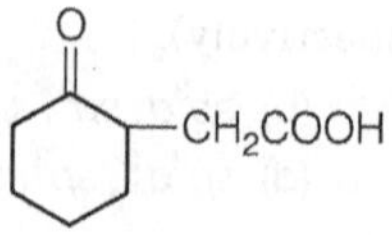 (d) None of these

40. The structure of N- propyl acetamide is
(a) $CH_3CH_2CONHCH_3$
(b) $CH_3CON(CH_2CH_2CH_3)_2$
(c) $CH_3CONHCH_2CH_2CH_3$
(d) $CH_3CH{=}NCH_2CH_2CH_3$

41. Which *A* gives red colour in the reaction

$$A \xrightarrow[\text{(ii) NaOH}]{\text{(i) HNO}_2} \text{red colour}$$

(a) $CH_3CH_2NO_2$ (b) $(CH_3)_2{-}CHNO_2$
(c) $(CH_3)_3CNO_2$ (d) 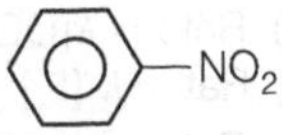

42. Alanine forms Zwitter ion which exists as
(a) $CH_3CH(\overset{\oplus}{N}H_3)COO^-$ in acidic medium
(b) $CH_3CH(\overset{\oplus}{N}H_3)COOH$ in a medium of pH = 4
(c) $CH_3CH(NH_2)COO^-$ in a medium of pH = 1
(d) $CH_3CH(NH_2)COO^-$ in a medium of pH = 2

43. Which gives only glucose by hydrolysis?
(a) Sucrose (b) Raffinose
(c) Maltose (d) Galactose

44. Which polysaccharide has α-glycoside linkage ?
(a) Amylose (b) Amylopection
(c) Cellulose (d) All of these

45. 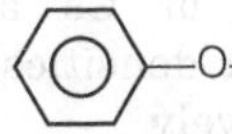 **—O—XH_3 + HI ⟶ Product (mixture)**

Mixture can be of

(a) 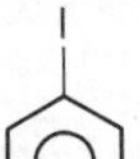, CH_3OH (b) 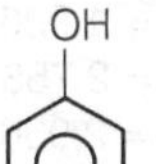, CH_3I
(c) Both (a) and (b) (d) None of these

46. *p*-cresol reacts with chloroform in alkaline medium to give the compound *A* which adds hydrogen cyanide to form the compound *B*. The latter on acidic hydrolysis gives chiral carboxylic acid '*C*' which is

(a) 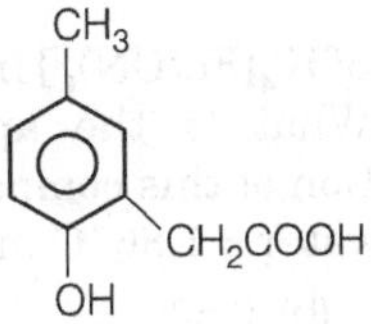 (b) 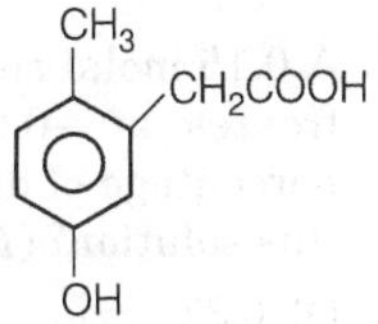
(c) 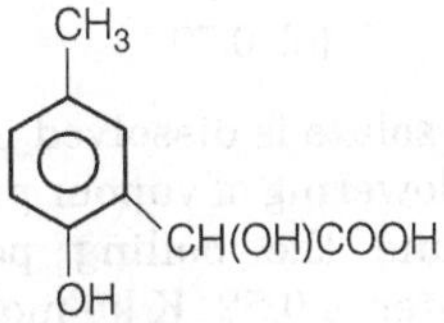 (d) 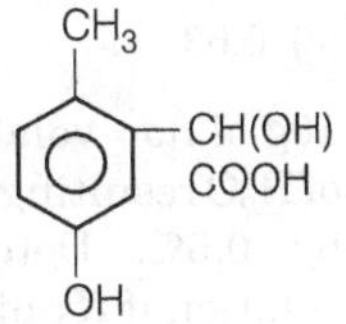

47. 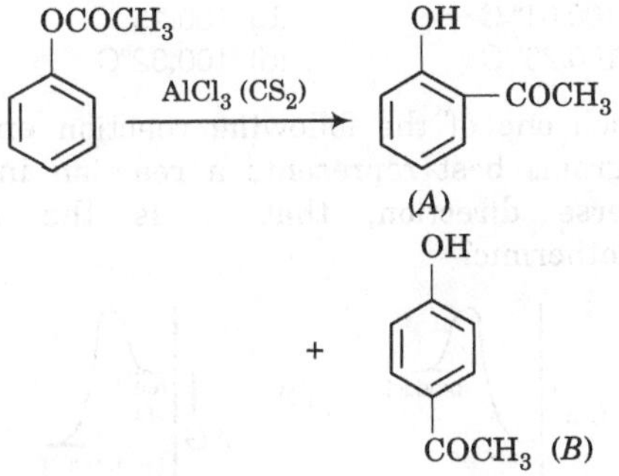

Choose the correct statement.
(a) *A* is more volatile than *B*
(b) *B* is more volatile than *A*
(c) *A* is formed more rapidly at a higher temperature
(d) *A* is formed in higher yield at a low temperature

48. Sodium extract of thioruea will be ...... colour in Lassaigne's test.
(a) blue (b) red (c) yellow (d) green

49. Consider the following redox reaction occurring in acidic medium

$$BrO_3^- \xrightarrow{E^\circ=1.5\text{V}} BrO^- \xrightarrow{E^\circ=1.6\text{V}} Br_2$$
$$BrO_3^- \xrightarrow{E^\circ=?\text{ V}} Br_2$$

The unknown standard reduction potential is
(a) –1.6 V (b) 1.6 V (c) –1.52 V (d) 1.52 V

50. The spontaneous reaction that takes place in this cell is
(a) $Zn + Ni \longrightarrow Zn^{2+} + Ni^{2+}$
(b) $Zn + Ni^{2+} \longrightarrow Zn^{2+} + Ni$
(c) $Ni + Zn^{2+} \longrightarrow Ni^{2+} + Zn$
(d) $Ni^{2+} + 2K \longrightarrow 2K^+ + Ni$

51. A 0.15 molal solution of $K_4[Fe(CN)_6]$ in water freezes at $-0.65°C$. What is the apparent percentage of dissociation of this compound in this solution? (K_f for water = $1.86°C\ mol^{-1}$)

(a) 0.33 (b) 0.52
(c) 0.63 (d) 0.79

52. 10g of non-volatile solute is dissolved in 180 g of H_2O resulting in lowering of vapour pressure by 0.5%. Determine the boiling point of solution if K_b of water is 0.52 K kg mol^{-1}.

(a) 100.01° C (b) 100.15° C
(c) 100.23° C (d) 100.32°C

53. Which one of the following reaction energy diagrams best represents a reaction in the reverse direction; that it is the most endothermic?

(a)
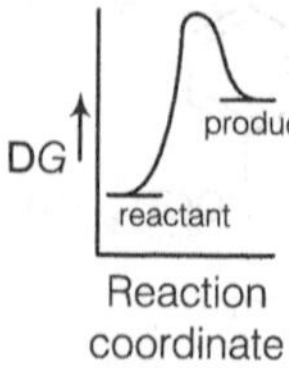

(b)
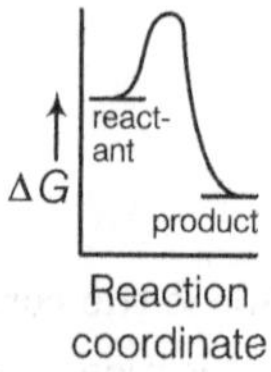

(c)
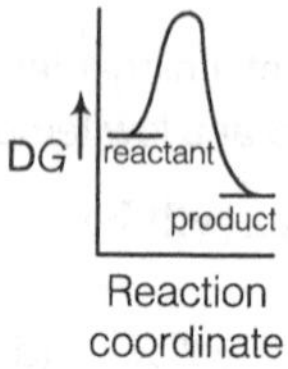

(b)
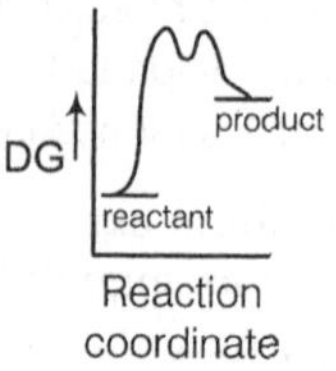

54. A certain reaction rate increases 1000 folds in the presence of a catalyst at 27°C. The activation energy of the original pathway is 98 kJ/mol. What is the activation energy of the new pathway?

(a) 80.77 kJ (b) 56.38 kJ
(c) 24.67 kJ (d) 90.43 kJ

55. Consider the following reaction in an aqueous solution

$$I^-(aq) + Cl^- \rightarrow IO^-(aq) + Cl^-(aq)$$

and the following initial concentration and initial rate data for this reaction.

Exp. No	$[I^-]$ M	$[OCl^-]$ M	Initial rate Ms^{-1}
1.	0.1000	0.0500	3.05×10^{-4}
2.	0.2000	0.0500	6.10×10^{-4}
3.	0.3000	0.0100	1.83×10^{-4}
4.	0.3000	0.0200	3.66×10^{-4}

Which of the following is the correct rate law for this reaction?

(a) Rate = $k\ [I^-]^2[OCl^-]$
(b) Rate = $k[OCl^-]$
(c) Rat = $k\ [I^-]^2$
(d) Rate = $k[I^-][OCl^-]$

56. Determine the pH of 0.024 M hydroxylamine hydrochloride solution. (K_b of hydroxyl amine is 10^{-8}).

(a) 8.4 (b) 6.2
(c) 3.8 (d) 5.6

57. Determine solubility of AgBr in a 0.1 M KCN solution. [K_f of $[Ag(CN)_2]^-$ = 5.6×10^8; K_{sp} of AgBr = 7.7×10^{-13}]

(a) 2×10^3 M (b) 4×10^{-6} M
(c) 2×10^{-6} M (d) 4×10^{-3} M

58. Which of the following forms cationic micelles above a certain minimum concentration?

(a) Sodium dodecyl sulphate
(b) Sodium acetate
(c) Urea
(d) Cetyl trimethyl ammonium bromide

59. Determine the relative size of Cs atom compared to a Li atom, if their densities are 1.87 g/cc and 0.53 g/cc respectively

(a) $r(Cs) = 1.753\ r(Li)$
(b) $r(Cs) = 1.936\ r(Li)$
(c) $r(Cs) = 2.753\ r(Li)$
(d) $r(Cs) = 2.936\ r(Li)$

60. Nylon threads are made of

(a) polyvinyl polymer
(b) Polyester polymer
(c) polyamide polymer
(d) polyethylene polymer

Mathematics

1. The least positive integer n, for which

$n! < \left(\frac{n+1}{2}\right)^n$ holds, is

(a) 1 (b) 2
(c) 3 (d) 4

2. The sum to n terms of the series:

$\frac{1}{1+1^2+1^4} + \frac{2}{1+2^2+2^4} + \frac{3}{1+3^2+3^4} + \dots$ is

(a) $\frac{n^2+1}{2(n^2+n+1)}$ (b) $\frac{n^2+n}{(n^2+n+1)}$
(c) $\frac{n^2+n}{2(n^2+n+1)}$ (d) None of these

3. The sum of n terms of the series $+1+\frac{4}{5}+\frac{7}{5^2}+\frac{10}{5^3}+\dots$ is

(a) $\frac{5}{4}+\frac{15}{16}\left(1-\frac{1}{5^{n-1}}\right)-\frac{(3n-2)}{4\cdot 5^{n-1}}$
(b) $\frac{5}{4}+\frac{1}{16}\left(1-\frac{1}{5^{n-1}}\right)-\frac{3n}{4\cdot 5^{n-1}}$
(c) $\left(1-\frac{1}{5^{n-1}}\right)-\frac{(3n+2)}{4\cdot 5^{n-1}}$
(d) None of the above

4. Sum of the series to n terms $5+7+13+31+85+\dots$ is

(a) 3^n+8n+1
(b) $\frac{1}{2}[3^n+8n-1]$
(c) $\frac{1}{2}(3^n+8n+1)$
(d) None of the above

5. If $b<0$, then the roots x_1 and x_2 of the equation $2x^2+6x+b=0$, satisfy the condition $\left(\frac{x_1}{x_2}\right)+\left(\frac{x_2}{x_1}\right)<K$, where K is equal to

(a) 2 (b) -2
(c) 0 (d) 4

6. If the roots of the equation $(a-1)(x^2+x+1)^2 = (a+1)(x^4+x^2+1)$ are real and distinct then the value of $a \in$

(a) $(-\infty, 3]$ (b) $(-\infty, -2) \cup (2, \infty)$
(c) $[-2, 2]$ (d) $[-3, \infty)$

7. If the sum of two of the roots of $x^3+px^2+qx+r=0$ is zero, then $pq=$

(a) $-r$ (b) $2r$ (c) $-2r$ (d) r

8. If m parallel lines in a plane are intersected by a family of n parallel lines, then the number of parallelograms that can be formed is

(a) $\frac{1}{4}mn(m-1)(n-1)$ (b) $\frac{1}{2}mn(m-1)(n-1)$
(c) $\frac{1}{4}m^2n^2$ (d) None of these

9. A person is permitted to select at least one and at most n coins from a collection of $(2n+1)$ distinct coins. If the total number of ways in which he can select coins is 255, then n equals

(a) 4 (b) 8 (c) 16 (d) 32

10. The value of x in the expression $(x+x^{\log_{10} x})^5$, if the third term in the expansion is 1,000,000, is

(a) $10, 10^{-3/2}$ (b) 100 or $10^{-3/2}$
(c) 10 or $10^{-5/2}$ (d) None of these

11. $\sum_{k=0}^{10} {}^{20}C_k$ is equal to

(a) $2^{19}+\frac{1}{2}\,{}^{20}C_{10}$ (b) 2^{19}
(c) ${}^{20}C_{10}$ (d) None of these

12. If the point (a, a) are placed in between the lines $|x+y|=4$ then

(a) $|a|=2$ (b) $|a|=3$
(c) $|a|<2$ (d) $|a|<3$

13. The number of rational values of m for which the y-coordinate of the point of intersection of the lines $3x+2y=10$ and $x=my+2$ is an integer is

(a) 2 (b) 4
(c) 6 (d) 8

14. A straight line cuts intercepts from the axis of coordinates the sum of the reciprocals of which is a constant K. Then it always passes through a fixed point

(a) (K, K) (b) $\left(\frac{1}{K}, \frac{1}{K}\right)$
(c) $(-K, -K)$ (d) $(K-1, K-1)$

15. If the line $\frac{x}{a}+\frac{y}{b}=1$ moves in such a way that $\frac{1}{a^2}+\frac{1}{b^2}=\frac{1}{c^2}$ where, c is a constant, then the locus of the foot of perpendicular from the origin on the straight line is

(a) Straight line (b) Parabola
(c) Ellipse (d) Circle

16. The pair of lines $\sqrt{3}x^2-4xy+\sqrt{3}y^2=0$ are rotated about the origin by $\frac{\pi}{6}$ in the anticlockwise sense. The equation of the pair in the new position is

(a) $x^2-\sqrt{3}xy=0$
(b) $xy-\sqrt{3}y^2=0$
(c) $\sqrt{3}x^2-xy=0$
(d) None of the above

17. The value of $\lim_{x\to 1}\frac{\left[\sum_{K=1}^{100}-x^K\right]-100}{x-1}$ is

(a) – 5050 (b) 0
(c) 5050 (d) None of these

18. The value of $\lim_{x\to\infty}\left\{\frac{x}{x+\frac{\sqrt[3]{x}}{x+\frac{\sqrt[3]{x}}{x+\sqrt[3]{x}}}\cdots}\right\}$ is

(a) 0 (b) 1 (c) 2 (d) 1/2

19. If $f(x)$ defined by $f(x)=\begin{cases}\frac{|x^2-x|}{x^2-x}, & x\neq 0,1\\ 1, & x=0\\ -1, & x=1\end{cases}$ then $f(x)$ is continuous for all

(a) x
(b) x except at $x=0$
(c) x except at $x=1$
(d) x excexpt at $x=0$ and $x=1$

20. Function $f(x)$ is defined as follows

$$f(x)=\begin{cases}ax-b, & x\le 1\\ 3x, & 1<x<2\\ bx^2-a, & x\ge 2\end{cases}$$

If $f(x)$ is continuous at $x=1$, but discontinuous at $x=2$ then the locus of the point (a,b) is a straight line excluding the point where it cuts the line

(a) $y=3$ (b) $y=2$
(c) $y=0$ (d) $y=1$

21. The maximum value of $(\cos\alpha_1)\cdot(\cos\alpha_2)\ldots(\cos\alpha_n)$. Under the restrictions $0\le\alpha_1,\alpha_2,\ldots\alpha_n\le\frac{\pi}{2}$ and $(\cot\alpha_1)\cdot(\cot\alpha_2)\ldots(\cot\alpha_n)=1$ is

(a) $\frac{1}{2^{n/2}}$ (b) $\frac{1}{2^n}$ (c) $\frac{1}{2n}$ (d) 1

22. The minimum value of the expression $\sin\alpha+\sin\beta+\sin\gamma$, where α,β,γ are real numbers satisfying $\alpha+\beta+\gamma=\pi$ is

(a) positive (b) zero
(c) negative (d) – 3

23. If θ be the angle between the unit vectors **a** and **b**, then $\cos\frac{\theta}{2}$ is equal to

(a) $\frac{1}{2}|\mathbf{a}-\mathbf{b}|$ (b) $\frac{1}{2}|\mathbf{a}+\mathbf{b}|$
(c) $\frac{|\mathbf{a}-\mathbf{b}|}{|\mathbf{a}+\mathbf{b}|}$ (d) $\frac{|\mathbf{a}+\mathbf{b}|}{|\mathbf{a}-\mathbf{b}|}$

24. If $\mathbf{a}\cdot\mathbf{i}=\mathbf{a}\cdot(\mathbf{j}+\mathbf{i})=\mathbf{a}\cdot(\mathbf{i}+\mathbf{j}+\mathbf{k})$, then **a** is equal to

(a) **i** (b) **k**
(c) **j** (d) (**i** + **j** + **k**)

25. If the scalar projection of the vector $x\mathbf{i}-\mathbf{j}+\mathbf{k}$ on the vector $2\mathbf{i}-\mathbf{j}+5\mathbf{k}$ is $\frac{1}{\sqrt{30}}$ then value of x is equal to

(a) $\frac{-5}{2}$ (b) 6 (c) – 6 (d) 3

26. If $\mathbf{a}\times\mathbf{b}=\mathbf{c}$, $\mathbf{b}\times\mathbf{c}=\mathbf{a}$ and a, b, c be the moduli of the vectors **a**, **b**, **c** respectively, then

(a) $a=1, b=1$ (b) $c=1, a=1$
(c) $\mathbf{a}\cdot(\mathbf{b}\times\mathbf{c})=1$ (d) $b=1, c=a$

27. A unit vectors coplanar with $\mathbf{i}+\mathbf{j}+2\mathbf{k}$ and $\mathbf{i}+2\mathbf{j}+\mathbf{k}$ and perpendicular to $\mathbf{i}+\mathbf{j}+\mathbf{k}$ is

(a) $\frac{\mathbf{j}-\mathbf{k}}{\sqrt{2}}$ (b) $\frac{\mathbf{i}+\mathbf{j}+\mathbf{k}}{\sqrt{3}}$
(c) $\frac{\mathbf{i}+\mathbf{j}+2\mathbf{k}}{\sqrt{6}}$ (d) $\frac{-\mathbf{j}+2\mathbf{k}}{\sqrt{5}}$

28. Let $A = \begin{bmatrix} 0 & \alpha \\ 0 & 0 \end{bmatrix}$ and $(A + I)^{50} - 50A = \begin{bmatrix} a & b \\ c & d \end{bmatrix}$, then the value of $a + b + c + d$ is

(a) 2 (b) 1
(c) 4 (d) None of these

29. For two unimodular complex numbers z_1 and z_2, $\begin{bmatrix} \bar{z}_1 & z_2 \\ \bar{z}_2 & z_1 \end{bmatrix}^{-1} \begin{bmatrix} z_1 & z_2 \\ -\bar{z}_2 & \bar{z}_1 \end{bmatrix}^{-1}$ is equal to

(a) $\begin{bmatrix} z_1 & z_2 \\ \bar{z}_1 & \bar{z}_2 \end{bmatrix}$ (b) $\begin{bmatrix} 1 & 0 \\ 0 & 1 \end{bmatrix}$
(c) $\begin{bmatrix} 1/2 & 0 \\ 0 & 1/2 \end{bmatrix}$ (d) None of these

30. If A is a square matrix of order n such that $|\text{adj}(\text{adj } A)| = |A|^9$, then the value of n can be

(a) 4 (b) 2
(c) either 4 or 2 (d) None of these

31. Coefficient of x in $f(x) = \begin{vmatrix} x & (1 + \sin x)^3 & \cos x \\ 1 & \log(1 + x) & 2 \\ x^2 & (1 + x)^2 & 0 \end{vmatrix}$ is

(a) 0
(b) 1
(c) – 2
(d) Cannot be determined

32. Let α_1, α_2 and β_1, β_2 be the roots of $ax^2 + bx + c = 0$ and $px^2 + qx + r = 0$ respectively. If the system of equations $\alpha_1 y + \alpha_2 z = 0$ and $\beta_1 y + \beta_2 z = 0$ has a non-trivial solution, then

(a) $b^2 pr = q^2 ac$ (b) $bpr^2 = qac^2$
(c) $bp^2 r = qa^2 c$ (d) None of these

33. The values of α for which the point $(\alpha - 1, \alpha + 1)$ lies in the larger segment of the circle $x^2 + y^2 - x - y - 6 = 0$ made by the chord whose equation is $x + y - 2 = 0$ is

(a) $-1 < \alpha < 1$ (b) $1 < \alpha < \infty$
(c) $-\infty < \alpha < -1$ (d) $\alpha \le 0$

34. The circles whose equations are $x^2 + y^2 + c^2 = 2ax$ and $x^2 + y^2 + c^2 - 2by = 0$ will touch each other externally if

(a) $\frac{1}{b^2} + \frac{1}{c^2} = \frac{1}{a^2}$ (b) $\frac{1}{c^2} + \frac{1}{a^2} = \frac{1}{b^2}$
(c) $\frac{1}{a^2} + \frac{1}{b^2} = \frac{1}{c^2}$ (d) None of these

35. The tangents to $x^2 + y^2 = a^2$ having inclinations α and β intersect at P. If $\cot\alpha + \cot\beta = 0$, then the locus of P is

(a) $x + y = 0$ (b) $x - y = 0$
(c) $xy = 0$ (d) None of these

36. An equilateral triangle SAB is inscribed in the parabola $y^2 = 4ax$ having its focus at S. If chord AB lies towards the lefit of S, then side length of this triangle is

(a) $2a(2 - \sqrt{3})$ (b) $4a(2 - \sqrt{3})$
(c) $a(2 - \sqrt{3})$ (d) $8a(2 - \sqrt{3})$

37. Minimum distance between the curves $y^2 = 4x$ and $x^2 + y^2 - 12x + 31 = 0$ is

(a) $\sqrt{5}$ (b) $\sqrt{21}$
(c) $\sqrt{28} - \sqrt{5}$ (d) $\sqrt{21} - \sqrt{5}$

38. If the line $lx + my + n = 0$ cuts the ellipse $\frac{x^2}{a^2} + \frac{y^2}{25} = 1$ in points whose eccentoric angles differ by $\frac{\pi}{2}$, then $\frac{a^2 l^2 + b^2 m^2}{n^2}$ is equal to

(a) 1 (b) 2 (c) 4 (d) $\frac{3}{2}$

39. If the tangent to ellipse $x^2 + 2y = 1$ at point $P\left(\frac{1}{\sqrt{2}}, \frac{1}{2}\right)$ meets the auxiliary circle at the points R and Q, then tangents to circle at Q and R intersect at

(a) $\left(\frac{1}{\sqrt{2}}, 1\right)$ (b) $\left(1, \frac{1}{\sqrt{2}}\right)$ (c) $\left(\frac{1}{2}, \frac{1}{2}\right)$ (d) $\left(\frac{1}{2}, \frac{1}{\sqrt{2}}\right)$

40. Which one of the following is independent of α in the hyperbola $(0 < \alpha < \pi/2)$ $\frac{x^2}{\cos^2\alpha} - \frac{y^2}{\sin^2\alpha} = 1$

(a) Eccentricity (b) Abscissa of foci
(c) Directrix (d) Vertex

41. If PQ is a double ordinate of the hyperbola $\frac{x^2}{a^2} - \frac{y^2}{b^2} = 1$ such that OPQ is an equilateral triangle, O bing the centre of the hyperbola, then the eccentricity e of the hyperbola satisfies.

(a) $1 < e < \frac{2}{\sqrt{3}}$ (b) $e = \frac{2}{\sqrt{3}}$
(c) $e = \frac{\sqrt{3}}{2}$ (d) $e > \frac{2}{\sqrt{3}}$

42. If $iz^4 + 1 = 0$, then z can take the value

(a) $\frac{1+i}{\sqrt{2}}$ (b) $\cos\frac{\pi}{8} + i\sin\frac{\pi}{8}$
(c) $\frac{1}{4i}$ (d) i

43. If $\cos\alpha + \cos\beta + \cos\gamma = \sin\alpha + \sin\beta + \sin\gamma = 0$, then the value of $\cos 3\alpha + \cos 3\beta + \cos 3\gamma$ is

(a) 0 (b) $\cos(\alpha + \beta + \gamma)$
(c) $3\cos(\alpha + \beta + \gamma)$ (d) $3\sin(\alpha + \beta + \gamma)$

44. If Q is real and z_1, z_2 are connected by $z_1^2 + z_2^2 + 2z_1z_2\cos\theta = 0$, then triangle with vertices 0, z_1 and z_2 is

(a) equilateral (b) right-angled
(c) isosceles (d) None of these

45. Let $f(xy) = f(x)\cdot f(y)$ for all $x, y \in R$. If $f'(1) = 2$ and $f(4) = 4$, then $f'(4)$ equal to

(a) 4 (b) 1
(c) $\frac{1}{2}$ (d) 8

46. If $y = \sqrt{(a-x)(x-b)} - (a-b)\tan^{-1}\sqrt{\frac{a-x}{x-b}}$, then $\frac{dy}{dx}$ is equal to

(a) 1 (b) $\sqrt{\frac{a-x}{x-b}}$
(c) $\sqrt{(a-x)(x-b)}$ (d) $\frac{1}{\sqrt{(a-x)(b-x)}}$

47. If $\sin^{-1}\left(\frac{x^2-y^2}{x^2+y^2}\right) = \log a$, then $\frac{d^2y}{dx^2}$ equals

(a) $\frac{x}{y}$ (b) $\frac{y}{x^2}$
(c) $\frac{y}{x}$ (d) 0

48. A man 1.6 m high walks at the rate of 30m/min away from a lamp which is 4 m above ground. How fast is the man's shadow lengthening?

(a) 22m/min (b) 20m/min
(c) 15m/min (d) 25m/min

49. The value P such that the length of subtangent and subnormal is equal for the curve $y = e^{Px} + Px$ at the point (0, 1) is

(a) $P = \pm 1$ (b) $P = \pm 2$
(c) $P = \pm\frac{1}{2}$ (d) None of these

50. AB is a diameter of a circle and C is any point on the circumference of the circle. then

(a) The area of ΔABC is maximum when it is isosceles
(b) The area of ΔABC is minimum when it is isosceles
(c) The perimeter of ΔABC is minimum when it is isosceles
(d) None of these

51. The value of $\tan\left\{\frac{1}{2}\cos^{-1}\left(\frac{\sqrt{5}}{3}\right)\right\}$ is

(a) $\frac{3+\sqrt{5}}{2}$ (b) $3+\sqrt{5}$
(c) $\frac{1}{2}(3-\sqrt{5})$ (d) None of these

52. The interval for which $2\tan^{-1}x + \sin^{-1}\frac{2x}{1+x^2}$ is independent of x is

(a) $|x| < 1$ (b) $|x| > 1$
(c) $|x| = 1$ (d) ϕ

53. The number of solutions of the equation $1 + \sin x \cdot \sin^2\frac{x}{2} = 0$ in $[-\pi, \pi]$ is

(a) zero (b) 1
(c) 2 (d) 3

54. If $x, y \in [0, 2\pi]$ then the total number of ordered pairs (x, y) satisfying $\sin x \cdot \cos y = 1$ is equal to

(a) 1 (b) 3
(c) 5 (d) 7

55. $\int x^x \log(ex)\,dx$ is equal to

(a) $x^x + c$ (b) $x\cdot\log x + c$
(c) $(\log x)^x + c$ (d) $x^{\log x} + c$

56. $\int\sqrt{1+\operatorname{cosec} x}\,dx$ is equal to

(a) $\pm\sin^{-1}(\tan x - \sec x) + c$
(b) $2\sin^{-1}(\cos x) + c$
(c) $\sin^{-1}\left(\cos\frac{x}{2} - \sin\frac{x}{2}\right) + c$
(d) $\pm 2\sin^{-1}\left(\sin\frac{x}{2} - \cos\frac{x}{2}\right) + c$

57. If $\int\frac{dx}{x^2(x^n+1)^{\frac{(n-1)}{n}}} = -[f(x)]^{1/n} + c$ then $f(x)$ is

(a) $1 + x^n$ (b) $1 + x^{-n}$
(c) $x^n + x^{-n}$ (d) None of these

58. The value of

$$\lim_{n \to \infty} \left\{ \frac{1}{na} + \frac{1}{na+1} + \frac{1}{na+2} + \dots + \frac{1}{n_b} \right\} \text{ is}$$

(a) log (ab) (b) log (a/b)
(c) log (b/a) (d) – log (a/b)

59. The value of $\int_a^b \frac{|x|}{x} dx$ is

(a) $|b| - |a|$ (b) $|a| - |b|$
(c) $|b| + |a|$ (d) $-|b| - |a|$

60. $y = \int_{1/8}^{\sin^2 x} \sin^{-1} \sqrt{t} dt + \int_{1/8}^{\cos^2 x} \cos^{-1} \sqrt{t}\, dt,$
$0 \le x \le x/2.$

(a) Is the equation of a straight line parallel to the x-axis
(b) Is the euqation of a straight line which is the bisector of first quadrant
(c) Is the equation of a straight line which is the bisector of second quadrant
(d) None of the above

61. The area of the region

$$[(x, y): x^2 + y^2 \le 1 \le x + y] \text{ is;}$$

(a) $\frac{\pi}{5}$ (b) $\frac{\pi}{4}$
(c) $\frac{\pi^2}{3}$ (d) $\frac{\pi}{4} - \frac{1}{2}$

62. The area of the figure bounded by the parabola $(y-2)^2 = x - 1$, the tangent to it at the point with the ordinate 3 and the x-axis is

(a) 3 (b) 6
(c) 9 (d) None of these

63. The degree of the differential equation

$$\left(\frac{d^2y}{dx^2}\right)^2 + \left(\frac{dy}{dx}\right)^2 = x \sin\left(\frac{d^2y}{dx^2}\right) \text{ is}$$

(a) 1 (b) 2
(c) 3 (d) None of these

64. The slope of the tangent at (x, y) to a curve passing through $\left(1, \frac{\pi}{4}\right)$ is given by $\frac{y}{x} - \cos^2\left(\frac{y}{x}\right)$ then the equation of the curve is

(a) $y = \tan^{-1}\left(\log\frac{c}{x}\right)$ (b) $y = x \tan^{-1}\left(\log\frac{x}{c}\right)$
(c) $y = x \tan^{-1}\left(\log\frac{c}{x}\right)$ (d) None of these

65. The solution of the equation $\frac{dy}{dx} = \frac{x(2\log x + 1)}{\sin y + y \cos y}$ is

(a) $y \sin y = x^2 \log x + \frac{x^2}{y} + c$
(b) $y \cos y = x^2 (\log x + 1) + c$
(c) $y \cos y = x^2 \log x + \frac{x^2}{2} + c$
(d) $y \sin y = x^2 \log x + c$

66. The curve for which the length of the normal is equal to the length of the radius vector, are

(a) only circles
(b) only rectangular hyperbolas
(c) either circles or rectangular hyperbolas
(d) None of the above

67. If $\frac{1}{x(x+1)(x+2)\dots(x+n)} = \frac{A_0}{x} + \frac{A_1}{x+1} + \frac{A_2}{x+2} + \dots + \frac{A_n}{x+n}$ then A_r is equal to

(a) $\frac{r!\,(1)^r}{(n-r)!}$ (b) $\frac{(-1)^r}{r!\,(n-r)!}$
(c) $\frac{1}{r!\,(n-r)!}$ (d) None of these

68. $\log_2(9 - 2^x) = 10^{\log(3-x)}$, solve for x.

(a) 0 (b) 3
(c) both (a) and (b) (d) 0 and 6

69. 3 numbers are in GP therefore, their logarithms are in

(a) GP (b) HP
(c) AP (d) None of these

70. In an equilateral triangle, the in-radius, circum-radius and one of the ex-radii are in the ratio

(a) 2 : 3 : 5 (b) 1 : 2 : 3
(c) 1 : 3 : 7 (d) 3 : 7 : 9

71. In a triangle, the length of the two larger sides are 24 and 22, respectively. If the angles are in AP, then the third side is

(a) $12 + 2\sqrt{3}$
(b) $12 - 2\sqrt{3}$
(c) $2\sqrt{3} + 2$
(d) $2\sqrt{3} - 2$

72. G is a set of all rational numbers except -1 and $*$ is defined by $a * b = a + b + ab$ for all $a, b \in G$, in the group $(G, *)$, the solution of $2^{-1} * x * 3^{-1} = 5$ is

(a) 71 (b) 68 (c) 63/5 (d) 72/5

73. If G is an abelian group then for all $a, b \in G$ $b^{-1} * a^{-1} * b * a$ is equal to

(a) $a * b$
(b) $a^{-1} * b^{-1}$
(c) e
(d) None of these

74. If $(G, *)$ is a group such that $(a * b)^2 = (a * a) * (b * b)$ for all $a, b * G$, then G is

(a) abelian
(b) finite
(c) infinite
(d) None of these

75. A graph which has no edges or node is known as

(a) Digraph
(b) Mixed graph
(c) Nullgraph
(d) Multigraph

76. If all edges are directed in a graph then it is called as

(a) Null graph
(b) Digraph
(c) Mixed graph
(d) None of these

77. Find the greatest value of xyz for positive values of x, y, z subject to the condition $xy + yz + zx = 12$.

(a) 64 (b) 8 (c) 16 (d) 32

78. a, b, c are prime numbers, x is an even number, y is an odd number. Which of the following is/are never true?

I. $a + x = b$ II. $b + y = c$
III. $ab = c$ IV. $a + b = c$

(a) I and II
(b) II and III
(c) Only III
(d) III and IV

79. $(p \wedge \sim q) \wedge (\sim p \wedge q)$

(a) A tautology
(b) A contradiction
(c) Both a tautology and a contradiction
(d) Neither a tautology nor a contradiction

80. $(\sim p \wedge q)$ is logically equivalent to

(a) $p \rightarrow q$
(b) $q \rightarrow p$
(c) $\sim(p \rightarrow q)$
(d) $\sim(q \rightarrow p)$

English

Directions (Q. Nos. 1-5) *In the following questions (a) part of the sentence is printed in bold. Below are given alternatives to the bold part at (a), (b) and (c), which may improve the sentence. Choose the correct alternative. In case no improvement is needed your answer is (d).*

1. No sooner **he reached** home than all the villagers gathered at his home to listen to his story.

(a) would he reach
(b) did he reach
(c) had he reached
(d) No improvement

2. Because of his mastery in this field, his suggestions **are wide accepted.**

(a) are widely accepted
(b) are wide acceptance
(c) have widely accepted
(d) No improvement

3. You are warned **against committing** the same mistake again.

(a) for committing
(b) against to commit
(c) to commit
(d) No improvement

4. The teacher gave the students some **advice**.

(a) advise
(b) advises
(c) advices
(d) No improvement

5. I shall **ring him** tomorrow in the afternoon.

(a) ring to him
(b) ring up to him
(c) ring him up
(d) No improvement

Directions (Q. Nos. 6-10) *In the following questions, out of the four alternatives, choose the one, which best expresses the meaning of the given word.*

6. RADIANT

(a) Bright
(b) Beautiful
(c) Plight
(d) Influential

7. PRUNE
(a) Lend (b) Reduce
(c) Expand (d) Prolong

8. DILETTANTE
(a) Opponent (b) Specialist
(c) Amateur (d) Expert

9. FOSTER
(a) Encourage (b) Fabricate
(c) Forment (d) Nurture

10. ENIGMA
(a) Elusive (b) Clear (c) Puzzle (d) Praise

Directions (Q. Nos. 11-15) *In the following questions, choose the word opposite in meaning to the given word.*

11. CONFORM
(a) Disappoint (b) Reform
(c) Deform (d) Dissent

12. ABORIGINAL
(a) Modern (b) Popular
(c) Current (d) Contemporary

13. AMENABLE
(a) Stubborn (b) Docile
(c) Obedient (d) Offensive

14. ACQUIT
(a) Confirm (b) Blame
(c) Punish (d) Indict

15. FORBIDDEN
(a) Allowed
(b) Prohibited
(c) Agreed
(d) Foresaken

Directions (Q. Nos. 16-20) *In the following questions, four alternatives are given for the idiom / phrase printed in bold in the sentence. Choose the alternative, which best expresses the meaning of the idiom / Phrase.*

16. Rahul **fought tooth and nail** to save his company.
(a) with weapons
(b) as best as he could
(c) using unfair means
(d) with strength and fury

17. Mr Roy is known as a **shop-lifter** in the city commercial centre.
(a) daily visitor
(b) buyer of all new things
(c) smuggler
(d) a thief in guise of customer

18. We should guard against our **green-eyed** friends.
(a) rich (b) jealous
(c) handsome (d) enthusiastic

19. This place affords **a bird's eye view** of the green valley below.
(a) beautiful view (b) general view
(c) narrow view (d) ugly view

20. I won't mind even if he **goes to dogs**.
(a) goes mad (b) is insulted
(c) is ruined (d) becomes brutal

Directions (Q. Nos. 21-25) *In the following questions out of the four alternatives, choose the one, which can be substituted for the given words / sentence.*

21. A person who lives alone and avoids other people.
(a) Ascetic (b) Recluse
(c) Unsocial (d) Agnostic

22. Explicit undertaking to do something.
(a) Agreement (b) Decision
(c) Settlement (d) Promise

23. Murder of a king.
(a) Matricide
(b) Genocide
(c) Regicide
(d) Homicide

24. A person interested in reading books and nothing else.
(a) Student (b) Book worn
(c) Scholar (d) Book-keeper

25. A book or picture produced merely to bring in money.
(a) Money-spinner
(b) Pot-hook
(c) Pot-boiler
(d) Blue-bird

Directions (Q. Nos. 26-30) *In the following questions, you have to rearrange the parts P, Q, R and S to produce a proper sentence.*

26. (P) took place
(Q) when militants opened fire on BSF men
(R) the encounter lasting over two hours
(S) in the town around 8.30 am
(a) QPRS (b) QSRP
(c) RPQS (d) RPSQ

27. They are plant eaters
(P) and various kinds of vegetation
(Q) browsing on grass
(R) and consume
(S) vast quantities of pasture
(a) QPRS (b) QSRP
(c) RSPQ (d) RSQP

28. (P) is becomes difficult
(Q) satisfying our desire
(R) but once we set about
(S) if not impossible to restrain them
(a) PQRS (b) QRPS (c) RQPS (d) RQSP

29. The captain
(P) when engaged against the enemy
(Q) who was himself a brave man
(R) never to lose heart
(S) advised the soliders
(a) PQSR (b) QPSR
(c) QSRP (d) SQRP

30. Some remarks
(P) put the police on the right scent
(Q) by a woman
(R) and they discoverd
(S) casually dropped
(6) the whole gang of brigands
(a) QSPR
(b) RPSQ
(c) RQSP
(d) SQPR

Directions (Q. Nos. 31-35) *In the following questions, some of the sentences have errors and some have none. Find out which part of a sentence has an error. The letter of that part is your answer. If there is no error, your answer is (d) i.e., 'No error'.*

31. Unless you do not listen to his advice (a)/I am not going (b)/ to help you. (c)/No error (d)

32. The teacher called Ravi (a)/and asked him (b)/to describe about the incident. (c)/No error (d)

33. Be smart (a)/not only in dress (b)/and also inaction. (c)/No error (d)

34. The reason for (a)/his failure is because (b)/he did not work hard. (c)/No error (d)

35. If I was you (a)/I would have (b)/terminated his services them and there. (c)/No error (d)

Directions (Q. Nos. 36-40) *In the following questions, groups of four words are given. In each group, one word is wrong spelt. Find the missplet word.*

36. (a) Laudable (b) Honourable
(c) Lovable (d) Honouraty

37. (a) Behaviour (b) Commend
(c) Mentenance (d) Appraise

38. (a) Focal
(b) Vocal
(c) Vehical
(d) Mystical

39. (a) Dairy
(b) Dafodil
(c) Dainty
(d) Damage

40. (a) Cureable
(b) Currency
(c) Campaign
(d) Chronicle

Answers

Physics

1. (d)	2. (c)	3. (a)	4. (d)	5. (c)	6. (d)	7. (a)	8. (a)	9. (a)	10. (b)
11. (a)	12. (a)	13. (d)	14. (a)	15. (c)	16. (c)	17. (c)	18. (b)	19. (b)	20. (c)
21. (c)	22. (c)	23. (b)	24. (d)	25. (d)	26. (a)	27. (a)	28. (a)	29. (c)	30. (b)
31. (b)	32. (d)	33. (c)	34. (c)	35. (a)	36. (c)	37. (a)	38. (a)	39. (c)	40. (c)
41. (a)	42. (a)	43. (c)	44. (a)	45. (a)	46. (a)	47. (a)	48. (b)	49. (d)	50. (a)
51. (c)	52. (b)	53. (d)	54. (d)	55. (c)	56. (a)	57. (a)	58. (c)	59. (c)	60. (c)

Chemistry

1. (a)	2. (d)	3. (a)	4. (b)	5. (b)	6. (d)	7. (a)	8. (b)	9. (d)	10. (d)
11. (c)	12. (b)	13. (d)	14. (d)	15. (d)	16. (a)	17. (c)	18. (b)	19. (c)	20. (a)
21. (d)	22. (a)	23. (b)	24. (b)	25. (d)	26. (c)	27. (d)	28. (a)	29. (c)	30. (b)
31. (a)	32. (b)	33. (c)	34. (c)	35. (b)	36. (a)	37. (d)	38. (a)	39. (c)	40. (c)
41. (a)	42. (b)	43. (c)	44. (a)	45. (b)	46. (c)	47. (a)	48. (b)	49. (d)	50. (b)
51. (a)	52. (b)	53. (b)	54. (a)	55. (d)	56. (c)	57. (a)	58. (d)	59. (a)	60. (c)

Mathematics

1. (b)	2. (c)	3. (a)	4. (b)	5. (b)	6. (b)	7. (d)	8. (a)	9. (a)	10. (c)
11. (a)	12. (c)	13. (c)	14. (b)	15. (d)	16. (c)	17. (c)	18. (b)	19. (c)	20. (a)
21. (a)	22. (c)	23. (b)	24. (a)	25. (a)	26. (d)	27. (a)	28. (a)	29. (c)	30. (a)
31. (c)	32. (a)	33. (a)	34. (c)	35. (c)	36. (b)	37. (a)	38. (b)	39. (a)	40. (b)
41. (d)	42. (b)	43. (c)	44. (c)	45. (d)	46. (b)	47. (d)	48. (b)	49. (c)	50. (a)
51. (c)	52. (b)	53. (a)	54. (b)	55. (a)	56. (d)	57. (b)	58. (c)	59. (a)	60. (a)
61. (d)	62. (c)	63. (d)	64. (c)	65. (d)	66. (c)	67. (b)	68. (a)	69. (c)	70. (b)
71. (a)	72. (a)	73. (c)	74. (a)	75. (d)	76. (b)	77. (b)	78. (c)	79. (b)	80. (d)

English

1. (b)	2. (a)	3. (d)	4. (d)	5. (c)	6. (a)	7. (b)	8. (c)	9. (d)	10. (c)
11. (d)	12. (a)	13. (a)	14. (d)	15. (a)	16. (d)	17. (d)	18. (b)	19. (b)	20. (c)
21. (b)	22. (d)	23. (c)	24. (b)	25. (c)	26. (c)	27. (a)	28. (c)	29. (c)	30. (d)
31. (a)	32. (c)	33. (c)	34. (b)	35. (a)	36. (d)	37. (c)	38. (c)	39. (b)	40. (a)

Hints & Solutions

Physics

1. Rate of change of displacement is known as velocity. The equation for displacment (y) of a body in SHM with angular velocity ω is given by

$$y = a\sin \omega t$$

where a is amplitude

velocity v = rate of change of displacement $= \frac{dy}{dt}$

$$v = \frac{dy}{dt} = \frac{d}{dt}(a\sin \omega t) = a\omega \cos \omega t$$

Given, $v_1 = v_2 = v_3$ and $a_1 = A_1, a_2 = A_2$.

$$a_3 = A_3$$

$$A_1\omega_1 = A_2\omega_2 = A_3\omega_3$$

2. The unit dyne/cm is CGS system of unit.

The CGS unit of force is dyne and SI unit is Newton. Also there are 10^5 dynes in one newton and 100 cm in 1 cm.

$$\therefore \quad \frac{70 \text{ dynes}}{\text{cm}} = \frac{70 \times 10^{-5}}{10^{-2}} = 7 \times 10^{-2} \text{ N/m}$$

3. The general gas equation is $\frac{p_1 V_1}{T_1} = \frac{p_2 V_2}{T_2}$ = gas constant

where p_1, V_1, p_2, V_2 are pressure and volume at temperatures T_1 and T_2 respectively.

Given $V_1 = 1500 \text{ m}^3, T_1 = 27 + 273 = 300 \text{ K}$,

$$p_1 = 4 \text{ atm}.$$

$$T_2 = -3°\text{C} = 273 - 3 = 270 \text{ K}, p_2 = 2 \text{ atm}.$$

$$\therefore \quad V_2 = \frac{p_1 V_1}{T_1} \times \frac{T_2}{p_2}$$

$$V_2 = \frac{4 \times 1500}{300} \times \frac{270}{2}$$

$$V_2 = 2700 \text{ m}^3$$

4. Work done against frictional force equals the kinetic energy of the body.

when a body of mass m, moves with velocity v, it has kinetic energy $k = \frac{1}{2}mv^2$, this energy is utilized in doing work against the frictional force between the tyres of the car and road.

$\therefore$ kinetic energy = work done against friction force

$$\frac{1}{2}m^2 = \mu\, mgs$$

where s is the distance in which the car is stopped and μ is coefficient of kinetic friction.

Given $\quad v = v_0$

$$s = \frac{v_0^2}{2\mu\, g}.$$

5. Average velocity is equal to ratio of total displacement to total time. Velocity is a vector quantity and average velocity is defined as the total displacement undergone by the body per unit time.

Displacement in first instance $= v_1 t$

Displacement in second instance $= v_2 t$

$$\therefore \quad \text{Average velocity} = \frac{v_1 t + v_2 t}{t + t} = \frac{v_1 + v_2}{2}$$

6. If the maximum kinetic energy of photoelectron emitted from the surface of a metal is E_K and W is the work function of the metal, the from Einstein's photoelectric equation we have

$$E_K = h\nu - W$$

where, $h\nu$ is the energy of the photon absorbed by the electron in the metal.

Also $\quad \nu = \frac{c}{\lambda}$

$$= \frac{\text{velocity}}{\text{wavelength}}$$

$$= E_K = \frac{hc}{\lambda} - \frac{hc}{\lambda_0} = hc\left(\frac{1}{\lambda} - \frac{1}{\lambda_0}\right)$$

Given, $\lambda = 3 \times 10^{-7}$ m,

$$\lambda_0 = 4 \times 10^{-7} \text{ m}.$$

$$\therefore \quad E_K = 6.6 \times 10^{-34} \times 3 \times 10^8 \times \left(\frac{1}{3 \times 10^{-7}} - \frac{1}{4 \times 10^{-7}}\right) \text{J}$$

$$= \frac{19.8 \times 10^{-19}}{12 \times 1.6 \times 10^{-19}} \text{ eV}$$

$$= 1.03 \approx 1 \text{eV}$$

7. In the equilibrium position energy is in the form of kinetic energy.

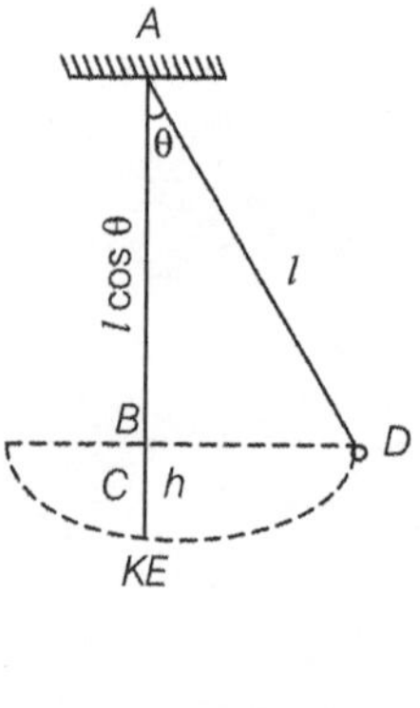

Height of bob at maximum angular displacement is

$$h = l - l\cos\theta$$

$$= l(1 - \cos\theta)$$

Also PE = KE

$$mgh = mgl(1 - \cos\theta)$$

8. Gravitational force provides the required centripetal force.

The gravitational force provides the required centripetal force in orbit of earth.

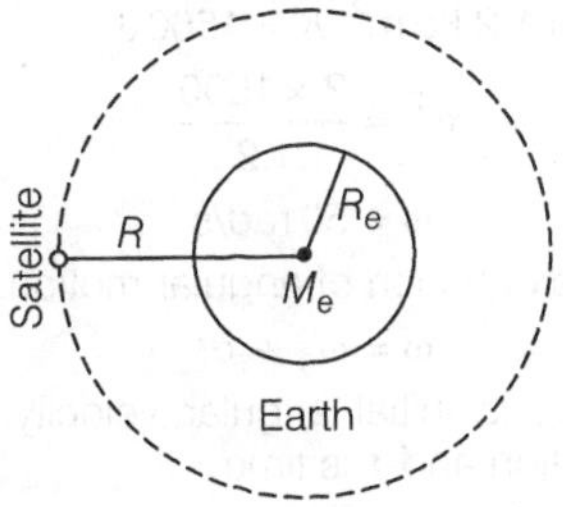

$$\therefore \quad \frac{GM_e M}{R^2} = \frac{mv_0^2}{R}$$

$$v_0 = \sqrt{\frac{GM_e}{R}}$$

$$\text{kinetic energy} = \frac{1}{2} mv_0^2$$

$$\therefore \quad KE = \frac{1}{2} m\left(\frac{GM_e}{R}\right)^{2/2}$$

$$= \frac{1}{2} \frac{mGM_e}{R}$$

$$KE \propto \frac{1}{R}$$

9. The surface tension (T) of a liquid is equal to the work (W) required to increase the surface area of the liquid film by unity at constant temperature.

As per key idea, tension

$$= \frac{\text{work done}}{\text{surface area}}$$

or $$T = \frac{W}{\Delta A}$$

Since soap bubble has two surface and surface area of soap bubble is $4\pi R^2$

where R is radius of bubble.

then $$W = T \times 2 \times 4\pi R^2$$

Given $$R = 2R$$

therefore $$W = T \times 2 \times 4\pi \times (2R)^2$$

$$W = 32\,\pi R^2 T$$

10. Form Einstein's mass energy relation

$$E = \Delta mc^2 \qquad \ldots(i)$$

where Δm is mass lost, c is speed of light. Also heat given by body is

$$E = mC\Delta\theta \qquad \ldots(ii)$$

where, C is specific heat.

Equating Eqs. (i) and (ii) we get

$$\Delta m = \frac{MC\Delta\theta}{C^2} = \frac{M \times 0.2 \times 100 \times 4.2 \times 10^3}{(3 \times 10^8)^2} \text{ J}$$

$$\frac{\Delta m}{M} = \frac{20 \times 4.2 \times 10^3}{(3 \times 10^8)^2}$$

% increase in mass

$$= \frac{\Delta m}{M} \times 100 = \frac{20 \times 4.2 \times 10^3}{(3 \times 10^8)^2} \times 100$$

$$= 9.3 \times 10^{-11}\%$$

11. Where the two coils are kept perpendicular to each other then resultant field in one coil is directed along the axis of coil and the other downwards, hence resultant field is

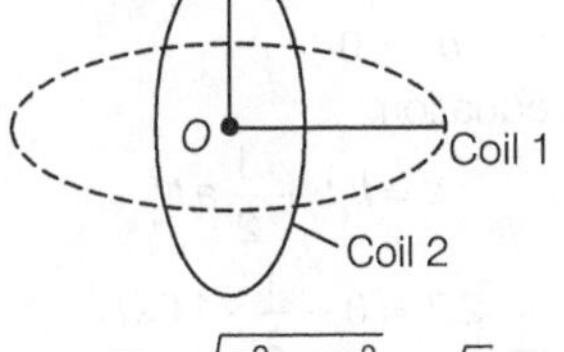

$$B' = \sqrt{B^2 + B^2} = \sqrt{2}\,B$$

Ratio of magnetic field due to one coil and resultant is $\dfrac{B}{B\sqrt{2}} = \dfrac{1}{\sqrt{2}}$

12. In the position of minimum deviation angle of incidence is equal to angle of emergence.

Let a ray of monochromatic light PQ be incident on face AB. $PQRS$ is path of light ray,

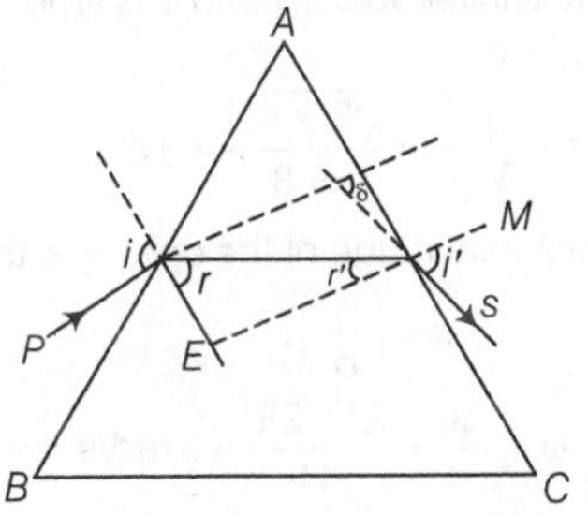

where i is angle of incidence r angle of refraction, r angle of incidence and i angle of emergence.

In position of minimum deviation

$$i' = i,\ r' = r,\ \delta = \delta m$$

$$2r = A \text{ or } r = \frac{A}{2}$$

Given $A = 60°$,

$$r = \frac{60°}{2} = 30°$$

Also from Snell's law

$$\mu = \frac{\sin i}{\sin r} = \frac{\sin i}{\sin 30^\circ}$$

$$\sqrt{2} = \frac{\sin i}{\sin 30^\circ}$$

$$\Rightarrow \quad \sin i = \sqrt{2} \times \frac{1}{2} = \frac{1}{\sqrt{2}}$$

$$\Rightarrow \quad i = 45^\circ$$

13. Net acceleration on the elevator car is more than acceleration due to gravity.

Since elevator car is ascending upwards, from Newton's second law, the net force is acting upwards hence resultant acceleration is

$$a_r = (g + a)$$
$$= (9.8 + 1.2) = 11 \text{ m/s}^2$$

Relative velocity of observer to elevator is

$$u_r = 0$$

from the equation

$$s = u_r t + \frac{1}{2} a_r t^2$$

$$2.7 = 0 + \frac{1}{2} \times 11 \times t^2$$

$$t^2 = \frac{5.4}{11}$$

$$t = \sqrt{\frac{5.4}{11}} \text{ s}$$

14. From equation of motion

$$h = ut + \frac{1}{2} gt^2$$

where u is initial velocity and t is time.

Since, $u = 0$

$$\therefore \quad t = \sqrt{\frac{2h}{g}} = \frac{\sqrt{2 \times 4.9}}{9.8} = 1 \text{ s}$$

The horizontal range of the drop = x, then

$$x = \left(\frac{v_t}{o}\right) t$$

Also, $\omega = \dfrac{\Delta\theta}{\Delta t} = \dfrac{21 \times 2\pi}{44} = 3 \text{ rad/s}$

Tangential speed $v_t = r\omega = 0.5 \times 3 \times 1.5$ m/s

$$x = 1.5 \times 1 = 1.5 \text{ m}$$

Locus of drop $= \sqrt{x^2 + r^2} = \sqrt{(1.5)^2 + (0.5)^2}$

$$= \sqrt{2.5} \text{ m}$$

15. Kinetic energy of rotation is half the product of the moment of inertia (I) of the body and the square of the angular velocity (ω) of the body.

Kinetic energy of rotation $= \frac{1}{2} \times$ moment of inertia $\times$ angular velocity

i.e., $$K = \frac{1}{2} I \omega^2$$

$$\omega^2 = \frac{2K}{I}$$

Given $I = 1.2 \text{ kg m}^2$, $K = 1500$ J

$$\omega^2 = \frac{2 \times 1500}{1.2}$$

$$\Rightarrow \quad \omega = 50 \text{ rad/s}$$

From the equation of angular motion, we have

$$\omega = \omega_0 + \alpha t$$

where ω_0 is initial angular velocity, α is angular acceleration and t is time

given $\omega_0 = 0$,

$$\omega = 50 \text{ rad/s},$$
$$\alpha = 25 \text{ rad/s}^2$$

$$t = \frac{\omega}{\alpha} = \frac{50}{25} = 2s$$

16. The kinetic energy of moving body is equal to half the product of the mass (m) of the body and the square of its speed (v^2).

kinetic energy $= \frac{1}{2} \times$ mass $\times$ (speed)^2

i.e., $$= \frac{1}{2} mv^2$$

Let mass of man is M kg and speed is v and speed of body is v_1, then

$$\frac{1}{2} Mv^2 = \frac{1}{2}\left\{\frac{1}{2}\frac{M}{2} v_1^2\right\} \quad \text{...(i)}$$

when man speed up by 1 m/s, than

$$v = v + 1$$

$$\therefore \quad \frac{1}{2} M (v + 1)^2 = \frac{1}{2}\left\{\frac{M}{2}\right\} \cdot v_1^2 \quad \text{...(ii)}$$

Dividing Eqs. (i) by (ii), we get

$$\frac{v^2}{(v + 1)^2} = \frac{1}{2}$$

$$\sqrt{2} v = v + 1$$

$$v = \frac{1}{\sqrt{2} - 1} \text{ m/s}$$

17. Larger the distance of planet from the sun, larger will be its period of revolution around the sun.

Kepler's third low planetary motion the square of the period of revolution of any planet around the sun is directly proportional to the cube of the semi-major axis of its elliptical orbit

$$\therefore \quad T^2 \propto R^3$$

$$\frac{T_s}{T_c} = \left(\frac{R_s}{R_c}\right)^{3/2}$$

given $R_s = 4R_c$

$$\frac{T_s}{T_c} = \left(\frac{4R_c}{R_c}\right)^{3/2} = 8$$

for $T_c = 1$ day

$T_s = 8$ days.

18. Force experienced by the surface is equal to change in momentum in one second.

From Newton's second law of motion, the rate of change of momentum of a body is equal to the external force applied on the body and the change in momentum always takes place in the direction of the force.

Momentum of ball before impact = mu

Momentum after impact = $-mu$

Change in momentum = $mu - (-mu) = 2mu$

Change in momentum due to n balls in one second is = $2mnu$.

Change of momentum = force experienced per second = $2mnu$.

19. When no external force acts on a body its angular moment remains conserved.

From law of conservation of angular momentum, we have

$$J = I\omega = \text{constant}$$

where I is moment of inertia of the body and ω is angular velocity.

Also $I = \frac{1}{2}MR^2$ (for disc) where M is mass and R is radius of disc.

$$J = \frac{1}{2}MR^2\ \omega = \text{constant}$$

Since, man moves towards the centre of the disc, distance of mass distribution R decreases hence, ω increases.

20. Use parallel axis theorem, from the theorem of parallel axis, the moment of inertia (I) of a body about given axis is equal to its moment of inertia I about its diameter, plus the product of the mass M of the body and square of perpendicular distance between the two axes. That is

$$I = I_d + MR^2 = \frac{1}{4}MR^2 + MR^2 = \frac{5}{4}MR^2$$

21. Value of acceleration due to gravity decreases on going below the surface of the earth.

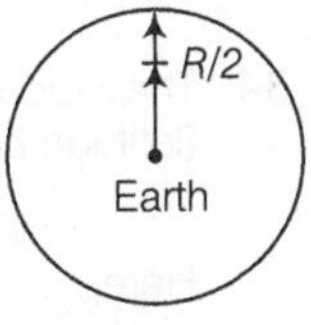

weight (w) of a body is defined as product of mass (m) and acceleration due to gravity (g)

$w = mg$

The value of g below the surface of earth at a distance h is given by

$$g' = g\left(1 - \frac{h}{R}\right)$$

$$\Rightarrow \quad W' = W\left(1 - \frac{h}{R}\right)$$

Given, $h = \frac{R}{2}$, $W = 500$ N

$$W' = 500\left(1 - \frac{1}{2}\right)$$

$$= \frac{500}{2} = 250 \text{ N}$$

22. At a crtain velocity of projection the body will go out of the gravitational field of the earth and will never return to the earth, this velocity is known as escape velocity.

$$v_e = \sqrt{\frac{2GM_e}{R_e}}$$

Given, $M_e = M_p$, $R_p = \frac{R_e}{4}$

$$\therefore \quad \frac{v_p}{v_e} = \sqrt{\frac{M_e}{M_e} \times \frac{R_e}{R_e/4}} = \sqrt{4} = 2$$

$$\Rightarrow \quad v_p = 2v_e = 2 \times 11.2$$

$$= 22.4 \text{ km/s}$$

23. Since liquid is incompressible there is no change in volume.

When strain is small the ratio of the normal stress to the volume strain is called the bulk modulus of material of the body.

$$B = \frac{\text{normal stress}}{\text{volume strain}}$$

= normal stress

change in volume/original volume

since, the liquid is incompressible change in volume is zero.

$$B = \frac{\text{normal stress}}{0} = \infty$$

Hence, bulk modulus is infinite.

24. The narrower the tube, the higher is the rise in water.

when a glass capillary

$$h = \frac{2T\cos\theta}{R\rho g}$$

$$h \propto \frac{1}{R}$$

$$\frac{h_1}{h_2} = \frac{R_2}{R_1}$$

$$\frac{3}{h_2} = \frac{1}{3}$$

$$h_2 = 9 \text{ mm}$$

25. $u = -20$ cm, $f = 20$ cm

From mirror formula

$$\frac{1}{f} = \frac{1}{v} + \frac{1}{u}$$

$$\frac{1}{20} = \frac{1}{v} + \frac{1}{-20}$$

$$\frac{1}{v} = \frac{1}{20} + \frac{1}{20}$$

$$\frac{1}{v} = \frac{2}{20} = v = 10 \text{ cm}$$

26. According to law of conservation of angular momentum if there is no torque on the system, then the angular momentum remains constant.

27. Equivalent focal length

$$\frac{1}{F} = \frac{1}{f_1} + \frac{1}{f_2}$$

$$= \frac{1}{20} + \frac{1}{30}$$

$$F = \frac{20 \times 30}{20 + 30} = \frac{600}{50} = 12 \text{ cm}$$

$$F = 12 \text{ cm}$$

28. Work done $W = \mathbf{F} \times \mathbf{d}$

$$= (6\,\mathbf{i} + 2\,\mathbf{j}) \times (3\,\mathbf{i} - \mathbf{j})$$

$$= 6 \times 3 + 2 \times -1$$

$$= 18 - 2 = 16 \text{ J}$$

$$W = 16 \text{ J}$$

29. From Einstein's theorem

$$\text{Relativistic momentum} = \frac{m_0 v}{\sqrt{1 - v^2/c^2}}$$

Putting $v = 2v$

$$\text{we get } p = \frac{m_0\,(2v)}{\sqrt{1 - (2v)^2/c^2}}$$

$$= 2\left(\frac{m_0 v}{\sqrt{1 - (2v)^2/c^2}}\right)$$

So the momentum becomes more than the double.

30. Position $x = Ka^m\, t^n$

Writing the dimension on the both sides.

$$[M^0\, L\, T^0] = [L\, T^{-2}]^m\, [T]^n$$

$$[M^0\, L^m\, T^{-2m+n}]$$

On comparing both sides.

$$m = 1$$

$$-2m + n = 0$$

$$n = 2m = 2 \times 1 = 2$$

31. We know that

$$\alpha = \left(\frac{\Delta i_C}{\Delta i_E}\right) \qquad \ldots(i)$$

and $$\beta = \left(\frac{\Delta i_C}{\Delta i_B}\right) = \frac{\Delta i_C}{\Delta i_E} \times \frac{\Delta i_E}{\Delta i_B}$$

$$= \alpha \frac{\Delta i_E}{\Delta i_B} \qquad \ldots(ii)$$

But $\Delta i_B = \Delta i_E - \Delta i_C$

from (ii)

$$\beta = \alpha \frac{\Delta i_E}{\Delta i_E - \Delta i_C} = \frac{\alpha}{1 - \dfrac{\Delta i_C}{\Delta i_E}}$$

$$\therefore \quad \beta = \frac{\alpha}{1 - \alpha}$$

32. $v_r^2 = v_m^2 - v^2$

$$v = \frac{320}{4} \text{ m/min} = 80 \text{ m/min}$$

$$v_m = \frac{5}{3} v_r$$

$$v_r^2 = t\frac{5}{3}(v_r)^2 - (80)^2$$

$$\frac{16}{9} v_r^2 = (80)^2$$

$$v_r = 60 \text{ m/min}$$

u^2, u_0 form canours linear

33. Let the radii of the thin spherical and the solid sphere are R_1 and R_2 respectively.

Then the moment of inertia of the spherical shell about their diameter

$$I = \frac{2}{3} MR_1^2 \qquad \ldots(i)$$

and the moment of inertia of the solid sphere is given by

$$I = \frac{2}{5} MR_2^2 \qquad \ldots(ii)$$

Given that the masses and moment of inertia for both the bodies are equal, then from Eqs. (i)and (ii)

$$\frac{2}{3} MR_1^2 = \frac{2}{5} MR_2^2 \Rightarrow \frac{R_1^2}{R_2^2} = \frac{3}{5}$$

$$\Rightarrow \quad \frac{R}{R_2} = \sqrt{\frac{3}{5}} \Rightarrow R_1 : R_2 = \sqrt{3} : \sqrt{5}$$

34. The apparent weight of person on the equator (latitude $\lambda = 0$) is given by

$$\omega' = w - mR\omega^2$$

Here, $\therefore \quad \omega' = (3/5)\, w = (3/5)\, mg \quad [\because \omega = mg]$

or $mR\omega^2 = mg - (3/5)\, mg = \left(\frac{2}{5}\right) mg$

or $\omega^2 = \frac{2g}{5R}$

$$\omega = \sqrt{\frac{2g}{5R}}$$

Here, $g = 9.8\,\text{ms}^{-2}$

and $R = 6400\,\text{km} = 6400 \times 10^3$ m

$$\therefore \quad \omega = \sqrt{\left(\frac{2}{5} \times \frac{9.8}{6400 \times 10^3}\right)}\ \text{rads}^{-1}$$

$$= 7.82 \times 10^{-4}\ \text{rads}^{-1}$$

35. From Bernoulli's theorem

$$p_1 + \frac{1}{2}\rho v_1^2 = p_2 + \frac{1}{\alpha}\rho v_2^2$$

$$\therefore \quad p_1 - p_2 = \frac{1}{2}\rho\,(v_2^2 - v_1^2)$$

$$\therefore \quad 10 = \frac{1}{2} \times 1.25 \times 10^3\,(v_2^2 - v_1^2)$$

$$\therefore \quad v_2^2 - v_1^2 = \frac{10 \times 2}{1.25 \times 10^3} = 16 \times 10^{-3} \qquad \ldots\text{(i)}$$

Also from equation of continuity

$$= A_1 v_1 = A_2 v_2$$

$$\pi r_1^2 v_1 = \pi r_2^2\, v_2$$

$$\therefore \quad \frac{v_1}{v_2} = \left[\frac{r_2}{r_1}\right]^2 = \frac{0.04}{0.1} = 0.4$$

$$v_1 = 0.4\, v_2 \qquad \ldots\text{(ii)}$$

Substituting this value in Eq. (i)

$$v_2^2 - (0.4\, v_2)^2 = 16 \times 10^{-3}$$

$$v_2 = 1.38 \times 10^{-1} = 0.138\,\text{ms}^{-1}$$

Rate of flow of glycrine $v = A_2 v_2$

$$= \pi r_2^2 v_2$$

$$= 6.93 \times 10^{-4}\ \text{m}^3\text{s}^{-1}$$

36. $E = \sigma T^4$

$$T_1 = T, \quad T_2 = 2T$$

$$\frac{E_1}{E_2} = \frac{T_1^4}{T_2^4} = \left(\frac{T}{2T}\right)^4 = \frac{1}{16}$$

$$E_2 = 16\,E$$

37. Since, police jeep and thief are moving in the same direction relative velocity is

$$v_\sigma = \mathbf{v}_1 - \mathbf{v}_2 = 153 - 45$$

$$= 108\ \text{km/h}$$

$$= 108 \times \frac{5}{18} = 30\ \text{min}$$

38. $E = \frac{-mz^2e^4}{8\,\varepsilon_0\, h^2} \cdot \frac{1}{n^2}$

$$Z_H = 1,$$

$$Z_{He} = 2$$

$$\frac{E_H}{E_{He}} = \frac{Z_H^2}{Z_{He}^2} = \frac{1}{4}$$

$$E_H = \frac{Z_H^2}{Z_{He}^2} = \frac{2}{4}$$

$$E_{He} = 4E_H$$

$$E_H = E_n$$

$$E_{He} = 4E_n$$

39. $W = \frac{hc}{\lambda}$

where h is planck's constant, c is speed of light, λ is wavelength

$$\lambda = \frac{hc}{W} = \frac{6.6 \times 10^{-34} \times 3 \times 10^8}{6.825 \times 1.6 \times 10^{-19}}$$

$$\lambda = 1800\ \text{Å}$$

40. A field effert transistor is a unipolar device since the conduction of current is due to majority charge carriers only.

41. The periodic-time of a simple pendulum is given by

$$T = 2\pi\sqrt{\frac{l}{g}}$$

$$T_1 = T,\ l_1 = l,\ l_2 = 9l$$

$$\frac{T_1}{T_2} = \sqrt{\frac{l_1}{l_2}} = \sqrt{\frac{1}{9}} = \frac{1}{3}$$

$$T_2 = 3T_1 = 3T$$

42. The free body diagram of the given situation is shown.

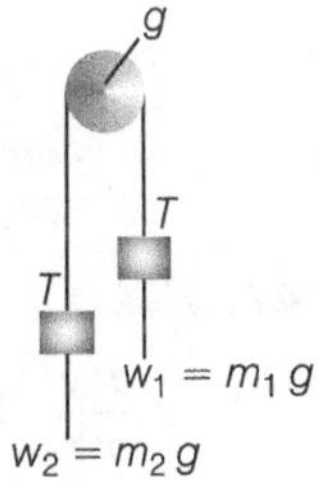

Taking the upward and downward motions respectively, we get

$$T_1 - mg = m_1\,(g + a)$$

$$T - 2mg = m_1 a$$

for second weight

$$m_2 g - T = m_2\,(g - a)$$

$$2m_2 g - T = m_2 a$$

$$T = \frac{4m_1 m_2 g}{(m_1 + m_2)}$$

$w_1 = m_1 g,\ m_2 = m_2 g$ hence switch

$$T = \frac{4\,w_1\,w_2}{(w_1 + w_2)}$$

43. The angle subtended at the eye becomes 10 times larger, this happens only when the tree appear 10 times nearer.

44. When the switch is closed, the current grows in the circuit but due to self-inductance of the switch inductor the growth is slow. On disclosing the switch the current suddenly decrease to zero on account of the infinite resistance of the circuit. Thus, rate cf growth of the current is lower as compared to the rate of decays

45. We know that $v = \sqrt{\frac{T}{m}}$

$$v \propto \sqrt{T}$$

$$\frac{v_1}{v_2} = \sqrt{\frac{T_1}{T_2}}$$

$$\frac{T_1}{T_2} = \left(\frac{v_1}{v_2}\right)^2$$

$$\frac{T_2 - T_1}{T_1} = \frac{v_1^2 - v_1^2}{v_1^2}$$

$$v_2 = v_1 + \frac{20}{100} v_1 = \frac{120}{100} v_1 = \frac{6}{5} v_1$$

$$= \frac{6}{5} \times 150$$

$$\frac{T_2 - T_1}{T_1} = \frac{(180)^2 - (150)^2}{(150)^2}$$

$$= \frac{30 \times 330}{150 \times 180} = 0.47$$

$$= \frac{T_2 - T_1}{T_1} \times 100 = 0.44 \times 100 = 44\%$$

46. $E - 40 \text{ GWh} = 40 \times 10^9 \text{J/s}$

$$(60 \times 60) = 1.44 \times 10^{14} \text{ J}$$

$$E = mc^2$$

$$m = \frac{E}{c^2} = \frac{1.44 \times 10^{14}}{(3 \times 10^8)^2} = 1 \times 10^3 \text{ kg}$$

$$m = 1.6 \text{ g}$$

47. $y = kt^2$

$$\frac{d^2y}{dt^2} = 2k$$

$$a = 2 \text{ m/s}^2$$

$$T_1 = 2\pi \sqrt{\frac{l}{g}}$$

$$T_2 = 2\pi \sqrt{\frac{l}{g + a_y}}$$

$$\frac{T_1^2}{T_2^2} = \frac{g + a_y}{g}$$

$$= \frac{10 + 2}{10} = \frac{6}{5}$$

48. $\mathbf{F} = -\frac{1}{4\pi\varepsilon_0} \frac{q_1 q_2}{r^2} = \mathbf{r}$

$$\mathbf{F} = -k \frac{q_1 q_2}{r^2} = \mathbf{r}$$

$$F = -k \frac{e.e}{r^2} \hat{\mathbf{r}} = -k \frac{e^2}{r^2} \hat{\mathbf{r}}$$

$$\hat{\mathbf{r}} = \frac{\mathbf{r}}{|\mathbf{r}|} = \frac{\mathbf{r}}{r}$$

$$F = -k \frac{e^2}{r^2} \cdot \frac{\mathbf{r}}{r} = -k \frac{e^2}{r^3} \mathbf{r}$$

49. Gauss's law

$$\phi = \frac{q}{\varepsilon_0}$$

$$\phi' = \frac{\phi}{6} = \frac{q}{6\varepsilon_0} = \frac{4\pi q}{6(4\pi\varepsilon_0)}$$

50. $\overline{A} \cdot \overline{B} = \overline{A + B}$

51. $F = mr\,\omega^2 = mr \times (2\pi mrn)^2 = 4\pi^2 mrn^2$

$$F \propto 2n^2$$

$$F = F_1 + F_2$$

$$n^2 = n_1^2 + n_2^2$$

$$n = \sqrt{n_1^2 + n_2^2}$$

$$n = \sqrt{(6)^2 + (8)^2}$$

$$n = \sqrt{100} = 10 \text{ Hz}$$

52. $K = K_0$ = total energy

As total energy remains conserved itself hence when U is maximum in SHM, $K = 0$

i.e., E is also equal to U_{max} *i.e.*,

$$U_{max} = E = K_0$$

53. Relative velocity of the parrot w.r.t. the train

$$[10 - (-5)] = (10 - (-5)) \text{ m/s} = 15 \text{ m/s}$$

Time taken by the parrot to cross the train

$$= \frac{150}{15} = 10 \text{ s}$$

54. Time taken by ice to grow a thickness of

$$t = \frac{eL}{2K\theta} y^2$$

Hence time interval to change thickness from 0 to y, from y to $2y$ and so on will be in the ratio

$$\Delta t_1 : \Delta t_2 : \Delta t_3$$

$$\therefore \quad = (1 - 0^2) : (2^2 - 1^2) : (3^2 - 2^2)$$

$$\Delta t_1 : \Delta t_2 : \Delta t_3 = 1 : 3 : 5$$

According to question = Δt_1 < terims

$$\Delta t_2 = 3\,\Delta H$$

$$= 3 \times 12 = 36 \text{ min}$$

55. As, $\sqrt{V} = a(Z-b)$

$$\sqrt{V} = a(Z-1)$$

$$V = a^2(Z-1)^2$$

$$\frac{C}{\lambda} = a^2(Z-1)^2$$

$$\lambda = \frac{c}{a^2(Z-1)^2}$$

$z = 43$, wavelength $= \lambda$

$$\lambda = \frac{c}{a^2(43-1)^2}$$

$$\lambda = \frac{c}{a^2 \times 42}$$

for $z = 29$, wavelength $= \lambda'$

$$\lambda' = \frac{C}{a^2(29-1)^2}$$

$$\lambda' = \frac{c}{28a^2}$$

$$\frac{\lambda'}{\lambda} = -\left(\frac{42}{28}\right)^2 = \left(\frac{3}{2}\right)^2$$

$$\lambda' = \left(\frac{9}{4}\right)\lambda$$

56. $\frac{1}{u} + \frac{1}{v} = \frac{1}{f}$

$$-\frac{1}{u^2}\frac{du}{dt} - \frac{1}{v^2}\frac{dv}{dt} = 0$$

$$= -\frac{1}{u^2}u_0 - \frac{1}{v^2}v_0 = 0$$

$$\frac{u_0}{v_0} = -\frac{u^2}{v^2} = -\frac{1}{m^2}$$

57. As, $I^2R = hA$

$$I^2\left(\frac{\rho L}{\pi r^2}\right) = h \times 2\pi rl, \; I \propto r^{3/2}$$

58. $f = \frac{\Delta V}{(\Delta Q_P)} = \frac{\Delta Q_V}{(\Delta Q_P)_p} = \frac{\mu\, Cv\, \Delta T}{\mu\, Cp\, \Delta T} = \frac{1}{T}$

$$f = \frac{7}{5}$$

$$f = \frac{5}{7}$$

59. In interference of light energy is neither created nor destroyed.

60. From colour triangle for pigment and dyes we observe that the given combination give black colour.

Chemistry

1. i for KCl = 2, i for for $CaCl_2$= 3

$$\Delta T_f \propto i$$

$$\frac{\Delta T_f(KCl)}{\Delta T_f(CaCl_2)} = \frac{2}{3}$$

$$\Delta T_f(CaCl_2) = \frac{3}{2} \times 2 = 3°C$$

Freezing point of $CaCl_2 = -3°C$

2. $\frac{r_1}{r_2} = \frac{p_1}{p_2}\sqrt{\frac{M_2}{M_1}}$

$$P_{He} = X_{He} \cdot p_{total}$$

$$= \frac{1}{4} \times 16 = 4 \text{ atom}$$

$$P_{N_2} = X_{N_2} \cdot p_{total}$$

$$= \frac{3}{4} \times 16 = 12 \text{ atom} = (p_{total}^{-} P_{He})$$

$$\frac{r_{He}}{r_{N_2}} = \frac{p_{He}}{p_{N_2}}\sqrt{\frac{M(N_2)}{M(He)}}$$

$$= \frac{4}{12}\sqrt{\frac{28}{4}} = 0.88$$

Hence, moles of He and N_2 effusing out initially are in the ratio 0.88 : 1.

3. Sc (21) = $1s^2\ 2s^2\ 2p^6\ 3s^2\ 3p^6\ 3d^1\ s^2$

19th electron goes into 4s

Hence, $n = 4$, $l = 0$,

$m = 0$, $m_s = +\frac{1}{2}$

4. (*A*) + tap water → white turbidity soluble in aq. NH_3.

Tap water has Cl^- and turbidity is soluble in $aq\, NH_3$.

Turbidity is of AgCl and thus (*A*) has Ag^+

(*A*) also gives ring test of NO_3^-

(*A*) has nitrate and thus (*A*) is $AgNO_3$

5. Greater the number of ions, greater the conduction hence, order of increasing conductivity is

$c<d<b<a$

(a) $[Pt(NH_3)_6]Cl_4 \rightleftharpoons [Pt(NH_3)_6]^{4+} + 4Cl^- = 5$ ions

(b) $[Cr(NH_3)_6]Cl_3 \rightleftharpoons [Cr(NH_3)_6] + 3Cl^- = 4$ ions

(c) $[Co(NH_3)_4Cl_2]Cl \rightleftharpoons [Co(NH_3)_4Cl_2] + Cl^- = 2$ ions

(d) $K_2[PtCl_6] \rightleftharpoons 2K + [PtCl_6]^{2-} = 3$ ions

6. EAN = $Z - $(O.N.) + $2x$ or $36 = 26 - 0 + 2x$ or $x = 5$

7. **For reaction *A***

$\Delta G° = 2\Delta G°_f\,(ZnO) + 2\Delta\, G°_f(SO_2)\ - 2G°_f\,(ZnS)$

$= 2\,[-318.2\ - 300.4 + 205.4]$

$= -826.4$ kJ ($\Delta G°_f$ (element) = 0)

For reaction *B*

$\Delta G° = \Delta G°_f\,(CO) - \Delta G°_f\,(ZnO)$

$= -137.3 + 328.2$

$= +180.9$ kJ

8. Ore- copper pyrites , $CuFeS_2$

Concentrated pyrites ore (froth- floatation process) is roasted below 800°C. FeS is converted to FeO while copper remains as sulphide.

Roasting $2FeS + 3O_2 \xrightarrow{\Delta} 2FeO + 2\,SO_2\uparrow$

Smelting $FeO + SiO_2 \longrightarrow \underset{\text{slag}}{FeSIO_3}$

(SiO_2 either is present ore or is added)

Bessemerisation

$2Cu_2S + 3O_2 \longrightarrow 2Cu_2O + 2SO_2$

$2Cu_2O + Cu_2S \longrightarrow 6Cu + SO_2$ (final step)

9. $CO_2 + H_2O \rightleftharpoons H_2CO_3 \rightleftharpoons H^+ + HCO_3^-$

$\underset{\text{yellow}}{2CrO_4^{2-}} + 2H^+ \longrightarrow \underset{\text{orange red}}{Cr_2O_7^{2-}} + H_2O$

Aqueous CO_2 is acidic and changes yellow CrO_4^{2-} into orange $Cr_2O_7^{2-}$.

10. Transition elements form complexers due to their smaller size, higher nuclear charge and presence of low energy vacant orbitals to accept lone pair of electrons donated by ligands.

11. ClO_2 reacts with KOH forming KCl_3 and $KClO_2$

$2KOH + 2ClO_2 \longrightarrow KClO_2 + KClO_3 + H_2O$

This reaction indicates that ClO_2 is a mixed anhydride of $HClO_2$ and $HClO_3$.

12. Alkene *A* is $CH_2{=}CH{-}CH{=}CH_2$

$CH_2{=}CH{-}CH{=}CH_2 + O_3 \longrightarrow$

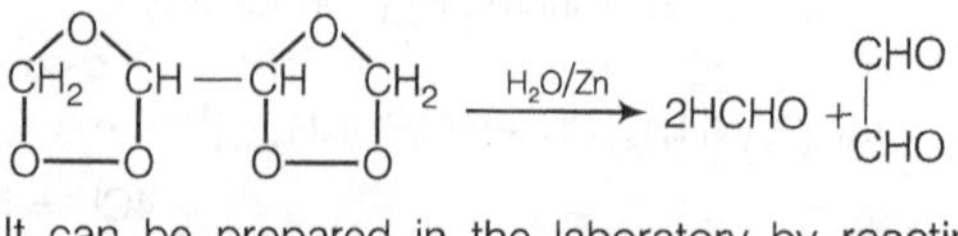

13. It can be prepared in the laboratory by reacting sodium, nitrite ($NaNO_2$) with a reducing agent such as Fe^{2+} in an acidic medium. NH_4NO_3 on heating at 250°C gives N_2O.

14. It absorbs CO_2 and releases O_2.

$4KO_2 + 2CO_2 \longrightarrow 2K_2CO_3 + 3O_2$

15. Stepwise formation of MgO involves

Step	Reaction	ΔH in kJ mol^{-1}
(i)	$Mg(s) \rightarrow Mg(g)$	147.7
(ii)	$Mg(g) \rightarrow Mg^{2+} + 2e^-$	2189.0
(iii)	$\frac{1}{2}O_2(g) \rightarrow O(g)$	498.4/2
(iv)	$O(g) + 1e^- \rightarrow O^-(g)$	−141.0
(v)	$O^-(g) te^- \rightarrow O^{2-}(g)$	q
(vi)	$Mg^{2+}(g) + O^{2-}(g) \rightarrow MgO(s)$	−3791.0

$Mg(s) + \frac{1}{2}O_2(g) \longrightarrow MgO(s)\,\Delta H = -1346.1 + q$

$Mg(s) + \frac{1}{2}O_2(g) \longrightarrow MgO(s)\,\Delta H_2 = -601.7$

By Born Herber cycle (based on Hes's law)

$\Delta H_1 = \Delta H_2$

$-1346.1 + q = -601.7$

$q = 744.4\,\text{Jmol}^{-1}$

16. $H_2S_2O_7 + H_2O \longrightarrow 2H_2SO_4$

SO_3 of $H_2S_2O_7$ is converted into H_2SO_4, hence SO_3 acts also as a dibasic acid.

Eq. wt. $(SO_3) = \frac{M}{2} = 40$

Let H_2SO_4 in fuming $H_2SO_4 = x$ g

$SO_3 = (1 - x)$ g

Equivalent of $H_2SO_4 = \frac{x}{49}$

$SO_3 = \frac{1-x}{40}$

Equivalent of NaOH used $= \frac{26.7 \times 0.8}{1000}$

$= 0.02136$

$\frac{x}{49} + \frac{1-x}{40} = 0.02136$

$x = 0.7927$ g H_2SO_4 in 1 g oleum.

Percentage of $H_2SO_4 = 79.27\%$

$SO_3 = 20.73\%$

17. $^{14}_{7}N + ^{1}_{0}n \longrightarrow ^{14}_{6}C + ^{1}_{1}H$

This reaction can be represented as

$^{14}_{7}N(n, p)^{14}_{6}C$

18. Ag^+ is reduced hence, it is oxidant Hydroquinone is oxidised hence, it is reductant.

19. N(7) and O^+ (8 – 1 = 7) = $1s^2\ 2s^2\ \underbrace{2p_x^1 2p_y^1 2p_z^1}_{\text{Maximum multiplicity most stable}}$

$$O = 1s^2, 2s^2\ 2p_x^2, 2p_y^1, 2p_z^1$$

Ionisation potential is dependent on the value of Z.

$$Z \text{ of } O^+ (8) > Z \text{ of } N(7)$$

Thus, ionisation potential of O^+ > N

Na with one valence electron has least ionisation potential.

20. Gd. (64) = $[Xe]4f^7 5d^1$

$4f^4$

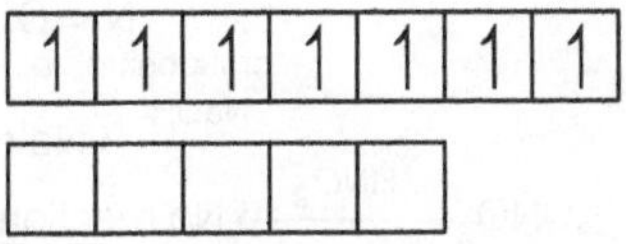

$5d^1$

All electrons are unpaired, hence greater stability

21. He^+, Li^{2+} and Be^{3+} each has one electron, hence no screening.

22. Symmetrical molecules will have net dipole zero while unsymmetrical will have dipoles. I is symmetrical and have no dipole while II and III have dipoles.

23. In PCl_5 No. of e^- pairs around central atom P = 5

No of lone pairs = 0 and hybridisation-sp^3d Shape = trigonal bipyramidal. In PCl_4^+– number. of e^- pairs = 4, lone pairs = 0 , hybridisation = sp^3 and shape – tetrahedral

In PCl_6^-– number of e^- pairs = 6, lone pairs = 0, hybridisation = sp^3d^2 and shape is octahedral

24. Repulsive interactions in decreasing order

$$= lp - lp > lp - bp > bp - bp$$

25. Based on Pauling's scale

$$(EN)_F - (EN)_{Cl} = k\sqrt{\Delta} = 0.208\sqrt{\Delta}$$

where, Δ = resonance energy

and k = conversion factor which is 0.208 for converting k cal into eV.

$$\Delta = (BE)_{Cl-F} - \sqrt{(BE)_{Cl-Cl}(BE)_{F-F}}$$
$$= 61 - \sqrt{58 \times 38}$$
$$= 61 - 46.95 = 14.05 \text{ kcal}$$
$$(EN)_{Cl} = (EN)_F - 0.208\sqrt{\Delta}$$
$$= 4.0 - 0.208\sqrt{14.05}$$
$$= 4.0 - 0.78 = 3.22 \text{ eV}.$$

26. Number of moles of O atoms = $\frac{1}{3}$ moles of H atoms

$$= \frac{1}{3} \times 3.18 = 1.06 \text{ mol}$$

27. In the compound

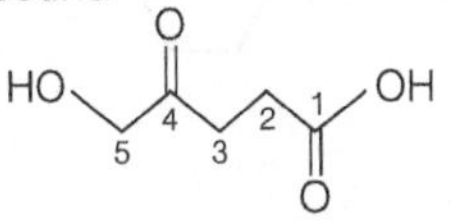

keto group is at 4C – atom

28. S_N1 reaction proceed through formation of carbocation which can form a pair of enantiomers. Thus, recemisation takes place (statement I). S_N2 reaction takes place forming pantavalent intermediate which loses Cl^- giving inverted alcohol. (Statement II) Hence, statements I and II both are correct.

29.

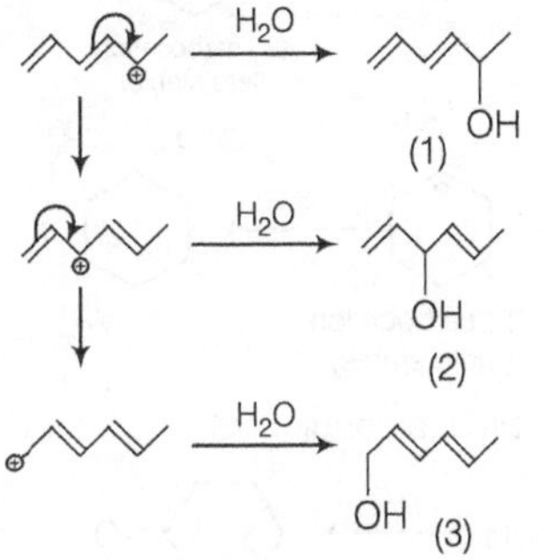

30. Electrophile $\overset{\oplus}{N}O_2$ attacks the following

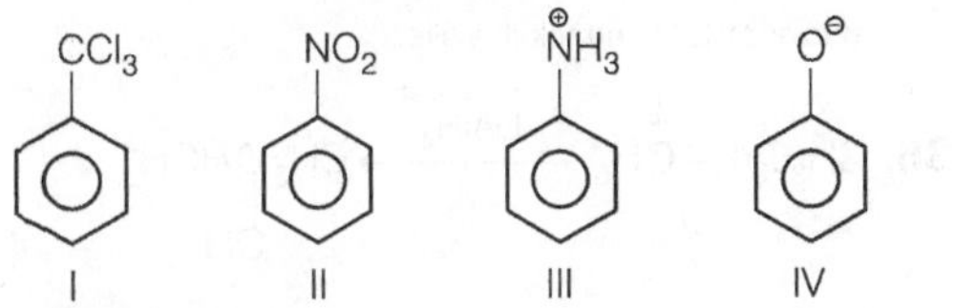

I II III IV

CCl_2, NO_2 and NH_3 groups are electron withdrawing groups. All deactivating and m-directing.

In negative charge on O is delocalised increasing electron density at o-and p-positions.

31. BE of C– Br bond < BE of C–Cl

Only C–Br is affected in Wurtz reaction.

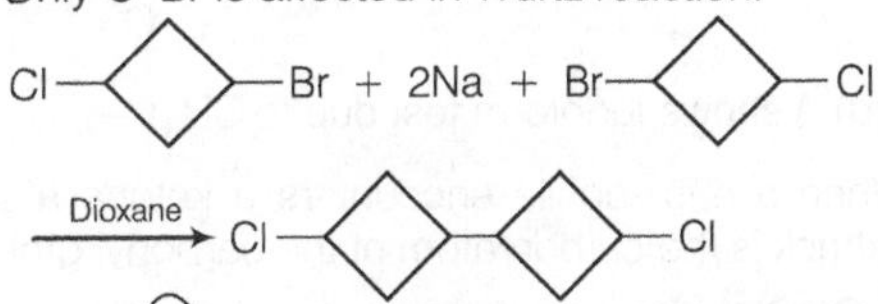

32. $CH_3CH{=}CH_2 \xrightarrow{D^-(NaBD_4)} CH_3\underset{\displaystyle D}{C}H\overset{\oplus}{C}H_2$

$\xrightarrow{H_2O_2/OH^-} CH_3\underset{\displaystyle D}{C}HCH_2OH$

33. 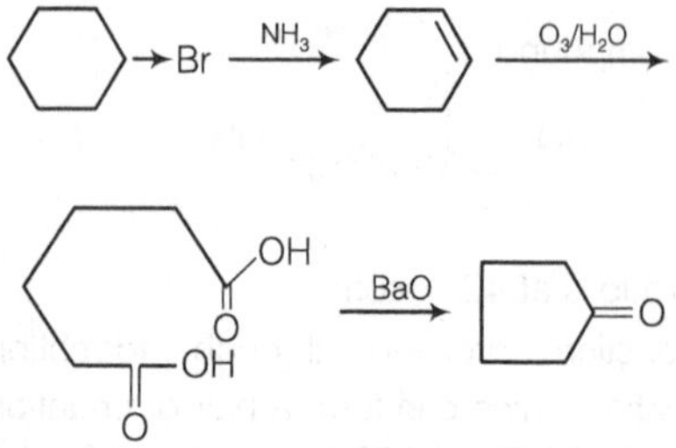

34. 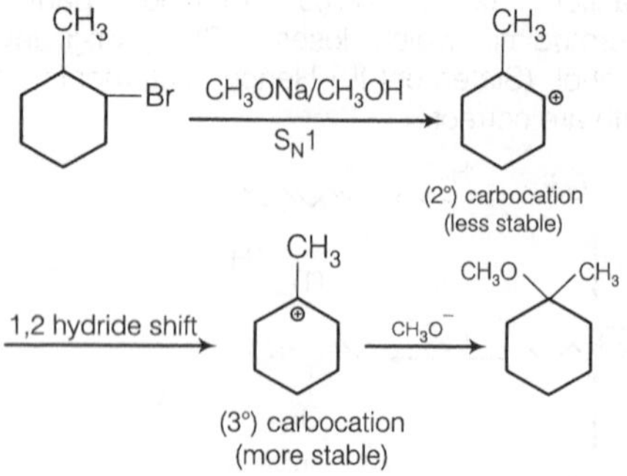

35. MnO_2 oxidises allylic alcohol

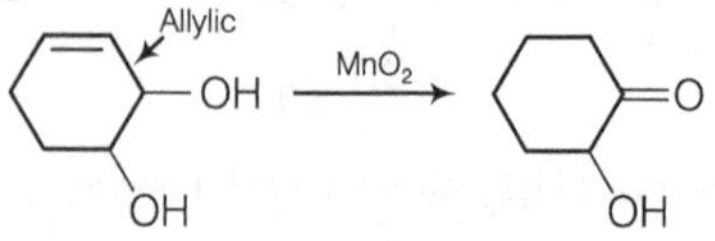

nucleophile H^- attack this site

36. 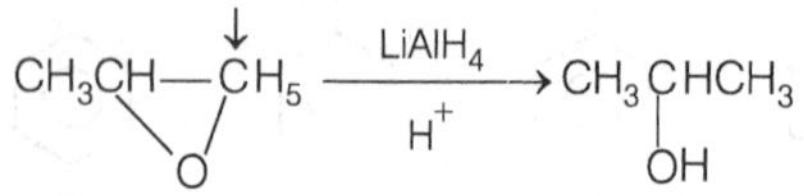

37. $CH_3\overset{O}{\overset{\|}{C}}OC_2H_5 + HCH_2\overset{O}{\overset{\|}{C}}OC_2H_5 \xrightarrow{C_2H_5ONa} CH_3\overset{O}{\overset{\|}{C}}CH_2\overset{O}{\overset{\|}{C}}OC_2H_5$ (A) Keto ester

(b) *A* forms oxime due to keto group

(c) *A* shows tautomerism

(d) *A* shows iodoform test due to $CH_3\overset{O}{\overset{\|}{C}}—$

38. When a nucleophile encounters a ketone a site of attack is the carbon atom of the carbonyl group.

39. —COOH attached to ring is eliminated due to presence of keto group at β – position.

40. $CH_3—\overset{O}{\overset{\|}{C}}—NHCH_2CH_2CH_3$

N- propylacetamide.

41. (a) $CH_3CH_2NO_2 \xrightarrow{HNO_2} CH_3\overset{NO_2}{\underset{N—OH}{C}}$

blue (nutroic acid)

$\xrightarrow{NaOH} CH_3—\underset{NONa}{\overset{}{C}}—NO_2$ (C=NONa) red

(b) $CH_3\overset{CH_3}{\overset{|}{C}}HNO_2 \xrightarrow{HNO_2} CH_3—\overset{CH_3}{\underset{N=O}{C}}—NO_2$

preudonitrol (blue)

$\xrightarrow{NaOH}$ No reaction blue

(c) $(CH_3)_3CNO_2 \xrightarrow{HNO_2}$ No reaction

42. $CH_3\underset{NH_2}{CH}COOH\ CH_3\underset{\oplus NH_3}{CH}COO^-$

$\xrightarrow[\text{acidic}]{H^+} CH_3\underset{\oplus NH_3}{CH}COOH$

(pH = 4, 1 or 2)

43. Sucrose has glucose and fructose, maltose has two units of glucose, raffinose has glucose, fructose and galactose. Galactose is a monosaccharide.

44. Amylose is a straight chain polymer have D- glucose units joined to gather by α – glycosidic linkage.

45. 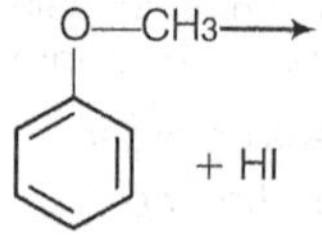

+ HI

This bond is cleaved due to greater stability of phenxide ion. Thus, C_6H_5OH and CH_3I is formed.

46. 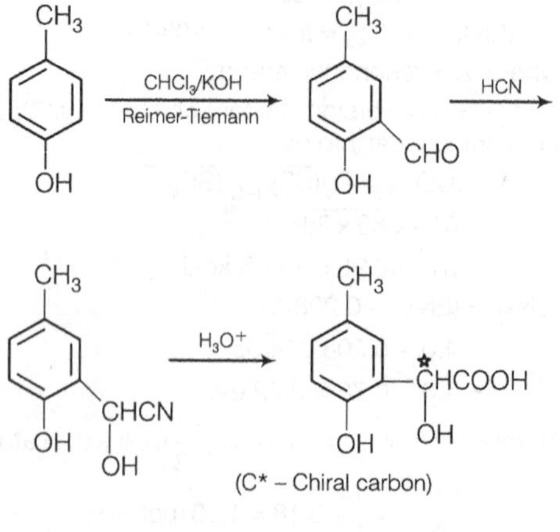

(C* – Chiral carbon)

47. 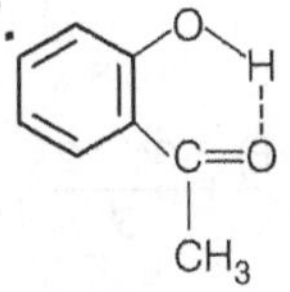

Boiling point of of A is lower than that of B. (In A, there is interamolecular H-bonding and in B, there is intermolecular H- bonding.) Thus, A is more volatile than B.

48. Sodium extract of thiourea will give red colour in Lassaigne's test due to the presence of both the elements nitrogen and sulphur.

49. The given half reactions are

	$E°$(inV)	$\Delta G° = -nFE°$
$BrO_3^- + 4e^- + 4H^+ \longrightarrow BrO^- + 2H_2O$	1.50	$-6F$
$BrO^- + e^- + 2H^+ \longrightarrow \frac{1}{2}Br_2 + H_2O$	1.60	$-1.6F$

Adding $BrO_3^- + 5e^- + 6H^+ \longrightarrow \frac{1}{2} Br_2 + 3H_2O$

$$\Delta G° = -7.6\,F = -5\,E°F$$

$$E° = \frac{7.6}{5} = 1.52\text{ V}$$

50. Zn is at lower potential, it will be oxidised and the cell reaction will be

$$Zn(s) + Ni^{2+}(aq) \longrightarrow Zn^{2+}(aq) + Ni(s)$$

51. $0.65 = i \times 1.86 \times 0.15$

$$i = 2.33 = 1 + 4\alpha$$

$$\alpha = -0.33$$

52. $p = p°x_1$

$$99.50 = 100\frac{n_1}{n_1 + n_2}$$

$$\frac{n_2}{n_1} = \frac{1}{191}$$

$$\Delta T_b = k_b m = K_b \frac{n_2}{n_1} \times \frac{1000}{m_1}$$

$$= 0.52 \times \frac{1}{191} \times \frac{1000}{18} = 0.15$$

$$T_b = 100.15°C$$

53. In reverse direction, product becomes reactant and *vice versa*. Hence, (b) is most endothermic in reverse direction

54. $\log \frac{k_2}{k_1} = \log 1000$

$$= \frac{E_1 - E_2}{RT}$$

$$= \frac{98000 - E_2}{8.314 \times 300}$$

E_2 (catalyst) = 80.77 kJ

55. From experiment no –3 and 4, doubling $[OCl^-]$ doubles the rate; *i.e.*, order with respect to OCl^- is one. Also, from experiment no 1 and 2, doubes the concentration of I^- double the rate, *i.e.* order with respect to I^- is also one.

rate $= k[I^-][OCl^-]$

56. $HON\overset{+}{H}_3 + H_2O \rightleftharpoons HONH_3OH + H^+$

$$K_h = \frac{K_w}{k_b} = \frac{10^{-14}}{10^{-8}} = 10^{-6}$$

$$[H^+] = \sqrt{K_h C}$$

$$= \sqrt{10^{-6} \times 0.024}$$

$$= 1.5 \times 10^{-4}$$

and pH = 3.8

57. $AgBr + 2CN^- \rightleftharpoons \underset{s}{Ag(CN)_2^-} + \underset{s}{Br^-}$

$0.1 - 2s$ $\quad K_{sp} = k_{sp} \cdot k = 4.3 \times 10^{-4}$

$$4.3 \times 10^{-4} = \left(\frac{s}{0.1 - 2s}\right)$$

$$s = 2 \times 10^{-3}$$

58. Cetyl trimethyl ammonium bromide forms cationic micelles above a certain minimum concentration.

59. $\frac{d(Cs)}{d(Li)} = \frac{1.87}{0.53} = \frac{133}{7}\left(\frac{r(Li)}{r(Cs)}\right)^3$

$$\frac{r(Cs)}{r(Li)} = 1.753$$

or $\quad r_{Cs} = 1.753\,r(Li)$

60. Nylon threads are made up of polyamide. Some common examples are nylon 6 and nylon 66.

Nylon 6 $H_2N-(CH_2)_5-COOH$

Monomer

$$-NH-(CH_2)_5-\overset{O}{\overset{||}{C}}-NH-(CH_2)_5-\overset{O}{\overset{||}{C}}-NH-$$

Polymer

Nylon 66 $\quad HO-\overset{O}{\overset{||}{C}}-(CH_2)_4COOH$

and $\quad H_2N-(CH_2)_6-NH_2$ are monomers

$$HN-\overset{O}{\overset{||}{C}}-(CH_2)_4-\overset{O}{\overset{||}{C}}-NH-(CH_2)_6-NH$$

nylon-66 (polymer)

$$-\overset{O}{\overset{||}{C}}-(CH_2)_4-\overset{O}{\overset{||}{C}}-$$

Mathematics

1. For $n = 1$, $1! < (1)^1$

that is, $1 < 1$ is not true.

For $n = 2$, $2! < \left(\frac{3}{2}\right)^1$

that is, $2 < \frac{9}{4}$ is true,

By induction, $n! < \left(\frac{n+1}{2}\right)^n \ \forall n \in N$ for $n > 2$.

$\therefore$ least $n = 2$.

2. Let T_r be the rth term of the given series.

Then, $T_r = \frac{r}{1 + r^2 + r^4}$, $r = 1, 2, 3, \ldots, n$

$$= \frac{r}{(r^2 + r + 1)(r^2 - r + 1)}$$

$$= \frac{1}{2}\left[\frac{1}{r^2 - r + 1} - \frac{1}{r^2 + r + 1}\right]$$

$\therefore$ Sum of the series $= \sum_{r=1}^{n} T_r$

$$= \frac{1}{2}\left\{\sum_{r=1}^{n}\left(\frac{1}{r^2 - r + 1} - \frac{1}{r^2 + r + 1}\right)\right\}$$

$$= \frac{1}{2}\left\{\left(1 - \frac{1}{3}\right) + \left(\frac{1}{3} - \frac{1}{7}\right) + \left(\frac{1}{7} - \frac{1}{13}\right) + \ldots + \left(\frac{1}{n^2 - n + 1} - \frac{1}{n^2 + n + 1}\right)\right\}$$

$$= \frac{1}{2}\left\{1 - \frac{1}{n^2 + n + 1}\right\} = \frac{n^2 + n}{2(n^2 + n + 1)}$$

3. Clearly, the given series is an arithmetico-geometric series whose corresponding AP and GP are respectively, 1, 4, 7, 10, ... and $1, \frac{1}{5}, \frac{1}{5^2}, \frac{1}{5^3}, \ldots$

The nth term of AP $= [1 + (n - 1) \times 3] = 3n - 2$

The nth term of GP $= \left[1 \times \left(\frac{1}{5}\right)^{n-1}\right] = \left(\frac{1}{5}\right)^{n-1}$

So, the nth term of the given series is

$$(3n - 2) \times \frac{1}{5^{n-1}} = \frac{3n - 2}{5^{n-1}}$$

Let, $S_n = 1 + \frac{4}{5} + \frac{7}{5^2} + \frac{10}{5^3} + \ldots + \frac{3n - 5}{5^{n-2}} + \frac{3n - 2}{5^{n-1}}$ ...(i)

$\frac{1}{5}S_n = \frac{1}{5} + \frac{4}{5^2} + \frac{7}{5^3} + \ldots + \frac{(3n - 5)}{5^{n-1}} + \frac{3n - 2}{5^n}$ ...(ii)

Subtracting (ii) from (i), we get

$$S_n - \frac{1}{5}S_n = 1 + \left\{\frac{3}{5} + \frac{3}{5^2} + \frac{3}{5^3} + \ldots + \frac{3}{5^{n-1}}\right\} - \frac{(3n - 2)}{5^n}$$

$$\Rightarrow \frac{4}{5}S_n = 1 + \frac{3}{5}\frac{\left\{1 - \left(\frac{1}{5}\right)^{n-1}\right\}}{\left(1 - \frac{1}{5}\right)} - \frac{(3n - 2)}{5^n}$$

$$\Rightarrow \frac{4}{5}S_n = 1 + \frac{3}{5}\frac{\left\{1 - \frac{1}{5^{n-1}}\right\}}{\left(\frac{4}{5}\right)} - \frac{(3n - 2)}{5^n}$$

$$\Rightarrow \frac{4}{5}S_n = 1 + \frac{3}{4}\left(1 - \frac{1}{5^{n-1}}\right) - \frac{(3n - 2)}{5^n}$$

$$\Rightarrow S_n = \frac{5}{4} + \frac{15}{16}\left(1 - \frac{1}{5^{n-1}}\right) - \frac{(3n - 2)}{4 \cdot 5^{n-1}}$$

4. The sequence of differences between successive terms is 2, 6, 18, 54, ...

Clearly, it is a GP. Let T_n be the nth term of the given series and S_n be the sum of its n terms.

Then,

$S_n = 5 + 7 + 13 + 31 + 85 + \ldots + T_{n-1} + T_n$ ...(i)

$S_n = 5 + 7 + 13 + 31 + \ldots + T_{n-1} + T_n$ ...(ii)

Subtracting (ii) from (i), we get

$$O = 5 + [2 + 6 + 18 + 54 + \ldots + (T_n - T_{n-1})] - T_n$$

$$\Rightarrow O = 5 + 2 \cdot \frac{3^{n-1} - 1}{3 - 1} - T_n$$

$$\Rightarrow T_n = 5 + (3^{n-1} - 1) = 4 + 3^{n-1}$$

$$\therefore S_n = \sum_{k=1}^{n} T_k = \sum_{k=1}^{n}(4 + 3^{K-1})$$

$$= \sum_{k=1}^{n} 4 + \sum_{k=1}^{n} 3^{k-1}$$

$$= 4n + (1 + 3 + 3^2 + \ldots + 3^{n-1})$$

$$= 4n + 1 \times \left(\frac{3^n - 1}{3 - 1}\right)$$

$$= 4n + \left(\frac{3^n - 1}{2}\right)$$

$$= \frac{1}{2}(3^n + 8n - 1)$$

5. The discriminant of the quadratic equation $2x^2 + 6x + b = 0$ is given by $D = 36 - 8b > 0$. Therefore, the given equation has real roots.

we have,

$$\frac{x_1}{x_2} + \frac{x_2}{x_1} = \frac{x_1^2 + x_2^2}{x_1 \cdot x_2} = \frac{(x_1 + x_2)^2 - 2x_1x_2}{x_1 \cdot x_2}$$

$$= \frac{(-3)^2 - 2(b/2)}{(b/2)} = \frac{18}{b} - 2 < -2 \quad [\because b < 0]$$

6. $x^4 + x^2 + 1 = (x^2 + 1)^2 - x^2$

$$= (x^2 + x + 1)(x^2 - x + 1)$$

$$x^2 + x + 1 = \left(x + \frac{1}{2}\right)^2 + \frac{3}{4} \neq 0 \qquad \forall\, x.$$

Now, $(a - 1)(x^2 + x + 1)^2 = (a + 1)(x^4 + x^2 + 1)$

$$(a - 1)(x^2 + x + 1) = (a + 1)(x^2 - x + 1).$$

$\Rightarrow$ $x^2 - ax + 1 = 0$ which has real and distinct roots

$\therefore \quad \Delta = a^2 - 4 > 0$

$\Rightarrow \quad a^2 > 4$

$\Rightarrow \quad a \in (-\infty, -2) \cup (2, \infty)$

7. Given that, $\alpha + \beta = 0$

$$\alpha + \beta + \gamma = -p$$

$\Rightarrow \quad \gamma = -p$

Substituting $\gamma = -p$ in the given equation,

$\Rightarrow -p^3 + p^3 - pq + r = 0$

$\Rightarrow \quad pq = r.$

8. Selection of 2 parallel lines from $m = {}^mC_2$

Selection of 2 parallel lines from $n = {}^nC_2$

Hence, number of parallelograms $= {}^mC_2 \cdot {}^nC_2$

$$= \frac{1}{4}(m - 1)(n - 1)mn$$

9. We have

$${}^{2n+1}C_1 + {}^{2n+1}C_2 + \ldots + {}^{2n+1}C_n = 255 \quad \ldots(i)$$

Also, the sum of binomial coefficients

$${}^{2n+1}C_0 + {}^{2n+1}C_1 + {}^{2n+1}C_2 + \ldots + {}^{2n+1}C_n + {}^{2n+1}C_{n+1} + + {}^{2n+1}C_{n+2} + \ldots + {}^{2n+1}C_{2n+1}$$

$$= (1 + 1)^{2n+1} = 2^{2n+1}$$

$$\Rightarrow {}^{2n+1}C_0 + 2\{{}^{2n+1}C_1 + {}^{2n+1}C_2 + \ldots + {}^{2n+1}C_n\} + {}^{2n+1}C_{2n+1} = 2^{2n+1}$$

$\Rightarrow \quad 1 + 2(255) + 1 = 2^{2n+1}$

$\Rightarrow \quad 1 + 255 = 2^{2n}$

$\Rightarrow \quad 2^{2n} = 2^8$

$\Rightarrow \quad n = 4$

10. $\log x$ is defined only when $x > 0$

Now, the 3rd term in the expansion

$T_{2+1} = {}^5C_2 \cdot x^{5-2} \cdot (x^{\log_{10} x})^2 = 1{,}000{,}000$ (given)

$\Rightarrow x^{3 + 2\log_{10} x} = 10^5.$

Taking logarithm of both sides, we get

$$(3 + 2\log_{10} x) \cdot \log_{10} x = 5$$

$\Rightarrow \quad 2y^2 + 3y - 5 = 0,$

where $\log_{10} x = y$

$\Rightarrow \quad (y - 1)(2y + 5) = 0$

$\Rightarrow \quad y = 1$ or $-5/2$

$\Rightarrow \log_{10} x = 1$ or $-5/2$

$\Rightarrow x = 10^1 = 10$ or $10^{-5/2}$

11. $\sum_{k=0}^{10} {}^{20}C_k = {}^{20}C_0 + {}^{20}C_1 + \ldots + {}^{20}C_{10}$

$$= \frac{1}{2}[2 \cdot {}^{20}C_0 + 2 \cdot {}^{20}C_1 + \ldots + 2 \cdot {}^{20}C_{10}]$$

$$= \frac{1}{2}[({}^{20}C_0 + {}^{20}C_{20}) + ({}^{20}C_1 + {}^{20}C_{19}) + \ldots + 2\,{}^{20}C_{10}]$$

$$= \frac{1}{2}[({}^{20}C_0 + {}^{20}C_1 + {}^{20}C_2 + \ldots + {}^{20}C_{19} + {}^{20}C_{20}) + {}^{20}C_{10}]$$

$$= \frac{1}{2}[2^{29} + {}^{20}C_{10}]$$

$$= 2^{19} + \frac{1}{2}\,{}^{20}C_{10}.$$

12. Lines $x + y = 4$ and $x + y = -4$ are parallel and points $(2, 2)$ and $(-2, -2)$ lie on these lines.

If point (a, a) lies between the lines then $a > -2$ and $a < 2$ *i.e.*,

$$-2 < a < 2 \Rightarrow |a| < 2$$

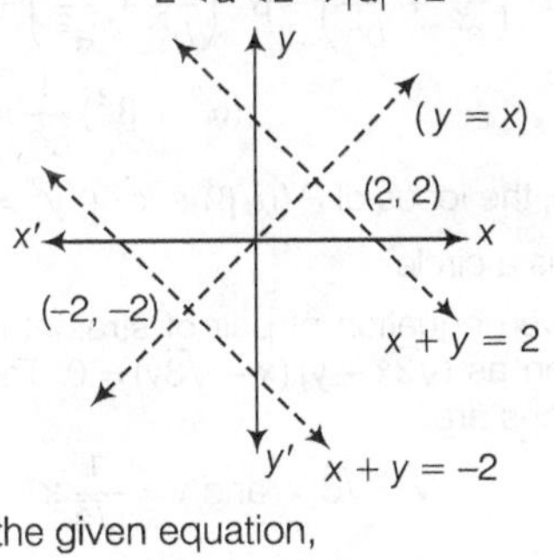

13. From the given equation,

$$3(my + 2) + 2y = 0$$

$\Rightarrow \quad y(3m + 2) = 4$

$\Rightarrow \quad y = \frac{4}{3m + 2},$

since, $y \in I \Rightarrow 3m + 2$

$\Rightarrow \quad \pm 1, \pm 2, \pm 4$

$\Rightarrow m = -1, -\frac{1}{3}, -\frac{4}{3}, 0, \frac{2}{3}, -2$

14. Let the equation of the line be $\frac{x}{a} + \frac{y}{b} = 1$ ...(i)

Its intercepts on y and x axes are b and a respectively.

According to the question,

$$\frac{1}{a} + \frac{1}{b} = \text{constant} = K \quad \text{(say)}$$

$$\therefore \quad \frac{1}{aK} + \frac{1}{bK} = 1$$

$$\Rightarrow \quad \frac{1/K}{a} + \frac{1/K}{b} = 1 \quad \text{...(ii)}$$

From (ii) it follows that line (i) passes through the fixed point $\left(\frac{1}{K}, \frac{1}{K}\right)$.

15. Variable line is $\frac{x}{a} + \frac{y}{b} = 1$ ...(i)

Any line perpendicular to (i) and passing through the origin will be

$$\frac{x}{b} - \frac{y}{a} = 0 \quad \text{...(ii)}$$

Now foot of the perpendicular from the origin to line (i) is the point of intersection (i) and (ii)

Let it be $P(\alpha, \beta)$, then

$$\frac{\alpha}{a} + \frac{\beta}{b} = 1 \quad \text{...(iii)}$$

and $$\frac{\alpha}{b} - \frac{\beta}{a} = 1 \quad \text{...(iv)}$$

Squarring and adding Eqs. (iii) and (iv), we get

$$\alpha^2\left(\frac{1}{a^2} + \frac{1}{b^2}\right) + \beta^2\left(\frac{1}{b^2} + \frac{1}{a^2}\right) = 1$$

$$\therefore \quad (\alpha^2 + \beta^2)\frac{1}{c^2} = 1$$

Hence, the locus of $P(\alpha, \beta)$ is $x^2 + y^2 = c^2$ which is a circle.

16. The given equation of pair of straight lines can be rewritten as $(\sqrt{3}x - y)(x - \sqrt{3}y) = 0$. Their separate equations are,

$$y = \sqrt{3}\,x \text{ and } y = \frac{1}{\sqrt{3}}x$$

$$\Rightarrow \quad y = \tan 60° x \quad \text{and} \quad y = \tan 30° x$$

After rotation, the separate equations are

$$y = \tan 90° x$$

and $$y = \tan 60° x$$

$\Rightarrow x = 0$ and $y = \sqrt{3} \cdot x$

$\therefore$ Combined equation in the new position is $x(\sqrt{3}x - y) = 0$ or $\sqrt{3}x^2 - xy = 0$

17. $$\lim_{x \to 1} \frac{\left[\sum_{K=1}^{100} x^K\right] - 100}{(x-1)}$$

$$= \lim_{x \to 1} \frac{(x + x^2 + x^3 + \ldots + x^{100}) - 100}{(x-1)}$$

$$= \lim_{x \to 1} \frac{(x-1) + (x^2-1) + (x^3-1) + \ldots + (x^{100}-1)}{(x-1)}$$

$$= \lim_{x \to 1} \left\{\left(\frac{x-1}{x-1}\right) + \left(\frac{x^2-1}{x-1}\right) + \left(\frac{x^3-1}{x-1}\right) + \ldots + \left(\frac{x^{100}-1}{x-1}\right)\right\}$$

$$= \lim_{x \to 1}\left(\frac{x-1}{x-1}\right) + \lim_{x \to 1}\left(\frac{x^2-1}{x-1}\right) + \lim_{x \to 1}\left(\frac{x^3-1}{x-1}\right) + \ldots + \lim_{x \to 1}\left(\frac{x^{100}-1}{x-1}\right)$$

$$= 1 + 2 + 3 + \ldots + 100$$

$$= \Sigma 100 = \frac{100 \times (100+1)}{2}$$

$$= 50 \times 101 = 5050 \quad \left\{\because \Sigma n = \frac{n(n+1)}{2}\right\}$$

18. Let $$y = \cfrac{x}{x + \cfrac{\sqrt[3]{x}}{x + \sqrt[3]{x} \ldots \text{infinity}}}$$

$$= \cfrac{x}{x + \cfrac{1}{x^{2/3}} \cdot \cfrac{x}{x + \sqrt[3]{x} \ldots \infty}} = \cfrac{x}{x + \cfrac{y}{x^{2/3}}}$$

$$\Rightarrow y = \frac{x^{5/3}}{x^{5/3} + y}$$

$$\Rightarrow y^2 + (x^{5/3}) \cdot y - x^{5/3} = 0$$

$$\therefore \quad y = \frac{-x^{5/3} \pm \sqrt{x^{10/3} + 4x^{5/3}}}{2}$$

$$= \frac{-x^{5/3} + \sqrt{x^{10/3} + 4x^{5/3}}}{2} \quad (\because y > 0)$$

$$= \frac{4x^{5/3}}{2(\sqrt{x^{10/3} + 4x^{5/3}} + x^{5/3})} = \frac{2}{\sqrt{\left(1 + \frac{4}{x^{5/3}}\right)} + 1}$$

$$\therefore \quad \lim_{x \to \infty} y = \frac{2}{\sqrt{1+0} + 1}$$

$$= \frac{2}{2} = 1.$$

19. We have; $f(x) = \begin{cases} \dfrac{x^2 - x}{x^2 - x} = 1, & \text{if } x < 0 \text{ or } x > 1 \\ -\dfrac{(x^2 - x)}{(x^2 - x)} = -1, & \text{if } 0 < x < 1 \\ 1, & \text{if } x = 0 \\ -1, & \text{if } x = 1 \end{cases}$

$= \begin{cases} 1, & \text{if } x \le 0 \text{ or } x > 1 \\ -1, & \text{if } 0 < x \le 1 \end{cases}$

Now, $\lim\limits_{x \to 0^-} 1 = 1$ and $\lim\limits_{x \to 0^+} f(x) = \lim\limits_{x \to 0} -1 = -1$

Clearly, $\lim\limits_{x \to 0^-} f(x) \ne \lim\limits_{x \to 0^+} f(x)$.

So, $f(x)$ is not continuous at $x = 0$. It can be easily seen that it is not continuous at $x = 1$ also.

20. Given; $f(x)$ is continuous at $x = 1$

$\therefore$ $f(1) =$ RHL

$\Rightarrow$ $f(1) = \lim\limits_{x \to 1} + f(x) \Rightarrow f(x) = \lim\limits_{h \to 0} f(1 + h)$

$\Rightarrow$ $a - b = \lim\limits_{h \to 0} 3(1 + h) \Rightarrow a - b = 3$ ...(i)

Again, given $f(x)$ is discontinuous at $x = 2$.

$\therefore$ LHL $\ne f(x)$

$\Rightarrow$ $\lim\limits_{x \to 2^-} f(x) \ne f(2) \Rightarrow \lim\limits_{h \to 0} f(2 - h) \ne f(2)$

$\Rightarrow$ $\lim\limits_{h \to 0} 3(2 - h) \ne 4b - a \Rightarrow 6 \ne 4b - a$ ...(ii)

Assume, $6 = 4b - a$

then from (i) and (ii), we get $b = 3$.

$\therefore$ locus $y = 3$

Which is impossible ($\because 6 \ne 4b - a$)

Hence, locus of (a, b) is $x - y = 3$ excluding the point when it cuts the line $y = 3$.

21. We are given that,

$(\cot \alpha_1) \cdot (\cot \alpha_2) \ldots (\cot \alpha_n) = 1$

$\Rightarrow$ $(\cos \alpha_1 \cdot (\cos \alpha_2) \ldots (\cos \alpha_n) = (\sin \alpha_1) \cdot (\sin \alpha_2) \ldots (\sin \alpha_n)$ ...(i)

Let $y = (\cos \alpha_1) \cdot (\cos \alpha_2) \ldots (\cos \alpha_n)$ (to be max.)

Squarring both sides, we get

$y^2 = (\cos^2 \alpha_1) \cdot (\cos^2 \alpha_2) \ldots (\cos^2 \alpha_n)$

$= \cos \alpha_1 \cdot \sin \alpha_1 \cdot \cos \alpha_2 \cdot \sin \alpha_2 \ldots \cos \alpha_n \cdot \sin \alpha_n$ [using (i)]

$= \dfrac{1}{2^n} [\sin 2\alpha_1 \cdot \sin 2\alpha_2 \ldots \sin 2\alpha_n]$

As $0 \le \alpha_1, \alpha_2, \ldots, \alpha_n \le \dfrac{\pi}{2}$

$\therefore$ $0 \le 2\alpha_1, 2\alpha_2, \ldots, 2\alpha_n \le \pi$

$\Rightarrow$ $0 \le \sin 2\alpha_1, \sin 2\alpha_2, \ldots, \sin 2\alpha_n \le 1$

$\therefore$ $y^2 \le \dfrac{1}{2^n} \cdot 1$ $\Rightarrow y \le \dfrac{1}{2^{n/2}}$

$\therefore$ Maximum value of y is $\dfrac{1}{2^{n/2}}$.

22. For $\alpha = -\dfrac{\pi}{2}, \beta = -\dfrac{\pi}{2}$ and $r = 2\pi$

$\sin\alpha + \sin\beta + \sin\gamma = -2$

$\Rightarrow$ minimum value of the expression is negative

23. $(\mathbf{a} + \mathbf{b}) \cdot (\mathbf{a} + \mathbf{b}) = |\mathbf{a}^2| + |\mathbf{b}^2| + 2\mathbf{a} \cdot \mathbf{b}$

$= 1 + 1 + 2(1)(1)\cos\theta = 2 + 2\cos\theta$

$\Rightarrow$ $|\mathbf{a} + \mathbf{b}|^2 = 2 \cdot 2\cos^2 \dfrac{\theta}{2}$

$\Rightarrow$ $\cos \dfrac{\theta}{2} = \dfrac{1}{2} |\mathbf{a} + \mathbf{b}|$

24. Let $\mathbf{a} = x\mathbf{i} + y\mathbf{j} + z\mathbf{k}$

then, $\mathbf{a} \cdot \mathbf{i} = (x\mathbf{i} + y\mathbf{j} + z\mathbf{k}) \cdot \mathbf{i} = x$

and $\mathbf{a} \cdot (\mathbf{i} + \mathbf{j}) = (x\mathbf{i} + y\mathbf{j} + z\mathbf{k}) \cdot (\mathbf{i} + \mathbf{j}) = x + y$

and $\mathbf{a} \cdot (\mathbf{i} + \mathbf{j} + \mathbf{k}) = (x\mathbf{i} + y\mathbf{j} + z\mathbf{k}) \cdot (\mathbf{i} + \mathbf{j} + \mathbf{k})$

$= x + y + z$

Given that $\mathbf{a} \cdot \mathbf{i} = \mathbf{a} \cdot (\mathbf{j} + \mathbf{i}) = \mathbf{a} \cdot (\mathbf{i} + \mathbf{j} + \mathbf{k})$

$\because$ $x = x + y = x + y + z$

Now, $x = x + y$ $\Rightarrow y = 0$

and $x + y = x + y + z$ $\Rightarrow z = 0$

Hence; $x = 1$

$\therefore$ $\mathbf{a} = \mathbf{i}$

25. Projection of $x\mathbf{i} - \mathbf{j} + \mathbf{k}$ on $2\mathbf{i} - \mathbf{j} + 5\mathbf{k}$

$= \dfrac{(x\mathbf{i} - \mathbf{j} + \mathbf{k}) \cdot (2\mathbf{i} - \mathbf{j} + 5\mathbf{k})}{\sqrt{4 + 1 + 25}} = \dfrac{2x + 1 + 5}{\sqrt{30}}$

But, given, $\dfrac{2x + 6}{\sqrt{30}} = \dfrac{1}{\sqrt{30}}$

$\Rightarrow$ $2x + 6 = 1$

$\Rightarrow$ $x = \dfrac{-5}{2}$

26. Since, $\mathbf{a} \times \mathbf{b} = \mathbf{c}$

$\therefore$ $(\mathbf{b} \times \mathbf{c}) \times \mathbf{b} = \mathbf{c}$

$\Rightarrow$ $(\mathbf{b} \cdot \mathbf{b})\mathbf{c} - (\mathbf{b} \cdot \mathbf{c})\mathbf{b} = \mathbf{c}$

$\Rightarrow$ $(\mathbf{b} \cdot \mathbf{b}) = 1,$ $\mathbf{c} \cdot \mathbf{b} = 0$

$\Rightarrow$ $b^2 = 1$ *i.e.*, $b = 1$

$\therefore$ **b** is a unit vector

$\therefore |\mathbf{c}| = |\mathbf{a}|$ $\Rightarrow c = a$.

27. Let $\mathbf{a} = t(\mathbf{i} + \mathbf{j} + 2\mathbf{k}) + s(\mathbf{i} + 2\mathbf{j} + \mathbf{k})$

$= (t + 3s)\mathbf{i} + (t + 2s)\mathbf{j} + (2t + s)\mathbf{k}$

given, $\mathbf{a} \cdot (\mathbf{i} + \mathbf{j} + \mathbf{k}) = 0$

$\therefore$ $t + s + t + 2s + 2t + s = 0$

$\Rightarrow$ $4(t + s) = 0$

$\Rightarrow$ $t + s = 0$

$\therefore$ $\mathbf{a} = t(\mathbf{i} + \mathbf{j} + 2\mathbf{k}) - t(\mathbf{i} + 2\mathbf{j} + \mathbf{k})$

$= t(-\mathbf{j} + \mathbf{k})$

But $|\mathbf{a}| = 1$

$\therefore$ $1 = 2t^2$

$$\Rightarrow \quad t = \pm \frac{1}{\sqrt{2}}$$

$$\therefore \quad \mathbf{a} = \pm \left(\frac{\mathbf{k} - \mathbf{j}}{\sqrt{2}}\right).$$

28. As, $A^2 = O, A^K = O \quad \forall \; K \geq 2$

Thus, $(A + I)^{50} = I + 50\,A$

$\Rightarrow \quad (A + I)^{50} - 50\,A = I$

$$\therefore \quad I = \begin{bmatrix} a & b \\ c & d \end{bmatrix}$$

$\Rightarrow a = 1, b = 0, c = 0, d = 1$

$\therefore \quad a + b + c + d = 1 + 0 + 0 + 1 = 2.$

29. $$\begin{bmatrix} \bar{z}_1 & -z_2 \\ \bar{z}_2 & z_1 \end{bmatrix}^{-1} \begin{bmatrix} z_1 & z_2 \\ -\bar{z}_2 & \bar{z}_1 \end{bmatrix}^{-1}$$

$$= \left\{\begin{bmatrix} z_1 & z_2 \\ -\bar{z}_2 & \bar{z}_1 \end{bmatrix}\begin{bmatrix} \bar{z}_1 & -z_2 \\ \bar{z}_2 & z_1 \end{bmatrix}\right\}^{-1}$$

$$= \begin{bmatrix} z_1\bar{z}_1 + z_2\bar{z}_2 & 0 \\ 0 & z_2\bar{z}_2 + z_1\bar{z}_1 \end{bmatrix}^{-1}$$

$$= \begin{bmatrix} |z_1|^2 + |z_2|^2 & 0 \\ 0 & |z_1|^2 + |z_2|^2 \end{bmatrix}^{-1}$$

$$= \begin{bmatrix} 2 & 0 \\ 0 & 2 \end{bmatrix}^{-1} = \begin{bmatrix} 1/2 & 0 \\ 0 & 1/2 \end{bmatrix}$$

30. We know that, in a square matrix of order n

$$|\text{adj } A| = |A|^{n-1}$$

$\Rightarrow |\text{adj (adj } A)| = |\text{adj } A|^{n-1} = |A|^{(n-1)^2}$

$\Rightarrow n^2 - 2n - 8 = 0$

$\Rightarrow \quad n = 4$ as $n = -2$ is not possible.

31. Coefficient x in $f(x)$ is equal to coefficient of x in

$$\begin{vmatrix} x & (1 + x - \frac{x^3}{3!} \ldots)^3 & 1 - \frac{x^2}{2!} \\ 1 & x - \frac{x^2}{2} & 2 \\ x^2 & 1 + x^2 & 0 \end{vmatrix}$$

$$= \text{coefficient of } x \text{ in} \begin{vmatrix} x & 1 & 1 \\ 1 & x & 2 \\ x^2 & 1 & 0 \end{vmatrix}$$

$= \text{Coefficient of } x \text{ in } [x(0 - 2) - 1(0 - 2x^2)$

$\qquad + 1(1 - x^3)]$

$= -2$

32. Given system has non-trivial solution then

$$\begin{vmatrix} \alpha_1 & \alpha_2 \\ \beta_1 & \beta_2 \end{vmatrix} = 0$$

$$\Rightarrow \quad \alpha_1\beta_2 = \alpha_2\beta_1$$

$$\Rightarrow \quad \frac{\alpha_1}{\beta_1} = \frac{\alpha_2}{\beta_2} = \frac{\alpha_1 + \alpha_2}{\beta_1 + \beta_2} = \sqrt{\frac{\alpha_1\,\alpha_2}{\beta_1\,\beta_2}}$$

$$\Rightarrow \quad \frac{-b/a}{-q/p} = \sqrt{\frac{c/a}{r/p}}$$

$$\Rightarrow \quad \frac{bp}{aq} = \sqrt{\frac{cp}{ar}}$$

$$\Rightarrow \quad b^2p^2ar = a^2q^2cp$$

$$\Rightarrow \quad b^2pr = q^2ac$$

33. The given circle

$$S(x, y) \equiv x^2 + y^2 - x - y - 6 = 0 \qquad \ldots\text{(i)}$$

has centre at $C \equiv \left(\frac{1}{2}, \frac{1}{2}\right)$.

According to the given conditions, the given point $P\,(\alpha - 1, \alpha + 1)$ must lie inside the given circle.

i.e., $\quad S\,(\alpha - 1, \alpha + 1) < 0$

$\Rightarrow (\alpha - 1)^2 + (\alpha + 1)^2 - (\alpha - 1) - (\alpha + 1) - 6 < 0$

$\Rightarrow \alpha^2 - \alpha - 2 < 0$ *i.e.*, $(\alpha - 2)(\alpha + 1) < 0$

$\Rightarrow -1 < \alpha < 2$

[using sign – scheme from algebra] $\qquad \ldots$(ii)

and also P and C must lie on the same side of the line.

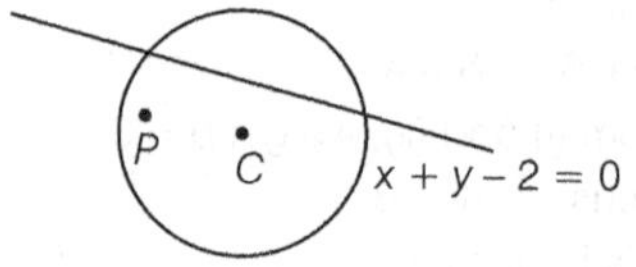

$$L\,(x, y) = x + y - 2 = 0$$

i.e., $L\left(\frac{1}{2}, \frac{1}{2}\right)$ and $L\,(\alpha - 1, \alpha + 1)$ must have the same sign.

Now, Isince $L\left(\frac{1}{2}, \frac{1}{2}\right) = \frac{1}{2} + \frac{1}{2} - 2 < 0$

therefore, we have

$$L\,(\alpha - 1, \alpha + 1) = (\alpha - 1) + (\alpha + 1) - 2 < 0$$

$$\Rightarrow \quad \alpha < 1 \qquad \ldots\text{(iv)}$$

Ineqalities (ii) and (iv) together give the permissible values of α as $-1 < \alpha < 1$.

34. The two circles are

$$x^2 + y^2 - 2ax + c^2 = 0$$

and $$x^2 + y^2 - 2by + c^2 = 0$$

centres : $C_1\,(a, 0) \qquad C_2\,(0, b)$

radii : $r_1 = \sqrt{a^2 - c^2}, \quad r_2 = \sqrt{b^2 - c^2}$

Since, the two circles touch each other externally, therefore

$$C_1C_2 = r_1 + r_2$$

$\Rightarrow \quad \sqrt{a^2+b^2} = \sqrt{a^2-c^2} + \sqrt{b^2-c^2}$

$\Rightarrow \quad a^2+b^2 = a^2-c^2+b^2-c^2 + 2\sqrt{a^2-c^2}\sqrt{b^2-c^2}$

$\Rightarrow \quad c^4 = a^2b^2 - c^2(a^2+b^2) + c^4$

$\Rightarrow \quad a^2b^2 = c^2(a^2+b^2)$

$\Rightarrow \quad \frac{1}{a^2} + \frac{1}{b^2} = \frac{1}{c^2}$

35. Let the coordinates of P be (h, k). Let the equation of a tangent from $P(h, K)$ to the circle $x^2+y^2=a^2$ be $y = mx + a\sqrt{1+m^2}$

Since, $P(h, K)$ lies on $y = mx + a\sqrt{1+m^2}$ therefore, $k = mh + a\sqrt{1+m^2} = (k-mh)^2 = a^2(1+m^2)$ this is a quadratic equation in m. Let the two roots be m_1 and m_2, then

$$m_1 + m_2 = \frac{2hK}{K^2-a^2}$$

But $\tan\alpha = m_1$, $\tan\beta = m_2$ and it is given that

$$\cot\alpha + \cot\beta = 0$$

$\therefore \quad \frac{1}{m_1} + \frac{1}{m_2} = 0$

$\Rightarrow \quad m_1 + m_2 = 0$

$\Rightarrow \quad \frac{2hK}{K^2-a^2} = 0$

$\Rightarrow \quad hK = 0$

Hence, the locus of (h, K) is $xy = 0$.

36. Let $A(at_1^2, 2at_2)$, $B(at_2^2, -2at_2)$.

We have,

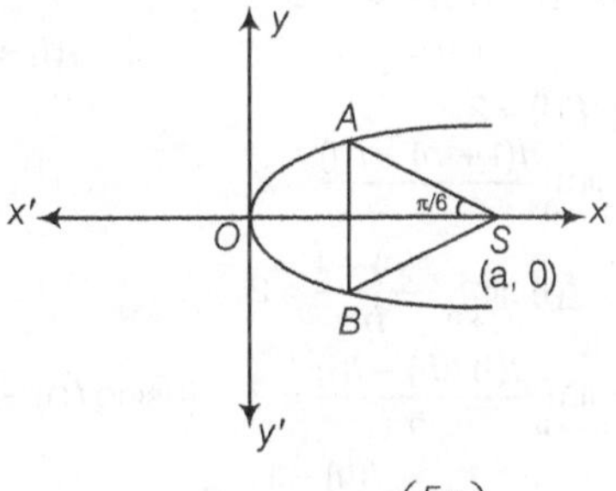

$$m_{AS} = \tan\left(\frac{5\pi}{6}\right)$$

$\Rightarrow \quad \frac{2at_1}{at_1^2 - a} = -\frac{1}{\sqrt{3}}$

$\Rightarrow \quad t_1^2 + 2\sqrt{3}t_1 - 1 = 0 \quad \Rightarrow t_1 = -\sqrt{3} \pm 2$

Clearly, $\quad t_1 = -\sqrt{3} - 2$ is rejected.

thus, $\quad t_1 = (2-\sqrt{3})$,

Hence, $\quad AB = 4at_1 = 4a(2-\sqrt{3})$.

37. Centre and radius of the given circle is $P(6, 0)$ and $\sqrt{5}$, respectively.

Now minimum distance between two curves always occurs along a line which normal to both the curves. Equation of normal to $y^2 = 4x$ at $(t^2, 2t)$ is

$$y = -tx + 2t + t^3.$$

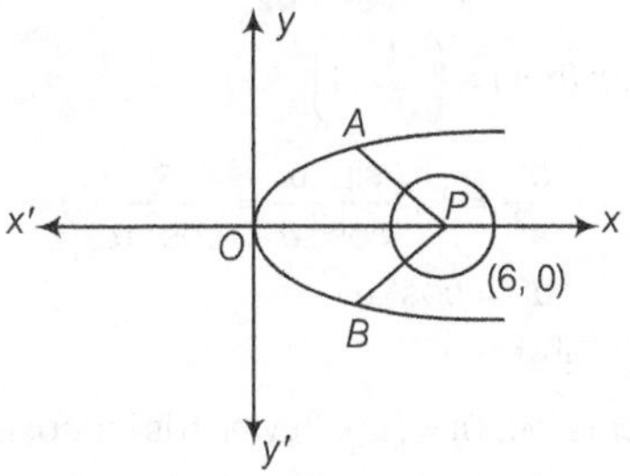

If it is normal to circle also, then it must pass though $(6, 0)$.

$\therefore \quad 0 = t^3 - 4t \quad \Rightarrow \quad t = 0 \quad$ or $\quad t = \pm 2$.

$\Rightarrow \quad A(4, 4)$ and $C(4, -4)$.

$\Rightarrow \quad PA = PC = \sqrt{20} = 2\sqrt{5}$.

$\Rightarrow$ Required minimum distance $= 2\sqrt{5} - \sqrt{5} = \sqrt{5}$.

38. Let the points of intersection of the line and the ellipse be $(a\cos\theta, b\sin\theta)$ and $\left\{a\cos\left(\frac{\pi}{2}+\theta\right), b\sin\left(\frac{\pi}{2}+\theta\right)\right\}$, since they lie on the given line $\quad lx + my + n = 0$.

$$la\cos\theta + mb\sin\theta + n = 0$$

$\Rightarrow \quad la\cos\theta + mb\sin\theta = -n$

and $\quad la\sin\theta + mb\cos\theta + n = 0$

$$la\sin\theta - mb\cos\theta = n$$

squarring and adding, we get,

$$a^2l^2 + b^2m^2 = 2n^2 \quad \Rightarrow \frac{a^2l^2 + b^2m^2}{n^2} = 2.$$

39. Equation of tangent to ellipse at given point is

$$x\left(\frac{1}{\sqrt{2}}\right) + 2y\left(\frac{1}{2}\right) = 1.$$

$\Rightarrow \quad x + \sqrt{2}y = \sqrt{2} \quad$ ...(i)

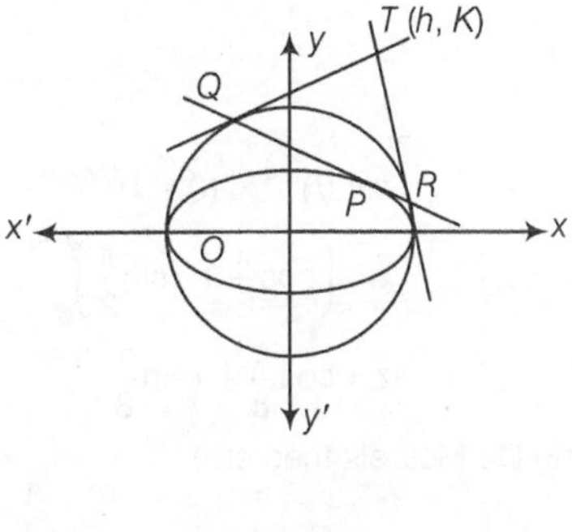

Now, QR is chord of contact of circle $x^2 + y^2 = 1$ with respect to point $T(h, K)$.

then, $QR \equiv hx + Ky = 1$ ...(ii)

Equations (i) and (ii) represent same straight line, then comparing ratio of coefficients we have

$$\frac{h}{1} = \frac{K}{\sqrt{2}} = \frac{1}{\sqrt{2}}$$

Hence; $(h, K) \equiv \left(\frac{1}{\sqrt{2}}, 1\right)$

40. $e^2 = 1 + \frac{b^2}{a^2} = 1 + \frac{\sin^2\alpha}{\cos^2\alpha} = \frac{1}{\cos^2\alpha}$

and $a^2 = \cos^2\alpha$

$\therefore$ $a^2e^2 = 1$

$\therefore$ *foci* $(\pm ae, 0) = (\pm 1, 0)$, which is independent of α.

41. Let the hyperbola be $\frac{x^2}{a^2} - \frac{y^2}{b^2} = 1$ and any double ordinate PQ be $(a\sec\theta, b\tan\theta)$, $(a\sec\theta, -b\tan\theta)$ and O is centre $(0, 0)$.

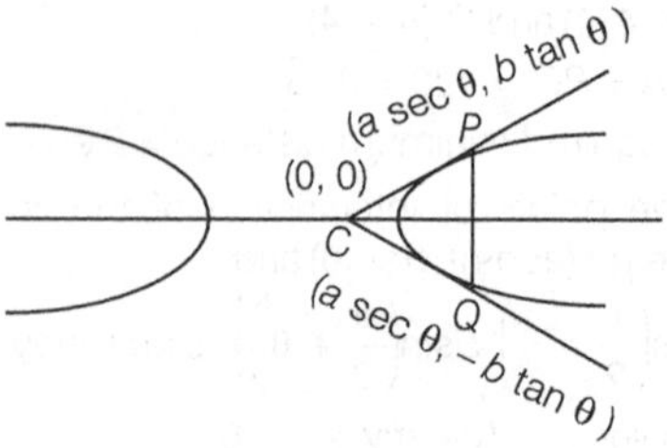

ΔOPQ being equilateral.

$\therefore$ $\tan 30° = \frac{b\tan\theta}{a\sec\theta}$

$\Rightarrow$ $3\cdot\frac{b^2}{a^2} = \text{cosce}^2\,\theta$

$\Rightarrow$ $3(e^2 - 1) = \text{cosce}^2\theta$

Now, $\text{cosec}^2\theta \geq 1$

$\therefore$ $3(e^2 - 1) \geq 1 \Rightarrow e^2 \geq \frac{4}{3}$

$\Rightarrow$ $e > 2/\sqrt{3}$

42. $iz^4 = -1$

$\Rightarrow$ $z^4 = -\frac{1}{i}$

$\Rightarrow$ $z^4 = i$

$\Rightarrow$ $z = (i)^{1/4} = (0 + i)^{1/4}$

$\Rightarrow$ $z = \left(\cos\frac{\pi}{2} + i\sin\frac{\pi}{2}\right)^{1/4}$

$\Rightarrow$ $z = \cos\frac{\pi}{8} + i\sin\frac{\pi}{8}$

(using De Moirrels theorem)

43. Let $a = \cos\alpha + i\sin\alpha$,

$b = \cos\beta + i\sin\beta$,

$c = \cos\gamma + i\sin\gamma$

then; $a + b + c = (\cos\alpha + \cos\beta + \cos\gamma) + i(\sin\alpha + \sin\beta + \sin\gamma)$

$= 0 + i0 = 0$

$\Rightarrow$ $a^3 + b^3 + c^3 = 3abc$

$\Rightarrow$ $(\cos 3\alpha + i\sin 3\alpha) + (\cos 3\beta + i\sin 3\beta) + (\cos 3\gamma + i\sin 3\gamma)$

$= 3[\cos(\alpha + \beta + \gamma) + i\sin(\alpha + \beta + \gamma)]$

$\Rightarrow$ $\cos 3\alpha + \cos 3\beta + \cos 3\gamma = 3\cos(\alpha + \beta + \gamma)$

44. $z_1^2 + z_2^2 + 2z_1z_2\cos\theta = 0$

$\Rightarrow$ $\left(\frac{z_1}{z_2}\right)^2 + 2\left(\frac{z_1}{z_2}\right)\cos\theta + 1 = 0$

$\Rightarrow$ $\left(\frac{z_1}{z_2} + \cos\theta\right)^2 = -(1 - \cos^2\theta) = -\sin^2\theta$

$\Rightarrow$ $\frac{z_1}{z_2} = -\cos\theta \pm i\sin\theta$

$\Rightarrow$ $\left|\frac{z_1}{z_2}\right| = \sqrt{(-\cos\theta)^2 + (\sin\theta)^2} = 1$

$\Rightarrow$ $|z_1| = |z_2|$

$\Rightarrow$ $|z_1 - 0| = |z_2 - 0|$

Thus, triangle with vertices 0, z_1, z_2 is iscosceles.

45. We have $f(xy) = f(x)f(y)$ for all $x, y \in R$.

Putting $x = y = 1$, we get

$f(1) = f(1)f(1)$

$\Rightarrow$ $f(1)[1 - f(1)] = 0$

$\Rightarrow$ $f(1) = 1$ $[\because f(1) \neq 0]$ ...(i)

Now, $f'(1) = 2$

$\Rightarrow$ $\lim_{h \to 0} \frac{f(1 + h) - f(1)}{h} = 2$

$\Rightarrow$ $f(1)\lim_{h \to 0} \frac{f(h) - 1}{h} = 2$

$\Rightarrow$ $\lim_{h \to 0} \frac{f(1)f(h) - f(1)}{h} = 2$ [using $f(1) = 1$] ...(ii)

$\Rightarrow$ $\lim_{h \to 0} \frac{f(h) - 1}{h} = 2$

Now $f'(4) = \lim_{h \to 0} \frac{f(4 + h) - f(4)}{h}$

$= \lim_{h \to 0} \frac{f(4)\cdot f(h) - f(4)}{h}$

$= \left\{\lim_{h \to 0} \frac{f(h) - 1}{h}\right\}\cdot f(4) = 2f(4)$ {from (i)}

$= 2 \times 4 = 8$

46. Let $x = a\cos^2\theta + b\sin^2\theta$

$\therefore \quad a - x = a - a\cos^2\theta - b\sin^2\theta = (a-b)\sin^2\theta$

and $x - b = a\cos^2\theta - b\sin^2\theta - b = (a-b)\cos^2\theta$

$\therefore \quad y = (a-b)\sin\theta\cdot\cos\theta - (a-b)\tan^{-1}\tan\theta$

$= \frac{a-b}{2}\sin 2\theta - (a-b)\theta$

$\therefore \quad \frac{dy}{dx} = \frac{dy/d\theta}{dx/d\theta} = \frac{(a-b)\cos 2\theta - (a-b)}{(b-a)\sin 2\theta}$

$= \frac{1-\cos 2\theta}{\sin\theta} = \tan\theta = \sqrt{\frac{a-x}{x-b}}$

47. We have $\sin^{-1}\left(\frac{x^2-y^2}{x^2+y^2}\right) = \log a$

$\Rightarrow \quad \frac{x^2-y^2}{x^2+y^2} = \sin(\log a)$

$\Rightarrow \quad \frac{1-\tan^2\theta}{1+\tan^2\theta} = \sin(\log a)$

(on putting $y = x\tan\theta$.)

$\Rightarrow \quad \cos 2\theta = \sin(\log a)$

$\Rightarrow \quad 2\theta = \cos^{-1}(\sin\log a)$

$\Rightarrow \quad \theta = \frac{1}{2}\cos^{-1}\{\sin(\log a)\}$

$\tan^{-1}\left(\frac{y}{x}\right) = \frac{1}{2}\cos^{-1}\{\sin\log a\}$

$\Rightarrow \quad \frac{1}{1+\frac{y^2}{x^2}}\cdot\frac{x\frac{dy}{dx}-y}{x^2} = 0$

$\Rightarrow \quad x\frac{dy}{dx} - y = 0$

$\Rightarrow \quad \frac{dy}{dx} = \frac{y}{x} \quad \ldots(i)$

$\Rightarrow \quad \frac{d^2y}{dx^2} = \frac{-y}{x^2} + \frac{1}{x}\cdot\frac{dy}{dx} = \frac{-y}{x^2} + \frac{1}{x}\left(\frac{y}{x}\right); \quad$ {from (i)}

$\Rightarrow \quad \frac{d^2y}{dx^2} = \frac{-y}{x^2} + \frac{y}{x^2} = 0.$

48. Let $PQ = 4m$ be the height of pole and, $AB = 1.6m$ be height of man. Let the end of shadow is R and it is at a distance of l from A when the man is at a distance x from PQ at some instant.

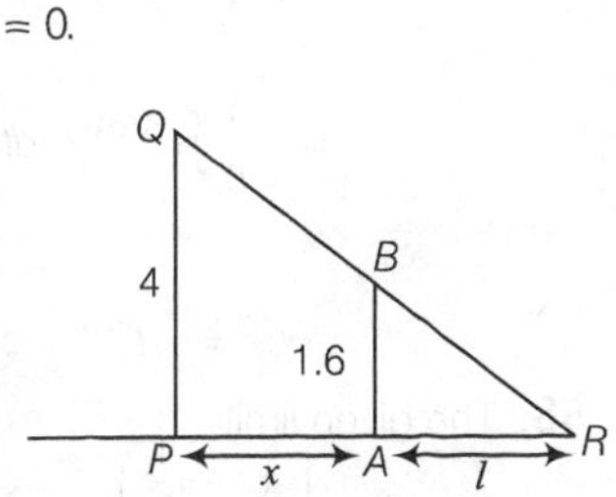

Since, ΔPQR and ΔABR are similar,

we have, $\frac{PQ}{AB} = \frac{PR}{AR}$

$\Rightarrow \quad \frac{4}{1.6} = \frac{x+l}{l}$

$\Rightarrow \quad 2x = 3l$

$\Rightarrow \quad \frac{2\,dx}{dt} = 3\cdot\frac{dl}{dt} \quad \left\{\text{given } \frac{dx}{dt} = 30 \text{ m/min}\right\}$

$\Rightarrow \quad \frac{dl}{dt} = \frac{2}{3}\times 30 \text{ m/min} = 20 \text{ m/min}.$

49. $\frac{dy}{dx} = Pe^{px} + P$

$\left(\frac{dy}{dx}\right)_{(0,1)} = 2P$

Subtangent $= \left|y\frac{dx}{dy}\right|$, subnormal $= \left|y\frac{dy}{dx}\right|$

Given; Subtangent = Subnormal

$\Rightarrow \frac{dy}{dx} = \pm 1 \Rightarrow 2p = \pm 1 \quad \Rightarrow \quad p = \pm\frac{1}{2}.$

50. Area of ΔABC, $A = \frac{1}{2}x\sqrt{d^2-x^2}$

for max/min $\frac{dA}{dx} = 0$

$\Rightarrow \quad \frac{1}{2}\sqrt{d^2-x^2} + \frac{1}{2}\times\left\{\frac{-2x}{2\sqrt{d^2-x^2}}\right\} = 0$

$\Rightarrow \quad \frac{d^2-x^2-x^2}{2\sqrt{d^2-x^2}} = 0$

$\Rightarrow \quad x = \frac{d}{\sqrt{2}}$

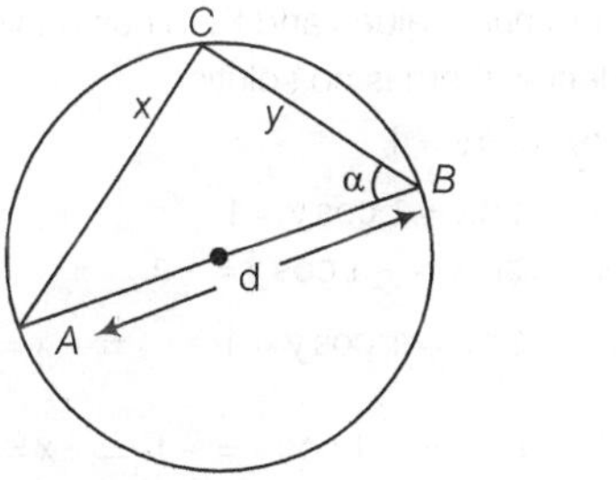

Also for $\quad x < \frac{d}{\sqrt{2}}, \frac{dA}{dx} > 0$

and for $\quad x > \frac{d}{\sqrt{2}}, \frac{dA}{dx} < 0.$

Hence; $x = \frac{d}{\sqrt{2}}$ is the point of maxima.

$\therefore$ A is max at $x = \frac{d}{\sqrt{2}}$, the area is maximum $y = \frac{d}{\sqrt{2}}$

$\therefore$ when Δ is isosceles,

51. Let $\cos^{-1}\left(\frac{\sqrt{5}}{3}\right) = \alpha$, then $\cos\alpha = \frac{\sqrt{5}}{3}$, where $0 < \alpha < \frac{\pi}{2}$.

Now, $\tan\frac{\alpha}{2} = \sqrt{\frac{1-\cos\alpha}{1+\cos\alpha}} = \sqrt{\frac{1-\sqrt{5}/3}{1+\sqrt{5}/3}}$

$= \sqrt{\frac{3-\sqrt{5}}{3+\sqrt{5}}} = \sqrt{\frac{(3-\sqrt{5})^2}{9-5}} = \frac{1}{2}(3-\sqrt{5})$

52. We know that;

if $|x| \le 1$, then $2\tan^{-1}x = \sin^{-1}\left(\frac{2x}{1+x^2}\right)$

and if $|x| > 1$, then $\pi - 2\tan^{-1}x = \sin^{-1}\left(\frac{2x}{1+x^2}\right)$

thus, if $|x| > 1$.

$2\tan^{-1}x + \sin^{-1}\frac{2x}{1+x^2} = 2\tan^{-1}x + \pi$

$- 2\tan^{-1}x = \pi$

which is independent of x.

53. $1 + \sin x \cdot \sin^2\frac{x}{2} = 0$

$\Rightarrow \quad 2 + 2\sin x \cdot \sin^2\frac{x}{2} = 0$

$\Rightarrow \quad 2 + \sin x(1-\cos x) = 0$

$\Rightarrow \quad 4 + 2\sin x(1-\cos x) = 0$

$\Rightarrow \quad 4 + 2\sin x - \sin 2x = 0$

$\Rightarrow \quad \sin 2x = 2\sin x + 4$

Above is not possible for any value of x as LHS has maximum value 1 and RHS has minimum value 2.

Hence, there is no solution.

54. $\sin x \cdot \cos y = 1$

$\Rightarrow \quad \sin x = 1, \cos y = 1$

or $\quad \sin x = -1, \cos y = -1$.

if $\quad \sin x = 1, \cos y = 1 \quad \Rightarrow x = \frac{\pi}{2}, y = 0, 2\pi$

if $\quad \sin x = -1, \cos y = -1 \Rightarrow x = \frac{3\pi}{2}, y = \pi$

Thus possible ordered pairs are

$\left(\frac{\pi}{2}, 0\right), \left(\frac{\pi}{2}, 2\pi\right), \left(\frac{3\pi}{2}, \pi\right)$.

55. Let $I = \int x^x (\log_e x)\,dx$

$= \int x^x (1 + \log x)\,dx$

let $t = x^x = e^{x\log x}$

$\Rightarrow \quad \frac{dt}{dx} = x^x\left\{x \cdot \frac{1}{x} + \log x\right\}$

$\Rightarrow \quad dt = x^x(1+\log x)\,dx$

$\therefore \quad I = \int t + c = x^x + c.$

56. Let $I = \int\sqrt{1+\operatorname{cosec} x}\,dx = \int\frac{\sqrt{1+\sin x}}{\sqrt{\sin x}}dx$

$= \int \pm \frac{\sin\frac{x}{2} + \cos\frac{x}{2}}{\sqrt{2\sin\frac{x}{2}\cdot\cos\frac{x}{2}}}\cdot dx$

$= \pm\int\frac{\sin\frac{x}{2}+\cos\frac{x}{2}}{\sqrt{1-\left(\sin\frac{x}{2}-\cos\frac{x}{2}\right)^2}}\cdot dx$

Put, $\quad \sin\frac{x}{2} - \cos\frac{x}{2} = t$

$\Rightarrow \quad \left(\cos\frac{x}{2} + \sin\frac{x}{2}\right)dx = 2dt$

$\therefore \quad I = \pm\int\frac{2\,dt}{\sqrt{1-t^2}} = \pm 2\sin^{-1}t + c$

$= \pm 2\sin^{-1}\left(\sin\frac{x}{2} - \cos\frac{x}{2}\right) + c$

57. We have;

$\int\frac{dx}{x^2(x^n+1)^{\frac{(n-1)}{n}}} = \int\frac{dx}{x^2\cdot x^{n-1}\left(1+\frac{1}{x^n}\right)^{\frac{(n-1)}{n}}}$

$= \int\frac{dx}{x^{n+1}(1+x^{-n})^{\frac{(n-1)}{n}}}$

Put, $\quad 1 + x^{-n} = t$

$\therefore \quad -nx^{-n-1}dx = dt$

$\Rightarrow \quad \frac{dx}{x^{n+1}} = \frac{-dt}{n}$

$\Rightarrow \quad \int\frac{dx}{x^2(x^n+1)^{\frac{(n-1)}{n}}} = -\frac{1}{n}\int\frac{dt}{t^{\frac{(n-1)}{n}}}$

$= -\frac{1}{n}\int t^{(1/n)-1}\,dt = -\frac{1}{n}\cdot\frac{t^{\left(\frac{1}{n}-1+1\right)}}{\left(\frac{1}{n}-1+1\right)} + c$

$= -t^{1/n} + c = -(1+x^{-n})^{\frac{1}{n}} + c$

58. The given limit

$L = \lim_{n\to\infty}\left\{\frac{1}{na} + \frac{1}{na+1} + \frac{1}{na+2} + \ldots + \frac{1}{na+n(b-a)}\right\}$

$$= \lim_{n \to \infty} \sum_{r=0}^{n(b-a)} \frac{1}{na + r}$$

$$= \lim_{n \to \infty} \frac{1}{n} \sum_{r=0}^{(b-a)n} \frac{1}{a + r/n}$$

$$= \int_0^{(b-a)} \frac{dx}{a + x} = [\log (a + x)]_0^{b-a} \qquad \left(\frac{r}{n} = x\right)$$

$$= \log b - \log a$$

$$= \log \left(\frac{b}{a}\right).$$

59. Case I. If $0 \le a < b$, then $\frac{|x|}{x} = 1$

$\therefore \quad I = \int_a^b 1 \cdot dx = b - a = |b| - |a|$

Case II : If $a < b \le 0$, then $|x| = -x$

$\therefore \quad I = \int_a^b \frac{-x}{x} dx = \int_a^b (-1) dx = [-x]_a^b$

$= -b - (-a) = |b| - |a|$

Case III If $a < 0 < b$.

then $|x| = -x$ when $a < x < 0$

and $|x| = x$ when $0 < x < b$

$$I = \int_a^b \frac{|x|}{x} dx = \int_a^0 \frac{|x|}{x} dx + \int_0^b \frac{|x|}{x} dx$$

$$= \int_a^0 \frac{-x}{x} dx + \int_0^b \frac{x}{x} dx$$

$$= \int_a^0 (-1) dx + \int_0^b (1) dx$$

$$= [-x]_a^0 + [x]_0^b$$

$$= a + b = b - (-a)$$

$$= |b| - |a|$$

Hence, in all cases,

$$I = \int_a^b \frac{|x|}{x} dx = |b| - |a|$$

60. Here, we have to prove that, y = constant or derivative of y with respect to x is zero.

$$y = \int_{1/8}^{\sin^2 x} \sin^{-1} \sqrt{t}\, dt + \int_{1/8}^{\cos^2 x} \cos^{-1} \sqrt{t}\, dt \quad \dots \text{(i)}$$

$\frac{dy}{dx} = \sin^{-1} \sqrt{\sin^2 x} \cdot 2 \sin x \cdot \cos x + \cos^{-1} \sqrt{\cos^2 x}$

$(-2 \sin x \cdot \cos x) = 2x \sin x \cdot \cos x - 2x \cdot \sin x \cdot \cos x$

$= 0$ for all x.

61. $x^2 + y^2 = 1$, $x + y = 1$ meet when

$$x^2 + (1 - x)^2 = 1$$

$\Rightarrow \quad x^2 + 1 + x^2 - 2x = 1$

$\Rightarrow \quad 2x^2 - 2x = 0$

$\Rightarrow \quad 2x(x - 1) = 0$

$\Rightarrow \quad x = 0, x = 1$

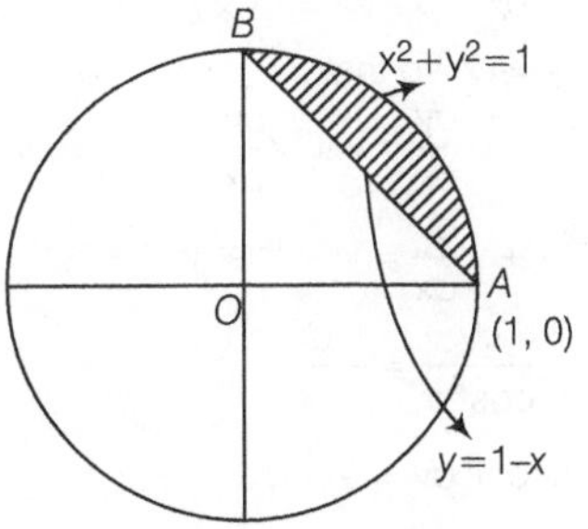

$\Rightarrow$ Meet at points A (1, 0), B (0, 1)

$\therefore$ Required area $= \frac{\pi}{4} - \frac{1}{2}(1) \cdot (1)$

$= \frac{\pi}{4} - \frac{1}{2}$ *i.e.*, [quad. $OAB - \Delta OAB$].

62. Given parabola, is

$$(y - 2)^2 = x - 1$$

$$\Rightarrow \quad \frac{dy}{dx} = \frac{1}{2(y - 2)}$$

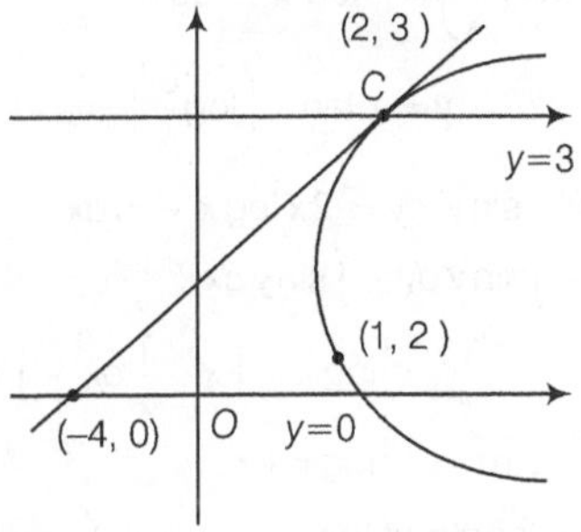

when, $y = 3$, $x = 2$

$$\therefore \quad \frac{dy}{dx} = \frac{1}{2(3 - 2)} = \frac{1}{2}$$

Tangent at $(x, 3)$ is

$$y - 3 = \frac{1}{2}(x - 2)$$

$\Rightarrow \quad x - 2y + 4 = 0.$

$\therefore$ Required area

$$\int_0^3 \{(y - 2)^2 + 1\} dy - \int_0^3 (2y - 4) dy$$

$$= \left[\frac{(y - 2)^3}{3} + y\right]_0^3 - [y^2 - 4y]_0^3$$

$$= \frac{1}{3} + 3 + \frac{8}{3} - (9 - 12) = 9.$$

63. Clearly, the given differential equation is not a polynomial indifferential coefficients. So, its degree is not defined.

64. We have, $\frac{dy}{dx} = \frac{y}{x} - \cos^2\left(\frac{y}{x}\right)$

Putting $y = vx$, so that

$$\frac{dy}{dx} = v + x\frac{dv}{dx},$$

$$\therefore \quad v + x\frac{dv}{dx} = v - \cos^2 v$$

$$\Rightarrow \quad \frac{dv}{\cos^2 v} = -\frac{dx}{x}$$

$$\Rightarrow \quad \sec^2 v\, dv = -\frac{1}{x}dx$$

on integration, we get

$$\tan v = -\log x + \log c$$

$$\Rightarrow \tan\left(\frac{y}{x}\right) = -\log x + \log c$$

this passes through $(1, \pi/4)$. therefore,

$$1 = \log c$$

$$\text{So;} \quad \tan\left(\frac{y}{x}\right) = -\log x + 1$$

$$\Rightarrow \quad \tan\left(\frac{y}{x}\right) = -\log x + \log c$$

$$\Rightarrow \quad y = x\tan^{-1}\left(\log\frac{c}{x}\right).$$

65. $(y\cos y + \sin y)\,dy = (2x\log x + x)\,dx$

$y\sin y - \int \sin y\, dy + \int \sin y\, dx$

$$= x^2\log x - \int x^2 \cdot \frac{1}{x}dx + \int x\,dx + c$$

$$\therefore \quad y\sin y = x^2\log x + c$$

66. We have length of the

normal = radius vector

$$\Rightarrow \quad y\sqrt{1+\left(\frac{dy}{dx}\right)^2} = \sqrt{x^2+y^2}$$

$$\Rightarrow \quad y^2\left\{1+\left(\frac{dy}{dx}\right)^2\right\} = x^2 + y^2$$

$$\Rightarrow \quad y^2\left(\frac{dy}{dx}\right)^2 = x^2$$

$$\Rightarrow \quad x = \pm y\frac{dy}{dx}$$

$$\Rightarrow \quad x = y\frac{dy}{dx} \quad \text{or } x = -y\frac{dy}{dx}$$

$$\Rightarrow \quad x\,dx - y\,dy = 0$$

$$\text{or} \quad x\,dx + y\,dy = 0$$

$$\Rightarrow x^2 - y^2 = c_1 \quad \text{or} \quad x^2 + y^2 = c_2$$

Clearly, $x^2 - y^2 = c_1$ represents a rectangular hyperbola and $x^2 + y^2 = c_2$ represents circle.

67. We have,

$$\frac{1}{x(x+1)(x+2)\ldots(x+n)} = \frac{A_0}{x} + \frac{A_1}{x+1} + \ldots + \frac{A_r}{x+r} + \ldots + \frac{A_n}{x+n}$$

$$\Rightarrow \quad 1 = A_0(x+1)(x+2)\ldots(x+n) + \ldots + A_r\, x(x+1)(x+2)\ldots(x+r-1)(x+r+1)\ldots(x+n) + \ldots + A_n x(x+1)\ldots(x+n-1)$$

Putting $x = -r$, we get;

$$1 = A_r(-r)(-r+1)(-r+2)\ldots(-3)(-2)(-1)(1)(2)\ldots(n-r)$$

$$\Rightarrow \quad 1 = A_r(-1)^r\{r(r-1)(r-2)\ldots(3\cdot2\cdot1\}(1\cdot2\cdot3\ldots(n-r))$$

$$\Rightarrow \quad A_r = (-1)^r / r!\,(n-r)!$$

68. $\log_2(9 - 2^x) = 10^{\log_{10}(3-x)}$

$$\Rightarrow \quad \log_2(9-2^x) = (3-x) \qquad [\because a^{\log_a b} = b]$$

$$\Rightarrow \quad 2^{3-x} = 9 - 2^x$$

$$\Rightarrow \quad \frac{2^3}{2^x} = 9 - 2^x$$

$$\Rightarrow \quad 8 = 2^x \times (9 - 2^x)$$

$$\Rightarrow \quad 2^{2x} - 2^x \times 9 + 8 = 0$$

Let $2^x = y$, then;

$$y^2 - 9y + 8 = 0$$

$$\Rightarrow \quad (y-8)(y-1) = 0$$

$$\Rightarrow \quad y = 8 \quad \text{or} \quad y = 1.$$

$$\Rightarrow \quad 2^x = 2^3 \quad \text{or} \quad 2^x = 2^0$$

$$\Rightarrow \quad x = 3 \quad \text{or} \quad x = 0.$$

But $x = 3$ does not satisfy the given equation, since log 0 is not defined.

69. a, b, c are in GP $\Rightarrow b^2 = ac$

taking log on both the sides, we get

$$2\log b = \log a + \log c$$

$$\Rightarrow \quad \log b = \frac{\log a + \log c}{2}$$

$\Rightarrow$ $\log a, \log b, \log c$ are in AP.

70. We have; $\Delta = \frac{\sqrt{3}}{4}a^2,\ s = \frac{3a}{2}$

$$\therefore \quad r = \frac{\Delta}{S} = \frac{a}{2\sqrt{3}},\ R = \frac{abc}{4\Delta}$$

$$= \frac{a^3}{\sqrt{3}\cdot a^2} = \frac{a}{\sqrt{3}}$$

$$\text{and} \quad r_1 = \frac{\Delta}{S-a} = \frac{\sqrt{3}/4\cdot a^2}{a/2} = \frac{\sqrt{3}}{2}\cdot a$$

$$\text{Hence;} \quad r : R : r_1 = \frac{a}{2\sqrt{3}} : \frac{a}{\sqrt{3}} : \frac{\sqrt{3}}{2}\cdot a = 1:2:3$$

71. Since the angles are in AP, therefore

$2B = A + C \quad \Rightarrow \quad 3B = A + B + C$

$\Rightarrow \; 3B = 180^\circ \Rightarrow \quad B = 60^\circ$

using the formula,

$\frac{\sin A}{a} = \frac{\sin B}{b}$, we get

$$\sin A = \frac{a}{b}\sin B = \frac{24}{22} \cdot \sin 60^\circ = \frac{6\sqrt{3}}{11}$$

Now; $\sin C = \sin(180^\circ - (A + B)) = \sin(A + B)$

$= \sin A \cdot \cos B + \cos A \cdot \sin B$

$$= \left(\frac{6\sqrt{3}}{11}\right)\left(\frac{1}{2}\right) + \sqrt{\left\{1 - \left(\frac{6\sqrt{3}}{11}\right)^2\right\}} \cdot \left(\frac{\sqrt{3}}{2}\right)$$

$$= \frac{1}{22}\left(\frac{\sqrt{3}}{2}\right)(12 + 2\sqrt{3})$$

Now, $c = \frac{b}{\sin B} \sin C$

$\Rightarrow \quad c = 12 + 2\sqrt{3}.$

72. $2^{-1} * x * 3^{-1} = 5$

$\Rightarrow \; x = 2 * (5 * 3) = 2 * (5 + 3 + 15)$

$= 2 * (23) = 2 + 23 + 46 = 71$

73. $b^{-1} * a^{-1} * b * a = (b^{-1} * a^{-1}) * (b * a) = (a * b)^{-1}$

$* (b * a) = (a * b)^{-1} * (a * b) = e.$

74. $(a * b)^2 = (a * a) * (b * b)$ for all $a, b \in G$

$\Rightarrow \; (a * b) * (a * b) = (a * a) * (b * b)$ for all $a, b \in G$

$a * (b * a) * b = a * ((a * b) * b$

for all $a, b \in G$

$\Rightarrow \; b * a = a * b$ for all $a, b \in G$

{by cancellation laws}

$\Rightarrow$ G is abelian.

77. By hit and trial method taking $x = y = z = 2$,

we get $xy + yz + zx = 12$

and $\quad xyz = 8.$

78. Since, c is a prime number so it can have only two factors 1 and c. $ab = c$ gives two more factors of c as a and b, which is not possible.

79. $(p \wedge \sim q)(\sim p \wedge q) = (p \wedge \sim p) \wedge (\sim q \wedge q)$

$= f \wedge f = f$

(By using associative laws and commutative laws.)

$\therefore \quad (p \wedge \sim q) \wedge (\sim p \wedge q)$ is a contradiction.

80. $\sim p \wedge q = \sim(q \rightarrow p)$

English

6. Radiant means giving a warm bright light.

7. Prune means to make something smaller by removing parts.

8. Dilettante means a person, who does something, but is not serious about it.

9. Foster means to encourage something to develop.

10. Enigma means a person or thing that is mysterious or difficult to understand.

11. Conform means to agree with or match something.

12. Aboriginal means relating to the original people, animals etc.

13. Amenable means easy to control.

14. Acquit means to state officially that somebody is not guilty of a crime.

15. Forbidden means not allowed.

31. Remove 'do not'.

32. Remove 'about'.

33. Use 'but' in place of 'and'.

34. Use 'that' in place of 'because'.

35. Use 'were' in place of 'was'.

36. The correct spelling is 'Honorary'.

37. The correct spelling is 'Maintenance'.

38. The correct spelling is 'Vehicle'.

39. The correct spelling is 'Daffodil'.

40. The correct spelling is 'Curable'.

Manipal
Engineering Entrance Exam
Solved Paper 2012

Physics

1. Oxygen is 16 times heavier than hydrogen. Equal volumes of hydrogen and oxygen are mixed. The ratio of speed of sound in the mixture to that in hydrogen is

(a) $\sqrt{8}$ (b) $\sqrt{2/17}$
(c) $\sqrt{1/8}$ (d) $\sqrt{32/17}$

2. A man, standing between two cliffs, claps his hands and starts hearing a series of echoes at intervals of one second. If the speed of sound in air is 340 ms^{-1}, the distance between the cliffs is

(a) 680 m (b) 1700 m
(c) 340 m (d) 1620 m

3. A beam of light of wavelength 600 nm from a source falls on a single slit 1 mm wide and the resulting diffraction pattern is observed on a screen 2 m away. The distance between the first dark fringes on either side of the central bright fring is

(a) 2.4 cm (b) 2.4 mm
(c) 1.2 mm (d) 1.2 cm

4. A thin plano-convex lens acts like a concave mirror of focal length 0.2 m when silvered from its plane surface. The refractive index of the material of the lens is 1.5. The radius of curvature of the convex surface of the lens will be

(a) 0.1 m (b) 0.75 m
(c) 0.4 m (d) 0.2 m

5. A balloon is rising vertically up with a velocity of 29 ms^{-1}. A stone is dropped from it and it reaches the ground in 10 s. The height of the balloon when the stone was dropped from it is ($g = 9.8\ ms^{-1}$)

(a) 400 m (b) 150 m
(c) 100 m (d) 200 m

6. In a Young's double slit experiment, the separation between the two slits is 0.9 mm and the fringes are observed one metre away. If it produces the second dark fringe at a distance of 1 mm from the central fringe, the wavelength of the monochromatic source of light used is

(a) 450 nm (b) 400 nm
(c) 500 nm (d) 600 nm

7. A count rate meter shows a count of 240 per minute from a given radioactive source. One hour later the meter shows a count rate of 30 per minute. The half-life of the source is

(a) 80 min (b) 120 min
(c) 20 min (d) 30 min

8. Two conductors of the same material have their diameters in the ratio 1 : 2 and their lengths in the ratio 2 : 1. If the temperature difference between their ends is the same, then the ratio of amounts of heat conducted per second through them will be

(a) 4 : 1 (b) 1 : 4 (c) 8 : 1 (b) 1 : 8

9. An ideal gas heat engine operates in a Carnot's cycle between 227°C and 127°C. It absorbs 6×10^4 J at high temperature. The amount of heat converted into work is

(a) 1.6×10^4 J (b) 1.2×10^4 J
(c) 4.8×10^4 J (d) 3.5×10^4 J

10. Effective capacitance between A and B in the figure shown is (all capacitances are in μF)

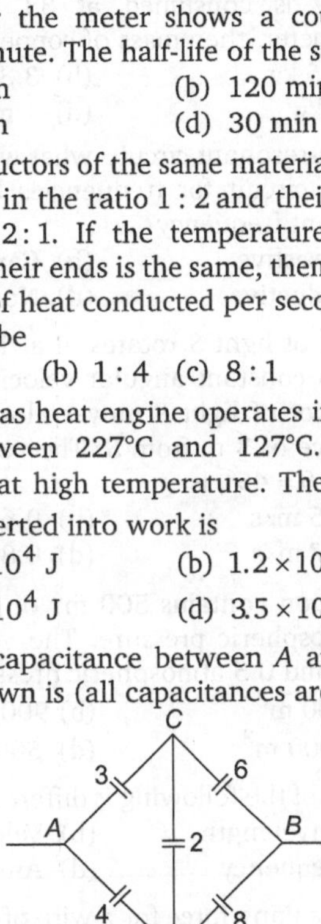

(a) $\frac{3}{14}$ μF (b) $\frac{14}{3}$ μF
(c) 21 μF (d) 23 μF

11. When the bob of a simple pendulum swings, the work done by tension in the string is
(a) > 0 (b) < 0
(c) zero (d) maximum

12. For a particle executing simple harmonic motion, the kinetic energy K is given by, $K = K_0 \cos^2 \omega t$. The maximum value of potential energy is
(a) K_0 (b) zero
(c) $K_0 / 2$ (d) not obtainable

13. A particle is moving along a circular path of radius 5 m with a uniform speed 5 ms^{-1}. What will be the average acceleration when the particle completes half revolution?
(a) Zero (b) 10 ms^{-2}
(c) 10π ms^{-2} (d) $\frac{10}{\pi}$ ms^{-2}

14. ECE of copper is 3.3×10^{-7} kg/C. If 100 kWh energy is consumed at 33 V in a copper voltameter, then mass of copper liberated is
(a) 3.6 kg (b) 3.3 kg
(c) 1 kg (d) 1 mg

15. In non-resonant circuit, what will be the nature of the circuit for frequencies higher than the resonant frequency?
(a) Resistive (b) Capacitive
(c) Inductive (d) None of these

16. A spot of light S rotates in a horizontal plane with a constant angular velocity of 0.1 rad/s. The spot of light P moves along the wall at a distance of 3 m from S. The velocity of spot P, where $\theta = 45°$, is
(a) 0.5 m/s (b) 0.6 m/s
(c) 0.7 m/s (d) 0.8 m/s

17. A balloon contains 500 m^3 of He at 27°C and 1 atmospheric pressure. The volume of He at −3°C and 0.5 atmospheric pressure will be
(a) 700 m^3 (b) 900 m^3
(c) 1000 m^3 (d) 500 m^3

18. Which of the following is different from others?
(a) Wavelength (b) Velocity
(c) Frequency (d) Amplitude

19. The breaking force for a wire of diameter D of a material is F. The breaking force for a wire of the same material of radius D is
(a) F (b) $2F$
(c) $\frac{F}{4}$ (d) $4F$

20. If pressure of a gas contained in a closed vessel is increased by 0.4% when heated by 1°C, its initial temperature must be
(a) 250 K (b) 250°C
(c) 2500 K (d) 25°C

21. In an electromagnetic wave, the electric and magnetising fields are 100 V/m and 0.265 A/m. The maximum energy flow is
(a) 26.5 W/m^2 (b) 36.5 W/m^2
(c) 46.7 W/m^2 (d) 765 W/m^2

22. A bullet of mass 10 g is fired from a gun of mass 1 kg with recoil velocity of gun is 5 m/s. The muzzle velocity will be
(a) 30 km/min (b) 60 km/min
(c) 30 m/s (d) 500 m/s

23. Work done by air when it expands from 50 L to 150 L at a constant pressure of 2 atmosphere is
(a) 2×10^4 J (b) 2×100 J
(c) 2×10^{11} J (d) 2×10^{15} J

24. In an electron gun, the control grid is given a negative potential, relative to cathode in order to
(a) accelerate the electrons
(b) decrease KE of electrons
(c) repel the electrons
(d) decelerate the electrons

25. In an LCR circuit having $L = 8$ H, $C = 0.5\mu$F and $R = 100\Omega$ in series, the resonance frequency in rad/s is
(a) 600 (b) 200
(c) $250/\pi$ (d) 500

26. Power of a lens is −4 D and for second lens, power is +2D; the total power of the couple is
(a) −2 D (b) 6 D
(c) −6 D (d) −8 D

27. In which one of the following regions of the electromagnetic spectrum will the vibrational motion of molecules give rise to absorption?
(a) Ultraviolet (b) Microwave
(c) Infrared (d) Radio wave

28. During inelastic collision of two particles
(a) $(KE)_{final} = (KE)_{initial}$
(b) $(KE)_{final}$ must be greater than $(KE)_{initial}$
(c) $(KE)_{final}$ must be less than $(KE)_{initial}$
(d) $(KE)_{final}$ must be greater or less than $(KE)_{initial}$

29. Two gases are at absolute temperatures 300 K and 350 K respectively. The ratio of average kinetic energy of their molecules is
(a) 6 : 7 (b) 36 : 49
(c) 7 : 6 (d) 343 : 216

30. The current in a coil changes from 0 to 2 A in 0.05 s. If the induced emf is 80 V, the self-inductance of the coil is,
(a) 1 H (b) 0.5 H
(c) 1.5 H (d) 2 H

31. Equal volumes of monoatomic and diatomic gases at the same temperature are given equal quantities of heat. Then,
(a) the temperature of diatomic gas will be more
(b) the temperature of monoatomic gas will be more
(c) the temperature of both will be zero
(d) nothing can be said

32. A copper wire and a steel wire of the same diameter and length are joined end to end and a force is applied which stretches their combined length by 1 cm. Then, the two wires will have
(a) the same stress and strain
(b) the same stress but different strains
(c) the same strain but different stresses
(d) different stresses and strains

33. The length of the second's pendulum is decreased by 0.3 cm when it is shifted to Chennai from London. If the acceleration due to gravity at London is 981cm/s^2, the acceleration due to gravity at Chennai is
(a) 981 cm/s^2 (b) 978 cm/s^2
(c) 984 cm/s^2 (d) 975 cm/s^2

34. How does, the electric field (E) between the plates of a charged cylindrical capacitor vary with the distance r from the axis of the cylinder?
(a) $E \propto \frac{1}{r^2}$ (b) $E \propto \frac{1}{r}$
(c) $E \propto r^2$ (d) $E \propto r$

35. A thin bar magnet of length $2L$ is bent at the mid-point so that the angle between them is 60°C. The new length of the magnet is
(a) $\sqrt{2}\,L$ (b) $\sqrt{3}\,L$
(c) $2\,L$ (d) L

36. A capacitor is connected to a cell of emf E and some internal resistance r. The potential difference across the
(a) cell is E (b) cell is $< E$
(c) capacitor is $< E$ (d) capacitor is $> E$

37. The radius of the smallest electron orbit in hydrogen-like ion is $(0.51 \times 10^{-10} / 4)$m, then it is
(a) hydrogen atom (b) He^+
(c) Li^{2+} (d) Be^{3+}

38. de-Broglie wavelength of a body of mass 1 kg moving with velocity of 2000 m/s is
(a) 3.32×10^{-27} Å (b) 1.5×10^7 Å
(c) 0.55×10^{-22} Å (d) None of these

39. A voltmeter of resistance 998 Ω is connected across a cell of emf 2 V and internal resistance 2Ω. The potential difference across the voltmeter is
(a) 1.99 V (b) 3.5 V (c) 5 V (d) 6 V

40. If a person with a spring balance and a body hanging from it goes up and up in an aeroplane, then the reading of the weight of the body as indicated by the spring balance will
(a) goes on increasing
(b) goes on decreasing
(c) first increase and then decrease
(d) remains the same

41. A mass $3m$, initially at rest at the origin explodes into three fragments of equal mass. Two of the fragments have speed v each and move perpendicular to each other. The third fragment will move with a speed
(a) $\frac{v}{\sqrt{2}}$ (b) $\frac{v}{2}$ (c) v (d) $v\sqrt{2}$

42. A nucleus X initially at rest, undergoes alpha decay according to the equation

$$_{92}X^A \rightarrow {}_ZY^{228} + \alpha$$

Then, the value of A and Z are
(a) 94, 230 (b) 232,90
(c) 190,32 (d) 230,94

43. The energy gap between the valence band and the conduction band for a material is 6 eV. The material is
(a) an insulator
(b) a metal
(c) an intrinsic semiconductor
(d) a superconductor

44. When the inputs of a two input logic gate are 0 and 0, the output is 1. When the inputs are 1 and 0 the output is 0. The logic gate is of the type
(a) AND (b) NAND
(c) NOR (d) OR

45. If a transistor biased in the common-emitter mode, the emitter current is
(a) much smaller than base curent
(b) much larger than base current
(c) nearly equal to the base current
(d) much smaller than the collector current

46. A sound has an intensity of 2×10^{-8} W / m^2. Its intensity level in decibels is ($\log_{10} 2 = 0.3$)
(a) 23 (b) 3
(c) 43 (d) 4.3

47. A ray reflected successively from two plane mirrors inclined at a certain angle undergoes a deviation of 240°. Then, the number of images observable is
(a) 3 (b) 5
(c) 7 (d) 9

48. If 1 A of current is passed through $CuSO_4$ solution for 10 s, the number of copper atoms deposite at the cathode will be about
(a) 1.6×10^{20} (b) 8×10^{19}
(c) 3.1×10^{19} (d) 6.2×10^{19}

49. Two points A and B are situated at distance x and $2x$ respectively from the nearer pole of a magnet 2 cm long. The ratio of magnetic field at A and B is
(a) 4 : 1 exactly (b) 4 : 1 approximately
(c) 8 : 1 exactly (d) 8 : 1 approximately

50. Electromagnetic waves travel only through
(a) material medium
(b) vacuum
(c) oscillating electric and magnetic fields
(d) oscillating electric and magnetic fields whose directions are perpendicular to each other

51. The ratio of magnetic moments of two short magnets which give null deflection in tan B position at 12 cm and 18 cm from the centre of a deflection magnetometer is
(a) 2 : 3 (b) 8 : 27
(c) 27 : 8 (d) 4 : 9

52. A 220 V main supply is connected to a resistance of 100 kΩ. The effective current is
(a) 2.2 mA (b) $2.2\sqrt{2}$ mA
(c) $\frac{2.2}{\sqrt{2}}$ mA (d) None of these

53. Two bodies of different masses m_1 and m_2 have equal momenta. Their kinetic energies E_1 and E_2 are in the ratio
(a) $\sqrt{m_1} : \sqrt{m_2}$ (b) $m_1 : m_2$
(c) $m_2 : m_1$ (d) $m_1^2 : m_2^2$

54. When the light enters from air to glass, for which colour the angle of deviation is maximum?
(a) Red (b) Yellow
(c) Blue (d) Violet

55. Two waves are represented by $y_1 = A \sin (kx - \omega t)$ and $y_2 = A \cos (kx - \omega t)$. The amplitude of resultant wave is
(a) $4A$ (b) $2A$
(c) $\sqrt{2}A$ (d) A

56. An atom of mass M which is in the state of rest emits a photon of wavelength λ. As a result, the atom will deflect with the kinetic energy equal to (h is Planck's constant)
(a) $\frac{h^2}{M\lambda^2}$ (b) $\frac{1}{2}\frac{h^2}{M\lambda^2}$
(c) $\frac{h}{M\lambda}$ (d) $\frac{1}{2}\frac{h}{M\lambda}$

57. If the angular momentum of any rotating body increases by 200%, then the increase in its kinetic energy is
(a) 400% (b) 800%
(c) 200% (d) 100%

58. A cylindrical tank has a hole of 1 cm^2 in its bottom. If the water is allowed to flow into the tank from a tube above it at the rate of 70 cm^2/s, then the maximum height up to which water can rise in the tank is
(a) 2.5 cm (b) 5 cm
(c) 10 cm (d) 0.25 cm

59. X-rays and γ-rays of the same energies may be distinguished by
(a) their velocity
(b) their ionising power
(c) their intensity
(d) method of production

60. Which of the following materials is non-crystalline?
(a) Copper (b) Sodium chloride
(c) Wood (d) Diamond

Chemistry

1. The vapour density of N_2O_4 at a certain temperature is 30. What is the percentage dissociation of N_2O_4 at this temperature?
(a) 46% (b) 53%
(c) 75% (d) 92%

2. Four different colloids have the following gold number, which one has the most effective action?
(a) 5 (b) 10
(c) 15 (d) 13

3. In which of the following case, increase in concentration of ion cause increase in E_{cell} ?
(a) $Pt(H_2)|H^+(aq)$
(b) $Pt|$ Quinhydrone $|H^+(aq)$
(c) $Ag|Ag^+(aq)$
(d) $Ag, AgCl|Cl^-(aq)$

4. The relative lowering of vapour pressure of an aqueous solution containing a non volatile solute is 0.0125. The molality of the solution is
(a) 0.70 (b) 0.30
(c) 0.125 (d) 0.07

5. Which of the following is known as Glauber's salt?
(a) $Na_2S_2O_3 \cdot 5H_2O$ (b) $Na_2SO_4 \cdot 10H_2O$
(c) $Na_2CO_3 \cdot 10H_2O$ (d) $Na(NH_4)HPO_4$

6. $PhCHClBr \xrightarrow{tBuO^-}$ Carbene
In the above reaction which of the following will be most probable?
(a) $\ddot{C}HBr$ (b) $\ddot{C}HCl$
(c) $\ddot{C}PhBr$ (d) $\ddot{C}PhCl$

7. Which of the following alcohol is unable to turn orange colour of chromic acid to green?
(a) 1° alcohol (b) 2° alcohol
(c) 3° alcohol (d) Allyl alcohol

8. Which of the following test is performed by carbohydrates?
(a) Biuret test (b) Nitroprusside test
(c) Ninhydrin test (d) Molisch's test

9. Which of the following is not used as propellant?
(a) Hydrazine (b) UDMH
(c) MMH (d) Phenyl hydrazine

10. Which of the following species have the same number of electrons in its outermost as well as penultimate shell?
(a) Cl^- (b) O^{2-}
(c) Na^+ (d) Mg^{2+}

11. Which of the following is the formula of Sorel's cement?
(a) $KCl \cdot MgCl_2 \cdot 6H_2O$
(b) $MgCl_2 \cdot 5MgO \cdot xH_2O$
(c) $CaSO_4 \cdot 2H_2O$
(d) $MgCO_3 \cdot CaCO_3$

12. $S_{(rhombic)} + O_2(g) \longrightarrow SO_2(g)$, $\Delta H = -297.5$ kJ
$S_{(monoclinic)} + O_2(g) \longrightarrow SO_2(g)$, $\Delta H = -300$ kJ
The data can predict that
(a) rhombic sulphur is yellow in colour
(b) monoclinic sulphur has metallic lusture
(c) monoclinic sulphur is more stable
(d) $\Delta H_{transition}$ of S_R to S_M is endothermic

13. Which of the following is a measurement of water pollution?
(a) PSC (b) PCB
(c) BOD (d) COD

14. The half-life period of a first order process is 1.6 min. It will be 90% complete in
(a) 5.3 min (b) 10.6 min
(c) 43.3 min (d) 99.7 min

15. Which of the following is tetrabasic acid?
(a) Orthophosphorus acid
(b) Orthophosphoric acid
(c) Metaphosphoric acid
(d) Pyrophosphoric acid

16. Which of the following order for bond angles is correct?
(a) $BF_3 < NF_3 < PF_3 < ClF_3$
(b) $BF_3 < NF_3 < PF_3 > ClF_3$
(c) $ClF_3 < PF_3 < NF_3 < BF_3$
(d) $BF_3 \approx NF_3 < PF_3 < ClF_3$

17. The triad of the nuclei that is isotonic, is
(a) ${}_6C^{14}, {}_7N^{14}, {}_9F^{19}$ (b) ${}_6C^{14}, {}_7N^{15}, {}_9F^{17}$
(c) ${}_6C^{14}, {}_7N^{14}, {}_9F^{17}$ (d) ${}_6C^{12}, {}_7N^{14}, {}_9F^{19}$

18. The oxidation number of cobalt in $K[Co(CO)_4]$ is
(a) −1 (b) −3 (c) +1 (d) +3

19. Out of Cu, Ag, Fe and Zn, the metal which can displace all others from their salt solution, is
(a) Zn (b) Cu
(c) Ag (d) Fe

20. Which of the following has magic number of neutrons?
(a) $_{13}Al^{27}$ (b) $_{26}Fe^{56}$
(c) $_{83}Bi^{209}$ (d) $_{92}U^{238}$

21. If K_a of HCN $= 4 \times 10^{-10}$, then the pH of 2.5×10^{-1} molar HCN(*aq*) is
(a) 1 (b) 2.5
(c) 4 (d) 5

22. A coffee cup calorimeter initially contains 125 g of water at a temperature of 24.2°C. After adding of 10.5 g of KBr, the temperature becomes 21.1°C. The heat of solution is
(a) 40 J/g (b) 117 J/g
(c) 167.7 J/g (d) 420.05 J/g

23. What product is formed when cyclohexanone is oxidised?
(a) $HOOC \cdot (CH_2)_4 \cdot COOH$
(b) CH_3CH_2COOH
(c) $CH_3(CH_2)_4 \cdot COOH$
(d) None of the above

24. The reaction
$$CH_3CN + 4H \xrightarrow{Na/C_2H_5OH} CH_3CH_2NH_2$$ is
known as
(a) Hofmann's bromamide reaction
(b) Mendius reaction
(c) Perkin's reaction
(d) Sabatier reaction

25. Action of Br_2 on cyclopentene gives
(a) cyclopentyl bromide
(b) 1,2-dibromo cyclopentane
(c) cyclopentyl dibromide
(d) no reaction

26. In a face-centred cubic arrangement of *A* and *B* atoms whose *A* atoms are at the corner of the unit cell and *B* atoms at the face-centre, one of *A* atom is missing from one corner in the unit cell. The simplest formula of the compound is
(a) $A_{24}B_7$ (b) A_7B_{24}
(c) A_7B_4 (d) AB_2

27. Beryllium shows diagonal relationship with
(a) B (b) Mg
(c) Al (d) Na

28. A compound has the molecular formula X_4O_6. If 10 g of X_4O_6 has 5.72 g *X*, then calculate atomic mass of *X*.
(a) 32 amu (b) 64 amu
(c) 67 amu (d) 98 amu

29. Which of the following is Rinmann's green?
(a) ZnO (b) $ZnS + BaSO_4$
(c) $ZnSO_4 \cdot 7H_2O$ (d) $ZnO \cdot CoO$

30. For the valence electron in copper, the four quantum numbers are
(a) $n = 4, l = 0, m = 0, s = \frac{1}{2}$
(b) $n = 4, l = 0, m = +2, s = \frac{1}{2}$
(c) $n = 4, l = 1, m = +2, s = \frac{1}{2}$
(d) $n = 4, l = 1, m = 0, s = \frac{1}{2}$

31. Geometry of BF_3 is
(a) trigonal planar (b) pentagonal
(c) trihedral (d) tetrahedral

32. Which of the following will represent the normality of 10% (*w/V*) acetic acid?
(a) 1.66 N (b) 16.6 N
(c) 3.32 N (d) 33.2 N

33. The oxidation number of sulphur in $H_2S_2O_8$ is
(a) +2 (b) +4
(c) +6 (d) +7

34. Which of the following reaction requires catalyst?
(a) $S + O_2 \longrightarrow SO_2$
(b) $C + O_2 \longrightarrow CO_2$
(c) $2SO_2 + O_2 \longrightarrow 2SO_3$
(d) All of the above

35. $N_2(g) + 3H_2(g) \rightleftharpoons 2NH_3(g) + 22$ kcal
The activation energy for the forward reaction is 50 kcal. What is the activation energy for the backward reaction?
(a) 72 kcal (b) 28 kcal
(c) −72 kcal (d) −28 kcal

36. Malonic acid and succinic acid are distinguished by
(a) heating (b) $NaHCO_3$
(c) both (a) and (b) (d) None of these

37. When chloroform is treated with chlorine in the presence of sunlight, it yields
(a) urotropine (b) pyrene
(c) chloropicrin (d) chloritone

38.
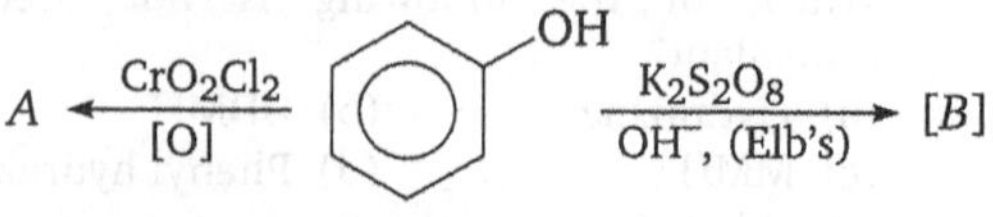

Which of the following are *A* and *B*?

(a) Both are 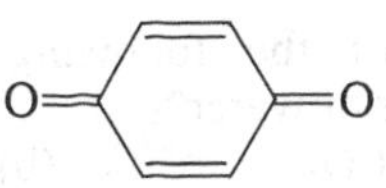

(b) Both are 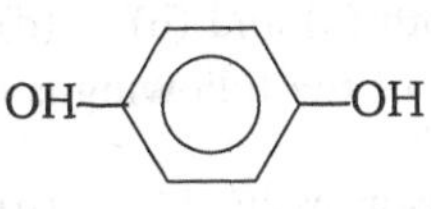

(c) 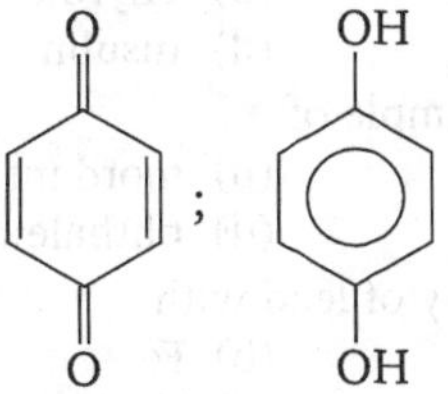 respectively

(d) 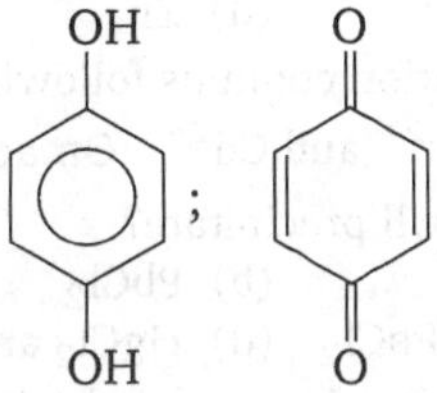 respectively

39. Which of the following will explain synthesis of RNA?
(a) Translation (b) Coagulation
(c) Transcription (d) Replication

40. Which of the following is a narrow spectrum antibiotics?
(a) Penicillin G (b) Dettol
(c) Aspirin (d) Analgin

41. When aniline reacts with oil of bitter almonds, condensation takes place and benzyl derivatives are formed. These are known as
(a) Millon's base (b) Schiff's reagent
(c) Benedict reagent (d) Schiff's base

42. 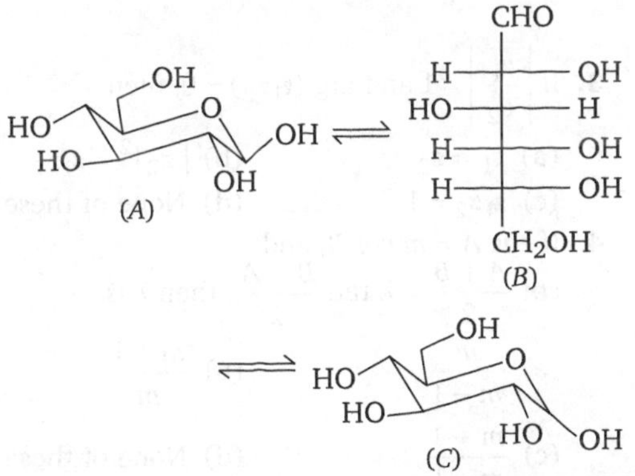

Which of the following is correct about (*A*), (*B*) and (*C*)?
(a) *A*,*B* and *C* are anomers
(b) *A* and *C* are anomers
(c) *A*, *B* and *C* are metamers
(d) None of the above

43. The IUPAC name of 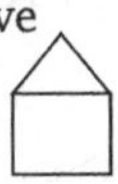 is
(a) bicyclo (2,1,0) pentane
(b) 1,2-cyclopropyl cyclobutane
(c) 1,2-methylene cyclobutane
(d) cyclopentane (4,3) annulene

44. In the conversion of ethanol into methanol which of the following reagents will be used?
(a) $K_2Cr_2O_7/H_2SO_4$ (b) $NaOH + CaO$
(c) Cl_2 + aq. KOH (d) All of these

45. PDI for natural polymers is generally close to
(a) zero (b) 1
(c) 10 (d) 100

46. Which of the following is diamagnetic?
(a) H_2 (b) H_2^- (c) H_2^+ (d) He_2^+

47. Which of the following is a property of non-metal?
(a) Electronegativity
(b) Basic nature of oxide
(c) Reducing property
(d) Low ionisation potential

48. One mole of ice is converted into water at 273 K. The entropies of $H_2O(s)$ and $H_2O(l)$ are 38.20 and 60.01 J mol^{-1} K^{-1} respectively. The enthalpy change for the conversion is
(a) 3 kJ mol^{-1} (b) 4 kJ mol^{-1}
(c) 5 kJ mol^{-1} (d) 6 kJ mol^{-1}

49. Complete the following reaction,

$$P_4 + NaOH + H_2O \longrightarrow ?$$

(a) $PH_3 + Na_2HPO_2$ (b) $PH_3 + NaH_2PO_2$
(c) $H_3PO_4 + NaO$ (d) $PH_3 + Na_2PO_4$

50. The elevation in boiling point would be highest for
(a) 0.08 M $BaCl_2$ (b) 0.15 M KCl
(c) 0.10 M Glucose (d) 0.06 M $Ca(NO_3)_2$

51. What is the magnetic moment of $[FeF_6]^{3-}$?
(a) 2 BM (b) 5.9 BM
(c) 7 BM (d) 35 BM

52. In the given reaction, identify compound *C*?

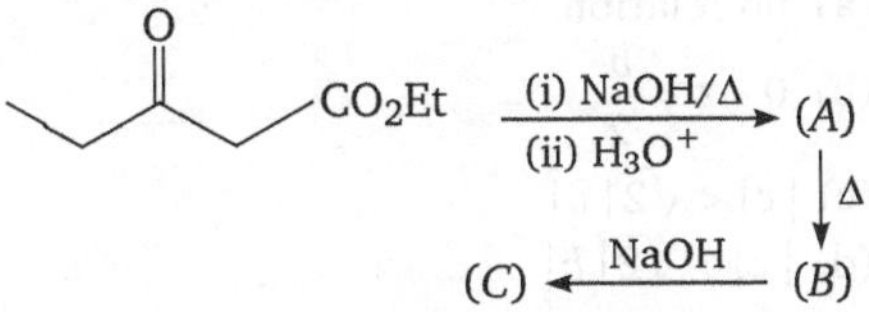

(a) 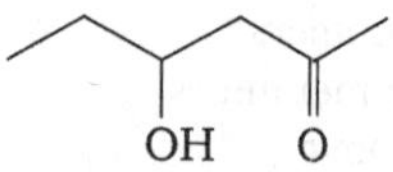

(b) 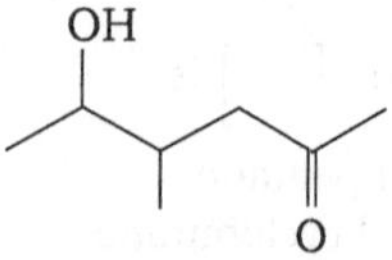

(c) 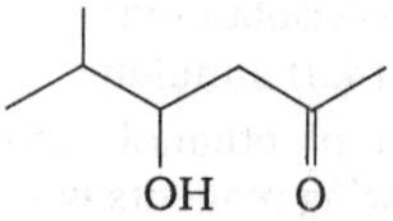

(d) 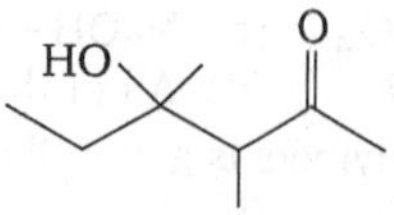

53. The ratio of rate of diffusion of hydrogen and helium gas is
(a) 1 : 1.4 (b) 1 : 1 (c) 1.4 : 1 (d) 1 : 2

54. The reaction,

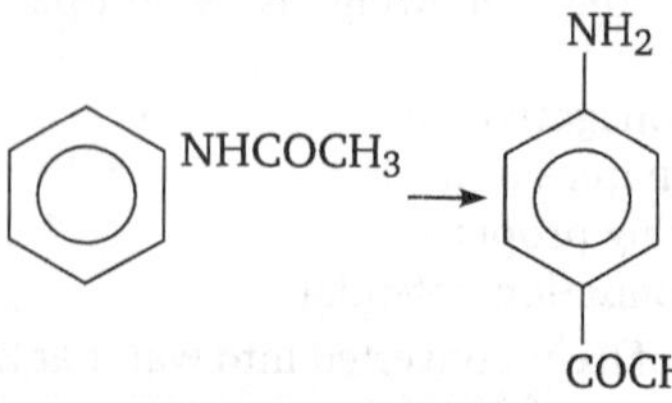

is known as

(a) substitution reaction
(b) addition reaction
(c) elimination reaction
(d) None of the above

55. Which of the following is responsible for hardness of water?
(a) $CaHCO_3$ (b) $CaCl_2$
(c) Both (a) and (b) (d) None of these

56. Which of the following is differ in nature from others?
(a) Vasopressin (b) Thyroid
(c) Oxytocin (d) Insulin

57. Indigo is an example of
(a) vat dye (b) mordant dye
(c) azo dye (d) phthalein dye

58. Solder is an alloy of lead with
(a) Cu (b) Fe
(c) Sn (d) Zn

59. An aqueous solution contains following ions : Hg_2^{2+}, Hg^{2+}, Pb^{2+} and Cd^{2+}. On adding HCl, which of these will precipitate?
(a) Hg_2Cl_2 (b) $PbCl_2$
(c) Hg_2Cl_2 and $PbCl_2$ (d) $HgCl_2$ and $PbCl_2$

60. Which of the following molecules/species are aromatic in nature?

(a) 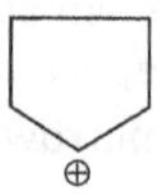

(b) 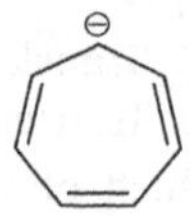

(c) 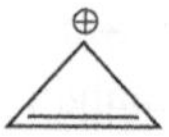

(d) 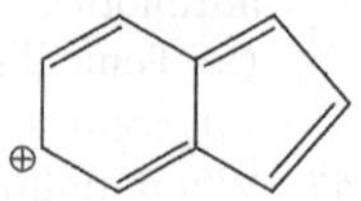

Mathematics

1. The sum of the series $1 + \frac{2^2}{2!} + \frac{3^2}{3!} + \frac{4^2}{4!} + \ldots$ is
(a) $2e$ (b) $3e$
(c) $2e - 1$ (d) None of these

2. If $f(x) = x^2 + 2bx + 2c^2$
and $g(x) = -x^2 - 2cx + b^2$
are such that min $f(x) >$ max $g(x)$, then relation between b and c is
(a) no relation
(b) $0 < c < \frac{b}{2}$
(c) $|c| < \sqrt{2}|b|$
(d) $|c| > \sqrt{2}|b|$

3. If $\left|\frac{z_1}{z_2}\right| = 1$ and $\arg(z_1 z_2) = 0$, then
(a) $z_1 = z_2$ (b) $|z_2|^2 = z_1 z_2$
(c) $z_1 z_2 = 1$ (d) None of these

4. If $\cos A = m \cos B$, and $\cot \frac{A+B}{2} = \lambda \tan \frac{B-A}{2}$, then λ is
(a) $\frac{m}{m-1}$ (b) $\frac{m+1}{m}$
(c) $\frac{m+1}{m-1}$ (d) None of these

5. In a triangle, the angle are in AP and the lengths of the two larger sides are 10 and 9, respectively, then the length of the third side can be

(a) $5 \pm \sqrt{6}$ (b) 0.7
(c) $15 - \sqrt{6}$ (d) None of these

6. The complex number $z = x + iy$ which satisfy the equation $\left|\frac{z - 5i}{z + 5i}\right| = 1$, lie on

(a) the x-axis
(b) the straight line $y = 5$
(c) a circle passing through the origin
(d) None of the above

7. If $x > 1, y > 1, z > 1$ are in GP, then $\frac{1}{1 + \ln x}$, $\frac{1}{1 + \ln y}$, $\frac{1}{1 + \ln z}$ are in

(a) AP (b) HP
(c) GP (d) None of these

8. If $\frac{e^x}{1 - x} = B_0 + B_1 x + B_2 x^2 + \ldots + B_n x^n + \ldots$, then $B_n - B_{n-1}$ equals

(a) $\frac{1}{n!}$ (b) $\frac{1}{(n-1)!}$
(c) $\frac{1}{n!} - \frac{1}{(n-1)!}$ (d) 1

9. If $f(x) = \begin{cases} \frac{\sin [x]}{[x]}, & [x] \neq 0 \\ 0, & [x] = 0 \end{cases}$

where $[x]$ denotes the greatest integer less than or equal to x, then $\lim_{x \to 0} f(x)$ equals

(a) 1 (b) 0
(c) -1 (d) None of these

10. If $y(t)$ is a solution of $(1 + t)\frac{dy}{dt} - ty = 1$ and $y(0) = -1$, then $y(1)$ equal to

(a) $-\frac{1}{2}$ (b) $e + \frac{1}{2}$
(c) $e - \frac{1}{2}$ (d) $\frac{1}{2}$

11. The locus of the mid-points of the chord of the circle $x^2 + y^2 = 4$ which subtends a right angle at the origin, is

(a) $x + y = 2$ (b) $x^2 + y^2 = 1$
(c) $x^2 + y^2 = 2$ (d) $x + y = 1$

12. From a point A common tangents are drawn to the circle $x^2 + y^2 = a^2/2$ and parabola $y^2 = 4ax$. The area of the quadrilateral formed by the common tangents to the chord of contact of the parabola is

(a) $\frac{15a^3}{4}$ sq units (b) $\frac{15a^2}{4}$ sq units
(c) $\frac{7a^2}{4}$ sq units (d) $\frac{a^2}{4}$ sq unit

13. Let $\omega = -\frac{1}{2} + i\frac{\sqrt{3}}{2}$, then value of the determinant $\Delta = \begin{vmatrix} 1 & 1 & 1 \\ 1 & -1 - \omega^2 & \omega^2 \\ 1 & \omega^2 & \omega \end{vmatrix}$ is

(a) 3ω (b) $3\omega(\omega - 1)$
(c) $3\omega^2$ (d) $3\omega(1 - \omega)$

14. The harmonic mean of the root of the equation $(5 + \sqrt{2})x^2 - (4 + \sqrt{5})x + 8 + 2\sqrt{5} = 0$ is

(a) 2 (b) 4
(c) 6 (d) 8

15. Domain of the function $\log |x^2 - 9|$ is

(a) R (b) $R - [-3, 3]$
(c) $R - \{-3, 3\}$ (d) None of these

16. Let $f : R \to R$ be such that $f(1) = 3$ and $f'(1) = 6$. Then, $\lim_{x \to 0} \left[\frac{f(1 + x)}{f(1)}\right]^{1/x}$ equals

(a) 1 (b) $e^{1/2}$
(c) e^2 (d) e^3

17. The value of $[\mathbf{a}\ \mathbf{a} + \mathbf{b}\ \mathbf{a} + \mathbf{b} + \mathbf{c}]$ is

(a) $[\mathbf{a}\,\mathbf{b}\,\mathbf{c}]^2$ (b) $[\mathbf{a}\,\mathbf{b}\,\mathbf{c}]$
(c) $2[\mathbf{a}\,\mathbf{b}\,\mathbf{c}]$ (d) $3[\mathbf{a}\,\mathbf{b}\,\mathbf{c}]$

18. If $\mathbf{a} = -\mathbf{i} + 2\mathbf{j} - \mathbf{k}$, $\mathbf{b} = \mathbf{i} + \mathbf{j} - 3\mathbf{k}$ and $\mathbf{c} = -4\mathbf{i} - \mathbf{k}$, then $\mathbf{a} \times (\mathbf{b} \times \mathbf{c}) + (\mathbf{a} \cdot \mathbf{b})\mathbf{c}$

(a) $5\mathbf{i} + 5\mathbf{j} - 15\mathbf{k}$ (b) 0
(c) $12\mathbf{j} + 4\mathbf{k}$ (d) $-3\mathbf{i} + 6\mathbf{j} - 3\mathbf{k}$

19. If $I(m, n) = \int_0^1 t^m (1 + t)^n \, dt$, then the expression for $I(m, n)$ in terms of $I(m + 1, n - 1)$ is

(a) $\frac{2^n}{m + 1} - \frac{n}{m + 1} I(m + 1, n - 1)$

(b) $\frac{n}{m + 1} I(m + 1, n - 1)$

(c) $\frac{2^n}{m + 1} + \frac{n}{m + 1} I(m + 1, n - 1)$

(d) $\frac{m}{m + 1} I(m + 1, n - 1)$

20. The ΔPQR is inscribed in the circle $x^2 + y^2 = 25$. If Q and R have coordinates (3, 4) and (–4, 3), respectively. Then, $\angle QPR$ is equal to

(a) $\frac{\pi}{2}$ (b) $\frac{\pi}{3}$

(c) $\frac{\pi}{4}$ (d) $\frac{\pi}{6}$

21. The equation of directrix of the parabola $y^2 + 4y + 4x + 2 = 0$ is

(a) $x = -1$ (b) $x = 1$

(c) $x = -\frac{3}{2}$ (d) $x = \frac{3}{2}$

22. If $a > 2b > 0$, then positive value of m for which $y = mx - b\sqrt{1 + m^2}$ is a common tangent to $x^2 + y^2 = b^2$ and $(x - a)^2 + y^2 = b^2$ is

(a) $\frac{2b}{\sqrt{a^2 - 4b^2}}$ (b) $\frac{\sqrt{a^2 - 4b^2}}{2b}$

(c) $\frac{2b}{a - 2b}$ (d) $\frac{b}{a - 2b}$

23. If $\alpha \in \left[0, \frac{\pi}{2}\right)$, then $\sqrt{x^2 + x} + \frac{\tan^2 \alpha}{\sqrt{x^2 + x}}$ is always greater than or equal to

(a) $2 \tan \alpha$ (b) 1

(c) 2 (d) $\sec^2 \alpha$

24. If an edge of a cube measure 2 m with a possible error of 0.5 cm. Find the corresponding error in the calculated volume of the cube

(a) 0.6 m^3 (b) 0.06 m^3

(c) 0.006 m^3 (d) 0.0006 m^3

25. Let f be twice differentiable function satisfying $f(1) = 1$, $f(2) = 4$, $f(3) = 9$, then

(a) $f''(x) = 2, \forall x \in R$

(b) $f'(x) = 5 = f''(x)$. For some $x \in (1, 3)$

(c) there exists atleast one $x \in (1, 3)$ such that $f''(x) = 2$

(d) None of the above

26. The order of the differential equation whose general solution is given by

$$y = (C_1 + C_2) \cos (x + C_3) - C_4 e^{x + C_5}$$

where, C_1, C_2, C_3, C_4, C_5 are arbitrary constants, is

(a) 5 (b) 4

(c) 3 (d) 2

27. Evaluate $\int (\sqrt{\tan x} + \sqrt{\cot x})\, dx$

(a) $\sqrt{2} \tan^{-1} (\sqrt{\tan x} - \sqrt{\cot x}) + C$

(b) $\frac{2}{\sqrt{2}} \tan^{-1} x + C$

(c) $\sqrt{2} \tan^{-1} \left(\frac{\sqrt{\tan x} - \sqrt{\cot x}}{\sqrt{2}}\right) + C$

(d) None of the above

28. In an election, number of candidates exceeds the number to be elected by 2. A man can vote in 56 ways. Then, the number of candidates is

(a) 5 (b) 6

(c) 7 (d) 8

29. The curve described parametrically by $x = t^2 + t + 1$, $y = t^2 - t + 1$ represents

(a) a pair of straight lines

(b) an ellipse

(c) a parabola

(d) a hyperbola

30. Tangent is drawn to ellipse $\frac{x^2}{27} + \frac{y^2}{1} = 1$ at $(3\sqrt{3} \cos \theta, \sin \theta)$, where $\left(\theta \in \left(0, \frac{\pi}{2}\right)\right)$. Then, the value of θ such that the sum of intercept on axes made by this tangent is minimum, is

(a) $\frac{\pi}{3}$ (b) $\frac{\pi}{6}$

(c) $\frac{\pi}{8}$ (d) $\frac{\pi}{4}$

31. Suppose a, b, c are in AP and a^2, b^2, c^2 are in GP. If $a < b < c$ and $a + b + c = \frac{3}{2}$, then the value of a is

(a) $\frac{1}{2\sqrt{2}}$ (b) $\frac{1}{2\sqrt{3}}$

(c) $\frac{1}{2} - \frac{1}{\sqrt{3}}$ (d) $\frac{1}{2} - \frac{1}{\sqrt{2}}$

32. The function $f(x) = \frac{\ln (1 + ax) - \ln (1 - bx)}{x}$ is not defined at $x = 0$. The value which should be assigned to f at $x = 0$ so that it is continuous at $x = 0$, is

(a) $a - b$ (b) $a + b$

(c) $\ln a + \ln b$ (d) None of these

33. Domain of definition of function $f(x) = \sqrt{\sin^{-1} (2x) + \frac{\pi}{6}}$ for real-valued of x, is

(a) $\left[-\frac{1}{4}, \frac{1}{2}\right]$ (b) $\left[-\frac{1}{2}, \frac{1}{2}\right]$

(c) $\left[-\frac{1}{2}, \frac{1}{9}\right]$ (d) $[0, \pi]$

34. If $x\,dy = y(dx + y\,dy)$, $y(1) = 1$ and $y(x) > 0$. Then, $y(-3)$ is equal to

(a) 3 (b) 2

(c) 1 (d) 0

35. Evaluate $\int \frac{(x-1)e^x}{(x+1)^3}\,dx$

(a) 0

(b) $\frac{(x+1)^2}{e^x} + C$

(c) $\frac{e^x}{(x+1)^2} + C$

(d) $\frac{e^x}{(x-1)^2} + (x+1)^2 + C$

36. If $a = \log_2 3$, $b = \log_2 5$, $c = \log_7 2$, then $\log_{140} 63$ in terms of a, b, c is

(a) $\frac{2ac+1}{2a+abc+1}$ (b) $\frac{2ac+1}{2a+c+a}$

(c) $\frac{2ac+1}{2c+ab+a}$ (d) None of these

37. Local maximum value of the function $\frac{\log_e x}{x}$ is

(a) 1 (b) e

(c) $\frac{1}{e}$ (d) None of these

38. $A + iB$ form of $\frac{(\cos x + i \sin x)(\cos y + i \sin y)}{(\cot u + i)(1 + i \tan v)}$ is

(a) $\sin u \cos v [\cos (x + y - u - v) + i \sin (x + y - u - v)]$

(b) $\sin u \cos v [\cos (x + y + u + v) + i \sin (x + y + u + v)]$

(c) $\sin u \cos v [\cos (x + y + u + v) - i \sin (x + y + u + v)$

(d) None of the above

39. In the set Q^+ of all positive rational numbers, the operation * is defined by the formula $a * b = \frac{ab}{6}$. Then, the inverse of 9 with respect to * is

(a) 4 (b) 3

(c) $\frac{1}{9}$ (d) $\frac{1}{3}$

40. If $\alpha + \beta + \gamma = \pi$, then the value of $\sin^2 \alpha + \sin^2 \beta - \sin^2 \gamma$ is equal to

(a) $2 \sin \alpha$ (b) $2 \sin \alpha \cos \beta \sin \gamma$

(c) $2 \sin \alpha \sin \beta \cos \gamma$ (d) $2 \sin \alpha \sin \beta \sin \gamma$

41. If in a ΔABC

$$\frac{2\cos A}{a} + \frac{\cos B}{b} + \frac{2\cos C}{c} = \frac{a}{bc} + \frac{b}{ac},$$

then $b^2 + c^2$ is equal to

(a) a^2 (b) ac

(c) bc (d) None of these

42. In any ΔABC, the least value of $\left(\frac{\sin^2 A + \sin A + 1}{\sin A}\right)$ is

(a) 3 (b) $\sqrt{3}$

(c) 9 (d) None of these

43. Let λ and α be real

$$\lambda x + \sin \alpha \cdot y + \cos \alpha \cdot z = 0,$$
$$x + \cos \alpha \cdot y + \sin \alpha \cdot z = 0,$$
$$-x + \sin \alpha \cdot y - \cos \alpha \cdot z = 0$$

has a non-trivial solution. If $\lambda = 1$, then all the values of α is equals

(a) $n\pi + \frac{\pi}{2}$ (b) $n\pi + \frac{\pi}{4}, n\pi$

(c) $n\pi \pm \frac{\pi}{2}$ (d) $n\pi + \frac{\pi}{4}, \frac{\pi}{4}$

44. If the roots of the equation

$$x^2 - 2ax + a^2 + a - 3 = 0$$

are real and less than 3, then

(a) $a < 2$ (b) $2 \le a \le 3$

(c) $3 < a \le 4$ (d) $a > 4$

45. If z_1 and z_2 are two complex numbers such that $|z_1| = |z_2| + |z_1 - z_2|$, then

(a) $\text{Im}\left(\frac{z_1}{z_2}\right) = 0$ (b) $\text{Re}\left(\frac{z_1}{z_2}\right) = 0$

(c) $\text{Re}\left(\frac{z_1}{z_2}\right) = \text{Im}\left(\frac{z_1}{z_2}\right)$ (d) None of these

46. The number of five digits numbers that can be formed without any restriction is

(a) 990000 (b) 10000

(c) 90000 (d) None of these

47. The middle term in the expansion of $\left(x^2 + \frac{1}{x^2} + 2\right)^n$ is

(a) $\frac{n}{[(n/2)!]^2}$

(b) $-\frac{(2n!)}{[(n/2)!]^2}$

(c) $\frac{1\cdot3\cdot5\ldots(2n+1)}{n!}2^n$

(d) $\frac{(2n)!}{(n!)^2}$

48. The value of sin (22°30′) is

(a) $\sqrt{\frac{\sqrt{2}+1}{2\sqrt{2}}}$ (b) $\sqrt{\sqrt{2}-1}$

(c) $\sqrt{\frac{\sqrt{2}-1}{2\sqrt{2}}}$ (d) None of these

49. If $\frac{\pi}{2}<\theta<\frac{3\pi}{2}$, then the value of $\sqrt{\frac{1-\sin\theta}{1+\sin\theta}}$ is equal to

(a) $-\sec\theta+\tan\theta$ (b) $\sec\theta+\tan\theta$

(c) $\sec\theta-\tan\theta$ (d) None of these

50. If $\int\frac{2^x}{\sqrt{1-4^x}}dx=k\sin^{-1}(2^x)+C$, then k is equal to

(a) log 2 (b) $\frac{1}{2}\log 2$

(c) $\frac{1}{2}$ (d) $\frac{1}{\log 2}$

51. The area of the region bounded by $y=-1$, $y=2$, $x=y^3$ and $x=0$ is

(a) $\frac{17}{4}$ sq units (b) $\frac{1}{4}$ sq unit

(c) 4 sq units (d) None of these

52. The mid-point of the line joining the common points of the line $2x-3y+8=0$ and $y^2=8x$ is

(a) (3, 2) (b) (5, 6)

(c) (4, −1) (d) (2, −3)

53. The value of $\cos^2 48°-\sin^2 12°$ is

(a) $\frac{\sqrt{5}+1}{8}$ (b) $\frac{\sqrt{5}-1}{8}$

(c) $\frac{\sqrt{5}+1}{16}$ (d) None of these

54. The number of solutions of $\sin^2\theta+3\cos\theta=3$ in $[-\pi,\pi]$, is

(a) 4 (b) 2

(c) 0 (d) None of these

55. The height of the chimney when it is found that on walking towards it 50 m in the horizontal line through its base, the angle of elevation of its top changes from 30° to 60°, is

(a) $25\sqrt{3}$ m (b) 25 m

(c) $25\sqrt{4}$ m (d) $\frac{25}{\sqrt{3}}$ m

56. If $\lim_{x\to\infty}\left(\frac{x^3+1}{x^2+1}-(ax+b)\right)=2$, then

(a) $a=1, b=1$ (b) $a=1, b=2$

(c) $a=1, b=-2$ (d) None of these

57. If $f(x)=\begin{cases}\frac{|x+2|}{\tan^{-1}(x+2)}, & x\neq-2\\ 2, & x=-2\end{cases}$, then $f(x)$ is

(a) continuous at $x=-2$

(b) not continuous at $x=-2$

(c) differentiable at $x=-2$

(d) continuous but not derivable at $x=-2$.

58. Let P be any point on the curve $x^{2/3}+y^{2/3}=a^{2/3}$. Then, the length of the segment of the tangent between the coordinate axes of length

(a) $3a$ (b) $4a$

(c) $5a$ (d) a

59. The motion of a particle along a straight line is described by the function $x=(2t-3)^2$, where x is in metre and t is in second. Then, the velocity of the particle at origin is

(a) 0 (b) 1

(c) 2 (d) None of these

60. The value of $\cos\frac{2\pi}{7}+\cos\frac{4\pi}{7}+\cos\frac{6\pi}{7}$ is

(a) 1 (b) −1

(c) $\frac{1}{2}$ (d) $-\frac{1}{2}$

61. The maximum value of $3\cos\theta+4\sin\theta$ is

(a) −5 (b) 5

(c) 25 (d) None of these

62. Which of the following is not a statement?

(a) Every set is a finite set

(b) 8 is less than 6

(c) Where are you going

(d) The sum of interior angles of a triangle is 180°

63. The least positive remainder when $123 \times 125 \times 127$ is divided by 124 is
(a) 4 (b) 120
(c) 121 (d) 130

64. The vertex connectivity of any tree is
(a) one (b) two
(c) three (d) None of these

65. A five digit number divisible by 3 is to be formed using the digits 0, 1, 2, 3, 4 and 5, without repetition. The total number of ways this can be done, is
(a) 216 (b) 240
(c) 600 (d) 3125

66. Given positive integers $r > 1$, $n > 2$ and the coefficient of $(3r)$ th and $(r + 2)$ th terms in the binomial expansion of $(1 + x)^{2n}$ are equal, then
(a) $n = 2r$ (b) $n = 2r + 1$
(c) $n = 3r$ (d) None of these

67. The number of solutions of $\log_4 (x - 1) = \log_2 (x - 3)$ is
(a) 3 (b) 1
(c) 2 (d) 0

68. The value of $\lim_{x \to 0} \frac{\int_0^{x^2} \cos^2 t \, dt}{x \sin x}$ is
(a) 2 (b) 1
(c) 0 (d) None of these

69. The value of $\int_2^3 \frac{\sqrt{x}}{\sqrt{5 - x} + \sqrt{x}} dx$ is
(a) 1 (b) 1/2
(c) 2 (d) None of these

70. The solution of differential equation $(e^x + 1) y \, dy = (y + 1) e^x \, dx$ is
(a) $(e^x + 1)(y + 1) = Ce^y$
(b) $(e^x + 1) |(y + 1)| = Ce^{-y}$
(c) $(e^x + 1)(y + 1) = \pm Ce^y$
(d) None of the above

71. The equation of the circle passing through (1, 1) and the points of intersection of $x^2 + y^2 + 13x - 3y = 0$ and $11x + \frac{1}{2} y + \frac{25}{2} = 0$ is
(a) $4x^2 + 4y^2 - 30x - 10y = 25$
(b) $4x^2 + 4y^2 - 30x - 13y - 25 = 0$
(c) $4x^2 + 4y^2 - 17x - 10y + 25 = 0$
(d) None of the above

72. The equation of the common tangent to the curves $y^2 = 8x$ and $xy = -1$ is
(a) $3y = 9x + 2$
(b) $y = 2x + 1$
(c) $2y = x + 8$
(d) $y = x + 2$

73. If k be the perimeter of the ΔABC, then $b \cos^2 \frac{C}{2} + c \cos^2 \frac{B}{2}$ is equal to
(a) k (b) $2k$
(c) $\frac{k}{2}$ (d) None of these

74. Let $f(x)$ be a positive function. Let

$$I_1 = \int_{1-k}^{k} x f\{x(1 - x)\} \, dx,$$

$$I_2 = \int_{1-k}^{k} f\{x(1 - x)\} \, dx,$$

where $2k - 1 > 0$, then $\frac{I_1}{I_2}$ is
(a) 2 (b) k
(c) $\frac{1}{2}$ (d) 1

75. If P is a variable point on the ellipse $\frac{x^2}{a^2} + \frac{y^2}{b^2} = 1$, with AA' as major axis. Then, the maximum value of the area of the $\Delta APA'$ is
(a) ab sq unit (b) $2ab$ sq unit
(c) $\frac{ab}{2}$ sq unit (d) None of these

76. In a ΔABC, if $\frac{\cos A}{a} = \frac{\cos B}{b} = \frac{\cos C}{c}$ and the side $a = 2$, then area of the triangle is
(a) 1 sq unit (b) 2 sq units
(c) $\frac{\sqrt{3}}{2}$ sq units (d) $\sqrt{3}$ sq units

77. The area bounded by the curves $y = |x| - 1$ and $y = -|x| + 1$ is
(a) 1 sq unit (b) 2 sq units
(c) $2\sqrt{2}$ sq units (d) 4 sq units

78. The value of the expression $\sqrt{3} \operatorname{cosec} 20° - \sec 20°$ is equal to
(a) 2 (b) $\frac{2 \sin 20°}{\sin 40°}$
(c) 4 (d) $\frac{4 \sin 20°}{\sin 40°}$

79. If in a ΔPQR, $\sin P, \sin Q, \sin R$ are in AP, then
(a) the altitudes are in AP
(b) the altitudes are in HP
(c) the medians are in GP
(d) the medians are in AP

80. A man and a woman appears in an interview for two vacancies in the same post. The probability of man's selection is $\frac{1}{4}$ and that of the woman's selection is $\frac{1}{3}$. What is the probability that none of them will be selected?
(a) $\frac{1}{2}$ (b) $\frac{1}{12}$
(c) $\frac{1}{4}$ (d) None of these

English

Directions (Q. Nos. 1-5) *In the following questions, some of the sentences have errors and some have none. Find out which part of a sentence has an error. The letter of that part is your answer. If there is no error, your answer is (d) i.e., No error.*

1. The whole country (a)/was suffering by (b)/a financial crisis. (c)/No error. (d)
2. I do not know where could he have gone (a)/so early (b)/in the morning. (c)/No error. (d)
3. I suggest that (a)/he goes (b)/to the doctor as soon as he returns from taking the examination. (c)/No error. (d)
4. The introduction of tea and coffee (a)/and such other beverages (b)/have not been without some effect. (c)/No error. (d)
5. Inspite of the roadblock (a)/the guards allowed us (b)/enter the restricted area to search for our friends (c)/No error. (d)

Directions (Q. Nos. 6-10) *In the following questions, out of the four alternatives, choose the one which best expresses the meaning of the given word.*

6. **Illicit**
(a) immoral (b) illegal
(c) ineligible (d) illegible
7. **Flair**
(a) talent (b) tendency
(c) bias (d) need
8. **Conservation**
(a) preservation (b) respiration
(c) correction (d) confusion
9. **Abysmal**
(a) sickening (b) gloomy
(c) sad (d) bottomless
10. **Salient**
(a) valient (b) variant
(c) prodent (d) prominent

Directions *(11-15) In the following questions, choose the word(s) opposite in meaning to the given word.*

11. **Suppress**
(a) stir up (b) rouse
(c) urge (d) incite
12. **Loosen**
(a) fasten (b) accelerate
(c) delay (d) paste
13. **Rebellion**
(a) forgiveness (b) retribution
(c) submission (d) domination
14. **Idiosyncrasy**
(a) insanity (b) sanity
(c) generality (d) singularity
15. **Sanguine**
(a) diffident (b) hopeless
(c) cynical (d) morose

Directions (Q. Nos. 16-20) *In the following questions, groups of four words are given. In each group, one word is wrongly spelt. Find the misspelt word.*

16. (a) prioratise (b) picturise
(c) hesitate (d) perforate
17. (a) mendacous (b) obnoxcious
(c) pernicious (d) ferocious
18. (a) pennetrate (b) irritate
(c) hesitate (d) perforate
19. (a) passagway (b) causeway
(c) subway (d) straightway
20. (a) rapport (b) support
(c) repport (d) purport

Directions (Q. Nos. 21-25) *In the following questions, four alternatives are given for the Idio/Phrase printed in* **bold** *in the sentence. Choose the alternative which best expresses the meaning of the Idiom/Phrase.*

21. Don't worry about the silly row. It was just a **storm in a tea cup.**

(a) important matter dealt with ease
(b) hot tea being served
(c) commotion over a trival matter
(d) confusion and chaos

22. The Rajput warriors **set their face against** the invader.
(a) became enemies (b) turned away from
(c) faced difficulty (d) opposed strongly

23. Syria is now **currying favour with** America.
(a) pleasing
(b) favouring
(c) obliging
(d) ingratiating itself with

24. Our principal is not a man **to mince matters.**
(a) to confuse issues
(b) to say something mildly
(c) to mix everything together
(d) to be very modest

25. We tend to **take for granted** the conveniences of modern life.
(a) to consider (b) to admit
(c) to accept readily (d) to care for

Directions (Q. Nos. 26-30) *In the following questions, out of the four alternatives, choose the one which can be substituted for the given words/sentence.*

26. Belief that God is in everything and that everything is God
(a) Atheism (b) Pantheism
(c) Scepticism (d) Animism

27. A picture of a person or a thing drawn in such a highly exaggerated manner as to cause laughter
(a) Cartoon (b) Cacography
(c) Cartography (d) Caricature

28. The state of being miserable be reft of all possessions
(a) Dependant (b) Complacent
(c) Destitute (d) Omnipresent

29. That which cannot be called back
(a) Irresponsible (b) Irrevocable
(c) Irredeemable (d) Incalculable

30. One who journeys from place to place
(a) Quack (b) Cannibal
(c) Itinerant (d) Courier

Directions (Q. Nos. 31-35) *In the following questions, a sentence has been given in Active Voice/Passive Voice. Out of the four alternatives suggested, select the one which best expresses the same sentence in Passive Voice/Active Voice.*

31. We have already done the exercise.
(a) Already, the exercise has been done by us
(b) The exercise has already been done by us
(c) The exercise had been already done by us
(d) The exercise is already done by us

32. The main skills we seek to develop include analysing, interpreting and evaluating ideas.
(a) The main skills sought by us to develop include analysing, interpreting and evaluating ideas
(b) The main skills that we are seeking to be developed include analysing, interpreting and evaluating ideas
(c) The main skills sought to be developed by us include analysing, interpreting and evaluating ideas
(d) The main skills include analysing, interpreting and evaluating kleas which are sought by us to develop

33. Who can question Gandhi's integrity?
(a) By whom Gandhi's integrity can be questioned?
(b) By whom can Gandhi's integrity be questioned?
(c) Gandhi's integrity can be questioned by whom?
(d) Who could have questioned Gandhi's integrity?

34. He presented me a bouquet on my birthday.
(a) A bouquet is presented to me on my birthday by him
(b) I was presented on my birthday a bouquet by him
(c) I was presented a bouquet on my birthday by him
(d) I will be presented a bouquet on my birthday by him

35. This surface feels smooth.
(a) This surface is felt smooth
(b) This surface is smooth when it is felt
(c) This surface when felt is smooth
(d) This surface is smooth as felt

Directions (Q. Nos. 36-40) *In the following questions, a part of the sentence is printed* ***bold.*** *Below are given alternatives to the* ***bold*** *part at (a), (b) and (c), which may improve the sentence. Choose the correct alternative. In case no improvement is needed, your answer is (d).*

36. In the desert, the sun is the master, all else **resigns** before its merciless rays.
(a) collapses (b) falls
(c) retires (d) No improvement

37. I intend **to learn** French next year.
(a) learning (b) learn
(c) have learnt (d) No improvement

38. The police **needed** him for armed robbery.
(a) liked (b) was after
(c) were looking to (d) No improvement

39. There is **no more room** for you in this compartment.
(a) no more seat
(b) no more space
(c) no more accommodation
(d) No improvement

40. It is easy to see why cities grew **on the river banks**
(a) along the river banks
(b) in the river banks
(c) upon the river banks
(d) No improvement

Answers

Physics

1. (b)	**2.** (c)	**3.** (b)	**4.** (d)	**5.** (d)	**6.** (d)	**7.** (c)	**8.** (d)	**9.** (b)	**10.** (b)
11. (c)	**12.** (a)	**13.** (d)	**14.** (a)	**15.** (c)	**16.** (b)	**17.** (b)	**18.** (d)	**19.** (c)	**20.** (a)
21. (a)	**22.** (d)	**23.** (a)	**24.** (c)	**25.** (c)	**26.** (a)	**27.** (a)	**28.** (b)	**29.** (a)	**30.** (d)
31. (b)	**32.** (a)	**33.** (b)	**34.** (a)	**35.** (d)	**36.** (a)	**37.** (d)	**38.** (a)	**39.** (a)	**40.** (b)
41. (d)	**42.** (b)	**43.** (a)	**44.** (c)	**45.** (b)	**46.** (c)	**47.** (b)	**48.** (c)	**49.** (d)	**50.** (d)
51. (b)	**52.** (a)	**53.** (c)	**54.** (d)	**55.** (c)	**56.** (b)	**57.** (b)	**58.** (a)	**59.** (d)	**60.** (c)

Chemistry

1. (b)	**2.** (a)	**3.** (d)	**4.** (a)	**5.** (b)	**6.** (b)	**7.** (c)	**8.** (d)	**9.** (d)	**10.** (a)
11. (b)	**12.** (d)	**13.** (c)	**14.** (a)	**15.** (d)	**16.** (c)	**17.** (b)	**18.** (a)	**19.** (a)	**20.** (c)
21. (d)	**22.** (c)	**23.** (a)	**24.** (b)	**25.** (b)	**26.** (b)	**27.** (c)	**28.** (a)	**29.** (d)	**30.** (a)
31. (a)	**32.** (a)	**33.** (c)	**34.** (c)	**35.** (a)	**36.** (a)	**37.** (b)	**38.** (c)	**39.** (c)	**40.** (a)
41. (d)	**42.** (b)	**43.** (a)	**44.** (d)	**45.** (b)	**46.** (a)	**47.** (a)	**48.** (d)	**49.** (b)	**50.** (b)
51. (b)	**52.** (d)	**53.** (c)	**54.** (d)	**55.** (c)	**56.** (b)	**57.** (a)	**58.** (c)	**59.** (c)	**60.** (c)

Mathematics

1. (a)	**2.** (d)	**3.** (b)	**4.** (c)	**5.** (a)	**6.** (a)	**7.** (b)	**8.** (a)	**9.** (d)	**10.** (a)
11. (c)	**12.** (b)	**13.** (b)	**14.** (b)	**15.** (c)	**16.** (c)	**17.** (b)	**18.** (a)	**19.** (b)	**20.** (c)
21. (d)	**22.** (a)	**23.** (a)	**24.** (b)	**25.** (c)	**26.** (c)	**27.** (c)	**28.** (b)	**29.** (c)	**30.** (b)
31. (d)	**32.** (b)	**33.** (a)	**34.** (a)	**35.** (c)	**36.** (d)	**37.** (c)	**38.** (a)	**39.** (a)	**40.** (c)
41. (a)	**42.** (a)	**43.** (b)	**44.** (a)	**45.** (a)	**46.** (c)	**47.** (d)	**48.** (c)	**49.** (a)	**50.** (d)
51. (a)	**52.** (b)	**53.** (a)	**54.** (d)	**55.** (a)	**56.** (c)	**57.** (b)	**58.** (d)	**59.** (a)	**60.** (d)
61. (b)	**62.** (c)	**63.** (c)	**64.** (a)	**65.** (a)	**66.** (a)	**67.** (b)	**68.** (b)	**69.** (b)	**70.** (a)
71. (d)	**72.** (d)	**73.** (c)	**74.** (c)	**75.** (a)	**76.** (d)	**77.** (b)	**78.** (c)	**79.** (b)	**80.** (a)

English

1. (b)	**2.** (a)	**3.** (b)	**4.** (c)	**5.** (c)	**6.** (b)	**7.** (a)	**8.** (a)	**9.** (d)	**10.** (d)
11. (d)	**12.** (a)	**13.** (c)	**14.** (c)	**15.** (b)	**16.** (a)	**17.** (b)	**18.** (a)	**19.** (a)	**20.** (c)
21. (c)	**22.** (d)	**23.** (d)	**24.** (a)	**25.** (c)	**26.** (b)	**27.** (d)	**28.** (c)	**29.** (b)	**30.** (c)
31. (b)	**32.** (b)	**33.** (b)	**34.** (c)	**35.** (a)	**36.** (d)	**37.** (d)	**38.** (c)	**39.** (d)	**40.** (a)

Hints & Solutions

Hints & Solutions

Physics

1. Let one mole of each gas has same volume as V. When they are mixed, then density of mixture is

$$\rho_{mixture} = \frac{\text{mass of } O_2 + \text{mass of } H_2}{\text{volume of } O_2 + \text{volume of } H_2}$$

$$= \frac{32+2}{V+V} = \frac{34}{2V} = \frac{17}{V}$$

Also, $\rho_{H_2} = \frac{2}{V}$

Now, velocity $v = \left(\frac{\gamma P}{\rho}\right)^{1/2}$ or $v \propto \frac{1}{\sqrt{\rho}}$

$$\frac{v_{mixture}}{v_{H_2}} = \sqrt{\left(\frac{\rho_{H_2}}{\rho_{mixture}}\right)}$$

$$= \sqrt{\left(\frac{2/V}{17/V}\right)} = \sqrt{\left(\frac{2}{17}\right)}$$

2. Let the distance between the two cliffs be d. Since, the man is standing midway between the two cliffs, then the distance of man from either end is $d/2$.

The distance travelled by sound (in producing an echo)

$$2 \times \frac{d}{2} = v \times t \Rightarrow d = 340 \times 1 = 340 \text{ m}$$

3. The distance between the first dark (minima) fringes on either side of the central maximum is twice the fringe width of either maximum or minimum.

i.e., $\beta = 2\frac{D\lambda}{d} = \frac{2 \times 2 \times 600 \times 10^{-9}}{1 \times 10^{-3}}$

$$= 2.4 \times 10^{-3} = 2.4 \text{ mm}$$

4. After silvering the plane surface, plano-convex lens behaves as a concave mirror of focal length

$$\frac{1}{F} = \frac{2}{f_{lens}}$$

But $F = 0.2 \text{ m}$

$\therefore$ $f_{lens} = 2F = 2 \times 0.2 = 0.4 \text{ m}$

Now, from lens maker's formula

$$\frac{1}{f_{lens}} = (\mu - 1)\left(\frac{1}{R_1} - \frac{1}{R_2}\right)$$

$\therefore$ $\frac{1}{0.4} = (1.5 - 1) \times \frac{1}{R_1}$ $[\because R_2 = \infty]$

$\Rightarrow$ $R_1 = 0.5 \times 0.4 = 0.2 \text{ m}$

5. For a stone which is thrown downwards from a balloon rising upwards, the equation of motion is

$$h = -ut + \frac{1}{2}gt^2$$

$$= -29 \times 10 + \frac{1}{2} \times 9.8 \times (10)^2$$

$$= -290 + 490$$

$$= 200 \text{ m}$$

6. The distance between two consecutive dark or bright fringes is recognised as β (fringe width) and that between central fringe and first dark fringe on either side is $\frac{\beta}{2}$.

Given, spacing between second dark fringe and central fringe $= \beta + \frac{\beta}{2}$

or $\frac{3\beta}{2} = 1 \text{ mm}$ (Given)

or $\beta = \frac{2}{3} \times 1 \text{ mm}$ or $\frac{\lambda D}{d} = \frac{2}{3} \text{ mm}$

$\therefore$ $\lambda = \frac{2}{3} \times 10^{-3} \times \frac{0.9 \times 10^{-3}}{1}$

$\Rightarrow$ $\lambda = 0.6 \times 10^{-6} \text{ m}$

$\therefore$ Wavelength of the monochromatic source of light

$$\lambda = 600 \times 10^{-9} \text{ m} = 600 \text{ nm}$$

7. The number of counts left after time t

$$N = N_0\left(\frac{1}{2}\right)^{t/T_{1/2}}$$

$\therefore$ $30 = 240\left(\frac{1}{2}\right)^{60/T_{1/2}}$

$[\because t = 1 \text{ h} = 60 \text{ min}]$

or $\frac{30}{240} = \left(\frac{1}{2}\right)^{60/T_{1/2}}$

or $\left(\frac{1}{2}\right)^3 = \left(\frac{1}{2}\right)^{60/T_{1/2}}$

Comparing the powers, we get

$$\therefore \quad \frac{60}{T_{1/2}} = 3$$

$$T_{1/2} = \frac{60}{3}$$

or $$T_{1/2} = 20 \text{ min}$$

8. The resistance of conductor

$$R = \frac{\rho l}{A} = \frac{\rho l}{\pi r^2}$$

or $$R \propto \frac{l}{r^2}$$

$$\therefore \quad \frac{R_1}{R_2} = \frac{l_1}{l_2} \times \frac{r_2^2}{r_1^2}$$

$$= \frac{2}{1} \times \left(\frac{2}{1}\right)^2 = \frac{8}{1}$$

Thermal potentials between the ends of the rods are same. So, heat conducted per second

$$H = \frac{Q}{t} \propto \frac{1}{R}$$

$$\therefore \quad \frac{H_1}{H_2} = \frac{R_2}{R_1} = \frac{1}{8} = 1:8$$

9. Using the relation

$$\frac{W}{Q_1} = \frac{Q_1 - Q_2}{Q_1}$$

or $$\frac{W}{Q_1} = 1 - \frac{Q_2}{Q_1}$$

or $$\frac{W}{Q_1} = 1 - \frac{T_2}{T_1} \qquad \left(\because \frac{Q_1}{Q_2} = \frac{T_1}{T_2}\right)$$

or $$W = Q_1\left(1 - \frac{T_2}{T_1}\right)$$

$$\therefore \quad W = 6 \times 10^4 \left[1 - \frac{127 + 273}{227 + 273}\right]$$

or $$W = 6 \times 10^4 \left(1 - \frac{400}{500}\right)$$

$$= 6 \times 10^4 \times \frac{100}{500}$$

$$= 1.2 \times 10^4 \text{ J}$$

10. The points C and D will be at same potenitals since, $\frac{3}{6} = \frac{4}{8}$. Therefore, capacitance of 2 μF will be unaffected. So, the equivalent circuit can be shown as the effective capacitance in upper arm in series, is given by

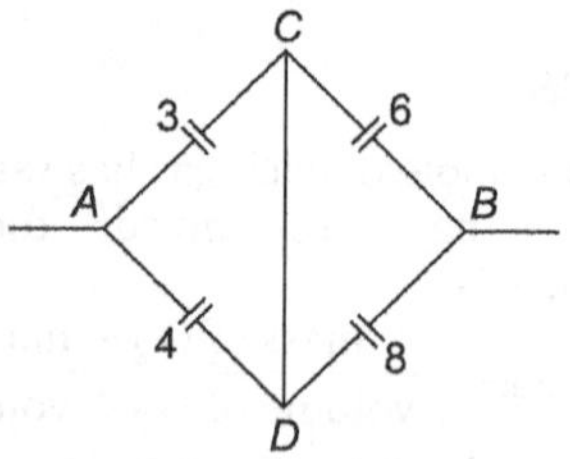

$$C_1 = \frac{3 \times 6}{3 + 6} = \frac{18}{9} = 2 \mu\text{F}$$

The effective capacitance in lower arm in series, is given by

$$C_2 = \frac{4 \times 8}{4 + 8} = \frac{32}{12} = \frac{8}{3} \mu\text{F}$$

Hence, the resultant capacitance in parallel is given by

$$C = C_1 + C_2 = 2 + \frac{8}{3} = \frac{14}{3} \mu\text{F}$$

11. Tension in the string is along the radius of circular path adopted by the bob; while displacement of the bob is along the circumference of the path. Hence, again **F** and **s** are at 90° and so, $W = 0$.

12. $K_{max} = K_0 =$ total energy

As total energy remains conserved in SHM, hence when U is maximum in SHM, $K = 0$, *i.e.*, E is also equal to U_{max}, *i.e.*, $U_{max} = E = K_0$.

13. When the particle completes half revolution, change in velocity

$$\Delta v = [5 - (-5)] \text{ m/s} = 10 \text{ m/s}$$

Time taken to complete the half revolution is

$$t = \frac{\pi r}{v} = \frac{\pi \times 5}{5} = \pi \text{ s}$$

Thus, average acceleration $= \frac{\Delta v}{t} = \frac{10}{\pi} \text{ m/s}^2$

14. Mass of copper liberated

$$m = zit = z\left(\frac{P}{V}\right)t$$

$$= (3.3 \times 10^{-7})\left(\frac{100 \times 36 \times 10^5}{33}\right)$$

$$= 3.6 \text{ kg}$$

15. At resonant frequency

$$X_L = X_C \qquad \left(\omega L = \frac{1}{\omega C}\right)$$

At frequencies higher than resonance frequencies

$$X_L > X_C$$

i.e., behaviour is inductive.

16. The situation is shown in figure.

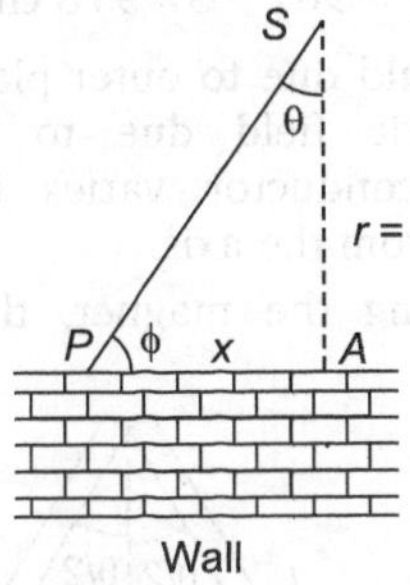

From figure,

$$x = r\tan\theta$$

$\therefore$ Velocity of P is

$$v = \frac{dx}{dt} = r\sec^2\theta\left(\frac{d\theta}{dt}\right)$$

where, $\frac{d\theta}{dt}$ = angular velocity of rotation of spot

$$= \omega$$

$$\therefore \quad v = \omega r\sec^2\theta$$

At $\phi = 45°$, so $\theta = 45°$

Hence, $v = 0.1 \times 3 \times \sec^2 45°$

$$= 0.1 \times 3 \times 2 = 0.6 \text{ m/s}$$

17. Using standard gas equation,

$$\frac{p_1V_1}{T_1} = \frac{p_2V_2}{T_2} \text{ or } V_2 = \frac{p_1V_1T_2}{p_2T_1}$$

$$\therefore \quad V_2 = \frac{1 \times 500 \times (273 - 3)}{0.5 \times (273 + 27)}$$

$$V_2 = \frac{1 \times 500 \times 270}{0.5 \times 300}$$

$$V_2 = 900 \text{ m}^3$$

18. Amplitude is independent of wavelength, velocity and frequency of oscillation.

19. Breaking stress $= \frac{\text{Breaking force}}{\text{Area}}$ = constant

$$\frac{F}{\left(\frac{\pi D^2}{4}\right)} = \frac{F'}{\pi D^2}$$

$$F' = 4F$$

20. $\frac{pV}{T}$ = constant; V = constant

$$\therefore \quad \frac{p}{T} = \text{constant}$$

$$\therefore \quad \frac{p}{T} = \frac{100.4p}{1000} \times \frac{1}{(T+1)}$$

$$\frac{1}{T} = \frac{100.4}{100} \times \frac{1}{(T+1)}$$

$$\frac{1}{T} = \frac{100.4}{100T + 100}$$

$$T = 250 \text{ K}$$

21. Maximum rate of energy flow, $S = E_0 \times H_0$

Given, $E_0 = 100$ V/m, $H_0 = 0.265$ A/m

$$\therefore \quad S = 100 \times 0.265 = 26.5 \text{ W/m}^2$$

22. Conservation of linear momentum gives,

$$m_1v_1 + m_2v_2 = 0$$

$$m_1v_1 = -m_2v_2$$

$$v_1 = \frac{-m_2v_2}{m_1}$$

Given, $m_1 = 10 \text{ g} = \left(\frac{10}{1000}\right)$ kg, $m_2 = 1$ kg,

$$v_2 = -5 \text{ m/s}$$

$$\therefore \quad v_1 = \frac{+1 \times 5}{10/1000}$$

or $\quad v_1 = 500$ m/s

23. Work done

$$W = p(V_2 - V_1)$$

Given, $p = 2 \times 10^5 \text{ N/m}^2$, $V_1 = 50 \times 10^{-3} \text{ m}^3$,

$$V_2 = 150 \times 10^{-3} \text{ m}^3$$

$$\therefore \quad W = 2 \times 10^5 (150 - 50) 10^{-3}$$

$$W = 2 \times 10^5 \times 100 \times 10^{-3}$$

$$W = 2 \times 10^4 \text{ J}$$

24. In an electron gun, the control grid is given a negative potential relative to cathode in order to repel the electrons so that a converging beam of electrons emerges from gun.

25. From the formula, $\omega = \frac{1}{\sqrt{LC}}$

Given, $L = 8$ H,

$$C = 0.5\,\mu\text{F} = 0.5 \times 10^{-6} \text{ F}$$

$$\therefore \quad \omega = \frac{1}{\sqrt{8 \times 0.5 \times 10^{-6}}}$$

$$\Rightarrow \quad \omega = \frac{1}{2\times10^{-3}} = 500$$

$$\Rightarrow \quad \omega = \frac{250}{\pi}\ \text{rad/s}$$

26. Total power

$$P = P_1 + P_2$$
$$P = -4D + 2D$$
$$P = -2D$$

27. Molecular spectra due to vibrational motion lies in the microwave region of electromagnetic spectrum. Due to Kirchhoff's law in spectroscopy the same will be absorbed.

28. During inelastic collision of two particles kinetic energy after collision is not equal to kinetic energy before collision. Here, KE appears in other forms. In some cases, $(KE)_{final} < (KE)_{initial}$ such as when initial KE is converted into internal energy of the product (as heat, elastic or excitation energy) while in other cases $(KE)_{final} > (KE)_{initial}$ such as when internal energy stored in the colliding particles is released.

29. From the relation,

$$\overline{KE} = \frac{3}{2}kT$$

$$\Rightarrow \quad \frac{(\overline{KE})_1}{(KE)_2} = \frac{T_1}{T_2} \quad \text{(become } k \text{ is constant)}$$

Given, $T_1 = 300\text{ K}, T_2 = 350\text{ K}$

$$\therefore \quad \frac{(KE)_1}{(KE)_2} = \frac{300}{350} = \frac{6}{7}$$

30. From the formula, $\varepsilon = -L\dfrac{dI}{dt}$

Take magnitude on both sides.

$$|\varepsilon| = \left|-L\frac{dI}{dt}\right| \quad \text{or} \quad \varepsilon = L\frac{dI}{dt}$$

$$\therefore \quad 80 = L\left(\frac{2}{0.05}\right) \Rightarrow 80 = L \times 40$$

$$\Rightarrow \quad L = \frac{80}{40} \Rightarrow L = 2\text{H}$$

32. As same force is applied to each wire having equal area of cross-section, so stress is same but extensions will be different for wires of different materials. So, same stress but different strains are there.

33. $L_1 = \dfrac{g_1T^2}{4\pi^2} = \dfrac{g_1}{\pi^2}$; $L_2 = \dfrac{g_2T^2}{4\pi^2} = \dfrac{g_2}{\pi^2}$

Since, length is decreased, g_2 is less than g_1.

$$\therefore \quad L_1 - L_2 = \frac{g_1 - g_2}{\pi^2}$$

or $\quad (L_1 - L_2)\pi^2 = g_1 - g_2$

or $\quad 0.3 \times 10 = g_1 - g_2$

$$\therefore \quad g_2 = 981 - 3 = 978\ \text{cm/s}^2$$

34. Electric field due to outer plate is zero and so the electric field due to inner cylindrical charged conductor varies inversely as the distance from the axis.

35. On bending the magnet, the length of the magnet

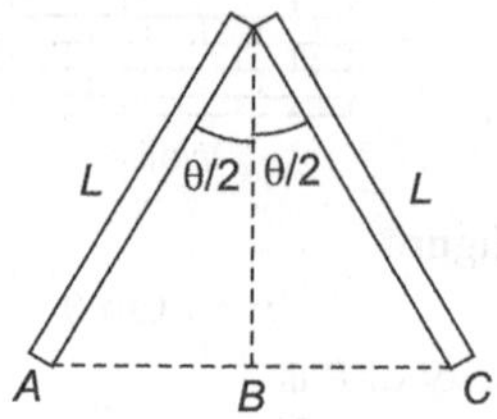

$$AC = AB + BC$$
$$= L\sin\left(\frac{\theta}{2}\right) + L\sin\left(\frac{\theta}{2}\right)$$
$$= 2L\sin\left(\frac{\theta}{2}\right) = 2L\sin 30°$$
$$= 2L \times \frac{1}{2} = L$$

36. When a capacitor is fully charged, then no current is drawn by it. When no current flows in the circuit, potential difference across the cell = emf of cell = potential difference across the capacitor.

37. For hydrogen like atom, the radius of nth orbit

$$r_n^Z = \frac{n^2}{Z}a_0$$

Here, $\quad a_0 = 0.51 \times 10^{-10}\ \text{m}$

$$\therefore \quad r_n^Z = \frac{0.51 \times 10^{-10}}{4}\ \text{m}$$

In the ground state, $n = 1$

$$\therefore \quad \frac{0.51\times10^{-10}}{4} = \frac{1^2}{Z} \times 0.51 \times 10^{-10}$$

$$\therefore \quad Z = 4$$

So, the atom is triply ionised beryllium (Be^{3+}).

38. $\lambda = \dfrac{h}{mv} = \dfrac{6.6 \times 10^{-34}}{1 \times 2000}$

$$= 3.3 \times 10^{-37}\ \text{m} = 3.3 \times 10^{-27}\ \text{Å}$$

39. From Ohm's law

$$i = \frac{E}{R + r} = \frac{2}{998 + 2} = 2 \times 10^{-3}\ \text{A}$$

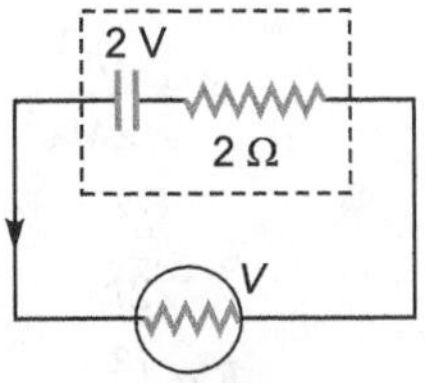

Potential difference across the voltmeter is

$$V = iR = (2 \times 10^{-3}) \times 998$$
$$= 1.996\ \text{V}$$

40. Initially, due to upward acceleration apparent weight of the body increases but then it decreases due to decrease in gravity.

41. From law of conservation of momentum,

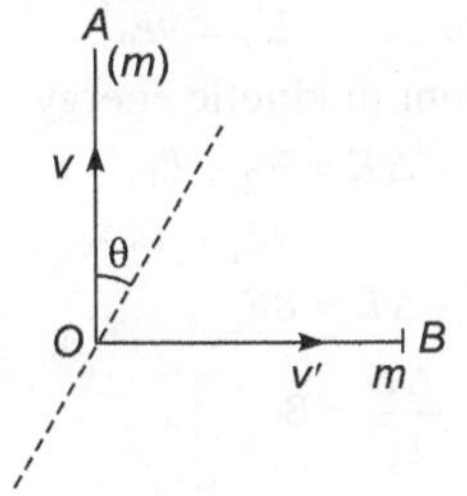

$$mv - mv' \cos\theta = 0 \quad \ldots(\text{i})$$
$$mv - mv' \sin\theta = 0 \quad \ldots(\text{ii})$$

From Eqs. (i) and (ii), we get

$$\theta = 45°$$
$$v' = \frac{v}{\cos\theta} = v\sqrt{2}$$

42. The decay equation is

$$_{92}X^A \longrightarrow {}_ZY^{228} + \alpha$$

α-particle is nucleus of $_2He^4$

$$_{92}X^A \longrightarrow {}_ZY^{228} + {}_2He^4$$
$$A = 228 + 4 = 232$$
$$Z = 92 - 2 = 90$$

43. Insulators have large energy gap between valence and conduction band (about 6 eV), while semiconductors have a smaller one and conductors, the smallest energy gap.

44. NOR gate is obtained when the output of OR gate is made as the input of NOT gate. Boolean expression for NOR gate is

$$Y = \overline{A + B}$$

Symbol of NOR gate

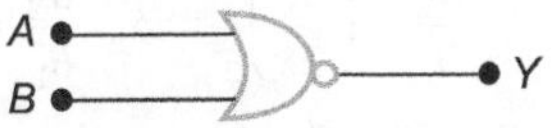

Truth Table

A	B	C
0	0	1
1	0	0
0	1	0
1	1	0

45. In common-emitter amplifier, the emitter current is much larger than the base current.

46.
$$L = 10 \log_{10}\left(\frac{I}{I_0}\right)$$

$$\therefore \quad L = 10 \log_{10}\left(\frac{2 \times 10^{-8}}{10^{-12}}\right)$$

$$(\because I_0 = 10^{-12}\ \text{W/m}^2)$$

$$L = 10 \log_{10}(2 \times 10^4)$$
$$L = 10\,[\log_{10} 2 + \log_{10}(10)^4]$$
$$L = 10\,(4.3)$$
$$L = 43 \text{ decibel}$$

47. $D = 360° - 2\theta \Rightarrow 240° = 360° - 2\theta$

$2\theta = 120°, \quad \therefore \theta = 60°$

Then, number of images observable

$$N = \frac{360°}{60°} - 1$$
$$N = 6 - 1$$
$$N = 5$$

48. Number of Cu ions liberated is

$$n = \frac{I \cdot t}{e\,(\text{valence electron})}$$

$$n = \frac{1 \times 10}{1.6 \times 10^{-19} \times 2}$$

$$n = 3.125 \times 10^{19}$$

$$\therefore \quad n \approx 3.1 \times 10^{19}$$

49. For a magnet $B = \frac{\mu_0}{4\pi}\frac{2M}{r^3}$ (nearly)

$$\Rightarrow \quad \frac{B_2}{B_1} = \left(\frac{r_1}{r_2}\right)^3$$

$$\Rightarrow \quad \frac{B_2}{B_1} = \left(\frac{x}{2x}\right)^3 \Rightarrow \frac{B_2}{B_1} = \frac{1}{8}$$

Thus, $B_1 : B_2 = 8 : 1$ approximately

50. Electromagnetic waves travel through oscillating electric and magnetic fields whose directions are perpendicular to each other.

51. According to question,

$$\frac{M_1}{d_1^3} = \frac{M_2}{d_2^3}$$

$$\Rightarrow \quad \frac{M_1}{M_2} = \left(\frac{d_1}{d_2}\right)^3 = \left(\frac{12}{18}\right)^3$$

$$\therefore \quad \frac{M_1}{M_2} = \frac{8}{27}$$

52. Effective current is the rms value. Here, 220 V is the labelled value of AC which is also the rms value. Hence,

$$I_{rms} = \frac{E_{rms}}{R}$$

$$I_{rms} = \frac{220}{100 \times 10^3}$$

$$I_{rms} = 2.2 \text{ mA}$$

54. The refractive index of glass is the greatest for violet colour and is smaller for red colour and as $\delta \propto (\mu - 1)$ therefore deviation for violet will be maximum.

55. Given, $\quad y_1 = A \sin (kx - \omega t)$

$$y_2 = A \cos (kx - \omega t)$$

or $\quad y_2 = A \sin \left(kx - \omega t + \frac{\pi}{2}\right)$

Phase difference of two waves $= \frac{\pi}{2}$

$\because$ Resultant amplitude

$$R = \sqrt{A^2 + A^2 + 2AA \cos \phi}$$

$$= \sqrt{A^2 + A^2 + 2A^2 \cos \frac{\pi}{2}}$$

$$= \sqrt{2A^2} \qquad \left(\because \cos \frac{\pi}{2} = 0\right)$$

$$R = \sqrt{2}A$$

56. Momentum of photon, $p = \frac{h}{\lambda}$

Kinetic energy of photon of mass M, $K = \frac{p^2}{2M}$

or $\quad K = \frac{h^2}{2M\lambda^2}$

57. $\quad E = \frac{L^2}{2I}$

$$\therefore \quad E \propto L^2$$

$$\frac{E_2}{E_1} = \left(\frac{L_2}{L_1}\right)^2$$

$$\frac{E_2}{E_1} = \left[\frac{L_1 + 200\% \text{ of } L_1}{L_1}\right]$$

$$= \left[\frac{L_1 + 2L_1}{L_1}\right]^2 = (3)^2$$

$$E_2 = 9E_1$$

Increment in kinetic energy

$$\Delta E = E_2 - E_1$$

$$= 9E_1 - E_1$$

$$\Delta E = 8E_1$$

$$\frac{\Delta E}{E_1} = 8$$

or percentage increase = 800%

58. The height of water in the tank becomes maximum when the volume of water flowing into the tank per second becomes equal to the volume flowing out per second.

Volume of water flowing out per second $= A\sqrt{2gh}$

Volume of water flowing in per second $= 70 \text{ cm}^3/\text{s}$

$$\therefore \quad A\sqrt{2gh} = 70$$

$$1\sqrt{2 \times 980 \times h} = 70$$

$$h = \frac{4900}{1960} = 2.5 \text{ cm}$$

59. The production of X-rays is an atomic property whereas the production of γ-rays is a nuclear property.

60. Wood is non-crystalline.

Chemistry

1. The reaction is

$$N_2O_4 \rightleftharpoons 2NO_2$$

Molecular weight of $N_2O_4 = 2\times14 + 4\times16 = 92$

$\therefore$ Vapour density (D) of N_2O_4

$$= \frac{\text{molecular weight}}{2} = \frac{92}{2} = 46$$

From, $\alpha = \frac{D-d}{(n-1)d}$

$$\alpha = \frac{46-30}{(2-1)30} = \frac{16}{30} = 0.533 = 53.3\%$$

2. Protective power of a lyophilic colloid is measured in terms of gold number. Lower the value of gold number, better is the protective power of a protective colloid. Thus, the order will be

(a) > (b) > (d) > (c)

3. (a) $\frac{1}{2}H_2(g) \longrightarrow H^+(aq) + e^-$

$$E_{cell} = E^\circ_{cell} - 0.0591 \log [H^+]$$

(b) 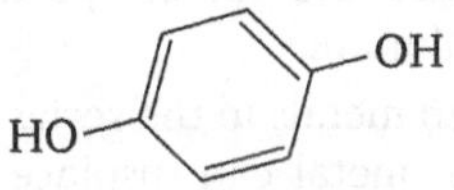 $\rightleftharpoons$

 $+ 2H^+ + 2e^-$

$$E_{cell} = E^\circ_{cell} - 0.0591 \log [H^+]^2$$

(c) $Ag(s) \longrightarrow Ag^+(aq)$

$$E_{cell} = E^\circ_{cell} - 0.0591 \log [Ag^+]$$

(d) $Ag(s) + Cl^-(aq) \longrightarrow AgCl(s) + e^-$

$$E_{cell} = E^\circ_{cell} - 0.0591 \log \frac{1}{[Cl^-]}$$

Thus, in case (d), the E_{cell} increases with increase in the concentration of ions.

4. According to Raoult's law, "Relative lowering of vapour pressure = mole fraction of solute"

$$(X_B) = 0.0125$$

$$\because \quad X_B = \frac{m\cdot M_A}{10000 + m\cdot M_A}$$

$$\therefore \quad 0.0125 = \frac{m\times18}{1000 + m\times18}$$

$$12.5 + 0.225m = 18m$$

$$17.775m = 12.5$$

$$m = \frac{12.5}{17.775} = 0.70$$

5.

Formula	Name
(a) $Na_2S_2O_3\cdot5H_2O$	hypo
(b) $Na_2SO_4\cdot10H_2O$	Glauber's salt
(c) $Na_2CO_3\cdot10H_2O$	Washing soda
(d) $Na(NH_4)HPO_4$	Microcosmic salt

6. $PhCHClBr \xrightarrow[-t\text{-}BuOH]{tBuO^-} PhC^-ClBr \longrightarrow Ph\ddot{C}Cl$

7. 3° alcohol offers resistance to oxidation. Hence, Cr^{3+} ions are not formed, and the solution does not turn green.

8. Molisch's test is a general test for carbohydrates, while ninhydrin test is given by all proteins. Nitroprusside test is given by proteins which contains —SH group. Due to the presence of —CONH linkage, alkaline solution of proteins give biuret test.

9. Rocket propellants are a combination of an oxidiser and fuel. Phenyl hydrazine can't act as propellant.

10. $Cl^- = 2, 8, 8$

$O^{2-} = 2, 8$

$Na^+ = 2, 8$

$Mg^{2+} = 2, 8$

Cl^- contains same number of electrons in outermost and penultimate shells.

11. Mixture of $MgCl_2$ and MgO is called Sorel's cement. It is $MgCl_2\cdot5MgO\cdot xH_2O$.

12. Substracting equation (II) from (I), we get

$$S_R \longrightarrow S_M, \qquad \Delta H = +2.5\text{ kJ}$$

Thus, S_R to S_M transition is an endothermic reaction.

13. (a) PSC (polar stratospheric clouds) are special type of clouds formed over Antarctica in winter that play important role in ozone layer depletion.

(b) PCB (poly chlorinated biphenyls) are toxic chemicals with high stability and resistance to oxidation. These are used for fluids in transformers and capacitors.

(c) BOD (biochemical oxygen demand) is a measure of dissolved oxygen that would be needed by the microorganisms to oxidise organic and inorganic compounds present in polluted water.

(d) COD (chemical oxygen demand) is the amount of oxygen, in ppm, that would be required to oxidise the contaminants.

14. For a first order reaction

$$k = \frac{2.303}{t} \log \frac{a}{a-x}$$

At half-time,

$$x = \frac{a}{2}$$

$$\therefore \quad k = \frac{2.303}{1.6} \times \log \frac{a}{a - \frac{a}{2}}$$

$$= \frac{2.303}{1.6} \times \log 2$$

$$= \frac{2.303}{1.6} \times 0.3010$$

$$= 0.4332$$

For 90% completion,

$$x = 0.9a$$

$$\therefore \quad k = \frac{2.303}{t} \log \frac{a}{a - 0.9a}$$

$$0.4332 = \frac{2.303}{t} \log \frac{a}{0.1\,a}$$

$$= \frac{2.303}{t} \log 10$$

$$= \frac{2.303}{t} \times 1$$

$$t = \frac{2.303}{0.4332}$$

$$= 5.3 \text{ min}$$

15. $H_4P_2O_7$ (pyrophosphoric acid) is tetrabasic because it gives four H^+ ions.

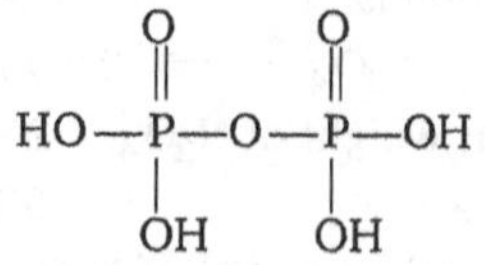

16. Bond angles of ClF_3, PF_3, NF_3 and BF_3 are 80°90′, 101°, 106° and 120° respectively.

17. Isotonic species have the same the number of neutrons.

No. of neutrons in

$${}_6C^{14} = 14 - 6 = 8$$

$${}_7N^{14} = 14 - 7 = 7$$

$${}_9F^{19} = 19 - 9 = 10$$

$${}_7N^{15} = 15 - 7 = 8$$

$${}_9F^{17} = 15 - 7 = 8$$

$${}_6C^{12} = 12 - 6 = 6$$

Hence, ${}_6C^{14}$, ${}_7N^{15}$, ${}_9F^{17}$ are isotonic.

18. $K[Co(CO)_4]$

$$+1 + x + 0 \times 4 = 0$$

$$\therefore \quad x = -1$$

19. The metal placed above in electrochemical series can displace the metals placed below from their salt solution.

The order of given metals in the series is Zn, Fe, Cu, Ag. Thus, Zn metal can displace all other three metals from their salt solutions.

20. Magic numbers are 2, 8, 20, 28, 50 or 82 protons or 2, 8, 20, 28, 50, 82 or 126 neutrons in the nucleus. These numbers impart stability to the nucleus.

(a) ${}_{13}Al^{27}$, no. of neutrons $= 27 - 13 = 14$

(b) ${}_{26}Fe^{56}$, no. of neutrons $= 56 - 26 = 30$

(c) ${}_{83}Bi^{209}$, no. of neutrons $= 209 - 83 = 126$

(d) ${}_{92}U^{238}$, no. of neutrons $= 238 - 92 = 146$

21. $[H_3O^+] = \sqrt{K_a \times C}$

$$= \sqrt{4 \times 10^{-10} \times 2.5 \times 10^{-1}}$$

$$= \sqrt{10^{-10}} = 10^{-5}\ M$$

$$\because \quad pH = -\log [H_3O^+]$$

$$\therefore \quad pH = -\log (10^{-5}) = 5$$

22. $\Delta H = ms\Delta T = (125 + 10.5) \times 1 \times 3.1$

$$= 420.05 \text{ cal} = \frac{420.05}{10.5} \text{ cal/g}$$

$$= 40 \text{ cal/g} = 167.7 \text{ J/g}$$

23. 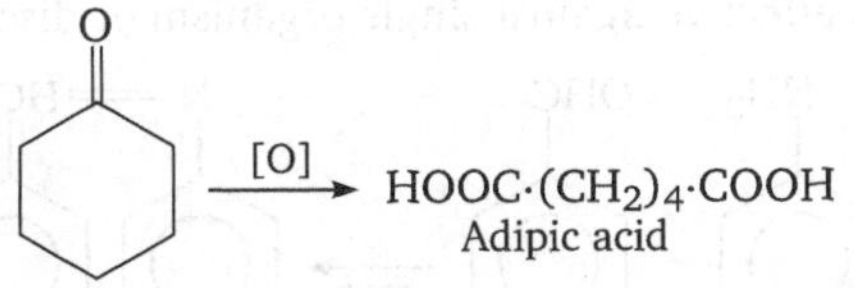

24. The conversion of —CN to —CH_2NH_2 by catalytic reduction is called Mendius reaction.

25. 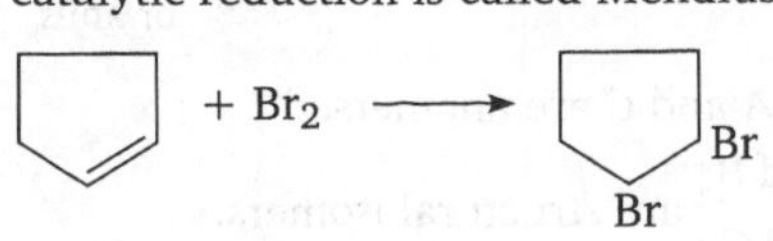

26. No. of atoms of A from corners of unit cell $= \frac{7}{8}$

No. of atoms of B from faces of unit cell = 3

$\therefore \quad A : B = \frac{7}{8} : 3 = 7 : 24$

Thus, formula of the compound is A_7B_{24}.

27. The elements of II period show similar properties as the elements of III period, which are diagonally placed to them. This is known as diagonal relationship.

II Period :	Li	Be	B	C	N
III Period :	Na	Mg	Al	Si	P

(Li ↘ Mg, Be ↘ Al, B ↘ Si, C ↘ P)

28. Molecular weight of $X_4O_6 = (4a + 96)$

$\because$ $(4a + 96)$g X_4O_6 contains $X = 4a$g

$\therefore$ 10 g X_4O_6 contains $X = \frac{4a \times 10}{4a + 96}$

$$5.72 = \frac{4a \times 10}{4a + 96}$$

$$22.88a + 549.12 = 40a$$

$$17.12a = 549.12$$

$$a = 32.07 \text{ amu.}$$

29.

	Formula	Name
(a)	ZnO	Philosopher's wool
(b)	$ZnS + BaSO_4$	Lithopone
(c)	$ZnSO_4 \cdot 7H_2O$	White vitriol
(d)	$ZnO \cdot CoO$	Rinmann's green

30. $_{29}Cu = 2, 8, 18, 1$

$$= 1s^2, 2s^2 2p^6, 3s^2 3p^6 3d^{10}, 4s^1$$

$\therefore$ Quantum numbers for $4s^1$ electron are

$$n = 4, \quad l = 0, \quad m = 0, \quad s = +\frac{1}{2}$$

31. $_5B$(G.S.) = 2s: [⇅] 2p: [↑][ ][ ]

B(E.S.) = 2s: [↑] 2p: [↑][↑][ ]

sp^2-hybridisation

Due to sp^2-hybridisation, BF_3 has trigonal planar structure.

32. Normality, $N = \frac{w}{E \cdot V}$

$$= \frac{10 \text{ g}}{60 \times \frac{100}{1000} \text{ L}}$$

$$= 1.66 \text{ N}$$

33. H—O—S(=O)(=O)—O—O—S(=O)(=O)—O—H

$\because$ 2O-atoms are in peroxide linkage

$\therefore$ Their oxidation state = −1

and, oxidation state of remaining 6O-atoms = −2

Oxidation state of H-atoms = +1

Let oxidation state of S atom = x

$$S_2 \quad H_2 \quad O_6 \quad O_2$$

$$\therefore \quad 2x + 2(+1) + 6(-2) + 2(-1) = 0$$

$$2x + 2 - 12 - 2 = 0$$

$$x = +6$$

34. Reaction (a) and (b) are exothermic and spontaneous so do not need catalyst. On the other hand, conversion of SO_2 into SO_3 need catalyst. Catalyst used for this reaction may be Pt or V_2O_5 (contact process) or NO (lead chamber process)

$$2SO_2 + O_2 \xrightarrow{V_2O_5} 2SO_3$$

35. $N_2(g) + 3H_2(g) \rightleftharpoons 2NH_3(g) + 22$ kcal

$\because$ Activation energy for the forward reaction = 50 kcal

$\therefore$ Activation energy for the backward reaction

$= 50 + 22 = 72$ kcal

36. $CH_2(COOH)_2 \xrightarrow{\Delta} CH_3COOH + CO_2$

Malonic acid

CH_2COOH–$CH_2COOH \longrightarrow$ $(CH_2CO)_2O$

Succinic acid

37. $CHCl_3 + Cl_2 \xrightarrow{h\nu} CCl_4 + HCl$

Carbon tetra chloride or pyrene

38.

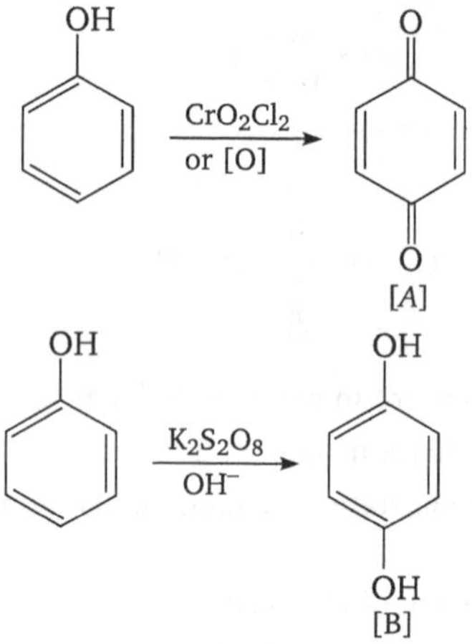

Mechanism

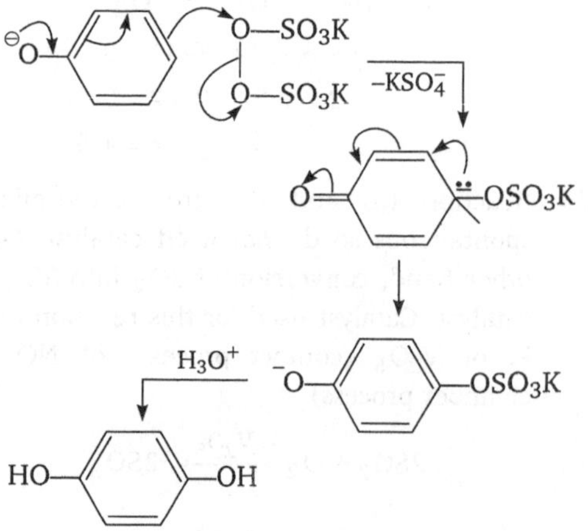

39. In transcription, synthesis of RNA takes place while in translation, synthesis of protein occurs in the cytoplasm of the cell.

40. Penicillin G is a narrow spectrum antibiotics. It is effective against single organism or disease.

41.

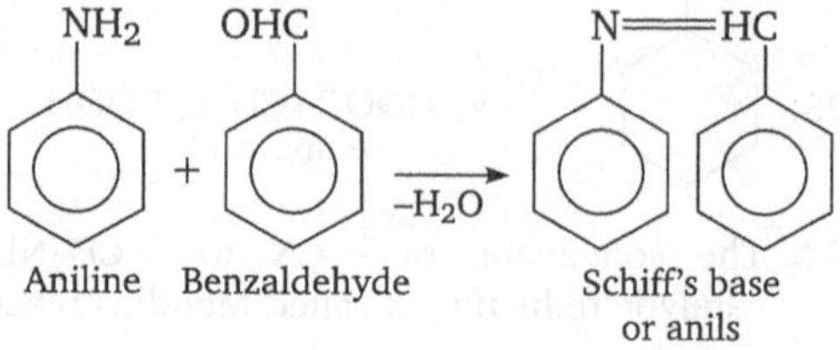

42. Only *A* and *C* are anomers.

A and *B*, *B* and *C* } are structural isomers.

43. According to IUPAC nomenclature  is bicyclo (2,1,0) pentane.

44. $CH_3CH_2OH \xrightarrow[K_2Cr_2O_7/H_2SO_4]{[O]} CH_3CHO$

$\xrightarrow[K_2Cr_2O_7/H_2SO_4]{[O]} CH_3COOH \xrightarrow[\Delta]{NaOH + CaO}$

$CH_4 \xrightarrow{Cl_2} CH_3Cl \xrightarrow[-KCl]{aq.\ KOH} CH_3OH$

45. PDI abbreviates as polydispersity index of polymer.

$$PDI = \frac{\overline{M}_w}{M_n}$$

For natural polymers, PDI = 1
and for synthetic polymers, PDI > 1

46. Diamagnetic species have all electrons paired, whereas paramagnetic species have unpaired electrons.

$H_2 = \sigma 1s^2$

No. of unpaired electron $= 0 \Rightarrow$ diamagnetic

$H_2^- = \sigma 1s, \sigma^* 1s^1$

No. of unpaired electron $= 1 \Rightarrow$ paramagnetic

$H_2^- = \sigma 1s^1$

No. of unpaired electron $= 1 \Rightarrow$ paramagnetic

$He_2^+ = \sigma 1s^2, \sigma^* 1s^1$

No. of unpaired electron $= 1 \Rightarrow$ paramagnetic

47. Non-metals are electronegative in nature.

48. $\because \quad \Delta G = \Delta H - T\Delta S$

At equilibrium, $\Delta G = 0$

$\therefore \quad \Delta H = T\Delta S$

$\Delta H = 273 \times (60.01 - 38.20)$

$= 5954.13 \text{ J mol}^{-1} = 5.954 \text{ kJ mol}^{-1}$

$\approx 6 \text{ kJ mol}^{-1}$

49. $\underset{\text{White phosphorus}}{P_4} + 3NaOH + 3H_2O \xrightarrow{\Delta} \underset{\text{Phosphene}}{PH_3} + \underset{\text{Sodium hypophosphate}}{3NaH_2PO_2}$

50. $\because \quad \Delta T_b = \dfrac{1000 \cdot K_b \cdot w}{m \cdot W} \times i$

$\therefore \quad \Delta T_b \propto i \cdot \dfrac{w}{m \cdot \dfrac{W}{1000}} \quad (\because K_b = \text{constant})$

or $\quad \Delta T_b \propto i \cdot M$

$\left(\because \text{Molality} = \dfrac{w}{m \cdot \dfrac{W}{1000}} \text{ and assuming, Molarity, } M = \text{molality, } m\right)$

Now, for the given solutions,

(a) For 0.08 M $BaCl_2$, $i \cdot M = 3 \times 0.08 = 0.24$

(b) For 0.15 M KCl, $i \cdot M = 2 \times 0.15 = 0.30$

(c) For 0.10 M glucose, $i \cdot M = 1 \times 0.10 = 0.10$

(d) For 0.06 M $Ca(NO_3)_2$, $i \cdot M = 3 \times 0.06 = 0.18$

Hence, ΔT_b is maximum for 0.15 M KCl solution.

51. No. of unpaired electrons in Fe = 5

Magnetic moment, $\mu = \sqrt{n(n+2)} = \sqrt{5(5+2)}$

$= \sqrt{35} = 5.92$ BM

52.

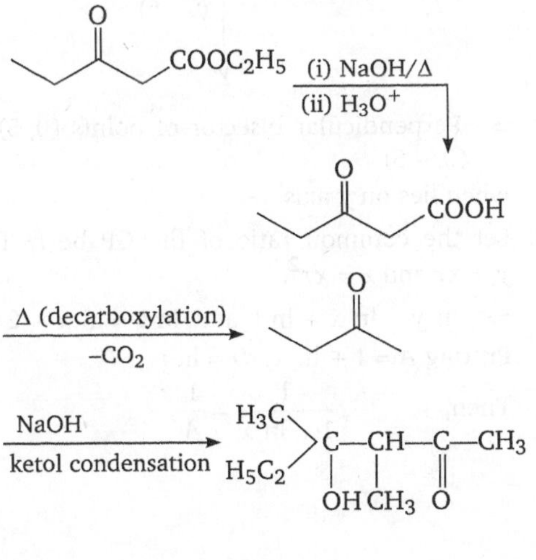

53. $\because \quad \dfrac{r_H}{r_{He}} = \sqrt{\dfrac{M_{He}}{M_H}}$

$= \sqrt{\dfrac{4}{2}} = \sqrt{2} = 1.4$

$\therefore r_H : r_{He} = 1.4 : 1$

54.

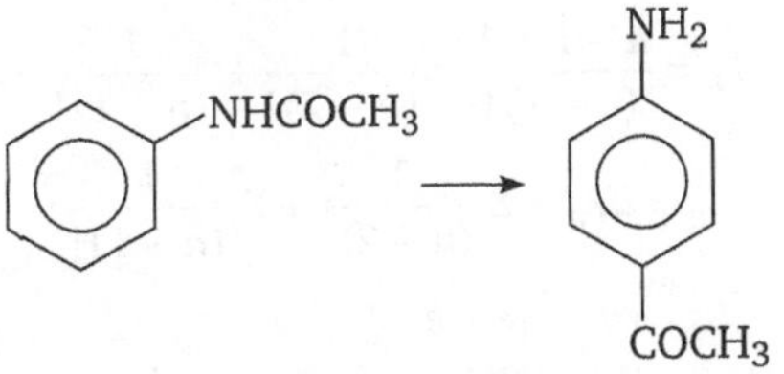

Above reaction is an example of rearrangement reaction.

55. Temporary hardness of water is due to the presence of bicarbonates of calcium and magnesium while permanent hardness is due to the presence of soluble chlorides and sulphates of calcium and magnesium.

56. Vasopressin, oxytocin, insulin, angiotensin, are peptide hormones whereas thyroid, adrenaline are amine hormones.

57. Indigo is a vat dye. Vat dyes are insoluble in water and cann't be applied directly.

58. Solder is an alloy of tin and lead. Its melting point is quite low, hence, it is very useful in stitching ICs in various electrical instruments.

59. Hg_2^{2+} and Pb^{2+} have been placed in I group of basic radicals as their K_{sp} are less than the II group ions (Hg^{2+}, Cd^{2+}). Hence, only these will precipitate.

60.

$2\pi e^- \Rightarrow 4 \times 0 + 2 = 4n + 2$

$\therefore$ According to Huckel rule, cyclopropenyl carbocation is aromatic in nature.

Mathematics

1. Let $S = 1 + \frac{2^2}{2!} + \frac{3^2}{3!} + \frac{4^2}{4!} + \ldots$

$\therefore \quad T_n = \frac{n^2}{n!} = \frac{n}{(n-1)!}$

$\Rightarrow \quad T_n = \frac{n-1+1}{(n-1)!} = \frac{1}{(n-2)!} + \frac{1}{(n-1)!}$

$\therefore \quad S_n = \Sigma T_n = \Sigma \frac{1}{(n-2)!} + \Sigma \frac{1}{(n-1)!}$

$= e + e$

$= 2e$

2. We have, $f(x) = (x+b)^2 + 2c^2 - b^2$

$\Rightarrow \quad \min f(x) = 2c^2 - b^2$

and $\quad g(x) = b^2 + c^2 - (x+c)^2$

$\Rightarrow \quad \max g(x) = b^2 + c^2$

Thus, $\quad \min f(x) > \max g(x)$

$\Rightarrow \quad 2c^2 - b^2 > b^2 + c^2$

$\Rightarrow \quad c^2 > 2b^2 \Rightarrow |c| > \sqrt{2}|b|$

3. Let $z_1 = \gamma_1 (\cos \theta_1 + i \sin \theta_1)$.

Given that, $\left|\frac{z_1}{z_2}\right| = 1$

$\Rightarrow \quad |z_1| = |z_2| \Rightarrow |z_1| = |z_2| = \gamma_1$

Now, $\arg(z_1 z_2) = 0 \Rightarrow \arg(z_1) + \arg(z_2) = 0$

$\Rightarrow \quad \arg(z_2) = -\theta_1$

Therefore, $z_2 = \gamma_1 (\cos(-\theta_1) + i \sin(-\theta_1))$

$= \gamma_1 (\cos \theta_1 - i \sin \theta_1) = \bar{z}_1$

$\Rightarrow \quad \bar{z}_2 = (\bar{\bar{z}}_1) = z_1 \Rightarrow z_2 \bar{z}_2 = z_1 z_2$

$\Rightarrow \quad |z_2|^2 = z_1 z_2$

4. We have, $\cos A = m \cos B$

$\Rightarrow \quad \frac{\cos A}{\cos B} = \frac{m}{1}$

On applying componendo and dividendo rule, we get

$\Rightarrow \quad \frac{\cos A + \cos B}{\cos A - \cos B} = \frac{m+1}{m-1}$

$\Rightarrow \quad \frac{2\cos\left(\frac{A+B}{2}\right)\cos\left(\frac{B-A}{2}\right)}{2\sin\left(\frac{A+B}{2}\right)\sin\left(\frac{B-A}{2}\right)} = \frac{m+1}{m-1}$

$\Rightarrow \quad \cot\left(\frac{A+B}{2}\right) = \left(\frac{m+1}{m-1}\right)\tan\left(\frac{B-A}{2}\right)$

But, $\cot \frac{A+B}{2} = \lambda \tan\left(\frac{B-A}{2}\right)$ (given)

$\Rightarrow \quad \lambda = \frac{m+1}{m-1}$

5. Let the angles be $A = x - d$, $B = x$, $C = x + d$,
Then, $x - d + x + x + d = 180°$

$\Rightarrow \quad 3x = 180° \Rightarrow x = 60°$

Therefore, two larger angles are B and C.
Hence, $b = 9$ and $c = 10$.
Now,

$\cos B = \frac{c^2 + a^2 - b^2}{2ac}$

$\Rightarrow \quad \frac{1}{2} = \frac{100 + a^2 - 81}{20a}$

$\Rightarrow \quad a^2 - 10a + 19 = 0$

$\Rightarrow \quad a = 5 \pm \sqrt{6}$

6. $\left|\frac{z - 5i}{z + 5i}\right| = 1 \Rightarrow |z - 5i| = |z + 5i|$

(Using definition $|z - z_1| = |z - z_2|$ gives Perpendicular bisector of z_1 and z_2.)

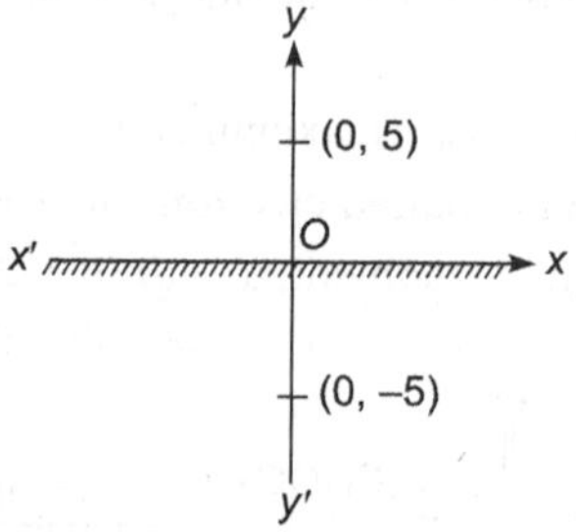

$\Rightarrow$ Perpendicular bisector of points (0, 5) and (0, – 5).
whch lies on y-axis.

7. Let the common ratio of the GP be r. Then, $y = xr$ and $z = xr^2$.

$\Rightarrow \quad \ln y = \ln x + \ln r$ and $\ln z = \ln x + 2\ln r$

Putting $A = 1 + \ln x$, $D = \ln r$

Then, $\quad \frac{1}{1 + \ln x} = \frac{1}{A}$,

$$\frac{1}{1+\ln y}=\frac{1}{1+\ln xr}=\frac{1}{1+\ln x+\ln r}=\frac{1}{A+D}$$

and $\frac{1}{1+\ln z}=\frac{1}{1+\ln x+2\ln r}=\frac{1}{A+2D}$

Here we see that A, $A+D$ and $A+2D$ are in AP.

$\therefore$ $\frac{1}{A}, \frac{1}{A+D}$ and $\frac{1}{A+2D}$ are in HP.

Therefore, $\frac{1}{1+\ln x}, \frac{1}{1+\ln y}, \frac{1}{1+\ln z}$, are in HP.

8. We have,

$$\frac{e^x}{1-x}=B_0+B_1x+B_2x^2+\ldots+B_nx^n+\ldots$$

$$\Rightarrow e^x(1-x)^{-1}=B_0+B_1x+B_2x^2+\ldots+B_nx^n+\ldots$$

$$\Rightarrow \left(1+\frac{x}{1!}+\frac{x^2}{2!}+\ldots+\frac{x^{n-1}}{(n-1)!}+\frac{x^n}{n!}+\ldots\right)\times$$

$$(1+x+x^2+\ldots+x^{n-1}+x^n+\ldots\infty)$$

$$=B_0+B_1x+B_2x^2+\ldots+B_nx^n+\ldots$$

On comparing the coefficients of x^n and x^{n-1} on both sides, we get

$$\frac{1}{n!}+\frac{1}{(n-1)!}+\ldots+\frac{1}{2!}+\frac{1}{1!}+1=B_n$$

and $\frac{1}{(n-1)!}+\frac{1}{(n-2)!}+\ldots+\frac{1}{2!}+\frac{1}{1!}+1=B_{n-1}$

$$B_n-B_{n-1}=\frac{1}{n!}$$

9. As, $f(x)=\begin{cases}\frac{\sin[x]}{[x]}, & [x]\neq 0\\ 0, & [x]=0\end{cases}$

$$\Rightarrow f(x)=\begin{cases}\frac{\sin[x]}{[x]}, & x\in R-[0,1)\\ 0, & 0\le x<1\end{cases}$$

$\therefore$ RHL at $x=0$

$$\lim_{x\to 0^+} f(x)=\lim_{x\to 0}\frac{\sin[0+h]}{[0+h]}=0$$

LHL at $x=0$

$$\lim_{x\to 0^-} f(x)=\lim_{h\to 0}\frac{\sin[0-h]}{[0-h]}$$

$$=\lim_{h\to 0}\frac{\sin(-1)}{-1}=\sin 1$$

Since, RHL $\neq$ LHL

$\therefore$ Limit does not exist.

10. Here, $\frac{dy}{dt}-\left(\frac{t}{1+t}\right)y=\frac{1}{1+t}$ and $y(0)=-1$

which represents linear differential equation of first order.

$$\text{IF}=e^{-\int\left(\frac{t}{1+t}\right)dt}=e^{-\int\frac{(t+1-1)}{1+t}dt}$$

$$=e^{-t+\log(1+t)}$$

$$=e^{-t}(1+t)$$

$\therefore$ Required solution is

$$y(\text{IF})=[\int Q\cdot(\text{IF})\,dt]+C$$

$$ye^{-t}(1+t)=\int\frac{1}{1+t}e^{-t}(1+t)\,dt+C$$

$$=\int e^{-t}\,dt+C=-e^{-t}+C$$

At $t=0, y=-1$

$\therefore$ $-1(1+0)=-e^0+C$

$\Rightarrow$ $C=0$

$\therefore$ $ye^{-t}(1+t)=-e^{-t}$

At $t=1$

$ye^{-1}(1+1)=-e^{-1}$

$\Rightarrow$ $y=-\frac{1}{2}$

11. As, we have to find the locus of mid-point of chord and we know perpendicular drawn from centre to the chord, it bisects the chord.

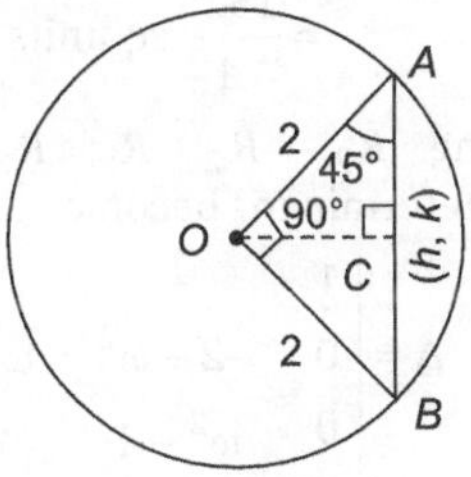

Given, $\angle OAB=90°$

Also, OC bisects the $\angle AOB$

$\therefore$ $\angle COA=\angle OAC=45°$

In ΔOAC, $\frac{OC}{OA}=\sin 45° \Rightarrow OC=\frac{2}{\sqrt{2}}=\sqrt{2}$

$\therefore$ $h^2+k^2=OC^2$

$\Rightarrow$ $h^2+k^2=(\sqrt{2})^2$

or $x^2 + y^2 = 2$ is required equation of locus of mid-point of chord subtending right angle at centre.

12. Equation of any tangent of the parabola, $y^2 = 4ax$ is $y = mx + \frac{a}{m}$. This line will touch the circle $x^2 + y^2 = \frac{a^2}{2}$.

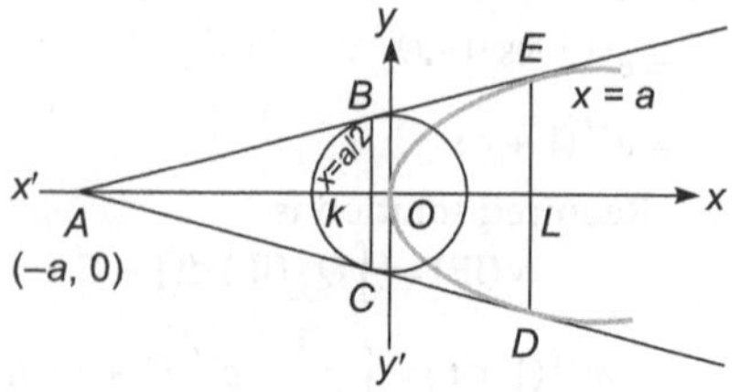

$$\therefore \quad BC = 2\sqrt{OB^2 - OK^2}$$
$$= 2\sqrt{\frac{a^2}{2} - \frac{a^2}{4}}$$
$$= 2\sqrt{\frac{a^2}{4}} = a$$

and we know that DE is the latusrectum of the parabola, so its length is $4a$.

Thus, area of trapezium $BCDE$

$$= \frac{1}{2}(BC + DE)(KL)$$
$$= \frac{1}{2}(a + 4a)\left[\frac{3a}{2}\right]$$
$$= \frac{15a^2}{4} \text{ sq units}$$

13. Applying $R_2 \to R_2 - R_1;\ R_3 \to R_3 - R_1$, the given determinant becomes

$$\Delta = \begin{vmatrix} 1 & 1 & 1 \\ 0 & -2-\omega^2 & \omega^2 - 1 \\ 0 & \omega^2 - 1 & \omega - 1 \end{vmatrix}$$

$$= 1[(\omega - 1)(-2 - \omega^2) - \{(\omega^2 - 1)(\omega^2 - 1)\}] - 1(0) + 1(0)$$
$$= (-2 - \omega^2)(\omega - 1) - (\omega^2 - 1)^2$$
$$= -2\omega + 2 - \omega^3 + \omega^2 - (\omega^4 + 1 - 2\omega^2)$$
$$= -2\omega + 1 + \omega^2 - (\omega + 1 - 2\omega^2)$$
$$= 3\omega^2 - 3\omega$$
$$= 3\omega(\omega - 1)$$

14. Let α, β be the roots of given quadratic equation. Then,

$$\alpha + \beta = \frac{4 + \sqrt{5}}{5 + \sqrt{2}} \quad \text{and} \quad \alpha\beta = \frac{8 + 2\sqrt{5}}{5 + \sqrt{2}}$$

Again, H be the harmonic mean between α and β, then $H = \frac{2\alpha\beta}{\alpha + \beta} = \frac{16 + 4\sqrt{5}}{4 + \sqrt{5}} = 4$

15. For $x = -3, 3; |x^2 - 9| = 0$

Therefore, $\log|x^2 - 9|$ does not exist at

$$x = -3, 3$$

Hence, domain of function is $R - \{-3, 3\}$.

16. Let $\quad y = \left[\frac{f(1 + x)}{f(1)}\right]^{1/x}$

$$\Rightarrow \quad \log y = \frac{1}{x}[\log f(1 + x) - \log f(1)]$$
$$\Rightarrow \lim_{x \to 0} \log y = \lim_{x \to 0} \frac{[\log f(1 + x) - \log f(1)]}{x}$$
$$\Rightarrow \lim_{x \to 0} \log y = \lim_{x \to 0}\left[\frac{1}{f(1 + x)} f'(1 + x)\right]$$

(using L' Hospital's rule)

$$= \frac{f'(1)}{f(1)} = \frac{6}{3} = 2$$
$$\Rightarrow \quad \lim_{x \to 0} y = e^2$$

17. Now, $[\mathbf{a} \quad \mathbf{a} + \mathbf{b} \quad \mathbf{a} + \mathbf{b} + \mathbf{c}]$

$$= [\mathbf{a} \times (\mathbf{a} + \mathbf{b})] \cdot (\mathbf{a} + \mathbf{b} + \mathbf{c})$$
$$= (\mathbf{a} \times \mathbf{b}) \cdot (\mathbf{a} + \mathbf{b} + \mathbf{c}) = (\mathbf{a} \times \mathbf{b}) \cdot \mathbf{c}$$
$$= [\mathbf{a}\ \mathbf{b}\ \mathbf{c}]$$

18. $\mathbf{a} = -\mathbf{i} + 2\mathbf{j} - \mathbf{k}, \mathbf{b} = \mathbf{i} + \mathbf{j} - 3\mathbf{k}, \mathbf{c} = -4\mathbf{i} - \mathbf{k}$

$$\therefore \quad \mathbf{a} \times (\mathbf{b} \times \mathbf{c}) + (\mathbf{a} \cdot \mathbf{b})\mathbf{c} = (\mathbf{a} \cdot \mathbf{c})\,\mathbf{b}$$
$$= (4 + 1)(\mathbf{i} + \mathbf{j} - 3\mathbf{k})$$
$$= 5\mathbf{i} + 5\mathbf{j} - 15\mathbf{k}$$

19. Here, $I(m, n) = \int_0^1 t^m (1 + t)^n \, dt$

[We apply integration by parts, taking $(1 + t)^n$ as first and t^m as second function]

$$I(m, n) = \left[(1 + t)^n \cdot \frac{t^{m+1}}{m + 1}\right]_0^1 - \int_0^1 n(1 + t)^{n-1} \cdot \frac{t^{m+1}}{m + 1} \, dt$$

$$= \frac{2^n}{m+1} - \frac{n}{m+1} \int_0^1 (1+t)^{n-1} \cdot t^{m+1}\, dt$$

$$\therefore \quad I(m, n) = \frac{2^n}{m+1} - \frac{n}{m+1} \cdot I(m+1, n-1)$$

20. Let O is the point on centre and P is the point on circumference. Therefore, angle QOR is double the angle QPR. So, it is sufficient to find the angle QOR.

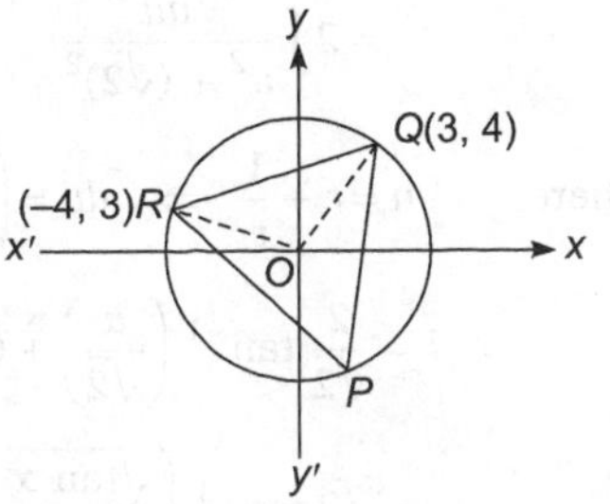

Now, slope of OQ, $m_1 = \frac{4-0}{3-0} = \frac{4}{3}$

Slope of OR, $m_2 = \frac{3-0}{-4-0} = -\frac{3}{4}$

Again, $m_1 m_2 = -1$

Therefore, $\angle QOR = 90°$

Which implies that $\angle QPR = \frac{90°}{2} = 45°$.

21. We have, $y^2 + 4y + 4x + 2 = 0$

$$\Rightarrow \quad (y+2)^2 + 4x - 2 = 0$$

$$\Rightarrow \quad (y+2)^2 = -4\left(x - \frac{1}{2}\right)$$

Replace, $y + 2 = Y,\ x - \frac{1}{2} = X$

We have, $Y^2 = -4X$

This is a parabola with directrix at $X = 1$

$$\Rightarrow \quad x - \frac{1}{2} = 1$$

$$\Rightarrow \quad x = \frac{3}{2}$$

22. Given, $y = mx - b\sqrt{1+m^2}$ touches both the circles.

So, distance from centre = Radius of both the circles

$$\therefore \quad \frac{|-b\sqrt{1+m^2}|}{\sqrt{m^2+1}} = b$$

and $$\frac{|ma - 0 - b\sqrt{1+m^2}|}{\sqrt{m^2+1}} = b$$

$$\Rightarrow \quad |ma - b\sqrt{1+m^2}| = |-b\sqrt{1+m^2}|$$

$$\Rightarrow \quad m^2a^2 - 2abm\sqrt{1+m^2} + b^2(1+m^2) = b^2(1+m^2)$$

$$\Rightarrow \quad ma - 2b\sqrt{1+m^2} = 0$$

$$\Rightarrow \quad m^2a^2 = 4b^2(1+m^2)$$

$$\Rightarrow \quad m^2(a^2 - 4b^2) = 4b^2$$

$$\Rightarrow \quad m = \frac{2b}{\sqrt{a^2 - 4b^2}}$$

23. Here, $\alpha \in \left[0, \frac{\pi}{2}\right) \Rightarrow \tan\alpha$ is (+ve)

As, we know, if $a, b > 0 \Rightarrow \frac{a+b}{2} \ge \sqrt{ab}$

i.e., AM ≥ GM

$$\therefore \quad \frac{\sqrt{x^2+x} + \frac{\tan^2\alpha}{\sqrt{x^2+x}}}{2} \ge \sqrt{\sqrt{x^2+x} \cdot \frac{\tan^2\alpha}{\sqrt{x^2+x}}}$$

(using AM ≥ GM)

$$\Rightarrow \quad \sqrt{x^2+x} + \frac{\tan^2\alpha}{\sqrt{x^2+x}} \ge 2\tan\alpha$$

24. Let x be the length of an edge of a cube and V be the volume of that cube.

$$\therefore \quad V = x^3$$

On differentiating w.r.t. x, we get

$$\frac{dV}{dx} = 3x^2$$

Let δV be error in V and corresponding error δx in x.

$$\therefore \quad \delta V = \frac{dV}{dx}\,\delta x = 3x^2\,\delta x$$

Given that, $x = 2$ m and $\delta x = 0.5$ cm $= \frac{0.5}{100}$ m

$$\therefore \quad \delta V = 3(2)^2\left(\frac{0.5}{100}\right)$$

$$= \frac{12 \times 0.5}{100} = \frac{6}{100} = 0.06\ \text{m}^3$$

25. Let $g(x) = f(x) - x^2$

$\Rightarrow$ $g(x)$ has atleast 3 real roots which are $x = 1, 2, 3$ (by mean value theorem)

$\Rightarrow$ $g'(x)$ has atleast 2 real roots in $x \in (1, 3)$

$\Rightarrow$ $g''(x)$ has atleast 1 real root in $x \in (1, 3)$

$\Rightarrow$ $f''(x) = 2$ for atleast one $x \in (1, 3)$

26. We have,

$$y = (C_1 + C_2)\cos(x + C_3) - C_4 e^{x + C_5} \quad \ldots(i)$$

$$\Rightarrow y = (C_1 + C_2)\cos(x + C_3) - C_4 e^x \cdot e^{C_5}$$

Now, let $C_1 + C_2 = A, C_3 = B, C_4\, e^{C_5} = C$

$$\Rightarrow \quad y = A\cos(x + B) - Ce^x \quad \ldots(ii)$$

On differentiating w.r.t. x, we get

$$\frac{dy}{dx} = -A\sin(x + B) - Ce^x \quad \ldots(iii)$$

Again, differentiating w.r.t. x, we get

$$\frac{d^2y}{dx^2} = -A\cos(x + B) - Ce^x \quad \ldots(iv)$$

$$\Rightarrow \quad \frac{d^2y}{dx^2} = -y - 2Ce^x$$

$$\Rightarrow \quad \frac{d^2y}{dx^2} + y = -2Ce^x \quad \ldots(v)$$

Again, differentiating w.r.t. x, we get

$$\frac{d^3y}{dx^3} + \frac{dy}{dx} = -2Ce^x \quad \ldots(vi)$$

$$\Rightarrow \quad \frac{d^3y}{dx^3} + \frac{dy}{dx} = \frac{d^2y}{dx^2} + y \text{ [from Eq. (v)]}$$

which is a differential equation of order 3.

27. Let $I = \int (\sqrt{\tan x} + \sqrt{\cot x})\, dx$

$$= \int \frac{\tan x + 1}{\sqrt{\tan x}}\, dx$$

Put $\tan x = t^2$

$$\Rightarrow \sec^2 x\, dx = 2t\, dt$$

$$\Rightarrow \quad dx = \frac{2t\, dt}{1 + \tan^2 t} = \frac{2t}{1 + t^4}\, dt$$

$$\therefore \quad I = \int \frac{t^2 + 1}{\sqrt{t^2}} \cdot \frac{2t}{t^4 + 1}\, dt$$

$$= 2\int \frac{t^2 + 1}{t^4 + 1}\, dt$$

$$= 2\int \frac{1 + \frac{1}{t^2}}{t^2 + \frac{1}{t^2} - 2 + 2}\, dt$$

$$= 2\int \frac{1 + \frac{1}{t^2}}{\left(t - \frac{1}{t}\right)^2 + (\sqrt{2})^2}\, dt$$

$$= 2\int \frac{du}{u^2 + (\sqrt{2})^2}$$

where, $\quad u = t - \frac{1}{t} \Rightarrow du = \left(1 + \frac{1}{t^2}\right) dt$

$$\Rightarrow \quad I = \frac{2}{\sqrt{2}} \tan^{-1}\left(\frac{u}{\sqrt{2}}\right) + C$$

$$= \sqrt{2}\tan^{-1}\left(\frac{\sqrt{\tan x} - \sqrt{\cot x}}{\sqrt{2}}\right) + C$$

28. Let number of candidates be n. Therefore, $n - 2$ are to be elected and so one can vote up to $(n - 2)$. Hence, number of ways in which one can vote

$$= {}^nC_1 + {}^nC_2 + \ldots + {}^nC_{n-2} = 56 \quad \text{(given)}$$

$$\Rightarrow \quad 2^n - ({}^nC_0 + {}^nC_{n-1} + {}^nC_n) = 56$$

$$\Rightarrow \quad 2^n - n = 58 \Rightarrow 2^n = 58 + n$$

Which is satisfied by $n = 6$ only.

29. We have, $\quad x = t^2 + t + 1 \quad \ldots(i)$

and $\quad y = t^2 - t + 1 \quad \ldots(ii)$

Now, $\quad x + y = 2(1 + t^2) \quad \ldots(iii)$

and $\quad x - y = 2t \quad \ldots(iv)$

Now, from Eqs. (iii) and (iv), we get

$$x + y = 2\left[1 + \left(\frac{x - y}{2}\right)^2\right]$$

$$\Rightarrow \quad x + y = 2\left[\frac{4 + x^2 + y^2 - 2xy}{4}\right]$$

$$\Rightarrow \quad x^2 + y^2 - 2xy - 2x - 2y + 4 = 0 \quad \ldots(v)$$

$\therefore$ On comparing with

$$ax^2 + 2hxy + by^2 + 2gx + 2fy + c = 0$$

We get, $a = 1, b = 1, c = 4, h = -1, g = -1, f = -1$

$\because \quad \Delta = abc + 2fgh - af^2 - bg^2 - ch^2$

Now, $\Delta = 1\cdot 1\cdot 4 + 2(-1)(-1)(-1) - 1\times(-1)^2 - 1\times(-1)^2 - 4(-1)^2$

$= 4 - 2 - 1 - 1 - 4$

$= -4$ therefore, $\Delta \neq 0$

and $ab - h^2 = 1\cdot 1 - (1)^2 = 1 - 1 = 0$

So, it is equation of a parabola.

30. Equation of tangent is drawn at a $(3\sqrt{3}\cos\theta, \sin\theta)$ to the curve

$$\frac{x^2}{27} + \frac{y^2}{1} = 1$$

is $\dfrac{x\cos\theta}{3\sqrt{3}} + \dfrac{y\sin\theta}{1} = 1$

Thus, sum of intercepts

$= (3\sqrt{3}\sec\theta + \operatorname{cosec}\theta) = f(\theta)$ (say)

$\Rightarrow f'(\theta) = 3\sqrt{3}\sec\theta\tan\theta - \operatorname{cosec}\theta\cot\theta$

$$= \frac{3\sqrt{3}\sin^3\theta - \cos^3\theta}{\sin^2\theta\cos^2\theta}$$

Put $f'(\theta) = 0$

$\Rightarrow \sin^3\theta = \dfrac{1}{3^{3/2}}\cos^3\theta$

$\Rightarrow \tan\theta = \dfrac{1}{\sqrt{3}}$

$\Rightarrow \theta = \dfrac{\pi}{6}$

31. Since, a, b, c are in AP.

$\therefore a = A - D, b = A, c = A + D$

Where, A is the first term and D is the common difference of an AP.

Given, $a + b + c = \dfrac{3}{2}$

$\Rightarrow (A - D) + A + (A + D) = \dfrac{3}{2}$

$\Rightarrow 3A = \dfrac{3}{2}$

$\Rightarrow A = \dfrac{1}{2}$

$\therefore$ The numbers are $\dfrac{1}{2} - D, \dfrac{1}{2}, \dfrac{1}{2} + D$

Also, $\left(\dfrac{1}{2} - D\right)^2, \dfrac{1}{4}, \left(\dfrac{1}{2} + D\right)^2$ are in GP.

$\Rightarrow \left(\dfrac{1}{4}\right)^2 = \left(\dfrac{1}{2} - D\right)^2\left(\dfrac{1}{2} + D\right)^2$

$\Rightarrow \dfrac{1}{16} = \left(\dfrac{1}{4} - D^2\right)^2$

$\Rightarrow \dfrac{1}{4} - D^2 = \pm\dfrac{1}{4}$

$\Rightarrow D^2 = \dfrac{1}{2}$ ($\because D = 0$ is not possible)

$\Rightarrow D = \pm\dfrac{1}{\sqrt{2}}$

$\therefore a = \dfrac{1}{2} \pm \dfrac{1}{\sqrt{2}}$

So, out of the given value $a = \dfrac{1}{2} - \dfrac{1}{\sqrt{2}}$ is the right choice.

32. For $f(x)$ to be continuous, we must have

$$f(0) = \lim_{x\to 0} f(x)$$

$\therefore \lim_{x\to 0}\dfrac{\log(1 + ax) - \log(1 - bx)}{x}$

$= \lim_{x\to 0}\dfrac{a\log(1 + ax)}{ax} + \dfrac{b\log(1 - bx)}{-bx}$

$= a\cdot 1 + b\cdot 1$ $\left[\text{using, } \lim_{x\to 0}\dfrac{\log(1 + x)}{x} = 1\right]$

$= a + b$

$\therefore f(0) = a + b$

33. Here, $f(x) = \sqrt{\sin^{-1}(2x) + \dfrac{\pi}{6}}$, to find domain, we must have,

$\sin^{-1}(2x) + \dfrac{\pi}{6} \geq 0$ $\left(\text{but } -\dfrac{\pi}{2} \leq \sin^{-1}\theta \leq \dfrac{\pi}{2}\right)$

$\therefore -\dfrac{\pi}{6} \leq \sin^{-1}(2x) \leq \sin\dfrac{\pi}{2}$

$\Rightarrow \sin\left(-\dfrac{\pi}{6}\right) \leq 2x \leq \sin\left(\dfrac{\pi}{2}\right)$

$\Rightarrow -\dfrac{1}{2} \leq 2x \leq 1$

$\Rightarrow -\dfrac{1}{4} \leq x \leq \dfrac{1}{2}$

$\Rightarrow x \in \left[-\dfrac{1}{4}, \dfrac{1}{2}\right]$

34. $x\,dy = y\,(dx + y\,dy), y > 0$

$\Rightarrow x\,dy - y\,dx = y^2\,dy$

$\Rightarrow \dfrac{x\,dy - y\,dx}{y^2} = dy$

$\Rightarrow \quad -d\left(\frac{x}{y}\right) = dy$

On integrating both sides, we get

$$\frac{x}{y} = -y + C \quad \ldots(i)$$

As $y(1) = 1 \Rightarrow x = 1, y = 1$

$\therefore \quad C = 2$

$\therefore$ Eq. (i) becomes

$$\frac{x}{y} + y = 2$$

Again, for $x = -3$

$\Rightarrow \quad -3 + y^2 = 2y$

$\Rightarrow \quad y^2 - 2y - 3 = 0$

$\Rightarrow \quad (y+1)(y-3) = 0$

As $y > 0$, we take $y = 3$, (neglecting $y = -1$)

35. Let $I = \int \frac{(x-1)e^x}{(x+1)^3} dx$

$\Rightarrow \quad I = \int \left\{\frac{x+1-2}{(x+1)^3}\right\} e^x \, dx$

$$= \int \left\{\frac{1}{(x+1)^2} - \frac{2}{(x+1)^3}\right\} e^x \, dx$$

$$= \int e^x \cdot \frac{1}{(x+1)^2} dx - 2\int e^x \frac{1}{(x+1)^3} dx$$

Applying integrating by parts, we get

$$= \left(\frac{1}{(x+1)^2} e^x - \int e^x \frac{(-2)}{(x+1)^3} dx\right) - 2\int e^x \frac{1}{(x+1)^3} dx$$

$$= \frac{e^x}{(x+1)^2} + C$$

Note We can use the formula

$$\int e^x [f(x) + f'(x)] \, dx = e^x f(x) + C$$

36. Now, $\log_{140} 63 = \log_{2^2 \times 5 \times 7} (3 \times 3 \times 7)$

$$= \frac{\log_2 (3 \times 3 \times 7)}{\log_2 (2^2 \times 5 \times 7)} = \frac{\log_2 3 + \log_2 3 + \log_2 7}{2\log_2 2 + \log_2 5 + \log_2 7}$$

$$= \frac{2a + \frac{1}{c}}{2 + b + \frac{1}{c}} = \frac{2ac + 1}{2c + bc + 1}$$

37. Let $f(x) = \frac{\log_e x}{x}$

On differentiating, w.r.t. x, we get

$$f'(x) = \frac{1}{x^2} - \frac{\log_e x}{x^2}$$

For maximum or minimum value of $f(x)$,

Put $\quad f'(x) = 0$

$\Rightarrow \quad \frac{1 - \log_e x}{x^2} = 0$

$\Rightarrow \quad \log_e x = 1$

$\Rightarrow \quad x = e$, which lies in $(0, \infty)$.

For $x = e$, $f''(x) = -\text{ve}$

Hence, y is maximum at $x = e$ and its maximum value $= \frac{\log_e e}{e} = \frac{1}{e}$

38. LHS

$$= \frac{(\cos x + i \sin x)(\cos y + i \sin y)}{(\cos u + i \sin u)(\cos v + i \sin v)} \sin u \cos v$$

$$= \sin u \cos v \, e^{ix} \cdot e^{iy} \cdot e^{-iu} \cdot e^{-iv}$$

$$= \sin u \cos v [\cos (x + y - u - v) + i \sin (x + y - u - v)]$$

39. Let e be the identity,

$\therefore \quad a * e = a$

$\Rightarrow \quad \frac{ae}{6} = a$

$\Rightarrow \quad e = 6$

$\because \quad a * a^{-1} = e$

$\therefore \quad a * a^{-1} = 6$

$\Rightarrow \quad \frac{aa^{-1}}{6} = 6$

$\Rightarrow \quad a^{-1} = \frac{6 \times 6}{a}$

$\therefore \quad 9^{-1} = \frac{6 \times 6}{9} = 4$

40. $\therefore \quad \sin^2 \alpha + \sin (\beta - \gamma) \sin (\beta + \gamma)$

$$= \sin^2 \alpha + \sin (\pi - \alpha) \sin (\beta - \gamma)$$

$$= \sin \alpha [\sin \alpha + \sin (\beta - \gamma)]$$

$$= \sin \alpha [\sin (\pi - (\beta + \gamma)) + \sin (\beta - \gamma)]$$

$$= \sin \alpha [\sin (\beta - \gamma) + \sin (\beta + \gamma)]$$

$$= 2 \sin \alpha \sin \beta \cos \gamma$$

41. We have, $\dfrac{2\cos A}{a} + \dfrac{\cos B}{b} + \dfrac{2\cos C}{c} = \dfrac{a}{bc} + \dfrac{b}{ac}$

On multiplying both sides by abc

$\Rightarrow 2bc\cos A + ac\cos B + 2ab\cos C = a^2 + b^2$

$$\Rightarrow (b^2 + c^2 - a^2) + \frac{(c^2 + a^2 - b^2)}{2} + (a^2 + b^2 - c^2) = a^2 + b^2$$

$$\Rightarrow (c^2 + a^2 - b^2) = 2a^2 - 2b^2$$

$$\Rightarrow b^2 + c^2 = a^2$$

42. $\dfrac{\sin^2 A + \sin A + 1}{\sin A} = \sin A + \dfrac{1}{\sin A} + 1$

$$= \left[\left(\sqrt{\sin A} - \frac{1}{\sqrt{\sin A}}\right)^2 + 3\right] \geq 3$$

43. The given system has a non-trivial solution, if

$$\begin{vmatrix} \lambda & \sin\alpha & \cos\alpha \\ 1 & \cos\alpha & \sin\alpha \\ -1 & \sin\alpha & -\cos\alpha \end{vmatrix} = 0$$

By expanding the determinant along (C_1), we get

$$\lambda(-\cos^2\alpha - \sin^2\alpha) - 1(-\sin\alpha\cos\alpha - \sin\alpha\cos\alpha) - 1(\sin^2\alpha - \cos^2\alpha) = 0$$

$$\lambda = \sin 2\alpha + \cos 2\alpha$$

For $\lambda = 1$, $\sin 2\alpha + \cos 2\alpha = 1$

$$\Rightarrow \frac{1}{\sqrt{2}}\sin 2\alpha + \frac{1}{\sqrt{2}}\cos 2\alpha = \frac{1}{\sqrt{2}}$$

$$\Rightarrow \cos\left(2\alpha - \frac{\pi}{4}\right) = \cos\left(2n\pi \pm \frac{\pi}{4}\right)$$

$$\Rightarrow 2\alpha = 2n\pi \pm \frac{\pi}{4} + \frac{\pi}{4}, \text{ } n \text{ being an integer.}$$

$$\Rightarrow \alpha = n\pi + \frac{\pi}{4}, n\pi$$

44. Let $f(x) = x^2 - 2ax + a^2 + a - 3 = 0$. Since, $f(x)$ has real roots both less than 3. Therefore, $D > 0$ and $f(3) > 0$

$\Rightarrow 4a^2 - 4(a^2 + a - 3) > 0$

and $a^2 - 5a + 6 > 0$

$\Rightarrow a < 3$ and $a < 2$ or $a > 3$

$\Rightarrow a < 2$

45. $\because |z_1| = |z_2| + |z_1 - z_2|$

$\Rightarrow |z_1| - |z_2| = |z_1 - z_2|$

$\Rightarrow (|z_1| - |z_2|)^2 = |z_1 - z_2|^2$

$\Rightarrow |z_1|^2 + |z_2|^2 - 2|z_1||z_2| = |z_1|^2 + |z_2|^2 - 2|z_1||z_2|\cos(\theta_1 - \theta_2)$

$\Rightarrow \cos(\theta_1 - \theta_2) = 1$

$\Rightarrow \theta_1 - \theta_2 = 0$

$\Rightarrow \arg(z_1) - \arg(z_2) = 0$

$\Rightarrow \dfrac{z_1}{z_2}$ is purely real.

$\Rightarrow \text{Im}\left(\dfrac{z_1}{z_2}\right) = 0$

46. Since, total number of numbers 1 to 5 digits are 99999 and total number of numbers 1 to 4 digits are 9999.

Hence, the total number of numbers of exactly 5 digits = 99999 − 9999 = 90000.

47. $\because \left(x^2 + \dfrac{1}{x^2} + 2\right)^n = \left[\left(x + \dfrac{1}{x}\right)^2\right]^n = \left(x + \dfrac{1}{x}\right)^{2n}$

The number of terms in the expansion of $\left(x + \dfrac{1}{x}\right)^{2n}$ is $2n + 1$ which is odd. Therefore, $(n + 1)$th term will be middle term.

$$\therefore T_{n+1} = {}^{2n}C_n (x)^{2n-n}\left(\frac{1}{x}\right)^n = {}^{2n}C_n x^n \cdot \frac{1}{x^n} = {}^{2n}C_n = \frac{(2n)!}{n!\,n!} = \frac{(2n)!}{(n!)^2}$$

48. We know,

$$\sin^2 A = \left(\frac{1 - \cos 2A}{2}\right)$$

$$\Rightarrow \sin^2(22°30') = \frac{1 - \cos 45°}{2} = \frac{\left(1 - \frac{1}{\sqrt{2}}\right)}{2} = \frac{\sqrt{2} - 1}{2\sqrt{2}}$$

$$\Rightarrow \sin(22°30') = \sqrt{\frac{\sqrt{2} - 1}{2\sqrt{2}}}$$

49. $\sqrt{\dfrac{1-\sin\theta}{1+\sin\theta}} = \sqrt{\dfrac{(1-\sin\theta)^2}{1-\sin^2\theta}}$

$= \dfrac{1-\sin\theta}{\sqrt{\cos^2\theta}} = \dfrac{1-\sin\theta}{|\cos\theta|}$

$= \dfrac{1-\sin\theta}{-\cos\theta} \qquad \left(\because \dfrac{\pi}{2} < \theta < \dfrac{3\pi}{2}\right)$

$= -\dfrac{1}{\cos\theta} + \dfrac{\sin\theta}{\cos\theta}$

$= -\sec\theta + \tan\theta$

50. Let $I = \int \dfrac{2^x}{\sqrt{1-4^x}}\,dx$

Let $2^x = t$

$\Rightarrow \quad 2^x \log 2\,dx = dt$

$\therefore \quad I = \dfrac{1}{\log 2}\int \dfrac{1}{\sqrt{1-t^2}}\,dt$

$= \dfrac{1}{\log 2}\sin^{-1}(t) + C$

$= \dfrac{1}{\log 2}\sin^{-1}(2^x) + C$

But $I = k\sin^{-1}(2^x) + C,$

$\Rightarrow \quad k = \dfrac{1}{\log 2}$

51. Required area $= \int_{-1}^{2} |x_1|\,dy$

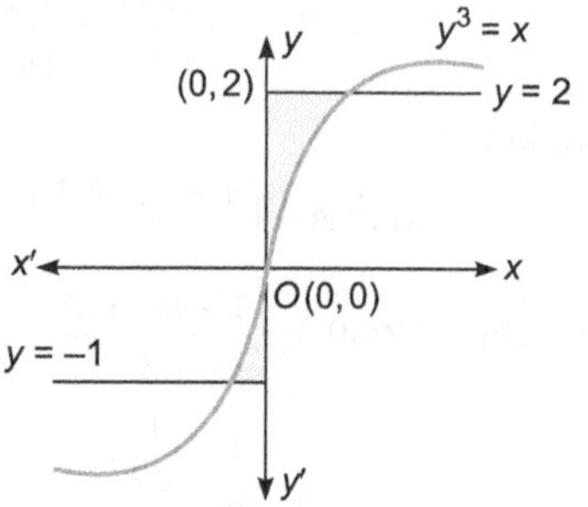

$= \int_{-1}^{0}(-x_1)\,dy + \int_0^2 x_1\,dy$

$= -\int_{-1}^{0} y^3\,dy + \int_0^2 y^3\,dy$

$= -\left[\dfrac{y^4}{4}\right]_{-1}^{0} + \left[\dfrac{y^4}{4}\right]_0^2 = \dfrac{1}{4} + 4$

$= \dfrac{17}{4}$ sq units.

52. Let (x_1, y_1) be the mid-point of the line joining the common points of the given line and the given parabola. Then, the equation of line is

$$yy_1 - 4(x + x_1) = y_1^2 - 8x_1$$

$\Rightarrow \quad 4x - yy_1 + y_1^2 - 4x_1 = 0$

This line and $2x - 3y + 8 = 0$ represents the same line.

$\therefore \quad \dfrac{4}{2} = \dfrac{-y_1}{-3} = \dfrac{y_1^2 - 4x_1}{8}$

$\Rightarrow \quad y_1 = 6,\ y_1^2 - 4x_1 = 16$

$\Rightarrow \quad y_1 = 6,\ x_1 = \dfrac{36-16}{4}$

$\Rightarrow \quad y_1 = 6,\ x_1 = 5$

$\therefore$ The required point is (5, 6).

53. $\cos^2 48° - \sin^2 12° = \dfrac{1}{2}(2\cos^2 48° - 2\sin^2 12°)$

$= \dfrac{1}{2}[(1 + \cos 96°) - (1 - \cos 24°)]$

$= \dfrac{1}{2}(\cos 96° + \cos 24°)$

$= \dfrac{1}{2}\left[2\cos\left(\dfrac{96° + 24°}{2}\right)\cos\left(\dfrac{96° - 24°}{2}\right)\right]$

$= \cos 60° \cos 36°$

$= \dfrac{1}{2} \times \left(\dfrac{\sqrt{5}+1}{4}\right)$

$= \dfrac{\sqrt{5}+1}{8}$

54. $\sin^2\theta + 3\cos\theta = 3$

$\Rightarrow \quad 1 - \cos^2\theta + 3\cos\theta = 3$

$\Rightarrow \quad \cos^2\theta - 3\cos\theta + 2 = 0$

$\Rightarrow \quad (\cos\theta - 1)(\cos\theta - 2) = 0$

$\therefore \quad \cos\theta = 1 \qquad (\because \cos\theta \neq 2)$

$\Rightarrow \quad \theta = 0°$ as $\theta \in [-\pi, \pi]$

$\therefore$ The number of solution is 1.

55. Let $PQ = h$ be the height of chimney. A and B are the two points 50 m apart.

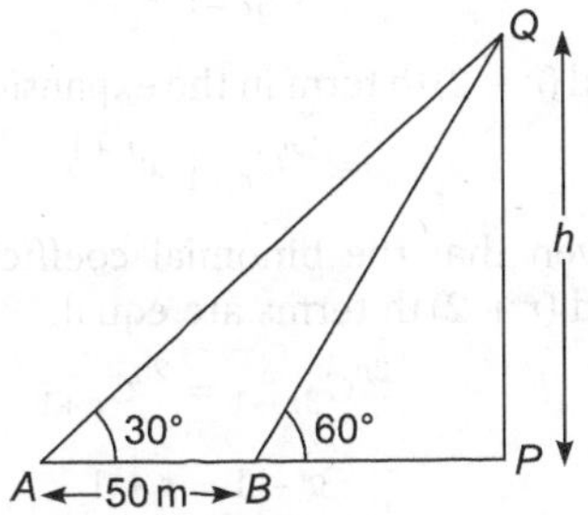

In ΔAPQ,

$$\tan 30° = \frac{h}{AP}$$

$$\Rightarrow \quad AP = h \cot 30° \quad \ldots(i)$$

and in ΔQBP,

$$\tan 60° = \frac{h}{BP}$$

$$\Rightarrow \quad BP = h \cot 60° \quad \ldots(ii)$$

$$\because \quad AP - BP = 50$$

$$\therefore \quad h(\cot 30° - \cot 60°) = 50$$

$$\Rightarrow \quad h = \frac{50}{\left(\sqrt{3} - \frac{1}{\sqrt{3}}\right)} = \frac{50\sqrt{3}}{3-1}$$

$$= \frac{50\sqrt{3}}{2} = 25\sqrt{3} \text{ m}$$

56. $$\lim_{x \to \infty} \left(\frac{x^3 + 1}{x^2 + 1} - (ax + b) \right) = 2$$

$$\Rightarrow \quad \lim_{x \to \infty} \frac{x^3(1-a) - bx^2 - ax + (1-b)}{x^2 + 1} = 2$$

$$\Rightarrow \quad \lim_{x \to \infty} \frac{x(1-a) - b - \frac{a}{x} + \frac{1-b}{x^2}}{1 + \frac{1}{x^2}} = 2$$

$$\Rightarrow \quad 1 - a = 0 \text{ and } -b = 2$$

$$\Rightarrow \quad a = 1 \text{ and } b = -2$$

57. $$f(x) = \begin{cases} \dfrac{|x+2|}{\tan^{-1}(x+2)}, & x \neq -2 \\ 2, & x = -2 \end{cases}$$

$$\therefore \quad \lim_{x \to -2^-} f(x) = \lim_{h \to 0} f(-2-h)$$

$$= \lim_{h \to 0} \frac{|-2-h+2|}{\tan^{-1}(-2-h+2)}$$

$$= \lim_{h \to 0} \frac{-h}{\tan^{-1} h} = -1$$

and $$\lim_{x \to -2^+} f(x) = \lim_{h \to 0} f(-2+h)$$

$$= \lim_{h \to 0} \left[\frac{|-2+h+2|}{\tan^{-1}(-2+h+2)} \right]$$

$$= \lim_{h \to 0} \frac{h}{\tan^{-1} h} = 1$$

$$\because \quad \lim_{x \to -2^-} f(x) \neq \lim_{x \to -2^+} f(x)$$

So, $f(x)$ is not continuous as well as not differentiable at $x = -2$.

58. Let the coordinates of the point P be (x_1, y_1). This point lies on the curve

$$x^{2/3} + y^{2/3} = a^{2/3} \quad \ldots(i)$$

$$\therefore \quad x_1^{2/3} + y_1^{2/3} = a^{2/3} \quad \ldots(ii)$$

On differentiating Eq. (i) w.r.t. x, we get

$$\frac{2}{3} x^{-1/3} + \frac{2}{3} y^{-1/3} \frac{dy}{dx} = 0$$

$$\Rightarrow \quad \frac{dy}{dx} = -\frac{y^{1/3}}{x^{1/3}}$$

$$\Rightarrow \quad \left(\frac{dy}{dx}\right)_{(x_1, y_1)} = -\left(\frac{y_1}{x_1}\right)^{1/3}$$

The equation of the tangent at (x_1, y_1) to the given curve is

$$y - y_1 = -\frac{y_1^{1/3}}{x_1^{1/3}}(x - x_1)$$

$$\Rightarrow \quad x x_1^{-1/3} + y y_1^{-1/3} = x_1^{2/3} + y_1^{2/3}$$

$$\Rightarrow \quad x x_1^{-1/3} + y y_1^{-1/3} = a^{2/3} \quad \text{[using Eq. (ii)]}$$

This tangent meets the coordinate axes at $A(a^{2/3} x_1^{1/3}, 0)$ and $B(0, a^{2/3} y_1^{1/3})$.

$$\therefore \quad AB = \sqrt{(0 - a^{2/3} x_1^{1/3})^2 + (a^{2/3} y_1^{1/3} - 0)^2}$$

$$= \sqrt{a^{4/3}(x_1^{2/3} + y_1^{2/3})} \quad \text{[from Eq. (ii)]}$$

$$= \sqrt{a^{4/3} \cdot a^{2/3}}$$

$$= \sqrt{a^2} = a$$

59. Given that, $x = (2t - 3)^2$

On differentiating w.r.t. x, we get

$$\frac{dx}{dt} = 2(2t - 3)(2)$$

At origin, $x = 0$

$$\therefore \quad 2t - 3 = 0 \Rightarrow t = \frac{3}{2}$$

$$\therefore \quad \text{Velocity} = \frac{dx}{dt} = 4\left(2 \times \frac{3}{2} - 3\right) = 0$$

60. $\therefore \cos\frac{2\pi}{7} + \cos\frac{4\pi}{7} + \cos\frac{6\pi}{7}$

$$= \frac{1}{2\sin\frac{\pi}{7}} \times \left[2\sin\frac{\pi}{7}\cos\frac{2\pi}{7} + 2\sin\frac{\pi}{7}\cos\frac{4\pi}{7} + 2\sin\frac{\pi}{7}\cos\frac{6\pi}{7}\right]$$

$$= \frac{1}{2\sin\frac{\pi}{7}} \times \left[\sin\frac{3\pi}{7} - \sin\frac{\pi}{7} + \sin\frac{5\pi}{7} - \sin\frac{3\pi}{7} + \sin\pi - \sin\frac{5\pi}{7}\right]$$

$$= -\frac{1}{2}$$

61. The maximum value of $3\cos\theta + 4\sin\theta$

$$= \sqrt{3^2 + 4^2}$$

$$= \sqrt{25} = 5$$

62. 'Where are you going' is not a statement.

63. Since, $123 = -1 \pmod{124}$

$$125 = 1 \pmod{124}$$

$$127 = 3 \pmod{124}$$

$$\therefore \quad 123 \times 125 \times 127 = (-1)(1)(3) \bmod (124)$$

$$= -3 \bmod (124)$$

$$= 121 \bmod (125)$$

64. The vertex connectivity of any tree is one.

65. Since, a five digit number is formed by using digits 0, 1, 2, 3, 4 and 5, divisible by 3 *i.e.*, only possible when sum of digits is multiple of three which gives two cases.

Case I Using digits 0, 1, 2, 4, 5 the number of ways $= 4 \times 4 \times 3 \times 2 \times 1 = 96$.

Case II Using digits 1, 2, 3, 4, 5 the number of ways $= 5 \times 4 \times 3 \times 2 \times 1 = 120$

$\therefore$ Total numbers formed $= 120 + 96 = 216$

66. $3r$ th term in the expansion of $(1 + x)^{2n}$

$$= {}^{2n}C_{3r-1}\, x^{3r-1}$$

and $(r + 2)$th term in the expansion of $(1 + x)^{2n}$

$$= {}^{2n}C_{r+1}\, x^{r+1}$$

Given that the binomial coefficients of $(3r)$th and $(r + 2)$th terms are equal.

$$\therefore \quad {}^{2n}C_{3r-1} = {}^{2n}C_{r+1}$$

$$\Rightarrow \quad 3r - 1 = r + 1$$

$$\text{or} \quad 2n = (3r - 1) + (r + 1)$$

$$\Rightarrow \quad 2r = 2 \text{ or } 2n = 4r$$

$$\Rightarrow \quad r = 1 \text{ or } n = 2r$$

$$\text{But} \quad r > 1$$

$$\therefore \quad n = 2r$$

67. $\log_4 (x - 1) = \log_2 (x - 3) = \log_{4^{1/2}} (x - 3)$

$$\Rightarrow \quad \log_4 (x - 1) = 2\log_4 (x - 3)$$

$$\Rightarrow \quad \log_4 (x - 1) = \log_4 (x - 3)^2$$

$$\Rightarrow \quad (x - 1) = (x - 3)^2$$

$$\Rightarrow \quad x^2 + 9 - 6x = x - 1$$

$$\Rightarrow \quad x^2 - 7x + 10 = 0$$

$$\Rightarrow \quad (x - 2)(x - 5) = 0$$

$$\Rightarrow \quad x = 2 \text{ or } x = 5$$

Hence, $x = 5$

[$\because x = 2$ makes $\log (x - 3)$ undefined.]

68. $\lim\limits_{x \to 0} \dfrac{\int_0^{x^2} \cos^2 t\, dt}{x \sin x}$

Using L' Hospital's rule, we get

$$= \lim_{x \to 0} \frac{\cos^2 (x^2) \cdot 2x - 0}{x\cos x + \sin x}$$

$$= \lim_{x \to 0} \frac{2\cos^2 (x^2)}{\cos x + \frac{\sin x}{x}}$$

$$= \frac{2}{2} = 1$$

69. Let $$I = \int_2^3 \frac{\sqrt{x}}{\sqrt{5 - x} + \sqrt{x}}\, dx \quad \text{...(i)}$$

$$\therefore \quad I = \int_2^3 \frac{\sqrt{5 - x}}{\sqrt{x} + \sqrt{5 - x}}\, dx \quad \text{...(ii)}$$

On adding Eqs. (i) and (ii), we get

$$2I = \int_2^3 \frac{\sqrt{x} + \sqrt{5-x}}{\sqrt{x} + \sqrt{5-x}}\, dx$$

$$= \int_2^3 1\, dx = 3 - 2 = 1$$

$$\Rightarrow \quad I = \frac{1}{2}$$

70. The given differential equation is

$$(e^x + 1) \cdot y\, dy = (y+1) e^x\, dx$$

$$\Rightarrow \quad \frac{y\, dy}{(y+1)} = \frac{e^x}{(e^x+1)}\, dx$$

$$\Rightarrow \quad \left(\frac{y+1-1}{y+1}\right) dy = \frac{e^x}{e^x+1}\, dx$$

$$\Rightarrow \quad dy - \frac{1}{y+1}\, dy = \frac{e^x}{e^x+1}\, dx$$

On integrating, we get

$$\Rightarrow \quad y - \log|y+1| = \log(e^x+1) + \log k$$

$$\Rightarrow \quad y = \log|(y+1)(e^x+1)|\, k$$

$$\Rightarrow \quad (y+1)(e^x+1) = e^y C$$

71. The required equation of circle is

$$(x^2 + y^2 + 13x - 3y) + \lambda\left(11x + \frac{1}{2}y + \frac{25}{2}\right) = 0 \quad \ldots\text{(i)}$$

This circle passes through (1, 1).

$$(1^2 + 1^2 + 13 - 3) + \lambda\left(11 + \frac{1}{2} + \frac{25}{2}\right) = 0$$

$$\therefore \quad 12 + \lambda(24) = 0$$

$$\Rightarrow \quad \lambda = -\frac{1}{2}$$

On putting this value of λ in Eq. (i), we get

$$x^2 + y^2 + 13x - 3y - \frac{11}{2}x - \frac{1}{4}y - \frac{25}{4} = 0$$

$$\Rightarrow \quad 4x^2 + 4y^2 + 52x - 12y - 22x - y - 25 = 0$$

$$\Rightarrow \quad 4x^2 + 4y^2 + 30x - 13y - 25 = 0$$

72. Tangent to the curve $y^2 = 8x$ is $y = mx + \frac{2}{m}$.

So, it must satisfy $xy = -1$

$$\Rightarrow x\left(mx + \frac{2}{m}\right) = -1 \Rightarrow mx^2 + \frac{2}{m}x + 1 = 0$$

Since, it has equal roots.

$$\therefore \quad D = 0$$

$$\Rightarrow \quad \frac{4}{m^2} - 4m = 0$$

$$\Rightarrow \quad m^3 = 1 \Rightarrow m = 1$$

So, equation of common tangent is $y = x + 2$.

73. $b\cos^2\frac{C}{2} + c\cos^2\frac{B}{2}$

$$= \frac{b}{2}(1 + \cos C) + \frac{c}{2}(1 + \cos B)$$

$$= \frac{b}{2} + \frac{c}{2} + \frac{1}{2}(b\cos C + c\cos B)$$

$$= \frac{b}{2} + \frac{c}{2} + \frac{a}{2} \quad (\because a = b\cos C + c\cos B)$$

$$= \frac{a+b+c}{2} = \frac{k}{2}$$

74. $I_1 = \int_{1-k}^{k} x f\{x(1-x)\}\, dx$

$$= \int_{1-k}^{k} (1-x) f\{(1-x)(1-(1-x))\}\, dx$$

$$\left(\because \int_a^b f(x)\, dx = \int_a^b f(a+b-x)\, dx\right)$$

$$= \int_{1-k}^{k} (1-x) f\{(1-x)x\}\, dx$$

$$= \int_{1-k}^{k} f\{x(1-x)\}\, dx - \int_{1-k}^{k} x f\{x(1-x)\}\, dx$$

$$\Rightarrow \quad I_1 = I_2 - I_1 \Rightarrow 2I_1 = I_2$$

$$\Rightarrow \quad \frac{I_1}{I_2} = \frac{1}{2}$$

75. The maximum area corresponds to when P is at either end of the minor axis.

$\therefore$ Required area $= \frac{1}{2}(2a)b = ab$ sq unit

76. We have, $\frac{\cos A}{a} = \frac{\cos B}{b} = \frac{\cos C}{c}$

$$\Rightarrow \quad \frac{\cos A}{k\sin A} = \frac{\cos B}{k\sin B} = \frac{\cos C}{k\sin C}$$

$$\Rightarrow \quad \cot A = \cot B = \cot C$$

$$\Rightarrow \quad A = B = C = 60°$$

$\therefore$ ΔABC is an equilateral triangle.

$\therefore$ Area of triangle $= \frac{\sqrt{3}}{4}a^2$

$$= \frac{\sqrt{3}}{4} \times 2^2 = \sqrt{3} \text{ sq units}$$

77. The region is clearly square with vertices at the points (1, 0), (0, 1), (–1, 0) and (0, –1). So, its area $= \sqrt{2} \times \sqrt{2} = 2$ sq units.

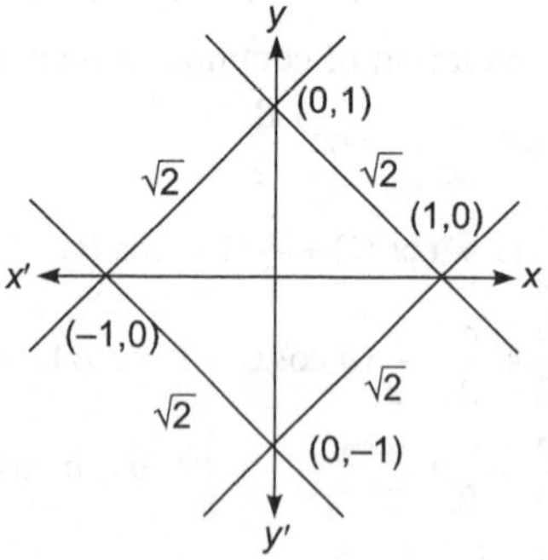

78. $\sqrt{3}$ cosec 20° – sec 20°

$$= \tan 60^\circ \operatorname{cosec} 20^\circ - \sec 20^\circ$$

$$= \frac{\sin 60^\circ \cos 20^\circ - \cos 60^\circ \sin 20^\circ}{\cos 60^\circ \sin 20^\circ \cos 20^\circ}$$

$$= \frac{\sin (60^\circ - 20^\circ)}{\cos 60^\circ \sin 20^\circ \cos 20^\circ}$$

$$= \frac{\sin 40^\circ}{\frac{1}{2} \sin 20^\circ \cos 20^\circ}$$

$$= \frac{2 \sin 20^\circ \cos 20^\circ}{\frac{1}{2} \sin 20^\circ \cos 20^\circ} = 4$$

79. By sine rule

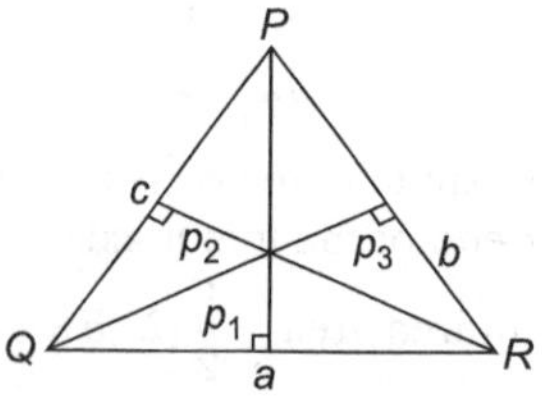

$$\frac{a}{\sin P} = \frac{b}{\sin Q} = \frac{c}{\sin R} = k \quad \text{(say)}$$

Also, $\frac{1}{2} a p_1 = \Delta$

$\Rightarrow \quad \frac{2\Delta}{a} = p_1$

$\Rightarrow \quad p_1 = \frac{2\Delta}{k \sin P}$

Similarly, $p_1 = \frac{2\Delta}{k \sin Q}$ and $p_3 = \frac{2\Delta}{k \sin R}$

Since, sin P, sin Q, sin R are in AP.

$\therefore \quad \frac{1}{\sin P}, \frac{1}{\sin Q}, \frac{1}{\sin R}$ are in HP

$\Rightarrow \quad \frac{2\Delta}{k \sin P}, \frac{2\Delta}{k \sin Q}, \frac{2\Delta}{k \sin R}$ are in HP

$\Rightarrow p_1, p_2, p_3$ are in HP.

80. Let E_1 be the events that man will be selected and E_2 the events that woman will be selected.

Then, $P(E_1) = \frac{1}{4}$

So, $P(\overline{E}_1) = 1 - \frac{1}{4} = \frac{3}{4}$

and $P(E_2) = \frac{1}{3}$

$\Rightarrow \quad P(\overline{E}_2) = \frac{2}{3}$

$\therefore$ E_1 and E_2 are independent events.

$\therefore \quad P(\overline{E}_1 \cap \overline{E}_2) = P(\overline{E}_1) \times P(\overline{E}_2)$

$$= \frac{3}{4} \times \frac{2}{3} = \frac{1}{2}$$

Manipal

Engineering Entrance Exam

Solved Paper 2011

Physics

1. Which of the following statements is wrong?
(a) Voltmeter should have high resistance
(b) Ammeter should have low resistance
(c) Ammeter is placed in parallel across the conductor in a circuit
(d) Voltmeter is placed in parallel across the conductor in a circuit

2. As the temperature of hot junction increases, the thermo emf
(a) always increases
(b) always decreases
(c) may increase or decrease
(d) always remains constant

3. A proton and an α-particle enter a uniform magnetic field perpendicularly with the same speed. If proton takes 25 μs to make 5 revolutions, then the periodic time for the α-particle would be
(a) 50 μs (b) 25 μs
(c) 10 μs (d) 5 μs

4. Two parallel conductors A and B of equal length carry currents i and $10i$, respectively, in the same direction. Then
(a) A and B will repel each other with same force
(b) A and B will attract each other with same force
(c) A will attract B, but B will repel A
(d) A and B will attract each other with different forces

5. An open pipe is suddenly closed at one end with the result that the frequency of third harmonic of the closed pipe is termed to be higher by 100 Hz than the fundamental frequency of the open pipe. The fundamental frequency of the open pipe is
(a) 200 Hz (b) 150 Hz
(c) 100 Hz (d) 250 Hz

6. Two slits 4 mm apart, are illuminated by light of wavelength 6000 Å. What will be the fringe width on a screen placed 2 m from the slits?
(a) 0.12 mm (b) 0.3 mm
(c) 3.0 mm (d) 4.0 mm

7. A thin prism P_1 of angle 4° and refractive index 1.54° is combined with another thin prism P_2 of refractive index 1.72 to produce dispersion without deviation. The angle of P_2 is
(a) 4° (b) 5.33°
(c) 2.6° (d) 3°

8. While viewing a distant object with a telescope suddenly a housefly sits on objective lens. The correct statement is that
(a) housefly will be seen enlarged in image
(b) housefly will be seen reduced in image
(c) intensity of image will be decreased
(d) intensity of image will be increased

9. A capacitor of capacitance 5 μF is connected as shown in the figure. The internal resistance of the cell is 0.5 Ω. The amount of charge on the capacitor plate is

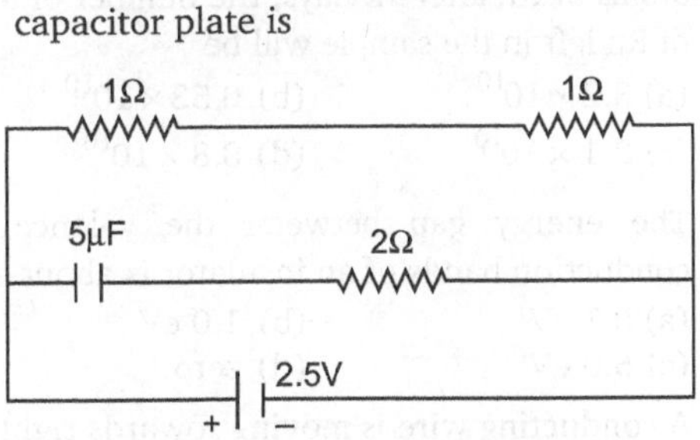

(a) 10 μC (b) 5 μC
(c) 6 μC (d) 10.2 μC

10. A charge of 10^{-9} C is placed on each of the 64 identical drops of radius 2 cm. They are then combined to form a bigger drop. Its potential will be

(a) 7.2×10^3 V (b) 7.2×10^2 V
(c) 1.44×10^2 V (d) 1.44×10^3 V

11. A particle moves in the *xy*-plane under the action of a force **F** such that the components of its linear momentum **p** at any time t are $p_x = 2\cos t$, $p_y = \sin t$. The angle between **F** and **p** at time t is

(a) 90° (b) 0°
(c) 180° (d) 30°

12. A fighter plane is moving in a vertical circle of radius r. Its minimum velocity at the highest point A of the circle will be

(a) $\sqrt{3gr}$ (b) $\sqrt{2gr}$
(c) $\sqrt{gr}$ (d) $\sqrt{gr/2}$

13. In Millikan's oil drop experiment, a charged drop of mass 1.8×10^{-13} kg is stationary between its plates. The distance between its plates is 0.9 cm and potential difference is 2000 V. The number of electrons in the drop is

(a) 500 (b) 50
(c) 5 (d) 10

14. Decay constant of radium is λ. By a suitable process its compound radium bromide is obtained. The decay constant of radium bromide will be

(a) λ (b) more than λ
(c) less than λ (d) zero

15. Rn decays into Po by emitting an α-particle with half-life of 4 days. A sample contains 6.4×10^{10} atoms of Rn after 12 days, the number of atoms of Rn left in the sample will be

(a) 3.2×10^{10} (b) 0.53×10^{10}
(c) 2.1×10^{10} (d) 0.8×10^{10}

16. The energy gap between the valence and conduction bands of an insulator is about

(a) 0.1 eV (b) 1.0 eV
(c) 5.0 eV (d) zero

17. A conducting wire is moving towards right in a magnetic field B. The direction of induced current in the wire is shown in the figure. The direction of magnetic field will be

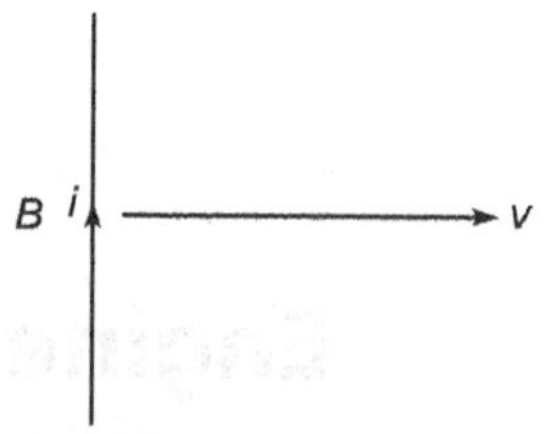

(a) in the plane of paper pointing towards right
(b) in the plane of paper pointing towards left
(c) perpendicular to the plane of paper and downwards
(d) perpendicular to the plane of paper and upwards

18. An electron moving with velocity 2×10^{-7} m/s, describes a circle in a magnetic field of strength 2×10^{-2} T. If $\frac{e}{m}$ of electron is 1.76×10^{11} C/kg. Then the diameter of the circle will be

(a) 11 cm (b) 1.1 cm
(c) 1.1 mm (d) 1.1 m

19. In an AC circuit the potential difference and current are represented respectively by $V = 100 \sin(100t)$ volt,
$I = 100 \sin\left(100t + \frac{\pi}{3}\right)$ milliampere. The power in the circuit is

(a) 2.5 W (b) 5 W
(c) 10 W (d) 10^4 W

20. Escape velocity from earth's surface is 11 km/s. If radius of a planet is double than that of earth and density is same as that of earth, then the escape velocity from this planet will be

(a) 5.5 km/s (b) 11 km/s
(c) 16.5 km/s (d) 22 km/s

21. The orbital angular momentum of a satellite revolving at a distance r from the centre is L. If the distance is increased to $16r$, then the new angular momentum will be

(a) $16L$ (b) $64L$
(c) $\frac{L}{4}$ (d) $4L$

22. The acceleration of a particle performing SHM is 12 cm/s^2 at a distance of 3 cm from the mean position. Its time period is

(a) 0.5 s (b) 1.0 s
(c) 2.0 s (d) 3.14 s

23. A soap bubble in vacuum has a radius of 3 cm and another soap bubble in vacuum has a radius of 4 cm. If the two bubbles coalesce under isothermal condition, then the radius of the new bubble is

(a) 2.3 cm (b) 4.5 cm
(c) 5 cm (d) 7 cm

24. A 10 cm^3 cube floats in water with a height of 4 cm^3 remaining above the surface. The density of the material from which the cube is made is

(a) 0.6 g cm^{-3} (b) 1.0 g cm^{-3}
(c) 0.4 g cm^{-3} (d) 0.24 g cm^{-3}

25. The heat flows through a rod of length 50 cm and area of cross-section 5 cm^2. Its ends are respectively at 25°C and 125°C. The coefficient of thermal conductivity of the material rod is 0.092 kcal/m°C. The temperature gradient of the rod is

(a) 2°C/ cm (b) 2°C/m
(c) 20°C/cm (d) 20°C/m

26. 5 mol of hydrogen gas is heated from 30°C to 60°C at constant pressure. Heat given to the gas is (given $R = 2$ cal/ mol deg)

(a) 750 cal (b) 630 cal
(c) 1050 cal (d) 1470 cal

27. A fish at a depth of 12 cm in water is viewed by an observer on the bank of a lake. To what height the image of the fish is raised?

(Refractive index of lake water $= \frac{4}{3}$)

(a) 9 cm (b) 12 cm
(c) 3.8 cm (d) 3 cm

28. The rest energy of an electron is 0.511 MeV. The electron is accelerated from rest to a velocity 0.5 c. The change in its energy will be

(a) 0.026 MeV (b) 0.051 MeV
(c) 0.079 MeV (d) 0.105 MeV

29. In Millikan's oil drop experiment, an oil drop carrying a charge Q is held stationary by a potential difference 2400 V between the plates. To keep a drop of half the radius stationary, the potential difference had to be made 600V. What is the charge on the second drop?

(a) $\frac{Q}{4}$ (b) $\frac{Q}{2}$
(c) Q (d) $\frac{3Q}{2}$

30. Band spectrum is characteristic of the

(a) atoms
(b) molecules
(c) amorphous solids
(d) crystalline solids

31. As a result of radioactive decay $_{92}U^{238}$ is converted into $_{91}Pa^{234}$. The particles emitted during this decay are

(a) a proton and a neutron
(b) a proton and two α-particles
(c) an α-particle and a β-particle
(d) two β-particles and a proton

32. A p-type semiconductor has acceptor levels 57 meV above the valence band. The maximum wavelength of light required to create a hole

(a) 57 Å (b) 57×10^{-3} Å
(c) 217105 Å (d) 11.61×10^{-33} Å

33. In an explosion a body breaks up into pieces of unequal masses. In this case

(a) both parts will have numerically equal momentum
(b) lighter part will have more momentum
(c) heavier part will have more momentum
(d) both parts will have equal kinetic energy

34. A force of 10 N acts on a body of mass 20 kg for 10 s. Change in its momentum is

(a) 5 kg m/s (b) 100 kg m/s
(c) 200 kg m/s (d) 1000 kg m/s

35. In which case does the potential energy decrease?

(a) On compressing a spring
(b) On stretching a spring
(c) On moving a body against gravitational force
(d) On the rising of an air bubble in water

36. A heavy mass is attached to a thin wire and is whirled in a vertical circle. The wire is most likely to break

(a) when the mass is at the highest point of the circle
(b) when the mass is at the lowest point of the circle
(c) when the wire is horizontal
(d) at an angle of $\cos^{-1}\left(\frac{1}{3}\right)$ from the upward vertical

37. The time taken by AC of 50 Hz in reaching from zero to the maximum value is
(a) 50×10^{-3} s (b) 5×10^{-3} s
(c) 1×10^{-3} s (d) 2×10^{-3} s

38. The molar heat capacity of rock salt at low temperature varies with temperature according to Debye's T^3 law

Thus $C = k \frac{T^3}{\theta^3}$ where $k = 1940$ Jmol^{-1}K^{-1}
$\theta = 281$ K

Calculate how much heat is required to raise the temperature of 2 moles of rock salt from 10 K to 50 K
(a) 800 J (b) 373 J
(c) 273 J (d) None of these

39. The velocity of an electron in the second orbit of sodium atom (atomic number =11) is v. The velocity of an electron in its fifth orbit will be
(a) v (b) $\frac{22}{5}v$
(c) $\frac{5}{2}$ (d) $\frac{2}{5}v$

40. A charged particle enters a magnetic field H with its initial velocity making an angle of 45° with H. The path of the particle will be
(a) a straight line (b) a circle
(c) an ellipse (d) a helix

41. A 10 μF capacitor is charged to a potential difference of 50V and is connected to another uncharged capacitor in parallel. Now the common potential difference becomes 20V. The capacitance of second capacitor is
(a) 10 μF (b) 20 μF
(c) 30 μF (d) 15μF

42. An alternating voltage $E = 200\sqrt{2}\sin(100\,t)$ V is connected to a 1 μF capacitor through an AC ammeter. The reading of ammeter is
(a) 10 mA (b) 20 mA
(c) 40 mA (d) 80 mA

43. In a step-up transformer the turn ratio is 1 : 2. A Leclanche cell (emf 1.5 V) is connected across the primary. The voltage across the secondary is
(a) 3.0 V (b) 0.75 V (c) zero (d) 1.5 V

44. If 10% of the current passes through a moving coil galvanometer of resistance 99 Ω. Then the shunt resistance will be
(a) 9.9 Ω (b) 11 Ω
(c) 10 Ω (d) 9 Ω

45. How much energy will necessary for making a body of 500 kg escape from the earth? ($g = 9.8$ m/s^2, radius of the earth $= 6.4 \times 10^6$ m)
(a) about 9.8×10^6 J (b) about 6.4×10^8 J
(c) about 3.1×10^{10} J (d) about 27.4×10^{12} J

46. In a neon discharge tube 2.9×10^{18} Ne^+ ions move to the right each second while 1.2×10^{18} electrons move to the left per second, electrons charge is 1.6×10^{-19} C. The current in the discharge tube is
(a) 1 A, towards right
(b) 0.66 A, towards right
(c) 0.66 A, towards left
(d) zero

47. Water is falling on the blades of a turbine at a rate of 100 kg/s. If the height of the fall is 100 m, the power transferred to the turbine will be approximately
(a) 100 kW (b) 10 kW (c) 1 kW (d) 100 W

48. A metre scale is standing vertically on the earth's surface on one of its end. It now falls on earth without slipping. Find the velocity with which the free end of the scale strikes the earth. ($g = 10$ m/s^2)

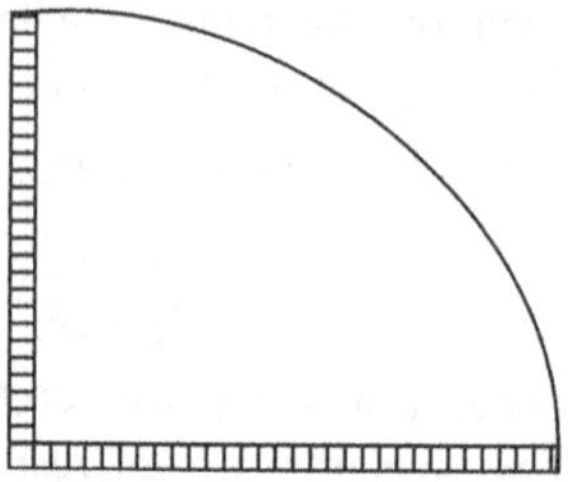

(a) 9.8 m/s (b) 5.47 m/s
(c) 4.5 m/s (d) 1 m/s

49. For germanium crystal, the forbidden energy gap in joule is
(a) 1.216×10^{-19} (b) 1.76×10^{-19}
(c) 1.6×10^{-19} (d) zero

50. An electron is accelerated through a potential difference of 200 V. If e/m for the electron be 1.6×10^{11} C/kg, the velocity acquired by the electron will be
(a) 8×10^6 m/s (b) 8×10^5 m/s
(c) 5.9×10^6 m/s (d) 5.9×10^5 m/s

51. At what temperature will the resistance of a copper wire become three times its value at 0°C? (temperature coefficient of resistance for copper $= 4 \times 10^{-3}$/°C)
(a) 400°C (b) 450°C
(c) 500°C (d) 550°C

52. A coil having an area of 2m^2 is placed in a magnetic field which changes from 1 Wb/m^2 to 4 Wb/m^2 in 2 s. The emf induced in the coil will be
(a) 4 V (b) 3 V
(c) 2 V (d) 1 V

53. A 10 m long copper wire while remaining in the east-west horizontal direction is falling down with a speed of 5.0 m/s. If the horizontal component of the earth's magnetic field $= 0.3 \times 10^{-4}$ Wb/m^2 the emf developed between the ends of the wires is
(a) 0.15 V (b) 1.5 V
(c) 0.15 mV (d) 1.5 mV

54. A current of 1.6 A is passed through a solution of $CuSO_4$. How many Cu^{++} ions are liberated in one minute?
(electronic charge $= 1.6 \times 10^{-19}$ C)
(a) 3×10^{20} (b) 3×10^{19}
(c) 6×10^{20} (d) 6×10^{19}

55. A solenoid has an inductance of 60 H and a resistance of 30 Ω. If it is connected to a 100 V battery, how long will it take for the current to reach $\frac{e-1}{e} = 63.2\%$ of its final value?
(a) 1 s (b) 2 s
(c) e second (d) $2e$ second

56. Three blocks of masses m_1, m_2 are connected by massless strings as shown on a frictionless table. They are pulled with a force $T_3 = 40$ N. If $m_1 = 10$ kg, $m_2 = 6$ kg and $m_3 = 4$ kg, the tension T_2 will be

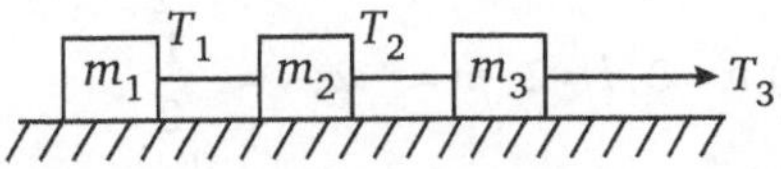

(a) 20 N (b) 40 N
(c) 10 N (d) 32 N

57. The force constant of weightless spring is 16 N/m. A body of mass 1.0 kg suspended from it is pulled down through 5 cm and then released. The maximum kinetic energy of the system (spring + body) will be
(a) 2×10^{-2} J (b) 4×10^{-2} J
(c) 8×10^{-2} J (d) 16×10^{-2} J

58. A mass M is moving with a constant velocity on a line parallel to the x-axis. Its angular momentum with respect to the origin or z-axis
(a) is zero
(b) remains constant
(c) goes on increasing
(d) goes on decreasing

59. Which of the following pairs is an isobar?
(a) ${}_1H^1$ and ${}_1H^2$ (b) ${}_1H^2$ and ${}_1H^3$
(c) ${}_6C^{12}$ and ${}_6C^{13}$ (d) ${}_{15}P^{30}$ and ${}_{14}Si^{30}$

60. The half-life of the isotope ${}_{11}Na^{24}$ is 15 h. How much time does it take for $\frac{7}{8}$th of a sample of this isotope to decay?
(a) 75 h (b) 65 h
(c) 55 h (d) 45 h

Chemistry

1. The largest number of molecules is in
(a) 36 g of water
(b) 28 g of CO_2
(c) 46 g of CH_3OH
(d) 58 g of N_2O_5

2. In the following reaction, which choice has value twice that of the equivalent weight of the oxidising agent?

$$SO_2 + 2H_2O \longrightarrow S + 2H_2O_2$$

(a) 16 (b) 48
(c) 64 (d) 32

3. Orbital is
(a) circular path around the nucleus in which electrons are revolves
(b) space around the nucleus where the probability of finding the electron is maximum
(c) amplitude of electrons wave
(d) None of the above

4. Which of the following sets of quantum number is not permitted?
(a) $n=3, l=3, m=0, s=+\frac{1}{2}$
(b) $n=3, l=2, m=+2, s=-\frac{1}{2}$
(c) $n=3, l=2, m=-2, s=-\frac{1}{2}$
(d) $n=3, l=0, m=0, s=+\frac{1}{2}$

5. An element M has an atomic mass of 19 and atomic number 9, its ions is represented by
(a) M^+ (b) M^{2+} (c) M^- (d) M^{2-}

6. In S_8 molecules, the type of hybridisation exhibited by sulphur is
(a) sp^2 (b) sp^3
(c) sp (d) sp^3d

7. In the oxyacids of chlorine, Cl—O bond contains
(a) $d\pi - d\pi$ bonding
(b) $d\pi - p\pi$ bonding
(c) $p\pi - p\pi$ bonding
(d) None of the above

8. Anhydrous $AlCl_3$ fume in air due to
(a) oxidation (b) hydrolysis
(c) reduction (d) hydrogenation

9. The nature of Fe_2O_3 is
(a) acidic (b) basic
(c) amphoteric (d) None of these

10. Concentrated nitric acid (HNO_3) oxidises phosphorus to
(a) H_3PO_4 (b) H_3PO_3
(c) $H_4P_2O_7$ (d) H_3PO_2

11. Which of the following has the highest nucleophilicity?
(a) F^- (b) OH^-
(c) CH_3^- (d) NH_2^-

12. Number of isomeric primary amines obtained from $C_4H_{11}N$ are
(a) 3 (b) 4
(c) 5 (d) 6

13. Which of the following reaction doesn't support the acidic nature of alkyne?
(a) Reaction with HBr
(b) Reaction with Grignard reagent
(c) Reaction with ammoniacal silver salt
(d) Reaction with metallic sodium

14. $CH_3-CH_2-\underset{\underset{CH_3}{|}}{N}-\overset{\overset{O}{\|}}{C}-H$;

The IUPAC name of the compound is
(a) N-formyl-N-methyl ethanamide
(b) N-ethyl-N-methyl methanamide
(c) N-methyl-N-oxo ethanamine
(d) None of the above

15. The product of the following reaction,

$C_6H_5COCHBr_2 \xrightarrow{OH^-} ?$ is

(a) C_6H_5COCHO (b) $C_6H_5COCOOH$
(c) C_6H_5CHO (d) $C_6H_5CHOHCOOH$

16. Which of the following alkenes will react fastest with H_2 under catalytic hydrogenation conditions?

(a) 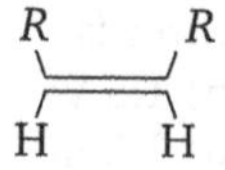

(b) 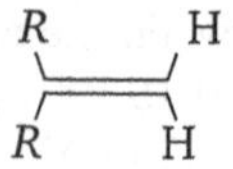

(c) 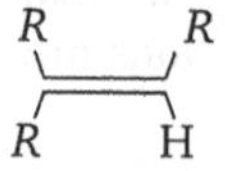

(d) 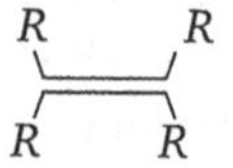

17. Glucose reacts with acetic anhydride to give pentaacetyl derivative. Which of the following is true about that?
(a) that can reduce Fehling or Tollen's reagent
(b) that's soluble in dil. NaOH solution
(c) that consumes one mole of HIO_4
(d) Br_2/H_2O can oxidise

18. Complete hydrolysis of cellulose gives
(a) D-fructose (b) D-ribose
(c) D-glucose (d) L-glucose

19. The saponification value of an oil or fat is measured in terms of
(a) NH_4OH (b) NaOH
(c) C_6H_5OH (d) kOH

20. N_2 gas is liberated when [HCl + $NaNO_2$] reacts with the following compounds;
A. $CH_3CH_2NH_2$ *B*. urea
C. CH_3CONH_2 *D*. $C_6H_5NH_2$
The answer is
(a) *A, B, C* (b) *B, C, D*
(c) *A, C, D* (d) *A, B, D*

21. Benzaldehyde condenses with N,N-dimethyl aniline in presence of anhydrous $ZnCl_2$ to give
(a) azo dye (b) malachite
(c) michlers ketone (d) buffer yellow

22. Which of the following reactions is given by only primary amines?
(a) Reaction with HNO_2
(b) Reaction with chloroform and alcoholic KOH
(c) Reaction with acetyl chloride
(d) Reaction with Grignard reagent

23. $CH_3COOH \xrightarrow{NH_3} \xrightarrow{\Delta} ?$
The product of the reaction is isomeric with
(a) $CH_2(NH_2)—CHO$ (b) $CH_3CH{=}NOH$
(c) $HCONH—CH_3$ (d) All of these

24. Which one of the following compound gives aspirin on reacting with acetic anhydride in presence of H_2SO_4?
(a) 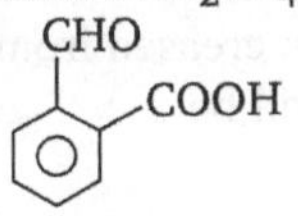
(b) 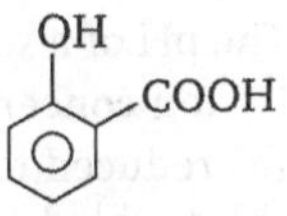
(c) 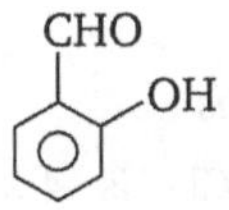
(d) 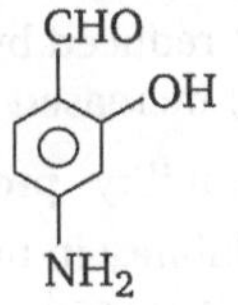

25. $R—C(=O)Z + Na^- \longrightarrow R—C(=O)Na + Z^-$
is fastest when Z is
(a) Cl (b) NH_2
(c) OC_2H_5 (d) $OCOCH_3$

26. Aldehyde not showing cannizzaro's reaction is
(a) paraldehyde (b) chloral
(c) formaldehyde (d) acetaldehyde

27. Both HCHO and CH_3CHO gives similar reactions with all the reagents except
(a) Schiff reagent
(b) Fehling solution
(c) ammoniacal $AgNO_3$
(d) ammonia

28. Which of the following statements is not correct?
(a) All alcohols are miscible with water
(b) Only lower alcohols are miscible with water
(c) All alcohols are not poisonous
(d) Methanol is poisonous

29. Cyclohexanol is
(a) phenol (b) 1° alcohol
(c) 2° alcohol (d) 3° alcohol

30. In the manufacture of ethanol from sugar the enzymes are
(a) diastase and zymase
(b) maltase and zymase
(c) diastase and invertase
(d) invertase and zymase

31. The action of chloral on chlorobenzene gives
(a) BHC (d) DDT
(c) gammexene (d) lindane

32. Which halide will be least reactive in respect to hydrolysis?
(a) vinyl chloride (b) allyl chloride
(c) ethyl chloride (d) *t*-butyl chloride

33. By Wurtz reaction, a mixture of methyl iodide and ethyl iodide gives
(a) butane
(b) ethane
(c) propane
(d) a mixture of above three

34. The cyanide process is used for obtaining
(a) Cu (b) Na
(c) Zn (d) Ag

35. Which of the following ore does not represent the ores of iron?
(a) Cassiterite (b) Limonite
(c) Haematite (d) Magnetite

36. van-Arkel method of purification of metals involves converting the metal to a
(a) volatile stable compound
(b) non-volatile stable compound
(c) volatile unstable compound
(d) None of the above

37. The first element of rare earth metal is
(a) cerium (b) cesium
(c) lanthanide (d) actinide

38. Which of the following transitions involves maximum amount of energy?
(a) $M^-(g) \longrightarrow M(g)$
(b) $M^-(g) \longrightarrow M^+(g)$
(c) $M^+(g) \longrightarrow M^{2+}(g)$
(d) $M^{2+}(g) \longrightarrow M^{3+}(g)$

39. Transition metal with low oxidation number will act as
(a) an oxidising agent (b) a base
(c) an acid (d) None of these

40. Chloride of which of the following element is coloured?
(a) Hg (b) Ag
(c) Co (d) Zn

41. Spiegeleisn is an alloy of
(a) Fe, Co and Cr (b) Fe, Co and Mg
(c) Fe, Mg, and C (d) Fe, C and Mn

42. Which of the following ions form most stable complex compound?
(a) Mn^{2+} (b) Ni^{2+}
(c) Fe^{2+} (d) Cu^{2+}

43. Coordination number of Fe in the complexes $[Fe(CN)_6]^{4-}$, $[Fe(CN)_6]^{3-}$ and $[Fe Cl_4]^-$ would be respectively
(a) 6, 4, 6 (b) 6, 6, 4
(c) 6, 3, 3 (d) 2, 3, 3

44. Which of the following is a wrong statement?
(a) $Ni(CO)_4$ has zero oxidation number for Ni
(b) $Ni(CO)_4$ has oxidation number +4 for Ni
(c) Ni is metal
(d) CO is gas

45. Rusting of iron is catalysed by which of the following?
(a) Fe (b) Zn
(c) O_2 (d) H^+

46. Standard electrode potential of NHE at 298 K is
(a) 0.05 V (b) 0.10 V
(c) 0.50 V (d) 0.00 V

47. For a cell reaction involving a two electron change, the standard emf of the cell is found to be 0.295 V at 25°C. The equilibrium constant of the reaction at 25°C will be
(a) 10 (b) 1×10^{10}
(c) 1×10^{-10} (d) 10×10^{-2}

48. The ionic conductance of Ba^{2+} and Cl^- are respectively 127 and 76 ohm^{-1}-cm^2 at infinite dilution. The equivalent conductance (inohm^{-1} cm^2) of $BaCl_2$ at infinite dilution will be
(a) 139.5 (b) 203
(c) 279 (d) 101.5

49. Multimolecular colloids are present in
(a) soap solution (b) sol of proteins
(c) sol of gold (d) All of these

50. In physical adsorption, the gas molecules are held by solid surfaces through
(a) strong chemical forces
(b) van der waals' forces
(c) metallic bonds
(d) gravitational forces

51. The osmotic pressure of a 5% (wt./vol.) solution of cane sugar at 150°C is
(a) 3.078 atm (b) 4.078 atm
(c) 5.078 atm (d) 2.45 atm

52. The mole fraction of water in 20% aqueous solution of H_2O_2 is
(a) $\frac{20}{80}$ (b) $\frac{80}{20}$
(c) $\frac{34}{77}$ (d) $\frac{77}{68}$

53. The pH of a solution is increased from 3 to 6, its H^+ ion concentration will be
(a) reduced to half
(b) doubled
(c) reduced by 1000 times
(d) increased by 1000 times

54. Solubility product of $BaCl_2$ is 4×10^{-9}. Its solubility in mol/L would be
(a) 1×10^{-3} (b) 1×10^{-9}
(c) 4×10^{-27} (d) 1×10^{-27}

55. An example of a Lewis acid is
(a) NaCl (b) $MgCl_2$
(c) $AlCl_3$ (d) $SnCl_4$

56. Radioactive decay is a
(a) first order reaction
(b) zero order reaction
(c) second order reaction
(d) third order reaction

57. The rate constant for the first order reaction is 60 s^{-1}. How much time will it take to reduce the concentration of the reactant to $\frac{1}{16}$ th value?
(a) 4.6×10^{-2} s
(b) 4.6×10^{4} s
(c) 4.6×10^{2} s
(d) 4.6×10^{-4} s

58. $(\Delta H - \Delta U)$ for the formation of carbon monoxide (CO) from its elements at 298 K is ($R = 8.314\ JK^{-1}\ mol^{-1}$)
(a) $-1238.78\ J\ mol^{-1}$ (b) $1238.78\ J\ mol^{-1}$
(c) $-2477.57\ J\ mol^{-1}$ (d) $2477.57\ J\ mol^{-1}$

59. An ideal gas at constant temperature and pressure expands, then its
(a) internal energy remains same
(b) internal energy decrease
(c) internal energy increases
(d) entropy first increases and then decreases

60. At which temperature nitrogen under 1.00 atm pressure has the same root mean square speed as that of CO_2 at STP
(a) 0°C (b) 27°C (c) – 99°C (d) – 200°C

Mathematics

1. The value of $\begin{vmatrix} a^2 & b^2 & c^2 \\ (a+1)^2 & (b+1)^2 & (c+1)^2 \\ (a-1)^2 & (b-1)^2 & (c-1)^2 \end{vmatrix}$ is

(a) $3\begin{vmatrix} a^2 & b^2 & c^2 \\ a & b & c \\ 1 & 1 & 1 \end{vmatrix}$ (b) $4\begin{vmatrix} a^2 & b^2 & c^2 \\ a & b & c \\ 1 & 1 & 1 \end{vmatrix}$

(c) $2\begin{vmatrix} a^2 & b^2 & c^2 \\ a & b & c \\ 1 & 1 & 1 \end{vmatrix}$ (d) None of these

2. If $a_1x + b_1y + c_1z = 0$, $a_2x + b_2y + c_2z = 0$, $a_3x + b_3y + c_3z = 0$ and $\begin{vmatrix} a_1 & b_1 & c_1 \\ a_2 & b_2 & c_2 \\ a_3 & b_3 & c_3 \end{vmatrix} = 0$, then the given system has
(a) one trivial and one non-trivial solution
(b) no solution
(c) one solution
(d) infinite solution

3. Let A be a square matrix all of whose entries are integers. Then, which one of the following is true?
(a) If det $(A) = \pm 1$, then A^{-1} exists but all its entries are not necessarily integers
(b) If det $(A) \neq \pm 1$, then A^{-1} exists and all its entries are non-integers
(c) If det $(A) = \pm 1$, then A^{-1} exists and all its entries are integers
(d) If det $(A) = \pm 1$, then A^{-1} need not exist

4. Which of the following statements is/are correct?
I. Adjoint of a unit matrix is a unit matrix.
II. $A(\text{adj } A) = (\text{adi } A)\ A = |A|\, I$
III. Adjoint of a symmetric matrix is symmetric.
IV. Adjoint of a diagonal matrix is a diagonal matrix.
(a) I
(b) II
(c) I or II
(d) All statements are correct

5. The value of $\int \frac{dx}{x\sqrt{1-(\log x)^2}}$ is
(a) $\cos^{-1}(\log x + C)$
(b) $x \log(1 - x^2) + C$
(c) $\frac{1}{2}\cos^{-1}(\log x + C)$
(d) $\sin^{-1}(\log x) + C$

6. $\int \frac{x^2}{(x^2+2)(x^2+3)}\, dx$ is equal to
(a) $-\sqrt{2}\tan^{-1}\left(\frac{x}{\sqrt{2}}\right) + \sqrt{3}\tan^{-1}\left(\frac{x}{\sqrt{3}}\right) + C$
(b) $\sqrt{2}\tan^{-1}\left(\frac{x}{\sqrt{2}}\right) + \sqrt{3}\tan^{-1}\left(\frac{x}{\sqrt{3}}\right) + C$
(c) $-\sqrt{2}\tan^{-1} x + \sqrt{3}\tan^{-1} x + C$
(d) None of the above

7. The degree and order of the differential equation of the family of all parabolas whose axis is x-axis, are respectively
(a) 2, 1 (b) 1, 2
(c) 3, 2 (d) 2, 3

8. The differential equation satisfied by the family of curve $y = ax\cos\left(\frac{1}{x} + b\right)$, where a, b are parameters, is

(a) $x^2 y_2 + y = 0$ (b) $x^4 y_2 + y = 0$
(c) $xy_2 - y = 0$ (d) $x^4 y_2 - y = 0$

9. Integrating factor of equation
$$(x^2+1)\frac{dy}{dx} + 2xy = x^2 - 1 \text{ is}$$
(a) $\frac{2x}{x^2+1}$ (b) $\frac{x^2-1}{x^2+1}$
(c) x^2+1 (d) None of these

10. The value of $\int_0^{\pi/2} |\sin x - \cos x|\, dx$ is
(a) 0 (b) $2(\sqrt{2}-1)$
(c) $\sqrt{2}-1$ (d) $2(\sqrt{2}+1)$

11. The equation of the circle which passes through the points (2, 3) and (4, 5) and the centre lies on the straight line $y - 4x + 3 = 0$, is
(a) $x^2 + y^2 + 4x - 10y + 25 = 0$
(b) $x^2 + y^2 - 4x - 10y + 16 = 0$
(c) $x^2 + y^2 - 4x - 10y + 25 = 0$
(d) None of the above

12. The number of solutions of $\log_4 (x-1) = \log_2 (x-3)$ is/ are
(a) 0 (b) 1
(c) 2 (d) 3

13. If α, β are the roots of $x^2 - 3x + 1 = 0$, then the equation whose roots are $\frac{1}{\alpha-2}, \frac{1}{\beta-2}$, is
(a) $x^2 + x + 1 = 0$ (b) $x^2 - x - 1 = 0$
(c) $x^2 + x - 1 = 0$ (d) None of these

14. If x is real, then the value of $\frac{x^2+34x-71}{x^2+2x-7}$ does not lie between
(a) 8 and –5 (b) –5 and 9
(c) 0 and 9 (d) 5 and 9

15. The square root of $\sqrt{50} + \sqrt{48}$ is
(a) $2^{1/4}(3+\sqrt{2})$ (b) $2^{1/4}(\sqrt{3}+\sqrt{2})$
(c) $2^{1/4}(2+\sqrt{2})$ (d) $2^{1/4}(\sqrt{3}+2)$

16. Let $I_1 = \int_\alpha^{\pi-\alpha} xf(\sin x)\, dx$, $I_2 = \int_{xf(\sin x),\, dx}^{\pi-\alpha} f(\sin x)\, dx$, then I_2 is equal to
(a) $\frac{\pi}{2} I_1$ (b) πI_1
(c) $\frac{2}{\pi} I_1$ (d) $2 I_1$

17. The area of figure bounded by $y = e^x, y = e^{-x}$ and the straight line $x = 1$ is
(a) $\left(e + \frac{1}{e}\right)$ sq unit (b) $\left(e - \frac{1}{e}\right)$ sq unit
(c) $\left(e + \frac{1}{e} - 2\right)$ sq unit (d) $\left(e + \frac{1}{e} + 2\right)$ sq unit

18. The domain and range of $f(x) = \sin^{-1}[x]$ are
(a) $[0,2), \left\{-\frac{\pi}{2}, \frac{\pi}{2}\right\}$
(b) $[-1,2), \left\{-\frac{\pi}{2}, 0, \frac{\pi}{2}\right\}$
(c) $[-1,2), \left\{-\pi, \frac{-\pi}{2}, 0, \frac{\pi}{2}, \pi\right\}$
(d) None of the above

19. If **a**, **b** and **c** are non-collinear vectors such that for some scalars x, y, z, $x\mathbf{a} + x\mathbf{b} + x\mathbf{c} = 0$, then
(a) $x = 0, y = 0, z = 0$
(b) $x \neq 0, y \neq 0, z = 0$
(c) $x = 0, y \neq 0, z \neq 0$
(d) $x \neq 0, y \neq 0, z \neq 0$

20. Let **a**, **b**, **c**, be three vectors such that $\mathbf{c} \neq 0$ and $\mathbf{a} \cdot \mathbf{b} = 2\mathbf{a} \cdot \mathbf{c}, |\mathbf{a}| = |\mathbf{c}| = 1, |\mathbf{b}| = 4$ and $|\mathbf{b} \times \mathbf{c}| = \sqrt{15}$, if $\mathbf{b} - 2\mathbf{c} = \lambda\alpha$. then, λ equals
(a) –1 (b) 1
(c) – 4 (d) 2

21. The image of the point with position vector $\mathbf{i} + 3\mathbf{k}$ in the plane $r \cdot (\mathbf{i} + \mathbf{j} + \mathbf{k}) = 1$ is
(a) $\mathbf{i} - 2\mathbf{j} + \mathbf{k}$ (b) $\mathbf{i} + 2\mathbf{j} - \mathbf{k}$
(c) $-\mathbf{i} - 2\mathbf{j} + \mathbf{k}$ (d) None of these

22. The vectors **c**, **a**, $= x\,\mathbf{i} + y\,\mathbf{j} + z\,\mathbf{k}$ and $\mathbf{b} = \mathbf{j}$ are such that **a**, **b**, **c** form a right handed system, then **c** is
(a) 0 (b) $y\,\mathbf{j}$
(c) $-z\,\mathbf{i} + x\,\mathbf{k}$ (d) $z\,\mathbf{i} - x\,\mathbf{k}$

23. The equation of plane passing through a point A (2, –1, 3) and parallel to the vectors $\mathbf{a} = (3, 0, -1)$ and $\mathbf{b} = (-3, 2, 2)$ is
(a) $2x - 3y + 6z + 25 = 0$
(b) $3x - 2y + 6z + 25 = 0$
(c) $2x - 3y + 6z - 25 = 0$
(d) $3x - 2y + 6z - 25 = 0$

24. $\lim_{x\to 1} \frac{(2x-3)(\sqrt{x}-1)}{2x^2+x-3}$ is equal to
(a) $-1/10$ (b) $1/10$
(c) $-1/8$ (d) None of these

25. Which is the correct order for a given number α in increasing order?
(a) $\log_2 \alpha, \log_3 \alpha, \log_e \alpha, \log_{10} \alpha$
(b) $\log_{10} \alpha, \log_3 \alpha, \log_e \alpha, \log_2 \alpha$
(c) $\log_{10} \alpha, \log_e \alpha, \log_2 \alpha, \alpha \log_3 \alpha$
(d) None of the above

26. If $x = \frac{\sqrt{5}+\sqrt{2}}{\sqrt{5}-\sqrt{2}}, y = \frac{\sqrt{5}-\sqrt{2}}{\sqrt{5}+\sqrt{2}}$, then $3x^2 + 4xy - 3y^2$ is equal to
(a) $\frac{1}{3}(56\sqrt{10}+12)$ (b) $\frac{1}{3}(56\sqrt{10}-12)$
(c) $\frac{1}{3}(56+12\sqrt{10})$ (d) None of these

27. If $x + y = 1$, then $\sum_{r=0}^{n} r^2 \cdot {}^nC_r\, x^r\, y^{n-r}$ is equal to
(a) nxy (b) $nx(x + yn)$
(c) $nx(nx + y)$ (d) None of these

28. If $Z = f(x + ay) + \phi(x - ay)$, then
(a) $Z_{xx} = Z_{yy}$ (b) $Z_{xx} = a^2 Z_{yy}$
(c) $Z_{yy} = a^2 Z_{xx}$ (d) None of these

29. The existence of the unique solution of the system $x + y + z = \lambda$, $5x - y + \mu z = 10$, $2x + 3y - z = 6$ depends on
(a) μ only (b) λ only
(c) λ and μ both (d) neither λ nor μ

30. Let $A = \begin{bmatrix} 1 & -1 & 1 \\ 2 & 1 & -3 \\ 1 & 1 & 1 \end{bmatrix}$ and $10B = \begin{bmatrix} 4 & 2 & 2 \\ -5 & 0 & \alpha \\ 1 & -2 & 3 \end{bmatrix}$. If B is the inverse of matrix A, then α is
(a) -1 (b) -2
(c) 2 (d) 5

31. The angle of intersection of ellipse $\frac{x^2}{a^2} + \frac{y^2}{b^2} = 1$ and circle $x^2 + y^2 = ab$ is
(a) $\tan^{-1}\left(\frac{a+b}{ab}\right)$ (b) $\tan^{-1}\left(\frac{a-b}{\sqrt{ab}}\right)$
(c) $\tan^{-1}\left(\frac{a+b}{\sqrt{ab}}\right)$ (d) $\tan^{-1}\left(\frac{a-b}{ab}\right)$

32. If $S = \sum_{n=0}^{\infty} \frac{(\log x)^{2n}}{(2n)!}$, then S is equal to
(a) $x + x^{-1}$ (b) $x - x^{-1}$
(c) $\frac{1}{2}(x + x^{-1})$ (d) None of these

33. The value of $\log 2 + 2\left(\frac{1}{5} + \frac{1}{3} \cdot \frac{1}{5^3} + \frac{1}{5} \cdot \frac{1}{5^5} + \dots\right)$, is
(a) $\log 2 + \log 3$ (b) $\log 2 + 2$
(c) $\frac{1}{2}\log 2$ (d) $\log 3$

34. The equation of parabola whose vertex and focus are (0,4) and (0,2) respectively, is
(a) $y^2 - 8x = 32$ (b) $y^2 + 8x = 32$
(c) $x^2 + 8y = 32$ (d) $x^2 - 8y = 32$

35. The line $x + 2y = 4$ is translated parallel to itself by 3 unit in the sense of increasing x and then rotated by 30° in the anti-clockwise direction about the point where the shifted line cuts the x-axis. The equation of the line in the new position is
(a) $y = \tan(\theta - 30°)(x - 4 - 3\sqrt{5})$
(b) $y = \tan(30° - \theta)(x - 4 - 3\sqrt{5})$
(c) $y = \tan(\theta + 30°)(x + 4 + 3\sqrt{5})$
(d) $y = \tan(\theta - 30°)(x + 4 + 3\sqrt{5})$

36. The distance between the foci of a hyperbola is double the distance between its vertices and the length of its conjugate axis is 6. The equation of the hyperbola referred to its axes as axes of coordinates are
(a) $3x^2 - y^2 = 3$
(b) $x^2 - 3y^2 = 3$
(c) $3x^2 - y^2 = 9$
(d) $x^2 - 3y^2 = 9$

37. If the equation
$$12x^2 + 7xy - py^2 - 18x + qy + 6 = 0$$
represents a pair of perpendicular straight lines, then
(a) $p = 12, q = -1$ (b) $p = -12, q = 1$
(c) $p = 12, q = 1$ (d) $p = 1, q = 1$

38. If the roots of the equation $\frac{\alpha}{x-\alpha} + \frac{\beta}{x-\beta} = 1$ be equal in magnitude but opposite in sign, then $\alpha + \beta$ is equal to
(a) 0 (b) 1
(c) 2 (d) None of these

39. If $2a + 3b + 6c = 0$, then at least one root of the equation $ax^2 + bx + c = 0$ lies in the interval
(a) (1, 2) (b) (0, 1)
(c) (2, 3) (d) (3, 4)

40. The ratio in which the line $3x + 4y + 2 = 0$ divides the distance between $3x + 4y + 5 = 0$ and $3x + 4y - 5 = 0$, is
(a) 7 : 3
(b) 3 : 7
(c) 2 : 3
(d) None of the above

41. If the circle $x^2 + y^2 + 2gx + 2fy + c = 0$ is touched by $y = x$ at P such that $OP = 6\sqrt{2}$, then the value of c is
(a) 36 (b) 72
(c) 144 (d) None of these

42. The coefficient of x^4 in the expansion of $(1 + x + x^2 + x^3)^n$ is
(a) ${}^nC_4$
(b) ${}^nC_4 + {}^nC_2$
(c) ${}^nC_4 + {}^nC_2 + {}^nC_4 \cdot {}^nC_2$
(d) ${}^nC_4 + {}^nC_2 + {}^nC_1 \cdot {}^nC_2$

43. $1 + \frac{1+x}{2!} + \frac{1+x+x^2}{3!} + \frac{1+x+x^2+x^3}{4!} + \dots$ is equal to
(a) $\frac{e^x+1}{x-1}$ (b) $\frac{e^x+1}{x+1}$
(c) $\frac{e^x-e}{x+1}$ (d) $\frac{e^x-e}{x-1}$

44. The inequation $n! > 2^{n-1}$ is true for
(a) for all $n \in N$
(b) for all $n > 1$
(c) for all $n > 2$
(d) None of the above

45. Let $A = \{(x, y) : y = e^x, x \in R\}$ and $B = \{(x, y) : y = e^{-x}, x \in R\}$. Then,
(a) $A \cap B = \phi$ (b) $A \cap B \neq \phi$
(c) $A \cup B = R^2$ (d) None of these

46. In ΔABC, $\frac{b-c}{r_1} + \frac{c-a}{r_2} + \frac{a-b}{r_3}$ is equal to
(a) 0 (b) abc
(c) $a + b + c$ (d) $ab + bc + ca$

47. $ABCD$ is a rectangular field . A vertical lamp post of height 12 m stands at the corner A. If the angle of elevation of its top from B is 60° and from C is 45°, then the area of the field is
(a) $48\sqrt{2}$ m^2 (b) $12\sqrt{2}$ m^2
(c) 48 m^2 (d) $12\sqrt{3}$ m^2

48. The function $f(x) = \max\{(1 - x), (1 + x), 2\}$, $x \in (-\infty, \infty)$ is equivalent to
(a) $f(x) = \begin{cases} 1+x, & x \le -1 \\ 2, & -1 < x < 1 \\ 1-x, & x \ge 1 \end{cases}$
(b) $f(x) = \begin{cases} 1-x, & x \le -1 \\ 2, & -1 < x < 1 \\ 1+x, & x \ge 1 \end{cases}$
(c) $f(x) = \begin{cases} 1-x, & x \le -1 \\ 1, & -1 < x < 1 \\ 1+x, & x \ge 1 \end{cases}$
(d) None of the above

49. The number of values of x in the interval $[0, 5\pi]$ satisfying the equation $3\sin^2 x - 7\sin x + 2 = 0$ is
(a) 0 (b) 5 (c) 6 (d) 10

50. $\left(1 + \cos\frac{\pi}{8}\right)\left(1 + \cos\frac{3\pi}{8}\right)\left(1 + \cos\frac{5\pi}{8}\right)\left(1 + \cos\frac{7\pi}{8}\right)$ is equal to
(a) $\frac{1}{2}$ (b) $\frac{1}{8}$
(c) $\cos\frac{\pi}{8}$ (d) $\frac{1+\sqrt{2}}{2\sqrt{2}}$

51. The number of solutions of the equation $2\sin^{-1}\sqrt{x^2 - x + 1} + \cos^{-1}\sqrt{x^2 - x} = \frac{3\pi}{2}$ is
(a) 0 (b) 2
(c) 4 (d) ∞

52. Each side of a square subtends an angle of 60° at the top of a tower h metre high standing in the centre of the square. If a is the length of each side of the square, then
(a) $2a^2 = h^2$ (b) $2h^2 = a^2$
(c) $3a^2 = 2h^2$ (d) $2h^2 = 3a^2$

53. The value of $\cos^{-1}\left(\cos\frac{7\pi}{6}\right)$ is
(a) $\frac{7\pi}{6}$ (b) $\frac{5\pi}{3}$
(c) $\frac{5\pi}{6}$ (d) $\frac{13\pi}{6}$

54. If $n \in N$, then $|\sin nx|$
(a) $\le n|\sin x|$ (b) $\ge n|\sin x|$
(c) $= n|\sin x|$ (d) None of these

55. In ΔABC, if $a = 30, b = 24, = 18$, then r_3 is equal to
(a) 15 (b) 18 (c) 36 (d) 12

56. The circumcentre of a triangle formed by the lines $xy + 2x + 2y + 4 = 0$ and $x + y + 2 = 0$, is
(a) $(0, -1)$ (b) $(-1, 0)$
(c) $(1, 1)$ (d) $(-1, -1)$

57. One diagonal of a square is along the line $8x - 15y = 0$ and one of its vertex is $(1, 2)$. Then, the equation of the sides of the square passing through this vertex are
(a) $23x + 7y = 9, 7x + 23y = 53$
(b) $23x - 7y - 9 = 0, 7x + 23y - 53 = 0$
(c) $23x - 7y + 9 = 0, 7x + 23y + 53 = 0$
(d) None of the above

58. The two curves $x^3 - 3xy^2 + 2 = 0$ and $3x^2 y - y^3 = 2$
(a) cut at right angles
(b) touch each other
(c) cut at an angle $\pi/3$
(d) cut at an angle $\pi/4$

59. If $\frac{2x+3}{(x+1)(x-3)} = \frac{a}{x+1} + \frac{b}{x-3}$, then $a + b$ is equal to
(a) 1 (b) 2
(c) 9/4 (d) $-1/4$

60. $\lim_{n \to \infty} \frac{1}{n}\left[\sec^2 \frac{\pi}{4n} + \sec^2 \frac{2\pi}{4n} + \ldots + \sec^2 \frac{n\pi}{4n}\right]$ is equal to
(a) $\frac{4}{\pi}$ (b) $\frac{2}{\pi}$
(c) $\frac{3}{\pi}$ (d) $\frac{5}{\pi}$

61. If $f(x) = |\cos x|$ and $g(x) = [x]$, then $gof(x)$ is equal to
(a) $|\cos [x]|$ (b) $|\cos x|$
(c) $[|\cos x|]$ (d) $|[\cos x]|$

62. If $\lim_{x \to \infty} \left(1 + \frac{a}{x} + \frac{b}{x^2}\right)^{2x} = e^2$, then
(a) $a = 1, b = 2$ (b) $a = 2, b = 1$
(c) $a = 1, b \in R$ (d) None of these

63. The area bounded by the curve $y^2 = 2x + 1$ and the straight line $x - y - 1 = 0$ is given by
(a) $\frac{9}{2}$ sq units (b) $\frac{43}{6}$ sq units
(c) $\frac{35}{6}$ sq units (d) None of these

64. The value of $f(0)$, so that the function $f(x) = \frac{2 - (256 - 7x)^{1/8}}{(5x + 3x)^{1/5} - 2}$, $x \neq 0$ is continuous everywhere is given by
(a) -1 (b) 1
(c) 2^6 (d) None of these

65. The set of points where the function $f(x) = x|x|$ is differentiable, is
(a) $(-\infty, \infty)$ (b) $(-\infty, 0) \cup (0, \infty)$
(c) $(0, \infty)$ (d) $[0, \infty)$

66. If $f(x) = |x|^{|\sin x|}$, then $f'\left(-\frac{\pi}{4}\right)$ is equal to
(a) $\left(\frac{\pi}{4}\right)^{1/\sqrt{2}} \left(\frac{\sqrt{2}}{2} \log \frac{4}{\pi} - \frac{2\sqrt{2}}{\pi}\right)$
(b) $\left(\frac{\pi}{4}\right)^{1/\sqrt{2}} \left(\frac{\sqrt{2}}{2} \log \frac{4}{\pi} + \frac{2\sqrt{2}}{\pi}\right)$
(c) $\left(\frac{\pi}{4}\right)^{1/\sqrt{2}} \left(\frac{\sqrt{2}}{2} \log \frac{\pi}{4} - \frac{2\sqrt{2}}{\pi}\right)$
(d) $\left(\frac{\sqrt{2}}{2} \log \frac{\pi}{4} + \frac{2\sqrt{2}}{\pi}\right)$

67. A circle touches the x-axis and also touches the circle which centre at $(0, 3)$ and radius 2. The locus of the centre of the circle is
(a) a parabola (b) a circle
(c) an ellipse (d) a hyperbola

68. The circle on focal radii of a parabola as diameter touches
(a) the x-axis
(b) the tangent at the vertex
(c) the directrix
(d) None of the above

69. The equation of the ellipse whose axes are parallel to the coordinate axes having its centre at the point $(2, -3)$ and focus at $(3, -3)$ and one vertex at $(4, -3)$ is
(a) $\frac{(x-2)^2}{4} + \frac{(y+3)^2}{3} = 1$
(b) $\frac{(x+2)^2}{3} + \frac{(y+3)^2}{4} = 1$
(c) $\frac{(x+2)^2}{4} + \frac{(y+3)^2}{3} = 1$
(d) None of the above

70. The diameter of $16x^2 - 9y^2 = 144$ which is conjugate to $x = 2y$ is

(a) $y = \frac{32x}{9}$ (b) $x = \frac{16}{9}y$

(c) $y = \frac{16}{9}x$ (d) None of these

71. If $\frac{2x+3}{(x+1)(x-3)} = \frac{a}{x+1} + \frac{b}{x-3}$, then $a + b$ is equal to

(a) 1

(b) 2

(c) 9/4

(d) –1/4

72. If $\frac{e^x + 2}{(e^x - 1)(2e^x - 3)} = -\frac{3}{e^x - 1} + \frac{B}{2e^x - 3}$, then B is equal to

(a) 1 (b) 3

(c) 5 (d) 7

73. $\frac{3\sqrt{2}}{\sqrt{6}+\sqrt{3}} - \frac{4\sqrt{3}}{\sqrt{6}+\sqrt{2}} + \frac{\sqrt{6}}{\sqrt{3}+\sqrt{2}}$ is equal to

(a) $5\sqrt{2}$ (b) $3\sqrt{2}$

(c) $2\sqrt{3}$ (d) 0

74. If $\log_3 2$, $\log_3 (2^x - 5)$ and $\log_3\left(2^x - \frac{7}{2}\right)$ are in AP, then x is equal to

(a) $1, \frac{1}{2}$ (b) $1, \frac{1}{3}$

(c) $1, \frac{3}{2}$ (d) None of these

75. If the roots of the cubic equation $ax^3 + bx^2 + cx + d = 0$ are in GP, then

(a) $c^3a = b^3d$ (b) $ca^3 = bd^3$

(c) $a^3b = c^3d$ (d) $ab^3 = cd^3$

76. In the group $G = (\{1, 2, 3, 4\}, \times_5)$ the solution of $2^{-1} \times (3 \times_5 x) = 4$ is

(a) 1 (b) 2

(c) 3 (d) None of these

77. Negation of "Ram is in Class X or Rashmi is in Class XII" is

(a) Ram is not in Class X but Rashmi is in Class XII

(b) Ram is not in Class X but Rashmi is in Class XII

(c) Either Ram is not in Class X or Rashmi is not in class XII

(d) None of the above

78. The inverse of the proportion $(p \wedge \sim q) \Rightarrow r$ is

(a) $\sim r \Rightarrow \sim p \vee q$ (b) $\sim p \vee q \Rightarrow \sim r$

(c) $r \Rightarrow p \vee \sim q$ (d) None of these

79. The least remainder when 17^{30} is divided by 5, is

(a) 2 (b) 1

(c) 4 (d) 3

80. The solution of differential equation $dy - \sin x \sin y \, dx = 0$ is

(a) $e^{\cos x} \cdot \tan \frac{y}{2} = C$ (b) $e^{\cos x} \cdot \tan y = C$

(c) $\cos x \cdot \tan y = C$ (d) $\cos x \cdot \sin y = C$

General English and Aptitude

Directions (Q. Nos. 1-4) *Choose the alternative which can be substituted for the given group of words.*

1. A person concerned with practical results and values

(a) plagiarist (b) realist

(c) pragmatist (d) fundamentalist

2. Member of a band of robbers

(a) dacoit (b) brigand

(c) thief (d) pirate

3. A person without manners or polish is

(a) rustic (b) naive

(c) boorish (d) barbarian

4. A speech by an actor at the end of a play

(a) epilogue (b) monologue

(c) duologue (d) prologue

Directions (Q. Nos. 5-8) *Choose the most suitable alternative to fill in the blank.*

5. The teacher ordered Kamal to leave the room and .... him to return.

(a) stopped (b) rufused

(c) forabade (d) challenged

6. I hope you must have .... by now that failures are the stepping stones to success.

(a) known (b) felt

(c) decided (d) realised

7. The tyrant .... anyone whom he regarded as a rival.
(a) massacred (b) killed
(c) exterminated (d) slaughtered

8. In a little publicised deal, Pepsi Cola has .... the entire soft drink market in Afghanistan.
(a) conquered
(b) swallowed
(c) captured
(d) occupied

Directions (Q. Nos. 9-12) *Choose the alternative which is nearest in meaning to the word given in capital letters.*

9. DILETTANTE
(a) Opponent (b) Specialist
(c) Amateur (d) Expert

10. FLAK
(a) Adventure (b) Advice
(c) Criticism (d) Praise

11. HOODLUM
(a) Pioneer (b) Criminal
(c) Devotee (d) Scholar

12. SPASMODIC
(a) Continuous (b) Gradual
(c) Intermittent (d) Spontaneous

Directions (Q. Nos. 13-16) *Choose the alternative which is opposite in meaning of the word given in capital letters.*

13. SHAME
(a) Glorify (b) Exalt
(c) Dignify (d) Enshrine

14. RESCUE
(a) Extricate (b) Waver
(c) Bind (d) Desert

15. AGONY
(a) Pleasure (b) Bliss
(c) Ecstasy (d) Fear

16. REQUISITE
(a) Dispensable (b) Random
(c) Inappropriate (d) Chaotic

Directions (Q. Nos. 17-20) *In each of the following questions, a sentence is given with a part of it italicised. Choose the alternative which is an improvement upon this italicised part. If the sentence is correct as it is, choose (d) as your answer.*

17. Old habits *die hardly*.
(a) die much hardly
(b) die hard
(c) die too hard
(d) No improvement

18. The teacher asked, "*Why you are late*?"
(a) Why you were late
(b) Why late you are
(c) Why are you late
(d) No improvement

19. It is ten years since I *have begun* living here.
(a) begun (b) had begun
(c) began (d) No improvement

20. The various practices and norms for bank's transactions *are laid down* by the Reserve Bank of India.
(a) are laid up (b) are led down
(c) are lead up (d) No improvement

Directions (Q. Nos. 21-25) *In the following questions, choose the option which shows common feature in the relationship given in each question.*

21. Knot : Watt : Fathom
(a) The terms are connected with rope
(b) They are units of measurement
(c) The terms are used by sailors
(d) The terms are used for installing electricity

22. Majlis : Diet : Knesset
(a) These are parliaments of countries
(b) These are old names of certain countries
(c) These are foreign languages
(d) These are the names of foods eaten in different countries

23. Kanha : Periyar : Dachigam
(a) These are mountain peaks
(b) These are famous lagoons
(c) These are hill stations
(d) These are animal sanctuaries

24. Stirrup : Anvil : Drum
(a) These are used by folk artists
(b) The items are used by riders
(c) They are parts of ear
(d) These are musical instruments

25. Kennedy : Indira : Palme
(a) They were Prime Ministers
(b) They were very popular among children
(c) They were Presidents.
(d) All of them were assassinated

Directions (Q. Nos. 26-30) *In the following questions, choose the group of words that shows the same relationship as given in the question.*

26. Chair : Door : Stick
(a) Book : Pen : Notebook
(b) Statue : Brick : Pitcher
(c) Mason : Carpenter : Cobbler
(d) Tomato : Potato : Brinjal

27. Music : Guitar : Performer
(a) Trick : Rope : Acrobat
(b) Dance : Tune : Instrument
(c) Food : Recipe : Cook
(d) Patient : Medicine : Doctor

28. Hair : Brush : Wig
(a) Cement : Brick : Building
(b) Paper : Pen : Pencil
(c) Iron : Hammer : Axe
(d) Bread : Butter : Milk

29. Village : City : Suburb
(a) Footpath : Road : Highway
(b) Pillow : Quilt : Bed
(c) Flowers : Garden : Park
(d) Pen : Pencil : Colour

30. Smile : Laugh : Cry
(a) Frown : Anger : Temper
(b) Sit : Sleep : Play
(c) Touch : Catch : Release
(d) Morning : Night : Day

31. Which river of India is called Vridha Ganga?
(a) Krishna (b) Godavari
(c) Cauveri (d) Narmada

32. The capital of Tanzania is
(a) Nairobi (b) Lusaka
(c) Kaupala (d) Dar-e-Salaam

33. During which decade did the population of India record a negative growth rate?
(a) 1921-31 (b) 1911-21
(c) 1941-51 (d) 1931-41

34. Marx belonged to
(a) Germany (b) Holland
(c) France (d) Britain

35. Who is the author of 'A River Sutra'?
(a) VS Naipaul (b) Nirad C Chaudhri
(c) Gita Mehta (d) Vikram Seth

36. Who is known as the 'Father of Indian Missile Technology'?
(a) Dr UR Rao
(b) Dr APJ Abdul Kalam
(c) Dr Chidambaram
(d) Dr Homi Bhabha

37. The art and science of map making is called
(a) remote sensing
(b) cartography
(c) photogrammetry
(d) mapping

38. WTO basically promotes
(a) financial support
(b) global peace
(c) unilateral trade
(d) multilateral trade

39. The forest in Sunderban is called
(a) scrub jungle (b) mangrove
(c) deciduous forest (d) tundra

40. Indian Standard Time relates to
(a) 75.5° E longitude
(b) 82.5° E longitude
(c) 90.5° E longitude
(d) 0° longitude

Answers

Physics

1. (c)	**2.** (b)	**3.** (c)	**4.** (b)	**5.** (a)	**6.** (b)	**7.** (d)	**8.** (c)	**9.** (a)	**10.** (a)
11. (a)	**12.** (c)	**13.** (b)	**14.** (a)	**15.** (d)	**16.** (c)	**17.** (c)	**18.** (b)	**19.** (a)	**20.** (d)
21. (d)	**22.** (d)	**23.** (c)	**24.** (a)	**25.** (a)	**26.** (c)	**27.** (d)	**28.** (c)	**29.** (b)	**30.** (b)
31. (c)	**32.** (c)	**33.** (d)	**34.** (b)	**35.** (d)	**36.** (b)	**37.** (b)	**38.** (c)	**39.** (d)	**40.** (d)
41. (d)	**42.** (b)	**43.** (c)	**44.** (b)	**45.** (c)	**46.** (b)	**47.** (a)	**48.** (b)	**49.** (a)	**50.** (a)
51. (c)	**52.** (b)	**53.** (d)	**54.** (a)	**55.** (b)	**56.** (d)	**57.** (a)	**58.** (b)	**59.** (d)	**60.** (d)

Chemistry

1. (a)	**2.** (d)	**3.** (b)	**4.** (a)	**5.** (c)	**6.** (b)	**7.** (b)	**8.** (b)	**9.** (c)	**10.** (a)
11. (c)	**12.** (b)	**13.** (a)	**14.** (b)	**15.** (a)	**16.** (a)	**17.** (b)	**18.** (c)	**19.** (d)	**20.** (a)
21. (b)	**22.** (b)	**23.** (c)	**24.** (b)	**25.** (a)	**26.** (d)	**27.** (d)	**28.** (a)	**29.** (c)	**30.** (d)
31. (b)	**32.** (a)	**33.** (d)	**34.** (d)	**35.** (a)	**36.** (a)	**37.** (a)	**38.** (d)	**39.** (a)	**40.** (c)
41. (d)	**42.** (d)	**43.** (b)	**44.** (b)	**45.** (d)	**46.** (d)	**47.** (b)	**48.** (a)	**49.** (c)	**50.** (b)
51. (c)	**52.** (c)	**53.** (c)	**54.** (a)	**55.** (c)	**56.** (a)	**57.** (a)	**58.** (a)	**59.** (a)	**60.** (c)

Mathematics

1. (b)	**2.** (d)	**3.** (c)	**4.** (d)	**5.** (d)	**6.** (a)	**7.** (a)	**8.** (b)	**9.** (c)	**10.** (b)
11. (c)	**12.** (b)	**13.** (b)	**14.** (d)	**15.** (b)	**16.** (c)	**17.** (c)	**18.** (b)	**19.** (a)	**20.** (c)
21. (c)	**22.** (d)	**23.** (c)	**24.** (a)	**25.** (b)	**26.** (a)	**27.** (c)	**28.** (c)	**29.** (a)	**30.** (d)
31. (b)	**32.** (c)	**33.** (d)	**34.** (c)	**35.** (a)	**36.** (c)	**37.** (c)	**38.** (a)	**39.** (b)	**40.** (b)
41. (b)	**42.** (d)	**43.** (d)	**44.** (c)	**45.** (b)	**46.** (a)	**47.** (a)	**48.** (b)	**49.** (c)	**50.** (b)
51. (b)	**52.** (b)	**53.** (c)	**54.** (a)	**55.** (d)	**56.** (d)	**57.** (b)	**58.** (a)	**59.** (b)	**60.** (a)
61. (c)	**62.** (c)	**63.** (d)	**64.** (d)	**65.** (a)	**66.** (a)	**67.** (a)	**68.** (b)	**69.** (a)	**70.** (a)
71. (b)	**72.** (d)	**73.** (d)	**74.** (d)	**75.** (a)	**76.** (a)	**77.** (d)	**78.** (b)	**79.** (c)	**80.** (a)

General English and Aptitude

1. (c)	**2.** (b)	**3.** (a)	**4.** (a)	**5.** (c)	**6.** (d)	**7.** (c)	**8.** (c)	**9.** (c)	**10.** (d)
11. (b)	**12.** (c)	**13.** (b)	**14.** (d)	**15.** (c)	**16.** (a)	**17.** (b)	**18.** (c)	**19.** (c)	**20.** (d)
21. (b)	**22.** (a)	**23.** (d)	**24.** (c)	**25.** (d)	**26.** (b)	**27.** (a)	**28.** (c)	**29.** (d)	**30.** (b)
31. (b)	**32.** (d)	**33.** (b)	**34.** (a)	**35.** (c)	**36.** (b)	**37.** (b)	**38.** (d)	**39.** (b)	**40.** (b)

Hints & Solutions

Physics

1. Ammeter is always connected in series with circuit.

3. Time period of proton $T_p = \frac{25}{5} = 5\,\mu s$

By using $T = \frac{2\pi m}{qB}$

$$\Rightarrow \quad \frac{T_\alpha}{T_p} = \frac{m_\alpha}{m_p} \times \frac{q_p}{q_\alpha} = \frac{4m_p}{m_p} \times \frac{q_p}{2q_p} = 2$$

$$\Rightarrow \quad T_\alpha = 2T_p = 10\,\mu s$$

4. The magnitude of force acting on length l of each conductor due to other is

$$\frac{F}{l} = \frac{\mu_0}{2\pi} \frac{i_1 i_2}{r}$$

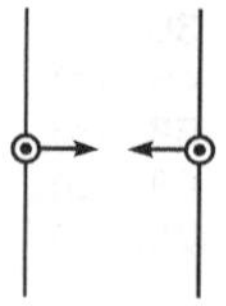

Given $\quad i_1 = i, i_2 = 10\,i$

$$\therefore \quad \frac{F}{l} = \frac{\mu_0}{2\pi} \frac{10 i^2}{r}$$

From Fleming's left hand rule, this is force of attraction.

5. Fundamental frequency of open pipe

$$n_1 = \frac{v}{2l}$$

Frequency of third harmonic of closed pipe

$$n_1 = \frac{3v}{4l}$$

$$\therefore \quad \frac{3v}{4l} - \frac{v}{2l} = 100$$

$$\frac{v}{4l} = 100$$

$$\frac{1}{2}\left(\frac{v}{2l}\right) = 100$$

$$\therefore \quad n_1 = 200 \text{ Hz}$$

6. Separation of slits $d = 4 \text{ mm} = 4 \times 10^{-3}$ m

Wavelength $\quad \lambda = 6000\text{ Å} = 6 \times 10^{-7}$ m

Distance of screen $D = 2$ m

Fringe width $\beta = \frac{\lambda D}{d}$

$$= \frac{6 \times 10^{-7} \times 2}{4 \times 10^{-3}}$$

$$= 3 \times 10^{-4} \text{ m}$$

$$= 0.3 \text{ mm}$$

7. Angle of deviation $\delta = (n - 1)\,A$.

Dispersion produced by both the prism's will be equal.

$$(n_1 - 1)\,A_1 = (n_2 - 1)\,A_2$$

$$\Rightarrow \quad A_2 = \frac{(n_1 - 1)\,A_1}{(n_2 - 1)} = \frac{(1.54 - 1)}{(1.72 - 1)} = 3°$$

Hence, the angle of prism P_2 is 3°.

8. If housefly sits on objective lens, then the intensity of image will decrease, because some of light rays will be stopped by housefly but housefly will not be seen.

9. We know that capacitor offers infinite resistance for DC source.

Current taken from the cell

$$i = \frac{E}{R + r}$$

$$\therefore \quad i = \frac{2.5}{1 + 1 + 0.5} = 1\text{ A}$$

Potential drop across capacitor

$$V = E - ir$$

$$= 2.5 - 1 \times 0.5 = 2\text{ V}$$

The current in 2 Ω resistor is zero.

Charge on capacitor

$$Q = CV$$

$$= 5 \times 2 = 10\,\mu C$$

10. Volume of bigger drop = volume of 64 small drops

i.e., $\frac{4}{3}\pi R^3 = 64 \times \frac{4}{3}\pi r^3$

or $R = 4r$

Potential on small drop

$$V_1 = \frac{1}{4\pi\varepsilon_0}\frac{q}{r}$$

Potential on bigger drop

$$V_2 = \frac{1}{4\pi\varepsilon_0}\frac{64q}{R}$$

$$V_2 = \frac{1}{4\pi\varepsilon_0}\frac{64q}{4r}$$

$$= \frac{9\times10^9 \times 64\times10^{-9}}{4\times2\times10^{-2}}$$

$$= 7.2\times10^3 \text{ V}$$

11. $p_x = 2\cos t$, $p_y = 2\sin t$,

$\therefore$ $\mathbf{p} = p_x\,\mathbf{i} + p_y\,\mathbf{j}$

$= 2\cos t\,\mathbf{i} + 2\sin t\,\mathbf{j}$

Force = rate of change of momentum

$= \frac{d\mathbf{p}}{dt}$

$\Rightarrow$ $\mathbf{F} = -2\sin t\,\mathbf{i} + 2\cos t\,\mathbf{j}$

$\therefore \cos\theta = \frac{\mathbf{F}\cdot\mathbf{p}}{|\mathbf{F}|\cdot|\mathbf{p}|}$

$$= \frac{(-2\sin t\,\mathbf{i} + 2\cos t\,\mathbf{j})(2\cos t\,\mathbf{i} + 2\sin t\,\mathbf{j})}{\sqrt{4\sin^2 t + 4\cos^2 t}\sqrt{4\cos^2 t + 4\sin^2 t}}$$

$\Rightarrow$ $\cos\theta = 0$

$\Rightarrow$ $\theta = 90°$

12. When a body tied to the end of a string is revolved in a vertical circle, the speed of the body is different at different points of the circular paths.

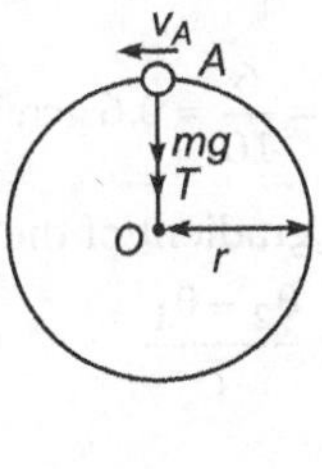

The resultant of tension (T) and weight (mg) provides the necessary centripetal force

i.e., $T + mg = \frac{mv_A^2}{r}$

Tension (T) becomes zero at the points of minimum velocity, therefore

$\therefore$ $0 + mg = \frac{mv_A^2}{r}$

$\Rightarrow$ $v_A = \sqrt{gr}$

13. Drop is stationary, i.e.,

weight of drop = electric force

i.e., $mg = qE$

or $mg = q\frac{V}{d}$

or $mg = ne\frac{V}{d}$

$\therefore$ $n = \frac{mgd}{eV}$

$$= \frac{1.8\times10^{-13}\times10\times0.9\times10^{-2}}{1.6\times10^{-19}\times2000}$$

$n = 50$

14. Many nuclei are radioactive. This means that they are unstable like radium and will eventually decay by emitting as particle transforming the nucleus into another nucleus. A chain of decays take place until a stable nucleus is reached. In the compound radium bromide, radium is radioactive while bromide is not, hence decay constant of compound is same as that of radium that is λ.

15. Remaing amount $N = N_0\left(\frac{1}{2}\right)^n$

$$N = 6.4\times10^{10}\left(\frac{1}{2}\right)^{12/4}$$

$= 0.8\times10^{10}$ atoms

16. An insulator has its valence band filled with electrons and its conduction band is empty. The energy gap is large approximately 5.0 eV, so it is not possible for ordinary process to provide enough energy for valence electrons to break out of their valence band and jump into the conduction band where they might carry electric current.

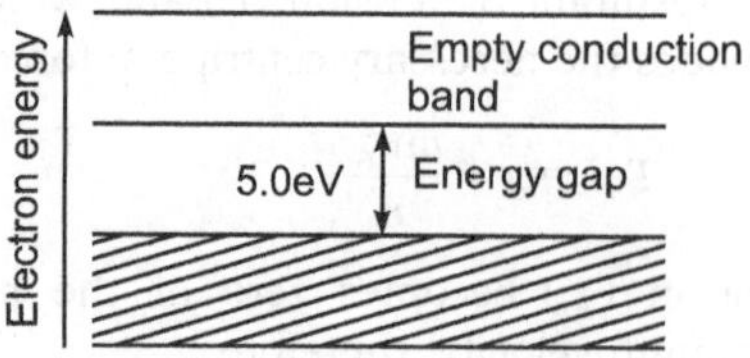

17. According to Fleming's right hand rule, magnetic field will be perpendicular to the plane of paper and act downward.

18. The force exerted on electron moving with velocity v is

$$F = evB \sin \theta$$

where $\theta = 90°$

$$\therefore \quad F_B = evB \quad \ldots\text{(i)}$$

Centripetal force is given by

$$F_C = \frac{mv^2}{r} \quad \ldots\text{(ii)}$$

Equating Eqs. (i) and (ii), we get

$$r = \frac{mv^2}{eB}$$

Diameter $d = 2r = \dfrac{2v}{(e/m)B}$

$$\Rightarrow \quad d = \frac{2 \times 2 \times 10^7}{1.76 \times 10^{11} \times 2 \times 10^{-2}}$$

$$= 1.1 \text{ cm}$$

19. Given, $V = 100 \sin 100t$

$$i = 100 \sin\left(100t + \frac{\pi}{3}\right)$$

Comparing with standard forms

$$V = V_0 \sin \omega t$$

$$I = I_0 \sin(\omega t + \phi)$$

$$\therefore \quad V_0 = 100 \text{ V}$$

$$I_0 = 100 \text{ mA} = 100 \times 10^{-3} \text{ A}$$

$$\phi = \frac{\pi}{3}$$

Power $P = \dfrac{V_0 I_0}{2} \cos \phi$

$$\therefore \quad P = \frac{100 \times 100 \times 10^{-3}}{2} \cos\frac{\pi}{3} = 2.5 \text{ W}$$

20. Escape velocity $v_e = \sqrt{\dfrac{2GM}{R}}$

$$= R\sqrt{\frac{8}{3}G\rho}$$

$$\therefore \quad v \propto R$$

We can say, if radius of a planet is double than that of earth then escape velocity from this planet will be 22 km/s.

21. $L = mvr = m\sqrt{\dfrac{GM}{r}}\, r = m\sqrt{GMr}$

$$\therefore \quad L \propto \sqrt{r}$$

$$\Rightarrow \quad \frac{L_2}{L_1} = \sqrt{\frac{r_2}{r_1}}$$

$$= \sqrt{\frac{16r}{r}} = 4$$

$$\Rightarrow \quad L_2 = 4L$$

22. Acceleration $a = \omega^2 y$

$$a = \left(\frac{2\pi}{T}\right)^2 y$$

$$\Rightarrow \quad T^2 = \frac{4\pi^2 y}{a}$$

$$\Rightarrow \quad T = 2\pi\sqrt{\frac{y}{a}}$$

$$= 2\pi\sqrt{\frac{3}{12}}$$

$$T = 2\pi \times \frac{1}{2} = 3.14 \text{ s}$$

23. $r = \sqrt{r_1^2 + r_2^2} = \sqrt{9 + 16} = \sqrt{25} = 5$ cm.

24. $\dfrac{\text{Volume of cube submerged}}{\text{Total volume}}$

$$= \frac{\text{density of cube material}}{\text{density of water}}$$

$$\Rightarrow \quad \frac{10 - 4}{10} = \frac{d}{1}$$

$$\therefore \quad d = \frac{6}{10} = 0.6 \text{ gcm}^{-3}$$

25. Temperature gradient of the rod

$$\frac{\Delta\theta}{\Delta x} = \frac{\theta_2 - \theta_1}{l}$$

$$\therefore \quad \frac{\Delta\theta}{\Delta x} = \frac{125-25}{50}$$

$$= \frac{100}{50} = 2°\text{C/cm}$$

26. $Q = nC_p\,\Delta T$

For hydrogen

$$C_p = \frac{7}{2}R$$

So, $Q = n\frac{7}{2}R\,\Delta T$

$$= 5 \times \frac{7}{2} \times 2 \times 30$$

$$= 1050 \text{ cal}$$

27. Apparent rise $= d\left(1 - \frac{1}{_a\mu_w}\right)$

$$= 12 \times \left(1 - \frac{3}{4}\right)$$

$$= 12 \times \frac{1}{4} = 3 \text{ cm}$$

28. According to Einstein's equation relation between rest energy and moving energy is

$$E = E_0\left(1 - \frac{v^2}{c^2}\right)^{-1/2}$$

Given, $v = 0.5\,c$

$$\therefore \quad E = E_0\left(1 - \frac{(0.5\,c)^2}{c^2}\right)^{-1/2}$$

$$E = E_0(1 - 0.25)^{-1/2}$$

$$= \frac{E_0}{\sqrt{0.75}}$$

Change in energy $\Delta E = E - E_0$

$$= \frac{E_0}{\sqrt{0.75}} - E_0$$

$$= E_0\left(\frac{1}{\sqrt{0.75}} - 1\right)$$

$$= 0.511\left(\frac{1}{\sqrt{0.75}} - 1\right)$$

$$= 0.079 \text{ MeV}$$

29. Force on charge, $F = QE$

or $\quad F = \frac{QV}{d} \quad (\because V = Ed)$

For drop to be stationary,

weight of drop = force due to charge

i.e., $\quad mg = \frac{QV}{d}$

For two drops, $\frac{Q_1}{Q_2} \cdot \frac{V_1}{V_2} = \frac{m_1}{m_2}$

or $\quad \frac{Q_1}{Q_2} \cdot \frac{V_1}{V_2} = \frac{\frac{4}{3}\pi r_1^3 \rho}{\frac{4}{3}\pi r_2^3 \rho}$

or $\quad \frac{Q_2}{Q_1} = \frac{r_2^3}{r_1^3} \times \frac{V_1}{V_2}$

$\therefore \quad \frac{Q_2}{Q} = \frac{(r/2)^3}{r^3} \times \frac{2400}{600}$

or $\quad Q_2 = \frac{Q}{2}$

30. Band spectrum is defined as a molecular spectrum consisting of numerous closely spaced, often unresolved emission lines or absorption lines which occur across a limited range of frequencies. Each line represents an increment of energy due to a change in the rotational state of the molecule. Bands caused by titanium oxide, zirconium and carbon compounds are characteristic of low temperature stars.

31. Decrease in mass number = 238 – 234 = 4

No. of α-particles emitted $= \frac{4}{4} = 1$

Decrease in atomic number due to emission of 1 α-particle is 2, but now atomic number is 91, hence number of β-particles = 91 + 2 – 92 = 1.

32. $E = 57 \text{ meV} = 57 \times 10^{-3} \text{ eV}$

$$\therefore \quad \lambda = \frac{12375}{E} = \frac{12375}{57 \times 10^{-3}}$$

$$= 217.105 \times 10^3 \text{Å}$$

$$\lambda = 217105 \text{Å}$$

33. From law of conservation of linear momentum *i.e.,*

final momentum = initial momentum

$$m_1v_1 + m_2v_2 = 0 \quad \text{...(i)}$$

where m_1 and m_2 are masses of pieces.

(initial momentum is zero because body is stationary)

From Eq. (i), $m_1v_1 = -m_2v_2$

So, numerical value of momentum of both pieces will be equal.

34. $dp = F \times dt$

$$= 10 \times 10 = 100 \text{ kg-m/s}$$

36. The tension at the highest point is zero, but tension at the lowest point is 6 times the weight of the body. Hence, wire is most likely to break at the lowest point.

37. The time taken by AC in reaching from zero to maximum value is

$$t = \frac{1}{4} = \frac{1}{4f}$$

$$= \frac{1}{50 \times 4} = 5 \times 10^{-3} \text{ s}$$

38. $dQ = nCdT$

$$dQ = nk\frac{T^3}{\theta^3}dT$$

$$Q = \frac{nk}{\theta^3}\int_{T_1}^{T_2} T^3 dT$$

$$= \frac{nK}{\theta^3}\left(\frac{T_2^4 - T_1^4}{4}\right)$$

$$= \frac{2 \times 1940 \times (50^4 - 10^4)}{(281)^3 \times 4}$$

$$= 272.79 \text{ J} = 273 \text{ J}$$

39. Velocity of electron in nth orbit

$$v_n = \frac{Ze^2}{2\varepsilon_0 nh}$$

$$\Rightarrow \quad v \propto \frac{1}{n}$$

$$v_1 = v, n_1 = 2, n_2 = 5$$

$$\therefore \quad \frac{v}{v_2} = \frac{n_2}{n_1}$$

$$\therefore \quad \frac{v}{v_2} = \frac{5}{2}$$

$$\Rightarrow \quad v_2 = \frac{2v}{5}$$

40. If charged particle enters in magnetic field making an angle less than 90°, then its path will be a helix.

41. $C_1 = 10\mu\text{F}, V = 50 \text{ V}$

Charge on first capacitor

$$q = C_1V$$

$$= 10 \times 50 = 500 \mu\text{C}$$

$$\text{Common potential } V = \frac{\text{total charge}}{\text{total capacitance}}$$

$$\therefore \quad 20 = \frac{500}{C_1 + C_2}$$

$$\text{or} \quad 20 = \frac{500}{10 + C_2}$$

$$\text{or} \quad C_2 = 15\mu\text{F}$$

42. Given $E = 200\sqrt{2}\sin(100t)$ ...(i)

Comparing Eq. (i) with

$$E = E_0 \sin \omega t$$

Peak voltage $E_0 = 200\sqrt{2}, \omega = 100$

$$C = 1\mu\text{F} = 1 \times 10^{-6} \text{ F}$$

$$\text{Reading of ammeter, } i = \frac{E_{rms}}{X_C}$$

$$= \frac{E_0\omega_C}{\sqrt{2}} = \frac{E_{rms}}{1/\omega C} = \frac{E_0\omega C}{\sqrt{2}}$$

$$= \frac{200\sqrt{2} \times 100 \times 1 \times 10^{-6}}{\sqrt{2}}$$

$$= 20 \times 10^{-3} \text{ A}$$

$$= 20 \text{ mA}$$

43. Transformer works only in alternating current. Primary coil is connected to Lechlanche cell *i. e.,* connected by DC source, so voltage across secondary will be zero.

44. Resistance of galvanometer $G = 99\ \Omega$

$$\text{Resistance of shunt } S = \frac{i_gG}{i - i_g}$$

$$= \frac{10 \times 99}{100 - 10} = 11\ \Omega$$

45. Given $m = 500$ kg, $g = 9.8\ m/s^2$

AND $R_e = 6.4 \times 10^6$ m

Energy required to scape

$$E = mg\ R_e$$

$$= 500 \times 9.8 \times 6.4 \times 10^6$$

$$= 3.1 \times 10^{10}\ J$$

46. Direction of current is opposite to the direction of flow of electron and in the same direction of positive ion.

Total current $I = I_{Ne} + I_{electron}$

$$= (2.9 \times 10^{18} + 1.2 \times 10^{18})\ e$$

$$= 4.1 \times 10^{18} \times 1.6 \times 10^{-19}$$

$$= 0.66\ A$$

Direction of current will be the direction of flow of Ne^+ ion *i. e.* towards right.

47. Power $= \dfrac{\text{work}}{\text{time}}$

$$= \frac{mgh}{t} = \left(\frac{m}{t}\right) gh$$

Given, $\dfrac{m}{t} = 100$ kg/s, $h = 100$ m, $g = 10\ m/s^2$

$\therefore$ Power $= 100 \times 10 \times 100$

$$= 100\ kW$$

48. Velocity of free end

$$v = \sqrt{3gh}$$

$$= \sqrt{3 \times 10 \times 1}$$

$$= 5.47\ m/s$$

49. Forbidden energy gap for germanium crystal

$$= 0.76\ eV$$

$$= 0.76 \times 1.6 \times 10^{-19}\ J$$

$$= 1.216 \times 10^{-19}\ J$$

50. $V = 200\ V, \dfrac{e}{m} = 1.6 \times 10^{11} C/kg$

Kinetic energy of electron $= eV$

i.e, $\quad \dfrac{1}{2} mv^2 = eV$

or $\quad v^2 = \dfrac{2e}{m} V$

or $\quad v = \sqrt{2\dfrac{e}{m}V}$

$$= \sqrt{2 \times 1.6 \times 10^{11} \times 200}$$

$$= 8 \times 10^6\ m/s$$

51. From definition of temperature resistance coefficient

$$R_t = R_0\ (1 + \alpha t)$$

$$3R_0 = R_0\ (1 + 4 \times 10^{-3}\ t)$$

$$3 = 1 + 4 \times 10^{-3}\ t$$

$$4 \times 10^{-3}\ t = 2$$

$$t = \frac{2}{4 \times 10^{-3}}$$

$$= 500°C$$

52. Induced emf $e = nA\ \dfrac{dB}{dt}$

A = area of coil $= 2m^2$

and $\quad n = 1$

$$e = \frac{1 \times 2 \times (4-1)}{2} = 3\,V$$

53. Given, $l = 10$ m, $v = 5$ m/s

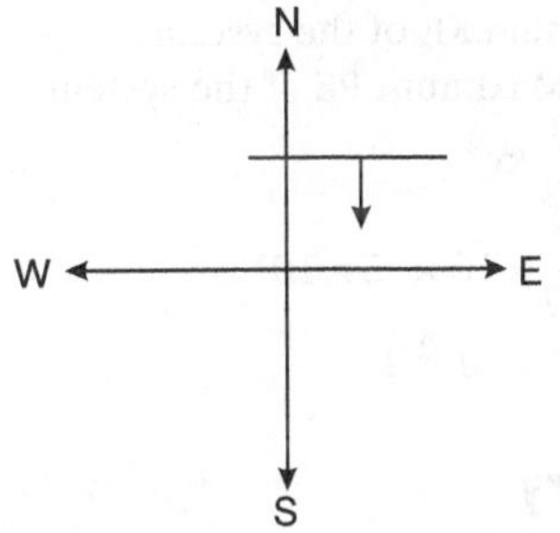

and $\quad H = 0.3 \times 10^{-4}\ Wb/m^2$

Induced emf $e = Hvl$

$$= 0.3 \times 10^{-4} \times 5 \times 10$$

$$= 1.5 \times 10^{-3}\ V$$

$$= 1.5\ m\,V$$

54. $i = 1.6$ A,$t = 1$ minute $= 60$ s

Charge on one $Cu^{++} = 2e$

Let n Cu^{++} are liberated from

$$Q = ne$$
$$it = n\,(2e)$$
$$1.6 \times 60 = n\,(2 \times 1.6 \times 10^{-19})$$
$$n = 3 \times 10^{20} \text{ ions}$$

55. Given, $L = 60$ H

$R = 30\ \Omega$

$V = 100$ V

In *L-R* circuit current to reach 63.2% of its final value in time constant.

Equation of current in *L-R* circuit

$$I = I_0\,(1 - e^{-Rt/L})$$

Required time = time constant

$$= \frac{L}{R} = \frac{60}{30} = 2\text{ s}$$

56. $$T_2 = \frac{(m_1 + m_2)T_3}{m_1 + m_2 + m_3}$$
$$= \frac{(10 + 6) \times 40}{10 + 6 + 4}$$
$$= \frac{16 \times 40}{20}$$
$$= 32\text{ N}$$

57. Maximum KE of the system

= Maximum PE of the system

$$= \frac{1}{2}kx^2$$
$$= \frac{1}{2} \times 16 \times (5 \times 10^{-2})^2$$
$$= 2 \times 10^{-2}\text{ J}$$

58. Angular momentum of particle w.r.t. origin

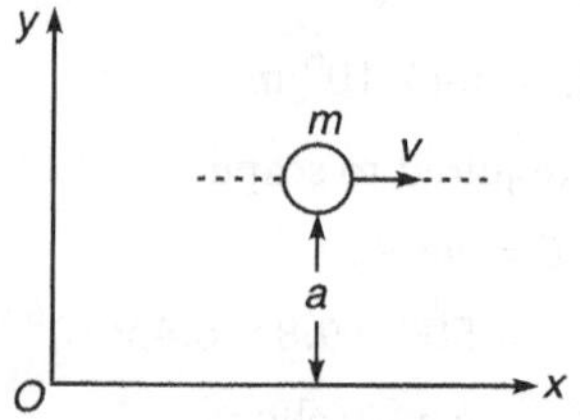

= linear momentum × perpendicular distance of line of action of linear momentum from origin

$= mv \times a = mva =$ constant

59. Isobars are the elements which have same mass number. So, ${}_{15}P^{30}$ and ${}_{14}Si^{30}$ are isobars.

60. Let initial amount be N_0.

Remaining amount $N = N_0 - \frac{7N_0}{8} = \frac{N_0}{8}$

$$\therefore \quad N = N_0\left(\frac{1}{2}\right)^n$$
$$\Rightarrow \quad \frac{N_0}{8} = N_0\left(\frac{1}{2}\right)^n$$
$$\Rightarrow \quad \left(\frac{1}{2}\right)^3 = \left(\frac{1}{2}\right)^n$$
$$\Rightarrow \quad n = 3$$
$$\therefore \quad \frac{t}{T_{1/2}} = 3$$
$$\Rightarrow \quad \frac{t}{15} = 3 \quad \text{or} \quad t = 15 \times 3$$
$$t = 45\text{ h}$$

Chemistry

1. 36 g of water $= \frac{36}{18} = 2$ mol

28 g of $CO_2 = \frac{28}{44} = 0.64$ mol

46 g of $CH_3OH = \frac{46}{32} = 1.44$ mol

58 g of $N_2O_5 = \frac{58}{108} = 0.54$ mol

2. $SO_2 + H_2O \longrightarrow 3S + 2H_2O$

Eq. wt. of SO_2 in the reaction

$$= \frac{\text{molar mass of } SO_2}{\text{change in oxidation state}} = \frac{64}{4} = 16$$

Twice of 16 will be 32.

3. An atomic orbital is a three dimensional region of definite shape around the nucleus where the probability of finding the electron is maximum. Therefore, an atom has both characteristic energy and a characteristic shape.

4. For n, $l = 0$ to $(n - 1)$

Therefore, when $n = 3$, $l \neq 3$

5. Configuration of an element ${}_9M = 1s^2, 2s^2, 2p^5$. Hence, its ion will be M^- because it can easily gain one electron to have octet.

6. In S_8 molecules, each sulphur is sp^3 hybridised with two lone pairs of electrons occupying two of the hybrid orbitals and reducing the S—S—S bond angle to 105°.

7. In oxyacids of chlorine, HClO, $HClO_2$, $HClO_3$ and $HClO_4$, the central Cl atom is sp^3 hybridised. The half filled unhybridised *d*-orbitals of Cl and *p*-orbitals of oxygen overlap to form π-bonds.

8. Anhydrous $AlCl_3$ is highly hygroscopic and hence gives fumes of hydrogen chloride by the hydrolysis of moist air.

$$AlCl_3 + \underset{\text{moist air}}{3H_2O} \longrightarrow Al(OH)_3 + 3HCl$$

9. Fe_2O_3 is amphoteric in nature as it reacts with both acid and alkali.

$$Fe_2O_3 + \underset{\text{acid}}{6HCl} \longrightarrow 2FeCl_3 + 3H_2O$$

$$Fe_2O_3 + \underset{\text{alkali}}{Na_2CO_3} \longrightarrow 2NaFeO_2 + CO_2$$

10. $P_4 + 20HNO_3 \longrightarrow 4H_3PO_4 + 20NO_2 + 4H_2O$

11. CH_3^- is the best nucleophile because carbon has the least electronegativity among key atoms of the given options.

12. Four isomeric primary amines are obtained from $C_4H_{11}N$ are possible.

(i) $CH_3CH_2CH_2NH_2$

(ii) 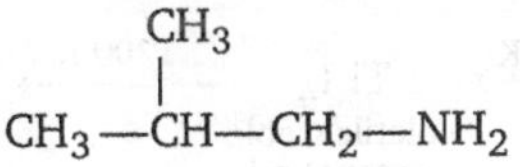

(iii) 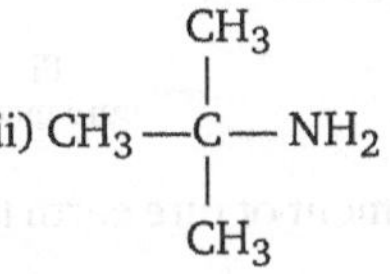

(iv) 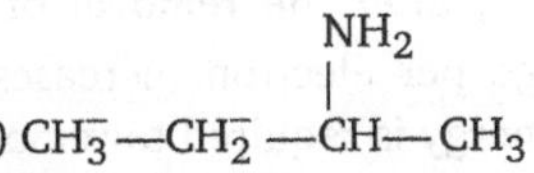

13. Reaction of alkyne with HBr supports it addition nature.

14. 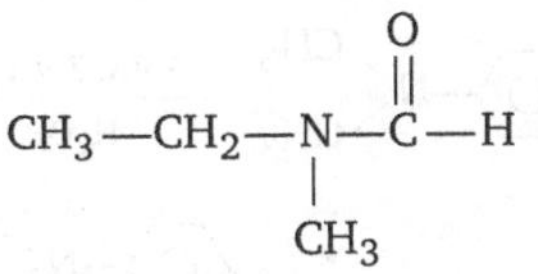

N-ethyl-N-methyl methanamide.

15. 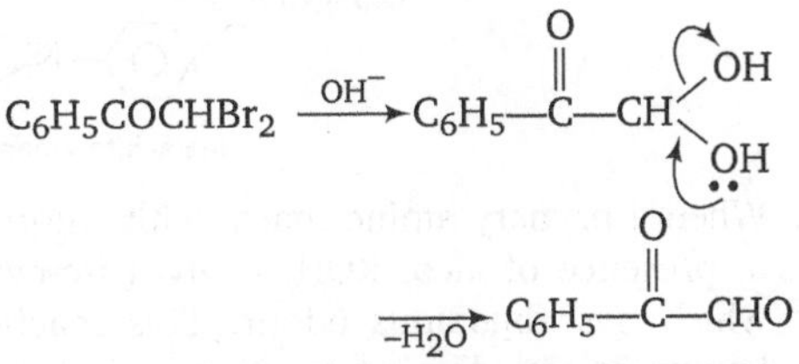

16. During catalytic hydrogenation, the hydrogens are transferred from the catalyst to the same side of the double bond. Evidently, smaller the number of *R* substituents, lesser is the steric hinderance and hence faster is the rate of hydrogenation. Thus, option (a) with two *R* groups on the same side of the molecule is correct.

17. Pentacetyl derivative can undergo basic hydrolysis.

18. Cellulose is a polysaccharide, composed of D-glucose units which are joined by β-glycosidic linkages. On complete hydrolysis cellulose produces D-glucose.

$$(C_6H_{10}O_5)_n + nH_2O \xrightarrow{H^+} \underset{\text{D-glucose}}{nC_6H_{12}O_6}$$

19. Saponification value is the number of mg of KOH required to neutralise the fatty acid resulting from the complete hydrolysis of 1 g of oil or fat.

20. $CH_3CH_2NH_2 \xrightarrow{NaNO_2 + HCl} CH_3CH_2OH + N_2 + H_2O + NaCl$

$NH_2CONH_2 \xrightarrow{NaNO_2 + HCl} 2N_2 + H_2O + CO_2 + NaCl$

$CH_3CONH_2 \xrightarrow{NaNO_2 + HCl} CH_3COOH + N_2 + H_2O + NaCl$

$C_6H_5NH_2 \xrightarrow{NaNO_2 + HCl} C_6H_5N_2^+Cl^- + H_2O + NaCl$

21. 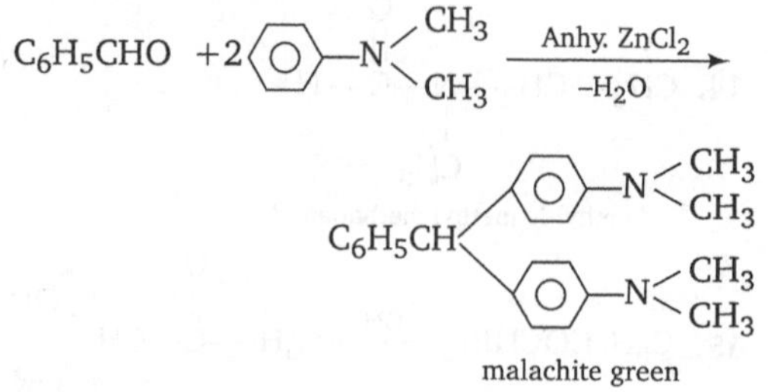

22. When a primary amine reacts with chloroform in presence of alco. KOH, it gives *iso*-cyanide which has abnoxious odour. This reaction is known as carbylamine reaction and it is given only by primary amines.

$$\underset{\text{primary amine}}{RHN_2} + CHCl_3 + \underset{\text{alco.}}{3KOH} \xrightarrow{\Delta} RNC + 3KCl + 3H_2O$$

23. $$\underset{\text{acetic acid}}{CH_3COOH} \xrightarrow{NH_3} CH_3COONH_4 \xrightarrow[-H_2O]{\Delta} \underset{\text{acetamide}}{CH_3CO\ NH_2}$$

The isomers of CH_3CONH_2 are

$$\underset{(I)}{NH_2CH_2CHO}, \underset{(II)}{CH_3—CH=NOH},$$

$$\underset{(III)}{H—CONH—CH_3}$$

24. Salicylic acid gives aspirin on reaction with acetic anhydride in presence of H_2SO_4.

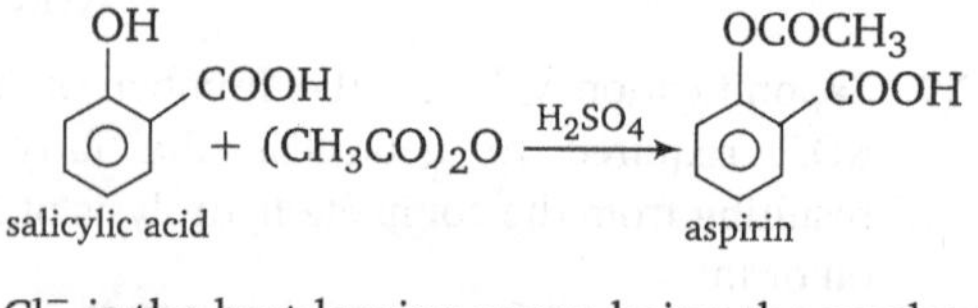

25. Cl^- is the best leaving group being the weakest nucleophile out of NH_2^-, Cl^-, $O^-C_2H_5$ and CH_3COO^-.

26. Acetaldehyde does not show cannizzaro's reaction due to the presence of α-atom.

27. HCHO and CH_3CHO give different reactions with NH_3.

$$\underset{\text{formaldehyde}}{6HCHO} + 4NH_3 \longrightarrow \underset{\text{Urotropine}}{(CH_2)_6N_4} + 6H_2O$$

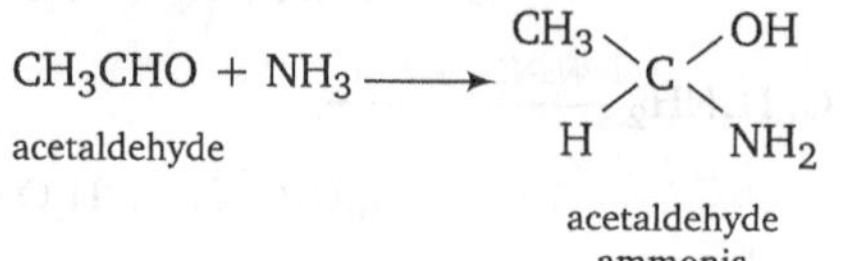

28. Solubility of alcohols in water decreases with increase in molecular masses.

29. 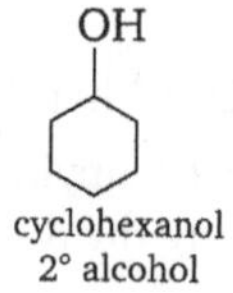

30. $$C_{12}H_{22}O_{11} + H_2O \xrightarrow{\text{Invertase}} \underset{\text{glucose}}{C_6H_{12}O_6} + \underset{\text{fructose}}{C_6H_{12}O_6}$$

$$C_6H_{12}O_6 \xrightarrow{\text{Zymase}} 2C_2H_5OH + 2CO_2$$

31. 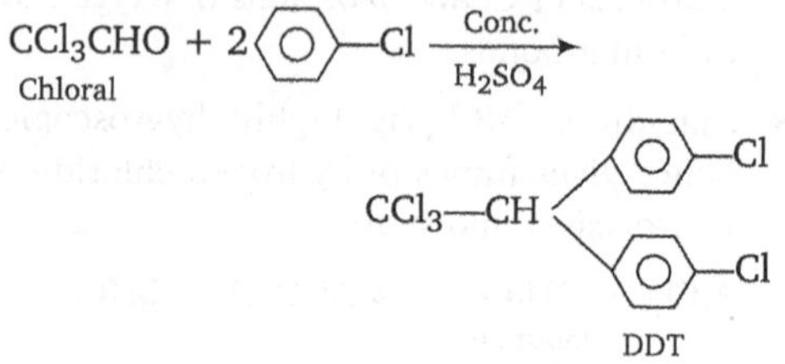

32. Vinyl chloride is least reactive towards hydrolysis because of non-reactivity of chlorine atom due to resonance stabilization.

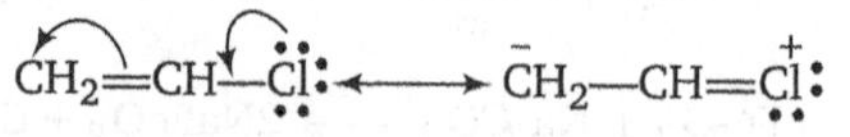

33. Wurtz reaction

$$CH_3I + 2Na + IC_2H_5 \xrightarrow[-2NaI]{\text{Dry ether}} C_2H_6 + C_3H_8 + C_4H_{10}$$

34. Cyanide process is used in the extraction of silver and gold.

35. SnO_2—Cassiterite, Fe_3O_4—Magnetite

Fe_2O_3—Haematite, $Fe_2O_3 \cdot 3H_2O$—Limonite

36. $$Ti + 2I_2 \xrightarrow{500\ K} \underset{\text{volatile stable compound}}{Ti\ I_4} \xrightarrow{1700\ K} \underset{\text{pure metal}}{Ti} + 2I_2$$

37. Cerium is the first element of rare earth metals.

38. $M^{2+} \longrightarrow M^{3+}$, after the removal of $2e^-$, the nuclear charge per electron increases, due to which high energy is required to remove $3e^-$.

39. Transition metal which have low oxidation number show the oxidising nature because of great tendency to lose the electron.

40. In $CoCl_3$, Co is in + 3 state

$Co^{3+} = 1s^2, 2s^2\ 2p^6, 3s^2\ 3p^6\ 3d^6, 4s^\circ$

4 unpaired electrons, so it will be coloured.

41. Spiegeleisn is an alloy of Fe, C and Mn.

42. The magnitude of stability constants for some divalent metal ions of the first transition series with oxygen or nitrogen donor ligands increases in the order

$Mn^{2+} < Fe^{2+} < Co^{2+} < Ni^{2+} < Cu^{2+} < Zn^{2+}$

43. Coordination number is equal to total number of ligands in a complex.

44. In $[Ni(CO)_4]$, Ni has zero oxidation number.

45. Rusting of iron is catalysed by H^+ ions.

46. At 298 K, standard electrode potential of NHE is 0.00 V.

47. $\Delta G^\circ = -nFE^\circ$

$$\Delta G^\circ = -2.303\,RT \log K_c$$

$$nFE^\circ = 2.303\,RT \log K_c$$

$$\log K_c = \frac{nFE^\circ}{2.303RT} = \frac{2 \times 96500 \times 0.295}{2.303 \times 8.314 \times 298}$$

$$\log K_c = 9.97$$

$$K_c = 1 \times 10^{10}$$

48. Equivalence conductance = conducting power of all the ion produced by one gram equivalent of an electrolyte.

Since, Eq. wt. of $BaCl_2 = \frac{\text{mol. wt.}}{2}$ then

$$\lambda^\infty_{BaCl_2} = \frac{1}{2}\lambda^\infty\, Ba^{2+} + \lambda^\infty Cl^-$$

$$= \frac{127}{2} + 76$$

$$= 139.5\ \text{ohm}^{-1}\text{cm}^2$$

49. In multimolecular colloids, colloidal size particles or molecules are built by the aggregation of ions or smaller particles. *eg*. gold sol.

50. Physical adsorption involves physical forces *i.e.*, van der Waals' forces.

51. $C = \frac{5}{342} \times \frac{1}{100} \times 1000 = \frac{50}{342}$ mol L^{-1}

$$\pi = \frac{50}{342} \times 0.082 \times 423$$

$$= 5.07 \text{ atm}$$

52. Mole fraction of $H_2O = \dfrac{80/18}{\frac{80}{18} + \frac{20}{34}} = \frac{34}{77}$

53. When pH = 3 then $[H^+] = 10^{-3}$ M. After that we increased the pH from 3 to 6 then $[H^+] = 10^{-6}$ M means reduced 1000 times.

54. $BaCl_2 \rightleftharpoons \underset{s}{Ba^{2+}} + \underset{2s}{2Cl^-}$ $K_{sp} = 4s^3$,

$$s^3 = \frac{4 \times 10^{-9}}{4} = 10^{-3}$$

$$s = 10^{-3} \text{ M}$$

55. $AlCl_3$ in a lewis acid because it is a electron pair acceptor, its central atom have a vacant *d*-orbital.

56. Radioactive decay is a first order reaction.

57. $t = \frac{2.303}{k} \log \frac{[A]_0}{[A]}$

$$= \frac{2.303}{60} = \log \frac{a}{\frac{a}{16}} = \frac{2.303}{60} \log 16$$

$$= \frac{2.303}{60} \times 1.204$$

$$= 0.0462 \text{ s} = 4.6 \times 10^{-2} \text{ s}$$

58. $C_{(s)} + \frac{1}{2}O_{2(g)} \longrightarrow CO_{(s)}$ $(\Delta H - \Delta U) = -\Delta nRT$

$$= -\frac{1}{2} \times 8.314 \times 298 = -1238.78$$

59. At constant temperature and pressure, internal energy of ideal gas remains unaffected.

60. Let RMS speed of nitrogen at T K be u and is equal to that of CO_2 at STP

$$u = \sqrt{\frac{3RT}{28}} = \sqrt{\frac{3R \times 273}{44}}$$

$$T = \frac{273 \times 28}{44} = 173.73 \text{ K} = -99.27^\circ\text{C}$$

Mathematics

1. Let $\Delta = \begin{vmatrix} a^2 & b^2 & c^2 \\ (a+1)^2 & (b+1)^2 & (c+1)^2 \\ (a-1)^2 & (b-1)^2 & (c-1)^2 \end{vmatrix}$

Applying $R_2 \to R_2 - R_3$

$$= 4 \begin{vmatrix} a^2 & b^2 & c^2 \\ a & b & c \\ (a-1)^2 & (b-1)^2 & (c-1)^2 \end{vmatrix}$$

Applying $R_3 \to R_3 - (R_1 - 2R_2)$

$$= 4 \begin{vmatrix} a^2 & b^2 & c^2 \\ a & b & c \\ 1 & 1 & 1 \end{vmatrix}$$

2. Given system has infinite solution.

3. As det $(A) = \pm 1$, A^{-1} exists and

$$A^{-1} = \frac{1}{\det(A)}(\text{adj } A) = \pm \text{ adj } A$$

All entries in adj (A) are integers.

$\therefore A^{-1}$ has integer entries.

4. All statements are correct.

5. Let $I = \int \frac{dx}{x\sqrt{1-(\log x)^2}}$

Put $\log x = t$

$\Rightarrow \frac{1}{x} dx = dt$

$\therefore I = \int \frac{dt}{\sqrt{1-t^2}} = \sin^{-1} t + C$

$= \sin^{-1}(\log x) + C$

6. $\int \frac{x^2}{(x^2+2)(x^2+3)} dx = \int \left(\frac{3}{x^2+3} - \frac{2}{x^2+2}\right) dx$

$= \frac{3}{\sqrt{3}} \tan^{-1}\left(\frac{x}{\sqrt{3}}\right) - \frac{2}{\sqrt{2}} \tan^{-1}\left(\frac{x}{\sqrt{2}}\right) + C$

$= \sqrt{3} \tan^{-1}\left(\frac{x}{\sqrt{3}}\right) - \sqrt{2} \tan^{-1}\left(\frac{x}{\sqrt{2}}\right) + C$

7. Let the equation of parabola whose axis is x- axis, is

$$y^2 = \pm 4a(x-h)$$

On differentiating w.r.t. x, we get

$2yy_1 = \pm 4a$

$\Rightarrow yy_1 = \pm 2a$

Again differentiating, we get

$(y_1)^2 + yy_2 = 0$

Hence, order and degree are respectively 1 and 2.

8. Given that, $y = ax \cos\left(\frac{1}{x} + b\right)$

On differentiating w.r.t. x, we get

$y_1 = a\left[\cos\left(\frac{1}{x}+b\right) - x \sin\left(\frac{1}{x}+b\right)\left(-\frac{1}{x^2}\right)\right]$

$\Rightarrow y_1 = a\left[\cos\left(\frac{1}{x}+b\right) + \frac{1}{x} \sin\left(\frac{1}{x}+b\right)\right]$

Again differentiating, we get

$y_2 = a\left[-\sin\left(\frac{1}{x}+b\right)\left(-\frac{1}{x^2}\right) - \frac{1}{x^2}\sin\left(\frac{1}{x}+b\right) - \frac{1}{x^3}\cos\left(\frac{1}{x}+b\right)\right]$

$\Rightarrow y_2 = -\frac{ax}{x^4}\cos\left(\frac{1}{x}+b\right)$

$\Rightarrow x^4 y_2 + y = 0$

9. Given differential equation can be rewritten as

$$\frac{dy}{dx} + \frac{2x}{1+x^2} y = \frac{x^2-1}{x^2+1}$$

Here, $P = \frac{2x}{1+x^2}$ and $Q = \frac{x^2-1}{x^2+1}$

$\therefore$ IF $= e^{\int \frac{2x}{1+x^2} dx} = e^{\log(1+x^2)}$

$= 1 + x^2$

10. $\int_0^{\pi/2} |\sin x - \cos x| \, dx$

$= \int_0^{\pi/4} -(\sin x - \cos x)\, dx$

$+ \int_{\pi/4}^{\pi/2} -(\sin x - \cos x)\, dx$

$= -[-\cos x - \sin x]_0^{\pi/4} + [-\cos x - \sin x]_{\pi/4}^{\pi/2}$

$= -\left[-\frac{1}{\sqrt{2}} - \frac{1}{\sqrt{2}} + 1\right] + \left[-0 - 1 + \left(\frac{1}{\sqrt{2}} + \frac{1}{\sqrt{2}}\right)\right]$

$= \sqrt{2} - 1 - 1 + \sqrt{2}$

$= 2(\sqrt{2} - 1)$

11. Let the centre of circle be (h, k). Then

$\sqrt{(h-2)^2 + (k-3)^2} = \sqrt{(h-4)^2 + (k-5)^2}$ ...(i)

and $k - 4h + 3 = 0$ ...(ii)

From Eq. (i)

$-4h - 6k + 8h + 10k = 16 + 25 - 4 - 9$

$\Rightarrow \quad h + k - 7 = 0$ ...(iii)

On solving Eqs. (ii) and (iii), we get the centre of circle (2, 5).

Now, radius $= \sqrt{(2-2)^2 + (5-3)^2} = 2$

$\therefore$ Required equation of circle is

$(x-2)^2 + (y-5)^2 = 2^2$

$\Rightarrow \quad x^2 + y^2 - 4x - 10y + 25 = 0$

12. Given that, $\log_4 (x-1) = \log_2 (x-3)$

$\therefore \quad x - 1 = (x-3)^2$

$\Rightarrow \quad x^2 - 7x + 10 = 0$

$\Rightarrow \quad (x-5)(x-2) = 0$

$\Rightarrow \quad x = 2, 5$

$\Rightarrow x = 5$, but $x = 2 \log_2(x-3)$ is not satisfied.

13. Since, α and β are the roots of $x^2 - 3x + 1 = 0$, then

$\alpha + \beta = 3$ and $\alpha\beta = 1$

Now, $\quad S = \frac{1}{\alpha - 2} + \frac{1}{\beta - 2}$

$= \frac{\alpha + \beta - 4}{\alpha\beta - 2(\alpha + \beta) + 4}$

$= \frac{3-4}{1 - 2\cdot 3 + 4} = 1$

and $\quad P = \frac{1}{\alpha - 2} \times \frac{1}{\beta - 2}$

$= \frac{1}{\alpha\beta - 2(\alpha + \beta - 2)}$

$= 1$

$\therefore$ Required equation is

$x^2 - Sx + P = 0$

$\Rightarrow \quad x^2 - x - 1 = 0$

14. Let $y = \frac{x^2 + 34x - 71}{x^2 + 2x - 7}$

$\Rightarrow \quad x^2(y-1) + 2(y-17)x + (71 - 7y) = 0$

For real values of x, its discriminant, $D \geq 0$

$\therefore \quad 4(y-17)^2 - 4(y-1)(71-7y) \geq 0$

$\Rightarrow \quad (y^2 - 34y + 289) - (71y - 7y^2 - 71 + 7y) \geq 0$

$\Rightarrow \quad y^2 - 14y + 45 \geq 0$

$\Rightarrow \quad (y-5)(y-9) \geq 0$

$\Rightarrow \quad y \leq 5$ and $y \geq 9$

Hence, y does not lie between 5 and 9

15. Now, $\sqrt{50} + \sqrt{48} = 5\sqrt{2} + 4\sqrt{3}$

$= \sqrt{2}[5 + 2\cdot\sqrt{2}\cdot\sqrt{3}]$

$= \sqrt{2}(\sqrt{3} + \sqrt{2})^2$

$\therefore \quad \sqrt{\sqrt{50} + \sqrt{48}} = 2^{1/4}(\sqrt{3} + \sqrt{2})$

16. Since, $I_1 = \int_a^{\pi - a} x f(\sin x)\, dx$

$= \int_a^{\pi - a} (\pi - x) f(\sin(\pi - x))\, dx$

$= \int_a^{\pi - a} (\pi - x) f(\sin x)\, dx$

$= \int_a^{\pi - a} \pi f(\sin x)\, dx - I_1$

$\Rightarrow \quad 2I_1 = \pi I_2$

$\Rightarrow \quad I_2 = \frac{2}{\pi} I_1$

17. Required area $= \int_0^1 (e^x - e^{-x})\, dx$

$= [e^x + e^{-x}]_0^1$

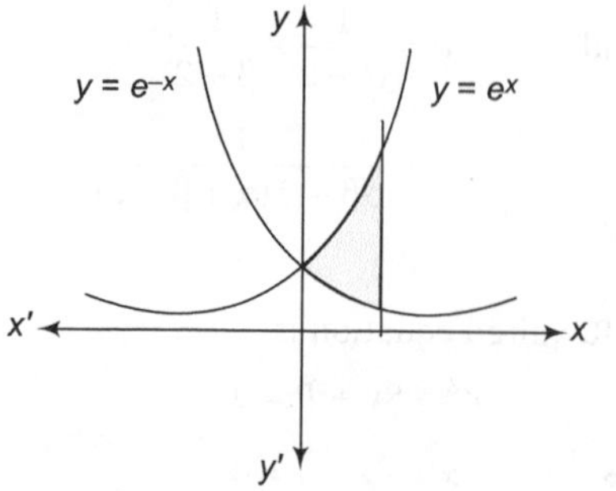

$$= \left(e + \frac{1}{e} - 2\right) \text{ sq unit}$$

18. $f(x) = \sin^{-1}[x]$ is defined, if $-1 \le [x] \le 1$

$\Rightarrow \quad [x] = \{-1, 0, 1\}$

$\therefore \quad x \in [-1, 2)$

$\therefore$ Range is $\{\sin^{-1}(-1), \sin^{-1} 0, \sin^{-1} 1\}$

$$= \left\{-\frac{\pi}{2}, 0, \frac{\pi}{2}\right\}$$

19. Since, $x\mathbf{a} + y\mathbf{b} + z\mathbf{c} = 0$

Also, **a**, **b** and **c** are non-collinear vectors, then $x = y = z = 0$

20. Let angle between **b** and **c** is α, then

$$|\mathbf{b} \times \mathbf{c}| = \sqrt{15} \text{ (given)}$$

$\Rightarrow \quad |\mathbf{b}||\mathbf{c}| \sin \alpha = \sqrt{15}$

$\Rightarrow \quad \sin \alpha = \frac{\sqrt{15}}{4}$

$\therefore \quad \cos \alpha = \frac{1}{4}$

Also, $\quad \mathbf{b} - 2\mathbf{c} = \lambda \mathbf{a}$

$\Rightarrow \quad (\mathbf{b} - 2\mathbf{c})^2 = \lambda^2 (\mathbf{a})^2$

$\Rightarrow \quad |\mathbf{b}|^2 + 4|\mathbf{c}|^2 - 4\mathbf{b} \cdot \mathbf{c} = \lambda^2 |\mathbf{a}|^2$

$\Rightarrow \quad 16 + 4 - 4\{|\mathbf{b}||\mathbf{c}| \cos\alpha\} = \lambda^2$

$\Rightarrow \quad \lambda^2 = 16$

$\Rightarrow \quad \lambda = \pm 4$

21. Let Q be the image of the point $P(\mathbf{i} + 3\mathbf{k})$ in the plane $\mathbf{r} \cdot (\mathbf{i} + \mathbf{j} + \mathbf{k}) = 1$. Then, PQ is normal to the plane. Since, PQ passes through P and in normal to the given plane, therefore equation of PQ is $\mathbf{r} = (\mathbf{i} + 3\mathbf{k}) + \lambda(\mathbf{i} + \mathbf{j} + \mathbf{k})$

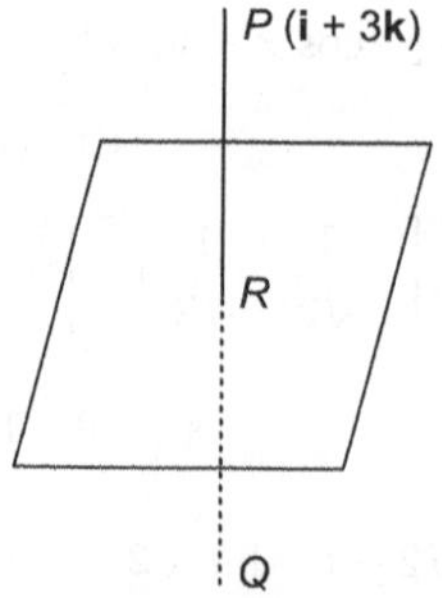

Since, Q lies on the line PQ, so let the position vector of Q be $(\mathbf{i} + 3\mathbf{k}) + \lambda(\mathbf{i} + \mathbf{j} + \mathbf{k})$ *i.e.*, $(1 + \lambda)\mathbf{i} + \lambda\mathbf{j} + (3 + \lambda)\mathbf{k}$.

Since, R is the mid-point of PQ, therefore position vector of R is

$$\frac{(1 + \lambda)\mathbf{i} + \lambda\mathbf{j} + (3 + \lambda)\mathbf{k} + \mathbf{i} + 3\mathbf{k}}{2}$$

$$= \left(\frac{\lambda + 2}{2}\right)\mathbf{i} + \frac{\lambda}{2}\mathbf{j} + \left(\frac{6 + \lambda}{2}\right)\mathbf{k}$$

$$= \left(\frac{\lambda}{2} + 1\right)\mathbf{i} + \frac{\lambda}{2}\mathbf{j} + \left(3 + \frac{\lambda}{2}\right)\mathbf{k}$$

Since, R lies on the plane

$$\mathbf{r} \cdot (\mathbf{i} + \mathbf{j} + \mathbf{k}) = 1$$

$$\therefore \quad \left[\left(\frac{\lambda}{2} + 1\right)\mathbf{i} + \frac{\lambda}{2}\mathbf{j} + \left(3 + \frac{\lambda}{2}\right)\mathbf{k}\right] \cdot (\mathbf{i} + \mathbf{j} + \mathbf{k}) = 1$$

$$\Rightarrow \quad \left[\frac{\lambda}{2} + 1 + \frac{\lambda}{2} + 3 + \frac{\lambda}{2}\right] = 1$$

$\Rightarrow \quad \lambda = -2$

Hence, the position vector of Q is $-\mathbf{i} - 2\mathbf{j} + \mathbf{k}$.

22. Since, **a**, **b**, **c** form a right handed system

$\therefore \quad \mathbf{c} = \mathbf{b} \times \mathbf{a}$

$= \mathbf{j} \times (x\mathbf{i} + y\mathbf{j} + z\mathbf{k})$

$= x(\mathbf{j} \times \mathbf{i}) + z(\mathbf{j} \times \mathbf{k})$

$= -x\mathbf{k} + z\mathbf{i} = z\mathbf{i} - x\mathbf{k}$

23. Since, plane is parallel to a given vector, it implies that normal of plane must perpendicular to the given vectors. Given point to which plane passes through is (2, – 1, 3). Let A, B and C are direction ratios of its normal.

$\therefore$ Equation of plane is

$$A(x-2)+B(y+1)+C(z-3)=0$$

Now, normal to plane $A\,\mathbf{i}+B\,\mathbf{j}+C\,\mathbf{k}$ is perpendicular to the given vectors

$$\mathbf{a}=3\mathbf{i}+0\,\mathbf{j}-\mathbf{k}$$

and $\quad \mathbf{b}=-3\mathbf{i}+2\mathbf{j}+2\mathbf{k}$

$\therefore \quad 3A+0B-C=0 \quad \ldots(\text{i})$

and $\quad -3A+2B+2C=0 \quad \ldots(\text{ii})$

From Eqs. (i) and (ii), we get

$$\frac{A}{2}=\frac{B}{-3}=\frac{C}{6}$$

$\therefore$ Equation of plane be

$$2(x-2)-3(y+1)+6(z-3)=0$$

i.e., $\quad 2x-3y+6z-25=0$

24. $\lim\limits_{x\to 1}\dfrac{(2x-3)(\sqrt{x}-1)}{2x^2+x-3}$

$$=\lim_{x\to 1}\frac{(2x-3)(\sqrt{x}-1)}{(2x+3)(x-1)}$$

$$=\lim_{x\to 1}\frac{(2x-3)}{(2x+3)(\sqrt{x}+1)}$$

$$=\frac{2-3}{(2+3)(\sqrt{1}+1)}=-\frac{1}{10}$$

25. Since, 10, 3, e, 2 are in decreasing order.

$\therefore \quad \log_{10}\alpha, \log_3\alpha, \log_e\alpha, \log_2\alpha$ are in increasing order.

26. Given that, $y=\dfrac{\sqrt{5}-\sqrt{2}}{\sqrt{5}+\sqrt{2}}$

and $\quad x=\dfrac{\sqrt{5}+\sqrt{2}}{\sqrt{5}-\sqrt{2}}$

$\Rightarrow \quad y=\dfrac{1}{x}$

$\Rightarrow \quad xy=1$

$\therefore \quad 3x^2+4xy-3y^2=3(x-y)(x+y)+4xy$

$$=3\left(\frac{\sqrt{5}+\sqrt{2}}{\sqrt{5}-\sqrt{2}}-\frac{\sqrt{5}-\sqrt{2}}{\sqrt{5}+\sqrt{2}}\right)\times\left(\frac{\sqrt{5}+\sqrt{2}}{\sqrt{5}-\sqrt{2}}+\frac{\sqrt{5}-\sqrt{2}}{\sqrt{5}+\sqrt{2}}\right)+4$$

$$=3\left[\frac{(\sqrt{5}+\sqrt{2})^2-(\sqrt{5}-\sqrt{2})^2}{(5-2)(5-2)}\right][(\sqrt{5}+\sqrt{2})^2+(\sqrt{5}-\sqrt{2})^2]+4$$

$$=\frac{1}{3}\cdot 4\sqrt{10}\cdot 2(5+2)+4$$

$$=\frac{56}{3}\sqrt{10}+4$$

$$=\frac{1}{3}(56\sqrt{10}+12)$$

27. $\sum\limits_{r=0}^{n} r^2\,{}^nC_r\,x^r y^{n-r}$

$$=\sum_{r=0}^{n}[r(r-1)+r]\,{}^nC_r\,x^r y^{n-r}$$

$$=\sum_{r=0}^{n} r(r-1)\,{}^nC_r\,x^r y^{n-r}+\sum_{r=0}^{n} r\,{}^nC_r\,x^r y^{n-r}$$

$$=\sum_{r=2}^{n-2} r(r-1)\frac{n}{r}\cdot\frac{n-1}{r-1}\,{}^{n-2}C_{r-2}\,x^2\cdot x^{r-2}y^{n-r}+\sum_{r=1}^{n} r\cdot\frac{n}{r}\,{}^{n-1}C_{r-1}\,x\cdot x^{r-1}y^{n-r}$$

$$=n(n-1)x^2\sum_{r=2}^{n-2}{}^{n-2}C_{r-2}\,x^{r-2}y^{(n-2)-(r-2)}+nx\sum_{r=1}^{n}{}^{n-1}C_{r-1}\,x^{r-1}y^{(n-1)-(r-1)}$$

$$=n(n-1)x^2(x+y)^{n-2}+nx(x+y)^{n-1}$$

$$=n(n-1)x^2+nx \qquad (\because x+y=1)$$

$$=nx(nx-x+1)$$

$$=nx(nx+y) \qquad (\because x+y=1)$$

28. Given that,

$$Z=f(x+ay)+\phi(x-ay)$$

$$Z_x=f'(x+ay)+\phi(x-ay)$$

$\Rightarrow \quad Z_{xx}=f''(x+ay)+\phi''(x-ay)$

and $\quad Z_y=af'(x+ay)-a\phi'(x-ay)$

$\Rightarrow \quad Z_{yy}=a^2f''(x+ay)-a^2\phi'(x-ay)$

Hence, $a^2Z_{xx}=Z_{yy}$

29. For unique solution of the given system, $\Delta \neq 0$

i.e., $\begin{vmatrix} 1 & 1 & 1 \\ 5 & -1 & \mu \\ 2 & 3 & -1 \end{vmatrix} \neq 0$

Hence, it depends only on μ.

30. Given, $\begin{bmatrix} 4 & 2 & 2 \\ -5 & 0 & \alpha \\ 1 & -2 & 3 \end{bmatrix} = 10A^{-1}$

$$\therefore \quad \begin{bmatrix} 4 & 2 & 2 \\ -5 & 0 & \alpha \\ 1 & -2 & 3 \end{bmatrix}\begin{bmatrix} 1 & -1 & 1 \\ 2 & 1 & -3 \\ 1 & 1 & 1 \end{bmatrix} = \begin{bmatrix} 10 & 0 & 0 \\ 0 & 10 & 0 \\ 0 & 0 & 10 \end{bmatrix}$$

$$\Rightarrow \quad \begin{bmatrix} 10 & 0 & 0 \\ -5+\alpha & 5+\alpha & -5+\alpha \\ 0 & 0 & 10 \end{bmatrix} = \begin{bmatrix} 10 & 0 & 0 \\ 0 & 10 & 0 \\ 0 & 0 & 10 \end{bmatrix}$$

$$\Rightarrow \quad -5+\alpha = 0$$

$$\Rightarrow \quad \alpha = 5$$

31. Given equation of curves are

$$\frac{x^2}{a^2} + \frac{y^2}{b^2} = 1 \text{ and } x^2 + y^2 = ab$$

$$\therefore \quad \frac{ab - y^2}{a^2} + \frac{y^2}{b^2} = 1$$

$$\Rightarrow \quad y^2\left(\frac{a^2 - b^2}{a^2b^2}\right) = \frac{a-b}{a}$$

$$\Rightarrow \quad y^2 = \frac{ab^2}{a+b}$$

and $\quad x^2 = \dfrac{a^2b}{a+b}$

$$\Rightarrow \quad x = a\sqrt{\frac{b}{a+b}}$$

and $\quad y = b\sqrt{\dfrac{a}{a+b}}$

Slope of tangent at ellipse $= \dfrac{-b^2x}{a^2y}$

$$= -\frac{b^2}{a^2}\sqrt{\frac{a}{b}}$$

Slope of tangent at circle $= -\dfrac{x}{y} = -\sqrt{\dfrac{a}{b}}$

$$\therefore \quad \theta = \tan^{-1}\left[\frac{\sqrt{\frac{a}{b}} - \frac{b^2}{a^2}\sqrt{\frac{a}{b}}}{1 + \frac{b^2}{a^2}\cdot\frac{a}{b}}\right]$$

$$\Rightarrow \quad \theta = \tan^{-1}\left(\frac{a-b}{\sqrt{ab}}\right)$$

32. Given that,

$$S = \sum_{n=0}^{\infty} \frac{(\log x)^{2n}}{(2n)!}$$

$$= \left(\frac{e^{\log x} + e^{-\log x}}{2}\right) = \frac{x + x^{-1}}{2}$$

33. $\log 2 + 2\left(\dfrac{1}{5} + \dfrac{1}{3}\cdot\dfrac{1}{5^3} + \dfrac{1}{5}\cdot\dfrac{1}{5^5} + \dots \infty\right)$

$$= \log 2 + \log\left(\frac{1 + \frac{1}{5}}{1 - \frac{1}{5}}\right)$$

$$= \log 2 + \log\left(\frac{3}{2}\right) = \log 3$$

34. Since, vertex is (0, 4) and focus is (0, 2).

$$\therefore \quad a = 2$$

$\therefore$ Equation of parabola is

$$(x-0)^2 = -4\cdot 2(y-4)$$

$$\Rightarrow \quad x^2 + 8y = 32$$

35. The equation of a line parallel to $x + 2y = 4$ is $x + 2y = k$. Since, the distance between these two lines is 3, therefore

$$\frac{k}{\sqrt{1+4}} - \frac{4}{\sqrt{1+4}} = 3$$

$$\Rightarrow \quad k = 4 + 3\sqrt{5}$$

This shifted line cuts x-axis at $(k, 0)$. After rotation the slope of the line is $\tan(\theta - 30°)$, where $\tan\theta =$ (slope of

$$(x + 2y = 4) = -\frac{1}{2}$$

∴The equation of the line in the new position is

$$y - 0 = \tan(\theta - 30°)(x - k),$$

$$\Rightarrow \quad y = \tan(\theta - 30°)(x - k)$$

where $k = 4 + 3\sqrt{5}$

36. According to given condition, $2ae = 2 \cdot 2a$

$$\Rightarrow \quad e = 2$$

and $\quad 2b = 6 \Rightarrow b = 3$

Hence, $a = \frac{3}{\sqrt{3}} = \sqrt{3}$

∴Required equation is

$$\frac{x^2}{3} - \frac{y^2}{9} = 1$$

$$\Rightarrow \quad 3x^2 - y^2 = 9$$

37. The second degree equation will represent a pair of perpendicular straight lines, if

$$\begin{vmatrix} a & h & g \\ h & b & f \\ g & f & c \end{vmatrix} = 0 \text{ and } a + b = 0$$

Given pair of line is

$$12x^2 + 7xy - py^2 - 18x + qy + 6 = 0$$

$$\therefore \quad \begin{vmatrix} 12 & 7/2 & -9 \\ 7/2 & -p & q/2 \\ -9 & q/2 & 6 \end{vmatrix} = 0$$

and $\quad 12 - p = 0$

$$\Rightarrow \quad p = 12$$

$$\therefore \quad \begin{vmatrix} 12 & 7/2 & -9 \\ 7/2 & -12 & q/2 \\ -9 & q/2 & 6 \end{vmatrix} = 0$$

$$\Rightarrow \quad 12\left(-72 + \frac{q^2}{4}\right) - \frac{7}{2}\left(-21 + \frac{9q}{2}\right)$$

$$-9\left(\frac{7q}{4} - 108\right) = 0$$

$$\Rightarrow \quad -864 - 3q^2 - \frac{147}{2} - \frac{63q}{4}$$

$$-\frac{63q}{4} + 972 = 0 \Rightarrow q = 1$$

38. Given equation can be rewritten as

$$x^2 - 2(\alpha + \beta)x + 3\alpha\beta = 0$$

Let roots be α' and $-\alpha'$.

$$\therefore \quad \alpha' + (-\alpha') = 2(\alpha + \beta)$$

$$\Rightarrow \quad \alpha + \beta = 0$$

39. Let $f(x) = ax^2 + bx + c$

Again, let $f(x) = \int f(x)\,dx$

$$= \frac{a}{3}x^3 + \frac{bx^2}{2} + cx$$

At $\quad x = 0,$

$$f(0) = 0$$

Agin, at $x = 1$

$$f(1) = \frac{a}{3} + \frac{b}{2} + c$$

$$= \frac{2a + 3b + 6c}{6} = 0$$

Hence, one root lies between (0, 1).

40. Since, the lines $3x + 4y + 2 = 0$ and $3x + 4y + 5 = 0$ are on the same side of the origin. The distance between these lines is $5 - 2 = 3$.

And the line $3x + 4y + 2 = 0$ and $3x + 4y - 5 = 0$ are opposite side of the origin, The distance between these lies is 7.

Hence, required ratio is 3 : 7.

41. Given equation of line $y = x$ is written in polar form is

$$\frac{x}{\cos\theta} = \frac{y}{\sin\theta} = r, \text{ where } \theta = \frac{\pi}{4}$$

For point P, $r = 6\sqrt{2}$. Therefore, coordinates of P are given by

$$\frac{x}{\cos\frac{\pi}{4}} = \frac{y}{\sin\frac{\pi}{4}} = 6\sqrt{2}$$

$$\Rightarrow \quad x = 6, y = 6$$

Since, point (6, 6) lies on

$$x^2 + y^2 + 2gx + 2fy + c = 0$$

$$\therefore \quad 72 + 12(g + f) + c = 0 \quad \text{...(i)}$$

Since, $y = x$ touches the circle, therefore the equation $2x^2 + 2x(g + f) + c = 0$ has equal roots.

$\Rightarrow \quad 4(g + f)^2 = 8c$

$\Rightarrow \quad (g + f)^2 = 2c \quad \ldots \text{(ii)}$

From Eq. (i)

$[(12(g + f)]^2 = [-(c + 72)]^2$

$\Rightarrow \quad 144(g + f)^2 = (c + 72)^2$

$\Rightarrow \quad 144(2c) = (c + 72)^2$ [from Eq. (ii)]

$\Rightarrow \quad (c - 72)^2 = 0$

$\Rightarrow \quad c = 72$

42. $(1 + x + x^2 + x^3)^n = (1 + x)^n (1 + x^2)^n$

$= (1 + {}^nC_1 x + {}^nC_2 x^2 + \ldots\ldots + {}^nC_n x^n)$
$\times (1 + {}^nC_1 x^2 + {}^nC_2 x^4 + \ldots\ldots + {}^nC_n x^{2n})$

$\therefore$ The coefficient of x^4

$= {}^nC_2 + {}^nC_2 \cdot {}^nC_1 + {}^nC_4$

$= {}^nC_4 + {}^nC_2 + {}^nC_1 \cdot {}^nC_2$

43. $T_n = \dfrac{1 + x + x^2 + \ldots + x^{n-1}}{n!}$

$= \dfrac{x^n - 1}{x - 1} \cdot \dfrac{1}{n!}$

$= \dfrac{1}{1 - x}\left(\dfrac{x^n}{n!} - \dfrac{1}{n!}\right)$

$\Rightarrow \quad \sum_{n=1}^{\infty} \dfrac{1}{1 - x}\left(\dfrac{x^n}{n!} - \dfrac{1}{n!}\right)$

$= \dfrac{1}{(x - 1)}\left[\sum_{n=1}^{\infty} \dfrac{x^n}{n!} - \sum_{n=1}^{\infty} \dfrac{1}{n!}\right]$

$= \dfrac{1}{(x - 1)}(e^x - e)$

44. Given inequation is

$n! > 2^{n-1}$

For $\quad n = 1, 2$

$1! \not> 1$

$\Rightarrow \quad 1 \not> 1$

and $\quad 2! \not> 2^1 \Rightarrow 2 \not> 2$

For $n = 3$

$3! > 2^2$

$\Rightarrow \quad 6 > 4$

Hence, it is true for $n > 2$.

45. Given sets are

$A = \{(x, y) : y = e^x, x \in R\}$ and
$B = \{(x, y) : y = e^{-x}, x \in R\}$

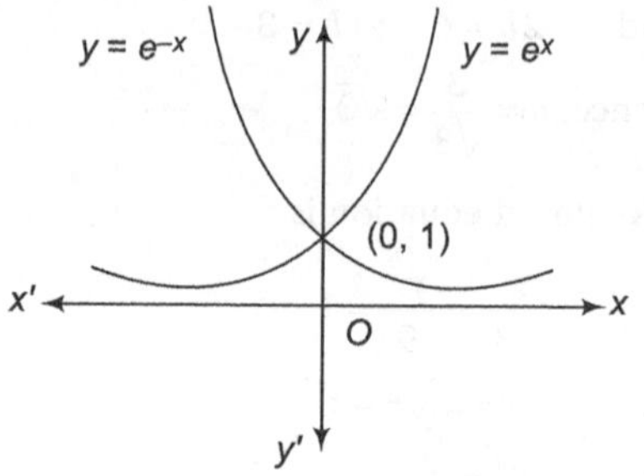

It is clear from the graph that two curves intersect at one point.

$\therefore \quad A \cap B \neq \phi$

46. Now, $\dfrac{b - c}{r_1} + \dfrac{c - a}{r_2} + \dfrac{a - b}{r_3}$

$= \dfrac{[(b - c)(s - a) + (c - a)(s - b) + (a - b)(s - c)]}{\Delta}$

$= 0$

47. Let AE is a vertical lamp post

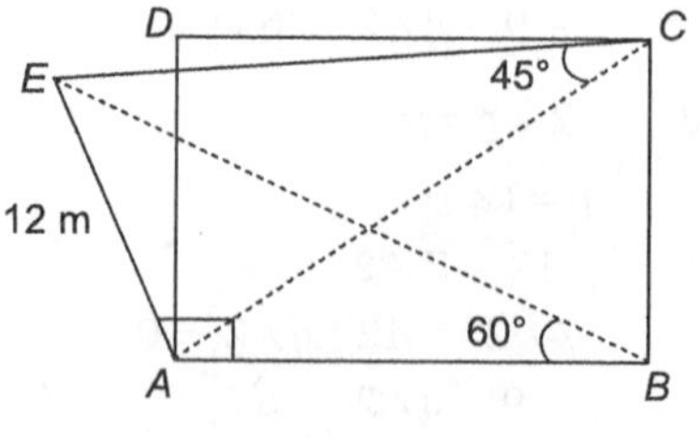

In ΔAEC,

$\tan 45° = \dfrac{AE}{AC}$

$\Rightarrow \quad AC = AE = 12 \text{ m}$

and in ΔABE,

$\tan 60° = \dfrac{AE}{AB}$

$\Rightarrow \quad AB = \dfrac{AE}{\sqrt{3}} = 4\sqrt{3}$

Now, $BC = \sqrt{AC^2 - AB^2}$

$= \sqrt{144 - 48} = \sqrt{96}$

$= 4\sqrt{6}$

∴ Area of rectangle $ABCD$

$= AB \times BC$

$= 4\sqrt{3} \times 4\sqrt{6}$

$= 48\sqrt{2}\ m^2$

48. Given that,

$$f(x) = \max\{(1-x), 2, (1+x)\}$$

For $x \le -1$, we find that $1 - x \ge 2$

and $1 - x \ge 1 + x$

$\therefore \max\{(1-x), 2, (1+x)\} = 1 - x$

For $-1 < x < 1$, we find that

$$0 < 1 - x < 2 \text{ and } 0 < 1 + x < 2$$

$\therefore \max\{(1-x), 2, (1+x)\} = 2$

For $x \ge 1$, we find that

$$1 + x \ge 2,\ 1 + x > 1 - x$$

$\therefore \max\{(1-x), 2, (1+x)\} = 1 + x$

Hence, $f(x) = \begin{cases} 1 - x, & x \le -1 \\ 2, & -1 < x < 1 \\ 1 + x, & x \ge 1 \end{cases}$

49. Given, $3\sin^2 x - 7\sin x + 2 = 0$

$\Rightarrow \quad (2\sin x - 1)(\sin x - 2) = 0$

$\Rightarrow \quad \sin x = \frac{1}{3} \qquad (\because \sin x \ne 2)$

Let $\sin^{-1}\frac{1}{3} = \alpha,\ 0 < \alpha < \frac{\pi}{2}$ are the solutions in $[0, 5\pi]$. Then,

$\alpha, \pi - \alpha, 2\pi + \alpha, 3\pi - \alpha, 4\pi + \alpha, 5\pi - \alpha$ are the solutions.

50. $\left(1 + \cos\frac{\pi}{8}\right)\left(1 + \cos\frac{3\pi}{8}\right)\left(1 - \cos\frac{3\pi}{8}\right)\left(1 - \cos\frac{\pi}{8}\right)$

$= \left(1 - \cos^2\frac{\pi}{8}\right)\left(1 - \cos^2\frac{3\pi}{8}\right)$

$= \left(\sin\frac{\pi}{8}\sin\frac{3\pi}{8}\right)^2$

$= \frac{1}{4}\left(2\sin\frac{3\pi}{8}\sin\frac{\pi}{8}\right)^2$

$= \frac{1}{4}\left(\cos\frac{\pi}{4} - \cos\frac{\pi}{2}\right)^2 = \frac{1}{8}$

51. Since, $\sin^{-1}\sqrt{x}$, $\cos^{-1}\sqrt{x}$ are defined for $x \le 1$ and $x \ge 0$.

$\therefore \quad 0 \le \sqrt{x^2 - x + 1} \le 1$

and $\quad 0 \le \sqrt{x^2 - x} \le 1$

$\Rightarrow \quad -1 \le x^2 - x \le 0$

and $\quad 0 \le x^2 - x \le 0$

$\Rightarrow \quad x^2 - x = 0$

$\Rightarrow \quad x = 0, 1$

Hence, number of solutions are two.

52. Let $ABCD$ be a square of each side of length a. It is given that $\angle OCP = 60°$.

Length of diagonal

$$AC = \sqrt{a^2 + a^2} = a\sqrt{2}$$

$\therefore \quad PC = \frac{AC}{2} = \frac{a}{\sqrt{2}}$

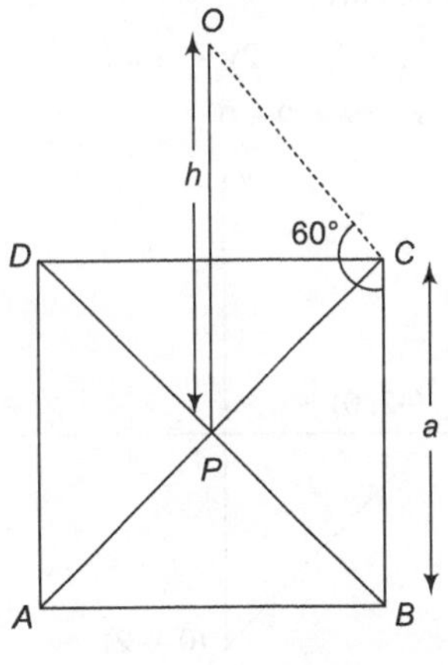

In $\quad \Delta OCP,\ \tan 60° = \frac{OP}{PC}$

$\Rightarrow \quad \sqrt{3} = \frac{h}{a/\sqrt{2}}$

$\Rightarrow \quad \sqrt{3}a = \sqrt{2}h$

$\Rightarrow \quad 3a^2 = 2h^2$

53. $\cos^{-1}\left(\cos\frac{7\pi}{6}\right)=\cos^{-1}\left(\cos\left(2\pi-\frac{5\pi}{6}\right)\right)$

$$=\cos^{-1}\left(\cos\left(\frac{5\pi}{6}\right)\right)=\frac{5\pi}{6}$$

54. Let $T=|\sin nx|$

Put $n=2$, then

$$T=|\sin 2x|$$
$$=|2\sin x\cos x|\le 2|\sin x|$$

55. Given sides of a triangle are

$$a=30,\ b=24,\ c=18$$

$\therefore$ $$s=\frac{a+b+c}{2}$$
$$=\frac{30+24+18}{2}=36$$

Now, $\Delta=\sqrt{s(s-a)(s-b)(s-c)}$

$$=\sqrt{36(36-30)(36-24)(36-18)}$$
$$=\sqrt{36\times 6\times 12\times 18}$$
$$=\sqrt{36\times 36\times 36}=216$$

$\therefore$ $$r_3=\frac{\Delta}{s-c}=\frac{216}{18}=12$$

56. Given lines are

$$xy+2x+2y+4=0 \quad \text{...(i)}$$

and $$x+y+2=0 \quad \text{...(ii)}$$

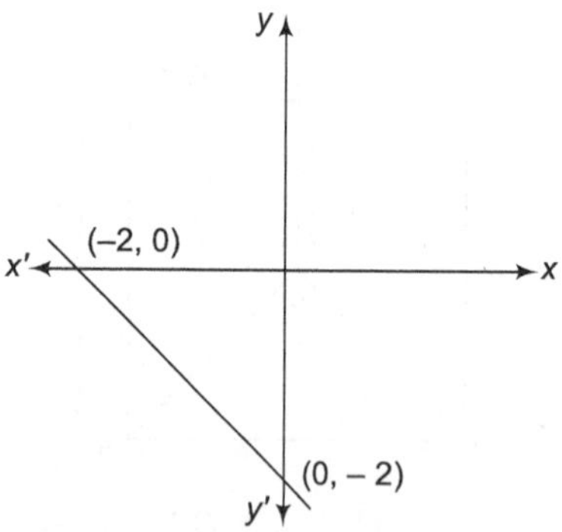

From Eqs. (i) and (ii), we get

$$xy=0$$

$\Rightarrow$ $$x=y=0$$

$\therefore$ Vertices of triangle are $(-2, 0)$, $(0, 0)$, $(0, -2)$.

In a right angled triangle circumcentre is the mid-point of hypotenuse.

$\therefore$ Point $(-1, -1)$ is the required coordinate of the circumcentre.

57. Slop of BD is $\frac{8}{15}$ and angle made by BD with AD and DC is 45°.

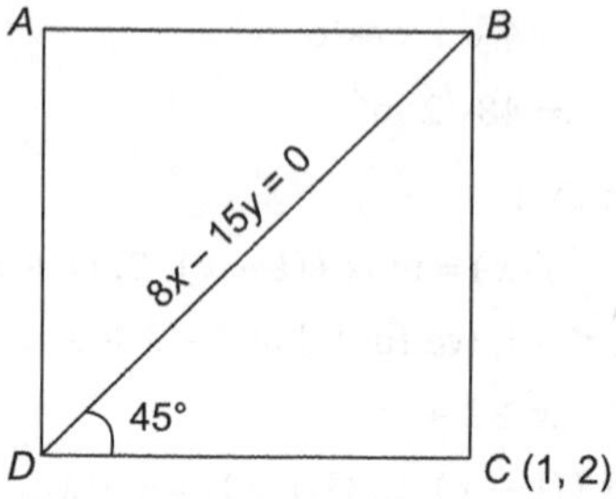

So, let slope of DC be m, then

$$\tan 45°=\pm\frac{m-\frac{8}{15}}{1+\frac{8}{15}m}$$

$\Rightarrow$ $$(15+8m)=\pm(15m-8)$$

$\Rightarrow$ $$m=\frac{23}{7}\text{ and }-\frac{7}{23}$$

Hence, the equation of DC and AD are

$$y-2=\frac{23}{7}(x-1)$$

$\Rightarrow$ $$23x-7y-9=0$$

and $$y-2=-\frac{7}{23}(x-1)$$

$\Rightarrow$ $$7x+23y-53=0$$

58. Given curves are

$$x^3-3xy^2+2=0 \quad \text{...(i)}$$

and $$3x^2y-y^3-2=0 \quad \text{...(ii)}$$

On differentiating Eqs. (i) and (ii), with respect to x, we get

$$\left(\frac{dy}{dx}\right)_{C_1}=\frac{x^2-y^2}{2xy}$$

and $$\left(\frac{dy}{dx}\right)_{C_2}=\frac{2xy}{x^2-y^2}$$

$$\left(\frac{dy}{dx}\right)_{C_1}\times\left(\frac{dy}{dx}\right)_{C_2}=-1$$

59. Given that,

$$\frac{2x+3}{(x+1)(x-3)} = \frac{a}{(x+1)} + \frac{b}{(x-3)}$$

$$\Rightarrow \quad (2x+3) = a(x-3) + b(x+1)$$

On comparing the coefficient of x and constant, we get

$$2 = a + b \text{ and } 3 = -3a + b$$

$$\Rightarrow \quad a = -\frac{1}{4} \text{ and } b = \frac{9}{4}$$

$$\therefore \quad a + b = -\frac{1}{4} + \frac{9}{4} = 2$$

60. $$\lim_{n \to \infty} \frac{1}{n}\left[\sec^2 \frac{\pi}{4n} + \sec^2 \frac{2\pi}{4n} + \ldots + \sec^2 \frac{n\pi}{4n}\right]$$

$$= \lim_{n \to \infty} \frac{1}{n} \sum_{r=1}^{n} \sec^2 \frac{r\pi}{4n}$$

$$= \int_0^1 \sec^2 \frac{\pi x}{4}\, dx$$

$$= \frac{4}{\pi}\left[\tan \frac{\pi x}{4}\right]_0^1 = \frac{4}{\pi}$$

61. Given that, $f(x) = |\cos x|$ and $g(x) = [x]$,

$$\therefore \quad gof(x) = g(|\cos x|)$$

$$= [|\cos x|]$$

62. Since, $$\lim_{x \to \infty}\left(1 + \frac{a}{x} + \frac{b}{x^2}\right)^{2x} = e^2$$

$$\Rightarrow \quad \lim_{x \to \infty}\left(1 + \frac{ax+b}{x^2}\right)^{2x} - e^2 \qquad \ldots(\text{i})$$

$$\Rightarrow \quad \lim_{x \to \infty} \frac{ax+b}{x^2} \text{ must be equal to zero.}$$

$$\Rightarrow \quad \lim_{x \to \infty} \frac{ax+b}{x^2} = 0$$

$$\Rightarrow \quad \lim_{x \to \infty}\left(\frac{a}{x} + \frac{b}{x^2}\right) = 0$$

$$\Rightarrow \quad a \text{ and } b \neq 0$$

$\therefore$ From Eq. (i)

$$e^{\lim_{x \to \infty} \frac{2(ax+b)}{x}} = e^2$$

$$\Rightarrow \quad e^{2a} = e^2$$

$$\Rightarrow \quad a = 1 \text{ and } b \in R$$

63. The point of intersection of curve $y^2 = 2x + 1$ and line $x - y - 1 = 0$ is (14, 3) and (0, – 1)

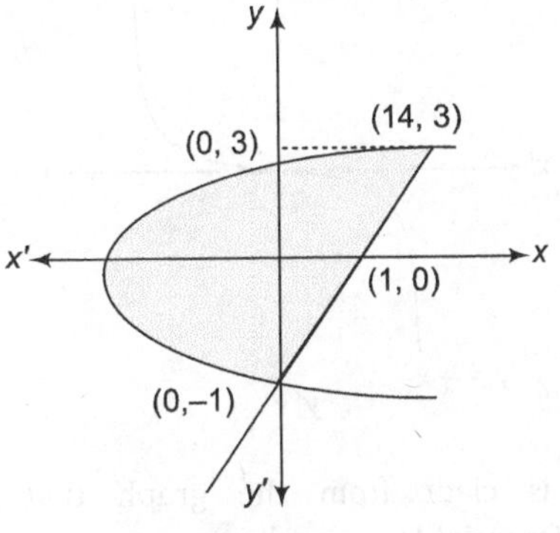

$$\therefore \text{Required area} = \int_{-1}^{3} (x_2 - x_1)\, dy$$

$$= \int_{-1}^{3} \left\{(y+1) - \frac{(y^2-1)}{2}\right\} dy$$

$$= \frac{1}{2}\int_{-1}^{3} (2y + 3 - y^2)\, dy$$

$$= \frac{1}{2}\left[y^2 + 3y - \frac{y^3}{3}\right]_{-1}^{3}$$

$$= \frac{1}{2}\left[(9 + 9 - 9) - \left(1 - 3 + \frac{1}{3}\right)\right]$$

$$= \frac{1}{2}\left(9 + \frac{5}{3}\right)$$

$$= \frac{16}{3} \text{ sq units}$$

64. For $f(x)$ to be continuous everywhere, we must have

$$f(0) = \lim_{x \to 0} f(x)$$

$$= \lim_{x \to 0} \frac{2 - (256 - 7x)^{1/8}}{(5x + 32)^{1/5} - 2}$$

$$= \lim_{x \to 0} \frac{\frac{7}{8}(256 - 7x)^{-7/8}}{(5x + 32)^{-4/5}}$$

$$= \frac{7}{8} \times \frac{2^{-7}}{2^{-4}} = \frac{7}{64}$$

65. Given that, $f(x) = x|x|$

$$\therefore \quad f(x) = \begin{cases} x^2, & x \geq 0 \\ -x^2, & x < 0 \end{cases}$$

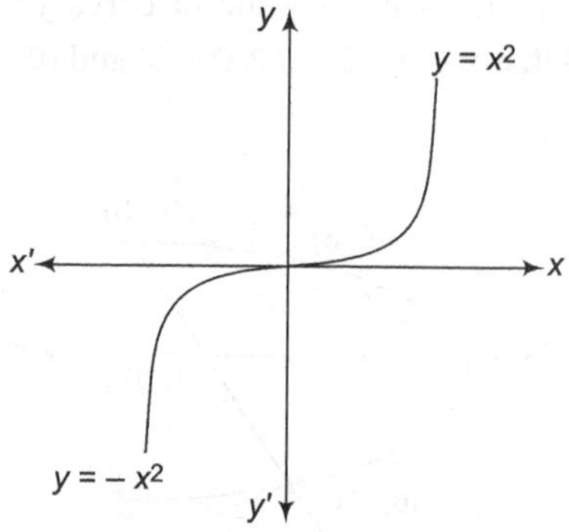

It is clear from the graph that $f(x)$ is differentiable everywhere.

66. Since, $f(x) = |x|^{\sin x}$

when $x < 0$

$$f(x) = (-x)^{(-\sin x)}$$

$$= e^{-\sin x \log(-x)}$$

$$\Rightarrow \quad f'(x) = e^{-\sin x \log(-x)} \left(-\cos x \log(-x) - \frac{\sin x}{x}\right)$$

$$\Rightarrow \quad f'(x) = (-x)^{-\sin x} \left(-\cos x \cdot \log(-x) - \frac{\sin x}{x}\right)$$

$$\Rightarrow \quad f'\left(-\frac{\pi}{4}\right) = \left(\frac{\pi}{4}\right)^{1/\sqrt{2}} \left(-\frac{1}{\sqrt{2}} \log \frac{\pi}{4} + \frac{4}{\pi}\left(-\frac{1}{\sqrt{2}}\right)\right)$$

$$= \left(\frac{\pi}{4}\right)^{1/\sqrt{2}} \left(\frac{\sqrt{2}}{2} \log \frac{4}{\pi} - \frac{2\sqrt{2}}{\pi}\right)$$

67. Let the centre of a circle be $C_1(h, k)$

Since, $\quad C_1C_2 = r_1 + r_2 \quad$ (given)

$$\Rightarrow \quad \sqrt{(h-0)^2 + (k-3)^2} = |k+2|$$

$$\Rightarrow \quad h^2 = 5(2k-1)$$

Hence, locus of a point is $x^2 = 5(2y-1)$ which represents a equation of parabola.

68. Let $S(a, 0)$ be the focus of the parabola $y^2 = 4ax$ and $P(at^2, 2at)$ be a point on it. Then, the equation of a circle on SP as diameter is $(x-a)(x-at^2) + (y-0)(y-2at) = 0$

It meets y-axis at $x = 0$

$$\therefore \quad y^2 - 2aty + a^2t^2 = 0$$

$$\Rightarrow \quad (y - at)^2 = 0$$

This shows that y-axis meets the circle in two coincident points. Hence, the circle touches the tangent at the vertex.

69. Let $2a$ and $2b$ be the major and minor axes of the ellipse. Then, its equation is

$$\frac{(x-2)^2}{a^2} + \frac{(y+3)^2}{b^2} = 1 \quad \text{...(i)}$$

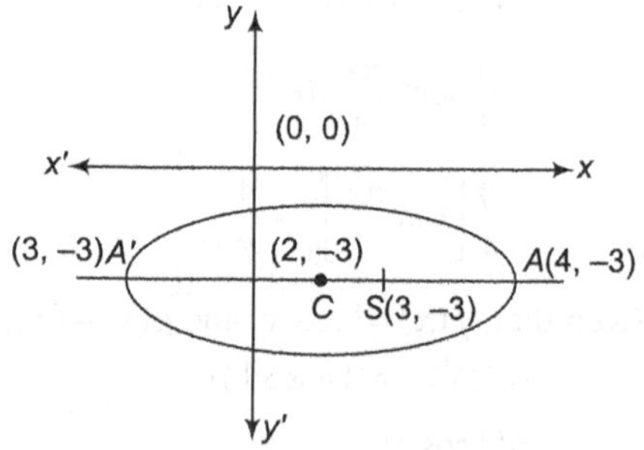

Here, semi-major axis $CA = a$

$$\Rightarrow \quad \sqrt{(4-2)^2 + (-3+3)^2} = a$$

$$\Rightarrow \quad a = 2 \quad \text{...(ii)}$$

Here, $CS = ae$

$$\Rightarrow \quad \sqrt{(2-3)^2 + (-3+3)^2} = ae$$

$$\Rightarrow \quad ae = 1 \quad \text{...(iii)}$$

From Eqs. (ii) and (iii), we get

$$e = \frac{1}{2}$$

Now, $\quad b^2 = a^2(1 - e^2)$

$$\Rightarrow \quad b^2 = 4\left(1 - \frac{1}{4}\right) = 3$$

On substituting the values of a and b in Eq. (i) we get

$$\frac{(x-2)^2}{4} + \frac{(y+3)^2}{3} = 1$$

70. The diameters $y = m_1 x$ and $y = m_2 x$ are conjugate diameters of the hyperbola

$$\frac{x^2}{a^2} - \frac{y^2}{b^2} = 1, \text{ if } m_1 m_2 = \frac{b^2}{a^2}$$

Given equation is

$$\frac{x^2}{9} - \frac{y^2}{16} = 1$$

Here, $a^2 = 9, b^2 = 16$

Also given, $x = 2y$

Then, $m_1 = \frac{1}{2}$

$$\because \quad \frac{1}{2} \times m_2 = \frac{16}{9}$$

$$\Rightarrow \quad m_2 = \frac{32}{9}$$

Hence, the required equation of diameter is

$$y = \frac{32x}{9}$$

71. $\therefore (2x + 3) = a(x - 3) + b(x + 1)$

Put $x = -1$ and $x = 3$ respectively, we get

$$1 = -4a \text{ and } 9 = 4b$$

$$\Rightarrow \quad a = -\frac{1}{4} \text{ and } b = \frac{9}{4}$$

$$\therefore \quad a + b = -\frac{1}{4} + \frac{9}{4} = 2$$

72. Here, $e^x + 2 = -3(2e^x - 3) + B(e^x - 1)$

On equating the coefficient of e^x and constant, we get

$$1 = -6 + B,$$

$$2 = 9 - B \Rightarrow B = 7$$

73. $$\frac{3\sqrt{2}}{\sqrt{6} + \sqrt{3}} - \frac{4\sqrt{3}}{\sqrt{6} + \sqrt{2}} + \frac{\sqrt{6}}{\sqrt{3} + \sqrt{2}}$$

$$= \frac{3\sqrt{2}(\sqrt{6} - \sqrt{3})}{6 - 3} - \frac{4\sqrt{3}(\sqrt{6} - \sqrt{2})}{6 - 3} + \frac{\sqrt{6}(\sqrt{3} - \sqrt{2})}{3 - 2}$$

$$= \sqrt{2}(\sqrt{6} - \sqrt{3}) - \sqrt{3}(\sqrt{6} - \sqrt{2}) + \sqrt{6}(\sqrt{3} - \sqrt{2})$$

$$= 0$$

74. Since, it is an AP.

$$\therefore \quad \log_3(2^x - 5) = \frac{\log_3 2 + \log_3\left(2^x - \frac{7}{2}\right)}{2}$$

$$\Rightarrow \quad 2\log_3(2^x - 5) = \log_3\left(2 \cdot \left(2^x - \frac{7}{2}\right)\right)$$

$$\Rightarrow \quad (2^x - 5)^2 = 2^{x+1} - 7$$

$$\Rightarrow \quad 2^{2x} - 12 \cdot 2^x - 32 = 0$$

$$\Rightarrow \quad x = 2, 3$$

But $x = 2$, $\log_3(2^x - 5)$ is not defined.

75. Let $\frac{A}{R}$, A, AR be the roots of the equation $ax^3 + bx^2 + cx + d = 0$, then

Product of roots

$$A^3 = -\frac{d}{a}$$

$$\Rightarrow \quad = A = -\left(\frac{d}{a}\right)^{1/3}$$

Since, A is a root of the equation.

$$\therefore \quad aA^3 + bA^2 + cA + d = 0$$

$$\Rightarrow \quad a\left(-\frac{d}{a}\right) + b\left(-\frac{d}{a}\right)^{2/3} + c\left(-\frac{d}{a}\right)^{1/3} + d = 0$$

$$\Rightarrow \quad b\left(\frac{d}{a}\right)^{2/3} = c\left(\frac{d}{a}\right)^{1/3}$$

$$\Rightarrow \quad b^3 \cdot \frac{d^2}{a^2} = c^3 \cdot \frac{d}{a}$$

$$\Rightarrow \quad b^3 d = c^3 a$$

76. The composition table is as given below

$\times_5$	1	2	3	4
1	1	2	3	4
2	2	4	1	3
3	3	1	4	2
4	4	3	2	1

Clearly, identity element = 1

$$\therefore \quad 2^{-1} \times_5 (3 \times_5 x) = 4$$

$$\Rightarrow \quad 3 \times_5 (3 \times_5 x) = 4$$

$$\Rightarrow \quad (3 \times_5 3) \times_5 x = 4$$

$\Rightarrow \quad 4 \times_5 x = 4$

$\Rightarrow \quad x = 4^{-1} \times_5 4 = 1$

77. Let p : Ram is in class x, q : Rashmi is in class XII

Given proposition is $p \vee q$.

Its negation is $\sim (p \vee q) = \sim p \wedge \sim q$

Ram is not in class X and Rashmi is not in class XII.

78. Inverse of $p \Rightarrow q$ is $\sim p \Rightarrow \sim q$

$\therefore$ Inverse of $(p \wedge \sim q) \Rightarrow r$ is

$$\sim (p \wedge \sim q) \Rightarrow \sim r$$

i. e., $\quad (\sim p \vee q) \Rightarrow \sim r$

79. $(17)^2 = 289 \equiv (-1) \pmod 5$

$\Rightarrow \quad (17^2)^{15} \equiv (-1)^{15} \pmod 5$

$\Rightarrow \quad 17^{30} \equiv - \pmod 5$

i. e., $\quad 17^{30} \equiv 4 \pmod 5$

80. Given equation can be rewritten as

$$\text{cosec } y \, dy = \sin x \, dx$$

On integrating both sides, we get

$$\log \tan \frac{y}{2} = -\cos x + \log C$$

$$\Rightarrow \quad \log \frac{\tan \frac{y}{2}}{C} = -\cos x$$

$$\Rightarrow \quad \frac{\tan \frac{y}{2}}{C} = e^{-\cos x}$$

$$\Rightarrow \quad e^{\cos x} \cdot \tan \frac{y}{2} = C$$

General English and Aptitude

21. Knot, Watt and Fathom, all are units of measurement.

22. Majlis, Diet and Knesset, all are parliament name of countries.

23. Kanha, Periyar and Dachigam, all are animal sanctuaries.

24. Stirrup, Anvil and Drum, all are parts of ear.

25. Kennedy, Indira and Palme, all of them were assassinated.

26. Chair, Door and Stick all are made of the same raw material.

27. As, a performer plays music on a guitar. Similarly, an acrobat performs tricks on a rope.

28. First is used to make the second and the third.

29. Second and third are bigger and more sophisticated forms than the first and second respectively.

30. Second is more intense form of the first, while third is opposite of second.

Manipal
Engineering Entrance Exam
Solved Paper 2010

Physics

1. In the figure shown, the magnetic field induction at the point O will be

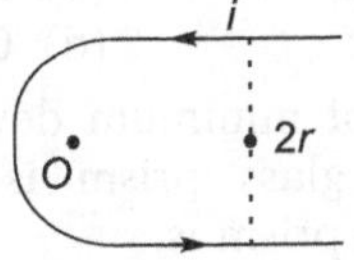

(a) $\frac{\mu_0 i}{2\pi r}$ (b) $\left(\frac{\mu_0}{4\pi}\right)\left(\frac{i}{r}\right)(\pi+2)$
(c) $\left(\frac{\mu_0}{4\pi}\right)\left(\frac{i}{r}\right)(\pi+1)$ (d) $\frac{\mu_0}{4\pi}\frac{i}{r}(\pi-2)$

2. In the electrical network shown in the figure, the potential difference across $3\,\Omega$ resistance will be

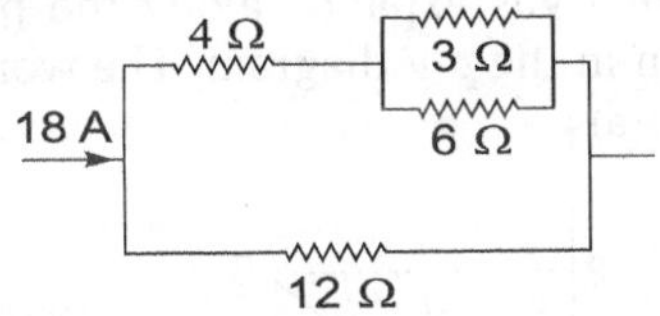

(a) 12 V (b) 2.4 V
(c) 24 V (d) 36 V

3. Three identical thermal conductors are connected as shown in figure. Considering no heat loss due to radiation, temperature at the junction will be

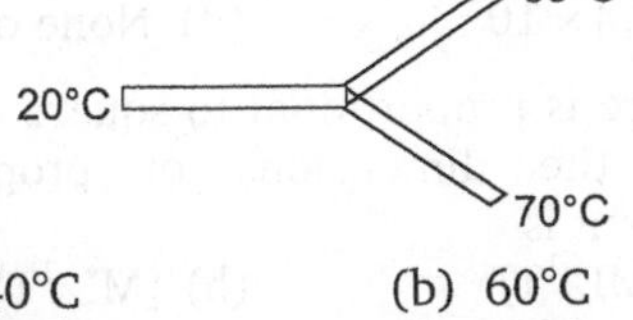

(a) 40°C (b) 60°C
(c) 50°C (d) 35°C

4. Surface tension vanishes at
(a) absolute zero temperature
(b) transition temperature
(c) critical temperature
(d) None of the above

5. A transistor is working in common emitter mode. Its amplification factor (β) is 80. If the base current is 250 μA, the collector current will be
(a) 1.25 μA (b) $\frac{250}{80}$ μA
(c) 430 μA (d) 250×80 μA

6. From an inclined plane two particles are projected with same speed at same angle θ, one up and other down the plane as shown in figure, which of the following statements is/are correct?

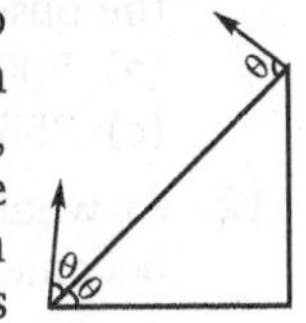

(a) The time of flight of each particle is the same
(b) The particles will collide the plane with same speed
(c) Both the particles strike the plane perpendicularly
(d) The particles will collide in mid air if projected simultaneously and time of flight of each particle is less than the time of collision

7. A battery of emf 10 V is connected to resistance as shown in figure. The potential difference $V_A - V_B$ between the points A and B is

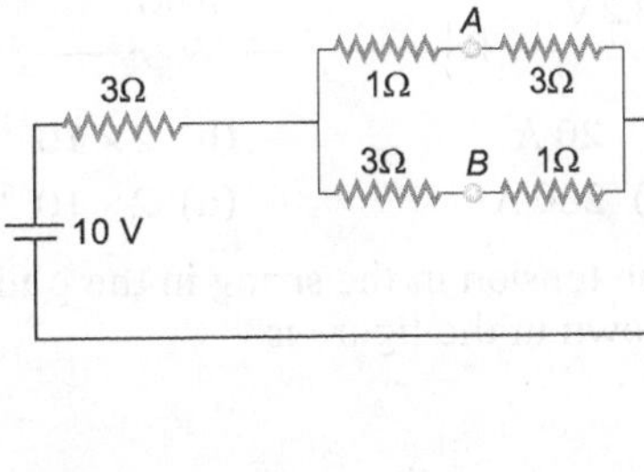

(a) -2 V (b) 2 V
(c) 5 V (d) $\frac{20}{11}$ V

8. Solar radiation is
(a) transverse electromagnetic wave
(b) longitudinal electromagnetic wave
(c) stationary wave
(d) None of the above

9. Fundamental frequency of an open pipe is f_o, Fundamental frequency when it is half filled with water is
(a) f_o (b) $f_o/2$
(c) $2f_o$ (d) $3f_o$

10. If the rms velocity of a gas is v, then
(a) v^2T = constant
(b) v^2/T = constant
(c) vT^2 = constant
(d) v is independent of T

11. A sounding source of frequency 500 Hz moves towards a stationary observer with a velocity 30 m/s. If the velocity of sound in air is 330 m/s, find the frequency heared by the observer.
(a) 500 Hz (b) 550 Hz
(c) 355 Hz (d) 55.5 Hz

12. At what height h above earth, the value of g becomes $g/2$? (R = Radius of earth)
(a) $3R$ (b) $\sqrt{2}R$
(c) $(\sqrt{2}-1)R$ (d) $\frac{1}{\sqrt{2}}R$

13. A freshly prepared radioactive source of half-life 2 h emits radiation of intensity which is 64 times the permissible safe level. Calculate the minimum time after which it would be possible to work safely with this source.
(a) 12 h (b) 24 h
(c) 6 h (d) 130 h

14. The current in the circuit shown in the figure, considering ideal diode is

(a) 20 A (b) 2×10^{-3} A
(c) 200 A (d) 2×10^{-4} A

15. The tension in the string in the pulley system shown in the figure is

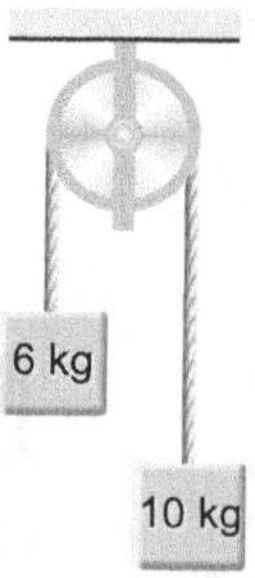

(a) 75 N (b) 80 N
(c) 7.5 N (d) 30 N

16. A glass flask having mass 390 g and an interior volume of 500 cm^3 floats on water when it is less than half filled with water. The density of material of the flask is
(a) 0.8 g/cc (b) 2.8 g/cc
(c) 1.8 g/cc (d) 0.28 g/cc

17. The angle of minimum deviation δ_m for an equilateral glass prism is 30°. Refractive index of the prism is
(a) $1/\sqrt{2}$ (b) $\sqrt{2}$
(c) $2\sqrt{2}$ (d) $1/2\sqrt{2}$

18. When the wavelength of sound changes from 1 m to 1.01 m, the number of beats heard are 4. The velocity of sound is
(a) 404 m/s (b) 4.04 m/s
(c) 414 m/s (d) 400 m/s

19. An ideal gas expands along the path AB as shown in the p-V diagram. The work done is

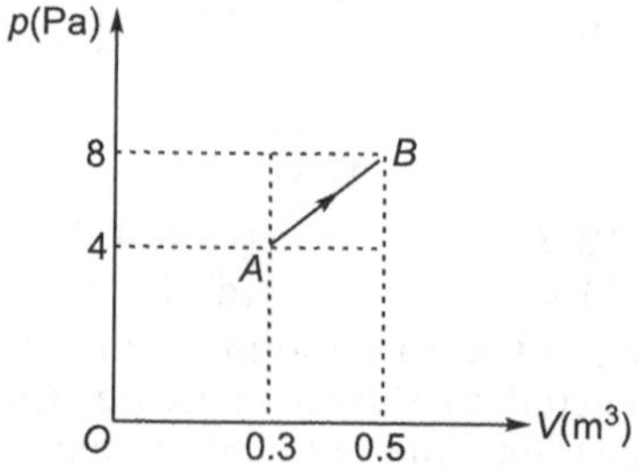

(a) 4×10^4 J (b) 1.2×10^5 J
(c) 2.4×10^5 J (d) None of these

20. If force is proportional to square of velocity, then the dimensions of proportionality constant is
(a) $[ML^{-1}T]$ (b) $[ML^{-1}T^0]$
(c) $[MLT^0]$ (d) $[M^0LT^{-1}]$

21. Two bodies A and B having temperatures 327°C and 427°C are radiating heat to the surrounding. The surrounding temperature is 27°C. The ratio of rates of heat radiation of A to that of B is

(a) 0.52 (b) 0.31
(c) 0.81 (d) 0.42

22. Two bulbs 40 W and 60 W and rated voltage 240 V are connected in series across a potential difference of 420 V. Which bulb will work at above its rated voltage?

(a) 40 W bulb (b) 60 W bulb
(c) Both will work (d) None of these

23. Two cars A and B move along a concentric circular path of radius r_A and r_B with velocities v_A and v_B maintaining constant distance, then $\frac{v_A}{v_B}$ is equal to

(a) $\frac{r_B}{r_A}$ (b) $\frac{r_A}{r_B}$
(c) $\frac{r_A^2}{r_B^2}$ (d) $\frac{r_B^2}{r_A^2}$

24. A body of mass 10 kg is moving with a constant velocity of 10 m/s. When a constant force acts for 4 s on it, it moves with a velocity 2 m/s in the opposite direction. The acceleration produced in it is

(a) 3 m/s^2 (b) -3 m/s^2
(c) 0.3 m/s^2 (d) 0.03 m/s^2

25. A beam of light is incident at 60° to a plane surface. The reflected and refracted rays are perpendicular to each other then refractive index of the surface is

(a) $\sqrt{3}$ (b) $\frac{1}{\sqrt{3}}$
(c) $\frac{1}{2\sqrt{3}}$ (d) None of these

26. Two wires of lengths l and $2l$, radii r and $2r$ respectively having same Young's modulus are hung with a weight mg. Net elongation is

(a) $\frac{3mgl}{\pi r^2 Y}$ (b) $\frac{2mgl}{3\pi r^2 Y}$
(c) $\frac{3mgl}{2\pi r^2 Y}$ (d) $\frac{3mgl}{4\pi r^2 Y}$

27. A ball rolls off the top of stairway with a horizontal velocity of magnitude 1.8 m/s. The steps are 0.20 m high and 0.20 m wide. Which step will the ball hit first?

(a) First (b) Second
(c) Third (d) Fourth

28. The peak value of an alternating emf E given by $E = E_0 \cos \omega t$ is 10 V and its frequency is 50 Hz. At a time $t = \frac{1}{600}$ s, the instantaneous value of the emf is

(a) 10 V (b) $5\sqrt{3}$ V
(c) 5 V (d) 1 V

29. A circuit area 0.01 m^2 is kept inside a magnetic field which is normal to its plane. The magnetic field changes from 2 T to 1 T in 1 ms. If the resistance of the circuit is 2 Ω. The amount of heat evolved is

(a) 0.05 J (b) 50 J
(c) 0.50 J (d) 500 J

30. A convex lens is placed between object and a screen. The size of object is 3 cm and an image of height 9 cm is obtained on the screen. When the lens is displaced to a new position, what will be the size of image on the screen?

(a) 2 cm (b) 6 cm
(c) 4 cm (d) 1 cm

31. A gas is suddenly expanded such that its final volume becomes 3 times its initial volume. If the specific heat at constant volume of the gas is $2R$, then the ratio of initial to final pressures is nearly equal to

(a) 5 (b) 6.5
(c) 7 (d) 3.5

32. Two pendulums have time periods T and $5T/4$. They start SHM at the same time from the mean position. What will be the phase difference between them after the bigger pendulum completed one oscillation?

(a) 45° (b) 90°
(c) 60° (d) 30°

33. A body is coming with a velocity of 72 km/h on a rough horizontal surface of coefficient of friction 0.5. If the acceleration due to gravity is 10 m/s^2, find the minimum distance it can be stopped.

(a) 400 m (b) 40 m
(c) 0.40 m (d) 4 m

34. A bullet comes out of the barrel of gun of length 2 m with a speed 80 m/s. The average acceleration of the bullet is

(a) 1.6 m/s^2 (b) 160 m/s^2
(c) 1600 m/s^2 (d) 16 m/s^2

35. A disc of radius 0.1 m is rotating with a frequency 10 rev/s in a normal magnetic field of strength 0.1 T. Net induced emf is

(a) $2\pi \times 10^{-2}$V (b) $\pi \times 10^{-2}$V
(c) $\frac{\pi}{2} \times 10^{-2}$V (d) None of these

36. 1 cm^3 of water at its boiling point absorbs 540 cal of heat to becomes steam with a volume of 1671 cm^3. If the atmospheric pressure $= 1.013 \times 10^5$ N/m^2 and the mechanical equivalent of heat = 4.19 J/cal, the energy spent in this process in overcoming intermolecular forces is

(a) 540 cal (b) 40 cal
(c) 500 cal (d) zero

37. A string fixed at both ends oscillates in 5 segments, length 10 m and velocity of wave is 20 m/s. What is the frequency?

(a) 5 Hz (b) 15 Hz
(c) 10 Hz (d) 2 Hz

38. When the amplitude of a body executing SHM becomes twice what happens?

(a) Maximum potential energy is doubled
(b) Maximum kinetic energy is doubled
(c) Total energy is doubled
(d) Maximum velocity is doubled

39. The time period of a geostationary satellite at a height 36000 km is 24 h. A spy satellite orbits earth at a height 6400 km. What will be the time period of spy satellite? [Radius of the earth = 6400 km]

(a) 5 h (b) 4 h
(c) 3 h (d) 12 h

40. A bomb of mass 9 kg explodes into two parts. One part of mass 3 kg moves with velocity 16 m/s, then the KE of the other part is

(a) 162 J (b) 150 J
(c) 192 J (d) 200 J

41. In the reaction

$$_7N^{14} + \alpha \longrightarrow {}_8X^{17} + {}_1p^1$$

identify X.

(a) O_2 (b) N_2
(c) He (d) Ar

42. Current in a coil changes from 5 A to 10 A in 0.2 s. If the coefficient of self-induction is 10 H, then the induced emf is

(a) 112 V (b) 250 V
(c) 125 V (d) 230 V

43. The force of interaction between two charges $q_1 = 6\mu C$ and $q_2 = 2\mu C$ is 12 N. If charge $q = -2\mu C$ is added to each of the charges, then the new force of interaction is

(a) 2×10^{-7} N (b) Zero
(c) 30 N (d) 2×10^{-3} N

44. The number of turns in primary coil of a transformer is 20 and the number of turns in the secondary is 10. If the voltage across the primary is 220 V, what is the voltage across the secondary?

(a) 110 V (b) 130 V
(c) 190 V (d) 310 V

45. An electron of an atom transits from n_1 to n_2. In which of the following, maximum frequency of photon will be emitted?

(a) $n_1 = 1$ to $n_2 = 2$ (b) $n_1 = 2$ to $n_2 = 1$
(c) $n_1 = 2$ to $n_2 = 6$ (d) $n_1 = 6$ to $n_2 = 2$

46. A rod of length L and mass M is bent to form a semicircular ring as shown in figure. The moment of inertia about XY is

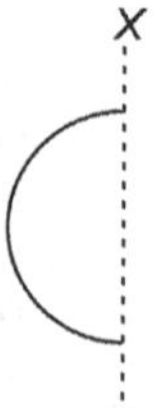

(a) $\frac{1}{4}\frac{ML^2}{\pi^2}$ (b) $\frac{2ML^2}{3\pi^2}$
(c) $\frac{ML^2}{2}$ (d) $\frac{ML^2}{2}$

47. In the figure shown, $m_1 = 10$ kg, $m_2 = 6$ kg, $m_3 = 4$ kg. If $T_3 = 40$ N, $T_2 = ?$

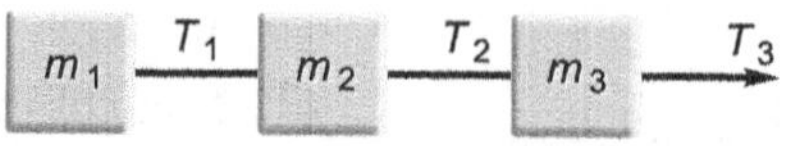

(a) 13 N (b) 32 N
(c) 25 N (d) 35 N

48. A stone is thrown at an angle θ to be horizontal reaches a maximum height H. Then the time of flight of stone will be

(a) $\sqrt{\frac{2H}{g}}$ (b) $2\sqrt{\frac{2H}{g}}$
(c) $\frac{2\sqrt{2H \sin \theta}}{g}$ (d) $\frac{\sqrt{2H \sin \theta}}{g}$

49. Two parallel plates of area A are separated by two different dielectrics as shown in figure. The net capacitance is

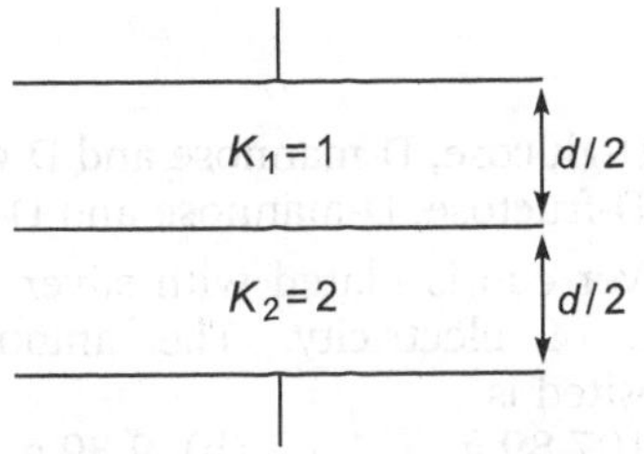

(a) $\frac{\varepsilon_0 A}{2d}$ (b) $\frac{\varepsilon_0 A}{d}$

(c) $\frac{3\varepsilon_0 A}{d}$ (d) $\frac{4\varepsilon_0 A}{3d}$

50. Two springs of force constants k_1 and k_2 are connected as shown. The effective spring constant k is

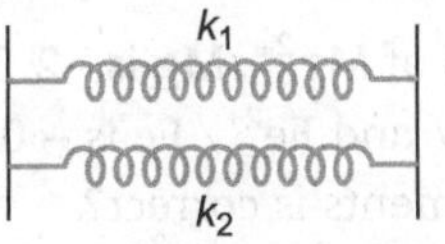

(a) $k_1 + k_2$ (b) $\frac{k_1}{k_2}$

(c) $k_1 k_2$ (d) $2k_1 k_2$

51. A body of weight 2 kg is suspended as shown in figure. The tension T_1 in the horizontal string (in kg-wt) is

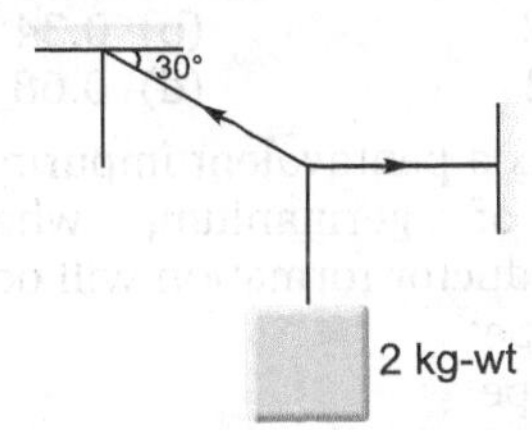

(a) $2/\sqrt{3}$ (b) $\sqrt{3}/2$

(c) $2\sqrt{3}$ (d) 2

52. If the momentum of a body is increased by 100%, then the percentage increase in the kinetic energy is

(a) 150% (b) 200%

(c) 225% (d) 300%

53. In Young's double slit experiment, slit separation is 0.6 mm and the separation between slit and screen is 1.2 m. The angular width is (the wavelength of light used is 4800 Å)

(a) 30 rad (b) 8×10^{-4} rad

(c) 12 rad (d) 70.5 rad

54. X-ray of wavelength $\lambda = 2$ Å is emitted from the metal target. The potential difference applied across the cathode and the metal target is

(a) 5525 V (b) 320 V

(c) 6200 V (d) 3250 V

55. Two identical masses m moving with velocities u_1 and u_2 collide perfectly inelastically. Find the loss in energy.

(a) $m(u_1 - u_2^2)$ (b) $\frac{m}{4}(u_1 - u_2)^2$

(c) $\frac{m}{2}(u_1 - u_2)^2$ (d) $m(u_1 - u_2)^3$

56. A constant torque acting on a uniform circular wheel changes its angular momentum from A_0 to $4A_0$ in 4 s. The magnitude of this torque is

(a) $\frac{3A_0}{4}$ (b) A_0

(c) $4A_0$ (d) $12A_0$

57. When a number of small droplets combines to form a large drop, then

(a) energy is absorbed

(b) energy is liberated

(c) energy is neither liberated nor absorbed

(d) process is independent of energy

58. With what minimum acceleration can a fireman slide down a rope while breaking strength of the rope is $\frac{2}{3}$ of the weight?

(a) $\frac{2}{3}g$ (b) g

(c) $\frac{1}{3}g$ (d) Zero

59. If two waves represented by $y_1 = 4\sin\omega t$ and $y_2 = 3\sin\left(\omega t + \frac{\pi}{3}\right)$ interfere at a point. The amplitude of the resulting wave will be about

(a) 7 (b) 6

(c) 5 (d) 3.5

60. Saturated vapour is compressed to half its volume without any change in temperature, then the pressure will be

(a) doubled (b) halved

(c) the same (d) Zero

Chemistry

1. The root mean square velocity of a gas is doubled when the temperature is
(a) increased four times
(b) increased two times
(c) reduced to half
(d) reduced to one fourth

2. Acidity of phenol is due to
(a) hydrogen bonding
(b) phenolic group
(c) benzene ring
(d) resonance stabilisation of its anion

3. In which one of the following pairs the radius of the second species is greater than that of the first?
(a) Na, Mg (b) O^{2-}, N^{3-}
(c) Li^+, Be^{2+} (d) Ba^{2+}, Sr^{2+}

4. The radius of the first Bohr orbit of hydrogen atom is 0.529 Å. The radius of the third orbit of H^+ will be
(a) 8.46 Å (b) 0.705 Å
(c) 1.59 Å (d) 4.79 Å

5. Both Co^{3+} and Pt^{4+} have a coordination number of six. Which of the following pairs of complexes will show approximately the same electrical conductance for their 0.001 M aqueous solutions?
(a) $CoCl_3 \cdot 4NH_3$ and $PtCl_4 \cdot 4NH_3$
(b) $CoCl_3 \cdot 3NH_3$ and $PtCl_4 \cdot 5NH_3$
(c) $CoCl_3 \cdot 6NH_3$ and $PtCl_4 \cdot 5NH_3$
(d) $CoCl_3 \cdot 6NH_3$ and $PtCl_4 \cdot 3NH_3$

6. When ice melts into water, the entropy
(a) becomes zero (b) remains same
(c) decreases (d) increases

7. Reduction of nitrobenzene in the presence of Zn/NH_4Cl gives
(a) azobenzene
(b) hydrazobenzene
(c) N-phenyl hydroxyl amine
(d) aniline

8. Which one of the following is not a protein?
(a) Wool (b) Nail
(c) Hair (d) DNA

9. Which of the following hexoses will form the same osazone when treated with excess phenyl hydrazine?
(a) D-glucose, D-fructose and D-galactose
(b) D-glucose, D-fructose and D-mannose
(c) D-glucose, D-mannose and D-galactose
(d) D-fructose, D-mannose and D-galactose

10. A silver cup is plated with silver by passing 965C of electricity. The amount of Ag deposited is
(a) 107.89 g (b) 9.89 g
(c) 1.0002 g (d) 1.08 g

11. The standard emf of a cell involving one electron change is found to be 0.591 V at 25°C. The equilibrium constant of the reaction is ($F = 96,500 \text{ C mol}^{-1}$)
(a) 1.0×10^1 (b) 1.0×10^5
(c) 1.0×10^{10} (d) 1.0×10^{30}

12. $E°$ values of Mg^{2+}/Mg is –2.37 V, of Zn^{2+}/Zn is –0.76 V and Fe^{2+}/Fe is – 0.44 V. Which of the statements is correct?
(a) Zn will reduce Fe^{2+}
(b) Zn will reduce Mg^{2+}
(c) Mg oxidises Fe
(d) Zn oxidises Fe

13. The maximum proportion of available volume that can be filled by hard spheres in diamond is
(a) 0.52 (b) 0.34
(c) 0.32 (d) 0.68

14. If we mix a pentavalent impurity in a crystal lattice of germanium, what type of semiconductor formation will occur?
(a) *p*-type
(b) *n*-type
(c) Both (a) and (b)
(d) None of the two

15. Which one of the following complexes is an outer orbital complex?
(Atomic numbers Mn = 25, Fe = 26, Co = 27, Ni = 28)
(a) $[Fe(CN)_6]^{4-}$ (b) $[Mn(CN)_6]^{4-}$
(c) $[Co(NH_3)_6]^{3+}$ (d) $[Ni(NH_3)_6]^{2+}$

16. The product of reaction between alcoholic silver nitrite with ethyl bromide is
(a) ethene (b) ethane
(c) ethyl nitrile (d) nitro ethane

17. *Iso*-propyl chloride undergoes hydrolysis by
(a) S_N1 mechanism
(b) S_N2 mechanisms
(c) S_N1 and S_N2 mechanisms
(d) Neither S_N1 nor S_N2 mechanism

18. *o*-toluic acid on reaction with Br_2 + Fe gives

(a) 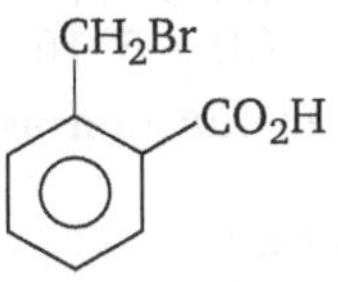
(b) 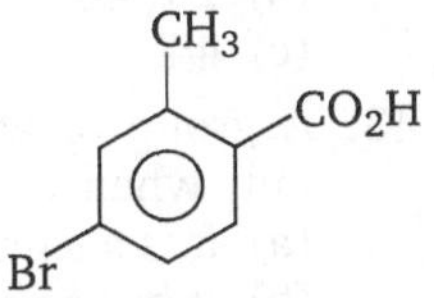

(c) 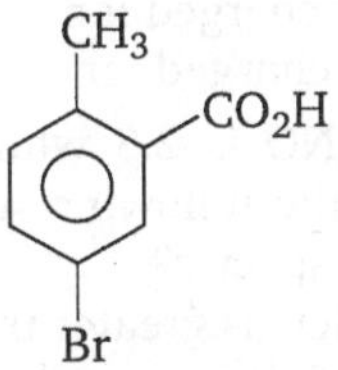
(d) 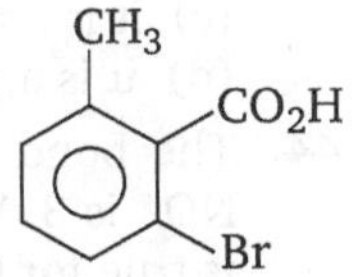

19. Consider the acidity of the carboxylic acids
I. PhCOOH
II. *o*-$NO_2C_6H_4COOH$
III. *p*-$NO_2C_6H_4COOH$
IV. *m*-$NO_2C_6H_4COOH$
Which of the following order is correct?
(a) I > II > III > IV (b) II > IV > III > I
(c) II > IV > I > III (d) II > III > IV > I

20. Which of the following does not answer iodoform test?
(a) *n*-butyl alcohol
(b) Acetophenone
(c) Acetaldehyde
(d) Ethylmethyl ketone

21. Which one of the following undergoes reaction with 50% sodium hydroxide solution to give the corresponding alcohol and acid?
(a) Phenol (b) Benzaldehyde
(c) Butanal (d) Benzoic acid

22. The IUPAC name of

$$CH_3—\underset{OH}{CH}—CH{=}\underset{CH_3}{\underset{|}{C}}—CHO \text{ is}$$

(a) 4-hydroxy-1-methylpentanal
(b) 4-hydroxy-2-methylpent-2-en-1-al
(c) 2-hydroxy-4-methylpent-3-en-5-al
(d) 2-hydroxy-3-methylpent-2-en-5-al

23. The activation energy of exothermic reaction $A \rightarrow B$ is 80 kJ mol^{-1}. The heat of reaction is 200 kJ mol^{-1}. The activation energy for the reaction $B \rightarrow A$ (in kJ mol^{-1}) will be
(a) 80 (b) 120
(c) 40 (d) 280

24. Which of the following electrolytes is least effective in coagulating ferric hydroxide solution?
(a) KBr (b) K_2SO_4
(c) K_2CrO_4 (d) $K_4[Fe(CN)_6]$

25. Of the following outer electronic configurations of atoms, the highest oxidation state is achieved by which one of them?
(a) $(n-1)d^8, ns^2$ (b) $(n-1)d^5, ns^1$
(c) $(n-1)d^3, ns^2$ (d) $(n-1)d^5, ns^2$

26. The relative lowering of vapour pressure of a dilute aqueous solution containing non volatile solute is 0.0125. The molality of the solution is about
(a) 0.70 (b) 0.50
(c) 0.90 (d) 0.80

27. When hydrogen peroxide is added to acidified potassium dichromate, a blue colour is produced due to formation of
(a) CrO_3 (b) Cr_2O_3
(c) CrO_5 (d) CrO_4^{2-}

28. In the following reaction,
$NaOH + S \longrightarrow A + Na_2S + H_2O$; A is
(a) Na_2SO_4 (b) Na_2SO_3
(c) Na_2S (d) $Na_2S_2O_3$

29. On igniting Fe_2O_3 at 1400°C, the product obtained is
(a) Fe_2O_3 melt (b) FeO
(c) Fe_3O_4 (d) metallic iron

30. Sulphuric acid has great affinity for water because
(a) it hydrolyses the acid
(b) it decomposes the acid
(c) acid forms hydrates with water
(d) acid decomposes water

31. Helium-oxygen mixture is used by deep sea divers in preference to nitrogen-oxygen mixture because
(a) helium is much less soluble in blood than nitrogen
(b) nitrogen is much less soluble in blood than helium
(c) due to high pressure deep under the sea nitrogen and oxygen react to give poisonous nitric oxide
(d) nitrogen is highly soluble in water

32. One mole of CO_2 contains
(a) 3 g atoms of CO_2
(b) 18.1×10^{23} molecules of CO_2
(c) 6.02×10^{23} atoms of O
(d) 6.02×10^{23} atoms of C

33. The equivalent weight of $KMnO_4$ for acid solution is
(a) 79 (b) 52.16
(c) 158 (d) 31.6

34. Which has the highest weight?
(a) 1 m^3 of water
(b) A normal adult man
(c) 10 L of Hg
(d) All have same weight

35. Which has the highest *e/m* ratio?
(a) He^{2+} (b) H^+
(c) He^+ (d) D^+

36. The ionisation potential order for which set is correct?
(a) $Cs < Li < K$ (b) $Cs > Li > B$
(c) $Li > K > Cs$ (d) $B > Li < K$

37. The oxidation state of Fe in Fe_3O_4 is
(a) +3 (b) 8/3
(c) +6 (d) +2

38. For a first order reaction, the concentration changes from 0.8 to 0.4 in 15 min. The time taken for the concentration to change from 0.1 M to 0.025 M is
(a) 30 min (b) 15 min
(c) 7.5 min (d) 60 min

39. When phenol is treated with excess of bromine water, it gives
(a) *m*-bromophenol
(b) *o*-and *p*-bromophenols
(c) 2, 4-dibromophenol
(d) 2, 4, 6-tribromophenol

40. Which reaction is suitable for the preparation of α-chloroacetic acid?
(a) Hell-Volhard Zelinsky reaction
(b) Nef reaction
(c) Stephen's reaction
(d) Perkin condensation

41. Which one of the following compounds will dissolve in an alkali solution after it has undergone reaction with Hinsberg reagent?
(a) CH_3NH_2 (b) $(CH_3)_3N$
(c) $(C_2H_5)_2NH$ (d) $C_6H_5NHC_6H_5$

42. Angle strain in cyclopropane is
(a) 24° 44′ (b) 9° 44′
(c) 44′ (d) – 5° 16′

43. A group of atoms can function as a ligand only when
(a) it is a small molecule
(b) it has an unshared electron pair
(c) it is a negatively charged ion
(d) it is a positively charged ion

44. The bond order in NO is 2.5 while that in NO^+ is 3. Which of the following statements is true for these two species?
(a) Bond length in NO^+ is greater than in NO
(b) Bond length in NO is greater than in NO^+
(c) Bond length in NO^+ is equal to that in NO
(d) Bond length is unpredictable

45. In a homonuclear molecule which of the following set of orbitals is degenerate?
(a) $\sigma 2s$ and $\sigma 1s$ (b) $\pi 2p_x$ and $\pi 2p_y$
(c) $\pi 2p_x$ and $\sigma 2p_z$ (d) $\sigma 2p_z$ and $\overset{*}{\pi} 2p_x$

46. The pH of a neutral water sample is 6.5. Then the temperature of water
(a) is 25°C
(b) is more than 25°C
(c) is less than 25°C
(d) can be more or less than 25°C

47. Which is Lewis acid?
(a) BF_3 (b) NF_3
(c) Cl^- (d) H_2O

48. The pH of 10^{-10} M NaOH solution is nearest to
(a) 4 (b) – 10
(c) 4 (d) 7

49. Which of the following is not true for carbanions?
(a) The carbon carrying the charge has eight valence electrons
(b) They are formed by heterolytic fission
(c) They are paramagnetic
(d) The carbon carrying the charge is sp^3 hybridised

50. Which among the following statements is correct with respect to the optical isomers?
(a) Enantiomers are non-superimposable mirror images
(b) Diastereomers are superimposable mirror images
(c) Enantiomers are superimposable mirror images
(d) *Meso* forms have no plane of symmetry

51. Inductive effect involves
(a) delocalisation of σ-electrons
(b) displacement of σ-electrons
(c) delocalisation of n-electrons
(d) displacement of π-electrons

52. The solubility of AgCl in 0.2 M NaCl solution is (K_{sp} of AgCl $= 1.20 \times 10^{-10}$)
(a) 6.0×10^{-10} M
(b) 0.2 M
(c) 1.2×10^{-10} M
(d) 0.2×10^{-10} M

53. Lipids are
(a) nucleic acids occurring in plants
(b) proteins occurring in animals
(c) carbohydrates occurring in plants
(d) fats of natural origin

54. The enzyme pepsin hydrolyses
(a) proteins to amino acids
(b) fats to fatty acids
(c) glucose to ethyl alcohol
(d) polysaccharides to monosaccharides

55. Carbon cannot be used in the reduction of Al_2O_3 because
(a) it is an expensive proposition
(b) the enthalpy of formation of CO_2 is more than that of Al_2O_3
(c) pure carbon is not easily available
(d) the enthalpy of formation of Al_2O_3 is too high

56. Oxalic acid when heated with conc H_2SO_4, gives
(a) H_2O_2 and CO_2
(b) CO and CO_2
(c) H_2O_2 and CO
(d) CO_2 and H_2S

57. At 25°C, the highest osmotic pressure is exhibited by 0.1 M solution of
(a) urea
(b) glucose
(c) KCl
(d) $CaCl_2$

58. Ammonia is dried over
(a) slaked lime
(b) calcium chloride
(c) phosphorus pentoxide
(d) quick lime

59. $C_6H_6 \xrightarrow[350\text{ K}]{H_2SO_4} A \xrightarrow[\text{Fusion}]{\text{Alkali}} B \xrightarrow[H_2O]{Br_2} C$

In the above sequence, C is
(a) o-bromophenol
(b) p-bromophenol
(c) m-bromophenol
(d) 2, 4, 6-tribromophenol

60. Collin's reagent is used to convert
(a) $>C{=}O \longrightarrow >CHOH$
(b) $-CH_2OH \longrightarrow -CHO$
(c) $-CHO \longrightarrow -COOH$
(d) $-CHO \longrightarrow -CH_2OH$

Mathematics

1. The range of the function $f(x) = \log_e(3x^2 - 4x + 5)$ is
(a) $\left(-\infty, \log_e \frac{11}{3}\right]$
(b) $\left[\log_e \frac{11}{3}, \infty\right)$
(c) $\left[-\log_e \frac{11}{3}, \log_e \frac{11}{3}\right]$
(d) None of the above

2. The domain of the function $f(x) = \frac{\sqrt{9 - x^2}}{\sin^{-1}(3 - x)}$ is
(a) (2, 3)
(b) [2, 3)
(c) (2, 3]
(d) None of these

3. The numbers a_n's are defined by
$$a_0 = 1, a_{n+1} = 3n^2 + n + a_n, (n \ge 0)$$
Then, a_n is equal to
(a) $n^3 + n^2 + 1$
(b) $n^3 - n^2 + 1$
(c) $n^3 - n^2$
(d) $n^3 + n^2$

4. Nishi has 5 coins each of the different denomination. The number different sums of money she conform
(a) 32
(b) 25
(c) 31
(d) None of these

5. There are P copies of n-different books. The number of different ways in which a non-empty selection can be made from them, is
(a) $(P + 1)^n - 1$
(b) $P^n - 1$
(c) $(P + 1)^{n-1} - 1$
(d) None of these

6. Out of 18 points in a plane no three are in the same straight line except five points which are collinear. The number of straight lines that can be formed joining them, is

(a) 143 (b) 144
(c) 153 (d) None of these

7. The term independent of x in $\left[\sqrt{\left(\frac{x}{3}\right)}+\sqrt{\left(\frac{3}{2x^2}\right)}\right]^{10}$ is
(a) 1 (b) ${}^{10}C_1$
(c) $\frac{5}{12}$ (d) None of these

8. The greatest coefficient in the expansion of $(1+x)^{2n}$ is
(a) ${}^{2n}C_n$ (b) ${}^{2n}C_{n-1}$
(c) ${}^{2n}C_{n-2}$ (d) None of these

9. The number of terms in the expansion of $(\sqrt{5}+\sqrt[4]{11})^{124}$ which are integers, is equal to
(a) 0 (b) 30 (c) 31 (d) 32

10. The constant term in the expansion of $(1+x)^m\left(1+\frac{1}{x}\right)^n$ is
(a) ${}^{m+n}C_{m-1}$ (b) ${}^{m+n}C_n$
(c) ${}^{m+n}C_{m-n}$ (d) None of these

11. If $a>1$, roots of the equation
$$(1-a)x^2+3ax-1=0 \text{ are}$$
(a) one positive and one negative
(b) both negative
(c) both positive
(d) both non-real complex

12. The number of values of the triplet (a, b, c) for which $a\cos 2x + b\sin^2 x + c = 0$ is satisfied by all real x, is
(a) 0 (b) 2
(c) 3 (d) infinite

13. If α, β, γ are such that $\alpha+\beta+\gamma=2$, $\alpha^2+\beta^2+\gamma^2=6$, $\alpha^3+\beta^3+\gamma^3=8$, then $\alpha^4+\beta^4+\gamma^4$ is
(a) 5 (b) 18
(c) 12 (d) 36

14. If a, b, c are three distinct positive real numbers of real roots of $ax^2+2b|x|-c=0$ is
(a) 4 (b) 2
(c) 0 (d) None of these

15. If A is a skew-symmetric matrix, then trace of A is
(a) 1 (b) -1
(c) 0 (d) None of these

16. If the matrix $A=\begin{bmatrix} y+a & b & c \\ a & y+b & c \\ a & b & y+c \end{bmatrix}$ has rank 3, then
(a) $y \neq (a+b+c)$
(b) $y \neq 1$
(c) $y=0$
(d) $y \neq -(a+b+c)$ and $y \neq 0$

17. For positive numbers x, y, z the numerical value of the determinant $\begin{vmatrix} 1 & \log_x y & \log_x z \\ \log_y x & 1 & \log_y z \\ \log_z x & \log_z y & 1 \end{vmatrix}$ is
(a) 0 (b) 1
(c) 2 (d) None of these

18. If A is an orthogonal matrix, then
(a) det A = not exist (b) det $A=0$
(c) det $A=\pm 1$ (d) None of these

19. Let a, b, c be real numbers with $a \neq 0$ and let α, β be the roots of the equation $ax^2+bx+c=0$, then $a^3x^2+abcx+c^3=0$ has roots
(a) $\alpha^2\beta, \beta^2\alpha$ (b) α, β^2
(c) $\alpha^2\beta, \beta\alpha$ (d) $\alpha^3\beta, \beta^3\alpha$

20. If n_1, n_2 are positive integers, then $(1+i)^{n_1}+(1+i^3)^{n_1}+(1+i^5)^{n_2}+(1+i^7)^{n_2}$ is a real number if and only if
(a) $n_1=n_2+1$
(b) $n_1=n_2$
(c) n_1, n_2 are any two negative integers
(d) n_1, n_2 are both any positive integers

21. The equation $|z+i|-|z-i|=k$ represent a hyperbola, if
(a) $-2<k<2$ (b) $k>2$
(c) $0<k<2$ (d) None of these

22. The period of the function $f(x)=|\sin 4x|+|\cos 4x|$ is
(a) $\pi/2$ (b) $\pi/8$
(c) $\pi/4$ (d) None of these

23. If $\cos x=\frac{2\cos y-1}{2-\cos y}$, where $x, y \in (0, \pi)$, then $\tan\frac{x}{2}\cdot\cot\frac{y}{2}$ is equal to
(a) $\sqrt{2}$ (b) $\sqrt{3}$
(c) $\frac{1}{\sqrt{2}}$ (d) $\frac{1}{\sqrt{3}}$

24. The value of $\cos\frac{2\pi}{15}\cdot\cos\frac{4\pi}{15}\cdot\cos\frac{8\pi}{15}\cdot\cos\frac{16\pi}{15}$ is equal to

(a) $\frac{1}{16}$ (b) $\frac{1}{32}$

(c) $\frac{1}{64}$ (d) $\frac{1}{8}$

25. The sum of all the solutions of the equation $\cos x\cdot\cos\left(\frac{\pi}{3}+x\right)\cdot\cos\left(\frac{\pi}{3}-x\right)=\frac{1}{4}$, $x\in[0, 6\pi]$ is

(a) 15π (b) 30π

(c) $\frac{110\pi}{3}$ (d) None of these

26. $2\tan^{-1}(\text{cosec}\tan^{-1}x-\tan\cot^{-1}x)$ is equal to

(a) $\cot^{-1}x$ (b) $\cot^{-1}\frac{1}{x}$

(c) $\tan^{-1}x$ (d) None of these

27. The value of

$$\sin^{-1}\left[\cos\left(\sin^{-1}\sqrt{\frac{2-\sqrt{3}}{4}}+\cos^{-1}\frac{\sqrt{12}}{4}+\sec^{-1}\sqrt{2}\right)\right]$$ is

(a) 0 (b) $\frac{\pi}{4}$ (c) $\frac{\pi}{6}$ (d) $\frac{\pi}{2}$

28. In a ΔABC, $\angle B=90°$, then $\tan^2\left(\frac{A}{2}\right)$ is

(a) $\frac{b-c}{b+c}$ (b) $\frac{b+c}{b-c}$

(c) $\frac{b-2c}{b+c}$ (d) None of these

29. In a triangle, if $r_1>r_2>r_3$, then

(a) $a>b>c$ (b) $a<b<c$

(c) $a>b$ and $b<c$ (d) $a<b$ and $b>c$

30. A and B are two points on one bank of a straight river and C, D are two other points on the other bank of river, if direction from A to B is same as that from C to D and $AB=a$, $\angle CAD=\alpha$, $\angle DAB=\beta$, $\angle CBA=\gamma$, then CD is equal to

(a) $\frac{a\sin\beta\cdot\sin\gamma}{\sin\alpha\cdot\sin(\alpha+\beta+\gamma)}$

(b) $\frac{a\sin\alpha\cdot\sin\gamma}{\sin\beta\cdot\sin(\alpha+\beta+\gamma)}$

(c) $\frac{a\sin\alpha\cdot\sin\beta}{\sin\gamma\cdot\sin(\alpha+\beta+\gamma)}$

(d) None of the above

31. In a trapezoid of the vector $\vec{BC}=\lambda\vec{AD}$. We will, then find that $\vec{P}=\vec{AC}+\vec{BD}$ is collinear with $\vec{AD}$. If $\vec{P}=\mu\vec{AD}$, then

(a) $\mu=\lambda+1$ (b) $\lambda=\mu+1$

(c) $\lambda+\mu=1$ (d) $\mu=2+\lambda$

32. If $P(\vec{p})$, $Q(\vec{q})$, $R(\vec{r})$ and $S(\vec{s})$ be four points such that $3\vec{p}+8\vec{q}=6\vec{r}+5\vec{s}$, then the lines PQ and RS are

(s) skew (b) intersecting

(c) parallel (d) None of these

33. Let $\vec{a}=2\hat{i}+\hat{j}-2\hat{k}$ and $\vec{b}=\hat{i}+\hat{j}$, if $\vec{c}$ is a vector such that $\vec{a}\cdot\vec{c}=|\vec{c}|$, $|\vec{c}-\vec{a}|=2\sqrt{2}$ and the angle between $\vec{a}\times\vec{b}$ and $\vec{c}$ is 30°, then $|(\vec{a}\times\vec{b})\times\vec{c}|$ is equal to

(a) $\frac{2}{3}$ (b) $\frac{3}{2}$

(c) 2 (d) 3

34. Consider a tetrahedron with faces F_1, F_2, F_3, F_4. Let $\vec{V}_1, \vec{V}_2, \vec{V}_3, \vec{V}_4$ be the vectors whose magnitudes are respectively equal to areas of F_1, F_2, F_3, F_4 and whose directions are perpendicular to these faces in outward direction, then $|\vec{V}_1+\vec{V}_2+\vec{V}_3+\vec{V}_4|$ equals

(a) 1 (b) 4

(c) $\vec{0}$ (d) None of these

35. If V is the volume of the parallelopiped having three coterminus edges as $\vec{a}$, $\vec{b}$ and $\vec{c}$, then the volume of the parallelopiped having three coterminus edge as

$$\vec{\alpha}=(\vec{a}\cdot\vec{a})\vec{a}+(\vec{a}\cdot\vec{b})\vec{b}+(\vec{a}\cdot\vec{c})\vec{c}$$

$$\vec{\beta}=(\vec{a}\cdot\vec{b})\vec{a}+(\vec{b}\cdot\vec{b})\vec{b}+(\vec{b}\cdot\vec{c})\vec{c}$$

$$\vec{\gamma}=(\vec{a}\cdot\vec{c})\vec{a}+(\vec{b}\cdot\vec{c})\vec{b}+(\vec{c}\cdot\vec{c})\vec{c}$$ is

(a) V^3 (b) $3V$

(c) V^2 (d) $2V$

36. Let us define the length of a vector $a\hat{i}+b\hat{j}+c\hat{k}$ as $|a|+|b|+|c|$. This definition coincides with the usual definition of length of a vector $a\hat{i}+b\hat{j}+c\hat{k}$, iff
(a) $a=b=c=0$
(b) any two of a, b and c are zero
(c) any one of a, b and c is zero
(d) $a+b+c=0$

37. Let $\vec{u}=\hat{i}+\hat{j}$, $\vec{v}=\hat{i}-\hat{j}$ and $\vec{w}=\hat{i}+2\hat{j}+3\hat{k}$, if $\hat{n}$ is a unit vector such that $\vec{u}\cdot\hat{n}=0$ and $\vec{v}\cdot\hat{n}=0$, then $|\vec{w}\cdot\hat{n}|$ is equal to
(a) 3 (b) 0
(c) 1 (d) 2

38. For two events A and B, it is given that $P(A)=P\left(\frac{A}{B}\right)=\frac{1}{4}$ and $P\left(\frac{B}{A}\right)=\frac{1}{2}$. Then
(a) A and B are mutually exclusive events
(b) A and B are dependent events
(c) $P\left(\frac{A}{B}\right)=\frac{3}{4}$
(d) None of the above

39. The least number of times a fair coin must be tossed so that the probability of getting atleast one head is atleast 0.8 is
(a) 7 (b) 6
(c) 5 (d) 3

40. If x follows a binomial distribution with parameters $n=100$ and $p=\frac{1}{3}$, then $p(X=r)$ is maximum when r equals
(a) 16 (b) 32
(c) 33 (d) None of these

41. If $\bar{x}_1$ and $\bar{x}_2$ are the means of two distributions such that $\bar{x}_1<\bar{x}_2$ and $\bar{x}$ is the mean of the combined distribution, then
(a) $\bar{x}<\bar{x}_1$ (b) $\bar{x}>\bar{x}_2$
(c) $\bar{x}=\frac{\bar{x}_1+\bar{x}_2}{2}$ (d) $\bar{x}_1<\bar{x}<\bar{x}_2$

42. If the axes be turned through an angle $\tan^{-1}2$. What does the equation $4xy-3x^2=a^2$ become?
(a) $X^2-4Y^2=a^2$ (b) $X^2+4Y^2=a^2$
(c) $X^2+4Y^2=-a^2$ (d) None of these

43. If t_1, t_2 and t_3 are distinct, the points $(t_1, 2at_1+at_1^3)$, $(t_2, 2at_2+at_2^3)$, $(t_3, 2at_3+at_3^3)$ are collinear, if
(a) $t_1t_2t_3=1$
(b) $t_1+t_2+t_3=t_1t_2t_3$
(c) $t_1+t_2+t_3=0$
(d) $t_1+t_2+t_3=-1$

44. The distance of the point (2, 3) from the line $2x-3y+9=0$ measured along a line $x-y+1=0$ is
(a) $4\sqrt{2}$ (b) $2\sqrt{2}$
(c) $\sqrt{2}$ (d) $\frac{1}{\sqrt{2}}$

45. The equation of the straight line which passes through the intersection of the lines $x-y-1=0$ and $2x-3y+1=0$ and is parallel to x-axis, is
(a) $y=3$ (b) $y=-3$
(c) $x+y=3$ (d) None of these

46. If the straight line $a_1x+b_1y+c_1=0$, $a_1x+b_1y+c_2=0$, $a_2x+b_2y+d_1=0$ and $a_2x+b_2y+d_2=0$ are the sides of rhombus, then
(a) $(a_2^2+b_2^2)(c_1-c_2)^2=(a_1^2+b_1^2)(d_1-d_2)^2$
(b) $(a_1^2+b_1^2)|d_1-d_2|=(a_2^2+b_2^2)|c_1-c_2|$
(c) $(a_2^2+b_2^2)(d_1-d_2)^2=(a_1^2+b_1^2)(c_1-c_2)^2$
(d) $(a_1^2+b_1^2)|c_1-c_2|=(a_2^2+b_2^2)|d_1-d_2|$

47. The equation $3x^2+7xy+2y^2+5x+5y+2=0$ represents
(a) a pair of straight lines
(b) a circle
(c) an ellipse
(d) a hyperbola

48. To the lines $ax^2+2hxy+by^2=0$ the lines $a^2x^2+2h(a+b)xy+b^2y^2=0$, are
(a) equally inclined
(b) perpendicular
(c) bisector of the angle
(d) None of the above

49. Consider four circles $(x\pm1)^2+(y\pm1)^2=1$, then the equation of smaller circle touching these four circle is
(a) $x^2+y^2=3-\sqrt{2}$
(b) $x^2+y^2=6-3\sqrt{2}$
(c) $x^2+y^2=5-2\sqrt{2}$
(d) $x^2+y^2=3-2\sqrt{2}$

50. The locus of the point of intersection of tangents to the circle $x = a\cos\theta,\ y = a\sin\theta$ at the points, whose parametric angles differe by $\frac{\pi}{2}$, is

(a) a straight line
(b) a circle
(c) a pair of straight line
(d) None of the above

51. The equation of the circle passing through (1, 0) and (0, 1) and having the smallest possible radius is

(a) $x^2 - y^2 - x - y = 0$
(b) $x^2 + y^2 - x - y = 0$
(c) $x^2 + y^2 + x + y = 0$
(d) $x^2 + y^2 - 2x - 2y = 0$

52. The length of the common chord of the two circles $(x-a)^2 + (y-b)^2 = c^2$ and $(x-b)^2 + (y-a)^2 = c^2$ is

(a) $\sqrt{4c^2 + 2(a-b)^2}$ (b) $\sqrt{4c^2 - (a-b)^2}$
(c) $\sqrt{4c^2 - 2(a-b)^2}$ (d) $\sqrt{2c^2 - 2(a-b)^2}$

53. If $\left(m_i, \frac{1}{m_i}\right)$ are four distinct points on a circle, then

(a) $m_1m_2m_3m_4 = 1$
(b) $m_1m_2m_3m_4 = -1$
(c) $m_1m_2m_3m_4 = \frac{1}{2}$
(d) $\frac{1}{m_1} + \frac{1}{m_2} + \frac{1}{m_3} + \frac{1}{m_4} = \frac{1}{4}$

54. If the normal to the parabola $y^2 = 4ax$ at the point $(at^2, 2at)$ cuts the parabola again at $(aT^2, 2aT)$, then

(a) $-2 \le T \le 2$
(b) $T \in (-\infty, -8) \cup (8, \infty)$
(c) $T^2 < 8$
(d) $T^2 \ge 8$

55. The number of real tangents that can be drawn to the ellipse $3x^2 + 5y^2 = 32$ passing through (3, 5) is

(a) 0 (b) 1
(c) 2 (d) infinite

56. If the equation $lx^2 + 2mxy + ny^2 = 0$ represents a pair conjugate diameter of the hyperbola $\frac{x^2}{a^2} - \frac{y^2}{b^2} = 1$, then

(a) $la^2 + nb^2 = 0$ (b) $la^2 = nb^2$
(c) $2la^2 = nb^2$ (d) None of these

57. If e is the eccentricity of the hyperbola $\frac{x^2}{a^2} - \frac{y^2}{b^2} = 1$ and θ is the angle between the asymptotes, then $\cos\left(\frac{\theta}{2}\right)$ is equal to

(a) $\frac{1}{e}$ (b) $\frac{-1}{e}$
(c) e (d) $\frac{2}{e}$

58. The tangent and normal to a rectangular hyperbola $xy = c^2$ at a point cuts off intercepts a_1 and a_2 on one axis and b_1, b_2 on the other, then $a_1a_2 + b_1b_2$ is equal to

(a) 1 (b) 2
(c) 3 (d) 0

59. The direction cosines of any normal to the xy-plane are

(a) 1, 0, 0 (b) 0, 1, 0
(c) 1, 1, 0 (d) 0, 0, 1

60. The locus of the equation $xy + yz = 0$ is

(a) a pair of straight lines
(b) a pair of parallel lines
(c) a pair of perpendicular planes
(d) None of the above

61. The reflection of the point $(2, -1, 3)$ in the plane $3x - 2y - z = 9$ is

(a) $\left(\frac{26}{7}, \frac{15}{7}, \frac{17}{7}\right)$ (b) $\left(\frac{26}{7}, \frac{-15}{7}, \frac{17}{7}\right)$
(c) $\left(\frac{15}{7}, \frac{26}{7}, \frac{-17}{7}\right)$ (d) $\left(\frac{26}{7}, \frac{17}{7}, \frac{-15}{7}\right)$

62. $\lim_{x \to 0} \frac{\sqrt{1 - \cos 2x}}{\sqrt{2} \cdot x}$ is

(a) 1 (b) -1
(c) zero (d) does not exist

63. If $f(x) = \begin{cases} ax^2 + b, & 0 \le x < 1 \\ x + 3, & 1 < x \le 2 \\ 4, & x = 1 \end{cases}$, then the value of (a, b) for which $f(x)$ cannot be continuous at $x = 1$ is

(a) (2, 2) (b) (3, 1)
(c) (4, 0) (d) (5, 2)

64. If $f(x) = \log_x(\log x)$, then $f'(x)$ at $x = e$ is

(a) $1/e$ (b) e
(c) $-1/e$ (d) 0

65. Radius of the circle

$$\vec{r}^2 + \vec{r}\cdot(2\hat{i} - 2\hat{j} - 4\hat{k}) - 19 = 0,$$
$$\vec{r}\cdot(\hat{i} - 2\hat{j} + 2\hat{k}) + 8 = 0, \text{ is}$$

(a) 5 (b) 4
(c) 3 (d) 2

66. If $x^y = e^{x-y}$, then $\frac{dy}{dx}$ is equal to

(a) $\frac{\log x}{(1+\log x)^2}$ (b) $\frac{x-y}{(1+\log x)}$
(c) $\frac{x-y}{(1+\log x)^2}$ (d) $\frac{1}{(1+\log x)}$

67. If $y = \sin(m \sin^{-1} x)$, then $(1-x^2)\,y'' - xy'$ is equal to

(a) m^2y (b) my
(c) $-m^2y$ (d) None of these

68. If $f(x) = (ax+b)\sin x + (cx+d)\cos x$, then the values of a, b, c and d such that $f'(x) = x\cos x$ for all x, are

(a) $b = c = 0, a = d = 1$
(b) $b = d = 0, a = c = 1$
(c) $c = d = 0, a = b = 1$
(d) None of the above

69. If a particle is moving such that the velocity acquired is proportional to the square root of the distance covered, then its acceleration is

(a) a constant (b) $\propto s^2$
(c) $\propto \frac{1}{s^2}$ (d) $\propto \frac{1}{s}$

70. The point in the interval $[0, 2\pi]$, where $f(x) = e^x \sin x$ has maximum slope, is

(a) $\frac{\pi}{4}$ (b) $\frac{\pi}{2}$
(c) π (d) $\frac{3\pi}{2}$

71. The function $f(x) = \frac{ax+b}{(x-1)(x-4)}$ has a local maxima at $(2, -1)$, then

(a) $b = 1, a = 0$ (b) $a = 1, b = 0$
(c) $b = -1, a = 0$ (d) $a = -1, b = 0$

72. If $z = \tan(y + ax) + \sqrt{y - ax}$, then $z_{xx} - a^2 z_{yy}$ is equal to

(a) 0 (b) 2
(c) $z_x + z_y$ (d) $z_x z_y$

73. If $\int f(x)dx = F(x)$, then $\int x^3 f(x^2)dx$ is equal to

(a) $\frac{1}{2}[x^2\{F(x)\}^2 - \int \{F(x)\}^2 dx]$
(b) $\frac{1}{2}[x^2F(x^2) - \int F(x^2)d(x^2)]$
(c) $\frac{1}{2}[x^2F(x) - \frac{1}{2}\int \{F(x)\}^2 dx]$
(d) None of the above

74. $\int \frac{x^2+4}{x^4+16}\,dx$ is equal to

(a) $\frac{1}{2\sqrt{2}}\tan^{-1}\left(\frac{x^2+4}{2x}\right) + c$
(b) $\frac{1}{2\sqrt{2}}\tan^{-1}\left(\frac{x^2-4}{2\sqrt{2}x}\right) + c$
(c) $\frac{1}{2\sqrt{2}}\tan^{-1}\left(\frac{x^2+4}{2\sqrt{2}x}\right) + c$
(d) $\frac{1}{2}\tan^{-1}\left(\frac{x^2-4}{2x}\right) + c$

75. Evaluate $\int \frac{1}{(x+1)\sqrt{x^2-1}}\,dx$

(a) $\sqrt{\frac{x+1}{x-1}} + c$ (b) $\sqrt{\frac{x-1}{x+1}} + c$
(c) $\sqrt{\frac{1}{x+1}} + c$ (d) None of these

76. The value of the integral $\int_{\pi/2}^{3\pi/2} [\sin x]\,dx$, where $[\cdot]$ denotes the greatest integer function, is

(a) $\frac{\pi}{2}$ (b) $-\frac{\pi}{2}$
(c) 0 (d) π

77. The area of the loop the curve $ay^2 = x^2(a-x)$ is

(a) $\frac{8a^2}{15}$ sq unit (b) $\frac{4a^2}{15}$ sq unit
(c) $\frac{2a^2}{15}$ sq unit (d) None of these

78. Solution of the differential equation

$$x = 1 + xy\frac{dy}{dx} + \frac{(xy)^2}{2!}\left(\frac{dy}{dx}\right)^2 + \frac{(xy)^3}{3!}\left(\frac{dy}{dx}\right)^3 + \dots \text{ is}$$

(a) $y = \log_e(x) + c$
(b) $y = (\log_e x)^2 + c$
(c) $y = \pm\sqrt{(\log_e x)^2 + 2c}$
(d) $xy = x^y + k$

79. If the solution of the differential equation $\frac{dy}{dx} = \frac{ax+3}{2y+f}$ represents a circle, then the value of a is

(a) 2 (b) −2
(c) 3 (d) −4

80. The approximate value of $\int_1^5 x^2 dx$ using trapezoidal rule with $n = 4$ is

(a) 41 (b) 41.5
(c) 41.75 (d) 42

General English and Aptitude

Directions (1-4) *In each of the following questions, choose the alternative which best expresses the meaning of the word given in capital letters.*

1. MOSQUERADE
(a) to provide support
(b) to go in disguise
(c) to mesmerise
(d) marathon race

2. PREPOSTEROUS
(a) careful (b) casual
(c) absurd (d) deterrent

3. SOLITUDE
(a) musical composition
(b) aloneness
(c) true statement
(d) single mindedness

4. PROPITIOUS
(a) favourable (b) clean
(c) nearby (d) patriotic

Directions (5-8) *In each of the following questions, choose the alternative which is opposite in meaning to the word given in capital letters.*

5. RELENTLESS
(a) merciless (b) yielding
(c) monotonous (d) incisive

6. FORBEARANCE
(a) patience (b) self-control
(c) intolerance (d) preference

7. PALTRY
(a) absolete (b) cautious
(c) random (d) plentiful

8. EXODUS
(a) influx (b) return
(c) home coming (d) restoration

Directions (9-12) *In each of the following questions, choose the most suitable alternative to fill in the blank.*

9. The........argument put forth for not disclosing the facts did not impress anybody.
(a) intemperate (b) spurious
(c) specious (d) convincing

10. Director, he said, would.......the matter at once.
(a) invigilate (b) explore
(c) investigate (d) survey

11. Everyone was.........by surprise when she announced her plan to marry that boy.
(a) moved (b) shaken
(c) taken (d) prevailed

12. Your case is so unique that I am not getting any........to support it.
(a) reason (b) help
(c) happening (d) precedent

Directions (13-16) *In each of the following questions, choose the alternative which can be substituted for the given sentence.*

13. The writing or compiling of dictionaries
(a) Lexicography (b) Numismatics
(c) Cytology (d) Demography

14. The study of Insects
(a) Chromatics (b) Dactylology
(c) Calligraphy (d) Entomology

15. Murder of one's brother
(a) Regiside (b) Foeticide
(c) Fratricide (d) Uxoricide

16. Government by one person
(a) Autonomy (b) Autocracy
(c) Plutocracy (d) Theocracy

Directions (17-20) *In each of the following questions, choose the alternative which can best improve the given sentence by substituting the italicised portion. If the sentence is correct as it is, your answer is (d).*

17. She cut a *sad figure* in her first performance on the stage.
(a) made a sorry figure
(b) cut a sorry face
(c) cut a sorry figure
(d) no improvement

18. *No sooner I saw* the tiger, than I ran away.
(a) as soon as I saw
(b) no sooner I had seen
(c) no sooner did I see
(d) no improvement

19. If he *had time* he will call you.
(a) would have (b) would have had
(c) has (d) no improvement

20. *All his answers* were correct.
(a) his all answers (b) his every answers
(c) all of his answers (d) no improvement

Directions (21-25) *In the following questions, choose the group of words that shows the same relationship as given at the top of every question.*

21. Jam : Jelly : Pickles
(a) Butter : Marmalade : Grapes
(b) Granite : Basalt : Coke
(c) Cow : Dry : Draft
(d) Bleat : Bray : Grunt

22. Mountain : Height : Climber
(a) River : Length : Water
(b) Land : Farmer : Crop
(c) College : Building : Student
(d) Sea : Depth : Diver

23. Head : Brain : Think
(a) Eyes : Lashes : See
(b) Skin : Sweat : Touch
(c) Feet : Dance : Toe
(d) Mouth : Teeth : Chew

24. Jute : Cotton : Wool
(a) Potato : Carrot : Bean
(b) Canada : Chile : Asia
(c) Liver : Heat : Blood
(d) Shark : Cod : Eel

25. Sial : Sima : Mantle
(a) Core : Asteroid : Comet
(b) Pneumonia : Tetanus : Hepatitis
(c) Calcite : Magnesium : Zinc
(d) Wrestling : Karate : Boxing

Directions (26-30) *In the following questions, choose the option which shows common feature in the relationship given in each question.*

26. Rourkela : Bokaro : Durgapur
(a) They have steel plants
(b) They have coal mines
(c) They have atomic power plants
(d) They have the best technical colleges

27. Chlorine : Fluorine : Iodine
(a) These are names of inert gases
(b) These are gases at room temperature
(c) These are transition elements
(d) These are halogens

28. Petrol : Phosphorus : Cooking gas
(a) They are fuels
(b) They are highly inflamable
(c) They can be sold without permit
(d) India has to import them

29. Green : Violet : Orange
(a) They are primary colours
(b) These colours occur together in a rainbow
(c) They are made by mixing other colours
(d) These colours are not found in butterflies

30. Supernova : Protostar : Red Giant
(a) These are kinds of stars
(b) These are members of galaxies
(c) These are stages in the life of a star
(d) These move about the Sun

31. The classical dance, Kathakali belongs to
(a) Tamil Nadu (b) Karnataka
(c) Orissa (d) Kerala

32. The highest civilian award in India is
(a) Paramveer Chakra
(b) Bharat Ratna
(c) Bhatnagar Award
(d) Kalinga Award

33. Who is called 'Nightingale of India'?
(a) Sarojini Naidu (b) Lata Mangeshkar
(c) Amir Khushro (d) Asha Bhosale

34. Nehru Trophy is related with
(a) Cricket (b) Volleyball
(c) Hockey (d) Badminton

35. The headquarters of NATO is situated at
(a) Brussels (b) Geneva
(c) Newyork (d) Paris

36. The first Indian woman to swim across the English channel is
(a) Bachendri Pal (b) Santosh Yadav
(c) Nirja Mishra (d) Aarti Saha

37. Which of the following river is related with Bhakra-Nagal Project?
(a) Ravi (b) Satluj
(c) Gandak (d) Mahanadi

38. Namdhapha National Park is located in
(a) Rajasthan
(b) Orissa
(c) Arunachal Pradesh
(d) Uttar Pradesh

39. Which of the following is known as 'evening star'?
(a) Venus (b) Mercury
(c) Saturn (d) Jupiter

40. Which of the following was the founder of Indian National Congress?
(a) Jawahar Lal Nehru
(b) Mahatma Gandhi
(c) Moti Lal Nehru
(d) A. O. Hume

Answers

Physics

1. (b)	2. (c)	3. (c)	4. (c)	5. (d)	6. (a)	7. (b)	8. (a)	9. (a)	10. (b)
11. (b)	12. (c)	13. (a)	14. (b)	15. (a)	16. (b)	17. (b)	18. (a)	19. (b)	20. (b)
21. (a)	22. (b)	23. (b)	24. (b)	25. (a)	26. (c)	27. (d)	28. (b)	29. (a)	30. (d)
31. (a)	32. (b)	33. (b)	34. (c)	35. (b)	36. (c)	37. (a)	38. (d)	39. (b)	40. (c)
41. (a)	42. (b)	43. (b)	44. (a)	45. (b)	46. (a)	47. (b)	48. (b)	49. (d)	50. (a)
51. (c)	52. (d)	53. (b)	54. (c)	55. (b)	56. (a)	57. (b)	58. (c)	59. (b)	60. (c)

Chemistry

1. (a)	2. (d)	3. (b)	4. (d)	5. (c)	6. (d)	7. (c)	8. (d)	9. (b)	10. (d)
11. (c)	12. (a)	13. (b)	14. (b)	15. (d)	16. (d)	17. (c)	18. (c)	19. (d)	20. (a)
21. (b)	22. (b)	23. (d)	24. (a)	25. (d)	26. (a)	27. (c)	28. (b)	29. (d)	30. (c)
31. (a)	32. (d)	33. (d)	34. (a)	35. (b)	36. (c)	37. (b)	38. (a)	39. (d)	40. (a)
41. (a)	42. (a)	43. (b)	44. (b)	45. (b)	46. (b)	47. (a)	48. (d)	49. (c)	50. (a)
51. (b)	52. (a)	53. (d)	54. (a)	55. (d)	56. (b)	57. (d)	58. (d)	59. (d)	60. (b)

Mathematics

1. (b)	2. (b)	3. (c)	4. (c)	5. (a)	6. (b)	7. (d)	8. (a)	9. (d)	10. (b)
11. (c)	12. (d)	13. (b)	14. (b)	15. (c)	16. (d)	17. (a)	18. (c)	19. (a)	20. (d)
21. (a)	22. (b)	23. (b)	24. (a)	25. (b)	26. (c)	27. (a)	28. (a)	29. (a)	30. (b)
31. (a)	32. (b)	33. (b)	34. (c)	35. (a)	36. (b)	37. (a)	38. (c)	39. (d)	40. (c)
41. (d)	42. (a)	43. (c)	44. (a)	45. (a)	46. (a)	47. (a)	48. (a)	49. (d)	50. (b)
51. (b)	52. (c)	53. (a)	54. (d)	55. (c)	56. (b)	57. (a)	58. (d)	59. (d)	60. (c)
61. (b)	62. (d)	63. (d)	64. (a)	65. (b)	66. (a)	67. (c)	68. (a)	69. (a)	70. (b)
71. (b)	72. (a)	73. (b)	74. (b)	75. (b)	76. (b)	77. (a)	78. (c)	79. (b)	80. (d)

General English and Aptitude

1. (b)	2. (c)	3. (b)	4. (a)	5. (b)	6. (c)	7. (d)	8. (c)	9. (c)	10. (b)
11. (d)	12. (c)	13. (a)	14. (d)	15. (c)	16. (b)	17. (c)	18. (c)	19. (c)	20. (d)
21. (d)	22. (d)	23. (d)	24. (d)	25. (d)	26. (d)	27. (d)	28. (b)	29. (b)	30. (c)
31. (d)	32. (b)	33. (a)	34. (c)	35. (a)	36. (d)	37. (b)	38. (c)	39. (a)	40. (d)

Hints & Solutions

Physics

1. Field due to a straight wire of infinite length is $\frac{\mu_0 i}{4\pi r}$ if the point is on a line perpendicular to its length while at the centre of a semicircular coil is $\frac{\mu_0 \pi i}{4\pi r}$.

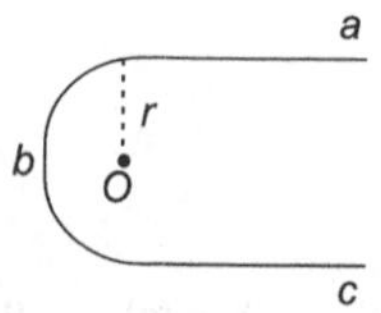

$$\therefore \quad B = B_a + B_b + B_c$$
$$= \frac{\mu_0}{4\pi}\frac{i}{r} + \frac{\mu_0}{4\pi}\frac{\pi i}{r} + \frac{\mu_0}{4\pi}\frac{i}{r}$$
$$= \frac{\mu_0}{4\pi}\frac{i}{r}(\pi + 2) \text{ out of the page}$$

2. As 3Ω and 6Ω resistances are in parallel their equivalent resistance will be 2Ω. Here 2Ω and 4Ω are in series, their equivalent resistance will be 6Ω.

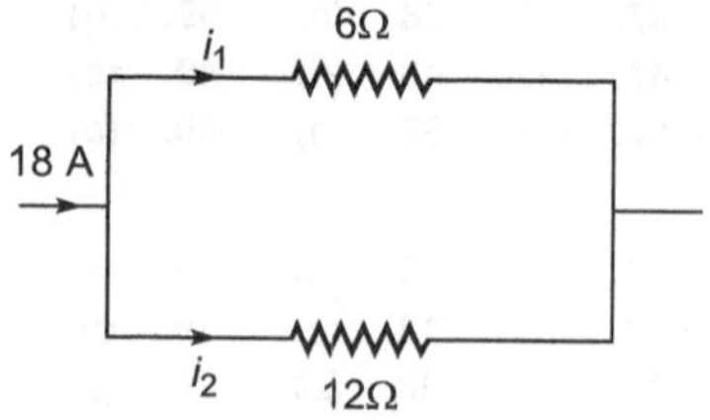

From current distribution law

$$i_1 = \frac{12 \times 18}{18} = 12\text{ A}$$
$$i_2 = \frac{6 \times 18}{18} = 6\text{ A}$$

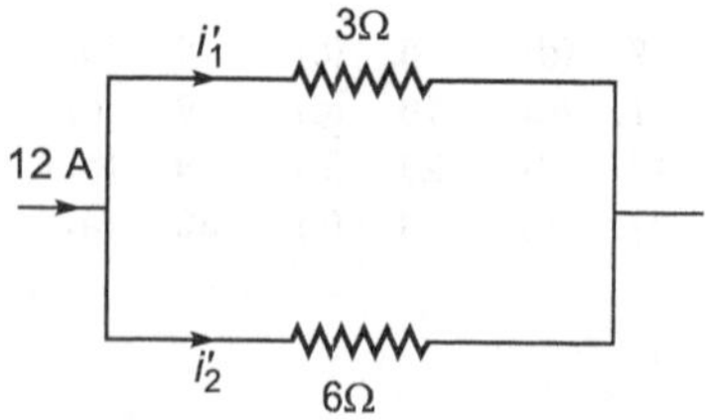

Now 12 A current is entering in parallel combination of 3Ω and 6Ω again from current distribution law

$$i_1' = \frac{6 \times 12}{9} = 8\text{ A}$$
$$i_2' = \frac{3 \times 12}{9} = 4\text{ A}$$

$\therefore$ Potential difference across 3Ω resistance

$$= 8 \times 3 = 24\text{ V}$$

3. Let the temperature of junction be θ. In equilibrium, rate of flow of heat through rod 1 is equal to sum of rate of flow of heat through rods 2 and 3, *ie*,

$$\left(\frac{dQ}{dt}\right)_1 = \left(\frac{dQ}{dt}\right)_2 + \left(\frac{dQ}{dt}\right)_3$$
$$\therefore \quad \frac{KA(\theta - 20)}{l} = \frac{KA(60 - \theta)}{l} + \frac{KA(70 - \theta)}{l}$$
$$\Rightarrow \quad \theta - 20 = 130 - 2\theta$$
$$\text{or} \quad 3\theta = 150$$
$$\therefore \quad \theta = 50°\text{C}$$

4. The surface tension of liquids decreases with increase of temperature. For small temperature difference it decreases almost linearly. The surface tension of a liquid becomes zero at a particular temperature, called the critical temperature of that liquid.

5. Current amplification factor, $\beta = \frac{i_C}{i_B}$

$$\therefore \quad i_C = i_B \times \beta$$
$$= 250 \times 80\,\mu\text{A}$$

6. Here, $\alpha = 2\theta$, $\beta = \theta$

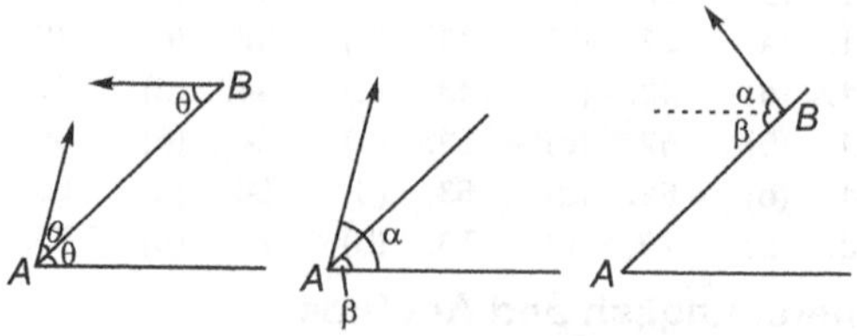

Time of flight of A is,

$$T_1 = \frac{2u \sin(\alpha - \beta)}{g \cos\beta}$$
$$= \frac{2u \sin(2\theta - \theta)}{g \cos\theta}$$
$$= \frac{2u}{g}\tan\theta$$

Time of flight of B is, $T_2 = \frac{2u \sin\theta}{g\cos\theta}$

$$= \frac{2u}{g}\tan\theta$$

So, $T_1 = T_2$. The acceleration of both the particles is g downwards. Therefore, relative acceleration between the two is zero or relative motion between the two is uniform. The relative velocity of A w.r.t. B is towards AB, therefore collision will take place between the two in mid air.

7. $R_{eq} = 5\Omega$, current $i = \frac{10}{5} = 2\text{A}$ and current in each branch = 1 A. Potential difference between C and A,

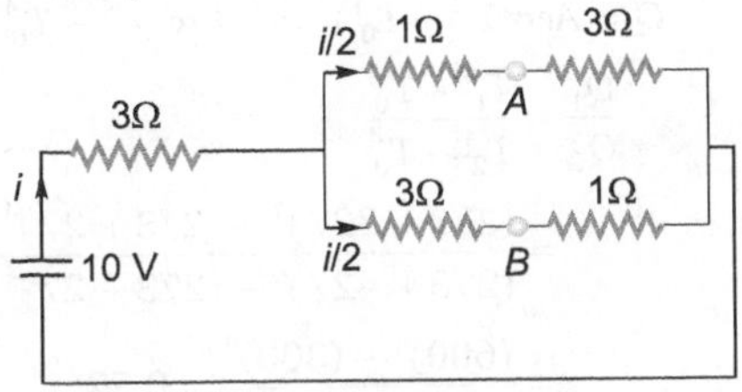

$$V_C - V_A = 1 \times 1 = 1\text{ V} \quad \text{...(i)}$$

Potential difference between C and B

$$V_C - V_B = 1 \times 3 = 3\text{V} \quad \text{...(ii)}$$

On solving Eqs. (i) and (ii) we get

$$V_A - V_B = 2\text{V}$$

8. Solar radiations are transverse electromagnetic waves. The central core of the sun emits a continuous electromagnetic spectrum.

9. If an open pipe is half submerged in water, it will become a closed organ pipe of length half that of open pipe as shown in figure. So, its frequency will become,

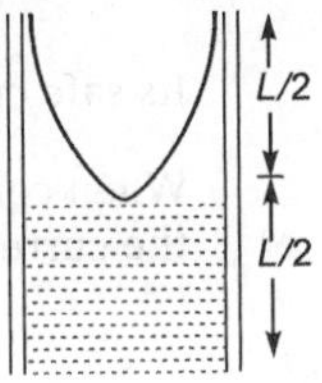

$$f_c = \frac{v}{4\left(\frac{L}{2}\right)} = \frac{v}{2L} = f_o$$

ie, equal to that of open pipe *ie*, frequency will remain unchanged.

10. Rms velocity is given by

$$v = \sqrt{\frac{3kT}{m}} \quad \text{or} \quad v^2 = \frac{3kT}{m}$$

For a gas, k and m are constants.

$$\therefore \quad \frac{v^2}{T} = \text{constant}$$

11.

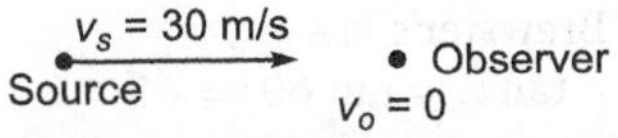

Here, $n = 500$ Hz, $v_o = 0$

$v_s = 30$ m/s, $v = 330$ m/s

$$\therefore \text{ From, } \quad n' = n\left(\frac{v - v_o}{v - v_s}\right)$$

$$= 500\left(\frac{330}{330 - 30}\right)$$

$$= 550 \text{ Hz}$$

12. The value of acceleration due to gravity at height h (when h is not negligible as compared to R)

$$g' = g\frac{R^2}{(R+h)^2}$$

Here, $$g' = \frac{g}{2}$$

$$\therefore \quad \frac{g}{2} = g\frac{R^2}{(R+h)^2}$$

or $$\frac{1}{2} = \frac{R^2}{(R+h)^2}$$

or $$\sqrt{\frac{1}{2}} = \frac{R}{R+h}$$

or $$R + h = \sqrt{2}R$$

$$\therefore \quad h = (\sqrt{2} - 1)R$$

13. For working safely, the activity must reduce to $\frac{1}{64}$.

$$\frac{N}{N_0} = \left(\frac{1}{2}\right)^n = \frac{1}{64}$$

$$\therefore \quad n = 6$$

Thus, $$t = nT = 6 \times 2 = 12\text{ h}$$

14. Potential difference across diode

$$= 3.2 - 3 = 0.2\text{ V}$$

$\therefore$ Current through diode

$$i = \frac{0.2}{100} = 2 \times 10^{-3}\text{ A}$$

15.

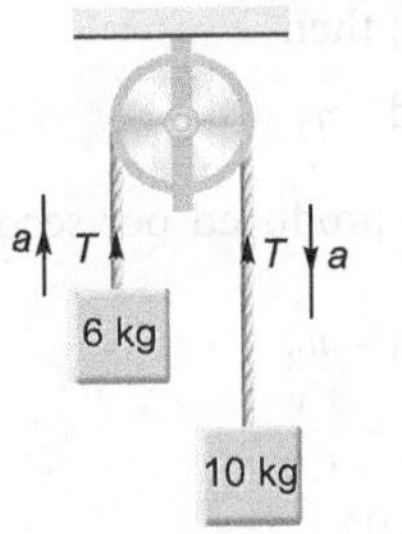

$$10g - T = 10a \quad \text{...(i)}$$

$$T - 6g = 6a \quad \text{...(ii)}$$

From Eqs. (i) and (ii), we get

$$T = 75\text{N}$$

16. As the flask floats in water when less than half filled with water, it will float just fully submerged when half filled. In this situation, mass of flask + mass of water in it = $V\sigma$

ie, $390 + 250 = V \times 1$ (as $\sigma = 1$ g/cc)

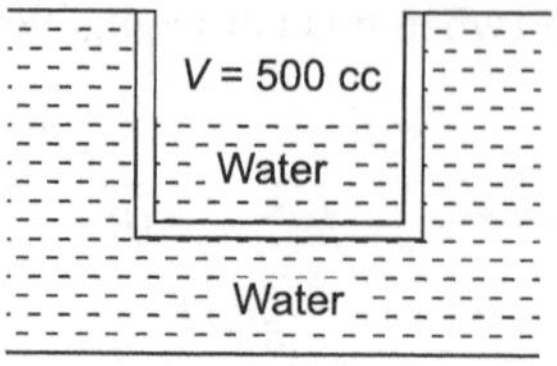

ie, outer volume of flask

$$V = 640 \text{ cc}$$

Now as inner volume of flask is given to be 500 cc, so the volume of the material of flask $= 640 - 500 = 140$ cc. But as mass of flask is 390 g, so density of material of flask

$$\rho = \frac{m}{V} = \frac{390}{140} = 2.8 \text{ g/cc}$$

17. $A = 60°, \delta_m = 30°$

As $$\mu = \frac{\sin\left(\frac{A + \delta_m}{2}\right)}{\sin\left(\frac{A}{2}\right)}$$

$\therefore$ $$\mu = \frac{\sin\left(\frac{60° + 30°}{2}\right)}{\sin\left(\frac{60°}{2}\right)} = \frac{\sin 45°}{\sin 30°}$$

$$= \frac{1/\sqrt{2}}{1/2} = \frac{2}{\sqrt{2}} = \sqrt{2}$$

18. Let n_1 and n_2 be the frequencies of the two notes of wavelengths λ_1 and λ_2.

Here, $\lambda_1 = 1$ m, $\lambda_2 = 1.01$ m

If v is velocity of sound, then

$$n_1 = \frac{v}{1} \quad \text{and} \quad n_2 = \frac{v}{1.01}$$

If the number of beats produced per second is b, then

$$b = n_1 - n_2$$

$$4 = \frac{v}{1} - \frac{v}{1.01}$$

$\therefore$ $$v = \frac{4.04}{0.01} = 404 \text{ m/s}$$

19. Work done = area of trapezium

$$= \frac{1}{2} \times (8 \times 10^5 + 4 \times 10^5) \times 0.2$$

$$= 1.2 \times 10^5 \text{ J}$$

20. As $F \propto v^2$

or $$F = kv^2$$

$\therefore$ $$k = \frac{F}{v^2}$$

$$[k] = \frac{[F]}{[v^2]} = \frac{[\text{MLT}^{-2}]}{[\text{L}^2\text{T}^{-2}]}$$

$$= [\text{ML}^{-1}\text{T}^0]$$

21. If temperature of surrounding is considered, then net loss of energy of a body by radiation

$$Q = Ae\sigma(T^4 - T_0^4) \Rightarrow Q \propto (T^4 - T_0^4)$$

$\therefore$ $$\frac{Q_1}{Q_2} = \frac{T_1^4 - T_0^4}{T_2^4 - T_0^4}$$

$$= \frac{(273 + 327)^4 - (273 + 27)^4}{(273 + 427)^4 - (273 + 27)^4}$$

$$= \frac{(600)^4 - (300)^4}{(700)^4 - (300)^4} = 0.52$$

22. Resistance of 40 W bulb $= \frac{240 \times 240}{40}$

$$= 1440\ \Omega$$

Its safe current $= \frac{240}{1440} = 0.167$ A

Resistance of 60 W bulb $= \frac{240 \times 240}{60}$

$$= 960\ \Omega$$

Its safe current $= \frac{240}{960} = 0.25$ A

When connected in series to 420 V supply, then the current

$$i = \frac{420}{1440 + 960} = \frac{420}{2400}$$

$$= 0.175 \text{ A}$$

Thus, current is greater for 40 W bulb, so it will fuse.

23. Angular velocity ω is constant.

$\therefore$ $$v \propto r \text{ or } \frac{v_A}{v_B} = \frac{r_A}{r_B}$$

24. According to equation of motion

$$v = u + at$$

$\Rightarrow$ $$-2 = 10 + a \times 4$$

$\therefore$ $$a = -3 \text{ m/s}^2$$

25. According to Brewster's law

$$\mu = \tan i_p = \tan 60° = \sqrt{3}$$

26. Young's modulus, $Y = \frac{mgl}{a_1 l_1}$

$\therefore$ $$l_1 = \frac{mgl}{Y\pi r^2} \quad \text{...(i)}$$

and $Y = \frac{mg(2l)}{a_2 l_2} = \frac{mg(2l)}{\pi(2r)^2 l_2}$

or $l_2 = \frac{mgl}{2Y\pi r^2}$ ...(ii)

From Eqs. (i) and (ii), we have

$\therefore \quad l_1 + l_2 = \frac{mgl}{Y\pi r^2} + \frac{mgl}{2Y\pi r^2} = \frac{3}{2}\frac{mgl}{Y\pi r^2}$

27. Given, $x = 0.20$ m, $y = 0.20$ m, $u = 1.8$ m/s.
Let the ball strikes the nth step of stairs,
Vertical distance travelled

$= ny = n \times 0.20 = \frac{1}{2}gt^2$

Horizontal distance travelled, $nx = ut$

or $t = \frac{nx}{u}$

$\therefore \quad ny = \frac{1}{2}g \times \frac{n^2x^2}{u^2}$

or $n = \frac{2u^2}{g}\frac{y}{x^2} = \frac{2 \times (1.8)^2 \times 0.20}{9.8 \times (0.20)^2} = 3.3 \approx 4$

28. $E = E_0 \cos \omega t = 10 \cos(2\pi \times ft)$

$= 10 \cos\left(2\pi \times 50 \times \frac{1}{600}\right)$

$= 10 \cos\left(\frac{\pi}{6}\right) = 10 \times \frac{\sqrt{3}}{2} = 5\sqrt{3}$ V

29. Induced emf produced in coil

$e = \frac{-d\phi}{dt} = \frac{-d}{dt}(BA)$

$\therefore \quad |e| = A\frac{dB}{dt} = 0.01 \times \frac{1}{1 \times 10^{-3}}$

$|e| = 10$ V

Current produced in coil,

$i = \frac{|e|}{R} = \frac{10}{2} = 5$ A

Heat evolved $= i^2Rt = (5)^2 \times (2) \times 1 \times 10^{-3}$

$= 0.05$ J

30. From displacement method size of object, $O = \sqrt{I_1 I_2}$

Here, $O = 3$ cm, $I_1 = 9$ cm,

$\therefore \quad 3 = \sqrt{9I_2}$

or $I_2 = 1$ cm

31. As gas is suddenly expanded so it is an adiabatic process,

ie, $pV^\gamma =$ constant

or $p_1V_1^\gamma = p_2V_2^\gamma$

Given $V_2 = 3V_1$, $C_V = 2R$

$\therefore \quad C_p = 2R + R = 3R$

$\Rightarrow \quad \gamma = \frac{C_p}{C_V} = \frac{3R}{2R} = 1.5$

$\therefore \quad \frac{p_1}{p_2} = \left(\frac{V_2}{V_1}\right)^\gamma = (3)^{1.5} = 5.1 \approx 5$

32. When bigger pendulum of time period $\frac{5T}{4}$ completes one vibration, the smaller pendulum will complete $\left(\frac{5}{4}\right)$ vibrations. It means the smaller pendulum will be leaded the bigger pendulum by phase $\frac{T}{4}$ sec $= \frac{\pi}{2}$ rad $= 90°$.

33. Given, $u = 72$ km/h $= 20$ m/s

$a = \mu g = 0.5 \times 10$ m/s^2

From $v^2 = u^2 - 2as$

$\therefore \quad (0)^2 = (20)^2 - 2 \times 0.5 \times 10 \times s$

$\therefore \quad s = \frac{20 \times 20}{2 \times 0.5 \times 10}$

or $s = 40$ m

34. Given, $s = 2$ m, $u = 80$ m/s, $v = 0$

From $v^2 = u^2 - 2as$

$\therefore \quad (0)^2 = (80)^2 - 2 \times a \times 2$

or $a = \frac{80 \times 80}{4} = 1600$ m/s^2

35. Net induced emf due to disc is

$\varepsilon = \frac{1}{2}B\omega l^2$

$= \frac{1}{2} \times 0.1 \times 2\pi \times 10 \times 0.1 \times 0.1$

$= \pi \times 10^{-2}$ V

36. According to first law of thermodynamics

$\Delta Q = \Delta U + \Delta W$

$\therefore \quad \Delta U = \Delta Q - \Delta W$

$= 540 - \frac{p(V_2 - V_1)}{J}$

$= 540 - \frac{1.013 \times 10^5 \times [(1671 - 1) \times 10^{-6}]}{4.2}$

$= 540 - 40 = 500$ cal

37. Frequency of vibration is given by

$n = \frac{p}{2l}\sqrt{\frac{T}{m}} \quad \left(\text{where } v = \sqrt{\frac{T}{m}}\right)$

$\therefore \quad n = \frac{pv}{2l} = \frac{5 \times 20}{2 \times 10} = 5$ Hz

38. $v_{max} = A\omega$

when A becomes twice v_{max} is also doubled.

39. From Kepler's third law of planetary motion

$$T^2 \propto R^3$$

$$\therefore \quad \frac{T_2^2}{T_1^2} = \frac{R_2^3}{R_1^3}$$

$$\text{or} \quad \frac{T_2^2}{(24)^2} = \left(\frac{6400 + 6400}{36000 + 6400}\right)^3$$

$$\text{or} \quad T_2^2 = (24)^2 \times \left(\frac{16}{53}\right)^3$$

$$\Rightarrow \quad T_2 = 4\,\text{h}$$

40. From the law of conservation of momentum

$$3 \times 16 + 6 \times v = 9 \times 0$$

$$\text{or} \quad v = -8\,\text{m/s}$$

$$\Rightarrow \quad v = 8\,\text{m/s} \quad \text{(numerically)}$$

Therefore, its kinetic energy

$$K = \frac{1}{2} \times 6 \times (8)^2 = 192\,\text{J}$$

41. In the reaction

$${}_7\text{N}^{14} + \alpha \longrightarrow {}_8X^{17} + {}_1p^1$$

8 is the atomic number of oxygen molecule.
So, here X is oxygen (O_2) molecule.

42. Induced emf in the coil is given by

$$e = \frac{Ldi}{dt}$$

$$\text{or} \quad e = 10 \times \frac{(10 - 5)}{0.2}$$

$$\Rightarrow \quad e = 250\,\text{V}$$

43. New force $F = \dfrac{1}{4\pi\varepsilon_0} \dfrac{(6-2)(2-2)}{r^2} \times 10^{-12}$

$$= 0$$

44. Given, $N_P = 20$, $N_S = 10$, $e_P = 220$ V

$\therefore$ Transformation ratio, k

$$\frac{e_S}{e_P} = \frac{N_S}{N_P}$$

$$\text{or} \quad e_S = \frac{N_S}{N_P} \times e_P$$

$$= \frac{10}{20} \times 220 = 110\,\text{V}$$

45. As $\quad E_1 > E_2$

$\therefore \quad \nu_1 > \nu_2$

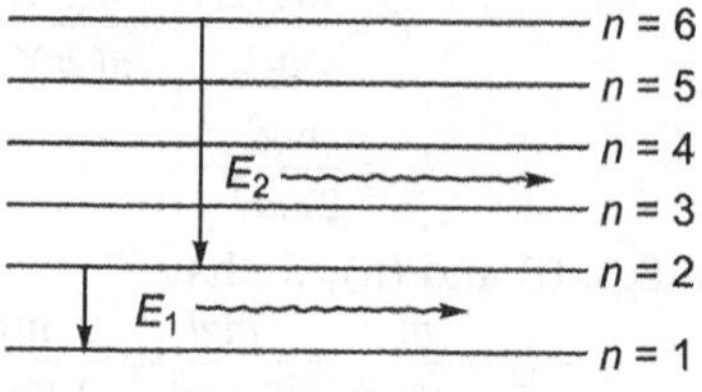

ie, photons of higher frequency will be emitted if transition takes place from $n = 2$ to $n = 1$.

46. Here,

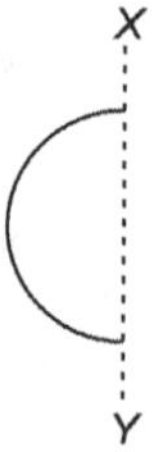

$$L = \pi R \text{ or } R = \frac{L}{\pi}$$

$\therefore$ Moment of inertia about XY

$$= \frac{1}{2}\left(\frac{1}{2}MR^2\right)$$

$$= \frac{1}{4}\frac{ML^2}{\pi^2}$$

47. Let a be the acceleration of each block. Then,

$$T_3 = (m_1 + m_2 + m_3)a \quad \ldots\text{(i)}$$

$$\text{and} \quad T_2 = (m_1 + m_2)a \quad \ldots\text{(ii)}$$

From Eqs. (i) and (ii), we get

$$T_2 = \left(\frac{m_1 + m_2}{m_1 + m_2 + m_3}\right) \times T_3$$

$$= \left(\frac{10 + 6}{10 + 6 + 4}\right) \times 40 = 32\,\text{N}$$

48. $H = \dfrac{u^2 \sin^2\theta}{2g}$ and $T = \dfrac{2u\sin\theta}{g}$

$$\Rightarrow \quad T^2 = \frac{4u^2\sin^2\theta}{g^2}$$

$$\therefore \quad \frac{T^2}{H} = \frac{8}{g}$$

$$\Rightarrow \quad T = \sqrt{\frac{8H}{g}} = 2\sqrt{\frac{2H}{g}}$$

49. The net capacitance $C = \dfrac{\varepsilon_0 A}{\dfrac{d_1}{K_1} + \dfrac{d_2}{K_2}}$

$$= \frac{\varepsilon_0 A}{\dfrac{d}{2}\left(\dfrac{1}{1} + \dfrac{1}{2}\right)} = \frac{4\varepsilon_0 A}{3d}$$

50. Effective spring constant of parallel combination

$$k_e = k_1 + k_2$$

51. $T \sin 30° = 2$ kg-wt

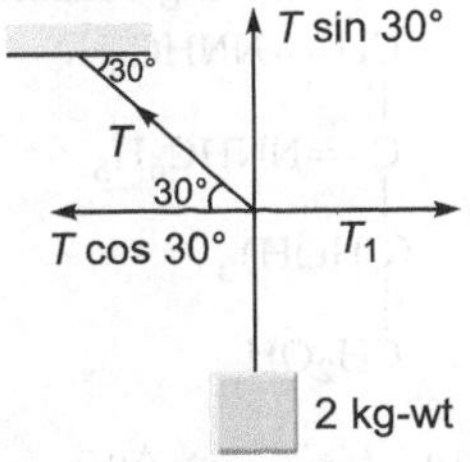

$\Rightarrow \quad T = 4$ kg-wt

$T_1 = T \cos 30°$

$= 4 \cos 30° = 2\sqrt{3}$

52. $\text{KE} = \frac{p^2}{2m} \Rightarrow \text{KE}_2 = \text{KE}_1 \left(\frac{p_2}{p_1}\right)^2 = \text{KE}_1 \left(\frac{2p}{p}\right)^2$

$\Rightarrow \quad \text{KE}_2 = 4\text{KE}_1$

$= \text{KE}_1 + 3\text{KE}_1 = \text{KE}_1 + 300\%$ of KE_1

53. Angular width $\theta = \frac{\beta}{D}$

$= \frac{\lambda}{d} = \frac{4800 \times 10^{-10}}{0.6 \times 10^{-3}}$

$= 8 \times 10^{-4}$ rad

54. Potential difference $V = \frac{hc}{e\lambda}$

$= \frac{6.6 \times 10^{-34} \times 3 \times 10^{8}}{1.6 \times 10^{-19} \times 2 \times 10^{-10}}$

$= 6200$ volt

55. Loss in kinetic energy

$= \frac{1}{2} \frac{m_1 m_2 (u_1 - u_2)^2}{(m_1 + m_2)}$

$= \frac{1}{2} \frac{m \cdot m (u_1 - u_2)^2}{(m + m)}$

$= \frac{m}{4} (u_1 - u_2)^2$

56. Torque $\vec{\tau} = \frac{d\vec{L}}{dt} = \frac{L_2 - L_1}{\Delta t} = \frac{4A_0 - A_0}{4} = \frac{3A_0}{4}$

57. When a number of small droplets coalesce to form a bigger drop surface energy is released because its surface area decreases.

58. If man slides down with some acceleration, then its apparent weight decreases. For critical condition rope can bear only 2/3 of his weight. If a is the minimum acceleration, then

tension in the rope $= m(g - a)$

$= $ breaking strength

$\Rightarrow \quad m(g - a) = \frac{2}{3} mg$

$\Rightarrow \quad a = g - \frac{2g}{3} = \frac{g}{3}$

59. Resulting amplitude of two interfering waves

$A = \sqrt{a_1^2 + a_2^2 + 2a_1 a_2 \cos \phi}$

Here, $\phi = \pi/3$, $a_1 = 4$, $a_2 = 3$

$\Rightarrow \quad A \simeq 6$

60. The saturation vapour pressure is the static pressure of a vapour when the vapour pressure of some material is in equilibrium with the liquid phase of that material. The saturation vapour pressure of any material is solely dependent on the temperature of that material. Since, temperature change does not occur in the given case, hence pressure will be the same.

Chemistry

1. $v_{rms} = \sqrt{\frac{3RT}{M}}$

$\therefore \quad v_{rms} \propto \sqrt{T}$

$\therefore$ At two different temperatures,

$\frac{v_{rms}}{v'_{rms}} = \sqrt{\frac{T}{T'}}$

Given, $\quad v'_{rms} = 2v_{rms}$

$\frac{1}{2} = \sqrt{\frac{T}{T'}}$

or $\quad \frac{1}{4} = \frac{T}{T'}$

$\therefore \quad T' = 4T$

$\therefore v_{rms}$ gets doubled, when the temperature is increased four times.

2. All the reaction proceed by stable ions. After the lose of H^+ ion, phenol forms phenoxide ion. The phenoxide ion is resonance stabilized, thus makes the phenol more acidic.

3. O^{2-} and N^{3-} both are isoelectronic but differ in the charge possessed by them. As the negative charge increases, the electrons are held less and less tightly by the nucleus, therefore ionic radii increases. Hence, ionic radii of N^{3-} is greater than O^{2-}.

In a period from left to right atomic radii decreases but in a group on moving downwards it increases.

4. According to Bohr model,
radius of hydrogen atom

$$(r_n) = \frac{0.529 \times n^2}{Z} \text{ Å}$$

(where, n = number of orbit, Z = atomic number)

$$r_3 = \frac{0.529 \times (3)^2}{1} = 4.761 \text{ Å}$$

5. Co^{3+} and Pt^{4+} have 6 coordination number.
$CoCl_3 \cdot 6NH_3$ and $PtCl_4 \cdot 5NH_3$

$$[Co(NH_3)_6]Cl_3 \xrightarrow{\text{In solution}} [Co(NH_3)_6]^{3+} + 3Cl^-$$

$$[PtCl(NH_3)_5]Cl_3 \xrightarrow{\text{In solution}} [PtCl(NH_3)_5]^{3+} + 3Cl^-$$

Number of ionic species are same in the solution of both complexes, therefore their equimolar solutions will show same conductance.

6. Entropy is the measure of randomness. In liquids randomness is more than solids.
∵ When ice melts, randomness increases, (solid $\longrightarrow$ liquid)
∴ Entropy increases.

7. Reduction of nitrobenzene by Zn and NH_4Cl gives N-phenyl hydroxylamine.

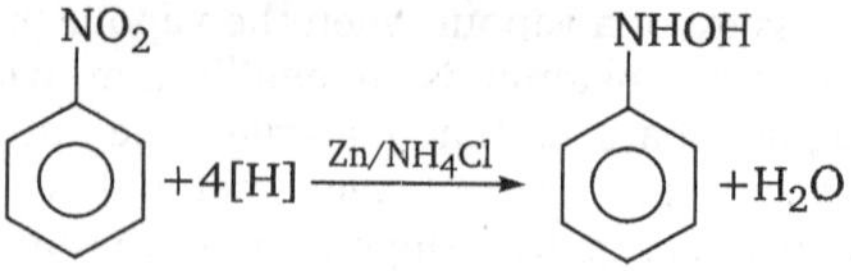

N-phenyl hydroxylamine

8. Proteins are the polymer of amino acids and contain polypeptide bonds. DNA, however, are the polymers of nucleotide and contain no polypeptide bond. Hence, DNA is not a protein.

9. D-glucose, D-fructose and D-mannose form the same osazone when treated with excess of phenyl hydrazine because they differ only in Ist and 2nd carbon atoms which are transformed to the same form.

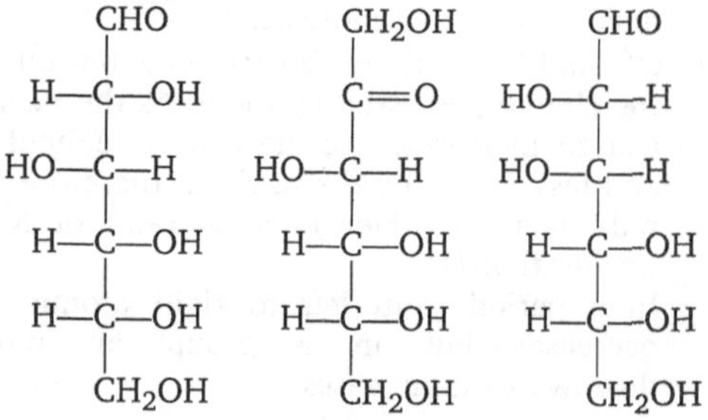

D-glucose D-fructose D-mannose

They form the following osazone

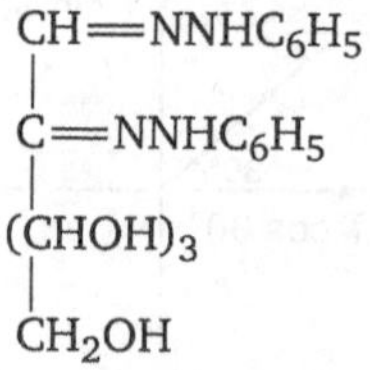

10. $$Ag^+ + e^- \longrightarrow Ag$$

∵ 96500 C are required to deposite Ag $= 108$ g

∴ 965 C are required to deposite Ag

$$= \frac{108}{96500} \times 965$$

$$= 1.08 \text{ g}$$

11. $$E°_{cell} = \frac{2.303\, RT}{nF} \log K_{eq}$$

$$E°_{cell} = \frac{0.0591}{n} \log K_{eq} \quad \text{(At 298 K)}$$

$$0.591 = \frac{0.0591}{1} \log K_{eq}$$

$$\therefore \log K_{eq} = 10$$

$$\therefore K_{eq} = 1 \times 10^{10}$$

12. Higher the negative value of $E°$, more is the reducing power. The order of $E°$ values (negative value) is

$$\underset{(Mg)}{-2.37} > \underset{(Zn)}{-0.76} > \underset{(Fe)}{-0.44}$$

∴ Mg can reduce both Zn^{2+} and Fe^{2+}. Zn can reduce Fe^{2+}, but not Mg^{2+}. Fe cannot reduce Mg and Zn but can oxidise them.

13. In diamond,
the maximum proportion of available volume that can be filled by hard spheres $= \frac{\pi\sqrt{3}}{16}$
$= 0.34$

14. On adding a pentavalent impurity with germanium, we get n-type of semiconductors because excess of electrons is responsible for conduction.

15. The complex in which nd orbitals are used in hybridisation, are called outer orbital complex.
(a) $[Fe(CN)_6]^{4-}$ =

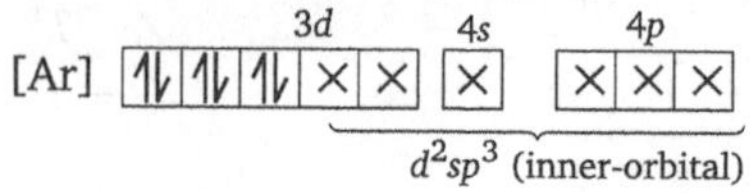

(b) $[Mn(CN)_6]^{4-}$ =

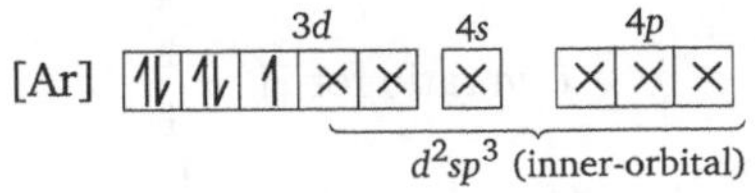

(c) $[Co(NH_3)_6]^{3+}$ =

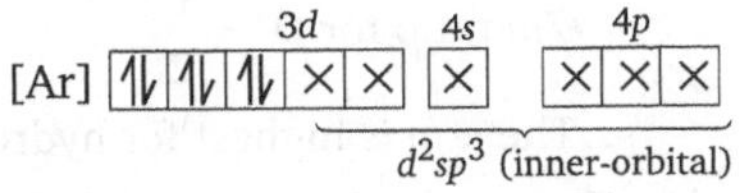

(d) $[Ni(NH_3)_6]^{2+}$ =

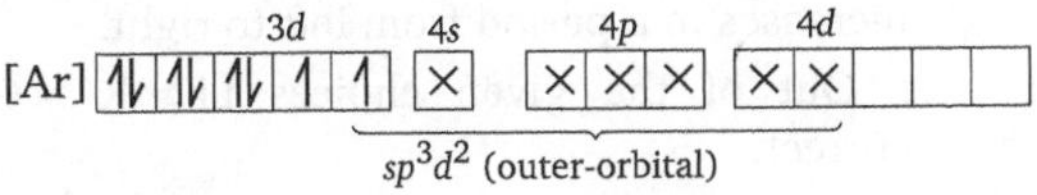

16. $C_2H_5Br + AgNO_2\ (alc.) \rightarrow C_2H_5NO_2 + AgBr$
nitro ethane

17. *Iso*-propyl chloride is a 2° halide and 2° halides can undergo hydrolysis either by S_N1 or S_N2 mechanism depending upon the nature of solvent used.

18.

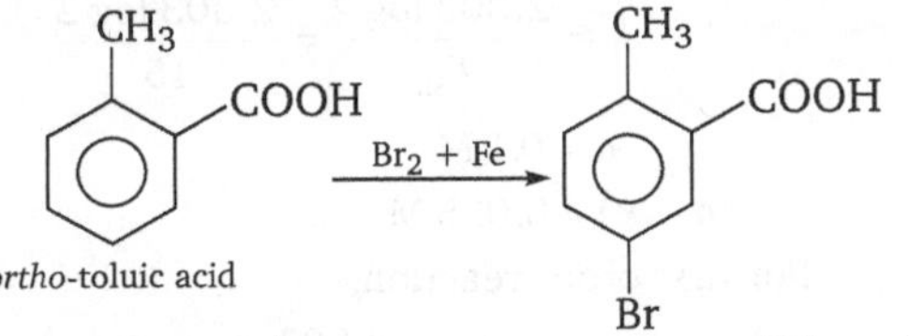

ortho-toluic acid

($\because$ In the product, —Br is *para* to —CH_3 and *meta* to —COOH.)

19. —NO_2 group at any position shows electron withdrawing effect, thus acid strength is increased. But *o*-nitro benzoic acid believed to have *ortho* effect. As a result, resonance gets prevented. Hence, its acid strength is maximum, thus, the order of acid strength (II) < (III) > (IV) > (I)
(The effect is more at *para* position than *meta*.)

20. Iodoform test is given by the compounds containing either CH_3CO—group or $CH_3\overset{|}{C}HOH$ group.
The structures of the given compounds are as
(a) $CH_3CH_2CH_2CH_2OH$
(b) $CH_3COC_6H_5$
(c) CH_3CHO
(d) $CH_3COC_2H_5$
$\therefore$ *n* butyl alcohol does not give iodoform test because it does not possess the CH_3CO—or $CH_3\overset{|}{C}HOH$ group.

21. $A + NaOH \longrightarrow$ alcohol + acid
Thus, it is Cannizaro reaction.
A is thus aldehyde without H at α-carbon.
(like C_6H_5CHO, HCHO)
$$\underset{\text{benzaldehyde}}{2C_6H_5CHO} + NaOH \longrightarrow C_6H_5CH_2OH + C_6H_5COONa$$

22. $\overset{5}{C}H_3—\underset{OH}{\overset{4}{C}H}—\overset{3}{C}H=\underset{CH_3}{\overset{2}{C}}—\overset{1}{C}HO$
4-hydroxy-2-methylpent-2-en-1-al

23. $E_a\ (A \rightarrow B) = 80\ \text{kJ mol}^{-1}$
Heat of reaction $(A \rightarrow B) = 200\ \text{kJ mol}^{-1}$
For $(B \rightarrow A)$ backward reaction,
$E_a\ (B \rightarrow A) = E_a\ (A \rightarrow B)$ + heat of reaction
$= 80 + 200 = 280\ \text{kJ mol}^{-1}$

24. $Fe(OH)_3$ is a positively charged sol, thus coagulated by negative ion (anion). Smaller the charge on anion, smaller is its coagulating power or higher is its flocculation value.
(a) $KBr \longrightarrow K^+ + Br^-$
(b) $K_2SO_4 \longrightarrow 2K^+ + SO_4^{2-}$
(c) $K_2CrO_4 \longrightarrow 2K^+ + CrO_4^{2-}$
(d) $K_4[Fe(CN)_6] \longrightarrow 4K^+ + [Fe(CN)_6]^{4-}$
$\because Br^-$ has smaller charge.
$\therefore$ KBr is least effective in coagulating $Fe(OH)_3$ sol.

25. Maximum oxidation state exhibited by *d* block elements (O.S) = no. of *ns* electrons + no. of $(n-1)\,d$ electrons.
(a) O. S. = 2 + 2 = 4 (b) O. S. = 5 + 1 = 6
(c) O. S. = 3 + 2 = 5 (d) O. S. = 5 + 2 = 7
$\therefore (n-1)d^5ns^2$ configuration will achieve the highest oxidation state.

26. Relative lowering of vapour pressure = mole fraction of solute **(Raoult's law)**
$$\frac{p - p_S}{p} = \chi_2$$
$$\frac{p - p_S}{p} = \frac{wM}{mW}$$
$$0.0125 = \frac{wM}{mW}$$
or $$\frac{w}{mW} = \frac{0.0125}{18} = 0.00070$$
Hence, $$\text{molality} = \frac{w}{mW} \times 1000$$
$$= 0.0007 \times 1000 = 0.70$$

27. H_2O_2 oxidises the acidified potassium dichromate solution into blue peroxide of chromium, CrO_5.

$$Cr_2O_7^{2-} + 2H^+ + 4H_2O_2 \longrightarrow 2CrO_5 + 5H_2O$$

28. $NaOH + S \longrightarrow \underset{\text{'A'}}{Na_2SO_3} + Na_2S + H_2O$

29. On igniting at 1400°C, Fe_2O_3 get reduced to metallic Fe.

$$3Fe_2O_3 + CO \longrightarrow 2Fe_3O_4 + CO_2$$
$$Fe_3O_4 + CO \longrightarrow 3FeO + CO_2$$
$$FeO + CO \longrightarrow Fe + CO_2$$

30. H_2SO_4 forms hydrate with water. That's why it has great affinity towards water.

31. In blood, He is much less soluble than nitrogen, hence He—O_2 mixture is used by deep sea divers in preference to N_2—O_2 mixture.

32. One mole = 6.02×10^{23} molecules/atoms/ions.

$\because$ CO_2 molecule has one C and two oxygen atoms.

$\therefore$ CO_2 contains 6.023×10^{23} atoms of C and 6.023×10^{23} molecules of O_2.

33. In acidic medium following reaction takes place.

$$8H^+ + 5e^- + MnO_4^- \longrightarrow Mn^{2+} + 4H_2O$$

$\therefore$ Equivalent weight of $KMnO_4$ in acidic medium $= \dfrac{\text{molecular weight of } KMnO_4}{5}$

$$= \frac{158}{5}$$
$$= 31.6$$

34. (a) Density of water = 1 g cm^{-3}

Mass of water = $1 \text{ m}^3 = 10^6 \text{ cm}^{-3}$

$$\text{Mass} = \text{volume} \times \text{density}$$
$$= 10^6 \text{ cm}^{-3} \times 1 \text{ g cm}^{-3}$$
$$= 10^6 \text{ g}$$
$$= \frac{10^6}{10^3} \text{ kg}$$
$$= 1000 \text{ kg}$$

(b) Mass of normal adult man = 65 kg

(c) Density of Hg = 13.6 g cm^{-3}

Volume of Hg = $10 \text{ L} = 10 \times 1000 \text{ cm}^{-3}$

$\therefore$ Mass of Hg $= 13.6 \times 10 \times 1000$
$$= 136000 \text{ g}$$
$$= 13.6 \text{ kg}$$

$\therefore$ Mass of 1 m^3 water is highest.

35. e/m ratio for $He^{2+} = \dfrac{2}{4}$

e/m ratio for $H^+ = \dfrac{1}{1}$

e/m ratio for $He^+ = \dfrac{1}{4}$

e/m ratio for $D^+ = \dfrac{1}{2}$

$\therefore$ The e/m is highest for hydrogen.

36. The ionisation potential decreases down the group (due to increase in size of atom) and increases in a period from left to right.

$\therefore$ Out of the given choices Li > K > Cs is correct.

37. Let the oxidation state of Fe in $Fe_3O_4 = x$

$\therefore$ $3x + 4 \times (-2) = 0$

or $3x - 8 = 0$

$\therefore$ $x = \dfrac{8}{3}$

38. $T_{50} = 15$ min

$$k = \frac{2.303 \log 2}{T_{50}} = \frac{2.303 \log 2}{15}$$

$$a = 0.1 \text{ M}$$
$$(a - x) = 0.025 \text{ M}$$

For first order reaction,

$$k = \frac{2.303}{t} \log \left(\frac{a}{a-x} \right)$$

$$\frac{2.303 \log 2}{15} = \frac{2 \times 2.303}{t} \log \frac{0.1}{0.025}$$
$$= \frac{2.303}{t} \log 4$$

$\therefore$ $\dfrac{2.303 \log 2}{15} = \dfrac{2 \times 2.303 \log 2}{t}$

$\therefore$ $t = 30$ min

39. 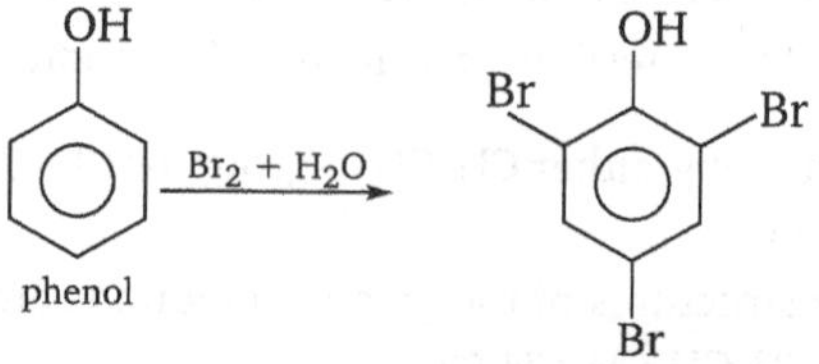

phenol

2,4,6-tribromophenol

40. $CH_3COOH \xrightarrow[\Delta]{Cl_2/\text{Red P}} \underset{\alpha\text{-chloroacetic acid}}{CH_2ClCOOH}$

This reaction is called Hell-Volhard-Zelinsky reaction.

41. $CH_3NH_2 + C_6H_5SO_2Cl \longrightarrow C_6H_5SO_2NHCH_3$

1° amine; Hinsberg's reagent; N-methyl benzene sulphonamide

$\xrightarrow[\text{(alkali)}]{\text{NaOH}} C_6H_5SO_2N(Na)CH_3$

soluble sodium salt

42. Angle strain, $\alpha = \frac{1}{2}[109°28' - \theta]$

In case of cyclopropane,

$$\theta = 60°$$

$$\therefore \quad \alpha = \frac{1}{2}(109°28' - 60°)$$

$$= 24°44'$$

43. An species or group of atoms can act as ligand only when it carries an unshared pair, *ie*, lone pair of electrons.

44. Bond order $\propto \frac{1}{\text{bond length}}$

BO of NO < BO of NO^+

∴ Bond length of NO is greater than the bond length of NO^+.

45. $\pi\ 2p_x$ and $\pi\ 2p_y$ or $\overset{*}{\pi}\ 2p_x$ and $\overset{*}{\pi}\ 2p_y$ orbitals have nearly equal energy and thus, are called degenerate orbitals.

46. pH varies inversely with temperature. Thus, if the pH of neutral water is 6.5, the temperature of water is more than 25°C.

47. Electron acceptors are Lewis acids. They are electron deficient compounds. BF_3 is Lewis acid because B has only 6 electrons in its valence shell and it can accept electrons. NF_3, Cl^- and H_2O have lone pair of electrons. Thus, they are electron donors and Lewis bases.

48. Given, concentration of NaOH = 10^{-10} M

$$\underset{10^{-10}}{NaOH} \longrightarrow \underset{10^{-10}}{Na^+} + \underset{10^{-10}}{OH^-}$$

$\therefore$ $[OH]^-$ from NaOH $= 10^{-10}$

$\therefore$ Total $[OH^-] = 10^{-7} + 10^{-10}$

$$= 10^{-7}(1 + 0.001)$$

$$= 10^{-7}\left(\frac{1001}{1000}\right)$$

$$= 10^{-10} \times 1001$$

$$pOH = -\log[OH^-]$$

$$= -\log(1001 \times 10^{-10})$$

$$= -3.004 + 10$$

$$= 6.9996$$

$\because$ pH + pOH = 14

$\therefore$ pH = 14 − 6.9996

$$= 7.0004 \approx 7$$

49. Carbanions contain even number of valence electrons and thus, show diamagnetic behaviour.

50. Enantiomers are non-superimposable mirror images. *eg*, lactic acid

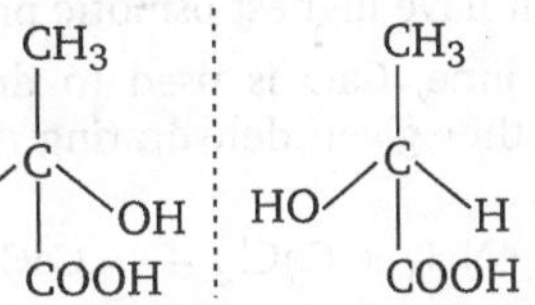

Diastereomers are non-superimposable and are not the mirror images of each other. Moreover, *meso* form has plane of symmetry.

51. Inductive effect involves only displacement (and not delocalisation) of σ-electrons.

52. Given, concentration of NaCl = 0.2 M

$$K_{sp}(AgCl) = 1.20 \times 10^{-10}$$

Let the solubility of AgCl in NaCl = x

$$AgCl \longrightarrow \underset{x}{Ag^+} + \underset{x}{Cl^-}$$

Solubility $\underset{x}{}$ $\quad \underset{0.2}{NaCl} \longrightarrow \underset{0.2}{Na^+} + \underset{0.2}{Cl^-}$

$\therefore$ $[Ag^+] = x$ and $[Cl^-] = (x + 0.2)$

$\therefore$ $K_{sp}(AgCl) = [Ag^+][Cl^-]$

$$= x(x + 0.2)$$

$$= x^2 + 0.2x$$

$\therefore$ $K_{sp} = 0.2x \ (x^2 << 1)$

or $1.2 \times 10^{-10} = 0.2x$

$\therefore$ $x = 6 \times 10^{-10}$

53. Naturally occurring fats are called lipids.

54. Pepsin hydrolyses proteins into amino acids as

$$\text{Proteins} \xrightarrow{\text{Pepsin}} \text{amino acids.}$$

55. The enthalpy of formation of Al_2O_3 is very high and hence, it is not possible to reduce it by carbon.

56. $\underset{\text{oxalic acid}}{\begin{matrix} COOH \\ | \\ COOH \end{matrix}}$ + conc. $H_2SO_4 \longrightarrow CO + CO_2 + H_2O$

Concentrated H_2SO_4 is a strong dehydrating agent.

57. Osmotic pressure is a colligative property *ie*, depends only upon the number of particles or ions in solution. More the number of ions in solution, more will be the osmotic pressure of solution.

(i) 0.1 M urea and 0.1 M glucose will have same number of molecules in solution as they don't ionise.

(ii) $KCl \longrightarrow K^+ + Cl^-$ (2 ions)

(iii) $CaCl_2 \longrightarrow Ca^{2+} + 2Cl^-$ (3 ions)

$\therefore CaCl_2$ produces maximum number of ions.

$\therefore$ It will have highest osmotic pressure.

58. Quick lime, CaO is used to dry ammonia, as with other given dehydrating agents ammonia reacts.

$$4NH_3 + CaCl_2 \longrightarrow CaCl_2 \cdot 4NH_3$$
$$4NH_3 + 2P_2O_5 \longrightarrow 4NH_4PO_3$$

$Ca(OH)_2$ is never used as dehydrating agent.

59.

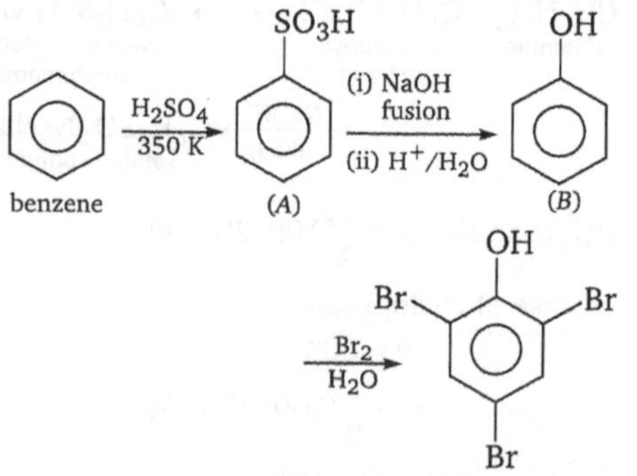

2,4,6-tribromophenol (C)

60. Collin's reagent is used to convert

$$—CH_2OH \longrightarrow —CHO.$$

Mathematics

1. $f(x)$ is defined, if $3x^2 - 4x + 5 > 0$

$$\Rightarrow \quad 3\left[x^2 - \frac{4}{3}x + \frac{5}{3}\right] > 0$$

$$\Rightarrow \quad 3\left[\left(x - \frac{2}{3}\right)^2 + \frac{11}{9}\right] > 0$$

which is true for all real x.

$\therefore$ Domain of $f = (-\infty, \infty)$

Let $y = \log_e(3x^2 - 4x + 5) \Rightarrow e^y = 3x^2 - 4x + 5$

$\Rightarrow 3x^2 - 4x + (5 - e^y) = 0$

For x to be real,

$$16 - 12(5 - e^y) \geq 0$$

$$\Rightarrow \quad 12e^y \geq 44$$

$$\Rightarrow \quad e^y \geq \frac{11}{3}$$

$$\Rightarrow \quad y \geq \log_e \frac{11}{3}$$

Range of $f = \left[\log_e \frac{11}{3}, \infty\right)$

2. $\sqrt{9 - x^2}$ is defined for

$$9 - x^2 \geq 0 \Rightarrow (3 - x)(3 + x) \geq 0$$

$$\Rightarrow \quad (x - 3)(x + 3) \leq 0 \quad \ldots\text{(i)}$$

$$\Rightarrow \quad -3 \leq x \leq 3$$

$\sin^{-1}(3 - x)$ defined for

$$-1 \leq 3 - x \leq 1$$

$$\Rightarrow \quad -4 \leq -x \leq -2$$

$$\Rightarrow \quad 2 \leq x \leq 4 \quad \ldots\text{(ii)}$$

Also, $\sin^{-1}(3 - x) \neq 0$

$$\Rightarrow \quad 3 - x \neq 0 \text{ or } x \neq 3 \quad \ldots\text{(iii)}$$

From Eqs. (i), (ii) and (iii), we get

The domain of

$$f = ([-3, 3] \cap [2, 4]) - \{3\} = [2, 3)$$

3. Given, $a_0 = 1$

and $\quad a_{n+1} = 3n^2 + n + a_n$

$$\Rightarrow \quad a_n = 3(n-1)^2 + (n-1) + a_{n-1}$$

Now, put $n = 1, 2, 3, \ldots, n$

We get,

$$a_1 = 0 + a_0 = 0 + 1 = 1$$

$$\Rightarrow \quad a_2 = 4 + a_1$$

$$\Rightarrow \quad a_3 = 14 + a_2$$

$$\Rightarrow \quad a_4 = 30 + a_3$$

....

....

....

$$\Rightarrow \quad a_{n-1} = 3(n-2)^2 + (n-2) + a_{n-2}$$

$$\Rightarrow \quad a_n = 3(n-1)^2 + (n-1) + a_{n-1}$$

On adding, we get

$$a_n = (1 + 4 + 14 + 30 + \ldots + 3(n-1)^2 + (n-1))$$

$$\Rightarrow \quad a_n = \Sigma\, 3(n^2 + 1 - 2n) + (n - 1)$$

$$\Rightarrow \quad a_n = \Sigma(3n^2 + 3 - 6n + n - 1)$$

$$\Rightarrow \quad a_n = \Sigma(3n^2 - 5n + 2)$$

$$\Rightarrow \quad a_n = 3\Sigma n^2 - 5\Sigma_n + 2\Sigma 1$$

$$\Rightarrow \quad a_n = \frac{3n(n+1)(2n+1)}{6} - \frac{5n(n+1)}{2} + 2n$$

$$\Rightarrow \quad a_n = \frac{n}{2}(n+1)\{2n+1-5\} + 2n$$

$$\Rightarrow \quad a_n = \frac{n}{2}(n+1)(2n-4) + 2n$$

$$= n(n+1)(n-2) + 2n$$

$$\Rightarrow \quad a_n = (n^2+n)(n-2) + 2n$$

$$= n^3 - 2n^2 + n^2 - 2n + 2n$$

$$\Rightarrow \quad a_n = n^3 - n^2 \qquad (\because n \geq 0)$$

4. Required number of ways

$$= {}^5C_1 + {}^5C_2 + {}^5C_3 + {}^5C_4 + {}^5C_5 = 2^5 - 1 = 31$$

5. Number of selections of any number of copies of a book $= P + 1$, (because copies of the same book are identical things). Similarly, for each book.

$\therefore$ Total number of selections

$$= (P+1)(P+1)\ldots \text{ to } n \text{ factors}$$

$$= (P+1)^n$$

But this includes a selections which is empty *ie,* zero copy of each is selected. Excluding this, the required number of non-empty selections

$$= (P+1)^n - 1$$

6. The number of straight lines

$$= {}^{18}C_2 - ({}^5C_2 - 1) = 144$$

7. The general term is

$$T_{r+1} = {}^{10}C_r \left(\sqrt{\frac{x}{3}}\right)^{10-r} \left(\sqrt{\frac{3}{2x^2}}\right)^r$$

$$= {}^{10}C_r \left(\frac{1}{3}\right)^{5-(r/2)} \left(\frac{3}{2}\right)^{r/2} \cdot x^{5-(3r/2)}$$

for term independent of x.

$5 - \frac{3r}{2} = 0 \Rightarrow r = \frac{10}{3}$, which is not a positive integer. Hence, there is no term independent of x.

8. $$\frac{\text{The coefficient of } t_{r+1}}{\text{The coefficient of } t_r} = \frac{{}^{2n}C_r}{{}^{2n}C_{r-1}}$$

$$= \frac{2n-r+1}{r}$$

The coefficient of $t_{r+1} \geq$ The coefficient of t_r, provided

$$\frac{2n-r+1}{r} \geq 1 \text{ or } 2n+1 \geq 2r$$

or $\quad r \leq \frac{2n+1}{2}$ or $r \leq n + \frac{1}{2}$

Hence, the greatest coefficient

= The coefficient of $(n+1)$th term

$$= {}^{2n}C_n = \frac{(2n)!}{(n!)^2}$$

9. The general term in the expansion of $(\sqrt{5} + 4\sqrt{11})^{124}$ is

$$T_{r+1} = {}^{124}C_r (\sqrt{5})^{124-r} (\sqrt[4]{11})^r$$

$$= {}^{124}C_r \cdot 5^{(124-r)/2} \cdot 11^{r/4}$$

For integer terms $r = 0, 4, 8, 12, \ldots, 124$

$\therefore$ Number of terms which are integers = 32

10. We have,

$$(1+x)^m \left(1 + \frac{1}{x}\right)^n = (1+x)^m \left(\frac{x+1}{x}\right)^n$$

$$= \frac{(1+x)^{m+n}}{x^n} = x^{-n}(1+x)^{m+n}$$

$\therefore$ Required term independent of x

= Coefficient of x^0 in $x^{-n}(1+x)^{m+n}$

= Coefficient of x^n in $(1+x)^{m+n}$

$$= {}^{m+n}C_n$$

11. Let α, β are the roots of the equation $(1-a)x^2 + 3ax - 1 = 0$, then $\alpha + \beta = \frac{3a}{a-1}$,

$$\alpha\beta = \frac{1}{a-1}$$

As $a > 1$, $\alpha + \beta > 0$ and $\alpha\beta > 0$ and $D = 9a^2 + 4(1-a)$

$$= 9\left(a^2 - \frac{4}{9}a + \frac{4}{9}\right) = 9\left\{a\left(a - \frac{4}{9}\right) + \frac{4}{9}\right\} > 0, \quad \text{as}$$

$a > 1$

$\therefore$ The equation has real and positive roots.

12. $a\cos 2x + b\sin^2 x + c = 0$

$$\Rightarrow \quad a(1 - 2\sin^2 x) + b\sin^2 x + c = 0$$

ie, $\quad (b - 2a)\sin^2 x + (a + c) = 0$

It is an identity, if $b - 2a = 0$, $a + c = 0$, so

$$\frac{a}{1} = \frac{b}{2} = \frac{c}{-1}$$

Thus, number of triplet (a, b, c) are infinite.

13. We have,

$$(\alpha + \beta + \gamma)^2 = \alpha^2 + \beta^2 + \gamma^2 + 2(\alpha\beta + \beta\gamma + \gamma\alpha)$$

$$\Rightarrow \quad 4 = 6 + 2(\alpha\beta + \beta\gamma + \gamma\alpha)$$

$$\Rightarrow \quad \beta\gamma + \gamma\alpha + \alpha\beta = -1 \qquad \ldots(\text{i})$$

Also, $\quad \alpha^3 + \beta^3 + \gamma^3 - 3\alpha\beta\gamma$

$$= (\alpha + \beta + \gamma)(\alpha^2 + \beta^2 + \gamma^2 - \alpha\beta - \beta\gamma - \gamma\alpha)$$

$\Rightarrow \quad 8 - 3\alpha\beta\gamma = 2(6+1)$

$\Rightarrow \quad 3\alpha\beta\gamma = 8 - 14 = -6$

or $\quad \alpha\beta\gamma = -2 \quad \ldots$(ii)

Now, $(\alpha^2+\beta^2+\gamma^2)^2 = \alpha^4+\beta^4+\gamma^4 + 2(\alpha^2\beta^2+\beta^2\gamma^2+\gamma^2\alpha^2)$

$= (\alpha^4+\beta^4+\gamma^4) + 2[(\alpha\beta+\beta\gamma+\gamma\alpha)^2 - 2\alpha\beta\gamma(\alpha+\beta+\gamma)]$

$\Rightarrow (\alpha^4+\beta^4+\gamma^4) = 36 - 2[(-1)^2 - 2(-2)(2)] = 18$

14. $ax^2 + 2b|x| - c = 0$

$\Rightarrow \quad a|x|^2 + 2b|x| - c = 0$

$\therefore \quad |x| = \dfrac{-2b \pm \sqrt{4b^2+4ac}}{2} = -b \pm \sqrt{b^2+ac}$

Since, a, b, c are positive. So,

$|x| = -b + \sqrt{b^2+ac}$

$\therefore x$ has two real values, neglecting

$|x| = -b - \sqrt{b^2+ac}$, as $|x| > 0$

15. Since, the diagonal elements of a skew-symmetric matrix are always zero.

$\therefore$ Trace of $A = 0$

16. Here, the rank of A is 3.

Therefore, the minor of order 3 of $A \neq 0$

$$\Rightarrow \begin{vmatrix} y+a & b & c \\ a & y+b & c \\ a & b & y+c \end{vmatrix} \neq 0$$

[Applying $C_1 \to C_1 + C_2 + C_3$ and taking $(y+a+b+c)$ common from C_1]

$$\Rightarrow (y+a+b+c)\begin{vmatrix} 1 & b & c \\ 1 & y+b & c \\ 1 & b & y+c \end{vmatrix} \neq 0$$

[Applying $R_2 \to R_2 - R_1, R_3 \to R_3 - R_1$]

$$\Rightarrow (y+a+b+c)\begin{vmatrix} 1 & b & c \\ 0 & y & 0 \\ 0 & 0 & y \end{vmatrix} \neq 0$$

Expanding along C_1

$\Rightarrow \quad (y+a+b+c)(y^2) \neq 0$

$\Rightarrow \quad y \neq 0$ and $y \neq -(a+b+c)$

17. We have, $\begin{vmatrix} 1 & \log_x y & \log_x z \\ \log_y x & 1 & \log_y z \\ \log_z x & \log_z y & 1 \end{vmatrix}$

$$= \begin{vmatrix} 1 & \dfrac{\log y}{\log x} & \dfrac{\log z}{\log x} \\ \dfrac{\log x}{\log y} & 1 & \dfrac{\log z}{\log y} \\ \dfrac{\log x}{\log z} & \dfrac{\log y}{\log z} & 1 \end{vmatrix}$$

$$= \frac{1}{\log x \cdot \log y \cdot \log z}\begin{vmatrix} \log x & \log y & \log z \\ \log x & \log y & \log z \\ \log x & \log y & \log z \end{vmatrix} = 0$$

[$\because$ all rows are identical]

18. Since, A is an orthogonal matrix, therefore,

$AA' = I \Rightarrow |A \cdot A'| = |I| \Rightarrow |A||A'| = 1$

$\Rightarrow \quad |A||A| = 1 \quad (\because |I| = 1)$

$\Rightarrow \quad |A|^2 = 1 \quad (\because |A'| = |A|)$

$\Rightarrow \quad |A| = \pm 1$

19. Dividing the equation,

$a^3x^2 + abcx + c^3 = 0$ by c^2, we get

$$a\left(\frac{ax}{c}\right)^2 + b\left(\frac{ax}{c}\right) + c = 0$$

$\Rightarrow \quad \dfrac{ax}{c} = \alpha, \beta$ are roots.

$\Rightarrow \quad x = \dfrac{c}{a}\alpha$ and $x = \dfrac{c}{a}\beta$

$\Rightarrow \quad x = \alpha^2\beta$ and $x = \alpha\beta^2$

[$\because \alpha\beta = \dfrac{c}{a}$ as α, β are the roots of $ax^2 + bx + c = 0$]

are roots of above equation.

20. Expression can be written as

$= (1+i)^{n_1} + (1-i)^{n_1} + (1+i)^{n_2} + (1-i)^{n_2}$

$$= 2^{n_1/2}\left(\frac{1}{\sqrt{2}} + i\frac{1}{\sqrt{2}}\right)^{n_1} + 2^{n_1/2}\left(\frac{1}{\sqrt{2}} - i\frac{1}{\sqrt{2}}\right)^{n_1} + 2^{n_2/2}\left(\frac{1}{\sqrt{2}} + i\frac{1}{\sqrt{2}}\right)^{n_2} + 2^{n_2/2}\left(\frac{1}{\sqrt{2}} - i\frac{1}{\sqrt{2}}\right)^{n_2}$$

$$= 2^{n_1/2}\left\{\left(\cos\frac{\pi}{4} + i\sin\frac{\pi}{4}\right)^{n_1} + \left(\cos\frac{\pi}{4} - i\sin\frac{\pi}{4}\right)^{n_1}\right\}$$

$$+ 2^{n_2/2}\left\{\left(\cos\frac{\pi}{4} + i\sin\frac{\pi}{4}\right)^{n_2} + \left(\cos\frac{\pi}{4} - i\sin\frac{\pi}{4}\right)^{n_2}\right\}$$

$$= n^{n_1/2}\left\{\cos\frac{n_1\pi}{4} + i\sin\frac{n_1\pi}{4} + \cos\frac{n_1\pi}{4} - i\sin\frac{n_1\pi}{4}\right\}$$

$$+ 2^{n_2/2}\left\{\cos\frac{n_2\pi}{4} + i\sin\frac{n_2\pi}{4} + \cos\frac{n_2\pi}{4} - \sin\frac{n_2\pi}{4}\right\}$$

$$= 2^{n_1/2}\cdot 2\cos\frac{n_1\pi}{4} + 2^{n_2/2}\cdot 2\cos\frac{n_2\pi}{4} = \text{real}$$

ie, n_1, n_2 are any two positive integers.

21. Let $z = x + iy$, then $|z + i| - |z - i| = k$ becomes

$\therefore \quad \sqrt{x^2 + (y+1)^2} - \sqrt{x^2 + (y-1)^2} = k \quad \ldots\text{(i)}$

or $\quad x^2 + (y+1)^2 - x^2 - (y-1)^2$

$$= k\{\sqrt{x^2 + (y+1)^2} + \sqrt{x^2 + (y-1)^2}\}$$

$\therefore \quad \sqrt{x^2 + (y-1)^2} + \sqrt{x^2 + (y+1)^2} = \dfrac{4y}{k} \quad \ldots\text{(ii)}$

From Eqs. (i) and (ii),

$$2\sqrt{x^2 + (y+1)^2} = k + \frac{4y}{k}$$

$$\Rightarrow \quad 4x^2 + 4y^2 + 8y + 4 = k^2 + \frac{16y^2}{k^2} + 8y$$

$$\Rightarrow \quad 4x^2 + \left(4 - \frac{16}{k^2}\right)y^2 = k^2 - 4$$

For an hyperbola, $\dfrac{4k^2 - 16}{k^2} < 0 \Rightarrow k^2 - 4 < 0$

$$\Rightarrow \quad |k| < 2 \text{ or } -2 < k < 2$$

22. Since, $|\sin x| + |\cos x|$ is a periodic function with period $\pi/2$.

$\therefore \quad f(x) = |\sin 4x| + |\cos 4x|$ is a periodic function with period $\dfrac{1}{4}\cdot\dfrac{\pi}{2}$ *ie*, $\dfrac{\pi}{8}$.

23. $\cos x = \dfrac{2\cos y - 1}{2 - \cos y}$

$$\Rightarrow \quad \frac{1 - \tan^2\frac{x}{2}}{1 + \tan^2\frac{x}{2}} = \frac{2\left(\dfrac{1 - \tan^2\frac{y}{2}}{1 + \tan^2\frac{y}{2}}\right) - 1}{2 - \left(\dfrac{1 - \tan^2\frac{y}{2}}{1 + \tan^2\frac{y}{2}}\right)}$$

$$\Rightarrow \frac{1 - \tan^2\frac{x}{2}}{1 + \tan^2\frac{x}{2}} = \frac{2 - 2\tan^2\frac{y}{2} - 1 - \tan^2\frac{y}{2}}{2 + 2\tan^2\frac{y}{2} - 1 + \tan^2\frac{y}{2}}$$

$$\Rightarrow \quad \frac{1 - \tan^2\frac{x}{2}}{1 + \tan^2\frac{x}{2}} = \frac{1 - 3\tan^2\frac{y}{2}}{1 + 3\tan^2\frac{y}{2}}$$

$$\Rightarrow 1 + 3\tan^2\frac{y}{2} - \tan^2\frac{x}{2} - 3\tan^2\frac{x}{2}\tan^2\frac{y}{2}$$

$$= 1 - 3\tan^2\frac{y}{2} + \tan^2\frac{x}{2} - 3\tan^2\frac{x}{2}\tan^2\frac{y}{2}$$

$$\Rightarrow \quad 6\tan^2\frac{y}{2} = 2\tan^2\frac{x}{2}$$

$$\Rightarrow \quad \tan\frac{x}{2}\cdot\cot\frac{y}{2} = \sqrt{3}$$

24. $\cos\dfrac{2\pi}{15}\cdot\cos\dfrac{4\pi}{15}\cos\dfrac{8\pi}{15}\cos\dfrac{16\pi}{15}$

$$= \frac{1}{2\sin\frac{2\pi}{15}}\cdot 2\sin\frac{2\pi}{15}\cdot\cos\frac{2\pi}{15}\cdot\cos\frac{4\pi}{15}\cdot\cos\frac{8\pi}{15}\cdot\cos\frac{16\pi}{15}$$

$$\Rightarrow \quad \frac{1}{2\sin\frac{2\pi}{15}}\cdot\sin\frac{4\pi}{15}\cdot\cos\frac{4\pi}{15}\cdot\cos\frac{8\pi}{15}\cdot\cos\frac{16\pi}{15}$$

$$\Rightarrow \quad \frac{1}{4\sin\frac{2\pi}{15}}\cdot\sin\frac{8\pi}{15}\cdot\cos\frac{8\pi}{15}\cdot\cos\frac{16\pi}{15}$$

$$= \frac{1}{8\sin\frac{2\pi}{15}}\cdot\sin\frac{16\pi}{15}\cdot\cos\frac{16\pi}{15}$$

$$= \frac{1}{16\sin\frac{2\pi}{15}}\cdot\sin\frac{32\pi}{15}$$

$$= \frac{1}{16\sin\frac{2\pi}{15}}\cdot\sin\left(2\pi + \frac{2\pi}{15}\right) = \frac{1}{16\sin\frac{2\pi}{15}}\cdot\sin\frac{2\pi}{15}$$

$$= \frac{1}{16}$$

25. We have, $\cos x\cdot\cos\left(\dfrac{\pi}{3} + x\right)\cos\left(\dfrac{\pi}{3} - x\right) = \dfrac{1}{4}$

$$\Rightarrow \quad \cos x\left(\frac{1}{4}\cos^2 x - \frac{3}{4}\sin^2 x\right) = \frac{1}{4}$$

or $\quad \dfrac{\cos x}{4}(4\cos^2 x - 3) = \dfrac{1}{4}$

$$\Rightarrow \quad 4\cos^3 x - 3\cos x = 1$$

$$(\because \cos 3x = 4\cos^3 x - 3\cos x)$$

or $\quad \cos 3x = 1$

$$\Rightarrow \quad 3x = 2n\pi$$

$\Rightarrow x = \dfrac{2n\pi}{3}$, where $n = 0, 1, 2, 3, 4, 5, 6, 7, 8, 9$

$(\because x \in [0, 6\pi])$

$\therefore$ The required sum $= \dfrac{2\pi}{3}\sum_{n=0}^{9} n = 30\pi$

26. The given expression

$$= 2\tan^{-1}[\operatorname{cosec}\tan^{-1}x - \tan\cot^{-1}x]$$

$$= 2\tan^{-1}\left[\operatorname{cosec}\left\{\operatorname{cosec}^{-1}\frac{\sqrt{1+x^2}}{x}\right\} - \tan\left\{\tan^{-1}\left(\frac{1}{x}\right)\right\}\right]$$

$$= 2\tan^{-1}\left[\frac{\sqrt{1+x^2}}{x} - \frac{1}{x}\right] = 2\tan^{-1}\left[\frac{\sqrt{1+x^2}-1}{x}\right]$$

$$= 2\tan^{-1}\left[\frac{\sec\theta - 1}{\tan\theta}\right] \quad [\text{putting } x = \tan\theta]$$

$$= \tan^{-1}\left[\frac{1-\cos\theta}{\sin\theta}\right]$$

$$= 2\tan^{-1}\left[\frac{2\sin^2\frac{\theta}{2}}{2\sin\frac{\theta}{2}\cdot\cos\frac{\theta}{2}}\right]$$

$$= 2\tan^{-1}\tan\frac{\theta}{2} = 2\cdot\frac{\theta}{2} = \theta = \tan^{-1}x$$

27. We have,

$$\sin^{-1}\left[\cos\left(\sin^{-1}\sqrt{\left(\frac{2-\sqrt{3}}{4}\right)} + \cos^{-1}\frac{\sqrt{12}}{4} + \sec^{-1}\sqrt{2}\right)\right]$$

$$= \sin^{-1}\left[\cos\left(\sin^{-1}\left(\frac{\sqrt{3}-1}{2\sqrt{2}}\right) + \cos^{-1}\frac{\sqrt{3}}{2} + \cos^{-1}\frac{1}{\sqrt{2}}\right)\right]$$

$$\left[\because \left(\frac{2-\sqrt{3}}{4}\right)^{1/2} = \frac{\sqrt{3}-1}{2\sqrt{2}}\right]$$

$$= \sin^{-1}[\cos(15^\circ + 30^\circ + 45^\circ)]$$

$$= \sin^{-1}[\cos 90^\circ] \quad \left(\because \sin 15^\circ = \frac{\sqrt{3}-1}{2\sqrt{2}}\right)$$

$$= \sin^{-1} 0$$

$$= 0$$

28. In given right angled triangle, we have

$$\cos A = c/b$$

$$\Rightarrow \frac{1-\tan^2\frac{A}{2}}{1+\tan^2\frac{A}{2}} = c/b$$

$$\Rightarrow \frac{\left(1+\tan^2\frac{A}{2}\right) - \left(1-\tan^2\frac{A}{2}\right)}{\left(1+\tan^2\frac{A}{2}\right) + \left(1-\tan^2\frac{A}{2}\right)} = \frac{(b-c)}{(b+c)}$$

(by componendo and dividendo rule)

Hence, $\tan^2\frac{A}{2} = \frac{b-c}{b+c}$

29. We have, $r_1 > r_2 > r_3$

$$\Rightarrow \frac{\Delta}{s-a} > \frac{\Delta}{s-b} > \frac{\Delta}{s-c}$$

$$\Rightarrow \frac{s-a}{\Delta} < \frac{s-b}{\Delta} < \frac{s-c}{\Delta}$$

$$\Rightarrow s-a < s-b < s-c$$

$$\Rightarrow -a < -b < -c \Rightarrow a > b > c$$

30. Clearly,

$\angle CDA = \beta$, $\angle DCB = \gamma$, $\angle ACB = \pi - (\alpha + \beta + \gamma)$

Using since rule in ΔCAB, we get

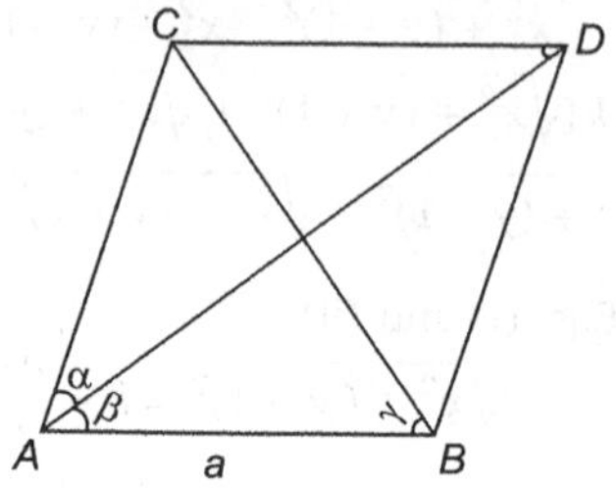

$$\frac{AB}{\sin(\pi - (\alpha+\beta+\gamma))} = \frac{CA}{\sin\gamma}$$

$$\Rightarrow CA = \frac{a\cdot\sin\gamma}{\sin(\alpha+\beta+\gamma)}$$

Now, using sine rule in ΔCAD,

we get $\frac{CA}{\sin\beta} = \frac{CD}{\sin\alpha}$

$$\Rightarrow CD = \frac{CA\cdot\sin\alpha}{\sin\beta} = \frac{a\sin\alpha\cdot\sin\gamma}{\sin\beta\cdot\sin(\alpha+\beta+\gamma)}$$

31. We have, $\vec{\mathbf{P}} = \vec{\mathbf{AC}} + \vec{\mathbf{BD}} = \vec{\mathbf{AC}} + \vec{\mathbf{BC}} + \vec{\mathbf{CD}}$

$$= \vec{\mathbf{AC}} + \lambda\vec{\mathbf{AD}} + \vec{\mathbf{CD}} = \lambda\vec{\mathbf{AD}} + (\vec{\mathbf{AC}} + \vec{\mathbf{CD}})$$

$$= \lambda\vec{\mathbf{AD}} + \vec{\mathbf{AD}} = (\lambda+1)\vec{\mathbf{AD}}$$

But, $\vec{\mathbf{P}} = \mu\,\vec{\mathbf{AD}}$

$$\therefore \quad \mu = \lambda + 1$$

32. Given, $3\vec{\mathbf{p}} + 8\vec{\mathbf{q}} = 6\vec{\mathbf{r}} + 5\vec{\mathbf{s}}$

$$\Rightarrow \frac{3\vec{\mathbf{p}} + 8\vec{\mathbf{q}}}{8+3} = \frac{6\vec{\mathbf{r}} + 5\vec{\mathbf{s}}}{6+5}$$

$\Rightarrow$ The point which divides PQ in ratio 8 : 3 in the same as the point which divides RS in the ratio 5 : 6. Hence, the line PQ and RS intersect.

33. We have, $\vec{\mathbf{a}}\cdot\vec{\mathbf{c}} = |\vec{\mathbf{c}}|$ and $|\vec{\mathbf{c}} - \vec{\mathbf{a}}| = 2\sqrt{2}$

$$\Rightarrow \vec{\mathbf{a}}\cdot\vec{\mathbf{c}} = |\vec{\mathbf{c}}| \text{ and } |\vec{\mathbf{c}}|^2 + |\vec{\mathbf{a}}|^2 - 2(\vec{\mathbf{a}}\cdot\vec{\mathbf{c}}) = 8$$

$$\Rightarrow |\vec{\mathbf{c}}|^2 + 9 - 2|\vec{\mathbf{c}}| = 8$$

$\Rightarrow \quad (|\vec{c}| - 1)^2 = 0$

$\Rightarrow \quad |\vec{c}| = 1$

$\Rightarrow \quad |(\vec{a} \times \vec{b}) \times \vec{c}| = |\vec{a} \times \vec{b}||\vec{c}| \sin 30°$

$= \frac{1}{2}|\vec{a} \times \vec{b}| = \frac{3}{2}$

$[\because \ \vec{a} \times \vec{b} = 2\hat{i} - 2\hat{j} + \hat{k}]$

34. We have,

$\vec{V}_1 = \frac{1}{2}(\vec{a} \times \vec{b}), \vec{V}_2 = \frac{1}{2}(\vec{b} \times \vec{c}), \vec{V}_3 = \frac{1}{2}(\vec{c} \times \vec{a})$

and $\vec{V}_4 = \frac{1}{2}\{(\vec{c} - \vec{a}) \times (\vec{b} - \vec{a})\}$

$\therefore \vec{V}_1 + \vec{V}_2 + \vec{V}_3 + \vec{V}_4 = \frac{1}{2}(\vec{a} \times \vec{b} + \vec{b} \times \vec{c} + \vec{c} \times \vec{a} + \vec{c} \times \vec{b} - \vec{c} \times \vec{a} - \vec{a} \times \vec{b}) = \vec{0}$

$|\vec{V}_1 + \vec{V}_2 + \vec{V}_3 + \vec{V}_4| = \vec{0}$

35. We have,

$|[\vec{a}\,\vec{b}\,\vec{c}]| = V$

Let V_1 be the volume of parallelopiped formed by the vectors $\vec{\alpha}$, $\vec{\beta}$ and $\vec{\gamma}$. Then,

$V_1 = |[\vec{\alpha}\,\vec{\beta}\,\vec{\gamma}]|$

$$\Rightarrow \quad V_1 = \begin{vmatrix} \vec{a}\cdot\vec{a} & \vec{a}\cdot\vec{b} & \vec{a}\cdot\vec{c} \\ \vec{a}\cdot\vec{b} & \vec{b}\cdot\vec{b} & \vec{b}\cdot\vec{c} \\ \vec{a}\cdot\vec{c} & \vec{b}\cdot\vec{c} & \vec{c}\cdot\vec{c} \end{vmatrix} [\vec{a}\,\vec{b}\,\vec{c}]$$

$\Rightarrow \quad V_1 = [\vec{a}\,\vec{b}\,\vec{c}]^2[\vec{a}\,\vec{b}\,\vec{c}]$

$\Rightarrow \quad V_1 = [\vec{a}\,\vec{b}\,\vec{c}]^3$

$\Rightarrow \quad V_1 = V^3$

36. We have, $|a\hat{i} + b\hat{j} + c\hat{k}| = |a| + |b| + |c|$

$\Rightarrow \sqrt{a^2 + b^2 + c^2} = |a| + |b| + |c|$

$\Rightarrow a^2 + b^2 + c^2$

$= a^2 + b^2 + c^2 + 2(|a||b| + |b||c| + |c||a|)$

$\Rightarrow |a||b| + |b||c| + |c||a| = 0$

$\Rightarrow ab = bc = ca = 0$

$\Rightarrow$ Any two of a, b and c are zero.

37. We have, $\vec{u}\cdot\hat{n} = 0$ and $\vec{v}\cdot\hat{n} = 0$

$\Rightarrow \quad \hat{n} \perp \vec{u}$ and $\hat{n} \perp \vec{v}$

$\Rightarrow \quad \hat{n} = \pm \frac{\vec{u} \times \vec{v}}{|\hat{u} \times \vec{v}|}$

Now, $\vec{u} \times \vec{v} = (\hat{i} + \hat{j}) \times (\hat{i} - \hat{j}) = -2\hat{k}$

$\hat{n} = \pm \hat{k}$

Hence, $|\vec{w}\cdot\hat{n}| = |(\hat{i} + 2\hat{j} + 3\hat{k})\cdot(\pm\hat{k})| = 3$

38. We have, $P(A) = P\left(\frac{A}{B}\right) = \frac{1}{4}$. This shows that A and B are independent events. So,

$P(B) = P\left(\frac{B}{A}\right) = \frac{1}{2}$

Now, $P\left(\frac{A}{B}\right) = \frac{1}{4} \Rightarrow P\left(\frac{\overline{A}}{B}\right) = 1 - \frac{1}{4} = \frac{3}{4}$

39. Suppose the coin is tossed n times. Let X be the number of the heads obtained. Then, X follows a binomial distribution with parameters n and $p = \frac{1}{2}$

Now, $P(X \geq 1) \geq 0.8 \Rightarrow 1 - P(X = 0) \geq 0.8$

$\Rightarrow \quad 1 - {}^nC_0 p^0 (1 - p)^n \geq 0.8$

$\Rightarrow \quad \left(\frac{1}{2}\right)^n \leq 0.2$ or $\left(\frac{1}{2}\right)^n \leq \frac{1}{5}$

$\Rightarrow \quad 2^n \geq 5$

$\Rightarrow \quad n \geq 3$

Hence, the least value of n is 3.

40. Since, $(n + 1)p = \frac{101}{3}$ is not an integer.

Therefore, $p(X = r)$ is maximum when $r = \left[\frac{101}{3}\right] = 33$

41. Let n_1 and n_2 be the number of observations in two groups having means $\overline{X}_1$ and $\overline{X}_2$ respectively.

Then $\quad \overline{X} = \frac{n_1\overline{X}_1 + n_2\overline{X}_2}{n_1 + n_2} - \overline{X}_2$

Now, $\overline{X} - \overline{X}_1 = \frac{n_1\overline{X}_1 + n_2\overline{X}_2}{n_1 + n_2} - \overline{X}_1$

$= \frac{n_2(\overline{X}_2 - \overline{X}_1)}{n_1 + n_2} > 0 \quad [\because \overline{X}_2 > \overline{X}_1]$

$\Rightarrow \quad \overline{X} > \overline{X}_1 \quad \ldots(i)$

And $\overline{X} - \overline{X}_2 = \frac{n(\overline{X}_1 - \overline{X}_2)}{n_1 + n_2} < 0 \quad [\because \overline{X}_2 > \overline{X}_1]$

$\Rightarrow \quad \overline{X} < \overline{X}_2 \quad \ldots(ii)$

From Eqs. (i) and (ii), $\overline{X}_1 < \overline{X} < \overline{X}_2$

42. Here, $\tan\theta = 2$,

So, $\quad \cos\theta = \frac{1}{\sqrt{5}}, \sin\theta = \frac{2}{\sqrt{5}}$

For x and y, we have

$$x = X\cos\theta - Y\sin\theta = \frac{X - 2Y}{\sqrt{5}}$$

and $$y = X\sin\theta + Y\cos\theta = \frac{2X + Y}{\sqrt{5}}$$

The equation $4xy - 3x^2 = a^2$ reduces to

$$\frac{4(X - 2Y)}{\sqrt{5}} \cdot \frac{(2X + Y)}{\sqrt{5}} - 3\left(\frac{X - 2Y}{\sqrt{5}}\right)^2 = a^2$$

$$\Rightarrow \quad 4(2X^2 - 2Y^2 - 3XY) - 3(X^2 - 4XY + 4Y^2) = 5a^2$$

$$\Rightarrow \quad 5X^2 - 20Y^2 = 5a^2$$

$$\Rightarrow \quad X^2 - 4Y^2 = a^2$$

43. The given points are collinear, if

$$\begin{vmatrix} t_1 & 2at_1 + at_1^3 & 1 \\ t_2 & 2at_2 + at_2^3 & 1 \\ t_3 & 2at_3 + at_3^3 & 1 \end{vmatrix} = 0$$

$$\Rightarrow \quad a\begin{vmatrix} t_1 & 2t_1 + t_1^3 & 1 \\ t_2 & 2t_2 + t_2^3 & 1 \\ t_3 & 2t_3 + t_3^3 & 1 \end{vmatrix} = 0$$

Applying $R_2 \to R_2 - R_1, R_3 \to R_3 - R_1$, we get

$$\begin{vmatrix} t_1 & 2t_1 + t_1^3 & 1 \\ t_2 - t_1 & 2(t_2 - t_1) + (t_2^3 - t_1^3) & 0 \\ t_3 - t_1 & 2(t_3 - t_1) + (t_3^3 - t_1^3) & 0 \end{vmatrix} = 0$$

$$\Rightarrow (t_2 - t_1)(t_3 - t_1) \begin{vmatrix} t_1 & 2t_1 + t_1^3 & 1 \\ 1 & 2 + t_2^2 + t_1^2 + t_2t_1 & 0 \\ 1 & 2 + t_3^2 + t_1^2 + t_3t_1 & 0 \end{vmatrix} = 0$$

$$\Rightarrow (t_2 - t_1)(t_3 - t_1)(t_3 - t_2)(t_3 + t_2 + t_1) = 0$$

$$\Rightarrow \quad t_1 + t_2 + t_3 = 0 \quad [\because t_1 \neq t_2 \neq t_3]$$

44. The slope of the line $x - y + 1 = 0$ is 1. So, it makes an angle of 45° with x-axis.

The equation of a line passing through (2, 3) and making an angle of 45°

$$\frac{x - 2}{\cos 45^\circ} = \frac{y - 3}{\sin 45^\circ} = r$$

Coordinates of any point on this line are

$$(2 + r\cos 45^\circ, 3 + r\sin 45^\circ)$$

or $$\left(2 + \frac{r}{\sqrt{2}}, 3 + \frac{r}{\sqrt{2}}\right)$$

If this point lies on the line $2x - 3y + 9 = 0$, then $4 + \sqrt{2}r - 9 - \frac{3r}{\sqrt{2}} + 9 = 0$

$$\Rightarrow \quad r = 4\sqrt{2}$$

Alternate Method

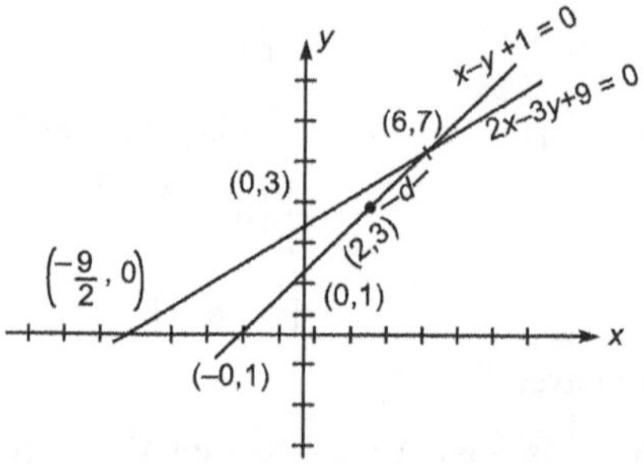

Since the point (2, 3) lies on the line $x - y + 1 = 0$. Therefore the distance from (2, 3) to the line $2x - 3y + 9 = 0$ along the line $x - y + 1 = 0$ is equal to the distance between the points (2, 3) and intersection point of $2x - 3y + 9 = 0$ and $x - y + 1 = 0$ *ie*, (6, 7). Hence required distance

$$d = \sqrt{(6 - 2)^2 + (7 - 3)^2} = \sqrt{32}$$

$$d = 4\sqrt{2}$$

45. The equation of any line through the point of intersection of the lines $x - y - 1 = 0$ and $2x - 3y + 1 = 0$ is

$$(x - y - 1) + \lambda(2x - 3y + 1) = 0$$

$$\Rightarrow \quad (2\lambda + 1)x - y(3\lambda + 1) + (\lambda - 1) = 0 \quad \text{...(i)}$$

The line in Eq. (i) will be parallel to x-axis, if it is of the form y = constant, therefore coefficient of x in Eq. (i) = 0

ie, $$2\lambda + 1 = 0 \Rightarrow \lambda = -\frac{1}{2}$$

On putting $\lambda = -\frac{1}{2}$ in Eq. (i), we get $y = 3$

This is the equation of the required line.

46. Since, given straight lines are the sides of a rhombus. Therefore, distance between the parallel lines

$a_1x + b_1y + c_1 = 0$, $a_1x + b_1y + c_2 = 0$ and $a_2x + b_2y + d_1 = 0$, $a_2x + b_2y + d_2 = 0$ must be equal

$$\therefore \quad \frac{|c_1 - c_2|}{\sqrt{a_1^2 + b_1^2}} = \frac{|d_1 - d_2|}{\sqrt{a_2^2 + b_2^2}}$$

$$\Rightarrow (a_2^2 + b_2^2)(c_1 - c_2)^2 = (a_1^2 + a_1^2)(d_1 - d_2)^2$$

47. The given equation is of the form $ax^2 + 2hxy + by^2 + 2gx + 2fy + c = 0$, on comparing the given equation with it

We obtain $a = 3$, $h = \frac{7}{2}$, $b = 2$, $g = \frac{5}{2}$, $f = \frac{5}{2}$ and $c = 2$

Now, $abc + 2fgh - af^2 - bg^2 - ch^2$

$$= 12 + 2 \times \frac{5}{2} \times \frac{5}{2} \times \frac{7}{2} - 3\left(\frac{5}{2}\right)^2 - 2\left(\frac{5}{2}\right)^2 - 2\left(\frac{7}{2}\right)^2$$

$= 0$

Hence, the given equation represents a pair of straight lines.

48. If the two pairs of straight lines have the same bisectors, then the two pairs are equally inclined.

The equation of the bisectors of the angle between the lines given by $ax^2 + 2hxy + by^2 = 0$ is

$$\frac{x^2 - y^2}{a - b} = \frac{xy}{h} \quad \ldots \text{(i)}$$

The equation of the bisectors of the angle between the lines given by $a^2x^2 + 2h(a + b)xy + b^2y^2 = 0$ is

$$\frac{x^2 - y^2}{a^2 - b^2} = \frac{xy}{h(a + b)}$$

$$\Rightarrow \quad \frac{x^2 - y^2}{a - b} = \frac{xy}{h} \quad \ldots \text{(ii)}$$

Clearly Eqs. (i) and (ii) are the same. The two pairs of straight lines are equally inclined.

49. $A_1B_1 = \sqrt{4 + 4} = 2\sqrt{2}$

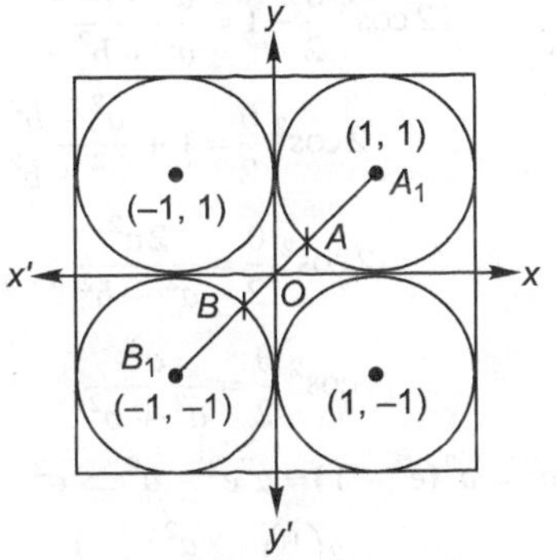

$$AB = 2\sqrt{2} - 2 = 2(\sqrt{2} - 1)$$

$\Rightarrow OA = OB = \dfrac{AB}{2} = \sqrt{2} - 1$ which is the radius of required circle.

Thus, equation of required circle is

$$x^2 + y^2 = (\sqrt{2} - 1)^2 = 3 - 2\sqrt{2}$$

50. We have, the given circle $x = a\cos\theta$, $y = a\sin\theta$

$$\Rightarrow \quad x^2 + y^2 = a^2 \quad \ldots \text{(i)}$$

Tangents to the circle (i) at θ and $\left(\theta + \dfrac{\pi}{2}\right)$ are

$$x\cos\theta + y\sin\theta = a$$

and $x\cos\left(\theta + \dfrac{\pi}{2}\right) + y\sin\left(\theta + \dfrac{\pi}{2}\right) = a$

or $\quad x\cos\theta + y\sin\theta = a$

and $\quad -x\sin\theta + y\cos\theta = a$

Squaring and adding, we get

$$x^2 + y^2 = 2a^2$$

which is a circle.

51. Let the equation of the required circle be

$$x^2 + y^2 + 2gx + 2fy + c = 0 \quad \ldots \text{(i)}$$

This passes through $A(1, 0)$ and $B(0, 1)$, therefore,

$$1 + 2g + c = 0 \text{ and } 1 + 2f + c = 0$$

$$\Rightarrow \quad g = -\left(\frac{c + 1}{2}\right) \text{ and } f = -\left(\frac{c + 1}{2}\right)$$

Let r be the radius of circle (i), then

$$r = \sqrt{g^2 + f^2 - c}$$

$$\Rightarrow \quad r = \sqrt{\left(\frac{c + 1}{2}\right)^2 + \left(\frac{c + 1}{2}\right)^2 - c}$$

$$\Rightarrow \quad r = \sqrt{\frac{c^2 + 1}{2}}$$

$$\Rightarrow \quad r^2 = \frac{1}{2}(c^2 + 1)$$

Clearly, r is minimum when $c = 0$ and the minimum value of r is $\dfrac{1}{\sqrt{2}}$.

For $c = 0$, we have

$$g = -\frac{1}{2} \text{ and } f = -\frac{1}{2}$$

On substituting the values of g, f and c in Eq. (i), we get,

$$x^2 + y^2 - x - y = 0$$

This is equation of the required circle.

52. The equations of two circles are

$$S_1 \equiv (x - a)^2 + (y - b)^2 = c^2 \quad \ldots \text{(i)}$$

and $\quad S_2 \equiv (x - b)^2 + (y - a)^2 = c^2 \quad \ldots \text{(ii)}$

The equation of the common chord of these circle is

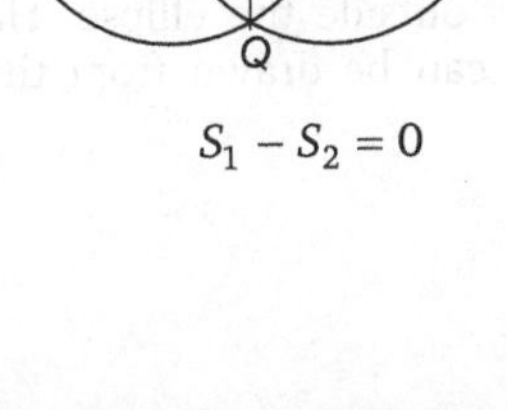

$$S_1 - S_2 = 0$$

$\Rightarrow (x-a)^2-(x-b)^2+(y-b)^2-(y-a)^2=0$

$\Rightarrow (2x-a-b)(b-a)+(2y-b-a)(a-b)=0$

$\Rightarrow 2x-a-b-2y+b+a=0$

$\Rightarrow x-y=0$

Let C_1 and C_2 be the centres of circles (i) and (ii) respectively. Then, the coordinates of C_1 and C_2 are (a, b) and (b, a) respectively.

$$C_1M=\left|\frac{a-b}{\sqrt{1+1}}\right|=\frac{|a-b|}{\sqrt{2}}$$

In right ΔC_1PM, we have,

$$PM=\sqrt{C_1P^2-C_1M^2}=\sqrt{c^2-\frac{(a-b)^2}{2}}$$

$$\therefore \quad PQ=2PM=2\sqrt{c^2-\frac{(a-b)^2}{2}}$$

$$=\sqrt{4c^2-2(a-b)^2}$$

53. Let the points $\left(m_i, \frac{1}{m_i}\right)$, $i=1, 2, 3, 4$ lie on the circle $x^2+y^2+2gx+2fy+c=0$

Then,

$$\Rightarrow m_i^2+\frac{1}{m_i^2}+2gm_i+\frac{2f}{m_i}+c=0, i=1, 2, 3, 4$$

$$\Rightarrow \quad m_i^4+2gm_i^3+cm_i^2+2fm_i+1=0,$$
$$i=1, 2, 3, 4$$

$\Rightarrow m_1, m_2, m_3$ and m_4 are the roots of the equation

$$m^4+2gm^3+cm^2+2fm+1=0$$

$$\Rightarrow \quad m_1m_2m_3m_4=\frac{1}{1}=1$$

54. Equation of the normal of the parabola $y^2=4ax$ at the point $(at^2, 2at)$ is

$$y+tx=2at+at^3 \quad \ldots(i)$$

$\because$ Eq. (i) cuts the parabola again at $(aT^2, 2aT)$.

Then, $\quad 2aT+taT^2=2at+at^3$

$$\Rightarrow \quad 2a(T-t)=-at(T^2-t^2)$$

$$\Rightarrow \quad 2=-t(T+t) \quad (\because t\neq T)$$

$$\Rightarrow \quad t^2+tT+2=0$$

$\because t$ is real, $\therefore T^2-4\cdot 1\cdot 2\geq 0$

$$\Rightarrow \quad T^2\geq 8$$

55. Since, $3(3)^2+5(5)^2-32>0$. So, the given point lies outside the ellipse. Hence, two real tangents can be drawn from the point to the ellipse.

56. Let $y=m_1x$ and $y=m_2x$ be the lines represented by $lx^2+2mxy+ny^2=0$, then

$$m_1m_2=\frac{l}{n} \quad \ldots(i)$$

But $y=m_1x$ and $y=m_2x$ are conjugate diameters of the hyperbola.

$$\therefore \quad m_1m_2=\frac{b^2}{a^2} \quad \ldots(ii)$$

From Eqs. (i) and (ii), we get

$$\frac{l}{n}=\frac{b^2}{a^2}\Rightarrow la^2=nb^2$$

57. The equations of the asymptotes of the hyperbola

$$\frac{x^2}{a^2}-\frac{y^2}{b^2}=1 \text{ are } y=\pm\frac{b}{a} \text{ ie,}$$

$$y=\frac{b}{a}x \text{ and } y=\frac{-b}{a}x$$

If θ is the angle between the two asymptotes,

$$\text{then, } \tan\theta=\frac{\frac{b}{a}+\frac{b}{a}}{1-\frac{b}{a}\cdot\frac{b}{a}}=\frac{2b}{a}\times\frac{a^2}{a^2-b^2}=\frac{2ab}{a^2-b^2}$$

From this, we get $\cos\theta=\dfrac{a^2-b^2}{a^2+b^2}$

$$\Rightarrow \quad 2\cos^2\frac{\theta}{2}-1=\frac{a^2-b^2}{a^2+b^2}$$

$$\Rightarrow \quad 2\cos^2\frac{\theta}{2}=1+\frac{a^2-b^2}{a^2+b^2}$$

$$\Rightarrow \quad 2\cos^2\frac{\theta}{2}=\frac{2a^2}{a^2+b^2}$$

$$\Rightarrow \quad \cos^2\frac{\theta}{2}=\frac{a^2}{a^2+b^2}$$

But $b^2=a^2(e^2-1)=a^2e^2-a^2\Rightarrow a^2+b^2=a^2e^2$

$$\therefore \quad \cos^2\left(\frac{\theta}{2}\right)=\frac{a^2}{a^2e^2}=\frac{1}{e^2}$$

$$\Rightarrow \quad \cos\left(\frac{\theta}{2}\right)=\frac{1}{e}$$

58. Let the point be $P\left(ct, \frac{c}{t}\right)$

The equation of the tangent to the hyperbola $xy=c^2$ at P is

$$x\frac{c}{t}+cty=2c^2$$

$$\Rightarrow \quad x+yt^2-2ct=0 \quad \ldots(i)$$

Then, putting $y = 0, x = 0$ successively in Eq. (i), we get $a_1 = 2ct$ and $b_1 = \frac{2c}{t}$

($\because$ intercepts of tangent on the axes are a_1 and b_1)

Again the equation of the normal to the hyperbola $xy = c^2$ at P is

$$xt^3 - yt - ct^4 + c = 0$$

As before, $a_2 = ct - \frac{c}{t^3}$ and $b_2 = -ct^3 + \frac{c}{t}$

$$\therefore\ a_1a_2 + b_1b_2 = 2ct\left(ct - \frac{c}{t^3}\right) + \frac{2c}{t}\left(-ct^3 + \frac{c}{t}\right)$$

$$= 0$$

59. The equation of xy-plane is $z = 0$

$\therefore$ Direction cosines of its normal are 0, 0, 1.

60. $xy + yz = 0 \Leftrightarrow (x + z)y = 0 \Leftrightarrow y = 0$

or $x + z = 0$

The equations $y = 0$ and $x + z = 0$ represents planes which are at right angles as their altitudes numbers are $< 0, 1, 0 >$ and $< 1, 0, 1 >$ respectively.

61. By image formula,

$$\frac{x_2 - x_1}{a} = \frac{y_2 - y_1}{b} = \frac{z_2 - z_1}{c} = -\frac{2(ax_1 + by_1 + cz_1 + d)}{a^2 + b^2 + c^2}$$

$$\Rightarrow \frac{x_2 - 2}{3} = \frac{y_2 + 1}{-2} = \frac{z_2 - 3}{-1} = -2\frac{(6 + 2 - 3 - 9)}{9 + 4 + 1}$$

$$\Rightarrow \frac{x_2 - 2}{3} = \frac{y_2 + 1}{-2} = \frac{z_2 - 3}{-1} = \frac{-2(-4)}{14}$$

$$\therefore\ x_2 = \frac{26}{7},\ y_2 = \frac{-15}{7},\ z_2 = \frac{17}{7}$$

62. $$\lim_{x \to 0} \frac{\sqrt{1 - \cos 2x}}{\sqrt{2} \cdot x} = \lim_{x \to 0} \frac{\sqrt{1 - (1 - 2\sin^2 x)}}{\sqrt{2} \cdot x}$$

$$= \lim_{x \to 0} \frac{\sqrt{2\sin^2 x}}{\sqrt{2} \cdot x} = \lim_{x \to 0} \frac{|\sin x|}{x}$$

Let $f(x) = \frac{|\sin x|}{x}$

Then,

$$f(0 + 0) = \lim_{h \to 0} \frac{|\sin(0 + h)|}{h} = \lim_{h \to 0} \frac{\sin h}{h} = 1$$

and

$$f(0 - 0) = \lim_{h \to 0} \frac{|\sin(0 - h)|}{-h} = \lim_{h \to 0} \frac{\sin h}{-h} = -1$$

$\because$ $f(0 + 0) \neq f(0 - 0)$

$\therefore$ The limit does not exist.

63. We have,

$$\lim_{h \to 0} f(1 - h) = \lim_{h \to 0} a(1 - h)^2 + b = a + b$$

$$\Rightarrow \lim_{h \to 0} f(1 + h) = \lim_{h \to 0} (1 + h) + 3 = 4$$

and $f(1) = 4$

$\therefore$ $f(x)$ will not be continuous at $x = 1$, if $a + b \neq 4$

64. We have, $f(x) = \log_x(\log x) = \frac{\log(\log x)}{\log x}$

$$\Rightarrow f'(x) = \frac{\log x \cdot \frac{1}{\log x} \cdot \frac{1}{x} - \log(\log x) \cdot \frac{1}{x}}{(\log x)^2}$$

$$= \frac{1 - \log(\log x)}{x(\log x)^2}$$

$$\therefore\ f'(e) = \frac{1 - \log(\log e)}{e(\log e)^2} = \frac{1}{e}$$

65. Given, circle is intersection of sphere.

$$x^2 + y^2 + z^2 + 2x - 2y - 4z - 19 = 0 \quad \ldots(i)$$

and plane $x - 2y + 2z + 8 = 0$ ...(ii)

Centre of sphere is $(-1, 1, 2)$.

P = length of the perpendicular from $(-1, 1, 2)$ upon Eq. (ii)

$$= \frac{-1 - 2 + 4 + 8}{\sqrt{1 + 4 + 4}} = \frac{9}{3} = 3$$

R = Radius of the sphere $= \sqrt{1 + 1 + 4 + 19} = 5$

Radius of the circle $= \sqrt{R^2 - P^2} = \sqrt{25 - 9}$

$$= \sqrt{16} = 4$$

66. We have, $x^y = e^{x-y}$

$$\Rightarrow y \log x = (x - y)\log e = x - y$$

$$\Rightarrow y = \frac{x}{1 + \log x}$$

$$\Rightarrow \frac{dy}{dx} = \frac{(1 + \log x) \cdot 1 - x \cdot \frac{1}{x}}{(1 + \log x)^2} = \frac{\log x}{(1 + \log x)^2}$$

67. We have, $y = \sin(m \sin^{-1} x)$

$$\frac{dy}{dx} = \cos(m \sin^{-1} x) \cdot \frac{m}{\sqrt{1 - x^2}}$$

$$\Rightarrow (1 - x^2)\left(\frac{dy}{dx}\right)^2 = m^2 \cos^2(m \sin^{-1} x)$$

$$= m^2(1 - y^2)$$

On differentiating w.r.t. x, again, we get

$$(1-x^2)2\cdot\frac{dy}{dx}\cdot\frac{d^2y}{dx^2}+\left(\frac{dy}{dx}\right)^2\cdot(-2x)$$

$$=-2m^2y\frac{dy}{dx}$$

$$\Rightarrow \quad (1-x^2)\frac{d^2y}{dx^2}-x\frac{dy}{dx}=-m^2y$$

[$\because$ Dividing by $2dy/dx$]

68. We have, $f(x)=(ax+b)\sin x+(cx+d)\cos x$

$f'(x)=a\sin x+(ax+b)\cos x+c\cos x$
$-(cx+d)\sin x$

But $f'(x)=x\cos x$ for all x (given)

$\therefore\ x\cos x=(a-d)$
$\sin x+(b+c)\cos x+ax\cos x-cx\sin x$

Equating the coefficients of $\sin x$, $\cos x$, $x\cos x$ and $x\sin x$, we get

$$a-d=0,\ b+c=0,\ a=1,\ c=0$$

$\therefore b=c=0$ and $a=d=1$

69. Let v be the velocity of the particle when the distance covered is s. Then,

$$v\propto\sqrt{s} \qquad \text{(given)}$$

$$\Rightarrow \quad v=\lambda\sqrt{s}$$

$$\Rightarrow \quad \frac{dv}{dx}=\frac{\lambda}{2\sqrt{s}}$$

$$\Rightarrow \quad v\frac{dv}{ds}=\frac{\lambda v}{2\sqrt{s}}=\frac{\lambda^2}{2}=\text{constant}$$

$\therefore$ The acceleration is constant

70. We have, $f(x)=e^x\sin x$

$$\Rightarrow \quad f'(x)=e^x\cos x+\sin x\cdot e^x$$

and $f''(x)=-\sin x\cdot e^x+\cos x\cdot e^x+\cos x\cdot e^x$
$+\sin x\cdot e^x$

Now, for maximum or minimum slope put $(f'(x))'=0\Rightarrow f''(x)=0$

$$\Rightarrow \quad 2\cos x\cdot e^x=0$$

$$\Rightarrow \quad \cos x=0\Rightarrow x=\frac{\pi}{2}$$

Also, $f'''(x)=-2\sin x\cdot e^x+2\cos x\cdot e^x=$ negative

$\therefore$ Slope is maximum at $x=\pi/2$.

71. Clearly, $f(2)=-1$

$$\Rightarrow \quad -1=\frac{2a+b}{(2-1)(2-4)} \Rightarrow 2a+b=2$$

Now, $f'(x)=\dfrac{4a+5b-2bx-ax^2}{(x-1)^2(x-4)^2}$,

$$f'(2)=0$$

$$\Rightarrow \quad b=0\Rightarrow a=1$$

$$\Rightarrow \quad f'(x)=-\frac{(x-2)(x+2)}{(x-1)^2(x-4)^2}$$

Clearly, for $x>2$, $f'(x)<0$ and for $x<2$, $f'(x)>0$.

Thus, $x=2$ is indeed the point of local maxima for $y=f(x)$

72. We have, $z=\tan(y+ax)+\sqrt{y-ax}$

$$\Rightarrow z_x=a\sec^2(y+ax)+\frac{(-a)}{2\sqrt{y-ax}}$$

$$\Rightarrow z_{xx}=2a^2\sec^2(y+ax)\tan(y+ax)$$
$$-\frac{a^2}{4}(y-ax)^{-3/2}$$

and $\quad z_y=\sec^2(y+ax)+\dfrac{1}{2\sqrt{y-ax}}$

$$\Rightarrow z_{yy}=2\sec^2(y+ax)\tan(y+ax)$$
$$-\frac{1}{4}(y-ax)^{-3/2}$$

$$a^2z_{yy}=2a^2\sec^2(y+ax)\cdot\tan(y+ax)$$
$$-\frac{a^2}{4}(y-ax)^{-3/2}$$

Now, $\quad z_{xx}-a^2z_{yy}=0$

73. We have, $\int f(x)\,dx=F(x)$

$$\therefore\ \int x^3f(x^2)\,dx=\frac{1}{2}\int x^2f(x^2)\,d(x^2)$$

$$=\frac{1}{2}[x^2F(x^2)-\int F(x^2)d(x^2)]$$

74. $\displaystyle\int\frac{x^2+4}{x^4+16}dx$

$$=\int\frac{1+\frac{4}{x^2}}{x^2+\frac{16}{x^2}}dx=\int\frac{d\left(x-\frac{4}{x}\right)}{\left(x-\frac{4}{x}\right)^2+8}$$

$$=\int\frac{dt}{t^2+(2\sqrt{2})^2},\ \text{where } t=x-\frac{4}{x}$$

$$=\frac{1}{2\sqrt{2}}\tan^{-1}\left(\frac{x^2-4}{2\sqrt{2}\,x}\right)+c$$

75. Let $I=\displaystyle\int\frac{1}{(x+1)\sqrt{x^2-1}}dx$

Put $\quad x+1=\dfrac{1}{t}$

$$\Rightarrow \quad dx=-\frac{1}{t^2}dt,\ \text{then}$$

$$I = \int \frac{1}{\frac{1}{t}\sqrt{\left(\frac{1}{t}-1\right)^2 - 1}} \left(\frac{-1}{t^2}\right) dt$$

$$\Rightarrow \quad I = -\int \frac{dt}{\sqrt{1-2t}} = -\int (1-2t)^{-1/2} dt$$

$$= -\frac{(1-2t)^{1/2}}{(-2)\left(\frac{1}{2}\right)} + c$$

$$\Rightarrow \quad I = \sqrt{1-2t} + c$$

$$\Rightarrow \quad I = \sqrt{1 - \frac{2}{x+1}} + c$$

$$\Rightarrow \quad I = \sqrt{\frac{x-1}{x+1}} + c$$

76. $\int_{\pi/2}^{3\pi/2} [\sin x] dx = \int_{\pi/2}^{\pi} [\sin x] dx + \int_{\pi}^{3\pi/2} [\sin x]\, dx$

$$= \int_{\pi/2}^{\pi} 0\, dx + \int_{\pi}^{3\pi/2} (-1) dx = -[x]_{\pi}^{3\pi/2}$$

$$= -\left(\frac{3\pi}{2} - \pi\right) = -\frac{\pi}{2}$$

77. The area of the loop of the curve $ay^2 = x^2(a-x)$ is given by

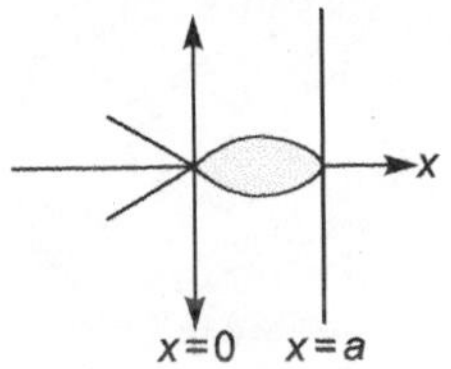

$$2\int_0^a y\, dx = 2\int_0^a x\sqrt{\frac{a-x}{a}} dx$$

$$= 4a^2 \int_0^{\pi/2} \sin^3\theta \cdot \cos^2\theta \cdot d\theta$$

[Let $x = a\sin^2\theta$

$\Rightarrow \quad dx = 2a \sin\theta \cdot \cos\theta \cdot d\theta$]

$$= 4a^2 \cdot \frac{2 \cdot 1}{5 \cdot 3 \cdot 1} = \frac{8a^2}{15} \text{ sq unit.}$$

78. The given differential equation can be written as

$$x = e^{xy\frac{dy}{dx}}$$

$$\Rightarrow \quad \log x = xy\frac{dy}{dx} \Rightarrow ydy = \frac{\log x}{x} dx$$

$$\Rightarrow \quad ydy = \log x\, d(\log x)$$

On integrating, we get

$$\frac{y^2}{2} = \frac{(\log x)^2}{2} + c$$

$$\Rightarrow \quad y^2 = (\log_e x)^2 + 2c$$

$$\Rightarrow \quad y = \pm\sqrt{(\log_e x)^2 + 2c}$$

79. We have, $\frac{dy}{dx} = \frac{ax+3}{2y+f}$

$$\Rightarrow \quad (ax+3)\,dx = (2y+f)dy$$

On integrating, we obtain

$$a\frac{x^2}{2} + 3x = y^2 + fy + c$$

$$\Rightarrow \quad -\frac{a}{2}x^2 + y^2 - 3x + fy + c = 0$$

This will represent a circle, if

$$-\frac{a}{2} = 1$$

($\because$ coefficient of x^2 = coefficient of y^2)

$$\Rightarrow \quad a = -2$$

80. $\int_1^5 x^2 dx$

$$= \frac{4}{4}\left[\frac{1}{2}f(1) + f(2) + f(3) + f(4) + \frac{1}{2}f(5)\right]$$

$$= \left[\frac{1}{2} + 4 + 9 + 16 + \frac{25}{2}\right] = 42 \text{ [using } f(x) = x^2]$$

